Primate Societies

JERYLLYNN

Primate Societies

Edited by Barbara B. Smuts

Dorothy L. Cheney

Robert M. Seyfarth

Richard W. Wrangham

Thomas T. Struhsaker

With 46 Contributors

The University of Chicago Press
Chicago and London

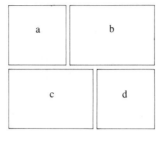

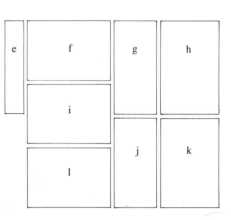

Key to cover photos and photographers

a) *Tarsius bancanus,* Patricial C. Wright

b) *Gorilla gorilla beringei,* Dorothy L. Cheney

c) *Pan troglodytes,* Joan B. Silk

d) *Colobus badius,* Lysa Leland

e) *Hylobates muelleri,* Peter Becker

f) *Alouatta seniculus,* Carolyn M. Crockett

g) *Lemur catta,* David Haring

h) *Propithecus verreauxi,* Alison F. Richard

i) *Brachyteles arachnoides,* John Robinson

j) *Erythrocebus patas,* Dana Olson

k) *Loris tardigradus,* David Haring

l) *Presbytis entellus,* Jim Moore/Anthrophoto

THE UNIVERSITY OF CHICAGO PRESS, CHICAGO 60637
THE UNIVERSITY OF CHICAGO PRESS, LTD., LONDON

© 1987 by The University of Chicago
All rights reserved. Published 1987
Printed in the United States of America

95 94 93 92 91 90 89 88 87   5 4 3 2 1

Library of Congress Cataloging-in-Publication Data

Primate societies.

   Bibliography: p.
   Includes index.
   1. Primates—Behavior. 2. Primates—Ecology.
3. Animal societies. 4. Social behavior in animals.
5. Mammals—Behavior. 6. Mammals—Ecology.
I. Smuts, Barbara B.
QL737.P9P6744   1987   599′.0524   86-7091
ISBN 0-226-76715-9
ISBN 0-226-76716-7 (pbk.)

Contents

Foreword. David A. Hamburg   vii

Preface   ix

1   The Study of Primate Societies
Dorothy L. Cheney, Robert M. Seyfarth,
Barbara B. Smuts, and Richard W.
Wrangham   1

PART I   EVOLUTION OF DIVERSITY   9

2   Lorises, Bushbabies, and Tarsiers:
Diverse Societies in Solitary Foragers
Simon K. Bearder   11

3   Malagasy Prosimians: Female
Dominance
Alison F. Richard   25

4   Tamarins and Marmosets: Communal
Care of Offspring
Anne Wilson Goldizen   34

5   Monogamous Cebids and Their
Relatives: Intergroup Calls and Spacing
John G. Robinson, Patricia C. Wright, and
Warren G. Kinzey   44

6   Howlers: Variations in Group Size and
Demography
Carolyn M. Crockett and John F.
Eisenberg   54

7   Capuchins, Squirrel Monkeys, and
Atelines: Socioecological Convergence
with Old World Primates.
John G. Robinson and Charles H.
Janson   69

8   Colobines: Infanticide by Adult Males.
Thomas T. Struhsaker and Lysa Leland   83

9   Forest Guenons and Patas Monkeys:
Male-Male Competition in One-Male
Groups
Marina Cords   98

10   Desert, Forest, and Montane Baboons:
Multilevel Societies
Eduard Stammbach   112

11   Cercopithecines in Multimale Groups:
Genetic Diversity and Population
Structure
Don J. Melnick and Mary C. Pearl   121

12   Gibbons: Territoriality and Monogamy
Donna Robbins Leighton   135

13   Orangutans: Sexual Dimorphism in a
Solitary Species
Peter S. Rodman and John C. Mitani   146

14   Gorillas: Variation in Female
Relationships
Kelly J. Stewart and Alexander H.
Harcourt   155

15   Chimpanzees and Bonobos: Cooperative
Relationships among Males
Toshisada Nishida and Mariko Hiraiwa-
Hasegawa   165

PART II   SOCIOECOLOGY   179

16   Life Histories in Comparative
Perspective
Paul H. Harvey, R. D. Martin, and
T. H. Clutton-Brock   181

17   Food Distribution and Foraging Behavior
John F. Oates   197

18   Interactions among Primate Species
Peter M. Waser   210

19   Predation
Dorothy L. Cheney and Richard W.
Wrangham   227

20   Demography and Reproduction
R. I. M. Dunbar   240

21   Dispersal and Philopatry
Anne E. Pusey and Craig Packer   250

22   Interactions and Relationships between
Groups
Dorothy L. Cheney   267

23   Evolution of Social Structure
Richard W. Wrangham   282

PART III   GROUP LIFE   297

24   Kinship
     Sarah Gouzoules and Harold Gouzoules   299

25   Conflict and Cooperation
     Jeffrey R. Walters and Robert M. Seyfarth   306

26   Social Behavior in Evolutionary Perspective
     Joan B. Silk   318

27   Infants, Mothers, and Other Females
     Nancy A. Nicolson   330

28   Infants and Adult Males
     Patricia L. Whitten   343

29   Transition to Adulthood
     Jeffrey R. Walters   358

30   Patterning of Sexual Activity
     Sarah Blaffer Hrdy and Patricia L. Whitten   370

31   Sexual Competition and Mate Choice
     Barbara B. Smuts   385

32   Gender, Aggression, and Influence
     Barbara B. Smuts   400

33   Can Nonhuman Primates Help Us Understand Human Behavior?
     Robert A. Hinde   413

34   Dynamics of Social Relationships
     Frans B. M. de Waal   421

PART IV   COMMUNICATION AND INTELLIGENCE   431

35   Communication by Sight and Smell
     Anne C. Zeller   433

36   Vocal Communication and Its Relation to Language
     Robert M. Seyfarth   440

37   Intelligence and Social Cognition
     Susan Essock-Vitale and Robert M. Seyfarth   452

38   Local Traditions and Cultural Transmission
     Toshisada Nishida   462

PART V   THE FUTURE   475

39   Conservation of Primates and Their Habitats
     Russell A. Mittermeier and Dorothy L. Cheney   477

40   Future of Primate Research
     Dorothy L. Cheney, Robert M. Seyfarth, Barbara B. Smuts, and Richard W. Wrangham   491

Appendix: The Order Primates: Species Names and a Guide to Social Organization   499
List of Contributors   507
Bibliography   511
Index   565

Foreword

Despite centuries of fascination with monkeys and apes, reflecting our intuitive recognition of their close biological relationship to the human species, little was known about the behavior of nonhuman primates in their natural habitats until recently. In the past two decades, the study of the behavior of nonhuman primates has rapidly developed into one of the most stimulating areas of research at the interface between the biological and social sciences. Indeed, these inquiries make clear why it is useful to take a broad view of the life sciences—unifying biology, behavior, and social organization in an evolutionary framework. Also in the last two decades the inherent fascination of the quest for human origins has been intensified by a surge of discoveries involving not only an enrichment of established lines of inquiry (such as stones and bones) but also the engagement of the physical, biological, and behavioral sciences through molecular biology, geology, and animal behavior.

Some of the most significant scientific opportunities for insight into human origins unfortunately are transient. Tropical forests are disappearing rapidly, and with them we are losing their populations of prosimians, monkeys, and apes. Similarly, the few remaining hunter-gatherer populations are disappearing or changing their life-styles. Within a short time, these ways of life may disappear from the earth. Research on crucial elements of the human and prehuman past is therefore truly urgent: it must be done now or never.

The contributors to this volume present information from primate field studies in an evolutionary context. The results are partly experimental; for example, they document playbacks of primate vocalizations that help crack the codes of communication within species. But the volume's principal aim is to provide reliable observational data on ecology and social behavior in more than one hundred species of primates. This body of information has allowed authors to draw systematic comparisons between species, which make it possible in turn to generate testable hypotheses.

It is remarkable that we now have so much information about wild primates, for only twenty years ago fewer than a dozen species had been studied intensively in the wild. Obviously more information provides a much firmer foundation for the development both of broad generalizations and of hypotheses that future research

can test. The same growth of knowledge is reflected in the upsurge of long-term field studies. These have shown, as we anticipated, that rare events, such as a single aggressive interaction between groups, can have long-lasting and biologically significant effects on behavior. More than anything, these complex and difficult studies, often sustained over many years at a single field site, have documented the profound importance of long-term social relationships for patterning social interactions.

One of the most fascinating features of this book is the light it casts on the richness and diversity of primate social behavior. For example, not only does the primate group typically include animals of different age, sex, kinship, and dominance status, but there are shifting subgroups, temporary alliances, and long-term associations or even "friendships" that cut across traditional categories. All this reflects the wide variety of ways in which social behavior contributes to strategies of survival and reproduction.

This volume contains lessons that may have profound significance for contemporary dilemmas: human behavioral tendencies grow deep in the soil of primate evolution. Consider the relationship between attachment and aggression. In primates, attachments to other group members are often developed and protected through aggression. One of the most frequent contexts of aggressive behavior is protection of close relatives and friends. For example, female macaques usually react with submission and deference to higher-ranking animals. But in defense of close kin, a female will threaten or even attack a more dominant animal; this, in fact, is the only context in which she will routinely do so. Thus, attachment and aggression are closely linked in the adaptive functioning of primate societies.

This relationship is seen also in interactions between groups. The planned, brutal, and male-dominated attacks between chimpanzees of adjacent communities provide the most striking example since this extreme form of aggression between communities is associated with particularly warm affiliative behavior among males in the same community. When the chimpanzee evidence is considered along with information on responses to strangers in other nonhuman primate species, it seems likely that the human tendency to react with fear and hostility to strangers has its roots in our nonhuman past. One of the exciting aspects of this volume, therefore, is the development of explanations for specific patterns of behavior, such as elements of aggressive behavior, that nonhuman primates share with the human species.

The evidence of progress in this and many other areas is all the more remarkable when one reflects on the extraordinary obstacles in the way of primate research. In the often remote areas where primates still survive in nature, there are truly formidable difficulties involving logistics, disease, language, culture, and even violence. Populations and their habitats are dwindling so fast that sometimes it takes months even to find a viable population; later the observers may find themselves working as hard to protect their subjects as to study them. Moreover, funding for research in this field has scarcely been abundant. One cannot fail to be deeply impressed with the remarkable dedication of the scientists who have made these observations.

Building on their hard-won, careful data, the authors of this volume provide valuable analyses. They link information and ideas from various disciplines, adopt common terms and concepts, and look to the future with vision and imagination. This book is a landmark in the scientific study of primate behavior and human origins.

David A. Hamburg, M.D.

Preface

Although C. R. Carpenter conducted several field studies before the Second World War, it was not until the 1950s, when the Japan Monkey Center was organized at Kyoto University and other research sites were established in Africa, Central America, and Asia, that the systematic study of free-ranging primates became firmly established. By the early 1960s, data on a variety of species, including chimpanzees, gorillas, macaques, and baboons, had become available for the first time (e.g., Southwick 1963), thus alerting social scientists to the potential significance of primate studies. To evaluate these results, David Hamburg and Sherwood Washburn organized a primate study group at the Center for Advanced Study in the Behavioral Sciences (CASBS), Stanford, California, in 1962–63. The group involved six primatologists who spent nine months as fellows at CASBS, along with a number of other consulting researchers. One result of their activity was the publication in 1965 of *Primate Behavior: Field Studies of Monkeys and Apes,* edited by Irven DeVore, the first major collection of papers on primate field research. This volume quickly became a classic for thousands of undergraduates. Even today it can be found not only in universities but on the bookshelves of lawyers, doctors, business executives, and others for whom it represented their first, and in many cases their only, exposure to the social behavior of our closest living relatives.

Although the naturalistic study of primate behavior expanded rapidly between 1955 and 1965, its growth in the last twenty years has been even more phenomenal. Research on prosimians, monkeys, and apes is now conducted throughout the world by hundreds of investigators. Most species have received some attention, and many have been studied repeatedly in different places. In several cases, groups of individually recognized individuals have been studied continuously for ten years or more. Each new study site generates new observations, just as long-term studies constantly bring surprises. The growth of information has forced the growth of new ideas, and long-term primate studies were among the first to show the critical importance in mammals of kinship, social relationships, and individual variations in behavior. Of equal significance, advances in evolutionary theory have revolutionized the interpretation and design of primate studies.

Such rapid progress brings with it special problems of dissemination and synthesis of results. Partly because advances have occurred on such a broad front, it is at present remarkably difficult for students, professionals in other fields, and even those within primatology to keep up. Because primatology has always involved researchers and concepts from several different disciplines (including biology, anthropology, psychology, and even sociology), new findings appear in a bewildering variety of dissertations, journals and books. Furthermore, there are few reviews that compile diverse information on the basic social behavior of particular species or species-groups. There is also a shortage of synthetic articles on particular aspects of behavior, such as kinship, mate choice, or relations between groups. As a result, discussions of primate behavior in anthropology, biology, and psychology textbooks are routinely dated and sometimes seriously misinformed. Students and professionals in related fields therefore have a difficult time integrating primatology's newest findings.

In 1981, with these concerns in mind, David Hamburg asked Barbara Smuts to organize a study group along lines similar to the 1962–63 project. The CASBS agreed once again to act as a host, and in 1983, twenty years after the first project, the editors of this volume gathered to spend nine months reviewing the current status of primate field research. The present volume is the result.

Primate Societies reviews current knowledge of all species that have been studied in the wild and highlights theoretical issues of central importance. It is designed for a broad audience with diverse needs. We hope that it will serve as a reference work for primatologists and other field researchers; as a convenient, up-to-date synthesis of new findings for social scientists and behavioral biologists; and as a text for advanced undergraduates and graduate students.

The book reflects three important decisions made when the volume was conceived. First, because we hoped the book would function as a valuable reference, we have reviewed all major topics in depth and, as far as possible, supported generalizations with facts and references. Second, although we have included information on ecology that contributes to understanding primate societies, the central topic is social behavior. This distinguishes the present volume from several recent and valuable books on primate ecology (Richard 1985; Chivers, Wood, and Bilsborough 1984; Rodman and Cant 1984). Third, the book emphasizes field studies. The focus on wild primates reflects both our own interests as field-workers and the urgent nature of primate field research due to worldwide destruction of primate habitats and other sources of human disturbance (chapter 39).

From the start it was clear that this book would depend on a large number of primatologists with a diverse set of backgrounds, interests, and research experience. At the same time, we felt that consistency of content, style, and purpose across chapters was essential. To the extent that we have met this goal, it reflects cooperative and diligent authors and a united group of editors. For this reason, emphases or gaps in particular chapters often reflect editorial policy as well as the authors' own preferences.

In a book of this size that attempts to review an entire field, the issue of overlap between chapters is a difficult one. On the one hand, we wished to avoid annoying repetition whenever possible. On the other hand, we wanted each chapter to stand alone so that those interested in particular chapters would not be forced to refer continually to other chapters for relevant data or theories. We have attempted to resolve these conflicting goals through the following compromise. Whenever possible, repetition is minimized by having one chapter serve as the primary source for a given set of data or a particular theoretical perspective. When it is necessary for other chapters to refer to the same material, a brief summary is generally provided, along with references to the chapter where more detailed treatment occurs.

The preparation of any book that involves five editors, forty chapters, and forty-six authors requires an extraordinary amount of support from many people. We are deeply indebted to David Hamburg for inspiring the primate project and the volume that came out of it. David Hamburg and Irven DeVore provided invaluable assistance during the organization of the primate study group. As editors we could not have completed this task had we, during 1983–84, not been pampered by the staff of CASBS—one of only a handful of institutions that provides a supportive base for cooperative intellectual enterprises of this kind. We are especially grateful to Gardner Lindzey and Bob Scott for making our stay at the center possible, and to Alan Henderson, Frances Duignan, Margaret Amara, Lynn Gale, and other members of the CASBS staff for their unfailing and cheerful support in even the most mundane areas. Special thanks go to Muriel Bell for her editorial assistance, and to the National Science Foundation, the Alfred P. Sloan Foundation, and the Exxon Education Fund, which provided financial support for the editors while we were fellows at CASBS.

To the contributors to this volume, we express heartfelt thanks. Our authors met the difficult challenge of synthesizing diverse findings with creative energy and personal commitment. We are deeply grateful for their remarkable perseverance, generosity, and patience in response to eighteen months of more or less continuous editorial pressures. This is their book, and we hope that they will agree with us that all the hard work has been worthwhile.

All of the authors have agreed to join the editors in donating their share of the royalties from this volume to a primate conservation fund (*Primate Societies* Conservation Fund, c/o Wildlife Conservation International, New York Zoological Society, Bronx, N.Y. 10460, USA, or P.O. Box 62844, Nairobi, Kenya). Similarly, the photographers whose pictures appear in the book and on the cover have agreed to waive their fees on behalf of this fund. We owe a special thanks to Nancy DeVore, who offered the services of her photographic agency, Anthrophoto, free of charge. We are deeply grateful to the many people who were willing to contribute time and money to help ensure the survival of the wild primates who are the subjects of this book.

We thank the many colleagues who improved the book tremendously by acting as external reviewers for different chapters: Robert Bailey, Nicholas Blurton Jones, Curt Busse, Carolyn Crockett, John Eisenberg, Lynn Fairbanks, Philip Gingerich, Tom Gordon, Kristen Hawkes, Sarah Hrdy, Kevin Hunt, Lysa Leland, James McKenna, Peter Marler, John Mitani, John Pollock, Trevor Poole, Carolyn Ristau, Charles Snowdon, Robert W. Sussman, and Claudia Thompson.

We owe a special debt to Elizabeth Ross for her painstaking drawing of figures and to John G. Robinson and Lysa Leland for generous assistance in planning the volume. We also thank Virginia Vitzthum for manuscript preparation and Deborah Middleton and Richard Connor for extensive help with the references. Janet Friedman prepared the index with assistance from Anne Keddy and Marc Hauser.

Authors would like to thank the following people who provided information and feedback for their chapters (the numbers of the chapters to which the acknowledgments apply are given in parentheses): E. M. B. Barrett, C. S. Harcourt, C. Niemitz, R. S. Pitts (2); Patricia Wright (4); Rolando Aquino, Tom Defler, Dorothy Fragasy, Sally Mendoza (5); A. Estrada, K. Milton, T. Pope, R. Rudran, R. Sekulic, A. Shoemaker, R. Thorington, Jr. (6); Robert Bailey, John Eisenberg, Gordon Orians, Kent Redford (7); E. Bennet, T. M. Butynski, M. Hauser, S. Hrdy, C. Marsh, J. F. Oates, J. P. Skorupa (8); E. J. Brennan, T. M. Butynski, J. Chism, S. H. Curtin, D. K. Olson, T. T. Struhsaker (9); M. Leighton, J. Moore (12); all personnel of the Mountain Gorilla Project (14); T. Hasegawa, T. Kano (15); David Chivers, Annie Gautier-Hion, W. Thomas Jones, Russell Mittermeier, John Oates, John Robinson, Peter Rodman, Thomas Struhsaker, Mary Sue Waser, (18); Tom Butynski, Joseph Skorupa (19);

Richard Connor (23); G. Hausfater, J. G. Vandenbergh (25); Robert Boyd (26); James McKenna, Marten de-Vries (27); E. O. Smith (28); J. Altmann, M. Perreira (29); Leo Berenstain, Martin Daly, Daniel Estep, E. B. Keverne, Donna Leighton, Martha McClintock, Peter Rodman, Joan Silk, Joseph Skorupa, Meredith Small, Thomas Struhsaker, Margo Wilson (30); R. W. Smuts, John Watanabe (31 and 32); Joseph Skorupa (39). Authors would also like to thank the following people and organizations for (a) assistance during fieldwork: Thomas and Cecilia Blohm (6); C. L. Darsono, Indonesian Institute of Sciences, Indonesian Ornithological Society, Pertamina Oil Company (13); D. Fossey, A. Nemeye, L. Rwelekana, and field assistants and researchers of the Mountain Gorilla Project too numerous to name individually, Office Rwandais du Tourisme et des Parcs Nationaux, Institut Zairois pour la Conservation de la Nature (14); Staff Sergeant Alfred Holmes and personnel of the Gibraltar regiment (35); (b) help in preparing manuscripts and illustrations: Gina Habets, Mary Peters (27); Pam Carpenter (30); Bob Dodsworth, Jackie Kinney, Mary Schatz (34); (c) research and bibliographic assistance: Susan McDonald (19); Yongmi Han (30); and (d) general support and encouragement: G. A. Doyle, R. D. Martin (2); R. Brooks (6); Orville A. Smith (13); R. Hinde, Gloria Stewart, and J. Stewart (14); John Watanabe (31 and 32); F. D. Burton (35).

Like many other sciences, the study of primate behavior and ecology has experienced its greatest growth in the Western, industrialized nations of the world. Unlike many other disciplines, however, primatology requires close links between scientists in industrialized countries, where modern equipment and techniques are available, and those in developing countries, where most nonhuman primates actually live. We and the authors would therefore like to thank the governments of all countries that have allowed and encouraged nonhuman primate field research, particularly when they have not always derived immediate, tangible benefits from such work: Bolivia, Botswana, Brazil, Burma, Burundi, Cambodia, Cameroon, People's Republic of China, Republic of China, Colombia, Congo, Costa Rica, Dahomey, Ethiopia, Gabon, Gambia, Guatemala, Guinea, Honduras, India, Indonesia, Ivory Coast, Japan, Kenya, Laos, Liberia, Madagascar, Malaysia, Mexico, Morocco, Namibia, Nepal, Nicaragua, Nigeria, Pakistan, Panama, Peru, Philippines, Rwanda, Saudi Arabia, Senegal, Sierra Leone, South Africa, Sudan, Tanzania, Thailand, Uganda, Venezuela, and Zaire. We also thank the University of Chicago Press for allowing us to donate two hundred copies of this book to institutions in primate host countries. We hope that volumes like the present one will contribute to growth of the biological sciences in these nations.

This book rests on the work of many individuals supported by many different institutions. Here we join our authors in citing the public and private organizations that have supported field research on primates over the years: the Boise Fund (Oxford); Edwin Edwards Scholarship Fund of the University of Michigan; Friends of the National Zoo; Harry Frank Guggenheim Foundation; Indonesian Ornithological Society; Japanese Ministry of Education, Science, and Culture and its subdivision, Special Project Research on Biological Aspects of Optimal Strategy and Social Structure; Joseph Henry Fund of the National Academy of Sciences; L. S. B. Leakey Foundation; Medical Research Council (Great Britain); National Academy of Sciences (United States); National Environmental Research Council (Great Britain); National Geographic Society; National Institutes of Health, National Institutes of Mental Health, and National Science Foundation (United States); New York Zoological Society; Oxford Polytechnic Research Committee; P. S. R. Fund for Graduate Research; Regents' Fellowships, University of California, Davis; Royal Society and Royal Society Leverhulme Trustees (Great Britain); Science Research Council (Great Britain); Scientific Research Society of Sigma Xi; Smithsonian Institution; T. C. Schneirla Research Fund; University of Michigan Block Grants; Wenner-Gren Foundation for Anthropological Research; World Wildlife Fund; W. T. Grant Foundation; and Zoological Society of London. Chapters 25 and 29 are publication numbers 9979 and 9624, respectively, of the Journal Series of the North Carolina Agricultural Research Service, Raleigh, North Carolina, and chapter 34 is publication number 24-007 of the Wisconsin Regional Primate Research Center.

Finally, we thank Susan Abrams, whose encouragement and enthusiasm have been extraordinary, and who has expressed empathy for editors and authors alike by persistently quoting John Donne: "to write is to war with idiots in the caves of the heart and mind."

1 | The Study of Primate Societies

Dorothy L. Cheney, Robert M. Seyfarth, Barbara B. Smuts, and Richard W. Wrangham

ORGANIZATION OF *Primate Societies* AND ITS INTENDED AUDIENCE

The many similarities between ourselves and nonhuman primates have made monkeys and apes the focus of research in a variety of scientific disciplines. Anthropologists, for example, study primates to seek clues to the social behavior of early humans. The biomedical community depends on primates because of their physical similarity to humans, while primates offer psychologists a unique opportunity to study intelligence in creatures whose brain, more than that of any other species, is like our own. Finally, for biologists primates mirror the diversity of mammals themselves, extending from solitary, nocturnal species whose social organization is relatively simple to those that live in large, diurnal groups whose kinship, long-term relationships, and forms of reciprocity pose special challenges for theories about the evolution of social behavior.

With this diverse audience in mind, *Primate Societies* reviews what is known about the social behavior of primates in their natural habitats. The focus of the book is on social organization and behavior, although some ecological data are presented to provide a necessary background.

Primate Societies is organized into five parts. The chapters in part 1, "Evolution of Diversity," are ordered taxonomically and according to social organization. They describe the behavior and ecology of the approximately 200 species that make up the order Primates. Each chapter also addresses a specific behavioral or ecological topic of special relevance to that particular group. Our aim in this section has been to provide a complete, authoritative description of species differences and intraspecific variation among nonhuman primates. Authors focus on well-documented results rather than on theoretical issues, illustrate the magnitude of variation within each taxonomic group, and illuminate correlations among ecology, social organization, and behavior. Taken together, the chapters in part 1 should specify what data are available for testing theories as well as point to the gaps in current knowledge. These gaps are particularly apparent for New World monkeys and prosimians, which have received comparatively less scientific attention than the Old World monkeys and apes.

To achieve a concise presentation of information, authors have summarized their data in tables wherever possible. This sort of quantitative approach entails risks, however, because results presented in numerical or tabular form often assume an aura of precision and infallibility that may not be justified. Readers should be sensitive to the fact that, due to the varying lengths of time that different species have been studied, the quality of the data varies tremendously, and results are not always as definitive as they might seem.

As each species is mentioned in part 1, both Latin and common English names are given. Thereafter, English names are used except in those cases where to do so might be ambiguous. For a complete taxonomy of the Primates, as well as a complete listing of Latin names and the English equivalents we have adopted, readers are referred to the appendix. Technical terms are listed in the index, where a separate entry gives the pages on which each term has been defined.

The chapters in parts 2 through 5 abandon the taxonomic approach in order to address particular issues that apply to many different species. In part 2, "Socioecology," eight chapters on habitats and foraging behavior, community ecology, predation, life histories, dispersal, intergroup relations, and the relation between ecology and social organization examine primates as part of larger ecological communities and describe social behavior at the local population level. Part 3, "Group Life," follows with eleven chapters on social behavior that review the current understanding of social interactions within primate groups. These are followed by four chapters in part 4, "Communication and Intelligence," that discuss communication, intelligence, learning, and cultural transmission. Part 5, "The Future," includes one chapter on conservation that documents the very real danger of extinction threatening most of the world's nonhuman primates. Part 5 concludes with a chapter highlighting important new areas of research.

Who Should Read This Book?

Students. Primate Societies is primarily intended for advanced undergraduates, graduate students, and professionals in related fields interested in the behavior and ecology of nonhuman primates. We have assumed that readers will be familiar with the basic concepts of evolutionary theory and behavioral ecology that are learned in an introductory course in animal behavior.

At the same time, *Primate Societies* is also intended as a useful, thorough reference for professional anthropologists, psychologists, biologists, and those from the biomedical community. Although scientists in these fields approach primatology from quite different perspectives, all share a need for up-to-date information on the behavior of different species under natural conditions and on the theories that have been used to organize and explain these data.

Anthropologists. Many of the early students of primate behavior were anthropologists who sought analogies for the social organization of early humans in the behavior of nonhuman primates. Traditionally, anthropologists have tended to select a single primate species (such as the chimpanzee because of its genetic similarity to humans, or the baboon because it inhabits the East African savanna where early humans are thought to have evolved) and imagine this species' behavior as representative of the ancestral human condition. As the chapters in part 1 reveal, however, there is extraordinary diversity in primate social organization, even within a given habitat. Any attempt to extrapolate from living primates to early humans should obviously take account of this diversity and, rather than searching for single species as models, should attempt instead to uncover some of the basic principles that link ecology and social structure. Some of these principles are described, to the best of our current knowledge, in parts 2 through 4.

Psychologists and the Biomedical Community. In the United States alone, thousands of nonhuman primates serve as subjects in laboratory experiments each year. Primates are often the preferred subjects for psychologists investigating learning, memory, and other complex cognitive processes and for biomedical scientists whose work may ultimately be applicable to human health and disease. In each of these fields investigators must conduct experiments within strictly controlled conditions and limited budgets. Much of this research involves monkeys and apes that have been removed from their natural habitats and housed in individual cages, which, while efficient, allow no opportunity for normal social interaction. Scientists themselves often know little about their subjects' history or their subjects' behavior under natural conditions.

In the past, these conditions were viewed simply as the price one had to pay for conducting carefully controlled experiments at reasonable cost. It has become increasingly clear, however, that ignorance of an animal's natural social behavior, in addition to leading to potentially inhumane treatment, may produce bad science. Consider, for example, the search by neuropsychologists for similarities in the way human and nonhuman primates perceive communicative sounds. When humans are asked to discriminate between different speech sounds, most individuals perform better with their right ear than with their left. This is explained by the existence in the left cerebral hemisphere of an area specialized for speech perception (e.g., Masterson and Imig 1984). Despite the many similarities between human and nonhuman primate brains, however, for years similar cerebral lateralization could not be demonstrated for nonhuman primates (e.g., Hamilton 1977). Finally it was recognized that while humans had been tested with biologically important speech sounds, monkeys and apes had been tested with biologically irrelevant stimuli such as tones and hisses. The investigators' ignorance of their subjects' natural behavior and vocal communication had led them to use inappropriate stimuli in their tests. In 1978, when experiments first examined how primates perceive their own species' calls, a right-ear advantage clearly emerged (Petersen et al. 1978). Further insights into the similarities between human and nonhuman primate brains quickly followed (Hefner and Hefner 1984).

This example is not an isolated case, but forms part of a growing body of evidence that laboratory investigators who remain ignorant of their subjects' natural behavior risk making important scientific errors (e.g., Seligman and Hager 1972; Hinde and Stevenson-Hinde 1973). For all those working with captive primates, then, we offer *Primate Societies* as a reference to the diverse and complex ways in which evolution has shaped their behavior.

Biologists. Finally, field biologists, including those of us who have edited this volume, study primates for still other reasons, to elucidate the principles that govern the evolution of social organization and behavior. In the past twenty years, with the publication of papers by Hamilton, Trivers, Maynard Smith, and others, there has been a theoretical revolution in evolutionary biology, leading to the reexamination of earlier work, the establishment of new journals, and a dramatic increase in studies designed to clarify how factors such as kinship, reciprocity, sexual selection, and life history affect the evolution of behavior. Somewhat surprisingly, however, primate research has rarely been in the forefront of this work (Richard 1981), and most general textbooks in animal behavior rely primarily on studies of insects, fish, birds, or nonprimate mammals (e.g., Krebs and Davies 1981; Alcock 1984). This raises two questions. First, why have primate studies remained apart from other work in animal behavior? And second, can the behavior of primates be explained in terms of the same general principles that explain the behavior of other animals?

One reason primates are rarely used to test evolutionary theories derives from problems associated with studying them in their natural habitats. Many primate species live in areas where they are difficult to observe or identify individually and where political and logistical problems can often entirely prevent research. More important, even when individuals can be followed and recognized, primates are so long-lived that it may take at least a decade before data on kinship and reproductive success begin to accumulate. Little wonder that scientists interested in reproductive life histories have turned to animals that are short-lived, accessible, and easy to observe. It has also proved far more difficult to conduct manipulative experiments on nonhuman primates than on birds or nonprimate mammals, and as a result, experimental tests of evolutionary theories on nonhuman primates have lagged far behind those on other animals.

It is, of course, too early to know whether the behavior of nonhuman primates, like that of humans, will pose special problems for theories designed to explain the behavior of other animals. No doubt much of primate behavior and ecology will be consistent with generalizations drawn from other species. Even at this early stage, however, at least three features of primate behavior suggest unusual levels of complexity when compared with other animals. We mention these here because they offer additional reasons why primates have often been ignored by behavioral ecologists seeking simple explanations for a given pattern of behavior, and why biologists unfamiliar with primate research may profit from the material contained in *Primate Societies*.

What Makes Primates Different from Other Animals?

1. *Primates have unusually varied and diverse ways of expressing themselves socially.* Although it is always dangerous to pretend to know the limits of animal communicative repertoires (see chap. 36), we can say that the touching, hugging, mouthing, mounting, lip smacking, vocalizing, greeting, and grooming of primates allow them many subtly different ways of expressing affinity and perhaps more means of developing complex social relationships than are found in most other species. Moreover, primates seem to move easily from one behavioral "currency" to another, often apparently "trading," for example, a mount for tolerance at a food source, or a bout of grooming for later support in an alliance. Such diverse interactions have two consequences. On the one hand, they increase the likelihood that primates, more than any other animals, will have evolved complex, reciprocal interactions (e.g., Trivers 1971). On the other hand, however, they increase the difficulty of documenting the existence of complex features like reciprocity (see chaps. 25, 26).

2. *The social organization of many primate species is unusually complex.* A typical primate group contains individuals of different ages, sexes, dominance ranks, and kinship. Although this in itself is not unusual, primates also form temporary alliances, subgroups, and even long-term associations that cut across such categories.

The result is (1) a complex network of interactions, with many alternative strategies for survival and reproduction, and (2) social groups in which individuals are likely to pursue a number of different strategies during their lifetimes. To cite just one example, baboon males can gain access to estrous females either by attaining alpha rank and intimidating rivals, by forming alliances with other males, or by establishing a long-term bond with particular adult females (see chaps. 31, 32). These different modes of achieving mating opportunities seem at first glance to be ideally suited for testing models of behavior that focus on alternative competitive strategies (e.g., Maynard Smith and Price 1973; Maynard Smith 1982; Parker 1984). However, most such models depend on a number of simplifying assumptions. For example, most assume that interactions are dyadic, that there are a very limited number of alternative behavioral strategies, and that knowledge of opponents is based directly on recent experience. These assumptions seldom apply to primates, whose social interactions often involve more than two individuals, numerous possible responses, and, at least in some cases, observational learning (chaps. 34, 37). As a result, theoretical models that have been valuable in explaining competitive interactions in other organisms are of only limited usefulness for primates. The wealth of detailed data on primate social interactions could, however, prove extremely useful in developing new models that are more relevant to species exhibiting complex social behavior (see Dunbar 1984b for an example).

3. *Primates form various kinds of long-term social relationships.* As is the case with other animal species, the function of some types of competitive and cooperative behavior in nonhuman primates can often be described in terms of their direct consequences for reproductive success. For example, aggressive behavior may allow a male to maintain a high dominance rank, thereby increasing his access to estrous females. Because primates are long-lived, intelligent creatures, however, there are also many more subtle, indirect ways in which a given pattern of behavior can affect fitness.

Ample evidence, for example, shows that primates groom or form alliances not only for their immediate benefits but also to establish and maintain particular social relationships. For instance, close kin groom one another after fighting, males sometimes groom and maintain close proximity to pregnant and lactating females, and female vervets and baboons often form alliances against animals that already rank lower than they do (Walters 1980; Cheney 1983a; Datta 1983a; Smuts 1985; de Waal, in press). In each of these and many other cases, the adaptive consequences of behavior are difficult to specify if we analyze social interactions only in terms of their immediate effects. The function of grooming, alliances, and other acts becomes clearer, however, if we think of such behaviors as contributing incrementally to social relationships that themselves have adaptive consequences (Hinde 1976; Seyfarth, Cheney, and Hinde 1978). The male baboon who associates with a female during pregnancy and protects her infant during its early months may reap no immediate reproductive benefits, but his behavior may nevertheless lead to a close, long-term relationship that persists when the female resumes sexual cycling (Seyfarth 1978b; Smuts 1985). When two female baboons or macaques form an alliance against a third, they may be acting both to reaffirm the existing dominance hierarchy (Walters 1980) and to solidify a social bond that benefits them in a subsequent, more costly interaction. Male chimpanzees who groom frequently, travel together, and reconcile after fights receive few obvious immediate reproductive benefits. Over long periods of time, however, their interactions may lead to a close relationship with consequences not only for their dominance ranks but also for their success in intergroup competition (e.g., Goodall et al. 1979; Nishida et al. 1985; chap. 15).

The widespread existence of long-term social relationships among primates has frequently forced primatologists to approach the evolution of behavior with a perspective that differs in its emphasis from that traditionally used in the study of birds and nonprimate mammals. For example, traditional ethological research has concentrated on interactions between individuals and has examined the function of single acts by measuring their immediate and long-term consequences. In primates, however, an interaction like grooming clearly has consequences beyond both its immediate function of ectoparasite removal and the longer-term function of making subsequent grooming bouts more likely, since grooming can also contribute to the maintenance of a relationship that may have important reproductive consequences. Primate research has revealed clearly that individual acts do not occur independently and that one type of behavior can affect another, contributing over time in an almost imperceptible way to a particular social bond (Hinde 1976; Seyfarth, Cheney, and Hinde 1978). The importance of long-term relationships among primates makes their behavior of particular interest to students of human behavior.

While primate studies have begun to document the importance of analyzing behavior at the level of social relationships (e.g., Hinde 1983a), this research has also raised some intriguing problems. One concerns the quantitative approach to data analysis. In primatology as in most behavioral studies, scientists attempt to test hypotheses with quantitative data drawn from as many interactions as possible. Typically, this involves summing data from all observed instances and giving equal weight to each one. However, such quantification may be misleading if behavior is in fact best analyzed in terms of long-term relationships because single interactions can have unique, pivotal effects on the future of a social bond

(Simpson 1973a). Consider, for example, the hypothesis that primates groom at least in part to increase the probability of subsequent support in an alliance. By this reasoning, animals should attempt to groom often with high-ranking individuals, since it is these individuals that potentially offer the greatest support in alliances (Seyfarth 1977). When the hypothesis is tested, however, it is often found that animals do indeed groom those of high rank often, but that high-ranking animals rarely reciprocate as predicted (e.g., Fairbanks 1980; Silk 1982). Does this mean that the hypothesis is disproven? Not necessarily, because high- and low-ranking animals do not benefit equally from forming alliances with each other. It is entirely possible that one crucial act of reciprocity from a high-ranking animal is sufficient to elicit grooming from low-ranking animals for days or months thereafter. Primatologists, and any others interested in social relationships, must thus steer an uneasy course between the demands of quantification and the need to recognize that each interaction can potentially make a unique contribution to a particular social relationship.

Primate Societies thus has two goals: to provide a compendium of facts, and to offer a conceptual framework that challenges many of the assumptions currently held by anthropologists, biologists, psychologists, and those in the biomedical community. As noted earlier, scientists are drawn to primates becase they are the animals most like ourselves. In this respect primates provide a unique opportunity to study, indirectly, the ways in which our own behavior—even those aspects of it that we think of as uniquely human—reflects our evolutionary heritage.

Danger of Extinction

The day is long past when a book on primate behavior could be written without some discussion of the very real danger facing many primate species in their natural habitat. At present, approximately 50% of all nonhuman primate species are threatened, and, unless there are major changes in human population growth and habitat use during the next 20 years, many primate species will be lost forever (chap. 39).

Although primates are fascinating in their own right, their preservation is not simply an aesthetic concern. Primate conservation is linked inextricably to the preservation of tropical forests, whose destruction threatens the economies and climates of countries the world over. The world's tropical forests are being eliminated an an ever increasing rate, and most will probably disappear before the end of this century. The adverse effect of habitat destruction on the economies of developing countries is already obvious, and erosion, soil depletion, drought, and food shortages are increasingly common. Even the developed countries of North America, Europe, and Asia, however, will soon feel the repercussions of tropical forest destruction. Not only do we depend on tropical forests as

FIGURE 1-1. A Kenyan assistant, Raphael Mututua, recording the behavior of yellow baboons in Amboseli National Park, Kenya. (Photo: Jeanne and Stuart Altmann)

a genetic source of domestic and medicinal plants, but there is also increasing evidence that the elimination of tropical forests will have a severe, and perhaps irreversible, effect on global climate and rainfall patterns. Ultimately, the preservation of tropical forests depends critically on large infusions of economic support from the developed countries of the world, support that currently is given reluctantly and that is hopelessly inadequate to the task. If *Primate Societies* goes some way toward helping to pique and maintain an interest in the conservation of primates and their habitats, it will more than have served its purpose.

FIELD RESEARCH

Due to limitations of space, we have chosen not to include a chapter on field methods in *Primate Societies*, but since methodological issues continue to play an important role in the field, we discuss them briefly here. (For a more detailed treatment, see Altmann 1974 and Dunbar 1976.)

Habituation and Recognition

At the beginning of any study the field primatologist faces a number of problems. He or she must locate subjects, habituate them to the presence of humans without disrupting their natural behavior, distinguish one group from another, and, if possible, learn to recognize ani-

mals individually. These early stages of fieldwork often represent a low point in morale and productivity. To begin with, habituating monkeys to close-range observation is extremely frustrating and dull. Little can be accomplished if your subjects vanish whenever you approach, although good work on vegetation, fecal samples, and warning cries has been conducted during this period. When the observer is finally lucky enough to get close to the subjects, attempts to recognize animals individually can create further frustration and anguish. In some studies, of course, the difficulties of individual recognition have been overcome by capturing and marking animals, while observers of species such as chimpanzees and gorillas report that their animals can be recognized individually as easily as humans, even after only one or two days' observation. For the most part, however, primatologists give thanks for cuts on the ear, misshapen noses, wounds that leave scars, and other abnormalities that facilitate recognition until, after months of frustration, individual monkeys are distinguished almost as easily as individual humans.

Observational Sampling

Once a group of primates has been habituated and the animals can be recognized individually, the observer is ready to begin sampling their behavior (fig. 1-1). Underlying such observations is a fundamental principle: one cannot record everything. Even in small groups, social behavior or feeding can occur in rapid sequences, and the observer will often want to gather carefully limited data on the durations of events, intervals between events, or sequences of events. From a large population of many behavioral acts, the observer must therefore sample a small proportion and design rules for sampling to yield data that comprise a representative, unbiased subset of all the events that have occurred.

Hinde (1973), Altmann (1974), and Dunbar (1976) discuss and compare the many different ways in which observers can sample behavior, and readers are referred to these papers for more detailed discussion of how different scientific questions or different field conditions may require different sampling methods. As an introduction to the data presented throughout *Primate Societies,* we briefly describe here perhaps the three most common methods of observational data collection.

Instantaneous Sampling. This method, also called point sampling, is used to record one or more classes of behavior at a predetermined instant in time. For example, suppose an observer wanted to determine whether individuals in a group of black-and-white colobus monkeys showed a particular pattern of feeding and resting throughout the day, perhaps feeding early in the morning, late in the afternoon, and resting at noontime. A test of this hypothesis using instantaneous sampling would proceed as follows. First, decide the time interval at which monkeys will be sampled, for example, one individual every minute. Second, produce a randomized list of the individual monkeys to be sampled. Third, beginning at the earliest light of dawn, locate the first monkey listed and record its activity at the instant that the minute changes or over a predetermined, brief interval (e.g., 5 sec) beginning on the minute. Then, repeating this procedure for the next monkey on the list, continue until all subjects have been sampled equally often for a specified length of time.

If this protocol is repeated for a number of days, and if sampling sessions are varied so that they include an equal number of sessions from dawn to dusk, the observer will find it fairly easy to compare, for each individual, the proportion of time spent in different activities at different times of day. Instantaneous sampling is often useful because it allows a large number of individuals to be sampled daily, and it minimizes the length of time between successive samples on the same individual. For example, if an observer spends 5 hours with a group of 30 animals, and carries out instantaneous sampling on all of them, he or she may well obtain 4 to 5 samples on each individual in a day, with no more than 1 hour between successive samples on a given animal. Thus short-term changes in behavior, such as those occurring during the female reproductive cycle, can be documented more readily by instantaneous sampling than by other methods. The value of instantaneous samples is limited, however, because it provides no information on sequences of interactions or their duration.

Focal Animal Sampling. Suppose one wanted to test the hypothesis that genetically related individuals were more likely than others to groom within 1 minute after an aggressive interaction. Instantaneous sampling would clearly be inadequate to address this question, but an appropriate method could be devised if the observer simply increased the duration of time during which data were recorded. For instance, an observer might locate the first animal on the list and then follow that individual, recording all its activities, not for an instant but for some predetermined time, say 5 minutes, 10 minutes, 1 hour, or even a day. Such long-term follows of a preselected individual are an example of focal animal sampling, perhaps the most widely used method in the study of primate behavioral ecology. Focal animal sampling is an extremely versatile technique because the data it produces can be used to answer a variety of questions about frequencies (how often a behavior occurs), rates (how often a behavior occurs per unit time), sequences (how often behavior A follows behavior B), and durations (how long a behavior typically lasts). In the example cited above, an observer could not only test the original hypothesis but could also determine whether rates of aggression were higher among kin than among nonkin, whether fighting bouts among nonkin lasted longer, or whether aggression among kin was preceded by different events than aggression among nonkin. One weakness of focal animal sampling, however, results from the time devoted to each sample. It is rarely possible to sample many animals on a given day, and considerable time may elapse between successive samples on the same individual. Short-term changes in individual behavior may thus come and go between samples and never appear in an observer's records during a particular period.

Ad libitum Sampling. A variety of conditions may conspire to make regular, ordered sampling of behavior by either instantaneous or focal animal sampling impossible. In dense vegetation, for example, it may prove extremely difficult to find "the next subject" on an observer's list, and all primatologists are familiar with the mysterious process that causes subjects to disappear from sight shortly after a focal sample on them has begun. More important, some extremely significant behaviors such as copulation may be so rare that most instances would go unrecorded even by regular observation. To overcome these limitations, a variety of ad libitum sampling techniques, each designed to supplement data gathered by more systematic methods, have been developed by different observers. Each method of ad-lib sampling has its own assets and limitations, and each reflects the different problems posed by work on different species.

For example, after 18 months in the field an observer may return saying that copulations were extremely rare, that they never appeared in her systematic samples, but that she recorded them whenever she saw them. What conclusions can be drawn from data gathered in this way? First, because copulation is not always conspicuous, the data do not allow her to measure copulation rates quantitatively, although she can say something about the frequency of copulation relative to other behaviors (it is rare). Even though the observer spent, for instance, 1,500 contact hours with a group and saw 67 copulations, we cannot conclude that copulations in this species occur at the rate of one every 22.4 hours. Since the observer probably missed many events, the true rate of copulation is almost certainly higher than her estimate. On the other hand, if the same data referred to prominent, conspicuous behavior such as intergroup fights, ad-lib data could be used to calculate rates of occurrence, since the observer probably obtained a complete record of the behavior every time it occurred.

Although data gathered by ad-lib sampling cannot always be used to estimate rates of behavior, they can be extremely useful in other sorts of analyses. For example, an observer may have been interested in the formation of alliances—when two individuals join together in directing aggression against a third. Alliances were rare during his study, but he noted the participants in every alliance he observed. If he was relatively certain that the likelihood of observing the alliance was unaffected by the identities of the participants, then ad-lib data could be used to show, for example, that alliances among kin were more successful than those among nonkin. The data could also be used to show that alliances among males occurred primarily in competition for females, while alliances among females occurred primarily in competition for food, since there is no reason to believe that the visibility of behavior in these two contexts would be different for the two sexes. Finally, the observer's data could also be used to show that, among females, most alliances were formed with the highest-ranking individual, provided that the observer could be confident that such alliances were not somehow more conspicuous than others and hence more likely to have been observed.

This brief discussion can only hint at the variety of ways in which ad libitum sampling can be used and the many precautions one must take in each case. Such relatively unsystematic sampling has played an important role in field primatology because, as described earlier, both primate habitats and the animals themselves do not always lend themselves to carefully planned, faultlessly executed sampling schedules. The careful use of less ideal methods such as ad-lib sampling is likely to become even more important as research increasingly focuses on those species that have traditionally received less attention because they are difficult to study.

Field Experiments

Field experiments have always played a major role in research on birds and nonprimate mammals, but until recently such experiments were rarely used by primatologists, for a number of different reasons. Most important, the same logistical factors that prevent observation at close range can discourage any thought of experimentation in the field. Accounts of the abnormal behavior that results when primates interact with humans in zoos or wild animal parks have encouraged the belief that behavior will quickly be distorted if the animals interact with their observers in any way. These considerations are certainly understandable and important, but it has also become clear during the past ten years that, with suitable precautions, well-controlled field experiments can be conducted on primates without distorting their behavior and that, as with other species, such experiments can provide new insights into behavior that cannot be obtained through observation alone.

Among primatologists, Hans Kummer and his colleagues were pioneers in the integration of observational and experimental field techniques (e.g., Nagel 1973; Bachmann and Kummer 1980; Sigg 1980; Sigg and Falett 1985). Beginning with detailed observational data on hamadryas baboons, these investigators formulated hypotheses about the mechanisms underlying food gathering, partner preferences, and social structure. They then tested such hypotheses by experimentally manipulating food availability or social competition, either among their free-ranging subjects or in large enclosures. A particularly ingenious experiment examined the mechanisms underlying social structure in olive and hamadryas baboons by transplanting males and females from one subspecies to another (Nagel 1973; see also chap. 10).

A second group of primatologists, following a long tradition of field experiments on bird vocalizations, has used portable tape recorders and loudspeakers to conduct playback experiments on free-ranging primates (e.g., Gautier 1974; Waser 1977b; Robinson 1979c; Seyfarth, Cheney, and Marler 1980a; Sekulic 1982c; Gouzoules, Gouzoules, and Marler 1984; Mitani 1985c). Such work was originally designed to provide new information on the use of vocalizations, and in this area it has made important contributions (chap. 36). Perhaps more important, however, vocal playbacks offer an opportunity to test hypotheses about an animal's social or ecological knowledge because they allow an observer to mimic the presence of certain individuals or the occurrence of certain events under specified conditions. Thus Seyfarth and Cheney (1984a), for example, used playback experiments to test the hypothesis that vervet monkeys remember past interactions and in some cases behave cooperatively only to those who have behaved cooperatively toward them in the past.

These are only a few examples of the ways in which field experiments can be used to supplement data collected by observations. Primatologists have also re-introduced to a group female baboons with timed-release hormone implants (Saayman 1968), introduced novel objects to test for learning and cultural transmission in macaques (Menzel 1966), and exposed chimpanzees (Kortlandt 1972) and vervet monkeys (Cheney and Seyfarth 1981) to stuffed predators. Thus although most research on primates relies almost entirely on observation, the methods that can be useful in the field are clearly limited only by the investigator's ingenuity.

PART I

Evolution of Diversity

When DeVore (1965b) reviewed the ecology and social organization of primates, data on group size, population density, and home ranges were available for just five species (mountain gorillas, lar gibbons, hanuman langurs, savanna baboons, and mantled howler monkeys). A glance at the chapters in part 1 shows the subsequent explosion in our knowledge of primates in the wild. Open-country species have now been studied in a variety of habitats, while many species that are harder to observe have been the subject of one or more long-term investigations. The surge of knowledge has been exciting but all too daunting for those attempting to keep up with the literature. The following chapters therefore offer a detailed review of ecology, reproduction, and social organization throughout the primate order.

Each chapter discusses a species or group of species that are closely related to each other. No attempt has been made to give equal space to all species. Some groups are discussed briefly because they still have not been studied extensively. Others, such as baboons and macaques, are given little space for the opposite reason: they are so well-known that material on them makes a substantial contribution to chapters in parts 2 through 4. The relationship between taxonomic groups is summarized in table 0.1. In addition to taxonomic considerations, the grouping of cercopithecine species of Old World monkeys in chapters 8, 9, and 10 takes similarities in social organization into account. For example, vervet monkeys *Cercopithecus aethiops* live in multimale groups. They are therefore discussed in chapter 10 along with other species living in multimale groups, rather than in chapter 8 with the majority of *Cercopithecus* species.

A word is needed on the scientific and common names used throughout the book. There are numerous systems of primate classification, each with their own merits. We therefore invited authors of the chapters in part 1 to use whichever scientific and common names they felt to be most appropriate for the particular species that they discuss. We then combined these names into a single taxonomy, which has been followed throughout the book and is shown in the appendix. Compared to other systems of nomenclature, our taxonomy tends to "lump" species, some of which may come to be firmly separated in the future.

In addition to its systematic review, each chapter in part 1 discusses a special topic. The special topic is given in the subtitle of each chapter, and in the text it is marked with a black diamond. Topics were selected because of their particular importance to the species under consideration, but all are relevant to other primates, as well as to other animals. Some topics highlight unusual types of social relationship, such as the unexplained tendency for female lemurs to dominate males (chap. 3) or the remarkable degree of cooperation among male chimpanzees (chap. 15). Others draw attention to general problems that are well studied in the group in question. Examples include intraspecific variations in grouping patterns (chap. 6) and the causes of infanticide (chap. 8). The special topics deal largely with important problems that we hope will be understood more clearly soon.

The compilation of results from so many field studies highlights not only the solid achievements of primatology over the last twenty years but also the gaps to be filled during the next few decades. Many of the chapters in part 1 refer to threatened species and habitats, and chapter 39 shows clearly why for many populations more field research is needed soon if it is to be done at all. As scientists we might prefer to tackle easily soluble problems rather than deal with the small sample sizes and uncertain observations of elusive primates. But as recorders of nature we must choose now whether to describe the unknown before it vanishes or let it remain unknown forever. It is extraordinary that serious studies of lowland gorillas are only now beginning; that a striking new species of monkey (*Cercopithecus salongo*) was discovered in Zaire in the 1970s; that there are numerous large species that have never been habituated, much less observed systematically, such as various colobines, guenons, baboons, macaques, and uakaris. Adventurous investigators will find even more challenges among the smaller and nocturnal primates.

The unknown must be explored if we are to put primatology on a firm foundation because what we have yet to learn will almost certainly lead to changes in our ideas about the evolution of behavior. The study of primates is a young field, and the following chapters show plainly what a large proportion of knowledge rests on weak foundations. For example, chimpanzees are one of the better-studied species, yet detailed data come from only two sites in the extreme east of their geographical range. It is clear that their social organization is different in other areas, but we do not yet know what aspects of the behavior are species-specific. Similar examples are legion. We therefore hope that the primate diversity and new research opportunities shown in part 1 will stimulate action as well as thought. Many species are threatened with extinction in our lifetimes. Now is the time to study them and to save them.

TABLE O-I The Nonhuman Primates

Suborder Infraorder Superfamily Family Subfamily		Number of Species
Prosimii	Prosimians	
Lemuriformes		22
Daubentonioidea		
Daubentoniidae	Aye-aye	1
Lemuroidea		
Cheirogaleidae	Mouse lemurs, dwarf lemurs, and forked lemur	7
Indriidae	Indri, sifakas, and woolly lemur	4
Lemuridae	Lemurs	7
Lepilemuridae	Gentle lemurs and sportive lemurs	3
Lorisiformes		13
Lorisidae		
Galaginae	Bushbabies	8
Lorisinae	Lorises	5
Tarsiiformes		3
Tarsiidae	Tarsiers	3
Anthropoidea	Monkeys and apes	
Platyrrhini	New World monkeys	47
Ceboidea		
Callimiconidae	Goeldi's marmoset	1
Callitrichidae	Tamarins and marmosets	15
Cebidae		
Alouattinae	Howler monkeys	6
Atelinae	Spider monkeys and woolly monkeys	7
Cebinae	Capuchins and squirrel monkeys	6
Pitheciinae	Owl monkeys, titi monkeys, uakaris, and sakis	12
Catarrhini	Old World monkeys and apes	81
Cercopithecoidea		
Cercopithecidae		
Cercopithecinae	Macaques, guenons, baboons, patas monkeys, and mangabeys	41
Colobinae	Colobines	27
Hominoidea		
Hylobatidae	Gibbons	9
Pongidae	Great apes (orangutans, gorillas, and chimpanzees)	4

2 | Lorises, Bushbabies, and Tarsiers: Diverse Societies in Solitary Foragers

Simon K. Bearder

The division of living primates into prosimians (lemurs, lorises, bushbabies, and tarsiers) and simians, or anthropoids (monkeys, apes, and humans), is a reflection of major differences in their sensory anatomy and physiology. The anthropoid primates have occupied diurnal niches since an early stage of their evolution, and their sense organs and perceptual abilities are adapted accordingly. In contrast, all the prosimians, including those lemurs that are now diurnal, show the hallmarks of a long history of adaptation to nocturnal conditions: they have relatively large eyes, sensitive nocturnal vision, large independently movable ears, elaborate tactile hairs (vibrissae), and a well-developed sense of smell. These sensory specializations are accompanied by differences in the organization of the brain and by a marked contrast in social relations and systems of communication compared to anthropoid primates (Charles-Dominique 1978). Thus, the primate order includes an unusually wide array of social systems, both diurnal and nocturnal.

Although prosimians share a number of physical and behavioral characteristics, it is not clear how closely they are related. Lorises and bushbabies (lorisids) can be grouped with the lemurs as strepsirhines (or moist-nosed primates), while tarsiers show affinities with the remaining primates (the haplorhines, or hairy-lipped species). In common with anthropoids, tarsiers have a muscular upper lip associated with facial expressions; fine independent control of the fingers; and the presence of a fovea in the eye. All these are lacking in the strepsirhines. It is therefore normally thought that tarsiers and anthropoids are derived from a common ancestral stock that was diurnal, tarsiers having become secondarily adapted to a nocturnal life (Martin 1978).

Lorises, bushbabies, and tarsiers coexist with monkeys and apes over a large area of Africa and Asia. Unlike most monkeys and apes, but in common with other small nocturnal mammals, they are not generally seen in groups. It was once thought that the strong tendency for solitary foraging (to which there are now known to be exceptions), was associated with a correspondingly simple pattern of social organization (Crook and Gartlan 1966). Detailed field studies, however, have revealed an extraordinary diversity and complexity of types of social organization within this superficially uniform adaptation to the small-bodied, nocturnal niche. The nature of these social systems, their taxonomic distribution, and their occurrence in different ecological circumstances are still not well understood, but they provide considerable potential for further analysis of the ecological and phylogenetic causes of different types of primate sociality. In this chapter I review present knowledge of the ecological and anatomical adaptations that distinguish the living lorisids and tarsiers and categorize the principal types of social systems among nocturnal primates in general. This provides a basis for understanding the probable condition of ancestral primate species.

Surprisingly, despite their nocturnal habits, it is possible to study these animals in the wild with a considerable degree of success. The first systematic field study, undertaken in 1966, employed the simple but highly effective technique of mounting a flashlight at eye level and using its red or white light to observe the animals. Under good conditions the eyeshine from lorisids can be detected from 100 m or more. The subjects are undisturbed by red light and soon ignore the presence of an observer, with the result that they can sometimes be followed at close range (1–10 m) throughout the night (Charles-Dominique and Bearder 1979). The tarsiers, which lack a reflective eyeshine (*tapetum lucidum*), have been observed under seminatural conditions (Niemitz 1979, 1984) or located by their calls and followed in the light of a headlamp (MacKinnon and MacKinnon 1980b). Subsequent use of radio tracking and efficient techniques of live capture and marking now make it possible to identify and locate these animals with ease.

Throughout this review, data are taken from studies that span at least 12 months and often considerably longer. Information for separate species in each of the tables is based on a single study site unless otherwise stated. Differences in emphasis between the available field studies include general ecology (Barrett 1984; Charles-Dominique 1977a; Doyle and Bearder 1977; C. S. Harcourt 1984; MacKinnon and MacKinnon 1980b; Niemitz 1979); diet (Bearder and Martin 1980; Charles-Dominique 1974b; Harcourt 1980); locomotion (Crompton 1983, 1984); communication (Charles-Dominique 1977b; Clark 1978a; Harcourt 1981; Nash 1982); reproduction (Nash 1983); and social relationships (Bearder and Martin 1979; Charles-Dominique 1977a; Clark 1978a).

TAXONOMY AND DISTRIBUTION

The Lorisiformes and Tarsiiformes include three sub-families (lorises, bushbabies, and tarsiers) and at least sixteen species (table 2-1). The lorises are found in Africa and Asia. There are two species in Africa, the potto (*Perodicticus*) and the angwantibo (*Arctocebus*), which are partly sympatric. The potto is three to four times the weight of the angwantibo. The number of Asian lorises (*Loris, Nycticebus*) is still open to question, but the 3 forms commonly recognized at present are allopatric. Like the African species, they include a substantial range of body sizes (190–1,675 g). Bushbabies (*Galago*) are confined to Africa, where there are 2 large (>700 g) and 6 to 9 smaller species (<400 g), with as many as 3 species living sympatrically. There are at least 3 species of tarsier (*Tarsius*), all small bodied and living in different areas of Southeast Asia (fig. 2-1).

TABLE 2-1   Distribution, Body Weights, and Habitat Preferences for Lorises, Bushbabies, and Tarsiers

Taxonomic Name	Common Name	Distribution	Mean Body Weights (g) (sample size)	Preferred Habitats
Family Lorisidae Subfamily 1: Lorisinae[a] (lorises)				
Nycticebus pygmaeus	Pygmy loris	Vietnam	190–230 (3)	
N. coucang	Slow loris	S. E. Asia, Assam to Borneo	M: 679 (56) F: 626 (44)	Tropical rain forest with continuous canopy
Loris tardigradus	Slender loris	S. India, Sri Lanka	227–355 (10)	Tropical rain forest and woodland
Arctocebus calabarensis	Angwantibo	W. Africa between Niger and Zaire rivers	150–270 (30)	Secondary growth in rain forest; mainly tree-fall zones
Perodicticus potto	Potto	W. and central Africa, Guinea to W. Kenya	850–1,600 (33)	Primary/secondary tropical forests
Subfamily 2: Galaginae[b] (bushbabies or galagos)				
Galago demidovii[c]	Demidoff's	W. and central Africa; Uganda	46–88 (66)	Primary/secondary tropical forests
G. zanzibaricus	Zanzibar	E. Uganda, Kenya, and Tanzania	M: 159 (12) F: 136 (14)	Lowland/coastal tropical forests
G. alleni	Allen's	W. Africa between Niger and Zaire rivers	188–340 (17)	Primary rain forests—understorey
G. senegalensis[d]	Lesser	W. and Sub-Saharan Africa, Somalia, E. and S. Africa	M: 210 (21) F: 193 (14)	Woodland-savanna, riverine bush, forest fringes
G. elegantulus	Needle-clawed	W. Africa between Niger and Zaire rivers	270–360 (39)	Primary/secondary rain forest
G. inustus	Lesser needle-clawed	E. Zaire, W. Uganda	170–250	Tropical relic forest
G. crassicaudatus	Thick-tailed	E. and southern Africa	M: 1,510 (8) F: 1,258 (9)	Subtropical/tropical forests; woodland, riverine, and coastal vegetation
G. garnettii	Greater	E. Africa	M: 822 (8) F: 721 (5)	Tropical, temperate, and coastal forests
Family Tarsiidae Subfamily: Tarsiinae (tarsiers)				
Tarsius bancanus	Western	Borneo, Sumatra	M: 128 (21) F: 117 (16)	Primary/secondary tropical forest with dense understorey
T. syrichta	Philippine	Philippines	M: 122 (4) F: 110 (4)	Primary/secondary tropical forest
T. spectrum[e]	Spectral	Sulawesi (Celebes)	Similar to other tarsiers	Primary/secondary, montane, and mangrove forests; coastal and thorn scrub

SOURCES: Petter and Petter-Rousseaux 1979. Additional data on body weights come from Napier and Napier 1967, Kingdon 1974, and references listed in table 2.2.

a. Speciation in the Lorisidae may be more extensive than is indicated here. The weights of slow lorises, for example, vary between regions, from 360 to 1,675 g, and different gestation lengths have been recorded within species (*N. coucang, A. calabarensis*).

b. Different groups of bushbabies have been given generic status, but the divisions remain unclear and the single genus *Galago* is used here.

c. Probably two species, *G. demidovii* and *G. thomasi*.

d. Possibly three species, *G. senegalensis, G. moholi,* and *G. gallarum* (Olson 1979).

e. Possibly two species, *T. spectrum* and *T. pumilus* (Niemitz 1984).

FIGURE 2-1. *Tarsius spectrum pumilus* in its natural habitat in Sulawesi. This species (now thought to be best classified as *T. pumilus*, distinct from *T. spectrum*) is easily distinguished from *T. bancanus* by its facial features. (Photo: Carsten Niemitz)

The confusion regarding the taxonomy of these primates is largely a consequence of their secretive, nocturnal habits. Early taxonomists could not rely on direct observation of living animals to provide a starting point for separating species, and it is now becoming clear that for all three groups certain species have been erroneously lumped together. Separate species may look superficially similar, but their behavior is usually distinctly different, and such differences are now being used to help refine the taxonomy. *Galago crassicaudatus* and *G. garnettii*, for example, were considered to be a single species until recently. Yet they exhibit important differences in their contact vocalizations, locomotion, litter size, and method of infant carriage (C. S. Harcourt 1984), as well as in morphology (Dixon and van Horn 1977; Olson 1979). Similar differences exist within other species groups, and their taxonomic revision is far from complete.

ECOLOGY

Activity Patterns

All the species considered here are nocturnal. They generally leave their sleeping places after nightfall and return before dawn at times that are usually related to critical levels of light intensity. Movement between sleeping sites during the daytime has been seen in response to heavy rainfall or extremes of temperature, and diurnal foraging may occur when food is in short supply. In South Africa, for example, on the coldest nights in midwinter (minimum −5°C), when insects and gums are hard to find, lesser bushbabies (*G. senegalensis*) may return to huddle at a sleeping site up to 7 hours before sunrise, and at these times they are frequently active during the day (Bearder and Martin 1980).

Locomotion

The two subfamilies of the Lorisidae have distinct styles of locomotion. Lorises are slow-moving, quadrupedal climbers, while bushbabies and tarsiers are vertical clingers and leapers. These generalizations obscure a considerable variety of locomotor behavior within and between species (table 2-2). Lorises are incapable of leaping; they are specialized for "bridging" between branches that may be set at varied angles (fig. 2-2). The limbs and trunk are elongated, and there is considerable freedom of movement at the joints. The tail is reduced to a stump. Lorises can maintain a grip while remaining completely immobile for long periods, due to a specialized network of blood vessels in the wrists and ankles. Such peculiarities have been interpreted as part of a cryptic strategy that helps lorises avoid detection by predators.

Bushbabies and tarsiers are active and agile, with a maximum horizontal leap of around 5 m in the most specialized species (*G. senegalensis, G. elegantulus, Tarsius* spp.). They have powerful, elongated hind legs and a long tail, yet several species are predominantly quadrupedal runners rather than leapers (*G. demidovii, G. zanzibaricus, G. crassicaudatus*), and even on the ground they tend to run rather than hop bipedally as in the majority of species. The hands and feet (and the undersurface of tarsiers' tails, which are used as a prop) carry highly efficient friction ridges (dermatoglyphs). Some species (*G. elegantulus, G. inustus, G. garnettii*) also have pointed nails that can dig into bark to achieve a grip on broad, smooth tree trunks.

The habit of urine washing of the hands and feet is common to a number of lorisids but absent in others (*G. elegantulus* and lorises, except for *Loris tardigradus*). It is also shown by some lemurs (*Microcebus murinus*) and New World monkeys (*Aotus, Saimiri*). Dampening of the palms and soles in this way may disseminate olfactory signals, and it also appears to improve grip in the absence of sweat glands on the extremities (Harcourt 1981).

Diet

Lorisids consume a relatively high-energy diet that may include fruits, gums, and animal prey. Either fruits or gums may be ignored, but all species eat some animal matter. Leaves are never eaten, although sometimes they are licked to provide moisture and exudates.

An important factor relating to the diet of lorises is their unusually low metabolic rate, such that they require less energy than expected on the basis of their body weight (chap. 16). The basal metabolic rate of the potto and the slow loris is about 40% below the expected value (Martin 1984a), with profound consequences for their behavior.

Tarsiers are unusual among nocturnal prosimians because they eat exclusively animal food (90% arthropods, 10% vertebrates—birds, bats and snakes in the case of *T.*

FIGURE 2-2. Slow loris. (Photo: David Agee/Anthrophoto)

FIGURE 2-3. *Tarsius bancanus* eating a neurotoxic snake (*Maticora intestinalis*). (Photo: Carsten Niemitz)

bancanus) (fig. 2-3). Niemitz (1979) has suggested that the western tarsier occupies the ecological niche of a small owl in dense undergrowth, where owls are unable to fly. Indeed, tarsiers show striking convergence with owls: large eyes, absence of a *tapetum lucidum;* the ability to rotate the head to look directly backward; and the use of hearing rather than sight as the primary means of locating prey. In common with owls, they feed on a similar range of species, their locomotion is completely noiseless, and they employ an ambush type of predation while moving fairly close to the ground (Niemitz 1979). The spectral tarsier (*T. spectrum*) from Sulawesi, using its hands, will also grab at insects in the foliage. In this respect it is more like a bushbaby (MacKinnon and MacKinnon 1980b).

Gum feeding is a conspicuous characteristic of several lorisids, and it is linked with adaptations of the digestive tract and teeth (Hladik 1979; Martin 1979; Bearder and Martin 1980). Gum feeders use a specialized tooth-scraper or tooth-comb (procumbent incisors and canines of the lower jaw) to scrape hardened gum from the surface of trees, and it appears that gums are mainly digested in an enlarged caecum through the action of symbiotic bacteria (Charles-Dominique 1977a; Hladik 1979).

The ability to utilize gums (complex polymerized sugars with protein and trace minerals) has probably enabled these small primates to survive seasonal shortages of fruits and insects. In many habitats, particularly those away from the equator, such foods are periodically un-

FIGURE 2-4. Lesser bushbaby *Galago senegalensis moholi* feeding on gum (Photo: Simon Bearder)

available due to long- or short-term fluctuations in climate. At these times the animals are able to subsist on gum alone (e.g., *G. senegalensis, G. crassicaudatus;* fig. 2-4). However, gums are eaten throughout the year, and it is likely that their significant calcium content supplements the low levels of calcium in fruits and insects (Bearder and Martin (1980). Gums also play a role in helping reduce interspecific competition. Exudate feeding, including the use of gums, is also common among the small-bodied marmosets and tamarins (chap. 4).

Defense

Environmental hazards, including adverse climatic conditions and the presence of predators, have a major effect on the behavior of lorisids and tarsiers. Protection during the day is achieved in a number of ways. The angwantibo and Allen's bushbaby (*G. alleni*) cannot tolerate exposure to sunlight or heavy rainfall; their sleeping places are invariably sheltered. Other species (e.g., *G. demidovii*) build spherical leaf nests or group together when the weather is cold. Predation is countered by a combination of crypsis (camouflage) and direct protection behind a screen of vegetation. Lesser bushbabies, for example, may sleep on a platform of leaves in trees that are highly thorny, thus gaining protection from below and above. If, however, an attack is imminent, the animals will either escape rapidly from a relatively open sleeping site or defend a more sheltered site by biting and spitting.

At night a similar correspondence exists between avoiding detection by predators and responding in an appropriate way if detected, but some tarsiers appear to be remarkably unafraid of other species at night, and predation is possibly infrequent. The specialized locomotor behavior of the lorises undoubtedly avoids attracting attention, but it leaves them vulnerable should they encounter a predator at close quarters. Not surprisingly, these species display elaborate means of defense in such circumstances: the head is tucked beneath the chest, and they either lunge toward an arboreal opponent (potto and slow loris) or they bite a terrestrial adversary from an unexpected angle (angwantibo). (The back of the neck in the potto is protected by a complex scapular shield with a thick skin, long guard hairs, and projections of the cervical vertebrae; Charles-Dominique 1977a). In dire emergencies, even species living in the canopy will let go of the branch and fall to the forest floor. Conversely, bushbabies rely on their agility to escape from danger; they often retreat to a safe vantage point and give alarm calls that may promote a mobbing response in conspecifics nearby.

Interspecific Competition

A final well-studied facet of the ecology of these species concerns interspecific competition (cf. chap. 18). In parts of Africa where two or more lorisids live sympatrically, there is a clear partitioning of resources. This

TABLE 2-2 Comparison of Ecological Niche and Population Density among Lorises, Bushbabies, and Tarsiers

Species	Characteristic Locomotion	Use of Strata/Supports When Active	Diet[a]—% Contribution			Sleeping Sites	Pop. Density per km²	Sources
			Fruits	Gums	Animal Prey			
Peninsular Malaysia								
Nycticebus coucang	Slow climbing	Interconnecting branches and vines in canopy and understorey (0–20 m)	50+[b]	10–[b]	30+[b]	Tangled vegetation; many sites	20	Barrett 1984
Gabon, West Africa								
Arctocebus calabarensis	Slow climbing	Secondary undergrowth, especially tree-fall zones (0–5 m)	14	0	85 mostly caterpillars	Foliage with dense cover	<7	Charles-Dominique 1977a Charles-Dominique and Bearder 1979
Perodicticus potto	Slow climbing	Interconnecting branches and vines; prefers canopy (5–30 m)	65	21	10 mostly ants	Tangled vegetation; many sites	8–10	Ibid.
Galago demidovii	Fast moving: mainly quadrupedal	Dense secondary undergrowth or vine-curtins in canopy (5–40 m)	19	10	70 small insects	Spherical leaf-nest or dense vegetation; few sites	50–80	Ibid.
G. alleni	Active leaping	Undergrowth (1–2 m); prefers vertical supports	73 fallen fruit	0	25	Tree hollows; few sites	15–20	Ibid.
G. elegantulus	Running, leaping, and climbing	Canopy (5–30 m); can negotiate broad, smooth supports	5	75	20 orthopterans beetles	Branch fork in shelter of foliage; few sites	15–20	Ibid.
Kenya, East Africa								
Galago zanzibaricus	Fast moving; mainly quadrupedal	Ground to canopy (0–13 m); prefers undergrowth	30	0	70	Tree hollows; few sites	169	C. S. Harcourt 1984
G. garnettii	Quadrupedal walking and running; bipedal hopping	Undergrowth to upper canopy (5–20 m); prefers canopy	50	0	50	Tangled vegetation; hollows rarely; many sites	38	Ibid.
South Africa								
Galago senegalensis	Active leaping; bipedal hopping	Ground to upper canopy; prefers lower strata (0–4 m)	0	48	52 no vertebrates	Flat leaf-nest, tree hollow or branch fork in a thorn tree; few sites	31	Bearder and Martin 1979 Harcourt 1980
G. crassicaudatus	Quadrupedal walking and running; some leaping	Ground to upper canopy; prefers upper strata (4–12 m)	33 0[c]	62 41[c]	5 59[c]	Tangled vegetation or flat leaf-nest; few sites		Bearder and Doyle 1974 Harcourt 1980
Borneo								
Tarsius bancanus	Active leaping	Ground and undergrowth (0–2 m); prefers vertical supports	0	0	100 invertebrates and vertebrates	Open branch shaded by foliage	40–115	Niemitz 1984
Sulawesi								
Tarsius spectrum	Active leaping (less specialized than above)	Ground to canopy (0–9 m); uses vertical supports below 3 m when hunting	0	0	100 no vertebrates	Tangled vegetation or tree hollow; multiple exits; few sites	300–800	MacKinnon and MacKinnon 1980b

a. Proportions of foods in the diet are estimations based on analysis of stomach contents (West African species); fecal analysis (East African species); and direct observation (remaining species).

b. Few observations were made on this species.
c. Second study site.

is based partly on anatomical and physiological specializations (body size, gut anatomy, limb morphology), but it is due mainly to behavioral mechanisms, including food preferences and use of different forest strata and types of support. Ecological separation is well illustrated for the five sympatric lorisids of Gabon (Charles-Dominique 1974b, 1977a). Three species that share the canopy of the primary forest (pottos, dwarf bushbabies and needle-clawed bushbabies) specialize on fruits, insects, and gums respectively (table 2-2), while those in the undergrowth feed mainly on fallen fruit (Allen's bushbabies) or caterpillars (angwantibos). In East Africa, where the dietary overlap of sympatric bushbabies is considerable, greater bushbabies (*G. garnettii*) spend

TABLE 2-3   Reproductive Parameters for Lorises, Bushbabies, and Tarsiers

Species[a]	Birth Seasonality	Gestation Length (days)	Modal Litter Size (maximum)	Interbirth Interval (months)	Age at Sexual Maturity (months)	Longevity (years)	Use of Nest or Hollow	Method of Infant Carriage
Galago demidovii	None	111–14	1 (2)	12	8–10	13	Yes	Mouth
Tarsius spectrum	May, Nov–Dec	?	1	?	?	8–12	Yes	Mouth
T. bancanus	Seasonal	90(?)	1	?	13	8–12	No	Fur
Galago zanzibanicus	Feb–Mar, Aug–Oct	120	1 (2)	8 and 4[b]	?	?	Yes	Mouth
Arctocebus calabarensis	None	131–36	1	4–5[b]	9–10	11	No	Fur
Galago senegalensis moholi	Oct–Nov, Jan–Feb	121–25	2	8 and 4[b]	9–12	16	Yes	Mouth
G. alleni	None	133	1 (2)	12	8–10	12	Yes	Mouth
Loris tardigradus	Apr–May, Oct–Nov	163–67	1	?	?	12	No	Fur
Nycticebus coucang	None	193	1	12	?	16	No	Fur
Galago garnettii	Aug–Nov	130–35	1	12	12–18	15	Yes	Mouth
Perodicticus potto	Aug–Jan	170	1	12	18	15	No	Fur
Galago crassicaudatus	Oct–Nov	135+(?)	2 (3)	12	18–24	15	Yes	Mouth and fur

SOURCES: Van Horn and Eaton 1979; also, see table 2-2.
   a. Species presented in order of body size; see table 0-1.
   b. A postpartum estrus may occur.

most time in the canopy while the Zanzibar bushbaby (*G. zanzibaricus*) inhabits the undergrowth (C. S. Harcourt 1984). A similar relationship holds true in South Africa between *G. crassicaudatus* and *G. senegalensis*. Body size is an important factor; in each case the smaller species is better suited to the lower arboreal zones (see Crompton 1984).

Home Range and Population Density

Reported measures of home range size and population density for nocturnal species are difficult to interpret. Alternative methods of estimation can produce wide discrepancies, and the figures also vary in relation to the length of the study. One useful measure is the relative size of home ranges for individuals in a single study population, which provides a guide to their social organization. The reliability of population densities in table 2-2 cannot yet be assessed; they are presented here purely to reflect current opinions.

Despite the lack of undisputed figures, there is little doubt that these small secretive species are among the least threatened of any primates. Nevertheless, they are vulnerable in the face of extensive destruction of their habitats. The slow loris (*Nycticebus coucang*) in peninsular Malaysia is known to be adversely affected by even moderate levels of selective logging (Barrett 1981), and, ironically, it is possible that some species will be forced to extinction even before they are recognized as distinct forms.

REPRODUCTIVE PATTERNS

The reproductive potential of a number of species is indicated in table 2-3. In most cases births generally coincide with warm/wet seasons, when food and cover are readily available. The shortest birth seasons occur at latitudes away from the equator, where birth peaks are a consequence of synchronized periods of estrus (e.g., 1 or 2 weeks in *G. senegalensis* and *G. crassicaudatus*). The breeding potential of males may also vary on a seasonal basis, with maximum testis size coinciding with estrus peaks (Charles-Dominique 1977a; Bearder 1984).

In lorisids, estrus is accompanied by rupture of the vulval membranes that normally remain sealed. Some species are able to breed more than once each year due to a postpartum estrus; when births are seasonal, the second birth is followed by a period of anestrus (e.g., *G. zanzibaricus, G. senegalensis*) or, alternatively, matings follow births without a break (*Arctocebus*). Olfactory signals promote an increased frequency of visits by males, and both *G. senegalensis* and *N. coucang* have mating chases in which females at the peak of estrus are followed by up to six males (Bearder and Martin 1979; Elliot and Elliot 1967). In most species the act of mating is brief, but copulations lasting more than 1 hour have been recorded for *G. crassicaudatus* and *G. demidovii*, which may represent a strategy of mate guarding (Clark 1984). Species with brief copulations have a copulatory plug, which may perform a similar function (e.g., *Arctocebus, T. bancanus;* van Horn and Eaton 1979).

As in general among mammals, the gestation length of lorisids increases with body size, but bushbabies have shorter gestation periods than lorises of the same size. Thus newborn bushbabies are poorly developed compared to lorises, and they are incapable of gripping firmly to the mother's fur. Oral transport of infants and the use of a nest or tree hollow occurs in all bushbabies (with the possible exception of *G. elegantulus*), but this is unnecessary in the case of lorises.

The reproductive characteristics of tarsiers are among the clues that suggest they share a distinct common ancestry with the anthropoids. At any given maternal weight, a tarsier or anthropoid produces a neonate three times heavier than a lemur or lorisid of comparable size (Martin 1978; chap. 16). The gestation length of tarsiers is uncertain, but there are indications of variations between species. Infant *T. bancanus* are carried on the fur, and they are not placed in a nest or hollow. *T. syrichta* and *T. spectrum,* on the other hand, exhibit oral transport of infants and use tree hollows, which suggests that their gestation periods may be shorter.

Contrary to the general rule among mammals (Clutton-Brock and Harvey 1983), litter size in this group does not decrease as body size increases. The largest species (*G. crassicaudatus*) may have triplets, while the smallest (*G. demidovii, G. zanzibaricus,* and *Tarsius* spp.) regularly have singletons. There is, however, an interesting relationship between reproductive output and environmental conditions. Those species with greater rates of reproduction (higher litter size and/or annual birth rate) live in habitats that appear to have irregular fluctuations in food supply and environmental hazards. These conditions are supposed to favor a high reproductive rate (r-selection) whereby individuals can take advantage rapidly of temporary resource abundance (Martin 1979, 1981a). Thus, species living in relatively dry or cold climates seem to have relatively high reproductive potential (Nash 1983). Similarly, the forest-dwelling *Arctocebus* is almost totally confined to tree-fall zones that are unpredictable in occurrence and relatively short-lived; it can give birth at least twice as often as sympatric lorisids. Where two species share the same unpredictable habitat, they both have a high reproductive potential (e.g., *G. crassicaudatus* and *G. senegalensis*). In this case the typical mammal rule is followed: the smaller species breeds faster, perhaps because it is more vulnerable to predators and climatic extremes.

SOCIAL ORGANIZATION OF *G. senegalensis*: A CASE STUDY

Reference is often made to social animals when what is actually meant is group-living or gregarious species. In fact all mammals exhibit social behavior in relation to mating and rearing, and those with solitary foraging hab-

its may require elaborate means of communication to coordinate their social activities. Without direct contact, nocturnal prosimians are able to monitor by sight (Pariente 1979), sound (Petter and Charles-Dominique 1979), and particularly by smell (Schilling 1979), details concerning conspecifics. In may cases they also meet at night, with bouts of mutual inspection, allogrooming, and play, or they sleep huddled together during the day.

The surprising complexity of social systems in solitary foragers is well illustrated by the lesser bushbaby (*G. senegalensis;* fig. 2-5). This species has been studied in an area of acacia savannah in South Africa where all individuals entering a 1 × 1 km site were marked, and some were followed from birth to adulthood with the aid of radio tracking (Bearder and Martin 1979). The maintenance and dynamics of the social system can be understood by tracing the social development of individuals.

Infants (usually twins) are born in a nest or in a hollow. At night the mother forages alone, but she regularly returns to nurse the young, which she may move from one retreat to another. After a few days the infants are "parked" in a tree near the nest, and the mother's visits become less frequent as the young mature.

Juveniles may follow an adult, but they usually forage alone. They gradually increase their range of movement within the area occupied by the mother, which she defends from encroachment by neighboring females and juveniles by scent marking, calling near boundaries, and fighting if necessary. Once the offspring reach maturity, their ranging patterns change in relation to their sex.

Maturing daughters (200 days) adopt territorial behaviors toward strange females and young, and their movements tend to become separate from those of the mother. However, they are often unable to establish separate territories because of the presence of aggressive female neighbors. They continue to have regular amicable contacts with the mother (sleeping together on 40% of days observed), and related females may rear offspring in the same nest. If, however, an adjacent territory becomes available (following the death of the resident), either the daughter or the mother will make an abrupt shift into the new area, and their relationship soon becomes one of mutual intolerance.

At puberty (300 days), males begin to dominate and court adult females; yet providing they remain subordinate, they are not excluded from the areas occupied by older males. Consequently at this stage males are free to extend their movements in any direction, and their ranges are unusually large (mean area 15.9 ha, compared to 6.7 ha for females). Most males make a prereproductive migration away from the area of their birth. These migrations are remarkably stereotyped and involve traveling either due east or due west for two or three nights before settling in a new area 1 or 2 km from their origin. The migrations are known to result in a two-way transfer

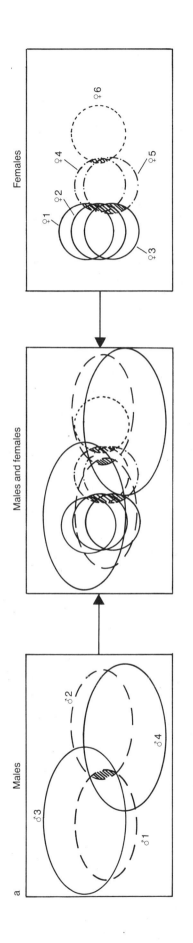

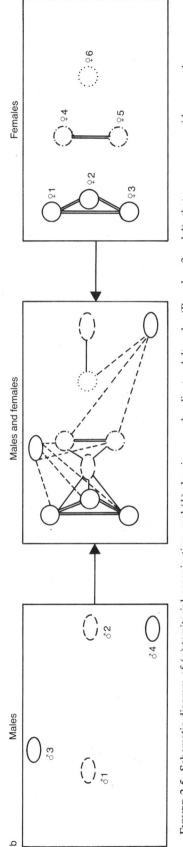

FIGURE 2-5. Schematic diagram of (*a*) territorial organization, and (*b*) sleeping associations in *Galago senegalensis moholi*. (*a*) Six females are depicted (circular territories), including two matriarchies (1-2-3, 4-5) in which related females frequently sleep together. Two A-males (1 and 2) exclude each other from territories within which they have prior access to females. They tolerate the presence of subordinate adult males (B-males, 3 and 4), that range over a wide area and occasionally sleep with females when the A-male is elsewhere (see Bearder and Martin 1979). (*b*) Two parallel lines = frequent association; single line = regular association; dashed line = occasional association.

of males between local populations, there is no aggression toward males prior to their departure, and incoming males are able to establish contacts with resident females—all of which suggests that an important function of migration is the avoidance of inbreeding (chap. 21).

The key to understanding relationships between males lies in the recognition of two classes of sexually mature adults: (1) older, heavier individuals (A-males; mean weight 226 g) that have preferential social contacts with females (sleeping associations; 54% alone; 46% in groups); and (2) younger, nonterritorial adults that are similar in terms of overall size, but lighter in weight (B-males; mean weight 211 g), and that have less frequent contact with females (sleeping associations: 84% alone; 16% in groups). The territories of A-males (mean area 11 ha) are centered over those of one or more females. Other A-males are excluded, but B-males are simply chased from the immediate vicinity. B-males, in turn, displace one another in an age-graded hierarchy. If an A-male disappears, he is quickly replaced by the highest ranking B-male, who becomes territorial.

The pattern of female territoriality ensures that single females or kin groups (matriarchies) are fairly evenly spaced out in suitable areas of vegetation. The size of their individual territories reaches a maximum where the density of the vegetation is low (12 ha at 29% cover,

compared to 5 ha at 55% cover). The distribution of male ranges is strongly influenced by the distribution of females, and male range size bears no obvious relationship to environmental resources. The fact that A-males exclude only males of a similar high rank confirms that they are not defending food resources or refuges. However, mutual exclusion between A-males prevents them from establishing privileged bonds with females in neighboring areas that appear to facilitate acceptance of a male by the female during estrus. Their territoriality can, therefore, be interpreted as a strategy that ensures mating access to females without wasting energy on subordinates. B-male may eventually become A-males; until then they attempt to find and mate with estrus females that are temporarily unguarded by the resident A-male.

SOCIAL SYSTEMS OF NOCTURNAL PRIMATES ♦

Measures of sociality among lorisids and tarsiers are compared in table 2-4. Nine well-studied species are divided into three types on the basis of differing patterns of range overlap among adults. Using these limited criteria, further categories are apparent among other nocturnal primates and are summarized to provide a background for discussing the social behavior of ancestral species (fig. 2-6).

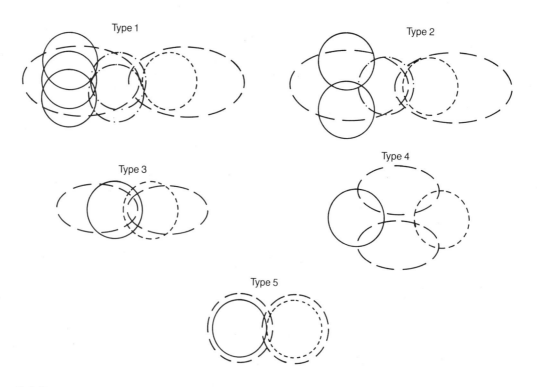

FIGURE 2-6. Patterns of territorial overlap in nocturnal primates: (type 1) *Galago alleni;* (type 2) *Perodicticus potto;* (type 3) *Tarsius bancanus;* (type 4) *Microcebus coquereli;* (type 5) *Lemur mongoz* (?). Circles represent female ranges, and ovals male ranges, except in type 5, where both sexes are portrayed with circles.

Type 1: Bushbabies, Excluding G. zanzibaricus. No two species listed in table 2-4 are the same in all aspects of their social behavior. Type 1 social systems, as illustrated above by *G. senegalensis,* are characterized by the fact that (1) adult male ranges are larger than those of females, and (2) matriarchies are present (related adult females with overlapping ranging areas). The predominant patterns common to these species are as follows (adapted from Charles-Dominique 1978):

1. The female territory is sufficiently large to allow feeding throughout the year. Where the habitat is homogeneous, each female has a range of approximately the same size. Territory ownership is signaled to conspecifics by scent marks and vocalizations; only adult females are repulsed. At first the adult daughter may share her mother's territory. Later there is less overlap, but related females may continue to sleep together and often share critical resources (e.g., highly productive feeding areas).

TABLE 2-4 Variability within Selected Parameters of Social Life among Lorisids and Tarsiers

Species	Study Habitat[a]	Range Size (hectares)	Pattern of Range Overlap[b] M/F	F/F	M/M	Matri-archy (size)	Social Contacts at Night	Adult Sleeping Association	Maximum Observed Sleeping Group and Composition	Mating Relationships	Sources
			TYPE 1 (male ranges larger than those of females; matriarchies present)								
Galago alleni	A	F: 8.0–16.0 M: 30.0–50.0	P	P	A	Yes (3)	Year-round visits by males to females	Males alone; females in groups	6—females and young	Polygynous	Charles-Dominique 1977a
G. senegalensis moholi	B	F: 4.4–11.7 M: 9.5–22.9	P	P	P	Yes (3)	Year-round visits by males to females; A and B males with overlapping ranges	Alone or in groups, males never together	6—male, females, and young	Polygynous; females may attract more than one male	Bearder and Martin 1979
G. demidovii	A	F: 0.6–1.4 M: 0.5–2.7	P	P	P	Yes (3)	As above; juveniles tend to follow adults while foraging	Males alone; females often in groups	10—females and young	As above (?)	Charles-Dominique 1977a
G. garnettii	C	F: 11.6 M: 17.9	P	P	P	Yes (2)	Year-round visits by males to females; A and B males with overlapping ranges	Alone	3—female and young	Polygynous	C. S. Harcourt 1984
G. crassicaudatus	D	F: 7.0 M: 10.0(+)	P	P	P	Yes (3)	As above; young often travel with mother	Alone or in groups	6—male, females, and young	Polygynous; females may attract more than one male	Bearder and Doyle 1974
			TYPE 2 (male ranges larger than those of females; matriarchies absent)								
Perodicticus potto	A	F: 6.0–9.0 M: 9.0–40.0	P	A	A	No	Little direct contact; no loud calls: indirect olfactory monitoring; young follow mother	Alone	2—female and young	Usually polygynous	Charles-Dominique 1977a
			TYPE 3 (male and female ranges coincide—nongregarious)								
Tarsius bancanus	E	F & M: 1.0–2.0	C	A	A	No	Little direct contact; pairs occasionally move together; shared territorial marking	Usually alone	2—female and young	Usually pairs; sometimes polygynous(?)	Niemitz 1984
Galago zanzibaricus	C	F: 1.6–2.6 M: 1.9–2.9	C	C (?)	A	? (2)	Males associate with single female or female pair; regular contact calls	Male and female(s) usually together	6—male, females, and young	Monogamy and bigamy indicated	C. S. Harcourt 1984
Tarsius spectrum	F	F & M: 0.9–1.1	C	P	P	Families	Pairs forage close together; auditory, visual, and tactile contact common; complex synchronized duets; pairs may tolerate adult offspring	Usually together	6—males, females, and young	Probably monogamous	MacKinnon and MacKinnon 1980b

a. Study Habitat: A = primary/secondary rain forest, Gabon, W. Africa; B = acacia savanna, Transvaal, S. Africa; C = coastal forest, E. Africa; D = subtropical riverine forest, Transvaal, S. Africa; E = primary/secondary rain forest, lowland Sarawak; F = coastal forest and scrub, N.E. Sulawesi.

b. A = overlap absent; P = overlap present; C = ranges coincide.

Cooperative defense by females has not been observed.

2. Male territories (A-males) overlap those of one or more females. There is no relationship between male territory size and dietary requirements.

3. Young males are nonterritorial (B-males). They usually emigrate at puberty and either become floaters or settle in a new area, thus gaining access to females but avoiding direct contact with other males.

4. A special bond based on familiarity links a territorial male to his female(s). Male stability and territoriality depends on the presence of females.

5. The A-male territory is signaled by scent and vocalizations. Only adult males are repulsed. Where B-males are tolerated, the degree of tolerance varies between species (reaching a maximum in *G. crassicaudatus*).

6. Territoriality is exhibited throughout the year, even when reproduction is seasonal.

7. Male-male and female-female territories (between matriarchies) have small zones of overlap (interaction zones) where territorial signals are exchanged.

8. Male-female relations are stable throughout the year, based on regular visits that involve direct contact, contact from a distance (7 to 15 distinct calls), and indirect contact (3 to 5+ scent marks).

9. During the night the animals are most often alone (solitary phase), but direct contact may be prolonged (several hours). By day, especially females and offspring may congregate at regularly used sleeping trees (gregarious phase). Adult males sleep alone more frequently.

10. The mother carries the infant in her mouth (*G. elegantulus* and *G. crassicaudatus* also exhibit fur carriage). Infants are deposited in the vegetation and recovered later or at dawn (baby parking). Special calls are associated with the retrieval of infants. Juveniles tend to follow the mother, or another adult, which provides opportunities for learning food habits, range use, and avoidance of dangers.

Type 2: Perodicticus potto and possibly other lorises. The social system of the potto is distinguished from the former type by the absence of matriarchies and by other factors that ensure that the animals remain inconspicuous and that energy expenditure is reduced to a minimum. Hence, vocal exchanges are limited; there are no alarm calls, and social interactions are generally indirect, with an emphasis on scent marking. Infants are invariably transported on the mother's fur, first ventrally and then dorsally. Baby parking is practiced, but pottos may also carry the infant or juvenile while foraging. Consequently, the mother does not have to travel far to retrieve her offspring, and at dawn she simply finds the nearest patch of dense vegetation in order to sleep. Juveniles follow the mother, and they undergo dietary conditioning whereby they learn to eat relatively unpalatable prey

(noxious and irritant forms) that are detected and captured with the minimum of movement. (If given a choice in captivity, such species are ignored—see Charles-Dominique 1977a.) Variations of type 2 systems appear to be shown by other relatively inactive species (e.g., *Arctocebus*), and they are clearly related to the unusual locomotor and metabolic peculiarities of the lorises.

Type 3: Tarsiers and G. zanzibaricus. Type 3 social systems are far from homogeneous, but they share the fact that male and female territories coincide (synterritorial). In some cases males may make brief excursions outside their territories to visit other females, but they usually restrict their attention to their synterritorial companions.

A male *G. zanzibaricus* usually shares his territory with one or two females. At night the adults move independently, but they maintain regular contact through calling, and they almost invariably sleep together during the day. The dispersal patterns of subadults are not fully known (C. S. Harcourt 1984; Harcourt and Nash, in press). It is not clear why *G. zanzibaricus* males associate more closely with females than do other bushbabies.

T. bancanus normally forms solitary ranging pairs, with subadults of both sexes leaving to establish territories elsewhere. Very little physical contact occurs between the male and female, either by day or at night. For example, they do not sleep or sit together, and they do not groom each other even when they mate. Seven distinct calls have been described, but this species is usually silent. Calls are never used in a territorial context. Olfactory marking is well developed; there are glands on the face, chest, and anogenital region, and the urine has a penetrating odor. Evidence of pairing includes the fact that both sexes scent mark at specific sites, and the female is known to cease marking following the death of the male (Niemitz 1979, 1984).

T. spectrum provides an interesting contrast. Here, physical contact is common and includes allogrooming, playing, tail intertwining, and huddling. Adults, juveniles, and mature offspring sometimes sleep together (maximum, four individuals), and they may forage close together and cooperate in defense of the territory. In addition to scent-marking behaviors, there are approximately 15 distinct vocalizations. The most striking are loud territorial songs, or highly synchronized duets analogous to those of gibbons, in which the male and female components are somewhat different (MacKinnon and MacKinnon 1980b, 1984). No reasons have been proposed to explain why *T. spectrum* is more gregarious than *T. bancanus*.

Types 4 and 5: Other Nocturnal Primates. The remaining nocturnal primates include the majority of lemurs (Cheirogaleidae, Daubentoniidae, Lepilemuridae, *Avahi laniger,* and *Lemur mongoz*—see chap. 3) and the night monkeys (*Aotus*). The social systems of many of these species are poorly known, but some of them fall within the broad categories described above (e.g., *Phaner furcifer,* type 3). Finer distinctions will undoubtedly be made as field studies progress, but at least two further categories are already apparent: type 4, in which male and female ranges appear to be largely separate from one another, as in the mouse lemurs, *Microcebus murinus* (Martin 1973a) and *Mirza coquereli* (Pages 1980), where limited resources *may* be a critical factor; and type 5 systems, where a male, female, and their offspring forage together as a cohesive group, as in *Lemur mongoz, Hapalemur griseus* (Charles-Dominique 1978), and *Aotus trivirgatus* (Wright 1978, chap. 5). Monogamous group foraging places these species apart from all other nocturnal forms. Interestingly, there are good reasons in each case to believe that in their evolutionary past their ancestors were at least partially diurnal (Charles-Dominique 1978; Martin 1979).

Reconstructing the Condition of Ancestral Mammals and Primates

There have been a number of attempts to reconstruct the probable structure and behavior of the common ancestor of modern mammals (Kay and Cartmill 1977; Lillegraven, Kielan-Jaworowska, and Clemens 1980; Eisenberg 1981; Martin 1973b; Charles-Dominique 1975). The consensus of opinion is that the ancestral placental mammals were probably small, shrewlike, terrestrial or semiterrestrial species, with clawed digits and nonopposable thumbs and first toes. The eyes were laterally placed and relatively small, but the olfactory apparatus was elaborate, and they probably had sensitive hearing and touch (vibrissae). Such creatures occupied the evolving tropical forests of the Cretaceous and Paleocene (90 to 65 million years ago), where they would have faced competition for insect prey from diurnal vertebrates (reptiles and primitive birds). It is therefore likely that they were nocturnal (Charles-Dominique 1975).

This reconstruction is based partly on the fragmentary fossil record (paleoanatomy and paleoecology) and partly on extrapolation from a broad comparision of living mammals. Particular weight is given to the smaller, forest-dwelling species in the tropics, which occupy what might be considered the niches of the Paleocene (Eisenberg 1981). In making generalizations about hypothetical ancestors, it is assumed that widespread characteristics of living representatives are more likely to represent archaic features retained from a common ancestor than those with a restricted distribution. From this it follows that the early placental mammals would have displayed the overall pattern of solitary foraging and a dispersed polygynous social network, which is retained by many of their descendants among diverse mammalian orders (Charles-Dominique 1978). Interspecific comparisons

have been used to infer other aspects of their behavior. Thus it is likely that urine and anogenital secretions played an important part in communication; ovulation may have been induced, in association with prolonged copulation; altricial infants would have been carried by mouth and protected in a nest; patterns of threat, grooming, vocalization, and parental care were, most probably, relatively simple (see Eisenberg 1981).

This provides a suitable starting point for considering the evolution of the earliest primates. By the latter half of the Eocene (40 to 50 million years ago), the lineages giving rise to the living Lorisiformes, Lemuriformes, Tarsiiformes, and Anthropoidea were established (Kay and Cartmill 1977). Their earlier common ancestors are thought to have differed from the mammalian forebears mainly in terms of a shift away from reliance on the olfactory and tactile sensitivity of the nose and muzzle toward increasing coordination between the eyes and hands. The typical primate characteristics of forward-facing eyes, grasping hands and feet, and flattened nails instead of claws cannot simply be a result of the selective influences of arboreal life, since many arboreal specialists are not adapted in this way (e.g., squirrels and tree shrews). More precisely, they probably evolved in relation to their visually directed predation on insects and other prey in the fine branches of the canopy and marginal growth of tropical forests (the visual predation hypothesis—see Cartmill 1972, 1974b, 1974c). Support for this interpretation is provided by convergent adaptations in the visual apparatus, typical of predators such as cats, and by similarities with the grasping abilities of other fine-branch dwellers, including climbing mice, opossums, and chameleons.

The living nocturnal prosimians fit this picture of the ancestral primate condition extremely well, but there is considerable uncertainty about the extent to which living species have remained physically and behaviorally conservative. The foregoing review of nocturnal primates reveals that their social systems fall into at least five general categories that bear some relationship to phylogenetic lineage and also to detailed differences in ecology. Perhaps the least specialized system, from which others may be derived, is that typical of bushbabies (type 1). This can be defined as a polygynous system including predominantly solitary foraging, direct and indirect social contact at night, partially gregarious sleeping habits, female philopatry, and male transfer between dispersed groups. Bushbabies lack the female dominance over males that is found in some nocturnal lemurs; they do not have the slow locomotion or unusually low metabolic rate of the lorises; and they show no signs of a previous history of diurnal activity, as in *Tarsius, Aotus,* and the gregarious nocturnal lemurs (Martin 1979; Charles-Dominique 1978). Furthermore, dispersed polygyny is common among other mammals that are thought to occupy "primitive" niches (e.g., Viverridae, Tragulidae, Manidae and Tenrecidae; Charles-Dominique 1978). Polygyny therefore seems more likely than monogamy to have been common in ancestral primates.

Similarities in the Social Systems of Nocturnal and Diurnal Species

The development of gregariousness and the movement into diurnal niches by the ancestors of anthropoid primates (and later by the diurnal lemurs) follow from an evolutionary increase in body size. Even among the nocturnal prosimians, gregarious foraging is most conspicuous in the larger representatives (*G. crassicaudatus, P. potto*) where the size of the mother relative to her offspring means that she is often able to carry them with her at night, and they later join her to forage as a cohesive unit. Such larger species eat a relatively high proportion of fruit, which allows grouping, as opposed to insects, which seem to demand a solitary foraging technique in nocturnal species. Charles-Dominique (1975) has argued convincingly that larger size and consequent changes in energy needs (including an increased ability to consume leaves) would have enabled arboreal mammals to overcome competition for diurnal niches from birds, whose maximum body weight is limited by the demands of flight. Increased size in diurnal primates would have helped reduce the risk of predation from birds of prey, and this may also partially explain why nocturnal primates are generally small—since they can more easily hide from visually oriented predators while asleep during the day.

Whatever their ancestry, modern primates differ radically depending on when they are active. Broadly speaking, nocturnal life is strongly associated with a small body size; a high-energy diet; poor manipulative abilities; solitary foraging techniques; and reliance on olfactory and auditory communication, correlated with a smaller brain size in relation to body weight. Conversely, diurnal life has resulted in primates that are generally larger; with a more varied diet (including folivory); improved manipulative abilities; gregarious foraging habits; and reliance on visual acuity and color vision; accompanied by greater expansion of the brain (anthropoids have relatively larger brains than prosimians; diurnal lemurs have relatively larger brains than nocturnal species; Martin 1975, 1979).

In spite of these differences, there are striking similarities between the two groups. Consider the possibility that the gregarious social systems of diurnal primates are largely parallel versions of the dispersed systems of nocturnal species. Contrasting attributes may result mainly from differences in body size and communication abilities. The patterns common to both groups include varying degrees of tolerance between adult females, reaching a maximum among female kin; various patterns of dispersal and transfer of subadults, with a strong tendency toward migration by males (chap. 21); varying degrees of tolerance between adult males, including the acceptance of immigrants, which are displaced according to an established dominance hierarchy; and various patterns of bonding between individual males and females that maintain contacts throughout the year. These relate to a diversity of mating systems, both polygamous and monogamous.

Finally, although the social systems of diurnal primates are generally gregarious and those of nocturnal species are dispersed, even this distinction is not absolute. Long-term field studies show that the movements of individuals away from a group are more common in diurnal species than was initially apparent, and gregarious tendencies are now known to exist in a number of nocturnal species. Further comparisons will surely prove rewarding.

3 | Malagasy Prosimians: Female Dominance

Alison F. Richard

A sea channel has separated Madagascar from the southeast coast of Africa for about 120 million years. Prosimians reached Madagascar and evolved on this island continent in isolation from monkeys and apes and from many of the competitors and predators facing primates elsewhere. Today there are five lemuriform families containing a total of 20 species (fig. 3-1), but four of the five also include species that have recently gone extinct (see table 39-2). About two thousand years ago, people colonized Madagascar. They hunted, opened up previously forested areas, and introduced new species, particularly cattle and goats. The Malagasy primates were apparently unable to respond effectively to the resulting pressures, and at least 14 species went extinct over the next 600 years, along with at least 2 other mammal species, a radiation of ostrichlike elephant birds, and 2 species of giant tortoise (Dewar 1984). This wave of extinctions eliminated all the largest animals on the island, including lemurs the size of goats (*Megaladapis* spp.) and birds that stood 3 m tall (*Aepyornis* spp.). Thus, when studying the living lemuriforms, it is important to remember that they cannot be considered as members of intact ecological communities.

The lemuriforms, like other prosimians, have retained a greater proportion of primitive characters (i.e., those resembling the ancestral condition) than have the anthropoid primates. In particular, they show less expansion of the brain, especially the neocortex, in relation to body size, and the area of the brain associated with olfaction is relatively large (Tattersall 1982). Unlike anthropoid primates, they have a rhinarium, a moist area of skin at the tip of the nose that is sensitive to scents, and olfactory signals play an important role in communication. They constitute the only prosimian group that contains large-bodied (i.e., >1 kg) diurnal species.

The Malagasy primates occupy a wide range of forests and woodlands. With the exception of one or two species, they are highly arboreal. Thus, while they occur the length of Madagascar in the east and the west, the largely deforested central plateau is devoid of primates. In coastal areas species ranges are disjunct, broken up by stretches of open, often degraded habitat. Outside Madagascar, two species (*Lemur fulvus* and *Lemur mongoz*) occur in the wild on the Comoro Islands, where they were introduced by people several hundred years ago.

The first major survey of lemur ecology and social behavior was done by Petter (1962). Twelve- to eighteen-month studies have since been done on 4 diurnal species (*Lemur catta, Lemur fulvus, Indri indri,* and *Propithecus verreauxi*) and on 1 nocturnal species (*Lepilemur mustelinus*), and another 5 species have been studied for shorter periods (*Phaner furcifer, Microcebus murinus, Mirza coquereli, Cheirogaleus medius, Lemur mongoz*). Still, there remain 10 species, including most of those living in the eastern rain forests, about which little or nothing is known.

ECOLOGY

The activity cycles and specializations of vision of the Malagasy primates present a curious problem. Most species have a retina with a reflective layer (the *tapetum lucidum*) for aiding night vision, yet several, including three indriids, two lepilemurids, and *Lemur catta,* are exclusively diurnal. Other members of the genus *Lemur* lack a *tapetum,* yet at least some populations of *L. mongoz* are reportedly nocturnal, and, at least in the Comoro Islands, *L. fulvus* is diel, that is, active by night as well as by day (Sussmann and Tattersall 1976; Tattersall 1979). A diel rhythm may represent a transitional stage between nocturnality and diurnality made possible in Madagascar (and, by extension, in the Comoros) by the extinctions of many diurnal species about 2000 years ago. Alternatively, Tattersall (1982) suggests that it is more widespread among species of *Lemur* than we think, and may even be the ancestral activity pattern for the genus as a whole. (See Wright 1978 for a related discussion of determinants of the activity pattern of a New World primate, *Aotus trivirgatus*).

Two general categories of locomotor behavior are found in the lemuriforms—vertical clinging and leaping and arboreal quadrupedalism (table 3-1). Vertical clingers and leapers travel by leaping between vertical supports and generally rest in an upright posture (fig. 3-2). They can reach almost any part of a tree, however, and feed in a wide range of suspensory postures. The arboreal quadrupeds move with agility among small branches and twigs on the periphery of tree crowns and leap wide gaps to cross from one crown to another. *L. catta* and, possibly, *Lemur coronatus,* are exceptional, habitually traveling on the ground and in the trees using

FIGURE 3-1. Representatives of the five lemuriform families: (*top left*) Daubentoniidae: aye-aye (the only surviving species in this family); (*bottom left*) Cheirogaleidae: western gray mouse lemur; (*top right*) Indriidae: indri; (*bottom right*) Lemuridae: ring-tailed lemur; (*top left on facing page*) another Lemuridae: brown lemur; (*bottom left on facing page*) Lepilemuridae: sportive lemur. (Photos: Russell Mittermeier/Anthrophoto)

FIGURE 3-2. White sifaka resting in the upright posture typical of vertical clingers and leapers. (Photo: Russell Mittermeier/Anthrophoto)

broad horizontal limbs in preference to thin, less stable branches (Sussman 1974).

The lemuriforms show varying degrees of dietary specialization, but taken together they eat a wide range of items including insects and vegetable goods. The Lemuridae, Lepilemuridae, and Indriidae are vegetarians. Except for *Hapalemur,* they rarely use their hands to manipulate food items but rather pull food-bearing branches to their mouth and feed from them directly. *Hapalemur* is reported to feed primarily on young bamboo shoots and leaves, and its anterior dentition is specialized for processing this food (Petter and Peyrieras 1970; Milton 1978). Shoots are detached with the incisors, clamped between the upper and lower canines and premolars, and pulled sideways with the hands, thereby stripping off the fibrous outer layer. The tender interior is then pushed back into the side of the mouth and chewed up. *Lepilemur,* the other extant genus in the Lepilemuridae, may exhibit a specialization of digestive physiology that is unique among primates. During the dry season in the arid forests of the south it has been observed to spend 91% of feeding time on leaves and 6% on flowers and fruit together. It has also been seen eating its own feces. Based on these observations, Hladik and Charles-Dominique (1974) have suggested that *Lepilemur* survives, at least seasonally, on a high-fiber diet by practising cecotrophy. Cecotrophs ferment food in the large intestine, excrete a proportion of the nutrients released by this process in their feces, eat the feces, and assimilate the nutrients therein. However, cecotrophy has not been seen in a subsequent study (Russell 1977), and the digestive physiology of this species remains in question.

TABLE 3-1   Activity Patterns and Spacing in Malagasy Primates

Species	Activity Cycle	Positional Behavior	Diet	Social Group Size	Home Range Size (ha)	Density (ha)
Lemur catta	Diurnal	SQ	Leaves, fruit, insects, flowers	17 (5–30)	6–23	1.5–3.5
L. mongoz	Nocturnal/ ?diel/diurnal	AQ	Nectar, flowers, fruit, leaves	2 (+ offspring)	1.15	
L. macaco	Diurnal	AQ	Fruit, leaves, flowers	10 (4–15)		0.58
L. fulvus	Diurnal/diel	AQ	Leaves, fruit, flowers, bark	9 (4–17)	0.75–1	0.4–12
Varecia variegata	Diurnal	AQ	Fruit, ?	2 (+ offspring)		
Lepilemur mustelinus	Nocturnal	VCL	Leaves, flowers	Solitary	0.1–0.3	2–8.1
Indri indri	Diurnal	VCL	Leaves, fruit, flowers, seeds	2 (+ offspring)	17.7–18	0.09–0.16
Propithecus verreauxi	Diurnal	VCL	Fruit, flowers, leaves, bark	5 (3–13)	2–8.5	1.1–1.5
Cheirogaleus medius	Nocturnal	AQ	Nectar, fruit, insects, gum	Solitary	4.0	3–4
Microcebus murinus	Nocturnal	AQ	Insects, flowers, fruit, leaves, sap	Solitary (nesting groups of 2–9)	0.07–2.0	4–26
Mirza coquereli	Nocturnal	VCL, AQ	Insects, fruit, gum, insect secretions, small vertebrates	Solitary	2.5–3.0	0.3
Hapalemur griseus	Crepuscular	VCL, AQ	Bamboo, ?	3–6	0.47	0.62

SOURCES: Pollock 1979b; Tattersall 1982; Sussman, Richard, and Ravelojaona 1985.

NOTES: SQ = semiterrestrial quadruped; AQ = arboreal quadruped; VCL = vertical clinger and leaper.

The proportions of leaves, flowers, and fruit eaten by lemurids and indriids vary between species and, within species, from region to region and from season to season (Richard 1978). Interspecific variations do not correspond closely with variations in tooth and gut morphology. For example, Sussman (1974) reported that in the dry season sympatric L. fulvus and L. catta fed, respectively, for 89% and 43% of the time on leaves. Yet experiments with captive animals have shown that L. catta grinds up and digests leaves more efficiently than L. fulvus (Sheine 1979). One explanation for this anomaly is that certain extant species exploit foods today from which they were previously excluded by now-extinct species. Again, it is important to remember that the living Malagasy primates do not live in intact ecological communities.

Daubentonia madagascariensis lives in the east coast rain forest. Several of its traits, including a particularly well-developed olfactory lobe, large ears, robust and continuously growing incisors, and a middle finger modified to form a thin probe, appear to be specializations for the exploitation of insect larvae hidden under the bark of trees (Petter 1977). Madagascar has no woodpeckers, and aye-ayes may have evolved to fill the predatory role played elsewhere by these birds (Cartmill 1974d).

Cheirogaleids, all small (<1 kg) animals, have been most intensively studied in the dry forests of western Madagascar (Charles-Dominique et al. 1980) where the summer diet of the four species present consists of a variety of fruits and insects. During the long, dry austral winter, Phaner furcifer and Mirza coquereli feed heavily on gums and insect secretions, whereas Cheirogaleus medius hibernates for at least 6 and perhaps as many as 7 months. During the summer, members of this species build up fat reserves for the winter, and mean adult body weight fluctuates from 217 gm at the beginning of the period of hibernation to 142 gm at the end. While hibernating, animals remain totally inactive. Three to five individuals may hibernate together, piled on top of one another inside a deep hole in a tree trunk. Microcebus murinus does not enter a state of profound torpor like C. medius, but during winter months these small primates (50 to 100 gm) may spend several days at a time inside a hollow trunk, their overall activity is much reduced, and they too gain weight in the summer and lose it in winter, although the cycle is less marked than in C. medius.

Reported population densities and home range sizes vary widely within and between species (table 3-1). Primates are not uniformly distributed in the forests of Madagascar, and some of this variation may be a result of extrapolating from very small samples. However, there is good evidence that certain populations achieve local densities higher than those reported for any other unprovisioned primate. For example, the density of Lemur fulvus in western gallery forest can reach 1,227 animals/km^2 (inferred from censuses of 110 individuals carried out between July and November 1970). Indri has the lowest reported density, at 9 to 16 animals/km^2.

All of the Malagasy primates are threatened by habitat destruction, whether through slash-and-burn cultivation,

TABLE 3-2 Reproductive Parameters for Malagasy Primates

Species	Gestation Length* (days)	Age of Females at Sexual Maturity[†] (months)	Interbirth Interval[‡] (months)	Peak Birth Period in the Wild
Lemur catta	130–35	21–24 30 at Berenty[a]	12 12 or 24 at Berenty[b]	Aug–Nov[c]
L. macaco	120	20–24	12	Aug–Nov[d]
L. fulvus	120	20–24	12	Aug–Nov[d]
Varecia variegata	99–103	20	12	Oct–Nov (captive animals in Tananarive)[d]
Indri indri		? 48–84[e]	? 24 or 36[e]	May[e]
Propithecus verreauxi	162	? 30 at Berenty[a] ? 42 at Hazafotsy[f]	12 24[a,f]	June–Aug[a,f]
Cheirogaleus medius	62	12	12	Nov–Mar[g]
Microcebus murinus	60–64	8–12	12	Nov–Mar[h]
Mirza coquereli	86	9	12	Jan–Feb[i]

SOURCES: Unless indicated otherwise, data are from the Duke University Primate Center (Pollock, pers. comm.). Other sources are as follows: (a) Jolly 1966a; (b) Jolly, pers. comm.; (c) Budnitz and Dainis 1975; (d) Tattersall 1982; (e) Pollock 1975, pers. comm.; (f) Richard 1978; (g) inferred from timing at DUPC; (h) Martin 1972; (i) Charles-Dominique et al. 1980.

*All values based on 50 animals, except *M. coquereli* (N = 5), at the DUPC.

[†]Values from DUPC based on 50 animals except *M. coquereli* (N = 5); values for the wild inferred from observations of size and behavior.

[‡]Values from DUPC based on 50 animals except *M. coquereli* (N = 5). Values for the wild based on census data and do not take account of miscarriages or infant deaths.

clearance for large-scale agriculture, cutting for firewood or construction materials, or the grazing and browsing of domesticated herds. Traditional taboos against hunting, which helped protect some species, are breaking down; this may constitute an additional threat for them. Probably the most endangered species are those living at low density with a limited geographical distribution. But for several species we do not have a clear idea of their distribution or abundance, so for the time being it is probably wisest to consider *all* the lemuriforms in jeopardy.

REPRODUCTION

The reproductive parameters of Malagasy primates in the wild are poorly known, and most values in table 3-2 come from captive animals. Still, a number of generalizations can be made. All species are seasonal breeders, the mating season ranging from a few days (*P. verreauxi*) to about 2 months (certain cheirogaleids). Gestation length is shorter in Malagasy prosimians than it is in most anthropoids (in relation to maternal body weight), and infants are relatively altricial (i.e., born at an early stage of development) (chap. 16). Several species (*Microcebus, Phaner, Lepilemur, Cheirogaleus, Varecia,* and *Daubentonia*) use nests or holes in tree trunks as sleeping places and as places in which to keep their newborns. In captivity, twins and triplets are common in several cheirogaleids and in *Varecia,* but the incidence of multiple births in the wild is unknown. In the wild, some or all the cheirogaleids may experience a postpartum estrus following a full-term gestation if the infant dies shortly after birth, though this has yet to be documented. Evidence from captive animals suggests that postpartum es-

trus is uncommon following the birth of a surviving infant: in seven breeding seasons at the Duke University Primate Center there have been only two instances (one in *C. medius* and one in *L. fulvus*) of a post-partum estrus and successful birth following a full-term gestation (Pollock, pers. comm.). Limited evidence suggests that the large species, including *Lemur* and the diurnal indriids, give birth every other year if the infant survives. If the infant dies, the female may conceive in the next mating season.

In *Microcebus murinus* and *Cheirogaleus medius,* estrus occurs after the entrance to the vagina, or vulva, has opened. The vulva is closed during the rest of the year. Opening is accompanied by swelling and pinkening of the skin around the vulva during a period of 4 to 6 days. Closure of the vulva outside estrus has been observed in only a few of the other species studied, notably *Varecia, Lepilemur,* possibly *Hapalemur,* and some individuals of *Lemur fulvus* and *Lemur coronatus.* In *L. catta* and *Propithecus verreauxi,* estrus is marked by swelling and pinkening of the external genitalia and by repeated presentation of the anogenital region to males.

An adult sex ratio biased toward males has been reported in two Malagasy primates. At Berenty, male sifakas (*Propithecus verreauxi*) have outnumbered females in five out of the seven censuses done between 1963 and 1980 (N = 42–58), and in one out of three other forests censused males outnumbered females (N = 86) (Richard, in press). Petter (1962) found a sex ratio of 0.71 among animals older than one year in 10 social groups of *Lemur macaco* (N = 96). The latter finding suggests that the imbalance may be present at birth, but further evidence is needed to establish the determinants of this unusual sex ratio in these two species and also whether it occurs in others.

SOCIAL ORGANIZATION

Even brief studies have yet to be done of some lemuriforms, and for none is there a long-term study of known individuals, but the findings are tantalizing nonetheless. On the one hand there is evidence of traits that are unusual among primates, such as female dominance over males. On the other hand, there are striking parallels with the anthropoids. For example, *L. catta* live in societies in which kinship plays an important part in structuring social relations and in which females and their female descendants form the core of the social group (Sussman and Taylor, pers. comm.). In these features they resemble both macaques and baboons (chap. 11).

Most of the nocturnal lemuriforms are predominantly solitary, like the lorisiforms described in chapter 2. They communicate vocally and interact at low rates with individuals with whom they share part or all of their home range. For example, heterosexual or female pairs of *Lepilemur* often share most or all of a tiny home range (0.3 ha for males and 0.18 ha for females). Meetings occur 1 to 3 times a night and last from 5 minutes to an hour, during which the pair moves, feeds, rests, and occasionally grooms together (Russell 1977). In *Microcebus murinus,* nonreproductive males range on the periphery of an area occupied by a few males and many females. The home ranges of the central males overlap with the home ranges of one or more females. The home ranges of these males and females contain preferred habitat— young forest with an abundant shrub layer—whereas areas occupied by peripheral males tend to be covered by more mature, less desirable forest. The central male regularly visits and interacts with the females whose home ranges overlap with his, and he mates with them during the breeding season. There are affiliative relations between the females themselves, which frequently sleep together during the day, usually in groups of 2 to 9 individuals, though as many as 15 may be found together (Martin 1972, 1973a). Among the cheirogaleids, relations between males and females appear closest in *Phaner furcifer.* Adult pairs call back and forth and are frequently in close proximity throughout the night. In one instance, the home range of a male encompassed those of two females, and he alternated between them in the course of each night (Charles-Dominique and Petter 1980).

The diurnal lemuriforms are gregarious, habitually ranging in groups of 2 to 30 individuals. In some species, groups are thought to be monogamous pairs with their young (*Indri, Lemur rubriventer, Varecia, Propithecus diadema, Hapalemur griseus*). *Avahi,* the one nocturnal indriid, is also believed to be monogamous. However, with the exception of *Indri,* this conclusion is based solely on observations of group size. Intermittent

observations of *Indri* over a 10-year period now suggest that groups may be less stable and their sexual composition more variable than we once thought (Pollock 1979b, pers. comm.).

Indri groups have exclusive use of their large home ranges, and, in "singing battles," defend range boundaries against other groups. No fights have been seen during these encounters, which are strictly vocal confrontations (Pollock 1979b). *Lemur mongoz* live in monogamous pairs in northwest Madagascar. These pairs and their young occupy overlapping home ranges, but intergroup encounters have not been seen (Sussman and Tattersall 1976). If these animals are indeed monogamous with overlapping home ranges, this combination is unusual among mammals, for most monogamous species are territorial (chaps. 12, 23). In the Comoro Islands, however, the composition of groups in certain populations is at variance with strict monogamy. One group, for example, contained six adult-size individuals, four males and two females, and one juvenile. Tattersall (1978) suggests that smaller dry-season groupings coalesce in the wet season in a fission-fusion pattern recalling the spider monkey (*Ateles* sp.) (chap. 7).

Groups containing more than one adult male and female characterize *P. verreauxi, L. catta, L. fulvus,* and *L. macao* (fig. 3-3). In *L. catta,* females form a temporally and spatially stable core of the group and males transfer frequently, perhaps annually, between groups. In one 12-month study, the male composition changed in all eight of the groups censused (Budnitz and Dainis 1975). Subordinate males are spatially and socially peripheral to the group most of the time (Jolly 1966a; K. C. Jones 1983; Budnitz and Dainis 1975). During the mating season, however, the male dominance hierarchy breaks down and males compete to mate with the females in their group.

In some species, the identity of the social unit is less clear, and it has been suggested that a neighborhood, or social network, spans several foraging groups (Jolly 1966a; Richard 1978; Pollock 1979b; Tattersall 1982). This suggestion is based on two kinds of observations. First, sleeping aggregations of separate foraging groups have been reported in *L. macaco, L. fulvus,* and *P. verreauxi* (Pollock 1979b). Second, during the mating season, *P. verreauxi* males leave their own groups and visit others containing estrous females. While the mating season is short, the timing of estrus seems to be staggered between groups so that a visiting male does not necessarily miss the opportunity to mate with females in his own group, although in neither of the two groups observed were such matings seen. The response to visiting males varies. Some are chased away, while others are sexually solicited by the females and mate without interference from the group males. Preliminary evidence suggests that animals in neighboring groups may maintain stable social relationships with one another (Richard

FIGURE 3-3. Black lemurs: (*above*) adult female on the right and subadult female on the left; (*below, right*) adult male. Black lemurs typically live in groups with two or more adults of each sex. Note the sexual dichromatism: although the males are solid black, the females are reddish-brown and have a white ruff around the face. (Photo: Russell Mittermeier/Anthrophoto)

1974b, in press). In some parts of the Comoro Islands, movement in and out of *L. fulvus* groups is frequent, but it is not known whether animals are moving between subgroups of a stable group or whether groups are unstable. The sex of transferring animals is also unknown (Tattersall 1978, 1979). Until studies have been done of individually recognized animals, the nature of the social and mating systems of this species will remain unresolved. Only in *L. catta* and *P. verreauxi* is there strong evidence that just males transfer, and even in these species the timing of transfer in relation to age and maturational state is poorly documented.

Spacing patterns can sometimes vary as much within as between the species discussed above. For instance, in the moist forests of northwestern Madagascar extensive overlap occurs between the home ranges of neighboring foraging groups of *P. verreauxi*, whereas in the dry forests of the south there is little or no overlap and groups defend the boundaries of their home ranges. This difference in the spacing behavior of the two populations may be attributable to a difference in the distribution of resources in the two habitats (Richard 1978).

◆FEMALE DOMINANCE

In all the lemuriforms for which we have appropriate data, females have been found to be individually dominant to males, particularly in the context of feeding. Female-dominant species include *Lemur catta, Propithecus verreauxi, Indri indri, Microcebus murinus,* and *Phaner furcifer.* Dominance relationships can be assessed by either approach-retreat interactions or the direction of aggression-submission. In species for which we have good data, the hierarchies that result from either measure are usually very similar. In several lemuriforms, submissive gestures are rare or difficult to see and dominance is determined according to approach-retreat interactions. In *Lemur catta,* the incidence of aggression within groups is high, particularly between males. Aggression takes the form of chasing, cuffing, scent marking, and, in males, stink fighting (Jolly 1966a). Acts of submission include retreat or cowering as a dominant animal approaches, and low-ranking males habitually walk with lowered head and tail carriage, lagging behind the group and generally avoiding other animals. Females are much less frequently aggressive than males, and the female dominance hierarchy is less easy to detect, although the few agonistic encounters observed suggest that it is stable. Yet, "at any time . . . a female may casually supplant any male or irritably cuff him over the nose and take a tamarind pod from his hand." (Jolly 1984, 198).

In contrast with *Lemur catta,* the frequency of all kinds of interaction in *P. verreauxi* and *I. indri* is low (Richard and Heimbuch 1975; Pollock 1979a). In both species, dominance can be determined by the direction of approach-retreat interactions and by who bites, threatens, or cuffs whom. *P. verreauxi* bare their teeth, roll their tails up between their legs, and hunch their backs to show submission. Most agonistic encounters occur in the context of feeding. Pollock (1979a) distinguishes between voluntary and imposed displacements in *I. indri.* A voluntary displacement consists of a swift advance by one animal and "the immediate, hasty, and sometimes poorly coordinated departure" of the other (p. 149). In an imposed displacement, one animal kicks or bites another, who immediate retreats. Pollock never saw an animal try to retaliate once it had been accosted in this way. Aside from the act of retreat itself, submissive gestures are not seen in *I. indri.* In both species, females have been found to be dominant to males with only one exception: Richard and Heimbuch (1975) report one *P. verreauxi* dyad in which a young adult male consistently displaced an adult female. The genealogical relationship, if any, between them was not known.

Phaner furcifer and *Microcebus murinus* spend much or all of the night foraging alone or at least at some distance from conspecifics, yet when a male and female are in proximity, the latter is apparently able to displace the former. Charles-Dominique and Petter (1980) report instances of deferral by male *P. furcifer* to females at favored gum trees. Perret (1982) reports female priority in captive *M. murinus.*

In summary, either female dominance has been observed or the situation is unclear in all lemuriforms studied to date, including species that live in groups containing several adult females and males, and species that may be monogamous. Female dominance is exhibited primarily in the context of approach-retreat interactions at feeding stations.

Female dominance is rare in other primates and in other orders of mammals (Hrdy 1981b). Among anthropoid primates, females are sometimes dominant to males in *Cercopithecus talapoin,* several macaque species, and *Cebus apella* (chap. 32). Some form of female dominance is also present in golden and Chinese hamsters (*Mesocricetus, Cricetulus* spp.), Maxwell's duikers (*Cephalophus maxwelli*), otters (*Amblonyx cinerea* and *Lutrogale perspicilla*), beavers (*Castor canadensis*), feral nutria (*Nutria coypu*) in the southern United States, spotted hyaenas (*Crocuta crocuta*), and dwarf mongooses (*Helogale parvula*) (Ralls 1976). Considering just this handful of mammals, however, a single definition of female dominance is inappropriate because both the nature and function of dominance vary from species to species. In *Cercopithecus talapoin* and *Macaca mulatta,* for example, single females are sometimes subordinate to single males, and only in coalition with other females are they dominant (chap. 32). In golden hamsters, the outcome of an agonistic interaction between a male and female depends strongly on the female's reproductive condition. In species in which males are usually dominant to nonpregnant females, a pregnant or lactating female is often dominant to males. If a female is neither pregnant nor lactating, the outcome is likely to depend on the stage of her estrous cycle. In captive Maxwell's duikers, Ralls (1976) found that context determined the outcome of agonistic interactions between males and females: males had priority at preferred resting places, for example, whereas females had priority of access to preferred pieces of food.

In short, it is unlikely that the phenomenon of female dominance is homologous (i.e., derived from a common ancestor) among mammals or, indeed, that it has evolved independently in several species to serve a similar function. Considering just the lemuriforms, however, female dominance appears to be represented by a distinctive and homogeneous set of behaviors, notably the consistent displacement of males by females at feeding sites.

In most primates, adult males have priority of access to food, and displace without challenge immatures and adult females at feeding sites. These males are usually larger and have more weaponlike canine teeth and a more aggressive temperament than females. This difference is usually explained in terms of more intense intrasexual competition and higher variance in reproductive success among males than among females. Male dominance over females is generally presumed to be an incidental consequence of the morphological features favored by intrasexual competition (e.g., Pollock 1979a; Hrdy 1981b; Jolly 1984).

Two explanations have been proposed to explain the few instances of female dominance seen in mammals. Each emphasizes different aspects of biology and behavior as keys to the phenomenon of female dominance:

1. In monogamous species, a male may benefit his present and future offspring if he defers to his mate, the mother of those offspring, and gives her priority of access to food. This argument is in keeping with more general explanations of the evolutionary logic of parental care (Trivers 1972).

2. In species that breed seasonally, with marked reproductive synchrony, females experience particular seasonal stress during pregnancy and lactation. It is in the interest of males who have mated to defer to females, at least during such periods, because they may have fathered the females' offspring. They regain weight and compete for mates during periods of plenty (Hrdy 1981b).

The first explanation can be appropriately applied to *Phaner furcifer* and probably to *Indri indri*. In these species males may be confident of paternity and may enhance their reproductive success by giving their mate priority of access to food. In most other lemuriforms, however, there is no evidence of monogamy and considerable evidence of promiscuity. Jolly (1984) has discussed the strengths and weaknesses of the second explanation in these cases. One general problem concerns the assumption that males are willing to suspend intrasexual competition during the lean season in favor of females, for one could equally argue that males should compete even at that time to ensure their own advantage at the start of the season of abundance (Jones 1981). Like most such debates, the resolution of this one must await improvements in our ability to assign costs and benefits to particular patterns of behavior, and it may well prove that neither explanation is correct.

Specific predictions of the second explanation are that female dominance should occur in polygynous species when (1) mating takes place during a brief period in the season of plenty and (2) lactating or gestating females experience particular seasonal stress. Extreme seasonality of mating is well documented in all lemuriforms studied so far, and *L. catta* and *P. verreauxi* mate toward or at the end of the season of abundance. In contrast, *M. murinus* mates in September when males weigh least after a period of torpor during the austral winter (Petter-Rousseaux 1980). Jolly (1984) suggests that prosimian females experience greater reproductive stress than anthropoid females in seasonal environments. Lemuriforms have a shorter gestation period in relation to maternal body weight, and their young are altricial, less developed at birth. It is probable that, like other altricial mammals, they also have a higher growth rate that imposes a heavy energy demand on the mother. This hypothesis is supported by data showing a relatively high postnatal brain growth index for the subfamily Lemurinae (chap. 16). The larger-bodied lemuriforms may be particularly stressed because, unlike the small species, they cannot encompass the cycle of gestation and lactation within a single season of abundance. Inevitably some portion of the cycle will fall in the season of scarcity. In sum, Jolly concludes, it is possible that prosimians invest relatively more energy than anthropoids in reproduction and that large-bodied lemuriforms experience particular stress during some phase of the reproductive cycle. Unfortunately, there are no data with which to test these ideas.

Our knowledge of the Malagasy prosimians is still so limited that any explanation of the phenomenon of female dominance is necessarily speculative at best. However, the fact of female dominance in these primates remains, and it is possibly linked to other features that are rare or absent among anthropoids. These include lack of sexual dimorphism in body size even in the largest-bodied species and an equal or male-biased sex ratio in at least some species. The causal links, if any, between these traits have yet to be explained.

The study of lemuriform biology and behavior is important not just to increase our understanding of these prosimians but also to increase our understanding of the anthropoid primates and, indeed, of other mammals. As the opportunity to work in Madagascar is renewed, the study and conservation of the lemuriforms must surely be a high priority.

4 | Tamarins and Marmosets: Communal Care of Offspring

Anne Wilson Goldizen

Marmosets and tamarins belong to two New World primate families: the Callitrichidae and the Callimiconidae. Callitrichids are distinguished from all other anthropoids by their small size (ranging from *Cebuella* at less than 160 g to *Leontopithecus* at 650 to 750 g), by having claws instead of fingernails, by their dental formulae, and by the fact that about 80% of the litters contain twins.

These characteristics of callitrichids were originally considered primitive traits (Hershkovitz 1977), but there now exist two competing views. One view, held by Leutenegger (1979) and others, is that these traits are the result of phyletic dwarfism. Leutenegger believes that the ancestors of the Callitrichidae were frugivorous and that a shift to an insectivorous diet caused selection for a smaller body size. Among primate species and placental mammals in general, as maternal weight decreases, total litter weight increases relative to maternal weight (Leutenegger 1973). When a primate infant's weight at birth exceeds 15% of the mother's weight, birth becomes prohibitively difficult. In callitrichids, total litter weights range from 14.1 to 23.5% of maternal weight (Leutenegger 1979). Leutenegger argues that selection favored twinning because two smaller infants are easier to give birth to than one large one. The callitrichids' single pair of nipples and the morphologies of the callitrichid uterus and placenta support the argument that callitrichids evolved from ancestors that had single young.

Others, claiming that callitrichids are not necessarily phyletic dwarfs, argue that each of these special callitrichid characteristics can be explained by forms of selection other than selection for small size (Sussman and Kinzey 1984). In favor of this view, callitrichids do not have relatively large brains (chap. 16), which are expected in phyletic dwarfs (Clutton-Brock and Harvey 1980). This controversy was reviewed by Sussman and Kinzey (1984), who conclude that the claws, dental formulae, and regular twinning of callitrichids are derived, rather than primitive, traits, but that whether they are the result of "dwarfing" is not clear.

Taxonomy and Distribution

The taxonomic status of many callitrichid species is still controversial. Researchers even disagree about whether Goeldi's marmoset (*Callimico goeldii*) belongs in its own family, the Callimiconidae, or in the Callitrichidae or the Cebidae. For convenience, Hershkovitz's taxonomy of the Callitrichidae is used in this paper (Hershkovitz 1977). The Callitrichidae include two genera of tamarins: *Saguinus* (ten species): *fuscicollis, nigricollis, imperator, mystax, labiatus, inustus, oedipus, leucopus, bicolor, midas*) and *Leontopithecus* (one species: *rosalia*); and two genera of marmosets: *Callithrix* (three species: *jacchus, argentata, humeralifer*) and *Cebuella* (one species: *pygmaea*). The common names of these species are listed in the appendix.

The genus *Saguinus* occurs throughout much of the neotropical lowland rain forest, from Panama to Bolivia to northeastern Brazil. *Leontopithecus*, now very rare in the wild, occurs in seasonal tropical forest in three small and separate areas near the coast of Brazil (fig. 4-1). *Callithrix* occurs throughout much of Brazil, mostly in savanna or savanna-forest habitats. *Cebuella* occurs in western Amazonia—in parts of Peru, Ecuador, and Brazil. *Cebuella* is more common in seasonally flooded or earlier successional habitats than in other types of forest. *Cebuella* is sympatric throughout its range with several

FIGURE 4-1. Adult lion tamarin. (Photo: Russell Mittermeier/Anthrophoto)

Saguinus species, and *Leontopithecus* coexists with *Callithrix jacchus,* but the two tamarin genera are entirely allopatric (geographically nonoverlapping), as are the two marmoset genera. *Callimico* inhabits lowland rain forest from southeastern Peru and northern Bolivia to southern Columbia, but appears to be rare throughout its range. Approximate range maps for all of the callitrichid and callimiconid species are available in Hershkovitz (1977) and Wolfheim (1983).

Field Studies

Only five long-term field studies of callitrichids have been published: three on *S. oedipus* (Dawson 1978; Neyman 1978; Garber 1980, 1984), one on *S. fuscicollis* and *S. imperator* (Terborgh 1983; Terborgh and Goldizen 1985), and one on *Cebuella pygmaea* (Soini 1982b). Data on the ecology of these and other callitrichids also exist from shorter studies. However, in only one of these studies was behavioral data regularly gathered on marked individuals (Terborgh and Goldizen 1985). The other field studies either involved unmarked individuals or marked individuals that were not sufficiently habituated to be observed at close range. Due to this paucity of behavioral data on wild callitrichids, the study of *S. fusci-collis* is by necessity emphasized in the behavioral section of this chapter (fig. 4-2). This study occurred in the Manu National Park in southeastern Peru, which will henceforth be referred to simply as Manu.

ECOLOGY
Habitat Selection

Sussman and Kinzey (1984) review callitrichid habitat selection. Several researchers have suggested that callitrichids prefer secondary forests and edge habitats to primary forests, especially mature open forests. However, other studies have found that in areas containing both of these types of forests, the callitrichid species present use all of the habitats. Perhaps most callitrichids do equally well in both young and old forests, and the suggested "preference" for secondary forest is an artifact of the fact that callitrichids have often been studied in secondary forests located near human populations. In such areas, the population sizes of the larger primate species are often greatly reduced due to hunting, and the callitrichid population densities may be abnormally high because of reduced competition with larger primates. Several authors have suggested that callitrichid home ranges typically include a variety of habitat types.

FIGURE 4-2. Juvenile saddleback tamarin, *Saguinus fuscicollis,* in primary forest in Manu National Park, Peru. Estimated age is 3 to 4 months. (Photo: Anne Goldizen)

Feeding

Sussman and Kinzey (1984) also review callitrichid feeding ecology. The callitrichids and *Callimico* are all diurnal and arboreal and eat insects, small vertebrates, fruit, tree exudates (gum and sap), and nectar. Marmosets and tamarins have distinct morphological specializations for feeding. The marmosets have lower incisors as long as their canines, which they use to gnaw holes in trees, causing exudate flow. The lower incisors of the tamarins are less than one-third the length of their canines, fitting the normal primate dentition pattern. Thus tamarins are unable to gnaw holes to cause exudate flow (Coimbra-Filho and Mittermeier 1978). Ramirez, Freese, and Revilla (1978) found that a *Cebuella* troop spent 32% of its daily activity budget eating exudates or chewing exudate holes, but only rarely ate fruits. *Callithrix* has not been well studied in the wild, and reports on the importance of exudate eating in this genus are contradictory. While *C. jacchus* is thought to eat a lot of exudates, Rylands (1981) reported that *C. humeralifer* ate exudates only infrequently. The two tamarin genera and *Callimico* eat far more fruits than exudates, though at least *Saguinus* occasionally eats exudates either by raiding holes made by *Cebuella* or by feeding on exudates that ooze from wounds on trees (Coimbra-Filho and Mittermeier 1973; Pook and Pook 1981; Garber 1980; Terborgh 1983; Izawa 1975).

Garber (1980, 1984) found that 14% of the feeding time of *S. oedipus* in Panama was spent eating exudates, principally from *Anacardium excelsum* trees. Analysis of the exudate of this species showed high ratios of calcium to phosphate content, similar to some other plant exudates. Garber therefore suggested that callitrichids eat exudates to complement a high phosphorus/low calcium ratio in the digestible parts of the insects that they eat, and the low calcium content of the fruits in their diet. He also suggested that calcium may be particularly important to lactating females because of the fast growth of their twin infants (Garber 1984). While this explanation may be appropriate for *Cebuella, Callithrix* species, and *S. oedipus,* which do seem to eat a lot of exudates, it does not account for the feeding habits of *S. fuscicollis* or *S. imperator.* Terborgh (1983) found that *S. imperator* spent only 1% of its feeding time on exudates in the wet season and 2% in the dry season, while the equivalent percentages for *S. fuscicollis* were 3% and 9%. At Terborgh's study site, most exudate eating occurred during the dry season, when fruits were scarce and when females do not lactate (Goldizen, pers. obser.). This suggests that the tamarins at this site obtain sufficient calcium from food items other than exudates.

Two *Saguinus* species have been studied in sufficient detail to document differences in their insect-foraging techniques. *S. fuscicollis* forages primarily by investigating knotholes and crevices on trunks and branches and catches mostly large insects (1 to 2 inches long), 89% of which are hidden before capture. *S. imperator* does most of its insect foraging on leaves. *S. imperator's* prey are roughly the same size as those of *S. fuscicollis,* but 79% are exposed prior to capture (Terborgh 1983).

Activity Patterns

The best data on callitrichid activity budgets are from Terborgh (1983). *S. fuscicollis* and *S. imperator* spent 20 to 21% of their daily time budgets traveling and 16 to 17% feeding on plant materials. They differed in the amounts of time spent resting and insect foraging; for *S. fuscicollis* these percentages were 44% and 16% while for *S. imperator* they were 25% and 34%. While resting or feeding on fruits or nectar, groups of these species are very cohesive, with group diameters generally less than 20 m. During travel, or travel combined with insect foraging, *S. fuscicollis* groups often spread out much more, with group diameters occasionally greater than 50 m, but all group members are always in regular vocal contact (Goldizen, pers. obser.).

Mixed-Species Associations

Several studies have reported close associations between two *Saguinus* species, including *S. fuscicollis* with *S. imperator, S. fuscicollis* with *S. labiatus,* and *S. fuscicollis* with *S. mystax* (Pook and Pook 1982; Terborgh 1983). In these cases, two troops of different species spend some time traveling together and some time apart, though often calling back and forth. These pairs of troops are even thought to defend a single territory (Terborgh 1983). Interestingly, in northwestern Bolivia, where *S. imperator, S. labiatus,* and *S. mystax* are sympatric, they are never found together in mixed troops, although each travels with *S. fuscicollis* (Izawa and Bejarano 1981). Competition between species in paired *Saguinus* troops may be avoided by the use of different insect-foraging techniques, or by preferences for different heights in the forest. For example, both Pook and Pook (1981) and Yoneda (1981) found that *S. fuscicollis* spends more time in lower forest layers than does *S. labiatus.*

Further work is needed to fully understand the benefits of mixed-species troops to the individuals involved, but Terborgh (1983) suggests three possibilities. The first possible benefit is improved predator detection due to larger group size. The second two are possible benefits gained from the knowledge of which fruit trees have recently been visited by the other species. Terborgh argues that this knowledge allows each tamarin group to avoid traveling to trees where the other group has fed recently, and which are thus unlikely to contain ripe fruits. This could either save the tamarins the energetic expenses of travel to unprofitable trees or reduce the risk of predation incurred by extra travel. Terborgh tentatively favors the latter explanation, but some recent anecdotal evidence from Manu supports the importance of the first explanation. During several months, an *S. imperator* group consisting of only one male and one female shared its territory with an *S. fuscicollis* group. The female *S. imperator* was quite habituated to the observer's presence, while the male was not. When the *S. fuscicollis* group was being observed, the female *S. imperator* normally traveled in the midst of the *S. fuscicollis* group, while her mate remained 50 to 100 m away, apparently too scared to approach more closely (Goldizen, pers. obser.). The preference of the female *S. imperator* for the company of five *S. fuscicollis* over that of her mate suggests predator detection benefits of the mixed-species groups. If either the second or third of Terborgh's explanations for the benefits of mixed-species troops were more important, the female could have remained with her mate 50 to 100 m from the *S. fuscicollis* and accrued the benefits via vocal communication.

Predation

The threat of predation appears important to understanding callitrichid behavior and ecology. At Manu, tamarins are known to have been eaten by *Spizaetus ornatus* (the ornate hawk eagle) as well as by ocelots (*Felis pardalis*). Other possible predators of callitrichids are harpy eagles (*Harpia harpyja*), crested eagles (*Morphnus guianensis*), other wild cats, tayras, and snakes. Tamarin groups were attacked by raptors approximately once every 1 or 2 weeks (Goldizen, pers. obser.).

Direct data on predation rates were hard to obtain (cf. chap. 19). Nonetheless, observations indirectly suggest that predation is a serious problem to callitrichids and that they are constantly alert to the danger. At Manu, members of *S. fuscicollis* groups rarely feed all at once; nearly always at least one adult, positioned in or near the feeding tree on an open branch, scans the surrounding area (Goldizen, unpub. data). Alarm calls are given more than once an hour, usually false alarms at harmless large birds or terrestrial noises. Other group members react to these false alarms—adults by immediately scanning the area, and juveniles by jumping to the ground or running for cover. When not feeding or foraging, the tamarins rest, generally well hidden in vines, and without movement or vocalizations. When young tamarins have play sessions, often involving much chasing and vocalizations, the adults of the group station themselves on different sides of the playing offspring, as if they were being especially alert for predators that could be attracted by the raucous young (Goldizen, pers. obser.). Caine (1984) reported experiments on the scanning behavior of captive *S. labiatus;* she suggested that this behavior has evolved to be an important way for tamarins to avoid predation. While more firm data on the importance of predation to callitrichids is necessary, anyone who watched wild tamarins would form the subjective impression that these little animals are constantly scared and watchful.

Population Densities

Reported population densities and home range sizes vary greatly between species (table 4-1). Some of the population density estimates listed in table 4-1 are probably unrealistically high. Since these monkeys do not occupy all habitats at any particular site, densities calculated from group sizes and home range sizes are likely to be misleading. For species other than *Cebuella*, most studies found home range sizes of about 10 to 50 ha. It is not yet clear why the home range sizes of *Saguinus* species are particularly large at Manu (Terborgh and Goldizen, unpub. obser.). Daily ranging distances of these species are generally between 1 and 2 km (table 4-1).

Cebuella appears to differ from the other genera in several ways: (1) troops have extremely small home ranges, usually centered around a few trees from which they harvest exudates, (2) troop members do not remain cohesive during the day but scatter widely throughout the small home range (Ramirez, Freese, and Revilla 1978), and (3) a troop occupies its home range for a few months to a few years and then moves as a group to a new range. Soini (1982b) found that 6 of his 14 study troops changed home ranges temporarily or permanently within a 1-year period. Such migratory troop movements are quite unusual among primates. Terborgh (1983) reported that during his 14-month study one group of *S. fuscicollis* and one group of *S. imperator* changed territories, but this does not seem to be common, as no *S. fuscicollis* groups ($N = 6$) changed territories at the same study site between 1977 and 1985 (Goldizen, pers. obser.).

Conservation

Leontopithecus rosalia and *Saguinus oedipus* are probably the most endangered callitrichid species, followed by *Callithrix jacchus*. In 1975 it was estimated that only 100 to 200 wild *L. rosalia* remained, due to habitat destruction (Magnanini 1978). Both *C. jacchus* and *Leontopithecus* inhabit very small remnants of forest along the heavily populated Atlantic forest region of Brazil. The Colombian population of *S. oedipus* is highly endangered, due both to extensive habitat destruction and to capturing for export (Neyman 1978). Details on the conservation status of these species are given in chapter 39.

REPRODUCTIVE PARAMETERS

Table 4-2 shows available information on several demographic and reproductive parameters for *Callimico* and the callitrichids. Known gestation periods range from approximately 128 days in *Leontopithecus* to approximately 148 days in *C. jacchus* (Sussman and Kinzey 1984). Age at sexual maturity is not known exactly for

TABLE 4-1   Population Density, Home Range Size, and Day Range Length for Callitrichids and *Callimico goeldii*

Species and Study Site	Population Density (ind/km^2)	Home Range Area (ha)	Day Range Length (m)	Sources
Saguinus oedipus				
Canal Zone, Panama	27–36	26–43	2,061	Dawson 1979
Sucre, Colombia		7.8–10		Neyman 1978
S. nigricollis				
Rio Caqueta, Colombia	10–21	ca. 30–50	ca. 1,000	Izawa 1978a
S. fuscicollis				
Mucden, Bolivia	10–20	26–40	1,140–1,590	Yoneda 1981
Manu Park, Peru	ca. 7–8	50–100		Terborgh and Goldizen, unpub.
Manu Park, Peru	16	30	1,220	Terborgh 1983
S. imperator				
Manu Park, Peru	ca. 4–5	ca. 50–100		Terborgh and Goldizen, unpub.
Manu Park, Peru	12	30	1,420	Terborgh 1983
S. labiatus				
Mucden, Bolivia	7–15	23–41		Yoneda 1981
S. mystax				
Northeastern Peru	25–32			Soini 1982a
Callithrix humeralifer				
Rio Aripuana, Brazil		12–14	740–1,500	Rylands 1981
C. pygmaea				
Rio Maniti, Peru	34–247	0.2–0.4		Soini 1982b
Rio Nanay, Peru		0.3		Ramirez, Freese, and Revilla 1978
C. goeldi				
Pando, Bolivia		30–60	ca. 2,000	Pook and Pook 1981

TABLE 4-2   Reproductive Parameters for Callitrichids and *Callimico goeldii*

Species	Gestation[a] Period (days)	Age at Sexual Maturity (months)	Interbirth Interval (months)	Peak Birth Period	Study Site
Saguinus fuscicollis	140–50	23[b]	ca. 12[b]	Aug–Jan[b]	Manu Park, Peru
				Sept–Dec[c]	Mucden, Bolivia
S. labiatus	140–50			Oct–Dec[c]	Mucden, Bolivia
S. mystax	140–50	F: 15–17[d]	11–12[d]	Nov–Feb[d]	Northeastern Peru
		M: 17–18			
Callithrix jacchus	148 ± 4.3[e]	F: 20–24[e]			
Leontopithecus rosalia	128.6[f] (125–32)			Sept–Mar[g]	Southern Brazil
Cebuella pygmaea			5–7[h]	Nov–Jan[h] May–June	Rio Maniti, Peru
Callimico goeldii			ca. 6[i]	Sept–Nov[j]	Pando, Bolivia

SOURCES: (a) references in Hershkovitz 1977; (b) Terborgh and Goldizen, unpub.; (c) Yoneda 1981; (d) Soini 1982a; (e) Hearn 1978; (f) Kleiman 1978a; (g) Coimbra-Filho and Mittermeier 1973; (h) Soini 1982b; (i) Masataka 1981; (j) Pook and Pook 1981.

any of the species, but studies suggest that it is between 18 and 24 months for most of them. The interbirth intervals and birth seasonalities listed in the table refer only to wild individuals. In captivity, all of the callitrichids appear capable of producing young twice each year, but *Cebuella* is the only species for which this is documented in the wild. Masataka (1981) suggests that *Callimico* may have young twice per year, but further evidence is needed to support this. All populations of callitrichids studied have exhibited one or two distinct birth peaks during the

year, although some births generally occur throughout the year.

Estrous cycles have been found to average 15 to 17 days, but their lengths are variable (Hearn 1978; Hampton and Hampton 1978). Callitrichids do not menstruate.

SOCIAL ORGANIZATION

Table 4-3 presents data on group sizes and compositions in wild callitrichids and *Callimico*. Group sizes are almost always between 2 and 10; larger groups are occa-

TABLE 4-3. Group Size and Composition of Wild Groups of Callitrichids and *Callimico goeldii*

Species and Study Site	Group Size	Aver. No. Adult Females/ Group	Aver. No. Adult Males/ Group	Solitary Individuals	Transient Groups	Source
Saguinus fuscicollis						
Manu Park, Peru	3–10 (incl. infants)	1.14	1.99	Only males seen	Two all-male groups seen	Terborgh and Goldizen, unpub.
Mucden, Bolivia	2–7 aver. = 4.7			Seen		Yoneda 1981
S. mystax						
Northeastern Peru	2–8 aver. = 5.3 (infants not incl.)	1.6	2.1	3 nulliparous females seen alone	Groups without infants seen	Soini 1982a
S. labiatus						
Mucden, Bolivia	2–6 aver. = 4.2			Seen		Yoneda 1981
S. oedipus						
Canal Zone, Panama	Aver. = 6.39 (infants not incl.)	2.1	2.4	None seen		Dawson 1978
Sucre, Colombia	3–13 (incl. infants)	1.8	3.0	Males and females seen	Groups without infants seen	Neyman 1978
Leontopithecus rosalia						
Southern Brazil	2–8					Coimbra-Filho and Mittermeier 1973
Callithrix humeralifer						
Rio Aripuaña, Brazil	4–13 (infants incl.)	2.5	2.5			Rylands 1981
Cebuella pygmaea						
Rio Maniti, Peru	2–9 aver. = 6 (infants not incl.)	>1	>1	Males and females seen	Groups without territories seen	Soini 1982b
Callimico goeldii						
Mucden, Bolivia	6 (1 group)	2	1			Masataka 1981
Pando, Bolivia	8–9 (1 group)					Pook and Pook 1981

sionally seen, but are assumed to be temporary associations of two separate troops. Several authors report transient groups of two or more adults without infants, seen wandering through territories held by other groups (Sussman and Kinzey 1984, pers. obser.). Solitary individuals of both sexes have also been reported for several species, although it is not known how long such individuals remain alone.

Resident groups consist of at least one adult of each sex and young of up to three or four different ages. As table 4-3 shows, groups commonly contain multiple adult males and adult females. Because wild groups are fairly small and have only one set of young at a time, and because captive individuals usually do best when kept in pairs, callitrichids have until recently been considered monogamous.

However, in a population of six groups of individually marked *S. fuscicollis* in Manu, which were monitored for a 4-year period, Terborgh and Goldizen (1985) found a high degree of variability in the groups' mating systems, both between groups and within groups over time. Groups had the following adult group compositions: one female and one male 22% of the time, one female and more than

one male 61% of the time, more than one adult of each sex 14% of the time, and adults of only one sex 3% of the time. When groups contained more than one adult female, only one female generally bred within a given breeding season, as reported in most other studies. However, Terborgh and Goldizen (1985) once observed two females breeding a few months apart (but during the same breeding season) while living together in the same group. In four out of five two-male groups in which copulations were observed, both males were seen to mate with the single adult female. Terborgh and Goldizen consider these four two-male groups to fit their two conditions for cooperative polyandry: (1) two or more males must mate with one female, with neither male monopolizing the female around the time that she ovulates, and (2) all of the males must help care for the female's young. Thus this population contained monogamous groups, polyandrous groups, and a group with two breeding females.

In a further study of the same *S. fuscicollis* population, five groups were studied over a 15-month period in 1984 and 1985 (Goldizen, unpub.). Of these five groups, three were thought to be polyandrous—all three groups having begun 1984 as newly formed trios of two adult males and one female, without young. None of the males were previous offspring of the females. One of these troops was studied intensively. During the 2-month period around the time of conception, one male was observed mating with the female 14 times, and the other male 10 times (Goldizen and Terborgh, in press). Both males also mated with the female during her pregnancy, during lactation, and after weaning of the young. The fourth group in the population contained one adult male and two adult females that both became pregnant, although one apparently lost her young immediately after birth. The mating system of the fifth group was not determined. The possible adaptive significance of the variable mating system of this *S. fuscicollis* population is discussed later in this chapter.

More long-term field studies of individually marked marmosets and tamarins are clearly needed in order to determine whether other populations of *S. fuscicollis*, and populations of other species, have social systems similar to that found in this *S. fuscicollis* population. Studies of marked wild callitrichids show that both sexes transfer between groups (Dawson 1978; Neyman 1978; Terborgh and Goldizen, unpub.), although insufficient data exist to know whether one sex emigrates more frequently than the other. While Terborgh and Goldizen have not intensively studied the social behavior of *S. imperator,* at Manu this species appears to have the same variable social system as does *S. fuscicollis,* with many of the groups beginning as trios of two adult males and one adult female. Although matings have not been seen in *S. imperator,* group compositions at least suggest that some groups may be polyandrous. *S. mystax* in northern Peru have group compositions similar to those of the *S. fuscicollis* at Manu, as well as adult sex ratios biased in favor of males, so this species may also have some cooperatively polyandrous groups (Garber, Moya, and Malaga 1984).

Almost nothing is known about the social organization of wild *Callimico,* although Masataka (1981) observed one group when it visited an artificial feeding platform over a 5-month period. The group had two adult females, each with infants, one male, and six juveniles and subadults. After the first 2 weeks of life, most infant carrying was done by nonmothers.

The land tenure system varies greatly both within and between *Saguinus* species. If territoriality is defined as defense of an area with stable boundaries, then *S. fuscicollis, S. imperator, S. labiatus, S. oedipus,* and *S. mystax* have all been reported to be territorial, at least at some localities (Terborgh 1983; Yoneda 1981; Dawson 1978; Neyman 1978; Soini 1982a). However, the degree to which adjacent home ranges overlap can vary greatly within species, for instance from less than 10% overlap in *S. fuscicollis* at Manu (Terborgh and Goldizen, un-

pub.), to 79% overlap between adjacent groups in *S. fuscicollis* at a Bolivian site (Yoneda 1981). *S. oedipus* shows similar variability: from almost complete overlap and no territorial defense in an upland troop, to territoriality and only 13% overlap in a nearby lowland troop (Dawson 1979).

Neyman (1978) and Terborgh and Goldizen (unpub.) report similar patterns of territorial defense in *S. oedipus* in Columbia and *S. fuscicollis* at Manu, with aggressive confrontations between some neighboring groups occurring every few days. At Manu, intergroup encounters last anywhere from 1 hour to over 1 day and involve intermittent loud vocalizations and aggression. The encounters tend to begin aggressively, often with actual physical fights and chases occurring between the adult males and subadults of both sexes of the two groups. The adult females and small young usually remain in the background, although they occasionally participate in chases. Encounters generally follow the same pattern, alternating between approximately half-hour periods during which the two groups are close together and actively interacting, and slightly longer periods during which the groups separate by 100 to 200 m and rest or feed. Aggressive interactions are greatly reduced after the first hour or two. After this, the interactive periods often involve play between the young from the two groups, while the adults of the two groups rest about 50 m apart (Goldizen, unpub. obser.). These interactions appear to have two purposes: first, enforcement of the boundary, and second, allowing subadults to check for possible breeding positions in the neighboring group, or possibly to look for an individual with whom they could emigrate in search of a new territory.

The home ranges of *Cebuella* troops at a site in northeastern Peru did not overlap, and overt interactions were never seen between adjacent groups (Soini 1982b). No information is available on the presence or absence of territoriality or on the nature of intergroup interactions in *Leontopithecus, Callithrix,* or *Callimico.*

HELPING BEHAVIOR
Nonmaternal Infant Care in Callitrichids

The pattern of infant care among callitrichids is clearly distinct from other primates because juveniles and subadults often provide substantial amounts of help. Furthermore, adult males provide extensive care, as occurs also in titi monkeys (*Callicebus moloch* and *Callicebus torquatus*) and owl monkeys (*Aotus trivirgatus*) (chap. 5), and siamangs (*Syndactylus symphalangus*: Chivers 1974), but with the difference that in callitrichids two or more adult males may help. In this section I describe the types and quantities of infant care given by different age-sex classes of callitrichids both in captivity and in the wild. Then I present possible explanations for these contributions to infant care. The help given by nonreproductive individuals is discussed in the context of theories developed for bird and mammal "helpers at the nest," while discussion of the help of adult males is combined with a discussion of the variable mating system found in *S. fuscicollis* and perhaps in other callitrichid species as well.

The Manu study of *S. fuscicollis* provides the only quantitative data on infant carrying among wild callitrichids. In 1980 two troops were observed over the first 3 months of the infants' lives. The infants were carried at least part of the time until they were about 3-months old. Twins were sometimes carried simultaneously by one individual and at other times were carried separately by two individuals. One group had four adult males for the first 6 or 7 weeks of its single infant's life and was then reduced to one adult male after the other three emigrated. The second group contained two adult males during the entire 3-month period. In both groups, all of the adult males carried infants; in one group the proportions of carrying by the two males were 18% and 58% respectively of the group's total carrying. The amount of carrying by yearlings (full or half-siblings of the infants) also varied. When the first group had four adult males, the yearling male and female were never observed carrying the group's one infant, but immediately after the males' emigrations the two yearlings began to carry the infant as much as, or more than, the adult female. During the same period, the twin infants in the second troop were carried by the two adult males and one 2-year-old male, and initially also by the adult female. Throughout the observation period, the yearling male in that group did virtually no carrying (Terborgh and Goldizen 1985).

In the second *S. fuscicollis* group during the infants' third month of life, all of the group members were observed sharing large insects (Orthoptera) with the infants (Terborgh and Goldizen 1985). In a typical interaction,

an infant saw an individual with an insect and ran to it, giving a specific vocalization. The older individual then passively allowed the infant to take the insect or sometimes held the prey item as the infant ate it. The same behaviors were observed with fruits that the infants were not capable of opening. Once the infants began to learn to catch insects for themselves, they still obtained insects from other group members, but they did so by snatching these from the "giver," apparently against its will. Brown and Mack (1978) report that not only do captive *Leontopithecus* adults donate food to infants, but they give "food calls" to "invite" newly weaned infants to take food items.

In the polyandrous group that I studied in 1984, the two males who both mated with the same female subsequently carried her two infants for roughly equal amounts of time. The group's two juveniles did not carry their younger siblings, perhaps because the juveniles were only 9- or 10-months old when the infants were born. However, once the infants began eating insects, all the group members, including the juveniles, were observed "giving" food to the infants.

Captive studies of three species have found that all group members carry infants. Table 4-4 presents data from five studies of infant care, which provide a rough idea of the relative amounts of carrying done by different age-sex classes and allow comparisons between the studies. Two points are of particular interest. First, adult males do substantially more infant carrying than do adult females. Second, even though younger individuals do less carrying than adults, their contribution is nonetheless significant.

Although these studies seem to suggest species differences, conclusions about interspecific variation are premature because individual differences within species are so large. For example, one study of *C. jacchus* found

TABLE 4-4   Relative Amounts of Infant Carrying by Age-Sex Class in Callitrichids

Species	Type of Study	Weeks Studied from Birth	Groups and Litters Studied	Adult Males	Adult Females	Young Males	Young Females	Sources
Saguinus fuscicollis	Captive	6	7 groups 8 litters	1.58	1.03	0.51	0.24	Epple 1975
	Captive	12	1 group 3 litters	2.37	0.58	0.98	0.52	Vogt, Carlson, and Menzel 1978
	Wild	9–12	2 groups 2 litters	1.60	0.70	0.40	0.80	Terborgh and Goldizen, unpub.
Leontopithecus rosalia	Captive	12	4 groups 7 litters	1.89	1.31	0.15	0.19	Hoage 1978
Callithrix jacchus	Captive	4	1 group 1 group	0.64	1.35	0.51	1.50	Box 1975a
Averages of all studies weighted by number of groups studied				1.65	1.05	0.43	0.40	

NOTES: To obtain the numbers presented, expected amounts of carrying by each age-sex class were calculated for each study from the group compositions, based on the assumption that each group member should do an equal amount of carrying. For each age-sex class, the table presents the actual amount of carrying done, divided by the amount of carrying expected. The calculations do not differentiate between carrying of one or two infants.

that the adult female carried infants more than twice as much as did the adult male during the first 4 weeks of the infants' lives (Box 1975a), while a second study of the same species found the opposite (Ingram 1977). In a study of seven *S. fuscicollis* groups, mothers were responsible for 1 to 69% of the carrying during the first 40 days, while dominant males performed 30 to 96% (Epple 1975). Although data are insufficient to determine whether the amount of carrying done by different age-sex classes differs between species, the last row of table 4-4 tentatively indicates that, in general, adult male callitrichids carry infants more than do adult females, who in turn carry more than do young animals.

There may be species differences in the ages at which infants begin to be carried by different age classes. In Hoage's (1978) study of four *L. rosalia* groups, only the mothers carried infants during the infants' first week of life, and they remained the principal carriers until the fourth week. In *S. fuscicollis* and *C. jacchus*, both parents carried infants from the day of birth, although in some studies nonparents did not carry the infants during the first few days. Again, more studies are needed to determine whether these patterns represent true species differences.

Helping Behavior in Other Animals

Helping behavior, defined as the care of offspring by individuals who are not their parents, occurs in birds, mammals, and insects. It has been reported in over 150 bird species (reviewed by Brown 1978 and Emlen 1984) and in over 30 mammal species (reviewed by Bertram 1983 and Emlen 1984). Most mammalian cooperative breeding involves canids. Some well-studied examples of mammalian cooperative breeders include the cape hunting dog (*Lycaon pictus*), the black-backed jackal (*Canis mesomelas*), the coyote (*Canis latrans*), the wolf (*Canis lupus*), and the dwarf mongoose (*Helogale parvula*). The most extreme form of helping behavior occurs in the social insects, in which the majority of females either become or are born sterile and spend their lifetimes helping to raise the offspring of their sisters (Wilson 1971).

The most striking aspects of helping behavior (not all exhibited by all helping species) are delayed reproduction by helpers (or permanent sterility), donation of food to young, significant loss of foraging or hunting time due to "baby-sitting" of young, and increased risk of predation or injury due to defense of young. These aspects of helping behavior are interesting because they are hard to explain in terms of individual selection; all at least potentially reduce the helper's reproductive success. The most commonly discussed benefits to helpers of helping behavior are increased inclusive fitness through kin selection, delayed benefits through reciprocal altruism, and benefits of gaining experience in parental care.

Adaptive Significance of Helping by Nonreproductive Callitrichids

One class of helpers in callitrichids includes nonbreeding individuals who help raise full or half-siblings. Helping by these individuals is probably explained by a combination of nepotistic gains and the benefits of becoming experienced at infant care. As long as their help does not decrease their own chances of future reproduction, helpers increase their inclusive fitness by increasing their younger siblings' chances of survival, and therefore of reproduction. It has also been found that captive adult callitrichids do not care for their young properly if they did not gain early experience at infant care by helping to raise their siblings (Epple 1975; Hoage 1978). Thus, by helping to care for their siblings, young callitrichids may increase their personal reproductive success. No evidence exists to date that reciprocal altruism occurs in callitrichids, but this is not likely to be detected without careful long-term studies.

To fully understand why mature callitrichids often remain in their natal group and help until they are 3 or 4-years old, we need to understand why they postpone their own breeding. In his "ecological constraints" model, Emlen lists four reasons why mature offspring in many cooperatively breeding species may have a very low chance of breeding on their own (Emlen 1982, 1984). These are (1) high risks of dispersal, (2) low probability of finding a territory, (3) low probability of finding a mate, and (4) low chance of reproducing successfully even if a mate and a territory are acquired. If any of these constraints were strong enough, a mature offspring would be better off remaining as a helper in its natal territory. The first two of these constraints are more likely than the last two to apply to callitrichid populations. The second might occur in territorial populations such as the Manu *S. fuscicollis* population, where in some years all the available forest seems to be "owned" by established tamarin groups (Goldizen, pers. obser.), as is the case in many cooperatively breeding birds (Emlen 1984). The first constraint may be a factor in all tamarin populations; a solitary tamarin dispersing in search of a territory is probably at high risk of being caught by a predator.

Adaptive Significance of Helping by Polyandrous Male Callitrichids

In Terborgh and Goldizen's *S. fuscicollis* population, polyandrous males could also be considered helpers. Although they do not fit the standard definition of helpers, these males presumably do not recognize their genetic relationship to the young. To explain the occurrence of polyandry in *S. fuscicollis*, Terborgh and Goldizen (1985) suggest that a lone reproductive pair without young helpers would have little chance of successfully raising twin offspring. At birth twin callitrichids together weigh from 14 to 25% of the mother's weight. By weaning, this percentage is probably close to 50%. Thus lactating females presumably suffer significant energetic stress and for this reason do relatively little infant carrying, especially after the first week or two. In a lone pair, then, the male would have to carry the young almost all day, and since individuals carrying babies rarely eat or forage (Goldizen, pers. obser.), this does not seem feasible. Thus, Terborgh and Goldizen argue that a pair should accept a second male in the group, and the males should share the reproduction of the female. Even if a male were not the father of every infant he helped to raise, over his lifetime he would raise enough of his own offspring to benefit from this helping behavior.

For this explanation of polyandry to be reasonable, neither male must be able to detect the timing of the female's ovulation, so that polyandrous males have roughly equal chances of paternity. Callitrichids show no morphological signs of estrus (e.g., sex skins), and reports are mixed as to whether there are behavioral indications of ovulation. Kleiman (1978a) suggests that *L. rosalia* copulate, groom, sniff, and approach each other more frequently around the time of ovulation, while Brand and Martin (1983) found that mating behavior was not reliably correlated with the females' estrogen levels in *S. oedipus,* and Hearn (1978) found that mating did not always increase at the time of ovulation in *C. jacchus.* More studies are needed on this question, but the possibility remains that females of at least some callitrichid species exhibit "concealed ovulation" (chap. 30).

Terborgh and Goldizen's proposed explanation of facultative polyandry in *S. fuscicollis* needs much further testing in wild callitrichids. A few of the more important types of data needed are the number of reproductively active males in groups with and without nonreproductive helpers, the patterning of copulations by polyandrous males with respect to the female's ovulation, and the costs of infant carrying. More studies are needed to determine whether cooperative polyandry is common in the Callitrichidae. Since all of the species generally have twins of high birth weight, it certainly seems possible that the facultative polyandry of *S. fuscicollis* could be general to the family.

While some forms of nonparental infant care do occur in other primates (chaps. 27 and 28), the types of help provided by nonreproductive helpers in these species seem to be substantially different from those found in callitrichids. Whereas nonreprodutive callitrichid helpers carry heavy infants for long distances and donate food to them, nonreproductive helpers in other primate species groom infants, carry them for short distances, "baby-sit" them, usually for short time periods, and occasionally defend them against predators. One possible exception is night monkeys (*Aotus trivirgatus*); in a captive group of *Aotus,* the juvenile carried its infant sibling about 15 to 25% of the time until the infant was 5 weeks old, when the juvenile ceased carrying it (Wright 1984). To compare the magnitudes of the help provided by helpers in different primate species, data are needed on the costs and benefits of the different kinds of infant care provided by the helpers.

In diurnal anthropoids, facultative polyandry and twinning are both unique to the callitrichids. It is likely that polyandry, and the extensive helping by nonreproductive helpers, is related to their frequent twinning. However, why callitrichids evolved large litters is still unresolved. Until this is better understood, the evolution of callitrichid social systems will remain somewhat of a puzzle.

5 | Monogamous Cebids and Their Relatives: Intergroup Calls and Spacing

John G. Robinson, Patricia C. Wright, and Warren G. Kinzey

The pitheciines illustrate a variety of social adaptations. A number of species are monogamous and groups occupy exclusive ranges, but species differ in the way that they maintain the pair-bond and defend space. Sexual dimorphism in size is not pronounced among pitheciines, but sex differences in pelage color and vocalizations are reported in a number of species. *Aotus* (the night or owl monkey) is the only nocturnal primate in the Neotropics. Despite these interesting social variations, good field data exist only for *Aotus* and two of the species of *Callicebus* (titi monkeys). Much less is known about the societies of *Pithecia* (sakis) and *Chiropotes* (bearded sakis). *Cacajao* (uakaris) is among the least studied of all primates.

NUMBER OF GENERA AND SPECIES

The five genera considered in this chapter have traditionally been separated into two or more subfamilies. Rosenberger (1981) united them into a single subfamily on the basis of morphological criteria. The appendix

presents the accepted species. *Aotus* has usually been considered a single species. Based on variability in karyotype, color, pelage, and susceptibility to malarial infection, Hershkovitz (1983) recently divided *Aotus* into at least nine allopatric species. Since additional karyotypes remain to be described, and cross-mating studies are necessary to determine which karyotypes are sexually incompatible, in this review we will consider *Aotus* as the single species, *A. trivirgatus*.

PHYSICAL CHARACTERISTICS

Sexual dimorphism in body size is slight in *Aotus, Callicebus,* and *Pithecia,* and not much greater in the larger-bodied genera *Chiropotes* and *Cacajao* (table 5-1). A sex difference in pelage color (sexual dichromatism), rare among primates, is found in all species of *Pithecia* (Hershkovitz 1979). In these species, the pelage of the face and crown of males is short and contrasts with the darker body color. This condition is most striking in *P. pithecia* (fig. 5-1), intermediate in *P. monachus,* and

FIGURE 5-1. Adult male white-faced saki (*Pithecia pithecia*) from Surinam. The short, light pelage of the face contrasts with the darker body color. (Photo: Russell Mittermeier/Anthrophoto)

FIGURE 5-2. Red uakari male (*Cacajao calvus*) showing piloerection in threat. Animals of both sexes lose the cranial and facial hair as adults. (Photo: Russell Mittermeier/Anthrophoto, courtesy of Monkey Jungle, Florida)

least in *P. hirsuta*. In females, this pelage is long and does not contrast with the general body color. Coat color is sexually dichromatic only in *P. pithecia:* males are blackish while females are brownish agouti. In all species, juveniles of both sexes resemble females. Age dichromatism also characterizes *Cacajao calvus rubicundus* (Fontaine 1981): on reaching adulthood, animals of both sexes lose their facial and cranial hair and the exposed skin develops a characteristic bright scarlet color. Adult males are most easily distinguished from females by the extreme prominence of the muscles overlying the frontal and parietal bones (fig. 5-2). These develop synchronously with other secondary sexual characteristics (Fontaine 1981). *Cacajao* is unusual in having a very short tail.

GEOGRAPHIC DISTRIBUTION

Aotus is the most widely distributed of the pitheciines; it is found from Panama to northern Argentina. Its distribution reaches the Pacific Ocean in Colombia and Ecuador, and the Atlantic in Amazonian Brazil. The distribution of *Callicebus* is less extensive. The genus occurs only as far north as the Llanos of Venezuela and Colombia and as far south as the Paraguayan Chaco. One species (*C. personatus*) occurs in the eastern Brazilian coastal forests. Neither species occurs in the Guyanas, and *Aotus* does not occur in eastern Brazil (Thorington and Vorek 1976). The three larger-bodied genera are restricted to the Amazon and Orinoco river basins (see Wolfheim 1983). Congeneric sympatry occurs in the

TABLE 5-1 Ecological Characteristics of Pitheciines

Species and Study Site	Body Weight (g) Male	Body Weight (g) Female	% Time[a] Foraging	% Foraging on Fruits	% Foraging on Leaves	% Foraging on Insects	% Foraging on Other	Sampling[b] Method	Study[c] Period	Sources
Aotus trivirgatus										
Peru	795	780	45	75	10	15	0	TF	L	Wright 1985; Aquino, pers. comm.
Paraguay				16	40	11	33	TF	S	Wright 1985
Captivity	945	895								Elliott, Sehgal, and Shalifoux 1976
Callicebus moloch										
Peru			26	48	40	12	0	TF	S	Terborgh 1983
Peru			41	54	28	17	1	TF	L	Wright 1985
Peru			24	70	26	1	3	TT	S	Kinzey 1978
Captivity	1,266	1,200								Robinson and Ramirez 1982
C. personatus										
Brazil			18	81	18	0	1	TT	S	Kinzey and Becker 1983
C. torquatus										
Peru			26	67	13	14	6	TT	S	Kinzey 1977
Pithecia hirsuta										
Peru				71	16	0	13	F	S	Happel 1982
P. pithecia										
Surinam	1,875	1,866		93	0	0	7	F	S	Fleagle and Mittermeier 1980; Mittermeier and van Roosmalen 1981
Chiropotes albinasus										
Brazil	3,175	2,518	90	<10	0	0		F	L	Ayres 1981
C. satanas										
Surinam	3,000	2,980	93	1	0	6		F	L	Fleagle and Mittermeier 1980; Mittermeier and van Roosmalen 1981
Cacajao calvus										
Captivity	4,100	3,550		+	+					Fontaine 1981

a. Includes time spent feeding on leaves, fruits, etc., and foraging for invertebrates.

b. Time-taking method (TT) records the proportion of time animals spend feeding on different items; frequency method (F) records the frequency that different items were taken; time-foraging method (TF) records the time allocated to foraging for and feeding on different items. Compared to the other methods, time foraging (TF) overestimates the relative importance of insects.

c. L = long-term studies, usually lasting at least a year; S = short-term studies.

subfamily: *Pithecia hirsuta* possibly occurs with *P. monachus* west of the Jurua River and with *P. albicans* east of the same river (Hershkovitz 1979). *Callicebus moloch* is sympatric with *C. torquatus* along both banks of the upper Guaviare River (Hernandez-Camacho, pers. comm.) and between the Amazon and Purus rivers (Kinzey and Gentry 1979).

ECOLOGY
Habitat Preferences

Most members of this subfamily have distinct habitat preferences. *Aotus* is the most versatile genus, occurring from seasonally deciduous scrub forest to high-altitude cloud forest up to the timber line at 3,000 m (Hernandez-Camacho and Cooper 1976; Napier 1976). *Callicebus* is found only in low-altitude forests (under 500 m), with the greatest population densities in areas characterized by forest openings and successional vegetation, such as swamps and the edges of rivers and lakes (Hernandez-Camacho and Cooper 1976; Kinzey 1981; Terborgh 1983; Wright 1985). Where *C. moloch* and *C. torquatus* occur in sympatry, the former is found in lower-slope forests growing on clays and loams, while the latter occurs in upper-slope white sand forests (Kinzey and Gentry 1979).

Aotus is found in all strata of the forest (Wright 1981). *Callicebus*, especially *C. moloch*, spends more time in the understory and lower canopy levels of the forest (Kinzey 1976, 1981). *Pithecia* is a canopy-dwelling species primarily found in highland and lowland nonflooded forests (Mittermeier and van Roosmalen 1981; Happel 1982). Unlike the three preceding genera, *Cacajao* and *Chiropotes* occur only in undisturbed forest. Both are seen mostly in the upper canopy. The former is specialized to flooded whitewater and blackwater forests, and the latter to upland forests (Pires 1973; Mittermeier and van Roosmalen 1981; Ayres 1981).

Diet

The smallest pitheciines, *Aotus* and *Callicebus*, have a very generalized diet that includes leaves, insects, flowers, and fruit (table 5-1). The leaves that are taken are immature, if from trees, and both mature and immature, if from lianas (Wright 1985). When seeds are ingested, they usually pass through the digestive tract intact. Within both genera, the diet varies considerably in different regions. For instance, *Aotus* in the subtropical Chaco of Paraguay, where the species is very diurnal, spend 40% of their foraging time feeding on leaves, while in lowland Amazon forest in Peru they allocate only an estimated 5 to 10% to this food (Wright 1985). *C. torquatus*, which resides in a habitat where the foliage has a high concentration of secondary compounds, takes more insects than *C. moloch* (Kinzey 1978). Diet also varies seasonally. For instance, *C. mo-*

loch in Peru spends 40% of its feeding time eating leaves, 43% eating fruits, and 15% foraging for insects in the dry season, but 23% of its time on leaves, 58% on fruit, and 18% on insects in the wet season when fruit and insects are abundant (Wright 1985). *Aotus* in Paraguay spends 47% of its feeding time eating leaves in the cold, dry season month of July, but only 15% of its time in September when flowers, fruits, and insects are abundant (Wright 1985).

The diets of the larger pitheciines, *Pithecia*, *Chiropotes*, and *Cacajao*, are made up largely of fruits and seeds. *Pithecia* and *Chiropotes* are major seed predators (Mittermeier and van Roosmalen 1981). In 55% of feeding observations, *P. pithecia* in Surinam ingested seeds (Buchanan, Mittermeier, and van Roosmalen 1981). In 66% of the feeding observations of *C. satanas*, seeds were masticated, often from unripe fruits of species eaten by other monkeys only when ripe, while in 28%, ripe fruit mesocarp but not seeds were used (van Roosmalen, Mittermeier, and Milton 1981). The comparable figures for *C. albinasus* were 36% and 54% of observations (Ayres 1981). This reliance on seeds varies seasonally: at the end of the drier season in Aripuana, Brazil, *C. albinasus* ingested seeds in 80% of feeding observations, while at the height of the wetter season, this figure was as low as 10% (Ayres 1981). *Chiropotes* obtain additional protein from insects (Ayres 1981; Ayres and Nessimian 1982). *Chiropotes* have specialized teeth with robust and broad canines and well-developed jaw muscles that allow them to crack open very hard nuts and seeds. *Pithecia* and *Cacajao* have a similar jaw and tooth morphology. Fontaine (1981) describes *Cacajao* as a frugivore-folivore in a seminatural habitat.

Movements and Use of Space

The smaller pitheciines, *Aotus*, *Callicebus*, and *Pithecia*, occur in small groups of rarely more than six animals (table 5-2). Groups have short day ranges of under 1 km. With the exception of *C. torquatus*, range sizes do not exceed 10 ha, perhaps because these small groups exploit densely distributed fruit resources that provide food for a long period of time. For example, Wright (1984, 1985) reported that 93% of the feeding trees of *C. moloch* were relatively common species with crown diameters of 10 m or less, which ripened a few fruits at a time over periods of 2 to 6 months. *Aotus* and *C. moloch* repeatedly return to specific sleeping sites, such as holes in hollow trees and dense vine tangles, but *C. torquatus* sleep on large exposed limbs high in the canopy (Kinzey 1981; Wright 1981). Groups of all three genera occupy relatively exclusive ranges and regularly encounter other groups at range boundaries.

The larger *Chiropotes* and *Cacajao* occur in large groups of up to 30 animals. Day ranges are probably correspondingly large: Ayres (1981) noted that *C. albinasus*

TABLE 5-2 Group Characteristics, Movements, and Use of Space among Pitheciines

Species and Study Site	Group Size	No. Groups	Group Composition				Home Range Size (ha)	Daily Path Length (km)	Population Density (#/km²)	Sources
			Adult Male	Adult Female	Juve-niles	Infants				
Aotus trivirgatus										
Peru	4.5	9	1.0	1.0	1.5	1.0	10.0	0.71	40	Wright 1985
Paraguay	3.1	21							14	Stallings 1984
Callicebus moloch										
Colombia	3.2	8	(2.0	)	0.5	0.6	0.5	0.57	400	Mason 1968
Colombia	3.1	18	(2.2	)	0.4	0.5	3.7		57	Robinson 1977
Peru	4.2	7	1.0	1.0	1.2	1.0	8.0	0.67	20	Wright 1985
Paraguay	2.8	26							17	Stallings 1984
C. personatus										
Brazil	3.7	15	1	1	0–3	0–1	4.7	0.70		Kinzey and Becker 1983
C. torquatus										
Colombia	4.0	5	(2.1	)	0.9	1.0	14.2		32	Defler 1983
Peru	3.9	15	1	1	0–3	0–1	18.0	0.82	20	Kinzey 1978; Easley 1982
Pithecia hirsuta										
Peru	4.5	4	1	1	1.3	1.3			>37.5	Happel 1982
P. pithecia										
Surinam	2.7						4–10		3.6	Buchanan, Mittermeier, and van Roosmalen 1981
Chiropotes albinasus										
Brazil	25	1	8	9	5	3	250–350	2.5–5	7–8	Ayres 1981
C. satanas										
Surinam	8–30						200–250	2.5	7–8	Van Roosmalen, Mittermeier, and Milton 1981
Cacajao calvus										
Brazil	5–30									Fontaine 1981
C. melanocephalus										
Brazil	15–25									Ibid.

groups move 2.5 to 3.5 km during the wetter season and 4.5 km during the drier season. Home ranges of three groups of this species varied from 80 to 200 ha in size. Overlap of these ranges did not exceed 10%. Ayres never observed intergroup encounters. Nothing is known about *Cacajao* range use.

Population Densities

There is considerable variation in the reported densities of the different species. While some of this variation might be accounted for by differences in forest productivity and the presence of food competitors, most of it probably results from differences in sampling methods and from varying amounts of habitat disturbance. Nevertheless, some trends are clear: First, within the subfamily there is an inverse correlation between body weight and population density (fig. 5-3), as found among primates generally (Clutton-Brock and Harvey 1977b). Second, *Callicebus torquatus* occurs at much lower densities than *C. moloch*, when they occur in the same general area, because *C. torquatus* inhabits forests of lower plant productivity and has a more insectivorous diet (Kinzey and Gentry 1979; Kinzey 1981). Finally, in

habitat islands created by human disturbance, some species can reach very high densities. *C. moloch* densities in Socay Forest, a 6.9 ha forest fragment in Colombia, reached 400/km² (Mason 1968).

CONSERVATION

Both *Aotus* and *Callicebus* can survive in highly disturbed secondary forests, because (*a*) their diets are catholic, (*b*) group ranges are small, and (*c*) they are infrequently hunted because they are small and cryptic. These species, however, are threatened locally. The northern Colombian *Aotus* is endangered as a result of extensive deforestation and removal of animals for malarial research. The forest habitat of *Callicebus moloch ornatus* in central Colombia is rapidly being converted to ranch land, and the subspecies might be endangered. Similarly, *C. personatus*, the eastern Brazilian endemic, is listed as vulnerable in the IUCN *Red Data Book*—primarily because the Atlantic forests have been decimated.

No species of *Pithecia* is endangered. However the genus occurs at relatively low densities even in relatively undisturbed forest, so large areas are required to maintain viable populations. *Pithecia* is hunted for meat and for its large bushy tail, which is sold as a duster in Peru and Brazil.

Chiropotes s. satanas and *C. s. utahicki,* which occur south of the Amazon River in an area of moderate human populations and extensive deforestation, are listed as endangered, and *C. albinasus,* whose range is traversed by highways of the trans-Amazon system and overlapped by a number of large colonization schemes, is listed as vulnerable in the IUCN *Red Data Book.* Both species of *Cacajao* are listed as vulnerable. There are indications that the flooded habitat of this genus can support intensive and sustained rice cultivation (e.g., Goodland 1980), and large-scale conversion of these forests to farmland is under consideration.

REPRODUCTIVE PARAMETERS

Gestation periods in this subfamily vary between 133 and 173 days (table 5-3). The ages of sexual maturity are poorly known and depend on the measure used. For *Aotus* and *Callicebus* we used the age at which animals leave their natal groups. For *Cacajao c. rubicundus,* for which there are good captive data on four individuals, we used age of first birth for females and age of testes descent in males (Fontaine 1981). In captivity, both males and females developed the characteristic scarlet facial skin at 3 years. Shortly after this, the female conceived and gave birth to an infant when she was 3 years 7 months. Two of the three males did not develop full adult male characteristics for an additional 2 to 3 years, at which time their testes descended and their size and muscularity increased. The third male developed adult characteristics directly at 3 years of age.

Data on estrus are also poor. Only *Chiropotes* shows morphological signs of estrus: the anogenital area and labia become deep red at times when *C. albinasus* females are receptive (Hick 1968a; Ayres 1981). Estrous cycles are known only for *Aotus* (16 days; Bonney, Dixson, and Fleming 1980) and *Callicebus* (17 to 20 days; Sassenrath et al. 1980). There are no reports of postpartum estrus following a normal gestation in the subfamily, even under captive conditions (*Aotus:* Hunter et al. 1979).

SOCIAL ORGANIZATION

The Pitheciinae includes several species that are consistently monogamous, in that within a social group only a single male and female breed at one time. Groups of *Aotus, Callicebus,* and *Pithecia* are usually made up of an adult pair, sometimes a single infant, and sometimes

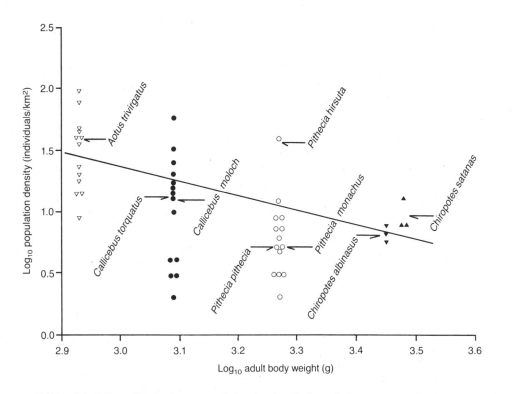

FIGURE 5-3. Effect of body size on population density. Each symbol represents a single density estimate, and arrows indicate mean densities for particular species. Linear regression of the mean density of each species illustrates trend. ▽ = *Aotus trivirgatus;* ● = *Callicebus* spp.; ○ = *Pithecia* spp.; ▼ = *Chiropotes albinasus;* ▲ = *Chiropotes satanas.* SOURCES: Wright 1981; Kinzey 1981; Buchanan, Mittermeier, and van Roosmalen 1981; van Roosmalen, Mittermeier, and Milton 1981; Fontaine 1981; and studies quoted in this chapter. Densities calculated from river surveys or from habitat islands not included.

juveniles born in preceding years (Wright 1981; Kinzey 1981; Buchanan, Mittermeier, and van Roosmalen 1981). Group size therefore normally varies between two and five. On occasion, larger groups with additional adults have been reported. Hernandez-Camacho and Cooper (1976) reported collections of several adult pairs of *Aotus* in the same nest site and a group of seven (*Callicebus moloch* in Colombia. Adult *C. moloch* sometimes join established pairs and their offspring (Robinson, pers. obser.). However, there are no reports of groups with more than one breeding female, and only a single report (Mason 1966) of more than one male mating with a female, so the presumption is that usually a single pair of adults is breeding.

In *Chiropotes* and *Cacajao*, groups of 10 to 30 are commonly recorded (van Roosmalen, Mittermeier, and Milton 1981; Fontaine 1981). Aggregations of more than 100 *Cacajao* have also been reported (Fontaine 1981;

Defler, pers. comm.). No information on group composition in *Cacajao* is available, but *Chiropotes* groups are composed of approximately equal numbers of adult males and females. In a group of 26 *C. albinasus*, Ayres (1981) counted 8 adult males, 9 adult females, 5 juveniles and 2 to 3 infants. Within this group, spatially segregated subgroups composed of a single male, a single female, and juveniles often moved and fed together. *Chiropotes* might represent another variation on the monogamous theme: large groups might be relatively permanent aggregations of monogamous subunits. If so, this would be the only known primate, other than *Homo sapiens*, characterized by this type of social organization.

Formation of New Groups

In *Aotus* and *Callicebus*, subadult males of 2.5 to 3.5 years of age leave their natal groups. No agonistic behavior has been observed at or immediately prior to their de-

TABLE 5-3 Demographic and Reproductive Parameters of Pitheciines

Species and Study Site	Birth Seasonality	Age at Sexual Maturity		Gestation Length (days)	Litter Size	Inter-birth Interval (months)	External Sign of Estrus	Sources
		Male	Female					
Aotus trivirgatus								
Peru	None (peak: Aug–Feb)	2.5	2.5	133	1 1	12	None	Wright 1985 Hunter et al. 1979
Captivity								
Callicebus moloch								
Colombia	Dec–Mar							Mason 1966
Peru	July–Nov	2.5	2.5	167	1		None	Wright 1985
Captivity				160	1			Fragaszy, pers. comm.
C. personatus								
Brazil	Aug–Oct				1		None	Kinzey and Becker 1983; Kinzey, pers. obser.; Kimura, pers. comm.
C. torquatus								
Colombia	Jan–Feb	2–2.5	2–2.5		1			Defler 1983
Peru	Dec–Jan	2–3.5	2–3.0		1	12	None	Kinzey 1981, unpub. data
Pithecia pithecia								
Surinam	Jan–Feb (?)				1			Buchanan, Mittermeier, and Milton 1981
Captivity				163, 176				Hick 1968a
Chiropotes albinasus								
Brazil	None (peak: Feb–Mar, Aug–Sep)			152–67	1		Anogenital area and labia become bright red	Ayres 1981 Van Roosmalen, Mittermeier, and Milton 1981
Captivity								
C. satanas								
Surinam	Dec–Jan							Ibid.
Captivity				152–67				Fontaine 1981
Cacajao calvus								
Captivity	None (peak: May–Oct)	5.5	3.6		1		None	Ibid.

NOTE: See text for definition of categories.

parture. The age at which subadult females leave is uncertain, as is the process of mate selection. In *Aotus*, both sexes have a monotonous "hoot" vocalization of 1 to 6 notes that is audible for over 500 m. There are two "hoot" types, gruff and pure tone, and it appears that males give the former and females the latter (Wright 1985). These calls are given repeatedly by solitary animals, often for 1 to 2 hours. These calls might attract mates, since other solitaries also hoot and move toward the caller (Wright 1985).

In *Callicebus*, both solitary males and females sometimes give short call sequences (Robinson 1979a). Wright (1985) observed group-living subadults chase same-sex solitaries from the home range. On occasion, solitaries may pair up and remain temporarily in an area (Robinson 1979a). Permanent pair formation and establishment of a home range have not been observed.

Social Interactions

Social interactions within groups of the monogamous species are primarily affiliative. In *Callicebus*, aggression is rare, and a dominance rank between the sexes cannot be discerned (Wright 1984). Grooming is the most frequent interaction in *Callicebus torquatus*: group members spend an average of 1 hour a day grooming one another. Bouts are distributed throughout the day, though most grooming occurs in the late afternoon after the group has arrived at its sleeping bough (Kinzey and Wright 1982).

Captive studies of *Callicebus moloch* have more fully described the pattern of interactions in this monogamous species. Males and females familiar with one another approached, followed, came into contact, groomed, and twined their tails more than did pairs that were strangers to one another (Mason 1971). In comparison to heterosexual cage mates of the polygamous cebid *Saimiri*, heterosexual *Callicebus* cage mates were spatially closer to one another, their activities were more synchronized, they spent more time in mutual and unilateral gazing at their cage mate, and they were more distressed when separated from that cage mate (Mason 1974; Cubicciotti and Mason 1975; Phillips and Mason 1976). In paired tests, they preferred to be in proximity to their former cage mate rather than with another opposite sex animal (Mason 1975). Finally, the frequency of agonistic behaviors during feeding was lower than for *Saimiri* cage mates (Fragaszy 1978).

There are indications of sex differences in behavior in both *Callicebus* species. On average, male and female *C. torquatus* groomed one another equally. On the sleeping bough, however, the male groomed each offspring more than he groomed the female, while the female partitioned her grooming more equally among group members (Kinzey and Wright 1982). The data on the first animal during group progressions are contradictory. Adult

females of *C. torquatus* generally led group progressions (Kinzey et al. 1977). *C. moloch* males in Colombia led most progressions (59%; $N = 79$), while females led only 34% (Robinson 1977). In Peru, Wright (1984) found that progressions of *C. moloch* groups were led by adult females on 60% of occasions ($N = 625$) and by adult males on only 19%.

Although also monogamous, *Aotus* does not appear to share the same suite of behaviors. In this genus, aggression is rare, but so is grooming. In captivity, Moynihan (1964) reports grooming only before copulation. At Wright's (1984) study site in Paraguay, grooming did not precede copulation on the three observed occasions, and the pair averaged only 1 minute of grooming a day.

In captive and semi-free-ranging *Cacajao calvus*, aggression is uncommon and clear dominance ranks are only discernible in caged animals. Allogrooming is common, reciprocal, and dyadic (Fontaine 1981). Adult females are the age-sex class most commonly involved in aggressive and affiliative interactions, but Fontaine's captive colony only included a single adult male.

Little is known of intragroup social behavior in the other members of the subfamily. *Pithecia* in captivity frequently groom one another (Buchanan, Mittermeier, and van Roosmalen 1981).

Parental Care

Callicebus and *Aotus* have extensive paternal care (Dixson and Fleming 1981; Fragaszy, Schwartz, and Schinosaka 1982). In captivity, the *Aotus* male carries the infant 70 to 81% of the time during the first month of life (Dixson and Fleming 1981; Wright 1984). In the wild, the *Callicebus* male carries the infant 80 to 90% of the time (Wright 1984). Male carrying in both species continues until the infant is about 4-months old (Dixson and Fleming 1981; Wright 1984) (fig. 5-4). Other caregiving behaviors by fathers include food sharing, playing, and grooming. Siblings carry infants 2 to 25% of the time in these genera, but only during the first 6 to 8 weeks of life. The infant is carried by the mother only 3% of the time in both *Callicebus* and *Aotus* between birth and weaning (6 to 7 months). In the other three genera, infant carrying is done by the female (Buchanan, Mittermeier, and van Roosmalen 1981; van Roosmalen, Mittermeier, and Milton 1981; Fontaine 1981), although Hick (1973) mentions male carrying in *Pithecia*.

INTERGROUP CALLS AND SPACING ◆

Groups of *Aotus*, *Callicebus*, and *Pithecia* all occupy relatively exclusive areas with clearly defined boundaries (Wright 1978; Mason 1968; Buchanan, Mittermeier, and van Roosmalen 1981). Overlap of home ranges is slight. How different species maintain that exclusive use of space varies within the subfamily.

FIGURE 5-4. Family group of owl monkeys (*Aotus trivirgatus*) from Peruvian Amazonia. The adult male is carrying an infant. (Photo: Russell Mittermeier/Anthrophoto)

Callicebus moloch has been described as "territorial" (Mason 1966, 1968) in that, in addition to groups occupying exclusive areas, spacing between groups is maintained by site-dependent aggression: the probability that a group will attack, rather than avoid, another group depends on the site at which the encounter takes place. In *C. moloch,* that probability is low at the center of a group's own range, increases the closer the group is to the boundary, and then drops off rapidly as the boundary is crossed (Robinson 1979a). The outcome of an aggressive encounter between two groups therefore varies with locality. Each group is more aggressive and therefore displaces others when it is within its own exclusive area. Groups are most aggressive close to but on their side of the boundary, a "doughnut"-shaped aggression field (see Waser and Wiley 1979) that results in the clear definition and reinforcement of the conventional location of the boundaries. Boundaries between adjacent ranges are defined and reinforced by frequent encounters between neighboring groups. Boundaries are very stable: one group studied in Peru maintained nearly identical boundaries for 9 years (Janson 1975; Terborgh 1983; Wright 1985).

Intergroup interactions usually begin with the males of different groups calling alone from near their boundaries in the early morning. These loud calls, which can be heard by human observers up to 1 km away, are composed of short "chirrup-pump" sequences and much longer "solo male call" sequences (Robinson 1979a). Very occasionally the female of a mated pair chirrup-pumps simultaneously with the male. These calls mark the location of the calling animal. The response to this calling depends on both the location and proximity of neighboring groups. If the groups are near a common boundary or in proximity to one another, the mated pairs frequently begin loud "duetting," long complex sequences of calls in which the male and female vocal contributions are tightly coordinated (Robinson 1979c). Neighboring pairs counterduet and move toward each other. This brings them together at the boundary. At the boundary, neighboring pairs countercall with duets, "chirrup-pant-pump" sequences, and chirrups. Members of the mated pair sit close together, give arch displays (Moynihan 1966), and sometimes twine their tails. Both male and female might chase their neighbors, but physical contact is rare. Following encounters, groups move away from boundaries and begin feeding (Robinson 1981b). This pattern of group interaction, a sequence of spatial and vocal responses, was experimentally replicated by playing back different recorded calls (to which the group always responded) from different locations within the range and at different proximities to subject groups (Robinson 1979c, 1981b). The frequency and length of intergroup interactions vary with weather, time of the year, and the density of groups, but typically a group will countercall with its neighbors on most mornings. Not all intergroup interactions lead to encounters at the boundary, and the progression of spatial and vocal responses might end at any point. A single solo male call or duet can last several minutes, and the whole interaction frequently lasts several hours.

Callicebus torquatus at some localities apparently defines boundaries in a similar way (Defler 1983). But in a *Callicebus torquatus* population in Peru, the mechanism that regulates spacing is proximity-dependent avoidance of neighboring groups. The location of groups is identified in the early morning by loud duets or solo male calls from near the center of the home range (Kinzey 1977). Experimental playback of solo male calls from locations near and inside the boundary of a subject group's range resulted in the group moving rapidly away, while duet playbacks resulted in counterduetting followed by slow movement away from or parallel to the playback speaker. Groups never approached the speaker or neighboring groups at boundaries. Exclusive use of space is maintained by mutual avoidance and possibly by restricting movements to familiar areas (Kinzey and Robinson 1983). Boundaries are not rigidly defined, and over a period of years they shift considerably (Easley and Kinzey 1985). The difference in the response to intergroup calls, both within and between species of *Callicebus,* might depend on the average range size in different areas. *C. moloch* at Robinson's site had ranges of 4 ha, *C. torquatus* at Defler's site had ranges of 14 ha, and at Kinzey's site, 20 ha. With a larger range, there is a higher cost, both in time and energy, of moving to the edge of a range and interacting with neighbors.

Aotus shows yet a different spacing mechanism. The species has no loud intergroup call. Instead it has a "resonant whoop" (Moynihan 1964; Wright 1981), a series of 10 to 20 notes given in coordination by the mated pair, but the call can only be heard for 50 m and is given only during close-range intergroup encounters. Playback of "resonant whoop" recordings results in the subject group approaching and giving whoops. All intergroup encounters ($N = 15$) observed by Wright (1985) took place when neighboring groups were discovered in fruit trees in border areas during nights when the moon was full and directly overhead. Once during a new moon, two groups met, but silently avoided each other. This suggests that active defense of fruit trees might only be possible when animals can see neighbors, although it is possible that groups range further on moonlit nights and are therefore more likely to encounter other groups in border areas (Wright 1985). Exclusive use of space apparently is maintained by defense of all fruiting trees, rather than space per se, within the range.

Pithecia, like *Callicebus* and *Aotus,* has duetting calls that are given during intergroup interactions. The *Pithecia monachus* "hoo-hoo" is a tonal call given by both sexes that is highly contagious between groups (Buchanan 1978). The "throat rattles" and "twitter growls" of *Pithecia pithecia* are low-pitched, rather atonal, long, pulsed roars, which are also contagious (Buchanan 1978). Although they are pitched higher, they are spectrographically rather similar to *Alouatta* roars (see chap. 6). How these calls are used in spacing is not clear. What is clear is that neighboring groups of *Pithecia* sometimes feed together in the same fruiting tree in zones of overlap (Happel 1982). These multigroup aggregations might be seasonal (Buchanan 1978).

Exclusive use of space, vocal duets, and monogamy are often found together in primates. Exclusive use of space and monogamy are frequently associated because (*a*) to maintain exclusive use of space, the diameter of the home range must be small relative to daily path length (Mitani and Rodman 1979); (*b*) small home ranges are only possible in species that live in small groups and exploit relatively dense resources that are predictable in time and space; and (*c*) monogamous primates occur in small groups. These characteristics are epitomized in *Aotus, Callicebus,* and *Pithecia.* Home ranges are small relative to daily path lengths (see table 5-2), groups are small, and a major part of both *Aotus* and *Callicebus* diet is uniformly dispersed, small fruit trees, with each tree supplying food over an extended period (Easley 1982; Wright 1984, 1985). Duets and monogamy are frequently associated because such complex, mutually coordinated displays can only develop in species with long-lasting bonds (Serpell 1981; Farabaugh 1982). Pair-bonds in these pitheciine species last many years (Robinson 1977; Kinzey 1981). Two further questions might be asked: Are duets important to the maintenance of the group's exclusive use of space or the pair-bond? Why is exclusive use of space advantageous?

Duets certainly communicate between neighboring groups in these species. The vocal and spatial responses to duets are often different from the responses to other loud calls, as illustrated by the *Callicebus* examples. They also depend on the vocal structure of the duet and not its volume. This is illustrated by the duets of *C. moloch* in which the female's calls, which affect the structure of the male's calls, are inaudible to neighboring animals (Robinson 1979c). In addition, duets communicate between mates. Duets are characterized by vocal and spatial coordination in the mated pair in at least *Aotus* and *Callicebus* (Robinson 1981b; Wright 1985). Duets ensure the close proximity of the mated pair during boundary encounters, the only time when they are close to conspecifics.

Exclusive use of space has at least two consequences. One is that it allows animals to know where resources are and to have exclusive access to them. A relationship between intergroup interactions and resource availability was reported in Wright's (1985) study of *Aotus* and *C. moloch.* She found that in both species, intergroup encounters only occurred when trees were fruiting in areas of overlap. However, in Robinson's (1977) study of *C. moloch,* the presence of resources and the probability of intergroup interactions were not clearly related. The other consequence of exclusive use of space is the exclusion of sexual competitors from the range. Exclusive access to mates allows monogamy. Intergroup interactions and sexual competition are related in *C. moloch.* Animals vocalized more strongly and approached more rapidly to calls from same-sex animals, and at the boundary, animals were more aggressive toward same-sex neighbors (Robinson 1981b). However, the observation that *Pithecia* groups sometimes coalesce in border areas

(Happel 1982) indicates that excluding sexual competitors from the home range is not the only mechanism that can isolate animals from opposite-sex conspecifics. Another mechanism is mate guarding, a behavior described during boundary encounters in *C. moloch* (Mason 1966).

The behavioral mechanisms regulating spacing and grouping therefore vary within the subfamily, within a genus, and even within a species. Duets affect the vocal and spatial responses of neighboring groups and of one's mate, but the manner in which they do so varies among species. Exclusive use of space can be maintained by (*a*) site-dependent aggression and regular definition of the conventional location of boundaries, (*b*) defense of an area's resources, or (*c*) site attachment and avoidance of the ranges of neighboring groups. One consequence of exclusive use of space (both living in one's own exclusive area and avoiding the ranges of other groups) is being certain of the location and abundance of resources. Another consequence is the isolation of one's mate from opposite-sex conspecifics. Maintaining an exclusive space however is not the only mechanism that has these consequences. For example, mate guarding also results in the sexual isolation of one's mate. Pitheciines are a splendid illustration of how different species can regulate superficially similar social systems in different ways.

6 | Howlers: Variations in Group Size and Demography

Carolyn M. Crockett and John F. Eisenberg

Howler monkeys, genus *Alouatta,* provide fresh insights and challenges for models of primate societies. Part of the importance of howler monkeys for our understanding of primate societies as a whole derives from the fact that the genus *Alouatta* is a relatively well-studied group. Howlers are most interesting, however, because they present us with numerous examples of intra- and interspecific similarities and differences that seemingly defy our efforts to make sense of them.

♦ TAXONOMY AND DISTRIBUTION

Alouatta is the single genus within the subfamily Alouattinae of the Neotropical family Cebidae (Napier and Napier 1967). The six species currently recognized by most authorities are distributed from southern Mexico to northern Argentina (Wolfheim 1983; Crockett 1986). The species (and alternative or former names), common names, and general areas of distribution are as follows:

Alouatta palliata (*A. villosa*): Mantled howler; southern Mexico; southern Guatemala, and south through the rest of Central America to the west coast of Colombia and Ecuador.

Alouatta pigra (*A. palliata/villosa pigra*): Guatemalan or black howler; southeastern Mexico, Belize, and northern Guatemala.

Alouatta seniculus: Red howler; northern South America, north of the Amazon River in the east and extending south into Bolivia in the west.

Alouatta belzebul: Red-handed howler; northeastern Brazil, south of the Amazon River.

Alouatta fusca (*A. guariba*): Brown howler; southeastern Brazil along the coast, extending into extreme northeastern Argentina.

Alouatta caraya: Black howler; interior of southern Brazil, eastern Bolivia, Paraguay, and northern Argentina.

The six species are found in essentially allopatric (nonoverlapping) distributions. However, small areas of sympatry have been reported for *A. palliata* and *A. pigra* in Mexico (Smith 1970), *A. palliata* and *A. seniculus* in Colombia (Hernández-Camacho and Cooper 1976), and *A. caraya* and *A. fusca* in northeastern Argentina (Crespo 1982).

Howler monkeys have the largest distribution and occupy the widest range of habitat types of any Neotropical primate (Eisenberg 1979; Wolfheim 1983). They are found from sea level to over 2,300 m, in wet evergreen to highly seasonal semideciduous habitats (Wolfheim 1983; Glander 1978; Milton 1980; Neville 1972b; Mittermeier and van Roosmalen 1981; Gaulin and Gaulin 1982; Eisenberg 1979). *Alouatta* exists in communities of up to 11 other primate species (Terborgh 1983) and is reported to occur in most areas where other primates are found (Freese et al. 1982). In some places *Alouatta* is the only primate species, reflecting its ability to adapt to many ecological conditions (Eisenberg 1979). Although *A. palliata* and *A. pigra* are seriously threatened, primarily because of forest destruction within their small geographic distributions, the widely distributed *A. seniculus* and *A. caraya* are among the few primates to receive a "safe" status rating by Wolfheim (1983).

OVERVIEW OF PHYSICAL AND BEHAVIORAL CHARACTERISTICS

Howlers are often referred to as the largest Neotropical primate. Actually, the differences among species and subspecies of howlers in body weight and measurements are such that members of the genera *Ateles, Brachyteles,* and *Lagothrix* are of similar size or larger (e.g., Milton, 1982, table 6; Hill 1962; Napier and Napier 1967). With adult weights of about 4 to 10 kg for all four genera, the largest Neotropical primates are modest in size compared to many Old World monkeys. Howlers are sexually dimorphic in body size, with males larger than females, but *A. palliata* is only about half as dimorphic in weight as *A. seniculus* and *A. caraya* (Thorington, Rudran, and Mack 1979; table 6-1). Furthermore, the scrotum of *A. palliata* males is not evident until the testes descend at puberty, and immatures cannot be sexed by human observers unless the animals are captured (Thorington, Rudran, and Mack 1979). Visible testis descent occurs in infancy or sometime prior to puberty in the other five species (table 6-1). Another intriguing interspecific difference is that one species and one subspecies are sexually dichromatic, with males changing color at puberty (table 6-1).

An outstanding anatomical peculiarity of the genus *Alouatta* is the possession of an enlarged hyoid bone, which acts as a resonator and amplifier when howler monkeys produce their characteristic long calls. The enlargement of the hyoid, while incipient in *Callicebus* and *Pithecia*, is most pronounced in *Alouatta* (Hill 1962). The hyoid is larger in the male than in the female, and its size varies across species—the largest hyoid being in *A. seniculus* and the smallest in *A. palliata* (table 6-1). Larger hyoids produce lower fundamental frequencies and affect the overall tonal quality of the long call or "roar" (Sekulic 1981). Roars are at times produced by males only but often involve a full troop chorus with adult females and juveniles participating (Baldwin and Baldwin 1976b; Chivers 1969; Sekulic 1982a, 1982b, 1982c). Howlers give these loud calls most often in the early morning, but the detection of another troop or extratroop animals can stimulate roaring at any time of day. The temporal patterning of roaring throughout the day may vary seasonally, with suppression of vocalizations during the midday heat of the dry season (Horwich and Gebhard 1983; Sekulic 1982b).

The dawn calls apparently function to announce location, although the effects of calling on movements of troops has not been clearly shown (but see Chivers 1969). Daytime howls are usually given in the context of interactions among neighboring troops whose home ranges in part overlap. The relationship between howling and ranging patterns will be discussed further later in this chapter. Since hyoids are sexually dimorphic in size, the age and sex of the howlers are somewhat revealed by their roars. Thus, it is possible that roaring may provide information about the composition of the troop. The presence of a relatively large number of males, as revealed by their vocalizations, may deter extratroop males

from attempting to invade. Male incursions have been reported for mantled howlers and red howlers (Clarke 1983; Crockett and Sekulic 1984; Rudran 1979; see also Sekulic 1982c).

Aside from the distinctive roaring vocalizations of howler monkeys, numerous other calls show homologies with other Neotropical species (Eisenberg 1976; Baldwin and Baldwin 1976b).

Another notable anatomical adaptation is the howler's prehensile tail, a feature shared with some other cebids (Grand 1978). The tail is used for support during feeding and for crossing gaps in vegetation, including "bridging" behavior in which one animal (often an adult female) holds branches of adjacent trees together while another animal (usually an infant or juvenile) crosses over (mantled howlers: Carpenter 1934; red howlers: Crockett and Eisenberg, pers. obser.). Howler locomotion is usually quadrupedal, on the tops of branches at a relatively slow pace (Carpenter 1934; Schön Ybarra 1984). While howlers generally are considered to be strictly arboreal (e.g., Carpenter 1934), they will come to the ground under some circumstances and in some habitat types, for example, to cross from one clump of vegetation to another (Glander 1978; Schön Ybarra 1984; Terborgh 1983). In taller forests, howlers are usually found in the middle to upper strata, but even in such habitats they spend some time at lower heights and in forest fringes (e.g., Mittermeier and van Roosmalen 1981).

Alouatta palliata is the only species whose behavior patterns have been described in detail (Carpenter 1934; Eisenberg 1976, Appendix IV). However, several of the

TABLE 6-1. Species and Sex Differences in Morphology of Howler Monkeys

Species	Age of Visible Testis Descent	Pelage Dichrom.; Color	A ♀ weight[a] % of A ♂	Hyoid Volume (ml)[b] Male	Female	♀/♂ (%)
Alouatta palliata	Puberty (ca. 3–4 yr)[a,c]	No; blackish or brownish w/ golden "mantle" on back[d]	84	4.5	2.0	44
A. pigra	Infancy (<4 mo)[e]	No; black[f]	—	—	—	—
A. seniculus	Infancy (ca. 1 mo)[g]	No; red-orange; golden dorsum, maroon ventrum[d,g]	69	69.5	12.5	18
A. belzebul	Before puberty[h]	No; black or blackish brown w/rufous to golden extremities[d]	—	56.5	12.5	22
A. fusca	Before puberty[h]		—	39.0	8.5	22
A. f. fusca		No; brown				
A. f. clamitans		Yes; immatures and most adult females brown; adult males rufous[i]				
A. caraya	Infancy (<4 mo)[j]	Yes; immatures blonde; adult females olive buff, yellowish; adult males black[k]	68	23.0	8.0	35

SOURCES: (a) Thorington, Rudran, and Mack 1979; (b) Sekulic 1981; (c) Froehlich, Thorington, and Otis 1981; (d) Hill 1962; (e) Horwich 1983b; (f) Smith 1970; (g) Crockett, pers. obser.; (h) Thorington, pers. comm.; (i) Kinzey 1982; (j) Shoemaker, pers. comm.; (k) Shoemaker 1978.

NOTES: Pelage description is simplified; consult original references for details.
Dichrom. = sexual dichromatism in pelage color.

more stereotyped patterns have been reported for three or four howler species, although their exact form may vary slightly from species to species. These behaviors include tongue flicking, chin-throat-chest rubbing, back rubbing, anogenital rubbing, and the arch display—an aggressive posture in which the back is arched and head lowered, sometimes accompanied by piloerection and branch breaking (Braza, Alvarez, and Azcarate 1981; Carpenter 1934; Eisenberg 1976; Horwich 1983a; Sekulic and Eisenberg 1983; Young 1982; Crockett, pers. obser.). Contrary to old accounts, howlers do not throw feces or purposefully defecate on humans. However, they sometimes move as a group to the same branch for defecation; this usually occurs in frequently used trees after a rest or sleep period, which results in the accumulation of large dung piles (Braza, Alvarez, and Azcarate 1981; Shoemaker 1979; Crockett and Eisenberg, pers. obser.). Some of these behaviors may be involved in olfactory communication. Neotropical primates have relatively well-developed olfactory systems compared to Old World monkeys, and many New World monkeys, including some *Alouatta* species, have odor-producing scent glands (Epple and Lorenz 1967; Epple and Moulton 1978; chap. 35). The arch display also is not unique to *Alouatta,* and a similar behavior is performed by *Leontopithecus* and some other Neotropical monkeys (Rathbun 1979).

DIET

Howlers clearly can be classified as vegetarians, but *folivore-frugivore* is a more appropriate label than *folivore*. Detailed studies have shown that howlers consume considerable amounts of fruit in addition to leaves; young leaves are typically eaten in greater amounts than mature ones (table 6-2). Although ripe fruits are generally preferred (table 6-2), howlers consume more immature fruits than do other sympatric primate species (e.g., Terborgh 1983). Some of the fruits eaten are highly fibrous, such as figs, which comprise a major part of *Alouatta*'s diet; others, like the winged areas of various leguminous fruits, are composed of material similar to leaves and may require similar digestive processes (Milton 1980,

pers. comm.; Milton et al. 1982). Nevertheless, in contrast to Old World colobines which have evolved complex sacculated stomachs and bacterial symbionts to digest a high-cellulose diet, howler digestive tracts are not highly specialized. Their dimensions overlap with some of the more frugivorous species, although the howler's large intestine is somewhat enlarged (Milton 1980; Chivers and Hladik 1980).

Interspecific differences in howler diets appear to be smaller than habitat and seasonal differences within species (table 6-2; Crockett 1986). In moister habitats, both red and mantled howlers spent similar amounts of feeding time ingesting leaves and fruit. In contrast, mantled howlers in a highly seasonal semideciduous habitat spent comparatively more feeding time eating leaves and flowers and less eating fruits (table 6-2). Lists of plant species eaten in the various study areas reveal many families, genera, and even species in common. Consumption of animal matter is insignificant and confined to a few insects (ants, termites, and wasps) that are mostly eaten incidentally.

ACTIVITY PATTERNS AND RANGING BEHAVIOR

Howlers, being diurnal, show little activity at night (Neville 1972a). They also display little activity during the day, sometimes taxing the patience of their human observers. A large part of the day is spent stationary—resting and napping, presumably digesting their bulky diet. Howlers spend 63 to 79% of the daytime at rest, with studies of *A. palliata, A. seniculus,* and *A. caraya* overlapping in these estimates (Milton 1980; Richard 1970; Smith 1977; Gaulin and Gaulin 1982). In contrast, the similar-sized *Ateles geoffroyi* spends only 40 to 54% of daytime at rest (Eisenberg and Kuehn 1966; Richard 1970).

Table 6-3 includes day range lengths and home range sizes from several studies. Average day ranges vary less than 600 m across sites and species, while within-site day-to-day variability is more pronounced. Seasonal differences in day range length have been reported, with

TABLE 6-2. Feeding Time Spent Eating Major Food Types by Howler Monkeys (percentage)

Species	Site	Leaves	(Young)	(Mature)	Fruits	(Unripe)	(Ripe)	Flowers	Sources
Alouatta palliata	1	49	(39)	(10)	51	(9)	(41)	<1	Estrada 1984b
	2	48	(>34)	(<14)	42	—	—	10	Milton 1980
	3	69[a]	(44)	(19)	13	—	—	18	Glander 1978
A. seniculus	4	53	(45)	(8)	42	(14)	(28)	5	Gaulin and Gaulin 1982

NOTES: Sites are as follows: (1) high evergreen rain forest, annual rainfall = 4,500 mm; (2) lowland moist "semideciduous" forest, 2,730 mm; although Barro Colorado Island has a short dry season and Leigh, Rand, and Windsor 1982 classify it as semideciduous, its rainfall and leaf fall pattern make it intermediate between true evergreen rain forest (Estrada 1984b) and true tropical dry forest (Glander 1978); (3) lowland tropical dry forest (semideciduous), 1,431 mm; (4) lower montane wet forest, 1,942 mm.

a. Includes items such as petioles, excluded from young and mature leaf categories.

daily travel farther, on average, during wet than dry season in drier semideciduous habitats (Sekulic 1982b; Glander 1978), and farther in dry than wet season in a less-seasonal forest with higher rainfall (Milton 1980).

Milton (1980) suggests that mantled howlers are travel minimizers, energetically constrained because the foliage part of their diet is low in ready energy. Mantled howlers traveled almost exclusively to reach food sources and did so in a very directed manner (Milton 1980; Estrada 1984b). The travel of red howlers in two semideciduous habitats is in this regard similar to mantled howlers, except that mean travel distances were shorter in a thorny open-woodland habitat than in a gallery forest with a more continuous canopy (Crockett, unpub. data). A more thorough analysis than is possible here will likely show that day-to-day, seasonal, and habitat differences in howlers' typical travel distance are largely explained by the distribution of food resources sought at a particular point in time (Estrada 1984b; Milton 1980; Glander 1978; Sekulic 1982b), as well as by differences in ease of travel imposed by the structure of the habitat type.

Home range size varies more within species than between them (table 6-3). This suggests that, as for day range length, the differences in home range size are a function of environment rather than species membership. However, home range size is more variable than day range, and home range size and average day range length are not significantly correlated (Spearman rank-order correlation = $-.13$, $N = 9$, $p > .05$; data from sources cited in table 6-3; *A. palliata*, *A. pigra*, and *A. seniculus*). Thus, the size of the supplying area on a long-term basis is not a direct reflection of its use on a daily basis. Energetic constraints may make very long day-travel impossible for a howler, while the size of the supplying area required by a troop on an annual basis

TABLE 6-3. Howler Monkey Population and Ranging Characteristics

Species and Study Site	Habitat	Troop Size Mean	Troop Size Range	Individuals per km²	Home Range (ha)	Day Range (m) $\overline{X}$ (min–max)	Sources
Alouatta palliata							
Mexico	EG	9.1	5–16	23	60	123 (11–503)	Estrada 1982, 1984b
Outside reserve		6.0					Ibid.
Costa Rica	SD	8.9–15.4	2–39	90	10	596 (207–1,261)	Heltner, Turner, and Scott 1976; Glander 1978; Clarke 1983
Panama BCI	MSD	8.0–23.0	2–45	16–90	31	443 (104–792)	Milton 1980; Collias and Southwick 1952
Coastal forest		18.9	7–28	1,040	3–7		Baldwin and Baldwin 1976
A. pigra							
Guatemala	EG	5.5	4–7	5, 13	125	250 (40–700)	Schlichte 1978a
Belize	SR	4.4–6.2	2–10	8–ca. 22	11–24		Bolin 1981; Horwich and Gebhard 1983
A. seniculus							
Venezuela	SD	6.3–7.6	3–13	25–54			Braza, Alvarez, and Azcarate 1981
Hato Mas.	SDOW	8.5–10.5	4–16	83–118	4–7	340–445 (20–840)	Neville 1972; Rudran 1979; Sekulic 1982b; Crockett 1984, 1985, unpub.
Hato Mas.	SDGF	5.9–8.3	4–14	<36 53	25	542 (290–655)	Crockett 1984, 1985, unpub.
Colombia	MEG	9.0	9,9	ca. 15	22	706	Gaulin and Gaulin 1982
Colombia	SD	5.8, 6.8	3–9	23, 27	13, 24		Defler 1981
Peru	EG	6.0	5–8	30	10–20		Terborgh 1983
Bolivia	EG	7.4	1–14	8, 120			Freese et al. 1982
Surinam	EG	4.3	2–8	17	6–11		Mittermeier, in Wolfheim 1983
A. caraya							
Argentina	SD	7.2–8.9	3–19	130, 131			Thorington, Ruiz, and Eisenberg 1984

NOTES: Habitat types: EG = evergreen; SD = semideciduous; MSD = moist semideciduous; SDGF = SD gallery forest; SDOW = SD open woodland; MEG = montane EG; SR = secondary riverine. For troop size, density, and ranging data, a dash separates ranges (e.g., over time) and a comma separates two estimates (e.g., two sites).

Home and day range data include estimates from single troops and averages of several troops (consult original reference for details). Densities preceded by *ca.* indicate that individuals per km² were not given in reference but estimated from information presented therein.

may vary enormously depending upon density and distribution of food sources as well as upon the presence of food competitors.

Home range size is inversely related to howler population density ($r_s = -.53$, $N = 12$, $p < .05$, one-tailed). This may not be simply a function of higher food densities permitting the same number of howlers to survive in a smaller supplying area (i.e., home range). There is some evidence that home ranges decrease in average size over time as population density increases within the same site (Chivers 1969). Also, home range overlap may increase as population density increases (Baldwin and Baldwin 1972a). Neither of these possibilities has been conclusively shown.

In fact, despite considerable speculation regarding the nature of howler ranging patterns vis-à-vis territorial defense, few data are available. Milton (1980) and Sekulic (1982b) argue that howlers cannot be regarded as territorial because home range overlap is sometimes considerable. Other authors (e.g., Horwich 1983a) maintain that howlers demonstrate territoriality despite some overlap. Only Sekulic (1982c) presents data showing clearly the location and outcome of intertroop interactions with respect to home range. When interactions occurred the

troops were within sight of each other, either in areas of overlapping use or in their respective ranges. During her year-long study, the home ranges of four red howler troops overlapped 28 to 63% with those of their neighbors, and the troops spent significantly more time than expected in areas of overlap than in areas of exclusive use (Sekulic 1982b). Intertroop interactions in red howlers primarily involve loud vocal displays, often characterized by countercalling, the two opposing troops alternating in their roars (fig. 6-1). Outright displacements are rare (Sekulic 1982c), but sometimes one troop will "charge" the opponent troop, which then retreats a short distance rather rapidly (Crockett, pers. obser.). These interactions are vigorous and are about the most energetically expensive thing that red howlers do. Sekulic (1982c) reported that in 22 of 27 intertroop interactions, the approached (defending) troop howled first; the troop that remained after the interaction called last at a retreating troop in 19 of 25 cases.

Whether mantled howlers differ from red howlers in their spacing patterns is not clear. Baldwin and Baldwin (1976b, 105) did not quantify duration, but indicated that "the troop that eventually retreated first usually roared more than the other troop" and that the troop that

FIGURE 6-1. Red howler monkeys in Venezuela begin the day with strident howls after confronting a neighboring troop. As usual, they roar in unison. (Photo: Carolyn Crockett)

had been using an area was more likely to retreat first and to vocalize more. Home range overlap in mantled howlers is variable in both space and time, with available estimates ranging from 5 to 94%, and even 100% for some individual troops (Carpenter 1934; Chivers 1969; Baldwin and Baldwin 1972a, Milton 1980). It has also been suggested that mantled howler troops tend to avoid each other and that the dawn calls are somehow used in this avoidance (Chivers 1969). However, as nicely put by Sekulic (1982b, 41), "there should be no advantage for a howler troop that occupies a small home range to advertise its movement to one area, because neighbors on the other side may learn that an area is unoccupied. Therefore, although informing neighbors may reduce energy expended in interactions one day, it could also reduce the resources available at the other side of the home range for the following day." Howlers may to some extent be energetically constrained from maintaining exclusive territories (Milton 1980; Sekulic 1982b). Whether territorial or not, however, howlers appear to be actively laying a claim to a piece of essential real estate and defending their right to be there.

REPRODUCTIVE PARAMETERS

Table 6-4 summarizes reproductive data for three of the six howler species. Howlers are not seasonal breeders, and births can occur during any month. However, the distribution of births varies considerably among different howler populations. Births sometimes occur in "clusters" that are not sufficiently predictable to warrant labeling "seasonal peaks" (e.g., A. palliata: Glander 1980; Clarke and Glander 1984). Our A. seniculus population at Hato Masaguaral, Venezuela, shows a distinct pattern of births, averaged over several years, with very few births in May through July (early wet season) com-

pared to all other months (Crockett and Rudran, in prep.). Interbirth intervals after surviving offspring average about 17 to 23 months with a normal range of 11 to 26 months (table 6-4; Crockett and Sekulic 1984; Glander 1980; Milton 1982).

Gestation length (total range 180 to 194 days) and estrous cycle length (averages ca. 16 to 20 days) show considerable interspecific similarity (table 6-4; Colillas and Coppo 1978; Crockett and Sekulic 1982; Glander 1980; Shoemaker 1979). From the available descriptions of changes in female genitalia correlated with estrus, it appears that A. palliata shows subtle cyclic swelling and color change, A. caraya does not, and A. seniculus shows subtle, inconsistent swelling with considerable individual variation (Glander 1980; Colillas and Coppo 1978; Crockett and Sekulic 1982).

Chemical cues of estrus are suspected to be important, although no pheromones have yet been identified. Howler males nuzzle and lick females' genitalia, and, in A. palliata at least, taste urine (Carpenter 1934; Crockett and Sekulic 1982; Eisenberg 1976; Glander 1980; Horwich 1983a). Sexual solicitations are performed by both sexes and often include tongue flicking. Typically females are receptive for 2 to 4 days (Carpenter 1934; Glander 1980; Crockett and Sekulic 1982). In general, a female's estrus is signaled to observers by proximity between the female and a male. This proximity can be sufficiently prolonged (several hours or days) that the label *consortship* seems appropriate. However, in A. seniculus close consortship did not occur between females and unfamiliar males who had recently entered their troops. Instead, females maintained a distance from a new male and intermittently approached and solicited copulations; they withdrew again after the mount was over (Crockett, pers. obser. of three females in two troops).

TABLE 6-4. Reproductive Parameters for Howler Monkeys

| Species and Study Site | Birth Seasonality | Age (mo) at Sex. Maturity | | Gestation Length in Days (range) | IBI in Months (range) | Estrous Cycle | | Sources |
		Male	Female			Cues (phys)	Length (days)	
Alouatta palliata								
Mexico	None							Estrada 1982
BCI	None				ca. 17[a]			Milton 1982
Costa Rica	None (clusters)	42[a]	36[a]	186[a] (180–94)	22.5[b] (18–25)	Subtle	16.3[c]	Glander 1980; Clarke and Glander 1984
A. seniculus								
Venezuela	None (peaks)	58–66[a]	43–54[b]	191[b] (186–94)	16.6[c] (10.5–26)	None reliable	17[b]	Crockett and Sekulic 1982, 1984; Crockett and Rudran, in prep.
A. caraya								
Captive (US and Argentina)	None	24–37[a]	35–42[a]	187[a]		None detected	19.7[b]	Shoemaker 1979, 1982 Colillas and Coppo 1978

NOTES: Litter size usually one. Age at sexual maturity = age at conception of first offspring; IBI = interbirth interval; phys. = physical cues of estrus.

a. $1 < N < 5$.
b. $5 < N < 20$.
c. $N > 20$.

Limited data suggest that species differences may exist in age of sexual maturity, although these may prove to be due to differing environmental (e.g., nutritional) conditions. Female *A. palliata* and *A. caraya* become fertile at about the same age (ca. 3 to 3.5 years), while *A. seniculus* are nearly a year older at maturity (ca. 3.5 to 4.5 years) (Glander 1980; Froehlich, Thorington, and Otis 1981; Shoemaker 1982; Crockett and Rudran, unpub.). There does not appear to be a dramatic sex difference in age of fertility, and a captive-born *A. caraya* male sired an offspring when he was 37-months old (Shoemaker 1982). Physical fertility can, of course, occur prior to the environmental or social conditions conducive to conception by females and fertilization by males. Males in particular may be socially inhibited from siring offspring for several years after they are fertile, for example, until they attain maximum weight. Most red howler females first give birth at about 5 years of age, while males are rarely fathers before the age of 7 years (Crockett and Rudran, unpub.). In general, males are 1 to 2 years older than females when first classified as "adult" (but see Glander 1980). Thus, social adulthood, especially in males, is not directly equivalent to sexual maturity as defined by fertility.

SOCIAL INTERACTIONS AND SPATIAL RELATIONSHIPS

Few studies of social interactions and intratroop spacing have been published, although several investigators have data under analysis. The available reports either do not identify individuals or have not sampled all subjects in an equivalent, unbiased manner. The following is a very brief and necessarily preliminary summary.

Allogrooming among howlers is an uncommon event when compared to the same activity in Old World monkeys (Horwich 1983a; Jones 1980b; Neville 1972a, 1983; Eisenberg 1976; Richard 1970). Less than 2% of a howler's daytime is spent grooming others, and bout durations average only about one minute. Most allogrooming is performed by females, especially adults, and the principal recipients are the females' own infants or adult males. Grooming by adult and immature males is very rare. Overt social interactions generally occur at such a low rate among howlers that spatial relationships probably provide the best quantitative data on associations of individuals within howler troops. In multimale troops, adult males differ in their average proximity to adult females, with the closest male being most likely to mate with the females (Sekulic 1983b; also see C. B. Jones 1982).

The infrequency of interactions may explain why agonistic rank has not been clearly quantified for howlers (fig. 6-2). Clarke and Glander (1984), based on access to preferred food and resting sites, refer to high-ranking and low-ranking mantled howlers, but the ranking data

have not been published. Jones (1980b) presents dominance rankings for two troops of mantled howlers, based on a combination of aggressive interactions and nonaggressive supplantations. Her limited data suggest that adult males ranked above adult females and that, within sexes, younger adults were of higher rank than older ones. In a study of mantled howlers in another site, Smith (1977, 123) writes, "I had the impression that the members of a troop treated each other as individuals and had a rank order. However, I was never able to document the social order of even the males of one troop in spite of my efforts to record the interactions of the four males." Comparable information has not been published for other *Alouatta* species.

INTRA- AND INTERSPECIFIC VARIATION IN SOCIAL ORGANIZATION ♦

Typical troop size and adult composition are important components of any description of social organization. Since *Alouatta* has been studied in a number of loca-

FIGURE 6-2. Adult male red howler, *Alouatta seniculus*. The slashed upper lip is the kind of injury typically attributed to male-male aggression. (Photo: Carolyn Crockett)

tions, this genus is particularly appropriate for assessing the nature of intra- and interspecific variation in these characters. Furthermore, because *Alouatta* species are essentially allopatric, we can rule out competition with sympatric species in the same genus as a source of major differences. We will examine data for four of the six species (little information is available for *A. belzebul* and *A. fusca*). Best studied are *A. palliata,* especially on Barro Colorado Island, Panama (BCI), and *A. seniculus,* especially at our research site, Hato Masaguaral, Venezuela.

Table 6-3 has already suggested that mean troop size shows both intra- and interspecific variation. Figures 6-3 and 6-4 and tables 6-5, 6-6, and 6-7 present additional information for discussing and evaluating intra- and interspecific variation in *Alouatta* social organization. Figure 6-3 shows changes in mean troop size over time for *A. palliata* and *A. seniculus.* Table 6-5 presents the data on population density, mean troop size, mean adult composition, and adult sex ratio upon which the linear regressions in table 6-6 and figure 6-4 are based. Table 6-7 summarizes the adult composition of *Alouatta* troops of four species.

Interspecific differences between *A. palliata* and *A. seniculus* were tested with Mann-Whitney *U* test (Siegel 1956), although it should be kept in mind that some numbers in table 6-5 represent repeated samples of the same population at different times. *Alouatta palliata* differs significantly from *A. seniculus* in having larger mean troop size, more adult males and adult females per troop, and more adult females per male ($U = 18, 22, 0,$ and $48,$ respectively; $N_1 = 16, 13, 13, 13; N_2 = 23; p < .001,$ one-tailed, for all comparisons). The two species do not differ significantly with respect to the population densities at which they are found ($U = 77; N_1 = 10; N_2 = 16; p > .05$). *Alouatta pigra* tends to have the lowest values for these variables while *A. caraya* is similar to *A. seniculus* (table 6-5; fig. 6-4).

Variation in Mean Troop Size

Figure 6-3 shows that mean troop size varies about as much in the same location over time as between sites, in different habitats. Reductions in troop size have apparently resulted from numerous factors; yellow fever epidemics, habitat destruction, abnormal rainfall, food-

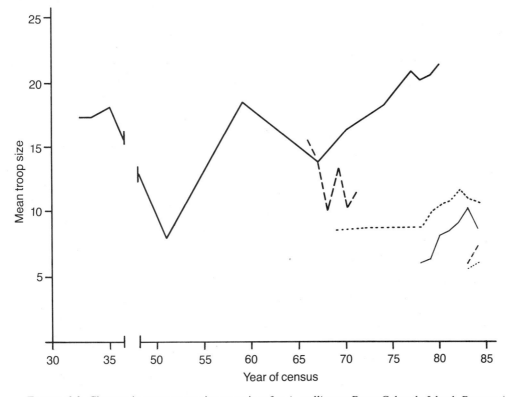

FIGURE 6-3. Changes in mean troop size over time for *A. palliata* at Barro Colorado Island, Panama (————) and Finca Taboga, Costa Rica (— — —); and for *A. seniculus* at Hato Masaguaral, Venezuela, in open woodland habitat (------) and gallery forest habitat (————). Connected data points for 1978 through 1984 are repeated counts of the same established troops over time. Means of new troops in 1983 and 1984 are presented for comparison (woodland — — —; forest ------) (table 6-5 presents overall mean troop size, new and established troops combined). SOURCES: Milton 1982; Heltne, Turner, and Scott 1976; Neville 1972b, 1976; Rudran 1979; and Crockett 1985, unpub.

crop failure, human predation, natural disasters such as hurricanes, and other causes of increased mortality (e.g., Collias and Southwick 1952; Heltne, Turner, and Scott 1976; Crockett 1985). On the other hand, conditions producing decreased mortality may account for some increases in troop size. For example, above average rainfall at Hato Masaguaral during the period 1979 to 1981 may partially account for increases in troop size during that period (fig. 6-3). Milton (1982) suggests that under stable conditions, the optimal troop size for *A. palliata* at BCI is around 19 individuals. In the long term, *A. seniculus* troops at Hato Masaguaral average around 8 to 9 individuals. Variation in troop size within sites at the same point in time may be related to different densities of food resources in troops' home ranges (e.g., Gaulin, Knight, and Gaulin 1980). Mortality and physical condi-

tion appear to be related to dietary quality (Froehlich, Thorington, and Otis 1981; Milton 1982).

Red howlers have been studied at the most locations, and smaller troops tend to occur in evergreen rain forests compared to semideciduous habitats (table 6-3). This is reflected in the finding that mean troop size increases with population density, and howler density is lower in evergreen forests (fig. 6-4a; table 6-3). Interspecific competition with other primate species may be involved; primate diversity and biomass are generally positively related to rainfall and inversely related to seasonal drought conditions typical of semideciduous habitats (Eisenberg 1979).

Complicating all of these generalizations is the revelation that recently formed troops are smaller than troops that have been in existence for several years or more

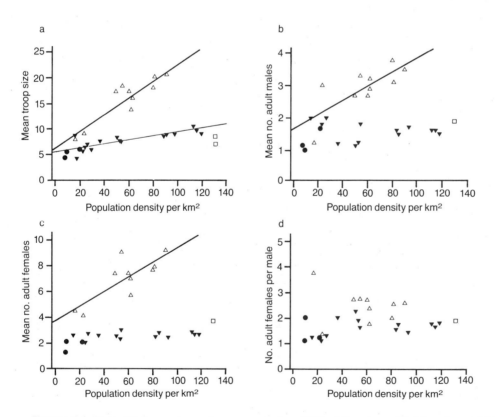

FIGURE 6-4. Plots of (*a*) mean troop size; mean number of (*b*) adult males and (*c*) females; and (*d*) adult females per male, each as a function of population density; mean "rest-of-troop" size as a function of mean number of (*e*) adult males and (*f*) females, where "rest of troop" = mean troop size minus mean number adult males(*e*) or adult

(Crockett 1985). The proportion of new troops in the census can therefore affect the overall mean troop size. Population growth involving new troop formation has been documented for red howlers at Hato Masaguaral (Crockett 1984) and inferred for mantled howlers by the increased number of troops on Barro Colorado Island (Milton 1982).

When Carpenter (1934) began his study of *A. palliata,* part of the island was in early second growth and unoccupied by howlers. By 1959 the second growth had matured to the point where howler troops were occupying areas from which they had been absent during the mid-1930s. With the exception of a crash during the 1940s from yellow fever, the total island population underwent a dramatic increase over the same period. The estimated number of troops islandwide increased from about 23 in 1932 to about 66 in 1974 (Milton 1982), although the manner in which the new troops formed was not documented.

Solitary individuals and small extratroop parties (unisexual and bisexual) have been reported for four species (e.g., Estrada 1982; Schlichte 1978a; Bolin 1981; Gaulin and Gaulin 1982; Rudran 1979; Thorington, Ruiz, and Eisenberg 1984). Some of these form the nucleus for new troops (Crockett 1984; Jones 1980b, 396). At Hato Masaguaral, 6% of the study population of about 500 red howlers is extratroop (Crockett 1985). Red howler population growth at Hato Masaguaral is partly due to an increase in woody vegetation and suitable habitat over the past 20 years. The advance in vegetative cover is related to altered drainage patterns resulting from construction of a major dam and highway in the 1950s as well as to the

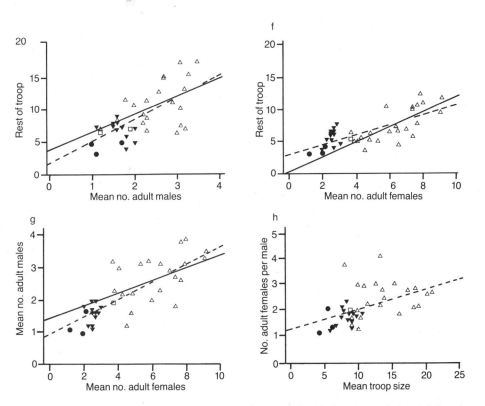

females (*f*); (*g*) mean number of adult males as a function of adult females; and (*h*) adult females per male as a function of mean troop size. Statistically significant linear regressions (table 6-6) are drawn: Thick line = *A. palliata* (△); thin line = *A. seniculus* (▼); dashed line = all four species, including *A. pigra* (●) and *A. caraya* (□)

TABLE 6-5. Population Density, Troop Size, Adult Composition, and Sex Ratio of Howler Monkeys

Species and Study Site	Year	Density per km^2	Mean Troop Size	Mean No. Adult Males	Mean No. Adult Females	A♀/A♂	Number of Troops	Sources
Alouatta palliata								
Mexico		23	9.1	3.0	4.1	1.37	17	Estrada 1982
Panama								
BCI[a]	1932	49	17.3	2.7	7.4	2.74	23	Milton 1982
	1933	60	17.4	2.7	7.4	2.74	28	Ibid.
	1951	16	8.0	1.2	4.5	3.75	30	Ibid.
	1959	54	18.5	3.3	9.1	2.76	44	Ibid.
	1967	62	13.8	2.9	7.0	2.41	27	Ibid.
	1970	62	16.2	3.2	5.8	1.81	6	Ibid.
	1974	80	18.2	3.8	7.7	2.03	6	Ibid.
	1977	90	20.8	3.5	9.2	2.63	13	Ibid.
	1978	81	20.2	3.1	7.9	2.55	13	Ibid.
Coastal		(1,040)[b]	18.9	3.9	8.0	2.06	8	Baldwin and Baldwin 1976a
Costa Rica								
La Pacífica	1967	—	13.2	1.8	7.4	4.11	5	Heltne, Turner, and Scott 1976
	1969	—	10.0	3.2	3.7	1.16	6	Ibid.
	1970	—	12.5	2.0	6.0	3.00	4	Ibid.
Taboga	1966	—	15.4	2.6	7.7	3.00	7	Ibid.
	1967	—	13.5	2.3	6.4	2.78	10	Ibid.
	2/1968	—	10.0	1.6	4.8	2.92	8	Ibid.
	7/1968	—	8.9	2.3	3.8	1.67	8	Ibid.
	1969	—	13.1	3.1	6.5	2.13	17	Ibid.
	2/1970	—	10.4	3.1	5.3	1.68	7	Ibid.
	7/1970	—	10.8	2.3	6.5	2.89	8	Ibid.
	4/1971	—	11.3	2.2	4.9	2.18	22	Ibid.
	6/1971	—	9.9	2.2	4.3	1.94	15	Ibid.
A. pigra								
Guatemala		9[c]	5.5	1.0	2.0	2.00	4	Schlichte 1978a
Belize	1979	8	4.4	1.1	1.2	1.07	13	Bolin 1981
	1981	22[d]	6.2	1.7	2.1	1.27	9	Horwich and Gebhard 1983
A. seniculus								
Venezuela								
Hato Masaguaral								
Gallery	1981	36	7.7	1.2	2.5	2.08	17	Crockett 1984
	1983	50	8.3	1.1	2.5	2.26	21	Crockett, unpub.
	1984	53	7.8	1.2	2.3	1.93	25	Ibid.
Open wd	1969	83[e]	8.5	1.6	2.5	1.56	26	Neville 1972b
	1975	85	8.7	1.5	2.7	1.80	10	Eisenberg 1979, unpub.
	1978	93[e]	8.9	1.7	2.5	1.47	20	Rudran 1979
	1981	112	10.5	1.6	2.9	1.81	26	Crockett 1984
	1983	115	9.7	1.6	2.7	1.70	29	Crockett, unpub.
	1984	118	9.1	1.5	2.7	1.83	29	Ibid.
El Frío		54	7.6	1.8	3.0	1.67	5	Braza, Alvarez, and Azcarate 1981
		25	6.3	—	—	—	141	Ibid.
Colombia								
Andes		15[d]	9.0	2.0	2.5	1.25	2	Gaulin and Gaulin 1982
Llanos I		23	5.8	1.8	2.0	1.14	4	Defler 1981
Llanos II		27	6.8	2.0	2.7	1.33	6	Ibid.
Peru		30	6.0	—	—	—	22	Terborgh 1983
Surinam		17	4.3	—	—	—	17	Mittermeier in Wolfheim 1983
A. caraya								
Argentina	1978	130	8.9	1.9	3.7	1.95	11	Thorington, Ruiz, and Eisenberg 1984
	1979	131	7.2	—	—	—	—	Ibid.

a. BCI = Barro Colorado Island; density estimates are calculated by dividing total population estimate (Milton 1982) by 15 km^2 (the size of the island) for all years except 1932 and 1933. For 1932 and 1933, population size was divided by 8.2 km^2, the approximate area of old forest and part of tall young forest (Foster and Brokow 1982) where howlers were found during Carpenter's (1934) study.

b. Density of 1,040 individual per km^2 reported by Baldwin and Baldwin 1976a is an order of magnitude larger than other density estimates and has been excluded from linear regressions (table 6-6, fig.

6-4). Based on a study of only 10-weeks duration in a habitat whose surrounding area was being cleared, it may not reflect population density on an annual basis.

c. Median of density estimates in two areas (5 and 13 per km^2).

d. Density calculated from information and maps presented in original references.

e. Density estimates calculated to be comparable to Crockett's estimates for same area (original sources estimated densities for subset(s) of larger study area). Open wd = open woodland habitat.

control of range fires and the creation of permanent ponds by the ranch's owner since about 1960 (also see San José and Fariñas 1983).

The formation of new, smaller troops by emigrants from existing, larger troops may account for some decreases in mean troop size, even though overall population density is increasing. This may partly account for the dips in mean troop size on BCI in 1967 and Hato Masaguaral in 1984 (fig. 6-3; table 6-5).

Male Composition

Howler monkey social organization has been described as unimale, multimale, and age graded (Eisenberg 1979). The age-graded system, as originally proposed by Eisenberg, Muckenhirn, and Rudran (1972), differs from other multimale systems in that only one male does most of the breeding, and he is thought to be related to the other, usually younger, males.

TABLE 6-6. Linear Regressions of Population Density, Troop Size, and Adult Composition and Sex Ratio in Howler Monkeys

Variables Independent × Dependent	% Variation Explained	F	d.f.	P
Density × Mean Troop Size				
Four species	11.5	3.77	1,29	NS
A. palliata	81.3	34.80	1,8	<.01
A. seniculus	61.1	21.94	1,14	<.01
Density × Adult Male Mean				
Four species	2.9	.74	1,25	NS
A. palliata	55.6	10.01	1,8	<.05
A. seniculus	4.3	.49	1,11	NS
Density × Adult Female Mean				
Four species	4.0	1.04	1,25	NS
A. palliata	63.4	13.87	1,8	<.01
A. seniculus	21.7	3.05	1,11	NS
Density × Adult Females per Adult Male				
Four species	.9	.22	1,25	NS
A. palliata	3.0	.24	1,8	NS
A. seniculus	9.4	1.14	1,11	NS
Adult Male Mean × Rest of Troop				
Four species	51.0	39.58	1,38	<.01
A. palliata	27.6	7.99	1,21	<.05
A. seniculus	12.2	1.53	1,11	NS
Adult Female Mean × Rest of Troop				
Four species	53.2	43.14	1,38	<.01
A. palliata	54.5	25.12	1,21	<.01
A. seniculus	14.2	1.82	1,11	NS
Adult Female Mean × Adult Male Mean				
Four species	60.6	58.34	1,38	<.01
A. palliata	22.1	5.95	1,21	<.05
A. seniculus	2.3	.26	1,11	NS
Troop Size × Adult Females per Adult Male				
Four species	23.7	11.83	1,38	<.01
A. palliata	1.6	.34	1,21	NS
A. seniculus	10.4	1.28	1,11	NS

SOURCE: Data in table 6-5.

Tables 6-5 and 6-7 provide evidence for variation in male composition. Between species, the percentage of troops that contain only one adult male ("unimale") varies from 0 to 100%, with the mean number of adult males ranging from slightly over one to nearly four. Intraspecific variation is also conspicuous. Variation, in male composition of A. palliata studied in the same site over time varied from 11 to 80% one-male troops (Heltne, Turner, and Scott 1976). However, in comparison to A. pigra, A. seniculus, and A. caraya, troops of A. palliata are more likely to be multimale, and these troops sometimes have three or more males. The other species' multimale troops are usually composed of two and occasionally three adult males (table 6-7).

One factor that partially accounts for troops of A. palliata having more adult males, on average, is that A. palliata males tend to be classified as adult as soon as their testes descend at puberty, while similar-aged males of the other species are classed as subadults for a year or so. However, much of the difference persists despite the problems of age classification. The larger number of males appears to be related to A. palliata's larger average troop size and the positive correlations among troop size, number of adult males, number of adult females, and adult females per male for this species (fig. 6-4, e–h).

Most multimale howler troops may be functionally unimale (cf. chap. 31). In multimale troops, all males do not have equal access to receptive females. Clarke (1983, 244) reported that for A. palliata, "the dominant male was the only male observed consorting or copulating with adult females in peak receptivity in 42 out of 49 female estrous cycles." Similarly, Sekulic (1983a, 198) reported that for A. seniculus in three two-male troops, "only three of the 26 copulations involved a male that did not have an obvious priority of access to females." Furthermore, preliminary paternity exclusion analysis of red howler blood proteins suggests that only one male in a multimale troop usually sires all or most of the offspring (T. Pope, in prep.).

As would be predicted for a polygynous, sexually dimorphic species, male-male competition is sometimes severe. Male invasions and associated infanticides have been reported for A. palliata and A. seniculus, and some males have died of injuries received while fighting other males (Clarke 1983; Crockett and Sekulic 1984; Rudran 1979; Sekulic 1983a). The adult-female-to-male ratio is typically highest in A. palliata (fig. 6-4h), and mortality among maturing males is suspected to be especially high in this species (Froehlich, Thorington, and Otis 1981). It is therefore puzzling that A. palliata has the lowest degree of sexual dimorphism. Socionomic sex ratio and sexual dimorphism are generally positively correlated since both reflect competition among males for females (Clutton-Brock and Harvey 1977a; chap. 13). However, since one red howler male may be more successful at

TABLE 6-7. Adult Composition of Howler Monkey Troops

Species and Study Site	Date	% with 1 A♂	Number of Troops with					Number of Troops with						Sources
			1♂	2♂	3♂	>3♂	Max #♂	1♀	2♀	3♀	4♀	>4♀	Max #♀	
A. palliata														
Mexico		6	1	4	7	5	(5)	0	2	4	4	7	(6)	Estrada 1982, 1984a
Panama														
BCI	1932	17	4	8	4	7	(6)	0	0	2	1	18	(14)	Carpenter 1934
	1951	80	24	6	0	0	(2)	1	3	9	3	14	(9)	Collias and Southwick 1952
Coastal		0	0	2	1	5	(6)	0	0	1	0	7	(13)	Baldwin and Baldwin 1976a
A. pigra														
Guatemala		100	4	0	0	0	(1)	1	2	1	0	0	(3)	Schlichte 1978a
Belize	1979	92	12	1	0	0	(2)	11	2	0	0	0	(2)	Bolin 1981
	1981	44	4	4	1	0	(3)	2	4	3	0	0	(3)	Horwich and Gebhard 1983
A. seniculus														
Venezuela		40	2	2	1	0	(3)	0	1	3	1	0	(4)	Braza, Alvarez, and Azcarate 1981
Hato Mas.														
OW	1978	40	8	10	2	0	(3)	1	10	7	2	0	(4)	Rudran 1979
	1981	44	12	15	0	0	(2)	0	8	15	4	0	(4)	Crockett 1984, 1985, unpub.
	1983	45	13	15	1	0	(3)	—	—	—	—	—	—	Ibid.
GF	1981	91	19	1	1	0	(3)	0	16	8	1	0	(4)	Ibid.
	1983	86	18	3	0	0	(2)	—	—	—	—	—	—	Ibid.
Colombia		30	3	5	2	0	(3)	2	3	4	1	0	(4)	Defler 1981
A. caraya														
Argentina		45	5	2	4	0	(3)	0	2	3	4	2	(7)	Thorington, Ruiz, and Eisenberg 1984

NOTES: Date = year of more than one composition estimate for the same site; Hato Mas. = Hato Masaguaral; OW = open woodland; GF = gallery forest.

monopolizing females in their smaller troops than the dominant male in a mantled howler troop, the actual number of females mated per male may be more similar between these two species than the adult sex ratios suggest.

Are multimale howler troops age-graded? Tooth wear aging of *A. palliata* males on BCI and *A. seniculus* males at Hato Masaguaral suggests that multimale troops are usually composed of different-aged males (Froehlich, Thorington, and Otis 1981; Thorington et al. unpub. data). Younger males tend to be higher ranking in mantled howler troops (Jones 1980b), but no such generalization can be made about red howlers at Hato Masaguaral; in some troops the elder is the breeding male while in others it is the younger. At Hato Masaguaral, about one-third of the multimale troops resulted when one male remained in his natal troop after reaching adulthood; in another third, two males immigrated together. In other cases, a single male invaded and co-resided with the original resident to produce a multimale composition. In short, multimale compositions result from many sources, of which the natal male maturing under the tolerance of his father (the original formulation of the "age-graded" system) is only one of several possibilities.

Even though howler troops sometimes (or in the case of *A. palliata,* usually) have more than one adult male, howler social organization shares similarities with unimale harem systems (chaps. 8, 9). Other species typically thought of as unimale are sometimes found in multimale troops (e.g., *Colobus guereza:* Dunbar and Dunbar 1976; see chap. 9). Dunbar and Dunbar (1976) and Eisenberg, Muckenhirn, and Rudran (1972) suggest that this variation reflects an ontogenetic sequence, with multimale troops resulting from the maturation of natal males over a period of time. Consistent with this idea, newly formed red howler troops are much more likely to be unimale than are long-established troops (Crockett 1985). However, as indicated earlier, this is not the only source of multimale troops.

Female Composition

Troops of *A. palliata* typically have more adult females than the other three species (*A. caraya* may be intermediate between *A. palliata* and *A. seniculus;* tables 6-5, 6-7; fig. 6-4g). Red howler troops generally have two or three and occasionally one or four adult females. At Hato Masaguaral, troops have had single females only after the death of a second female. Bolin (1981) found many *A. pigra* troops with single females, after a recent hurricane which may have resulted in high mortality. Two years later in the same area most troops had two or three females (Horwich and Gebhard 1983).

The mean number of adult females is a somewhat better predictor of the size of the rest of the troop than is the mean number of adult males; this is especially so for *A. palliata* (table 6-6; fig. 6-4f vs. 6-4e). Since there are more adult females than adult males per troop, adult females naturally comprise a larger proportion of troop size. Thus, in order to fairly evaluate the contribution of each sex to variation in troop size, each adult class was compared with the size of the "rest of the troop," defined as mean troop size minus the mean number of adults of that sex. Mean values for *A. seniculus* do not vary much, and the only significant relationship in table 6-6 is between mean troop size and population density (fig. 6-4a). However, when individual red howler troop counts were used (Crockett 1984), the number of adult females explained 26% of the variation in the size of the rest of the troop ($p < .01$) and 24% of the variation in the number of immatures ($p < .01$); in contrast, adult males explained less than 7% of the variation in the rest of the troop (*NS*), the number of immatures (*NS*), and the number of females (*NS*).

The number of immatures accounted for 92% of the variation in red howler troop size (Crockett 1984). Since most immatures are natal, it is the adult females that produce them. Thus, it follows that troop members can directly influence troop size by somehow limiting the number of females allowed to reproduce. This is the most likely proximate mechanism to account for "typical" troop sizes reported for many species. It has been suggested for both red howlers and mantled howlers that females compete among themselves for opportunities to breed (Crockett 1984; Jones 1980b), and thus it could be said that females directly determine troop size. Virtually all adult female howlers in troops produce offspring, and there is no evidence for complete reproductive suppression of troop females once they start breeding. Rather, it appears that maturing females compete for troop membership itself. Recall that mantled howler troops are consistently larger, with more females than the other species. These interspecific differences in troop size and the number of adult females have a proximate explanation in the nature of female dispersal.

It turns out that maturing individuals of both sexes emigrate from mantled and red howler troops (e.g., Crockett 1984; Clarke and Glander 1984). In polygynous primates, male emigration is typical while female emigration is uncommon (Wrangham 1980; Chap. 21). Female transfer has been reported for both mantled and red howlers at several sites (Estrada 1982; Jones 1980b; Crockett 1984; Gaulin and Gaulin 1982). It appears that

female emigration is a regular feature of both species' societies and that female-female competition is a mediating factor. Some females of both species do succeed in staying in their natal troops and breeding there. However, of those that leave, red howler females are much more likely to form new troops with other extratroop males and females than to immigrate into existing troops, where resident females offer much opposition (Crockett 1984; Sekulic 1982a). In marked contrast to red howlers, young mantled howler females commonly immigrate into established troops, where they sometimes oust older, reproductive females, who in turn occasionally form new troops (Jones 1980b). In general, more maturing females leave than stay, but the proportion could vary tremendously depending on mean troop size, carrying capacity, and opportunities for new troop formation. For example, at Hato Masaguaral, about 60% of primiparous births occurred in natal troops in one habitat compared to only 16% in another habitat where most primiparous births occurred in newly formed troops (Crockett 1984).

These data raise several evolutionary questions: (1) Why should females emigrate? (2) Why should it be in the interests of females (rather than males) to determine troop size? and (3) Why should howler species differ in troop size?

Maturing red howler females may emigrate because the number of reproductive positions seems to be limited (presumably by other females) to about four (Crockett 1984). This argument does not extend directly to mantled howlers, whose troops may have as many as 14 adult females, perhaps because it is easier for new females to join. Nevertheless, intrasexual competition appears to be the best explanation for female dispersal in howlers (Crockett 1984; Jones 1980b; chap. 21). Opportunities for colonization also may have favored emigration by females of both species (although this is unlikely to be the primary cause; chap. 21). Howlers are pioneer species capable of living in seemingly marginal habitats (Eisenberg 1979), which perhaps is related to their vegetarian diet and the fact that howlers do not need to drink water.

Incest avoidance does not account satisfactorily for female emigration in howlers since males also transfer and some females stay in their natal troops. Furthermore, male tenure is usually sufficiently short, compared to female age at maturity, so fathers would rarely have the chance to breed with their daughters (Crockett 1984; see also Moore and Ali 1984).

Any evolutionary argument favoring female emigration in combination with male transfer must address the disadvantages of unlimited female recruitment: What limits reproductive female group size, and over what resource are the females competing? Why, in the ultimate sense, females should behave so as to affect troop size stems from different natural selection pressures on the two sexes, rooted in differences in parental investment (Trivers 1972). Current theory suggests that females should be selected to maximize food intake while males compete among themselves for access to those females (Clutton-Brock, Guinness, and Albon 1982; Wrangham 1980). Thus, if there is such a thing as an "optimal troop size," measured for example in foraging efficiency and perhaps involving defense of a suitable home range, then it would be in the greater reproductive interests of females rather than males to affect troop size. Here the *Alouatta* data pose some problems. The degree of interspecific variation in the average number of adult females and correlated troop size traditionally would be explained in terms of differences in foraging strategy or habitats occupied. Or, contrasting predation pressures might favor different troop sizes (van Schaik 1983). However, such interspecific differences between mantled and red howler foraging ecology or predation pressures, if they exist, are not conspicuous.

As a consequence of bisexual emigration and immigration, adults may have few or no adult relatives in the same troop (Clarke and Glander 1984; Crockett 1984). Relatedness in red howler troops is highly variable. In some troops two or more adult females are related (usually as mother-daughter); in other cases males are known to be related, or are possibly so because they emigrated from the same troop. Most newly formed troops are composed of unrelated individuals. Thus members of a howler troop, by roaring in unison at competitors for food, space, and even troop membership, cooperate with kin and nonkin alike in this activity. This suggests that while kinship can be an important correlate of cooperation, it is not an essential component (Wasser 1982; Wrangham 1982).

CONCLUSION

We have presented a rather complicated picture of variation in the social organization of howler monkeys. Even so, we have been unable to address all of the interesting questions posed by *Alouatta*, such as the sexual dichromatism of *A. caraya* and *A. f. clamitans* (but see Crockett 1986). The main interspecific difference among *Alouatta* species is in the typical number of adult females. There is some evidence that troop size and to some extent mean number of adult males track the number of females. The main factor limiting the number of adult females per troop appears to be female reproductive competition. However, the ultimate factors producing species differences in average female group size are open to speculation. Perhaps there is some significant difference between the foraging strategy or resource base of *A. palliata* compared to the other howler species, although such a difference is not conspicuous given present data. On the other hand, *A. palliata* differs from the other species in several aspects of demography and dimorphism, thus suggesting that generalizing from mantled howlers to all howlers should be done with discretion. Finally, *Alouatta* illustrates the oversimplification created by dichotomies such as folivore-frugivore, unimale-multimale, and territorial-nonterritorial, for howlers can be characterized as all of these.

7 | Capuchins, Squirrel Monkeys, and Atelines: Socioecological Convergence with Old World Primates

John G. Robinson and Charles H. Janson

The Cebinae and the Atelinae include both the smallest and the largest cebid monkeys. Members of the subfamily Cebinae include squirrel monkeys (genus *Saimiri*), which weigh under 1 kg, and capuchin monkeys (genus *Cebus*), which weigh from 2.5 to 4 kg. Members of the subfamily Atelinac include spider monkeys (genus *Ateles*), woolly monkeys (genus *Lagothrix*), and the muriqui or woolly spider monkey (genus *Brachyteles*). *Ateles* weigh about 8 kg, *Lagothrix* about 10 kg, and *Brachyteles* weigh about 12 kg. Sexual dimorphism in size is not pronounced in *Saimiri* nor in *Ateles*. Males are larger than females in *Cebus*, *Lagothrix*, and *Brachyteles*. With the increase in body size, males develop larger forelimbs and tail, while the importance of the hindlimbs is reduced (Napier 1976; Hershkovitz 1977). *Saimiri* are quadrupedal and, except in newborns, lack a prehensile tail. *Cebus* have arms that are nearly as long as their legs, opposable thumbs, and great dexterity (Grand 1978). Although their tails are prehensile, they are fully furred, not very dexterous, and only able to support an adult weight for short periods. *Ateles*, *Brachyteles*, and to a lesser extent *Lagothrix* show great finger prehensibility and a reduction or elimination of the thumb—adaptations associated with an ability to semibrachiate. The prehensile tail in *Lagothrix*, *Ateles*, and *Brachyteles* is large, and the distal third of the ventral surface is naked. The tail is able to fully support the body for extended periods and is also capable of precise and agile movements. These adaptations allow these larger genera to gain increased access to fruit and leaves on terminal branches.

TAXONOMY AND DISTRIBUTION

Saimiri (squirrel monkeys) include at least two allopatric species (but see Herskkovitz 1984), the Central American *S. oerstedii* and the South American *S. sciureus*. *Cebus* (capuchin or sapajou monkeys) are split into four species: three allopatric "nontufted" (referring to the absence of hair tufts on the crown) species, *C. capucinus* in Central America, *C. albifrons* and *C. olivaceus* (= *nigrivittatus*) in South America, and the "tufted" *C. apella*, which is sympatric with the last two. The monospecific *Brachyteles* (muriqui) are restricted to the Atlantic coastal forest of Brazil. *Lagothrix* (woolly monkeys) are separated into *L. flavicauda*, which has a restricted range in northern Peru, and *L. lagothricha*, which is distributed widely throughout South America. No *Lagothrix* occur northeast of the junction of the Amazon and Negro rivers. Four allopatric species of *Ateles* (spider monkeys) are recognized—*A. geoffroyi* in Central America, *A. fusciceps* on the Pacific slopes of the Andes in Colombia and Ecuador and in the Darien of Panama, and two restricted to South America, *A. belzebuth* and *A. paniscus*. The taxonomy follows Honacki, Kinman, and Koeppl's (1982) list of mammalian species.

ECOLOGY
Susceptibility to Predation

The smaller genera, *Saimiri* and *Cebus*, are more vulnerable to predators and give alarm calls to large carnivorous mammals, boas, and almost any large flying bird (Robinson 1981a; Terborgh 1983). Terborgh noted attacks on both *Saimiri* and *Cebus* by eagles and hawk eagles in Peru. In Guyana, a pair of harpy eagles brought a *Cebus* monkey to the nest every 25 days on average (Rettig 1978). The larger atelines are less vulnerable to raptors, although *Ateles* do give alarm calls to harpy eagles (Janson, pers. obser.). There are no published reports of attacks or predation on atelines.

Diet

The two cebines, *Saimiri* and *Cebus*, are frugivore-insectivores, while the three atelines, *Ateles*, *Lagothrix*, and *Brachyteles*, are frugivore-folivores (table 7-1). *Ateles* is the most frugivorous, with fruits making up over 80% of items taken (van Roosmalen 1980). *Brachyteles* is the most folivorous, with animals devoting an average of about two-thirds of their foraging time to leaves (Fonseca et al., in press). For the other genera, fruits make up about half of their diet. The smaller *Saimiri* and *Cebus* supplement their diet with invertebrates. *Saimiri*, as leaf gleaners, snatch mostly cryptic, immobile prey, such as caterpillars, from the surfaces of leaves (Thorington 1968). In contrast, *Cebus*, more manipulative foragers, take advantage of their size and strength to reach prey hidden inside the bases of palm

TABLE 7-1. Diet Composition in Capuchins, Squirrel Monkeys, and Atelines

Species and Study Site	% Time		% Foraging on				% Fruit Ripe	Sampling Method	Study Period	Sources
	Foraging	Moving	Fruits	Leaves	Other	Insects				
Time-Taking (TT), Frequency (F), and Items Ingested (II) Methods										
Cebus albifrons Colombia			(	80	)	20		TT	L	Defler 1979a
C. capucinus BCI, Panama	25	46	65	15	0	20		II	L	Freese and Oppenheimer 1981
C. olivaceus Venezuela	47	22	49.1	7.6	1.5	35.1	91.1	F	L	Robinson, in press *b*
			46.7	7.8	1.4	37.0		TT	L	
Ateles belzebuth Colombia	22	15	83	7	10	0	99	TT	L	Klein and Klein 1977
A. geoffroyi BCI, Panama	11	28	80	20	0	0		II	S	Richard 1970; Hladik and Hladik 1969
A. paniscus Surinam			82.9	7.9	6.4	0.2	96.0	F	L	Van Roosmalen 1980
Lagothrix lagothricha Colombia			++	++	+	+		F	S	Kavanagh and Dresdale 1975
Peru			>80	+	+			TT	S	Ramirez, in press
L. flavicauda Peru			++	+	+			F	S	Leo Luna 1981
Brachyteles arachnoides Brazil	25	42	39.0	57.1	4.0	0		TT	S	Fonseca 1983
Brazil	18	27	26.1	66.1	7.6	0		TT	S	Young 1983
Brazil	27	10	19.5	68.2	12.4	0		TT	L	Milton 1984
Time-Foraging (TF) Method										
Saimiri sciureus Surinam			(	28.0	)	72.0		TF	S	Mittermeier and Van Roosmalen 1981
Peru			12.3	0	5.7	82.0		TF	L	Terborgh 1983
Cebus albifrons Peru	61	21	24.6	2.1	9.6	63.6		TF	L	Ibid.
C. capucinus Costa Rica	(	70–80	)			30–51		TF	S	Freese and Oppenheimer 1981
C. olivaceus Venezuela	47	22	31.1	5.2	1.0	55.4	91.1	TF	L	Robinson, in press *b*
Surinam			(	66.7	)	33.3		TF	S	Mittermeier and Van Roosmalen 1981
C. apella Peru	66	21	17.6	2.1	4.7	75.4		TF	S	Terborgh 1983
Surinam			(	52.7	)	47.3		TF	S	Mittermeier and Van Roosmalen 1981

NOTES: Plus (+) = less important; double plus (++) = more important. The time-taking method (TT) records the proportion of time animals spend feeding on different items; the frequency method (F) records the frequency with which different items were taken; the items-ingested method (II) visually estimates the weight of different items ingested; the time-foraging method records the time allocated to foraging for and feeding on different items. Robinson (in press *b*) discusses the influence of different methods on results. Methods measuring the time-taking individual items, the frequency that each item is taken, and the numbers of each item that are ingested give similar results. Recording the time foraging on different items gives different results because the capture success on invertebrates is low.

L = long-term studies, usually lasting at least a year; S = short-term studies.

fronds, dead twigs, branches, and so forth (Freese and Oppenheimer 1981; Terborgh 1983; Robinson, in press b). Both cebines take only small quantities of leafy material. Because foraging for invertebrates is time-consuming, cebines spend 70 to 80% of their day foraging and moving between foraging sites, and they allocate more time to foraging for invertebrates than feeding on fruits. The larger atelines take invertebrates only on occasion. Instead they supplement their diet primarily with flowers and both young and mature leaves. Because harvest rates and digestion time for these items are much higher than those for invertebrates, atelines only spend about half their daylight hours foraging and moving.

When fruits are seasonally scarce, frugivores must either move to areas of greater fruit abundance or consume food items (usually leaves or invertebrates) with lower energetic returns per unit foraging effort. While these responses are not mutually exclusive, seasonal move-ment is generally adopted by the "ripe-fruit specialists": *Saimiri* in Peru (Terborgh 1983; fig. 7-1), the nontufted *Cebus* species (Freese and Oppenheimer 1981; Robinson, in press a), and *Ateles* (Van Roosmalen 1980). These species move to different locations within their large ranges in order to exploit seasonally fruiting trees. In contrast, *Saimiri* in Colombia (Bailey, pers. comm.) shift almost exclusively to insects, while *Cebus apella* in Peru switch from fruit to *Astrocaryum* palm nuts, *Scheelea* frond pith, and more invertebrates (Janson, pers. obser.).

Movements, Use of Space, and Intergroup Relations

Mean day range generally is longest in *Ateles*, which rely on ripe fruits, a patchily distributed resource. It is intermediate in *Cebus* and *Saimiri*, which rely on both fruits and insects, and shortest in *Lagothrix* and *Brachyteles*, which take fruit and leaves (table 7-2).

TABLE 7-2. Group Size, Use of Space, Population Density, and Intergroup Overlap in Capuchins, Squirrel Monkeys, and Atelines

Species and Study Site	Group Size	Home Range (ha)	Home Range per Ind	Mean Day[a] Range (km)	Day[a] Range (km)	Population Density (#/km)	Area per Ind	Over-lap[b] Index	Inter-group[c] Relations	Sources
Saimiri oerstedii										
Panama	23				2.5–4.2	130	0.8	Great		Baldwin and Baldwin 1981
S. sciureus										
Peru	35	>250	>7.1			50	2.0	>3.6	+	Terborgh 1983
Colombia	25–35	65–130				50–80	1.5	2.2		Klein and Klein 1975
Colombia, S. Sofia	42	33	1.3	1.5	0.7–2.3	175	0.6	2.2	+	Bailey, pers. comm.
	54	44	1.2							
Cebus albifrons										
Colombia	35	115	3.3							Defler 1979a
Peru	15	>150	>10.0	1.8	1.5–2.2	24	4.2	>2.4	−	Terborgh 1983
C. capucinus										
BCI, Panama	11–15	80	5.9			18–24	4.8	1.2	−	Freese and Oppenheimer 1981
Costa Rica	15–20	50	2.9	2	1–3	5–7	16.6			Freese 1976
C. olivaceus										
Venezuela	20	257	12.9	2.1	1.0–3.6	25	4.0	3.2	−	Robinson, in press b
C. apella										
Colombia	16	>260	>16.3		1–3			>2		Izawa 1980
Peru	10	125	12.5	2.1	1.6–2.6	40	2.5	5.0	−/+	Janson, in prep.
Ateles belzebuth										
Colombia	18	259–389	18.0		0.5–4.0	15–18	6.1	3.0	−	Klein and Klein 1976
A. geoffroyi										
BCI, Panama	15	100–115	7.2							Dare 1974
A. paniscus										
Surinam	18	220	12.2		0.5–5.0	8	12.2	1.0	−	Van Roosmalen 1980
Lagothrix lagothricha										
Colombia	23–43	>400	>13.3	1		12	8.3		+	Nishimura and Izawa 1975
Brachyteles arachnoides										
Brazil	22	300				5				Fonseca 1983
Brazil	20	150		0.8	0.5–1.6					Young 1983
Brazil				0.6	0.4–1.4					Milton 1984

a. For *Ateles* and the *Brachyteles* population studied by Milton 1984, these figures are for followed subgroups or parties.

b. Home range overlap measured using $O = HR \times D/GS$, where HR is average home range size, GS is average group size, and D is population density. This provides a measure of the average number of group ranges completely overlapping at any one point in space (Terborgh 1983).

c. Plus (+) = tolerant; plus or minus (+/−) = tolerant in some circumstances; minus (−) = intolerant.

FIGURE 7-1. (*top*) A young adult female wedge-capped capuchin (*Cebus olivaceus*) feeding on figs; (*left*) a squirrel monkey eating a guava. (Top photo: John Robinson; left photo: Robert Bailey/Anthrophoto)

The species that rely most heavily on patchily distributed fruit trees have the largest ranges. In Peru, both *Saimiri* and *Cebus albifrons* have much larger ranges than *C. apella* (Terborgh 1983). Comparisons of species in different areas do not reveal the same relationship, however, because of variations in the spatial distribution of resources and the density of animals.

Range overlap among groups depends on population density and home range size, and thus intraspecific variation at different study areas is as great as interspecific variation at a single site (table 7-2). For instance, in Colombia, *C. albifrons* groups share only about 20% of their ranges with neighboring groups (Defler 1979a), while in Peru intergroup overlap is complete (Terborgh 1983). Overlap of neighboring *Ateles* groups is 20 to 30% in Colombia (Klein and Klein 1975) but virtually nonexistent in Surinam (van Roosmalen 1980).

Intergroup relations in *Saimiri* and *Lagothrix* seem to be generally amicable, and groups can coalesce into large aggregations for extended periods (Baldwin and Baldwin 1981; Durham 1975; Nishimura and Izawa 1975). In contrast, relations between groups of *Cebus, Ateles,* and *Brachyteles* are usually hostile (Terborgh 1983; Klein and Klein 1975; Young 1983; see also chap. 22). In *Ateles,* conflicts typically begin with males calling, often accompanied by one or more females. As other animals are drawn to the area, conflicts may escalate to include intense visual and vocal displays by members of both sexes (Klein 1974; van Roosmalen 1980). In *C. olivaceus,* conflicts typically consist of running and chasing

skirmishes between the adult and subadult males, and sometimes also include the high-ranking females of interacting groups (Robinson, in press *b*). Only in *Ateles* is there convincing evidence that groups defend fixed ranges (van Roosmalen 1980). In *Cebus,* with the possible exception of *C. capucinus* (Freese and Oppenheimer 1981), groups defend resources or the area where the group is located. *Cebus albifrons* groups defend the resource patch that the group is exploiting (Terborgh 1983). *Cebus olivaceus* groups do likewise, and playback experiments have shown that the probability of intergroup conflict is higher when animals are feeding in concentrated food patches such as fruit trees rather than on more scattered food items such as invertebrates (Robinson, in press *a*). Intergroup conflicts in *C. apella* also occur when groups are near or in fruit trees. In other contexts, or if the fruit trees are very large, groups will tolerate one another (Terborgh 1983). The outcome of interactions is site-dependent in *Ateles* and possibly in *C. capucinus;* groups displace others in more central regions of their ranges (van Roosmalen 1980; Freese and Oppenheimer 1981). In *C. olivaceus,* the outcome depends largely on the size of the interacting groups (Robinson, in press *b*). Intergroup tolerance does not appear to be related to home range size or the extent of range overlap (see table 7-2).

Population Densities

Despite the variation in population densities at different locations, there is an inverse correlation between body size and population densities when species of similar diet are compared. Furthermore, when body size is held constant, the more folivorous species achieve greater population densities than do the more insectivorous ones (fig. 7-2). These correlations are probably related to energetic and trophic factors respectively: a given resource can support more small animals than large, and within the forest the biomass of leaves greatly exceeds that of insects (Leigh, Rand, and Windsor 1982; see also chap. 17).

CONSERVATION

Both cebines and atelines are restricted to arboreal habitats, so their survival is threatened by the increasing destruction of the forest by human encroachment (see chap. 39). The most vulnerable species are those with restricted ranges in areas of high human density, those susceptible to human disturbance, and those that occur at low population densities. *Brachyteles* are restricted to the Atlantic coast of Brazil, where forests have been decimated. Population density is naturally low, and animals are also hunted for food, two factors that exacerbate the effects of habitat destruction. The total population of this genus may be no more than about 400 animals (Fonseca

1983). The other two atelines, *Lagothrix* and *Ateles,* are also hunted heavily in areas near human habitation because they are large and considered savory eating (Castro, Revilla, and Neville 1975; Yost and Kelley 1983). *Lagothrix flavicauda,* because of its limited distribution in northern Peru, is listed as endangered in the IUCN *Red Data Book* (1982) (fig. 7-3). All the other species of *Ateles* and *Lagothrix* are listed as vulnerable or indeterminate. No *Cebus* species is currently threatened, although in certain areas animals are hunted heavily (Hawkes, Hill, and O'Connell 1982). *Saimiri oerstedii,* while locally abundant, is restricted to a small area in western Panama and southern Costa Rica and is highly endangered.

REPRODUCTIVE PARAMATERS

Larger cebids have lower reproductive rates than smaller ones (Eisenberg 1978); rates range from one birth every 3 to 4 years in *Ateles* up to about one birth every year in *Saimiri* (table 7-3). This trend is not caused by differences in litter size, which is one, nor by the duration of gestation, which varies from 150 to 160 days in the cebins up to 225 to 230 days in *Ateles* (based on captive data). Instead, most of the difference is accounted for by variation in the period of infant dependency, which ranges from 5 months in *Saimiri* up to 3 years in *Ateles* (fig. 7-4). Females do not appear to resume cycling until their infants either are weaned or have died. *Cebus,* for

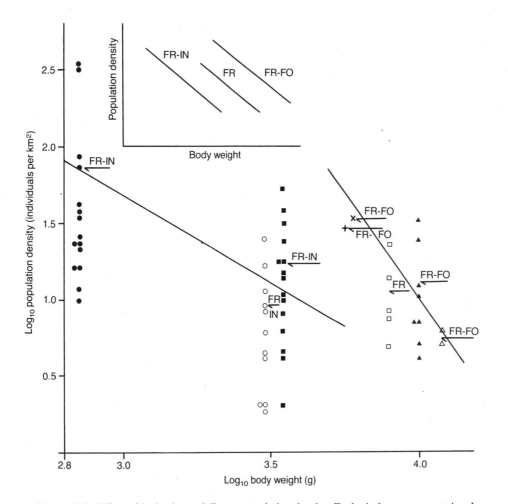

FIGURE 7-2. Effect of body size and diet on population density. Each circle, square, or triangle represents a single density estimate. Arrows indicate mean densities of the following: ● = *Saimiri* spp.; ○ = *Cebus* spp. (nontufted); ■ = *Cebus apella* (tufted); □ = *Ateles* spp.; ▲ = *Lagothrix lagotricha*; △ *Brachyteles arachnoides.* Mean densities of *Alouatta seniculus* (+) and *A. palliata* (X) are also included because they are large frugivore-folivores often included in the subfamily Atelinae (Rosenberger 1981). Linear regression lines through the means of frugivore-insectivores (FR-IN) and through the means of frugivore-folivores (FR-FO) are for visual comparison of trends. Inset shows expected relationship. SOURCES: Milton and de Lucca 1984; Ramirez, in press; Robinson and Ramirez 1982; chap. 7. Densities calculated from river surveys or habitat islands are not included (modified from Robinson and Ramirez 1982).

FIGURE 7-3. The yellow-tailed woolly monkey (*Lagothrix flavicauda*) is in danger of extinction due to its limited distribution and human hunting. (Photo: Andrew Young)

FIGURE 7-4. Black-handed spider monkey (*A. geoffroyi*) female with infant. (Photo: Russell Mittermeier/ Anthrophoto)

example, produce infants about every 2 years if the infant survives, but every year if the infant is lost before the next breeding season (Janson and Robinson, unpub. data). The degree of birthing synchrony varies greatly: births in *Saimiri* occur in a clearly defined season, in *Cebus* mostly within a period of about 3 months, and in *Ateles* throughout the year.

The age of maturity is difficult to assess precisely. For females, one measure is the age at which they first give birth. This occurs at 2.5 to 3 years in *Saimiri*, 4 to 7 years in *Cebus*, 5 years in *Ateles*, and 6 years in *Lagothrix* (table 7-3). Females may show some estrous behavior

and may even copulate years before their first conception. For males, there is no accepted measure. In *Cebus olivaceus*, for example, males may copulate with females when they are 3-years old, although these copulations are usually outside the breeding season. They emigrate from their natal troops usually between the age of 2 and 8, but do not appear to reach full adult size until they are about 15. Even then they may remain reproductively inactive if they have not assumed breeding control of a group (Robinson, unpub. data). In *C. capucinus*, Opphenheimer (1969) reports that males reach full adult size at about 8 years of age.

TABLE 7-3.   Reproductive Parameters for Capuchins, Squirrel Monkeys, and Atelines

Species and Study Site	Birth Seasonality	Age at Sexual[a] Maturity (years)		Gestation Length (days)	Interbirth Interval (months)	External Signs of Estrus[b]	Sources
		Male	Female				
Saimiri sciureus							
S. Sofia, Colombia	Feb.–Mar.					PT[c]	Bailey, pers. comm.
Panama	July	2.5–4	2.5	152–68	14?	PT	Baldwin and Baldwin 1981
Cebus albifrons							
Colombia	None (peak: Feb–July)				14?	AA	Defler 1979b
Peru	None (peak: Aug–Jan)	(?/3)			18	AA	Janson, pers. obser.
C. capucinus							
BCI, Panama	None (peak: Dec–Apr)	8	4	157–67	19	AA	Freese and Oppenheimer 1981; Wolf et al. 1975
C. olivaceus							
Venezuela	May–Aug	15 (3/2)	6 (2/6)		26	AA	Robinson, unpub. data
C. apella							
Peru	Oct–Jan	(2/6)	(3/5)	149–58	22	AA	Janson, unpub. data; Wright and Bush 1977
Ateles belzebuth							
Colombia	None				>36	AA	Klein 1971
A. fusciceps							
captivity	None?	4.8	4.3	226	23–31	AA	Eisenberg 1973, 1976
A. geoffroyi						AA	Klein 1971; Dare 1974;
Panama and captivity	None	5	5	226–32	22–46 17–46		Eisenberg 1973, 1976
A. paniscus							
Surinam	None (peak: Nov–Feb)	4.2–5.4	4.2–5.4		46–50	AA	Van Roosmalen 1980
Lagothrix lagothricha							
captivity	Inadequate information	8 (3/?)	8 (?/6)	225	18–24	AA	Wolf et al. 1975; Williams 1967
Brachyteles arachnoides							Fonseca 1983
Brazil	Inadequate information						

NOTE: Litter size for all is one.

a. Numbers in parentheses are other specified measures of sexual maturity: for males, entry preceding slash is earliest reported age of first copulation with estrous females, age after slash is earliest reported age of emigration from natal group; for females, entry preceding slash is earliest reported age of first estrous behavior, entry after slash is earliest reported age of first pregnancy.

b. PT = passive tolerance of males; AA = active approach to

males. Males also appear sensitive to olfactory cues of estrus in the female's urine in *Saimiri* (Candland et al. 1980), *Cebus* (Robinson 1979b), *Ateles* (Klein and Klein 1971; Eisenberg 1976), and *Brachyteles* (Milton 1984). No data are available for *Lagothrix*.

c. In *Saimiri*, females are apparently sensitive to olfactory cues in male urine associated with male breeding condition (Candland et al. 1980).

SOCIAL ORGANIZATION
Migration between Groups

Migration of males is the rule in cebines, although a low level of movement by adult females has been recorded in most species (table 7-4). In *Cebus,* migrants move directly from one group into another, and solitary animals are rarely observed (Oppenheimer 1969; Robinson and Janson, unpub. data). Little information is available for the atelines. In *Ateles,* females often visit neighboring groups, especially when they are carrying newborn infants, and young females may emigrate permanently (van Roosmalen 1980). Whether male *Ateles* emigrate or remain in their natal groups is not known.

Group Size and Composition

The largest social groups occur in *Saimiri,* in which stable groups of over 50 have been reported (Baldwin and Baldwin 1981). Groups are generally smaller in the other genera (table 7-2). *Cebus* group sizes range from 10 to 35, with the nontufted species forming the largest groups (Freese and Oppenheimer 1981). *Lagothrix flavicauda* live in stable groups of between 4 and 10 individuals (Graves and O'Neill 1980; Leo Luna 1981). Estimates of group size in *L. lagotricha* are based on counts rather than long-term studies— a problem if groups coalesce. The available evidence indicates that groups usually contain between 10 and 30 animals (Kavanaugh and Dresdale 1975; Nishimura and Izawa 1975; Ramirez, in press). Groups of *Brachyteles* range in size from 15 to 25 individuals (Young 1983; Fonseca 1983). At some sites, these groups move as single, widely dispersed units (Young 1983; Fonseca 1983), while at others groups fragment into subgroups or parties that move independent of one another (Milton 1984). *Ateles* groups also range from 15 to 25 individuals and invariably fragment into subgroups of varying size and composition. Most subgroups are composed of 3 or fewer individuals, and solitary animals are common, but subgroups are larger in months when resources are plentiful (Klein and Klein 1977; van Roosmalen 1980). Subgroup composition in *Ateles* is highly variable and temporary. The most permanent association is that of a mother and her dependent offspring. The most frequently observed composition is an adult male, an adult female, and her dependent offspring (Izawa, Kimura, and Nieto 1979; van Roosmalen 1980).

Multimale, multifemale groups are the norm in cebines and atelines, although a single male may occur with several females when groups are small. In most species there are more adult females than males (table 7-4), although the sex ratio is even in *Cebus apella* in Peru. In *Cebus,* unbalanced sex ratios are partly the result of the later maturation of males, since the sex ratio bias disappears nearly completely when both adults and subadults are counted (Robinson and Janson, unpub. data). In *Ateles,* however, females outnumber males by two to one (van Roosmalen 1980). In cebines, the ratio of juveniles to adults is nearly even, whereas atelines appear to have fewer juveniles than adults, perhaps because of their low reproductive rate.

Breeding System

All genera are potentially polygamous. *Saimiri* has a restricted mating season, when males fatten and become reproductively active (DuMond and Hutchinson 1967). Males aggressively compete for females by displaying and sometimes seriously wounding one another. While there is no strong relation between dominance and mating success among breeding males, subadult or subordinate males are often peripheralized and do not breed. Female *Saimiri* are promiscuous, with mounts initiated by either sex (Baldwin and Baldwin 1981).

Male-male competition is less obvious in *Cebus.* Usually a single female is in estrus at a time, and males rarely fight over access to her (Janson 1984). Although the dominant male does not guard the estrous female, he is the only one to achieve substantial mating success. In *C. apella,* this success results from female choice, since the estrous female follows and solicits copulations from the dominant male almost continuously for 3 to 4 days in the middle of her 4 to 6 days of mating activity (Janson 1984). On the remaining 1 to 2 days, usually at the end of her behavioral estrus, a female solicits and copulates with up to six subordinate males in a day. These matings are likely to be infertile, however, since ovulation probably occurs in the middle of estrus (Wright and Bush 1977).

In *Ateles,* aggression between males is also infrequent, and estrous females actively choose their mating partners. Although some females form consortships with specific males that last for up to 3 days, they have also been observed to mate with a number of different males on a single day. Females initiate sexual activity, and the face-to-face copulation may last for up to 25 minutes (Eisenberg 1976; van Roosmalen 1980). In *Brachyteles* also, successive copulations with many male partners has been reported (Aguirre 1971).

SOCIAL INTERACTIONS
Aggression and Dominance

Dominance hierarchies, based upon the outcome of agonistic interactions, occur in cebine groups, but they vary in structure and stability. Dominance ranks of all the animals within *Cebus* groups are linear, and rank reversals rare. In Colombia, only 9.1% of all interactions in a *C. apella* group involved a rank reversal (Izawa 1980). In another group in Peru, reversals accounted for only 0.7% of aggressive interactions, and nearly all changes in the hierarchy during the 7-year study resulted from growth and maturation of juveniles (Janson, unpub. data). *C.*

olivaceus and *C. albifrons* also have stable hierarchies (rank reversals = 10.8% and 11.6% of all interactions respectively (Robinson 1981a; Janson, unpub. data). When only females are considered, dominance hierarchies are even more stable (for *C. olivaceus*, reversals = 6.9%). In contrast, *Saimiri* adult females cannot be ranked, and adult male ranks are very unstable and can change drastically every few weeks (Baldwin and Baldwin 1981). In *Ateles,* aggression is rare, but adult males can be clearly ranked; there are also dominance relations among some females (van Roosmalen 1980).

In all well-studied species, at least some adult males are always dominant to adult females in dyadic interactions. However, both *Saimiri* and *Cebus* females have been observed to form successful coalitions against males (Baldwin and Baldwin 1981; Robinson 1981a). In addition, the highest-ranking female in *Cebus* is dominant to all but the highest-ranking male (Robinson 1981a). Thus, in both *Saimiri* and *Cebus,* spatial relationships within the group and access to resources can be largely regulated by females. In contrast, in *Ateles* and *Brachyteles*

(Klein 1974; Fonseca et al., in press), adult males are invariably dominant to females, and adult male *Ateles* also exhibit more aggressive behaviors than do females (Fedigan and Baxter 1984). Female coalitions have never been observed, but males form coalitions against females (Klein and Klein 1971; Klein 1974). Adults are usually dominant to juveniles, but in *Cebus,* subadults and juveniles of both sexes frequently dominate some adult females (Robinson 1981a; Janson, unpub. data).

Most aggression in *Saimiri* occurs during the breeding season, when males form coalitions against one another. At other times of the year, aggression is most commonly directed against adult males by coalitions of females (Baldwin and Baldwin 1981). In all other species, aggression occurs most frequently over food, foraging sites, or water (Klein 1974; Freese and Oppenheimer 1981; Janson and Robinson, unpub. data). In *Cebus,* the dominant male and female are the two highest-ranked animals in the group, and most agonistic interactions involve either this male (*C. apella* in Peru: Janson, unpub. data; *C. albifrons* in Columbia; Defler 1979b), the female (*C. ap-*

TABLE 7-4.   Social Organization of Capuchins, Squirrel Monkeys, and Atelines

Species and Study Site	Group[a] Size	Group Composition[b]			
		Adult ♂	Adult ♀	Juveniles	Infants
Saimiri sciureus					
Colombia	22	3	5	10	4
S. Sofia, Colombia	41.7	3	12.3	17	9.3
Panama	25	2	6.5	11	5.5
Cebus albifrons					
Colombia	35	7	10	10	8
Peru	14.5	2.5	4.5	3.5	4
C. capucinus					
BCI, Panama	15	1	5	5	4
C. olivaceus					
Venezuela	>18.3	1.4	6.1	8.6	2.2
C. apella					
Peru	10	3	3	2.3	1.7
Colombia	16.3	3.7	5.7	5.3	1.7
Ateles belzebuth					
Colombia	>19.5	4	11.5	4	>0
A. paniscus					
Surinam	18	3	8	5	2
Lagothrix flavicauda					
Peru	9.1	3.0	4.0	0.8	1.2
L. lagothricha					
Colombia	>23.4	9	7.5	3.8	2.8
Brachyteles arachnoides					
Brazil	22	6	9	2	5
Brazil	25	8	8	4	5

NOTES: Ad = adult; SA = subadult. M = male; F = female.

a. Mean group sizes are underestimates in studies in which some of the larger groups were incompletely counted.

b. Counts combine subadult and juvenile age classes.

ella in Columbia: Izawa 1979; *C. albifrons* in Peru: Janson, pers. obser.), or both (*C. olivaceus* in Venezuela: Robinson 1981a).

As in most other primate species (see chap. 24), kinship appears to influence the distribution of female aggression. In all cebines and atelines studied, females show little aggression toward their dependent offspring and are often relatively tolerant of their older juveniles as well. Males may also show reduced aggression toward potential offspring. In one group of *Cebus apella* in Peru, the dominant male was tolerant of the juveniles that he may have sired, but aggressive toward those he could not have fathered (Janson 1984).

Affiliative Behaviors

In all cebine and ateline species, the relationship characterized by the most frequent affiliative interactions is that of the mother and her youngest offspring. The species differ, however, in which other affiliative relationships are most conspicuous. In *Saimiri*, it is the "F-class" relationship—that between two or three specific adult fe-

males (Vaitl 1978). In *Cebus*, the most common affiliative behaviors are directed by adult females to other females and to the adult male of the group (Robinson, unpub. data). In *Ateles*, it is the males who are the most affiliative, both to other males and to adult females (Fedigan and Baxter 1984).

The frequency of grooming varies widely among species. It is virtually absent between adults in *Saimiri* (Baldwin and Baldwin 1981), but it is the principal activity of adults during rest periods in *Cebus* (Janson and Robinson, unpub. data). In all species, mothers regularly groom their infants. Among *Cebus* species, dominant individuals are usually involved in a disproportionately high percentage of grooming bouts and often receive far more grooming than they give. For instance, the dominant male and female in a group of *C. apella* were involved in 63% ($N = 60$) of the bouts among eight adults and received nearly twice as much grooming as they gave (Janson, unpub. data). In *Ateles*, adult males and females with dependent offspring are involved in grooming bouts more than juveniles and females without offspring,

TABLE 7-4.   *(continued)*

Solitary Animals % Pop.	Migration		Breeding System		Sexual Dimorphism (♂ : ♀, kg)	Sources
	Sex	Age (yr)	♂ – ♂ Agress.	♀ Choice		
<5	M	Ad				Thorington 1968
<5	M	SA	High	Low		Bailey, pers. comm.
<5	M	SA			0.9:0.6	Baldwin and Baldwin 1981
					2.4:1.6	Defler 1979b
<1	M	3+				Janson, unpub. data
Rare	M	Ad	Med.		2.5:1.6	Freese and Oppenheimer 1981
	F	Ad				
Rare	M	2+	Med.	High		Robinson, unpub. data
	F	old				
<5	M	5+	Low	High	3.5:2.9	Janson 1984, unpub. data
	F	Ad				
Rare						Izawa 1980
Common					:5.8	Klein and Klein 1977
Common	F	Ad	Low	High	7.9:7.7	Van Roosmalen 1980
Rare						Leo Luna 1982
					11:7	Nishimura and Izawa 1975
						Fonseca 1983
Common	F	Ad			15:12	Young 1983

and give more grooming than they receive (Eisenberg and Kuehn 1966; Eisenberg 1976).

Alloparenting is common among the cebines and includes carrying juveniles, "baby-sitting," active or passive sharing of food, and even nursing. Subadult and adult female *Saimiri* commonly carry and play with young infants other than their own. Such "aunting" is usually performed by only one or two particular individuals (Baldwin and Baldwin 1981). In *Cebus*, adult females and juveniles of both sexes show great interest in infants from birth until about 2 months of age and will attempt to touch or pull an infant off its mother's back (Freese and Oppenheimer 1981). Infants aged over 4 months frequently move independent of their mothers, but must be carried when the travel route is difficult or when the group is moving rapidly. In one *C. olivaceus* group, alloparents, usually juveniles, were responsible for 52% ($N = 519$) of these carries. Alloparents showed a preference for same-sex infants, but females carried more frequently and for longer periods of time than did males (Robinson, unpub. data).

In *Cebus*, baby-sitting is performed in two ways. In *C. apella* and occasionally also in *C. albifrons*, infants (at least 2 months old) are left with the dominant male, who allows them to huddle or play near or on him (Janson, pers. obser.). In *C. albifrons* and *C. olivaceus*, older infants (over 4 months of age) aggregate in the group center (Robinson 1981a), often in the company of one or two adult or subadult females. In *Saimiri*, older infants are also found in peer play groups in the group center, but without obvious presence of any adults (Baldwin and Baldwin 1981).

Reports of food sharing and allomaternal nursing exist only for *Cebus*. In one *C. olivaceus* group, infants over 4 months of age nursed from allomothers on 40 of 165 occasions (Robinson, unpub. data). Janson has recorded allomaternal nursing of the dominant female's infant and the sharing of a partially consumed tree frog with an orphaned juvenile in *C. apella*.

SOCIOECOLOGICAL CONVERGENCE WITH OLD WORLD PRIMATES ♦

Although the cebines have different social systems, many aspects of their spatial structure and social organization bear at least superficial resemblances to those of a number of Old World monkeys and apes. Crook and Gartlan (1966) were the first to ask whether diet and susceptibility to predation determine the social system of a primate species, and Jolly (1972) has suggested that the three major phyletic groups of primates (lemurs, catarrhines, and platyrrhines) have each radiated into a similar array of ecological niches. Are there ecological and social parallels between the Neotropical (New World) cebines and atelines and Paleotropical (Old World) species?

Two forest primate communities in Africa, in Gabon and Uganda, have been well studied, and they include species with ecological similarities to the atelines and cebines. Four comparisons are particularly instructive: (1) *Cercopithecus talapoin* with *Saimiri*; (2) *Cercopithecus cephus* and the closely related *C. ascanius* with *Cebus*; (3) *Cercocebus albigena* with *Ateles*; and (4) *Colobus badius* with *Brachyteles*. Each of these pairs is comprised of genera with similar body weights and ecological characteristics (diets, day ranges, and home range area) (Table 7-5). Are they also similar in their spatial structure and social organization?

Spatial structure can be described at two levels: (1) the tendencies of groups to coalesce into multigroup aggregations and to fragment into subgroups (defined here to be parties of animals that move independent of one another and occupy different regions of the group range),

TABLE 7-5. Ecological and Social Characteristics of Some African Forest Primates

Species	Adult Weight (kg) Male	Female	% Foraging on Fruits	Leaves	Insects	Sampling Method	Home Range (ha)	Day Range (km)	Group Size	Sources
Cercopithecus talapoin	1.3	1.1	43	2	36	SC	120	2.3	70	Gautier-Hion 1978a
C. cephus	4.1	2.9	79	8	10	SC	20–50	0.5–1.0	10	Ibid.; Gautier-Hion and Gautier 1979
C. ascanius	4.0	2.9	44	16	22	F	30	1.4	33	Struhsaker 1978; Struhsaker and Leland 1979
Cercocebus albigena	9.0	6.5	81	6	6	SC				Gautier-Hion 1978a
	10.5	7.0	59	5	11	F	410	1.3	16	Struhsaker 1978; Struhsaker and Leland 1979
Colobus badius	10.5	7.0	6	74	3	F	71	0.6	20	Struhsaker 1978; Struhsaker and Leland 1979
Pan troglodytes	49	41	68	28	4	II				C. M. Hladik 1978
			58	20	—	TT	1250	2.0–6.4	(50)	Wrangham 1977

NOTE: Stomach contents (SC) method records the % dry weight of stomach contents. For other methods (F, TT, and II), see table 7-1.

and (2) the pattern of spatial associations among individuals within groups. Social organization is reflected in the group size, the group composition, and the breeding system.

C. talapoin, like *Saimiri*, are small and not strongly sexually dimorphic (see chap. 11). Both genera have a diet mainly composed of fruits and invertebrates and take only small amounts of leafy material. They have similar day ranges and range sizes, and only rare intergroup encounters (Gautier-Hion 1971b). Both live in large groups. Spatial structure of the two genera differ however. *Saimiri* groups sometimes coalesce into large aggregations of several hundred animals (Baldwin and Baldwin 1981), possibly in response to superabundant food resources. *Saimiri* groups also fragment into subgroups: groups on St. Sofia Island in Colombia, for instance, during the dry season break into dispersed subgroups that move independent of one another and rarely come together (Bailey, pers. comm.). This might be a response to the even spatial distribution of their food (mainly insects) at this time of year or, since the dry season coincides with the mating season, a consequence of competition between males. *Saimiri* groups also exhibit sexual segregation—the spatial association of same-sex adults (Baldwin and Baldwin 1981). This segregation is partly a consequence of aggression directed toward males by females (Baldwin and Baldwin 1981) and partly a consequence of strong associations between specific females (Mason 1971; Coe and Rosenblum 1974; Vaitl 1978). *C. talapoin*, in contrast, do not form independently moving subgroups nor multigroup aggregations (Gautier-Hion 1971b; Rowell 1973). While *C. talapoin* do exhibit sexual segregation, and same-sex associations are also maintained by aggression toward males by females, female-female associations are not as pronounced, and the most common association observed is that of a female and her offspring (Gautier-Hion 1971b). Like *Saimiri*, *C. talapoin* have a circumscribed mating season characterized by male-male conflict, but there appears to be no clear dominance hierarchy among males (Gautier-Hion 1971b).

Cercopithecus cephus and *C. ascanius* are similar to *Cebus* in weight and degree of sexual dimorphism (see chap. 9). The relative proportions of fruit, leafy material, and insects in their diets parallel those of *Cebus*. Day ranges and home ranges are smaller in the *Cercopithecus* species, but groups of both genera are similarly intolerant of other groups. Group sizes in both *Cebus* and the *Cercopithecus* species are variable, but range from about 10 to over 30. Neither genera show any tendency to form subgroups or multigroup aggregations. Sexual segregation is absent in both (Struhsaker and Leland 1979; Robinson 1981a; Janson, unpub. data). Group compositions, however, are not identical. In the two *Cercopithecus* species, each group usually contains only a single male (Gautier-Hion 1978a; Struhsaker and

Leland 1979; but see chap. 9). In contrast, although small *Cebus* groups contain only a single male, larger groups contain a number of males. Typically, however, only one of these males is central and apparently responsible for most, if not all, of the breeding.

Cercocebus albigena are similar in weight to *Ateles* and, like *Ateles*, have been described as specialists on widely spaced fruit trees. They are not quite as frugivorous as *Ateles* and take more invertebrates. Like *Ateles*, *Cercocebus* ranges are large, and groups are intolerant and avoid one another (Waser 1976). Group sizes of the two genera are similar. In neither species is there sexual segregation, and spatial association between adult males and females in both are more common than expected by chance (van Roosmalen 1980; Chalmers 1968a). *Cercocebus* and *Ateles* differ in at least two ways. First, *Cercocebus* groups do not fragment into independently moving subgroups (Waser 1977a). Second, the migratory sex in *Cercocebus* appears to be the males, while in *Ateles* it might be the females. Groups of both genera contain a number of fully adult males. In *Cercocebus*, one male performs more copulations than the others, but, as in *Ateles*, there is no clear relation between dominance rank and male reproductive success, and a number of males may copulate with a female during a single estrous period (Struhsaker and Leland 1979).

Colobus badius are only slightly smaller than *Brachyteles* (see chap. 8). Both species, highly folivorous, take mature and immature leaves (Struhsaker 1978; Fonseca 1983; Milton 1984). Compared to *Brachyteles*, *C. badius* groups have a smaller home range, but day ranges are of similar length. In contrast to *Brachyteles* groups, which do not tolerate one another, relations between *C. badius* groups vary from tolerance to aggression (Struhsaker and Leland 1979). Group sizes in the two species are similar. However, *C. badius* groups do not fragment into subgroups, in contrast with the spatial structure reported from at least some populations of *Brachyteles*. *C. badius* groups are characterized by female migration (Struhsaker and Leland 1979; Marsh 1979b), and the same may be true for *Brachyteles* (Fonseca et al., in press). In both species, groups frequently include a number of males and females, and an estrous female may copulate with a number of males in rapid succession (Struhsaker and Leland 1979; Aguirre 1971; Milton 1984).

In some respects, the social structure of *Ateles* and perhaps also of *Brachyteles* more closely resembles the fission-fusion society of the chimpanzee *Pan* (Cant 1977; van Roosmalen 1980; see also chap. 15). All three genera have the following common characteristics (van Roosmalen 1980; Fedigen and Baxter 1984; Milton 1984): subgroups form that move independent of one another and rarely, if ever, aggregate into the larger social unit; size and membership of subgroups vary; individual adults occupy different centers of activity or core areas within the group range, with males ranging over greater areas than females; males cooperate in intergroup interactions and may be related. In *Pan* (Wrangham 1979b), and possibly also in *Ateles* and *Brachyteles,* females are the more migratory sex. The chimpanzee "community" might be equivalent to the *Ateles* and *Brachyteles* "group." Are there ecological parallels between the genera?

Although chimpanzees have been described as fruit specialists, they are not the specialists that *Ateles* are. Neither are they as folivorous as *Brachyteles,* although leafy material makes up a large part of their diet. Nothing has been published on the seasonal variation in subgroup size in *Brachyteles,* but in *Ateles* and *Pan,* subgroups are larger when resources are more available (Wrangham 1977; van Roosmalen 1980). This suggests that in these two genera, subgroups are a response to the average food availability at food sources (patch size). Although the home range of the chimpanzee community is much larger, their use of space parallels that of *Ateles* and possibly *Brachyteles* groups: animals occupy different core areas within the total range, but group range is patrolled and defended (Goodall et al. 1979).

Although there are social and ecological parallels between selected Neotropical and Paleotropical genera, there are also many differences. Atelines and cebines differ socially from their ecological counterparts in the Paleotropics in their spatial structure: *Saimiri, Ateles,* and *Brachyteles* all form subgroups that independently move through the group range; *Saimiri* (and *Lagothrix*) groups coalesce into large multigroup aggregations, a grouping pattern that has not been observed in any Old World cercopithecine or colobine. In diet, *Ateles* and *Brachyteles* differ ecologically from their social counterpart *Pan,* and other differences may be revealed by further research. The failure to find more precise parallels probably results from how we measure the social and ecological characteristics of these species. Is the relative proportion of different food items consumed too simple a measure of the diet of a species? Are group size and composition important measures of social structure? Identifying the appropriate measures depends on an understanding of the mechanism relating social and ecological characteristics. The next step of comparative primatology is to describe these mechanisms.

8 | Colobines: Infanticide by Adult Males

Thomas T. Struhsaker and Lysa Leland

This subfamily of Old World monkeys is best known for its peculiar digestive anatomy and physiology. The large, four-chambered stomach is structured such that ingesta are initially separated in the fore chambers, which are less acidic than the distal portion. The higher pH (5 to 7) in the fore chambers permits fermentation by anaerobic, cellulytic bacteria. The polygastric condition distinguishes this subfamily from all other primates who have single-chambered (monogastric) stomachs. This ruminantlike digestion allows the colobines to exploit leaf diets generally unavailable to other primates and may also reduce their requirements for water (Bauchop 1978). As will be apparent throughout this chapter, the implications of this adaptation for social behavior and ecology are considerable.

The reduced thumb and unusually long fingers of the Colobus species may enhance their skills at limited brachiation. More striking than this, however, are the exceptionally long feet of some species. For example, Colobus badius tephrosceles have hind feet that are similar in length to both the tibia and femur: in other words, the hind foot represents somewhat more than one-third of the entire hind limb, including the foot (Struhsaker, unpub.). This anatomical feature permits great leaps between trees as well as a saltatory mode of locomotion (Struhsaker 1975).

TAXONOMY AND DISTRIBUTION

The colobines range in size from about 4 to 23 kg and are comprised of 5 to 7 genera (Presbytis, Rhinopithecus, Pygathrix, Nasalis, Simias, Colobus, Procolobus) and 24 to 30 species (Napier and Napier 1967; Thorington and Groves 1970; Struhsaker 1981b; Oates and Trocco 1983). Speciation in most cases is attributed to isolation during the extreme climatic fluctuations of the Pleistocene. In the case of Africa, these fluctuations apparently resulted in fragmentation of the forests into relatively small patches or ecological islands that acted as refugia (Kingdon 1974; Hamilton 1982; Struhsaker 1981b). In much of Malaysia and Indonesia, isolation conducive to speciation in the Pleistocene has been attributed to recurrent and intermittent changes in sea level, which in turn affected connections between islands and with the mainland (Medway 1970).

Where confusion exists concerning species affinities and phylogenetic relationships, it can usually be attributed to reliance on characteristics showing extreme variability, such as coat color or craniology. In the case of Colobus monkeys, much of the taxonomic confusion has been reduced through detailed analysis of vocalizations, which seem to be far more conservative features than coat color and perhaps many skeletal features as well (Struhsaker 1981b; Oates and Trocco 1983; also see Wilson and Wilson 1975 and Ruhiyat 1983 on Presbytis).

ECOLOGY

The colobines occur in tropical Africa (all Colobus species and Procolobus) and Asia (all other genera), and are primarily adapted and restricted to rain forests. A few species, such as Colobus guereza in Africa and Presbytis entellus in India and Sri Lanka, are more flexible. They can thrive in much drier habitats, usually by occupying riverine or ground-water forests within the savanna woodland or dry deciduous forest. The proboscis monkey (Nasalis larvatus), in contrast, is adapted to life in mangrove swamps, though populations also extend along rivers far inland. At the other extreme, some populations of the gray langur (Presbytis entellus) and most, if not all, of the golden snub-nosed monkey (Rhinopithecus roxellanae) live at relatively high altitudes (2,000 to 3,000 m) in mountain forests where they experience extreme seasonal variations in weather, including snowy winters. The gray langur is the only species of colobine known in which some populations have adapted to a coexistence with humans in towns and cities, where they actually derive much of their food from people.

Although all colobines are diurnal, limited activity does occur at night in some species, such as red colobus (C. badius), who, based on distinct vocalizations, apparently engage in limited sexual and aggressive interactions after nightfall (Struhsaker 1975). The great majority of forest species are arboreal (fig. 8-1), rarely coming to the ground except to feed on soil and aquatic plants or to drink (e.g., C. guereza and C. badius). In contrast, species and populations living in woodland or forest that has been severely fragmented frequently come to the ground to travel between trees or groves. Although the majority of colobines tend to be found higher in the forest canopy than cercopithecines, this vertical stratification seems to be primarily the result of differences in food habits and in the distribution of these foods in the forest rather than a direct response to interspecific competition for the same resource (Struhsaker 1978).

The colobine diet always contains a large proportion of leaves (about 35 to 75% of frequency data), with most species showing a definite preference for young leaves that are generally high in protein, low in lignin and condensed tannins, and highly digestible. Some species, such as the red colobus, select the petiole (leaf stalk) rather than the blade of mature leaves, apparently because of greater digestibility. Seed eating has not been generally recorded among the colobines, although the black colobus (*Colobus satanas*: McKey et al. 1981) of Cameroon, West Africa, feed largely on seeds (53.2%). Because we do not know if the seeds are digested or not, some of the high fruit consumption by Asiatic species, such as 45% for *P. entellus* (Hladik and Hladik 1972), 42 to 60% for *Presbytis melalophos* (Johns 1983), and 32 to 56% for *Presbytis obscura* (MacKinnon and MacKinnon 1978; Curtin 1980) may in fact constitute additional examples of seed eating. These species stand in marked contrast to *Presbytis aygula* of W. Java (Ruhiyat 1983) and the African *Colobus* living in rain forests that have been studied to date (*C. badius* and *C. guereza*). Their fruit-seed diet is generally less than 13%, or less than half that of the above Asiatic species. Two exceptions exist in red colobus populations living outside of rain forests. One is in the gallery forests and savanna woodland of Senegal (Gatinot 1978), and the other is in the riverine, ground-water forest of the Tana River, Kenya (Marsh 1981a). These red colobus consume much larger quantities of fruit and seed (36% and 25%) than those of the wetter forests, though still considerably less than *C. satanas* and the Asiatic species. Regardless of proportions, in all detailed studies the great majority of fruit eaten by colobines is unripe rather than, and in preference to, ripe fruit.

The distance traveled each day by a group of colobines averages about 500 to 600 m, with daily variation between 220 and 1,360 m. It is striking that, in spite of considerable differences in group size, population densities, and ecology, both African and Asian species should be so similar in daily travel distance (table 8-1). In the Kibale Forest of Uganda, for example, groups of red colobus varying in size from 9 to 68 individuals, and including some with overlapping home ranges, showed no significant correlation between group size and daily travel distance (table 8-2; see also chap. 20). This suggests that colobine ranging cannot be explained by relatively simple optimal foraging models because one would predict larger groups to travel further each day than smaller groups (see Krebs and Davies 1981 for discussion of foraging models).

FIGURE 8-1. (*above*) Red colobus monkey (*Colobus badius*) in the Kibale Forest, Uganda; (*right*) black-and-white colobus (*C. guereza*) also in Kibale Forest. (Photos: (*above*) Lysa Leland; (*right*) John Oates)

Home range size, on the other hand, seems to be more variable; it ranges from estimates of 2.5 ha for the *Presbytis senex* of Polonnaruwa in Sri Lanka to more than 100 ha for the red colobus of Gombe, Tanzania. Much of this can probably be attributed to ecological variables. For example, in Kibale, Uganda, red colobus have more diversified diets and larger home ranges than do both the black-and-white colobus there (Struhsaker and Oates 1975) and the red colobus living in the species-poor forest along the Tana River, Kenya (Marsh 1981b). Furthermore, differences in field and analytical methodology alone can lead to appreciable variation (see Struhsaker 1975).

Population and biomass densities are also variable; they range from as few as 12 individuals per km² for *C. guereza* in some parts of the Kibale Forest of western Uganda to as many as 300/km² for red colobus in the same forest. Corresponding biomass densities are 65 and 1,760 kg/km², respectively. The latter figure represents the highest known biomass density for any anthropoid (Struhsaker 1975). In general, the colobines travel less on a daily basis, but have both higher numerical and bio-

mass densities than do the cercopithecines (Struhsaker 1978, 1980). Food availability probably accounts for much of these differences, since edible leaf matter is usually more abundant than fruit and insects.

CONSERVATION

The majority of colobines, being highly dependent on old, mature rain forest, are threatened with appreciable population reductions, if not with extinction, due to the rapid and irreversible deforestation occurring throughout most of the tropics. Particularly vulnerable in Africa are those species or subspecies restricted to one or a few forests, such as *C. badius preussi* of Cameroon, *C. b. pennanti* of Fernando Po, *C. b. tephrosceles* in Uganda and Tanzania, *C. b. gordonorum* of Southern Tanzania, *C. b. kirkii* of Zanzibar, *C. satanas* of Cameroon and Equatorial Guinea, and the very rare *C. b. rufomitratus* of the Tana River, Kenya, certainly the most endangered colobine in Africa. Two black-and-white colobus (*Colobus polykomos* and *Colobus vellerosus*) of West Africa are also threatened by hunting and habitat destruction. The olive colobus, *Colobus verus* of West Africa, is poorly known, but it too is likely vulnerable. In most cases, deforestation is due to human populations expanding at the

rate of 2.5 to 4% annually and demanding more land for cultivation. In other cases, however, multinational companies may be the most immediate threat. For example, oil explorations and planned drilling in the Douala-Edea Reserve of Cameroon threaten one of the few remaining large and viable populations of *C. satanas*. Once an area of forest is opened by exploitation such as this or by timber extraction, agricultural settlement and hunting soon follow. Hunting in most of West and central Africa poses a severe threat to colobines because they tend to be less wary and thus more vulnerable than other primates. Local extinction from such development probably accounts for much of the patchy distribution of many colobines. Similar pressures jeopardize the future of Asiatic colobines, particularly *Presbytis johnii* of the Western Ghats in India, *Presbytis geei* of Bhutan, *Presbytis pileata* of Bangladesh, Assam, and West Burma, as well as all of the *Rhinopithecus* and *Pygathrix nemaeus* of subtropical China and north Indochina (Marsh, pers. comm.).

REPRODUCTION

Generalizations about colobine reproduction are difficult to make because few field studies have lasted more than

TABLE 8-1.   Colobine Day Ranges

Species	Day Range (m)	Group Size	Population Density (#/km²)	Habitat	Sources
Colobus badius tephrosceles	557	9–68 (x̄ = 34)	300	Kibale, Uganda; medium-altitude rain forest	Struhsaker 1975, unpub.
C. b. rufomitratus	603	18–22	200–39	Tana River, Kenya; riverine, ground-water, seasonal forest	Marsh 1981b
C. guereza	535	12	10–100	Kibale, Uganda; medium-altitude rain forest	Oates 1977c; Struhsaker 1978
C. satanas	459	15–16	<26	Douala-Edea, Cameroon; lowland rain forest	McKey and Waterman 1982
Presbytis entellus	360 (60–1,300)	15–23	16–135	Dharwar, India; dry, deciduous forest	Yoshiba 1967; Sugiyama 1965
P. johnii	500	9	—	S. India; monsoon forest	Tanaka 1965; Poirier 1968
P. cristata	200–500	32	—	W. Malaysia; mixed rain forest and city park	Bernstein 1968
P. obscura	950[a]	10.3	31	Krau, W. Malaysia; lowland rain forest	MacKinnon and MacKinnon 1978
P. melalophos	1,150[a]	9.3	74	" "	MacKinnon and MacKinnon 1978
	682	15	—	" "	Bennett 1983
	614	14	47.6	Sungai Tekam, W. Malaysia; primary lowland rain forest	Johns 1983

a. Perhaps unusually high due to short duration of study and relatively poor habituation of study groups.

TABLE 8-2.   Group Size and Daily Travel Distance among *Colobus badius* of Kibale, Uganda, 1970–82

Mean Group Size	Mean Daily Travel Distance (m)	No. Days
9	578.3	24
11	576.0	4
20	652.4	53
25	546.4	11
33	533.3	5
36	473.5	4
48	576.2	28
60	487.0	46
68	592.5	90

SOURCES: L. Isbell, L. Leland, J. Skorupa, and T. T. Struhsaker, unpub. data.
NOTE: $r = -0.295$, $p > 0.05$.

one or two consecutive years, and there are few records of colobine reproduction in captivity. Allowing for the possible exception of gray langurs (*P. entellus:* Hrdy 1977) who live high in the Himalayas and *Rhinopithecus roxellanae* in the mountains of Szechwan, no definite case of strict birth seasonality has been described for any Colobinae. Birth peaks, however, are indicated for at least three species: *P. senex, P. entellus,* and *Colobus badius* (table 8-3; fig. 8-2). The temporal pattern of these peaks tends to vary with climate. In some cases, such as *C. badius,* it appears bimodal, with peaks tending to coincide with rainy months. In contrast, the single birth peak for *P. s. senex* at Polonnaruwa, Sri Lanka, was coincident with the dry season. However, *P. s. monticola* at Horton Plains, also in Sri Lanka, had no apparent birth peak during 2 years. There the fluctuations in climate and food abundance were far less pronounced than at Polonnaruwa (Rudran 1973b).

Seasonality of births may sometimes be confused with intragroup birth synchrony (Sugiyama, Yoshiba, and Parthasarathy 1965). For example, Rudran (1973a) suggested that infanticide, which sometimes follows male replacement in one-male groups of *P. senex,* could result in highly synchronized births among females of the same group. Unless the sample covers several different social groups or otherwise allows for this variable, distinction between the two phenomena can be difficult (see discussion of the reproductive consequences of infanticide later in chapter).

An analogous problem arises in determining the age of sexual maturity for nonhuman primates under field con-

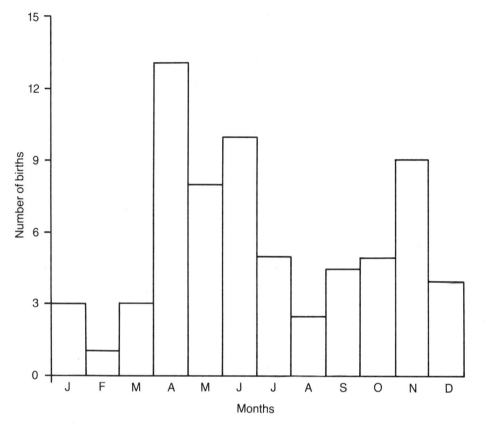

FIGURE 8-2. Monthly distribution of 68 births in the CW group of red colobus, Kibale Forest, Uganda, during 1971–83.

ditions. Because many individuals with known birth dates migrate and disappear before maturation, both long-term data and studies of several groups are required.

Perineal swellings related to estrus have been described only for *C. badius, C. satanas,* and *C. verus* of Africa, which generally live in multimale groups (Struhsaker and Leland 1979). Swellings associated with pregnancy have only been described for *C. badius* of Kibale, Uganda (Struhsaker and Leland 1985). No colobines studied to date who usually live in one-male groups have swellings. *P. entellus* females, however, who live in both one-male and multimale groups, have definite behaviors correlated with estrus. These consist of simultaneously shaking the head back and forth, dropping the tail to the ground, and presenting the perineum to the male (Jay 1965). There are no firm data on olfactory cues associated with estrus, but we do have behavioral evidence that these may be quite important for red colobus. Males often muzzle the perineum of a swollen female before mounting her (Struhsaker 1975).

The length of gestation has been established for only one species, *P. entellus,* and this under captive conditions (200 ± 10 days, from L. J. Neurater in Hrdy 1977). Gestation length is difficult to determine under field conditions because, even if one did have daily

samples of female sexual behavior, the estimations would be confounded by postconception estrus, now known to occur in the two best-studied species, *P. entellus* and *C. badius.* In the latter species, copulation can occur 101 days prior to parturition or about half way through the probable gestation period (Struhsaker and Leland 1985).

Interbirth intervals are not well documented for the Colobinae because, again, most studies have been too short. The most detailed data are from 16 females of one group of *C. badius* during a 13-year study (fig. 8-3; Struhsaker, unpub.). They show considerable variation, but with a mean interval of approximately 2 years between births (mode 23 mo). This typical interval can be shortened by more than 50% with the death of an infant ($N = 8$; $\bar{X} = 12.1$; range 9.2 to 17.6 mos.; 7 females). At the other extreme, interbirth intervals can be very long, exceeding 43 months. Some of these longer intervals may be an artifact of incomplete sampling, such as would occur if a birth was missed and the infant died soon thereafter. However, certain females rather consistently had longer intervals than others (by as much as some 10 mo.), and in these cases an alternative explanation seems warranted (Struhsaker, unpub.).

Although we cannot yet explain these apparent interindividual variations in interbirth interval, limited data for both *P. senex* and *P. entellus* suggest that interpopulation differences may be attributable to nutrition, as reflected by habitat and seasonal extremes. At Horton Plains, with its less variable climate and more abundant food supply throughout the year, *P. s. monticola* had an interbirth interval of only 16.5 months. This was roughly 25 to 30% shorter than *P. s. senex* ($\geq$ 22 to 25 mo) at Polonnaruwa where food was less abundant and more seasonally variable (Rudran 1973b). It certainly seems true that nutrition affects interbirth intervals of humans (Frisch and McArthur 1974; Frisch 1982; Bongaarts 1982), chimpanzees (Coe et al. 1979), Japanese macaques (A. Mori

TABLE 8-3. Reproductive Parameters for Colobines

Species and Study Site	Birth Seasonality	Age at Sexual Maturity (mo)		Gestation Length	Interbirth Interval	Ext. Signs Estrus	Sources
		Male	Female				
Colobus badius							
Kibale, Uganda	No (peaks: Apr–June and Nov)[b]	x̄ $\geq$ 46.5[a] $\geq$35.4–58[c]	~38–46[d,e]	NA	25.5 ± 5.1[b]	Yes, perineal swelling[b]	Struhsaker, unpub.
C. guereza							
Tigoni, Kenya (captivity)	No	—	—	NA[f]	17.5 ± 6.39[d,g]	No	Rowell and Richards 1979
Kibale, Uganda	No[b]	NA	NA	NA	25.2[a]	No	Oates, unpub.
C. verus							
Tiwai Is., Sierra Leone	Nov?–Feb?	NA	NA	5–6 mo?	?	Yes, perineal swelling	Ibid.
Presbytis entellus							
Central India	No (peak: Apr–May— no data	NA	42—no data	6—no data	24—no data	No (behavioral)	Jay 1965
Dharwar, India	No (peak: Nov–May)[b]	NA	NA	6–7—no data	>12 $\leq$ 24— no data	No (behavioral)	Sugiyama 1966, 1967; Sugiyama, Yoshiba, and Parthasarathy 1965
Mt. Abu, India	No	NA	NA	6.5 ± 0.3[h]	25.4[d,i]	No (behavioral)	Hrdy 1974, 1977
Jaipur, Rajasthan, India	Perhaps (peaks: Mar–May and Oct)[b]	NA	NA	NA	NA	NA	Prakash 1962
Jodhpur, Rajasthan, India	No?	NA (60)[l]	46.5–47[a,j]	6.6[a,k]	15[d]	NA	Vogel and Loch 1984
Rajaji, N. India	Perhaps (peak: Mar–July, but maybe also in Jan, Feb though no sample)[c]						Laws and Vonder Haar Laws 1984
P. johnii							
Kakachi, S. India	?	NA	NA	NA	>22[a]	No	Oates, unpub.
P. s. monticola							
Horton Plains, Sri Lanka	No regular peaks[b,m]	NA	NA	6.5–7.25[d,m]	16.5[a]	No	Rudran 1973b
P. s. senex							
Polonarruwa, Sri Lanka	No (peak: May–Aug)[b,m]	NA	NA	6.5–7.25[d,m]	$\geq$22–25[a]	No	Ibid.

a. $1 < N < 5$.

b. $N > 20$.

c. $N = 4$, 2 of which were based on rough estimates of birth dates.

d. $5 < N < 20$.

e. $N = 10$ extrapolations based on estimated age of juvenilehood plus 12 to 18 months to first pregnancy.

f. Estimated at 170 days by Rowell and Richards 1979, but no basis or data given.

g. Rowell and Richards 1979 give 16 intervals from 4 females over a 13-year period (all captive data). The range was 8 to 32 months. Data following 2 stillbirths and for 3 intervals of 8, 9, and 9 months have been excluded here. The latter were excluded because it seems unlikely the female(s) concerned experienced only a 1- to 2-month period of lactational amenorrhea without the loss of the suckling infant.

h. From captive animals; see Hrdy 1977, pers. comm. from L. J. Neurater.

i. $N = 5$, very rough estimates.

j. One female only, age of first birth (menarche in one captive was 41 months) (Ramaswami 1975).

k. Based on behavioral data only and assumes that no postconception estrus occurred.

l. No data, estimation only.

m. Most birth dates based on estimates derived from assessed ages.

n. Rough approximation based on behavioral data, which could be misleading due to postconception estrus.

1979a), vervets (Cheney et al., in press), and perhaps olive baboons (Strum and Western 1982). The same may also be so for the Colobinae.

SOCIAL ORGANIZATION

Group size and composition of bisexual groups vary considerably throughout the Colobinae. The majority of species studied to date tend to live in small social groups numbering about 10 to 15 and having only 1 fully adult male: *Presbytis senex* (Rudran 1973b), *P. cristata* (Bernstein 1968; Wolf and Fleagle 1977), *P. johnii* (Poirier 1969), *P. obscura* (Marsh and Wilson 1981a; L. Bennett, pers. comm.), *P. aygula* (Ruhiyat 1983), *P. melalophos* (Bernstein 1967b; Johns 1983; L. Bennett, per. comm.), *P. pileata* (Islam and Husain 1982), *P. thomasi* (Rijksen 1978), and *Colobus guereza* (Dunbar and Dunbar 1976; Oates 1977c). Exceptions have been reported for both *P. obscura* and *P. melalophos* (Curtin 1980), where groups have contained as many as four adult males. On the infrequent occasions when more than one adult male has been seen in *C. guereza* groups, this has been temporary and is believed associated either with male replacements

(Oates 1977c) or young males maturing in their natal groups (Dunbar and Dunbar 1976).

Although many populations of *P. entellus* live in small (ca. 15) one-male social groups (e.g., Sugiyama, Yoshiba, and Parthasarathy 1965), several others have large, multimale social groups, sometimes numbering even more than 100 (Jay 1965; Ripley 1967; Hrdy 1977; Yoshiba 1968; Laws and Vonder Haar Laws 1984). No satisfactory explanation has been advanced for these striking demographic differences. In some cases, particularly the very large counts of 100 or more, the observers may have mistaken temporary associations of two or more groups for one. However, in other cases, groups of 50 or so seem well documented (summarized in Hrdy 1977).

Large, multimale groups of more than 80 individuals, but averaging 50, seem to be the rule for most *C. badius* subspecies (Struhsaker 1975). However, in more seasonal and marginal habitats of Senegal and Kenya, groups are less than half this size (Struhsaker 1975; Gatinot 1978), and along the Tana River of Kenya they usually live in one-male groups (Marsh 1979a). Even in popula-

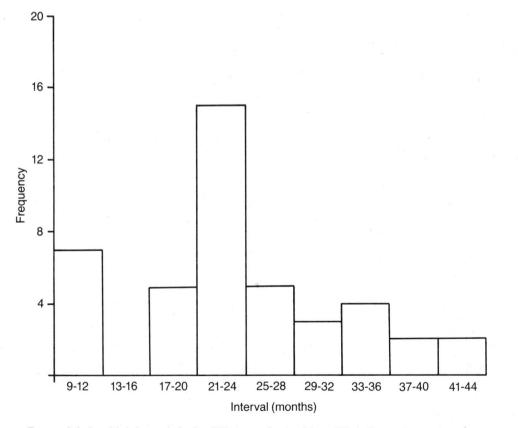

FIGURE 8-3. Interbirth intervals in the CW group of red colobus, Kibale Forest, Uganda, during 1971–83. N=43, $\bar{x}$=24.5 mo, s.d. ± 8.6; excluding 8 shortest $\bar{x}$=27.3 mo, s.d. ± 6.8; excluding 8 shortest and 4 longest $\bar{x}$=25.6 mo, s.d. ± 5.1; 8 shortest intervals followed the death of infants ranging in age from 1 week to 4 mo and belonging to 7 different mothers. Mean of 8 shortest = 12.1 mo.

tions typically having large groups, such as Kibale, Uganda, one can find groups as small as 8 individuals (Leland, pers. comm.).

The black colobus represents yet another variant; it lives in small groups of about 15 (5 to 30), with up to 3 adult males per group (Sabater Pi 1973; McKey 1978; McKey and Waterman 1982). Less well known are the large aggregations of more than 100 *Colobus angolensis* in Rwanda (M. Storz, pers. comm.; Struhsaker and Leland, unpub.) and several hundred *Rhinopithecus roxellanae* in China, which can fragment into smaller units of 50 or so (Schaller, pers. comm.). Whether these represent aggregations of several social groups (as sometimes occurs with *C. badius* in Uganda) or a single fusion-fission society (as in hamadryas baboons of Ethiopia, chap. 10) remains to be determined. However, the frequent formation of smaller aggregations of *C. angolensis* in Kenya apparently represents temporary and nonaggressive fusions of two multimale groups. These associations always lasted at least several hours and did not involve any distinctive behavior or social interactions (Moreno-Black and Bent 1982).

Proboscis monkeys (*Nasalis larvatus*) have not been studied in great detail, but preliminary observations indicate that they are extremely variable in foraging party size (2 to 63), which range from small groups with only one adult male to large groups with several males (Kern 1964; Kawabe and Mano 1972). It has been suggested that foraging parties vary in size according to patterns of food dispersal (MacDonald 1982).

Monogamy is extremely rare in Old World monkeys, but two clear cases have been described for the colobines of Siberut Island off the west coast of Sumatra (*Presbytis potenziani*: Tilson and Tenaza 1976; Watanabe 1981; *Simias* or *Nasalis concolor*: Tilson 1977; Watanabe 1981). *P. pontenziani* seems to be monogamous throughout its range on Siberut and shows several of the characteristics typical of other monogamous species, such as monomorphy, vocal duetting, and mutual visual displays between the male and female (Tilson and Tenaza 1976). There is evidence, however, that with higher population densities the *S. concolor* groups are polygynous, with more than half the groups containing two to four adult females and only one adult male (Watanabe 1981). *Presbytis aygula* in west Java is also variable, with one population appearing primarily monogamous and another polygynous (Ruhiyat 1983).

Groups containing only adult and/or subadult males have been described for relatively few colobines. They are commonly seen only among *P. entellus* and *P. senex* (Sugiyama, Yoshiba, and Parthasarathy 1965; Jay 1965; Rudran 1973b; Laws and Vonder Haar Laws 1984; Hrdy 1977). There are rare accounts of all-male groups in *P. melalophos* (Curtin 1980) and *C. guereza* (Oates 1977c), but like most other colobine species, extragroup males

are generally seen as solitary individuals. Nowhere are solitary males common, which suggests that extragroup males may suffer higher mortality than females.

Transfer of animals between social groups is rarely observed, and consequently, conclusions regarding patterns of dispersal are usually based on partial or indirect evidence. For example, the prevalence of one-male groups and the male majority of extragroup adults and solitaries clearly suggests that it is primarily the males who transfer from their natal groups. More direct evidence comes from observations of adult male replacement in one-male groups. For the few colobines studied in any detail, adult female membership in social groups is far more stable than that of the adult males (*C. guereza*: Oates 1977c; *C. satanas*: McKey and Waterman 1982; *P. entellus*: Hrdy 1977; *P. melalophos*: Curtin 1980; *P. senex*: Rudran 1973b; *P. johnii*: Poirier 1969; *P. cristata*: Bernstein 1968; Wolf and Fleagle 1977). Thus, the available evidence strongly suggests that among the colobines studied to date it is usually the males who transfer between groups, while females tend to remain as the social nucleus in their natal group. Clearly, however, more long-term studies are required to fully understand the details of colobine dispersal patterns (see Moore 1984 for a different view, which emphasizes the exceptional cases rather than the norm).

The only regular exceptions to this rule are most populations of *C. badius,* where apparently all females transfer to other groups near the onset of puberty. Most males either remain in their natal group or disappear. Although some young male *C. badius* leave their natal groups at puberty, few are successful at immigrating into new groups and only during adulthood. Only 3 male immigrations have been seen during some 27 group-years of study in Kibale, Uganda. The group studied longest and in greatest detail had no male immigrations in 14 years of study, while 3 adult males entered 2 other groups under study for 6 and 7 years respectively, but stayed for less than 1 year. In contrast, several adolescent female immigrations occurred every year (Struhsaker and Leland, unpub. data). The Tana River *C. badius rufomitratus* provide an interesting exception: it appears that most males as well as females leave their natal groups, for here they live in one-male groups (Marsh 1979a).

In summary, the colobines provide us with a wide array of social systems, including one-male matrilineal groups, multimale matrilineal and patrilineal groups, monogamous families, and enormous aggregations that may prove to be fusion-fission societies. (Following *Webster's Third New International Dictionary* 1965, we define *matrilineal* and *patrilineal* as designating descent, kinship, or derivation through the mother or father, respectively. Emphasis here is on the predominant sex that remains in its natal group.) Breeding within these groups is usually promiscuous when the groups contain several males (Jay 1965; Struhsaker 1975) or polygynous in the one-male groups (Hrdy 1977; Oates 1977b, 1977c) (see chap. 31). In multimale groups of some species such as red colobus (Struhsaker 1975), dominant males clearly perform the majority of copulations.

SOCIAL BEHAVIOR

Aggression (chases and fights) and dominance relations in the Colobinae are most apparent and intense among the males, who, in the presence of adult females, are generally intolerant of one another. Indeed, in most species, one-male social groups are common, and males aggressively resist the immigration of additional males. In those species or populations having multimale groups, such as *C. badius* (Struhsaker 1975; Struhsaker and Leland 1979) and some populations of *P. entellus* (Jay 1965), usually a distinct linear dominance hierarchy exists among males. The frequency of agonistic interactions, however, seems less than in multimale groups of cercopithecines (Struhsaker and Leland 1979). The same applies to aggression in all-male groups of *P. entellus* (Yoshiba 1968). In *C. badius*, the only colobine regularly found in multimale groups, male-male aggression generally involves harassment during copulation (Struhsaker and Leland 1985).

Aggression among females is usually infrequent, and dominance hierarchies are either weakly developed or not apparent (e.g., *C. badius:* Struhsaker 1975; *C. guereza:* Oates 1977c). In *P. johnii* agonistic encounters among females are rare, but a linear hierarchy has been described (Poirier 1969, 1970). Although reports conflict for *P. entellus,* females of this species appear to have more aggressive interactions among themselves than the other colobines studied to date, but not so frequently as macaque and baboon females. Most workers have concluded that dominance hierarchies among female *P. entellus* are poorly defined, unstable, nonlinear, and with relatively frequent reversals (Jay 1965; Ripley 1965; Sugiyama 1967; Yoshiba 1968; see Hrdy 1977 for review). In contrast, Hrdy's (1977) study of *P. entellus* at Mt. Abu concludes that a stable linear hierarchy existed among females on a short-term basis, though rank reversals were generally more frequent than in macaques and baboons (2.6% versus 0.4 to 1.0% of encounters). Nearly 70% of the agonistic encounters upon which the female hierarchy at Abu was established were induced through food provisioning by humans (Hrdy 1977). Under more natural conditions one might expect a lower frequency of occurrence of female-female agonistic encounters and perhaps a reduction in the importance of a hierarchy.

Grooming patterns among the four species of colobines (*P. entellus:* Jay 1965; Hrdy 1977; *P. johnii:* Poirier 1969; *C. badius:* Struhsaker 1975; Struhsaker and Leland 1979; *C. guereza:* Oates 1977c). in which this behavior has been studied suggest the following: females do most of the grooming and often, but not invariably, receive grooming more than expected by chance. Males in general are groomees more than or as much as expected and groomers less than expected on the basis of their proportional representation in the group. When they do groom, it is generally directed to adult females. This pattern prevails, regardless of ecology, group size, and composition, not only throughout the colobines, but in most, if not all, cercopithecines as well (Struhsaker and Leland 1979). Although grooming has been studied in few colobine species, *C. badius* appear exceptional in several ways. For example, adult females were both groomees and groomed one another less than expected based on their proportional representation in the group. Adult males, though not common groomers, groomed one another more and adult females less than expected. It has been suggested that these differences can be related to the social organization of *C. badius,* which, in contrast to other colobines, is patrilineal rather than matrilineal (Struhsaker and Leland 1979). The coalition of related males may be strengthened through grooming. Support for this idea comes from analogous grooming relations in two other patrilineal societies: in gorillas where females rarely groom one another unless closely related (chap. 14) and in chimpanzees where males of a community often groom one another, but adult females groom each other proportionately less (chap. 15).

INTERGROUP RELATIONS

Adjacent social groups in the majority of colobines studied to date generally interact aggressively (*P. entellus:* Ripley 1967; Sugiyama 1967; Hrdy 1977; *P. cristata:* Bernstein 1968; *P. aygula:* Ruhiyat 1983; *P. johnii:* Poirier 1968; *P. senex:* Rudran 1973b; *C. guereza:* Marler 1969; Oates 1977b, 1977c). Adult males are usually the most aggressive and frequent participants in these intergroup encounters, though in most populations of *P. entellus,* adult females (Ripley 1967) as well as younger age classes (Hrdy 1977) may play prominent roles in these fights. In at least one population of *P. entellus,* however, neighboring groups were extremely tolerant of one another (Jay 1965).

Territoriality has been described for most populations of *P. entellus,* whether they live in one-male or multimale social groups (Ripley 1967; Sugiyama 1967; Hrdy 1977), as well as for *P. senex* (Rudran 1973b), *P. johnii* (Poirier 1968), *P. cristata* (Bernstein 1968), and *C. guereza* (Marler 1969). Oates (1977c), however, has shown that at least in some groups of *C. guereza* there is extensive range overlap ($\geq$ 74%) and little, if any, exclusive use of an area in spite of defended core areas. In other words, even though a group supplanted other groups from its core area, exclusive use of this area did not result. Adult male loud calls, particularly among forest-living *Presbytis* and *C. guereza,* may be important in intergroup spacing. Extragroup males, including all-male groups, regularly trespass the territories of some *P. entellus* bisexual groups and apparently do not defend territories of their own (Hrdy 1977).

Once again, *C. badius* are exceptional among the Colobinae in that groups in most populations are not territorial and can show extreme tolerance for other groups. Complete overlap in home ranges of three or more groups is typical of *C. badius* in Uganda. Fights occur between groups, but these are almost exclusively between adult males. No boundaries are defended, and the outcome of these encounters is extremely variable, apparently depending to a large extent on the number and fighting capabilities of the males involved. During these intergroup conflicts, the cohesion within the male coalitions is most prominent (Struhsaker 1975). It is believed that these male coalitions defend females as reproductive resources against other male coalitions, analogous to the hypothesis offered for chimpanzee male coalitions (e.g., Wrangham 1979a).

♦ INFANTICIDE BY ADULT MALES

Infanticide by adult males occurs in several diverse orders of mammals: for example, Rodentia (Svare and Mann 1981; Huck, Soltis, and Coopersmith 1982; vom Saal and Howard 1982), Carnivora (Packer and Pusey 1983, 1984), and Primates (Angst and Thommen 1977; Hrdy 1979; Leland, Struhsaker, and Butynski 1984). Typically, a male kills infants of females he has not previously copulated with. These females soon come into estrus and usually mate with the infanticidal male. Correspondingly, the average interbirth interval is often reduced. In primates, infanticide was first described under field conditions for the gray langur, *Presbytis entellus* (Sugiyama 1965). When similar descriptions of this behavior were reported in other langur populations, Hrdy (1974) proposed the sexual selection hypothesis to explain this phenomenon. The hypothesis predicts that males, on average, gain a reproductive advantage by killing unrelated infants. Other researchers, who had not witnessed infanticide in their populations of langurs, claimed that this behavior was aberrant and occurred in overcrowded areas due primarily to human disturbance or interaction (Dolhinow 1977; Curtin and Dolhinow 1978; Boggess 1979, 1984). A third hypothesis, resource competition, was derived from the study of another langur, *Presbytis senex,* in Sri Lanka (Rudran 1973a).

So far, infanticide has been observed in three Colobinae species (table 8-4) and suspected in three others: *Presbytis senex* (Rudran 1973b); *Colobus guereza* (Oates 1977c); and *Colobus badius rufomitratus* (Marsh 1979a). Social organization in these species includes the more prevalent one-male groups as well as matrilineal and patrilineal multimale groups (see table 8-4). Consequently, much of the material for generating hypotheses and expanding our understanding of this complex phenomenon has been derived from comparison of species and social systems within this subfamily of Old World monkeys (see Hausfater and Hrdy 1984).

Nature of the Data Base

The evidence for infanticide is of variable quality. In gray langurs, for instance, infanticide has been attributed to the death or disappearance of nearly 50 infants. In fact, researchers have witnessed only seven of the infanticides (table 8-4). The other cases have been inferred from observed chases by males against mothers and their clinging infants, observed woundings of infants by males, fresh wounds on infants, dead infants killed by biting, disappearances of infants soon after male replacements, and reports by local informants (see Vogel and Loch 1984). Yet, because the act of infanticide is not only uncommon and relatively unpredictable but can occur in a matter of seconds, the chances of a researcher actually observing the event are unlikely.

To minimize the problematic issues concerning the cause of death, we give greatest emphasis in this overview to cases of infanticide or fatal wounding documented in the literature as actually observed by researchers in the field under nonmanipulative conditions. This very conservative treatment is, however, representative of all available information. The data base is broadened by including cases of infanticide from all nonhuman primates, not just the Colobinae.

Summary of Observed Infanticides

The 24 cases of observed infanticide span 10 species, 3 families of anthropoids, and 4 subfamilies: 37.5% Colobinae, 33.3% Cercopithecinae, 8.3% Alouattinae, and 20.8% Ponginae (table 8-4). If one includes the 112 or so additional cases of suspected infanticide that comprise an additional 7 species (table 8-4), then observed infanticide represents about 18% of all cases.

Social Structure. Depending on how one classifies the social systems of gorillas and howlers, 45.8 to 66.7% of the 24 observed cases occurred in species typically living in one-male social groups. However, only 41.7% of the groups actually had only a single adult male at the time of the infanticide because in some cases the infanticidal male was a recent immigrant and coexisted in the group with the previous resident male.

Circumstances. The overwhelming majority (91.3%) of observed infanticides (*N* = 23) was committed by immigrant males (73.9%) or males who did not belong to the infant's social group (17.4%). Only 2 occurred within the infanticidal male's natal group. The pattern of infanticide is most strongly influenced by the type of social organization (table 8-4). Of 23 cases, 56.5% followed almost immediately after male replacement or an immigrant rising in rank over the resident male in matrilineal one-male groups. One of these includes an inferred status change.

In matrilineal multimale groups of baboons, 3 out of 4 infanticides were committed by recent immigrant males. The fourth case (table 8-4, no. 17) was unusual because it occurred within the infanticidal male's natal group. This group, however, did not include his mother or most of his sibs who had fissioned from it 16 months earlier (Collins, Busse, Goodall 1984).

The remaining 6 cases (25%) involved males in patrilineal social organizations. Two of these males were exceptional for patrilineal societies: one, who killed an infant in his social group, was an immigrant male (no. 24); the other (no. 9), who committed infanticide in his *natal* group, was almost certainly not conceived in this group. In both of these cases, the males' attainment of sexual maturity and dominance status coincided with infanticide. The last 4 cases (2 chimpanzees and 2 gorillas) differ from the rest in table 8-4 in that the infanticidal males did not live in the same group as the infants they killed.

Relatedness of Infanticidal Males and Infant Victims. Establishing paternity for infant victims is certainly the most difficult problem to resolve under field conditions. However, it is often possible to make reasonable inferences from indirect evidence. In 95.7% of the observed cases ($N = 23$; table 8-4), the infanticidal male presumably did not sire the infant he killed because, at the time of conception, he was not in the group, he was sexually immature (no. 9), or, due to his low dominance status, he did not have priority of sexual access to females in the group (no. 24).

There is no verified case of a male killing his own offspring or sib. Cases that might possibly be viewed as exceptions invariably involve incomplete data sets and are, therefore, equivocal. There is only one observed infanticide where a male may have killed his own infant (no. 19). However, because the infanticidal male temporarily left the group around the time of its conception, a status change was inferred, suggesting that the infanticidal male was not the father.

The most equivocal cases of paternity concern suspected infanticides, especially when evidence is based only on disappearance (e.g., see Boggess 1984; Hrdy 1984 reply). In two suspected infanticides where fathers may have killed their own infants, it is of interest that the males were separated from the females they copulated with for much of the gestation period and birth (howlers: Crockett and Sekulic 1984; chimpanzees: Kawanaka 1981). Data from several studies suggest that males may respond differentially to females they had copulated with

and thus avoid killing their subsequent infants (e.g., Hrdy 1977; Struhsaker and Leland 1985). Separation from the female in time and space, however, may alter this discrimination.

Of the observed infanticides, only one case involved a male killing infants in the same group as his mother (no. 9). Although this male red colobus killed or is suspected of killing or wounding all infants under the age of 6.5 months, he did not attack his mother's infant. Furthermore, because he was probably conceived outside his natal group, he is unlikely to have killed a close relative. In howlers (Clarke 1983) there is also evidence of a male who avoided killing his own siblings during cases of suspected infanticide.

Infant Victims. All 24 infants in table 8-4 were still suckling and younger than the age at which weaning occurs and before the mother usually resumes sexual cycling. There was no apparent discrimination in terms of the sex of the infant victim.

Subsequent Matings with the Mother and Effect on Interbirth Interval. In 75% of the 12 cases where there is information, the infanticidal male copulated with the victim's mother following the killing. The infanticidal male could have sired the subsequent infant of the victim's mother in 50% of the 14 cases for which sufficient information is available. Not only did the infanticidal male gain sexual access to the victim's mother, but there was an appreciable shortening of the interbirth interval by as much as 66% following infanticide in all 4 cases where this could be determined.

The Response of Group Members to Infanticidal Attacks. Response of group members depends on the social organization and, in turn, most probably on the degree of genetic relatedness. In one-male groups where females are presumably closely related to one another, they often form aggressive coalitions against the infanticidal male (Hrdy 1977; Struhsaker 1977; Butynksi 1982b) (fig. 8-4). Coalitions involving both males (possibly sires) and females (relatives) may be formed against males posing a threat to infants in matrilineal multimale groups (e.g., baboons: Busse and Hamilton 1981; Collins, Busse, and Goodall 1984). In contrast, coalitions against infanticidal males in patrilineal multimale groups where females are not closely related to one another (e.g., red colobus) involved only adult males (possible sires or relatives) (Leland, Struhsaker, and Butynski 1984; Struhsaker and Leland 1985). Because infanticide has rarely been seen within multimale groups (in spite of greater observation effort on them than one-male groups), coalitions involving males or both males and females may be the most effective defense. In addition to forming defensive coalitions, females with infants threatened by in-

TABLE 8-4. Observed Cases of Infanticide among Primates

1. Subfamily	2. Species	3. Approx. No. Suspected Infanticides (total)	4. No. of Observed Infanticides (total)	5. Site/Source	6. Social Organization	7. Case Number	8. Circumstance
Colobinae		(>50)[a]	(9)				
	Presbytis entellus	38	1	Dharwar/ Sugiyama 1965	One-male matrilineal	1	After takeover
			3	Jodhpur/ Mohnot 1971	One-male matrilineal	2	After takeover
						3	After takeover
						4	After takeover
			3	Jodhpur Vogel and Loch 1984	One-male matrilineal	5	After takeover
						6	After takeover
						7	After takeover
	P. cristata	3	1	Kuala Selangor/ Wolf 1980; Wolf and Fleagle 1977	One-male breeding groups, matrilineal[e]	8	After intrusion and rise in rank
	Colobus badius	2	1	Kibale/Struhsaker and Leland 1985	Multimale patrilineal	9	Sexual maturity and rise in rank in natal group
Cercopithecinae		(26)[b]	(8)				
	Cercopithecus ascanius		2	Kibale/Struhsaker 1977	One-male matrilineal	10	After takeover
						11	After takeover
	C. mitis	2	1	Kibale/Butynski 1982b	One-male matrilineal	12	After takeover
	Papio cynocephalus ursinus	7	1	Moremi/Collins, Busse, and Goodall 1984	Multimale matrilineal	13	No data
	P. c. anubis	11	4	Gombe/Collins, Busse, and Goodall 1984	Multimale matrilineal	14	After immigration
						15	After immigration
						16	After immigration
						17	In fissioned natal group
Alouattinae		(25)[c]	(2)				
	Alouatta seniculus	21	1	Hato Masaguaral/ Rudran 1979	One-male breeding groups,[e] matrilineal	18	After intrusion
			1	Hato Masaguaral/ Crockett and Sekulic 1984	One-male breeding groups,[e] matrilineal	19	After inferred status change
Pongidae		(12)	(5)				
	Pan troglodytes	6[d]	2	Gombe/J. Goodall 1977	Multimale patrilineal	20	Strange female encountered
						21	Strange female encountered
	Gorilla gorilla beringei	6	3	Karisoke/Fossey 1979, 1984	One-male breeding units,[e] patrilineal	22	Intergroup encounter
						23	Intergroup encounter
						24	Rise in rank (immigrant male)

NOTE: Information in columns 4 through 14 refers only to observed cases.

a. See Vogel and Loch 1984 for a review of *P. entellus;* see introduction to this chapter for other colobine species.

b. See Collins, Busse, and Goodall 1984 for a review of baboon species; for *C. campbelli,* see Galat-Luong and Galat 1979.

c. See also Clarke 1983 for *A. palliata.*

d. See also Kawanaka 1981.

e. Adult and subadult males vary from 1 to 3. Only one male usually breeds.

f. Infanticidal male mounted but was not observed to copulate.

g. Not observed but possible.

h. Not observed but probable.

i. Not observed but unlikely.

j. Observed copulation at probable time of conception, but female also mated with other males.

TABLE 8-4. *(continued)*

9. Age/Sex of Infant Victim	10. Infanticidal Male Mates	11. Infanticidal Male Possible Sire	12. Comments on Infanticidal Male (I♂) Mate/Sire	13. Average Interbirth Interval (IBI) (mo)	14. IBI of Victim's Mother or Mean IBI of Infant-deprived Females in General (mo)
11.6 mo	Yes	Yes[h]	Conception during I♂ tenure	24	18
2.5 mo/male	Yes[f]	No data	Females presented to I♂ but he did not always copulate		
2.5 mo/male	?[g]	No data			
2.0 mo/female	Yes	No data			
1.0 mo/male	Yes	No data	Study ended	14.9	x̄ = 9.85
3.6 mo/male	Yes	No data			
1.5 mo/female	Yes	No data			
9 days	?[g]	No data	I♂ dominant until study ended; but matings not observed	No data	No data
1–1.5 mo/male	Yes	?Yes[j]	Female promiscuity	24–26	8
7 days	Yes[h]	Yes[h]	Females not identified; probable conception during I♂ tenure	11.5–28 bimodal	No data
1 day	Yes[h]	Yes[h]			
6 mo/female	No[i]	No[i]	I♂ replaced 4 days after infanticide	No data	No data
8 mo	No data	No data	Group not monitored	23.5	x̄ = 13.5
6.4 mo	Yes	?Yes[j]	(14) Female promiscuity	19	x̄ = 12.8
8.5 mo	No data	No data	(15) Group not monitored		
7.2 mo	No data	No[i]	(16) & (17) I♂ not observed to consort w/♀ at time of conception		
8.4 mo	No data	No[i]			
No data	?[g]	No	I♂ lethally wounded before conception (Hrdy 1979)		
1 day	Yes	Yes[h]	I♂ dominant at time of conception	16.6	x̄ = 10.5
1.5–2 yr	No[i]	No[i]	Females did not transfer into I♂♂'s community		
1.5–2 yr	No[i]	No[i]			
11 mo	Yes[h]	Yes[h]	Female transferred to I♂	39	25
1 day	No data	No data	Female disappeared wounded		—
3 mo	Yes	No[i]	Female emigrated from I♂		14

FIGURE 8-4. Two older adult female gray langurs charge an adolescent male to retrieve an infant he has snatched and severely wounded. (Photo: Sarah Blaffer Hrdy/Anthrophoto)

fanticidal males actively avoided these males and sometimes sought the proximity of possible sires (Hrdy 1977; Struhsaker and Leland 1985).

Observations of gray langurs (Hrdy 1977) and red colobus (Struhsaker and Leland 1985) clearly indicate that postconception estrus and promiscuity are effective ways in which females may confuse paternity recognition and thereby reduce the probability of attack by infanticidal males on subsequent offspring. Red colobus females pregnant at the time of attacks on other infants in the group copulated more frequently and later into their term of pregnancy than did females either before or after these attacks. A large proportion of their postconception copulations were with the infanticidal male, who did not attack their subsequent infants (Struhsaker and Leland 1985).

Hypotheses

How can the available information on infanticide by male nonhuman primates be interpreted from a theoretical standpoint? We present here three prevalent hypotheses on infanticide and discuss them in terms of their predictions and the basic data needed to test them. (For a more detailed treatment, see Hausfater and Hrdy 1984; and for a review of other hypotheses, see Hausfater and Vogel 1982).

First, the sexual selection hypothesis proposes that infanticide is an adaptive strategy whereby males kill the infants of other males in order to bring females into estrus sooner and thus increase their own reproductive success. Specifically, it predicts that (1) an infanticidal male

will typically not be the father or sibling of any infant he kills; (2) the mother will, on average, become sexually receptive and conceive earlier than if her infant had lived; (3) the infanticidal male will, on average, have greater reproductive success by mating with the mother sooner, than will a male who did not commit infanticide under the same conditions; and (4) the reproductive gain to the killer will be inversely correlated with the age of the infant at death and the average time an adult male has priority of access to ovulating females (see also Hrdy and Hausfater 1984). The available evidence for infanticide by male nonhuman primates most closely fits this hypothesis.

The sexual selection hypothesis is also consistent with the observation that infanticide occurs more often in one-male breeding groups than in multimale groups. There is more pressure to reproduce in a short amount of time because intense competition from extragroup males affects tenure length. In addition, infanticidal males are less likely to encounter severe counterattack from group members and are more likely to have exclusive sexual access to the mother afterward (e.g., Leland, Struhsaker, and Butynski 1984). Furthermore, the development of female counterstrategies, particularly promiscuity and postconception estrus, lends support to the suggestion that we are dealing with an evolutionary process in which males and females are both behaving in ways that improve their reproductive success.

Second, the resource competition hypothesis predicts that the death of an infant will result in increased access to resources for the killer and his descendants and a gain in their fitness (Rudran 1973a; reviewed in Hrdy 1979; Leland, Struhsaker, and Butynski 1984). It requires data showing that (1) competition for a resource is limiting the reproductive success and inclusive fitness of the infanticidal animal; and (2) the infanticidal animal or its offspring do in fact gain in fitness (in terms of increased resources) from infanticide. In no case of infanticide in nonhuman primates has adequate data been provided to support these predictions. It must be emphasized that an infanticidal male must first reproduce before any competitive advantage is gained by his direct descendants. If infanticide does have any effect on resource competition, it will be secondary to the primary effect of infanticide on reproductive success (Hrdy 1977).

Furthermore, if one assumes that resources are at least as critical, if not more so, for females, then we might expect females as well as males to commit infanticide. Yet, there are only three cases of observed infanticide by female primates, all in the same community and by the same mother and daughter chimpanzees (J. Goodall 1977, 1983). In other mammals, however, where resources such as nesting sites are severely limited, infanticide by females is more common (e.g., ground squirrels: Sherman 1981).

On the other hand, resource competition may be involved in rare cases of suspected and observed infanticide during intergroup encounters in matrilineal social organizations (e.g., yellow baboons: Shopland 1982; vervets: Hauser, pers. comm.). In patrilineal social organizations, however, infanticide by an extragroup male may be a means of attracting the female to transfer and mate with him (e.g., gorillas: Fossey 1983) and would best fit the sexual selection hypothesis.

Third, the social pathology hypothesis predicts that (1) infanticide will not be advantageous to the perpetrator, and (2) it will result from "enforced proximity" due primarily to human disturbance (Dolhinow 1977; Curtin and Dolhinow 1978; Boggess 1979, 1984; also see Hrdy 1979, 1984, and Leland, Struhsaker, and Butynski 1984 for rebuttal).

This hypothesis is not supported by the available information on primates. In the majority of unequivocal and clearly observed cases, the infanticidal male gained an opportunity for a reproductive advantage. Furthermore, because *enforced proximity* and *human disturbance* have no operational definitions (Boggess 1984), this hypothesis cannot adequately be tested. The prediction also does not fit populations where infanticide has been observed but human disturbance is negligible (e.g. Kibale, Uganda; Moremi, Botswana; and Hato Masaguaral, Venezuela (see references in table 8-4).

CONCLUSIONS

Infanticide by male nonhuman primates appears to be best explained by the sexual selection hypothesis. Although data from field studies are often sparse and incomplete, they demonstrate several important points: (1) infanticide does occur under natural, undisturbed conditions in the wild; (2) it is a widespread phenomenon that is more common than originally believed; (3) it typically occurs under a general set of conditions that are predicted by the sexual selection hypothesis; and (4) certain female behaviors may be explained as adaptive counterstrategies to the potentially great impact of infanticide on their reproductive success.

9 | Forest Guenons and Patas Monkeys: Male-Male Competition in One-Male Groups

Marina Cords

The genus *Cercopithecus* comprises more species than any other genus of African primates. If the closely related, monotypic genera *Allenopithecus, Erythrocebus,* and *C. (Miopithecus)* are included, there are at least 18 species in this group. The taxonomy of cercopithecine monkeys has been an area of controversy. Early classifications were based solely on comparative morphology (Hill 1966), but recent studies have also integrated information on DNA and protein similarities, as well as vocalizations (Ruvolo 1983). Both fossil and molecular evidence suggest that the guenon group (tribe: Cercopithecini) diverged from the baboon, macaque, and mangabey group (tribe: Papionini) about 10 million years ago. The radiation of modern cercopithecines began about 6 million years ago (Ruvolo 1983). Observations of hybrid animals in the wild and in captivity suggest that the speciation process is still under way since reproductive isolation between many species has not yet been established.

In this chapter I have adopted the species classification of Wolfheim (1983), except that the eastern and western forms of the *mona* group, *Cercopithecus mona* and *Cercopithecus campbelli,* are treated as separate species (Struhsaker 1970), while the subspecies *denti* and *wolfi* are included as *Cercopithecus pogonias* on the basis of vocalizations and distributional patterns respectively (Struhsaker, unpub.). The discussion includes the 16 species that do not routinely live in groups with more than one adult male (i.e., not *Cercopithecus aethiops* or *Cercopithecus (M.) talapoin;* see table 9-1). All are small- to medium-sized animals weighing approximately 2 to 8 kg as adults, except patas, which are larger (7 to 13 kg; Haltenorth and Diller 1980). Many of the forest species have brightly colored fur, especially on the tail and neck, and distinctive facial markings (Kingdon 1980). All species are diurnal.

In spite of the large number of species involved, this group of primates is little known relative to other African genera. The few long-term field studies have covered only about half of the species. This is because most *Cercopithecus* monkeys inhabit dense forest and are regularly hunted by humans, so observation is difficult; the more terrestrial patas are also often difficult to observe when they are dispersed in tall grasses whose color matches their own.

DISTRIBUTION

Most guenons inhabit the forests of West and central Africa. Species ranges typically fall into one of three biogeographic zones (Table 9-1): west of the Dahomey Gap, between the Dahomey Gap and the Congo and Ubangi rivers, or east of these rivers. *Erythrocebus patas,* the only nonforest monkey under consideration, occurs widely in the savannas and woodlands of the Sahel, from Senegal to Kenya.

Habitat preferences are variable both within and between forest-dwelling species (Wolfheim 1983; table 9-1). Some species are restricted to particular forest types while others occur in two or more kinds of habitat. *Cercopithecus mitis* is particularly noteworthy as it occupies a great diversity of habitats ranging from primary forest, to montane bamboo forest, coastal scrub, and relatively dry woodland. Patas monkeys typically occur in open habitats in steppe and woodland fringes. The habitat of patas differs from that of forest-dwelling cercopithecines in that it is highly seasonal.

The significance of variation in habitat preferences is not presently understood for several reasons. First, even in areas where a species' presence has been confirmed, population densities are mostly unknown. Second, detailed ecological studies have been made in only five sites on a total of seven species. In the absence of data on basic demography and community relationships in different geographical areas, it is difficult to assess the importance of specific ecological variables in determining distribution and abundance. Finally, most species are hunted by man and populations are declining as a result (Wolfheim 1983). Where hunting occurs, this important cause of mortality may obscure biological interactions that otherwise act to limit populations. Consequently, our interpretation and comparison of density estimates must be made with care.

In spite of these gaps in current knowlege, several findings emerge: (1) in forest habitats, two or more species are usually sympatric. In the Idenau Forest in Cameroon, for example, five species of *Cercopithecus* coexist (Gartlan and Struhsaker 1972). This degree of congeneric sympatry among primates is unequaled in Southeast Asia or South America. (2) Most forest species are primarily arboreal and are quick and agile in the trees; they may occasionally feed on the ground. Only *Cercopithe-*

cus neglectus, C. l'hoesti, C. mona, and perhaps *Allenopithecus nigroviridis* (Verheyen 1963) and *C. hamlyni* (J. Hart via R. Wrangham, pers. comm.) are often at ground level. Patas monkeys move along the ground, but spend a significant part of their time during daylight hours in trees, where much feeding occurs (Chism and Olson, unpub.). (3) Although some species—a minority in any given location—are mainly found in particular habitat types or strata, most niche separation among the forest-dwelling monkeys does not rely on gross habitat differences. Mixed-species groups of *Cercopithecus,* where participating species synchronize activity and location, are common and conspicuous (Struhsaker 1981a; Cords 1984b). Patas monkeys, on the other hand, are seldom seen with other sympatric primates because of differences in habitat preference. (4) The conservation status of several species is critical because of their already small population sizes, their limited geographic range, and the current destruction of their habitat (Wolfheim 1983). Particularly in West Africa, there is also considerable pressure from human hunting.

ECOLOGY

Feeding Behavior

The feeding and ranging ecology of a few species has been studied in detail (table 9-2); only *Cercopithecus ascanius* and *C. mitis* are known sufficiently well from different locations to allow interpopulational compari-

sons. Different methods have been used to quantify diet composition. Data from Gabon on *C. cephus, C. neglectus, C. nictitans,* and *C. pogonias* come from stomach contents expressed as percentage of dry weight. Other studies, such as those on *C. ascanius, C. diana, C. mitis,* and *E. patas,* have used a variety of feeding frequency scores. Unfortunately no algorithm exists for conversion of these measures, so the figures are not strictly comparable. Also, floristic variation between forests, or even different parts of the same forest, may increase apparent variation between and within species. In spite of these caveats, we know that the forest-dwelling species feed largely on fruit (fig. 9-1). Invertebrate prey (mainly insects) and leaves are also important for most species. Larger species are generally more folivorous and less insectivorous than smaller ones, and the degree of folivory increases when fruits are less frequently eaten (Gautier-Hion 1980; Rudran 1978; Struhsaker 1978; Cords 1984b). When leaves are eaten, young ones are preferred (Struhsaker 1978; Cords 1984b; Curtin, unpub.).

There is considerable overlap in the vegetable diets of sympatric guenons. Most plant species are shared by sympatric *Cercopithecus* species (e.g., 68% Makokou, Gabon; 87% Kakamega, Kenya). However, species may eat different amounts or different items from the same plant species, or may use shared plant foods at different times of year. Dietary overlap is consequently reduced if

TABLE 9-1. Geographical Distribution and Habitats of Forest Guenons

	Geography			Habitat							
Species	West of Dahomey Gap	Between Dahomey Gap and Congo/Ubangi	South or East of Congo/Ubangi	Primary Lowland Rain Forest	Montane Forest (>2,000 m)	Bamboo Forest	Flooded or Swamp Forest[a]	Secondary Forest	Riverine Forest	Coastal Forest	Dryer Woodland
Allenopithecus nigroviridis		+	+++				**				
Cercopithecus ascanius			+++	*			*	*	*		*
C. campbelli	+++			*				*	*		*
C. cephus		+++	+	*			*	*	*		
C. diana	+++			**				*	*		
C. erythrogaster		+++		*				*	*	*	
C. erythrotis		+++		*			*	*			
C. hamlyni			+++	*	**	*					
C. l'hoesti		+	+++	*	*			*	*		*
C. mitis			+++	*	*	*	*	*	*	*	*
C. mona	+	+++		*			*	*		*	
C. neglectus		++	++		*	*	**	*	*		
C. nictitans	++	++		**			*	*	*		
C. petaurista	+++			*			*	*	*	*	
C. pogonias		++	++	**			*	*			

SOURCES: Wolfheim 1983; Oates 1982; Struhsaker, pers. comm.

NOTES: Plus (+) ≤10% of range; double plus (++) = 10 to 90% of range; triple plus (+++) = ≥90% of range; * = found in; ** = typical of, or limited to.

a. Includes papyrus and mangrove swamps, and seasonally flooded forest.

TABLE 9-2. Feeding, Ranging, Population Density and Group Size of Patas Monkeys and Forest Guenons

Species and Study Site	Diet				Ranging[b]		Population Density (ind/km²)	Group Size[c]	Sources
	Fruit (%)	Leaves[a] (%)	Gum (%)	Inverte-brates (%)	Home Range (ha)	Day Range (m)			
Cercopithecus ascanius									
Kakamega, Kenya	61.2	7.2	2.8	25.0	55, 30, 27 (N = 3)	1,543 ± 296 (N = 84 d, 1 gp)	72	22, 23, 34 (N = 3)	Cords 1984b
Kanyawara, Kibale, Uganda	43.7	15.5	0	21.8	24 (N = 1)	1,447 ± 253 (N = 34 d, 1 gp)	140	30–35 (N = 1)	Struhsaker and Leland 1979
Ngogo, Kibale, Uganda	47.0	16.4	0	28.6	—	1,692 ± 377 (N = 48 d, 3 gp)	—	35–40 (N = 1)	Struhsaker, unpub.
Bangui, C.A.R.	✔	✔	✔	✔	15 (N = 1)	—	117	17–23 (N = 1)	Galat-Luong 1975
C. campbelli									
Adiopodoume, Ivory Coast	✔	✔	✔	0	3 (N = 1)	—	—	8–13 (N = 1)	Bourlière, Hunkeler, and Bertrand 1970
C. cephus									
Makokou, Gabon	81.3	6.1	0	12.6	52, 19 (N = 2)	1,295–1,980 (N = ? d, 2 gp)	—	8–15, 8–15+ (N = 2)	Gautier-Hion and Gautier 1974; Gautier-Hion, Emmons, and Dubost 1980; Gautier-Hion, Quris, and Gautier 1983
C. diana									
Bia Park, Ghana	41.4	6.2		24.5	189 (N = 1)	1,892 ± 573 (N = 50 d, 1 gp)	—	14, 20+, 30–40 (N = 3)	Curtin, unpub.
C. l'hoesti									
Kibale, Uganda	✔	✔	0	✔	—	—	—	17.4 ± 8.0 (N = 25 counts)	Butynski, unpub.
C. mitis									
Kakamega, Kenya	54.6	18.9	1.9	16.8	23 ± 9 (N = 5)	1,136 ± 228 (N = 80 d, 1 gp)	169	32.6 ± 8.9 (21–45) (N = 5)	Cords 1984b
Kanyawara, Kibale, Uganda	21.7–42.7	20.6–28.9	0	19.8–37.7	50.6 ± 14.8[d] (N = 4)	1,298 ± 345 (N = 65 d, 1 gp)	42	18.7 ± 8.5 (11–33) (N = 4)	Rudran 1978; Struhsaker 1978, Butynski 1982b, unpub.
Ngogo, Kibale, Uganda	30.1	22.4	0	35.9	350 (N = 1)	1,419 (676–2,446) (N = 90 d, 1 gp)	4–5		Butynski, unpub.
Budongo, Uganda	~60–70	✔	0	✔	8 (N = 4)	~150–1,000	185	13.6 ± 1.2 (12–17) (N = 6)	Aldrich-Blake 1970
Lake Kivu, Zaire	46	7	0	13	~25 (N = 2)	~594 ± 198 (N = 9 d, 1 gp)	—	10, 11, 16 (N = 3)	Schlichte 1978b
Muguga, Kenya	✔	✔	0	0	~16 (N = 1)	—	—	12+ (N = 1)	DeVos and Omar 1971
Nyeri, Kenya	—	—	—	—	13.2, 13.8, 14.2 (N = 3)	—	118	15, 16, 18 (N = 3)	Ibid.
C. neglectus									
Makokou, Gabon	77.3	9.4	0	4.9	≥6.6 ± 2.2[e] (N = 6)	530 (250–1,010) (N = 24 d, 1 gp)	0.3–0.5	4.0 ± 0.7 (N = 6)	Gautier-Hion and Gautier 1978; Gautier-Hion, Emmons, and Dubost 1980
Makokou, Gabon	—	—	—	—	~13 (N = 11)	—	28	5, 6 (N = 2)	Quris 1976
C. nictitans									
Makokou, Gabon	72.0	17.0	0.2	9.6	174 (N = 1)	1,825–1,980 (N = ? d, 1 gp)	—	13–20 (N = 1)	Gautier-Hion and Gautier 1974; Gautier-Hion, Emmons, and Dubost 1980; Gautier-Hion, Quris, and Gautier 1983

TABLE 9-2.   *(continued)*

Species and Study Site	Diet Fruit (%)	Leaves[a] (%)	Gum (%)	Inverte- brates (%)	Ranging[b] Home Range (ha)	Day Range (m)	Population Density (ind/km²)	Group Size[c]	Sources
C. pogonias Makokou, Gabon	82.5	1.2	0	16.1	174 (N = 1)	1,825–1,980 (N = ? d, 1 gp)	—	13–18 (N = 1)	Gautier-Hion and Gautier 1974; Gautier-Hion, Emmons, and Dubost 1980; Gautier-Hion, Quris, and Gautier 1973
Erythrocebus patas Laikipia, Kenya					3,200 (N = 1)	4,330 ± 1,520 (N = 55 d, 4 gp)	~1.5	35.5 ± 13.5 (15–73) (N = 10)	Chism and Olson, unpub.
Murchison Falls, Uganda	✔	✔	✔	✔	5,200 (N = 1)	mode: 2,000–2,500 (700–11,800) (N = 58 d, 4 gp)	0.035	20.6 ± 7.4 (5–30) (N = 8)	Hall 1965a

NOTE: When quantitative date are not available, tics (✔) indicate that a dietary item is consumed. D = days; gp = groups.

a. Includes both young and mature leaf blades and leaf buds. Samples from Gabon include all fiber.

b. For sample sizes ≤3, actual values are given. For sample sizes ≥4, means and standard deviations (along with sample sizes) are given when available. Alternatively, approximations are quoted (~).

c. For sample sizes ≤3, actual values or ranges (per group) are given. For sample sizes ≥4, means, standard deviations and ranges (in parentheses) are given where available. Approximations are indicated by a ~.

d. The three smallest home ranges are derived from only 12 to 19 days of observations.

e. Only 382 hours in contact with monkeys. The group that was studied longest (320 hr) had a home range of 7 ha.

calculated on the basis of plant-specific food items or on a monthly basis rather than as overlap in species use for the year as a whole (Struhsaker 1978; Cords 1984b). Overlap between species is higher for fruits than for less popular items such as leaves (Gautier-Hion 1980; Cords 1984b). It is therefore not unexpected that overlap is greatest when fruits are especially important in the diet (Gautier-Hion 1980; Cords 1984b; Struhsaker 1978). This observation suggests that interspecific competition is an important determinant of ecological segregation at certain times of year when resources are in short supply.

Species diverge in the invertebrate diet by using different prey items (Gautier-Hion 1980) or different capture methods and microhabitats (Struhsaker 1978; Cords 1984b). There are no data on seasonal variations in the overlap of their invertebrate prey, and the degree to which species compete for this important source of protein remains an open question.

Variation in diet also occurs within species. Table 9-2 shows that the proportions of the most important components in the diet may be quite different between study areas (compare *C. ascanius* and *C. mitis* in Kakamega and Kibale). Even within a given study area, the diets of different social groups (Rudran 1978) or of different age-sex classes (Gautier-Hion 1980; Rudran 1978; Cords, unpub.) may differ considerably. For example, Gautier-Hion found that adult males and females of the same species have different proportions of fruits and seeds, leaves and fiber, and animal matter in the diet over the course of the year, and females of one species may resemble males of another more closely than conspecific males, even during the dry season when dietary segregation is most extreme.

Ranging

Ranging patterns are of course closely related to diet. Patas, living in a poorer, drier, more seasonal and unpredictable environment, have larger home ranges and longer and more variable daily path lengths than forest guenons (table 9-2). Within the latter group, *C. neglectus* stands out with the smallest home ranges and shortest day ranges. Interpopulational comparisons between *C. ascanius* and *C. mitis* in Kakamega and Kibale are interesting in that day ranges are similar in both forests, whereas home ranges are not (N.B., even the *relative* sizes of home ranges are reversed). Home range data from other populations of *C. mitis* contribute to the variability. It seems that home range size is not a simple function of body size or diet. For these two species, at least, there is a negative relationship between home range size and population density (table 9-2; Nyeri *mitis* ignored since the density estimate includes areas of exotic plantations that the animals did not use). On the other hand, home range size for a particular social group may remain constant over time despite considerable fluctuations in group size (*C. mitis*: Cords, in press; *E. patas*: Olson, pers. comm.).

FIGURE 9-1. An adult male redtail monkey (*C. ascanius*) from Kibale Forest, Uganda, eating fruit. (Photo: Lysa Leland)

REPRODUCTION

Receptive females solicit males through conspicuous postures and facial expressions in some species (*C. mitis:* Tsingalia and Rowell 1984; *E. patas:* Rowell and Hartwell 1978; *C. ascanius:* Cords, unpub.), but commonly females simply approach males and remain in proximity (Cords, unpub.; Struhsaker, unpub.). With the exception of *Allenopithecus,* which has sexual swellings (Hill 1966), none of the guenons are characterized by obvious visual indicators of estrus, although olfactory cues cannot be ruled out. Because there are no external signs of estrus, gestation lengths are generally not known. Furthermore, *C. mitis* and *E. patas* females, whose gestation has been measured in captivity (table 9-3), are known to undergo postconception estrus (Rowell 1970; Rowell and Hartwell 1978), so changes in females' be-

havior cannot be used reliably to estimate gestation lengths in free-ranging populations of these or other species (Cords 1984a). Females in all species give birth to single offspring. All species show some seasonality of births, even though they may occur in most months (table 9-3). In many populations, births coincide with the dry season, but in others there is no relationship between rainfall and births. *E. patas* have the most distinct birth seasons, coincident with their highly seasonal habitat (fig. 9-2).

There are few data from natural populations on age at sexual maturity because of the lack of long-term records on known individuals. Data on males are especially difficult to assess since both social and physiological factors may influence the onset of breeding activity. Estimates based on behavior and morphology of captive animals are included in table 9-3 to give a rough idea of the age at which males are physiologically ready to mate (*C. cephus, C. neglectus, C. nictitans,* and *C. pogonias*). Figures for wild populations of *C. ascanius, C. campbelli, C. mitis,* and *E. patas* are based on estimated ages of the youngest males observed to complete copulations, although it could not be determined if these males were capable of impregnating females. For females, age at first birth is recorded. It should be noted that for all species except *C. campbelli* and *E. patas* females in Laikipia, ages are estimated based on observed changes in body size over time: the real age of first reproduction is not known for a single individual. In the seven forest guenons, females seem to give birth at around 4 to 5 years; males, who are larger as adults, apparently become sexually mature at about 5 to 6 years, or perhaps older. In contrast, patas development is relatively accelerated: females give birth at age 3, and males may first mate at 4 to 4.5 years, though it is known that they are capable of impregnating females at 3 to 3.5 years in captivity. Patas, being annual breeders also have the shortest interbirth intervals. Except for *C. campbelli* (for which the sample is very small), interbirth intervals in the forest guenons (three species) is around 2 years or more. Note that these data are biased toward shorter intervals: for example, in a 4-year study of Kakamega *C. mitis,* 3 females had 2-year interbirth intervals. At the end of 4 years, there remained 11 that had not given birth in 2 years or more. The exact interbirth intervals for these individuals are not known, but they will surely increase the average.

SOCIAL STRUCTURE AND BEHAVIOR

All 15 species apparently live in social groups of which only females are permanent members; males at puberty leave their natal groups and subsequently may be solitary, part of an all-male association (confirmed only in *E. patas* and one population of *C. mitis*) or part of a new

TABLE 9-3 Reproductive Parameters for Patas Monkeys and Forest Guenons

Species and Study Site	Birth Seasonality	Age at Sexual Maturity[a]		Gestation[b] (days)	Interbirth Interval[c] (mo)	Sources
		Male	Female			
Cercopithecus ascanius						
Kakamega, Kenya	Sep–Feb; peak Dec–Jan	6 yr[d]	—	—	>24[e]	Cords, unpub.
Kibale, Uganda	Most months; peak Nov–Feb	—	4–5 yr[d]	—	17.8 ± 6.2[e] (bimodal: 12 and 24 mo)	Struhsaker, unpub.
Bangui, C.A.R.	Aug–Sep	—	—	—	—	Galat-Luong 1975
C. campbelli						
Adiopodoumé, Ivory Coast	Peak Dec–Jan; some births in Aug, Sept	4.5 yr[d]	3 yr[d]	—	12[d]	Hunkeler, Bourlière, and Bertrand 1982; Galat-Luong and Galat 1979
C. cephus						
Makokou, Gabon	Peak Dec–Jan	5+ yr[d]	4 yr[d]	—	—	Gautier-Hion and Gautier 1976
C. l'hoesti						
Kibale, Uganda	Peak Dec–Feb	—	—	—	—	Butynski, unpub.
C. mitis						
Kakamega, Kenya	Most months; peak Jan–Mar	6+ yr[d]	5–6 yr[d]		>22[e]	Cords, unpub.
Kibale, Uganda	Peak Dec–Jan	—	—	140[d,h]	~21[d]	Rudran 1978; Butynski, unpub.
Aberdares, Kenya	Nov–Mar; peak Feb–Mar	—	—		—	Omar and DeVos 1971
Budongo, Uganda	Most months; peak Jan, Jul–Aug	—	—		—	Aldrich-Blake 1970
Captive, Limuru, Kenya	All months	—	53–97 mo[e]		20.2 ± 12.0[e] med: 18	Rowell and Richards 1979
C. neglectus						
Makokou, Gabon	Nov–Apr	5–6 yr[d]	3.5–4 yr[d]	177–87[g,i]	—	Gautier-Hion and Gautier 1976, 1978
Captive, Limuru, Kenya	Most months	—	48–94 mo[e]	~170[g,j]	27.4 ± 14.1[e] med: 23	Rowell and Richards 1979
C. nictitans						
Makokou, Gabon	Peak Dec–Jan	5–7 yr[d]	4 yr[d]	—	—	Gautier-Hion and Gautier 1976
C. pogonias						
Makokou, Gabon	Dec–Feb (one in Apr)	5–6 yr[d]	4 yr[d]	—	—	Ibid.
Erythrocebus patas						
Laikipia, Kenya	Dec–Jan	4–4.5 yr[d]	3 yr[d]		—	Chism, Rowell, and Olson 1984
Murchison, Uganda	Jan–Feb	4 yr[g]	—		—	Hall 1965a
Waza, Cameroon	Nov–Jan	—	—	167[f,k]	—	Struhsaker and Gartlan 1970
Captive, Limuru, Kenya	Most months	—	32–73 mo[d]		14.6 ± 4.7[e] med: 13	Rowell and Richards 1979
Captive, California, USA	Spring	3–3.5 yr[d]	25–46 mo[e] med: 35		med: 11.8[e]	Chism, Rowell, and Olson 1984

a. Values for all species except captive *E. patas* and Laikipia female *E. patas* are estimates; see text.

b. Gestation lengths for each species are from captive studies; see notes for each figure quoted.

c. Cases where the first of two infants died are excluded.

d. $0 < N \leq 5$.

e. $5 < N \leq 20$.

f. $N > 20$.

g. Sample size unavailable.

h. Rowell 1970, captive study.

i. Haltenorth and Diller 1980, estimate.

j. Rowell and Richards 1979, estimate.

k. Sly et al. 1983, captive study.

FIGURE 9-2. Two adult female patas monkeys (*Erythrocebus patas*) with nursing infants. (Photo: Dana Olson)

heterosexual group (fig. 9-3). For the most part, this pattern of events is inferred on the basis of group compositions, most of which report only a single adult male per group. Although data are available for most species, they vary considerably in quality; group enumerations are particularly difficult in forests. The problems are exacerbated if animals are not habituated (e.g., as those studied in Cameroon and Gabon) or are dispersed: in Kibale and Kakamega, *C. ascanius* and *C. mitis* are often spread over more than 100 m (Struhsaker and Leland 1979; Cords 1984b), although *C. cephus* groups in Gabon are evidently more compact (Quris et al. 1981). Patas may be spread out over 500 m (Chism, Rowell, and Olson 1984).

Most guenons live in groups of about 10 to 40 members (table 9-2; Wolfheim 1983). From long-term observations we know that, due to births, transient male membership, and mortality, there can be much fluctuation in size within a single group. Differences between conspecific groups within and between study areas can also be striking (table 9-2). Coupled with few data on most species, this variability severely limits the nature of interspecific comparisons.

C. neglectus, however, appears exceptional. In Gabon, *C. neglectus* lives in small family groups with one adult male, one adult female, plus young (Gautier-Hion and Gautier 1978). Kingdon (1974) reports much larger groups in East African populations, but his impression is not supported by Brennan's (1985) recent 3-month survey in Kenya, where groups contained two to six members. *C. neglectus* groups in Kenya may also differ from those in Gabon in that the Kenyan groups appear to contain more than one adult female per male. The social structure of *C. neglectus* is clearly quantitatively different from that of other cercopithecine species.

Intergroup Relations

Relations between groups of conspecifics are usually unfriendly. Some clearly territorial species show both site-specific defense of territorial boundaries and exclusive use of parts of the home range (*C. campbelli, C. mitis, C. ascanius:* Galat-Luong and Galat 1979; Struhsaker and Leland 1979; Cords 1984b). Patas groups may avoid one another (Hall 1965a) or engage in aggressive interactions (Struhsaker and Gartlan 1970; Chism, Rowell, and Olson 1984). Intergroup encounters may occur fre-

FIGURE 9-3. A subadult male blue monkey (*C. mitis*) from the Kibale Forest, Uganda. (Photo: Lysa Leland)

quently in this species despite large home ranges because several groups drink at the same water hole (Struhsaker and Gartlan 1970) or because range overlap is extensive (Chism and Olson, unpub.). The large size of the patas home range apparently prohibits effective range defense. Two neighboring groups of *C. cephus* in Gabon occasionally participated in aggressive encounters with one another, but at other times they moved and even slept together as part of a larger mixed-species group (Gautier and Gautier-Hion 1983). Data on other forest guenons are limited: observed intergroup contacts of *C. nictitans* and *C. pogonias* in Gabon are confined to vocal exchanges between males (Gautier-Hion, Quris, and Gautier 1983), but it is not clear whether groups are actively avoiding one another. Of the seven forest guenons whose ranging behavior has been studied, only *C. neglectus* ap-

pears to be nonterritorial, occasionally peacefully intermingling with individuals from neighboring groups and communally exploited space (Gautier-Hion and Gautier 1978).

When intergroup conflicts occur, in at least three species females and young are more actively involved than males (*C. ascanius, C. mitis,* and *E. patas*). Males may give loud calls (Gautier and Gautier-Hion 1983; Cords, unpub.). In Kenya, *C. mitis* and *E. patas* males were seen to join females in showing antagonism toward members of other groups only during the mating season (Chism, Rowell, and Olson 1984; Cords, unpub.), but *E. patas* males in Cameroon participated in intergroup conflicts at other times as well (Struhsaker and Gartlan 1970). In keeping with known residence patterns, females appear to defend the resources in their home range against neighboring conspecific females. From existing data, it is difficult to assess the relative costs and benefits of being territorial in these species: clearly larger home ranges should be more difficult to defend, and accordingly patas seem unable to defend them. But one is left with the enigmatic *C. neglectus* in its tiny home ranges.

Intragroup Relations

Compared to baboons or macaques, guenons interact seldom and subtly. Social organization within the group is not well understood for most species because in all but four studies (*C. ascanius, C. campbelli, C. mitis,* and *E. patas*), individuals were not recognized. In these few studies, dominance hierarchies were not apparent among females and juveniles. However, a network of dyads characterized by more-than-average affiliative behavior or proximity indicates that some sort of patterning of social relationships exists. From the field studies it is apparent that such dyadic relationships are especially marked among animals of similar age or sex, and possibly among kin; a study of social organization in captive patas confirms these impressions (Rowell and Olson 1983).

Females and juveniles interact socially mostly with one another. Allomaternal care is extensive in patas (Chism, Rowell, and Olson 1984) and occurs also in *C. campbelli, C. ascanius,* and *C. mitis* (Bourlière, Hunkeler, and Bertrand 1970; Struhsaker and Leland 1979). The ontogeny and functional aspects of social relationships have been little studied in the field (but see Chism, Rowell, and Olson 1984 for a discussion of patas development).

Most attention has been paid to adult males who are spatially and socially peripheral to the rest of the troop most of the time in at least five species (Bourlière, Hunkeler, and Bertrand 1970; Stuhsaker and Leland 1979; Quris et al. 1981; Chism and Olson, unpub.). Adult males are conspicuous troop members because of their loud rallying and alarm calls (Gautier and Gautier-

FIGURE 9-4. (*top*) Juvenile male patas monkeys harass an adult male who is copulating with an adult female. (*bottom*) An adult male threatens a juvenile male (not shown) away from an estrous female. Harassment of copulations by immature males is not uncommon in patas monkeys, but the functional significance of this behavior is unclear. (Photos: Dana Olson)

Hion 1983). In *E. patas, C. cephus,* and in at least one population of *C. mitis,* adult males are often near subadult males (Quris et al. 1981; Chism and Olson, unpub.; Cords, unpub.) (fig. 9-4). Males and females interact infrequently outside of the mating season.

◆ MALE-MALE COMPETITION IN ONE-MALE GROUPS

Since Struhsaker's review of cercopithecine social organization in 1969, forest guenons and patas monkeys have been seen as exemplars of a one-male-group type of social structure. From there, it seems a small step to infer that the typical mating system is "harem" polygyny, where only one adult male, the resident male, has reproductive priority to a group of females. Competition between males for access to breeding females should be intense in such polygynous species since differences in reproductive success between resident and extragroup males are potentially enormous; in other polygynous mammals such as elephant seals, for example, 14 to 35% of the males in residence during any one breeding season perform all copulations, and the five most active males account for over half of them (LeBoeuf 1974). Variance in reproductive success of males may be reduced over a lifetime, however, if each male assumes more than one role (i.e., resident or extragroup male) during his reproductively mature years. Given a certain-sized group of females, the lifetime reproductive advantage of a resident male thus depends on two things: (1) the degree to which he has exclusive access to fertile and receptive females, and (2) the length of his tenure(s) in a group.

Polygynous One-Male Groups

There are three kinds of evidence for polygynous one-male mating in forest guenons and patas monkeys. Two of them are indirect, in that mating has not necessarily been observed. First, most group counts, of which there have been about 100 with sufficient detail (for all species combined), include only one adult male. Data from 13 of the 16 species are available from short-term censuses (Struhsaker 1969; Struhsaker and Gartlan 1970; Quris 1976), longer studies where individual males were not recognized (Hall 1965a; Aldrich-Blake 1970; Gautier-Hion and Gautier 1974), or extended studies where recognized individuals are known to have persisted in particular social groups for periods up to several years (Hunkeler, Bourlière, and Bertrand 1972; Struhsaker and Leland 1979; Cords 1984a; Tsingalia and Rowell 1984).

Second, males in groups are known to be intolerant of other male intruders, who are chased and threatened (Struhsaker and Gartlan 1970; Galat-Luong and Galat 1979; Struhsaker and Leland 1979; Cords 1984a; Tsingalia and Rowell 1984). On the other hand, *extra*group males of some species (*C. mitis* in Kakamega, Tsingalia

and Rowell 1984; *E. patas:* Struhsaker and Gartlan 1970; Gartlan 1975; Chism and Olson, unpub.) may form loose, apparently ephemeral, affiliations, feeding and/ or ranging in proximity to one another. More obvious friendly interactions between members of such affiliations are rare.

Although the preceding information for patas and forest *Cercopithecus* stands in contrast to that for other cercopithecine species regularly living in multimale groups, conclusions about mating systems based on group enumerations should be interpreted carefully for several reasons: (1) Data from census-type studies may be of poor quality in that some group members are overlooked. From long-term studies where specific groups are contacted repeatedly, we know that complete group counts are rarely achieved on the first attempt. As mentioned previously, adult males of several species are often spatially peripheral to the rest of their groups. In at least some groups where more than one male was present (e.g., *C. mitis:* Tsingalia and Rowell 1984; *C. ascanius:* Cords 1984a), a similar pattern was observed, particularly for occasional residents, who nevertheless mated. Therefore the chances of missing males may be especially high. (2) Even if group counts are reliable, censuses and even long-term studies when groups are contacted only intermittently provide information on social structure for only one or a few points in time. If additional males join a group for brief periods of time, they are unlikely to be detected. As will be seen later, there is evidence that extragroup males join groups at particular, limited times (i.e., during the breeding season). Census counts that occur at other times of year may well indicate the presence of only a single male. (3) In any case, social structures and mating systems are not the same things. The mere presence of one or several males in a group most of the time does not mean that those males are breeding. When group counts are used as evidence for mating patterns, an inference is being made.

The best evidence for a polygynous one-male mating system in guenons comes from long-term studies of known social groups that individual males have monopolized during times when offspring were conceived. Table 9-4 presents selected data that include only males who persisted in groups for at least 12 months. In many cases, these males were observed to mate (Cords, unpub.; Struhsaker 1977, unpub.). Some were known to have been joined by other males, some of whom also mated, during certain breeding seasons, but in most breeding periods there is no evidence of multiple males joining the residents. Even though all of these groups were monitored intermittently to some degree, the pattern of events during recognized multimale influxes (described later) suggests that such influxes would be difficult to overlook if they occurred, even with intermittent sampling. Considering additionally the timing of influxes relative to births, it seems likely that most of the males whose tenures are given in table 9-4 monopolized females reproductively and that they probably sired the offspring conceived during their residence in a group. Furthermore, there are similar data from different groups or different times when, although the identity of the group males changed more often than once a year, only one was known to be in a group at a time (Butynski 1982b; Cords, unpub.; Struhsaker, unpub.), including times when infants were conceived.

Multimale Influxes and Promiscuous Mating

While the data discussed above indicate that the monopoly of a group of females by a single male is common, promiscuous mating has been directly observed in several study groups as well. In *C. ascanius* (Struhsaker 1977, unpub.; Cords 1984a), *C. mitis* (Tsingalia and Rowell 1984; Butynski, unpub.; Cords, unpub.), and *E. patas* (Olson et al., unpub.), a common pattern of events observed during certain periods of mating or conceptions was a rapid turnover of adult males, with simultaneous and sequential membership of 2 to 19 males in one social group. In some cases, the previous resident male persisted throughout such influxes (Struhsaker 1977; Tsingalia and Rowell 1984; Cords, unpub.), but in others he was replaced (Cords 1984a; Olson et al., unpub.; Struhsaker, unpub.). The tenures of newly joined males varied from less than a day to several months. Most males were observed to copulate, and both sexes mated

TABLE 9-4 Tenures of Recognized Males That Persisted for at Least One Year in Social Groups of Forest Guenons

Species and Study Site	Tenure Length (months)[b]	Source
Cercopithecus ascanius		
Kibale, Uganda	39, 28, 24[a], 23, 23[a], 21[a], 17[a], 15[a], 13[a]	Struhsaker, unpub.
Kakamega, Kenya	39	Cords, unpub.
C. campbelli		
Adiopodoume, Ivory Coast	34	Hunkeler, Bourlière, and Bertrand 1972
C. mitis		
Kibale, Uganda	62[a], 62, 60, 23, 22, 17	Butynski, unpub.; Rudran, 1978
Kakamega, Kenya	49[a], 46[a], 22	Cords, unpub.

a. Males known to have been joined by other males at some time(s) when females were either seen to mate or inferred to have mated because of births that occurred subsequently. The timing, duration, and reproductive consequences of these influxes were variable (see text).

b. Figures show minimum values, since males were often still in their groups at the end of the study. Data on all species except *C. campbelli* come from studies where observations were intermittent.

promiscuously (table 9-5). In the Kakamega Forest (*C. ascanius* and *C. mitis*) and Laikipia region (*E. patas*) of Kenya, male influxes were limited to peak breeding periods when most conceptions (for the years in question) occurred. In the Kibale Forest, Uganda, influxes also occurred outside of peak breeding periods, but were nonetheless accompanied by copulations (Struhsaker, unpub.).

Relatively frequent male-male agonistic interactions during multimale influxes reflected intense competition for mates (Struhsaker 1977; Cords 1984a, unpub.; Tsingalia and Rowell 1984). The reproductive consequences of this intermale competition could not be measured directly, but were estimated by inferring probable paternity based on observed mating behavior in relation to birth dates. For *C. ascanius*, this approach is complicated by the lack of precise information on gestation length. Furthermore, as Stern and Smith (1984) have shown with captive rhesus macaques, sexual activity may be a poor indicator of paternity.

In spite of these difficulties, it appears that males resident in social groups before and after multimale influxes frequently do not have a reproductive advantage relative to those who join groups as part of a multimale influx, at least in years when such influxes occur. In the Kakamega *C. mitis* and Laikipia *E. patas* groups, for example, resident males probably sired few, if any, of the offspring conceived during multimale influxes when they were either absent or not seen to mate. Resident males may have impregnated the occasional female who came into estrus in what is normally the birth season (Chism, Rowell, and Olson 1984; Tsingalia and Rowell 1984; Cords, unpub.). In Kakamega and Kibale *C. ascanius*, resident males may have sired some of the offspring conceived during multimale influxes, but other males were observed mating concurrently (Struhsaker 1977; Cords 1984a) and are also possible fathers. On the other hand, in the Kibale *C. ascanius* groups where influxes did not coincide closely with conceptions, at least half of conceptions for the years in which there were influxes occurred during times when only one male was known to be present (Struhsaker, unpub.).

In addition to the detailed studies discussed above, observations of other groups and other populations suggest that multimale influxes are widespread. In Kakamega, multimale influxes occurred in as many as three neighboring *C. mitis* groups simultaneously, with individual males moving from group to group over the course of weeks or hours. Males resident in one troop even mated with females in neighborhing groups (Tsingalia and Rowell 1984; Cords, unpub.). Other patas groups at Laikipia were also seen with more than one male (Chism, pers. comm.). In central and West African populations of *C. ascanius*, *C. diana*, and *C. campbelli*, the presence for several months of more than one adult or sexually

TABLE 9-5   Events Associated with Male Influxes during Mating Periods in Redtail, Blue, and Patas Monkeys

Species, Group, Year, Observation Months	Months When Mating/ Conceptions Occurred	Number of ♂♂ in Group during Mating Periods	Number of ♂♂ Seen in Group Simultaneously (min–max)[a]	Estimated ♂ Tenure Length (days); Median, Range[b]	Number of ♂♂ Seen to Mate	Number of Breeding ♀♀ in Group[c]	Number of Receptive ♀♀	Maximum Number of Receptive ♀♀ per Day	Number of Births Resulting from MCP and Months When They Occurred
KAKAMEGA									
Cercopithecus ascanius									
L group, 1980 (Jan–Dec)	Feb–Sep (227 d) (Dec)[d]	8	1–4 x̄ = 2.2	77 (1–153)[e] (N = 8)	7[f]	9–10	10	3	7 Sep, Oct, Dec. 1980; Jan 1981
C. mitis									
T group, 1981 (Jun–Nov)	Jun–Nov (~170 d)	16	2–10 x̄ = 4.2	(5–~105)[g] (N = 15)	9	18	6	4	3 Nov, Dec 1980; Jan, Mar 1981
C. mitis									
T group, 1983 (Jun–Aug)	Jun–Aug (≥54 d)	19	2–11 x̄ = 5.9	9 (1–44)[g] (N = 18)	12	18	13	8	Unknown
LAIKIPIA									
Erythrocebus patas									
M1 group, 1980 (Jan–Dec)	Jul–Aug (62 d)	3	1–3	3, 6, 62	2	9	9	≥3	8[h] Dec 1980; Jan 1981
E. patas									
M2 group, 1980 (Jan–Dec)	Jun–Sep (87 d) (Feb–Mar)[i]	8	1–7	20.5 (1–49) (N = 8)	6	19	16	≥4	15[h] Dec 1980; Jan, Feb 1981

NOTES: MCP = mating/conception period. Since patas conceptions were limited to a shorter period than were matings, and since most changes in male membership in groups occurred around conception periods, figures are given with reference to conception periods in this species. In *Cercopithecus*, either females were not recognized during the mating period (*C. mitis* 1981), birth dates are not known (*C. mitis* 1983), or gestation length is unknown (*C. ascanius*), so periods of conception cannot be identified. Given that subsequent births were spread over 5 months in each species, it is probable that conceptions were scattered throughout periods of mating during these years. Therefore figures for these species include the entire period in which mating occurred.

a. Both minimum and maximum values are conservative estimates of the real number of males present since some individuals may not have been detected. Means were calculated using probable tenures as described in note b.

b. Estimates include gaps in observations if males were present before and after such gaps. In *C. ascanius* and *C. mitis*, gaps were typically a few days long; in the M2 group of *E. patas*, there were three

gaps of about 2 weeks each as well. In all species, some tenures continued outside of the MCP, but values given here include only those fractions within MCPs.

c. In *C. mitis* and *E. patas*, includes females not fully grown if they mated. Some of these females later conceived.

d. One female that lost her last infant copulated in December. Her December estrus period is not included in the table.

e. Excludes natal subadult male who was in L group throughout the study, even though he mated. Maximum tenure length differs from Cords (1984a) because tenure calculations here are limited to time between first and last matings of the season.

f. Only four were seen to ejaculate, but entire series mounts are unlikely to be seen.

g. Excludes male resident in group before and after influx.

h. One additional female was obviously pregnant after the MCP, but disappeared before giving birth.

i. Two females may have lost infants from the previous year conceived in February and March. This conception period is not included in the table.

mature male has also been noted (Bourlière, Hunkeler, and Bertrand 1970; Galat-Luong 1975; Curtin, unpub.).

Evolutionary Consequences

The picture that emerges from these results is that guenon mating systems are tremendously variable. Specifically, several reproductive strategies for males are available, and over the short term, being a resident male of a one-male group is not necessarily the most advantageous of these. However, in order to assess the evolutionary consequences of different patterns of male behavior, it is necessary to know a male's reproductive success not only over a short period but also over an entire lifetime. From

the perspective of evolutionary theory, one must know the relative costs and benefits of attempting to maintain exclusive status in a one-male group and of competing through other means as a participant in multimale influxes into such groups during the breeding season.

The contrast between the reproductive consequences of these two kinds of strategies has been noted in a variety of animals (Rubenstein 1980). To date, however, no study of cercopithecine primates has gone on long enough to address this issue, though there have been a few relevant observations. For example, a few *C. mitis* and *E. patas* males, and one *C. ascanius* male, have been monitored as they switched between in-group and extragroup life, over the course of days to 2 or 3 years. Males may alternate between groups of females on an almost hourly or greater basis. Males resident in a group may steal copulations in neighboring groups. Clearly there are many ways to achieve lifetime reproductive success. Many of them do not rely on ability to defend groups of females against rival males. For example, the copulation frequencies of visiting males in the main study group of *C. mitis* in the Kakamega Forest (1981) were not correlated with their agonistic ranks (Tsingalia and Rowell 1984). Some visiting males were regularly chased by other males in the group, including the resident male, but still succeeded in copulating more than some of their higher-ranking rivals. In the Kakamega *C. ascanius* study group, a consistently low-ranking male had the second highest scores of copulation over the entire mating season (Cords 1984a).

In other cases, males have actively and successfully defended their priority of access to breeding females, at least in *C. ascanius* and *C. mitis*. Resident males have been observed to chase rivals from their groups, thereby keeping those groups for themselves (Cords 1984a; Tsingalia and Rowell 1984). Even during a multimale influx, both resident and newly joined males might, in theory, be able to ensure their reproductive priority because of their high rank. At present, however, few data support this hypothesis. Only in the study group of *C. ascanius* in Kakamega has a positive correlation between rank and sexual activity during multimale influxes been demonstrated for some parts of the mating period (Cords 1984a).

Other social factors may also influence a male's lifetime reproductive success. Males are likely to encounter one another year after year; their social relationships may make sense with respect to differential reproduction only in this larger context. Genetic relatedness between males might further complicate the evolutionary significance of those social relationships. Finally, female choice may be important, though the criteria on which it is based and its consequences in terms of paternity (and therefore sexual selection in males) are not known. At present only anecdotal evidence exists for female choice. Females have been observed soliciting and mating with particular males while concurrently disregarding the attentions given them by others (Cords, unpub.). Also, many, if not most, copulations are initiated by females approaching males (Chism, Rowell, and Olson 1984; Cords, unpub.; Struhsaker, unpub.).

Ecological Correlates

Like the evolutionary consequences, the ecological and demographic correlates of variation in cercopithecine mating patterns are not yet well understood. Clearly two important variables are the intensity of intermale competition for mates and the ease with which access to females can be monopolized. Intermale competition is indicated not only by the frequency of multimale influxes and number of participants, as described earlier, but also by the rate of replacement of males, each of whom is the only male resident in a group at a given time. For example, in an area having a high ratio of adult males to females, Butynski (1982b) reported six replacements in 21 months in a group of *C. mitis*, when no males were known to be in the group simultaneously. Males mating with females of other species, who produce fertile hybrid offspring, and infanticide by newly established resident males have been offered as further indications of how intense male-male competition can be (Struhsaker 1977; Butynski 1982b).

Various demographic factors might be expected to influence the degree of competition between males in different populations. The ratio of reproductive males to groups of females is obviously critical and may depend on several variables, including adult sex ratio, population density, and group size. These in turn may reflect the distribution and availability of resources. For example, the patas population in Laikipia, Kenya, where the two study groups were found to undergo multimale influxes in most breeding seasons, lived on a cattle ranch with water permanently available in many dispersed troughs. In contrast, water was severely limited for the patas in the Waza Reserve, Cameroon; males outside of heterosexual groups appeared to suffer especially, since they were denied access to the few water holes controlled by those groups (Gartlan 1975). The apparent differences in male mortality in these two populations, which could affect adult sex ratio, might explain why the one-male group structure persists in Waza, but not (continually) in Laikipia, where male-male competition is probably more intense.

Ecologically based arguments are less complete for *C. mitis*. Butynski (1982b) found a higher rate of replacement of males in one group in a part (Ngogo) of the Kibale Forest where the ratio of males to groups was seven times higher than elsewhere in the same forest. The reasons behind this difference in population structure are not clear: the *C. mitis* population density at Ngogo is four to five times lower than elsewhere (Struhsaker 1981a), possibly reflecting increased interspecific competition for food (Struhsaker, pers. comm.). However, if such competition occurs, it evidently affects heterosexual groups and extragroup males differently, the former being reduced in density and the latter increased relative to other parts of the same forest. When the Kibale population is compared to that in Kakamega, where multimale influxes have been observed more frequently, different ecological factors appear to be involved. *C. mitis* in Kakamega have a higher population density than in Kibale, again maybe reflecting lower interspecific competition for food. Groups in Kakamega are both larger and denser, and solitary males have been seen more frequently. The relative males-to-groups ratios in the two study areas are not known. Some groups in Kakamega may be more difficult to defend because they are larger. However, a series of male replacements and multimale influxes has been observed in one Kakamega group whose size (21 members) is similar to that of groups in Kibale.

For *C. ascanius,* there is little evidence to date for the importance of ecologically based population parameters in determining the degree of intermale competition. The incidence of multiple males in groups in three study populations is not correlated with population density or group size (Cords 1984a). The ratio of males to groups also does not correlate with the incidence of multimale groups, if one assumes identical sex ratios in the three populations. However, sex ratios are unknown.

It is important to remember that ecologically based comparisons of different study areas are still premature and that all specific conclusions as to the significance of such comparisons are highly tentative. One reason is that the density of adult males is not accurately known, nor easily calculated, in any population. Even assuming a 1:1 sex ratio at birth, subsequent mortality of males and females has not been measured. Furthermore, the long-term studies of *C. ascanius, C. mitis,* and *E. patas* in Kenya suggest that male density fluctuates widely on a local scale so that estimates for the population as a whole, which are based on a few home ranges at certain times of year, are likely to be misleading. Short but larger-scale censuses do not offer much hope of improvement since solitary animals are hard to detect. A second problem is that of quantifying the incidence of male membership changes and multimale influxes in different populations: variability within single groups may be sufficiently great from year to year (and perhaps even on a larger time scale) to encourage skepticism of estimates based on studies of one to two years or less. Variation between groups in one study area may be equally large (Cords, unpub.; Struhsaker, unpub.), so characterization of populational mating patterns based on a few groups might be misleading.

10 | Desert, Forest and Montane Baboons: Multilevel-Societies

Eduard Stammbach

TAXONOMY AND DISTRIBUTION

This chapter discusses four species that share a peculiarity in their social life: they form one-male units (OMU), which means that the basic social unit of their societies consists of one male, several females, and their dependent offspring. The OMUs of all four species combine to form higher-level social units that are structured differently in the respective species.

The hamadryas baboon (*Papio hamadryas*) belongs to the genus *Papio*. Of all the *Papio* species, the hamadryas baboon is the only one to form OMUs, although Dunbar and Nathan (1972) found a tendency for *Papio cynocephalus papio* males to associate with a small number of females. Hamadryas baboons inhabit wooded or subdesert steppe. During the night they use steep cliffs and rock ledges as sleeping quarters. They are confined to the arid zones of Ethiopia, Sudan, Somalia, and the adjacent coast of southern Arabia (Kummer 1968; Kummer et al. 1981). Adult hamadryas baboon males have dull pink faces and develop heavy silvery whiskers and profuse silvery shoulder mantles. The callosities are surrounded by rosy red skin.

The mandrill *Papio sphinx* and the drill *Papio leucophaeus* inhabit the lower levels near the ground of primary rain forest (Jouventin 1975a; Gartlan 1970; Hoshino et al. 1984), situated between the Niger and Congo rivers. In male mandrills both face and penine region are impressively colored bright red, blue, and white; even the hair shows differently tinted brown, silvery, and black areas. They possess an extensive glandular area beneath the pectoral hair tuft. Females exhibit no similar coloration. The closest relatives of the mandrill, the drills, have black faces.

Theropithecus gelada, the gelada baboon, is restricted to the highlands of Ethiopia (Eritrea, Semien), where it is found in montane grassland between 2,000 and 5,000 m. Like hamadryas baboons, geladas prefer steep rock cliffs as sleeping sites. Both adult male and female gelada baboons have a naked, triangular chest patch whose size and color is affected by cyclical hormonal changes in females. Additionally, gelada males wear whiskers and brown hairy mantles.

All four species are characterized by a distinct sexual dimorphism, the males being up to twice as large as the females (see table 10-1).

TABLE 10-1 Body Weight of Adult Male and Female Hamadryas, Gelada, Mandrills, and Drills

	Males (k)	Females (k)
Hamadryas	(16.4–19.5)	(10.5–11.5)
Gelada	20	16
	20.5	13.6
Mandrills	(21–28)	(11–12)
Drills	20	—

SOURCES: Hill 1970; second entry for gelada, Napier and Napier 1967.
NOTE: Range of values are given in parentheses.

ECOLOGICAL NICHE

Every morning, the hamadryas baboon troop leaves its sleeping cliff and subdivides into bands for a march through their habitat for the rest of the day (Kummer 1968; Sigg and Stolba 1981). This daily march will lead them to well-known feeding places and to water holes, distributed throughout their range. In the dry season only a few of the water holes are permanent. Frequent feeding stops will interrupt the march. The dry leaves, flowers, and beans or berries of *Acacia* and *Dobera* trees, *Grewia* bushes, and other different kinds of fruit provide the main food. During the rainy season, *Acacia* flowers and grass seeds are the preferred food. Additionally, roots and bulbs, dug out with great diligence, complete their diet. Animal food, for instance, insects or even small mammals such as dik-dik or hares, will be taken occasionally. At noon the baboons often rest near a water hole. After a short siesta, they return to their sleeping cliffs. On their way back, the bands subdivide again for foraging. A hamadryas baboon band will undertake a daily walk of up to 10 km (a maximum length of 32 km has been observed) each day, since food is scarce. The home range they utilize may range up to 28 km^2 (Sigg and Stolba 1981). Home ranges of neighboring bands may overlap by about 50%. Population densities reported for the hamadryas baboon range from 1.8 per km² (Kummer 1968) to 3 per km² (Nagel 1973).

The geographical neighbors of the hamadryas baboons, the gelada, clearly differ in their food habits: they feed mainly on grass seeds (Gramineae) found in the rich grassy plains of the Ethiopian highlands. Like the hamadryas baboon, geladas are terrestrial and leave their cliff in the morning to forage. As food availability is much greater than in the steppe, geladas move for only 1 or 2 km per day. In geladas, time spent feeding and moving averages 50 to 70% of their daily activity and may represent up to 90%. Their routes pass through sites where water is available. However, as water is fairly abundant, the waterholes are not as decisive for route choice in geladas as they are for the hamadryas. The ecological niche of gelada baboons may be compared with that of an ungulate (Kawai 1979; Dunbar and Dunbar 1975). The densities of geladas, ranging from 63 per km² (see Kawai 1979) to 77.6 per km² (Dunbar and Dunbar 1975), are much higher than those of hamadryas.

Much less is known about the behavior and ecology of the mandrill and drill. Both live in the dense primary rainy forest and, because of reduced visibility, are far more difficult to study. Nevertheless Sabater Pi (1972), Jouventin (1975a), and recently Hoshino et al. (1984) succeeded in making some few and interesting observations on the mandrill, as did Gartlan (1970) on the drill. The species not only live in similar habitats but also seem to have a nearly identical social organization. Mandrills range widely in the forests. Their feeding marches are comparable to those of the hamadryas baboon. Jouventin reports that mandrills move 5 to 15 km per day. Different groups use the same home range, which may be 40 to 50 km². Mandrills are mainly vegetarians; their diet consists of fruits, seeds, nuts, and shoots, but they will take fish, crabs, and insects on occasion. A detailed list of feeding plants was published by Jouventin (1975a). The large males remain mainly on the ground, while the females and young are semiarboreal and inhabit the undergrowth and middle layer of the forest. Mandrills exploit the lower levels of the forest, which are the least productive parts of the rain forest ecosystem (Jouventin 1975a). Jouventin therefore hypothesizes that mandrills live in a comparatively food-poor zone, like the hamadryas baboon. This indeed might explain the extended daily marches of the species.

CONSERVATION STATUS

Hamadryas baboons seem to be under slight pressure in Ethiopia, where they are regarded as crop raiders. Recently their habitat was affected by the Ethiopian civil war. Hamadryas baboons are trapped for commercial reasons: some populations have been found to lack juveniles and subadults (Wey, pers. comm.). The status of

geladas provides little cause for concern at the present moment. The habitats include the Simien Mountain National Park, founded partly to protect the Walia ibex (*Capra ibex walia*). At present, the conservation status of mandrills and drills is uncertain. Both species are hunted for meat, and drills are found in small, isolated patches of forest (IUCN *Mammal Red Data Book* 1982).

REPRODUCTIVE PARAMETERS

The only long-term studies concerning reproductive parameters of the four species have been conducted on hamadryas baboons (Sigg et al. 1982) and geladas (Dunbar and Dunbar 1975; Dunbar 1980a). Studies of the other species have lasted several months at most. Table 10-2 presents the available data. Data on the mandrill are unavailable, but the values should be comparable to those of other species.

Both hamadryas baboons and geladas lack a clear birth seasonality. Dunbar (1980a) reports two fairly distinct birth peaks in geladas that are correlated with seasonal rainfall: births occur most frequently during the dry season. Birth seasonality is reported from mandrills (Jouventin 1975a).

Females in all four species reach sexual maturity at about 4 to 5 years, males at 5 to 7 years. Sigg et al. (1982) reported adequate records on first estrus for 13 hamadryas females and could determine descent of the testes for hamadryas males by using photographs. Gestation length varied between 5 to 6 months, and interbirth intervals were estimated to be 2 years. In all four species, females show clear external signs of estrus: hamadryas, mandrill, and drill females have genital swellings; gelada females develop beadings at the border of the naked area on the chest and, additionally, a protruding vulva.

SOCIAL ORGANIZATION

Table 10-3 summarizes the studies from which data on social organization are available. As noted earlier, the basic social unit is the one-male unit (OMU), which consists of one or two males, several females, and their offspring. The OMUs represent the most basic level of the social structures of all four species. The higher levels of social structure, however, differ between genera. Table 10-4 summarizes the sizes of the different groupings at different organizational levels.

Hamadryas Baboons

In hamadryas baboons, two or three OMUs are strongly associated with one another and form the next higher level of social unit, the *clan* (Abegglen 1984). A clan is characterized by three aspects: (1) Social interactions occur more frequently within a clan than between clans. (2) The cohesion within a clan is apparent on the forag-

TABLE 10-2 Reproductive Parameters for Hamadryas, Gelada, Mandrills, and Drills

Species and Study Site	Birth Seasonality	Age at Sexual Maturity		Gestation Length	Interbirth Interval	Source
		Male	Female			
Hamadryas						
Erer-Gota	Peaks in May/June Nov/Dec			(5.5–5.8) captivity data		Kummer 1968
Erer-Gota	None	(57.5–81.5)	51.5 (48–60)		24 (12–36)	Sigg et al. 1982
Gelada						
Semien, Sankabar	Peaks in June/July Nov/Dec			(5–6) captivity data	24 (12–30)	Dunbar 1984b
Gich Plateau	None, inadequate information			(5–6) captivity data		Kawai 1979
Mandrills	Dec to April					Jouventin 1975b

NOTE: The mean values and/or range of values (in parentheses) are given in months.

ing march: clan members stay together and a clan may detach as a temporarily autonomous foraging subunit from the *band*. (3) The most surprising fact is that the mature males of a clan share an obvious morphological similarity. This led Abegglen (1984) to conclude that males of a clan are close genetic relatives. Stolba (1979) confirmed this observation later on in other bands. Thus several OMUs, several clans, and some single males form a band (Kummer 1968; Abegglen 1984; Sigg et al. 1982; Kummer et al. 1981). Bands are very stable and exclusive associations of individuals. Band membership and clan membership of males, at least in the most closely observed band in the study area, persisted from 1972 until the end of the study period in 1977 (details later). Social interactions were almost completely restricted to members of the same band. A band represented an autonomous unit of foraging. Sigg and Stolba (1981) observed that the main study band occupied one single sleeping rock (408 of 410 nights). Another band visited two rocks with almost equal frequencies. Up to three or four bands may congregate at one rock in one night in variously composed *troops*.

Geladas

In geladas, OMUs form clusters of closely associated OMUs. Dunbar and Dunbar (1975) and Kawai et al. (1983) refer to these clusters as *bands*. In contrast to the hamadryas baboon, the membership of gelada OMUs in a certain band is not constant, since OMUs may visit other bands or even pass their day alone. In addition to OMUs, males who are not associated with OMUs form all-male groups (AMG) that sometimes travel separately from bands.

A third level of organization in geladas is the *herd*, which, like the hamadryas baboon troop, is not reported to be a social group in the strict sense, but rather a tem-

porary congregation of OMUs at sites where grazing conditions are favorable. Herds are largest during dry season when the availability of food is reduced.

Mandrills and Drills

In the mandrill and drill, several OMUs may congregate, especially in the dry season, like those of gelada baboons. The few direct observations suggest that mandrills and drills occasionally meet in large groups. Adult male mandrills and drills often live alone if they are not a member of an OMU (Jouventin 1975a; Gartlan 1970; Hoshino et al. 1984). They do not form AMGs.

Interband Relations

Interband relationships in hamadryas baboons differ from those in geladas. Abegglen (1984) observed only brief interactions among members of different bands. Attempts of infants to interact across band units were suppressed, in contrast to the few encounters among juveniles. Adults seemed to avoid interband encounters completely. Very rarely intense physical aggression may occur, with the OMU leaders as the main fighters. Females remain in the "attack shadow" of their males. Kummer (1968) induced these "battles" when he emptied corn in a pile near the sleeping rock.

Ohsawa (see Kawai 1979) reports that band membership in geladas is not as consistent as in hamadryas baboons. A band may accept other bands or unfamiliar OMUs and merge into large mixed bands. Ohsawa thinks that geladas can form congregations without recognition among individuals. Mandrill groups are similarly compatible, and they may utilize the same feeding areas.

Stability and Formation of OMUs

Studies carried out on the hamadryas baboon (Abegglen 1984; Sigg et al. 1982) suggest that the membership of

TABLE 10-3 Studies of Social Behavior in Hamadryas, Gelada, Mandrills, and Drills

Species	Site	Person	Source	Period
Hamadryas	Erer-Gota	H. Kummer	Kummer 1968	Nov 1960–Oct 1961
	Erer-Gota	J. J. Abegglen	Abegglen 1984	May 1971–July 1972
		H. U. Müller	Müller 1980	
	" "	J. J. Abegglen	Sigg et al. 1982	Apr and May 1973
				Jan 1974
	" "	H. Sigg	Sigg and Stolba 1981	Feb 1974–June 1975
		A. Stolba		
	" "	J. J. Abegglen		Jan 1976–Feb 1977
		R. Wey	Sigg et al. 1982	
	Arabia	H. Kummer	Kummer et al. 1981	Jan + Feb 1980
Gelada	Semien			
	Sankaber	R. and P. Dunbar	Dunbar and Dunbar 1975	July 1971–Mar 1972
	Bole Valley	R. and P. Dunbar	Dunbar and Dunbar 1975	May 1972–Oct 1972
	Gich Plateau	M. Kawai	Kawai 1979	June 1973–Mar 1974
Mandrills	Gabon	P. Jouventin	Jouventin 1975b	Feb 1973–Mar 1973
				May 1974–July 1974
	Guinea	J. Sabater Pi	Sabater Pi 1972	Feb 1967–Oct 1968
	Cameroon	J. Hoshino	Hoshino et al. 1984	Aug 1979–Jan 1973
		A. Mori		
		H. Kudo		
		M. Kawai		
Drills	Cameroon	J. S. Gartlan	Gartlan 1970	Oct 1967–Dec 1968

TABLE 10-4 Size of Social Units of Hamadryas, Gelada, Mandrills, and Drills

Species and Study Site	Animals per OMU	Males per OMU	Females per OMU	Animals per Clan	Animals per Band	Animals per Troop (herd)	Source
Hamadryas							
Erer-Gota		(1–2)	1.9		68	ca. 120	Kummer 1968
					(42–82)		
Arabia		1.2					Kummer et al. 1981
		(1–2)					
Erer-Gota	7.3	(1–3)	(1–9)	2–3 OMGs	(61–69)	up to 236	Sigg et al. 1982
	(2–23)						
Gelada	10	(1–2)	3.7		(30–300)	up to 600	Dunbar 1984b
Sankabar			(1–10)				
Gich Plateau	9.9	1.2	3.5				Kawai 1979
		(1–3)	(1–8)				
Mandrills	(20–25)	(1–2)	(5–10)		up to 250		Jouventin 1975b
	13.9[a]				up to 250		Hoshino et al. 1984
Drills	(15–20)				up to 179		Gartlan 1970

NOTES: Mean range given in parentheses. OMU = one-male unit.
a. Number of animals per male.

adult males in an OMU remains stable for at least 3 years. The period of membership is shorter for females: 70% of the females of a well-studied band changed their OMU within 3 years. Since the whole study period lasted only 6 years, no upper limits can be given, but 70% of the male-female pair-bonds persisted for the whole study period.

Transfer of individuals between OMUs happens primarily within the band. Although juvenile and subadult males frequently travel around with other clans or even other bands for a certain time, they will return to their natal band at adulthood. Here they most often form the first stage of an OMU, an *initial unit* (IU), by establish-ing a relationship with a juvenile or subadult female of one of the OMUs of their clan (fig. 10-1). Alternatively they may attach themselves to an OMU as a *follower*. Later, they will attempt to take adult females away from the old OMU leader, usually during one of the rare but severe battles among adult males of the band. In the course of these battles, the females of one or more OMUs may be dispersed to new, different OMUs. Females change OMUs two or three times in their life, but tend to transfer into units that already contain females with whom they formerly lived in an OMU (Sigg et al. 1982). Abegglen (1984) reports that transferred females attempted to approach animals of their former OMU and

FIGURE 10-1. An adult male hamadryas baboon huddles with a juvenile female he has "adopted." (Photo: Hans Kummer)

tended to establish grooming relationships with unfamiliar females of the OMU rather slowly. He thus suggests that the bonds among females of an OMU contribute to group cohesion.

Defeated OMU leaders may live without females, but remain associated with their clan and continue to interact with their offspring. In contrast to the males that stay in their natal band and even in their natal clan for life, females not only change from one OMU to another, but may also change from clan to clan or even from band to band when forced by their new leader. In hamadryas baboons, males remain in the natal group and the females are responsible for the majority of migration. In this way, hamadryas baboons differ from most other cercopithecines (chap. 11).

In geladas, subadult males may adopt the "follower strategy," as do the hamadryas males, and build up close relationships with young females that have a weak grooming relationship with the leader male. Subsequently, the follower male and these females will attempt to form a new, independent OMU (Dunbar and Dunbar 1975). In the early stage, a new OMU is characterized by strong control of females by the male, whereas the bonds among the females themselves are of relatively greater importance in maintaining unit cohesion in "mature" OMUs.

A. Mori (1979c) and Mori and Dunbar (1985) report another male strategy for becoming an OMU leader: A male member of an AMG may attack an OMU leader who is persumed to be weak. If he wins, he will take over the OMU and become its new leader (fig. 10-2). The former leader will usually stay within the OMU as a second male. Mori also suggests that OMUs, having lost their leader, remain intact and are taken over as a group by an AMG male or merge entirely with another OMU. In contrast to the hamadryas baboons, therefore, bonds among the females of each OMU are so strong that a splitting up of the females among different OMUs does not occur. Female bonds in geladas are largely based on kinship (Dunbar 1979a). As in many cercopithecines (chaps. 9, 11), gelada females tend to stay together with their mat-

FIGURE 10-2. An adult male gelada baboon attacks the leader of a one-male unit in an attempt to take over the unit. (Photo: Richard Wrangham/Anthrophoto)

rilineal relatives for life (Dunbar 1979b; Dunbar 1984b).

No direct observations on unit formation and stability exist for the mandrills and drills. Taking into account that solitary males exist in both species, and extrapolating from captive studies, their social organizations would appear to be more geladalike than hamadryaslike (Emory 1975, 1976).

The similar basic social organization of all four species discussed in this chapter, namely the formation of OMUs, is associated with a convergent breeding system. Estrous females are mated nearly exclusively by their OMU leaders. Sometimes, younger males are able to "steal" a copulation behind the back of the leader. This kind of breeding system reduces the reproductive life of a male largely to the period when he is leading an OMU. Abegglen (1984) estimated this period to be only 3 to 6 years out of the approximately 20-year life of a hamadryas male. Sigg et al. (1982), however, suggest that the reproductive period of the male is longer because Abegglen may have underestimated the true age of the older OMU leaders. Dunbar (1984b) estimated that gelada males hold the breeding position within the OMU for 3.5 to 7 years and that male life expectancy is 12.3 years.

SOCIAL INTERACTIONS

The multileveled organization of the four species discussed here is expressed in a different pattern of interactions specific to each level. I will first deal with affiliative tendencies.

Grooming is generally limited to members of the same OMU. In hamadryas baboons it is the male leader of the OMU that is the focus of grooming by females, and centripetal females compete for access to him (Kummer 1968). Females groom the male more often than they are groomed by him. Grooming between the females of an OMU is roughly symmetrical, with each female grooming and being groomed approximately equally often (Sigg 1980). Grooming between members of different OMUs may occur but is limited to members of the same clan. In particular, the single adult males of the same clan will groom each other quite frequently (Abegglen 1984).

Gelada OMUs are characterized by a greater importance of bonds among the females. Grooming sociograms of geladas appear less centripetal; the interactions of the male, who is not the focal center of the unit, are limited primarily to only one or two of his females. The remaining females interact with each other: a female's affiliative interactions are confined to one or two close relatives (Dunbar 1983a). The male is rather peripheral socially, as compared to the hamadryas leader (Dunbar 1983b). Another interesting fact confirms this statement: an estrous hamadryas female will increase her time spent interacting with her leader, whereas among geladas, female estrus has minimal disruptive effects (Dunbar

1983d). Gelada OMU leaders seldom interact with each other (Dunbar 1983d).

Relevant information on mandrills is available only from captive studies (Emory 1975, 1976) and indicates that visual attention and social orientation toward the OMU leader are weaker in mandrills than in geladas. To summarize: affiliative interactions are highly centered around the male unit leader in hamadryas baboons, less so in geladas, and least in mandrills.

Agonistic interactions within OMUs are rare in all four species. One peculiarity of hamadryas social life has to be reported in this context. The OMU leaders may actively keep their females in their OMU by *herding* them. If a female moves away too far, the male will often chase and punish her with a neck bite. Gelada males do not herd their females in a comparable way. The frequency of herding is much lower, and it is relatively ineffective. The male may even be chased by two or more females after an attempt of herding (Dunbar 1983b). If a male tries to take a female back to the OMU, he will behave rather defensively by emitting appeasement vocalizations. Herding behavior is reported from drills (Gartlan 1970), but this behavior is not restricted to the leader male, who may be supported by subadult or juvenile males too.

Linear dominance-rank orders are established among females of each OMU (Sigg 1980; Dunbar 1983a). In geladas, the dominance rank of a female is matrilineally determined. The most dominant females of each matriline on their side determine the relative ranks of the matrilines: the more aggressive the female, the higher will be the rank of her matriline and thus the ranks of her relatives. Since most social interactions are restricted to members of an OMU, dominance relationships across OMUs are difficult to observe. However, these are present in males of a hamadryas band (Stolba 1979) and absent in gelada males (see Kawai 1979), which is consistent with the view that hamadryas males are part of a stable social unit, the band, whereas gelada males do not have regular and similarly intense social contacts.

Especially in hamadryas baboons, the multileveled organization of society can be recognized by an observer at the first glance: Spatial patterns clearly distinguish OMUs, less clearly clans, while the bands are clearly detached and easily distinguishable. Similar spatial patterns are recognizable in geladas too, but not as easily as in hamadryas baboons.

♦ MULTILEVEL SOCIETIES

One of the aims of primate ethology is to specify the relation between ecology and social structure. Primates invest a considerable amount of their time constructing and maintaining complex social systems. One of the most exciting questions is, What may be the use of a certain social organization in a particular ecological context? (chap. 23). Thus far, this chapter has presented three types of OMU societies found in four species of cercopithecine monkeys. Further discussion will now deal with evolutionary and functional aspects of OMUs. Relatively more information is available about hamadryas and gelada baboons, while relatively less is known about mandrills and drills.

The most striking fact concerning the four species discussed here is that a superficially identical form of social organization is developed and maintained by different processes. Hamadryas OMUs are held together by leader males whose aggressive behavior forces their females to stay within the OMU. Other males respect the close male-female bonds of an OMU as Kummer et al. (1973) were able to prove experimentally. If two hamadryas males and a hamadryas female are placed in a large outdoor cage, the males fight for the female. But if one of the males is allowed to observe the other male and the female interacting, no fight will take place when he is released into the cage with the other two individuals.

Studies on the social behavior of male baboon hybrids (*Papio cynocephalus anubis* × *P. hamadryas*) demonstrate the importance of the male's role in maintaining the hamadryas OMU (Müller 1980). Both *Papio* species share a basic behavioral repertoire that includes behaviors necessary to shape and maintain OMUs: males may aggressively herd females and exclude potential rivals, and females may focus their attention on only one male in the formation of an exclusive grooming. Müller found that hybrid males of the multimale anubis and the OMU-forming hamadryas baboon diverged in their social behavior from both parent species. Hybrids tried to maintain lasting relationships with nonestrous females as do hamadryas baboons, but lacked the efficient herding behavior that is typical for hamadryas males. Hybrids that are more hamadryaslike morphologically tended to prefer the hamadryas pattern of aggressive herding, whereas the more anubislike hybrids maintained close bonds with anestrous females by following and grooming, then assumed the behavior of anubis males during consortship with an estrous female (chap. 11). According to Sugawara (1979), hamadryaslike hybrids succeeded in forming and maintaining OMUs, whereas anubislike hybrids behaved according to an anubis type of social structure. Hamadryas baboons seem, during the course of evolution, to have applied behavioral components already present in their ancestors, in order to shape and maintain OMUs.

Geladas appear to have followed a different evolutionary path: The stability of gelada OMUs is due to strong female-female bonds (Dunbar 1983a). The importance of female relationships is underlined by the fact that even after loss of the leader male, the females of a gelada OMU remain together. A gelada OMU may have been derived from an organization that was characterized by distinct clusters of female relations. These clusters detached and, joined by a male, developed into OMUs (Ohsawa, see Kawai 1979).

The basic differences between the male-based type of OMU in hamadryas and the female-based type of OMU in gelada become even clearer if the behavior of the two species is compared in an ecological context. A hamadryas band, leaving its sleeping rock for its daily march, has to decide which route should be followed and which water hole should be reached at noon. Stolba (1979) studied this process of "decision making" in two particular bands. He found that the decision is apparently made mainly by the adult and subadult males of the band: a male leader, followed by the other members of his OMU, moves some steps in one direction. The other males may support this initiation by moving in the same direction, they may initiate movement in another direction, or they may not move at all. If an initial move is made by a male of the same clan, another male is more likely to agree. If the majority of the OMU leaders agrees with an initial movement, the band will move on.

Detailed analysis revealed that this decision does not necessarily determine the first part of the route, but rather will indicate the location of the water hole that is to be reached at noon. After leaving the rock, the band may now split: clans or even single OMUs may search for feeding places on their own and so may move away from the other band members, up to distances that exclude visual or auditory communication. At noon, however, the whole band usually converges at the water hole that was denoted by their initial direction of movement, or they will meet earlier at a rich feeding site. Since water holes are visited by nomads or by other hamadryas bands too, the band as a whole is strong enough to claim a place for drinking. These observations of coordination of travel suggest that the members of a band, or at least the males that support the route decisions, must possess some kind of mental map of their habitat (Sigg and Stolba 1981). The possibility of splitting up into smaller groups that is allowed by the multileveled organization, the knowledge about their habitat, and the communication in the "route-decision conference" allow hamadryas baboons to exploit the scarce and patchy food sources of their steppe habitat in an optimal way: the smaller the food patch, the smaller will be the size of the social unit that feeds in it. Kummer (1968) suggested that an OMU may exploit a single acacia tree, whereas the more extended feeding areas where the baboons dig for roots and bulbs provide food enough for the whole band.

Within the OMUs themselves there is also a differentiation of social position and an ecological division of labor. Sigg (1980) distinguished a *central female* (CF) that is more active socially than the other females, has a

closer social bond to the leader male, and is higher ranking than other females. She is, however less experienced in ecological tasks than the *peripheral female* (PF), who is less active in social affairs but concentrates on ecological aspects of daily march. About half of the feeding places exploited by OMUs that were closely observed by Sigg (1980) were first utilized by the PF and thereby denoted to the other members of the OMU. The PF approached areas that might hide a predator significantly more often than the other OMU members and thus assumed the role of reconnoitering potential dangers. Sigg confirmed this differentiation of positions in enclosure experiments: a PF was more successful than others in pointing out artificial food and water sites that had been shown to her alone when she was returned to the enclosure again together with the rest of her OMU.

Geladas appear to select routes more opportunistically, as they move from one feeding stop to the next. In addition, this process seems to involve only a single OMU. Gelada herds are more fluid than hamadryas bands, with the members of OMUs intermingled. Route decision processes in geladas thus differ from those of hamadryas baboons. Initiation of movement, that is, proposing a certain direction, is done mainly by lactating females. Dunbar (1983c) suggests that this occurs because of their largely increased need for food intake: the only means of increasing the amount of food is to increase progression speed. The proposed directions of lactating females are generally accepted, especially by their subordinate grooming partners. The final "decision" is very often made by the alpha-female, assisted by the male. The gelada male leader has to pay attention mainly to the alpha-female, who is responsible for maintaining group cohesion. Less frequently the male himself or even another individual of the OMU will be able to direct the group in a particular direction.

The comparison of processes of group coordination in hamadryas and gelada baboons reveals that a superficially similar social organization results from strong male-male bonds and the male's control of the other unit members in one species, and by strong female-female bonds and the prevalent influence of the dominant female in the other one. The two species thus have *analogous* and not *homologous* social structures.

Coordination of mandrill groups is maintained by both visual communication and vocalization (Jouventin 1975a; Hoshino et al. 1984). Jouventin (1975b) suggested that the brightly colored hindquaters of the adult males serve as signals for facilitating group progression in the dense vegetation of rain forests. The male stays mainly in the center of the group. The curious and mobile juvenile and subadult males, usually walking in front of the group, take the role of "route proposers." Communication by "contact calls" is important for group cohesion. In addition, the male uses a special "rallying call" to keep his OMU together in dangerous situations (Jouventin 1975a). In case of danger, however, he sometimes walks in front of his OMU (Gartlan 1970).

Relationships between OMUs seem to be reduced to regular meetings in large troops. Jouventin believes that intense feeding competition among the large leader males is responsible for the weak inter-OMU relationships. Leader males seldom climb trees, and they therefore feed on scarcely distributed food items including those that fall from other OMU members in the trees. Jouventin hypothesizes that the striking coloration of mandrill male faces has evolved in the context of male-male competition.

11 | Cercopithecines in Multimale Groups: Genetic Diversity and Population Structure

Don J. Melnick and Mary C. Pearl

More is known about the behavior of cercopithecines that live in multimale groups than about any other class of primates. Most theoretical discussions of the structure and function of kinship bonds, dominance interactions, and social relationships have been derived, either explicitly or implicitly, from studies of baboons, macaques, and vervet monkeys. The social behavior of these cercopithecines has to a large degree provided the conceptual framework within which other species have been described and evaluated, and this bias has had a profound effect on our perception of other primates.

Savanna baboons, macaques, vervets, mangabeys, and talapoins are discussed as a unit in this chapter because their social organizations are similar: most groups consist of more than one adult male, a number of adult females, and their offspring. Since so much is known about baboons, macaques, and vervets, many of the later chapters in this book will of necessity focus on these species' social behavior. As a result, their behavior will only be summarized here. What little is known about the social behavior of mangabeys and talapoins will be described in more detail.

TAXONOMY AND DISTRIBUTION

The 25 to 30 species considered here are all members of the Old World monkey subfamily Cercopithecinae. Their taxonomy does not enjoy unanimity among systematists and zoogeographers. The outline presented here is based on the taxonomies of Napier and Napier (1967), Thorington and Groves (1970), Szalay and Delson (1979), Fooden (1980), and Groves (1980).

There is disagreement about the taxonomic classification of savanna baboons. Some regard savanna baboons as conspecific with hamadryas baboons and classify them as *Papio hamadryas.* Here, however, we follow the suggestion of Thorington (Thorington and Groves 1970) and consider all savanna baboons as members of one species, *Papio cynocephalus,* and separate from hamadryas baboons. Savanna baboons include *cynocephalus* (yellow baboons), *ursinus* (chacma), and *anubis* (olive). (Another subspecies of *P. cynocephalus, P. c. papio,* is discussed in chap. 10.) Closely related to the baboons are the mangabeys, *Cercocebus,* which are generally divided into five species: *torquatus* (white-collared or sooty), *atys*

(sooty), *galeritus* (agile or Tana River), *albigena* (gray-cheeked), and *atterrimus* (black). The Asian cercopithecines, the macaques, belong to one very diverse genus, *Macaca,* consisting of 16 to 19 species (cf. Fooden 1980; Groves 1980). (Note that the appendix lists only one Sulawesi macaque, *Macaca nigra,* rather than dividing this group into four or seven species as advocated by Groves 1980 and Fooden 1980, respectively. This is done because nothing is known about the behavior or ecology of these species and because the taxonomy of this group is still not understood.)

Two members of the genus *Cercopithecus, C. aethiops* (vervets) and *C. talapoin* (talapoins), are also included in this chapter because, unlike other members of their genus, they are typically found in multimale groups. Talapoins are also known as *Miopithecus talapoin* (but see Napier and Napier 1967).

If abundance, breadth of distribution, and diversity of habitat are measures of evolutionary success, then the cercopithecines described here are indeed successful. Savanna baboons are distributed throughout most parts of sub-Saharan Africa and occupy a wide range of vegetational types, including semidesert, savanna, woodlands, some rain forest, and some high-altitude and coastal areas (Wolfheim 1983). Mangabeys range across central Africa from the Atlantic coast to western Uganda (Wolfheim 1983), with remnant populations in Kenya and Tanzania (Homewood 1978; Homewood and Rodgers 1981). In this broad equatorial zone, mangabeys occupy coastal plain, gallery, and inland swamp forests in the west, and montane, wet evergreen flood-plain, and semideciduous tropical lowland forest in the east (Wolfheim 1983).

Macaques, the most widely distributed of any primate genus, range across Asia from eastern Afghanistan to Japan, China, Taiwan, Southeast Asia, and Indonesia. Relic populations survive in Morocco and Algeria (Taub 1977). Within this geographical range, macaques occupy habitats ranging in altitude from sea level to 4,000 m, including tropical rain forest, monsoon forest, mangrove swamps, temperate forests, open woodland, grassland and dry scrub zones, and rocky cliffs and beaches in Taiwan and Japan.

The distribution of the vervet monkey roughly parallels that of the savanna baboon, and the two species, occupying many of the same habitat types, often occur sympatrically. The talapoin, in contrast, is more restricted; it occupies flooded primary forests, mangrove swamps, and gallery forests in West and central Africa along the Atlantic coast from southern Cameroon to Angola (Wolfheim 1983).

ECOLOGY AND FEEDING BEHAVIOR

Savanna baboons have often been described as generalized feeders (e.g., Hamilton, Buskirk, and Buskirk 1978). Throughout their range, baboons feed on grass, tubers, bulbs, corms, rhizomes, flowers, fruits, leaves, seeds, tree gum, insects, and eggs, and this eclectic diet is no doubt at least partially responsible for their wide geographical distribution (Harding 1976; Ransom 1981; Post 1982). Baboons (particularly adult males) also supplement this diet with meat, preying on hares, vervets, and infant gazelles (Stoltz and Saayman 1970; Hausfater 1975; Strum 1975). Baboons differ from other primates, however, in their ability to subsist almost exclusively on grass and subterranean corms, rhizomes, and bulbs, an adaptation that allows them to exploit open grassland habitats not normally utilized by other primates. Indeed, in some areas of Kenya such plants constitute over 80% of the baboon's diet (Harding 1976).

The patchy distribution of food resources in savanna environments, coupled with the baboon's relatively large size, means that baboon groups must often move over large areas in order to find enough to eat. This is manifested in both the distance traveled on any given day and the area used over an extensive period of time (i.e., the home range). In the arid savanna habitat of Amboseli, Kenya, for example, baboons typically travel 5.9 km per day, and home ranges are approximately 24 km² in size (Altmann and Altmann 1970). Population densities for baboons in open savanna habitats are consistently low across study sites, probably due to the baboons' similar dependence on grass and roots (table 11-1). In contrast, in forest habitats where fruit constitutes the largest proportion of their diet, baboons inhabit smaller home ranges and live at considerably higher population densities (e.g., Ransom 1981; see also table 11-1).

Vervets occur sympatrically with baboons in many areas of Africa and also exploit many of the same patchily distributed species of plants (Whitten 1983). Vervets, however, are more restricted than baboons to woodland areas and seldom venture into the open savanna. Their hands seem poorly adapted for digging up roots, and they rely far less than baboons on grass (Whitten 1983; Andelman, pers. comm.). Vervets also have much smaller day and home ranges and live at higher population densities than baboons (table 11-1). These differences are consistent with the general trends evident in cercopithecoids (Old World monkeys), where body size is positively correlated with home range but negatively corre-

lated with population density (Clutton-Brock and Harvey 1979).

Macaques occupy an enormous geographical range of great ecological diversity. The rhesus macaque (*Macaca mulatta*), which has the largest range of any macaque, lives in habitats that range from dry evergreen, mixed deciduous subtropical forests in Thailand (Eudey 1980) to Himalayan moist temperate forests in Pakistan (Goldstein 1984). Understandably, there is wide dietary variation among rhesus populations in different parts of their range. For example, Lindburg (1977) found that 65 to 70% of the diet of rhesus macaques in the moist tropical deciduous Siwalik Forest of north India was composed of fruit. By sharp contrast, fruit accounted for less than 9% of the food of rhesus living in the Himalayan foothills of Pakistan, where the bulk of the diet consisted of grass, clover, and other ground herbs (Goldstein 1984). Since the Pakistan rhesus macaques occupy a seasonal habitat that receives a great deal of snow in the winter, their diet changes dramatically during the course of the year. This is also true of Japanese macaques (*Macaca fuscata*), which, during the winter, shift from a diet of leaves, shoots, and fruit to one composed primarily of bark (Suzuki 1965; Iwamoto 1982).

Interspecific differences in the diets of macaques parallel those within species, especially across habitat types (e.g., Hladik 1975; Wheatley 1980; Iwamoto 1982). Even within the same geographical area, some interspecific differences in feeding behavior are evident. In Sumatra, for example, sympatric pigtailed macaques (*Macaca nemestrina*) and long-tailed macaques (*Macaca fascicularis*) exploit different forest strata and food types (Crockett and Wilson 1980). Differences in habitat and diet, both within and between species, are correlated with large differences in population density (table 11-1). Group size and composition, however, vary little, even across widely disparate habitats (table 11-1; see also later text).

While all mangabeys occupy forest habitats, some are semiterrestrial and obtain much of their food from the lower strata of the forest (e.g., agile and sooty magabeys: Jones and Sabater Pi 1968; Homewood 1978). These species also tend to live in larger groups than the strictly arboreal gray-cheeked mangabey. However, all studies of mangabeys report that fruits and seeds are the major (70 to 75%) component of the diet (Chalmers 1968b; Quris 1975; Waser 1975; Homewood 1978), which may explain gross similarities in home range size across species (table 11-1). Even within species, however, population densities may differ dramatically, depending upon specific local conditions (Waser 1975).

The most ecologically puzzling species in the group reviewed here is the talapoin money. This species is found primarily in rain forests, is completely arboreal, and is very small in body size (e.g., adult males weigh

Table 11-1 Group Size, Home Range, and Population Density in Savanna Baboons, Vervets, Macaques, Mangabeys, and Talapoins

Species	Location and Habitat Type	No. of Groups	X̄ Group Size	± Duration of Study (mo)	Home Range Size (ha)	No. of Days of Observation	Mean (m)	Range (m)	Pop. Density (per km²)	Sources
Papio c. anubis	Queen Eliz. Park, Uganda Gallery forest and savanna	2	45	22	470(390–500)					Rowell 1966a
	Nairobi Park, Kenya Savanna	5	39(17–76)	11	2,325(520–4,010)				3.9	DeVore and Hall 1965
	Metahara, Ethiopia Arid thorn scrub and gallery forest	1	87	2	430	12	5,800			Aldrich-Blake et al. 1971
	Gombe, Tanzania Forest	2	97	24	(390–520)				37.2	Ransom 1981
	Gilgil, Kenya Savanna	1	50	12	1,968		5,000			Harding 1976
P. c. cynocephalus	Amboseli, Kenya Semiarid savanna	1	36–45	11	2,408		5,900			Altmann and Altmann 1970
	Amboseli, Kenya Semiarid savanna	15	80(12–185)						9.7	DeVore and Hall 1965
P. c. ursinus	Transvaal, S. Africa Semiarid savanna	6	47.2(30–77)	16	1,735(1,295–2,331)	31 23	6,400 10,461	(3,219–9,656) (2,414–14,484)	4.0	Stoltz and Saayman 1970
	Cape Peninsula, S. Africa	3	45(20–80)	12	1,100(910–3,370)	32	4,667	(1,609–8,047)	2.4	DeVore and Hall 1965; Hall 1962a
Macaca mulatta	Asarori-Siwaliks, India Moist deciduous forest	3	41.3(9–80)	12		38	1,428	(350–2,820)		Lindburg 1971
	Kathmandu, Nepal Terai forests	8 10	32(20–51) 33.2(29–138)	2 60					9.3	Teas et al. 1980
M. fuscata	Japan Deciduous forest	14	36.6		530					Takasaki 1984
M. fiscata	Japan Evergreen forest	11	54.3		101					Takasaki 1984
M. nemestrina	Sumatra Mangrove	3	18.3(16–21)	14					17.9	Crockett and Wilson 1980
M. sinica	Polonnaruwa, Sri Lanka Semideciduous forest	18	24.8(8–43)	36	17–115				100	Dittus 1975, 1977
M. radiata	Southern India	4	34.5(6–58)	9	518					Simonds 1965
M. fascicularis	Sumatra Mangrove forest	4	27(22–30)	14	125	35	1,900		54.4	Wheatley 1980
Ceropithecus aethiops	Samburu, Kenya Savanna woodland	2	40	24	59–103				39	P. L. Whitten 1982
	Assirik, Senegal Savanna woodland	1		12	178			(943–2,087)	14.3	Harrison 1983a
	Amboseli, Kenya Semiarid savanna	3	(11–30)	78	28(22–49)				49–96	Cheney et al., in press
	Amboseli Semiarid savanna	9	24(7–35)	12	41.8(18.4–96)	90 24	1,440 950	(330–2,580) (135–1,550)	16.9–153.7	Struhsaker 1967b, 1967c, 1967d
C. (M.) talapoin	N. E. Gabon Inundated forest	4	112	21	122		2,323	(1,500–2,950)	92	Gautier-Hion 1973
Cercocebus albigena	Moira Bujuko, Uganda Semideciduous tropical rain forest	5	16.8(7–25)	22	21(13–26)				77	Chalmers 1968b
	Kibale, Uganda Montaine rain forest	5	14.4(6–28)	13	410	104	1,270	(835–1,465)	10	Waser 1975; Wolfheim 1983
	Equatorial Guinea	3	17.3(10–24)		200–300				22.1	Jones, pers. comm. in Wolfheim 1983
C. galeritus	Tana River, Kenya Evergreen flooding forest	4	26(17–36)	12	35(17–53)				50–300	Homewood 1978
	N. E. Gabon	3	10.5(7–18)		200				7.7–12.5	
C. torquatus	Rio Muni, W. Africa	Several	14–23		Large				133	Jones and Sabater Pi 1968
	Equatorial Guinea	5	35(20–60)							Jones, pers. comm.

about 1.4 kg; Gautier-Hion 1973). Unlike other forest-dwelling *Cercopithecus* species, however, talapoins live in large multimale groups (65 to 115 members), travel several kilometers per day, and occupy large home ranges (1.2 km²; Gautier-Hion 1973). Talapoins also occur at low population densities in relation to their body size (table 11-1). Gautier-Hion (1973) believes that large home ranges may be related to a heavy reliance on insects to supplement a frugivorous diet.

GROUP SIZE

Generally, primates that are large, terrestrial, and subsistent on relatively frugivorous diets tend to have larger home ranges and live in larger groups than smaller, ar-

boreal, or more folivorous species (Milton and May 1976; Clutton-Brock and Harvey 1977b, 1979; Takasaki 1981). However, these broad patterns do not always explain the characteristics of particular species or populations (Richard 1981). Indeed, it is not uncommon to find substantial variation in group size even within a particular study area (table 11-2). Is such variation so great that it obscures any intra- or interspecific differences?

To examine this question, simple one-way analyses of variance (Sokal and Rohlf 1981) were performed on species for which data on the group size of at least three populations were available (table 11-2). Within each of the seven subspecies or species analyzed, no statistically significant differences could be found (table 11-3). In-

TABLE 11-2   Group Size Variation within and between Species of Savanna Baboons, Vervets, Macaques, and Mangabeys

Species	Population	Location	No. of Groups	X̄ Group Size	Standard Deviation	Range	Source
Macaca mulatta	1[a]	Asarori, N. India	6	32.8	30.3	6–90	Makwana 1978
	2	Haldrani, N. India	5	41.0	25.2	16–77	Neville 1968b
	3	Dunga Gali, N. Pakistan	7	41.6	19.1	18–65	Melnick 1981
	4	Forest groups, U.P. India	7	47.7	12.5	32–68	Southwick, Beg, and Siddiqi 1965
M. fascicularis	1	Peninsular Malaysia	4	35.75	15.7	13–47	Furuya 1965
	2	Timur, Indonesia	12	18.2	4.8	10–30	Kurland 1973
	3	Peninsular Malaysia	4	29.8	27.0	14–70	Bernstein 1967a
	4[a]	E. Borneo	4	27.0	3.6	22–30	Wheatley 1980
M. radiata	1[a]	S. India	4	34.5	22.7	6–58	Simonds 1965
M. silenus	1[a]	S. India	4	21.0	9.6	12–34	Green and Minkowski 1977
M. sylvanus	1[a]	Morrocco	6	18.3	5.4	12–25	Deag and Crook 1971
M. assamensis	1[a]	W. Thailand	9	21.8	12.5	10–50	Fooden 1971
M. nemestrina	1	Sumatra	3	18.3	2.5	16–21	Crockett and Wilson 1980
Cercopithecus aethiops	1[a]	Amboseli, S. Kenya	9	17.9	6.7	9–29	Struhsaker 1973
	2	Awash, Ethiopia	7	17.7	7.5	9–27	Turner 1981
	3	Samburu, Kenya	5	22.8	5.2	19–31	Dracopoli et al. 1983
Cercocebus albigena	1[a]	Uganda	5	16.8	6.6	7–25	Chalmers 1968b
	2	Equatorial Guinea	3	17.3	7.0	10–24	Jones, pers. comm., in Wolfheim 1983
	3	Gabon	2	13.0	5.7	9–17	Gautier and Gautier-Hion 1969 Quris 1975
Papio cynocephalus anubis	1	Nairobi Park, Kenya	9	42.1	26.9	12–87	DeVore and Hall 1965
	2	Murchison Falls, W. Uganda	8	29.9	13.5	14–48	Hall 1965b
	3	C. Ethiopia	6	47.8	27.9	14–87	Aldrich-Blake et al. 1971
	4	W. Uganda	3	40.3	9.9	29–47	Rowell 1972
P. c. cynocephalus	1	Mesalani, Kenya	4	28.3	22.3	8–60	Maxim and Buettner-Janusch 1963
	2[a]	Amboseli, Kenya	20	56.0	40.8	18–196	Altmann and Altmann 1970
	3	W.C. Tanzania	4	81.75	43.0	27–130	Rasmussen 1981
P. c. ursinus	1[a]	N.E., S. Africa	6	47.2	17.8	30–77	Stoltz and Saayman 1970
	2	N. Botswana	7	79.0	43.4	7–128	Hamilton, Buskirk, and Buskirk 1976
	3	S.W., S. Africa (Capetown)	3	45.0	31.2	20–80	Hall 1963b

a. These populations are included in the interspecific comparisons in table 11-3.

TABLE 11-3 One-way Analysis of Variance of Intraspecific and Interspecific Differences in Group Size in Savanna Baboons, Vervets, Macaques, and Mangabeys

	INTRASPECIFIC				
Species	No. of Populations	No. of Groups	F-Ratio	p	r^2
Papio cynocephalus anubis	4	26	F (3, 22) = .803	.505	.099
P. c. cynocephalus	3	28	F (2, 25) = 1.855	.177	.129
P. c. ursinus	3	16	F (2, 13) = 1.830	.199	.220
Cercopithecus aethiops	3	21	F (2, 18) = 1.063	.366	.106
Macaca mulatta	4	25	F (3, 21) = .489	.694	.065
M. fascicularis	4	24	F (3, 20) = 2.334	.105	.259
Cercocebus albigena	3	10	F (2, 7) = .297	.752	.078

	INTERSPECIFIC				
Species	No. of Species	No. of Groups	F-Ratio	p	r^2
Macaques	7	36	F (6, 29) = .798	.579	.142
Macaques and vervets	8	45	F (7, 37) = 1.047	.416	.165
Macaques, vervets, and mangabeys	9	50	F (8, 41) = 1.110	.376	.178
Macaques, vervets, mangabeys, and baboons	12	85	F (11, 73) = 2.741	.005	.292

deed, variation in group size within populations was frequently as great as variation between populations. Perhaps more surprising, when baboons were excluded from the analysis, species and genera also failed to differ significantly from each other (table 11-3). For example, even though the average bonnet macaque (*Macaca radiata*) group in south India is almost twice the size of the average Barbary macaque (*Macaca sylvanus*) group in Morocco, intraspecific variation in group size obscures this apparent difference. Only when baboons were included in the analysis did significant interspecific differences in group size emerge ($F_{11,73}$ = 2.741, p = .005; table 11-3).

Intraspecific fluctuations in group size no doubt result in large part from the effects of variation in food abundance and predation pressure on mortality and fecundity (see chap. 20). However, historical processes may also exert some influence. For example, new cercopithecine groups are occasionally formed through the fission of a larger social group, a process that frequently results in two "daughter" groups of differing size (e.g., rhesus macaques: Chepko-Sade and Sade 1979). Short-term variation in offspring sex ratio, migration patterns, and intergroup relations can also cause marked fluctuations in group size (Dunbar 1979a; chap. 20).

Between species, similarities in group size probably occur in part because most of the cercopithecines included in the analysis share many of the same ecological adaptations. They are all diurnal, most are terrestrial or semiterrestrial, and none are folivorous. The differences seen in baboons may be related to their more complete terrestriality, larger body size, and exploitation of large resource patches (e.g., grass)(Andelman 1986).

REPRODUCTION

As is true of most primates, cercopithecine females give birth to one infant at a time. The average length of gestation in the species described here varies little, ranging from 160 to 180 days (table 11-4). This is surprising, given the positive correlation between body size and gestation length in primates (Eisenberg 1981; chap. 16) and the enormous difference in body size between an adult female talapoin (body weight 1.1 kg, gestation period 165 days; Gautier-Hion 1973) and an adult female yellow baboon (approximately 12.0 kg, gestation period 175 days; Altmann et al. 1977).

Age at sexual maturity is often difficult to define in wild primate populations, and, like all life-history parameters, it shows great intraspecific variation, depending upon both local food availability and social factors. For example, on Cayo Santiago Island, where rhesus macaques are provisioned, most females become reproductively mature by 3.5 to 4.5 years of age (Koford 1965; Sade et al. 1976). In contrast, in the temperate forests of the Himalayan foothills, rhesus females attain reproductive maturity at 5.5 years of age (Melnick 1981). Similarly, some studies have shown that high-ranking females become reproductively mature before low-ranking females, perhaps as a result of differential access to food resources and reduced stress and harassment (e.g, captive and provisioned rhesus macaques: Drickamer 1974; Wilson, Gordon, and Bernstein 1978; Wilson, Walker, and Gordon 1983; toque macaques [*Macaca sinica*]: Dittus 1977; yellow baboons: Altmann, Altmann, and Hausfater 1986; see also chap. 26). In general, reproductive maturity occurs at older ages among males than among females (table 11-4). Males also usually become reproductively mature long before they are able to compete successfully for females. Male yellow baboons, for example, between the ages of 4 and 7 become capable of fertilizing females, but they do not reach full adult size until they are approximately 10 (Altmann et al. 1977; see also chap. 20).

Most macaques, vervets, and talapoins show some seasonality in births, while baboons and mangabeys do not (table 11-4). Among seasonally breeding species, the timing of births appears to be primarily determined by environmental factors, such as photoperiod, warming temperatures, and increased rainfall. In northern Japan and the Himalayas, macaque parturition is concentrated during a short birth "pulse" of 2 months in the spring (Kawai, Azuman, and Yoshiba 1967; Melnick 1981). This ensures that infants are fully weaned and heavy when they face the rigors of the next winter (Pearl 1982). More typically, birth seasons are 3 or 4 months in length (table 11-4). The fact that baboons, which show no birth seasonality, are sympatric with vervets, which do, suggests that the presence or absence of birth seasonality is not a simple function of increased rainfall or temperature change and that subtle variations in diet, food distribution, and range size are probably equally important.

Interbirth intervals are also affected by both environmental and social factors. In provisioned groups, or in areas where food is abundant, interbirth intervals are shorter than in harsher habitats. At Cayo Santiago, for example, interbirth intervals average 12 months in length, while in the Himalayas rhesus females give birth at intervals of from 24 to 36 months (Pearl 1982). Similarly, among vervets in Amboseli, females living in groups with access to permanent waterholes give birth annually, while females in groups without surface water give birth biannually (Cheney et al., in press). Interbirth intervals are also affected by the survival of previous offspring; females whose previous infants survive usually experience longer interbirth intervals than those whose infants die before being weaned (e.g., baboons: Altmann et al. 1977; see also chap. 20). Another social factor potentially affecting interbirth intervals is female dominance rank, since some studies have documented shorter birth intervals in high-ranking, as opposed to low-ranking, females (e.g., captive vervets: Fairbanks and McGuire 1984; see also chap. 26).

Female baboons, mangabeys, talapoins, and some species of macaques exhibit pronounced sexual swelling during the follicular phase of their menstrual cycle (table 11-4; see also chap. 30). Other species of macaques, however, experience only a reddening of the perineum, while clear visual manifestations of ovulation are lacking in vervets. The precise function of sexual swellings is not yet known (see chap. 30). The presence or absence of sexual swellings is not obviously correlated with differences in social structure or birth seasonality, since all of these species live in multimale groups, and both seasonally and non–seasonally breeding species exhibit sexual swellings.

Females in multimale groups usually mate with a number of males during their estrous period. Most mating is concentrated at around the time of ovulation, although in some species mating sometimes occurs during other periods of the sexual cycle, including early pregnancy (chap. 30). The number of males with whom each female mates while in estrus varies greatly, both across species and across individuals. Male and female baboons typically form temporary, exclusive pair bonds, or sexual consortships, around the time of ovulation. These may last from as little as a few minutes to as long as 5 or 6 days, depending upon the dominance rank of the male, the intensity of male-male competition, and the sexual partners' social relationship (e.g., Hausfater 1975; Seyfarth 1978a, 1978b; Collins 1981; Rasmussen 1983b; Smuts 1985; see also chaps. 30, 31). Sexual consortships also occur among macaques and mangabeys, although the partners' movements and behavior appear to be less closely coordinated than among baboons (e.g., rhesus macaques: Kaufman 1965; Lindburg 1971; Japanese macaques: Enomoto 1978; Takahata 1982a; gray-cheeked mangabeys: Wallis 1983). Consortships are generally lacking in vervets and talapoins, and copulation usually occurs in the absence of other affinitive behavior, such as grooming (Rowell and Dixson 1975; Andelman 1985).

SOCIAL BEHAVIOR

Although social groups of mangabeys, baboons, vervets, macaques, and talapoins usually consist of a number of adult males, adult females, and offspring, the sex ratio of adults is often skewed toward females (table 11-5). Skewed adult sex ratios result at least in part from the fact that females usually mature several years before males and are thus considered to be adults at younger ages (Rowell 1969). Also, because males emigrate from the natal group, they may be more subject than females to predation, disease, and injury (Dittus 1975; see chaps. 19–21). Adult males are usually dominant to adult females in dyadic interactions, apparently because of their larger size (chaps. 25, 32). Talapoins may be an exception to this generalization; observations of both captive and free-ranging groups have suggested that females can supplant males even in the absence of the threat of a coalition (Gautier-Hion 1971b; Wolfheim 1977a; chap. 32).

Cercopithecine groups typically move as an integrated unit and do not regularly split into consistent subgroups. Talapoins again seem to be anomalous in this respect, since males and females habitually travel in distinct subgroups except during the breeding season (Gautier-Hion

TABLE 11-4   Reproductive Parameters for Savanna Baboons, Macaques, Vervets, Mangabeys, and Talapoins

Species and Study Site	Gestation Pd.[a] (days)	Sex. Maturity (years) Male	Female	Interbirth Interval (months) $\overline{X}$	Range	Peak Birth Pd.	External Signs of Estrus[b]	Sources
Papio c. anubis								
Uganda					(12–18)	None[c]	Perineal swelling	Rowell 1969; Chalmers and Rowell 1971
Gilgil, Kenya	180	5–7	4.5		(18–34)	None	Perineal swelling	Nicolson 1982; Scott 1984; Smuts, pers. comm.
P. c. cynocephalus								
Amboseli, Kenya	175	4–7	5–6	21[d]		None	Perineal swelling	J. Altmann et al. 1977
P. c. ursinus								
S. Africa					(18–24)	None	Perineal swelling	DeVore and Hall 1965; Stolz and Saayman 1970
Macaca mulatta	164(146–80)	4.5	3.5			Mar–June	Perineal reddening	Napier and Napier 1967
M. mulatta mulatta								
Uttar Pradesh, India						Mar–June	Perineal reddening	Southwick, Beg, and Siddiqi 1965
Cayo Santiago			3.5–4	12		Jan–June	Perineal reddening	Koford 1965
Dehra Dun				12		Mar–May	Perineal reddening	Lindburg 1971
M. mulatta villosa								
Pakistan		6–7	4.5–5.5	26	(24–36)	Mar–Apr	Perineal reddening	Melnick 1981; Pearl 1982
M. fuscata	170–80	4.5	3.5			Mar–Aug	Perineal reddening	Napier and Napier 1967
M. fuscata								
Takasakiyama			3.5			May–Sept		Lancaster and Lee 1965
Japan	180		6.5	23.2		Apr–July		A. Mori 1979a; Hiraiwa 1981
M. nemestrina								
Captive	170(162–86)						Perineal swelling	Napier and Napier 1967
M. sinica								
Sri Lanka		6–7	5–6	18	(8–24)	Feb–Apr	Perineal reddening	Dittus 1979, pers. comm.
		5–7	4.5–5.5					
M. radiata								
Madras	163(153–69)	6	4	12		Jan–May	Perineal reddening	Simonds 1965
M. fascicularis								
Malaysia						May–July		Kavanagh and Laursen 1984
Captive	167(153–79)	4.2	4.3					Napier and Napier 1967
Cercopithecus aethiops								
Amboseli, Kenya	163	5	4.5	16	(11–24)	Oct–Dec	None	Cheney et al., in press
Lolui Island, Uganda		5+	4+				None	Hall and Gartlan 1965
Cercocebus albigena								
Uganda	185	5–7	3+	33	(18–48)	None	Perineal swelling	Chalmers 1968a; Chalmers and Rowell 1971; Wallis 1983; Waser 1974; Leland, pers. comm.
Cercopithecus (M.) talapoin								
N.E. Gabon	165					Nov–Apr	Perineal swelling	Gautier-Hion 1971b, 1973; Rowell and Dixson 1975

a. Some gestation lengths are based on captive data (e.g., *C. aethiops*).

b. More references on perineal swellings can be obtained from chapter 30.

c. In savanna baboons, birth peaks are slight and mating takes place year round.

d. Based on 9 females whose infants survived the first year of life. If one based the calculation of IBI on a calculus that includes infant death, infant survival, and probability of survival, the figure is closer to 19.1 months.

TABLE 11-5 Demographic Structure of Social Groups of Savannah Baboons, Macaques, Vervets, Mangabeys, and Talapoins

Species and Study Site	No. of Groups	$\bar{X}$ Group Size	$\bar{X}$ No. Adult Males	$\bar{X}$ No. Adult Females	Adult Females/ Adult Males	$\bar{X}$ Juveniles and Infants	Sources
Papio c. anubis							
Nairobi, Kenya	5	25.8	3	7.4	1:2.5	15.4	DeVore and Hall 1965
Uganda	3	55.7	10.7	15.3	1:1.4	29.7	Rowell 1969
P. c. cynocephalus							
Amboseli, Kenya	1	39	8.4	12.6	1:1.5	18	Altmann and Altmann 1970
P. c. ursinus							
Cape Peninsula, S. Africa	3	42.7	4	13.3	1:3.3	25.3	Hall and DeVore 1965
Transvaal, S. Africa	6	47.2	13	20.5	1:1.6	13.7	Stoltz and Saayman 1970
Mt. Zebra Pk., S. Africa	1	24	2	8	1:4.0	14	Seyfarth 1976
Macaca mulatta							
Uttar Pradesh, India	7	49.8	5.6	19.2	1:3.4	25	Southwick, Beg, and Siddiqi 1965
Asarori, India	6	32.8	3.3	9.2	1:2.8	20.3	Makwana 1978
Rara Lake, Nepal	1	39	3	17	1:5.7	19	Teas et al. 1980
Dunga Gali, Pakistan	7	41.6	6.3	11.7	1:1.9	23.6	Melnick 1981; Pearl 1982
M. fuscata							
Kaminyu, Japan	1	81	7	24	1:3.4	50	Kurland 1977
M. sinica							
Sri Lanka	18	24.8	2.7	6.2	1:2.3	15.9	Dittus 1975
M. sylvanus							
Morocco	1	39	7	9	1:1.3	23	Taub 1980b
Morocco	6	18.3	2.7	5.2	1:1.9	10.5	Deag and Crook 1971
M. fascicularis							
Sumatra	1	29.1	5.7	9.9	1:1.7	15.5	Wheatley 1980
M. silenus							
S. India	4	21	2	7.8	1:3.9	11.3	Green and Minkowski 1977
M. nemestrina							
Lima Belos, Malaysia	1	50	2	18	1:9	30	Caldecott 1981
M. radiata							
Mysore, India	6	27.8	5.5	7.6	1:1.4	14.7	Simonds 1974
Dharwar, India	12	30	8.0	9.7	1:1.2	12.3	Sugiyama 1971
Cercopithecus aethiops							
Amboseli, Kenya	3	14–28	2–4	4–8	1:0.75– 1:7	10–13	Cheney et al., in press, pers. comm.
Samburu, Kenya	2	40.1	9.3	9.5	1:1.02	10.9	P. L. Whitten 1982
Assirik, Senegal	1		3–6	7			Harrison 1983a
Amboseli, Kenya	2	17	2	4	1:2	12	Struhsaker 1967b
Lolui Island		10	2	4	1:2	5	Gartlan and Brain 1968
S. Africa	3	19.7	1.7	5.7	1:3.4	12.3	Henzi and Lucas 1980
Cercocebus albigena							
Uganda (Mbira Bujoku)	2	21	4.5	8	1:1.8	8.5	Chalmers 1968b
Kibale, Uganda	1	14–16	3	6	1:2	5–7	Waser 1975
Cercopithecus (M.) talapoin							
N.E. Gabon	1	115	13	27	1:2.1	75	Gautier-Hion 1971a

1971b; Rowell and Dixson 1975). The New World squirrel monkeys (*Saimiri sciureus*) exhibit similar sexual segregation (chap. 7). In neither species, however, has this pattern of subgrouping been studied in any detail and its cause remains unknown.

Relationships among Females

In all of the species described here, females usually remain in their natal groups throughout their lives, while males transfer to other groups at around sexual maturity (chap. 21). As a result of this sex-biased dispersal pattern, females form the stable core of the social group and maintain close bonds with their maternal relatives throughout their lives (fig. 11-1). Within each group, adult females form linear dominance hierarchies, based on the direction of approach-retreat interactions (white cheeked magabeys: Chalmers and Rowell 1971; sooty

mangabeys: Bernstein 1976a; talapoins: Wolfheim 1977a; chap. 25). Among baboons, macaques, and vervets, dominance ranks remain relatively stable over considerable lengths of time, with daughters assuming ranks similar to those of their mothers (chaps. 25, 29). Rank acquisition appears to involve both active intervention by maternal kin in agonistic interactions and differential treatment by unrelated animals. From infancy, the offspring of high-ranking females receive less aggression, more support, and more affinitive gestures than do the offspring of low-ranking females (e.g., rhesus macaques: Sade 1972a, 1972b; Berman 1980a; pigtailed macaques: Massey 1977; Japanese macaques: Kurland 1977; stumptailed macaques: Gouzoules 1975; toque macaques: Dittus 1977, 1979; baboons: Cheney 1977a, 1978a; Altmann 1980; Walters 1980; vervets: Cheney 1983a; chaps. 25, 29).

FIGURE 11-1. An adult female vervet monkey and her two daughters sit together as they look at another group. (Photo: Patricia Whitten/Anthrophoto)

FIGURE 11-2. Grooming within matrilines. (*above*) A 2-year-old female rhesus monkey grooms her 1-year-old sister, who is leaning against their mother. The 1-year-old is touching the leg of her infant brother, who is nursing. (*right*) An adult female Japanese macaque grooms her juvenile daughter. Her youngest offspring and an older daughter are also nearby. (Photos: (*above*) Barbara Smuts/Anthrophoto; (*right*) Sarah and Harold Gouzoules)

A number of studies have documented reproductive benefits to females of high rank, including reduced interbirth intervals, reduced mortality during times of food scarcity, and increased offspring survival (see chap. 26). These benefits seem to accrue to high-ranking females as a result of their preferential access to scarce resources and the lack of harrassment from others. Rank is not correlated with reproductive success in all populations, however, at least in part because of the influence of selective factors that are less affected by female rank, such as predation (e.g., Cheney et al., in press).

Genealogical matrilines vary considerably in size and composition. For example, a wild group of rhesus macaques from the harsh Himalayan environment of northern Pakistan contained 5 matrilines averaging under 5 members (Pearl 1982). In contrast, provisioned groups of the same species on Cayo Santiago, Puerto Rico, contain as many as 11 matrilines that number up to 41 members (Datta 1981). Such differences in size clearly can affect the frequency of social interactions, within and between matrilines. Nevertheless, it is noteworthy that a

direct comparision of two groups of greatly different size from Cayo Santiago and Pakistan revealed quite similar patterns of interaction (Pearl and Schulman 1983). In fact, in all cercopithecine groups studied to date, most grooming and alliances occur among related females, despite variation in group and matriline size (see chap. 24) (fig. 11-2).

Although females associate preferentially with their maternal kin, such bonds nevertheless reveal complexity and change. Among Japanese macaques, for example, a female's interactions with her mother and siblings decrease in frequency with age, as her interactions become focused increasingly on her own offspring (Grewal 1980a). Regardless of their frequency, however, the quality of interactions with kin remains different from those with nonkin. Following fights, kin appear to reconcile more rapidly than do nonkin (e.g., Japanese macaques: Kurland 1977; rhesus macaques: de Waal and Yoshihara 1983), and are also more likely to aid each other in potentially costly disputes (e.g., baboons: Cheney 1977a; rhesus macaques: Datta 1983a). Kinship ties

may even extend for more than two generations. For example, a baboon grandmother was observed to harass an adult male who attacked her 7-month-old granddaughter (Collins, Busse, and Goodall 1984), and a wild rhesus grandmother was seen to hold her granddaughter during the temporary absence of the mother (Pearl and Melnick 1983).

Affinitive interactions among females are not restricted to close maternal relatives, however, and almost all individuals interact at least occasionally with other females in the group. In particular, females interact with animals of adjacent rank (Sade 1965; Seyfarth 1976) and also compete to establish bonds with high-ranking females. As a result, high-ranking females usually receive more grooming and alliances than do low-ranking females (chap. 25). A number of benefits to bonds with high-ranking females have been proposed, including reduced harrassment (Silk, Samuels, and Rodman 1981), increased support in aggressive interactions, and increased access to limited resources (Seyfarth 1977, 1983). In addition to benefiting from high rank as indi-

viduals, the members of dominant matrilines, while maintaining greater proximity and supporting each other in disputes at higher rates (Yamada 1963; Cheney 1977a, 1983a), often appear more cohesive than the members of subordinate matrilines.

Although relationships among females of differing dominance rank and genetic relatedness often vary greatly, they are almost always less hostile than those among females from different groups. Females rarely transfer to other groups and seldom interact affinitively with extragroup females (see chaps. 21, 22). Female vervets and macaques frequently participate in aggressive intergroup interactions and are usually particularly hostile toward members of their own sex. Such aggression is less common among baboons, but female baboons are nevertheless intolerant of females from other groups (chap. 22).

Relationships among Males

The cercopithecines discussed in this chapter are all distinguished by the fact that groups commonly include more than one male. Thus intrasexual competition does not occur in the form of attempts to control a bisexual group, as is the case, for example, among many forest-dwelling *Cercopithecus* species (chap. 9). Instead, male-male competition for females occurs within the group, after a male has already joined it. Relationships among males are usually more aggressive and less affinitive than those among females, probably both because most males are unrelated and because male reproductive success is largely determined by competition for females (gray-cheeked mangabeys: Struhsaker and Leland 1979; talapoins: Rowell and Dixson 1975; chaps. 25, 26). The outcome of male-male competitive interactions is typically a dominance hierarchy, based on the direction of approach-retreat interactions and aggressive disputes. Perhaps because most males emigrate from their natal groups, adult male dominance rank in macaques, vervets, and baboons cannot usually be predicted from maternal rank. Instead, male dominance rank is determined by such factors as age, size, and fighting ability (e.g., baboons: Hall and DeVore 1965; Hausfater 1975; Packer 1977, 1979b; vervets: Struhsaker 1967d; Japanese macaques: Sugiyama 1976; see chap. 25).

While access to females is sometimes correlated with male dominance rank (e.g., baboons: Hausfater 1975; Packer 1979b; rhesus macaques: Kaufman 1965; Chapais 1983c; captive talapoins: Keverne 1982; some groups of gray-cheeked mangabeys: Struhsaker and Leland 1979; see also Berenstain and Wade 1983), factors such as length of tenure in the group, the presence of alliance partners, and female choice can also influence mating success (Sugiyama 1976; Packer 1977, 1979b; Meikle and Vessey 1981; Strum 1982; Chapais 1983c; Fedigan 1983; Takahata 1982a, 1982b). Among baboons, males who establish close bonds (as measured by grooming, alliances, and proximity) with particular adult females often gain preferential access to such females when they are in estrus, even when the males are low ranking (Seyfarth 1978b; Collins 1981; Smuts 1985; chap. 31). Male-female relationships are less obviously associated with mating behavior in macaques and vervets. Moreover, male and female vervets seldom form long-term grooming relationships, and female choice of mates appears to be unrelated to past affinitive interactions (Andelman 1985).

It seems possible that the importance of male-female bonds in baboons, macaques, and vervets depends to some extent on the difference in body size between males and females. Baboon females weigh only approximately 50% as much as males and seldom form coalitions against them (Clutton-Brock and Harvey 1979; Packer and Pusey 1979). As a result, females may need to form bonds with males in order to protect themselves and their offspring against aggression from other males (Smuts 1985; chap. 31). In contrast, most female macaques,

vervets, and talapoins weigh at least 70% as much as males, and this reduction in the degree of sexual dimorphism apparently permits females in these species to form successful coalitions against males (Gautier-Hion 1971b; Rowell and Dixson 1975; Clutton-Brock and Harvey 1979; Packer and Pusey 1979; Cheney 1983b).

Although male-male interactions are usually more aggressive than those among females, this is not to say that males never cooperate to increase their mating success. Male vervets and macaques frequently emigrate in the company of brothers or natal group peers (see chap. 21), and rhesus macaque brothers have been observed to form alliances in their adopted groups that increase or maintain their dominance ranks (Meikle and Vessey 1981; see also Mizuhara 1964 for alliances among male Japanese macaques). Alliances among male baboons more often involve unrelated males, probably because baboons are less likely to emigrate with brothers or natal group peers (see chaps. 21, 31).

Grooming among adult male macaques, vervets, and talapoins is usually less common than among females, although it occurs at low rates, particularly during the nonbreeding season (e.g., rhesus macaques: Kaufmann 1967; Japanese macaques: Sugiyama 1976; bonnet macaques: Sugiyama 1971; vervets: Cheney, pers. comm.; talapoins: Wolfheim 1977a). Among Japanese and rhesus macaques, high-ranking males are groomed at higher rates than low-ranking males, as is true also for females (Kaufmann 1967; Sugiyama 1976). In contrast, grooming among male bonnet macaques appears to be unrelated to rank and also occurs at higher rates than in other macaque species (Sugiyama 1971). Male baboons and mangabeys groom each other only rarely (Saayman 1971b; Struhsaker and Leland 1979; Smuts 1985). Male baboons do, however, engage in "greeting behavior," a display that is characterized by stereotypic vocalizations, facial expressions, mounts, and touches. Such behavior may be related to the establishment and maintenance of alliances (Smuts and Watanabe, in prep.). Similar greeting displays occur among male bonnet macaques (Sugiyama 1971).

GENETIC DIVERSITY AND POPULATION STRUCTURE ◆

As described earlier in this chapter, male cercopithecines almost always migrate from their natal group at around sexual maturity and may subsequently migrate several more times during their lives. Females, on the other hand, remain in their natal groups throughout their lives. This pattern of dispersal affects both the social and genetic structure of a population. The proximate and ultimate consequences of dispersal are described in chapter 21. The focus here will be on the effects of dispersal on genetic differentiation within and between groups. Most of the data necessary to explore these effects come from studies of rhesus and Japanese macaques. Together with more limited data on vervets and baboons, they form the basis of the following discussion.

Dispersal patterns should affect the distribution of genetic variation within and between local populations, the degree of genetic similarity among a local population's social groups, and the level of inbreeding in both social groups and the population as a whole. Although there has been much theoretical discussion about the effects of dispersal on the distribution of a population's genetic diversity (e.g., Wright 1943, 1965), biochemical techniques for measuring such diversity in wild primates have only been applied since the 1960s, when it became possible to assay plasma and red cell proteins electrophoretically for phenotypic variation (e.g., Barnicot, Jolly, and Wade 1967).

Underlying the use of electrophoresis is the assumption that phenotypic variation is coded for by polymorphic genetic loci. The search for loci that will allow individuals, groups, or populations to be compared is essentially a sampling process. Typically, 50 to 90% of a population's members are screened for variation at 8 to 35 loci, of which only between 10 to 20% are ever found to be polymorphic (see following references). Individuals or groups are judged similar or different genetically depending on whether or not they possess the same alleles or similar frequencies of alleles at these polymorphic loci. There are a number of limitations to this type of analysis (Melnick, Jolly, and Kidd 1984; Melnick and Kidd 1985). Most important is the fact that electrophoresis measures only variation in proteins and may therefore fail to uncover genetic variation if different alleles code for what is electrophoretically the same protein phenotype (Ramshaw, Coyne, and Lewontin 1979). Moreover, since the analysis uses only soluble blood proteins, the extent of variation among all of an individual's primary gene products remains unknown (Brown and Langley 1979; McConkey, Taylor, and Phan 1979). Finally, genetic variation is not evenly distributed across all loci, nor is it the same for populations defined in different ways (e.g., social group versus regional population; Melnick, Jolly, and Kidd in prep.). Within these limitations, however, comparisons of individuals, groups, and populations can be made, provided that the same level of subdivision and the same loci are compared.

Distribution of Genetic Variation

Migration tends to reduce genetic differentiation among groups that exchange individuals and their genes. The greater the level of gene flow and the more random its distribution, the more likely that genetic diversity within social groups will approach the genetic diversity of the population as a whole (Nei 1977). Indeed, in a number of studies of nonhuman primates, individual social groups have been found to include over 90% of the ge-

netic diversity of the local population, which suggests that intergroup migration occurs at high rates (Cayo Santiago rhesus macaques: Duggleby 1978; Pakistan rhesus macaques: Melnick, Jolly, and Kidd 1984; Japanese macaques: Nozawa et al. 1982; vervets: Turner 1981; Dracopoli et al. 1983). Among both Japanese macaques and long-tailed macaques in Indonesia, groups become more genetically different with increasing geographical distance. Neighboring groups within local populations are genetically similar, while groups separated by more than 100 km are less so (Nozawa et. al 1982; Kawamoto, Ischak, and Supriatna 1984). This genetic structure corresponds closely to male migration patterns, since individuals are more likely to transfer to neighboring groups than to migrate over large distances (Kawanaka 1973).

Male vervets and rhesus macaques also tend to transfer to neighboring groups (chap. 21), and thus groups in these species might also be expected to show greater genetic differentiation as they become more separated geographically. Curiously, however, much of the genetic diversity in vervets throughout Kenya (Dracopoli et al. 1983) and rhesus macaques throughout Pakistan (Melnick 1983) can also be found in any local population or its constituent social groups. In fact, 73% of the genetic diversity of the entire rhesus species can be found among the individuals of any social group (Melnick, in prep.). These patterns may be the result of strong stabilizing selection, that is, selection that operates against the extremes of variation in the population. A more likely explanation, however, is that intra- and interpopulation migration are sufficiently high to keep genetic differentiation among groups low. Among vervets (Cheney and Seyfarth 1983) and rhesus macaques (Melnick, Pearl, and Richard 1984), at least some males always migrate to nonadjacent groups, and it is probable that even a few such long-distance migrants are sufficient to reduce genetic differentiation among widely separated groups (Bodmer and Cavalli-Sforza 1968). Interpopulation genetic differentiation may also be elevated in Japan and Indonesia as the result of islands and urban areas, which restrict male migration (Nozawa et al. 1982; Kawamoto, Ischak, and Supriatna 1984).

Genetic Similarities among Groups

Despite the fact that neighboring groups usually share many of the same alleles, groups may nevertheless differ significantly from their neighbors in the relative frequency with which these alleles are represented (baboons: Brett et al. 1977; Byles and Sanders 1981; rhesus macaques: Duggleby 1978; Melnick, Jolly, and Kidd 1984; see also Melnick, Jolly, and Kidd 1984 for a description of the statistical techniques used for examining genetic differentiation within and between groups). This occurs because groups are composed of clusters of relatives, and these nonrandom samples of individuals frequently exhibit distinctive allele frequencies. Indeed, many populations show more intergroup genetic heterogeneity than might be expected given the high rate of male migration (Cayo Santiago rhesus macaques: Duggleby 1978; Ober, Olivier, and Buettner-Janusch 1980; Pakistan rhesus macaques: Melnick, Jolly, and Kidd 1984; Melnick, Pearl, and Richard 1984; Japanese macaques: Nozawa et al. 1982; vervets: Turner 1981). These discrepancies appear to result from a number of behavioral factors.

First, male migration from the natal group is seldom completely random. In many species, males often migrate in the company of maternal half brothers or peers who may be paternal half brothers (e.g. rhesus macaques: Boelkins and Wilson 1972; Meikle and Vessey 1981; Japanese macaques: Sugiyama 1976; chacma baboons: Cheney and Seyfarth 1977; vervets: Cheney and Seyfarth 1983; see also chap. 21). Moreover, in some populations of vervets, rhesus, and Japanese macaques, males are more likely to migrate to particular neighboring groups than to others, and two groups may also exchange males at high rates (Koyama 1970; Cheney and Seyfarth 1983; Meikle, Tilford, and Vessey 1984). Such nonrandom, kin-structured migration should result in higher levels of differentiation between groups than might be expected if migration were random (Fix 1978). There is no evidence, however, that nonrandom migration results in an accumulation of inbreeding and its genetic effects, probably because non-natal males migrate more randomly than natal males, some natal males migrate alone and to more distant groups, and skewed migration patterns between specific pairs of neighboring groups do not persist over very long periods of time (Cheney and Seyfarth 1983; Melnick, Pearl, and Richard 1984). To date, no genetic studies of natural populations have produced any evidence of inbreeding. In fact, groups usually show more heterozygosity than would be expected by chance, or if males were mating with close relatives (Melnick 1982; Melnick, Pearl, and Richard 1984).

A second potential source of increased genetic differentiation between groups is group fission, a process by which a parent group divides into two distinct daughter groups. Group fission has been observed in several cercopithecine species, often under circumstances when groups are provisioned or expanding (Cayo Santiago rhesus macaques: Chepko-Sade and Sade 1979; Pakistani rhesus macaques: Melnick and Kidd 1983; Indian rhesus macaques: Southwick, Beg, and Siddiqi 1965; Japanese macaques: Furuya 1969; olive baboons: Nash 1976). On Cayo Santiago, groups divide according to maternal re-

latedness, a process that both elevates degrees of relatedness within groups and increases genetic differentiation between groups (Cheverud, Buettner-Janusch, and Sade 1978; Buettner-Janusch et al. 1983). In contrast, even though one small group in Pakistan also apparently divided along genealogical lines, the resulting genetic differences between the two daughter groups were the same as if the group had divided at random (Melnick and Kidd 1983). The differing consequences of group fission in the two populations appear to be due to differences in demographic structure and mating patterns. Unlike the Cayo Santiago population, the Pakistan population was not expanding in size, and groups were composed of many matrilines containing relatively few individuals. Since each fission group probably contained a number of small matrilines, the overall average kinship coefficients remained small. Regardless of whether fission results in higher kinship coefficients in the daughter groups, however, it should accelerate intergroup differentiation and founder-group divergence (Melnick and Kidd 1983).

Nonrandom mating provides a final source of genetic differentiation between groups, and even between matrilines within groups. On Cayo Santiago, genetic differentiation between matrilines appears to be accentuated as a result of the tendency of males to mate with females from the same matriline (MacMillan and Duggleby 1981). In Pakistan, in contrast, small group size and the ability of dominant males to monopolize the majority of matings may result in high kinship coefficients across matrilines (i.e., paternal half-siblings) and among individuals of similar age (Melnick and Kidd 1983). Such mating patterns may also cause social groups to become more rapidly differentiated from their neighbors than if mating were less restricted.

Current data on male dispersal and group fission allow several generalizations about genetic differentiation among multimale cercopithecine groups. First, genetic diversity is relatively evenly distributed throughout any local population of groups, apparently as a result of male migration. Second, because they contain clusters of relatives, social groups are distinctive in their specific frequencies of genes. Third, sex-biased dispersal appears to prevent inbreeding both in any one social group and in the population as a whole. Finally, in many cases the degree of genetic differentiation among groups is greater than might be expected if migration patterns were completely random. A number of behavioral factors, including nonrandom migration, group fission, and nonrandom mating, appear to contribute to such genetic differentiation.

12 | Gibbons: Territoriality and Monogamy

Donna Robbins Leighton

Gibbons (Hylobatidae) are the smallest of the apes. Most species weigh about 5 kg, although *Hylobates hoolock* weigh slightly more (6 to 8 kg), and the siamang (*Hylobates syndactylus*) is twice as large (10 to 12 kg). Gibbons are prototypical brachiators, with long arms and hands and flexible forelimb joints. Scientists are still debating whether these morphological features originally evolved for efficient food collection (i.e., hang-feeding and climbing on small branches; Fleagle 1976) or for efficient travel between food sources (Temerin and Cant 1983). The small size and suspensory locomotion of gibbons permit them to move more easily and directly through the rain forest canopy than sympatric orangutans and macaques (Temerin and Cant 1983; Grand 1984).

Most scientists now recognize the family Hylobatidae, the genus *Hylobates*, three subgenera—*Hylobates, Symphalangus* (*H. syndactylus*), and *Nomascus* (*H. concolor*)—and nine species of gibbons, but gibbon taxonomy remains a dynamic field. Some prefer to recognize only five gibbon species, and differences between *H. hoolock* and other *Hylobates* may warrant recognition of a fourth subgenus (Prouty et al. 1983). Disagreement also remains over the taxonomic implications of interbreeding between gibbon species (e.g., *moloch* and *syndactylus:* Myers and Shafer 1979), and classifications are still challenged by discoveries of populations with separate features that are characteristic of different species. The gibbons of southwestern Borneo, for example, sing like *H. agilis,* but have morphological features identical to *H. muelleri* (Groves 1984).

Gibbons are restricted to tropical evergreen and less seasonal parts of semievergreen rain forests in Asia. *H. agilis* occur in Sumatra, West Malaysia, southwestern Borneo, and southern Thailand; *concolor* in Vietnam, Laos, Cambodia, and southern China; *hoolock* in Assam, Bangladesh, and Burma; *klossii* in the Mentawai Islands west of Sumatra; *lar* in Thailand, peninsular Malaysia, and northern Sumatra; *moloch* in western Java; *muelleri* in all of Borneo but the southwestern corner; *pileatus* in southeastern Thailand and western Cambodia; and *syndactylus* in Sumatra and peninsular Malaysia. The siamang is sympatric with *lar* and *agilis.* All other species are allopatric, except for slight overlap between *lar* and *pileatus* in Thailand, between *lar* and *agilis* in West Malaysia, and between *agilis* and *muelleri* in Borneo. The coexistence of siamangs with smaller gibbons is attributed to differences in feeding patterns associated with larger body size (MacKinnon 1977; Gittins and Raemaekers 1980).

ECOLOGY AND CONSERVATION

Compared to many other primates (e.g., baboons or langurs), the behavior and ecology of gibbons appear to be remarkably consistent both within and between species. However, although field studies have been completed on all gibbons except *H. concolor,* nearly all have involved intensive observations of only one gibbon group. Because of variation in individual groups over time (*syndactylus* and *lar:* Chivers and Raemaekers 1980) and in different groups of the same species (table 12-1; see also MacKinnon 1977), more studies on each species must be conducted before any conclusive statements about apparent interspecific differences can be made.

Gibbons are frugivores. They east mostly ripe, sugar-rich, juicy fruits (Gittins and Raemaekers 1980) and also large quantities of figs (*Ficus* fruits) (table 12-1). Fruit comprises about 60% and leaves 30% of the diet (table 12-1) for most of the smaller gibbons, but monthly proportions of fruit sometimes exceed 90% or fall below 30%. The larger siamang (*H. syndactylus*) is more folivorous, *klossii* and *pileatus* may be more frugivorous, and *klossii* apparently obtains protein primarily from insects instead of young leaves (Whitten 1984).

At two study sites in West Malaysia, siamang diets average slightly more leaves than fruit, but reach 60% fruit in some months (Chivers 1974; MacKinnon 1977). In the fig-rich forest at Ketambe, Sumatra, siamangs may be more consistently frugivorous (West 1981). Siamang and *lar* groups with overlapping ranges have diets more similar to each other than to neighboring groups of the same species, but these similarities belie important differences in feeding ecology (MacKinnon 1977; Gittins and Raemaekers 1980). Siamangs eat leaves at a faster rate and eat more leaves that are mature or nearly mature. They visit fewer food patches (i.e., trees or lianas) each day, have longer feeding bouts, travel less far between patches, and appear to rely more on larger-sized fruit patches (Gittins and Raemaekers 1980).

TABLE 12-1 Gibbon Group Size, Feeding, and Ranging

Species	Study Site	No. Mo. Sampled	Group Size Mean (N)	Group Density (gps/km²)	Home Range Size (ha) Mean (N)	Range
Hylobates agilis	Sungai Dal, W. Malaysia	10	4.4 (7)	4.3	29 (1)	
H. hoolock	Hollongapar, Assam, India	2	3.2 (17)	2.2	22 (7)	18–30
H. klossii	Sirimuri, Siberut Is., Indonesia		3.8 (15)[a]			7–12
	Paitan, Siberut Is., Indonesia	5	3.7 (10)	2.8	33 (3)	31–35
H. lar	Tanjong Triang, W. Malaysia	15	3.3 (4)	2	44 (4)	20–46
	Kuala Lompat, W. Malaysia	12	3.5 (2)	0.7	54 (2)	50–58
	Khao Yai, Thailand			6.5	16 (5)	
H. moloch	Ujong Kulon, Java, Indonesia	12		2.7	17 (6)	12–22
H. muelleri	Kutai, E. Kalimantan, Indonesia	15	3.4 (7)	3	36 (4)	33–43
H. pileatus	Khao Soi Dao, Thailand	4	3.7 (12)	6.4	36 (1)	
H. syndactylus	Ulu Sempan, W. Malaysia	3	4 (4)	1.5	18 (4)	15–25
	Kuala Lompat, W. Malaysia	13	5 (1)	1	34 (1)	
	Kuala Lompat, W. Malaysia	12	3.0 (2)	0.8	47	
	Ketambe, Sumatra, Indonesia	2	3.8 (5)	1.8	35 (1)	

a. Revised estimate of Tilson 1981, in Brockelman and Srikosamatara 1984.

b. Whitten's estimate includes foraging time that was not spent traveling.

Gibbons have fixed home ranges that average about 34 hectares (table 12-1), with exclusive portions (territories) averaging 75% of this area. Siamang home ranges are smaller than those of sympatric *lar* groups, but similar in size to those of many other gibbon species. Presumably, the siamang's more folivorous diet permits smaller territories than expected for its larger body size (Gittins and Raemaekers 1980). Within species, home range size and gibbon densities can vary as much as fivefold between study sites (e.g., *klossii* and *lar:* table 12-1). There is no single, obvious explanation for these differences, although Marsh and Wilson (1981b), surveying 11 sites in peninsular Malaysia, found tree family diversity to be the best indicator of *lar* densities.

Muelleri, pileatus, and *syndactylus* travel about 800 to 900 m a day; all other species average 1,000 to 1,500 m (table 12-1). *Agilis* often take only 10 to 15 minutes to cross from one side of their territory to the other (Gittins 1980). Gibbons use their ranges unevenly, but tend to visit most sectors every few days. *Lar* and *agilis* normally take 2 to 3 days to cover their ranges completely, while *syndactylus* take 6 days (Gittins and Raemaekers 1980). Food is an important determinant of range use (Ellefson 1974). There is no strong evidence for patrolling behavior in gibbons (Gittins 1980), and boundary disputes normally do not affect ranging patterns in *agilis* (Gittins 1980). However, simulated intrusions using playbacks of gibbon duets have elicited immediate, temporary monitoring of the territory by siamangs (Chivers and MacKinnon 1977) and, less explicitly, by *muelleri* (Mitani 1985a).

Gibbons are active from 8 to 10 hours a day on average. Activity usually starts at dawn and stops well before sunset. Adult males and offspring become active sooner and often stay active later than females. Singing and the most intense feeding occurs in the morning, but compared to most other primates, gibbons show little change in activity over the day. Gibbons spend most of their day foraging in the main canopy (ca. 20 to 35 m high in primary forest) (fig. 12-1). Emergent trees are used primarily to rest, sleep, and sing. Little time is spent in the lower canopy, and this mostly to visit small food trees (Gittins and Raemaekers 1980).

Gibbon populations are currently threatened most by loss of habitat to cultivation and commercial logging. Gibbon densities may decline minimally and temporarily where few trees are extracted from a forest (Johns 1981), but in most timber operations, logging is intensive and appears to increase forest destruction caused by agricultural encroachment, flooding, wind storms, and forest fires. *Pileatus, concolor, moloch,* and possibly *hoolock* are currently endangered. Their future depends on the maintenance of the forest reserves to which they are not primarily or entirely restricted (see conservation chapters in Preuschoft et al. 1984; see also chapter 39).

REPRODUCTION

The limited demographic data available on gibbons show no birth seasonality in most study areas, and most samples are too small to determine the existence of birth peaks. In some studies peaks appear unlikely or at most weak (*lar:* Carpenter 1940; *klossii:* Tenaza 1975; Tilson 1981), but in more seasonal environments birth peaks may exist. All of three *hoolock* births observed by Tilson (1979) at Hollongapar Forest Reserve, Assam, India, occurred within a 2.5-month period (mid November to late

TABLE 12-1 *(continued)*

Day Range (m)		Monthly Dietary Proportions (mean and range)					
Mean	Range	Fruit	Fig	Leaves	Flowers	Insects	Sources
1,217	650–2,200	58 38–100	17	39 0–60	3 0–30	1 0–2	Gittins and Raemaekers 1980; Whitten 1984
		67		32	(leaves + flowers)	0	Tilson 1979; Chivers 1984 Chivers 1984
1,514	885–2,150	70	23	2	0	25[b]	Whitten 1984
	ca. 1 mile	67		33			Ellefson 1968a
1,490	450–2,900	50 36–60	22	29 14–53	7 3–8	13 6–24	Gittins and Raemaekers 1980; Whitten 1984 Chivers 1984
1,400		61 49–68		38 30–50	1 0–2	0	Kappeler 1984a; Chivers 1984
	850 350–1,890	62 27–90	24	32 8–73	4 0–16	2 0–6	Leighton, unpub.; Chivers 1984
	833 450–1,350	71	26	13	15	1	Srikosamatara 1984
	778 485–1,390	47 37–59	41	50 38–59	2 0–3	1 0–1	Chivers 1974
	969 320–2,860	32 25–36	24	58 16–61	9 5–15	2 0–4	Ibid.
	738 200–1,700	36 21–49	22	43 19–69	6 0–23	15 8–20	Gittins and Raemakers 1980
	933 150–1,550	59 56–61	42	24 14–33	4 1–6	2[c]	West 1981

c. An additional 11% of feeding time was spent searching (for insects) in leaves.

January), which supports McCann's (1933) earlier observation of a possible *hoolock* winter birth peak. At Kuala Lompat, Malaysia, three of four births for one female siamang and both of two *lar* births occurred between November and January (Chivers and Raemaekers 1980). Tilson (1979) and Chivers and Raemaekers (1980) speculate that birth peaks in gibbons may be timed to synchronize the commencement of weaning with periods of high fruit availability and may result from most conceptions occurring in seasons of high fruit abundance, but the data are insufficient to test this hypothesis.

Male and female gibbons mature at similar rates and appear to pair at 8 to 10 years of age (table 12-2). Females first menstruate at around 8 years of age (Carpenter 1940). They do not have sexual swellings; however, Carpenter (*lar:* 1940) and Chivers (*syndactylus:* 1974) both describe changes, presumably associated with ovulation, in the color and turgidity of female genitalia. In the typical 1-to-2-year field study, sexual behavior is not observed. Females cycle for only a few months (ca. 5), at intervals of 2 or more years between births (Ellefson 1974), and copulations during these periods of sexual activity occur at a rate of less than once a day (*lar:* Ellefson 1974; *syndactylus:* Chivers and Raemaekers 1980).

Few matings have been described. Carpenter (1940) observed that female *lar* were more aggressive than males in soliciting copulations, sometimes crouching before the male and pulling him toward themselves, but Ellefson (1974) found sexual solicitation by *lar* females to be rare. These observations as well as some on *klossii* (Tilson 1981), *muelleri* (Leighton, unpub.), and *syndactylus* (Chivers 1974) indicate that it is more common for

the male to initiate matings. If the female is willing to be mounted, she will bend forward to accommodate him; otherwise, she will ignore the male or move away.

SOCIAL ORGANIZATION

Gibbons are invariably monogamous and territorial, and defend their territories through regular loud morning songs and occasional encounters with neighbors and intruders. Outside of these intergroup activities, gibbon families lead a relatively subdued social life. In addition to the adult pair, a gibbon group potentially includes one infant (0 to 2 or 2.5 yr), one juvenile (2 to 4 yr; defined by locomotion independent of the mother), one adolescent (4 to 6 yr; not yet adult sized), and one subadult (6+ yr; fully grown, but unmated). Mean group size is normally about four (table 12-1). A mated pair produces an average of five to six offspring over a reproductive lifetime of 10 to 20 years (Carpenter 1940; Tilson 1981). Occasionally, a postreproductive adult is observed accompanying a gibbon family, and, in areas where *lar* and *pileatus* meet, groups containing two adult females of different species have been seen (Brockelman and Srikosamatara 1984). No stable groups have been observed with more than one female carrying an infant.

Given that gibbons rest for 20% (*lar:* Gittins and Raemaekers 1980) to 51% (*klossii:* Whitten 1984) of the period between their leaving and entering sleeping sites (the activity period), and sometimes sit for hours in the late afternoon before falling asleep, they spend surprisingly little time socializing with one another. Singing takes up about 4% of the activity period, grooming and social play usually less (table 12-3). The lack of social partners may contribute to such low levels of interaction.

FIGURE 12-1. A whitehanded gibbon (*H. lar*) feeds on the terminal branches of a tree. (Photo: David Chivers/ Anthrophoto)

This seems particularly likely for older infants and juveniles, which, compared to the young of most other primates, have no same-aged playmates and spend a negligible part of their day in social play (e.g., 2% of a young *lar*'s activity period: Ellefson 1974). Intragroup interactions increase in frequency, intensity, and aggressiveness during periods of sexual activity and as offspring near maturity and themselves show interest in mating and territorial behavior.

Female gibbons weigh nearly the same as males and have similar-sized canines. *H. concolor, hoolock,* and *pileatus* have sexually dimorphic coat colors, and all species but *hoolock* have sexually dimorphic vocal repertoires (Chivers 1977; Haimoff 1984). Males and females without sex-specific coloration are hard to distinguish by sight if the female is not carrying an infant or if, as is often the case, the observer cannot get a close and clear enough view to spot enlarged nipples or parapenal hair tufts.

Sexual monomorphism in gibbons is correlated with a comparatively high degree of behavioral and social equality between adult males and females. Partners in long-established pairs normally interact in a relaxed, tolerant,

and well-coordinated manner. In contrast to many species of primates, female gibbons are often "codominant" (Carpenter 1940) with males (*lar, agilis,* and *syndactylus:* Gittins and Raemaekers 1980). In one of two *lar* groups observed by Ellefson (1974), the male was dominant to the female; in the other dominance varied unpredictably until, in late pregnancy and early postpartum, the female became dominant to the male in feeding contexts. In two other species (*pileatus:* Srikosamatara 1980; *muelleri:* Leighton, unpub.), the female of the one group studied was more aggressive than the male, groomed him much less frequently than he groomed her, and, for *muelleri* at least, successfully displaced him at preferred feeding sites.

As with dominance, for gibbons there are relatively few clear, consistent differences in the social roles assumed by either sex. Several observers have suggested that the adult male is the principal protector against human predators (*klossii:* Tenaza 1976; *lar:* Carpenter 1940; *muelleri:* Leighton, unpub.) and the more active aggressor in intergroup encounters, but other roles are less predictably assumed by one sex or the other. In some pairs, the adult female almost always travels first in

TABLE 12-2 Gibbon Reproductive Parameters

Species	Birth Seasonality (N)	Age at Maturity (years)	Gestation Length (months)[a]	Interbirth Interval (months) Mean (N)	Ext. Signs of Ovulation	Study Area	Sources
Hylobates hoolock	Peak? 3					Hollongapor, Assam	Tilson 1979
H. klossii	None 8			40 (6)		Sirimuri, Siberut Is.	Tilson 1981
H. lar	None	8–10	7		See text	Soi Dao, Thailand	Carpenter 1940
H. muelleri	None? 5			Ca. 36 (4)[b]		Kutai, E. Kalimantan	Leighton, unpub.
H. syndactylus	Peak? 4+		7.5	42 (2)	Slight	Kuala Lompat, W. Malaysia	Chivers 1974;
				Ca. 36 (3)[c]	Color change[d]		Chivers and Raemaekers 1980

a. Ardito 1976.
b. Includes one known interval (33 months) + 3 intervals known to within ± 3 months.

c. Includes one infant that died.
d. Unclear if changes are associated with ovulation or menstruation.

TABLE 12-3 Gibbon Singing, Grooming, and Play

Species	Immatures in Group Studied	Percentage of Activity Period[a]			Sources
		Sing	Groom	Play[b]	
Hylobates agilis	A, I	5	<1	<1	Gittins and Raemaekers 1980
H. klossii	J	4	0	0	Whitten 1984
H. lar	S, A	3	3	<1	Gittins and Raemaekers 1980
H. muelleri	I	4	<1	<1	Leighton, unpub.
H. pileatus	A, J	4	5	3	Srikosamatara 1984
H. syndactylus	S, J	1	1	1	Gittins and Raemaekers 1980

NOTES: S = subadult; A = adolescent; J = juvenile; I = infant.

a. Most of these percentages represent the average for group members, with immatures generally underrepresented. The *klossii* data are for the adult male only, and the *muelleri* data are for the adults only. Individual percentages of time spent in social grooming or play for immatures other than infants (available in several theses) are slightly

higher than those of adults in some of these groups, but do not exceed the averages given by more than 1 or 2%.

b. Though most studies do not state whether these percentages exclude self-grooming and lone play, the percentages for *muelleri* do, and those for the other species are presumably accurate or slightly high for the social form of these activities.

group progressions (*klossii:* Tenaza 1975; *syndactylus:* Chivers 1974). In others, the male tends to lead (*lar:* Carpenter 1940), but travel order often varies greatly within a group. Similarly, in at least three studies (*muelleri:* Leighton, unpub.; *pileatus:* Srikosamatara 1980; *syndactylus:* Chivers 1974), adult males spent more time than their mates playing and grooming with their young, but too few groups have been observed intensively to determine if this pattern is common. Siamang males appear to be especially active fathers. They are the only gibbon males that help carry infants, and they appear to be more active than the females in both socializing and eventually evicting offspring (Chivers 1974).

BEHAVIORS MAINTAINING TERRITORIALITY IN GIBBONS

Territoriality in gibbons is routinely maintained through loud morning song bouts and ritualized encounters with neighbors at range boundaries. Exclusive spacing is also enforced by aggression toward solitary intruders and by gradual exclusion of maturing offspring from the territory. Because behaviors such as calling or chasing of nonresidents simultaneously restrict access to a gibbon's home range and to its mate, it is not clear whether they function primarily to defend mates or space (Brockelman and Srikosamatara 1984; Raemaekers, Raemaekers, and Haimoff 1984).

Morning songs

Gibbons have species-specific and, except for *hoolock,* sex-specific loud songs that can be heard clearly at a distance of up to 1 km. In all species but *concolor, hoolock,* and *syndactylus,* adult males sing long solos, usually in the darkness just prior to sunrise. The solos of unmated males tend to be longer and temporally separated from those of mated males, and apparently function to attract unmated females (*klossii:* Tilson 1981; *agilis, lar,* and *syndactylus:* Gittins and Raemaekers 1980).

In all species but *klossii* (and possibly *moloch:* Haimoff 1984; Kappeler 1984b), the female's song is sung as a duet with the male. Solitary, transient females almost never call, and widowed females tend to sing irregularly if at all. The female's loud song, therefore, seems to advertize her paired status as well as her presence on a territory. Gibbon pairs normally duet anytime between dawn and ten o'clock in the morning. Duets begin spontaneously or in response to calls by neighboring groups. They last about 15 minutes on average and include several "great calls" by the female, produced as she moves quickly about the crown of a tall tree. Great calls are introduced by and interspersed with exchanges of notes between the male and female. Different species show varying degrees of male participation in duets (Haimoff 1984). Other family members may also contribute vocally and behaviorally. Infants often squeal during a mother's great call, and juveniles may join duets with im-

perfect, sex-appropriate sounds. As they pass through adolescence, male offspring seem to be increasingly inhibited from singing with their parents. While subadult females usually give great calls simultaneously with their mothers, subadult males (except for siamangs) rarely sing in the presence of the adult male.

Most species sing duets almost daily. However, duets by siamang, like the solo songs of *klossii,* occur less frequently (ca. 25% of days versus 70 to 90% for other species; see references in table 12-1). Singing is less frequent on rainy or windy days, or following rainy nights, and is positively correlated with fruit abundance (*klossii:* Tenaza 1976; A. J. Whitten 1982; *lar* and *syndactylus:* Chivers and Raemaekers 1980), mating activity (Chivers and Raemaekers 1980), and territorial expansion (*klossii:* Tilson 1981). Song duration also increases with intruder pressure (*muelleri:* Mitani 1985a).

Recent playback experiments by Mitani (*muelleri:* 1984, 1985a, 1985c) and Raemaekers and Raemaekers (*lar:* 1985) have helped to elucidate the role of singing in mate and territory defense and to clarify the separate roles of male and female gibbons in maintaining monogamy. All subject groups in playback experiments responded frequently and aggressively to songs played from range centers, but the behaviors observed varied between species and with the sex of the "caller" and respondent. Female *muelleri* initiated duets and led joint approaches in response to playbacks of female solos or duets, while male *muelleri* led silent, joint approaches to male solos. In *lar,* females initiated almost all vocal responses to playbacks, but called consistently only in reaction to female solos and sometimes did not join their mate's approaches toward a male solo or duet (Raemaekers and Raemaekers 1985). Male *lar* led or co-led approaches to all playbacks except two of eight female solos. In a nonexperimental study, Kappeler (1984b) found that female *moloch* called alone in response to female solos and in all of four actual intrusions observed. Apparently, male *moloch* approach intruding pairs alone while their mates sing.

Playback experiments have confirmed the importance of intrasexual aggression as a proximate factor maintaining monogamy in gibbons. Furthermore, they suggest that intersexual aggression also helps enforce this mating system, at least in some species. However, since unmated, transient females generally do not sing, it is possible that a male gibbon interprets a female solo playback as an intrusion by a pair and is actually responding to the caller's silent mate. Nevertheless, one might expect a mated gibbon to resist the displacement of a proven mate by a stranger.

Intergroup Encounters

In his early account of territorial behavior in *lar,* Ellefson (1974) reported that 1-hour-long intergroup conflicts occurred almost daily and "occupied a substantial portion of a male gibbon's life" (p. 144). More recent studies show that intergroup conflicts are not always so frequent or time-consuming for males. Rates of intergroup encounters vary from once every two days in *lar* (Ellefson 1974) and *agilis* (Gittins 1980) to as low as once a month in *muelleri* (Mitani 1985a; Leighton, unpub.). In all species many encounters involve only duetting. Otherwise, opposing males typically sit or hang and stare at each other from the exposed limbs of tall trees and occasionally vocalize or chase each other. Male vocalizations in this context vary between species, from occasional soft "hoos" (*muelleri:* Leighton, unpub.) to loud, nearly continuous screaming (*moloch:* Kappeler 1984b). Fighting is rare. In 126 intergroup encounters involving *lar,* Ellefson (1974) observed fighting only once and suspected fighting in only 10 other cases. The incidence of facial scars, broken canines, and wounds on male gibbons also varies between sites, from none in *agilis* (Gittins 1980) to half of the males observed in *lar* (Ellefson 1974). There is consistently less evidence of fighting injuries among females than among males (Ellefson 1974). These observations suggest that male gibbons may enhance their infants' survival and the reproductive effort of their mates by assuming the riskier and more energetically demanding aggressive role of territorial encounters.

Female participation in border disputes is normally limited to calling. *Moloch* females, as the extreme, scream or great call throughout an encounter (Kappeler 1984b). In less stable social circumstances, however, females may become more aggressive. Newly paired females and females with weakened partners have been observed to chase adults of both sexes (e.g., *lar:* Chivers and Raemaekers 1980), and widowed females (*pileatus:* Brockelman and Srikosamatara 1984; *klossii:* Tilson 1981) have been seen to defend their territories both vocally and physically against solitary males and intruding neighbors.

Subadult and adolescent males often participate in intergroup conflicts and join their fathers in chasing a neighboring male (*syndactylus:* Chivers 1974; *lar:* Ellefson 1974; *agilis:* Gittins 1980; *klossii:* Tilson 1981). Indeed in two *lar* groups, a subadult male gradually took over the male role in territory defense from an aging father (Chivers and Raemaekers 1980; Raemaekers and Raemaekers, in press 1985). Similarly, one immature *pileatus* female was observed to defend her mother against an aggressive lone female, and a second immature female aided her parents in challenging a neighboring widow (Brockelman and Srikosamatara 1984). Though gibbon pairs can normally defend their territory without a subadult's assistance, a subadult's extra strength undoubtedly helps discourage intrusions and may occasionally be necessary to prevent a takeover (e.g., *lar:* Chivers and Raemaekers 1980). Furthermore, it may be that subadults have a better chance of obtaining a territory with their family's support than by themselves.

DISPERSAL AND FAMILY FORMATION

Since gibbon groups contain only one paired male and female, to become breeders maturing animals must either form new groups in usurped or unoccupied areas or replace or displace a same-sexed adult in their own or another group. Beginning when they are juveniles, young gibbons experience increasing adult aggression, most often in feeding contexts and primarily from their same sexed parent. Emigration from the natal territory seems to result at least in part from a subadult's desire to obtain a mate and territory of its own (Ellefson 1974), but events coinciding with the emigration of some maturing offspring suggest that parents also sometimes evict offspring that become potential mate or food competitors. For example, 2 weeks following the first observed sexual interactions between a subadult female *klossii* and her father, aggression between the female and her mother increased from 3.5 to 26.6 bouts per day (Tilson 1981). In *muelleri,* mother-daughter aggression increased with the commencement of mating between a female and her father and is suspected to have resulted in the subadult's departure within 3 days of her brother's birth (Leighton, unpub.). Similarly, a male siamang that was resuming sexual activity with his mate simultaneously became more antagonistic toward his subadult son (Chivers and Raemaekers 1980).

That optimal habitat is saturated with breeding pairs in most gibbon populations can be inferred from (1) the absence of unoccupied suitable forest, (2) the presence of transient, solitary animals (Chivers and Raemaekers 1980; Tilson 1981), (3) the quick invasion of territories following the disappearance of resident adults (Whitten 1980; Tilson 1981; Brockelman and Srikosamatara 1984), and (4) the readiness of transient females to join groups when tolerated by the adult female (Brockelman and Srikosamatara 1984).

Though more long-term studies are needed to determine rates of territory turnover in undisturbed gibbon populations, the apparently long tenure of territory occupants probably causes openings to occur infrequently. One siamang female, for example, is known to have held her breeding position for over 16 years (Chivers and Raemaekers 1980), and at least three *muelleri* pairs at one Bornean site remained stable for over 7 years (Leighton, unpub.). In contrast, the relatively frequent availability of mates and territories observed in studies of *klossii* were apparently due to human predation (Tenaza 1975; Tilson 1981).

If suitable breeding habitats are usually saturated with long-lived gibbon pairs, it is likely that mortality rates are high among dispersing subadults. Although no direct evidence supports this assumption, fewer solitary animals are usually observed than might be expected from the number and apparently high survival rate of infants and juveniles. Subadults faced with high dispersal risks and saturated breeding habitats might therefore be expected to attempt to remain in their natal group until a breeding opportunity arose. The feeding or mating costs to parents of a subadult's delayed dispersal may be partially offset by the fact that parents and subadults appear to aid each other in contesting or maintaining territories.

Current observations of dispersal in gibbons suggest that most individuals attempt to settle in or near their natal territory, but the proportion that successfully does so is unknown (table 12-4). Some observers have suggested that females tend to remain longer than males in or near their natal territories and that they then attract transient males or join local, unmated, territorial males (Ellefson 1974; Tilson 1981; Brockelman and Srikosamatara 1984). In support of this suggestion, transient males are often reported to be more numerous than transient females (e.g., *klossii:* Tenaza 1975), and, unlike solitary males, solitary subadult females do not call and have not been observed to defend a territory.

However, if males are more likely than females to establish a breeding area and then attract a mate, they too might benefit from delayed dispersal to a familiar area. In addition to the benefits of settling near kin, who are presumably less hostile (Waser and Jones 1983), subadult gibbons also potentially benefit from familial assistance in obtaining territories. Tilson (1981) describes three cases in which *klossii* groups temporarily expanded their ranges and subsequently left their offspring either alone (two males) or with a mate (one female) in the new area. Such parental aid in the establishment of an offspring's territory is similar to the "territorial budding" in the monogamous Florida scrub jay (Woolfenden and Fitzpatrick 1978) and may make territorial defense by more than one individual imperative.

Of the four subadults that were known to form new groups in Tilson's study, three received some form of parental assistance, but this proportion may be higher than in other populations (table 12-4). Even in long-term studies, the sample of observations on gibbon dispersal and family formation is very small and is complicated by (1) the unknown fates of subadults that emigrated from study areas, (2) the inferred identies of some study subjects (Chivers and Raemaekers 1980), (3) the influence of hunting on many study groups (Whitten 1980; Tilson 1981; Brockelman and Srikosamatara 1984), and (4) the presence of cleared land, poor forest, or large rivers, limiting dispersal within and beyond many study areas (Chivers and Raemaekers 1980).

Although the precise importance of parental assistance will probably remain unclear for a long time to come, its success no doubt varies with the relative strength of a given gibbon group and the subadult's ability to defend its new territory alone or with a mate. Indeed, one of the males assisted by his parents in Tilson's (1981) study subsequently failed to retain his territory. Variation in individual and group dominance relations as well as in local demography and territory quality presumably restricts the number of subadults that become breeders in this way, and there may be a number of different means by which family groups are formed. Carpenter (1940) and Ellefson (1974) have hypothesized that very limited breeding opportunities could make incest the easiest and thus the most common mode of obtaining mates and territories.

Tilson (1981) observed one definite and two possible cases in which *klossii* males inherited their natal territories by pairing with their widowed mothers (table 12-4). Mother-son pairs have also been observed in one group of siamangs (Chivers and Raemaekers 1980) and one group of *lar* (Raemaekers and Raemaekers, in press), but neither of these pairs lasted or, apparently, bred. The siamang son was displaced by a solitary male, and the *lar* son left to displace a neighboring male. In another *lar* study, a son returned to his aging parent's territory and was observed to duet with his sister, which indicated that he might eventually inherit the territory and perhaps mate with his sister (Chivers and Raemaekers 1980).

While data on family formation in gibbons indicate that incestuous pairs may be common, they probably usually involve mates of widely disparate ages whose pair-bond does not last long enough to result in much inbreeding (Tilson 1981). However, matings between kin may also be high if neighboring groups tend to be closely related, as has been suggested by several field studies (table 12-4). Although it seems reasonable to assume that animal species with characteristically severe competition for breeding positions may incur modest levels of inbreeding (Moore and Ali 1984), it is not yet known if monogamous, territorial primates are more inbred and show less inbreeding avoidance than do polygynous primates (see chapter 21).

TERRITORIALITY AND MONOGAMY ♦

Only 3% of mammals and 14% of primates are monogamous (Rutberg 1983). Some species (e.g., *Hylobates* spp.) are invariably monogamous, while others (e.g., *Simias concolor*) adopt this mating system only under specific environmental conditions (see Kleiman 1981). Monogamy is maintained either through the exclusion of reproductive adults from breeding groups (e.g., *Hylobates, Aotus,* and *Callicebus* spp.) or through the reproductive suppression of subordinate group members (Callitrichidae: see chap. 4).

Monogamy in primates is hypothesized to result from the inability of males to monopolize more than one breeding female (Rutberg 1983; van Schaik and van Hooff 1983). They are prevented from doing so by (1) the solitary nature, low density, and even dispersion of females in suitable breeding habitats, and/or (2) stringent requirements for male parental care (Kleiman 1981). For gibbons both conditions are thought to be important, with males obligated to mate monogamously as a result of the female's need for aid in territorial defense and, in the case of siamangs, perhaps also for infant carrying. However, the role of male territorial defense and paternal care in the evolution of monogamy in gibbons and several other primate species remains unclear (*Hylobates* spp.: Brockelman and Srikosamatara 1984; *Aotus* and *Callicebus* spp.: Wright 1984; Callitrichidae: chap. 4). For example, unless there are ecological constraints on group size, a male's territorial efforts could, in theory, be shared by several females, and nonbreeding adults could also accompany and help carry and protect the infants of females mated with polygynous males. Thus gibbon monogamy and territoriality should probably be considered in terms of the ecological factors that underly female dispersion patterns.

Explanations of female dispersal in nonhuman primates have concentrated on predation and food supply. The relative importance of these selective factors appears to vary greatly, however, even among monogamous species. Current data and hypotheses on the influence of each of these factors in the evolution of gibbon monogamy are examined below.

Role of Predation

Compared to the small arboreal primates of Africa and South America, gibbons appear to have few natural preda-

tors. Gibbons usually travel and sleep too high in the canopy to encounter leopards or tigers, and safe sleeping trees appear to be abundant (Chivers 1984; but see Tenaza 1975). The small size of gibbon groups and the often variable spatial distribution of individuals within groups do not seem well adapted for defense against the gibbons' most likely natural predators—raptors or felids. Although young gibbons are probably small enough to be taken by large raptors, their vulnerability could probably be reduced by larger group size (van Schaik and van Hooff 1983). Furthermore, while group members often sleep together in the same tree, they may also sleep in trees that are separated by over 100 m, and thus show no standard grouping response to potential nocturnal predators like felids or pythons. Finally, although human predation has been hypothesized to have influenced group size in *H. klossii*, the evidence that human hunting has constrained the evolution of polygyny in this and two other Mentawaian primate species (*Simias concolor* and *Presbytis potenziani*) is equivocal (see van Schaik and van Hooff 1983; chap. 19).

Role of Food Resources

Most recent theories on the evolution of monogamy in primates have tended to link selection for monogamy with selection for small group size and territoriality and have argued that these are largely determined by forag-

TABLE 12-4   Gibbon Dispersal and Family Formation Patterns

Species	Attempted Pattern of Family Formation						Fate of Subadults by Pattern of Dispersal and Family Formation								Sources
	L	A1	A2	A3	S1	S2	A1	A2	A3	S1	U1	U2	U3	U4	
Hylobates klossii		2/1					1/1						1/0		Tilson 1981
			3/1			1/0[a]		0/1			2/0	1/0			
					1/0				1/0						
	1/0		0/1										1/0	0/1[b]	Whitten 1980
H. lar	2/0			1/0[c]				1/0					2/0		Raemaekers, Raemaekers, and Haimoff 1984 Raemackers and Raemaekers, in press
	1/0					1/1[d]						1/1	1/0		Chivers and Raemaekers 1980
H. muelleri	0/1													0/1	Leighton, unpub.
H. pileatus	1/0			0/1				0/1					1/0		Srikosomatara 1984
H. syndactylus	0/1[e]		2/0		1/0		2/0						0/1	1/0	Chivers and Raemaekers 1980
Total (Male/female)	5/2	2/1	5/2	1/1	2/0	2/1	1/1	2/1	1/1	1/0	2/0	2/1	6/1	1/2	

NOTES: L = leave the (study) area directly from the natal group; A = settle adjacent to the natal territory (1) with parental assistance, (2) without parental assistance, (3) level of assistance not clear from available information; S = attempt to inherit the natal territory by taking the place of a missing or aging parent (1) confirmed attempts, (2) apparent attempts; U = fate unknown, but at last report (1) remaining peripheral to natal territory, (2) on natal territory with opportunity to pair with mother/sib, (3) not seen again, (4) probably not in vicinity of natal territory. Numbers = males/females.

a. Individual is also included under A2.

b. Female left the area after the male she had recently paired with was shot.

c. Male displaced an adjacent male after 6 months of staying on the natal territory with his widowed mother.

d. Male and female are brother and sister.

e. Female was presumably evicted from her natal territory by a strange male that subsequently paired with her widowed mother (Chivers, pers. commun).

ing constraints on females (Wrangham 1979a; Gittins and Raemaekers 1980; Rutberg 1983; van Schaik and van Hooff 1983). It is hypothesized that the frequent use of small, scattered, and perhaps also long-lasting food patches augments the value of territory defense, thus allowing gibbons efficiently to maintain a diet of high-quality, easily digested food. The critical feature of these "small" patches is that they are large enough for one female and her offspring to use regularly, but small enough that if two females and their offspring visited the same patches, the dominant could not be assured a sufficient meal or high feeding rates. Any antipredator or other benefit that might derive from female sociality is therefore outweighed by the costs of feeding competition.

There are a number of ways in which the systematic exploitation of small, scattered fruit patches might contribute to small group size and territoriality in gibbons. For example, small fruit patches occur at higher densities than large patches and probably provide more evenly dispersed fruit supplies; this may enable female gibbons to subsist in relatively small home ranges. Furthermore, in the process of feeding on and monitoring small, scattered fruit patches, female gibbons may cross and cover their ranges at a sufficiently high rate to permit effective patrolling (Wrangham 1979a). Increased group size would presumably force females to feed in more patches each day and to search more often for newly available patches. It would probably also demand increased range size and more range overlap, which could result in less efficient exploitation and patrolling of patches. Small group size, territoriality, and the selective use of small, scattered fruit patches therefore appear to be mutually dependent.

The gibbons' relatively small body size apparently enables them to acquire good-sized meals from small food patches. Small body size is usually associated with a greater reliance on more nutritious and easily digestible food items (Jarman 1974). Perhaps as a result, gibbons appear to have a narrower range of potential fruit and seed types, and hence a lower density of food patches in general, than do sympatric orangutans and macaques, which feed selectively on larger patches (MacKinnon 1977; MacKinnon and MacKinnon 1980a; Leighton and Leighton, unpub.). Gibbons also typically occupy primary forest, where fruit patches tend to occur at lower densities than in riverine habitats, where the similarly sized *Macaca fascicularis* obtain much of their food (Rodman 1978; Wheatley 1980). Finally, because they are bracheators and capable of rapid arboreal travel, gibbons can move quickly between scattered food patches (Temerin and Cant 1983) and have the potential rapidly to confront intruders detected from a distance. Thus, they may be better adapted to defend and exploit the food resources in a small area than to track larger, more widely dispersed fruit.

Despite these theoretical considerations, available data on gibbon foraging behavior and food availability are not adequate to allow a direct test of the hypothesis that gibbons feed selectively on small, scattered fruit patches. Raemaekers's (1977: referenced in Gittins and Raemaekers 1980) study of feeding in *lar* and *syndactylus* fails to show that a high proportion of the fruit patches used by gibbons on any particular day are disproportionately small or at low density. Moreover, Raemaekers's indexes of patch size may not reflect the actual amount of harvestable nutrients within fruit patches, and his sample of available patch sizes includes unripe fruits, which, he states, are rarely eaten. Nevertheless, the hypothesis that the selective use of small, scattered food patches leads to gibbon monogamy is supported indirectly by evidence that other monogamous primates also appear to employ such foraging strategies (two *Saguinus* spp.: Terborgh 1983; *Callicebus moloch:* Wright 1984; *Tarsius banacus:* MacKinnon and MacKinnon 1984) and by the fact that sympatric polygynous species, such as macaques, rely more than gibbons on large, densely clumped fruit patches.

Whether monogamy with joint defense of a single territory is the most advantageous mating system for both males and females is difficult to determine from current data. Males may in fact be energetically capable of defending an area large enough to provide two females with a joint territory. However, the ubiquity of female-female hostility (Brockelman and Srikosamatara 1984), female territoriality (Tilson 1981), and the lack of male dominance (Carpenter 1940) all suggest that females may prevent males from adopting this strategy, probably because polygyny would reduce female fitness.

Similarly, it is not known why male gibbons' ranges do not overlap with the separate ranges of more than one female, as is the case in orangutans (chap. 13). It is possible, however, that such a mating strategy might depress the fitness of males as well as their mates. For instance, males that failed to travel with a mate and aid her exclusively in territorial defense might be more likely than monogamous males to be cuckolded. Their mates might also produce fewer offspring than monogamous females, either due to the energetic cost of defending a territory alone or because they could not compete successfully against mated pairs for access to food patches. Ellefson's (1974) observation that the relative dominance of neighboring males can determine the access of groups to contested fruiting trees and Tilson's (1981) observation of pairs helping their offspring to usurp neighboring territories both suggest that lone females would not be able to compete with females that had the continuous aid of a mate. Moreover, if territory inheritance and budding are important strategies for obtaining a territory, then offspring of males whose territories overlapped those of several females might not only be vulnerable to expansionist groups when young but also less likely to become breeders once mature.

In conclusion, although gibbon females no doubt benefit from male aid in territorial defense, it seems unlikely that it is the sole cause of monogamy in gibbons, if only because it is easy to imagine how females in polygynous species might also benefit from increased male aid. Thus, gibbon monogamy probably also depends ultimately on ecological factors that simultaneously select for small group size and territorial defense by females. Until more data become available, however, the precise role of ecological factors in influencing female dispersion patterns and monogamy will remain a subject of conjecture.

13 | Orangutans: Sexual Dimorphism in a Solitary Species

Peter S. Rodman and John C. Mitani

In 1965 when Irven DeVore published *Primate Behavior,* the orangutan was virtually unstudied (Schaller 1965). The most extensive account of the orangutan's natural habits appeared in Wallace's (1869) report of his collecting expedition through the Malay Archipelago, which included description of his work in the Kuching River area of Sarawak where he observed and collected numerous orangutans. The orangutan was known to be predominantly solitary, but prevailing opinion was that this pattern, which is rare among diurnal primates and which seemed bizarre for an animal so closely related to humans, would prove to be a consequence of heavy human influence. It was expected that the normal social pattern would turn out to be quite chimpanzeelike. In 1967 D. A. Horr (1975) initiated the first long-term study of orangutans in a natural habitat. He has been followed by J. R. MacKinnon (1974), P. S. Rodman (1973, 1977, 1979, 1984), B. M. F. Galdikas (1978, 1981, 1983), and J. C. Mitani (1985b, 1985d) in various sites in Borneo, and by H. Rijksen (1978) and C. Schurmann (1982) at Ketambe in north Sumatra. The study by Galdikas, which began in 1971 and is still in progress, provides the most extensive long-term information on this great ape. The many recent studies of orangutans provide much comparative data with which to consider their social system and its relation to their striking sexual dimorphism.

BACKGROUND

Orangutans occurred widely in East and Southeast Asia during the Pleistocene, but today they are limited to forests of northern Sumatra and Borneo. The single living species *Pongo pygmaeus* comprises two subspecies. In northern Sumatra, *P. p. abelii* inhabits an area bounded approximately by 2° and 5° north latitude and by 95° and 99° east longitude (Rijksen 1978). The dispersion of *P. p. pygmaeus* in Borneo is not well known, but orangutans occupy forests in all parts of that island. There are pockets of absence, probably as a result of human population growth or unsuitable forest.

The normal habitat of orangutans is the tropical rain forest typical of much of Southeast Asia. Emergent trees of the family Dipterocarpaceae dominate this forest, where orangutans use primarily the middle canopy 10 to 25 m above ground. Orangutans also occupy less typical forest such as the peat swamp forest at Tanjung Puting (central Kalimantan; Galdikas 1978).

Orangutans are the largest living arboreal mammals. At maturity, females weigh an average of 37 kg, and males an average of 83 kg (Rodman 1984). Unlike gorillas and chimpanzees, which have relatively short, black hair, orangutans have coarse, long, straight hair that ranges in color from near black to orange in shade, but which is best described as "red." Young animals wear white "spectacles" of pale skin around their eyes, which darken to the chocolate color of adult skin as they mature. Unlike chimpanzees, female orangutans have no external signs of ovulation, but they develop pale labial swellings during pregnancy. Adult males have several intriguing secondary sexual characteristics that are described later in this chapter.

Orangutans contrast morphologically in important ways with the other apes. The anterior dentition differs sharply from that of the African apes in the large size of the central upper incisors and the small size of the lateral upper incisors. Their molar teeth have thickened enamel like human molars, unlike those of chimpanzees and gorillas. These conditions resemble closely the dentition of Miocene hominoids of Africa and Asia classified as *Sivapithecus* and *Ramapithecus* (Pilbeam 1984). In postcranial structure, orangutans have a highly mobile hip and a more fully opposable big toe than is found in the African apes. These latter differences reflect the orangutan's highly arboreal locomotion as compared with the more terrestrial habits of chimpanzees and particularly gorillas.

Broad/Gross Ecological Niche

Activity Patterns. In eastern Kalimantan orangutans awaken at approximately 0630 hours and retire (normally on a sleeping platform constructed by bending branches and leaves together) at approximately 1800 hours (Rodman 1979). During this 11.5 hour day, they spend an average of 46% of the time feeding, 39% resting, 11% traveling between feeding and resting sites, and the last waking minutes constructing the sleeping platform (Rodman 1979). Individuals travel 50 to 1,000 m each day, with only occasional longer day ranges. This distance is consistent among study areas, though the central tendency within this range varies from approximately 300 m at the Mentoko (Rodman 1977) to approximately 800 m at Tanjung Puting (Galdikas 1978).

Females travel nearly exclusively through the middle canopy of the forest, where their improvisational arboreal locomotion is best described as "quadrumanual clambering." Adult males are also arboreal, but all observers note that when they travel longer distances adult males normally come to the ground. On the ground, orangutans "fist walk," resting on curled toes and fingers in a ponderous quadrupedal pattern; occasionally they "crutch walk," drawing the hind limbs through the supporting forelimbs. Both arboreal and terrestrial locomotion are slow; their average speed in all travel, arboreal and terrestrial, is about 0.35 km/hr, which is approximately one-tenth the average speed of chimpanzees (Rodman 1984). Their slow speed on the ground seems best explained by a primary adaptation to arboreal life at the expense of quadrupedal agility. Their slow arboreal travel, in spite of primary adaptation for the arboreal milieu, is best explained as a consequence of their large body size. Although they are fully as arboreal as gibbons in all other respects, large size limits their agility and requires relatively greater caution in choosing supports and pathways.

Diets. Orangutans feed predominantly on fruit, though they also feed rather heavily on leaves and bark, with a small part of their feeding time devoted to ants and termites. This diet is similar in proportions of fruit and leaves to the diet of chimpanzees in East Africa (Rodman 1984). All field studies have shown the same dietary pattern in orangutans, with small variations dependent on local conditions. For example, the forest at Ketambe (North Sumatra; Rijksen 1978) contains a higher proportion of large fruiting fig trees, and orangutans there feed more heavily on figs than elsewhere. Figs are rare in the peat swamp forest of Tanjung Puting (Galdikas 1978), where they constitute only a small part of the orangutans' diet. Leaves form a variable proportion of the diet. Feeding time on leaves varies inversely with feeding time on fruit, and feeding time on fruit varies directly with the availability of suitable fruit, which suggests that orangutans prefer fruit when it is available. The orangutans' strength and dental and manual dexterity also permit them to feed on the bark of some dipterocarp species, a food resource that is not exploited by other forest animals, when fruit is scarce. The inner cambium of the bark is gnawed away with the incisors, and the fibrous mass is chewed and discarded. Orangutans also feed on some seeds, such as the unripe seeds of durian fruits and acorns of several oak species.

Population Densities

Population densities of orangutans vary from a low of 1 to 2 per km² in the upper Segama River area in 1969–70 (MacKinnon 1974) to approximately 5 per km² at Ketambe in 1973–76 (Rijksen 1978). Galdikas (1978) estimated density at 2 per km² in the Tanjung Puting study area, Horr (1975) estimated a similar density in the Lokan area in 1967–69, and Rodman (1973) reported a density of 3 per km² at the Mentoko site in 1970–71. The latter figure was approximately the same from 1975 through 1983.

Conservation Status

Although the orangutan population is larger than was estimated in early reports, the last 14 years have seen dramatic destruction of orangutan habitat. The Lokan and Segama study sites of Horr and MacKinnon have been logged, and one-third to one-half of the Kutai Reserve (Mentoko) has been logged since 1970. Forests surrounding these areas have been treated at least as harshly. In northern Sumatra the area occupied by orangutans has declined dramatically since the 1930s (Rijksen 1978). This decline has resulted from the expansion of the human population and from timber extraction, which has accelerated the decline of the orangutan's habitat.

The prognosis for orangutan populations, as for all tropical forest animals, is poor at best. Economic development will continue to remove habitat, and human populations in Borneo and Sumatra will continue to expand. Large national parks such as the Gunung Loeser Park of north Sumatra will shelter natural populations to the extent that they are carefully protected, but the long-term effects of removal of surrounding forests may ultimately alter even these refuges irrevocably.

Reproductive Parameters

Long life spans, long interbirth intervals, and short field studies relative to interbirth intervals have limited observations of mating and reproduction in orangutans despite long years of research. The age at which wild orangutans reach sexual maturity is unknown. Galdikas (1981) estimated that one female first gave birth at age 15. This estimate depends on the accuracy of her estimate of the age of the female at the beginning of the study, and her report of that estimate does not explain how she arrived at it (Galdikas 1981). In captivity, female orangutans have given birth as early as 7 years of age (D. M. Jones 1981), and one captive male orangutan only 6.5 years old has successfully fathered an offspring (Kingsley 1982, citing pers. comm. from Everts). Gestation lengths for wild orangutans have not been reported, but captive studies suggest gestations of 227 to 275 days, typically about 245 days (Martin 1981). Interbirth intervals in orangutans appear to be among the longest of any primate spe-

FIGURE 13-1. Orangutan mother and infant. (Photo: David Agee/Anthrophoto)

cies. Based on a sample of four females, Galdikas (1981) reported a minimum interval of 5 years and suggested a 6- to 7-year average. Her direct observations of one female indicated an interbirth interval of 8 to 9 years. These extremely long interbirth intervals reduce the availability of female mates and skew the operational sex ratio (the ratio of fertilizable females to breeding males) heavily toward males.

The lack of any external signs of estrus complicates evaluation of other reproductive variables in wild orangutans. Nadler (1977) reports that captive females show increased sexual responsiveness, as indicated by sexual solicitations and masturbation, during midcycle. Wild nulliparous females occasionally solicit adult males sexually, but whether this behavior corresponds to ovulation is not known (Galdikas 1978; Schurmann 1982). The behavior of nulliparous females contrasts sharply with that of parous females, who rarely solicit males sexually in the wild (Galidikas 1981; Mitani 1985d).

SOCIAL ORGANIZATION
General Patterns

Orangutans are the least gregarious of all diurnal primates. Although observations from different field stud-
ies vary in specifics, all studies concur that the primary units of orangutan populations are (1) solitary adult males; (2) adult females with offspring (fig. 13-1); and (3) solitary subadults. These units do not join in permanent social groups like those found in other diurnal anthropoid primates. Occasional larger groupings of animals occur when two or more primary units aggregate at a common food source. Some associations, however, last longer. Mitani (1985b) observed repeated associations between subadult males and adult females at the Mentoko, during which mating occurred over periods lasting up to 28 days. In contrast, adult males and adult females form consortships that may last as long as 5 days (Mitani 1985b). Two adult females may occasionally travel together for more than 1 day (Galdikas 1978).

Ranging Patterns and Dispersal

All field studies but MacKinnon's (1974) reveal that local populations of orangutans include resident animals, who are seen regularly through the study period, and nonresidents, who are seen briefly and irregularly. Estimates of the range sizes of residents vary. Rodman (1973) reported that females at the Mentoko occupy 0.4 to 0.6 km² ranges, but Mitani's more recent observations at the same site for a longer time revealed that female ranges exceeded 1.5 km². Galdikas (1978) reported 5 to 6 km² ranges for females of the Tanjung Puting Reserve for the period 1971 to 1975. Differences in estimates may result from differences in habitats, but they may also result from differences in sampling, since no published accounts are based on systematic, comparable surveys. For example, Galdikas (1978) reported the total range used by females for 4 years without indicating variation from year to year to allow comparison with shorter studies in other places. No study has produced a reliable estimate of males' range sizes because of the complexity of their pattern of ranging and dispersal.

There are no direct, continuous observations of dispersal of orangutans from their natal ranges, but it is probably gradual (Rodman 1973; Galdikas 1978; Rijksen 1978) and begins when juveniles first travel and forage independently from their mothers. Young animals spend progressively more time alone over the following years, occasionally rejoining their mothers for variable durations. The age when the young leave their natal ranges permanently is unknown. Based on ranging patterns observed for only a short period of 15 months, Rodman (1973) hypothesized that females settle near their mothers, probably in adjacent ranges, while males disperse farther before competing with others for residence within an area occupied by females. More recent reports based on longer studies (particularly Galdikas 1978) appear to agree with this interpretation of female behavior, but the male pattern is more complex. Judging by the large number of new males observed in a study area per year, some males wander over long distances. Others occupy stable ranges for variable lengths of time, but may depart from those ranges temporarily or permanently (Galdikas 1979). The male pattern probably depends on whether the local females are currently capable of conception or are reproductively preoccupied with pregnancy or lactation (Mitani 1985b).

Mating System

Most field studies have found that adult female orangutans occupy overlapping ranges nested within larger male ranges (Rodman 1973; Horr 1975; Galdikas 1978; Rijksen 1978; Mitani 1985d). The relatively large size of each female's home range, coupled with orangutans' slow movement, limits the ability of males to defend their ranges and maintain exclusive access to females (Rodman 1984). Mating therefore tends to be promiscuous, with each female mating with more than one male and each male mating with more than one female (Galdikas 1981; Schurmann 1982; Mitani 1985b).

Male Mating Strategies. Intrasexual competition apparently forces males of different ages and sizes to adopt alternative mating strategies (Mitani 1985b). Early field investigations suggested that subadult males mate with females during brief interactions (MacKinnon 1974; Galdikas 1979). More recent observations at the Mentoko, however, show that subadult males engage in sexual associations that last an average of 8 days (Mitani 1985b). In these associations the females rarely show proceptive sexual behavior and frequently resist mating attempts by subadults. Males nevertheless stay with females and continue to copulate forcibly with them until a larger male terminates the association. Galdikas (1985) suggests that subadult males at Tanjung Puting may follow a similar reproductive tactic.

The mating strategies of adult males are more complex. Some males may attempt to establish residency in an area that incorporates the ranges of two or more ovulating females (Rodman 1973; Rijksen 1978; Galdikas 1979; Mitani 1985b). Other males may wander over greater regions and seek matings opportunistically as they encounter females, although they face active competition from resident males (Mitani 1985b). Galdikas's long-term observations indicate that residency is not permanent, since a regularly observed male left the study area when nearby females gave birth. This male returned several years later when females began to cycle again, which suggests that male-male competition for fertile females determines the movements of adult males.

Field-workers disagree about the relative reproductive success of subadult and fully grown adult males. MacKinnon (1974) hypothesized that subadult males are reproductively more successful than adult males, since subadults mate with females more frequently than adults. It is not known, however, if subadult males are as likely as fully grown adult males to copulate with females at the time of ovulation. Galdikas (1985) infers from her observations that they are not. Moreover, adult males are always dominant to smaller subadult males, which suggests that large body size confers a competitive advantage when conflicts occur over access to females. As a result, most other field-workers assume that fully grown adult males father the majority of offspring (Rodman 1973; Horr 1975; Rijksen 1978; Galdikas 1981; Schurmann 1982).

One intriguing alternative not yet considered is that both subadult and adult males are reproductively successful at rates maintained by frequency-dependent selection acting on variations in male morphology and mating strategies (e.g., see Brockmann, Grafen, and Dawkins 1979). This could imply that not all males develop large body size and other secondary sexual characteristics as adults. Small body size may benefit some males by reducing the aggression they receive from larger males. Although fully grown males are almost invariably aggressive toward each other, they usually tolerate subadult males. Small body size may permit subordinate males to travel through areas controlled by larger resident males and to gain access to females without competing directly with larger, fully grown males.

In this context, it is interesting that Wallace (1869) collected a female-sized, but skeletally mature, male orangutan (i.e., with closed cranial sutures and fully mature dentition) that did not have the expected cheek pads and throat sac. Recently Kingsley (1982) has reported that when two male orangutans are caged together in zoos, the presence of a male with developed cheek pads retards development of cheek pads in the other for 2 to 3 years. Both observations suggest that not all small, sexually active male orangutans are truly subadults.

We currently lack paternity determinations of infants born in the field, and—inferential reasoning aside—in the absence of these data, it is not possible to assess the reproductive contributions of resident adults, nonresident adults, and subadult or small adult males. Sorting out these variations in male mating strategies is one of the challenges of continuing research on orangutans.

Female Resistance to Forced Matings. Females often resist forced matings by males, with the result that such copulations have been described as "rapes" (MacKinnon 1974; Galdikas 1981, 1985). Matings with small or subadult males are resisted more frequently than matings with adult males (MacKinnon 1974; Rijksen 1978; Galdikas 1981, 1985; Mitani 1985b), but resistance is rarely successful. Mitani (1985b) observed females struggle free only 13 times during 170 forced copulations; resistance was not more successful with subadult males than with adult males.

Given the low rate of success in deterring mating attempts, why do females resist males at all? Females may attempt to reduce the rate of male copulation attempts because mating interferes with other important activities, such as feeding and foraging or suckling of young. A female's resistance may also reflect the risk of premature pregnancy that could compromise investment in her current offspring. These hypotheses emphasize the indirect selective costs associated with unnecessary or ill-timed mating, but there may be direct costs of resistance as well. Females may resist subadult males more strenuously than adult males because females are more likely to injure themselves or their dependent offspring if they attempt to resist adult males that are much larger and stronger than subadult males and with whom the probability of successful resistance is low (Mitani 1985b). Alternatively, females may benefit in some as yet unspecified way from mating with adult males and therefore attempt more vigorously to avoid copulating with smaller males. This hypothesis suggests that female orangutans exercise choice in mating, but, as we discuss later, whether female orangutans do in fact choose among mates is currently debated.

Social Interactions and Relationships

As mentioned earlier, adult male and female orangutans spend most of their time alone. Mothers and their immature offspring are the only consistent social units. Observations of adult males meeting are rare (MacKinnon 1974; Galdikas 1978; Rijksen 1978; Mitani 1985b), in part because males occupy extremely large ranges and because they do not tolerate one another. Their encounters are usually aggressive, and chases and physical fights take place. Although aggressive encounters are seldom observed, many adult males have scars and broken fingers that may plausibly be attributed to injuries sustained in male-male fights (MacKinnon 1974; Galdikas 1979). Adult males tolerate, but can easily displace, subadult males, and subadult males are rarely aggressive toward each other (Galdikas 1979; Mitani 1985b). Resident adult males can be ranked according to the outcomes of their interactions (Galdikas 1979; Mitani 1985b). At Tanjung Puting (Galdikas 1979) and at the Mentoko (Mitani 1985b), the highest-ranking males were residents, while subordinate males were transients. Physical confrontations between males appear to be mediated by "long calls," which are given exclusively by fully grown adult males (MacKinnon 1974). Galdikas (1983) reported some observations suggesting that these calls mediate spacing among males, and Mitani (1985d) describes results from playback experiments that confirm Galdikas's observations. Long calls may therefore function to advertize a male's presence and may deter other males from approaching the area.

Interactions among adult females, although infrequent, appear to be relatively amicable. Rodman (1973) reported that one pair of adult females with their offspring met occasionally and that when they did, one female followed the other. Galdikas (1978) reported that adult females traveled together for up to "the better part of three days." In Galdikas's study, adolescent females were the most sociable age-sex class. They groomed adult males with whom they consorted, and they groomed one another when together. The behavior of one adolescent female changed abruptly at the birth of her first infant, when she became intolerant of others with whom she had previously had friendly relations (Galdikas 1978).

SEXUAL DIMORPHISM IN A SOLITARY SPECIES
Theory of Sexual Dimorphism

Two general hypotheses account for sexual dimorphism in animals. We may refer to these as *sexual selection* (Darwin 1871; Fisher 1958; Trivers 1972) and *ecological selection*. Sexual selection operates in two ways, either through intrasexual competition for mates, which favors characters that lead to success in such competi-

FIGURE 13-2. Adult male (*above*) adult female (*top right*) and subadult male (*bottom right*) orangutans. Note the size dimorphism between the adult male and female. (Photos: Peter Rodman)

tion, or through mate choice, which favors characters that are attractive to the opposite sex.

Ecological selection is also thought to produce sex differences in morphology or behavior in at least two ways. The first is reduction of feeding competition between males and females by niche divergence (Selander 1972). The second, suggested by Demment (1983), is based on his study of energetics, gut size, and dietary diversity in baboons and other herbivores. Demment argues that reproductive success of females of some species (e.g., savanna baboons) is limited largely by nutritional factors, in particular the energetic demands of pregnancy and lactation. Consequently, when compared with males, females have the smallest possible body sizes to minimize their own metabolic requirements relative to the requirements of gestation and lactation. Demment's analysis shows that at these small body sizes, females must choose relatively high-quality foods; that is, females must choose foods of low fiber content whose nutrients are rapidly assimilated by a small gut during a short period of digestion. Such foods are rare, however. In contrast, large body size is favored in males because increased gut capacity and increased digestive retention time allow them to exploit relatively abundant, low-quality (i.e., high-fiber) foods. Demment argues that in baboons, male size is limited by competition with the smaller ruminants, which are more efficient than nonruminant baboons at digesting fiber.

Sexual selection and ecological selection are not mutually exclusive. Size dimorphism resulting from sexual selection may have ecological consequences, and sexual dimorphism resulting from ecological selection may modify patterns of sexual selection. Consequently the causes and effects of sexual dimorphism are difficult to sort out precisely (fig. 13-2).

Sexual Dimorphism of Orangutans in Comparative Perspective

The extreme sex difference in body size of adult orangutans, similar in degree to that of gorillas and baboons, is repeated in other secondary sexual differences. At full maturity males have striking "cheek pads" consisting of subcutaneous accretions of fibrous tissue between the eyes and ears. Galdikas (1983) suggested that these pads may function to help locate the long calls of other males. Although interesting, this idea seems to us unlikely since the cheek pads lie in front of the ears. Alternatively the pads may assist in focusing the energy of the long call in a single direction, thus increasing its range. No other function of these odd structures has been proposed. Adult males also possess a large laryngeal sac that is often inflated when they give their loud "long call," as well as a sternal gland that is an accumulation of sebaceous glands on the chest. Observations of natural behavior have not suggested a function for this structure, which is unique among the Anthropoidea. Embryological precursors of cheek pads and throat sac are present in female fetuses, but rarely develop (Schultz 1940).

Among primates and other animals that live in one-male groups, the degree of sexual difference in size is usually directly related to variation in the number of females in a male's group (Alexander et al. 1979). Sexual dimorphism is also generally related to the socionomic sex ratio (ratio of breeding females to breeding males) in primates (Clutton-Brock, Harvey, and Rudder 1977). Male orangutans, however, do not defend clusters of females, as do gorillas and some other highly sexually dimorphic primates; their sexual dimorphism in size may therefore be puzzling.

Since the social pattern of orangutans is unusual among the primates, appropriate comparisons are better sought outside the order, among large mammals that have similar social systems. In table 13-1 we present comparative data for a number of essentially solitary bears and cats. The sex difference in body size of orangutans fits reasonably among these carnivores, although it is somewhat greater than in most species. Given other rather large contrasts between orangutans and such carnivores, it is probable that their shared breeding system is the source of the similarity in sexual dimorphism. This observation suggests that sexual dimorphism is normal among solitary species and that sexual dimorphism is related to a breeding system characterized by strong male-male competition regardless of whether or not animals live in groups.

Consequences and Correlates of Dimorphic Body Size in Orangutans

Body size has direct effects on metabolic rates, total nutrient requirements, gut capacities, food transport rates, and costs of locomotion (Kleiber 1975; Tucker 1975; Demment 1983; van Soest 1982). We might therefore expect male and female orangutans to differ in diet and ranging patterns simply because of differences in body size. On the other hand, motherhood imposes on females demands and constraints not faced by males, and these

TABLE 13-1 Sexual Dimorphism in Body Weight of Solitary, Polygynous Carnivores

Common Name	Species	Male (kg)	Female (kg)	Male/ Female
BEARS				
Spectacled bear	*Tremarctos ornatus*	158	61	2.6
Asiatic black bear	*Ursus thibetanus*	130	78	1.7
American black bear	*U. americanus*	193	116	1.7
Alaskan brown bear	*U. arctos*	139	95	1.5
Polar bear	*U. maritimus*	410	320	1.3
TIGERS	*Panthera tigris*			
	P. t. altaica	243	134	1.8
	P. t. tigris	219	130	1.7
	P. t. balica	95	73	1.3
	P. t. sumatrae	120	93	1.3
LEOPARD	*Panthera pardus*	64	44	1.5
ORANGUTAN	*Pongo pygmaeus*	83	37	2.2

SOURCE: Nowak and Paradiso 1983.

demands may interact in important ways with body size to decrease or increase male-female differences in feeding and ranging.

Observations suggest that males and females do differ in feeding and ranging. First, as noted previously, all observers agree that male orangutans come to the ground and travel on the ground more frequently and for more time than adult females. This may be a consequence of the large males' greater difficulty moving through the trees, or it may be that larger males face less risk from predation on the ground than smaller adult females and their vulnerable dependents. Regardless of the cause of this behavioral difference, males feed on and near the ground more than females (Rodman 1977; Galdikas 1978). Second, Rodman (1979) reports that one mature adult male fed more and traveled less than females observed at the same time and in the same area. Neither MacKinnon (1974) nor Galdikas (1978) found such differences, although MacKinnon lumped observations of mature males with observations of female-sized subadult males, with the result that any effects of body size were obscured. Galdikas suggested that Rodman's male subject was "past prime," like one of her subjects who was less active and fed more than her female subjects, but no other characteristics of the male suggested that he might be aged (Rodman, pers. obser.). The same adult male fed more on bark and leaves and less on fruit than females feeding in the same area at the same time (Rodman 1977). Galdikas found that her female subjects fed on more different types of food per day than did adult males, while Rodman (1979) found no difference in the mean number of food sources for one male and four female subjects.

Feeding differences between adult males and females may be consequences of differences in body size (Rodman 1979). According to this hypothesis, males are constrained by the energetic demands of their size to travel less and feed more than females. In contrast, females are permitted to be more selective of food sources by their greater mobility and (possibly) reduced food intake. The demands of pregnancy and lactation complicate the relationship, however, and may force females to feed on higher-quality food than males. Smaller female size may be selected in part to compensate for the costs of motherhood (Demment 1983).

Regardless of the functional origin of the sex differences in body size, a final consequence of the difference is that males are more than twice the size of females. This may account for females' relatively quick compliance with sexual advances of fully grown males compared with their resistance to copulations with smaller or subadult males (MacKinnon 1974; Mitani 1985b), although there is some evidence that nulliparous females may prefer adult (thus larger) males as consort partners (Galdikas 1985).

Sexual Selection Theory in Relation to Sexual Dimorphism in Orangutans

There are feeding differences between male and female orangutans that may be related to body size differences, and one potentially beneficial consequence for both males and females may be some reduction of competition. We have three reservations about concluding that there is a causal relationship between niche divergence and sexual dimorphism, however. First, the feeding differences reported do not seem large enough to suggest competition reduction as the principal selective factor in body size difference. Differences in intake of high-fiber leaves might be allowed by the male's larger gut capacity, but this difference in feeding seems too small to be the primary result of such large size difference. Second, such feeding differences as differential use of tree levels need not entail sexual dimorphism, since behavioral differentiation would be sufficient. Third, other sexually dimorphic characters such as the males' calls, the cheek pads, and the throat sac cannot be accounted for by ecological selection. The clearest effect of large male body size is in sexual relations. Large adult males displace smaller subadults from females, and large males may benefit from greater mating cooperation by females.

Research on a variety of animals has confirmed the importance of sexual selection for the evolution of sexual dimorphism (Selander 1972; Clutton-Brock, Harvey, and Rudder 1977; Alexander et al. 1979), and there is consensus among field-workers that this selective process played a critical role in the evolution of sexual dimorphism in orangutans (Rodman 1973, 1984; Horr 1975; Galdikas 1978, 1979; Rijksen 1978; Schurmann 1982; Mitani 1985b, 1985d). There is, however, disagreement concerning the mode of sexual selection under which dimorphic traits of orangutans evolved. Are sex differences between orangutans a result of mating competition among males or of female choice of mates, or both?

Among orangutans female parental investment greatly exceeds male investment, and operational sex ratios are heavily male biased because females conceive infrequently (Rodman 1973; Mitani 1985b). Under these conditions, theory predicts intense competition among males and the evolution of traits that increase a male's competitive success. Several authors (Horr 1975; Galdikas 1979; Rodman 1984; Mitani 1985b) have hypothesized that large body size confers a mating advantage on some males by helping them defeat others in direct contests for females. During a recent study, Mitani (1985b) used observations of interactions between males to test this hypothesis. He found that the majority of male-male interactions occurred in the presence of females, and he showed that large males disrupt mating by other males. The highest-ranking adult male consistently displaced other males, and all adult males supplanted the smaller subadult males. These results support the hypothesis that male-male competition for sexual access to females influenced the evolution of large male body size in orangutans.

Some field-workers propose, without substantial evidence for support, that the evolution of large male size results from female choice of mates (Rijksen 1978; Schurmann 1982). Galdikas (1985) reports that "Female-initiated consortships invariably involved adult males," which suggests some choice by females favoring larger, or at least more mature, males. Since observations of parous females who selectively choose large males as mates are rare, we feel it is premature to accept female choice as a primary selective factor in this case. Similarly, although Rijksen (1978) and Rodman (1973) described behavior suggesting that some females may approach calling males, their evidence is sparse. Moreover, no studies have shown that females selectively mate with males who call. At this time, the evidence suggests that male long calls have evolved as a result of male-male competition rather than female choice.

SUMMARY AND CONCLUSION

Two of three anatomical differences between male and female orangutans can be explained as consequences of selection on male calling effectiveness: the throat sac may contribute to the long-distance transmission of the male long call, and the cheek pads may act as reflectors to focus energy of the call in one direction. Since the long call is best explained as an adaptation for male-male competition, we conclude that the influence of female choice in the evolution of male orangutan morphology and behavior is currently best treated as secondary to the influence of competition among males for mates.

Sexual selection acting on male-male competition rather than female choice accounts for the sex difference in body size of orangutans. As in most aspects of natural organismal biology, however, we need quantitative information. Observations of adult females as they near conception and multiyear observations of adult males will fill several of the important gaps in the discussion we have presented here.

14 | Gorillas: Variation in Female Relationships

Kelly J. Stewart and Alexander H. Harcourt

TAXONOMY AND DISTRIBUTION

The gorilla is the largest living primate, yet the species was not known to science until the mid-nineteenth century (Savage and Wyman 1847). Although it is sometimes placed in the same genus as the chimpanzee, *Pan,* which it resembles anatomically and biochemically, we here take the more conventional approach and classify it as *Gorilla.* Within the genus, there is one species, *G. gorilla.* This is split into three races or subspecies that exist in two separate regions of equatorial Africa. The western lowland gorilla (*G. g. gorilla*) is the smallest of the three and occurs in Congo Brazzaville, Gabon, Equatorial Guinea, Cameroon, and the Central African Republic. The eastern lowland gorilla (*G. g. graueri*) occurs in eastern Zaire, while the rarest subspecies, the mountain gorilla (*G. g. beringei*) is found only in the mountains of southwest Uganda, northwest Rwanda, and eastern Zaire (Groves 1970; Harcourt, Fossey, and Sabater Pi 1981).

Very little is known about the western lowland gorilla, although it is by far the most numerous and widespread race. Most data come from the two eastern subspecies, especially from one population of mountain gorillas in the Virunga Volcanoes of east central Africa. There, at the Karisoke Research Center in Rwanda (Fossey 1983), several groups have been studied intensively since 1968, and periodic censuses have been conducted of nearly the whole population. The behavioral data described in this chapter are derived primarily from four bisexual and one all-male group in the Virunga Volcanoes, all of which have been observed for at least 2, and some for over 15, years.

ECOLOGY

Gorillas are found from sea level in West Africa to 3,400 m in the Virunga Volcanoes. Preferred habitats are forests with an open canopy and dense ground and shrub strata, where gorillas find most of their food. Such habitats include forest edges, regenerating secondary forest, montane and riverine forest, and, in some seasons, bamboo stands. Gorillas tend to avoid the dense canopy and bare floor of primary tropical forests (Schaller 1963; Tutin and Fernandez 1984).

Both eastern subspecies are primarily folivorous and eat the roots, leaves, stems, and pith of herbs, vines, shrubs, and bamboo (Casimir 1975; Fossey and Harcourt 1977; A. G. Goodall 1977). In the Virunga Volcanoes, for example, fruit makes up only 1.7% of all feeding records. The highly foliverous diet of the eastern population is due in part to the scarcity of fruit in the environment. In many areas of West Africa, fruit is far more common and also more frequently eaten (Sabater Pi 1977), although the precise figures are not known. Unlike chimpanzees, wild gorillas have never been seen to eat meat or to use tools to "fish" for ants or termites, but they do occasionally eat invertebrates (Fossey and Harcourt 1977; Tutin and Fernandez 1983).

Across Africa, the size of a gorilla group's yearly range varies from about 5 km² to about 35 km², depending on the region and also to some extent on group size (Jones and Sabater Pi 1971; Fossey and Harcourt 1977; A. G. Goodall 1977). Day journey length is from 0.5 to 1 km, which is small in relation to range size. Correlated with this small day range is a lack of territoriality. Indeed, in all areas for which enough data have been gathered, yearly ranges, and sometimes even core areas, overlap with those of other groups (Fossey 1974a; Fossey and Harcourt 1977).

Within regions, population densities obviously vary according to habitat (e.g., Tutin and Fernandez 1984), but across Africa density normally ranges from 0.35 to 0.75 animals/km². Biomass ranges from 0.3 to 0.65 kg/ha (Harcourt, Fossey, and Sabater Pi 1981).

CONSERVATION STATUS

The gorilla is vulnerable in most parts of its range, especially in East Africa, where populations are often split into tiny isolated pockets surrounded by human habitation (Schaller 1963). The least endangered of the subspecies is the western lowland gorilla, especially in Gabon, where the human population is low and vast tracks of forest still exist. Indeed, Gabon may contain as much as half the entire western population. Estimates of gorilla numbers in West Africa are rough guesses everywhere except in Gabon, where a recent census indicated a population of about 35,000 gorillas, plus or minus

7,000 (Tutin and Fernandez 1984). It is estimated that between 3,000 and 5,000 eastern lowland gorillas remain in eastern Zaire. The mountain gorilla, which has been more accurately censused, numbers only around 400, of which about 250 live in the Virunga Volcanoes (IUCN 1982a).

Forest clearance for commercial purposes or agriculture is the main threat to gorillas, but hunting also takes its toll. Gorillas are killed for food and as crop pests, especially in West Africa. In addition, infants are captured throughout Africa for sale to foreign zoos, and in East Africa, poachers sell gorilla skulls to tourists as souvenirs (IUCN 1982a; see also chap. 39).

REPRODUCTIVE PARAMETERS

One cause of the vulnerability of gorilla populations is the animal's long reproduction time. Table 14-1 presents information on several reproductive parameters, with all wild data coming from the mountain gorilla population in the Virunga Volcanoes. As the table shows, there is a period of "adolescent sterility" in females between first estrus at around 7.5 years and first conception at around 10. Only nulliparous females show sexual swellings, but these are very slight and inconspicuous. However, all females exhibit behavioral signs of receptivity, involving

characteristic soliciting approaches and postures toward the male (Harcourt et al. 1980). Sexual maturity in wild males is very difficult to determine, but, because of the competition they receive from older animals, males are unlikely to start breeding before about 15 years of age (Harcourt et al. 1980). It is not until this age that males develop their complete secondary sexual characteristics, in particular the silver saddle of hair on the back, that has earned fully adult males the name "silverback."

Across gorilla populations in East Africa, reproductive output for females is about one surviving offspring every 8 years, where *surviving* is defined as reaching breeding age. Thus, given a reproductive life span of approximately 25 years, a female may expect to produce about three surviving offspring in her lifetime (Harcourt, Fossey, and Sabater Pi 1981).

GROUP SIZE, GROUP COMPOSITION, AND MOVEMENT BETWEEN GROUPS

In all areas of their range, gorillas live in relatively stable bisexual groups, with solitary animals (usually fully adult males) making up about 10% of most populations (Harcourt, Fossey, and Sabater Pi 1981). Table 14-2 summarizes various group parameters from East and West Africa. In addition to the age-sex classes defined in this table, the other classes recognized for gorillas are infants, 0 to 3 years; juveniles, 3 to 6 years; subadults, 6 to 8 years; and black-backed males, 8 to ca. 13 years. Group size in West Africa is significantly smaller than in eastern populations, with the difference due to the number of females and immatures and not to the number of males: in all areas for which there are data, 60% of groups have only one adult male (Harcourt, Fossey, and Sabater Pi 1981). It might be expected that groups with more adult males would have more females, but this is not the case ($r_s = 0.14$; $N = 22$; $p>0.1$).

The gorilla is one of the few group-living primates in which females as well as males habitually emigrate from the group in which they were born (Harcourt 1978b). In the Virunga Volcano population, emigrating females transfer directly to either another group or to a solitary silverback and never range by themselves. Transfer appears voluntary, and females are not abducted by males (Harcourt 1978b; chap. 21). Female transfer normally occurs when two groups or a group and a solitary male

TABLE 14-1   Reproductive Parameters for Gorillas

Reproductive Variable	Data from Wild Population
Birth season	None
Age of sexual maturity	
Males	180[a] (captivity: 96, 72–114)
Females	89[b] (78–98)
	119[c] (114–130)
Gestation length	8.6 (0.23)
Litter size	1
Interbirth interval	47 (6.3)
External signs of estrus	None[d]

NOTES: Data are from the Virunga Volcanoes (Rwanda and Zaire). All measures are given in months.

   a. Age at sexual maturity for males cannot be judged in the wild, so figure given is estimated age of first breeding. Figures in parentheses show means and ranges for age of sexual maturity of males in captivity (Dixson 1981).

   b. Mean age when females first mate with fully adult males.

   c. Mean age of first conception.

   d. Very slight perineal swellings occur in nulliparous females.

TABLE 14-2   Demographic Parameters for Gorillas

		Group Size		AM/Group		AF/Group		Imm/Group	
	N	M	Max	M	Max	M	Max	M	Max
West	29	5	12	1	3	3	4	2	5
East	64	9	37	1	4	3	12	3	16

NOTES: Data from West Africa are largely from Rio Muni. Those from East Africa are largely from the Virunga Volcanoes (Rwanda and Zaire), Kahuzi-Biega (Zaire), and Bwindi Forest (Uganda). AM = fully adult male (>15 years); AF = fully adult female (>8 years); Imm = immature (<8 years); M = median; Max = maximum.

come into close proximity (usually tens of meters). While the males display and sometimes fight with each other, the female moves away from her group, approaches the new male, and then follows him when the encounter ends. Most females transfer at least once in their lives, usually before breeding (table 14-3).

Male gorillas also tend to emigrate from their natal groups before they have bred. However, unlike females, they very rarely join breeding groups. Of 17 animals of known sex who transferred at least once into bisexual groups, only one (a blackback) was a male, and no silverback has ever been known to join a breeding group (Harcourt 1978b). Instead, emigrant males either wander alone, are joined temporarily by females, or form all-male groups. In the Virungas, all-male groups may make up as much as 10% of all groups of two or more animals.

Since males rarely transfer to breeding groups, it seems likely that in the 40% of cases where groups include more than one silverback, the males are closely related. Indeed, not all males emigrate from their natal groups. Of 10 adult males first observed as immatures or as young, nonbreeding silverbacks, 3 remained in their presumed natal group to breed, 5 emigrated, and the fate of 2 others could not be determined. The factors likely to influence whether or not a male emigrates include the number of adult males in the group and their relative ages, both of which may affect the intensity of competition for females (Harcourt 1979a). The past history of the males' relationships also seems important (Harcourt and Stewart 1981).

It is often years before emigrant males permanently acquire females. Four males observed intermittently for 2 to 5 years after emigrating were usually seen alone. Occasionally, they associated temporarily with females and less often with another male (Caro 1976; Harcourt 1978b). The only consistently monitored emigrant male spent at least 3 years continually alone.

As mentioned above, males acquire females by attracting them away from other silverbacks during encounters with bisexual groups. Such encounters are characterized by intense male-male competition. In the Virunga Volcanoes, 80% of 19 encounters involved vigorous threat displays (e.g., chest beating [figs. 14-1, 14-2], foliage slapping, runs), and 50% of 16 encounters for which there was enough evidence included fights with physical contact (Harcourt 1978b; chap. 22). Such competition between males for females has probably contributed to the gorilla's extreme sexual dimorphism. Females weigh about 95 kg and are only about half as large as the 160 kg males. Males, but not females, also possess a number of morphological features associated with fighting ability, such as large canines (Harvey, Kavanagh, and Clutton-Brock 1978b).

Male and female emigration, female transfer, and male-male competition during interunit encounters have

TABLE 14-3 Parity of Female Gorillas Who Transferred between Breeding Units

	First Observed as:	
	Immatures	Adults
Nulliparous	9	—
Primiparous	3	—
Multiparous	—	2
No transfer	3	4

NOTES: The first column includes all females who were first observed as immatures (<8 years) and who reached breeding age (10 years) during the study. The second column presents data for females who were first observed as parous adults and who subsequently lived for at least 4 years.

also been observed in a population of *G. g. graueri* in eastern Zaire (Yamagiwa 1983).

DAILY ACTIVITY

A gorilla group's day typically has early morning and late afternoon peaks of travel and feeding, separated by a midday rest period of 1 to 2 hours (Schaller 1963; Harcourt 1978a). Feeding periods, which take up about 45% of an adult's day (Harcourt and Stewart 1984), are interspersed with intervals of relatively rapid, concerted travel. During most of the day, however, animals spread out in the thick vegetation, moving slowly and feeding as they go.

Intragroup aggression is generally mild and usually consists of aggressive vocalizations (cough-grunts). Among adults, aggression is more common during feeding than at other times. For example, in two Virunga Volcano groups eight of the nine adults who were involved in more than 5 aggressive incidents were more aggressive to other adults during feeding periods than during resting periods. The rate of aggression of the median adult during feeding was 2.25 incidents per hour within 5 m of others, 15 times higher than during rest periods. Thus, even in an area where most of the gorillas' food is abundant and evenly distributed, overt feeding competition apparently occurs. Interindividual distances during feeding periods are also greater than during rest periods (Fossey and Harcourt 1977), when group members cluster, often near the silverback (fig. 14-3). Although much of the rest period is spent dozing, it is also the time when rates of affinitive interactions, such as grooming, are highest.

SOCIAL RELATIONSHIPS WITHIN GROUPS
Adult Males and Females

The timing of group activity and the direction of travel are often determined by the dominant adult male. He ranks above all other group members and may hold this tenure (and thus priority of access to fertile females) for longer than a generation (10 years). The cohesion of go-

FIGURES 14-1 and 14-2. An adult male silverback mountain gorilla hoots and beats his chest during a threat display. (Photos: Alexander Harcourt/Anthrophoto)

rilla groups appears to depend primarily on the relationship of the dominant silverback with the adult females (Harcourt 1979c). Indeed, females tend to spend more time near the silverback than near any other adult. The proximity of females to the adult male is inversely proportional to the age of their offspring; with some new mothers the increase in proximity to the male is evident immediately after parturition (Harcourt 1979c). Close proximity between the dominant silverback and females with dependent offspring is a pattern also found in eastern lowland gorillas (Yamagiwa 1983).

The frequency and direction of grooming between parous females and adult males differ markedly between individuals and groups. In three of the four Virunga Volcano groups, the silverback groomed regularly with at least one parous female, whereas the females groomed each other only rarely. However, in one of these groups the male was responsible for most of the grooming, while in the other two groups the females groomed the males (Harcourt 1979c). In the fourth group, there was almost no grooming between the silverback and the females, although some females groomed each other.

Immatures and the Dominant Male

Immature mountain gorillas are strongly attracted to the dominant male, as has also been reported for eastern lowland gorillas (Yamagiwa 1983). In fact, the only other adult with whom immatures have an equal or stronger relationship is their mother (fig. 14-3). An attraction to the silverback becomes apparent between the second and third year of life, when the immatures regularly begin to seek the silverback's proximity, often leaving their mothers to do so (Stewart 1981). This attraction is especially obvious in infants and young juveniles that have lost their mothers. Indeed, orphans treat the silverback as their "caretaker" even when older siblings are present (Fossey 1979; Stewart 1981).

On the whole, the immatures are responsible for maintaining proximity with the silverback, and he does not actively seek their contact or carry them. His behavior toward immatures is usually simply one of tolerance, although during threatening situations he becomes extremely protective. This applies to threats from both outside (e.g., humans) and within the group. For example, silverbacks intervene in fights between group members

FIGURE 14-3. Adult females, juveniles, and infants cluster near the silverback male during a rest period. (Photo: Alexander Harcourt/Anthrophoto)

(by cough-grunting and running at the participants) three times more often than the median adult. When silverbacks support any one combatant, it is almost always the younger (89% of 33 incidents). Thus immatures are protected by the dominant male from escalated aggression.

Eight of 14 juveniles and subadults in three study groups also groomed regularly with the dominant silverback. As was the case for grooming between the male and adult females, in one group the silverback groomed the immatures, while in the other two the reverse was true. However, immatures whose mothers groomed with the male did not necessarily groom with him themselves (Stewart 1981).

Relationships among Immatures

Social interactions among immatures occur at a higher rate than those among adults. The most common type of interaction is play, the frequency of which is inversely related to age from the start of the second year. Grooming also occurs, with older immatures grooming more than younger ones. There is some evidence that maternal siblings groom each other preferentially. Thus, in two groups

for which data are available, the three juveniles with younger siblings gave a greater proportion of their grooming to their siblings than to other immatures (Stewart 1981).

Dominance Status

Among immatures, dominance rank (based on the direction of approach-retreat interactions) is positively related to age and not dependent upon sex or maternal rank. After puberty, however, sex differences become apparent. As males become larger than females, at about 11 years of age, they begin to dominate females in aggressive interactions and eventually come to outrank all adult females. Stable rank differences among nulliparous and parous females are generally not apparent (Harcourt 1979b), but the ranks of adult males are far more obvious. In both all-male and bisexual groups, older males are usually dominant to younger males (see also Schaller 1963). For example, in two bisexual groups for which data are available, the silverback male aggressively prevented two maturing males from mating with receptive females (Harcourt et al. 1980). However, in one case the presumed son gradually became dominant to the older male, and, over several years, both males mated with females.

♦ VARIATION IN FEMALE RELATIONSHIPS

In most group-living mammals, closely related females form the stable core of the social unit, while males transfer between groups (Packer 1979a; Greenwood 1980; chap. 21). Although some female dispersal occurs in many primate species (Moore 1984), gorillas are unusual in that females regularly emigrate from their natal groups before breeding and seldom interact with their close female kin as adults. Other primates so far studied in which females habitually transfer between breeding groups are red colobus (*Colobus badius:* Marsh 1979b; Struhsaker and Leland 1979), hamadryas baboons (*Papio hamadryas:* Sigg et al. 1982), and chimpanzees (*Pan troglodytes:* Pusey 1979; Goodall 1983; Kawanaka 1984). Female howler monkeys (*Alouatta palliata* and *A. seniculus*) also emigrate, but they do not always transfer between breeding groups (Jones 1980a; Crockett 1984; chap. 21).

Three separate questions with interdependent answers can be posed. Why do females live in groups? Why are these groups composed of kin? And why are some species characterized by female dispersal? The last is the question we are mainly concerned with here.

It is likely that a primary selective pressure in the evolution of primate kin groups has been the benefit gained, through individual and kin selection, from cooperation with one's relatives (Hamilton 1964; Bertram 1982; Wrangham 1980, 1982; chap. 23). Wrangham (1980) has suggested that a major difference between female-transfer and female-resident primate species lies in the importance of cooperation in feeding competition. Thus, when food is abundant and evenly distributed, as it is for mountain gorillas, cooperative defense will not be favored and a female-emigrant system may evolve. This idea can be considered from the perspective of female-female relationships. If kin-based alliances do increase individual competitive abilities, then relationships among close relatives should be characterized not only by coalitions in agonistic interactions but also by affinitive interactions, such as grooming, that help establish and maintain cooperative relationships. Grooming and coalitions appear to occur at higher rates in female-resident species than in species in which females transfer to other breeding groups (see chaps. 11, 25). This does not necessarily imply, however, that the need for cooperation among kin is the cause of female philopatry, since such cooperation could be the result of different emigration patterns rather than the cause of them. For example, if the costs of emigration for females are great (e.g., Clutton-Brock and Harvey 1976), then female residence could lead to high levels of cooperation simply as a result of kin selection (Hamilton 1964).

If resource competition among females has favored the evolution of kin-based alliances, then female-transfer species should differ from female-resident species in two important respects. First, female-transfer species should be characterized by relatively little competition. We are concerned here with direct or "interference" competition, when allies might be useful, rather than with indirect or "exploitative" competition, which occurs when animals simply avoid each other (Miller 1968). Second, even if close kin in female-transfer species do occasionally end up in the same group as adults, nepotism should be much less obvious than in female-resident species. These predictions are best examined in a female-transfer species in which some female kin remain together as adults. The Virunga Volcano population of mountain gorillas is ideally suited to such an investigation because some adult females do in fact live in the same group as their close female relatives.

We therefore start this section with two main questions: first, Is there evidence of significant interference competition among adult female gorillas? and second, Do kin cooperate? We attempt to address these questions by examining rank relationships among females and by comparing the interactions of female kin with those of nonkin. These results indicate that, although direct feeding competition is probably less important for gorillas than for many female-resident primates, it does nevertheless occur; it is reflected to some extent in females' dominance relationships. Furthermore, kinship influences the nature of females' interactions in much the same way that it does in cercopithecines. Kin do cooperate with each other, which suggests potential benefits to females in remaining with their relatives. We therefore go on to consider why female gorillas do not usually remain with their kin, and why, since they emigrate, they do not travel on their own to minimize feeding competition, as do chimpanzees (Wrangham and Smuts 1980).

Sources of Data

The data on intragroup relations come primarily from three Virunga Volcano groups, BV, UB, and NK (table 14-4). BV was observed in two distinct periods separated by a 4-year gap; because its composition differed markedly in the two periods, it is here considered as two separate groups, BV1 and BV2. Both BV and NK contained known adult relatives. In BV1 there was one kin pair—a breeding female and her nulliparous adult daughter. BV2 included a multiparous female (different from the mother in BV1) and her breeding adult daughter and a nulliparous adult daughter, both of whom had been born in the group and were full sisters. A third breeding female in BV2 had also been born in the group but had

TABLE 14-4   Number of Parous Females, Adult Nulliparous Females, and Adult Females with Close Female Kin in Four Virunga Volcano Study Groups

Study Group	Parous Females	Nulliparous Females	Females with Known Kin
UB	4	0	0
BV1	3	1	2
BV2	4	1	3
NK	6	0	2
TOTAL	17	2	7

NOTES: Close female kin included mothers and daughters and full sisters; $r = 0.5$. Study groups are coded according to the names of the dominant silverback in the group: UB: Uncle Bert (group 4); BV1: Beethoven (group 5) (before 1977); BV2: Beethoven (group 5) (after 1981); NK: Nunkie.

no close adult female relatives. In group NK there was one mother-daughter breeding pair. Like all other females in NK, both individuals had immigrated from another group, the mother 4 years before the daughter. Throughout this section, *close kin* refers only to mothers, daughters, or full sisters; *nonkin* are animals with a coefficient of relatedness of less than 0.5.

Is There Direct Competition among Adult Females?

Even in the Virunga Volcanoes, where food is abundant and evenly distributed, initial indications are that adult females compete directly as well as indirectly for food. For example, not only do they tend to be farther apart when feeding than when resting, but aggression during feeding is also more frequent than during resting. If interference competition has long-term significance (i.e., of possible influence on female reproductive success), then differences in competitive ability should be important to females, and there should be evidence of a dominance hierarchy (Wrangham 1980).

In three of four groups, the rank relationships among adult females were generally unclear, partly because the frequency of supplants was so low. For example, in group NK, the median number of supplants per dyad was 1.5 for every 100 hours of focal observation ($N = 15$ dyads). In these three groups, a difference in dominance rank could be detected in only 9 of 27 female dyads. The lack of clear rank differences between the others could have been due in part to the low rate of supplants. However, in 4 of the 12 dyads in which supplants were observed, both females supplanted each other, and in 2 of these dyads each female was responsible for an equal number of supplants. In other words, the females themselves seemed not to recognize differences in dominance.

Perhaps because the females spent more time in close proximity, supplant rates in BV2 were higher than in other groups, and rank differences were also clearer. Among the 6 dyads there was a median of 6.5 supplants

per dyad for every 100 hours of focal observation, and there was also a linear dominance hierarchy among the four adult females. As was the case also for the 9 dyads in the other groups, this dominance hierarchy was related to age; no younger animal dominated an older one. Watts (1985) obtained similar results for dominance relationships in this population.

The fact that daughters did not acquire ranks similar to those of their mothers suggests that the advantages of high rank, in particular its influence in ensuring offspring access to resources (Chapais and Schulman 1980), are not as important for gorillas as for some cercopithecine species (see also chaps. 11, 25, 26). In other words, success in direct feeding competition does not appear important. However, there is at least one other explanation for this observation. The dominant silverback, who is the father of most or all of the group's offspring, could in theory prevent some females from consistently supporting their offspring in disputes and keep some juveniles from benefiting at the expense of others. For example, a female that interferes in a fight on behalf of her offspring may in turn be attacked by the adult male. Thus, small group size, combined with extreme sexual dimorphism, might allow male gorillas to exert more influence over female interactions than do males in other primate species.

In summary, supplant frequencies and rank relationships suggest that interference competition does occur in gorillas, but at a relatively low level.

Do Kin Cooperate?

Relationships among female kin in female-resident species are typically more affinitive than those among nonkin (see chap. 11). When related female gorillas live in the same group as adults, they also treat kin differently from nonkin. In groups containing both related and unrelated adults, female gorillas spent more time in proximity with their kin both when resting and feeding. They also groomed their kin more, were less aggressive to them, and aided them more in agonistic interactions. Consanguinity did not, however, appear to affect rates of supplanting (table 14-5). Different measures of female affinity were used in different study periods, but rough values for median kin and nonkin dyads were as follows: proximity records within 5 m—kin 25%, nonkin 10%; grooming when animals were within 2 m of each other— kin 25%, nonkin 0%; aggressive incidents for every 10 hours that animals were within 5 m—kin 5, nonkin 15; aid in contests against dominant animals—kin 4%, nonkin 0%; supplants for every 10 hours that animals were within 5 m—kin 2.5, nonkin 3.5. Only one adult female spent more time near an unrelated female than near her close relatives. Furthermore, while 4 of 5 kin dyads groomed, only 2 of 20 nonkin dyads did so, and all of the 11 observed coalitions involved kin against nonkin. Re-

TABLE 14-5 Comparison of Adult Female Behavior with Kin and Nonkin

	More	Equal	Less
Proximity	7	—	—
Groom	7	—	—
Aggression	—	1	6
Supplant	2	1	4
Aid	4	—	—

NOTES: Numbers refer to the number of adult females whose behavioral scores with kin were more than, equal to, or less than their median scores with nonkin. Kin includes mothers and daughters and full sisters; all others are considered nonkin. See text for discussion of behavioral measures.

Kinship is defined as $r > 0.5$. Only four females aided any other adult female.

lated adult female gorillas were therefore friendlier and more tolerant of each other than of nonkin and sometimes cooperated in competitive interactions against nonkin. The immediate benefits of such cooperation were most obvious when gorillas fed on one of the rare clumped foods in their diet, such as tree roots or a log. Then, close kin would feed together to the exclusion of others.

Why Do Females Emigrate?

These data indicate that, although direct competition may exist at a relatively low level, there are still potential benefits to gorillas in remaining with kin. Why, then, do females leave their natal group and often forego the opportunity of kin-based alliances (cf., Wrangham 1980)? In fact, even in the absence of competition or cooperation, female emigration is difficult to explain since there are likely to be costs associated with dispersal into new areas and groups.

We suggest that female gorillas may have to leave their natal group to avoid inbreeding (Harcourt 1978b; see also Baker 1978; Greenwood 1980). It is not unlikely that, when a female reaches sexual maturity, the only breeding male in the group will be her father. This is because male immigration and the takeover of bisexual groups are extremely rare in mountain gorillas. Although the mean length of an adult male's tenure in a bisexual group is not known, at least some males retain the monopoly of fertile females for over 8 years, the time it takes a female to reach sexual maturity. The only three breeding silverbacks observed for at least as long as this period were all dominant males for over 10 years.

There is abundant genetic evidence from other primate species that inbreeding is deleterious (chap. 21). Behavioral evidence from gorillas suggests that females may avoid mating with males who were dominant when the females were young juveniles, (that is, their potential fathers). However, the females appear to have less inhibition regarding males with whom they matured and who are likely to be half-siblings. There are no data for full

siblings. Of seven females whose natal groups were known, the four who emigrated before breeding (and whose transfers were not precipitated by poacher attacks; Watts 1985), left groups that had only one silverback, and he was their likely or known father. In contrast, the three females who did not emigrate before breeding lived in groups that included at least one other younger adult male. In one case, the female is known to have been unrelated to the male, since he was the only male ever observed to immigrate into a bisexual group. Furthermore, of four young adult females who were observed to mate (but not necessarily breed) in their natal groups, three copulated only with the younger subordinate male, who was at most a half-sibling. Only one female was observed to copulate with her presumed father, and she appeared to be much less sexually attracted to him than to the younger male. Thus, as long as male immigration remains infrequent in gorillas, female philopatry is unlikely to become a predominant tactic.

In chimpanzees (Pusey 1980; Goodall 1983; Kawanaka 1984) and some red colobus (Struhsaker and Leland 1979), males rarely transfer to other groups, and the lack of unrelated males may also be one cause of dispersal by nulliparous females in these species (see also chap. 21).

Although inbreeding avoidance may account for the emigration of female gorillas from their natal groups, it cannot explain subsequent transfers, and it may also not be relevant in some other species. For example, in howler monkeys and one population of red colobus, some females emigrate even though males often transfer between groups (Marsh 1979b; Jones 1980a; see also chap. 21). In these populations, female emigration seems at least partially to result from competition with other females. Group life inevitably imposes some costs, which are likely to increase with group size (Alexander 1974). In the case of mountain gorillas, for example, Watts (1985) found that individuals in larger groups were involved in more supplants per other animal than those in smaller groups.

In female-resident species, females must presumably balance the disadvantages of group life with the benefits of cooperating with kin. In gorillas and other female-emigrant species, the balance may be different if there is less ecological pressure favoring kin-based alliances. Furthermore, once female emigration is established (perhaps due to inbreeding avoidance), adult females will typically have few allies in the social group and may do better by transferring to a group with fewer individuals. Emigration of subordinate females to smaller groups is therefore to be expected (e.g., Gauthreaux 1978; Rubenstein 1978).

Since dominance among gorillas is related to age, and since emigrating females are usually young (table 14-3), most emigrants are also likely to be relatively subordinate. Moreover, the only 3 breeding females that were ever observed to emigrate (when dispersal was not associated with the death of the dominant male) were all subordinate to at least one other female in their group. Furthermore, of 13 emigrants in the Virunga population for whom there are enough data, only 2 transferred to groups that already contained more than one female, even though the expected number, based on the distribution of group size in the population, was 6. The other 11 females joined either small groups or solitary males ($X^2 = 8.42$; $df = 1$; $p < 0.01$). This result contradicts the suggestion that females in female-emigrant species form groups in order to improve predator detection (Moore 1984); if this were the case, they should join larger, not smaller, groups.

There is some evidence from other species that group size affects female reproductive success (Downhower and Armitage 1971; Hoogland 1981; Clutton-Brock, Guinness, and Albon 1982; Dunbar 1984b). For the entire Virunga population, an inverse relation exists between group size and the number of infants and juveniles per female (0.75 versus 1.5 immature offspring per female in groups that were larger than as opposed to equal to or smaller than the median group size; Mann-Whitney U test, $U = 3$, $N = 4, 8$; $p < 0.02$; $r_s = -0.68$; $N = 12$; $p < 0.02$) (see also van Schaik 1983; Moore 1984). This could be due to the greater costs of feeding competition in larger groups.

Despite the possible influence of group size on reproductive success, we have no evidence that females who emigrate are forcibly evicted through the aggression of other females or that immigrants receive aggression from resident females (Harcourt 1979b). It is possible that the silverback prevents females from attacking immigrants, although the paucity of data on immigrant females precludes any definitive conclusions.

Why Live in a Group?

If group life does impose constraints on feeding through increased competition, it might be expected that females who emigrated would travel alone rather than joining males or other bisexual groups (Wrangham 1979a). We suggest that female gorillas associate with males in order to gain protection against predation and aggression from other gorillas (Wrangham 1979a). Although gorillas have few natural predators, leopards have occasionally been reported to prey on gorillas (Schaller 1963), and in all parts of their range gorillas are hunted by humans. Infants are also vulnerable to attacks by unrelated males. Indeed, over a 15-year period, 38% of all infant mortality in the Virunga Volcano groups was directly attributable to infanticide (Fossey 1984).

Further evidence that females require the protection of males is that female transfer from one male to another almost always occurs when the silverbacks are in close proximity (Harcourt 1978b), which suggests that it is dangerous for females to travel alone. Moreover, even un-

mated males sometimes travel together. All-male groups occur, even though such associations presumably increase feeding competition (Fossey and Harcourt 1977; Watts 1985) and possibly also decrease the probability of acquiring females. Finally, females may live together in groups instead of monogamously both because more than one are attracted to the same male and because male-male competition allows some males to monopolize mates at the expense of others (Wrangham 1979a).

Although similar arguments might be applied to female chimpanzees, chimpanzees feed primarily on fruit that is usually distributed in patches too small to support groups. These dietary constraints may prevent females from traveling in permanent associations (Wrangham 1979a), even though they are also subject to intraspecific aggression and infanticide (chap. 15).

If the protection of a male is essential to female reproductive success, then the quality of the male partner may be more important than the presence of female kin (see also Wrangham 1979a). There is some evidence that females make choices about particular males with whom to reside. For example, five of eight females followed after their initial transfer are known to have transferred at least once more before settling down to breed (Harcourt 1978b). Some of these movements were between lone silverbacks or between groups with similar numbers of females, which suggests that the females were searching for the right male and not just responding to group size. The bases on which females choose males is not known, although in some cases they seem to prefer an older male who has been a full silverback for several years. In three of four cases when a dominant male died in a group that also included a younger silverback or a silvering blackback, all females emigrated and joined an older male (Fossey 1983; Watts 1985; Stewart and Harcourt, pers. obser.). An apparent preference for older and possibly more experienced males also has been suggested for female eastern lowland gorillas (Yamagiwa 1983). Once a female finds a suitable male, she may remain with him for many years. The mean length of time that females remain with particular males is not known. However, all nine females observed as breeding adults for 8 or more years (long enough to raise two offspring to weaning age) stayed with the same male for at least that period of time and produced successive offspring with him. One other female and a male remained together for at least 16 years (Fossey 1983). The maintenance of a close bond with a high-quality male may be more important to a female's reproductive output than her bonds with female kin. Thus, male-female, rather than female-female, associations may be the initial basis for gorilla groups. Causes and consequences are difficult to disentangle, but females in female-emigrant species often seem to have closer bonds with males than with other females (gorillas: Harcourt 1979b; hamadryas baboons:

Kummer 1968; red colobus: Struhsaker and Leland 1979; chimpanzees; Wrangham and Smuts 1980).

Emigration with Kin

Theoretically, females who transfer with kin have the opportunity both to choose a protective male and to remain with their relatives. To some extent, this seems to occur. In the Virunga Volcano groups, 7 of 16 females transferred at least once in the company of females from their previous group. In some cases, such pairs remained together through several successive transfers. Two females who had matured together (and who may have been half or even full sisters) transferred together to three different males before settling to breed. Similarly, after the death of one dominant silverback, a mother-daughter pair transferred twice before settling together with another silverback. Females may even join particular groups because they contain relatives who transferred there previously (cf., Cheney and Seyfarth 1983 for male vervets), although the extent to which this occurs is difficult to determine. In other female-transfer primates there is also some evidence that kin remain together when possible and that associations among female kin are perhaps more common than was previously supposed. For example, some chimpanzee females emigrate to breed, but then return to their natal community (Pusey 1980, Goodall 1983). Female hamadryas baboons also tend to transfer into groups containing females likely to be close relatives (Sigg et al. 1982; chap. 10).

Given the social structure of gorillas groups, however, emigration with kin is unlikely to be the predominant pattern. A breeding female's loyalty to her mate will probably prevent her from joining a daughter who emigrates in search of unfamiliar mates. Perhaps as a result, mother-daughter transfer has been observed only following the death of a dominant silverback. Furthermore, since sisters will on average be 8 (and never less than 3) years apart in age, simultaneous transfer by full sisters will probably be uncommon. The number of half sisters or less closely related peers who are of roughly the same age, and thus likely to emigrate at the same time, will be constrained by small group size. Finally, subsequent transfers to join female kin may be constrained by the fact that multiple transfers can take females out of the range of their female relatives.

CONCLUSION

The frequency and nature of agonistic interactions among adult females suggest that direct competition for re-

sources may occur at lower levels in gorillas than in some female-resident primates. Thus, the benefits of kin-based alliances, at least in feeding competition, may be less important to female gorillas than, say, to female macaques or baboons (Wrangham 1980). Nevertheless, this is not enough to explain female emigration and transfer. Given that some interference competition does occur, gorillas could in theory benefit from cooperative alliances with their kin. Indeed, when female relatives occur in the same group as adults, they favor each other in much the same way as do kin in female-resident species. It appears that in any system in which females live in permanent groups there will always be a tendency for animals to associate with their relatives.

Various aspects of the mating system constrain the formation of female kin groups in gorillas. Male strategies seem important in preventing female philopatry. Females may be forced to emigrate initially to avoid breeding with their fathers, since silverbacks often retain tenure for more than one generation and prevent the immigration of unrelated males. The influence of a powerful male in a small group also appears to inihibit the development of kin-based alliances within the group, effectively limiting the usefulness of living with kin. When the costs of group life (likely to involve increased feeding competition) outweigh the benefits, females may do better by moving elsewhere. Thus, group size may be one basis on which females decide whether or not to emigrate and which group to join.

If competition resulting from group life constrains female reproductive success, it might be expected that female gorillas would travel alone rather than in groups. However, we suggest that the risk of predation and intraspecific aggression causes females to associate permanently with dominant, protective males. Groups may occur both because more than one female chooses the same male and because male-male competition allows some males to monopolize females at the expense of others. The optimal tactic for females might be for close relatives to choose the same male. Once again, however, constraints within the system may inhibit this. Long interbirth intervals, successive transfers, and the loyalty of breeding females to particular males decrease the likelihood that close relatives will be able to transfer simultaneously or at different times to the same silverback.

Although long-term bonds between males and females may be the social bases of gorilla groups, attraction among female kin is likely to be an important factor determining the extent and timing of female transfer.

15 | Chimpanzees and Bonobos: Cooperative Relationships among Males

Toshisada Nishida and Mariko Hiraiwa-Hasegawa

Chimpanzees have been known since the early 1960s to associate in temporary parties, unlike the stable groups of most large primates. In 1968 Nishida showed that these parties are formed within stable "unit-groups" (Nishida 1968), or communities (Goodall 1973), whose members share a common home range (see also Kawanaka 1984; Nishida et al. 1985). [*Unit-group* was the term chosen by Nishida (1968) when data first revealed that chimpanzees form a closed social network. Although *unit-group* therefore has priority, *community* is used here because it has become the standard term among anglophone primatologists since its introduction by Goodall (1973)—ED.] Bonobos are now known to be similar to chimpanzees in forming temporary parties within communities. However, since the late 1970s it has emerged that bonobo social behavior differs in many important ways from that of chimpanzees. A consistent tendency to form temporary parties within long-term communities is found in only one other genus of primates, the spider monkeys (chap. 7).

This chapter reviews the social behavior of the two species of *Pan* so as to highlight their similarities and differences. Special attention is paid to male-male relationships because male chimpanzees are unusually cooperative, and both species are remarkably tolerant compared to other great apes. They are discussed in the second half of the chapter.

SPECIES, DISTRIBUTION, AND HABITAT

There are three subspecies of the common chimpanzee *Pan troglodytes*, found in western, central and eastern Africa north of the Zaire River. The bonobo or pygmy chimpanzee *P. paniscus* forms a single group south of the Zaire River (Allen 1939). Bonobos are distinguished from chimpanzees by their slender frame, long hind limbs, short clavicle, and small molars (Johnson 1981). They are also commonly thought to be smaller and less sexually dimorphic than chimpanzees. However, although bonobos are indeed smaller than central African chimpanzees (*P. t. troglodytes*), present data suggest that they are the same weight as the East African subspecies (*P. t. schweinfurthii*) (Jungers and Susman 1984). Furthermore, the few published weights of wild adults show little difference in sexual dimorphism, with males being

around 30% heavier in both species (Jungers and Susman 1984).

Chimpanzees are primarily forest-living animals, but they also inhabit forest-savanna mosaic, woodland, and even dry savanna containing, for example, patas monkeys and baobab trees (*Adansonia digitata*) (Kano 1972; McGrew, Baldwin, and Tutin 1981; Baldwin, McGrew, and Tutin 1982). Nevertheless, they invariably rely on the presence of a small amount of forest for seasonal foods. Bonobos, by contrast, live only in lowland rain forest and swamp forest (Kano 1983, 1984).

Long-term studies of chimpanzees have been carried out in Gombe National Park (Goodall 1983) and the Mahale Mountains (Nishida 1979), both in Tanzania. In Mahale, long-term observations have been made on two neighboring communities, K group (30 individuals) and M group (100). Other populations have been studied in Tanzania (Suzuki 1969; Izawa 1970), Uganda (Reynolds and Reynolds 1965; Sugiyama 1968; Ghiglieri 1984, 1985), Equatorial Guinea (Sabater Pi 1979), Ivory Coast (Boesch and Boesch 1981), Guinea (Albrecht and Dunnett 1971; Sugiyama 1981), Liberia (Anderson, Williamson, and Carter 1983), and Senegal (Tutin, McGrew, and Baldwin 1983). Bonobos are confined to Zaire, where they have been studied in Wamba (Kano 1982a) and Lomako (Badrian, Badrian, and Susman 1981), with short studies also at Yalosidi (Kano 1983) and Lac Tumba (Horn 1980).

In Gombe, Mahale, and Wamba artificial feeding has been used to habituate communities. This can influence behavior at the artificial feeding area by increasing average party size and frequency of aggression (Wrangham 1974). No systematic effects on behavior have been found away from the feeding area, however (Wrangham and Smuts 1980; Goodall 1983). Without artificial feeding, individuals tend to be both shy and difficult to find. The most detailed descriptions of social behavior are therefore from Gombe, Mahale, and Wamba, which provide the majority of data for this chapter.

ECOLOGY

Both species are diurnal. So far their activity budgets have been confirmed only from Gombe, where male chimpanzees spend 46 to 60% of their time feeding, 8 to

TABLE 15-1   Some Ecological Comparisons of Chimpanzees and Bonobos

	Pan troglodytes	*Pan paniscus*
Major food type	Fruits	Fruits
Plant food types	Fruits, leaves, pithes, shoots, seeds, barks, flowers, wood, resin	Fruits, pithes, shoots, leaves, seeds, flowers, barks
Time spent feeding	46–60% (Gombe)	Ca. 30% (Wamba)
Vertebrate food	Primates, bovids, suids, rodents	Rodents, bovids, insectivores, snakes
Invertebrate food	Ants, termites, bees, caterpillars, beetles	Ants, termites, bees, crickets, caterpillars, beetles, earthworms, millipeds, snails
Minerals	Termite clay, rock	Termite clay
Subsistence tool use	Common and variable	Not observed, except for rain cover
Community range size	Densely wooded regions: 5–38 km^2 ($N = 8$) Sparsely wooded regions: 25–560 km^2 ($N = 5$)	Ca. 22 km^2 ($N = 1$, Lomako) 40–50 km^2 ($N = 3$, Wamba)
Population density	0.09–5/km^2	2–3/km^2

20% traveling, and 25 to 39% resting or grooming (Wrangham 1977). Preliminary data suggest that Gombe females spend more time feeding than males (Wrangham and Smuts 1980), whereas at Ngogo (Uganda) males appear to feed for longer (62%) than females (52%) (Ghiglieri 1985). There is some seasonal variation: for instance more time is spent feeding in the dry than the wet season.

Chimpanzees are omnivorous, with plant foods invariably predominant in their diet (table 15-1). Wrangham (1977), for example, found that over the year the proportion of time spent eating different foods was 56 to 71% for fruits, 18 to 21% for leaves, and 11 to 23% for other items including meat. Similarly, chimpanzees released onto an island in Gabon had a diet composed of 68% fruits (wet weight), 28% leaves, bark, and stems, and 4% animals (Hladik 1977). Seasonal changes in the distribution and availability of plant foods appear to influence grouping and ranging patterns. In particular there is evidence that average party size is larger when food is more abundant (Wrangham 1977).

Chimpanzees eat a variety of animal foods (table 15-1). In almost all sites they consume ants and termites, often with the aid of tools (Teleki 1974). They also occasionally hunt or scavenge vertebrates as prey, chiefly monkeys and ungulates (Nishida and Uehara 1983; Hasegawa et al. 1983). Both at Gombe and Mahale males have been found to eat more meat than females, whereas females eat more insects than males (McGrew 1979; Uehara 1984). Sex differences have also been found in plant diets. In the Ivory Coast both sexes crack open hardshelled fruits with stones or wooden clubs, but females are more frequent and efficient nut crackers than males (Boesch and Boesch 1981).

Bonobos have diets similar to those of chimpanzees, but "fibrous foods" (mostly herbs, shoots, pith, and stems of ground plants) are more important as an alternative food for bonobos when fruit is scarce (Kano 1983; Badrian and Malenky 1984). They also eat a wider range of invertebrates, including earthworms and millipedes (Kano 1983; Badrian and Malenky 1984; Kano and Mulavwa 1984). Bonobos have not been seen using tools to obtain invertebrates (Kano 1982a).

Community ranges vary widely in size (table 15-1). Ranges are larger (*a*) for larger communities and (*b*) for those living at low population density, presumably because food density is low also. Chimpanzees in arid low-density areas, such as Filabanga (Tanzania) or Mt. Assirik (Senegal), migrate seasonally within enormous ranges that include a variety of habitat types (over 200 km^2; Kano 1972; Baldwin, McGrew, and Tutin 1982).

Within chimpanzee communities the annual range of females and males is similar, since all individuals can use the entire area. Both in Gombe and Mahale, however, males have been shown to travel farther per day than females (Nishida 1979; Wrangham and Smuts 1980). In Gombe, for instance, the median day-range was 4.1 km for adult males versus 2.8 km for anestrous females. The sex difference occurs partly because males spend their time more evenly over the entire community range.

CONSERVATION

Although chimpanzees and bonobos are both protected by international legislation, they are in high demand for medical research, and hunting bans are not always enforced. In many countries they are heavily poached, even where they are legally protected. Occasional surveys have been made of chimpanzee populations (Tutin and Fernandez 1984), but the distribution and status of both species is poorly known. The spread of agriculture and timber exploitation is a severe threat throughout their ranges.

REPRODUCTIVE PARAMETERS

Reproductive parameters, given in table 15-2, are based primarily on field data given by Hasegawa and Hiraiwa-Hasegawa (1983; Mahale) and Tutin and McGinnis (1981; Gombe). No birth seasonality is apparent for ei-

TABLE 15-2   Reproductive Parameters for Chimpanzees and Bonobos

	Pan troglodytes	*Pan paniscus*
Age at sexual maturity		
Male: first ejaculation	9–10 years	?
suspected first fertilization	13–14 years	?
Female: menarche	10–11 years	8–9 years ?
first birth	10–18 years	13–14 years ?
Gestation period	228 days (*N* = 49, range 205–48)[a]	?
Birth interval	4–8 years	3–7 years ?
Adult estrous cycle	Regular	Irregular
	36 days (*N* = 46, range 25–84, in Gombe)	35–40 days (Wamba)
	31.5 days (*N* = 44, range 27–39, in Mahale)	46 days[b]
Adolescent estrous cycle	Irregular	Irregular (maximal and nearly maximal swelling phase consisted of 92% of observational days in Wamba)
Period of maximal swelling	9.6 days (*N* = 37, range 7–17, in Gombe)	14 to more than 20 days (Wamba)
	12.5 days (*N* = 27, range 8–19, in Mahale)	22.4 days[b]
Postpartum nonreceptivity	3–6 years	Less than 1 year

SOURCES: (a) Nissen and Yerkes 1943; (b) Dahl, pers. comm., cited in Thompson-Handler, Malenky, and Badrian 1984.

TABLE 15-3   Features of Social Units and Grouping Patterns of Chimpanzees and Bonobos

	Pan troglodytes	*Pan paniscus*
Social units	Multimale, bisexual community, or "unit-group"	Multimale, bisexual community, or "unit-group"
Size of community	20–105 (*N* = 10)	50–120 (*N* = 4)
Patterns of subgrouping	Party of any size and composition	Multimale bisexual parties consisting of matrifocal subunits are most common
Party size	Small (range 1–77); parties of less than 6 members are more common	Large (range 1–70); parties of more than 6 members are more common
Ratio of mature/immature	1:0.5–1.2 (*N* = 10)	1:1 (*N* = 1)
Male/female ratio	1:1.0–3.5 (*N* = 6)	1:1 (*N* = 1)
Socionomic sex ratio	1:0.7–3.6 (*N* = 8)	1:1.2 (*N* = 1)
Transfer between communities	Usually females, not males	Perhaps females, not males
Intercommunity relation	Antagonistic	Antagonistic

ther species. Bonobos have not yet been studied long enough to provide data on age at sexual maturity or birth interval; table 15-2 therefore gives provisional estimates.

Females of both species have conspicuous midcycle sexual swellings. Female bonobos are more sexually active than chimpanzees because their period of maximal tumescence is longer and they resume sexual swellings more rapidly after a birth. As yet, however, precise data are not available on bonobo sexual cycles.

SOCIAL ORGANIZATION OF CHIMPANZEES
Composition of Communities and Parties

Chimpanzee communities contain all age-sex classes and number anywhere from 20 to more than 100 individuals (table 15-3). Sex ratios (including immatures) have been

calculated for six communities, and in all cases females outnumbered males. The number of adult females is larger than that of adult males in eight communities but smaller in two communities (Hiraiwa-Hasegawa, Hasegawa, and Nishida 1984). In spite of the uneven sex ratio, there are no observations of all-male groups or lone males living outside communities.

Parties are temporary associations lasting from a few minutes to several days and varying in size from 1 to 77 (Hiraiwa-Hasegawa, Hasegawa, and Nishida 1984). The only long-term party is a mother with her dependent offspring. Most parties are small. Thus parties of six or less accounted for 82% of observations in Gombe (*N* = 498) (Goodall 1968), 62% in Budongo (*N* = 215) (Reynolds and Reynolds 1965), and 55% in Mahale (*N* = 218) (Nishida 1968). Community sizes for these three populations were estimated to be 60 to 80, 80, and 27 respectively.

Any combination of age-sex classes may be found together, but there is a clear sex difference in grouping patterns: adult males are consistently more sociable than adult females (Nishida 1968). In Gombe, for instance, the amount of time spent with at least one other adult is more than twice as much for males (73%) as it is for anestrous females (35%). Females also spend longer continuous periods alone, and shorter periods in parties, which means that they meet others less often than males do, and they tend to leave parties sooner (Wrangham and Smuts 1980). Figure 15-1 illustrates the sex difference in a small Mahale community by showing that the "familiarity index" is higher for males than females.

Female Relationships

Unrelated adult females show little evidence of mutual attraction (Nishida 1979). They rarely engage in long grooming bouts, for instance, and aggressive coalitions are uncommon. However, in Ngogo (Uganda) mothers with infants tend to travel with each other more than with adult males (Ghiglieri 1984). Lactating mothers are also often accompanied by nulliparous females who interact frequently with their infants (Nishida 1983a). Female dominance relationships have not been described in detail. The hierarchy is certainly less clear-cut than among males, but older females are generally dominant to younger ones. Newly immigrated females have the lowest status (Nishida 1979).

The mother-offspring relationship lasts for several years beyond weaning (Goodall 1975; Pusey 1983). Sons continue to travel with their mothers until adolescence and may transfer with them to new communities as juveniles ($N = 4$ cases at Mahale, 2 at Gombe) (Goodall 1983; Nishida et al. 1985). During adolescence sons begin to spend more and more time with adult males. By the time they reach adult size, males are dominant to all females and are typically integrated into the male hierarchy. Sons nevertheless continue to travel with their mothers occasionally and may groom with them for long periods (Pusey 1983). Association between adult daughters and their mothers is less common, but it occurs when females do not transfer at puberty. Mother-daughter association occurs primarily when the daughter is anestrus. During estrous periods she travels with adult males, and her mother may or may not join her.

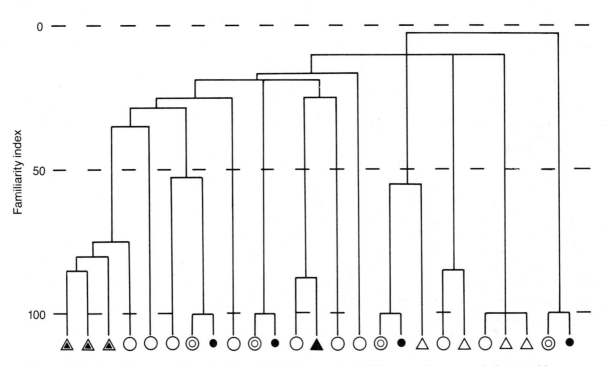

FIGURE 15-1. Association pattern based on hierarchical cluster analysis of K-group chimpanzees in January–May 1976, indicating strong bonds among adult males and between mother and offspring ($N=524$ parties). Figures represent "familiarity index": $100c/(a + b + c)$. *For two individuals A and B,* $a=$number of parties including A but not B; $b=$number of parties including B but not A; $c=$number of parties including both A and B. ▲=adult males; ○=adult cycling females; ◉=lactating females; ▲=adolescent male; △=adolescent or juvenile females; ●=infants. SOURCES: NISHIDA 1968, 1981.

Dispersal

Female chimpanzees show a strong tendency for migration between communities (chap. 21). In Mahale, for example, all K-group females ($N = 10$) who survived to early adolescence emigrated before their first birth. Although this suggests that adolescent females consistently emigrate, there is variation between communities. Thus during 15 years at Gombe only 2 females emigrated from the main study community, while at least 4 natal females remained and gave birth (and 12 females immigrated) (Goodall 1983). Similarly, in M-group between 1979 and 1983 only 3 natal females were presumed to have emigrated, while 2 remained and gave birth (and 7 immigrated from communities other than K group). Possibly females are more likely to remain in large than small communities.

Unlike adolescents, parous females do not normally transfer. However, between 1979 and 1982 every sexually cycling female in K-group ($N = 9$) regardless of age or parity, transferred to M-group. This high migration rate followed the disappearance of the adult males from K-group, apparently as a result of intercommunity aggression. It was therefore certainly an unusual case (Nishida et al. 1985).

Females transfer when they are in estrus and are attracted to high-ranking males, especially the alpha male (Nishida 1979). Males differ in their attitude toward immigrants: some protect them eagerly, while others are aggressive (Nishida and Hiraiwa-Hasegawa 1985). Resident females usually have few interactions with new immigrants, but sometimes the resident females show hostility, and make threats and displays or bark at them (Pusey 1980; Nishida 1981).

Mating Relationships

No evidence has been found for long-term relationships between particular females and particular males. Short-term sexual relationships occur in three patterns. In opportunistic matings a male copulates freely in the presence of other males; in the possessive pattern a male (typically the alpha) uses threats or aggression to prevent other males from mating; consortships occur when a male (not the alpha) and a female travel together away from other members of the community and maintain an exclusive mating relationship for a few days or weeks (Tutin 1979).

Matings occur principally during the period of maximal tumescence (table 15-2; fig. 15-2). The relative fre-

FIGURE 15-2. An adult male chimpanzee inspects a female by touching her vulva and sniffing his finger. Such genital inspections are common when females do not have sexual swellings, but may occur at any time. (Photo: Toshikazu Hasegawa)

quency of mating patterns appears to depend on the number of adult males present in a community. In a large community, 70 to 90% of all matings are opportunistic, with no indication of male-male competition. These occur especially during the early part of the estrus period. Ovulation occurs within one day of detumescence (Graham 1981). As the probable day of ovulation approaches, the proportion of possessive matings increases, especially by the alpha male. The alpha male thereby achieves a high mating success around the time of ovulation. It is still uncertain whether other high-ranking males have any advantage over low-ranking males in mating competition (Tutin and McGinnis 1981; Hasegawa and Hiraiwa-Hasegawa 1983). Tutin (1980) suggested that consortships, which occur at a higher frequency in Gombe than Mahale, may lead to a higher rate of conception than other mating patterns. Among consorting males no relationship has been found between dominance rank and frequency of consortships.

Intergroup Aggression and Infanticide

Consortships take place in the border area between neighboring community ranges. As a result they probably involve significant risk because chimpanzees have aggressive and dangerous intercommunity relationships. Meetings occur partly because parties meet by chance in overlap areas, but also because adult males occasionally "patrol" the border area of their communal territory (Bygott 1979; Goodall et al. 1979; Nishida 1979). The outcome is determined partly by calling. Small parties tend to avoid large parties, which leads to consistent dominance relationships by large communities over small (Nishida 1979).

Encounters between different communities carry a risk of severe aggression, except when a party encounters an estrous female in the process of transfer. This was seen most dramatically after the fission of Gombe's study community in 1970–71. Adult males of the main community repeatedly invaded the territory of the branch community and deliberately killed at least three adult males and one adult female. They were suspected of killing two additional adult males (Goodall et al. 1979). Similarly at Mahale, at least two adult males of K-group were thought to have been victims of intercommunity aggression (Nishida et al. 1985).

In addition to adult killings, infanticides (and subsequent cannibalism) have been reported in interactions both between and within communities (J. Goodall 1977; Kawanaka 1981). All intercommunity infanticides have followed the same pattern—adult males killing infants of unfamiliar females ($N = 3$ at Gombe, 2 at Mahale).

The cases of infanticide within communities, however, fall into two classes. At least three Gombe infants were killed by an unrelated adult female and her adolescent daughter. Possibly this benefited the adolescent by re-

pelling the victims' mothers and allowing her to secure a core foraging area near her mother (Pusey 1983).

The cases at Mahale, by contrast, were of familiar infants killed by resident males. They are therefore puzzling because the infants were apparently conceived within the killers' community (M-group), where the killers were possible fathers. The victims were all infants of

FIGURE 15-3. An adult male chimpanzee (Rodolf) reassures an adolescent female (Gilka) by touching her hand (Gombe National Park, 1973). The pair had been traveling together for 2 days, mostly without other companions. Shortly before the photograph was taken, Rodolf climbed into the tree (*Sterculia quinqueloba*) to eat seeds. Gilka watched, grunted softly, climbed up to the position shown, and extended her hand toward Rodolf. After he reached down to touch her hand, Gilka climbed past him to feed. If she had climbed past him directly, it is possible that he would have behaved aggressively. (Photo: Richard Wrangham/Anthrophoto)

females who had immigrated from K-group. Nine such females gave birth a total of 16 times after transferring to M-group: 4 of these infants were killed and eaten, and a fifth infanticide was strongly suspected. During the same period more than 30 infants were born to resident females (i.e., M-group members since 1972), none of whom, apparently, were victims of intracommunity infanticide. Since the mothers of the infanticide victims tended to occupy the periphery of the M-group's territory (i.e., the overlapping area of K-group and M-group ranges), M-group males may have recognized that K-group males were possible fathers (Nishida and Kawanaka 1985).

Communication

Aggressive behavior does not normally lead to physical violence. Chimpanzees have complicated systems of gestural and vocal communication, such as kissing, embracing, and touching, used for greeting, appeasement, reassurance, and display (fig. 15-3). These behaviors are performed in the context of reunions, dominance interactions, reconciliation, excitement, fear, pleasure, and grief (Goodall 1968; Nishida 1970; de Waal 1982). For example, adult males make frequent "charging displays," including branch shaking, branch dragging, stamping, slapping, and so forth (Goodall 1968). Subordinate individuals respond to such displays by giving "pant-grunts," that is, a series of soft or loud grunts. Dominant individuals, in their turn, may acknowledge the behavior of subordinates with reassurance signals such as soft biting or touching.

The most conspicuous vocalization is the "pant-hoot," which is a series of loud calls rising and falling in pitch, often ending in a scream (fig. 15-4). Males give pant-hoots more frequently than females, especially at rich feeding sites, at reunions of parties separated for a few days, in responses to distant calls, and when others give charging displays. Pant-hoots were given more often in a season of food abundance than of food scarcity. Therefore no evidence exists that they increase foraging efficiency (Wrangham 1977). The supposed functions of pant-hoots include demonstrating the location of individuals, announcing the presence of large food supplies, and threatening individuals in the same or other communities (Goodall 1968; Wrangham 1977). Chimpanzee communication has not been studied in detail in the wild

FIGURE 15-4. Two adult male chimpanzees, the alpha (*left*) and gamma, give pant-hoots (Mahale). Partial erection of hair indicates their state of arousal. (Photo: Toshisada Nishida)

but appears to be unusually complex and presumably related to the fission-fusion pattern of their social organization (see Marler 1976b).

BONOBO SOCIAL ORGANIZATION: CONTRASTS WITH CHIMPANZEES
Subgrouping within the Community

Bonobos also live in communities that split into temporary parties. However, unlike chimpanzees, "mixed parties" are predominant (i.e., including males and mothers; Kuroda 1979; Badrian and Badrian 1984), and parties are comparatively large, at least at Wamba. Thus 96% were mixed parties (Kano 1982a), compared to 30 to 50% mixed parties usual for chimpanzees (Tutin, McGrew, and Baldwin 1983); parties of six or more members accounted for 94% of all parties observed (N = 180, range 1 to 37, community size = 63).

Social interactions have been studied in detail only in one community, the E-group of Wamba. E-group was originally reported to be divided into northern and southern subcommunities (Kano 1982a; Kitamura 1983). Recent observations suggest that the community was then in the process of fission. Patterns of subgrouping were similar before and after the community divided. Kano (pers. comm.) has summarized his recent findings as follows.

Four types of subunit are typical: matrifocal units, male bands, male singletons, and female singletons. First, matrifocal units consist of mothers and their offspring, including mature sons. These are relatively stable associations. For instance, adult males are normally not separated from their mothers for more than a day. Six northern males and four southern males belong to matrifocal units. Second, male bands consist of adult males, of unknown relatedness. Five northern and three southern males form two male bands respectively, and the males of each band usually move together. Third, male singletons are old or seriously handicapped. Three northern and one southern male belong in this category. Finally, female singletons are nulliparous females who are presumed to be immigrants.

Subgrouping of communities (or, during the period of fission, subcommunities) is usually based on associations of the above subunits. Aggregations of matrifocal units are the most common parties. Matrifocal units also regularly join male bands, and these associations give rise to the largest parties. Male and female singletons associate with each other as well as with male bands and matrifocal units. There are no indications of affinity among these young adult females.

The male dominance hierarchy is linear, at least within the five-male band of the northern subcommunity. All of these males were dominant to adult males in matrifocal units. Adult males of matrifocal units benefit greatly from their mothers' assistance in dominance interactions

(Kano, in press). It is unknown whether adult males of matrifocal units ever join a male band before their mothers die.

Unlike chimpanzees, few sex differences are known in association patterns. Thus, the mean sex ratio both in small and large parties is equal to the sex ratio in the community (ca. 1:1). Party size is not changed by the presence of estrous females, which is not surprising since almost all parties contain estrous females. Thus there is no obvious tendency for males to associate with each other more than with females.

Female Relationships

Females appear to leave their natal communities in early adolescence. At least four adolescent females have been seen to interact with two different communities (Kano 1982a), but the details of transfer are not yet known.

Striking differences between bonobos and chimpanzees occur in grooming and food sharing. In bonobos, grooming among adults occurs most frequently between males and females, then among females, and least among males (Kano 1980; Kuroda 1980). This contrasts sharply with chimpanzees, for whom grooming is most common among males, then between males and females, and least among females. Similarly, in bonobos, food sharing among adults is most common among females, who share with unrelated individuals as well as with their own infants, but rare among males (Kuroda 1980); whereas in chimpanzees, male-male food sharing is common (Nishida 1970), while female chimpanzees almost never share food with individuals other than their infants (McGrew 1975).

These contrasts reveal the highly sociable nature of female bonobos, which is further illustrated by their genital-rubbing behavior. This is performed by adult or adolescent females embracing each other ventro-ventrally, both making rapid laternal pelvic thrusts. Genital rubbing occurs most often before competitive interactions over food or mates and apparently functions to resolve tensions between the participating females (Kano 1980; Kuroda 1980). Female chimpanzees, by contrast, have no special mechanisms for mutual appeasement or greeting. They avoid competition primarily by dispersing themselves widely in the community range.

Sex and Aggression

Bonobo sexual behavior also differs from that of chimpanzees. They have abundant, mutual precopulatory and copulatory signals, variable copulatory postures, and more prolonged copulations (table 15-4). In captivity numerous signals of posture and gesture are used to communicate desired changes of position, and mutual gazing occurs prior to and during nearly all copulatory bouts (Savage-Rumbaugh and Wilkerson 1978). The fact that ventro-ventral copulations occur regularly among bonobos (30%; Kano, pers. comm.) is in marked contrast to its virtual absence in chimpanzees. These behaviors may function to strengthen male-female bonds in bonobos.

Despite these remarkable differences in sexual behavior, no differences have been found in the interbirth interval or the age of weaning, nor is fecundity significantly greater in bonobos than chimpanzees. Bonobos have more anovulatory cycles than chimpanzees, and their more elaborate sexual behavior is apparently used to manipulate relationships rather than to increase reproductive rates (Kano, pers. comm.).

Finally, a series of differences occurs in aggressive behavior (table 15-5). The most striking difference is that bonobos have no pant-grunt sounds: instead, faint ⟨ku:ku:ku:⟩ sounds are given. Their charging displays contain fewer components, their aggressive behavior is milder than that of chimpanzees, and adult males are not always dominant to adult females in dyadic interactions (Kano, in press). These characteristics may be related to their comparatively stable parties and low levels of competition among individuals. Little is yet known of relationships between communities. Intercommunity en-

TABLE 15-4   Comparison of Sexual Behavior between Chimpanzees and Bonobos

	Pan troglodytes	*Pan paniscus*
Prolonged mutual gaze to communicate readiness to copulate	None	In nearly all bouts
Facial and vocal signals to communicate desired copulatory positions	None	In nearly all bouts[a]
Position signal	None	Very common[a]
Initiation of copulation	Mostly males	Mutual
Copulatory posture	Ventro-dorsal (virtually 100%)	Ventro-dorsal (70–74%) Ventro-ventral (26–30%)
Timing of copulation	Confined to maximal swelling phase	Maximal and submaximal swelling phase
Temporal distribution of copulation	Highest peak in the morning Second-highest peak in the evening	Highest peak in the morning Second-highest peak in the evening
Duration of copulation	7 sec (*N* = 45, range 3–15 in Gombe) 8.27 sec (*N* = 284 in Mahale)	15.3 sec (*N* = 121, range 1–52, in Wamba) 12.2 sec (*N* = 51, range 1.5–45, in Lomako)
Number of thrusts	8.8 (*N* = 1084, range 3–30 in Gombe)	43.8 (*N* = 16, range 14–78, in Wamba)
Copulation rate	0.62/hr (Mahale)	1.6/hr (Wamba)
Copulatory gaze	Virtually nonexistent	Very common
Copulatory call	One type of squeal	3 types of squeal[a]
Sexual interference	Mostly males	Mostly males
Male copulatory right	Dominance	Dominance

SOURCE: (a) Savage-Rumbaugh and Wilkerson 1978.

TABLE 15-5   Behavioral Contrasts between Chimpanzees and Bonobos

	Pan troglodytes	*Pan paniscus*
Male-male association	Very strong	Moderate
Male-female association	Weak	Strong
Female-female association	Very weak	Strong
Mother-son association	Up to adolescence	Up to adulthood
Grooming	Male-male most, female-female least	Male-female most, male-male least
Food sharing	Male-male most, female-female virtually absent	Male-female most, male-male least
Female genital-contact	Absent	Quite common
Rump contact in males	Very rare	Common
Long call	Pant-hoot with climax phase	Simple and without climax phase
Submissive greeting	Pant-grunt	No equivalent of pant-grunt, but faint [ku:ku:ku:]
Scream	[Kya:kya:]	[Nbi:nbi:] or [ngi:ngi:]

counters are sometimes avoided, but when they occur they appear antagonistic and may be accompanied by adult males giving branch-dragging displays.

Ecological Differences

Differences in social organization and behavior between the two species may be related ultimately to differences in food-patch size and patterns of food production between their habitats (Badrian and Badrian 1984). Unsystematic observations suggest that in Zaire large trees with numerous feeding sites are particularly common and that fruit production is more stable and continuous in Zairean than in Tanzanian forests. The forest structure may also differ in other ways. For example, there are no fig species in the food repertoire of the Wamba bonobos, and only three species of figs are eaten by bonobos at Lomako. Many different species of figs are eaten by chimpanzees, however, and they appear to be the most important food type throughout the range of chimpanzee habitats. Moreover, even in continuous primary forest the sizes of feeding parties of chimpanzees almost never exceed 15 (Ghiglieri 1984), whereas an aggregation of 30 to 50 bonobos in a single party is not uncommon (Kuroda 1979). Bonobos may also use terrestrial foods more than chimpanzees. These tend to occur in large patches and could be responsible for reduced feeding competition (Wrangham, 1986).

◆ COOPERATIVE RELATIONSHIPS AMONG MALES
Male Relationships among Old World Monkeys and Apes

In most monkeys with multimale groups, tolerant or cooperative relationships among males are rare or unknown. Male-male grooming, for example, is virtually nonexistent in rhesus monkeys (Lindburg 1973), Japanese macaques (Hasegawa and Hiraiwa 1980), and gray-cheeked mangabeys (Struhsaker and Leland 1979). Moreover, if grooming ever occurs, it is given entirely by subordinates to dominant males (Furuya 1957), unlike the more reciprocal system in chimpanzees. As another example, Watanabe (1979) studied alliance formation among Japanese macaques. Out of 905 cases only 4 alliances were between adult males. Relationships between males in these groups are thus primarily competitive.

This is explicable partly because males in multimale groups are often unrelated to each other, having immigrated as adolescents or young adults (chap. 21). As a result they are unfamiliar, and there are no direct genetic advantages to helping each other. Still, in some species cooperation can occur regularly even among unrelated males. It has been documented most frequently in savanna baboons, where reciprocal relationships among males are now known from a number of different sites

(Packer 1977; Rasmussen 1980; Collins 1981; Smuts 1985). It is not known why male relationships differ among cercopithecines living in multimale groups.

In the few cases where males are known to be related, cooperative behavior occurs regularly. For example, in red colobus only natal males appear to be accepted into the adult male subgroup, whose membership can be stable for several years at a time (Struhsaker and Leland 1979; chap. 21). Adult males spend much of their time close to each other and cooperate in aggression against neighboring groups. Two adult males may even be in physical contact while feeding from the same branch (Struhsaker 1975).

Among the great apes, orangutans have solitary adult males, while gorillas form one-male groups. Although more than one adult male may live in the same gorilla group, few interactions occur between them (Harcourt 1979a). Both in the wild and in captivity, males may live peaceably in all-male groups, but the introduction of females to captive males can lead to fighting (de Waal 1982). In wild groups two silverback males may cooperate to protect females and young against other groups or human poachers. The dominant silverback leads the group and decides the timing and direction of travel, while the subordinate brings up the rear. Cooperative pairs of this type appear to be close relatives, for example, father-son (Fossey 1983).

Only a few observations give direct evidence that alliances occur among male bonobos (Kano, in press). For example, when the two lowest-ranking males of the northern male band at Wamba were separated from the rest of the band, they showed subordinate behavior (grimacing, screeching, and avoidance) to males that they usually threatened. This implies that their usual companions acted as allies in aggressive encounters. Thus male-male bonds almost certainly occur among bonobos, but they are expressed remarkably little compared to those formed by chimpanzees.

Male Bonds among Chimpanzees

Male chimpanzees spend more time in parties than females, as discussed earlier, and as a result the highest levels of association between individuals occur among males (fig. 15-1). In addition to spending more time with each other than with females, males have frequent interactions in a variety of contexts.

Grooming is clearly affiliative, for example, and in Mahale Nishida (1979) found that grooming among adult males was observed more than 4 times as frequently as among adult females. (This does not take into account the amount of time individuals spent together.) Greeting by pant-grunting or kissing was observed among males 20 times as much as among females, and more than 80% of all the observed greetings by embracing were performed by adult males (Hasegawa, unpub. data). In Gombe, adult males greeted one another by pant-grunting about 9 times as frequently as did adult females (Bygott 1974).

Similarly, meat sharing is observed frequently among adult males, but rarely among adult females (Teleki 1973a, 1973b). The possessor sometimes breaks off small pieces of meat and hands it to a beggar or allows the beggar to nibble for meat. Chimpanzees also show considerable male-male tolerance in the context of mating. Females sometimes copulate with several males in rapid succession, which means that high-ranking males may connive at the copulations of lower-ranking ones. Other cooperative behavior includes joint charging displays and food calling.

Such examples stress the high degree of tolerance among male chimpanzees. Their relationships are also characterized by intense competition, however. Males appear continuously preoccupied with raising or maintaining their dominance rank (Goodall 1975), and this can lead to rapid changes in relationships. For example, while a gamma male was competing for status with a beta male, the beta male was attacked by the alpha. The gamma immediately took the opportunity to attack and outrank the beta (Nishida 1979). The tensions generated by such competition serve both to favor and to disrupt cooperative alliances, with complex results that will be discussed later.

Dominance Rank and Alpha Status. The dominance relationship between two males is detectable from the direction of pant-grunting, which is never reversed over the short term (de Waal 1982), as well as by other interactions such as an attack by the dominant individual (Bygott 1979). These signals are excellent predictors of the ability to win competitive interactions.

Age is the best single predictor of dominance rank. Adolescent males are dominated by all adult males. As they mature they gradually dominate an increasing number of males, until they achieve high-ranking or alpha status around the age of 20 to 25 years. Their status typically declines when they pass their physical prime, that is, by 30 to 40 years. Besides age, other important factors are size, physical condition, and personality traits, including the ability to form cooperative coalitions with other males (Goodall 1975; Bygott 1979). (Although mothers sometimes support their offspring in fights, there is no evidence that this has influenced the rank of any adult males; Bygott 1979.)

One adult male is typically recognized as the alpha since he never pant-grunts to others, while all pant-grunt to him. The alpha male characteristically shows high levels of aggression toward others (Riss and Busse 1977). Traveling with his hair slightly erect and shoulders fre-

quently hunched, he appears larger than his real size. In open ground he walks with marked confidence. The alpha male's tenure has varied between 3 and 10 years at Gombe ($N = 4$; Goodall 1975) and 3 to >7 years at Mahale ($N = 4$). Not all males achieve the top rank.

Alpha status confers several advantages. For instance, the male frequently takes meat from those that have caught prey and may supplant others from rich food sites. Reproductive benefits are presumably the most important, however, and are particularly striking in a small community. Thus Nishida (1983b) found that an alpha male obtained 80% of the copulations (cf. de Waal 1982.) Even in a large community a powerful alpha male can achieve disproportionate mating success by being possessive with different estrous females as each approaches the probable day of ovulation (Tutin and McGinnis 1981; Hasegawa and Hiraiwa-Hasegawa 1983).

Coalition and Manipulation. Males in small communities form a linear dominance hierarchy (Nishida 1979; Sugiyama and Koman 1979a), but if there are 10 males or more their dyadic relationships are less clear. For example, Bygott (1979) was able to distinguish only four classes of adult males in a 15-male community at Gombe: 1 alpha, 3 high-ranking, 6 middle-ranking, and 5 low-ranking males. The reason dominance relationships are difficult to describe in large communities is that males rarely meet on their own, so most interactions are affected by the presence of others. The relation between a male's dominance when alone and when with allies is complex, but is partly understood.

Dominance relationships can be classified into two categories according to whether interactions occur with or without the presence of a "third party." These have been termed "formal" and "real" relationships (de Waal 1982). "Formal" dominance is detected by the direction of pant-grunting or by attacks between two chimpanzees on their own. "Real" dominance is the result of conflicts where the outcome is influenced partly by other individuals. Reversals in real dominance interactions do not necessarily herald a permanent change in the formal dominance relationship, whereas a reversal in the formal dominance relationship invariably signals a long-term change. Males vulnerable to a change in formal dominance (perhaps because of repeated reversals in real dominance) therefore manipulate potential allies and rivals.

In M-group at Mahale, for instance, the alpha male has a close bond with three others (the oldest male, the second oldest male, and a prime male). The beta male, on the other hand, has an alliance with a young prime male. The relationship between the alpha and beta male is typically tense, especially in the first few minutes after a reunion. To prevent his rival from gaining new allies, the alpha male occasionally uses charging displays or chases to interrupt grooming between a member of his own clique and one of the other clique. Perhaps because of his top status, the alpha male is particularly solicitous of his allies. He grooms his allies more frequently and longer than he himself is groomed by them (see also Simpson 1973b) and shares his meat more extensively than males in other cliques.

Low-ranking males can use the insecurities of high-ranking males to their own advantage. The status of the M-group's oldest male, for example, is particularly interesting. The alpha male grooms him often, even though his grooming is hardly ever returned. He also lets the old male take large portions of meat, though the old male never shares with the alpha (Nishida and Hiraiwa-Hasegawa, pers. obser.). This curious asymmetry perhaps arises from the unique influence of the oldest male in real dominance interactions. Wrangham (1975) also showed that the two oldest, low-ranking males of Gombe were the most successful at obtaining and keeping meat, though the reason he suggested was that meat was a particularly valuable resource for them.

The duration of particular alliances varies widely. Goodall (1968, 1971) described four examples of alliances lasting several months or years between two adult males. In each case the pair was suspected to be brothers, and it was the younger member who benefited most (Bygott 1979). Likewise, Riss and Goodall (1977) described how an adult male attained alpha status partly as a result of consistent cooperation by his crippled elder brother. Alliances are not formed only by brothers, however, nor are they necessarily persistent. For example, Nishida (1981, 1983b) observed a young male taking over alpha status with the aid of an old gamma male who was probably unrelated to either competitor. The gamma male's subsequent strategy ("allegiance fickleness") was remarkable. By shifting his support (six times in 3 months) he maintained an unstable relationship between the two dominant males. Since both rivals depended on cooperation in order to win, neither could afford to show aggression to the gamma even when he tried to mate. As a result he achieved 50% of all matings, while the dominant males continued competing with each other.

Competition for alpha status has also been observed in captivity at Arnhem Zoo (de Waal 1982). Similar but much more elaborate social maneuvers than those seen in the wild occurred among three unrelated males. De Waal described these sometimes Machiavellian strategies in fine detail, and it is clear that the basic tactics of alliance formation are the same at Arnhem and in the wild. The only major difference is that at Arnhem females formed aggressive alliances with males and thereby substantially influenced male dominance relationships. The influence of females is trivial at Gombe and Mahale by comparison.

The dynamics of dominance competition among lower ranking males have not yet been described in the wild or captivity.

Alliance relationships may well not only result from but also be responsible for patterns of association. For instance, the fact that some males never meet on their own is probably not an accident. It suggests that one or both of them travels with others in order to avoid meeting his rival without support. This is probably a partial explanation for why males are more sociable than females.

Ambivalent Relationships. In spite of the tensions by male-male relationships, male chimpanzees are conspicuously sociable, even with their rivals. In M group, for instance, males of the different cliques usually interact peacefully with each other. Wrangham (1979a, 1979b) suggested that male-male bonding benefits each male by providing allies in defense of the community range (chap. 23). Certainly males from the same community occasionally cooperate to chase or attack males of other communities (Nishida 1979; Goodall et al. 1979), so intercommunity interactions are probably a key factor in the maintenance of male-male bonds.

Why, then, do males of the same community also fight each other occasionally? Competition occurs over access to estrous females, food, and other resources, and alpha males have priority of access: achieving alpha status therefore seems the best strategy for maximizing fitness. This means that both allying with an alpha male and defeating him are adaptive strategies. Males in the same community therefore have ambivalent relationships with one another. The result is a complex interweaving of strategies that is not yet understood.

For instance, it is uncertain what is responsible for community fission. Observations repeatedly indicate that within the male network the most tense relationship is that between the alpha and beta males (Riss and Busse 1977; de Waal 1982; Nishida 1983b). Sixteen males is the maximum number known in a community (Nishida 1979), and at Gombe fission occurred when there were 15, with particular tension between the 2 males who became alpha in the new communities (Bygott 1979). Presumably if there are too many males, competition becomes excessive. At the same time the community range may become too large to defend effectively.

There are many other problems awaiting resolution. For instance, Bygott (1979) suggested that overtly selfish activities by a dominant male would be penalized by the failure of other males to cooperate in defending the community range. But an alternative argument can be made. Since it is in the interests of both the alpha male and the subordinates to defend communally, the optimal strategy of the subordinates might be to cooperate even with a selfish alpha so long as he remains powerful, while waiting to usurp the alpha rank when the chance arises. Alpha's strategy, in turn, would be to form coalitions with old males and to be wary of the beta male, who is the most likely to defeat him.

Further observations are needed to resolve such debates. They will contribute not only to understanding the subtley of chimpanzee political maneuvers, but also to the analysis of chimpanzee society as a model for the evolution of human behavior (Itani 1980; McGrew 1981; de Waal 1982).

PART 2

Socioecology

Chapters in part 2 discuss aspects of the relationship between ecology and the social organization of primates—a problem of central concern for primatology since 1966, when Crook and Gartlan argued that impressive correlations occur between primate grouping patterns and ecological factors, including their diets, antipredator strategies, and habitats. A number of these correlations suggested that behavior is adapted to environmental pressures. For example, savanna baboons live in large social groups and forage in open areas, where they have few opportunities to escape from predators by climbing. Their large groups probably help them detect and deter predators. Hence, the tendency to live in large groups may have been favored by predator pressure. Crook and Garlan (1966) used examples such as this to suggest that the social behavior of primates would prove broadly adapted to ecological pressures, just as Crook (1965) had argued that bird societies were adapted to the environment. These analyses of bird and primate societies were two of the earliest attempts to understand ecological influences on species differences in social behavior, and they suggested that the major problems of social evolution would be solved rather easily.

Primatology was thus in the forefront of socioecology; it continues to be an important area of investigation in behavioral ecology because of the opportunities it presents for understanding the evolution of complex social behavior. Yet the early excitement generated by Crook and Gartlan's analysis has produced disappointingly few firm explanations of species differences in social behavior. Many of the facts of Crook and Gartlan's analysis are now known to be wrong (Eisenberg, Muckenhirn, and Rudran 1972; Clutton-Brock 1974), as are some of their conclusions and theoretical concepts (Clutton-Brock and Harvey 1976). Despite extensive later analyses, most of the fundamental problems that engaged the first investigators remain unsolved, such as why many primates live in groups, or why species vary in the number of males per group (chap. 23). Meanwhile new problems have been generated by data pouring in from field studies. Why do species differ in the pattern of intergroup dispersal, or dominance relationships, or mating patterns? Only tentative answers are available, but questions such as these, together with theories derived from studies of other animals, have stimulated new explana-

tory concepts. The chapters in part 2 are explorations of this exciting frontier. They review the current state of knowledge, define the important problems, and propose where answers are most likely to emerge.

Part 2 is concerned more with differences between than within species. This reflects the current state of the art, but it is somewhat unfortunate because, by isolating a few variables whose relationships can be examined in detail, intraspecific analyses are in theory likely to prove particularly helpful. For example, the only savanna baboons known to have territorial relationships between groups also live at a remarkably high population density, which suggests that crowding causes territoriality (Hamilton, Buskirk, and Buskirk 1976). For two reasons, however, analysis of intraspecific variation has not proved very rewarding to date. First, ecological factors cannot be isolated easily. For example, differences in population density can also be associated with differences in resource distribution. Either population density or distribution of resources may therefore be responsible for characteristics such as territoriality (chap. 22), or group size in howler monkeys (chap. 6). Second, many aspects of social behavior do not lend themselves easily to intraspecific analysis, either because they vary little within species (e.g., dominance relationships within macaques: chap. 25), or because their pattern of variation bears no apparent relationship to environmental variables (e.g., group size in baboons: chap. 11). In general, therefore, much intraspecific variation in social behavior has not yet been explained in relation to ecology. The chapters in part 2 accordingly focus on comparisons between species, but do not forget that intraspecific variation must be explained too.

Part 2 begins by asking how we can explain species differences in life-history variables, which are crucial to understanding the demographic composition of populations. The general answer given in chapter 16 is that body size and brain size are good predictors of patterns of individual growth and reproduction, whereas ecological and behavioral variables are not. A crucial question for understanding demographic patterns, accordingly, is why species differ in body and brain sizes. As chapter 16 shows, the evolution of these characteristics is still a puzzle but is beginning to be understood. For example, relatively large brains appear possible only in species with relatively high metabolic rates. Ultimately, therefore, ecological adaptations underlie species differences in brain size, and hence in life-history variables also.

Three ecological pressures are then discussed. First, we take as a given that primate diets are severely constrained by each species' morphology and physiology, a topic pursued in depth by several chapters in Chivers, Wood, and Bilsborough (1984). As in many animals, differences in food dispersal lead to differences in foraging patterns, which in turn affect social behavior directly. Chapter 17 shows that the foods eaten by most primates are distributed in particularly complex ways and illustrates some of the effects on behavior. Second, chapter 18 reviews the curious and often contradictory evidence on relationships between different species of primates. In theory the reasons different species associate could be similar to the reasons for grouping within species, but Waser shows that this is unlikely because associations between species are more fluid and variable than those within species. Present evidence suggests that relations with other species have had only minor effects on primate social organization, but firm conclusions are not possible until interspecies interactions are better understood. Finally, chapter 19 evaluates the widespread view that predator pressure has structured the evolution of primate groups. This is the first review of predation on primates, and it shows the need for caution before jumping to conclusions.

Intraspecific processes are analyzed next. Chapter 20 shows that rates of mortality and fertility respond sensitively to environmental changes. Demographic structure is therefore subject to short-term environmental influence. Group composition with respect to kinship, however, is clearly a largely species-specific trait. This is demonstrated in chapter 21, where Pusey and Packer summarize intergroup migration patterns in far more detail than previously available. Dispersal patterns appear to be important to social processes, not only because they determine kin relatedness within groups, but also because they are associated with species differences in the quality of relationships between neighboring groups (chap. 22). Chapter 22 stresses that there is substantial intraspecific variation also in intergroup relationships and shows that differences in food distribution are at least partly responsible.

The final chapter in this section reviews species differences in social organization by examining the different reproductive strategies of females and males and how they affect one another. Chapter 23 cautions us not to assume that groups are invariably adapted directly to their environment. Instead, groups may be formed because individuals who forage alone suffer from intraspecific aggression. This stresses just how complex the interplay of ecological and social variables can be and therefore how important it is to keep a broad perspective when analyzing environmental pressures on social evolution.

16 | Life Histories in Comparative Perspective

Paul H. Harvey, R. D. Martin,
and T. H. Clutton-Brock

In the game of life an animal stakes its offspring against a more or less capricious environment. The game is won if offspring live to play another round. What is an appropriate tactical strategy for winning this game? How many offspring are needed? At what age should they be born? Should they be born in one large batch or spread out over a long lifespan? Should the offspring in a particular batch be few and tough or many and flimsy? Should parents lavish care on their offspring? Should parents lavish care on themselves to survive and breed again? Should the young grow up as a family, or should they be broadcast over the landscape at an early age to seek their fortunes independently? (Horn 1978).

While the animal kingdom supplies a bewildering variety of answers to Henry Horn's questions, the range of variation within mammals is limited. For example, mammalian mothers invariably suckle their relatively

FIGURE 16-1. Primates vary in weight from mouse lemurs (ca. 60 gm) to adult male mountain gorillas (ca. 160 kg). Spider monkeys (ca. 8 kg) are one of the largest species in the New World. (Drawing: Teryl Lynn)

few young during early postnatal development, and there is always the potential for more than one litter during a lifetime. Yet, even within a single mammalian order such as Primates, the variety of life-history patterns, though limited, remains considerable. To take extremes, compare the mouse lemur (*Microcebus murinus*) with the gorilla (*Gorilla gorilla*) (fig. 16-1). Adult female mouse lemurs can probably produce one or two litters of two or three offspring each year, and the young can be parents themselves within a year of their own birth. On the other hand, adult female gorillas produce a single offspring every 4 or 5 years, and the young do not breed until they are about 10-years old.

Such differences between species have presumably evolved as adaptations for exploiting different ecological niches. Each niche is associated with a particular optimum body size, dictated in part by an animal's ability to garner and process available food supplies. Primates can be active at night or day; they can feed on insects, fruit, or leaves; and they can live in trees or on the ground. Diurnal, terrestrial folivores tend to be the largest species—the gorilla is over one thousand times heavier than the nocturnal, arboreal, omnivorous mouse lemur. In turn, differences in body size tend to be highly correlated with most aspects of life-history variation. For example, larger species take longer to grow to adult size. But how much longer? In fact, as we have seen, the thousandfold difference in size between the gorilla and the mouse lemur is reduced to a tenfold difference in age at first breeding. And there is merely a fourfold difference in gestation length (256 compared with 62 days). The picture is further complicated by the finding that some life-history variables, in particular the maximum recorded life span, are more highly correlated with species differences in brain size than with body size.

For decades biologists have been fascinated by the relationships between particular life-history variables and body size. Despite this interest and the numerous publications that have resulted, the selective forces responsible for the relationships are poorly understood. For example, it is well known that gestation length increases with body weight raised to the 0.13 power (approximately) across species within a number of mammalian orders (Kihlström 1972; Western 1979), but as far as we know, no one has suggested an evolutionary or functional reason why that particular value should be favored by

181

natural selection. More recently, several research papers have reported more extensive analyses in which a variety of life-history variables have been related to each other and to body size across various groups of birds and mammals (Millar 1977, 1982; Western 1979; Western and Ssemakula 1982; Stearns 1983). This has resulted in an increasing number of life-history profiles for different vertebrate groups, but a theoretical understanding of the relationships remains elusive.

One of our purposes in this chapter is to place the available primate data in the context of other studies. To that end we have based large parts of the discussion on our analysis of an extensive data base, but for brevity we have swept aside methodological and technical detail (contained in Harvey and Clutton-Brock 1985). After introducing the data, we examine and attempt to interpret some associations between body size and ecological niche differences across the primates. We then discuss briefly the idea of allometry, which will be used to facilitate the interpretation of changes in other variables, which are themselves related to body size. The first case of an allometric relationship considered is that between brain and body size. Brain size is introduced as a variable at this point in our analysis because several authors have noted that some life-history variables are more highly correlated with brain size than with body size. It has been argued that the brain may act as a pacemaker in development because neural tissue grows very slowly in the fetus, and that adult brain size may determine the maximum life span a species can attain because neurons that die are not replaced. Thus, it is desirable to investigate the relationship between brain and body size before these two variables are considered as correlates of life-history variation. We go on to discuss how a number of life-history variables change with both body and brain size. This leads us into a concluding section on precociality, revealing a number of evolutionary trade-offs that may constrain life-history variation in primates.

VARIABLES USED

The data are from 139 species representing 55 of the 56 extant primates genera listed by Corbet and Hill (1980). We have preferred accurate measures from animals living in natural populations, but these have been supplemented with observations made on captive animals. The full data set, together with further details on data sources, definitions of variables, criteria for exclusion from our sample, and methods used for deriving species' values can be found elsewhere (Rudder 1979; Harvey and Clutton-Brock 1985).

The variables used include adult and neonatal brain and body weight; litter size; gestation length; age at weaning, maturity, and first breeding; interbirth interval; maximum recorded life span; behavior patterns (nocturnal/diurnal, arboreal/terrestrial); breeding system; and diet. The data are summarized in tables 16-1 and 16-2.

We have used subfamilies as the level for discussion and analysis throughout. (See note at end of chapter.) Our subfamily classification differs slightly from that given in the appendix: we have subsumed the Lepilemuridae into the Lemuridae, and we have taken the Aotinae out of the Pitheciinae. Subfamilies were chosen because species from the same subfamily were found to be very similar on most measures. In fact, using a nested analysis of variance, we were able to show that for most measures about 85% of the original variation remains when subfamily values are considered (i.e., only 15% of the variance was located within subfamilies: Harvey and Clutton-Brock 1985). If we had used a lower taxonomic level for analysis, say the species, then species-rich genera could have biased our conclusions. For example, there are seven gibbon species in our data set. The gibbons are all monogamous frugivores, yet both monogamy and frugivory may only have arisen once in the common ancestry of contemporary gibbons. To count the gibbons as seven independent sources of data might inflate sample sizes to the point that misleading but significant results are obtained. However, there are also dangers associated with using too high a taxonomic level. If we had used the family instead of the subfamily, a lot of detail would have been lost; for example, we would have been left with only 68% of the original variation in gestation length.

BODY SIZE VARIATION

Even quite simple analyses reveal that body size is related to aspects of primate behavioral ecology, as can be seen from tables 16-1 and 16-2 (see also Clutton-Brock and Harvey 1977a, 1977b). Nocturnal primates tend to be small, while diurnal ones can be either small or large. Insectivorous primates tend to be small, and folivorous ones large. And terrestrial primates tend to be large. Interpretation of these patterns is complicated by associations between behavior patterns. For example, insectivorous primates are nocturnal and arboreal, while both folivorous and terrestrial primates are diurnal.

Diet is a major constraint on body size. Basal (or resting) energy needs increase with body weight, but not in direct proportion (Kleiber 1961). In fact, basal metabolic needs per unit time increase with the 0.75 power of body weight. Large species therefore need fewer calories per unit body weight than do small species. One result of this allometric relationships is that larger species can subsist on less nutritious food. If we assume that primates feed on the poorest quality food able to sustain them, then the findings that insectivorous primates are the smallest and folivorous ones the largest would be expected. But why do larger primates not maintain an energy-rich diet, for example, by spending more time searching for insects? Two separate studies of sympatric primate species (Charles-Dominique 1977a, five lorisid

TABLE 16-1 Species Averages for Primate Life-History Variables

	(1)	(2)	(3)	(4)	(5)	(6)	(7)	(8)	(9)	(10)	(11)	(12)	(13)	(14)
LEMUROIDEA														
LEMURIDAE														
Lemur catta	2.50	2.90	135	88.2	1.2	105	39	30.0	—	27.1	511	30.0	8.8	25.6
Lemur fulvus	1.90	2.50	118	81.4	1.0	135	—	27.8	10.0	30.8	547	23.3	10.7	25.2
Lemur macaco	2.50	2.50	128	100.0	1.0	—	33	—	24.0	27.1	—	—	—	25.6
Lemur mongoz	1.80	1.80	128	—	1.1	—	37	—	—	—	—	—	—	21.8
Lemur rubriventer	—	—	—	—	—	—	—	—	—	—	—	—	—	27.2
Varecia variegatus	3.10	3.60	102	107.5	2.0	90	30	23.5	5.2	—	365	—	10.6	34.2
Hapalemur griseus	2.00	2.00	140	48.0	1.0	—	—	28.6	—	12.1	—	24.0	—	14.7
Lepilemur m. mustelinus	0.64	0.61	135	34.5	1.0	75	—	—	21.0	—	—	21.0	2.9	9.5
Lepilemur m. ruficaudatus	0.75	0.75	—	—	—	—	—	—	—	—	—	—	—	—
CHEIROGALEIDAE														
Cheirogaleus major	0.40	0.40	70	18.0	2.0	—	30	—	—	8.8	—	—	—	5.9
Cheirogaleus medius	0.18	0.18	—	19.0	1.0	—	—	—	—	9.0	—	—	—	2.9
Microcebus coquereli	0.30	0.30	89	12.0	1.5	—	—	—	—	—	—	—	—	—
Microcebus murinus	0.08	0.08	62	6.5	1.9	40	50	11.50	9.5	15.5	312	—	—	1.8
Phaner furcifer	0.40	0.44	—	—	—	—	—	—	—	—	—	—	—	7.3
INDRIIDAE														
Indri indri	10.50	10.50	160	300.0	1.0	365	—	—	—	—	912	—	—	34.5
Avahi laniger	1.30	1.30	—	—	1.0	150	—	—	—	—	365	—	—	10.0
Propithecus diadema	7.50	7.50	—	—	—	—	—	—	—	—	—	—	—	37.0
Propithecus verreauxi	3.50	3.70	140	107.0	1.0	180	—	—	30.0	—	360	30.0	—	27.5
DAUBENTONIIDAE														
Daubentonia madagascariensis	2.80	2.80	—	—	1.0	—	—	—	29.0	—	912	—	—	45.2
LORISIDAE														
LORISINAE														
Loris tardigradus	0.26	0.29	163	12.7	1.6	—	40	—	13.0	12.0	—	—	2.7	6.7
Nycticebus coucang	1.20	1.30	193	49.3	1.0	90	40	—	—	14.5	—	—	4.0	10.0
Arctocebus calabarensis	0.31	0.32	134	25.2	1.0	115	39	—	9.5	9.5	—	—	2.3	7.7
Perodicticus potto	1.08	1.02	193	46.5	1.1	150	39	25.0	18.0	10.0	349	18.0	—	14.3
GALAGINAE														
Galago alleni	0.27	0.23	—	24.0	1.3	—	—	—	8.0	—	—	—	—	6.1
Galago crassicaudatus	1.26	1.42	135	47.4	1.1	90	—	16.5	12.0	15.0	360	—	4.0	11.8
Galago demidovii	0.62	0.63	111	7.5	1.2	45	—	30.8	8.0	14.0	330	9.0	1.2	2.7
Galago elegantulus	0.28	0.29	135	—	—	—	—	—	—	—	—	—	—	5.8
Galago senegalensis	0.21	0.24	124	11.5	1.6	75	—	10.8	6.7	16.0	200	—	2.3	4.8
TARSIIDAE														
Tarsius spectrum	0.20	0.20	157	30.0	1.0	68	24	17.0	14.0	12.0	152	—	—	3.8
Tarsius syrichta	0.12	0.13	180	26.2	1.0	—	28	—	—	—	—	—	—	4.0
CALLITRICHIDAE														
Callithrix humeralifer	0.30	0.30	—	—	—	—	—	—	—	—	—	—	—	—
Callithrix jacchus	0.29	0.31	148	28.0	2.1	63	16	17.0	12.0	12.0	157	16.7	4.4	7.9
Saguinus fuscicollis	0.37	0.42	149	40.0	1.5	90	—	24.1	—	—	242	—	—	9.3
Saguinus o. geoffroyi	0.51	0.50	—	50.0	1.9	55	—	—	—	—	243	—	—	10.5
Saguinus midas midas	0.53	0.60	127	36.0	2.0	70	16	24.0	20.0	13.0	240	—	—	10.4
Saguinus nigricollis	0.46	0.47	—	43.5	1.9	80	—	—	—	—	—	—	—	8.9
Saguinus oedipus oedipus	0.51	0.45	145	43.2	1.9	—	16	—	18.0	13.0	280	—	4.9	9.0
Saguinus midas tamarin	0.30	0.26	—	—	—	—	—	—	—	—	—	—	—	10.6
Leontopithecus rosalia	0.55	0.56	129	53.6	1.8	90	—	35.6	18.0	14.0	304	28.7	—	12.9
Cebuella pygmaea	0.14	0.16	136	16.0	2.1	90	—	24.0	24.0	10.0	154	24.0	—	4.2
CALLIMICONIDAE														
Callimico goeldii	0.53	0.65	154	48.6	1.0	65	27	15.8	8.5	9.0	167	16.5	5.8	10.8
CEBIDAE														
CEBINAE														
Cebus albifrons	2.60	2.60	—	234.0	—	270	—	—	43.1	—	—	—	—	82.0
Cebus apella	2.10	2.86	160	248.0	1.0	—	18	42.0	—	40.0	—	56.0	—	71.0
Cebus capucinus	2.70	3.80	—	230.0	1.0	—	—	—	—	—	—	—	29.0	79.2
Cebus nigrivittatus	2.30	2.90	—	—	1.0	—	—	—	—	—	—	—	—	80.8
Saimiri oerstedii	0.58	0.75	—	—	1.0	—	—	—	—	—	—	—	—	25.7
Saimiri sciureus	0.58	0.75	170	195.0	1.0	—	18	46.3	—	21.0	414	—	—	24.4

TABLE 16-1 *(continued)*

	(1)	(2)	(3)	(4)	(5)	(6)	(7)	(8)	(9)	(10)	(11)	(12)	(13)	(14)
ALOUATTINAE														
Alouatta caraya	5.70	6.70	—	—	1.0	—	—	—	—	—	—	60.0	—	56.7
Alouatta palliata	5.70	7.40	187	480.0	1.0	630	16	45.0	45.0	13.0	675	—	30.8	55.1
Alouatta seniculus	6.40	8.10	—	—	1.0	—	—	—	—	—	—	—	—	57.9
ATELINAE														
Ateles belzebuth	5.80	6.20	—	—	1.0	—	—	—	—	—	760	—	—	106.6
Ateles fusciceps	9.10	8.90	226	—	1.0	365	26	58.5	51.0	20.0	1095	57.0	—	114.7
Ateles geoffroyi	5.80	6.20	229	426.0	1.0	—	26	—	48.0	20.0	870	—	64.0	110.9
Ateles paniscus	5.80	6.60	—	480.0	1.0	—	—	—	—	—	880	—	—	109.9
Lagothrix lagothricha	5.80	6.80	225	450.0	1.0	315	25	—	98.0	12.0	720	—	—	96.4
Brachyteles arachnoides	9.50	9.50	—	—	—	—	—	—	—	—	—	—	—	120.1
AOTINAE														
Aotus trivirgatus	1.00	0.92	133	98.0	1.0	75	—	—	—	12.6	220	—	10.1	18.2
Callicebus moloch	1.05	1.10	—	—	1.0	—	—	—	—	—	365	—	—	19.0
Callicebus torquatus	1.10	1.10	—	—	1.0	140	—	48.0	—	—	365	—	—	22.4
PITHECIINAE														
Pithecia monachus	—	—	—	—	—	—	—	—	—	—	—	—	—	38.1
Pithecia pithecia	1.40	1.60	163	—	1.0	—	—	—	—	13.7	365	—	—	31.7
Chiropotes chiropotes	3.00	3.00	—	—	—	—	—	—	—	15.0	—	—	—	58.2
Chiropotes satanas	—	—	—	—	—	—	—	—	—	—	—	—	—	53.0
Cacajao calvus	—	—	—	—	—	—	—	—	—	—	—	—	—	73.3
Cacajao c. rubicundus	—	—	—	—	1.0	—	—	—	—	—	1095	—	—	75.2
CERCOPITHECIDAE														
CERCOPITHECINAE														
Macaca fascicularis	4.10	5.90	162	346.0	1.0	420	28	46.3	—	—	390	—	—	69.2
Macaca fuscata	9.10	11.70	173	503.0	1.0	—	28	60.0	—	—	—	—	—	109.1
Macaca maurus	5.10	9.50	163	—	1.0	—	—	—	—	—	—	—	—	—
Macaca mulatta	3.00	6.20	167	481.0	1.0	—	29	43.3	34.0	21.6	360	38.0	54.5	95.1
Macaca nemestrina	7.80	10.40	167	473.0	1.0	365	—	47.3	35.0	26.3	405	—	66.0	106.0
Macaca radiata	3.70	6.60	162	404.0	1.0	—	28	—	—	—	—	—	—	76.8
Macaca silenus	5.00	6.80	—	—	1.0	—	—	—	—	—	—	—	—	85.0
Macaca sinica	3.40	6.50	—	—	1.0	—	—	—	—	—	—	—	—	69.9
Macaca arctoides	8.00	9.20	175	485.0	1.0	—	29	—	—	30.0	525	—	—	104.1
Macaca sylvanus	10.00	11.20	—	—	1.0	—	—	46.0	46.0	—	945	—	—	93.2
Macaca nigra	6.60	10.40	176	455.0	1.0	—	36	66.0	49.0	18.0	540	—	—	94.9
Cercocebus albigena	6.40	9.00	177	425.0	1.0	210	28	72.0	48.0	21.0	510	—	—	99.1
Cercocebus atys	5.50	10.20	167	—	1.0	—	—	—	—	18.0	—	—	—	—
Cercocebus galeritus	5.50	10.20	171	—	1.0	—	30	78.0	—	19.0	—	—	—	114.7
Cercocebus torquatus	5.50	8.00	171	—	1.0	—	33	78.0	32.0	20.5	390	—	—	109.6
Papio c. cynocephalus anubis	12.00	21.00	180	1068.0	1.0	420	31	—	—	—	420	—	—	175.1
Papio c. cynocephalus	15.00	20.00	175	854.0	1.0	—	31	73.0	51.0	—	630	73.8	73.5	169.1
Papio c. hamadryas	9.40	21.50	172	—	1.0	—	—	—	—	35.6	—	—	—	142.5
Papio c. papio	13.00	26.00	184	—	1.0	—	—	—	—	—	423	—	—	165.3
Papio c. ursinus	16.80	20.40	187	—	1.0	—	—	—	38.0	—	—	60.0	—	214.4
Papio leucophaeus	10.00	17.00	176	—	1.0	—	33	60.0	42.0	28.6	450	—	—	152.7
Papio sphinx	11.50	25.00	173	613.0	1.0	—	35	60.5	—	29.1	523	—	—	159.4
Theropithecus gelada	13.60	20.50	170	464.0	1.0	450	34	54.0	49.5	—	525	—	—	131.9
Cercopithecus aethiops	3.56	4.75	163	314.0	1.0	—	33	47.7	30.0	31.0	365	—	—	59.8
Cercopithecus ascanius	2.90	4.20	—	—	1.0	180	—	—	—	—	—	—	—	66.5
Cercopithecus campbelli	3.60	3.60	—	—	1.0	—	—	—	40.0	—	—	—	—	65.8
Cercopithecus cephus	2.90	4.10	—	—	1.0	—	—	—	—	—	—	—	—	63.6
Cercopithecus diana	—	—	—	450.0	1.0	—	—	—	—	34.8	—	—	—	77.3
Cercopithecus erythrotis	—	—	—	—	1.0	180	—	—	—	—	—	—	—	65.2
Cercopithecus l'hoesti	4.70	8.50	—	—	1.0	—	—	—	—	—	—	—	—	76.0
Cercopithecus mitis	4.40	7.60	140	402.0	1.0	—	30	55.5	62.0	—	413	—	—	75.0
Cercopithecus mona	2.50	4.40	—	284.0	1.0	—	—	—	—	—	—	—	—	66.0
Cercopithecus neglectus	3.96	7.00	182	260.0	1.0	—	—	53.5	48.0	20.0	600	72.0	—	70.8
Cercopithecus nictitans	4.20	6.60	—	—	1.0	—	28	—	—	—	—	—	—	78.6
Cercopithecus pogonias	3.00	4.50	—	—	1.0	—	—	—	—	—	—	—	—	71.1
Cercopithecus aethiops pygerythrus	3.00	5.40	—	325.0	1.0	—	—	—	—	—	—	—	—	62.7

TABLE 16-1   *(continued)*

	(1)	(2)	(3)	(4)	(5)	(6)	(7)	(8)	(9)	(10)	(11)	(12)	(13)	(14)
Cercopithecus talapoin	1.10	1.40	162	180.0	1.0	180	36	53.0	48.0	22.3	365	114.0	—	37.7
Erythrocebus patas	5.60	10.00	163	—	1.0	—	—	36.0	33.0	20.2	420	42.0	—	106.6
Allenopithecus nigroviridis	—	—	—	—	—	—	—	—	—	—	—	—	—	62.5
COLOBINAE														
Presbytis aygula	6.20	6.30	—	—	1.0	—	—	—	—	—	—	—	—	80.3
Presbytis cristatus	8.10	8.60	—	—	1.0	—	—	—	—	—	—	—	—	64.0
Presbytis entellus	11.40	18.40	168	—	1.0	—	22	51.0	42.0	20.0	—	—	—	135.2
Presbytis geei	8.10	8.60	—	—	1.0	—	—	—	—	—	—	—	—	81.3
Presbytis johnii	12.00	14.80	—	—	1.0	—	—	—	—	—	—	—	—	84.6
Presbytis melalophos	6.60	6.70	—	—	1.0	—	—	—	—	—	—	—	—	80.0
Presbytis obscura	6.50	8.30	150	485.0	1.0	—	—	—	—	—	—	—	—	67.6
Presbytis potenziani	6.40	6.50	—	—	1.0	—	—	—	—	—	—	—	—	—
Presbytis rubicundus	6.30	6.30	—	—	1.0	—	—	—	—	—	—	—	—	92.7
Presbytis senex	7.80	8.50	—	360.0	1.0	225	—	—	—	—	569	—	—	64.9
Pygathrix nemaeus	—	—	165	—	1.0	—	—	—	—	—	495	—	—	108.5
Rhinopithecus roxellanae	—	—	—	—	—	—	—	—	—	—	—	—	—	121.7
Nasalis larvatus	9.90	20.30	166	450.0	1.0	—	—	—	—	—	—	—	—	94.2
Colobus angolensis	9.00	10.70	—	—	1.0	—	—	—	—	—	—	—	—	73.5
Colobus badius	5.80	10.50	—	—	1.0	—	—	—	—	—	520	—	—	73.8
Colobus guereza	9.25	11.80	—	445.0	1.0	390	—	55.0	—	—	365	—	—	82.3
Colobus polykomos	8.40	10.40	170	597.0	1.0	—	—	102.0	—	26.0	380	—	38.5	76.7
Colobus satanas	9.50	12.00	195	—	1.0	480	—	—	—	—	—	—	—	80.2
Colobus verus	3.60	3.80	—	—	1.0	—	—	—	—	—	—	—	—	57.8
HYLOBATIDAE														
Hylobates agilis	5.70	6.00	—	—	1.0	—	—	—	—	—	—	—	—	110.0
Hylobates concolor	5.80	5.60	—	—	1.0	—	—	—	—	—	—	—	—	131.7
Hylobates hoolock	6.50	6.90	—	—	1.0	700	28	—	84.0	—	—	—	55.0	108.5
Hylobates klossii	5.90	5.70	210	—	1.0	330	—	—	—	—	—	—	—	91.1
Hylobates lar	5.30	5.70	205	410.5	1.0	730	27	111.7	108.0	31.5	969	78.0	50.1	107.7
Hylobates moloch	5.70	6.00	—	—	1.0	—	—	—	—	—	—	—	—	113.7
Hylobates pileatus	—	—	—	—	—	—	—	—	—	—	—	—	—	114.2
Hylobates syndactylus	10.60	10.90	231	517.0	1.0	—	—	—	—	—	—	—	—	121.7
PONGIDAE														
Pongo pygmaeus	37.00	69.00	260	1728.0	1.0	1095	30	128.0	84.0	50.0	1025	115.5	170.3	413.3
Pan troglodytes	31.10	41.60	228	1756.0	1.0	1460	36	138.0	118.0	44.5	1825	156.0	128.0	410.3
Gorilla gorilla	93.00	160.0	256	2110.0	1.0	1583	28	118.2	78.0	39.3	1460	120.0	227.0	505.9
HOMINIDAE														
Homo sapiens	40.10	47.90	267	3300.0	1.0	720	28	232.0	198.0	60.0	1440	—	384.0	1250.0

NOTES: Column headings refer to (1) female weight in kg; (2) male weight in kg; (3) gestation length in days; (4) weight of individual neonates in g; (5) number of offspring per litter; (6) weaning age in days; (7) length of estrous cycle in days; (8) age at first breeding for females in months; (9) age at sexual maturity for females in months; (10) maximum recorded life span in years; (11) interbirth interval in days; (12) age at sexual maturity for males in months; (13) neonatal brain weight in grams; (14) adult brain weight in grams.

species in Gabon; Terborgh 1983, five New World monkey species in Peru) have provided evidence that the absolute quantity of insect food obtainable increases only marginally with increasing body size of the primate predator, such that larger-bodied primates are obliged to eat a greater proportion of plant food.

Large body size is also a distinct disadvantage for arboreal frugivores and folivores. Smaller species can move more freely through dense vegetation. And many nutritious food items such as young leaves and some fruits are often located on thin terminal twigs that are unable to bear the weight of larger species. This body size constraint is less important for terrestrial primates, which find much of their food at ground level and tend to be larger (Clutton-Brock and Harvey 1977a).

Antipredator defense mechanisms are also broadly correlated with body size. This topic is discussed elsewhere in this volume (see chap. 19).

More extensive discussions of factors selecting for small and large body size in primates can be found elsewhere (Clutton-Brock and Harvey 1977a, 1977b, 1983; Terborgh 1983). It should, however, be apparent that many selective forces are involved and that these are likely to differ in importance between habitats. Furthermore, the end result of evolution by natural selection is a constellation of functionally interrelated traits. This means that although we may find it convenient to write about, say, antipredator defense mechanisms being influenced by body size, it is equally plausible to argue that body size has been influenced by antipredator defense

TABLE 16-2   Behavior and Ecology of Primate Species of Different Subfamilies

Subfamily	N	D	A	T	FO	FR	IN	SI	M	G	SO	NS
Lemurinae	1	1	1	1	1	1	0	1	1	0	1	9
Cheirogaleinae	1	0	1	0	0	1	0	0	0	0	1	5
Indriinae	1	1	1	0	1	1	0	0	1	1	0	4
Daubentoniidae	1	0	1	0	0	0	1	0	0	0	1	1
Lorisinae	1	0	1	0	0	1	1	0	0	0	1	4
Galaginae	1	0	1	0	0	1	1	0	0	0	1	5
Tarsiidae	1	0	1	0	0	0	1	0	0	0	1	2
Callitrichidae	0	1	1	0	0	1	0	0	0	1	0	10
Callimiconidae	0	1	1	0	0	1	0	0	0	1	0	1
Cebinae	0	1	1	0	0	1	0	0	1	0	0	7
Alouattinae	0	1	1	0	1	0	0	0	1	0	0	3
Atelinae	0	1	1	0	0	1	0	0	1	0	0	7
Aotinae	1	1	1	0	0	1	0	0	0	1	0	3
Pitheciinae	0	1	1	0	0	1	0	0	1	1	0	7
Cercopithecinae	0	1	1	1	0	1	0	1	1	1	0	39
Colobinae	0	1	1	1	1	0	0	1	1	1	0	20
Hylobatidae	0	1	1	0	1	1	0	0	0	1	0	8
Pongidae	0	1	1	1	1	1	0	1	1	0	1	3
Hominidae	0	1	0	1	0	1	0	—	—	—	—	1

NOTES: Most subfamilies contain several species, some of which differ in their habits. In such cases we have indicated the presence of each variable represented in at least one species of the subfamily. 1 = present; 0 = absent; — = not easily classified. N = nocturnal; D = diurnal; A = arboreal; T = terrestrial; FO = folivorous; FR = fru- givorous; IN = insectivorous; SI = lives in single-male or harem breeding groups; M = multimale breeding groups; G = monogamous pairs; SO = solitary; NS = number of species present in the sample from each subfamily.

mechanisms: within a population of cryptic individuals, smaller animals may be at a selective advantage over larger ones.

ALLOMETRIC ANALYSIS

A common finding of comparative studies is that morphological, behavioral, and life-history variables tend to be highly correlated with body size. The relationships are often well described by a power function or "allometric" equation of the form

$$y = a(x)^b,$$

where y is the life-history variable, x is body size, and a and b are constants; a is usually referred to as the allometric coefficient, and b the allometric exponent. When this equation is logarithmically transformed, it plots as a straight line with slope b:

$$\log(y) = \log(a) + b \cdot \log(x).$$

We give an illustrative example of one allometric relationship, that between neonatal body weight and adult female body size, in figure 16-3.

The value of the allometric exponent can provide important clues to the adaptive significance of the allometric relationship. For example, it was thought for many years that the allometric exponent linking brain to body size across mammals was 0.67. Since surface areas are linked to body size with an allometric exponent of 0.67 (see R. D. Martin 1981b), it was widely held that brain size variation was intimately linked to surface area differences among species (Jerison 1973; Gould 1975). It has since become apparent that the exponent is not 0.67, but approximately 0.75. Basal energy needs increase with body size raised to the 0.75 power, and more recent interpretations of the allometric relationship between brain and body size suggest that brain size is tied to metabolic turnover.

Before performing an allometric analysis, it is often important to ensure that the comparison made is biologically meaningful. For example, mammals can be divided into precocial and altrical species. The former are born at a relatively advanced stage, with their eyes and ears open and with a covering of hair (see Eisenberg 1981). For any particular adult body size, it takes longer to gestate a precocial neonate, and since precocial mammals tend to be larger than altricial mammals as adults, a single allometric line placed through all mammals has a steeper slope than lines calculated for altricial or precocial mammals separately (Martin and MacLarnon 1985). Fortunately, the various orders of mammals tend to be composed of either altricial species or precocial species, but not both (though there are exceptions). Primates are moderately precocial, and this particular problem should not affect our analyses within the order, although we would expect the more altricial species belonging to other mammalian orders to have smaller allometric coefficients when gestation length or neonatal weight is plotted against adult body weight.

Unfortunately, deciding on which statistical method to use when determining the allometric exponent is not an easy matter. Three methods are in quite common use. They are model I regression, major axis, and reduced major axis analysis. For reasons discussed elsewhere (Harvey and Mace 1982; Seim and Saether 1983), we have used major axis analysis to define the line of best fit throughout this study. For each bivariate analysis, the major axis line reveals the allometric exponent as its slope and the logarithm of the allometric coefficient as its intercept. Allometric exponents relating the different life-history variables to body size are given in table 16-3.

The correlation matrix between the different continuously distributed variables used in this study is given in table 16-4. There are two obvious patterns. First, the variables tend to be highly correlated with each other. Second, the correlation coefficients are positive, except for those involving litter size, which are negative. Male and female adult body weights are so highly correlated with each other (0.996) that it is not worthwhile to distinguish between them, and for subsequent discussion only relationships with adult female body weight are used (fig. 16-2).

Two further points can be made about the discussion of allometric relationships that follow. First, when we refer to the "relative" value of a variable, we actually mean the value of that variable *after body size effects have been removed*. Relative values are measured as deviations from lines of best fit, taken as major axes in this study (table 16-5). If we say that a subfamily is typified by a large relative brain size, we therefore mean that its brain size is larger than would be predicted for a subfamily of that body size from the line of best fit. The second point is that we occasionally refer to partial correlations between life-history variables. These can be viewed in two ways. They are, in fact, the correlations between relative values of the variables concerned. In other words, they are the correlations between two life-history variables after the effects of body size have been removed. So, for example, if we mention the partial correlation between gestation length and neonatal body weight after the effects of body size have been removed, this can be interpreted as the correlation between relative gestation length and relative neonatal body weight.

ALLOMETRY OF BRAIN SIZE

Adult brain size is highly correlated with adult female body weight across primate subfamilies ($r = 0.96$). However, the allometric exponent of 0.94 is appreciably higher than that found in other mammalian orders (see Bennett and Harvey 1985): values do not usually exceed 0.75. In fact, the high exponent for primates is largely attributable to the effect of the single point for *Homo*. When the analysis is repeated with *Homo* removed, the exponent is reduced such that 0.75 lies within its 95% confidence limits (see table 16-3). *Homo* is clearly an extreme case (an outlier in relation to the best-fit line), and inclusion of this genus obviously distorts the slope of the line.

It has been pointed out elsewhere that folivorous primates tend to have relatively small brains (Clutton-Brock and Harvey 1980), and this analysis confirms that find-

TABLE 16-3   Relationships between Life-History Variables and Adult Body Weight in Primates

Life-History Variable	*a*	*b*	95% c.l. of *b*
Adult brain size	0.026	0.94	0.82–1.08
Adult brain size (-*Homo*)	0.045	0.86	0.74–0.99
Neonatal brain size	0.014	0.93	0.76–1.12
Neonatal brain size (-*Homo*)	0.019	0.88	0.70–1.11
Neonatal body weight	0.12	0.93	0.81–1.07
Gestation length	59.6	0.13	0.07–0.19
Weaning age	2.71	0.56	0.45–0.69
Age at maturity (female)	17.9	0.51	0.38–0.65
Age at maturity (male)	29.6	0.47	0.31–0.64
Age at first breeding (female)	43.5	0.44	0.35–0.54
Interbirth interval	29.1	0.37	0.27–0.47
Life span	780.9	0.29	0.17–0.40

NOTES: Best-fit major-axis regression lines were used to estimate allometric coefficients (*a*), allometric exponents (*b*), and the 95% confidence limits of the latter using the relationship log (life-history variable) = log (*a*) + (*b*) * log (adult female body weight). All weights have been converted to grams and times to days in order to facilitate comparison with studies from other groups.

ing: 6 of the 19 subfamilies contain folivorous species (table 16-2), and none ranks in the top ten relative brain sizes ($t = 5.32$; d.f. = 17; $p < 0.001$). However, as we have mentioned above, the 0.75 exponent across mammalian species (or orders) has led to the idea that metabolic rates are linked to brain size (Martin 1981; Armstrong 1983). This leads to the prediction that subfamilies with high metabolic rates will have relatively large brains. Unfortunately, as yet we cannot test this prediction directly because there are very few data on metabolic rates for primates species. We shall return to this discussion below.

Martin (1981) has argued that since the brain is such an energetically expensive organ to produce, neonatal brain size is determined (in evolutionary terms) by maternal metabolic turnover and should increase with maternal body size raised to the 0.75 power. The exponent linking neonatal brain size to maternal body size across primates is 0.93, which is almost identical to that linking adult brain size to adult body size (0.94). As with adult brain size, human neonatal brain sizes are very large and, when *Homo* is removed, the exponent is reduced so that 0.75 lies well within its 95% confidence limits (table 16-4). With humans removed, our analysis accords with the predicted exponent of neonatal brain size on mother's body weight of 0.75. Furthermore, adult brain size scales within a very similar exponent. This means that the relationship between adult and neonatal brain size is approximately isometric (i.e., the exponent is 1; see also Martin 1983). In other words, there is a fairly constant ratio between adult brain weight and neonate brain weight across primates (excluding the aberrant case of *Homo*), with the brain increasing by a factor of about 2.3 from birth to adulthood.

Because of the generally consistent relationship between adult brain weight and neonatal brain weight, subfamilies containing folivores score low on both counts. There is, however, one unusual case where adult and neonatal brain size do not follow this pattern. The Lorisinae have a much smaller relative neonatal brain size than we might expect from their relative adult brain size. The Lorisinae also have very low metabolic rates (see

Martin 1983). The small relative neonatal brain size for the Lorisinae, no species of which is folivorous, would be predicted by Martin's hypothesis that neonatal brain size is tied to maternal metabolic rate, but not by the hypothesis that neonatal brain size is related to diet in the way envisaged by Clutton-Brock and Harvey (1980).

ALLOMETRY OF LIFE-HISTORY VARIATION

Several recent studies have discussed allometric relationships between particular life-history variables and body size across a number of vertebrate groups (e.g., Western

TABLE 16-4   Product-Moment Correlation Matrix for Life-History Variables in Primates

	(2)	(3)	(4)	(5)	(6)	(7)	(8)	(9)	(10)	(11)	(12)	(13)
(1)	0.996	0.74	0.97	−0.52	0.91	0.92	0.89	0.78	0.86	0.89	0.95	0.96
(2)		0.73	0.97	−0.51	0.92	0.92	0.89	0.78	0.85	0.91	0.95	0.96
(3)			0.82	−0.61	0.84	0.81	0.81	0.62	0.63	0.84	0.84	0.80
(4)				−0.56	0.94	0.95	0.94	0.80	0.87	0.95	0.99	0.98
(5)					−0.56	−0.49	−0.47	−0.30	−0.41	−0.44	−0.51	−0.50
(6)						0.90	0.92	0.70	0.89	0.93	0.89	0.91
(7)							0.97	0.87	0.88	0.94	0.93	0.96
(8)								0.83	0.85	0.96	0.95	0.94
(9)									0.72	0.78	0.82	0.85
(10)										0.83	0.86	0.86
(11)											0.97	0.96
(12)												0.99

NOTES: The coefficients are based on the subfamily values given in table 16-2. Because of missing variables in some subfamilies, not all sample sizes are the same. Bracketed numbers refer to (1) female weight; (2) male weight; (3) gestation length; (4) weight of individual neonates; (5) number of offspring per litter; (6) weaning age; (7) age at first breeding for females; (8) age at sexual maturity for females; (9) maximum recorded life span; (10) interbirth interval; (11) age at sexual maturity for males; (12) neonatal brain weight; (13) adult brain weight.

TABLE 16-5   Relative Values of Various Life-History Variables and Postnatal Brain Growth Index for Primate Subfamiles

Subfamily	(1)	(2)	(3)	(4)	(5)	(6)	(7)	(8)	(9)	(10)	(11)
Lemurinae	−0.65	−0.74	0.27	−0.24	−0.44	−0.67	−0.78	−0.30	−0.35	−0.02	0.03
Cheirogaleinae	−0.55	—	—	−0.55	−0.28	−0.47	−0.06	—	−0.38	0.32	0.11
Indriinae	−0.42	—	—	−0.17	−1.07	−0.33	−0.30	−0.47	—	−0.28	—
Daubentoniidae	—	—	—	—	0.07	—	−0.13	—	—	0.54	—
Lorisinae	−0.38	−0.54	0.43	0.21	−0.09	0.43	−0.12	−0.05	0.07	0.16	−0.14
Galaginae	−0.15	−0.08	0.19	0.02	0.14	0.08	−0.16	−0.38	0.06	0.25	0.36
Tarsiidae	0.82	—	—	0.38	0.33	0.40	0.62	—	0.27	−0.18	0.30
Callitrichidae	0.27	0.49	−0.18	0.09	0.35	0.12	0.49	0.46	0.31	−0.14	0.10
Callimiconidae	0.02	0.03	−0.12	0.10	−0.05	−0.48	−0.62	−0.20	−0.45	−0.62	−0.41
Cebinae	0.57	1.08	−0.34	0.09	0.80	0.61	0.70	0.74	0.32	0.07	0.59
Alouattinae	0.23	−0.35	−0.16	0.00	−0.47	0.55	−0.07	0.06	−0.37	−0.04	−0.67
Atelinae	0.02	0.21	−0.25	0.17	0.01	0.02	0.28	—	−0.19	0.07	−0.54
Aotinae	0.27	0.15	−0.10	−0.11	0.10	−0.28	—	—	0.46	−0.26	−0.20
Pitheciinae	—	—	—	0.01	0.46	0.59	—	—	—	0.84	−0.27
Cercopithecinae	0.15	0.43	−0.43	−0.10	0.44	−0.29	−0.12	0.09	−0.14	−0.41	0.01
Colobinae	−0.11	−0.43	0.13	−0.14	−0.27	−0.34	−0.30	—	−0.20	−0.48	−0.20
Hylobatidae	−0.07	−0.08	0.02	0.12	−0.02	0.27	0.54	0.19	0.42	0.22	0.14
Pongidae	−0.36	−0.57	0.16	0.01	−0.37	0.15	−0.41	−0.14	−0.24	−0.07	−0.03
Hominidae	0.38	0.40	0.38	0.11	0.83	−0.40	0.45	—	0.43	0.02	0.83

NOTES: Relative weights or timings describe deviations from the double-logarithmic major-axis regression of the relevant variable on female body weight. Bracketed numbers refer to (1) relative neonatal body weight; (2) relative neonatal brain size; (3) postnatal brain growth index; (4) relative gestation length; (5) relative adult brain size; (6) relative weaning age; (7) relative age at maturity (female); (8) relative age at maturity (male); (9) relative age at first breeding (female); (10) relative interbirth interval; (11) relative life span.

1979; Western and Ssemakula 1982). One important finding is that similar allometric exponents often describe the same relationship in different groups (for example, that between neonatal weight and adult body weight). This could indicate that the same selective forces are acting in similar ways and within similar constraints in the different taxa. One of our primary objectives in this section is to describe these same life-history allometric relationships for primates, thus placing the primate data in a broader comparative context. In order to help interpret the relationships, we also compare body size and brain size as correlates of variation with the different variables. We have done this because some authors (notably Sacher 1959; Sacher and Staffeldt 1974) have argued that since neural tissue grows very slowly, it acts as a limiting factor in development, and that since the brain is an important homeostatic organ, its ultimate size determines maximum possible life span.

Our final objective in this section was originally to relate relative values of different life-history variables (given in table 16-5) to variation in the ecological and behavioral subfamily categories. The exercise proved

FIGURE 16-2. The degree of sexual dimorphism in body size varies substantially in primates, but is not sufficient to influence the analyses described in this chapter. Sexual dimorphism in body size is sometimes, but not always, correlated with the mating system. For example, monogamous species, such as gibbons (*right*) (chap. 12), tend to be monomorphic. This is not always true, however, as shown by de Brazza's monkeys (*left*) (chap. 9). In this species males are substantially larger than females, yet the social system is (at least sometimes) monogamous (chap. 9). (Drawing: Teryl Lynn)

fruitless: we could find no significant relationships. Although this finding is negative, it is actually very revealing and important. It means that the life-history variation and the ecological or behavioral variation we have managed to document cannot be related to each other independent of body size. It does, of course, remain possible that a more fine-grained or otherwise appropriate classification of primate ecology or behavior would reveal patterns that we have missed in the data.

Neonatal body weight is very highly correlated with adult body weight ($r = 0.97$), and the allometric exponent linking the two is 0.93 (see fig. 16-3). This is about the same as that found across species in the class Mammalia (e.g., Millar 1977; Stearns 1983; Western and Ssemakula 1982). There is no obvious explanation for this particular value of the exponent, although we shall return to it in a later section where we discuss variation in precociality among primates.

Despite the high correlation between neonatal and adult body weights, there is sufficient variation in relative neonatal body weight to merit attention. One factor that might influence relative neonatal body weight in mammals overall is litter size. Among mammals generally, litter size tends to be larger in smaller species, and large litter size may require the production of relatively small individual young. In the primates, however, litter size does not seem to be a complicating factor, especially since most species have only a single offspring: as noted by Leutenegger (1973), neonatal body weights tend to be relatively small in the strepsirhine primates (that is, the Lemuridae (including the Lepilemuridae) Cheirogaleidae, Indriidae, Daubentoniidae, Lorisinae, and Galagi-

nae), but there is no relation between relative neonatal body weight and litter size across strepsirhine subfamilies (see table 16-5). Furthermore, among the haplorhines (i.e., all the subfamilies that do not belong to the strepsirhines), twinning is the rule in the Callitrichidae, but relative neonatal body weight is fairly high, while the Colobinae do not twin but do bear relatively small young. Thus, while there are significant differences among primate subfamilies in relative neonatal weight, litter size does not account for them.

Two other suggestions have been made to account for variation in relative neonatal body weight among the primates. First Leutenegger (1973, 1976) has argued that the morphology of placentation has acted as an important selective constraint on the size of neonate that a mother can nourish. He suggests that since invasive (hemochorial) placentas are apparently superior to non-invasive (epitheliochorial) types with regard to coordination of maternal-fetal circulation and nutrition of the fetus, this allows species with hemochorial placentas to nourish the developing fetus until it is larger. Among the primates, the strepsirhines (which do indeed have epitheliochorial placentas) produce relatively smaller neonates than the haplorhines with their villous hemochorial placentas. However, the association between type of placentation and relative neonatal body weight does not hold across other mammalian orders (Martin 1984b). Indeed, the largest relative neonatal body weights are found in ungulates and cetaceans (dolphins and whales), which have noninvasive (epitheliochorial) placentas. The second suggestion is due to May and Rubenstein (1983), who argue that monogamous primates should give birth

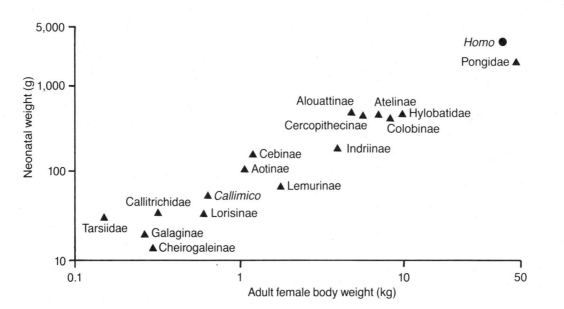

FIGURE 16-3. An example of an allometric relationship. When neonatal body weight is plotted against adult female body weight across 17 primate subfamilies, the relationship is linear when both axes are logarithmically scaled.

to relatively large young because the demands of lactation and postpartum care can be shared with a mate. We have been unable to confirm their prediction. For example, the Hylobatidae (and the Callitrichidae and *Callimico* in as far as they are monogamous—see chap. 4) do not produce relatively larger young than other haplorhine subfamilies (table 16-5). Thus, the reasons for variation in relative neonatal weight remain unclear.

Gestation length is not as strongly correlated with adult female body weight ($r = 0.74$) as it is with adult brain weight ($r = 0.80$). Not surprisingly, it is even more highly correlated with the neonatal size variables—neonatal body weight ($r = 0.82$) and neonatal brain weight ($r = 0.84$). We shall discuss these other relationships later.

The exponent relating gestation length to body weight (0.13) is similar to that found in other cross-species studies covering a variety of mammalian orders (e.g., Kihlström 1972; Western 1979). However, when species from different mammalian orders are lumped for analysis, the exponent is higher (ca. 0.23: Millar 1982, Stearns 1983). The exponent value obtained for mammals overall is higher because there is a general tendency (by no means so apparent within orders) for larger-bodied species to produce better-developed neonates requiring a relatively longer gestation period (e.g., artiodactyls, perissodactyls, pinnipeds, cetaceans, proboscideans).

The same line of reasoning that led to the suggestion that relative neonatal body weight should be related to placentation has led Leutenegger (1976) to the conclusion that relative gestation length is similarly affected. However, as Martin (1975) has pointed out, the primate data do not fully support this simple interpretation. As can be seen from table 16-5, although three of the strepsirhine subfamilies do have relatively short gestation lengths (Lemuridae, Cheirogaleidae, Indriidae), the other two for which we have data (Lorisinae and Galaginae) do not.

Weaning age is slightly more highly correlated with neonatal body weight ($r = 0.94$) than with adult female body weight ($r = 0.91$). The partial correlation between weaning age and neonatal body weight with the effect of mother's weight removed is 0.46. This correlation is not quite significantly positive ($0.05 < p < 0.1$), but the surprise is that it is not negative. We might have expected that high investment in the young before they are born would be compensated for by a lower postnatal investment. In fact, however, some species clearly invest more before birth and for longer after birth than others.

Weaning age scales with adult female body weight to the 0.56 power across primate subfamilies. There have been several studies reporting the scaling factor for other mammalian orders. The most comprehensive of these studies (Millar 1977) contains several important errors that have been discussed elsewhere (Clutton-Brock and

Harvey 1984). The others, which use more reliable measures and deal with higher correlation coefficients than Millar's, tend to report an exponent of about 0.25 (Blaxter 1971; Mace 1979; Reiss 1982; Russell 1982), which is well below the 95% confidence limits of our 0.56 value for primates. This means that weaning age increases more rapidly with increasing female body weight in primates than in other mammals that have been studied. We can think of no hypothesis that might explain this marked difference between primates and other mammals, but the consistency of the relationship between weaning age and adult female body weight across primates subfamilies suggests that some regular process is at work.

Age at maturity for males and females and *age at first breeding* for females all scale on adult body size with about the same exponent (0.47). Furthermore, their relative values tend to be highly correlated with each other (partial correlation between age at maturity for males and for females with the effect of adult body size removed is 0.84, and between age at maturity and age at first breeding for females with the effect of adult body size removed is 0.83). All three variables are more highly correlated with adult brain size (r about 0.95) than with adult body size (r about 0.90), and the partial correlations between them and adult brain size with the effects of body size removed remain significant and positive (r about 0.75). By contrast, the partial correlations between the variables and adult body size when the effects of adult brain size have been removed are nonsignificant and negative (about -0.20). This suggests, although the direction of causality cannot be determined, that postnatal brain development, learning skills, and social maturity may be intimately linked with physical maturity.

Interbirth interval is reasonably well correlated with adult body size ($r = 0.86$) and scales with adult body weight to the 0.37 power. For primates living in even moderately seasonal environments, however, we would expect further fine-tuning of this measure so that the period of maximum investment in the young coincides with the time of maximum food availability (e.g., Altmann 1980).

Life span is measured as maximum recorded life span, and most values are derived from captive specimens. It is a difficult measure to interpret biologically because those species that adapt well to zoos and that are commonly kept in captivity will be those for which most data on life span are available. Since life span is a maximum measure, larger sample size will tend to be associated with longer life spans. Nevertheless, we have included this measure because it has aroused so much interest following Sacher's (1959) discussion of the higher correlation of life span with brain size than with body size. Sacher's finding holds across the primate subfamilies analyzed here (table 16-3), with the partial correlation between life span and brain size being positive and reasonably high after the effects of body size have been removed ($r = 0.55$). However, for at least two reasons this does not necessarily mean life span is directly related to brain size. First, this correlation could merely result from brain size being less variable across individuals than body size. Another measure less variable than body size, the weight of the adrenal gland, is a better predictor of life span than is brain size (Economos 1980). The second reason why brain size may not be directly related to life span is that some intervening variable might be involved. For example, we have already seen that brain size is intimately linked to the age of female maturity. In fact, the correlation between life span and age at maturity ($r = 0.87$) is slightly higher than that between brain size and life span ($r = 0.85$). When the effects of age at sexual maturity are removed, the partial correlation between brain size and life span is very low ($r = 0.19$) and not significantly different from zero. It is conceivable that, in evolutionary terms, brain size and age at sexual maturity are related to each other, while life span is linked to age at sexual maturity. Of course, it is also possible that brain size is independently related to both life span and age at sexual maturity. Correlational analyses such as the ones performed here are unlikely to distinguish between these possibilities.

Across primates, life span increases with the 0.25 power of body size. Humans are, of course, characterized by relatively long lives and large brains. When humans are removed, the 0.55 correlation between relative life span and relative brain size mentioned in the last paragraph drops somewhat to 0.51.

LIFE-HISTORY COMPENSATION AND EVOLUTIONARY TRADE-OFFS

Investments are expected to reap rewards. However, all mammals have limited resources, and the extent and timing of investments are doubtless critical. One might therefore expect that investments would balance out overall, with different species spreading or concentrating investments. With life-history patterns we might expect a high investment early in life to result in benefits (or lower costs) later in life. For example, we might expect mothers investing in relatively large young to wean them at a relatively early age. But, as we have already seen, this does not appear to be the case. There is no evidence of compensation of heavy initial investment by later saving; to the contrary, subfamilies producing relatively large young tend to wean them relatively later. This means that subfamilies in which mothers invest heavily to produce relatively large neonates also invest heavily in the postnatal development of their young. Our main aim in this section is to establish whether high investment is a continuing feature of certain life-history patterns or whether compensatory effects occur. Our focus is pre-

natal versus postnatal development: How does the development of young at birth correlate with their later degree of development?

We need first to develop measures of the state of development of the young at birth. Primates are moderately precocial mammals: they are usually born with their eyes and ears open and with a covering of hair (see Eisenberg 1981). However, life-history variation within the order does not provide any single measure of differences in precociality. Development of the young can be defined in at least two ways using the variables given in table 16-1: first, by relative neonatal body weight, and second, by the degree of development of the brain at birth. We shall consider these measures in turn.

Body Weight Precociality

We have already assessed some of the suggested reasons for variation in relative neonatal body weight, but here we expand our discussion to show how this measure correlates with the relative values of other life-history variables (see table 16-6). The lack of compensation is clear: primate subfamilies that give birth to relatively large young have relatively long gestation lengths, advanced ages at weaning and maturity, and relatively long life spans. They also have relatively large neonatal and adult

TABLE 16-6 Product-Moment Correlation Coefficients between Derived Life-History Variables in Primates

	Relative Neonatal Body Weight	Postnatal Brain Growth Index
Relative gestation length	0.62	−0.11
Relative weaning age	0.46	−0.21
Relative age at maturity (M)	0.80	−0.63
Relative age at maturity (F)	0.74	−0.31
Relative age at first breeding (F)	0.56	0.06
Relative interbirth interval	−0.22	0.35
Relative life span	0.27	−0.27
Relative brain size	0.71	−0.14
Relative neonatal body weight	—	−0.59
Relative neonatal brain size	0.87	−0.63 (−0.86)

NOTES: Relative timings or sizes refer to deviations from major axis regressions of that particular variable on female body weight (see table 16-5). When two relative values are correlated, the correlation coefficient produced is the same as the partial correlation coefficient between the original variables with the effect of female body weight removed. Two correlations are given between relative neonatal brain size and the postnatal brain growth index. The one in parentheses is the correlation when *Homo*, a clearly aberrant point (see fig. 16-2), is removed.

brain sizes. As an example, we plot the relationship between relative neonatal body weight and relative age at maturity across the various subfamilies in figure 16-4.

We might have expected a positive relationship between relative neonatal body weight and relative gestation length because it will take longer to gestate a larger neonate. However, it is quite unexpected to find that those subfamilies that are relatively precocial (in the sense that they produce large young) also take relatively longer to mature.

Brain Size Precociality

The degree of development of the brain at birth provides two separate measures of precociality. The first is relative neonatal brain size: how large the neonatal brain is compared with the mother's body weight. The second is postnatal brain growth: how much the brain grows between birth and maturity.

Neonatal brain size is very highly correlated with neonatal body size ($r = 0.99$), and relative neonatal brain size is therefore very highly correlated with relative neonatal body weight ($r = 0.87$). The correlations between both measures of precociality and relative life-history variables are essentially the same, and we shall not repeat them here.

Although there is apparently no neuronal division after birth, the brain grows by expansion of existing neurons and by the addition of glial cells. Across subfamilies this growth is isometric (see also Martin 1983), with the exponent linking adult brain size to neonatal brain size being 1.02 and with brain size increasing by a factor of about 2.3 from birth to adulthood in primates generally. Subfamily deviations from the best-fit line of adult brain size on neonatal brain size provide a comparative measure of how much the brain grows after birth; we refer to them as postnatal brain growth indexes. It is apparent from figure 16-5 that relative neonatal brain size (a measure of how much the brain grows before birth) is negatively correlated with postnatal brain growth across primate subfamilies. There appears to be compensation or an "evolutionary trade-off," and thus we can define two extreme strategies for brain growth—a high rate of either prenatal or postnatal brain development. The same relationship was found in a study of primates by Martin (1983) and in studies of other mammals and of birds by Bennett and Harvey (1985).

However, there is one striking exception to the pattern in figure 16-5: *Homo*. Humans' relatively large brains are achieved by producing relatively large-brained neonates coupled with pronounced postnatal brain growth. The high rate of brain growth is reduced soon after birth among primates in general, while in humans it is maintained for about a year after birth. The dimensions of the mother's pelvis dictates that birth must take place after a

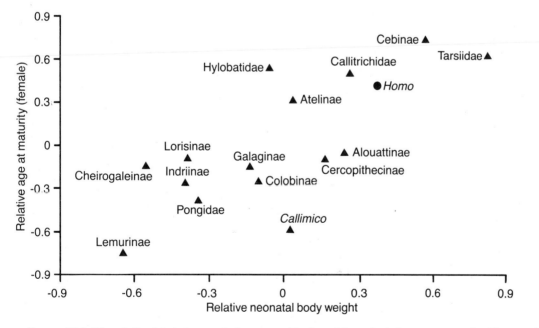

FIGURE 16-4. The relationship between relative neonatal body weight and relative age at maturity. The correlation between the two measures is the same as the partial correlation between neonatal body weight and age at maturity with the effects of body weight removed. Surprisingly, the relationship is positive. This means that those subfamilies investing in relatively large neonates also invest relatively heavily in their young. We might have expected compensation so that relatively large neonates matured relatively early.

9-month gestation in humans. Particularly high investment by the mother (and sometimes by both parents) is necessary during the first year of life when the human infant's brain is still developing rapidly and the child remains unable to walk.

So far in this chapter, the only significant behavioral correlate of a variable with the effects of body size removed is that between relative brain size and diet: subfamilies containing folivores have relatively small brains. Which phase in the ontogeny of the brain results in this association? The answer seems to be prenatal brain development: the four subfamilies containing folivores (Lemuridae, Pongidae, Colobinae, and Alouattinae) are much more strongly associated with a low relative neonatal brain size than with a low postnatal brain growth index (see fig. 16-5). It is interesting to note that the small relative neonatal brain size of the Lorisinae (which contains no folivorous species) is somewhat compensated for by a much higher postnatal brain development than is typical of the subfamilies containing folivores. This finding differs from Martin's (1983) supposition that the "gestation period is relatively longer in lorisines and allows for more foetal brain growth, thus offsetting the mother's low metabolic turnover." Indeed, the relative gestation length is long, but the resultant relative neonatal brain size is still small (see table 16-5). The Lorisinae are characterized by considerable postnatal brain growth rather than by an average neonatal brain size.

The final surprise to come from this survey is seen in table 16-6. On first thought we might have expected that

a late age at maturity would be associated with a greater degree of brain development after birth. To the contrary, those subfamilies that have more postnatal brain growth are the ones with a relatively early age at maturity, particularly among males. In fact, a relatively late age at maturity *is* associated with a relatively large brain. But the relatively large brain results from prenatal rather than postnatal brain growth.

CONCLUSION

G. Evelyn Hutchinson once wrote that priorities for ecological research should focus on the questions How big is it and how fast does it happen? As we have seen, when comparing life-history patterns, these questions appear to be two sides of the same coin: small animals have high rates of reproduction. To return to the extreme examples mentioned in the introduction, the smallest living primate, female mouse lemurs, can probably produce between three and six young in the year following their own birth, and they can go on doing that every year until they die (potential life span of about 15 years). The female gorilla, by contrast, waits 10 years before she can produce her first single young, and she can only produce others every 4 or 5 years until she dies (life span of about 40 years). In terms of potential reproductive success it pays to be the size of a mouse lemur rather than a gorilla.

Why, then, are some primates larger than others? We have suggested a number of possible reasons in the section on body size, though we should stress that selection acts on a whole suite of characters related to reproduction and not only body size alone. Selection for, say,

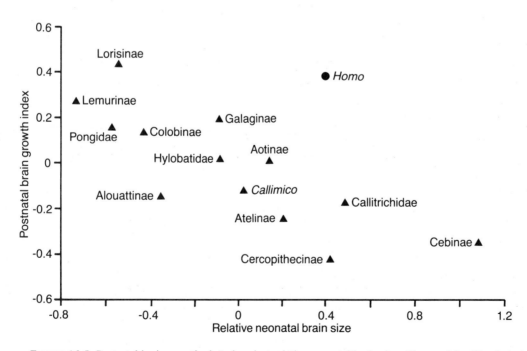

FIGURE 16-5. Postnatal brain growth plotted against relative neonatal brain size. Those subfamilies that give birth to relatively large-brained neonates are characterized by low levels of postnatal brain development.

longer life span or larger young might result in increased body size rather than vice versa.

The ways in which life-history variables scale with body size have served as a focus for this chapter. Three major conclusions have emerged. The first is that the scaling factors involved are generally very similar to those found in other orders of mammals (Millar 1979, 1982; Western 1979; Western and Ssemakula 1982) and in birds (Western and Ssemakula 1982). The second conclusion is that the balance of selective forces causing such tightly constrained and evolutionarily parallel relationships are little understood. The interpretation presents a major challenge for comparative biologists. Finally, deviations from the allometric relationships do not correlate with the broad behavioral and ecological subfamily classifications we have used. This may be a fault of the classifications, but it may simply indicate that primate life-histories can, in large part, be ordered along a simple size-time axis regardless of behavioral or ecological variables.

Despite the lack of ecological and behavioral correlates of life-history variables after the effects of body size have been removed, there is a ray of hope. It relates to brain size. The exponent linking brain to body size is not only repeatable within and among mammalian order (Martin and Harvey 1984), but it also has a familiar value: 0.75. That is the exponent linking basal metabolic needs to body size, and it suggests an energetic interpretation of differences in brain size (R. D. Martin 1981b, 1983; Armstrong 1983). The brain is an energetically expensive organ, and comparisons of brain to body size relationships in mammals, birds, and reptiles led R. D. Martin (1981b) to suggest that the size of the neonatal brain that a mother can nourish sets a constraint on adult brain size. The results reported in this chapter add support to Martin's ideas.

There are at least two further reasons for considering differences in brain size when comparing primate life-history variation. First, brain size is more closely correlated with several life-history variables than is body size. Neural tissue is slow to grow and may be the pacemaker of development (Sacher and Staffeldt 1974). The brain is also partly composed of nonrenewable neural tissue and is used in maintaining bodily homeostasis. Thus, brain size may be more highly correlated with life span than is body size (Sacher 1959). However, we must be careful when arguing that high correlation coefficients are necessarily associated with closer relationships. Such a statement may seem paradoxical because the correlation coefficient is in large part an assessment of the strength of a relationship. The reason is that body size can vary considerably during adult life, whereas brain size is more nearly constant. This means that body sizes may be measured with more statistical error than are brain sizes, thus leading to lower correlation coefficients. For example, Economos (1980) pointed out that another organ

less variable in weight than the body, the adrenal gland, is a better predictor of life span than is brain size.

The final reason for paying particular attention to differences in brain size is that deviations from the allometric relationship of brain on body size relate to diet. This suggests that something associated with diet may give a clue about factors influencing differences in brain size. It has been suggested that because folivores have small home ranges, they also need to store and process less information and therefore have smaller brains than frugivores (Clutton-Brock and Harvey 1980; Milton 1981). However, the metabolic needs hypothesis for differences in brain size might also explain the association. If folivores have lower metabolic needs than frugivores of the same size, the dietetic differences in relative brain sizes might be expected. Some preliminary evidence suggests that may be the case (see Harvey and Bennett 1983).

The ontogeny of the brain did, in fact, provide the one example of "life-history compensation" discussed in this chapter: primates that produce relatively large-brained neonates have relatively little brain growth after birth. Humans are the single exception. Our other searches for compensation came up with quite different findings: primates that have slow prenatal development or give birth to relatively large young are also those that wean and mature relatively late in life. Similar findings were reported by Lack (1968) in his comparative studies of life-history patterns in birds: among closely related birds of similar body size, those that produced larger eggs also took longer to fledge after hatching. If weaning and fledging weights are near adult body weights, primates and birds that produce relatively large young would need to invest less energy while rearing those young to weaning or fledging. Why that reduced investment is spread over a longer time span, thus adding increased maintenance costs to the investment in growth alone, and why the young develop relatively slowly remain open questions. In general, "how large it is and how fast it happens" seem to go hand in hand; also, how fast things happen early in life is related to how fast they happen later in life. As yet, the overall explanation of primate life-history patterns remains an elusive goal, but the systematic nature of the relationships discussed above indicates that continued research will eventually answer the questions raised by Horn (1978) in the quotation that opened this chapter.

SUMMARY

This chapter describes species differences in life-history patterns among primates. Most species belonging to the same subfamily tend to be similar on the various measures of life-history, such as gestation length, age at weaning, age at maturity, litter size, and life span, as well as on brain and body size. However, considerable differences among species belonging to different major subfamilies. Comparisons among subfamilies reveal clear allometric relations between life-history variation and body size. Adaptive explanations for the forms of the allometric relations are poorly developed. There are clear correlates of differences in behavior and ecology with body size and life-history variation, but not with relative life-history variation (life-history variation after the effects of body size have been removed). Nor is there evidence of compensation among life-history traits. For example, contrary to what might be expected, those primates giving birth to relatively large young wean those young relatively later. However, there is evidence of compensation in the development of the brain. Primates that give birth to small-brained young are characterized by extensive postnatal brain development.

NOTE: When a subfamily is the only one in its family, we refer to it by the family name (e.g. Hylobatinae is termed Hylobatidae). The Pongidae are treated as a single subfamily.

17 | Food Distribution and Foraging Behavior

John F. Oates

Since animals must get food if they are to maintain themselves and reproduce, the search for food is a crucial part of primates' lives and affects almost everything else they do. All primates have the same general need to acquire energy, amino acids, minerals, vitamins, water, and certain fatty acids, but their specific individual requirements vary and are met in a great variety of ways. No one species has a diet identical to another, and even within a species there is usually variation in diet between individuals, social groups, and populations. This is partly a result of spatial and temporal variation in what the environment provides (a topic pursued in more detail later) and partly a result of physiological and anatomical variation within and between primate species. Lactating females, for instance, have higher protein and mineral requirements than nonlactating females or similar-sized males. As animals get larger, they generally require more food, although their needs per kilogram of body weight may decrease (chap. 16). On the other hand, large primates cannot reach food items on the edge of tree crowns as readily as small individuals can, and they may not be agile enough to catch certain prey.

Limb structure as well as size will determine what foods a primate can reach and gather, while jaws, teeth, and guts will play a major part in determining what items can be processed once they are gathered. The strong jaws and teeth of gray-cheeked mangabeys allow them to open tough fruits that are unavailable to other monkeys in Uganda's Kibale Forest, while the forestomach fermentation chambers of colobus monkeys in the same forest allow them to derive a significant part of their energy requirements from the cell walls of leaves. The colobus can therefore exploit food resources unavailable to mangabeys, and vice versa. The microbes in their forestomach may also allow the colobus to cope with chemical components of leaves and unripe fruits that would be toxic to mangabeys and other cercopithecines with simple stomachs.

These differences in structure and function, combined with differences in habitat, produce a wide range of diets. This variety has been well summarized by Harding (1981), who shows that eclecticism is the rule in primate diets. The great majority of species eat a combination of fruits, leaves, and flowers, and most also eat some animal material, at least occasionally. Roots, bark, seeds, and gum also feature strongly in many diets. These foods are taken from a wide variety of sources; in most long-term, careful studies of feeding in wild primate populations, plant foods have been recorded to come from over 50 species. But despite this eclecticism, three broad categories of diet, emphasizing animal material, fruit, or leaves, can be recognized. This has been demonstrated by Chivers and Hladik (1980), who call the primates with such diets *faunivores, frugivores,* and *folivores,* respectively. In general, faunivores and frugivores get a major part of their protein requirements from insects, while folivores get most of their protein from leaves.

Food acquisition and processing and their relationship to diet are explored at length in Chivers, Wood, and Bilsborough (1984) and will not be given further attention here. It will be taken as a given that different primate species have different diets and are therefore seeking different sets of food items in their environment. This chapter, focusing on the behavioral aspects of the search, will pay particular attention to how patterns of movement in time and space are affected by the temporal and spatial distribution of food. Before these activity and ranging patterns are examined, some general aspects of food distribution will be reviewed.

THE FOOD SUPPLY
Tropical Environments

The great majority of nonhuman primates inhabit land areas with tropical climates; these areas are approximately bounded to north and south by the Tropics of Cancer and Capricorn at 23.5° latitude (see fig. 17-1). Only two of the 170 species listed in the appendix have their geographical ranges entirely outside these latitudes; these are the Barbary macaque (*Macaca sylvanus*) and the Japanese macaque (*M. fuscata*). Two other species, the Formosan macaque (*M. cyclopis*) and the golden langur (*Rhinopithecus roxellanae*), have geographical ranges that only just enter the tropics. All other nonhuman primate species have their entire range, or a substantial part of it, within the tropics.

Except at high altitudes, the tropics have warm air temperatures year-round, typically with a greater temperature range within any one 24-hour period than is found at the same time of day across different seasons. Variations in soil conditions and water supply (from both

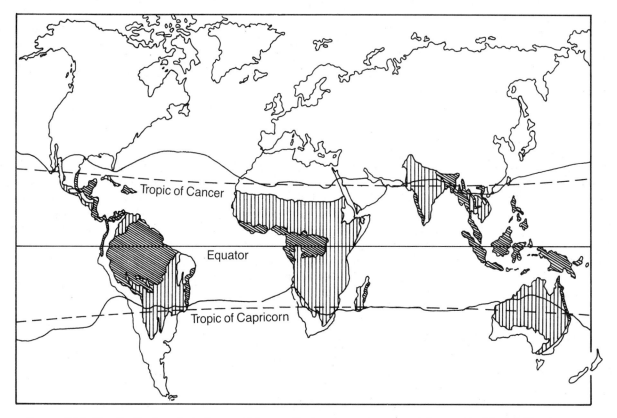

FIGURE 17-1. The distribution of rain forest and the woodland-savanna zones around the equator. Diagonal hatching shows tropical evergreen rain forests. Vertical hatching shows tropical and subtropical semievergreen and deciduous forests, dry woodlands, and natural savannas. The climatic boundaries of the tropics were defined by Nieuwolt 1977 as the 18°C sea-level isotherm for the coldest month and are shown oscillating about the Tropics of Cancer (Northern Hemisphere) and Capricorn (Southern Hemisphere). The vegetation map is from Walter 1973, adapted with permission from Springer-Verlag, New York.

precipitation and surface flow), rather than in temperature, are the major causes of variation in the vegetation from which primates directly or indirectly obtain their food. (Even those species that have a major animal component to their diet mostly eat herbivorous insects.) With the exception of the gelada baboon (*Theropithecus gelada*) and some populations of savanna baboons (*Papio* species) that feed heavily on grasses and herbs (Dunbar 1977; Iwamoto 1979; Post 1982), most primates get the majority of their food from woody-stemmed plants—trees especially, but also lianas. Some knowledge of tropical tree ecology is therefore necessary for a proper understanding of patterns of primate feeding behavior.

There are thousands of species of tropical trees, and they vary tremendously in size; in trunk and crown form; in both the form and the temporal and spatial production pattern of leaves, flowers, and fruits; and in chemistry. The particular set of tree species found in any one place, their density and dispersion pattern, and the other plant forms growing in association with them also vary greatly. This variation results from differences in the water supply, soil conditions, animal community, degree and type of human disturbance, prevalence of fire (which is usu-

ally caused by humans), and presence of regional barriers to dispersal (which will determine whether or not particular species can reach a given area). As a result, no two primate environments will have identical vegetation, and simple classifications of these environments can obscure important local variations.

Nevertheless, some broad categories of tropical wooded ecosystem can be recognized; they correspond to plant and animal assemblages that confront the primates living in them with broadly similar problems and opportunities. These broad categories, whose distribution is mapped in figure 17-1, are evergreen high forest ("tropical rain forest"), deciduous forest ("tropical dry forest"), and savanna. In this chapter I will describe the features of these environments only in very general terms. More detailed descriptions may be found in Hopkins (1974), Janzen (1975), Leigh, Rand, and Windsor (1982), Richards (1952), and Whitmore (1984).

Rain Forest Ecology

Rain forest typically grows close to the equator in lowland areas that have at least 1,500 mm of annual rainfall and no more than 4 consecutive months with less than

FIGURE 17-2. Rain forest in Manu National Park, Peru, continuous except where broken by rivers. (Photo: Anne Goldizen)

100 mm of rain. In this forest, the crowns of broad-leaved and mostly evergreen trees form a thick, continuous multilayered canopy (see fig. 17-2). Some of the trees in, or emerging from, the upper canopy reach great heights (often 40 m, sometimes 50 to 60 m). Tree species diversity is very high; it is common to find 50 to 100 species per hectare. However, a small number of species generally accounts for a major part of the plant biomass, and many species are represented by very few individuals. Different species vary in their adult size and crown shape and tend to show different patterns of clumping, sometimes in relation to soil conditions. The trees are exploited by abundant epiphytes (plants that attach themselves to the surfaces of stems and leaves without parasitizing them) and by large woody-stemmed climbers (lianas); these plants use the trees as supports to get themselves closer to the light. The spatial structure of the forest is therefore very complex in three dimensions, both in the size and arrangement of potential supports for arboreal animals and in the distribution of potential foods.

Patterns of leafing, flowering, and fruiting in rain forest are also complex. The timing of these phases may be roughly synchronized across a few species, but they are never synchronized across the whole community, and individuals of the same species often show different phasic patterns. However, community peaks of new leaf, flower, and fruit production typically occur. For instance, peaks of new leaf and flower production often occur in or at the end of a dry period, to be followed by one or more fruiting peaks in the wet season (Frankie, Baker, and Opler 1974; Whitmore 1984). Although the precise pattern varies from site to site, seasonal peaks and troughs in the abundance of particular primate foods are therefore an important feature of rain forest habitats, despite their relative stability. Superimposed on this pattern of fairly predictable seasonal change are occasional major perturbations resulting from unpredictable events, such as extremely wet episodes, or the failure of seasonal rains. Foster (1982) reported a fruit famine on Barro Colorado Island in 1970–71, apparently resulting from unusually heavy and prolonged dry-season rainfall. This famine had a major impact on the animal community (including the primates); some animals experienced changes in diet and feeding behavior, lowered reproductive output, and heavy mortality.

FIGURE 17-3. Mixed savanna and woodland, dominated by *Acacia xanthophloea*, in Amboseli National Park, Kenya. Baboons *P. c. cynocephalus* forage together with Burchell's zebra *Equus burchelli*, Grant's gazelle *Gazella granti*, and a bustard *Otis* sp. (Photo: Irven DeVore/Anthrophoto)

Rain forest leaves and fruits vary greatly in their size and architecture, and seeds are often protected by thick coats, husks, or pods that many primates cannot penetrate. In addition, a great array of chemical products are present in rain forest vegetation; each plant species has a distinct chemical profile, and there is often considerable chemical variation between members of the same species (as well as in the same leaf or fruit at different ages). Many of these chemical products are potentially toxic to primates (such as alkaloids) or can lower the efficiency of digestion (as do tannins and lignin), thus rendering the plant parts in which they are found less suitable as food

(see Waterman 1984 for a review). As a result, plant parts are often much less edible than is suggested by their superficial appearance.

Occurring partly in response to variation in the plant community, variations in the rain forest animal community place further constraints on primate feeding. For instance, many herbivorous insect species are specialists, feeding on a narrow range of plant species and parts. Different tree species will therefore tend to support different insect communities. Insect reproduction is often strongly influenced by climate and by the availability of particular plant phases, so insect numbers usually fluctuate considerably through the year. Many insects show obvious circadian patterns of activity, as do potential competitors for fruits, such as fruit bats, marsupials, squirrels, and frugivorous birds. In other words, animals, like plants, are a primate food source with a complex patterns of distribution in space and time.

If we define a food item as the edible portion of a plant or animal, discrete from other parts (in the case of a small animal this may be the whole body), then we can see that a primate seeking a particular set of food items in a rain forest will generally find that these items are patchily distributed. The lush appearance of the forest can be deceptive; items that fill nutritional needs, and that are readily reachable and edible, may be few and far between.

Savanna Zone Ecology

Tropical deciduous forests and savannas typically grow in areas with between 500 and 1,500 mm of annual rainfall and have at least one prolonged dry season (longer than 4 months). One result of a lower and more seasonal rainfall is a tree flora of lower diversity, with few or no very large individuals and a high frequency of deciduous species that drop all their leaves at some time during the year (often early in the dry season). More light therefore reaches the ground, where grass typically flourishes. When trees are abundant and their crowns are more or less in contact, this vegetation can be called *deciduous forest;* when there are large gaps in the canopy, the term *woodland* may be used; and when grass predominates and trees are scattered, the vegetation may be called *savanna* (see fig. 17-3). These drier-zone vegetation types usually grade into one another and are difficult to define precisely, but some generalizations can be made about them as primate environments. Compared with rain forest they have a simpler three-dimensional structure, though there are more frequently large gaps between trees, which must be crossed by primates seeking food in trees. Seasonal patterns of leaf, flower, and fruit production are more pronounced, and these phases are often quite closely synchronized so that some classes of food item fluctuate very greatly in abundance through the year. The animal community as a whole varies greatly from place to place, but usually contains more large terrestrial herbivores and carnivores and fewer dietary specialists than does the rain forest. Primates therefore face a very different array of potential competitors and predators. In these drier zones, variations in surface water distribution produce striking variations in the plant community; trees are most dense near rivers, lakes, and water holes, where the abundance of soil water may produce vegetation similar to rain forest (such forest along rivers in savanna areas is often called *gallery forest*). These areas of dense tree growth are important to primates as

sources of both food and sleeping sites, and the water itself may also be an essential resource at certain times of year. Although in some respects deciduous forests and savannas are less complex environments for primates than are rain forests, the food resources they offer to primates are even more distinctly clumped in space and time. Particular classes of food fluctuate greatly in abundance, and distinct "bottlenecks" of availability occur more frequently; for instance, at certain times of year, very few fruits can be found. Dietary specializations are rarely a successful strategy for primates in these environments, and flexibility is selected for.

Patchy Resources

The nature of primate environments is obviously such that primates' food is distributed in a patchy fashion, with areas of high food concentration separated by areas of low concentration. The size of the patches, the separation of the patches in space and time, and the density of items within patches all vary according to the particular food type and the particular environment. Since each species uses a characteristic set of items as food, different primate species sharing the same environment will each encounter different patterns of food-patch distribution. For instance, given that most trees (and other plants) produce more leaves than flowers or fruits, that they bear foliage for much longer periods than they bear reproductive parts, and that the mature phase of a leaf's life is usually considerably longer than the growth phase, patches of mature leaves will be more frequently encountered in space and time in a forest than patches of other plant items, and within patches leaves will usually have a higher density than will flowers or fruits. Insects and other animals typically have lower overall biomasses and densities than plant items. Therefore, a forest primate able to satisfy a large part of its nutrient needs from mature foliage will be able to find an adequate and predictable food supply within a smaller area than will a similarly sized primate whose diet is dominated by fruits and insects. However, although it is now widely assumed by primatologists that food-patch size and distribution are significant determinants of patterns of feeding (and, consequently, of social interaction), their measurement is difficult and time-consuming. Hence few accurate measurements of these variables have yet been made.

PATTERNS OF FEEDING
Introduction

An array of some 170 primate species interacting with a great variety of complicated food distribution patterns produces a picture of primate feeding behavior that is extremely complex. The complexity is further increased by the milieu in which feeding takes place—one in which predators must be avoided, competitors and weather coped with, social relationships maintained, and reproduction pursued. Primates are therefore faced with frequent decisions between conflicting pressures on their allocation of time.[1] However, since food is such a crucial resource, the actions needed to find and gather it are usually the major determinants of patterns of primate activity in space and time. The rest of this chapter examines these patterns. In this examination a distinction is made between *foraging,* which can be defined as searching for food, and actual *feeding,* the actions involved in picking (or catching), preparing, ingesting, and chewing food once it has been located.

Active Period

The great majority of primates, either strictly diurnal or nocturnal, rest throughout the night or day at carefully chosen sites. Activity, apparently cued by changing light levels, usually starts near to dawn or dusk. However, the precise timing of the start varies across species and populations and may be modified by environmental factors, in particular, ambient temperature (Richard 1978). The subsequent duration of the active (or "alert") period and the extent to which it is punctuated by periods of rest also vary greatly. Total duration is affected by latitude; at the equator, days and nights are each 12-hours long throughout the year, but at a latitude of 20° there is a difference of about 3 hours between the longest and shortest days of the year: a substantial difference in the amount of time available for feeding and social activities. In the same habitat, the consequences of diurnal or nocturnal activity will be very different; in each of these light phases, different sets of prey animals, competitors, and predators are active, and particular sense organs have different effectiveness.

Temporal Patterning of Foraging and Feeding

Primates wake up with empty stomachs, and a major feeding bout usually occurs early in their active period. Feeding is sometimes preceded by a short bout of social interaction, and may be delayed in very wet conditions or where animals spend some time in sunning behavior before other activity begins (e.g., black-and-white colobus, *Colobus guereza:* Oates 1977b; sifaka, *Propithecus verreauxi:* Richard 1978). The typical object of the first feeding bout of the day is a highly digestible, high-energy food that can be harvested at a rapid rate, such as a concentration of ripe fruits or young leaves. What follows this intense early-morning bout depends very much

on the nature of the primate and the distribution of food in its environment. A general distinction can be drawn between *foragers,* much of whose diet comes from small, highly dispersed food patches (particularly insects, but also certain fruits), and *banqueters* (Ripley 1970), much of whose food occurs in patches that are large relative to the animal's needs (such as leaves). Primates that get much of their protein requirement from insects may shift to foraging after an early bout of fruit feeding, but in species that get most of their protein (and much of their energy) from leaves, an early feeding bout that fills the stomach must be followed by a period of digestion before another banquet can be ingested.

Given the relative ease with which leaves may be located and harvested compared to other primate food items, and given the needs of digestion, a negative relationship exists across primate species between the proportion of the active period spent moving and the proportion of foliage in the diet (see fig. 17-4). A specific example of this relationship, and of the difference between foragers and banqueters, is provided by a comparison of redtail and red colobus monkeys in the Kibale Forest, Uganda. Struhsaker (1980) has shown that redtail monkeys (*Cercopithecus ascanius*) spend 21% of their day scanning the vegetation, apparently in search of the agile insects and ripe fruits that form a major part of their diet, while red colobus (*Colobus badius*), which feed predominantly on foliage, spend only 2% of the day scanning. Redtails spend 17% of the day traveling (during which they search for food), while red colobus travel for only 9% of the day. But red colobus spend more time actually ingesting and chewing food—41% of the day, compared with 34% for redtails. As a result of these different food-harvesting patterns, redtails spend only 10% of the day resting, while red colobus rest for 32% of the time. However, both species spend about the same amount of time (5% of the day) in social grooming. "One might think of the red colobus schedule as a repeated sequence consisting of a large meal followed by a long rest," says Struhsaker; he contrasts this with a redtail schedule consisting of a repeated series of "short move, scan and, finally, the ingestion of a single item." Similar differences have been found in the primates of the Manu National Park, Peru, by Terborgh (1983). Here, brown capuchins (*Cebus apella*) and squirrel monkeys (*Saimiri sciureus*) are estimated to spend about half their waking lives foraging for insects, and the time they devote to resting and social behavior is very brief. In contrast, the largely herbivorous dusky titis (*Callicebus moloch*) spend more than half their time resting. The titi has a very high percentage of leaves in its diet.

As in many biological classifications, the distinction between banqueters and foragers is not an absolute one. Rather, the two terms represent the ends of a spectrum. A folivorous colobine monkey can readily be classified

---

[1] Krebs 1978 has provided a useful general review of how variation in food distribution and food quality may influence the decisions made by animals as they harvest food. He points out that the use of the word *decision* need not imply conscious thought on the part of the animal. On the other hand, when considering animals with brains as large and complex as those of primates, we cannot exclude the possibility of awareness of action and its consequences, as cogently argued by Griffin 1976.

as a banqueter, and a highly insectivorous prosimian as a forager, but many species pursue somewhat intermediate strategies. For instance, on a single day a group of agile gibbons (*Hylobates agilis*) in the Malay Peninsula will typically combine several distinct fruit-feeding bouts at large food sources (small banquets) with periods of foraging for more scattered foods (including insects) and periods of traveling and resting (Gittins 1982).

Influence of Weather on Feeding Patterns

Primate feeding bouts are often reported to be most intense and prolonged at the beginning and the end of the active period. In many cases this is probably related not only to the need to build up energy reserves after and before a long sleep period but also to high ambient midday temperatures. Early afternoon ambient temperatures often reach 30 to 35°C in the lowland tropics. Combined with conditions of high relative humidity, such temperatures must cause considerable problems for primates attempting to regulate their body temperature at 36 to 38°C and would favor a reduction in the level of heat-generating activity. Captive studies by Müller, Kamau, and Maloiy (1983) have shown that the blue monkey (*Cercopithecus mitis*) and the black-and-white colobus (*Colobus guereza*) both become uncomfortable and restless at an am-

bient temperature above 28°C and that they start to lose thermoregulatory ability between 30 and 35°C. Clutton-Brock (1977) has pointed to evidence from several primate field studies that midday rest periods are most obvious at the hottest time of year, which suggests that weather conditions can influence activity patterns.

Conversely, under some conditions low temperatures may act to increase time spent feeding. In a comparison of gelada baboon populations at different altitudes, Iwamoto and Dunbar (1983) relate partly to temperature differences an increase in time spent feeding and a decrease in resting time, with cold temperatures at high altitude producing increased energy requirements for thermoregulation.

Rainy weather can also affect primate activity patterns, but the same conditions can produce very different responses in different species. Raemaekers (1980) reports that heavy showers generally cause siamangs (*Symphalangus syndactylus*) to halt until the rain abates, whereas lar gibbons (*Hylobates lar*) halt only briefly, and leaf monkeys (*Presbytis* spp.) seem not to halt at all.

Seasonal Changes in Feeding Behavior

The seasonal changes in food supply that occur to a greater or lesser extent in all primate environments have

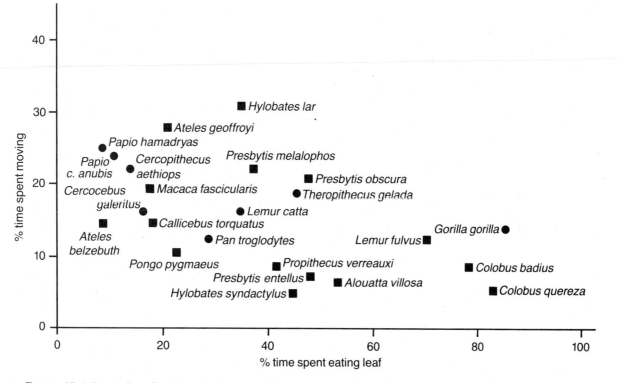

FIGURE 17-4. Proportion of active period spent moving versus proportion of feeding time spent eating foliage. Each point represents data from a single study, and no species is shown more than once. Only diurnal species, ranging in size from *Lemur catta* to *Gorilla gorilla*, are represented. Squares are arboreal, circles are terrestrial species. SOURCE: Redrawn from Clutton-Brock and Harvey 1977b with permission from Academic Press.

major consequences for primate feeding behavior. They produce changes in what is eaten and changes in the pattern of activities in time and space. Most studies of at least one year's duration that have collected data both on food availability and on feeding have shown that there are periods when preferred, high-quality food items are scarce and that at these times lower-quality items are eaten and harvesting patterns change. Although data from long-term studies are still quite sparse, the available evidence indicates that species relying heavily on foliage tend to decrease their level of activity at times of year when high-quality food is scarce, while species that rely heavily on fruits or insects increase the amount of time spent in looking for and processing food. For example, Richard (1978) reports that Verreaux's sifakas increased the proportion of the daily time budget allocated to resting by more than 10% in the dry season, despite a considerably shorter day length. At the height of the dry season, preferred food items such as young leaves, flowers, and fruits were in short supply, and mature leaves (which require more digestion time) became a major food source. This is the strategy of an energy economist. In contrast, the fruit and insect-eating *Cebus apella* in Manu National Park compensates for a scarcity of dry-season fruit by increasing the time devoted to searching for insects, thus increasing total feeding and foraging time (Terborgh 1983). *Cebus albifrons* shows a similar response to fruit scarcity by devoting long periods first to selecting hard-shelled *Astrocaryum* palm nuts that have been partially eaten by bruchid beetles and then to opening them by bashing them against tree branches (Terborgh 1983).

However, seasonal patterns are not always so clearcut. Observations made in Cameroon by Kavanagh (1978) suggest that two populations of the West African vervet monkey (*Cercopithecus aethiops tantalus*) responded to a dry-season decline of fruit (a major food item) in different ways. At one site, Buffle Noir, total feeding time increased, diet diversity decreased, and invertebrates accounted for more than 50% of feeding records; at Kalamaloue, a drier site, total feeding time decreased, diet diversity increased, and flowers and invertebrates together accounted for more than 65% of the diet.

An extreme case of behavioral and physiological adaptation to a seasonal environment is provided by some of the small nocturnal lemurs of western Madagascar. Hladik, Charles-Dominique, and Petter (1980) report that in the Marosalaza Forest, the fat-tailed dwarf lemur (*Cheirogaleus medius*) hibernates for 6 to 8 months during the dry portion of the year when its main food sources (especially fruit, but also insects and nectar) are scarce. *C. medius* lays down extensive fat deposits under the skin and in its tail during the wet season and uses this energy reserve to carry it through the hibernation period. Gray mouse lemurs (*Microcebus murinus*) in the same forest become lethargic during the dry season, but do not truly hibernate.

Seasonal changes in food distribution are probably an important factor in many of the instances where primates are found to have distinct breeding seasons (Lancaster and Lee 1965). Such breeding seasons may have pervasive influences on feeding and social behavior. For instance, squirrel monkeys in the Manu National Park have a concentrated mating season, when adult males largely give up their usual foraging activities (which normally take up about 50% of the active period) to engage in agonistic interactions and sexual displays. This behavior may explain the need for males to lay down extensive fat deposits prior to the mating season (see DuMond and Hutchinson 1967), a phenomenon paralleling the hibernation strategy of the dwarf lemur.

Feeding Synchrony, Competition, and Cooperation

A primate's feeding behavior varies through its active period; feeding often occurs in distinct bouts, separated by periods of travel and rest. If a number of animals is to move and function as a cooperative group, despite differences in food requirements based on age, sex, and reproductive condition, some synchronization of individual bouts must occur. However, although members of a primate social group generally travel in a more or less coherent fashion, there is rarely absolute synchronization in feeding. Clutton-Brock (1974) found that even at the height of a red colobus feeding bout, approximately half the group would be inactive, and Kavanagh (1978) found that tantalus monkey feeding was very highly synchronized only when the group was exploiting a preferred food source occurring in a distinct clump. Synchronization seems to be a probabilistic effect; social primates are more likely to feed when other group members are feeding and more likely to feed if they find themselves at a source of preferred food. This results in approximately synchronized bouts. The degree of synchronization varies with the type of primate, group size, and the food being exploited. Green (1978) presents evidence of less feeding synchrony in a large group of lion-tailed macaques (*Macaca silenus*) than in a smaller group and argues that this is the result of increased competition for individual items in a species in which the major sources of food are fruits, flowers, and insects. Such *interference competition* (direct interaction between competitors) is regarded by ecologists as distinct from *exploitation competition,* which arises from competitors using up part of a common resource without directly meeting (Pianka 1974).

Harcourt (1978a) has pointed out that the synchronization of group activity into nonrandom bouts may produce biases in the measurement of social behavior, which is likely to vary in both frequency and pattern with the predominant activity. For instance, 2-year-old gorillas prefer to associate closely with the silverback male during periods of travel and feeding, but are more likely to be close to their mother during rest periods (fig. 17-5). A misleading picture of social and spatial relationships is therefore obtained if social behavior is studied predominantly during particular activity bouts, such as rest periods.

The synchronization of feeding activity inevitably tends to increase interference competition for food items between individuals. Species that get much of their food by foraging tend to spread out while moving and feeding, thus reducing competition for each item (e.g., *Cercopithecus ascanius:* Struhsaker 1980; *Cebus nigrivittatus:* Robinson 1981a). In these circumstances, forest species often use "contact calls," which apparently serve to maintain coordination in the absence of visual contact. Despite the high dispersion typical of a foraging group, low-ranking individuals may still be supplanted frequently by dominant animals, except where the food items being exploited are very small and very highly dispersed. For instance, Post, Hausfater, and McCuskey (1980) found that adult male yellow baboons (*Papio cynocephalus*) at Amboseli, Kenya, had a significantly lower proportion of their feeding bouts interrupted than did juvenile males or females. However, dominance relationships can strongly influence feeding success even in the absence of direct supplantations. In a study of female vervet monkeys at Samburu-Isiolo, Whitten (1983) found that although supplantation rates over clumped food items were not significantly greater than those over randomly distributed foods, high-ranking females had a higher intake of food from clumped sources than did low-ranking females. This was apparently the result of low-ranking females avoiding food clumps where high-ranking females were feeding.

Feeding interference seems to be less common in banquet feeders than in foragers, especially when banqueters are exploiting large patches of abundant food items, such as mature leaves. Banqueters typically move in a closely coherent, goal-directed fashion from one major food source to another (e.g., mantled howlers: Milton 1980). On reaching a source, such as a tree crown, the animals spread themselves out in such a way that feeding interference is minimized (e.g., hanuman langurs: Ripley 1970). McKenna (1979b) has argued that the lower frequency of competition involved in the feeding strategy of colobines such as langurs is the fundamental reason for the lesser importance of dominance relationships in their social systems, as compared to the societies of cercopithecines. Social systems in which dominance relationships are particularly significant in determining individual behavior are characteristic of cercopithecines such as baboons and macaques that inhabit savanna and open woodland ecosystems, where much of the food supply is typically scattered in small clumps.

The opposing forces of competition and cooperation are brought together in those cases where individual primates cooperate with each other to deny access by other

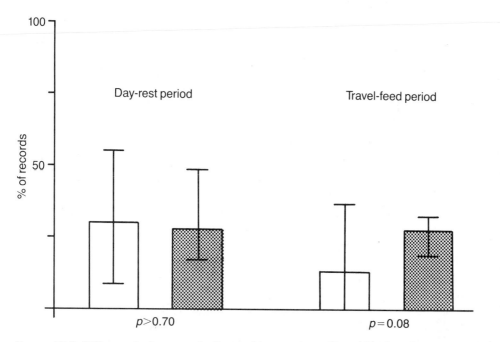

FIGURE 17-5. Difference in time spent by 2-year-old mountain gorillas within 5 m of mother (open histograms) and silverback male (shaded histograms) during periods when the group was (*a*) resting, or (*b*) foraging. Histograms show the median and bars show the range of the proportion of scan records for each of seven individuals. SOURCE: Redrawn from Harcourt 1978a with permission from E. J. Brill, Leiden, Netherlands.

individuals to particular food patches. This has been observed within groups (chaps. 25, 26) and between groups (chaps. 22, 23). Cooperation between individuals may also be used to increase feeding success in joint efforts to capture mobile prey. Because of its presumed parallel to a significant component of human food-gathering behavior, particular attention has been given to the role of active cooperation in the hunting of large prey animals by chimpanzees and baboons. Teleki (1973a, 1973b) has argued that the capture by chimpanzees of mammals such as monkeys at Gombe in Tanzania often involves plainly cooperative behavior. Busse (1978) has presented evidence, however, that individual hunting success decreases with increasing group size, and argues that episodes of apparent cooperation are more parsimoniously explained as instances of selfish individuals attempting to capture the same prey item at the same time. Reviewing observations of hunting by anubis baboons at Gilgil, Kenya, Strum (1981) describes individual adult males watching the hunting behavior of other males closely, and describes hunters apparently chasing their target purposefully in the direction of other baboons. However, Strum's data indicate that single males were about equally represented in both successful and unsuccessful hunts of gazelle groups. Whether or not there is significant cooperation by hunting chimpanzees and baboons, it seems that the highly goal-directed cooperative hunting of large-bodied prey, which is seen in many social carnivores and in man, is not typical of nonhuman primates in general.

Feeding and the Use of Space

Since food patches have a complex distribution in space as well as in time, primates are faced with decisions not only on when to feed but also on where to feed. This involves decisions on the direction, the distance, and the speed of movement. Factors in addition to food must also be taken into account. In savanna environments, where water holes and sleeping sites may be limited, their distribution in space and time strongly influences patterns of movement.

Several field studies have found evidence that primates can make highly directed movements to particular patches of food and other resources in a way strongly suggestive of goal direction and the existence of mental maps of the environment (e.g., Altmann and Altmann 1970, for yellow baboons; Marsh 1981b, for red colobus; Milton 1980, for howler monkeys; Sigg and Stolba 1981, for hamadryas baboons; Wrangham 1977, for chimpanzees). It has been argued that familiarity with an area and travel through it along relatively fixed pathways, rather than at random, results in a more efficient pattern of resource exploitation and allows important food sources to be monitored. Goal direction and mental maps are hard to test rigorously, but there is considerable evidence that

the patterns of primates' movements in space are strongly influenced by the spatial distribution pattern of resources, especially food resources.

The use of space, or "ranging" behavior, may be studied in a number of ways. We can examine in two or three dimensions the path followed by an individual primate or primate group during a particular activity period (a day, for instance) and measure both its length and its configuration (such as straight line, circle, or "random walk"). And we can take one or more paths and consider their areal distribution. Waser and Wiley (1979, 160) use the term *activity field* to describe "the distribution of an individual's time as a function of location." Primates, like most mammals, usually show a strong attachment to one particular area (they are "philopatric"), and hence the term *home range* is often applied to their long-term activity field. Jewell (1966, 103) has defined this as "the area over which an animal normally travels in pursuit of its routine activities."

In general, the wider the area over which a primate population's food resources are spread, the greater is an individual's daily path length and annual home range size. Thus, terrestrial species tend to have longer daily path lengths and larger annual home ranges than arboreal ones, and frugivores tend to follow longer paths and have larger annual ranges than folivores (Milton and May 1976; Clutton-Brock and Harvey 1977b). These general relationships are complicated in individual cases by such factors as body weight, metabolic rate, social group size, degree of local competition for food, nocturnal or diurnal activity, and the density and seasonal fluctuation of food resources at any given site. Clutton-Brock and Harvey (1977a) show that "population group weight" (the total weight of the number of animals in a population which regularly associate together and share a common home range) is positively correlated with home range size (fig. 17-6), probably because a greater weight of animals needs a larger food supply. However, some species with similar group weights have very different home range sizes (e.g., mantled howler monkeys, *Alouatta palliata,* versus talapoins, *Cercopithecus talapoin*). The causes for this difference seem to lie not only in the more folivorous diet of the larger howlers, but also in their much lower metabolic rate, which gives them lower food requirements per unit of body weight (Martin 1981a). The relationship between body size and metabolic rate (the larger an animal, the smaller its metabolic energy requirement per unit of body weight) probably also contributes to the larger range sizes of the nocturnal species versus the diurnal species shown in figure 17-6. The small nocturnal frugivorous prosimians not only have high metabolic rates but also include more insects in their diets than do large diurnal frugivores. As pointed out earlier in this chapter, insects are less densely distributed than plant food items. However, Martin (1981a) shows that even when metabolic energy requirements are taken into account, larger-bodied species still turn out to have greater-than-expected home range sizes. He argues that increased interspecific competition or some other factor probably obliges large-bodied species to exploit larger areas per individual. Martin has also shown that although terrestrial species have longer day range lengths than arboreal species, day range length does not increase consistently with increasing body size. This suggests that increased constraints are placed on locomotor expenditure as body size increases.

Several comparative studies provide interesting specific examples of relationships between some of the variables just considered. Raemaekers (1979) has compared the diet and patterns of both activity and range use in sympatric groups of lar gibbons and siamang in the Krau Game Reserve, Malaya. Each group contained one mated adult pair, a subadult male, and a juvenile. In 1 year, the group of siamang (average adult weight about 11 kg) spent 36% of their feeding time on fruit and 43% of their time on leaves, while the group of lar (average adult weight about 6 kg) spent 50% of their time on fruit and 29% of their time on leaves. The remaining 21% of each group's feeding time was spent on flowers and insects. Relying more on food items (leaves) that occur in larger clumps that are closer together, the siamang on average

moved less far each day than the smaller lar (738 m as against 1,490 m), moved more slowly, and fed for a longer part of the day (5.2 hr against 3.6 hr). The siamang group took an average of 6 days to cover all four quarters of its range, compared to 2.5 days for the lar group, and covered an area of 47 ha in one year, against the lar's 57 ha.

Similar relationships are demonstrated by comparative studies in the Kibale Forest (Struhsaker 1978) and in the Manu National Park (Terborgh 1983). Terborgh provides some striking examples of seasonal differences in range use by *Cebus* monkeys. During a period of food scarcity when their preferred foods became widely scattered, groups of both *C. apella* and *C. albifrons* ranged over wider areas and spent less time in any one area of their range than during the time of maximum fruit abundance (fig. 17-7).

We might expect ranging behavior and food supply to be related not only to group weight across species but also to group weight (and therefore size) within species. Clutton-Brock and Harvey (1977a) summarize evidence suggesting that, other things being equal, large groups tend to occupy larger ranges than small groups, and that home ranges are smallest where food is most abundant. However, given the complexity of food distribution in time and space and the range of factors influencing pat-

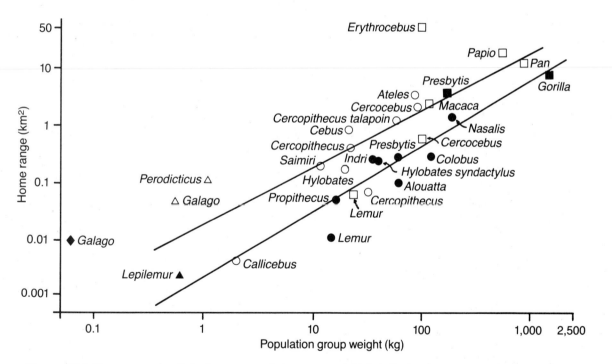

FIGURE 17-6. Home range size (in km) versus population group weight (kg). Each point represents the mean value for a genus. Open figure=frugivore; solid figure=folivore; diamond=nocturnal, arboreal insectivore; triangle=nocturnal, arboreal; circle=diurnal, arboreal; square=diurnal, terrestrial. Genera with species in more than one category are represented separately for each category (i.e., *Galago, Lemur, Cercocebus, Presbytis,* and *Hylobates*). Separate regression lines are shown for frugivores (*upper*) and folivores (*lower*). SOURCE: Redrawn from Clutton-Brock and Harvey 1977a with permission from Academic Press, New York.

terns of both movement and social organization, these relationships are rarely displayed in a simple fashion. For instance, Waser (1977a) found that the distance moved per day by gray-cheeked mangabeys (*Cercocebus albigena*) in the Kibale Forest increased with increasing group size, but apparently not in a linear fashion: a group of 28 individuals moved very much further than expected each day, compared with five groups of 6 to 16 individuals. Waser interprets these observations as indicating

that, up to a group size of 12 to 15, the energetic costs of adding a new member are relatively slight because the size of typical food patches allows several individuals to feed together in one patch without undue competition. When the group increases above about 15 individuals, all of them may have to move very much further per day to find an adequate number of sufficiently large patches. This should lower the efficiency of individual foraging and so constrain group size. This is an attractive argument, although it is based on a small sample of observations (the large mangabey group was followed for only 5 days). Similar constraints on group size have been suggested for black-and-white colobus by Oates (1977c) and for mantled howlers by Milton (1980).

Territoriality

Individual primates are affected in their use of space by competition between groups as well as within groups. Although in a few cases different groups have been observed to largely ignore each other or to intermingle without aggression, active avoidance or the aggressive defense of at least part of the home range is usual. In some species, such as gibbons, most of the home range is actively defended. Intergroup encounters are discussed in chapter 22 and therefore will not be considered in detail here, but some comments on relationships between range defense and diet are appropriate.

Brown (1964) has argued that territoriality—the defense of an area—will evolve where the resources that individuals require are limited but occur within an area that can be economically defended. Mitani and Rodman (1979) have examined this hypothesis in light of data from wild primate populations through the use of an *index of defendability*, which is average daily path length divided by the diameter of a circle with an area equal to the observed home range. This index was found equal to or greater than 1.0 in 13 territorial populations and less than 1.0 in 14 nonterritorial populations. Six nonterritorial populations also had an index greater than 1.0. Mitani and Rodman found that for a given group weight and diet, territorial species do not range further each day than do nonterritorial species. They interpret these findings to mean that in populations in which the normal daily movements of groups in search of food allow them to monitor the boundaries of their range without major extra energy expenditure, territoriality is feasible, but not a necessary consequence.

However, the monitoring of a territory does not necessarily imply frequent boundary patrols and clashes. Terborgh (1983) observes that although groups of tamarins (*Saguinus imperator* and *S. fuscicollis*) in the Manu National Park have no range overlap and defend territories, they concentrate their activities in small areas near the center of their ranges and visit boundary areas relatively infrequently. Terborgh suggests that, since the territories

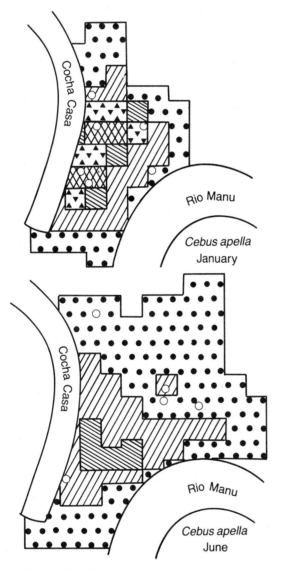

FIGURE 17-7. Home range use patterns of brown capuchins *Cebus apella* in Manu National Park, Peru, in two different seasons, January (*top*) and June (*bottom*). Grid size=1 ha. Sample lengths were 21 days for January and 20 days for June. Open circles show the locations of trees that collectively accounted for >50% of the feeding time for the respective samples. ⊡ 1–4 days; ▨ 5–8 days; ▩ 9–12 days; ▣ 13–16 days; ⊠ 17–18 days. From *Five New World Primates: A Study in Comparative Ecology,* J. Terborgh. © 1983 by Princeton University Press. Adapted with permission.

are only 400 to 800 m in diameter, animals in the center of a territory may be able to hear the vocal challenges that other groups make in approaching a border. Presumably, groups can act on such information in an economical fashion if moving to the boundary results in little further traveling than is normally involved in searching for food. The hypothesis that the ability to economically monitor and *reach* a boundary, rather than the ability to patrol it actively, is the more important factor in the development of territoriality would explain why Martin (1981a) did not find a strong relationship between territoriality and a ratio of day range length to the length of the home range perimeter.

The Manu tamarins live in groups of 2 to 10 individuals and specialize in exploiting fruit from plants that produce small numbers of ripe fruit over a long period of time. A small group of small-bodied monkeys can rely on finding these resources within a small area throughout the year, and hence this area may be profitably defended. In the same habitat occupied by the tamarins, Terborgh (1983) found that groups of 30 to 40 squirrel monkeys (*Saimiri sciureus*) had huge ranges (greater than 250 ha) that overlapped extensively with each other, mingled without obvious aggression, and shared food patches. Squirrel monkeys are fruit and insect eaters that rely heavily on large fig trees in the dry season when total numbers of fruits and insects in the forest are at their lowest point. Large fig trees with ripe crops are very widely scattered in space and time, but where they occur they can often provide enough food for more than one group. In these circumstances it is probably energetically unprofitable, if not impossible, for a squirrel monkey group to exploit its highly dispersed food resources and at the same time to defend them from exploitation by others. This is a good example of the pervasive influences that the spatial and temporal patterns of food distribution have on the behavior patterns of primates.

SUMMARY AND CONCLUSIONS

Most primates live in wooded environments in the tropics. These environments often appear highly productive, and in some seasons they are. Much of what is produced is not edible to primates, however, and what is edible by any given species is almost always patchily distributed in both time and space. Periods of food shortage occur frequently, although not necessarily predictably. Harvesting patterns are then liable to change, with folivores tending to be less active and frugivores more active. However, responses to food shortage vary greatly, even within a species.

Diet type has a major influence on feeding behavior. In general, species that rely heavily on leaves are banquet feeders, eating food from large patches in distinct bouts, spending much time resting between bouts, and displaying little intragroup competition. On the other hand, primates that rely more on highly dispersed fruits and insects forage more continuously, rest less, and compete with each other more. However, the particular temporal patterning of behavior shown by individual primates is affected by numerous other factors, including time of day, weather, and the behavior of conspecifics.

The spatial distribution of food strongly influences primates' patterns of movement. Although relationships between food distribution and ranging are complicated by factors such as body size and the form of locomotion employed, in general primates that exploit highly dispersed and unpredictable food supplies travel further each day and cover a larger area in the course of a year than do species feeding on dense resources that are more evenly distributed and predictably available. The dispersion of food patches and the costs of travel may interact with the size of individual patches to place upper limits on the size of primate social groups. If food-patch size remains constant, the number of patches that must be used in a given time will increase with group size, but energetic constraints will limit the number of different patches that can be visited.

If the food supply exploited by a primate population is distributed in such a way that social groups can economically defend an area that will supply their food requirements throughout an annual cycle, territorial behavior often occurs.

While available evidence indicates that the size and distribution of food patches have pervasive influences on the behavior of wild primates, the measurement of these parameters in complex tropical ecosystems is a difficult task for which primate ecologists are only just beginning to develop appropriate techniques. This is an important area for future research. When better measurements of food availability become available, we may have to modify some of the existing adaptive scenarios of primate behavior, which have often been constructed on quite flimsy evidence.

18 | Interactions among Primate Species

Peter M. Waser

Members of other vertebrate taxa may exert significant influences on primates, especially as competitors. Sloths influence the numbers of primate folivores in the Neotropics (Eisenberg and Thorington 1973). Potential frugivorous competitors include marsupials, rodents, ungulates, birds, bats, procyonids, and viverrids, which together outweigh primates in some localities (Emmons, Gautier-Hion, and Dubost 1983; Terborgh 1983). Nevertheless their size, habits, and numbers guarantee that primates of different species are generally each other's closest interactants, with the potential to act both as proximate (ecological) and as ultimate (evolutionary) influences shaping each other's behavior.

Primates have been reported to avoid, displace, approach, groom, forage, play, and mate with members of other primate species. Most live in communities of which other primate species are a significant component, whatever measure of importance one uses: number of species (up to 16), population density (up to 546 individuals/km^2), or biomass density (up to 2,652 kg/km^2) (see table 18-1).

In this chapter I review how primate species can modify each other's numbers, distributions, diets, and behavior (predatory effects are discussed in chap. 19). Since reports of interspecies interaction are scattered and often anecdotal, I draw when possible on data from six sites with diverse, relatively undisturbed, and rather completely described primate faunas (table 18-1): the Krau Game Reserve, Malaysia (Chivers 1980); the Kutai Nature Reserve, Borneo (Rodman 1978); the Kibale Forest Reserve, Uganda (Struhsaker 1975, 1978); Makokou, Gabon (Charles-Dominique 1977a; Gautier-Hion 1978a, 1980); the Raleighvallen-Voltzberg National Park, Surinam (Mittermeier and van Roosmalen 1981); and the Manu National Park, Peru (Terborgh 1983). Influences can be either negative or positive, and I discuss these under the rubrics of competition and mutualism in the first two sections of this chapter. In the final section, I discuss the ways in which primate species have influenced each other in evolutionary, rather than ecological, time.

In the interests of generalization, I have reanalyzed certain tabular data in comparable ways across species. In particular, as measures of dietary overlap I calculate either the summed percentages of specific food types shared by two different species (Holmes and Pitelka

TABLE 18-1   Species, Population Densities, and Biomass Densities for Six Primate Communities

Species	Adult Weight (kg)	Population density (ind/km^2)	Group Density (grp/km^2)	Biomass Density (kg/km^2)
Krau Game Reserve, Malaysia				
Hylobates syndactylus	10.0	5	1.3	39
Presbytis obscura	6.5/7.4	31	2.4	172
P. melalophos	6.9/6.7	74	6.8	406
Hylobates lar	5.4/5.0	6	1.9	30
Macaca fascicularis	3.1/3.3	39	1.2	89
(+ *M. nemestrina, Nycticebus coucang*)				
TOTAL 7 species		~156	~13	~746
Kutai Nature Reserve, Borneo				
Pongo pygmaeus	37.8/83.6	4	2.2	160
Presbytis aygula	6 /6.2	20	3.6	81
Macaca nemestrina	3.8/6.9	5	<0.4	<22
Hylobates muelleri	4.7/5.7	15	2.6	56
Macaca fascicularis	3.0/5.2	15	0.4	16
(+ *Nasalis larvatus, Presbytis rubicunda, P. frontata,*				
Nycticebus coucang, Tarsius bancanus)				
TOTAL 10 species		~61	~8	~335

TABLE 18-1   *(continued)*

Species	Adult Weight (kg)	Population density (ind/km²)	Group Density (grp/km²)	Biomass Density (kg/km²)
Kibale Forest Reserve, Uganda				
Pan troglodytes	45/45	3	0.2	30
Colobus guereza	7.0/10.5	58	6.5	317
C. badius	7.0/10.5	300	6.0	1760
Cercocebus albigena	7.0/10.5	9	0.6	60
Cercopithecus mitis	3.5/6.0	41	1.7	127
C. l'hoesti	3.0/6.0	5	0.5	13
C. ascanius	3.0/4.0	130	4.0	328
(+ *Papio anubis, Perodicticus potto, Galago inustus, G. demidovii*)				
TOTAL 11 species		~546	~19	~2652
Makokou, Gabon				
Papio sphinx	11.5/25.0	~5	?	?
Cercopithecus neglectus	4.0/7.0	28–38	6.0	110
C. nictitans	4.2/6.6	20–40	0.9	100
C. pogonias	3.0/4.5	20–25	0.9	60
C. cephus	2.9/4.1	20–30	2.0	80
C. talapoin	1.1/1.3	40–90	2.0	60
(+ *Gorilla gorilla, Pan troglodytes, Cercocebus albigena, Cercocebus galeritus, Colobus guereza, Perodicticus potto, Galago elegantulus, Galago alleni, G. demidovii, Arctocebus calabarensis;* several of these species occur in the general area but have not been sighted in the M'passa study area per se. Density estimates are approximate because of the patchy distribution of some species.)				
TOTAL 16 species		~148	~10	~410
Raleighvallen-Voltzberg National Park, Surinam				
Ateles paniscus	7.8/7.8	10		
Alouatta seniculus	6.0/8.5	17		
Cebus apella	3.0/3.9	13		
Saimiri sciureus	0.7/0.7	28		
Saguinus midas	0.5/0.5	23		
(+ *Cebus nigrivittatus, Chiropotes satanas, Pithecia pithecia*)				
TOTAL 8 species		106		
Manu National Park, Peru				
Ateles paniscus	8.0	25	?	175
Alouatta seniculus	8.0	30	5.0	180
Cebus apella	3.0	40	4.0	104
C. albifrons	2.8	35	2.3	84
Saimiri sciureus	0.9	60	1.7	48
Aotus trivirgatus	0.8	40	10.0	28
Callicebus moloch	0.8	24	8.0	17
(+ *Lagothrix lagotricha, Pithecia monachus, Callimico goeldii, Saguinus imperator, S. fuscicollis, Cebuella pygmaea*)				
TOTAL 13 species		275	~31	650

SOURCES: Krau Game Reserve: MacKinnon and MacKinnon 1980a; Chivers, pers. comm.; note however that longer-term censuses have shown significant variation in some figures, see Chivers 1980. Kutai Nature Reserve: Rodman 1978, pers. comm. Kibale Forest Reserve: Struhsaker 1975, 1978, 1981a. Makokou: Gautier-Hion 1978a, pers. comm.; Charles-Dominique 1977a. Raleighvallen-Voltzberg National Park: Mittermeier 1977; Mittermeier and Van Roosmalen 1981. Manu National Park: Terborgh 1983.

NOTES: Data are taken or calculated from most recently published figures, rounded to an equivalent number of significant digits. Weights are for females/males captured locally where data are available. In each site, species attaining at least 10 kg/km² biomass density are listed in descending order of individual weight.

1968) or the percentages of food species in one species' diet that are shared by another. These overlap indexes are readily calculable from published data, but they are subject to several limitations: (a) the values of both are influenced by how finely one subdivides "food types"; values are lower when finer distinctions are made; (b) dietary data for species being compared are generally collected in different seasons, years, or locations, which lowers the "overlap" values; (c) techniques by which diet is determined sometimes differ between species and study sites; and (d) data for the same species in the same site often vary between studies, thus reflecting sampling error and interobserver inconsistencies. In addition, there is no guarantee that the niche dimensions along which overlap can be calculated are those most important to primates. Because of these limitations, overlap indexes should be viewed as qualitative and relative despite their "quantitative" appearance.

COMPETITION

Stimulated by the truism that "complete competitors cannot coexist," investigators of multispecies primate communities have made the description of niche partitioning a primary goal. Similarities in resource use can be interpreted as indicating that species compete, differences as indicating that interspecific competition has been an important force in the past.

Competition, in the ecological literature, refers not to behavioral contest but rather to the reduction in availability of a resource to one individual by another. Species can influence each other through competition if some resource is in short supply and one species either aggressively excludes another from the limiting resource (interference competition) or simply eats it first (exploitation competition) (Miller 1967). Does interspecific competition occur between sympatric primates? Data consistent with this possibility are widespread in the literature and are of two sorts. The potential for interference competition exists wherever interspecific aggression occurs; exploitation competition is possible whenever niche overlap occurs.

Competition by Interference

The majority of primates interact aggressively with other species at least occasionally. *Cercocebus albigena* illustrates a common pattern: "in most cases, single mangabeys or small coalitions chased single individuals of other species, and the chases were generally brief. Once, a 15-minute series of chases, fights, and counterchases by most members of a blue monkey and mangabey group drove the mangabeys from a fruiting fig tree (Waser 1980, 66). In general, the outcome of aggressive interactions is determined by individual size. Interspecific dominance hierarchies therefore exist, with reversals occur-

ring where sizes are similar or intraspecific size variation is large (i.e., males of small species may displace females or subadults of larger species). Rarely, group size can reverse the effect of individual size (e.g., *Cebus albifrons* displaces *Cebus apella*, *Macaca spp.* can displace *Hylobates muelleri* and *Pongo pygmaeus*).

Unfortunately, neither the rate of interspecific aggressive interactions nor their effectiveness in excluding competitors has been widely documented. Many authors comment on the frequency with which different species share feeding trees *without* aggression: "competition was actually observed in the same tree at the same time, with *Chiropotes* and *Ateles* feeding on fruits of different stages of ripeness, but not interacting directly with one another" (Mittermeier and van Roosmalen 1981, 19). In Uganda, rates of aggression are higher between than within species, but remain very low in absolute terms. The number of hours two species spend together per aggressive interaction ranges from a minimum of about 3 to a maximum of nearly 160 (Struhsaker 1981a). Observations in several study sites indicate that even when aggression occurs, it may be ineffective at excluding anyone, particularly where the resource at issue is a large fruiting tree.

Where the rate of interspecies aggression is low, it seems unlikely to have a major impact on access to feeding sites. Theoretical arguments (Waser and Case 1981) suggest that, where food patches are difficult to find and where a group cannot stay indefinitely in a single feeding patch—as would be the case for most tropical forest frugivores—evicting other species from food sources used in common is an ineffective means of increasing one's share of food. On the other hand, if aggressive interactions provoke interspecies avoidance, even infrequent aggression could have important competitive effects; it seems quite likely that interspecies avoidance occurs, though hard evidence is difficult to obtain (but see fig. 18-3).

Competition by Exploitation?

Like interspecific aggression, niche overlap between species is ubiquitous, both for specific dietary items and along niche axes such as fruit size and foraging height. Overlap values for some species pairs and for some parts of the diet are remarkably high: as many as 96% of the fruit species used by one *Cercopithecus* species in Gabon can be used by another; shared percentages of prey-foraging substrates can total 94% in Peru; overlaps in foraging heights reach 92% in Borneo (tables 18-2, 18-3, 18-4, 18-5).

An illuminating indicator of the potential importance of interspecific competition can be acquired by comparing interspecific with intraspecific overlap. Such comparisons strongly imply that whether or not past inter-

TABLE 18-2   Dietary Overlap in Four Primate Communities: Plant Food Types

Krau Game Reserve, Malaysia[a]

Species A	Species B				
	Hylobates syndactylus	*Presbytis obscura*	*Presbytis melalophos*	*Hylobates lar*	*Macaca fascicularis*
H. syndactylus		14.3	12.0	48.5	24.4
P. obscura			40.2	20.2	29.9
P. melalophos				19.6	36.0
H. lar					35.2
M. fascicularis					

Kibale Forest Reserve, Uganda[b]

Species A	Species B				
	Colobus guereza	*Colobus badius*	*Cercocebus albigena*	*Cercopithecus mitis*	*Cercopithecus ascanius*
C. guereza		7.1	13.1	14.4	13.8
C. badius		24.3	5.1	14.1	4.7
C. albigena				26.2	21.2
C. mitis				16.0	33.8
C. ascanius					12.3

Makokou, Gabon[c]

Species A	Species B		
	Cercopithecus nictitans	*Cercopithecus pogonias*	*Cercopithecus cephus*
C. nictitans	—	89	93
C. pogonias	94	—	96
C. cephus	77	75	—

Manu National Park, Peru[d]

Species A	Species B					
	Cebus apella	*Cebus albifrons*	*Saimiri sciureus*	*Callicebus moloch*	*Saguinus imperator*	*Saguinus fuscicollis*
C. apella	—	25	33	04	10	10
C. albifrons	30	—	30	04	11	11
S. sciureus	35	26	—	04	10	10
C. moloch	27	22	27	—	19	08
S. imperator	21	19	21	06	—	26
S. fuscicollis	22	21	21	02	28	—

SOURCES: Krau Game Reserve: MacKinnon and MacKinnon 1980a; Kibale Forest Reserve: Struhsaker 1978; Makokou: calculated from Gautier-Hion 1980, fig. 4; Manu National Forest: calculated from Terborgh 1983, table 5.3.

NOTES: Species listed in descending order of weight. In Uganda, within-species overlap figures are for overlap between months. Most data are from groups not followed simultaneously in the same area, which results in overlap underestimates; samples from adjacent *C. mitis* groups in Uganda give a mean of 52.8% overlap within months, while simultaneous samples of overlapping *C. mitis* and *C. ascanius* groups give a mean overlap of 28.4% (Rudran 1978; Struhsaker 1981a).

a. Holmes-Pitelka overlap, diets broken down by plant species and food type; calculations based on % foraging time from system-atic behavioral observations; 376 species-specific plant food types distinguished.

b. Holmes-Pitelka overlap, diets broken down by plant species and food type; calculations based on % feeding time from systematic behavioral observations; >80 species and 16 plant food types distinguished.

c. % of fruit species in primate species A's diet shared by primate B; data from 52 to 100 stomachs per species; 93 plant species distinguished.

d. % of fruit species in primate species A's diet shared by primate B; data from opportunistic behavioral observations; 162 plant species distinguished.

specific competition has led to niche divergence, the potential for continued exploitation competition between some species pairs remains as great or greater than that for competition within species. The average overlap in plant food types between months for Ugandan *Cercopithecus mitis* is lower than that between *C. mitis* and *C. ascanius* (table 18-2). Malaysian *Hylobates lar* and *H. syndactylus* diets overlap less between adjacent conspecific groups than between species (MacKinnon 1977). In Gabon, male and female *C. cephus* diets resemble those of *C. nictitans* and *C. pogonias,* respectively, more than they do each other. In Peru, intraspecific overlaps "are reassuringly [i.e., not much?] higher than inter-specific overlaps except for *C. apella*" (Terborgh 1983, 101).

Studies of niche partitioning are, however, plagued by a number of problems. On the one hand, many interspecies differences are irrelevant to competition. Some primates specialize on eating seeds, others on the pulp surrounding them; a fruit used by one is nevertheless

gone for the other. Species with quite different locomotory capabilities may use them to gain access to the same food items. On the other hand, potentially critical niche dimensions have been incompletely described. If one primate takes fruit from main branches and the other takes it from twigs, they do not compete, yet the location from which food items are taken has rarely been a subject of study.

Another problem is that simple overlap measures do not detect asymmetries in competition. Species that eat flowers or immature fruit do not thereby cease to compete with those that eat the same species ripe, but competition does become one sided: species that eat ripe fruit are adversely affected by those that eat it immature, but not vice versa. If a rare and a common species feed on the same items, the common species should have much stronger effects on the rare one than vice versa.

Available overlap measures are often of limited precision. Annual overlap indexes likely conceal important seasonal fluctuations in diet. Overlap values calculated

TABLE 18-3   Dietary Overlap in Two Primate Communities: Prey Types

Makokou, Gabon[a]				
	Cercopithecus nictitans	*Cercopithecus pogonias*	*Cercopithecus cephus*	
C. nictitans	—	77	47	
C. pogonias		—	65	
C. cephus			—	

Manu National Park, Peru[b]					
	Cebus apella	*Cebus albifrons*	*Saimiri sciureus*	*Saguinus imperator*	*Saguinus fuscicollis*
C. apella	—	85	38	38	36
C. albifrons		—	42	42	39
S. sciureus			—	70	51
S. imperator				—	73
S. fuscicollis					—

SOURCES: Makokou: calculated from Gautier-Hion 1980, fig. 5; Manu National Park: calculated from Terborgh 1983, table 6.5.

NOTES: Species listed in descending order of weight. Holmes-Pitelka overlap; diets broken down by prey type.

a. Calculations based on % dry weight in stomachs. Eight prey types distinguished (caterpillars, orthoptera, ants, spiders, etc.).

b. Calculations based on % captures of identified prey during behavioral observations. Seven prey types distinguished (vertebrates, orthoptera, lepidoptera, hymenoptera, etc.).

TABLE 18-4   Dietary Overlap in the Manu Primate Community: Fruit Size

	Cebus apella	*Cebus albifrons*	*Saimiri sciureus*	*Callicebus moloch*	*Saguinus imperator*	*Saguinus fuscicollis*
C. apella	—	67	55	43	30	27
C. albifrons		—	65	30	23	21
S. sciureus			—	54	43	40
C. moloch				—	89	85
S. imperator					—	81
S. fuscicollis						—

SOURCE: Calculated from Terborgh 1983, table 5.4.

NOTES: Species listed in descending order of weight. Holmes-Pitelka overlap; diet broken down by fruit diameter; calculations based on % of foraging time during systematic behavioral observations, 6 size categories distinguished.

from groups in different years or slightly different locations underestimate the interaction of species coexisting in overlapping home ranges. As noted above, more precise measures of *intra*specific overlap, for instance overlap between sexes, age classes, or groups, would provide a valuable calibration of the importance of interspecific similarity.

Perhaps most important, for competition to occur, dietary overlap must involve a limiting resource. Where overlaps are high, an alternative interpretation is that shared resources are *not* in short supply.

What Resources, if Any, Are in Short Supply? Much circumstantial evidence suggests that food limits primate population densities. Food density and distribution have powerful effects on group size, home range size, and day range length (chap. 17). Group ranging patterns are often dictated by the density and distribution of food sources and, more specifically, of favored fruiting trees; intensity of use is often greater, and home ranges smaller or more tightly clustered where favored fruiting trees are concentrated. Across species, larger primates tend to have

lower population densities (fig. 18-1; see also Clutton-Brock and Harvey 1977b). Fruiting seasonality may be greater in the Neotropics than the Old World, implying lower food levels during seasonal bottlenecks, and primate biomass densities there are lower (Terborgh 1983; table 18-1).

While these data make a reasonable inferential case for the importance of some aspect(s) of food, direct evidence is rare. It is suggestive that several frugivorous species (*Cercocebus galeritus*, perhaps *Macaca fascicularis* and *Pongo pygmaeus*) attain considerably higher densities in forest types argued, because of their successional nature, to be more productive of fruits (Homewood 1978; Rodman 1978). The clearest answer to this question would come from cross-area comparisons of both primate density (or reproductive rate) and levels of putative limiting resources. This approach would also clarify what aspects of food are critical—whether fruit or arthropods, particular species or overall abundance, particular nutrients or specific plant defenses.

Though isolating potential limiting factors seems a daunting task, it is surprising that the approach closest to

TABLE 18-5   Dietary Overlap in Three Primate Communities: Foraging Height

	Kutai Nature Reserve, Borneo[a]				
	Pongo pygmaeus	*Presbytis aygula*	*Macaca nemestrina*	*Hylobates muelleri*	*Macaca fascicularis*
P. pygmaeus	—	87	30	84	62
P. aygula		—	20	92	94
M. nemestrina			—	20	24
H. muelleri				—	61
M. fascicularis					—

	Kibale Forest Reserve, Uganda[b]				
	Colobus guereza	*Colobus badius*	*Cercocebus albigena*	*Cercopithecus mitis*	*Cercopithecus ascanius*
C. guereza	—	60	69	76	70
C. badius		—	73	60	77
C. albigena			—	74	75
C. mitis				—	75
C. ascanius					—

	Raleighvallen-Voltzberg National Park, Surinam[c]				
	Ateles paniscus	*Alouatta seniculus*	*Cebus apella*	*Saimiri sciureus*	*Saguinus midas*
A. paniscus	—	72	43	27	39
A. seniculus		—	59	37	60
C. apella			—	66	86
S. sciureus				—	86
S. midas					—

SOURCES: Kutai Nature Reserve: Rodman 1978, fig. 10; Kibale Forest Reserve: Struhsaker 1978; Raleighvallen-Voltzberg National Park: Mittermeier and Van Roosmalen 1981, fig. 7.

NOTE: Species listed in descending order of weight.

a. Holmes-Pitelka overlap in height at first sighting during censuses; 7 height categories distinguished.

b. Holmes-Pitelka overlap in feeding heights during systematic behavioral observations; 13 height categories distinguished.

c. Holmes-Pitelka overlap in height at first sighting during censuses; 6 height categories distinguished.

an experimental test—comparing primate densities in undisturbed and adjacent felled forest patches—has been virtually unexploited. In Malaysia and Uganda, selective felling has markedly decreased group sizes and population densities of most, if not all, species (Struhsaker 1981a; Marsh and Wilson 1981b).

In summary, primates have the clear potential to exert significant competitive effects on each other, both through exploitation and through interference. We do not, however, know how often this potential is realized. To gaps in data and data analysis must be added the perennial question Does overlap in resource use involve resources that are in fact limiting? Future research may reap quicker answers by searching for the consequences of competition, rather than solely documenting niche partitioning. Below, I examine whether or not competition directly affects primates' numbers, distribution, diets, or ranging patterns.

Population Density. A classic indicator of competition is density compensation, an increase in one species' population density when a putative competitor is absent. Census data might readily be used to investigate this possibility, particularly since most primates are now confined to reserves or forest patches, the sort of "island" on which density compensation is most easily detected. Struhsaker (1975) inferred from surveys that *Cercopithe-*

cus mitis density is higher in Ugandan forest patches where potential competitors, particularly *Colobus badius* and *Cercocebus albigena,* are absent. Other possible cases of density compensation are summarized in table 18-6.

Distribution. In the extreme, competition can reduce the less efficient competitor's population size to zero. Evidence for local extinction might be sought in mosaic distributions of species that have generalized habitat requirements and similar diets. The patchy distribution of *Lagothrix* and *Ateles* in Peru could have such a cause, as could the interdigitation of *Callicebus moloch* and *Callicebus torquatus* home ranges (Kinzey and Robinson 1983; Terborgh 1983). A particularly clear case involves *Hylobates lar* and *H. agilis;* where they are sympatric, these two species are interspecifically territorial (Chivers 1977).

Diet. Interspecific competition may also be detected by a niche shift or expansion when a competitor is removed. If competition occurs, it should be manifest in seasonal environments by seasonal changes in niche overlap and breadth. To detect competition via niche shifts thus requires seasonal estimates of availability of the presumptive limiting resource (e.g., fruit, leaves, insects, all food

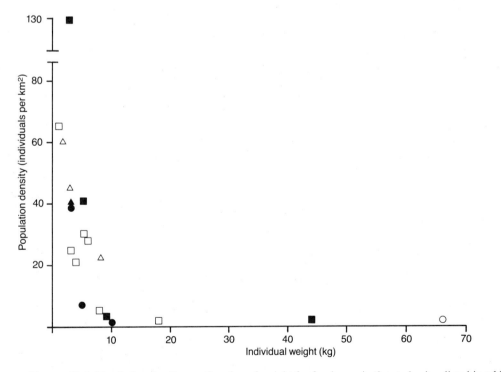

FIGURE 18-1. Population density as a function of weight for frugivores in the study sites listed in table 18-1. Since these species have broadly similar diets, the observed inverse relationship suggests that food limits population size. ●=Krau, Malaysia; ○=Kutai, Borneo; ■=Kibale, Uganda; □=Makokou, Gabon; ▲=Raleighvallen-Voltzberg, Surinam; △=Manu, Peru.

types together), as well as of interspecific overlap in use of that resource. Phenological asynchrony and year-to-year climatic variation produce bottlenecks and bonanzas to differing degrees and at differing times each year. Qualitatively, resources are often most scarce during the dry season (fruit in Gabon, Malaysia, Peru; young leaves in Gabon and Malaysia; insects in Gabon).

In Gabon, the proportions of fruits, leaves, and animal material in stomachs of three *Cercopithecus* species are most divergent during the main dry season, as would be expected if interspecific competition forces diets to diverge during periods of food scarcity. Decreased dietary overlap during seasonal periods of scarcity is substantial in Malaysia and striking in Peru (table 18-7).

In Peru, dry season overlap in fruit species decreases between *Cebus apella* and *Cebus albifrons,* but not between *Cebus* and *Saimiri;* conversely, overlap in prey-foraging microhabitat decreases between *Cebus, Saimiri,* and *Saguinus,* but not between *Cebus* species. Foraging tree height overlaps show no clear-cut seasonal changes, but overlaps in tree size as measured by canopy diameter sharply decrease during the dry season (tables 18-7, 18-8, 18-9). These comparisons, though qualitative, indicate what could be done with more precise seasonal comparisons—and support the idea that competition is occurring.

A little-exploited alternative to seasonal comparisons involves comparing primate diet overlaps and breadths in and out of mixed-species associations. It might be argued that the effect of a competitor would be most severe when it forages in the same place at the same time. Some

bird species overlap less in diet when they forage in mixed-species flocks, as would be expected if foraging together increases competition (Morse 1969). However, for primates the only published data indicate that diet diversity, foraging height overlap, and dietary overlap *increase* when species are in association (Gautier-Hion, Quris, and Gautier 1983).

Ranging Patterns. A final negative influence of competitors could be to increase the area or distances over which a group must forage to obtain adequate food. Where more species are sympatric, home ranges are often larger (table 18-6).

On the other hand, data from mixed-species associations show no obvious locomotor cost to associating with potential competitors: rates of movement for associations among *Cebus* and *Saimiri,* or different *Saguinus* species, or different *Cercopithecus* species are intermediate between those for the same species foraging alone (Gautier-Hion, Quris, and Gautier 1983; Terborgh 1983). The increased rate of movement for otherwise slower species (*Cebus apella, Saguinus imperator, Cercopithecus cephus*) presumably costs them something energetically. But surprisingly, movement rates for mixed-species associations are close to the mean rates of their constituent species moving independently. Why should species in association "split the difference" in their speeds? If competition increases when species join, one would expect that mixed-species groups would have to move farther to harvest the same amount of food than any species alone.

TABLE 18-6 Possible Cases of Density Compensation in Primates

Species	Density Comp. Present	Density Comp. Absent	Home Range Size (ha) Comp. Present	Home Range Size (ha) Comp. Absent	Potential Primate Competitors	Sources
Hylobates spp.[a]	2.3–24.5	42			*Hylobates syndactylus*	MacKinnon 1977
Macaca fascicularis[b]	0.0–2.6	5.8			*Hylobates* spp.	Marsh and Wilson 1981a
Pan troglodytes[b]	0.2	0.9			*C. mitis*	Struhsaker 1981a
Cercocebus albigena[c]	9	80	410	13	*Colobus badius,* *C. mitis*	Struhsaker 1978; Chalmers 1968b
C. albigena[b]	0.6	1.1	410	~200	*C. mitis*	Struhsaker 1981a, pers. comm.
Cercopithecus mitis[c]	35	185	55	8	*Cercocebus albigena,* *Colobus badius*	Rudran 1978
Alouatta seniculus[a]	180	622	20–30	3.2	*Cebus* spp., *Callicebus, Ateles*	Terborgh 1983; Rudran 1979
Saimiri sciureus[a]	33	>250	175	50	*Cebus* spp., *Callicebus, Saguinus* spp.	Chap. 7
Cebus albifrons			150	120	*Ateles, Callicebus, Cebus apella*	Terborgh 1983; Defler 1979a
Callicebus moloch[a]			6–12	0.3–4.2	*Cebus* spp., *Alouatta, Ateles*	Kinzey and Robinson 1983

a. Biomass density = kg/km^2.
b. Group density = groups/km^2.
c. Population density = individuals/km^2.

MUTUALISM

Do primate species always influence each other negatively? The answer is no, but inferences about positive effects come from a line of research quite unrelated to niche separation—from studies of mixed-species associations (fig. 18-2) (Gautier and Gautier-Hion 1969; Gautier-Hion, Quris, and Gautier 1983; Klein and Klein 1973; Struhsaker 1981a; Waser 1980). Primates can, of course, influence in positive ways each others' diets, ranging patterns, and numbers, even if they do not travel together: witness the doubling of *Cercopithecus talapoin* densities around *Homo sapiens* settlements (Gautier-Hion 1971a). Just as two species can compete by exploitation even when they forage independently in the same area, so primates might benefit from each other's presence at a distance. For instance, even groups several hundred meters apart might use each others' alarm calls as warnings of hunting eagles or cue on other vocalizations to locate food sources. Nevertheless, positive interactions have forced themselves on observer's attentions only in the context of mixed-species associations.

For reasons just discussed, comparison of behaviors in and out of associations will not detect all the effects two species have on each other. But it is surprising that such comparisons have not been more widely attempted, for they have several advantages. They come as close as is likely to be possible to controlling for habitat differences; when the same animals in the same place change their behavior while associating with another species, the changes are convincingly related to the other species' presence. In addition, where associations can be shown to involve behavioral attraction between species, positive effects must be strong; behavioral attraction is prima facie evidence that selection has favored associating.

Are Associations Necessarily Beneficial?

Data from mixed-species associations must be examined for positive interspecies influences with several caveats

TABLE 18-7   Seasonal Changes in Dietary Overlap in Three Primate Communities: Food Types

Krau Game Reserve, Malaysia[a]				
Hylobates syndactylus	*Presbytis obscura*	*Presbytis melalophos*	*Hylobates lar*	*Macaca fascicularis*
H. syndactylus —	64/68 (+)	56/67 (++)	83/80 (−)	72/71 (−)
P. obscura	—	70/96 (++)	59/88 (++)	49/95 (++)
P. melalophos		—	44/87 (++)	33/98 (++)
H. lar			—	89/91 (+)
M. fascicularis				—

Makokou, Gabon[b]		
Cercopithecus nictitans	*Cercopithecus pogonias*	*Cercopithecus cephus*
C. nictitans —	61/96 (++)	82/94 (++)
C. pogonias	—	79/97 (++)
C. cephus		—

Manu National Park, Peru[c]				
Cebus apella	*Cebus albifrons*	*Saimiri sciureus*	*Saguinus imperator*	*Saguinus fuscicollis*
C. apella —	80/100 (++)	67/99 (++)	43/98 (++)	17/97 (++)
C. albifrons	—	56/99 (++)	45/98 (++)	19/97 (++)
S. sciureus		—	50/97 (++)	25/96 (++)
S. imperator			—	70/98 (++)
S. fuscicollis				—

SOURCES: Krau Game Reserve: calculated from MacKinnon and MacKinnon 1980a, table 6.4; phenological/climate data from Chivers 1980; Makokou: calculated from Gautier-Hion 1980, table 3; phenological/climate data from Gautier-Hion 1980 and A. Hladik 1978; Manu National Park: calculated from Terborgh 1983, table 5.1; phenological/climate data from Terborgh 1983.

NOTES: Wet seasons are periods of increased fruit, leaf, and perhaps insect abundance. A plus sign (+) means that dietary overlap increased during the wet season, a double plus (++) that it increased more than 10% (as would be expected if interspecific competition forces diets to diverge during periods of food scarcity); a minus sign (−) indicates that overlap decreases during the wet season. In Gabon, comparisons between the main dry and minor wet seasons produce similar results.

a. Represents Jan.–Feb. (dry) versus July (wet); Holmes-Pitelka overlaps calculated from % feeding time during systematic behavioral observations. Food types distinguished: fruit, flowers, leaves, arthropods.

b. Represents June–Aug. (main dry) versus Aug.–Oct. (main wet); Holmes-Pitelka overlaps calculated from % dry weight in stomachs. Food types distinguished: fruit/seeds, leaves/fiber, animal material.

c. Represents dry versus wet season; Holmes-Pitelka overlaps calculated from % feeding time during systematic behavioral observations. Food types distinguished: fruit, seeds, sap, nectar, pith, flowers, other (animal matter excluded).

in mind. First, there are likely costs, as well as benefits, to association. Gautier-Hion, Quris, and Gautier (1983) have suggested that *C. cephus* sacrifices insect foraging efficiency by joining mixed-species groups, but increases fruit-foraging efficiency. When threatened from the ground, *C. cephus* abandons associations; groups are more cryptic and presumably safer from terrestrial predators when alone, but since they forage low in the canopy, association with more arboreal species may increase their safety from aerial predators.

Second, costs of associating can be as great or greater than benefits for some species; interactions between primate species can be commensal or parasitic. Mittermeier and van Roosmalen (1981) and Terborgh (1983) believe that the net effect of association on *Cebus* is zero or slightly negative, with net gains accruing only to *Saimiri* (but see Klein and Klein 1973). Many of the postulated effects of association listed below are one way.

Finally, conspicuous associations can occur even in the absence of net benefits to either species. Species might meet only at shared food resources, and, particularly where population densities are high, "associations" can occur simply because groups of finite size necessarily cross each others' paths (see fig. 18-3).

Though associations are not inevitably mutualistic or even commensal, there are nevertheless many positive ways that species in association could influence each other. Groups of a few primate species are found in association more often than out (table 18-10). In Gabon, Peru, and Uganda, solitary males as well as groups associate with other species. In some cases, associations are brief but frequent; in others, association is virtually a permanent condition. Many or all of the positive effects I now list can act together in any particular association, and attempts to determine their relative importance have only begun.

FIGURE 18-2. A polyspecific association among emperor tamarins (*upper left*), saddleback tamarins (*right*), and dusky titis (*lower left*). In Manu National Park it is not unusual to see such associations, though individuals rarely interact socially except with their own species. (Drawing: Teryl Lynn)

TABLE 18-8   Seasonal Changes in Dietary Overlap in the Manu Primate Community: Feeding Tree Size

	Cebus apella	*Cebus albifrons*	*Saimiri sciureus*
C. apella	—	26/62 (++)	27/64 (++)
C. albifrons		—	16/73 (++)
S. sciureus			—

SOURCE: Calculated from Terborgh 1983, table 5.7.

NOTES: Represents May–June (early dry) versus Nov.–Feb. (midwet). Holmes-Pitelka overlaps calculated from % feeding time; 6 feeding tree canopy diameter classes distinguished. Species listed in order of descending size. Fruit levels peak during the midwet season, are lowest in the early dry (Terborgh 1983). See table 18-7 for explanation of symbols.

TABLE 18-9   Seasonal Changes in Dietary Overlap in the Manu Primate Community: Foraging Substrate

	Cebus apella	*Cebus albifrons*	*Saguinus imperator*	*Saguinus fuscicollis*
C. apella	—	87/82 (−)	56/71 (++)	38/39 (+)
C. albifrons		—	68/78 (+)	35/37 (+)
S. imperator			—	5/17 (++)
S. fuscicollis				—

SOURCE: Calculated from Terborgh 1983, table 6.10.

NOTES: Represents Aug.–Sept. (late dry) versus Oct.–Dec. (early wet); Holmes-Pitelka overlaps calculated from % prey-foraging time. Substrate types distinguished: leaves, branches, palms, other. Species listed in order of descending weight. Insect levels probably peak during the early wet season, are lowest in the late dry (Terborgh 1983). See table 18-7 for explanation of symbols.

How Could One Species Benefit Another?

By forming associations, species gain access to otherwise unavailable food. In a number of mixed-species associations, smaller species gain access to food types too well protected for them to exploit alone (Struhsaker 1981a; Terborgh 1983). *Cercopithecus ascanius* and sometimes *C. mitis* follow the larger, heavier-jawed *Cercocebus albigena* to feed on leftover mesocarp of tough, heavy *Monodora* fruit, which *C. ascanius* cannot open alone. *Saimiri sciureus* gains a similar benefit from *Cebus apella* with hard, tightly packed clusters of *Scheelea* nuts. Terrestrial species use fruit dropped by arboreal ones, for example, Ugandan *Papio anubis* under *Cercopithecus* species.

Moving monkeys stir up insects; following a large group of any species could increase insect capture rate. Several cases are known in which birds gain this advantage by following monkeys, but reported capture rates of flushed prey by other monkeys are so low that this appears an insignificant effect of association for most primates (Gautier-Hion, Quris, and Gautier 1983; Struhsaker 1981a; Terborgh 1983).

Frugivorous primates with large home ranges can potentially exploit other species to locate new fruiting trees. Circumstantial evidence suggests that this effect is widespread. Species with larger home ranges (*Pan troglodytes, Cercocebus albigena, Cercopithecus mitis*) are reported to initiate associations with *Cercopithecus ascanius*, which uses a smaller area more intensively and thus can better locate newly ripening fruit crops (Struhsaker 1981a). *Cebus albifrons* and *Saimiri sciureus*, species with larger home ranges, tend to initiate associations with *Cebus apella*, with smaller ranges *Cebus albifrons* may then displace *Cebus apella* from fruiting trees they have discovered (Terborgh 1983).

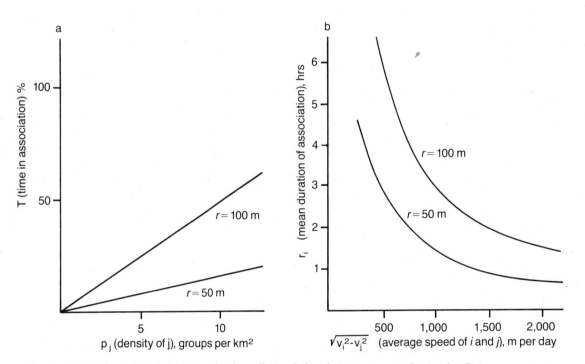

FIGURE 18-3. Expected characteristics of "chance" associations between two species *i* and *j*. Two groups moving independently of each other would be expected to initiate "association" at a rate $Z = 2\sqrt{\bar{v}_i^2 + \bar{v}_j^2}\, \rho_j r$; the mean duration of associations would be $\tau = (\pi^2 r) / (4\sqrt{\bar{v}_j^2 + \bar{v}_i^2}$; and the mean percentage of time in association would be $T = Z\tau = (\pi^2/2)\rho_j r^2$. (In these equations, v_i and v_j are the mean speeds of groups of *i* and *j*; ρ_j is the density of groups of *j*, and *r* is the criterion distance that defines *association;* see Waser 1982a, 1984 for derivation and discussion of assumptions).

Fig. 18-3a plots the expected percentage of time a focal group of species *i* should be seen in association with groups of *j* as a function of ρ and *r*. (*r* is equal to the sum of r_i [the mean radius of a group of species *i*], plus r_j [the radius of a group of *j*], plus *d* [the distance between the two groups' outmost animals beyond which the observer does not consider the groups to be associating].) For example, Struhsaker 1981a measured the radius of *Colobus badius* groups to be approximately 50 m, that of *Cercopithecus ascanius* groups to be approximately 25 m, and defined species to be in association if any two individuals were within 20 m of each other. Thus $r = r_i + r_j + d = 95$ m. For most species and studies, the value of *r* probably lies between 50 and 100 m.

Fig. 18-3b plots the expected duration of such "chance" associations as a function of the mean speeds of the two groups and of *r*. Observed percentage of time in association or association duration greater than these values would suggest that species are attracted to each other; less than these values, that species would avoid each other.

Food finding and predator avoidance can interact in complex ways. Struhsaker (1981a) has speculated that Ugandan *C. ascanius* associating with *Colobus badius* gain access to fruiting trees occupied by *Pan troglodytes,* which otherwise exclude (and may prey on) them. In Uganda, *Colobus badius* sometimes aggressively mob and displace *Pan;* it would be interesting to know if *C. ascanius* also associates with *Colobus* where *Pan* is a significant *Colobus* predator. In Gabon, small predation-prone species such as *C. cephus* and *C. talapoin* have been suggested to use association with other *Cercopithecus* species to gain access to less dense forest types that are more productive of fruit, but from which they would otherwise be excluded because they are too vulnerable to predation there (Gautier-Hion, Quris, and Gautier 1983).

Associating species might use food more systematically and thus efficiently. Cody (1971) suggested that flocking might allow birds to avoid feeding in areas recently exploited by competitors (if resources are not renewing) or (if they are) to regulate their return rates to maximize food renewal between visits. Cody's renewing-resource model has been questioned because of its extreme sensitivity to certain of its numerical assumptions (Pyke, Pulliam, and Charnov 1977). The question remains, Are there conditions under which an individual foraging with competitors saves enough search time to outweigh its preemption from some food sources? A formal analysis of this possibility would be desirable (e.g., Clark and Mangel 1984), as such an effect of foraging in mixed-species associations has been postulated for *Saguinus* in Peru and *Cercopithecus* in Gabon (Gautier-Hion, Quris, and Gautier 1983; Terborgh 1983).

Association increases intraspecific competitive ability. Over 5 years' observation in Gabon, *C. cephus* in a permanent mixed-species group steadily increased its home range size at the expense of an adjacent monospecific group (Gautier-Hion, Quris, and Gautier 1983). Primates in semipermanent mixed-species groups seem likely to benefit from increased success in intergroup encounters, assuming success in encounters increases with "group" size. *Saguinus* species in Peru, defending joint territories, represent the extreme of such a phenomenon (chap. 22).

Association might increase safety from predators. Other species could "dilute" in several ways the chances of an individual's being taken by a predator. If predators take at most a few animals per attack, and if their rate of attack

TABLE 18-10 Time Spent in Mixed-Species Associations in Three Primate Communities

Krau Game Reserve, Malaysia[a]					
Focal Species	*Hylobates syndactylus*	*Presbytis obscura*	*Presbytis melalophos*	*Hylobates lar*	*Macaca fascicularis*
H. syndactylus	—	2.9	3.2	6.4	<0.5

Kibale Forest Reserve, Uganda[b]							
Focal Species	*Pan troglodytes*	*Colobus guereza*	*Colobus badius*	*Cercocebus albigena*	*Cercopithecus mitis*	*C. l'hoesti*	*C. ascanius*
C. guereza	0.0	—	11.0	3.1	7.2	0.1	8.9
C. badius	0.1	0.7	—	1.8	11.6	0.0	24.1
C. albigena	0.7	2.5	8.4	—	8.7	0.1	15.4
C. mitis	0.0	0.2	7.8	0.5	—	0.1	12.9
C. ascanius	0.1	2.6	30.6	3.8	27.6	0.2	—

Makokou and Vicinity, Gabon[c]					
Focal Species	*Cercocebus albigena*	*Cercopithecus nictitans*	*C. pogonias*	*C. cephus*	*C. talapoin*
C. albigena	—	70	50	30	0
C. nictitans	14	—	40	55	4
C. pogonias	15	58	—	42	21
C. cephus	7	46	30	—	37
C. talapoin	0	5	16	40	—

SOURCES: Krau Game Reserve: Chivers, pers. comm.; Kibale Forest Reserve: Rudran 1978; Struhsaker 1981a; Waser 1980; Makokou and vicinity: Gautier and Gautier-Hion 1969.

a. % time in association (individuals in "close proximity") during systematic behavioral observations. Here and in other Asian sites, mixed-species associations are sighted extremely rarely, though no quantitative data are available (Struhsaker 1981a).

b. % time in association (individuals within 20 m, 50 m for *C. badius*) during systematic behavioral observations.

c. % time in association (individuals intermixed) when first sighted during censuses.

per group is independent of group size, then predatory risk goes down as group size goes up (Emlen 1973). The first assumption is met for most predators of primates, but data bearing on the second are lacking. Census results indicate that mixed-species associations are more conspicuous, at least to humans, than single-species groups (Struhsaker 1981a).

Individuals in mixed-species associations could increase their safety even in a group under attack by effectively interposing more individuals between themselves and a predator. In particular, animals that forage lower in the canopy could increase their safety from aerial predators by joining species that tend to move and feed above them; smaller or quieter animals might increase their safety from predators that cue on size or sound by immersing themselves in a group of a larger or noisier species (Gautier-Hion, Quris, and Gautier 1983; Waser 1982a).

Joining other species might increase the chance that predators are detected and their attacks therefore unsuccessful. Primates almost always respond to other species' alarm calls, even those of nonprimates. Such an advantage might be particularly strong for solitary individuals, for example, dispersing males. The advantage would also be stronger if the species joined is in a better position to detect predators (larger groups with more eyes, canopy foragers for aerial predators, understory foragers for terrestrial predators), and if it is characteristically more alert. (It has been suggested that males in unimale groups show greater vigilance because they have a known genetic stake in the group's vulnerable juveniles; Terborgh 1983; chap. 19).

Animals in mixed-species associations might better deter predators. Adults of larger species have been seen to rush at and mob potential predators (*Cercocebus albigena* and eagles, *Colobus badius* and chimpanzees, *Cercopithecus* species and cats, many species and humans). Many primate "alarm" calls actually have the conspicuous, easily localized acoustic structure usually associated with mobbing vocalizations.

Rates of observed predation are so low (chap. 19) that the importance of antipredatory benefits for mixed-species associations is generally argued on inference: most association types occur year-round and persist even when animals are not foraging (Gautier-Hion, Quris, Gautier 1983; Struhsaker 1981a); association frequency is geographically correlated with predator density, especially aerial predator density (both are most common in Africa, least in Asia (Struhsaker 1981a); smaller and presumably more vulnerable species are the most common participants in associations (table 18-10); terrestrial species, at little risk from aerial predators, are virtually never found in associations (Gautier-Hion, Quris, and Gautier 1983).

The infrequency of successful predatory attempts does not mean that predation risk is insignificant. But it does mean that differences in predatory risk for individuals with and away from other species will probably not be demonstrated by following monkey groups. Direct evidence in support of any specific antipredatory mechanisms would more efficiently be acquired by following predators. Two other approaches to documenting antipredator effects of interspecies association would also benefit from following predators. One would correlate the strength of association tendencies and the characteristics of associating species with their vulnerability to predation given knowledge of predator's hunting tactics. The other would compare rates of attempted and successful predation, rates of predator detection, and differences in time spent vigilant or in "risky" behaviors, for a species alone versus in association.

Association might have social benefits. Many primates play, groom, and otherwise interact nonaggressively with members of other species, though the rates of "friendly" interaction per hour of opportunity are typically much lower than they are within species (Struhsaker 1981a). While such behaviors may be adaptively neutral, they need not be. In Uganda, juveniles of other species play with *Colobus badius* juveniles at rates inversely proportional to their group size, for example, more frequently if they have fewer conspecific playmates. As another example, some individuals may receive real benefits from interspecific grooming. Solitary males are particularly likely to solicit grooming from other species and, at least in Uganda, not infrequently receive it. Such males may join other species to be groomed.

A final benefit potentially attainable from other species is reproductive. Conservatism or convergence in species-specific signals may account for some interspecific sexual interactions—for instance, *Colobus badius* investigating *Cercocebus albigena* sexual swellings (Waser 1980). But hybrids occur in mixed-species *Hylobates* groups (Chivers 1977) and are not uncommon in *Cercopithecus*. *C. mitis* groups in the southern Kibale Forest, Uganda, are rare. Dispersing males, however, are fairly common, mate with *C. ascanius,* produce viable offspring, and even backcross. Under these conditions, interspecific mating is a viable reproductive strategy, even if it is a male's option of last resort (Butynski 1982b).

COEVOLUTION

The preceding sections have explored how other species affect primate behaviors and numbers now. How important have other species been in the evolutionary past? Data and speculations are most extensive in two areas. Has interspecific competition been strong enough to limit the kinds of primate species that can coexist? Have the benefits of interaction been large enough to select for behavioral attraction between species?

Have Niches Diverged?

Does the structure of primate communities reflect the influence of past competitive interactions? Although the view that interspecific competition provides the primary limits to the number and kinds of species in a community has been virtually an article of faith among primatologists, the primacy of competition in structuring other communities has recently come into question (Connell 1983). The problem is a serious one for primate communities; some of their characteristics are indeed such that competition dogma might be expected to fail. In particular, the complex seasonal and interseasonal variation in plant phenology that characterizes many tropical forests provides precisely the conditions argued to relegate competition to an occasional and minor role (Wiens 1977).

To demonstrate niche divergence one might logically ask first whether species partition resources, and second whether differences in resource use, are greater than expected in the absence of competition. Most primate faunas can be broken qualitatively into three subsets with broadly different diets: large folivores, large frugivores,

and small insect, nectar, and tree-exudate feeders. Within these three guilds, how do primates partition resources (fig. 18-4)?

Habitat. In most primate communities, a minority of species are habitat specialists. Often these use "edge" or secondary habitats; sometimes swamp, riparian, or seasonally flooded forest is the habitat of specialization. A common feature of most of these habitat types is their relatively low stature and richness in vines, small trees, and other thin supports. Perhaps as a result, many habitat specialists tend to be small (for instance, *Cercopithecus talapoin, Arctocebus,* and *Galago demidovii* in Gabon; *Cebuella, Callimico, Callicebus,* and *Saguinus* in Peru). However, large habitat specialists exist, witness *Colobus guereza* as an "edge" species, or *Macaca fascicularis, Cercocebus galeritus,* and *Cercopithecus neglectus* as riparian forest denizens.

Foraging height. Species of edge and riparian habitat often forage more terrestrially than those of primary forest. In addition, most primary forest communities con-

FIGURE 18-4. A mother-son pair of chimpanzees sits in an oil-palm tree (*Elaeis guineensis*) after being supplanted from their feeding sites by two adult male olive baboons. Competition for food was clearly responsible for this association, which lasted only 10 minutes. (Photo: Richard Wrangham/Anthrophoto)

tain one or two predominantly terrestrial species, usually large: *Cercopithecus l'hoesti* and *Pan troglodytes* in Uganda; *Macaca nemestrina* in Malaysia and Borneo; *Pan* and *Papio sphinx* in Gabon. Terrestrial species often climb trees to feed, or feed on fallen fruit, so their dietary separation from arboreal species is not complete. But it is undoubtedly greater than that among arboreal species. While upper-, middle-, and lower-canopy feeders are sometimes distinguished in the literature, the great height overlap reported for most species pairs (table 18-5) makes the ecological importance of such distinctions equivocal.

Support size and locomotory style. Anatomy and size clearly differentiate species by the types of support they can reach and efficiently move on, and these differences may be at the root of habitat and foraging height specializations. But as yet, data on support size and locomotion are of limited value in understanding niche divergence, since what one species can gather by traversing fine branchlets, another may reach by breaking off the branch. The question that must be addressed is From what substrate is the food item taken?

Activity period. Most prosimians and *Aotus* are nocturnal, other primates are almost exclusively diurnal. Differences in activity period could conceivably reduce overlap in insect prey types, but are unlikely to reduce competition for fruit, leaves, or other plant parts.

Prey types and substrates. Investigators in Gabon, Peru, Surinam, and Uganda have independently noted that larger primates not only take fewer insects than do smaller primates but also tend to take them by different methods. Large species, such as *Cercocebus* in Uganda and *Cebus* in Peru, tend to take relatively immobile forms by destructive foraging, breaking off dead bark, splitting hollow vines or sticks, and tearing apart rotten wood. Small species, such as *Saguinus* in Peru, *C. talapoin* in Gabon, and *C. ascanius* in Uganda, tend to catch mobile prey with rapid grabs from bare surfaces or even from the air. Especially where primates are similar in size, other differences in prey foraging also occur, both in foraging substrate (e.g., among *Saguinus* species, between *Perodicticus potto* and *Galago* species) and in prey taxonomy (e.g., *Cercocebus albigena* stomachs contain primarily ants; *Cercopithecus pogonias* mainly orthoptera) (tables 18-3, 18-9).

Plant secondary compounds. Particularly among folivores, species apparently differ in their ability to digest and detoxify plant antiherbivore defenses. Where colobines coexist, two strategies recur: selecting for a wide variety of relatively less-protected young leaves and leaf parts, as practiced by *Colobus badius* in Uganda and

Presbytis melalophos in Malaysia; or concentrating on a species-poor diet often dominated by mature leaves, as practiced by *Colobus guereza* and *P. obscura* in the same two areas. Ability to detoxify particular classes of chemical defense may also separate species in some frugivores as well (chap. 17).

Fruit size and hardness. In the only study documenting species-specific use of fruit by size, Terborgh (1983) found a slight but nevertheless definite tendency for larger primates to eat larger fruit. The close relationship between monkey size differences and fruit size overlap (table 18-4) would make this niche dimension worth further investigation. Another clear qualitative trend is for larger or heavier-jawed species (*Chiropotes, Cebus apella, Cercocebus*) to eat thicker, tougher fruits.

Fruit dispersion. Perhaps the most universally noted dimension along which sympatric primates—especially sympatric frugivores—differ is that of patch size. On all continents there are primates that specialize on tree species, such as figs, characterized by large, evanescent fruit crops. "Large-patch" specialists, such as *Hylobates, Pongo, Pan, Cercocebus,* and *Ateles,* have diets dominated by large, fleshy, ripe fruits harvestable quickly in large quantities because of their spatial concentration. Other primates use these fruits too, but their diets are supplemented by other items, often fruit at earlier stages of ripeness and from trees that bear smaller fruit crops. As a general rule, large synchronously ripening fruit (or flower or young leaf) crops do not last long, and trees bearing them are widely spaced; thus large-patch specialists must travel long distances and maintain large, sometimes enormous, home ranges. In most cases, larger animals exploit larger patches, despite the fact that the infrequency of such patches in most habitats limits them to lower biomass (and population) densities than small-patch users (table 18-11).

Are all these differences adequate to demonstrate niche divergence, that is, do they show that coexisting species are more different than they would have been in the absence of interspecific competition? The simple answer is no. Even during the dry season, overlap indexes do not cluster around some constant level, as they should if interspecific competition had posed a limit to similarity (tables 18-2–18-5, 18-7–18-9). Primate communities show none of the regular spacing of body or feeding apparatus size that reflects niche overdispersion in certain other communities (Gautier-Hion and Gautier 1979; table 18-1).

A more reasoned answer to the question, however, is "not yet." Whether niches are overdispersed when multiple niche dimensions are taken into account cannot be calculated from available data, but primate niches are surely multidimensional. Moreover, some of the data

TABLE 18-11   Niche Partitioning by Patch Size among Frugivorous Primates

Species	Rank by				
	Patch Size	Home Range Size	Body Size	Population Density	Biomass Density
Malaysia					
Hylobates syndactylus	1	3	1	3	2
Hylobates lar	2	2	2	2	3
Macaca fascicularis	3	1	3	1	1
Borneo					
Pongo pygmaeus	1	1	1	3	1
Hylobates muelleri	2	3	2	1.5	2
Macaca fascicularis	3	2	3	1.5	3
Uganda					
Pan troglodytes	1	1	1	4	4
Cercocebus albigena	2	2	2	3	3
Cercopithecus mitis	3	3	3	2	2
Cercopithecus ascanius	4	4	4	1	1
Peru					
Ateles paniscus	1	1	1	4	1
Saimiri sciureus	2	2	4	1	4
Cebus albifrons	3	3	3	3	3
Cebus apella	4	4	2	2	2
Saguinus (both spp.)	5	5	5	5	5

SOURCES: Malaysia: Chivers 1980, esp. MacKinnon and MacKinnon 1980a; Borneo: Rodman, pers. comm., and Leighton, pers. comm.; Uganda: Struhsaker 1978; Peru: Terborgh 1983.

suggest that competition structures primate communities, but not in the traditional way. In this alternate view, competition has led to niches that are nested, not regularly spaced, along important niche dimensions.

Under what conditions could competition, a phenomenon usually perceived as working to reduce niche overlap, lead to niches that are nested? Communities of many other organisms—nectar-feeding birds and insects, seed-eating rodents—contain sets of species with nested niches, all species preferring similar high-quality or high-density resources but some able to subsist on lower density or poorer quality foods. In such communities, the requirement for stability is that the species with the included niche be able to control the preferred resource (Colwell and Fuentes 1975). For primates, the niches of large-patch specialists are included in those of other frugivores that, usually because of their smaller size, can make do on small as well as large patches. Large-patch specialists might in some cases be more efficient at exploiting them—the most likely mechanism would be increased locomotor efficiency during the long trips between patches—but in most cases the evidence suggests that they do, indeed, "control" the preferred resource through interference. Species that use larger patches are, with the exception of Saimiri, either larger or otherwise ag-

gressively dominant to species that are more generalized (table 18-11).

Has Selection Produced Interspecific Affinities?

Much discussion of mixed-species associations has been predicated on the assumption that species are actively attracted to each other, and thus implicitly that association is an "evolved" tendency. Subjective evidence to this effect can be overwhelming: some associations, such as those of Peruvian Saguinus species and Gabonese Cercopithecus, are virtually permanent. But as Klein and Klein (1973) first noted, the situation is not always so clear. They pointed out that associations between Colombian Ateles belzebuth and Alouatta seniculus, though they are common, long-lasting, and involve olfactory investigation, grooming, and play, are based solely on "mutual attraction to specific fruit trees or . . . contacts occurring on travel routes" (Klein and Klein 1973, 649). Are mixed-species associations based on behavioral attraction the exception or the rule?

One can start to answer this question by comparing observed association durations and their rates of initiation with those expected by "chance," if groups move independently (fig. 18-3). Several association types (in addition to those that are semipermanent) emerge as occurring far too much of the time to be "chance" phenomena: *Cercopithecus ascanius* with *C. mitis* in Uganda, *Cercocebus albigena* with *C. pogonias* and *nictitans* as well as *C. cephus* with *Miopithecus talapoin* in Gabon, *Cebus* species with *Saimiri* in several South American sites.

Beyond these clear-cut cases, there are a number of examples of association at levels slightly above that expected by chance, for example, *Cercopithecus ascanius* with *Cercocebus albigena* and *Colobus badius* (compare fig. 18-3 and table 18-10). In many of these, however, the involvement of interspecific attraction cannot yet be assessed because, among other things, association at levels slightly greater than "chance" might result from independent attraction to common food sources. In Uganda, Gabon, and Peru, the most common associations indeed involve species whose diets are most similar. Distinguishing attraction to a common food source from attraction to another species is much more difficult than determining whether movements are independent; one approach to this problem predicts the expected percentage of time fruiting trees would be used by primate species jointly versus singly if species independently discover and exploit them (Waser 1982a).

In summary, by no means all primate mixed-species associations form because one or both species are attracted to the other. Nevertheless, the evidence is clear that some do, and these are cases in which the advantages of past association have evolutionarily altered today's behavior.

CONCLUSION

Despite the many person-years devoted to acquiring the data reviewed here, a high-resolution picture of the ways primate species influence each other has not yet emerged. Detailed studies of resource levels and of multiple species' use of them simultaneously and in the same location are still required. But at the same time, in many areas—for instance, the use of survey data to detect density compensation—small efforts would reap large rewards in understanding. Such studies must be pursued to ensure that primates are not overwhelmed by the effects of the most significant interspecific interaction of all, that involving *Homo sapiens*.

SUMMARY

Primate species can in theory interfere with each other or aid each other. At present it is difficult to infer the importance of either of these relationships. At a proximate level, aggression over food is seen widely. However some species feed together peacefully, and the net effects of interspecific aggression on food availability are unknown. Competition can be inferred from the occurrence of interspecific territoriality and from higher population densities when competitors are absent. On the other hand, it is yet to be demonstrated that species have more divergent diets or greater travel distances when together than when alone.

Many species travel or feed together, sometimes for weeks at a time. Probable benefits of these associations include the use of complementary feeding techniques, an increased ability to supplant conspecific groups, and increased opportunities for social interaction including hybridization. Improved defense against predators is also possible, though data are scanty. Whatever the benefits, they are sufficient to have had effects over evolutionary time, since there is clear evidence that some primate species are strongly attracted to others.

The evolutionary effects of competition are less clear. Niche differentiation is known with respect to a number of variables (e.g., habitat use, animal prey type, digestive physiology, and fruit type), though it has not been established so clearly for others (e.g., foraging height, support size, locomotor style, and activity period). Where niche differences occur, it remains to be demonstrated that they result from an evolutionary process of niche divergence.

In general it is puzzling that many sympatric primates are remarkably similar ecologically. One solution may be that species with similar diets have nested niches: the larger species are specialists capable of dominating access to the preferred foods, while the smaller are generalists that use alternate foods when supplanted from their primary food patches.

19 | Predation

Dorothy L. Cheney and Richard W. Wrangham

As is often the case when there are few or no data, the influence of predation on primate behavior has generated much theoretical controversy. Predation has been argued to have exerted strong selective pressure on the evolution of sexual dimorphism (e.g., DeVore and Hall 1965), multimale groups (Hamilton, Buskirk, and Buskirk 1978), and even sociality itself (Alexander 1974; Terborgh 1983; van Schaik et al. 1983). Other authors attribute these characteristics to different selective pressures, such as sexual selection and feeding competition (e.g., Struhsaker 1969; Wrangham 1980). Yet predation is rarely observed, and the effects of predation on behavioral evolution or population regulation have not been accurately estimated in any species. There is an urgent need for better information on the effects of predation on primates living in undisturbed habitats.

In this chapter, we are concerned solely with predation by other species; intraspecific infanticide and cannibalism are discussed elsewhere (chaps. 8, 15). We first describe antipredator responses and the ways in which predators affect ranging and feeding behavior. These are issues for which the greatest amount of data exist, although only a few species have yet been studied systematically. Second, we investigate the demographic effects of predation. Although this also lends itself to empirical research, in practice it has proved difficult to measure predation rates and to compare predation with other factors that potentially limit populations, such as food supply and disease. Finally, we discuss the importance of predation as a selective pressure acting on morphology and social structure.

BEHAVIORAL ADAPTATIONS TO PREDATION
Defense against Predators

Cross-species comparisons of antipredator behavior in primates are complicated by two factors. First, few studies provide data on the relative frequency with which animals attack or flee from a given species of predator. Second, even within the same species or social group, a predator that usually causes flight may occasionally evoke attack, depending on the context of the encounter. Nevertheless, a few general trends are evident.

Most accounts of aggressive defense against predators concern larger primate species that typically live in multi-male groups. Baboons (*Papio cynocephalus*) have been observed to kill domestic dogs (Stoltz and Saayman 1970) and to mob and chase smaller carnivores such as cheetah and jackals (Altmann and Altmann 1970). They have also been reported to attack and even fight to the death with leopards and lions (Marais 1939; Altmann and Altmann 1970; Sapolsky, pers. comm.), although their more common response to such predators is to flee. Rhesus macaques (*Macaca mulatta*) have been seen mobbing a tiger (Lindburg 1977), while chimpanzees (*Pan troglodytes*), orangutans (*Pongo pygmaeus*), and gorillas (*Gorilla gorilla b.*) have all been observed threatening and even attacking leopards and lions (Schaller 1963; Rijksen and Rijksen-Graatsma 1975; Gandini and Baldwin 1978; Nishida, pers. comm.). Similarly, both baboons and chimpanzees will attack and display at stuffed leopards placed near the animals for experimental purposes (e.g., Kortlandt 1972).

In predator encounters involving baboons, adult males have been reported to be the primary defenders of the group and the last to retreat from attack (DeVore and Washburn 1963). Such aggressive defense, however, seems to depend very much on the predator in question; when particularly dangerous predators, such as people or lions are spotted, males are sometimes the first to flee, leaving behind mothers with infants (Hall 1963b; Rowell 1966a; Stoltz and Saayman 1970). This suggests that the protective behavior of males against less-threatening predators such as cheetah may sometimes be due more to the males' relative lack of vulnerability than to a high motivation to defend the group. Male attacks on predators, however, can act as effective deterrents. At Gombe, Tanzania, for example, adult male red colobus (*Colobus badius*) chased chimpanzees on 12 of the chimpanzees' 33 unsuccessful predation attempts on their group. In contrast, only 2 of 31 successful attacks involved active defense by males (Busse 1976).

Male defense against predators seems to be less common in one-male than in multimale groups, perhaps because a single male cannot often successfully drive off a predator. Hall (1965a), for example, reported that the vigilance and "distraction displays" of adult male patas monkeys (*Erythrocebus patas*) were the most striking feature of male behavior and interpreted them as antipredator behavior. Similar watchfulness has been reported for the adult male in bisexual groups of gray langurs (*Presbytis entellus*), although it is likely that such vigilance is also directed at least in part toward potential rival males (Hrdy 1977).

Although males are often reported to be the most active in aggressive antipredator defense, it is important to emphasize that females will also attack predators, especially if offspring are threatened. For example, at Gombe there were several recorded instances when female red colobus attempting to defend their offspring against chimpanzees were themselves killed (Wrangham and Bergmann-Riss, in prep.).

The antipredator behavior of small primates (< 1 kg body weight) consists largely of concealment, vigilance, and flight rather than attack, although many species are reported to mob snakes. Two patterns of flight are found. Rapid escape occurs in some fast diurnal and nocturnal species such as squirrel monkeys (*Saimiri* spp.: Terborgh 1983) and bushbabies (*Galago* spp.: Charles-Dominique 1977a; chap. 2). In contrast, antipredator strategies among the diurnal callitrichids, which are vulnerable to numerous predators, include a motionless and huddled sleeping style and a high rate of rapid surveillance (Terborgh 1983; Caine 1984). These species are apparently adapted to avoiding, rather than challenging, predators. Slow but silent avoidance also occurs in arboreal lorises (*Nycticebus* spp.) and pottos (*Periodicticus potto*). When confronted by snakes, pottos simply fall to the ground, run a few meters, and lie still. When attacked by carnivores, however, pottos fight back with their hands and bony neck region, and can knock opponents three times their size to the ground (Charles-Dominique 1977a).

Alarm Calls

Most birds and mammals give alarm calls to predators, and primates are no exception. Almost all primate species have been reported as having at least one type of alarm call, and many species whose vocalizations have been more systematically studied appear to have different alarm calls for different kinds of predators (e.g., vervet monkeys [*Cercopithecus aethiops*]: Struhsaker 1967a; Seyfarth, Cheney, and Marler 1980a, 1980b; red colobus: Struhsaker 1975; Goeldi's marmosets [*Callimico goeldii*]: Masataka 1983; pygmy marmosets [*Cebuella pygmaea*]: Pola and Snowdon 1975; cottontop tamarins [*Saguinus oedipus*]: Cleveland and Snowdon 1982; see also chap. 36). Vervet monkeys, for example, have at least six acoustically different alarm calls; they give different ones to leopards, smaller carnivores, eagles, snakes, baboons, and unfamiliar humans. Each call, which appears to be adapted to the hunting strategies of different predators, elicits different responses from those nearby (fig. 19-1).

The use of predator-specific alarm calls suggests that predators have exerted strong selection pressure on at least some aspects of behavior. However, whether differentiated alarm calls are consistently associated with vulnerability to a high diversity of predator species is not yet known. There is anecdotal evidence that the alarm calls of monkeys are more highly differentiated than those of the great apes, who appear to be less vulnerable than monkeys to many potential predators. For example, the alarm calls of vervets are distinguished easily by untrained humans, whereas no observers have noticed dif-

FIGURE 19-1. Vervet monkeys alarm-call at a python, one of their primary predators. (Photo: Richard Wrangham/Anthrophoto)

ferences in the alarm calls of chimpanzees to pythons, leopards, or unfamiliar humans.

Function of Antipredator Defense and Alarm Calls

There is considerable debate about whether antipredator behavior has evolved as a product of kin selection, to protect the actor's close relatives, or as a product of individual selection, to protect the actor itself (Maynard Smith 1965; Trivers 1971; Harvey and Greenwood 1978). In many cases where birds or nonprimate mammals have been found to attack predators, observations have suggested that the attacks function to protect close kin (see reviews by Kruuk 1972; Wittenberger 1981). Most research on alarm calls also supports the nepotism hypothesis. In prairie dogs and ground squirrels, for example, females with offspring and collateral kin alarm-call more than females without kin, and juvenile males who have not yet left their natal area alarm-call more than immigrant adult males (Sherman 1977; Dunford 1977; Hoogland 1983).

Few comparable data on individual differences in predator defense or alarm calling are available for primates. Although attacks on predators by male baboons or red colobus potentially function to defend offspring, no study has yet determined whether such antipredator behavior occurs most among those males who have fathered the most offspring. Moreover, reports of the alarm-calling behavior of males in multimale groups do not consistently support the nepotism hypothesis. For example, some observers of baboons have described peripheral males as the most likely to give alarm barks (Hall and DeVore 1965). Since these males are often recent immigrants (Busse 1984a; Collins 1984), their alarm calls probably do not function to warn offspring or other kin of danger. Similarly, although the highest-ranking male in one study of chacma baboons gave more alarm calls than other males, this male did not copulate the most (Stoltz and Saayman 1970; Saayman 1971a).

In one-male groups, alarm calls by long-established males probably have the effect of helping resident offspring. This supports the nepotism hypothesis. By contrast, males of shorter tenure probably have fewer close kin to benefit. Differences in rates of alarm calls between immigrant and established males might therefore help reveal the importance of kinship in maintaining alarm calls. Tenaza and Tilson (1977) have suggested that the loud alarm calls given by Kloss's gibbons (*Hylobates klossii*) help protect offspring even after they have left their parents' territory; it might be possible to test this hypothesis by determining if the amplitude of different species' alarm calls varies according to the distance over which offspring normally disperse. Other species differences in calling rates may also eventually help clarify the function of alarm calls. For example, although the adult male in *Cercopithecus pogonias* groups gives more alarm calls than adult females, the reverse is true for the sympatric *C. nictitans* (Struhsaker 1969; Gautier and Gautier-Hion 1983).

Experiments on captive vervets have shown that adult females alarm-call more when in the presence of kin than when with nonkin, which suggests that alarm calls function at least in part to protect genetic relatives (Cheney and Seyfarth 1985c). When data on free-ranging vervets are considered, however, it becomes more difficult to explain alarm calling entirely in terms of kin selection. In Amboseli, Kenya, high-ranking males and females alarm-call significantly more often than low-ranking individuals (Cheney and Seyfarth 1981, 1985c). However, since no correlation exists between rank and reproductive success (Cheney, Lee, and Seyfarth 1981; Cheney et al., in press), it seems unlikely that high-ranking animals alarm call most because they have more kin than low-ranking individuals. The members of high- and middle-ranking families are more susceptible to predation than low-ranking animals, a factor that might cause them to alarm-call at higher rates if calls functioned to protect potentially vulnerable kin. However, the predators that evoke most alarm calls are those to which the caller, rather than the caller's kin, are most vulnerable. Juveniles and infants, for example, are more vulnerable than adults to predation by baboons and eagles, while adults are more vulnerable to leopards. Adults give fewer alarm calls to baboons and eagles than to leopards, even though calls to the former species would be of very little cost to themselves and of great potential benefit to their offspring (Cheney and Seyfarth 1981).

It also seems unlikely that the reason why high-ranking vervets give more alarm calls than low-ranking animals is because they spot predators more often. For example, there is no consistent tendency for high-ranking animals to lead group progressions. Similarly, high- and low-ranking animals scan the area around them at similar rates, which suggests that they have equal opportunities to spot predators (Cheney and Seyfarth 1981, 1985c). These data suggest that low-ranking animals detect predators as often as high-ranking animals, but that they may occasionally withhold their alarm calls. The possible selective use of silence may be an effective means to mislead others about the presence of danger and may provide low-ranking individuals with a way to compete with their rivals (Cheney and Seyfarth 1985c).

Intragroup Spacing and Progression Order

Early observers, struck by the apparent tendency of savanna baboon groups to travel in a protective formation, suggested that consistent progression orders were adaptations against predation (e.g., Hall and DeVore 1965). Today there is conflicting evidence regarding this widely publicized idea. The original claim, based only on qualitative observations, was that young adult males kept to the edge of the group while vulnerable mothers and infants traveled with the protective dominant males at the center (DeVore and Washburn 1963; Hall and DeVore 1965). Numerical data from eight different study sites now show that although different age-sex classes may tend to occupy different positions, progressions are only weakly structured at best. Some observers have revealed a slight tendency for males to be in the front and on the edge of the group more than females (Harding 1977; Rhine and Westlund 1981; Rowell 1966a). Busse (1984a) and Collins (1984) also found that pregnant and low-ranking females were more peripheral than lactating or high-ranking females. In contrast, Altmann (1979b) concluded that progression order in a group of baboons in Amboseli, Kenya, was essentially random, even though predation pressure in this area was high.

Nor is there any consensus about which males are peripheral. In support of the DeVore hypothesis, some studies have shown that low-ranking and recent immigrant males were more likely to be in the front and rear of group progressions (Busse 1984a; Collins 1984). Rhine and Westlund (1981), however, reported that dominant males were the most likely to lead the group into potentially dangerous areas such as water holes. The original hypothesis was that peripheral males behave altruistically by exposing themselves as "sentinels," vulnerable to predation but ready to alert the group to impending danger (Hall and DeVore 1965). This interpretation now appears unnecessary, however, because the tendency for males to be peripheral is more easily explained with reference to social factors. The peripheralization of recent immigrant and maturing natal males is a widespread phenomenon in species in which males transfer (chap. 21), and no special explanations are needed for its occurrence in baboons. Peripheral males may alert the group to danger simply because they encounter predators often, not because they search for them. At present, however, there is no evidence that peripheral males are more vulnerable or alert the group to predators more than others. Indeed, even the relevance of progression order is now in some doubt. Since most carnivores hunt by night, predation risk during the day may not be as great as was first thought, and thus there may be no differential vulnerability between central and peripheral individuals (Collins 1984).

Polyspecific Associations

It has often been suggested that polyspecific associations (associations between two or more different species) occur as adaptations against predation (see chap. 18). Although several types of benefits from such associations are possible, the antipredator hypothesis has not yet been confirmed for any species. This is partly because it must first be demonstrated that polyspecific associations occur as a result of attraction rather than simply by chance (Waser 1984; chap. 18). The requisite evidence has been collected for only a few cases, so there are correspondingly few opportunities to test the antipredator hypothesis against the alternative suggestion that associations are the result of mutualistic feeding strategies. Attempts to evaluate the antipredator hypothesis have so far relied largely on inference, as illustrated by two cases in which feeding advantages from polyspecific association appear unlikely (see chap. 18 for a more detailed review).

For example, in Kibale, Uganda, redtail monkeys (*Cercopithecus ascanius*) associate frequently with red colobus. Waser (1984) argued that these are chance associations, but Struhsaker (1981a) considered them to be adaptive partly because red colobus have been observed to groom redtails, possibly in an effort to prolong the association. Moreover, the two species eat quite different food items and interact without aggression. This association contrasts with the association between redtails and blue monkeys (*Cercopithecus mitis*), which involve aggressive competition and seem to occur as a result of attraction to the same food sources. Because no feeding advantages can be detected for the associations of redtails and red colobus, Struhsaker (1981a) suggested that they function to increase the rate of predator detection. This example stresses that, where feeding benefits seem unlikely, whether or not associations occur by chance is a critical issue.

In a few cases the antipredator hypothesis is supported further by behavioral observations. Thus in their study of *Cercopithecus cephus*, *C. nictitans*, and *C. pogonias* in Makokou, Gabon, Gautier-Hion, Quris, and Gautier (1983) noted several effects of polyspecific association, including a more varied diet, more efficient food searching, and, at least for *C. cephus*, a larger home range (chap. 18). They also suspected an antipredator function, however, partly because there is no seasonal variation in the frequency of polyspecific associations, as might be expected if increased foraging efficiency were the only cause. Furthermore, the antipredator skills of the different species seem complementary. The three species exploit different parts of the forest, the smaller *C. cephus* foraging in the lower strata, while *C. nictitans* and *C. pogonias* forage in the higher levels. *C. cephus* alarm-calls primarily at terrestrial predators, while *C. pogonias* alarm-calls primarily at eagles (Gautier and Gautier-Hion 1983). *C. nictitans* and *C. pogonias* appear to derive increased protection from terrestrial predators in *C. cephus*'s presence, while *C. cephus* gains increased protection from eagles. Supporting this hypothesis is the fact that *C. cephus* exploits higher areas of the forest when associating with the larger two species. Moreover, of four observed predations by eagles on *C. cephus*, three occurred in monospecific groups. This case is perhaps the strongest for inferring that polyspecific associations represent antipredator strategies, and it illustrates the kind of data needed to provide a valid test.

The theoretical rationale for viewing large groups as adaptations against predation is usually applied to polyspecific associations in the same way as to relatively closed multifemale groups. One important difference between the two, however, is that most polyspecific associations are temporary (tamarins provide the only known exception; Terborgh 1983), whereas most multifemale groups are permanent. This means that any theory that attributes grouping tendencies to predation pressure needs to be formulated differently for associations between, as opposed to within, species. To date, most discussion has centered around the benefits of grouping per se rather than grouping with particular individuals (Bertram 1978; Pulliam and Caraco 1984; but see Helfman 1984). Such theories are therefore more easily applied to systems involving flexible grouping patterns, such as polyspecific associations, than to those with fixed groups.

PREDATION RATES
Estimating Predation Rates

Reliable estimates of predation rates can only be obtained if the predators themselves are systematically observed. Normally, however, accounts of predation on primates, as on other species of birds and mammals, are based on anecdotal observations. Most predators tend to be fearful of humans and to avoid groups that are being studied. Many predators also hunt at night, when observation conditions are poor. As a result, observers are usually confronted simply with the disappearance of one of their subjects. Depending on the age and sex of the animal, its condition, and the time elapsed since it was last seen, predation can be assumed, but it is usually impossible to confirm. Since few illnesses result in sudden death, it is often valid to conclude that healthy individuals who disappear overnight are preyed upon, at least among those age-sex classes that do not normally disperse. The overwhelming reliance on this assumption, however, underscores the difficulty of measuring predation rates, especially since sudden deaths can occur as a result of accidents or intraspecific aggression.

Few published studies report predation rates, the frequency with which animals of different age and sex are taken, or the types of predators to which populations are vulnerable. In order to assess the importance of predation on primates living in different habitats and types of social groups, we distributed questionnaires to a variety of workers at long-term sites. Results of the survey (table 19-1) reveal clearly that data on predation are woefully lacking. For most species, data are limited to only one study site, and many other species and even some genera could not be surveyed at all. Although we included only data from studies where some predation might be expected to occur, few observers expressed confidence in their estimates, and almost all commented on the difficulty of measuring predation accurately. The following

results should therefore be taken as extremely preliminary. Indeed, their major function is a heuristic one, to highlight the need for more systematic information. We exclude apes from our analysis, since the only reports of nonhuman predation on apes are anecdotal; no long-term studies have ever recorded successful predations (Schaller 1963; MacKinnon 1979; Fossey 1983; Chivers, pers. comm.; Nishida, pers. comm.).

Unfortunately, it proved impossible to assess the importance of human predation in areas where primates are hunted for food. Humans in these areas are likely to constitute the primary source of predation (chap. 39), which is a severe limitation. Another limitation derives from the fact that, even in many free-ranging populations, the eradication of natural predators by humans has resulted in negligible predation rates (e.g., hamadryas baboons [*Papio hamadryas*]; Japanese macaques [*Macaca fuscata*]; gray langurs). In such cases, humans or their agents (e.g., vehicles or electrocution for gray langurs; J. Moore, pers. comm.) are the most important "predators," even when most deaths are accidental.

Types of Predators

A huge array of species is known through anecdotal accounts to prey on primates, including crocodiles, pythons, a variety of raptors, and numerous carnivores. Since predation is seldom observed, however, primates at any given study site tend to have relatively few confirmed predators, even though many other species are suspected to be important. For example, the only known predators of vervets in Amboseli are leopards, martial eagles (*Polemaetus bellicosus*), baboons, and pythons, but many smaller carnivores (e.g., caracals), one other eagle species (the crowned hawk eagle, *Stephanoaetus coronatus*), and even the marsh mongoose are suspected predators (Cheney et al., in press). Similarly, although eagles, leopards, tigers, and golden cats are suspected to prey on long-tailed macaques (*Macaca fascicularis*) in Ketambe, Sumatra, only pythons have actually been observed to do so (van Schaik and van Noordwijk, pers. comm.).

Despite the paucity of information about types of predators, some general trends are evident. First, carnivores and raptors are the most common predators and in the survey account for 24 of the 30 different species of confirmed nonhuman predators. Second, smaller species are more likely than larger species to be taken by raptors. In the survey 9 primates weighing less than 2 kg (female body weight) were hunted by a total of 5 confirmed and 9 suspected bird species, whereas the 8 species weighing more than 5 kg had only 1 confirmed and 2 suspected bird predators. Third, larger primates are preyed upon by a larger variety of different carnivore species than are smaller primates. There were 3 confirmed and 4 suspected carnivore predators for smaller primates, but 10

TABLE 19-1 Estimated Predation Rates on Various Species of Primates

Q	Species	Y	PS	P	O	S	EPR	I	J	AF	AM	Sources
1	Aotus trivirgatus Manu, Peru	1.5	40	0	0		0	—	—	—	—	Wright 1984
1	Theropithecus gelada Sankaber, Ethiopia	1.5	270	0	0		0	—	—	—	—	R. Dunbar
1	Alouatta seniculus Hato Mas., Venezuela	7	375	0	0	4	<1					C. Crockett
1	Macaca sinica Polonnaruwa, Sri Lanka	8	450	2	13		<1	38	38	15	8	W. Dittus
1	Presbytis entellus Jodhpur, India	6	950	1	2	<10	<1	*	*	—	—	C. Vogel
1	P. entellus Abu, India	1	200	1	1		1					J. Moore
2	P. entellus Ranthambhore, India	0.5	125	1	0	9	10	*	—	*	*	J. Moore
2	Saimiri sciureus St. Sofia, Colombia	2	125	2	2		1–5	—	*	—	—	R. Bailey
1	Cercopithecus ascanius Kibale, Uganda	5	75	0	0		1–3					T. Butynski
2	C. ascanius Kakamega, Kenya	2.5	25	1	0		10					M. Cords
1	C. mitis Kibale, Uganda	6	75	0	0		1–3					T. Butynski
2	C. mitis Kakamega, Kenya	2.5	25	1	0		10					M. Cords
1	Cebus olivaceus Hato Mas., Venezuela	7	175	0	0	15	3	70	20	5	5	J. Robinson
1	Papio c. anubis Mara, Kenya	2	75	3	3	>4	3		*		*	R. Sapolsky
1	P. c. anubis Gombe, Tanzania	18	125	2	29		1	77	23	—	—	J. Goodall
1	P. c. cynocephalus Amboseli, Kenya	13	45	3	8	30	4–8	39	30	13	23	J. Altmann, S. Altmann, and G. Hausfater

confirmed and 11 suspected such predators for larger primates. Fourth, arboreal primates seem to be more vulnerable to raptors and less vulnerable to carnivores than terrestrial species.

Although these trends have long been suspected, their significance is still not clear. Different predator species presumably impose different types of selection pressure, but the extent to which grouping patterns are adapted to particular types of predators is not known. For instance, polyspecific associations may be more beneficial against eagles than against carnivores. Systematic data on the hunting strategies of predators are therefore needed.

Primates also fall prey to other species of primates. The most important nonhuman primate predator is certainly the chimpanzee, which has been observed to hunt and kill red colobus, black-and-white colobus (Colobus guereza), baboons, redtail monkeys, blue monkeys, and bushbabies (Teleki 1973a; Busse 1976; McGrew, Tutin, and Baldwin 1979b; Ghiglieri 1984; Takahata, Hasegawa, and Nishida 1984; fig. 19-2). Approximately 15% of the red colobus population at Gombe is estimated to be killed by chimpanzees each year, and such predation appears to be the major cause of red colobus mortality at this site (table 19-1). The only other confirmed primate predators are baboons, which regularly kill vervets (Hausfater 1976; Cheney et al, in press), and blue monkeys, which kill galagos (Butynski 1982a). In the New World, the only suspected primate predators are brown (Cebus apella) and white-fronted capuchins (C. albifrons), both of which commonly chase smaller primates, including night monkeys (Aotus trivirgatus) and dusky titi monkeys (Callicebus moloch). The chases resemble hunts, and capuchins are known to eat other small mammals such as squirrels, opossums, and anteaters (Wright 1984).

Although hunting rates are difficult to estimate, human predation no doubt accounts for the greatest number of primate deaths, even in areas where hunting methods are still primitive. For example, the Waorani of eastern Ecuador hunt large primates in preference to smaller ones, and in 867 monitored day hunts they killed 562 woolly (Lagothrix lagotricha) and 246 howler (Alouatta seniculus) monkeys (Yost and Kelley 1983; see also chap. 39). Similar hunting of relatively large monkeys occurs among the Efe in the Ituri Forest of Zaire (Bailey 1985) as well as among numerous tribes in Western Africa and Southeast Asia (Wolfheim 1983). The regular hunting of primates by hunter-gatherers suggests that humans were important predators of nonhuman primates long before

TABLE 19-1 *(continued)*

Q	Species	Y	PS	P	O	S	EPR	I	J	AF	AM	Sources
2	*P. c. cynocephalus* Mikumi, Tanzania	4	125	1	3		3–5	—	*	*	—	R. Rhine and G. Norton
1	*P. c. ursinus* Moremi, Botswana	3	125	2	10	28	9	18	18	50	13	Busse 1980; S. Smith
1	*Callicebus moloch* Manu, Peru	1.5	25	0	0	4	>4	*	*	*	—	Wright 1984
1	*Cercopithecus aethiops* Samburu, Kenya	2.2	75	2	3	8	6	64	27	0	9	P. Whitten
1	*C. aethiops* Amboseli, Kenya	7	75	4	12	59	15	48	15	20	17	D. Cheney et al.
1	*Colobus badius* Kibale, Uganda	1.3	60	1	1		1–2			*		J. Skorupa
1	*C. badius* Gombe, Tanzania	6	400	1	59		15	70	15	15	0	R. Wrangham and E. Bergmann-Riss
2	*Cercopithecus cephus* Makokou, Gabon	1	25	1	4		10					Gautier-Hion, Quris, and Gautier 1983
1	*Erythrocebus patas* Laikipia, Kenya	2.5	75	1	1	8	>10	38	0	50	13	J. Chism and D. Olson
1	*Macaca fascicularis* Ketambe, Sumatra	3	75	1	1	46	<11	35	40	15	10	C. van Schaik and M. van Noordwijk
1	*Cebus apella* Manu, Peru	5	30	0	0	10	13	40	30	20	10	J. Terborgh
1	*Galago senegalensis* Transvaal, South Africa	2	75	2	3	7	>15	M	—	*	*	S. Bearder and R. D. Martin
1	*Saguinus fuscicollis* Manu, Peru	10	35	2	2	>10	>15	M				J. Terborgh
1	*S. imperator* Manu, Peru	10	35	2	2	>10	>15	M				J. Terborgh

NOTES: Q = quality of observation (1 = predation rate estimated by extrapolation from suspected number of predations; 2 = predation rate roughly guessed); Y = number of study years; PS = population size; P = number of known predator species; O = number of observed predations; S = number of suspected predations; EPR = estimated predation rate (% of population per year); I, J, AF, AM = % of deaths estimated to have occurred in each age-sex class (* = 1 or more deaths recorded; M = many deaths estimated). Species are listed in ascending order of predation rate. All studies of the same species listed together. Unless otherwise stated, sources are personal communications.

the advent of firearms and that human hunting may have exerted an influence on the evolution of antipredator behavior and even social structure (Tenaza and Tilson 1985).

It should be emphasized that in some species, intraspecific killing accounts for more deaths per year than does predation. Infanticide appears to occur at a higher rate than predation among some populations of red howler monkeys (*Alouatta seniculus:* Crockett and Sekulic 1984), gray langurs (Hrdy 1977; Vogel and Loch 1984), blue monkeys (Butynski 1982b), mountain gorillas (*Gorilla g. beringei:* Fossey 1984), and chimpanzees (Nishida, pers. comm.). In such species, much of the behavior of males and females in bisexual groups may have evolved in response to the need to defend infants against potentially infanticidal immigrant males (see chap. 8). In some species, intraspecific killing is not restricted to infants, but also includes adult males and females (e.g., chimpanzees: Goodall 1983; gorillas: Fossey 1983).

Relation between Predation Rates and Social Structure

Hypotheses relating predation pressure to social behavior or morphology are difficult to test because some of the costs of predation pressure are not easily measured.

In addition to actual mortality, these costs include the effort expended to avoid detection and attack by predators. Thus, two species suffering the same predation rate might nevertheless experience different selective pressures on behavior depending on the rate at which they were detected or attacked. In some circumstances, attack rates are known to be high. For example, in one study eagles attacked a group of brown capuchins once every 2 weeks (Janson 1984); it seems likely that much of the capuchins' behavior was adapted to detecting and avoiding such attacks. Because of the shortage of systematic data on all predation costs, we limit the discussion here to actual mortality rates.

Strictly speaking, the data presented in table 19-1 do not warrant further analysis for three reasons. First, most predation rates were only roughly estimated or even guessed and thus cannot be considered reliable. Second, some species and genera are represented more than others, a factor that should preclude any statistical tests (e.g., Clutton-Brock and Harvey 1984). Third, it is impossible to evaluate the importance of past predation pressure on the evolution of behavior or morphology. For example, although hamadryas baboons do not now suffer high rates of predation, it is possible that predation pres-

FIGURE 19-2. Young adult male chimpanzee at Gombe (Evered) holds the partially broken skull of an adult red colobus killed and eaten by chimpanzees earlier in the day. (Photo: Patrick McGinnis)

sure in the past contributed to the tendency of one-male units to coalesce into large bands (chap. 10). Despite these caveats, there has been so much speculation on the relation between predation and social structure that we have decided to examine the data for any patterns that might emerge. The results are extremely preliminary, and our primary purpose is to encourage future research rather than to validate or reject any specific hypothesis.

In our sample of populations inhabiting relatively undisturbed areas, predation rates are negatively correlated with body size. Even when apes are excluded from the analysis, small primates appear to suffer higher predation rates than large ones do ($r = -.347$; $N = 30$; $p = .061$). There is no relation, however, between predation rates and habitat use or social structure. Thus, terrestrial species do not seem more vulnerable than arboreal species (fig. 19-3; $F_{(1,28)} = 1.146$, $p = .294$). For example, while some populations of baboons and vervets are subject to high predation rates, predation is also estimated to be high in such arboreal species as tamarins, bushbabies, and red colobus. Similarly, there are no differences in predation rates between species that typically live in multimale, as opposed to one-male, groups (fig. 19-3; $F_{(1,28)} = 1.122$, $p = .299$). High predation rates are re-

ported for vervets, which live in multimale groups, and for tamarins, which live in small family groups. Finally, predation is not consistently higher or lower among primates that live in female-bonded or non-female-bonded groups (see chap. 23)

Even within species showing the same social structure, predation rates vary across habitats. For example, although a high proportion of vervet monkeys in both Samburu and Amboseli appear to die of predation each year, the rate at Amboseli is twice that at Samburu. Similar variation is evident among different populations of baboons and red colobus (table 19-1).

At present, so little is known about predation that only very crude cross-species comparisons are possible. Nevertheless, what little data exist suggest no consistent differences in predation rates according to habitat use or social structure. Moreover, since infanticide can sometimes account for a higher proportion of deaths than does predation, in some species intraspecific killing may have exerted at least as strong a selective pressure on social structure and behavior (Hrdy and Hausfater 1984; chap. 8).

Relation between Predation and Demographic Structure

The effect of predation on population growth depends not only on mortality rates but also on population size and the age of the primary prey animals. Predation is most likely to limit populations or influence the composition of the population when it is concentrated among animals of high reproductive value. Thus predation will exert stronger selective pressure when it affects juvenile and young reproductive females rather than infants and males (Caughley 1977). Among most of the populations surveyed, infants seem to be the most vulnerable age class. In such cases, populations should be limited by predation only when infant mortality is unusually high (e.g., tamarins and Gombe red colobus).

Data in table 19-1 also suggest, however, that a considerable proportion of prey animals in some populations consists of adult females and that predation usually occurs at equal or higher rates among adult females than among adult males, even in species where females do not normally disperse. In some circumstances this may strongly influence population growth. For example, adult female vervets in Amboseli suffer high predation rates, and the high frequency of female mortality appears to have contributed to the population's decline, from 104 animals per km² in 1963 to approximately 49 animals per km² in 1983 (Struhsaker 1967c, 1976; Cheney et al., in press). Seventy percent of all vervet mortality is estimated to be due to predation on healthy animals, and it is therefore possible that predation holds population size below the limit set by the food supply. However, since the period between 1963 and 1983 has been characterized by both high predation rates and habitat deterioration, it is not clear if the population's decline has been due primarily to an increase in predation, a decrease in food supply, or both. Food supply is clearly important because there is significant variation in fecundity among females living in different habitats (Cheney et al., in press). Moreover, since the decrease in the number of mature trees on the open plains has apparently forced the monkeys to forage in more dangerous swampy areas, habitat deterioration may ultimately have contributed to an increase in predation rates (Cheney et al., in press). Similarly, Janson (1984) has argued that poor feeding conditions may increase the risk of predation among brown capuchins. In order to maintain adequate food intake, subordinate individuals are forced to forage away from the main group, with a possibly increased rate of predation.

In many primates, mortality is particularly high among migrant males (chap. 20). Little of this mortality, however, seems due to predation. In most of the species studied, mortality is high even when predators are rare and appears to be behaviorally induced through wounds or starvation (e.g., rhesus macaques: Koford 1965; Dittus 1975; howler monkeys: Otis, Froehlich, and Thorington 1981).

Unless all group members are monitored on a daily basis, it is usually impossible to determine whether predators are taking healthy or unhealthy animals. Such data, however, are relevant to any study of predation and population growth because if predators prey primarily on individuals that are sick, old, or likely to die anyway, the predators will have less effect on the population than if they take healthy animals (e.g., Schaller 1972; Kruuk 1972). In the Amboseli vervet population, for example, predation occurs at the highest rate in those months when animals are least likely to die of illness, which suggests that predators are killing healthy animals that would not otherwise die (Cheney et al., in press). The effect of predation on this population's growth is probably exacerbated as a result.

In summary, predation exerts the greatest influence on populations when it occurs at high rates, when it affects healthy juvenile and adult females, and when population size is small. Predation may therefore have a strong effect on population growth in some species, such as tamarins, vervets, and red colobus. Species that are regularly hunted by humans are probably the most likely to be held below carrying capacity, simply because human predation often occurs at higher rates than nonhuman predation.

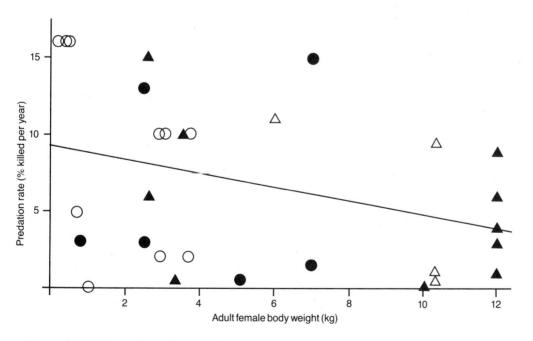

FIGURE 19-3. Estimated annual predation rates for arboreal and terrestrial primates living in one-male and multi-male groups. The regression line shows the relationship between predation rate and female body weight. Arboreal species are represented by circles, terrestrial species by triangles. Open circles or triangles represent one-male groups, closed circles or triangles multimale groups. SOURCE: Data are derived from table 19-1. See text for more details. Female body weights were obtained from the following sources: chaps. in part 1; Bailey 1985; Dittus 1979; Haddow 1952; Hrdy and Hrdy 1976; Napier and Napier 1967; Struhsaker 1980; P.L. Whitten 1982.

PREDATION AND THE EVOLUTION OF MORPHOLOGY AND SOCIAL STRUCTURE
Sexual Dimorphism

Sexual selection is generally regarded as the principal cause of the difference in size between males and females, especially in larger-bodied species (Darwin 1871; Crook 1972; Alexander et al. 1979; Popp 1983; Gaulin and Sailer 1984; chap. 13). A variety of secondary selective pressures have also been proposed, however, including interspecific food competition (Demment 1983), degree of arboreality (Clutton-Brock, Harvey, and Rudder 1977), incidental effects of increasing body size (Clutton-Brock, Harvey, and Rudder 1977), and predation (Leutenegger and Kelly 1977). Predation is thought to have favored large male body size because aggressive defense against predators seems to vary with the degree of sexual dimorphism. Thus monomorphic species tend to flee, while in highly dimorphic species such as baboons, males tend to attack (Leutenegger and Kelly 1977).

The correlation between size dimorphism and predator defense, however, can also be explained in other ways. In particular, since the degree of sexual dimorphism increases with body size (Clutton-Brock, Harvey, and Rudder 1977), males in highly dimorphic species may attack predators simply because they are large. Similarly, although predator pressure could in theory explain why terrestrial species tend to be more dimorphic than arboreal species for a given body size, this has been interpreted in other ways. For example, Clutton-Brock, Harvey, and Rudder (1977) suggest that it results from a relaxation of constraints on male body size otherwise imposed by feeding on thin branches.

Sexual dimorphism in canine size has also been attributed principally to sexual selection. Canine size, however, is not a simple function of dimorphism in body size. Although the reasons for the erratic relationship between canines and body size are unknown, the strength of sexual selection (Harvey, Kavanagh, and Clutton-Brock 1978b) and predator defense are two possible factors. It is hypothesized that predation is important for two reasons. First, males in species with more dimorphic teeth tend to be more aggressive toward predators (Leutenegger and Kelly 1977). Similar comments apply here as apply to the argument relating body size and predation: if large canines have evolved as a result of intrasexual competition, then aggressive responses to predators may be an effect, rather than a cause, of dimorphic teeth.

Second, terrestrial species have larger canines than expected on the basis of sexual dimorphism in body size (Harvey, Kavanagh, and Clutton-Brock: 1978b). Since the difference has not been explained by differences in the intensity of male-male competition in polygynous breeding systems, it has been ascribed to the increase in predation pressure that is thought to result from terrestriality. As both figure 19-3 and Harvey, Kavanagh,

and Clutton-Brock (1978b) note, however, terrestriality is only an approximate predictor of predation pressure. Moreover, the hypothesis fails to explain why unusually large male canines sometimes occur in species, including both terrestrial and arboreal, that do not show aggressive defense against predators (e.g., patas monkeys: Hall 1965a; howler monkeys: Leutenegger and Kelly 1977). Such exceptions suggest that predation, compared to sexual selection, has trivial effects on sexual dimorphism (Struhsaker 1969; Wrangham 1980).

Monogamy and Solitary Ranging

Many theories have argued that monogamy occurs when resources are defensible or when offspring require high rates of parental investment (Clutton-Brock and Harvey 1977a; Kleiman 1977; Wittenberger and Tilson 1980; Hrdy 1981b; Rutberg 1983; van Schaik and van Hooff 1983; chap. 12). Two additional suggestions have also been made concerning the role of predation.

First, monogamy may be favored when predation pressure is exceptionally low (Rutberg 1983; van Schaik and van Hooff 1983). This argument stems from the theory that larger groups form in response to high predator pressure and hence that smaller groups must reflect the absence of serious predation. In its support, several monogamous species, including gibbons (Chivers and Raemaekers 1980) and, until recently, the Mentawai langur *Presbytis potenziani,* have low rates of predation. Furthermore, there are no solitary or monogamous species in open country, where predators have been hypothesized to be most dangerous. To date, however, no way has been found to test the relative importance of alternative selective pressures, such as feeding competition and mate defense, in any of these species (chap. 12).

Arguing against this theory is the fact that many other primates living in small groups do appear to suffer high predation rates, especially smaller species such as the callitrichids (Terborgh 1983; table 19-1). Furthermore, the theory predicts that females will form multifemale groups if predation pressure increases. However, although predator pressure appears to vary in many species, no monogamous primate is known to become polygynous as a result of high hunting pressure. Indeed, the opposite appears to occur in the simakobu (*Simias concolor*).

The second suggestion is the opposite of the first and applies only to species that use cryptic antipredator tactics best employed in small groups. In these species monogamy or solitary ranging is hypothesized to evolve when predation pressure is exceptionally high because predation places a premium on successful evasion (Clutton-Brock and Harvey 1977a; Gautier-Hion and Gautier 1978; Hrdy 1981b; Terborgh 1983).

This theory has been proposed for two kinds of species. The first are small primates, including both diurnal species such as tamarins and nocturnal species that are mostly monogamous or solitary (Clutton-Brock and Harvey 1977a; Terborgh 1983; van Schaik and van Hooff 1983). In many of these species, males play the major role in carrying infants to safety in time of danger, and it seems possible that such paternal care occurs at least in part as a response to high predation pressure (Wright 1984; chaps. 4 and 5). However, the hypothesis that predation contributes to small group size and increased paternal investment cannot be applied to all small species. For example, squirrel monkeys, which also weigh less than 1 kg, live in large groups in which direct paternal care appears absent (chap. 7). It has been argued that the dependence of squirrel monkeys on figs forces them to forage rapidly over large areas and prevents them from adopting concealment as a defense against predation (Terborgh 1983). Large groups are therefore favored because they increase predator detection. This hypothesis may well explain why squirrel monkeys differ from other small primates, but the example underscores the fact that predation pressure does not predict similar grouping patterns in all small species.

The second group of primates for which high predation pressure has been hypothesized to have favored monogamy are the monogamous Old World monkeys, the two Mentawai colobines (the Mentawai langur and the simakobu), and possibly also de Brazza's monkey (*Cercopithecus neglectus*) (Gautier-Hion and Gautier 1978). The most interesting of these is the simakobu, which exhibits a clearer relationship between predation pressure and group size than any other primate (Tilson and Tenaza 1977; Watanabe 1981). In areas of low population density (7 to 8 animals per km²), groups are entirely monogamous. Where population density increases to 300 animals per km², groups include an average of two to three females. Low population densities appear to result from human hunting pressure because there are clear differences in hunting pressure across study sites, even though habitats seem similar ecologically. Furthermore, within high-density areas, monogamous groups are found along roads, where hunting pressure is likely to be high. Thus there is good evidence that intense predator pressure favors monogamy in this species (Watanabe 1981).

Curiously, however, over evolutionary time monogamy in the Mentawai Islands has been correlated with low, not high, predator pressure. Until 2,000 years ago, the Mentawai Islands were distinguished by a complete lack of mammalian and avian predators. Yet these islands also include the Mentawai langur, which, unlike the simakobu, is monogamous even at high densities. (De Brazza's monkey is monogamous in Gabon [Gautier-Hion and Gautier 1978] but possibly not in East Africa [Kingdon 1974; Brennan 1985]. The importance of predator pressure is not yet known, but it appears to be the only *Cercopithecus* species lacking an alarm call.)

The Mentawai colobines are therefore paradoxical. In the Mentawai langur, monogamy seems to have evolved

in the absence of predation pressure, but excessive predation apparently causes monogamy among simakobu (perhaps because it reduces population density below carrying capacity; Watanabe 1981). Resolution of this paradox may contribute to a better understanding of the relation between predation and monogamy in primates as a whole. In the meantime, a number of questions remain unresolved. For example, if predation does favor small group size, it might be expected to favor solitary behavior rather than monogamy. Moreover, while predation pressure can be argued to favor small group size, it cannot explain why most monogamous species are territorial (chap. 22). Indeed, territoriality seems to be more generally explicable in terms of the distribution and defensibility of food rather than of predation pressure (chap. 12). At present, it remains unclear whether predation has had any effects over evolutionary time in modifying or reversing the effects of other factors, such as the defensibility of food or mates (chap. 23).

Large Groups

The relation between predation pressure and monogamy is merely one aspect of the more general relation between predation and group size, since high predator pressure has also been invoked to explain the evolution of large groups. Large groups are considered beneficial because they reduce the rate of predation and the cost of vigilance (Crook 1972; Alexander 1974; Clutton-Brock and Harvey 1977a; van Schaik and van Hooff 1983; Terborgh 1983). This theory is compelling partly because large groups are difficult to explain in other ways (Alexander 1974; Terborgh 1983; Moore 1984; chap. 23).

There are three types of data supporting the predation theory. First, van Schaik (1983) used indirect evidence to determine whether individuals were safer in larger groups. He compared the relation between number of juveniles and group size in populations with and without predators. In populations that were free from predation, juveniles died at higher rates in large groups than in small groups. In contrast, where predators were present, juvenile mortality was unrelated to group size. Hence, juvenile mortality in small groups was relatively high only when predators were present. Van Schaik (1983) warned that this result is not conclusive because of difficulties in classifying the data (culled from 13 species), but the approach is clearly useful. It is the only evidence to suggest that, within species, primates in small groups are indeed likely to suffer higher rates of predation than those in large groups.

Second, predator detection has been shown to be more effective in larger groups of primates, as it is in other animals (van Schaik et al. 1983). However, this observation provides only incidental support for the predation theory because larger groups might be expected to provide greater vigilance, regardless of their adaptive significance (Pulliam and Caraco 1984).

Third, several interspecific differences in group size appear correlated with predation risk. Thus terrestrial and savanna species tend to live in groups larger than those of arboreal and forest species (Crook and Gartlan 1966; Clutton-Brock and Harvey 1977b). Similarly, chimpanzees, orangutans, and spider monkeys (*Ateles* spp.) which appear comparatively invulnerable to predation because of their large body size, forage in small groups (Rutberg 1983; Terborgh 1983). However, if group size is constrained by intragroup competition for food, such results could merely reflect differences in food distribution. Terrestrial primates, for example, use large food patches (Terborgh 1983; Andelman 1986), and this may favor large group size. In contrast, feeding competition may constrain group size in chimpanzees and orangutans (Wrangham 1986).

Multimale and All-Male Groups

Baboons, macaques, and vervets are largely terrestrial species and live in multimale groups. This observation has led a number of authors to argue that multimale groups may have evolved for protection against predators (DeVore 1963; Crook 1970, 1972; Busse 1976; Leutenegger and Kelly 1977; Hamilton, Buskirk, and Buskirk 1978). It now appears, however, that the number of males in social groups is significantly related to the number of females (Terborgh 1983; Andelman 1986), a relationship that is generally considered a result of reproductive competition rather than predation (chap. 23).

Red colobus and vervets, however, are habitually found in multimale groups, even though other members of their genera live primarily in one-male groups. Since both red colobus and vervets are also particularly vulnerable to predation, it seems possible that multimale groups in these species may have evolved because males provide protection against predation (Busse 1976).

This hypothesis can be tested by comparing the mean number of males and females per group in different habitats. If multimale groups have evolved as a response to predation pressure, groups in areas of high predation should contain more males than do groups in areas of low predation. This hypothesis, however, is not supported by available data. Over 7 years in Amboseli, the ratio of males to females in three vervet groups varied from 1:0.75 to 1:7, and variation in the adult sex ratio was not related to predation rate. The mean number of females per male at Amboseli (1.7) was more than at Samburu (1.0), even though the predation rate at Amboseli was double that at Samburu (table 19-1; Cheney, unpub. data). Similarly, the mean number of females per male in red colobus groups at Gombe is 2.5 (Marsh 1979a). In contrast, at Kibale, where predation rates are far lower (table 19-1), the figure is 2.0 (Skorupa, pers. comm.). Thus, there are slightly fewer males per female at Gombe than in an area where predation pressure is less intense.

More important, if multimale groups in red colobus have evolved in response to chimpanzee predation, it is not clear why multimale groups have not also evolved in other sympatric prey species, such as redtail monkeys and black-and-white colobus.

There is also no evidence that the efficacy of antipredator defense in vervets or red colobus always increases with the number of males. Male vervets seem relatively ineffective in deterring predators. Unlike baboons, vervets do not attack or actively defend themselves against predators, and their primary responses to predators are alarm calls and flight. Adult males, however, give most alarm calls to carnivores, the species to which they themselves are most vulnerable. Males seldom alarm-call at predators to which juveniles and infants are vulnerable. Thus the presence of more than one adult male does not necessarily decrease predation rates in the group as a whole. Similarly, although red colobus males provide effective deterrents against chimpanzees (Busse 1976; Ghiglieri 1984), they are probably less successful against other important predators such as the crowned hawk eagle (Struhsaker, pers. comm.).

It might also be argued that multimale groups in vervets have evolved because high predation rates prevent males from living alone, as do males in other species of their genus (chap. 9). This argument does not, however, explain why vervet males fail to form all-male groups, as do patas monkeys and langurs (chap. 21). There is evidence that all-male groups occur more often in terrestrial species than in arboreal species, and it seems possible that such groups have evolved at least in part as a defense against predation. The data are not completely consistent, however, since all-male groups also occur in some populations that today are subject to little or no predation (e.g., gelada baboons; Japanese and rhesus macaques; chap. 21).

It therefore seems unlikely that multimale groups evolved solely as a result of predation pressure. Even in species that are subject to heavy predation rates, the number of males can be accounted for by the size of the female group. Moreover, although the presence of more than one male may provide some deterrence against predation in some species, in others there is no evidence that males do in fact decrease the group's vulnerability. It seems probable that other selective factors, such as intergroup and mate competition, have been equally important in the evolution of multimale groups.

CONCLUSION

The development and evaluation of the theory that predation has shaped the evolution of social structure is still at an early stage. Although predation clearly affects behavior and ranging patterns, it fails to account for such factors as the prevalence of intergroup hostility, sex-biased dispersal patterns, variance in group size, and interspecific differences in social relationships. In particular, it has not been possible to explain in terms of predation pressure the presence or absence of dominance hierarchies and close female bonds. If future research can demonstrate that the predation theory explains these patterns, the theory will be strongly vindicated. If not, other theories may prove more useful, not only in explaining these and other aspects of grouping behavior but also in accounting for species differences in average group size.

SUMMARY

Little is known about the influence of predation on behavior and social structure. Many species have evolved means of antipredator defense, including active aggression and alarm calls. But the function of these behaviors is still poorly understood since there is no consistent correlation between the frequency of predator defense or alarm calls and the number of offspring or collateral kin. Predation is seldom observed, and few studies have been able to provide reliable estimates of predation rates. Across species, there is a negative correlation between predation pressure and body size. At present, however, there appears to be little relation between predation pressure and terrestriality, degree of sexual dimorphism, group size, mating system, or the number of males per group. Predation no doubt affects behavior, and in some populations predation may be as or more important than food supply in regulating or limiting population size. Current evidence on species differences in susceptibility to predation suggest, however, that predation has not played the primary role in the evolution of social structure or morphological traits.

20 | Demography and Reproduction

R. I. M. Dunbar

Demography is the study of population structure and composition. It is usually divided into two related but distinct aspects: life-history phenomena (birth and death rates, migration rates, etc.) and demographic structure (the composition of groups or populations in terms of different age and sex classes). The second is in part a consequence of the first, and between them they provide the link between the behavior of individual animals and the population-level phenomena that give rise to evolutionary change.

Life-history variables are usually presented in the form of a life table, which gives, for each year of an animal's life, an easy summary of the probability of survival from birth, the probability of dying during any interval, and the probability of giving birth during the interval. These variables are often called the population's vital statistics.

The study of demographic and life-history variables is important for at least three reasons. First, information of this kind can be used to assess the state of wild populations for conservation purposes and to predict their prospects of future survival. Second, knowledge of life-history variables is essential for an understanding of how evolutionary processes are likely to be influenced by the behavior of individual animals. Finally, because an animal's social behavior is influenced by the choice of potential interactees that it has available, it is important to understand how the demographic structure of groups is likely to change over time.

Although population biologists have always emphasized the fundamental importance of demographic data of these kinds, primatologists have generally been slow to appreciate their significance. Few field studies have done more than describe the composition of the groups that have been studied. During the past decade, however, a number of studies have begun to collect life-history data as a matter of routine.

This chapter reviews the few data that are available. I shall try to draw some concrete conclusions about how demographic processes are influenced by environmental variables and how the demographic processes themselves then determine the demographic structure of the population. Finally, I shall briefly discuss some ways in which the demographic structure of a population can affect the behavior of individual animals.

STUDIES

Data on demographic structure are available for almost all species. Group size and composition are conspicuous aspects of a species' biology that are usually easy to quantify in all but the most densely forested habitats. Even the earliest field studies tried to say something about these features. Data on life-history variables, however, require long-term studies on animals that can easily be censused on a regular basis as well as on large sample sizes if the estimates are to be at all reliable. Perhaps because they are easier to observe than are more arboreal species, species that live in relatively open habitats (mainly baboons and macaques) provide most of the data on primate life-history variables. Since the conclusions in this chapter are based largely on a single taxonomic group, the extent to which they are generally valid remains uncertain. However, a great deal of work has been done on the theory of demographic processes (see Caughley 1977), and many of the key relationships between demographic and environmental variables have been documented for species other than primates. Thus, at the very least, the data currently available allow us to compare a few primate species with many other animal species, even though in the process we may underestimate the variability within the order Primates.

Life-table data are normally obtained by following up the fate of an entire cohort of individuals (i.e., all the animals born in a given year). Naturally, this requires a study of the same length as the life span of the longest-lived member of the cohort. To date, only two species of primates have been studied for the requisite length of time (more than 20 years). These are the rhesus macaque (*Macaca mulatta*) population on Cayo Santiago in the Caribbean and the populations of Japanese macaques (*Macaca fuscata*) at Arashiyama, Takasakiyama, and Koshima Islet in Japan. A number of other species have been studied for shorter periods, so it has been possible to draw up partial life tables. These species include the *Papio hamadryas* population at Erer Gota, Ethiopia, and the *Cercopithecus aethiops* and *Papio cynocephalus*

cynocephalus populations at Amboseli, Kenya. A 20-year study has been carried out on the chimpanzees (*Pan troglodytes*) of the Gombe Stream Park, Tanzania, and a partial life table of this long-lived species has been drawn up.

Life tables can, however, also be constructed from shorter-term studies by determining the mortality rates within specified age classes over a limited period of time and then using these data to reconstruct a life table. Such a life table is termed an instantaneous life table, in contrast to the longitudinal or cohort life table discussed above. Whereas a cohort life table represents the history of one cohort, the instantaneous life table gives the demographic characteristics of a population at a specific point in time. Both kinds of life table suffer from the fact that environmental conditions fluctuate from one year to the next, so birth and death rates obtained from any one year may be different from those obtained in another. Thus, unless a very large number of cohorts are used to construct a longitudinal life table, it must be remembered that all life tables are population- and time-specific.

Instantaneous life tables have been constructed for populations of gelada (*Theropithecus gelada*) in Ethiopia and toque macaque (*Macaca sinica*) in Ceylon. Similar data are available for a small number of other species.

LIFE-HISTORY VARIABLES

Table 20-1, based on data given by Sade et al. (1976), gives a (compressed) life table for female rhesus macaques living on Cayo Santiago in the Caribbean. These data in fact derive from analyses of the age-specific birth and death rates during 2 years (1973 and 1974); the life table is thus an instantaneous one. The table shows for various age classes the proportion of animals born that survive to enter each interval (l_x), the probability of dying during each interval (q_x), and the life expectancy of an animal who enters a given interval (e_x). Life expectancy is correctly given as the mean age at death for animals entering a given interval, but it is also sometimes expressed as the number of additional years that an animal entering a given class can expect to live. Table 20-1 gives the first of these. The final key variable is the probability of giving birth to a female offspring during any given interval, usually known as the fecundity rate (m_x): note that this is not the same thing as the birth rate, which is the probability of giving birth to an infant of either sex (b_x) and is usually twice the fecundity rate.

The first three variables, (l_x, q_x, e_x) are all closely related: knowing any one of them allows one to determine any of the others. Usually we determine either l_x or q_x and estimate the others on this basis; the rhesus example shown in table 20-1 in fact is based on q_x. When multiplied together and summed across all age classes, l_x and m_x give an index of the demographic "health" of the population known as R_o, the net reproductive rate (penultimate column of table 20-1). This is equivalent to the number of female offspring born to each female during her lifetime and can be interpreted as the rate at which a female replaces herself. R_o is greater than 1 in expanding populations, equal to 1 in stable populations, and less than 1 in declining populations. In the rhesus case given in the table, $R_o = 1.876$, indicating that the population is expanding at a rapid rate.

The survival and fecundity rates can be used to calculate another index that is often of interest in behavioral studies: the female's reproductive value at any given age, v_x. This is the expected number of female offspring that will be born to the average female during the remainder of her life, given that she has reached any specific age. It is usually given by the formula:

$$v = \frac{e^{rx}}{l_x} \sum_{y=x}^{\infty} e^{-ry} l_y m_y,$$

where r is the Malthusian parameter (the natural rate of population growth).

The reproductive values of the various age classes of the Cayo Santiago rhesus are given in the final column of table 20-1. Reproductive value usually increases to reach a peak during the female's early reproductive years and then declines steadily with age. The lower values in the early years of life reflect the fact that some females die before reaching reproductive age; these obviously contribute $v_x = 0$ to the average value for all females of that particular age class.

A comparable index can be derived for males (Charlesworth 1980), although its value depends on how closely male reproductive rates are tied to their age. This may vary from one social system to another. Except in monogamous species, we cannot yet determine male reproductive rate for any wild primate. However, data from the Amboseli *P. c. cynocephalus* population suggests that, because the numbers of offspring fathered by males decline with the length of time a male is resident in any group (Altmann et al., in press), the relationship may be similar to that for females.

The mathematics of these and other indexes are described in detail by Caughley (1977).

MORTALITY AND SURVIVAL

The life tables of most large mammals, including primates, follow a general pattern in which mortality is initially high, decreases during the juvenile period, and then increases steadily with age. Life tables are usually computed separately for the two sexes: although survivorship and mortality are often similar during the first few years, they invariably differ considerably after puberty. Males often suffer higher mortality rates than females at any given age and therefore tend not to live so long. In the gelada, for example, the expectation of life at birth is 12.3 years for males compared to 13.8 years for females (Dunbar 1980a) (fig. 20-1). Similar differences in the mortality rates of the two sexes have been reported for baboons (Altmann and Altmann 1970), vervet monkeys (Cheney et al., in press), and macaques (Dittus 1975; Sugiyama 1976). Field studies of a number of species indicate that, despite an even sex ratio at birth, there are more females than males among the adults. However, at least part of this difference can be attributed to the fact that in sexually dimorphic species, males take as many as 2 years longer than females to reach physical maturity.

One reason males suffer higher mortality rates than females is that, in most species, it is the males that migrate between groups. As a result, males are exposed to higher risk of predation while moving alone between groups or during solitary periods. Males are also more likely to receive fatal injuries as a result of fighting, either to enter new groups or for access to reproductive females during the mating season (see for example Wilson and Boelkins 1970; Dittus 1977).

One exception to this general rule is the chimpanzee (*Pan troglodytes*). Data given by Teleki, Hunt, and Pfifferling (1976) yield life expectancies at birth of 14.3 years for males and only 10.9 years for females. The chimpanzee is also one of the few primate species in which the males stay in the natal home ranges and the females migrate out. However, that life expectancies at an age of 8 years are similar for the two sexes (15.9 versus 16.2 years, respectively) suggests that the difference is due largely to higher rates of mortality among females during the first few years of life. Because the sample size is small (33 known deaths), it may be that unusually high mortality due to a respiratory disease epidemic in 1968 may have biased these data.

Table 20-2 compares the life expectancies at birth for females of various species for which there are data. Also shown is the proportion of animals surviving from birth

TABLE 20-1 Life Table for Female *Macaca mulatta* on Cayo Santiago

Age (yr)	Survivorship l_x	Mortality Rate q_x	Life Expectancy e_x	Fecundity Rate m_x	$l_x m_x$	Reproductive Value v_x
0.5–2.4	1.000	0.196	7.5	0.000	0.000	1.88
2.5–4.4	0.804	0.165	10.5	0.064	0.051	2.36
4.5–6.4	0.671	0.146	12.6	0.633	0.425	2.72
6.5–8.4	0.573	0.129	13.5	0.800	0.458	2.44
8.5–10.4	0.499	0.188	15.9	0.645	0.322	1.89
10.5–12.4	0.405	0.153	16.4	0.712	0.288	1.53
12.5–14.4	0.343	0.166	17.9	0.587	0.201	0.97
14.5–16.4	0.286	0.336	18.5	0.000	0.000	0.46
16.5–18.4	0.190	0.247	19.2	0.500	0.095	0.69
18.5–20.4	0.143	0.503	19.5	0.000	0.000	0.25
20.5–22.4	0.071	1.000	22.0	0.500	0.036	0.05
22.5–24.4	0.000	—	—	—	—	—

SOURCE: Based on data given in Sade et al. 1976, table 1.
NOTE: $R_o = 1.876$.

TABLE 20-2 Life-History Data for Females from Selected Primate Populations

Species	Population	Life Expectancy at Birth (e_o) (years)	Survival to 4 Years (l_4)	Sources
Macaca mulatta	Cayo Santiago, Caribbean	7.5	0.73	Sade et al. 1976
M. fuscata	Takasakiyama, Japan	4.5	0.54	Masui et al. 1976
M. sinica	Polonnaruwa, Ceylon	0.6	0.20	Dittus 1975
Cercopithecus aethiops	Amboseli, Kenya	1.6	0.27	Cheney et al., in press
Theropithecus gelada	Gich, Semien Mts., Ethiopia	10.3	0.80	Ohsawa and Dunbar 1984
	Sankaber, Semien Mts., Ethiopia	13.8	0.88	Dunbar 1980a
Papio cynocephalus	Amboseli, Kenya	4.0	0.50	Altmann 1980
P. hamadryas	Erer Gota, Ethiopia	—	0.74	Sigg et al. 1982
Pan troglodytes	Gombe, Tanzania	10.9	0.46[a]	Teleki, Hunt, and Pfifferling 1976

a. Survival to 9 years (roughly equivalent in developmental terms to 4 years in baboons and macaques).

to the end of the juvenile period (taken to be 48 months, except for chimpanzees where 108 months is approximately equivalent). These two indexes give a rough guide to the demographic health of the population in cases where one of the more conventional indexes such as R_o cannot easily be calculated.

The wide variance in the life histories of animals in these populations is unlikely to reflect genetic differences between the species since most of them are taxonomically closely related. Rather, it probably reflects the marked differences in the environments of the populations concerned. Those with high mortality rates are generally characterized either by poor habitat conditions, high predation risk and declining populations (Amboseli, Polonnaruwa), or very high densities in limited space (Takasakiyama). The Cayo Santiago rhesus population, though living at high densities in a confined area, are heavily provisioned: this almost certainly accounts for their high rates of survival. Note, however, that even though provisioned, their survival rate is lower

than those for at least two naturally occurring populations—the gelada and the hamadryas baboon.

Figure 20-2 shows the full survivorship curves for females from as many as the populations in table 20-2 as provide sufficient data. These graphs highlight further aspects of life-history patterns. Thus, infant mortality rates are high in both hamadryas and the Amboseli baboons, but this rate slows after the first year in the hamadryas whereas it continues at a high level for a further year at Amboseli. This single difference is enough to result in a difference of 50% in the number of animals reaching maturity.

Data for Dittus's toque monkeys are not included on this graph because his estimates of survivorship are suspect. They are calculated from the observed age distribution by a method that assumes the population is demographically stationary (a technical term describing the condition under which a stable age distribution exists because births exactly balance deaths and the birth rate remains constant). Although Dittus (1975) presents three

FIGURE 20-1. Two adult male gelada baboons fighting. As in many other primate species, adult males suffer higher mortality rates than adult females, at least in part because of injuries sustained during fights such as this one. (Photo: Robin Dunbar).

indexes to show that the population is stationary, two of these derive from the l_x estimates that were calculated on the assumption the population was stationary; the third is subject to a number of assumptions of uncertain validity. In addition, it seems unlikely that the birth rate would have remained constant for the requisite period of time given the known fluctuations in environmental parameters. Caughley (1977) discusses this problem at some length. Thus, although Dittus's (1975) study is important, the data should be treated with some caution until further analyses become available.

There is a considerable body of evidence to suggest that, when mortality increases as a result of adverse environmental conditions, it does not fall evenly on the population as a whole. Juveniles and infants are generally found to suffer higher mortality than other age classes (*Cercopithecus aethiops:* Struhsaker 1973; *Alouatta palliata:* Milton 1982). In some cases, juvenile females take the brunt of the mortality because they are least able to defend themselves against aggression by other members of their group (*M. sinica:* Dittus 1977; captive *Macaca radiata:* Silk 1983). Studies of the Amboseli population of *C. aethiops* have shown that low-ranking animals are most likely to die of disease, while high- and middle-ranking animals are more likely to die

of predation (Cheney, Lee, and Seyfarth 1981; Cheney et al., in press). A. Mori (1979a) reported that, during a period of food shortage, the offspring of high-ranking *M. fuscata* females had higher growth curves than those of low-ranking females, as a result they appeared to have better chances of surviving to maturity. This advantage was due to the fact that high-ranking mothers were able to ensure that their offspring had access to the best food sources.

FECUNDITY

Fecundity rates vary considerably over the course of a female's life (table 20-1). All species for which there are data show fecundity increasing to a peak during the female's physical prime and declining steadily thereafter (for further examples, see Dittus 1975; Dunbar 1980a; Strum and Western 1982; Cheney et al., in press). This pattern is typical of most mammals.

Birth rates vary considerably, both within and between species. Given optimal environmental conditions, the females of many cercopithecine species are probably capable of giving birth every year. Most, however, fail to achieve such high rates in practice. Dunbar and Sharman (1983) found that the annual birth rate varied from 0.36 to 0.89 in a sample of 18 populations of baboons (genus

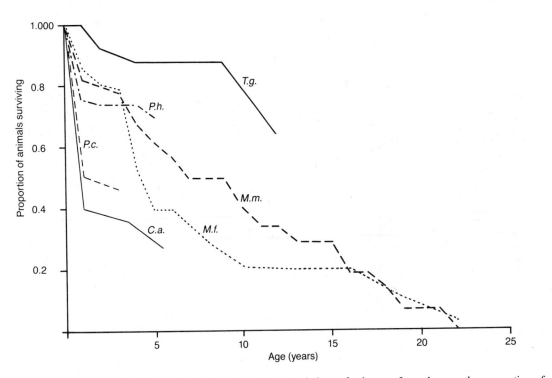

FIGURE 20-2. Survivorship graphs for females of five populations of primates. In each case, the proportion of animals still alive (l_x) is plotted at successive ages. *M.m.* = *Macaca mulatta,* Cayo Santiago, Puerto Rico; *M.f.* = *M. fuscata,* Takasakiyama, Japan; *T.g.* = *Theropithecus gelada,* Sankaber, Semien Mts., Ethiopia; *P.h.* = *Papio hamadryas,* Erer Gota, Ethiopia; *P.c.* = *P. cynocephalus,* Amboseli, Kenya; *C.a.* = *Cercopithecus aethiops;* Amboseli, Kenya. SOURCES: Sade et al. 1976; Masui et al. 1975; Dunbar 1980a; Sigg et al. 1982; Altmann 1980; Cheney et al., in press.

Papio) living in a wide variety of habitats. In addition, the variance in birth rates over time within populations is often considerable (table 20-3).

This variance in the birth rate may be due to a number of causes. Habitat conditions are especially likely to influence birth rates because they determine the condition of the animals; this in turn is known to influence fertility in many species of mammals (Sadleir 1969). A. Mori (1979a), for example, found that the birth rate of the Koshima population of *M. fuscata* fell from 0.67 per female per year during the period when the population was provisioned to 0.32 per year when provisioning was stopped and the animals had to fend for themselves. This fall in the birth rate was associated with a fall in the mean female body weight. It was also associated with a delay in the mean age at first reproduction from 6.2 to 6.8 years, possibly because females took longer to achieve the required minimum body weight to undergo puberty when food was limited. Correlations have also been found between the birth rate and the productivity of the habitat within populations over time (Strum and Western 1982), within one population across groups in different habitats (Cheney et al., in press), and between populations (Dunbar and Sharman 1983).

The overall birth rate may also be influenced in a number of different ways by demographic parameters. Since fecundity varies with a female's age (table 20-1), the overall birth rate for a group of females will depend on the age structure of the group. Birth rates will be lower if there is a high proportion of older females and higher if there are many younger ones. So far, only one study has examined age structure as a possible determinant (Strum and Western 1982); although this study found that age structure did not predict the observed fecundity rate, the correlation between observed and expected values only just fails to reach significance ($r_s = 0.569$; $t_7 = 1.831$; $p = 0.11$ 2-tailed test), which suggests that age structure probably was a contributory cause, even if it was not the main one.

It has also been hypothesized (Dunbar 1980b, 1984b; Wasser and Barash 1983) that birth rate may decline with declining dominance rank due to the physiologically disruptive effects of harassment by higher-ranking individuals. As a result, the mean birth rate for groups of females may decline as the group size increases (Dunbar 1980b). Dunbar and Sharman (1983) also found that birth rate declined as the adult sex ratio became increasingly biased in favor of females. In this case, the size of the female cohort could be ruled out as a possible factor, and the effect seemed due primarily to competition among the females for access to males (probably as protectors against other troop members rather than for sexual reasons; e.g., Smuts 1985). Van Schaik (1983) analyzed data from 16 primate species and found that the regression equation for number of infants per female on number of females in the group was negative in 22 of 27 populations. Van Schaik assumed that this effect was mainly a consequence of increasing competition for food as group size increases. Even if this were not the case, however, the social stress inherent in group living may be sufficient to generate the observed relationship. It is equally possible that social stress is the proximate cause of reproductive suppression that has ultimately been set in train by competition for access to food.

Altmann, Altmann, and Hausfater (in press) found that, in their population of *P. c. cynocephalus*, the daughters of high-ranking females underwent menarche (or puberty) up to a year earlier than daughters of low-ranking females, and that they conceived their first infant 6 months or more earlier as a result.

SEX RATIO

In most species of primates, the male-female sex ratio at birth approximates 50:50 (although very large samples may reveal sex ratios that are slightly, but significantly, male biased). Only one species of primate is known to have a sex ratio that deviates significantly from 50:50 even in small samples: *Galago crassicaudatus* has a

TABLE 20-3   Variation in the Annual Birth Rate over Time in Selected Primate Populations

| Species | Population | Birth Rate/Female | | Sample | | Sources |
		Min	Max	Groups	Years	
Macaca mulatta	Chhatari, Aligarh, India	0.727	1.000	2	15	Southwick and Siddiqi 1977
	Cayo Santiago, Caribbean	0.410	0.790	8	10	Drickamer 1974
M. fuscata[a]	Koshima Isle, Japan	0.150	0.750	1	12	A. Mori 1979a
	Arashiyama, Japan	0.380	0.870	1	17	Koyama, Norikoshi, and Mano 1975
Theropithecus gelada	Sankaber, Seimen Mts., Ethiopia	0.143	0.666	5	8	Dunbar 1980a
Papio anubis	Gilgil, Kenya	0.427	0.711	1	9	Strum and Western 1982
P. hamadryas	Erer Gota, Ethiopia	0.263	0.950	1	5	Sigg et al. 1982
Pan troglodytes	Gombe, Tanzania	0.034	0.227	1	10	Teleki, Hunt, and Pfifferling 1976

a. Data from period prior to artificial feeding only.

sex ratio of 57:43 males:females (Clark 1978b). One hamadryas band, however, managed to produce a sex ratio of 32:68 in favor of females in 82 births over a 7-year period (Sigg et al. 1982), a distribution that differs significantly from 50:50 ($X^2 = 10.976$; $p < 0.001$).

Because primates live in relatively small groups, there may be considerable variance around the average sex ratio, both between groups and within groups from one year to another, as a result of small sample biases (i.e., statistical sampling effects). Dunbar (1980a), for example, found sex ratios at birth as extreme as 25:75 and 67:33 in different years for one band of gelada, even though the average sex ratio was exactly 50:50. Similarly, Dittus (1975) reported neonatal sex ratios as extreme as 36:64 and 60:40 in *M. sinica,* and A. Mori (1979a) found sex ratios varying from 25:75 to 75:25 in one troop of *M. fuscata* over a 7-year period. These variations are almost entirely due to the fact that, when sample sizes are small, extreme values are significantly more likely to occur.

The adult sex ratio may also undergo considerable fluctuation within a group of monkeys over time (fig. 20-3). The number of adult females per adult male varied from 1.8 to 3.8 over a 16-year period in a semiprotected population of rhesus macaques in India (Southwick and Siddiqi 1977), while in the Arashiyama population of Japanese macaques it varied from 1.5 to 4.0 over an 18-year period (Koyama, Norikoshi, and Mano 1975). Several likely causes of fluctuations in the adult sex ratio include (*a*) variations in the neonatal sex ratio working

their way through the system (Dunbar 1979a; see also Altmann and Altmann 1979); (*b*) variations in the birth rate (Ohsawa and Dunbar 1984); (*c*) environmentally induced differences in the mortality rates of the two sexes (Dittus 1977; see also Dunbar and Sharman 1983); and (*d*) variations in the dispersal rates of males and females over time. Chapter 21 discusses in detail dispersal in nonhuman primates.

LONG-TERM DEMOGRAPHIC TRENDS
Population Growth

Long-term trends in population dynamics vary considerably across the order Primates. Overall, primate populations are in dramatic decline (chap. 39). Nonetheless, some populations of the more adaptable species such as baboons and macaques are able to coexist with humans, and these populations may be growing steadily. However, even those populations that have positive growth rates are probably subject to periodic catastrophic declines as a result of natural changes in environmental conditions. The *P. c. cynocephalus* and *C. aethiops* populations at Amboseli, Kenya, for example, underwent dramatic declines during the 1970s as a result of major long-term changes in the ecology of their environment (Altmann and Altmann 1979; Struhsaker 1973; Cheney et al., in press). In other cases, adverse conditions may be more transient, but their effect on primate populations may be just as serious. Dittus's (1977) population of toque monkeys at Polonnaruwa underwent a sudden decline in 1974

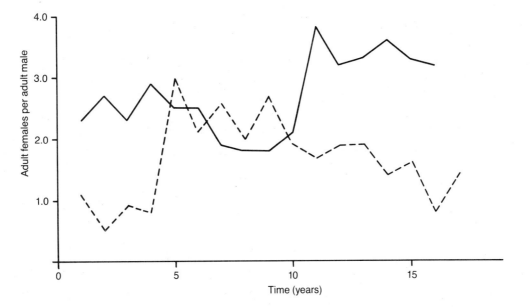

FIGURE 20-3. Variance in the adult sex ratio (females per male) over time in two free-ranging, protected populations of macaques. Solid line = *M. mulatta,* Chhatari, Aligarh, India; dashed line = *M. Fuscata,* Arashiyama, Japan. SOURCES: *M. mulatta:* Southwick and Siddiqi 1977; *M. fuscata:* Koyama, Norikoshi, and Mano 1975.

as a consequence of the worst drought in more than 44 years, although prior to that the population seems to have been fairly stable for at least a few years. Milton (1982) has suggested that periodic shortfalls in the availability of key dietary items may have caused the population crashes that occur from time to time in the Barro Colorado *Alouatta palliata*.

On the whole, it is unlikely that many populations of primates will achieve a stationary state (i.e., where $R_o = 1$, see p. 000). This is mainly because serious environmental catastrophes such as droughts or floods occur at intervals shorter than the typical life spans of most large mammals. In East Africa, for example, severe droughts occur on approximately a 10-year cycle: most larger primates that survive to sexual maturity can probably expect to live at least that long. The resulting impact on birth and mortality rates will generally be sufficient to cause major disruptions of the population's demographic trajectory, the reverberations of which may be detected a decade or more after the environmental conditions have returned to normal (see Dunbar 1979a).

Traditionally, food availability and predation pressure have been considered the two most important factors limiting the size of animal populations. Food availability seems to be the main factor limiting the size of the *M. sinica* populations in Ceylon (Dittus 1977) and the *A. palliata* populations on Barro Colorado (Milton 1982). Clear evidence from provisioned populations of *M. mulatta* in the Caribbean and *M. fuscata* in Japan shows that the provision of supplementary food allows the population to increase, often dramatically. When supplementary food is withdrawn, the population crashes (see A. Mori 1979a).

Access to other resources may also impose limits on natural populations. Harding (1976) found that the *P. c. anubis* population at Gilgil, Kenya, increased tenfold after cattle water troughs were built in the area: these apparently greatly increased the area accessible to the baboons, thereby increasing the gross amount of food available. One group in the *C. aethiops* population at Amboseli likewise seems to have difficulty obtaining water, at least during the dry season: severe mortality may result during unusually dry years, particularly when some plant species fail to fruit (Wrangham 1981; Cheney, Lee, and Seyfarth 1981; Cheney et al., in press). On balance, water is probably a less limiting resource than food.

Predation, on the other hand, can have a significant impact on natural populations. Of 68 deaths that occurred in the Amboseli *C. aethiops* population when an observer was present, 19 were due to illness, 12 were due to predation, and a further 37 were probably due to predation (the animals disappeared in a state of good health; Cheney et al., in press). In this case, predation seems to be the main cause of death, although it need not be the main factor that limits the size of the population. The effect of predation on nonhuman primate mortality is discussed further in chapter 19.

Two further factors that can limit the growth of a population are parasites and temperature stress. However, it is unlikely that either of these factors can do more than slow the rate of growth of a population, except in exceptional cases. Gelada baboons living at very high altitudes in the Simen Mountains of Ethiopia appear limited by the effects of low temperatures on birth and death rates (Ohsawa and Dunbar 1984). At lower altitudes, temperature can still be seen to have a significant effect on birth rates, but this is insufficient to prevent the population growing at very high rates. In this case, food density seems to be limiting the size of the local population, and group fission and emigration result in a fairly steady population size (Ohsawa and Dunbar 1984). Altitude is known to influence birth rates in humans, with low oxygen tension and the energy stress of maintaining body temperature the most likely causes (see Clegg and Harrison 1971).

Dittus (1977) has argued that most primate populations have built-in regulatory mechanisms that prevent the population rising to excessively high densities. He suggests that competition for scarce food resources results in high rates of mortality for animals with the lowest competitive abilities. Since immature females are particularly at risk, the birth rate will soon begin to fall because of reduced recruitment to the cohort of breeding females. This will cause the rate of growth of the population to slow down and, if adverse conditions persist, even to reverse. Note that this argument does not necessarily involve group selection: it is simply a consequence of the fact that immature females are often the animals that are least able to defend themselves. Rudran (1979) has suggested that infanticide may serve to regulate population numbers in *Alouatta* under similar circumstances, although this has been questioned (Milton 1982).

Group Structure

Most of the foregoing points apply equally to whole populations and to individual groups. However, because groups are discrete components of a population, it is important to remember that they are subject to small-sample bias effects in a way that whole populations very seldom are. This has the effect of making many extreme demographic events more likely to occur. As a result, neighboring groups of the same species may differ radically in their demographic structure for no reason other than chance. These differences may have profound demographic and behavioral consequences. Thus, it may be necessary, when seeking explanations for differences in behavior, to distinguish purely statistical effects from differences that are genuinely due to environmental factors.

Very few field studies have involved observation of more than one group. Even where wide censuses have been carried out, these are often limited to the recording of simple demographic variables such as group size and composition. In cases where it has been possible to study several groups in the same habitat in some detail, differences in social structure have been found that reflect differences in demographic structure. *T. gelada,* for example, live in one-male reproductive units that vary in size from 1 to 12 females. As the number of females in the group increases, the harem holder is able to spend less and less time grooming with each of them, and some he may not groom with at all. This results in the units becoming increasingly fragmented socially, and therefore becoming less cohesive; as a result, as their size increases they are more easily taken over by males from all-male groups (Dunbar 1984b).

Groups of different size and composition can often be formed into what seems to be a natural demographic sequence (Kummer 1968; Dunbar and Dunbar 1976). Although groups will in theory grow in a manner specified by the birth and death rates for their population, individual groups may have their growth trajectories disrupted by small-sample bias effects. A series of births of the same sex may result in marked deviations from the normal group composition, and these deviations in turn will affect social behavior.

Fission

If the growth rate of a population is positive, its groups will continue to grow in size until some environmental or demographic factor prevents them from doing so. If the growth rate is not held down by increased mortality or reduced birth rates (or both) once a group reaches a size at which it exceeds its food supply, then the group must either increase the size of its home range or part of the group must migrate to a new range elsewhere. The fissioning of groups has been documented in the Cayo Santiago and Japanese macaque populations (Koyama, Norikoshi, and Mano 1975; Sade et al. 1976). It has also been observed in free-living populations of baboons (Nash 1976; Ohsawa and Dunbar 1984).

Evidence from the Cayo Santiago macaque population, in which matrilineal relationships are well known, suggests that, when a group does undergo fission, it splits along the line of "least genetic resistance"—in other words, animals closely related to each other tend to stay together, so the division occurs in a way that separates the individuals least closely related (Chepko-Sade and Sade 1979). However, Melnick and Kidd (1983) have argued that this may only be true of expanding populations where matrilines can be very large. In natural populations whose growth is severely limited, fission may result in divisions that do not show any matrilineal correlations. They suggest that in these cases animals

may be more closely related through the paternal line and that patrilineality may be a relevant basis for determining which females remain together. As yet, the data on natural fissions are too few for any firm conclusions to be drawn either way.

DEMOGRAPHY AND BEHAVIOR

A number of behavioral consequences of demographic structure have already been mentioned. In order to underline the fundamental importance of demography to the study of behavior, I shall conclude by drawing attention to some more specific instances in which demographic structure has been shown to affect the frequencies of particular behavior patterns.

Altmann and Altmann (1979) drew attention to the often dramatic effects that small sample biases can have on the tenor of social life in a group of primates. Their study troop of *P. c. cynocephalus* baboons averaged some 35 to 40 individuals over a period of several years. The number of infants born each year was therefore small, and as many as 50% of these died before reaching sexual maturity (see fig. 20-1). As it happened, no female infants survived from the 1971 and 1972 birth cohorts; consequently, the female dominance hierarchy remained remarkably stable for a period of several years. The striking stability of female dominance relations led the Altmanns to characterize the female component as the stable core of the baboon troop. However, of the seven infants that survived from the 1973 cohort, six were female. When these matured into the adult hierarchy in 1976, they began to challenge the adult females for rank: there was a dramatic increase in the number of fights between females, and the female hierarchy was in constant turmoil with frequent rank changes and persistent instability. They were led to comment that "neither the remarkable peacefulness of the group before this time, nor the chaos since then makes much sense without knowing the demographic history of the group" (Altmann and Altmann 1979, 58).

Hausfater (1975) found that the frequency of wounding and the frequency of rank changes among males in the same group of baboons increased when one female came into estrus, but then decreased progressively as more females came into estrus at the same time. In other words, competition for access to cycling females increased the amount of serious fighting among males, though in the absence of any object over which to compete there was relatively little fighting. Data given by de Waal (1976) for two captive groups of *Macaca fascicularis* show that 69% of the aggressive acts exchanged between members of the same sex were between males when the adult male-female ratio was 1:1.3; however, when the sex ratio was 1:3.0, males accounted for only 28%. These data clearly suggest that fighting was most common among whichever sex had the more limited access to members of the opposite sex. Note that sexual access need not have been the only cause of aggression: other factors, such as competition for coalition or social partners, might have been more important in this case (see chaps. 31, 34).

Demographic variables may also have important implications for many aspects of behavioral ecology. A good example is the positive correlation between group size and day journey that has been reported for a number of species (*Cercocebus albigena:* Waser 1977a; *Papio* spp.: Sharman and Dunbar 1982; *T. gelada:* Iwamoto and Dunbar 1983, but see also chap. 8 for contrasting data on *Colobus badius*). This suggests that we need to take group size into account when comparing data from different habitats in order to identify the ecological determinants of behavior.

SUMMARY

Among primates, long-term demographic and life-history data are primarily derived from a relatively small number of studies of baboons, macaques, and vervets. As in many other mammals, mortality in these species is initially high, decreases during the juvenile period, then steadily increases with age. Fecundity increases during a female's physical prime and declines steadily thereafter. Both mortality and fecundity are strongly affected by habitat quality. Mortality is also affected by predation in some study areas, while fecundity may also be affected by female-female competition. Where mortality is high, it typically affects infants, juveniles, and those of low dominance rank more than others.

Demography can exert an important influence on social behavior. Variations in group size, infant survival, and adult sex ratio can affect factors as diverse as the length of day journeys, rates of aggression, and patterns of grooming within groups. Thus knowledge of the underlying life-history processes is often essential for a proper understanding of the behavior of both individual animals and whole populations. Although the data relevant to life-history variables are often difficult to obtain, it is important that the effort to do so be made.

21 | Dispersal and Philopatry

Anne E. Pusey and Craig Packer

Early field-workers were so struck by the apparent permanence of group membership in social primates that it was widely believed that primate groups were essentially closed genetic units. As the duration of field studies increased, however, it became apparent that there is actually considerable movement of individuals between groups (Itani 1972). It is now clear that although members of one sex often remain in their natal group throughout their lives, members of the other sex usually emigrate at sexual maturity. Among solitary primates, many animals also spend their entire lives near the site of their birth, but a proportion of individuals emigrate from their natal area (Waser and Jones 1983). Primates therefore resemble most other higher vertebrates in showing high rates of dispersal by one or both sexes (Greenwood 1980).

Examples of intergroup migration by primates can be spectacular, as in the case of two male Japanese macaques that traveled 60 km to a new group, or another male that left his natal group, traveled 17 km through urban Kyoto, and joined a group at the opposite side of the city (Norikoshi and Koyama 1975). Dispersing animals are likely to face increased risk of mortality from predation, starvation, or hostility from strange conspecifics (e.g., Gartlan 1975; Dittus 1977; Otis, Froehlich, and Thorington 1981). Why, then, do individuals ever transfer to other groups or disperse to other areas? In this chapter, we review patterns of dispersal in primates and then discuss the evolutionary explanations for this behavior.

DEFINITIONS

We divide dispersal into two components, emigration and transfer, and we separate natal from secondary dispersal. We define *natal emigration* as emigration from the natal group or range and *natal transfer* as emigration from the natal group followed by immigration into a group containing breeding members of the opposite sex. Note that an animal can emigrate without transferring. We do not consider the fusion of two preexisting breeding groups as transfer, nor the fission of a group into several new breeding units as emigration (chaps. 11, 22).

Secondary emigration/transfer, sometimes called "breeding dispersal" (e.g., Greenwood 1980), is any further movement after natal emigration/transfer. *Natal males* or *females* are individuals still residing in their natal group or range. *Immigrants,* sometimes called "transferred" animals (e.g., Packer 1979a), are individuals that have entered the group from elsewhere. Individuals show *natal philopatry* when they remain in the natal group or range as adults.

PATTERNS OF DISPERSAL IN PRIMATES

Although field studies of only a few months' duration sometimes document dispersal by one or more individuals, an accurate estimate of the rate of dispersal by each sex can only be gained from long-term data on known individuals. Table 21-1 provides data from the best-studied species on the extent of sex bias in dispersal, the proportion of individuals of each sex that disperse by full size, and the mean age at which natal dispersal typically occurs.

Gregarious Species

In most species that live in groups containing more than one breeding female, males show a much greater tendency than females both to emigrate from the natal group and to transfer between groups (table 21-1). However, a few gregarious species show dispersal by both sexes, and some show female-biased dispersal.

Male-biased dispersal occurs in all the Old World monkey species that have been well studied, except hamadryas baboons, red colobus, and possibly black colobus (*Colobus satanas*), and also appears to be the rule in social prosimians (Tattersall 1982; chap. 3). Little is known about dispersal in most species of New World monkeys, but male-biased dispersal has been described in some species of capuchins and squirrel monkeys (Terborgh 1983; chap. 7). In contrast, female emigration and transfer in all these species are very rare, although they do occur under certain circumstances (table 21-1).

In most of these species, almost all males emigrate by the time they have reached full size (table 21-1), and it is noteworthy when a male stays to breed in his natal group (e.g., Chapais 1983d). Males in many species also show considerable secondary emigration and transfer. For example, in some groups of Japanese macaques, complete replacement of males has occurred in as few as 2 years (Sugiyama 1976), and individual male olive baboons have been observed to join as many as five different

TABLE 21-1 Sex Differences in the Frequency and Age of Intergroup Transfer and Emigration

Species	No. Observed to Transfer or Immigrate		% Known to Emigrate by Full Size		Median Age (yr) at First Emigration		Age at Puberty	Age at Full Size/Birth	Sources
	Male	Female	Male	Female	Male	Female			
Macaca mulatta (Rhesus macaques)									
Cayo Santiago	137	14							Koford 1966
Cayo Santiago	—	—	96 (133)	0[a]	4 (3.5–5)		3–3.5	6.5	Colvin 1983b; Chepko-Sade, pers. comm.
La Parguera	45	0[a]	100 (16)	0[a]	4				Drickamer and Vessey 1973
Nepal	>9	0	100 (6)	0					Melnick, Pearl, and Richard 1984
M. fuscata (Japanese macaques)									
Hakone	24	0[a]	92 (26)[b]	0[a]					Fukuda et al. 1974 (quoted in Sugiyama 1976)
Arashiyma	41	0	89 (28)	0	4 (4)		4	8	Norikoshi and Koyama 1975
Ryozen	55	0	100 (13)	0	4.5 (4–5)				Sugiyama 1976; Sugiyama and Ohsawa 1982a
Shiga	13	0[a]	83 (12)	0					Sugiyama 1976
M. sinica (Toque macaques)	>12	0	75 (16)[b]	0	5.5			7–8	Dittus 1975, 1977
M. sylvanus (Barbary macaques) (Captive group)	53	0	58 (91)	0	4 (3–5)		3–5	7	Paul and Kuester 1985
M. fascicularis (Long-tailed macaques)	32	5	100 (>8)	0 (9)	5–6 (5–6)		5	9	Van Noordwijk and van Schaik 1985, pers. comm.
Papio cynocephalus anubis (Olive baboons)									
Gombe	86	1	98 (43)	0 (42)	7.5 (6–9)		5–6	7	Packer 1979a, unpub. data
Gilgil	12	0	85 (13)[c,e]	0			5–6	9–10	Smuts, Sluijter, and Noë, pers. comm.
P. c. cynocephalus (Yellow baboons)									
Mikumi	69	10	97 (37)[c]	0	Subadult/young adult				Rasmussen 1981; Rhine, pers. comm.
Amboseli	25	0	70 (10)	0	8–10 (8–10)		5–6	9–10	Altmann, Altmann, and Hausfater, in press; Altmann et al. 1977
P. c. ursinus (Chacma baboons)	18	0	100 (10)[e]	0	6.7		4	6.5–7	S. Smith, pers. comm.
P. hamadryas (Hamadryas baboons)									
Unit.	0	26	100 (13)	100 (26)	2	1.5–3.5			Sigg et al. 1982
Clan:	0	12	0 (7)	>23 (26)			M: 4.8–6.8 F: 4.3	10.3 6.1	
Cercopithecus aethiops (Vervet monkeys)									
South Africa	12	0[a]							Henzi and Lucas 1980
Amboseli	46	0[d]	93 (15)	0	5.5 (4.5–5.5)		5	5.5	Cheney, pers. comm.
Erythrocebus patas (Patas monkeys)	11	0	100 (7)	0 (10)	3		3	>5	Chism, Rowell, and Olson 1984; Rowell 1977; Chism, pers. comm.
Colobus guereza (Black-and-White Colobus)	>8	0							Oates 1977c
C. badius (Red colobus)									
Kibale	0	>20	60 (5)[e]	100 (>15)[c]	Old juv./ subadult	Old juv./ adolescent			Struhsaker and Leland, pers. comm.
Tana River	2	>12							Marsh 1979a

TABLE 21-1 Sex Differences in the Frequency and Age of Intergroup Transfer and Emigration

Species	No. Observed to Transfer or Immigrate		% Known to Emigrate by Full Size		Median Age (yr) at First Emigration		Age at Puberty	Age at Full Size/Birth	Sources
	Male	Female	Male	Female	Male	Female			
Presbytis entellus (Gray langurs)									
Mount Abu	8	0							Hrdy 1977
Jodphur	>11	2							Makwana and Advani 1981
Jodphur	6	1	100 (7)	0 (5)	Juvenile				Winkler et al. 1984, pers. comm.
P. senex (Purple-faced langurs)	10	3							Rudran 1973a
Lemur catta (Ring-tailed lemurs)	12	0			>2.5		2.5	2.5	K. C. Jones 1983
Saimiri sciureus (Squirrel monkeys; captive group)	>6	0							Scollay and Judge 1981
Alouatta seniculus (Red howlers)	12	0[f]							Rudran 1979
	32[g]	25[g]	92 (13)	43 (14)	4–6	2–3	M: 5 F: 4	6–7 5	Crockett 1984; Crockett and Rudran, pers. comm.
A. palliata (Mantled howlers)	3	8	86 (7)[c]	93 (14)[c]	2–4	2–4	M: 2.5–4 F: 3	4 3.8	Glander 1980, 1984
Gorilla gorilla (Gorillas)	1	16	63 (8)	63 (8)	11–12	8 (8)	M: ? F: 8	11.5 9–10	Harcourt 1978b; chap. 14
Pan troglodytes (Chimpanzees)									
Gombe	0	16	0 (9)	100 (8)[e]		11.5	10	13	Pusey 1979, 1980; Goddall 1983, 1986, pers. comm.
Mahale	0	13	0 (2)	100 (6)[e]		11	10	13	Nishida 1979; Kawanaka 1981

NOTES: The first pair of columns gives the number of individuals of each sex observed to transfer between groups or to immigrate into study groups from outside. These numbers exclude dependent animals that transferred with female kin and animals known to be returning to their natal group. The second pair of columns shows the proportion of each sex known to emigrate either permanently or temporarily (see footnote e) within a year or two of reaching full size or first parturition. Except where noted by footnote (c), data on natal emigration exclude disappearances. This means that data on the proportion emigrating are sometimes different from published figures. Numbers in parentheses are the number of individuals that could have reached full size. Data are taken only from studies at least 1-year long in which at least six individuals of one sex were observed to transfer or emigrate. They are also restricted to studies and study groups within studies in which approximately equal numbers of males and females could be individually identified. Numbers in parentheses following median age at first emigration show the age range over which the middle 50% of individuals

left. Unless otherwise noted, age at puberty and age at full size or first birth refer to the predominantly dispersing sex.

a. Authors imply that none of the sex transferred or emigrated.

b. Sample includes some individuals that had not yet reached full size by the end of the study, thus the percentage emigrating could be an underestimate.

c. N includes an unknown number of individuals that disappeared.

d. Does not include 5 females that joined other groups following the deaths of all the other females in their natal group since these are regarded as group fusions.

e. Includes one or more individuals that eventually returned and bred in their natal group (see text).

f. Does not include 2 female transfers that occurred after the census period.

g. Twenty males and 1 female transferred into established groups (containing breeding individuals of each sex) (Crockett 1984); the rest formed new groups with other dispersers (Crockett, pers. comm.)

groups (Packer 1979a). However, although males in many species may transfer several times, they almost never return to their natal group (e.g., olive baboons: Packer 1979a; yellow baboons: Rhine, pers. comm.; rhesus macaques: Drickamer and Vessey 1973; Japanese macaques: Sugiyama 1976; vervets: Cheney 1983b; toque macaques: Dittus 1979; long-tailed macaques: van Noordwijk and van Schaik, pers. comm.).

Species in which dispersal by both sexes is common include red howlers, mantled howlers, gorillas, hama-

dryas baboons, and some populations of red colobus. However, although both sexes show high rates of natal emigration in several of these species, there is often a sex bias in transfer. For example, in one population of red howlers, a considerable proportion of both males and females emigrated from their natal groups, but although males regularly transferred into established groups, females rarely did so. Instead, 24 of 25 transferring females formed new groups with solitary males and other dispersing females (table 21-1). Crockett (1984) observed 10 such new groups form during a 3-year study of 52 groups.

Although mountain gorillas and hamadryas baboons show natal emigration by both sexes, only females transfer between groups (table 21-1). In gorillas, some males take over their natal group when their father dies. Most, however, emigrate to become solitary or to join all-male groups and are eventually joined by emigrating females (chap. 14). Most females leave their natal group before breeding and transfer directly to other groups or solitary males (table 21-1). Hamadryas males leave their natal unit at about 2 years, but usually remain in their natal clan throughout their lives (see chap. 10). They acquire females either by gradually "adopting" juvenile females from the non-natal units of their natal clan or by challenging the males of other units and taking over one or more of their females. All females leave their natal units about 1 to 3 years before menarche, and most are initially taken over by males of their natal clan. However, females are usually taken over by other males later in life, and some end up in different clans and even bands. Thus, although both sexes leave their natal unit, females are more likely to leave their natal band.

Mantled howlers have been claimed to show male-biased dispersal (e.g., Scott, Malmgren, and Glander 1978; Froehlich, Thorington, and Otis 1981), but the only long-term study of one group revealed no obvious sex bias in either emigration or transfer (table 21-1). There is some evidence that both sexes also disperse at similar rates in black colobus (Harrison, pers. comm.). During a 9-month study of one group, two of three males disappeared and a neighboring group was later encountered that contained tame males. Also, one of seven adult females transferred to an adjacent group with her infant, and two adult females joined the group. In two other species, dispersal by both sexes has been observed, but more data are required to determine the extent of the sex bias. In sifakas (*Propithecus verreauxi*), male transfer is considered common (Richard 1974a; Jolly et al. 1982), but the only quantitative data on rates of male and female transfer come from a 6-month study in which seven males and three to five females transferred between groups following the death of an adult male (Jolly et al. 1982). In bonnet macaques (*Macaca radiata*), Simonds (1973) recorded several examples of male transfer and concluded that male-biased dispersal was a common feature of their social organization. However, in another study the only observed case of transfer was by a female (Rahaman and Parthasarathy 1969), and in a third study three females and three males transferred following a cyclone and another two males were also observed to transfer (Ali 1981).

In some populations of red colobus, transfer by both sexes is common, while in others transfer is strongly female biased (table 21-1). In rain forest areas, groups are

large and always include a number of males, whereas in more seasonal forests groups are smaller and include only one or two males (Struhsaker and Leland 1979; Marsh 1979a; Starin 1981; chap. 8). In the multimale groups, some males emigrate to become solitary, but male transfer is very rare. In contrast, all females transfer as adolescents (Struhsaker and Leland, pers. comm.). When groups contain only one or two males, however, there is both adult male replacement and female transfer (Marsh 1979a, 1979b; Starin 1981). Secondary transfer by adult females occurs in the one-male groups but is rare in the large multimale groups.

Finally, chimpanzees also show female-biased dispersal. All females transfer, at least temporarily, to other groups as adolescents (table 21-1). In many cases such transfer is permanent, but in one group at Gombe six of nine females later returned and bred in their natal group (Pusey 1979; Goodall 1983, pers. comm.). However, two of these females permanently changed subgroups (which later split into two different groups), and the first infants of two others were conceived outside the natal group (Pusey 1980). Transfer by adult females also occurs, and in some cases is probably secondary transfer, but it is less common than transfer by adolescents (Nishida 1979; Pusey 1979; Goodall 1983; chap. 15). In contrast to females, males usually remain in their natal groups (table 21-1), although dependent male juveniles have been observed to accompany female relatives to a new group (Pusey 1979; Itani 1980; Kawanaka 1981; Goodall 1983). A possible case of temporary male transfer occurred during a 6-month study in Guinea, when two males joined a group for a number of days (Sugiyama and Koman 1979a). However, male chimpanzees often spend considerable time away from their companions (e.g., Riss and Goodall 1977), and thus these males may already have been group members.

We know of no group-living primate in which there is a complete absence of intergroup transfer. It has been suggested that bonnet and Barbary macaques may be more highly inbred than other primates because many males are thought to remain in their natal groups (Wade 1979; Taub 1980b; Ali 1981; Moore and Ali 1984). However, field studies of each species have been of relatively short duration, and some male transfer has been observed in both. Moreover, in a semi-free-ranging population of Barbary macaques, most males eventually left their natal group (Paul and Kuester 1985). Gaps of over 2 years between migrations into a single group have been observed in the Gombe population of olive baboons, where 98% of males transferred (Packer 1979a; table 21-1), which suggests that only extensive long-term data can reveal whether transfer is rare or absent.

In summary, in many species characterized by male-biased dispersal, groups are composed almost entirely of female kin and immigrant males, while the converse is often true in species characterized by female dispersal. In these species, the like-sexed kin group may persist for generations (e.g., rhesus macaques: Chepko-Sade and Sade 1979). However, in the few species where both sexes transfer (e.g., mantled howlers), both males and females may be unrelated to like-sexed group members.

Monogamous Species

In monogamous gibbons, and tamarins that live in groups containing only one breeding female, both sexes apparently disperse at a similar rate. In one population of Kloss's gibbons (*Hylobates klossi*), subadults of both sexes left their natal groups. Two females and one male successfully paired with mates in territories adjacent to their parents' territory, two males lived at the periphery of their parents' territory, one male left the area, and one male eventually returned to replace his dead father (Tenaza 1975; Tilson 1981; chap. 12). Subadult male and female whitehanded gibbons (*Hylobates lar*) and siamangs (*Hylobates syndactylus*) also leave their natal groups (Chivers and Raemaekers 1980). Solitary individuals of both sexes are seen in all three species, and since groups contain only one adult of each sex, it appears that subadults of both sexes always emigrate unless a like-sexed parent dies.

In Geoffroy's tamarins (*Saguinus geoffroyi*), Dawson (1978) observed 18 immigrations by roughly equal numbers of males and females. Similar numbers of males and females also emigrated or disappeared. Immigrants were of all ages, but animals that emigrated or disappeared included a significantly higher proportion of immatures than adults. Adult male and female cottontop tamarins (*Saguinus oedipus*) have also been observed both to emigrate from and immigrate into established groups (Neyman 1978). Thus in tamarins both sexes apparently emigrate and transfer at similar rates.

Solitary Species

Solitary primate species typically show male-biased dispersal. In several species of bushbabies (*Galago senegalensis, G. crassicaudatus, G. demidovii*) and pottos (*Periodictus potto*), most females settle close to their natal range, while most males leave at puberty (see reviews by Waser and Jones 1983; chap. 2). This pattern of dispersal also appears to occur in orangutans (*Pongo pygmaeus:* Galdikas 1984; chap. 13).

SOLITARY ANIMALS AND NONBREEDING GROUPS

In many gregarious species, dispersing animals transfer directly from one bisexual group to another, but in others individuals may become solitary or join nonbreeding (usually all-male) bands for variable periods (fig. 21-1). Table 21-2 lists those species in which extragroup males appear to be a regular feature of that species' social organization. It does not include species in which solitary males are observed only rarely (e.g., baboons: Slatkin and Hausfater 1976; Hamilton and Tilson 1980).

In most cases, extragroup males are not individually identified, and nothing is known of the length of time they spend outside bisexual groups or of their relationships with such groups. The limited available data suggest great inter- and intra specific variation, ranging from a few weeks spent as a "semisolitary" male outside (but within the ranges of) bisexual groups (e.g., long-tailed macaques: van Noordwijk and van Schaik 1985) to many months spent as a member of an independent all-male band (e.g., geladas [*Theropithecus gelada*]: Dunbar and Dunbar 1975; U. Mori 1979a; gray langurs: Hrdy 1977). In some seasonally breeding species, extragroup males temporarily join bisexual groups during the mating season and subsequently leave again (e.g., patas: Chism and Olson 1982; blue monkeys [*Cercopithecus mitis*]: Tsingalia et al. 1984; chap. 9; some rhesus and Japanese macaques: Drickamer and Vessey 1973; Sugiyama 1976).

In several species, extragroup males form all-male bands (table 21-2). Such bands may include males from both the same and different natal groups (e.g., Japanese macaques, geladas, gray langurs; see references above). All-male bands are typically subordinate to bisexual groups and often occupy suboptimal ranges (e.g., patas: Gartlan 1975; purple-faced langurs: Rudran 1973a; gray langurs: Sugiyama 1967). Sometimes, however, all-male bands can be highly aggressive to males in bisexual groups.

Extragroup females are much less common than males, but they do occur in red howlers (Rudran 1979; Sekulic 1982a), mantled howlers (Glander 1980), purple-faced langurs (Rudran 1973a), and more rarely in gray langurs (Hrdy 1977). In addition, several female Japanese macaques have been observed to become solitary within the range of their natal group for varying periods (Burton and Fukuda 1981; Sugiyama and Ohsawa 1982a).

PROXIMATE CAUSES OF DISPERSAL

It is not always easy to determine why an individual emigrates at a particular time. In some species, emigration occurs as a consequence of coercive or aggressive behavior. In others, individuals emigrate as the result of an attraction to extragroup individuals. Sometimes both of these factors appear to operate simultaneously.

Abduction

In several species characterized by female emigration, it has been suggested that females are "abducted" by males outside their natal group. This appears common only among hamadryas baboons, however, where males adopt juvenile females or herd older females away from their defeated male rivals. Adoption usually occurs gradually;

the male shows protective behavior toward a female and prevents her from returning to her natal unit for successively longer periods (Abbegen 1984).

In chimpanzees, the actual transfer of females from one group to another has only rarely been observed. Males have twice been observed to lead females away from adjacent groups and attack them if they did not follow (Goodall 1983), but females are not continuously sequestered by males, and at least three females are known to have joined adjacent groups voluntarily (Pusey 1979, pers. obser.). Similarly, although Fossey (1983) suggests that male mountain gorillas sometimes attempt to force females into following them, Harcourt (1978b) concluded that six of seven observed transfers were voluntary.

Eviction

In a number of species, individuals are forced to emigrate as the result of intense aggression from conspecifics. The most obvious cases of eviction involve species living in one-male groups, and occur when an adult male

or an all-male band enters a bisexual group and drives out the previous breeding male (see chaps. 8, 9). Male replacements often involve fierce fights between the previous breeding male and the intruders and can result in severe wounds. Aggressive replacements have been observed in gray langurs (e.g., Sugiyama 1967; Newton 1984), purple-faced langurs (Rudran 1973a), blue monkeys (Butynski 1982b), redtail monkeys (*Cercopithecus ascanius:* Struhsaker 1977), some red colobus (Marsh 1979a), red howlers (Rudran 1979; Crockett and Sekulic 1984), and geladas (Nathan 1973; Dunbar 1984b). One-male groups seem more vulnerable to such incursions than multimale groups. For example, red howler groups often contain more than one adult male, but six of seven male incursions were into groups with only one resident male (Rudran 1979).

Clear cases of male eviction from multimale groups are less common but do occasionally occur. Male emigration in vervets is usually voluntary (Henzi and Lucas 1980; Cheney 1983b), but in one population four out of five males that had fallen in rank to other males were

FIGURE 21-1. A gray langur all-male band. (Photo: Sarah Blaffer Hrdy/Anthrophoto)

TABLE 21-2 Species in Which Extragroup Males Are a Common Feature of Social Structure

Species	Single or Multimale	Habitat	Size of Male Groups[a]	Comments	Sources
Japanese macaques	M	Terrest./forest	1–several	Adults usually solitary	Sugiyama 1976
Pigtailed macaques	M	Terrest./forest	1		Crockett and Wilson 1980; Rijksen 1978; Bernstein 1967a
Long-tailed macaques	M	Arboreal	1	Semisolitary	Van Noordwijk and van Schaik 1985
Rhesus macaques					
Puerto Rico	M	Terrest./forest	1–40	Groups are peripheral subgroups of bi-sexual groups	Boelkins and Wilson 1972; Drickamer and Vessey 1973
Nepal	M	Terrest./forest	2–12 (4–6)	Groups distinct from bisexual groups	B. Marriott, pers. comm.
Patas monkeys					
Kenya	S	Terrest./open	1–9	Adults usually solitary; immatures in groups	Chism and Olson 1982; Chism, pers. comm.
Cameroon	S	Terrest./open	1–8 (3.38)	Groups contain adults	Gartlan 1975
Gelada baboons	S	Terrest./open	4–13 3–13 (7.8)		U. Mori 1979a Dunbar and Dunbar 1975
Black-and-white colobus	S (usually)	Arboreal	1–2	Regularly seen but not common	Struhsaker and Leland 1979
Red colobus					
Kibale	M	Arboreal	1	Regularly seen but not common	Struhsaker and Leland 1979
West Africa	M	Arboreal	1		Struhsaker 1969
Tana River	S	Arboreal	1		Marsh 1979a
Redtail monkeys	S	Arboreal	1		Struhsaker 1969; Struhsaker and Leland 1979; chap. 9
Blue monkeys	S	Arboreal	1		Struhsaker 1969; chap. 9
Spot-nosed guenon	S	Arboreal	1		Struhsaker 1969
Crowned guenons	S	Arboreal	1		Ibid.
Mona monkeys	S	Arboreal	1		Ibid.
Moustached guenon	S	Arboreal	1		Ibid.
Red-eared nose-spotted monkeys	S	Arboreal	1		Ibid.
L'Hoest's monkeys	S	Arboreal	1		Ibid.
Mandrills	S	Terrest./forest	1		Ibid.; Sabater Pi 1972
Gray langurs					
Abu	S	Terrest./open	1–60		Hrdy 1977
Junbesi	M	Terrest./open	1–2		Boggess 1980
Dharwar	S	Terrest./open	2–59		Sugiyama 1967
Purple-faced langurs	S	Terrest./forest	(7.5)	Male groups sometimes include juvenile females	Rudran 1973a
Gorillas	S or M	Terrest./ thick veg.	1–9		Harcourt 1978b; Fossey 1983; chap. 14
Squirrel monkeys	M	Arboreal	1–5		Terborgh 1983
Brown capuchins	M	Arboreal	1–3		Ibid.
Red howlers	M or S	Arboreal	1–4	Male groups sometimes include females	Rudran 1979
Mantled howlers	M	Arboreal	1–?		Scott, Malmgren, and Glander, 1978

a. Mean group sizes, when known, are given in parentheses.

consistently harassed by the victor until they left (Henzi and Lucas 1980). Similarly, three male toque macaques disappeared following the establishment of a new dominant male, and one was severely wounded before his disappearance (Dittus 1975). Another ten males emigrated following a drop in dominance rank, but it is not clear if their departure was preceded by an increase in aggression.

The emigration of adult males in these cases is likely to be secondary dispersal, but in at least two species immature males are also evicted from their natal group. Sugiyama (1967) observed a male gray langur evict the breeding male from a group and then chase and bite all male juveniles until they eventually emigrated (see also Boggess 1980). In another gray langur population, adult males consistently harassed and prevented juvenile males from approaching their mothers until finally the juveniles became very peripheral and then joined all-male bands (Mohnot 1978). Similarly, following five cases of male replacements in purple-faced langurs, all males, juvenile females, and infants either emigrated or disappeared (Rudran 1973a). In one case, most emigrants (including two juvenile females) bore severe wounds. From these observations, Rudran concluded that juvenile females and adult and immature males in predominantly all-male bands had probably been evicted from bisexual groups. Subadult male black-and-white colobus also disappeared at the time of male replacements, but the precise cause of their departure is not known (Oates 1977c).

In these cases, individuals are presumably evicted by males that are not their father. In at least one species living in one-male groups (Lowe's guenons [*Cercopithecus campbelli lowei*]: Bourliere, Hunkeler, and Bertrand 1970), males are not evicted by their father, but emigrate voluntarily if their father is still present in the group when they mature. However, in other species relatives do sometimes evict immatures. In multimale groups of red colobus, immature males receive considerable aggression from other, presumably related males. Although most males nevertheless remain in the natal group, a few emigrate temporarily or even permanently as a result of this aggression (Struhsaker and Leland 1979, pers. comm). In various gibbon species, eviction by the like-sexed parent appears to be the primary cause of emigration by subadults, although attraction to extragroup animals may also be involved. In siamangs, whitehanded, Kloss's, and hoolock (*Hylobates hoolock*) gibbons, the increasing peripheralization of subadult males occurs as a direct result of the attacks, chases, and displays of their fathers. In addition, adult female Kloss's gibbons have been observed to chase and threaten their maturing daughters (Chivers and Raemaekers 1980; Tilson 1981; chap. 12).

There is also evidence that female aggression causes the emigration of some female howler monkeys. Crockett (1984) found that the emigration of juvenile females was often associated with wounding of the juvenile and some of the adult females in the group. In one case, the mother of the juvenile and the only other adult female in the group were observed in intense aggressive interactions at the time of the juvenile's departure. This juvenile eventually rejoined her mother in her natal group following the death of the other adult female. In mantled howlers, an adult female also emigrated after being chased away by two other females (Jones 1980b). Both Crockett and Jones suggest that breeding positions for females within howler groups are limited and that intense competition exists between females for group membership.

Young males in some species encounter increasing rates of aggression from resident adult males. This has been assumed to cause their eventual emigration although their departure was not observed (e.g., blue monkeys: Rudran 1978; redtail monkeys and black-and-white colobus: Struhsaker and Leland 1979). However, such aggression does not inevitably lead to emigration. Subadult males also receive higher rates of aggression in species in which males usually do not emigrate, such as red colobus and chimpanzees (Pusey 1978). Moreover, immature female toque macaques receive more aggression than immature males, but it is the males, rather than the females, that emigrate (Dittus 1977).

In many species, natal animals receive less aggression from other group members than do immigrants, and they also emigrate without any noticeable increase in aggression from other group members (e.g., Japanese macaques: Sugiyama 1976; olive baboons: Packer 1979a; vervets: Henzi and Lucas 1980; Cheney 1983b; long-tailed macaques: van Noordwijk and van Schaik 1985; rhesus macaques: Colvin 1983b; ring-tailed lemurs: K. C. Jones 1983; Lowe's guenon [*Cercopithecus campbelli lowei*]: Bourliere, Hunkeler, and Bertrand 1970; redtail monkeys: Struhsaker and Leland, pers. comm.; chimpanzees: Pusey 1980). Indeed, in one group of Japanese macaques where there were no adult males, immature males nevertheless continued to emigrate (Sugiyama 1976). Thus although it is sometimes suggested that emigration usually results from the coercive behavior of more dominant animals (e.g., E. O. Wilson 1975; Moore and Ali 1984), dispersal in many species is often voluntary.

Attraction to Extragroup Individuals

Sexual attraction to animals in other groups appears to be a common proximate cause of transfer. In many seasonally breeding species, male transfer occurs primarily during the mating season, which suggests that it is motivated by sexual attraction (e.g., rhesus macaques: Lindburg 1969; Boelkins and Wilson 1972; Drickamer and Vessey 1973; toque macaques: Dittus 1977; Japanese macaques: Sugiyama 1976; Barbary macaques: Paul and Kuester 1985; vervets: Henzi and Lucas 1980; Cheney 1983b; sifakas: Richard 1974a). Similarly, extragroup males often join bisexual groups during the mating season. Sexual attraction also seems to be an important cause of transfer in nonseasonally breeding species. The first visits of adolescent female chimpanzees to adjacent groups occur during their estrous periods, and almost all adult female transfers take place when the female is sexually cycling (Pusey 1979, 1980; Nishida 1979; Goodall 1983; Nishida and Hiraiwa-Hasegawa 1984). Male immigration in baboons is also most frequent when the group contains many estrous females (Packer 1979a; Rasmussen 1979; Manzolillo 1984). Even males in such predominantly solitary species as bushbabies move temporarily to areas where there are estrous females (Bearder and Martin 1979).

Natal emigration often appears motivated in particular by attraction to unfamiliar mates. As in many other mammals (e.g., Hill 1974), there is considerable evidence from primates that individuals that were familiar during immaturity (often close kin) show reduced sexual attraction as adults (e.g., humans: Shepher 1971; Wolf and Huang 1980; rhesus macaques: Sade 1968; Japanese macaques: Enomoto 1974; Barbary macaques: Paul and Kuester 1985; olive baboons: Packer 1979a; chimpanzees: Goodall 1968; Pusey 1980; Coe et al. 1979). In olive baboons, the initial interactions of transferring males are directed primarily toward females (Packer 1979a). These males subsequently compete at much higher levels than they had in their natal group for access to estrous females, and they also receive more solicitations from females in the new group. Natal transfer is not to groups containing greater numbers of females, but is rather to any group composed of unfamiliar females. Natal female chimpanzees (Pusey 1980), gorillas (chap. 14), and natal male vervets (Henzi and Lucas 1980; Cheney 1983b), long-tailed macaques (van Noordwijk and van Schaik 1985), and toque macaques (Dittus 1977) also show a greater attraction to mates outside their natal group.

In some species, natal emigration by males appears to result from an attraction to extragroup males or to emigrating group mates. In patas (Gartlan 1975) and geladas (Dunbar and Dunbar 1975; U. Mori 1979a), young males temporarily leave their natal units to interact with male peers or the members of all-male groups; this attraction is thought to contribute to eventual emigration. Juvenile male Japanese macaques also show great interest in solitary or peripheral males, and in some cases they eventually emigrate with these males (Sugiyama 1976). Similarly, immature male rhesus macaques initially join all-male subgroups associated with adjacent groups (Boelkins and Wilson 1972; Drickamer and Vessey 1973; Colvin 1983b). Young males in several species sometimes emigrate with peers or older brothers, and it is possible that the emigration of a group mate hastens their own emigration.

Other Contexts of Dispersal by Females

The only observed cases of female transfer in Japanese and rhesus macaques occurred between groups that had recently divided by group fission (Norikoshi and Koyama 1975; Chepko-Sade, pers. comm.); presumably these cases were the result of a continuing realignment of individuals following the fission. A recent troop fission may also have preceded the only cases of female transfer in yellow baboons at Mikumi, although it is also possible that these females had departed temporarily during the process of habituation to human observers (Rasmussen 1981).

Females have also been observed to disperse following habitat destruction or periods of high female mortality. For example, three female bonnet macaques transferred following a cyclone (Ali 1981), and two female long-tailed macaques emigrated from groups whose home ranges had been logged (van Schaik and van Noordwijk, pers. comm.). Similarly, a number of adult female vervets joined other groups following the deaths of all the other adult females in their previous group (Hauser, Cheney, and Seyfarth, in press).

In some species, adult females with infants occasionally emigrate following the entry to their group of a potentially infanticidal new male. Three female gray langurs were observed to emigrate temporarily in this context (Hrdy 1977), and a female purple-faced langur that left her group with an ousted male eventually joined a new group (Rudran 1973a). Similarly, two female long-tailed macaques with newborn infants transferred to another group following the immigration of a male that severely attacked one of them (van Schaik and van Noordwijk, pers. comm.). The entry of a new male also sometimes causes a group to fission, when some females leave with the previous breeding male (e.g., long-tailed macaques: Wheatley 1982; gray langurs: Sugiyama 1967).

AGE OF NATAL DISPERSAL

There is considerable inter- and intraspecific variation in the precise age and stage of maturation at which natal emigration takes place (table 21-1). In most species, the majority of males emigrate at puberty, but male hamadryas leave their natal unit before puberty and male savanna baboons remain for several years after puberty. In species showing regular natal emigration by females, this usually occurs after females have commenced sexual cycling, but female hamadryas are adopted by their first male when they are still juveniles, and at least some female howlers appear to emigrate before their first estrus (Glander 1980; Jones 1980a; Crockett 1984; table 21-1). Thus even though individuals of the dispersing sex do occasionally remain and breed in their natal group (e.g., baboons: Smuts 1985; rhesus macaques: Chapais 1983d), the great majority do not breed until they have left their natal group or area.

There appears to be considerably more variability in the relative size of dispersing animals. Males in some species typically leave when they are still only just over half grown (e.g., macaques, langurs, patas), while others do not usually leave until they are almost full sized (e.g., baboons, vervets, ring-tailed lemurs, gorillas). These differences are partly but not entirely due to differences in the relative size at which puberty occurs (table 21-1), and are not obviously correlated with the presence or absence of a solitary stage, whether dispersal is forced or voluntary, or with the extent of sexual dimorphism. In some groups of macaques, male rank (and probably reproductive success) increases with the length of tenure in the group, and males that transfer early may ultimately achieve higher reproductive success (e.g., Japanese macacques: Norikoshi and Koyama 1975; Sugiyama 1976; rhesus macaques: Drickamer and Vessey 1973). In contrast, emigrating male vervets and baboons often rapidly attain high dominance rank (Packer 1979a; Henzi and Lucas 1980; Ransom 1981; Strum 1982). Since male mating success and seniority are only occasionally correlated in baboons and vervets (Packer 1979a; Strum 1982; Smuts 1985; Andelman 1985), there may not be a similar advantage to early transfer in these species.

Several factors influence intraspecific variation in the age of natal emigration. In some species, the sons of high-ranking females emigrate at older ages than other males (e.g., rhesus and Japanese macaques: Colvin 1983b); in a few cases they may remain and breed for a few years in their natal group (Chapais 1983d). Provisioning (e.g., Sugiyama and Ohsawa 1982b) and a paucity of adjacent groups (e.g., Drickamer and Vessey 1973; Sugiyama 1976) also retard male emigration in Japanese and rhesus macaques. There is some indication that group size also affects the age that males leave. Some male Japanese macaques in abnormally large provisioned groups stay longer in their natal group than any males born in smaller provisioned groups (Sugiyama 1976). Similarly, male olive baboons born in large groups (Gilgil) remain longer than those born in small groups (Gombe) (Smuts, pers. comm.; Strum, pers. comm.; Packer 1979a; table 21-1). Finally, males whose mothers have died emigrate earlier in some populations of vervets (Cheney 1983b).

FACTORS INFLUENCING THE OCCURRENCE OF SOLITARIES AND ALL-MALE BANDS

Extragroup males are most common in species that typically live in one-male groups, although solitary or semisolitary males do occur in some macaques and howlers that live in multimale groups (table 21-2). In some of these species, the presence of extragroup males ob-

viously results from eviction during male replacements. In others, however, males emigrate voluntarily (e.g., patas: Olson, pers. comm.).

In some cases, species that are otherwise very similar show a striking difference in the tendency of males to transfer or to become solitary. For example, olive baboons, vervets, and Japanese macaques have a similar social organization (chap. 11), and in all these species virtually all males voluntarily leave their natal group. However, male baboons and vervets usually move directly from one group to another (Packer 1979a; Henzi and Lucas 1980; Cheney 1983b), whereas many male Japanese macaques spend up to 3 years as solitaries before entering another group (Sugiyama 1976). Packer and Pusey (1979) found that immigrating male Japanese macaques received considerably more aggression from adult females in the new group than did male baboons and proposed that this could account for the presence of solitary males in Japanese macaques. However, young male vervets also receive high levels of aggression from the females of new groups, but nevertheless do not become solitary. An additional reason for the difference may be that Japanese macaques have fewer natural predators than baboons or vervets and can better afford to avoid female aggression by remaining outside groups (see also chap. 19).

The risk of predation may also influence whether extragroup males are solitary or form all-male bands. Struhsaker (1969) pointed out that all-male bands are more common in terrestrial species that live in open habitats (e.g., some macaques, geladas, langurs, patas) than in arboreal species that live in forest (e.g., forest cercopithecines) (table 21-2) and suggested that male gregariousness is an adaptation to predation in open areas. In some species, immature males form all-male bands, whereas adults are usually solitary (e.g., Japanese macaques: Sugiyama 1976; one population of patas: Chism and Olson 1982). It is likely that smaller animals are more vulnerable to predation.

Another advantage to males from joining all-male bands may be the formation of alliances that increase their chances of entering bisexual groups. In lions (*Panthera leo*), groups of males are more successful than single males in taking over and then jointly defending groups of females (Bygott, Bertram, and Hanby 1979), and single males often join up with unrelated partners to form cooperative coalitions (Packer and Pusey 1982). Similar alliances also occur in red howlers, where pairs of males have been seen to oust resident males and then remain together in the group (Crockett and Sekulic 1984). In most primates, however, the members of all-male bands do not seem to cooperate with each other. Even though all-male bands in geladas and gray langurs sometimes attack bisexual groups and attempt to oust the resident male, the extent of cooperation between the males is usually only slight (U. Mori 1979a; Nathan 1973; Dunbar 1984b; Sugiyama 1965; Hrdy 1977). In these species it is rare for more than one male from an all-male band to remain together in a bisexual group following a takeover (e.g., Moore 1982).

SOLO OR PEER MIGRATION

In several species, individuals often emigrate in pairs or groups (rhesus macaques: Boelkins and Wilson 1972; Drickamer and Vessey 1973; Barbary macaques: Paul and Kuester 1985; vervets: Cheney and Seyfarth 1983; Japanese macaques: Sugiyama 1976; long-tailed macaques: van Noordwijk and van Schaik 1985; patas: Olson, pers. comm.; ring-tailed lemurs: K. C. Jones 1983; sifakas: Richard 1974a; chap. 11). In most of these species immature males are much more likely to emigrate in the company of group mates than are adult males. In contrast, there are other species in which individuals usually emigrate singly (e.g., male and female red howlers: Crockett 1984; female chimpanzees: Nishida 1979; Pusey 1979; female gorillas: Harcourt 1978b; male yellow baboons: Altmann and Altmann 1970; Rasmussen 1981; male olive baboons: Packer 1979a; male chacma baboons: S. Smith, pers. comm., but see Cheney and Seyfarth 1977). One difference between these two sets of species is that the former all breed seasonally while the latter do not. In seasonally breeding species, cohorts of individuals of the same age are likely to reach the same physical stage at the same time as well as experience the same external stimuli.

By transferring together, individuals may benefit both from increased vigilance against predators during the period of emigration and from the presence of a known individual that could be a potential ally in the new group (Cheney and Seyfarth 1983; van Noordwijk and van Schaik 1985). It is likely that younger, inexperienced animals are more vulnerable to predation and aggression than are adults (Cheney and Seyfarth 1983).

RECEPTION IN THE NEW GROUP

Immigrants frequently face greater hostility from members of their new group than from their old group. Transferring male baboons are often chased away from the new group by resident males, but can easily return to their prior group until they transfer successfully (Packer 1979a). The intensity of aggression toward immigrants is apparent from the fact that previously unscathed males in a number of species have been observed to receive wounds upon entering a new group (olive baboons: Packer 1979a; vervets: Henzi and Lucas 1980; Harrison 1983b; chap. 22; rhesus macaques: Lindburg 1969; toque macaques: Dittus 1977). Similarly, two adolescent female chimpanzees returned wounded after an absence from their natal group (Goodall 1986). There is a peak in male mortality in toque macaques (Dittus 1977) and mantled

howlers (Otis, Froehlich, and Thorington 1981) at the age at which dispersal takes place. At least some mortality is due to aggression from conspecifics, but in the case of mantled howlers it is not known whether the aggressors were members of the old or new group (see also chap. 20). Male toque macaques also often lose weight following transfer, presumably because they are excluded from food sources (Dittus 1977). Most aggression in the new group comes from like-sexed individuals, and in olive baboons immigrating males receive the most intense aggression from males of the same age (Packer 1979a). Migrating ring-tailed lemurs and sifakas are often vigorously repulsed by males of the new group (K. C. Jones 1983; Richard 1974a), as are male vervets, rhesus, and toque macaques (see earlier references). Similarly, immigrating female chimpanzees receive more aggression from resident females in their new group than in their natal group (Pusey 1980), and female red howlers that attempt to enter established groups receive intense aggression from resident females (Sekulic 1982a; Crockett 1984). Serious aggression against immigrating female red colobus, however, has not been observed (Marsh 1979a; Struhsaker and Leland, pers. comm.).

Immigrating animals may also face aggression from individuals of the opposite sex, and in species where sexual dimorphism is slight, female aggression can even deter males from entering groups. Female Japanese macaques often chase solitary males away from the group (Sugiyama 1976; Packer and Pusey 1979), and female vervets have been observed to wound young males attempting to join their group (chap. 22) (fig. 21-2). Female langurs and blue monkeys sometimes assist the resident male in chasing out intruding males (Hrdy 1977; Tsingalia and Rowell 1984), and a female Kloss's gibbon with an adolescent son but no mate consistently chased solitary males out of her territory (Tilson 1981).

FIGURE 21-2. Two adult female vervet monkeys threaten a male who is attempting to transfer into their group. (Photo: Dorothy Cheney)

Females may also influence whether or not a male is able to join their group by nonaggressive means. For example, female geladas showed more affiliative behavior toward one male that was trying to take over their unit than toward several others, and this male later became resident (U. Mori 1979a). Similarly, Smuts (1985) found that male olive baboons who were avoided least by females were the most likely to remain in the group.

In many cases, at least some group members welcome immigrants of the opposite sex. For example, female olive baboons showed more affiliative behavior to and preferred to mate with immigrant males rather than natal males, and sexually receptive females were most attracted to males that had recently entered the group (Packer 1979a). Similarly, Henzi and Lucas (1980) found that all male vervet immigrants received affiliative gestures from adult females within a few days of entering the group. One lineage of female Japanese macaques also showed a preference for mating with immigrant males rather than with males of their lineage (Enomoto 1974), and receptive female Japanese macaques and langurs have been observed to mate with solitary males (Sugiyama 1976; Hrdy 1977; Burton and Fukuda 1981). Finally, male chimpanzees (Nishida 1979; Goodall et al. 1979; Goodall 1983) and red colobus (Struhsaker and Leland, pers. comm.) show affiliative behavior toward immigrating cycling females, and male chimpanzees protect such females from the aggression of resident females. The friendliness of male chimpanzees toward cycling females is particularly striking, given their occasional severe attacks on anestrous females from other groups (Goodall et al. 1979; Nishida 1979; Pusey 1980; Wolf and Schulman 1984).

One salient feature of all-male bands is the peaceful and affiliative nature of interactions between males, which are in marked contrast to their hostile interactions with resident males in bisexual groups. In gray langurs (Mohnot 1978), purple-faced langurs (Rudran 1973a), and patas (Gartlan 1975), the members of all-male bands greet the approaches of immature males with friendly gestures, and new immigrants are readily integrated into the band. The reception of immigrants is more variable in geladas; some all-male bands welcome newcomers, while others resist them (Dunbar and Dunbar 1975; U. Mori 1979a). In Japanese and rhesus macaques, immature immigrant males also often form affiliative bonds with the members of peripheral male subgroups before attempting to enter the main bisexual group (Boelkins and Wilson 1972; Drickamer and Vessey 1973; Sugiyama 1976; Colvin 1983b).

CHOICE OF GROUP
Adjacent versus Nonadjacent Groups

Because transferring animals probably often have information on the composition of neighboring groups as a result of intergroup encounters (see chap. 22), most ani-

mals might be expected to transfer to adjacent groups. In species showing direct transfer for which data on several contiguous groups are available, almost all individuals transfer into adjacent groups (e.g., olive baboons: all cases of natal and secondary male transfer [Packer 1979a; unpub. data]; vervets: 96% of 23 natal males and 89% of 23 secondary transfers [Cheney and Seyfarth 1983, pers. comm.]; rhesus macaques: 82% of 11 males [Melnick, Pearl, and Richard 1984]; ring-tailed lemurs: all males [K. C. Jones 1983]; gorillas: all females [chap. 14]). Similarly, 77% of 26 natal transfers but only 32% of 19 secondary transfers by hamadryas females were to males of the same clan (Sigg et al. 1982). Although most individuals transfer to adjacent groups in these species, many eventually breed in groups far from their natal group because secondary transfer is common and return to the natal group is rare.

Species showing a solitary stage might be expected to show less tendency to settle in an adjacent group, since solitary individuals often range independently of a specific group. In long-tailed macaques, where males are sometimes semisolitary, only 6 of 21 males transferred into adjacent groups, but young males were significantly more likely than older males to do so (van Noordwijk and van Schaik 1985). In Japanese macaques, the percentage of males that disappeared from their natal group and were later found in or near an adjacent group averaged about 50% over five sites, and varied from 82% of 16 males at Shiga to 29% of 21 males at Ryozen (Sugiyama 1976). Males from three of these sites were found as far as 20 to 60 km away from their natal group.

Aggressiveness of New Group

Natal male olive baboons transfer to groups containing the fewest males of similar age to themselves, these being the males that would be most hostile to them (Packer 1979a). They also sometimes transfer to groups where the adult males are relatively less aggressive due to low levels of consorting activity (Manzolillo 1984). Similarly, natal male vervets transfer to groups from which they receive the least aggression from both males and females during intergroup encounters (Cheney 1983b).

Presence of Individuals from the Natal Group

Among vervets, the majority of males transferring for both the first and second time join groups that have previously received males from their former group (Cheney and Seyfarth 1983; Cheney, pers. comm.). Similar observations have been recorded for natal male long-tailed macaques (van Noordwijk and van Schaik 1985) and rhesus macaques in Puerto Rico (Boelkins and Wilson 1972; Drickamer and Vessey 1973; Meikle and Vessey 1981; Colvin 1983b). However, male rhesus in Nepal did not show this tendency (Melnick, Pearl, and Richard 1984). Cheney (1983b) has suggested that natal groupmates are potential allies, and that their presence is particularly important for the successful transfer of males not yet fully grown and subject to female as well as male aggression. The best data on the formation of alliances among males from the same natal group come from one study of rhesus macaques in which males tended to join maternal siblings in new groups. These males supported each other in alliances against other males, thereby improving their rank and probably their reproductive success (Meikle and Vessey 1981). In long-tailed macaques, although males that achieved top rank in their new group invariably did so without help from other males, the only males that succeeded were those that had been resident in the group for at least 6 months (van Noordwijk and van Schaik 1985). Van Noordwijk and van Schaik suggest that knowing the characters of the other males in the group is crucial for success in achieving high dominance rank. Thus there might be some advantage for young males (the only age class involved in top dominance takeovers) to immigrate into groups with known males. In contrast to these species, male olive baboons at Gombe often received the most intense aggression from males born in their natal group and were thus unable to join the same group (Packer 1979a).

Number and Quality of Mates

Whereas attraction to unfamiliar mates is a predominant feature of natal dispersal in many species, attraction to groups providing increased mating opportunities also occurs and seems to be of particular importance in the case of secondary transfer. Male olive baboons transferring for a second or subsequent time showed a significant tendency to enter groups that had more estrous females than the males' previous group (Packer 1979a). Most other studies have not separated natal from secondary transfer, but males have been observed to move to groups with greater adult-female-to-male sex ratios in ring-tailed lemurs (K. C. Jones 1983), Barbary macaques (Paul and Kuester 1985), and rhesus macaques at La Parguera (Drickamer and Vessey 1973). This is not the case, however, for rhesus macaques at Cayo Santiago (Boelkins and Wilson 1972), vervets (Henzi and Lucas 1980; Cheney 1983b), or non-natal long-tailed macaques (van Noordwijk and van Schaik 1985). The relative intensity of male-male competition in different neighboring groups may also influence immigrants' movements. For example, 10 of 13 adult male vervets in Amboseli improved their ranks and possibly also their reproductive success when they transferred; it seems likely that such individuals may have monitored male dominance relations in other groups and timed their transfers accordingly (Cheney and Seyfarth 1982b, 1983).

While males often move to groups with more estrous females or with a more skewed sex ratio than their previous group, female gorillas seem to prefer joining solitary males or those with only a few other females (Harcourt 1978b). Female gorillas, chimpanzees, and red colobus also seem to prefer males that are able to defend them successfully against other males. For example, 5 of 16 female gorillas transferred following the death of the dominant male silverback of their previous group (chap. 14). Similarly, when all but one of the adult males in one group of chimpanzees disappeared, all cycling females transferred to a larger neighboring group (Nishida and Hiraiwa-Hasegawa 1984). In red colobus, 10 females transferred to a group that had been taken over by a male whose subsequent tenure was 21 months, while 6 females emigrated from a group following the takeover of a male that remained for only 4 months (Marsh 1979a). Since male replacement carries the risk of infanticide, Marsh suggests that females may be able to assess the ability of males to maintain control of a group and transfer to those males that are best able to protect them. Extragroup female purple-faced langurs have also been observed to enter a group immediately after a takeover by a new male (Rudran 1973a).

Habitat Quality

Although it seems likely that an individual's choice of group or range might be influenced by the quality of environmental resources, data are largely lacking on this point, except in cases of habitat destruction. It seems likely that the quality of environmental resources would be more important to dispersing females than to males, but there are no relevant data from species showing female-biased dispersal. However, some evidence shows that range quality is unimportant to dispersing males in at least some species. Access to permanent water sources results in lower mortality rates in vervets, but several males transferred from groups that had access to permanent water sources to groups that did not (Cheney and Seyfarth 1983). Similarly, over 90% of male Japanese macaques leave heavily provisioned groups for groups or areas where there is no provisioning (Sugiyama 1976).

EVOLUTIONARY EXPLANATIONS FOR DISPERSAL

The proximate causes of dispersal obviously differ in different primate species and among individuals of different age or sex within the same species. For example, a male langur that is evicted from his natal group following a male takeover emigrates for quite a different reason than does an adolescent female chimpanzee. Consequently, several evolutionary forces are likely to have contributed to the broad patterns of dispersal in primates. Three major explanations for dispersal have been proposed: that individuals disperse to seek better or unexploited resources (Howard 1960; Lidicker 1962), that dispersal is a consequence of intrasexual competition (e.g., Waser and Jones 1983; Moore and Ali 1984), and that individuals disperse to avoid the deleterious genetic consequences of close inbreeding (e.g., Darling 1937; Itani 1972; Wilson 1975; Packer 1979a; chap. 11).

Because most primate populations appear to be at or near carrying capacity (e.g., Dittus 1977; Chivers and Raemaekers 1980), it is unlikely that dispersing primates can regularly find sufficiently unexploited resources to offset the cost of emigration. Although there are examples of dispersal resulting from severe habitat degradation (bonnet and long-tailed macaques), and during population expansion into previously marginal habitats following an improvement in ecological conditions (red howlers: Rudran 1979; Crockett 1984), such examples are unusual. Dispersal that results from attraction to more or better mates seems more common. However, in no species does this factor appear to be the predominant cause of natal (as opposed to secondary) dispersal, nor is it sufficient to account for the dispersal of most or all individuals of one sex. If increased reproductive opportunities were the primary cause of dispersal, individuals might be expected to forego emigration whenever their natal group contained a large number of animals of the opposite sex.

Intrasexual competition appears to be a major cause of dispersal for males in species that typically live in one-male groups, as well as for female howlers and both male and female gibbons. In addition, a proportion of secondary transfers by males in multimale species appears to result from attraction to groups where they can compete more successfully (vervets and possibly toque macaques). It is also possible that voluntary emigration by immatures to an extragroup phase sometimes forestalls inevitable eviction in such species as patas and geladas, and thus that dispersal results indirectly from competition. However, competition cannot explain dispersal in species where individuals voluntarily leave their natal group and directly enter new groups in spite of receiving high levels of aggression (male Japanese, rhesus, and toque macaques; olive baboons, vervets, and ring-tailed lemurs; female chimpanzees).

Voluntary natal transfer, especially in species where dispersing animals are attracted to unfamiliar mates (male olive baboons, vervets, long-tailed and toque macaques; female chimpanzees and gorillas), is more consistent with the inbreeding avoidance hypothesis. There is considerable evidence from captive primates that the offspring of close genetic relatives suffer from reduced viability (Ralls and Ballou 1982a) and that the severity of inbreeding depression is similar to that found in other groups of mammals (Ralls, Brugger, and Ballou 1979; Ralls and Ballou 1982b). Data on the severity of inbreeding depression in wild populations has proved difficult to collect because emigration greatly reduces the probability of close inbreeding and few field studies have been able to document the effect of inbreeding on offspring. The only quantitative data come from Packer's (1979a,

1985) study of olive baboons, where the eight presumed offspring of a single male breeding with females related to him by an estimated mean of 0.10 suffered 25 to 40% higher mortality prior to 1 month of age than offspring of males that were less closely related to their mates. In this population less than 17% of males could have died while transferring; as a result, the costs of inbreeding probably outweighed the costs of dispersal (see also Bengtsson 1978).

Even in the absence of inbreeding depression, another disadvantage of close inbreeding may be an increased susceptibility to disease. For example, cheetah (*Acinonyx jubatus*) show extremely high levels of homozygosity and reduced fecundity typical of inbred strains of domesticated mammals and are therefore assumed to have gone through an extreme population bottleneck (O'Brien et al. 1983). Apparently as a result of high levels of homozygosity, cheetah are more susceptible to infection and are more severely affected by particular pathogens than are more outbred felids (O'Brien et al. 1985).

Genetic studies of free-ranging primates have shown that levels of heterozygosity within social groups are very high, which has been cited as evidence that close inbreeding is avoided (see chap. 11). However, high levels of heterozygosity can result from high levels of dispersal, regardless of the causes of dispersal.

Two features of dispersal may have evolved as inbreeding avoidance mechanisms. First, dispersal is highly sex biased (table 21-1). Across polygynous species there is an inverse relationship between the extent of dispersal by each sex (fig. 21-3). Thus in most species the probability that members of one sex will disperse depends on the likelihood that the breeding members of the opposite sex are their close relatives. Note, however, that this pattern is not found in howlers where both sexes disperse at high rates. Second, even though secondary transfer is common in many species, individuals seldom return to the natal group. In addition, natal dispersal occurs before breeding (table 21-1). As a result, close genetic relatives of the opposite sex seldom reside in the same group as adults.

Although these factors decrease the probability of close inbreeding, they do not eliminate it entirely. Most individuals initially transfer to adjacent groups (Shields 1982), and individuals also may preferentially join groups with which there has been an exchange of members in the past (Cheney and Seyfarth 1983; see also chap. 11). However, the new group will nevertheless contain a lower proportion of close genetic relatives than the natal group, and thus any genetic costs of nonrandom transfer may be outweighed by the benefits of joining familiar and adjacent groups (Cheney and Seyfarth 1983).

In some cases, the costs of dispersal or the benefits of philopatry may be particularly high for certain individuals, with the result that they do not disperse at all. In rhesus and Japanese macaques, sons of very-high-ranking females occasionally remain in their natal groups, and there is evidence from rhesus macaques that female kin can assist such males in attaining a higher rank in their natal group than they are likely to achieve immediately in another group (Chapais 1983d). Such males do not mate with females of their matriline (Enomoto 1974; Chapais 1983c), which limits the number of females with whom they mate (Packer 1985); they may also incur some costs of close inbreeding by mating with patrilineal siblings. However, their high reproductive success may offset these costs. It should be emphasized, however, that even in these species most sons of high-ranking females emigrate (Sugiyama 1976; Colvin 1983b). Similarly, male gibbons and siamangs have been observed to remain in or return to their parents' territory following their father's death. Both Tilson (1981) and Chivers and Raemaekers (1980), however, emphasize that it is very difficult for individuals in these species to establish territories and that an incestuous union may represent a transitional stage during which adults that are nearing the end of their re-

productive lives pass on their territory to one of their offspring (see chap. 12).

Finally, there may be intraspecific variation in costs of dispersal or of inbreeding resulting from habitat differences. Male yellow baboons at Amboseli show a greater tendency to remain in their natal group than those at Mikumi (table 21-1); this is associated with a much lower population density at Amboseli (Altmann 1980; Rasmussen 1981). Costs of dispersal from predation may be higher in Amboseli because the distance between adjacent groups is greater. In Japanese macaques and olive baboons there is a tendency for males in larger groups to remain longer in their natal group (table 21-1). Because average coefficients of relatedness decrease with increasing group size (Bertram 1976), such males are likely to have a much greater range of suitable mates in their natal group.

Although the inbreeding avoidance hypothesis can account for sex-biased natal dispersal (fig. 21-4), it makes no predictions about which sex should disperse, nor has inbreeding avoidance been shown to be important in sec-

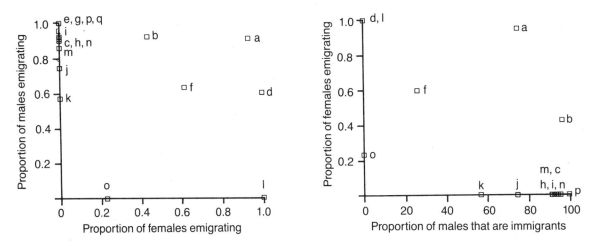

FIGURE 21-3 (*above left*). Extent of male versus female emigration in polygynous species. There is an inverse relationship between the extent of natal emigration by each sex in polygynous species (across species: $r_s = -0.613$, $N = 17$, $p < 0.02$; across genera: $r_s = -0.836$, $N = 9$, $p < 0.01$). a = *Alouatta palliata*; b = *A. seniculus*; c = *Cercopithecus aethiops*; d = *Colobus badius*; e = *Erythrocebus patas*; f = *Gorilla gorilla*; g = *Macaca fascicularis*; h = *M. fuscata*; i = *M. mulatta*; j = *M. sinica*; k = *M. sylvanus*; l = *Pan troglodytes*; m = *Papio c. anubis*; n = *P. c. cynocephalus*; o = *P. c. hamadryas*; p = *P. c. ursinus*; q = *Presbytis entellus*. SOURCES: Data are taken from all studies in table 21-1 for which there are data on both the number of transfers and the proportion of each sex that emigrate. Data for each species and subspecies are weighted means across study sites. Correlations are also calculated from genus averages since most species of *Macaca* and *Papio* showed similar patterns.

FIGURE 21-4 (*above right*). Female emigration versus availability of unrelated mates in natal group. The probability that females emigrate from their natal group is highest where the proportion of immigrant males in that group is lowest (across species $r_s = -0.666$, $p < 0.005$; across genera: $r_s = -0.920$, $p < 0.002$). The proportion of males that are immigrants is the number of male immigrants divided by the total number of male immigrants plus males that remained in or returned to their natal group as adults. The proportion of male immigrants in a female's natal group is highly correlated with the extent of male emigration ($r_s > 0.95$). However, in species where emigration is not followed by transfer into another group (e.g., male gorillas and red colobus at Kibale), the extent of emigration overestimates the proportion of immigrant males. See figure 21-3 for key. SOURCES: See figure 21-3.

ondary dispersal. Secondary dispersal sometimes coincides with the sexual maturation of offspring (Itani 1972; Henzi and Lucas 1980), and Henzi and Lucas have shown that average male tenure length is shorter than the age of maturation of females in several species. However, there is no direct evidence that secondary dispersal results from avoiding daughters, and Packer (1979a) found that father-daughter matings in olive baboons were avoided without dispersal.

WHICH SEX DISPERSES?

Most primates resemble other mammals and differ from birds in showing male-biased dispersal (Greenwood 1980). Since male-biased dispersal is strongly linked to polygyny (monogamous species show unbiased dispersal), we will first discuss various explanations for this trend and then discuss polygynous species that show female-biased dispersal.

In most polygynous species, males provide little direct paternal care. Female reproductive success appears limited primarily by nutritional constraints, while male reproductive success seems limited more by the numbers of females they can inseminate (Trivers 1972; Bradbury and Vehrencamp 1977; Wrangham 1980). Studies of various mammals suggest that foraging efficiency is increased by familiarity with the area and the phenology of food sources and that such familiarity is best gained by remaining in the natal area (Waser and Jones 1983). As a result, females may benefit more than males from philopatry. Wrangham (1980; chap. 23) has suggested that another advantage of female philopatry may be the cooperative defense of food sources by female kin. However, since female philopatry also characterizes most solitary mammals, it is possible that philopatry preceded gregariousness in the evolution of sociality (Waser and Jones 1983).

There are two additional reasons why males in polygynous species might be more likely to disperse than females. First, male reproductive success depends more than female reproductive success on access to mates, particularly when male parental care is slight. Males may therefore be more likely to benefit from moving to groups or areas where there are large numbers of mates (Clutton-Brock and Harvey 1976; Packer 1979a). Second, intrasexual competition for mates is probably more intense among males than among females in polygynous species (Trivers 1972); males may therefore be more likely to be evicted and to have to disperse to find mates (Clutton-Brock and Harvey 1976; Greenwood 1980; Dobson 1982).

If female primates usually benefit from remaining in their natal area, and if males are more likely to benefit from dispersal or be forced to disperse, then the exceptions to male-biased dispersal require explanation. In polygynous species in which females choose males on the basis of the quality of their territories, philopatry

may be important to male reproductive success (Greenwood 1980). While such "resource defense polygyny" is common in birds, it has not yet been described in any primate species, and thus this explanation for female-biased dispersal does not seem applicable. Wrangham (1980; chap. 23) has argued that species characterized by female-biased dispersal are ones whose food is distributed in such a way that it is not easily defended by groups of females. For these species, therefore, the benefits of female philopatry may not be great, and this may provide males with the opportunity to remain in their natal area. Nevertheless, there is evidence that females in some of these species may prefer to remain in their natal group or area. Both female gorillas (Harcourt 1978b) and chimpanzees (Pusey 1979, 1980; Goodall 1983) return to their natal group and area with greater frequency than do males in male-dispersing species. Females in both species also sometimes form cooperative coalitions against other females (Pusey 1983; chap. 14), and three female chimpanzees with high-ranking mothers are known to have settled in their natal range (Pusey 1983; Goodall 1986). Even female hamadryas baboons often rejoin their initial female companions (Sigg et al. 1982; Abegglen 1984).

There are five well-studied polygynous species that show female-biased or unbiased dispersal, and in three of these this pattern seems imposed on females by the reproductive strategies of males. In chimpanzees, overlapping female ranges may provide males with unusually high benefits from remaining with their kin and cooperatively defending a large joint range (Wrangham 1979a; chap. 15). The advantage to males from this cooperation may exceed the advantages of philopatry to females and "force" females to emigrate to avoid close inbreeding (Clutton-Brock and Harvey 1976; Pusey 1979; fig. 21-3). That male cooperation is related to female dispersal is supported by a comparison of chimpanzees with orangutans. Female orangutans closely resemble female chimpanzees in diet and ranging behavior. Unlike chimpanzees, however, orangutans appear to be characterized by greater male than female dispersal and a notable lack of male-male cooperation (chap. 13). A major difference between the two species seems to be in the energetic costs of locomotion, which may prevent groups of male orangutans from patrolling large areas as effectively as chimpanzees (Wrangham 1979a). It is therefore possible that female orangutans are able to show higher levels of natal philopatry than female chimpanzees because of lower benefits of philopatry to male orangutans.

Male hamadryas cooperate with males of their clan to prevent nonclan males from abducting their females (Sigg et al. 1982). The advantages to males of remaining near male kin may therefore again be high, and females, because they are abducted from their natal units as juveniles, may have little choice over whether or not to disperse. In gorillas, female-biased transfer may result at least in part from the long tenure length of males (chap. 14). Females always emigrate rather than mate with their father, but sometimes remain if the group contains a half-sibling or unrelated male (Fossey 1983; chap. 14). Although female dispersal in hamadryas and gorillas may occur as a result of male breeding strategies, it is not yet clear why male strategies in these species differ from those in similar species (e.g., hamadryas compared with geladas; gorillas compared with any other species living in one-male groups).

In at least two other species showing female-biased or unbiased dispersal—red colobus and mantled howlers— no complete explanation exists for female dispersal. Both species are folivorous, and in many folivorous mammals female philopatry is less common than in other mammals (Waser and Jones 1983; Moore 1984). Possibly because folivorous food sources are abundant, evenly distributed, and easy to exploit, there may be less advantage to remaining in the natal area or to cooperating with kin to defend food patches (Wrangham 1980). However, a number of folivorous primates show male-biased dispersal (e.g., black-and-white colobus, gray langurs, purple-faced langurs). Moreover, although red colobus males cooperate (Struhsaker and Leland 1979), it is not clear whether males in multimale groups are philopatric because of the advantages of cooperation or whether male cooperation is a by-product of female dispersal, which would allow males to remain in their natal group without incurring the costs of close inbreeding (see also Marsh 1979a).

CONCLUSION

The past two decades of research have greatly increased our knowledge of the causes and consequences of dispersal and intergroup transfer in primates. There is strong evidence that the overall patterns of primate dispersal result primarily from the avoidance of close inbreeding and from intrasexual competition. However, although the relationship between philopatry by one sex and dispersal by the opposite sex in the same species is now clear, existing hypotheses do not satisfactorily explain the distribution of female philopatry across species. Future studies could provide considerable insights by measuring the relative costs and benefits of philopatry and dispersal to each sex.

SUMMARY

In most primate species the majority of males permanently leave their natal group or area, whereas females typically remain. In a small number of species both sexes disperse, and in a few, females disperse rather than males. Some individuals may emigrate several times, but return to the natal group is rare. There are several proximate causes of dispersal. Abduction of females by males is only common in hamadryas baboons. Eviction by like-sexed individuals occurs in monogamous species and by males in species living in one-male groups. Sexual attraction to unfamiliar mates is common in many species. In most species, natal emigration typically coincides with sexual maturation, although there is considerable intraspecific variation. In some gregarious species, emigrating individuals transfer directly into other groups, but in others they spend some time as solitaries or as members of nonbreeding groups. All-male bands are more common in terrestrial species and may be a response to predation in open areas. Individuals transferring to other groups often receive more aggression in the new group. Several factors, including proximity, the presence of individuals from the natal group, and the number or quality of mates, can influence the choice of group in different species.

We examine three evolutionary explanations of dispersal: that individuals disperse to find better resources, to avoid intrasexual competition, or to avoid close inbreeding. Intrasexual competition is clearly involved in some natal dispersal and in secondary dispersal. Inbreeding avoidance accounts for the relationship between dispersal by one sex and philopatry of the opposite sex, and the rarity of return to the natal group. The predominant male bias in dispersal is similar to that found in other mammals and probably results because of greater advantages to males of dispersal and greater advantages to females of philopatry. The factors causing a reversal of this pattern in some species are not well understood.

22 | Interactions and Relationships between Groups

Dorothy L. Cheney

Most discussions of primate behavior focus on interactions that occur within the social group. Groups of primates do not exist in isolation from one another, however, and encounters between neighboring groups within local populations can be as complex and subtle as any that occur within groups. The relative friendliness or aggressiveness of such encounters varies widely, both within and between species, and it is difficult to generalize about their causes or the nature of their participants. Consider, for example, the intergroup encounters of vervet monkeys, gibbons, mangabeys, and gorillas—species whose behavior illustrates the diversity of interactions between primate groups.

East African vervet monkeys live in relatively small home ranges, which they aggressively defend against incursions by other groups. When another group is spotted at the border of the range, adult females often utter a particular call, alerting fellow group members to the other group's approach, while the alpha male displays by leaping through the trees. As the two groups come closer, juvenile males from the two groups may approach the border and play with each other. Occasionally, such friendly interactions also involve juvenile females and adults. More often, however, the encounter rapidly becomes aggressive. The groups' adult females, males, and juveniles line up against each other and chase, hit, and even bite one another (Cheney 1981, unpub. data).

Similar range defense occurs among most monogamous groups of gibbons. Gibbon pairs, using loud calls, displays, and occasional aggressive chases and physical contact defend their territories against incursions by neighboring groups (chap. 12). Calling and countercalling between neighboring pairs usually occur daily and attract both pairs to the boundary of their ranges, where aggressive chases may occur.

The behavior of these territorial species contrasts sharply with that of mangabeys, a forest cercopithecine that ranges over much larger areas than do vervets or gibbons. The ranges of mangabey groups overlap almost completely and are not defended. When two groups begin to approach each other, the adult males of each group often utter a characteristic loud call, which causes the groups to move away from each other. Aggressive intergroup encounters are rare. When they do occur, they usually concern access to a desirable resource such as a

fruiting tree, and they involve members of both sexes (Waser 1976).

In some species, adult males are the primary participants in intergroup interactions. When a group of mountain gorillas, for example, encounters another group or a solitary male, the resident silverback male often initiates an aggressive display that involves chest beating, branch breaking, and running charges. If a fight breaks out, only the adult males are involved; females and young actively avoid involvement in the interaction (Harcourt 1978b; chap. 14).

Clearly, the intergroup interactions of nonhuman primates are highly variable, and generalizations about such factors as the nature of resource defense and the relative roles of males and females are not immediately obvious. Attempts at cross-species comparisons are further confounded by a surprising lack of research on intergroup relations in many species, which is due primarily to the methodological difficulties of simultaneously studying social behavior in more than one group. This is particularly true for species such as baboons, whose intragroup behavior has been well studied, but whose large group size and home range make simultaneous observations of more than one group logistically difficult. As a result, accounts of intergroup encounters are often only descriptive or anecdotal.

Even when intergroup encounters have been described, definitions of what constitutes an encounter are often either vague or nonexistent and also differ considerably between study sites. For example, in species whose home ranges are large, intergroup encounters are often defined in terms of the distance separating two groups. Thus in their study of baboons, Cheney and Seyfarth (1977) defined an intergroup encounter as any approach of one group to within 500 m of another. This definition would clearly be inappropriate in the case of vervets, whose small home ranges cause them to be within 500 m of another group during most hours of the day. Perhaps as a result, intergroup encounters in species whose home ranges are small are usually defined in terms of physical or vocal contact (e.g., Cheney 1981).

The observer's definition of what constitutes an intergroup interaction clearly affects the rates at which groups are said to encounter each other as well as the frequency with which encounters are reported to involve physical contact. As a result, it is almost impossible to make cross-species comparisons about the relative aggressiveness of intergroup relations. Rates of encounters are also affected by observation conditions. In forest habitats, observers are frequently unable to determine when one group has approached to within 100 m of another, particularly when the majority of intergroup encounters take the form of long-range avoidance (e.g., see Kano's 1982a study of bonobos).

A final factor complicating the analysis of intergroup encounters concerns the fact that few descriptions differentiate among individuals. This affects assessment of both the rate and the function of intergroup interactions. For example, in many Old World monkey species, young males are the most frequent participants in intergroup encounters. Such interactions often precede the males' migration from the natal group and have different causes and consequences than interactions involving adult females, who do not migrate.

Cross-species comparisons of intergroup interactions are therefore hampered by both paucity of data and lack of comparable measures. The absence of standard methods and systematic studies seem particularly surprising given the ubiquity of intergroup conflicts in human and nonhuman primates.

In this chapter, I attempt to evaluate the causes and consequences of intergroup encounters in monkeys and apes. I begin with a description of the ways in which intergroup encounters are influenced by range size and the nature of resource defense. I then discuss sex differences in intergroup behavior that are apparently related to patterns of dispersal and resource defense. Finally, I discuss individual variation in intergroup interactions as well as the factors that affect relations between particular neighboring groups. Throughout, the term *range* is used to describe the area within which a group moves during a year and which may or may not be defended. *Territory* is reserved to describe the area that a group consistently attempts to defend against incursions by others. *Aggressive encounters* are defined as those involving threats, chases, hits, or bites between the members of different groups. The discussion concerns only group-living species. The interactions of solitary species are discussed in chapters 2 and 13.

VARIATION IN THE NATURE OF INTERGROUP ENCOUNTERS

Table 22-1 presents data (when reported) on average home range size, degree of range overlap, the frequency of intergroup encounters, and the presence or absence of range defense for a number of different species. Data on population densities may be obtained from chapters in part 1 (see also table 22-2). The list is not exhaustive, and it does not claim to include data from all sites. It should also be emphasized that when data are summarized in this manner, they frequently underestimate vari-

ability both within and between species. Nevertheless, a few generalizations emerge.

Range Defense

Range size is influenced by both dietary requirements and the distribution of food in space and time (chap. 17). If a group is able to find all the food it requires within a small geographical area, it often forages over a relatively fixed range. In contrast, when food resources are more patchily distributed, ranges tend to be larger and more variable (see Clutton-Brock and Harvey 1977b). Mitani and Rodman (1979) have shown that the presence or absence of territoriality is correlated with a group's ability to patrol its range on a regular basis. When a group is able to cover most of its range daily, defense is more likely than when home range size is larger than the average length of the daily foraging route.

Perhaps because large ranges are difficult and uneconomical to patrol, they often overlap extensively with those of other groups (>50%) (table 22-1). When overlap is great, aggressive encounters occur at low rates, and those that do occur usually concern access to a clumped, desirable resource, such as a fruiting tree, rather than a range boundary (e.g., capuchins, red howlers, baboons, mangabeys; see table 22-1 for references). Among many macaque species, aggressive intergroup encounters are most common in the areas of human provisioning. Away from temples or food bins, where resources are less concentrated, encounters generally involve passive avoidance (e.g., Southwick, Beg, and Siddiqi 1965; Kawanaka 1973; Lindburg 1977).

In contrast, in species whose ranges are small relative to the length of the daily foraging route, range overlap tends to be less than 50%. Intergroup encounters concern not just access to specific food resources, but defense of the entire range (e.g., ring-tailed lemurs, tamarins, dusky titis, redtail monkeys, blue monkeys, gibbon spp.; see table 22-1). Groups of vervet monkeys in Amboseli, Kenya, for example, defend all parts of their ranges against other groups. At any given time of year, however, the precise location of intergroup encounters along each group's border is determined by the location of water and particular fruiting or flowering plants.

Intergroup Dominance

When home ranges overlap extensively and are not defended, the aggressive defense of a particular resource may be more costly than the simple avoidance of other groups (Mitani and Rodman 1979). In such cases, intergroup competition is often mediated by the relative dominance ranks of the groups involved. Dominance is usually a function of group size and the number of adult males (e.g., baboons, macaque spp.; see table 22-1). Occasionally, however, intergroup dominance depends on more subtle factors, such as past relations among the male members of different groups (e.g., rhesus ma-

TABLE 22-1   Characteristics of Intergroup Encounters in a Number of Primate Species

Species	Study Site	No. Grps.	Range Size (km²)	% Overlap	No. Enc.	Rate of Enc.	Range De-fense?	Inter-grp. Dom?	Inter-grp. Call?	% Enc. Aggr.	Males Aggr? (%)	Females Aggr.? (%)	Occ. Friendly Enc? (%)	Sources
Lemur Catta														
(Ring-tailed lemur)	Madagascar	1	.06	Slight			+	−	+		+	+	+	Jolly 1966a; Klopfer and Jolly 1970
	Madagascar	12	.06–.23	Slight			+	−	+		−	+		Budnitz and Dainis 1975
L. fulvus														
(Brown lemur)	Madagascar	2	.01	Extensive			−	−		Rare			Mingling	Pollock 1979b
Indri indri														
(Indri)	Madagascar	2	.17	42		1/333 h	+	−	+	Rare	−	−		Pollock 1975, 1979b; chap. 3
Propithecus verreauxi														
(White sifaka)	Madagascar	5	.01–.03	31–100			Some	−			+	+		Jolly 1966a
	Madagascar	2	.07–.09	49–76	88		Some	−		31–69	+	+		Richard 1977
Saguinus imperator														
(Emperor tamarin)	Peru	4	.3	Slight		1/7 d	+	−	+ (a)		+	+	−	Terborgh 1983 (e)
S. fuscicollis														
(Saddleback tamarin)	Peru	3	.3	Slight		1/7 d	+	−	+ (a)		+	+	−	Ibid.
Callicebus moloch														
(Dusky titi monkey)	Colombia	3	.005	7–20		1/.9 d	+	−	+ (a)		+	+	−	Mason 1968
	Colombia	3	.03–.04	0–16		1/.8–2.6 d	+	−	+ (a)		+	+	−	Robinson 1979c, 1981b (a)
C. torquatus														
(Yellow-handed titi monkey)	Peru	1	.2				−		+ (r)				−	Kinzey and Robinson 1983
Saimiri sciureus														
(Squirrel monkey)	Peru	4	>2.5	Extensive			−		−				Mingling	Terborgh 1983
Alouatta seniculus														
(Red howler)	Venezuela	4	.04–.07	28–63	69	1/2.5 d	−	−	(r)		+	+	+	Sekulic 1982a, 1982b, 1982c (a, d)
A. palliata														
(Mantled howler)	Barro Colorado	5	.11	63			−		+ (r)					Chivers 1969
	Barro Colorado	2	.31	100			−	+	+ (r)		+			Milton 1980
Ateles belzebuth														
(Long-haired spider monkey)	Venezuela	3	2.6–3.9	20–30			+		+		+	−		Klein and Klein 1975
Lagothrix lagotricha														
(Humboldt's woolly monkey)	Colombia	5	4.0	Extensive										Nishimura and Izawa 1975
	Peru	4		Extensive			−				+	+	+	Durham 1975
Cebus olivaceus														
(Wedge-capped capuchin)	Venezuela	9	2.75	100		1/26 h	−	+	+		+	+		Robinson, in press *b* (c)
C. apella														
(Brown capuchin)	Peru	5	.8	Extensive		Daily	−		+	Rare			Mingling	Terborgh 1983
C. albifrons														
(White-fronted capuchin)	Peru	4	>1.5	Extensive			−		+ (r)	Rare				Ibid.
Papio cynocephalus														
(Baboon)	Transvaal, S. Africa	2	13–23.3	Extensive			−				+	−		Stoltz and Saayman 1970
	Transvaal, S. Africa	4	25.5–29	Extensive		1/13 d	−	−		Rare	+	−	+	Anderson 1981
	Craddock, S. Africa	1	3.0	100	174	1/1.3 d	−	+			+	Rare	+ (7)	Cheney and Seyfarth 1977 (c)
	Amboseli, Kenya	1	24.0	100			−	−			+	+	+	Altmann and Altmann 1970; Shopland 1982
	Nairobi, Kenya	8	5.2–40.2	Extensive			−			Rare	+			DeVore and Hall 1965
	Gombe, Tanzania	3	3.9–5.2	Extensive			−	+			+	−	+	Packer 1979a (c); Ransom 1981
	Namibia	3	4.0–9.4	Slight	48		+	−		65	+ (65)	−	+ (5)	Hamilton, Buskirk, and Buskirk 1975a, 1976
	Okavango, Botswana	6	2.1–6.5	10–18	21		+	+		100	+ (100)	−	+ (29)	Ibid.

TABLE 22-1 *(continued)*

Species	Study Site	No. Grps.	Range Size (km²)	% Overlap	No. Enc.	Rate of Enc.	Range Defense?	Inter-grp. Dom?	Inter-grp. Call?	% Enc. Aggr.	Males Aggr.? (%)	Females Aggr.? (%)	Occ. Friendly Enc? (%)	Sources
Macaca mulatta (Rhesus macaque)	Dehra Dun, India	9	1–15	40–100	142	1/1 h– 1/2.3 d	Some	+	–	37	+ (24)	+ (14)	+ (7)	Lindburg 1971, 1977
	Uttar Pradesh, India	3		80–90		1/3.5 d	Some	+	–					Southwick, Beg, and Siddiqi, 1965 (e)
	Nepal	15	.03–.2	20				+			+	+	Mingling	Teas et al. 1980
	Cayo Santiago	1		Extensive		1/1.3 h	–	+	–	35–45	+	+	+	Hausfater 1972 (e)
	La Cueva	3		Extensive		1/2.6 h	–	+	–	48	+	+	+	Vessey 1968; Gabow 1972
M. fuscata (Japanese macaque)	Japan	13	.24–26.7	10–50	95			–	+	–	7	+	+	+	Kawanaka 1973; Taka-saki 1981
M. radiata (Bonnet macaque)	India	1	.4	Extensive				–	+	–	Rare	+			Sugiyama 1971
M. fascicularis (Long-tailed macaque)	Bali	6		Extensive				–	+			+	+	+	Angst 1975
M. sylvanus (Barbary macaque)	Morocco	1			30	1/21 h			+	–	50	+ (50)	–	+	Deag 1973 (b)
Cercopithecus aethiops (Vervet)	Amboseli, Kenya (1964)	5	.18–.96	11–33		1/5 d	+	–	+		+	+		Struhsaker 1967b, 1967d
	Amboseli (1977–78)	3	.22–.34	26–42	234	1/1.8 d	+	–	+	47	+ (38)	+ (15)	+ (4)	Cheney 1981, unpub. (d, e)
	Amboseli (1980)	3	.22–.34	26–42	114	1/1.9 d	+	–	+	63	+ (45)	+ (36)	+ (36)	Ibid.
	Senegal	1	1.78	10	27	1/4 d	+	–	+	48	+ (48)	+ (11)	–	Harrison 1983b (d, e)
	Bakossi, Cameroon	2	.12–.15	0	2	1/43 h	+	–	+		+	–	–	Kavanagh 1981
	Buffle Noir, Cameroon	1	1.0	6			+	–	+		+	–	–	Ibid.
	Kalamahoue, Cameroon	1	.56	18			–		+	Rare	+	–	Mingling	Ibid.
C. ascanius (Redtail monkey)	Kibale, Uganda	1	.2–.3	Slight			+	–	+		+	+		Struhsaker and Leland, 1979; Struhsaker 1980
	Kakamega, Kenya	3	.27–.55	Slight			+	–	+		+	+		Chap. 9
C. mitis (Blue monkey)	Kibale, Uganda	5	.61	Slight			+	–	+		+	+		Rudran 1978; Struhsaker and Leland 1979
	Kakamega, Kenya		.23	Slight			+	–	+		+	+		Chap. 9
C. neglectus (De Brazza's monkey)	Gabon	6	.04–.10	Slight						Rare			Mingling	Gautier-Hion and Gautier 1978
C. talapoin (Talapoin)	Gabon	1	4.0	Slight	0	0								Gautier-Hion 1971b
Cercocebus albigena (Gray-cheeked mangabey)	Kibale, Uganda	3	4.1	72		1/120 d	–	–	+ (r)	Rare	+	+		Waser 1976; Struhsaker and Leland 1979
Erythrocebus patas (Patas monkey)	Laikipia, Kenya	2	38.5	Extensive			Some	+			+	+		Chism, Rowell, and Olson 1984; chap. 9
Colobus guereza (Black-and-white colobus)	Kibale, Uganda	1	.15	74	65	1/1.4 d	–	+	+	14	+	+	+	Oates 1977c (a)
	Chobe, Uganda	1	.10	Slight			+	–	+					Ibid.
	Budongo, Uganda	5	.14	Slight			+	–	+		+	+		Marler 1972
C. badius (Red colobus)	Kibale, Uganda	3	.35–.71	Extensive		1/2 d	–	+	+	7–36	+ (7–36)	–		Struhsaker 1975, 1980; Isbell 1984 (a)
	Tana, Kenya	1	.09	33	15	1/11.7 d	–	+		27	+ (27)	–		Marsh 1979a
Presbytis senex (Purple-faced langur)	Srilanka	56	.03–.07	Slight			+		+		+			Rudran 1973a

TABLE 22-1   *(continued)*

Species	Study Site	No. Grps.	Range Size (km²)	% Overlap	No. Enc.	Rate of Enc.	Range Defense?	Inter-grp. Dom?	Inter-grp. Call?	% Enc. Aggr.	Males Aggr? (%)	Females Aggr? (%)	Occ. Friendly Enc? (%)	Sources
P. entellus														
(Gray langur)	Orcha, India	3	3.9	Slight			−	+	+	0	−	−	Mingling	Jay 1965; Hrdy 1977
	Kaukori, India	1	7.8				−	+	+	0	−	−	Mingling	Jay 1965; Hrdy 1977
	Dharwar, India	6	.10−.33	Slight		Daily	Some	Some	+		+	+	+	Yoshiba 1968
	Mt. Abu, India	2	.3	Slight	26	1/30 h	+	−	+	Most	+	+ (58)	+	Hrdy 1977 (e)
	Sariska, India	2	.1−1.0	Extensive	6	1/28 h	−		+	33	+ (33)			Vogel 1975; Hrdy 1977
	Jodhpur, India	25	.6−1.3	Extensive			−				+	+		Mohnot 1980; Hrdy 1977
	Srilanka	4	.32−1.3	Slight	31	1/10 h	+	Some	+	84	+	+	+	Ripley 1967
P. aygula														
(Leaf monkey)	West Java	9	.14−.38	Slight	47		+	−	+	13−69	+ (13−69)	+ (3)		Ruhiyat 1983 (d, e)
P. johnii														
(Nilgiri langur)			.64−2.6	Slight			+		+		+		+	Poirier 1968
P. potenziani														
(Mentawai langur)	Pagai Islands	9	.15−.22	Slight			+	−	+		+	+		Tilson and Tenaza 1976 (d, e)
Hylobates klossi														
(Kloss's gibbon)	Siberut Islands	13	.05−.08	Slight		1/2.5 d	+	−	+ (a)		+	+		Tenaza 1975 (e); Tilson 1981
H. lar														
(Whitehanded gibbon)	Tanjong Triang, Malaysia	4	.2−.4	Slight	126	1/2 d	+	Some	+ (a)		+	−	−	Ellefson 1974 (e)
	Kuala Lompat, Malaysia		.54			1/8.4 d	+		+ (a)		+	+		Chivers and Raemaekers 1980 (e)
H. agilis														
(Agile gibbon)	Malaysia	7	.29	24		1/1.8 d	+	−	+ (a)		+	+ (10)	−	Gittins 1980 (e)
H. syndactylus														
(Siamang)	Malaysia	2	.32	30	7	1/7.3 d	+	−	+	28	+ (28)	−		Chivers 1974
H. moloch														
(Moloch gibbon)	Java	5	.17	8		1/2.2 d	+		+	8	+ (8)	−		Kappeler 1984b (d)
H. pileatus														
(Pileated gibbon)	Thailand	4	.36	25	21	1/5.6 d	+		+		+	+ (5)		Brockelman and Srikosamatara 1984 (e)
Pan troglodytes														
(Chimpanzee)	Gombe, Tanzania	2	10.0−13.0	Slight			Some	+	+		+	Rare	+	Wrangham 1977; Goodall et al. 1979
	Mahale, Tanzania	2	10.0−13.0	33−50			−	+	+		+	−	+	Nishida 1979
P. paniscus														
(Bonobo)	Wamba, Zaire	4	30.0−60.0	40−64	3		−	+	+					Kano 1982a
Gorilla gorilla b.														
(Mountain gorilla)	Rwanda	2	4.9−8.1	Extensive	35	1/8.2 d	−		+	48	+ (48)	−		Fossey and Harcourt 1977; Harcourt 1978b, pers. comm.; Fossey 1979

NOTES: Rate of Enc.; d = days, h = observer (not daylight) hours; Intergrp. Call: a and r indicate that calls function to attract or repel other groups; % Enc. Aggr.: no number is given if encounters are defined in terms of the presence of aggression. In Males Aggr. and Females Aggr., numbers indicate the proportion of encounters in which males or females showed aggressive behavior. Plus (+) and minus (−) indicate that authors noted the presence or absence of aggression but did not provide specific figures. Definitions of intergroup encounters are listed after each reference: (a) = approach to within 50 m or less; (b) = approach to within 100 m; (c) = approach to within 200 m; (d) = vocal or visual display; (e) = aggressive or friendly interactions. Letters are given only when authors provide specific definitions; most other authors imply that they define encounters in terms of aggressive behavior.

caques: Gabow 1972; red colobus: Struhsaker and Leland 1979).

Intergroup dominance hierarchies are less common when groups are territorial. In such cases, the outcome of intergroup encounters is typically influenced by the location of the encounter, with the resident group usually driving out the intruding one (e.g., vervets: Struhsaker 1967b, 1967d; Cheney 1981). This is not to say, however, that the boundaries of ranges are immutable. Groups occasionally expand their ranges at the expense of their neighbors', and in such cases the success of such expansion may be determined by the relative sizes of the groups or the dominance or fighting ability of a particular male or female (vervets: Cheney, unpub. gibbon spp.: Ellefson 1974; Tilson 1981).

Among species characterized by male dispersal, one-male groups (e.g., gray langurs, black-and-white colobus) are typically territorial, while multimale groups are not (e.g., baboons, mangabeys, macaque spp.; vervets are an exception to this generalization). It has been hypothesized that multimale groups have evolved in nonterritorial species because they allow females to compete more successfully for dominance with other groups (Wrangham 1980). Thus, for example, female baboons may be relatively unaggressive toward extragroup males in part because the recruitment of males aids females in intergroup competition. Across a number of species, however, the mean number of males per group is significantly related to the mean number of females (Terborgh 1983; Andelman 1986). This relation is generally considered to result primarily from mate competition, and indeed, in many species male migration patterns are influenced not just by the behavior of females but also by intergroup variation in the adult sex ratio (see chap. 21).

Aggressive and Friendly Interactions

Observers of nonterritorial groups often comment upon the relative lack of intergroup aggression, even when groups are in close proximity (e.g., bonnet macaques: Sugiyama 1971; baboons: DeVore and Hall 1965; Anderson 1981). Indeed, groups are occasionally observed to mingle peacefully in the same feeding tree, water hole, or sleeping site, even though long-term associations are rare (e.g., squirrel monkeys, woolly monkeys, brown capuchins, baboons, rhesus macaques, Barbary macaques; see table 22-1).

In contrast, when groups defend all or part of their home ranges, most intergroup interactions are characterized by aggression rather than by mutual avoidance. The absence in most studies of quantitative data on rates of intergroup aggression, however, precludes a direct comparison of rates of aggression between territorial and nonterritorial groups.

In many territorial species, females are frequently aggressive during intergroup interactions (e.g., ring-tailed lemurs, emperor and saddleback tamarins, vervets, red-tail monkeys, blue monkeys, Kloss's gibbons: table 22-1). Female aggression is more variable, however, in species that only infrequently defend ranges. In some of these, female aggression is common (e.g., macaque spp., red howlers, capuchins). In other species, however, males are the primary antagonists, perhaps because encounters more often concern mate, rather than food, defense (e.g., baboons, mountain gorillas).

Friendly interactions (play, copulation, grooming, touching, and mounting) between the members of different groups occur in both territorial and nonterritorial species. The primary participants are usually juvenile males, although members of all age-sex classes have been observed to engage in such interactions. Again, however, the lack of comparable quantitative data hinders any evaluation of the frequency of friendly interactions in territorial and nonterritorial groups.

Intergroup Calls

Some nonterritorial species, such as macaques, seem to lack specialized intergroup calls. Others, however, have evolved loud calls that aid in the regulation of intergroup spacing and cause groups to avoid each other. For example, the dawn chorus of roaring by mantled howler monkeys appears to cause groups to initiate their daily moves in the opposite direction of the nearest neighboring group (Chivers 1969; see also Terborgh 1983 for similar behavior in capuchins). These observations have also received experimental support. Playback experiments on mangabeys, siamangs, and yellow-handed titis have all shown that male loud calls cause neighboring groups to move away from the area in which the calls are played (Waser 1976; Chivers and MacKinnon 1977; Kinzey and Robinson 1983).

In contrast, most territorial species have specialized intergroup calls that function to attract, rather than to repel, neighboring groups, and are associated with aggressive range defense (table 22-1). Robinson (1981b) has demonstrated experimentally that intergroup calls cause dusky titis to approach the boundaries of their range, and playback experiments on gibbons and vervets have produced similar results (Chivers and MacKinnon 1977; Cheney and Seyfarth 1982b; Raemaekers, Raemaekers, and Haimoff 1984; Mitani 1985c; chap. 12). The relative frequency with which males and females call, however, is not obviously related to social structure. Males have been reported to call more than females in some monogamous species (e.g., whitehanded gibbons: Ellefson 1974; agile gibbons: Gittins 1984), as well as in some species that live in one-male, female-bonded groups (e.g., black-and-white colobus: Marler 1972; blue monkeys: Rudran 1978). In other monogamous species, though, females call as much as or more than males (e.g., dusky titis: Robinson 1981b; Kloss's gibbons: Tenaza 1975; moloch gibbons: Kappeler 1984b). Similarly, in multimale female-bonded groups of vervet monkeys, intergroup calls are given almost exclusively by females and juveniles (Cheney and Seyfarth 1982b).

Intraspecific Variation

While this discussion has focused primarily on interspecific differences, table 22-1 reveals almost as much intra- as interspecific variation in behavior. For example, at different study sites gray langurs have variously been described as having primarily peaceful and primarily aggressive intergroup encounters (Jay 1965; Hrdy 1977).

Variation in range size, population density, and frequency of intergroup aggression has also been reported for baboons, red colobus, black-and-white colobus, and rhesus macaques. Considerable variation in these factors may even exist over small geographical distances. Lindburg (1971, 1977), for example, reports that in a crowded forest reserve in India, intergroup encounters among rhesus macaques occurred at a rate of one per hour. In the nearby Siwalik Hills, however, where population density was lower, groups encountered each other at a rate of only once every 2.3 days.

Table 22-2 presents data on the correlation between population density, range size, and frequency of intergroup encounters for baboons, gray langurs, and vervets. Generally speaking, in areas of high population density groups range over smaller areas than in areas of low population density. Similarly, as among rhesus macaques, population density in baboons and langurs is generally inversely related to both the frequency of intergroup encounters and the proportion of encounters that are aggressive. This is not true, however, for vervets. For ex-

ample, over a number of different years the frequency of intergroup encounters in Amboseli, Kenya, was higher than in either Bakossi or Kalamoue, Cameroon, even though population density in Amboseli was considerably lower than in the latter two sites. Thus population density is unlikely to be the only factor affecting intraspecific variation in the frequency of intergroup encounters. Instead, encounters are also influenced by other, more complex, factors, such as the distribution and availability of food and water and the demographic composition of groups.

SEX DIFFERENCES

While ecological variables influence the economic costs of defending a range against other groups, other factors also affect the nature of intergroup encounters. In particular, the behavior of males and females often differs substantially, and these differences seem related to the types of resources that each sex attempts to defend.

In most mammalian species, females contribute more energy to the production and rearing of offspring than do

TABLE 22-2 Species Variation in Population Density, Range Size, and Frequency of Intergroup Encounters in Three Old World Monkey Species

	Population Density (km^2)	Aver. Range Size (km^2)	Rate of Enc.	% Enc. Aggr.
Baboons				
Transvaal	3.2	20.5–29.0	1/13 d	Rare +
Nairobi	3.9	22.9		Rare +
Namibia	5.3	6.7		65
Amboseli (1970)	6.3	24.0		
Okavango	24.0	4.5		100
		$r_s = -.700$		$r_s = .949$
Gray langurs				
Kaukori	2.7	7.8		0
Orcha	2.7–6.0	3.9		0
Jodhpur	18.0	0.6–1.3		
Mt. Abu	50.0	0.3	1/30 h	Most +
Dharwar	84.0–133.0	0.1–0.3	Daily +	
Sariska	104.0	0.1–1.0	1/28 h	33
Srilanka	100.0–200.0	0.25	1/10 h +	84 +
		$r_s = -.821*$	$r_s = .632$	$r_s = .791$
Vervets				
Senegal	14.3	1.78	1/4 d +	
Buffle Noir	18.0	1.0		
Amboseli (1980)	68.0	0.28	1/1.9 d	63
Amboseli (1977–78)	96.0	0.28	1/1.8 d	47
Amboseli (1964)	104.0	0.41	1/5 d	
Bakossi	113.0	0.12–0.15	1/43 h +	
Kalamoue	149.0	0.56	Rare	Rare
		$r_s = -523$	$r_s = -.638$	

SOURCES: See table 22-1. Additional data on baboons and langurs are derived from Anderson 1981 and Hrdy 1977. The rate of encounters in Amboseli (1964) is estimated from Struhsaker 1967b and 1967d.

NOTES: Spearman rank correlations refer to the correlation between population density and range size, the rate of intergroup encounters, or the frequency of aggressive encounters. Only study sites for which population density is known are shown. Plus (+) indicates scores that are tied with each other. For definition of (h) and (d), see table 22-1.
* = $p < .05$ (two-tailed).

males. Since female reproductive success appears limited primarily by energetic and nutritional constraints, it has been hypothesized that female grouping patterns are influenced primarily by the distribution of food. Thus, females should remain in their natal groups and cooperate with kin when kin-based alliances increase access to food patches (Wrangham 1980). In contrast, since the reproductive success of males is limited primarily by the availability of females, male grouping patterns are hypothesized to be dependent upon the distribution of females (Trivers 1972; Emlen and Oring 1977; Wrangham 1980; chap. 23).

The attempt to relate female grouping patterns to food distribution has led to several predictions about female responses to other groups (Wrangham 1980). First, when females remain in their natal groups with their kin, they should respond aggressively to other groups. Thus, females in most species of Old World monkeys are expected to be hostile toward females in other groups. Female aggression should occur at particularly high rates when groups inhabit relatively small ranges since such areas should be more economical to defend than larger ranges (Mitani and Rodman 1979). Females are predicted to be more aggressive toward females than toward males, both because their reproductive success is mediated by competition with other females for access to limited food resources and because males may aid females in dominating other groups. The latter factor may be particularly important in nonterritorial, multimale groups, where intergroup dominance is at least partially influenced by the number of males in each group.

Similar aggressive range defense is predicted for most monogamous species, where females range alone over relatively small areas (chap. 12). In contrast, when natural selection appears to have favored female dispersal over relatively large ranges, they should not attempt to defend food resources (Wrangham 1980). Females who form temporary aggregations with other females, as do chimpanzees and gorillas, are expected to ignore encounters with other groups and to take no part in range defense.

Finally, the intergroup behavior of males should primarily involve attempts to defend females from other males. Regardless of female association patterns, therefore, males may be expected to respond aggressively toward extra group males and to resist the immigration of additional males into their groups. Male aggression should be directed primarily against other males. Even when males participate in encounters involving territorial or resource defense, aggression toward females should occur at lower rates than aggression toward males.

In summary, a variety of theoretical arguments predicts that while the behavior of females during intergroup encounters will vary according to their dispersal and foraging patterns, the behavior of males will primarily concern the defense of females. In the following sections I discuss sex differences during intergroup encounters among primates of varying social structure and patterns of dispersal.

Species Characterized by Female Dispersal

Behavior of Females. In those primate species characterized by female dispersal, females often join groups that include few or no close kin and, as adults, interact at only low rates with other female group members (chimpanzees: chap. 15; gorillas: chap. 14; red colobus: chap. 8). Females in these species tend not to participate in intergroup interactions. Female gorillas, red colobus, and hamadryas avoid such encounters and do not respond aggressively to other groups (Harcourt 1978b; Struhsaker and Leland 1979; Fossey 1979; Abegglen 1984). Similarly, although female chimpanzees have been observed to attack females from other communities, such aggression appears rare, and most females are less hostile to extragroup females than males are to extragroup males (Goodall et al. 1979).

Behavior of Males. In contrast to females, male chimpanzees, red colobus, and gorillas are usually hostile toward the members of other groups, especially the members of their own sex. Such hostility seems to be ultimately related to the defense of females.

Among chimpanzees in Tanzania, and perhaps also among bonobos in Zaire, intergroup interactions appear to involve contests for dominance, with the members of smaller parties and communities giving way to larger ones (Goodall et al. 1979; Nishida 1979; Kano 1982a). Occasionally, males even seem to seek opportunities to attack unfamiliar males. At Gombe, for example, parties of males have been observed to go on raiding patrols into neighboring communities' ranges, and their attacks on males during such patrols have resulted in severe injury and even death (Goodall et al. 1979). Up to 25% of adult male mortality at Gombe is estimated to occur as a result of intercommunity hostility (Goodall 1983). Most such aggression is directed toward other males, and in at least one case males from a dominant community are believed to have killed all the males of a more subordinate community (Goodall 1983). Females have been observed to transfer to a more dominant group following the defeat or death of males in their prior community, which suggests that intercommunity conflicts for dominance are ultimately products of male-male competition for females (Nishida et al. 1985). Although males occasionally attack and even kill old, pregnant, or lactating females during raiding patrols, they have not been observed to attack young or estrous females, further supporting the hypothesis that intercommunity aggression is related primarily to the defense and acquisition of females (Wolf and Schulman 1984).

Intergroup aggression in male mountain gorillas also seems related to the acquisition and defense of females. While gorilla groups in the Virunga volcanoes of Rwanda do not defend fixed ranges, males do respond aggressively toward extragroup males, particularly when they are solitary (Harcourt 1978b). Approximately 48% of intergroup encounters escalate from chest-beating and branch-breaking displays into physical fights that can result in severe wounds. A male's successful defense of his group against other males not only potentially serves to deter rivals, but may also influence female choice, since females appear to prefer males who are able to defend them and their offspring against attacks by other males (chaps. 14, 31). Indeed, female transfer followed three of the four cases of cross-group infanticide documented in the Virunga population (Fossey 1983).

Male red colobus respond aggressively to other groups by means of loud calls and chases, whether or not they are defending fixed ranges (Struhsaker 1975; Marsh 1979a; Isbell 1984). At Kibale, Uganda, for example, ranges overlap extensively, and the outcome of intergroup encounters depends on the number, ages, and fighting abilities of each group's males. In contrast, red colobus ranges at the Tana River, Kenya, overlap only slightly, and intergroup dominance appears to be site-specific (Marsh 1979a; table 22-1).

The intergroup interactions of hamadryas baboons present a complex and intriguing problem because their multileveled society makes it difficult to define the group. Although males aggressively defend their females against other males, evidence shows some degree of cooperation among one-male units, clans, and even bands. Males from the same clan, who are suspected to be close relatives, travel and rest together and also occasionally groom each other (chap. 10; see also Kummer 1968; Sigg et al. 1982; Abegglen 1984). Clans in the same band have also been observed to cooperate in aggressive interactions against clans from other bands. Cooperative alliances may even occur among bands, even though the members of different bands seldom interact affinitively with each other. During Abegglen's (1984) study of three bands that shared a sleeping cliff, interband aggression was directed primarily at foreign bands that occasionally attempted to use the sleeping cliff. Five such instances were observed in 500 hours, and in four cases males from the resident bands drove away the foreign band. While the precise extent to which the resident bands acted in concert could not be determined, it appeared to the observer that the males were cooperating with each other in the defense of their females (Kummer, in Abegglen 1984). At the level of the one-male unit, the clan, and possibly even the band, therefore, hamadryas males form cooperative al-

liances with others. In no other primate species have similar interunit alliances been reported.

Monogamous Species

Most species that live in monogamous groups or groups containing only one breeding female are territorial. The successful production and rearing of offspring appears to depend on the exclusion of competitors from the territory's resources. Perhaps as a result, both the male and the female cooperatively defend their range against incursions by others.

Gibbons, tamarins, marmosets, and titi monkeys defend their ranges against other groups through calls, displays, and occasional chases and physical fights at the boundary of their territories (chaps. 4, 12). Such intergroup conflicts sometimes result in the creation of new territories. Tilson (1981), for example, has reported cases in which a pair of Kloss's gibbons slowly expanded their range at the expense of their neighbors', apparently in order to establish their offspring in adjacent territories (see also chap. 12).

The frequency of aggressive intergroup interactions among monogamous primates is highly variable (table 22-1). In some species, males appear to be more aggressive than females, although females do participate in intergroup interactions through calling (e.g., dusky titis: Robinson 1981b; moloch gibbons: Kappeler 1984b; cottontop tamarins: chap. 4). In other species, however, female aggression is more common. For example, in agile and Kloss's gibbons, females have been observed to attempt to expel individuals that intrude into their territories (Tenaza 1975; Gittins 1980; Tilson 1981). Most studies of monogamous species have suggested that animals tend to be most aggressive toward individuals of their own sex, perhaps because such individuals represent potential mate competition. In gibbons, this intrasexual aggression is sometimes even extended to maturing like-sexed offspring (chap. 12).

Only a few species of Old World monkeys are thought to be monogamous. Mentawai langurs, like gibbons and New World monkeys, defend their territories through loud calls and aggressive chases by the members of both sexes (Tilson and Tenaza 1976). The intergroup encounters of the de Brazza's monkey, however, may differ fundamentally from those of other species. Little is known about the behavior of this monkey, which appears to be monogamous in Gabon but not in East Africa (Kingdon 1974; Brennan 1985). In Gabon, monogamous groups have been observed to mingle peacefully in the same tree, which suggests that range defense may be rare or nonexistent (Gautier-Hion and Gautier 1978).

In Peru, two species of tamarins have been observed to cooperate in the defense of a joint territory (Terborgh 1983). Although the precise function and causal mecha-

a.   *b.*

c.

nisms of such cooperation are not known, it seems possible that joint territory defense may enable each species to control a larger area than it might in the absence of such associations. A similar correlation between polyspecific associations and increased range size has been observed among forest cercopithecines in Gabon (Gautier-Hion, Quris, and Gautier 1983).

Species Characterized by Male Dispersal

Behavior of Females. Female hostility toward other groups occurs in almost all Old World monkey species characterized by female philopatry (table 22-1; fig. 22-1). Antagonism toward extragroup females occurs even among gelada baboons, which regularly forage and travel in the close proximity of females from other one-

male units (U. Mori 1979b). However, although females seldom engage in friendly interactions or grooming with females in neighboring groups, they are not consistently aggressive toward males. Copulation and grooming between males and females from different groups have been reported, for example, in vervets (Cheney 1981), rhesus macaques (Lindburg 1971; Hausfater 1972), and baboons (Cheney and Seyfarth 1977). Moreover, females do not always respond aggressively to migrant males, even when the successful takeover of their group by a migrant could result in the infanticide of their offspring. For example, although female langurs at Abu were sometimes observed to aid the group's resident male in driving away potential migrants, they also frequently ignored such individuals (Hrdy 1977).

d.

FIGURE 22-1. In many Old World monkey species, adult females are active participants in intergroup interactions, and they are often particularly aggressive toward other females. Here, females in four different species threaten females, males, and juveniles in neighboring groups: (*a*) vervet monkeys; (*b*) redtail monkeys; (*c*) gray langurs; (*d*) gelada baboons. (Photos: (*a*) Dorothy Cheney; (*b*) Lysa Leland; (*c*) Sarah Blaffer Hrdy/Anthrophoto; (*d*) Robin Dunbar)

At least two factors appear to influence the rate of female aggression in intergroup encounters. First, females are most aggressive when intergroup encounters involve the defense of a territory or a clumped, defendable resource. For example, encounters among groups of Amboseli vervets and provisioned rhesus macaques are characterized by high rates of female aggression (fig. 22-2). In contrast, when most intergroup encounters involve supplants or avoidance, as is frequently the case among baboons and long-tailed, bonnet, and Japanese macaques, female aggression occurs at lower rates (table 22-1).

Second, when their aggression is directed toward extragroup males, interspecific variation in the frequency of female aggression seems to be affected by the degree of sexual dimorphism. While female aggression toward migrant males is common in macaques and vervets, it occurs less often in baboons (Packer and Pusey 1979; see also table 22-1). Macaque and vervet females are approximately 70 to 85% the weight of males, while female baboons weigh only about half as much as males (Napier and Napier 1967; P. L. Whitten 1982). Perhaps as a result of these differences in sexual dimorphism, fe-

male coalitions against males occur more frequently in macaques and vervets than in baboons, and females in these species may exert considerable influence over whether or not a male is able to transfer into their group (Packer and Pusey 1979; Cheney 1983b).

Unfortunately, cross-species comparisons of rates of female aggression during intergroup interactions are presently impossible since studies seldom provide quantitative data on this measure. Moreover, in their descriptions of such aggression, authors frequently fail to distinguish between aggression directed toward females and aggression directed toward males. This distinction is important, however, because female-female aggression frequently concerns defense of territorial boundaries or food resources, while female aggression toward males more often involves resistance to migrants (e.g., gray langurs: Hrdy 1977). The frequency of female aggression is therefore affected not just by competition for resources, but also by social and demographic factors. For example, natal male migration among Amboseli vervets is often preceded by a period of several months in which subadult males approach and attempt to interact with

FIGURE 22-2. An aggressive intergroup encounter between rhesus monkeys on Cayo Santiago. The females on the left with their backs to the camera belong to Group Q, a small, subordinate group. Those on the right belong to group F, a much larger, higher-ranking group, which won this encounter. (Photo: Barbara Smuts/Anthrophoto)

adult females and juveniles in other groups. The hostility of females toward such advances often precipitates aggressive intergroup interactions involving all the members of both groups. Thus when the rate of male migration among Amboseli vervets doubled between 1977–78 and 1980, there was a correspondent increase in both the proportion of aggressive encounters and the proportion of encounters involving female aggression (table 22-1).

Although intergroup encounters in Old World monkeys seldom escalate into physical attacks, females occasionally bite and injure the members of other groups. Hausfater (1972) reports that female rhesus macaques on Cayo Santiago were frequently wounded in intergroup encounters, although he does not mention the sex of their antagonists. Among gray langurs at Jodhpur, the cross-group "kidnapping" of infants by females resulted in the infants' death in 14% of cases, and such kidnapping often escalated into aggressive intergroup chases and fights (Mohnot 1980). Shopland (1982) observed a number of adult female baboons attack and wound a low-ranking female from another group, while an adult male

and a high-ranking adult female repeatedly bit and eventually killed her infant. Similarly, in 114 encounters observed during 1980 in Amboseli, female vervets bit and injured juvenile males attempting to migrate into their group on 8 occasions (7% of encounters).

Individual variation in the behavior of females. In many species of Old World monkeys, high-ranking females have preferential access to food and water (chap. 26), which suggests that such individuals may benefit more than others from defending their group's resources. Most studies do not comment upon the rank of female participants in intergroup encounters. Occasionally, however, preferred access to resources does seem to be associated with individual differences in the frequency and intensity of intergroup aggression. For example, over a number of different years, high-ranking female vervets in Amboseli were more aggressive during intergroup encounters than low-ranking females (Cheney, Lee, and Seyfarth 1981). Similar observations have been made for wedge-capped capuchins (Robinson, in press *a*) and one group of rhesus macaques on La Cueva Island (Vessey 1968). Hausfater (1972), however, found no correlation between rank and the frequency of female aggression in a group of rhesus macaques on Cayo Santiago.

In contrast to high-ranking females, low-ranking vervets in Amboseli were more likely to engage in affinitive interactions with the members of other groups (Cheney, Lee, and Seyfarth 1981). Similar affinitive interactions by females have been reported in baboons and rhesus macaques, although those studies have not commented upon the ranks of the individuals involved (Hausfater 1972; Cheney and Seyfarth 1977). It is worth noting, however, that when groups of rhesus and Japanese macaques have been observed to fission, it is generally the lower-ranking genealogies that split off from the larger group (Koyama 1970; Chepko-Sade and Olivier 1979). In addition to increasing average degrees of relatedness among individuals in the two new groups (chap. 11), such fissions also improve the ranks of some previously subordinate females.

Individual differences in the behavior of females during intergroup encounters therefore appear ultimately related to attempts to maintain or increase access to food resources. In the case of high-ranking females, such attempts seem to involve defense of existing resources and social rank. In contrast, the interactions of low-ranking females with extragroup animals potentially lead to disruptions in the stability of the group, which in some cases may help increase individual ranks or competitive abilities.

Behavior of Males. The reproductive success of males depends on the ability to monopolize or compete successfully for mates. Perhaps as a result, male behavior during intergroup encounters usually involves attempts either to gain access to groups of females or to exclude other males from such groups. Although data on rates of male and female aggression are rarely presented, males are usually reported to be more aggressive during intergroup encounters than females, and those studies that do present rates support this contention (table 22-1). Most intergroup aggression is directed toward other males.

In species such as black-and-white colobus, gray langurs, redtails, and blue monkeys, groups usually contain only one male (see chaps. 8, 9). In such cases, the resident male aggressively resists any attempts by solitary males or individuals from all-male groups to approach or take over his group. Male-male interactions often involve loud calls, which seem to function as advertisements of the resident male's presence and his willingness to fight to maintain possession of his group (table 22-1).

Male-male competition in one-male groups is often intense. Particularly in areas of high population density, a male's length of tenure in a bisexual group is often short, and challenges from solitary males or all-male groups

occur at high rates (Hrdy 1977; chaps. 8, 9). Among gray langurs at Abu, for example, resident males in bisexual groups always responded aggressively toward potential migrant males. In contrast, they were far less aggressive toward males in other bisexual groups, presumably because such males did not represent a challenge to their tenure (Hrdy 1977).

In species characterized by multimale groups, most competiton for adult females occurs within the group. However, perhaps because the intensity of male-male competition is affected by the number of males in the group, males often respond aggressively toward potential migrants and are more aggressive toward extragroup males than toward extragroup females (e.g., baboons: Hamilton, Buskirk, and Buskirk 1975a; Cheney and Seyfarth 1977; Packer 1979a; rhesus macaques: Hausfater 1972; Meikle and Vessey 1981; Japanese macaques: Sugiyama 1976; vervets: Cheney 1981; Cheney and Seyfarth 1983; Harrison 1983b).

Although few studies provide data on rates of fighting, male intergroup aggression seldom results in physical contact. As with intragroup fights, however, serious injuries can result when conflicts escalate. For example, in 114 encounters in 1980 among vervets in Amboseli, adult males bit potential migrant males on only 5% of occasions. All such bites, however, produced large, partially immobilizing wounds. Similarly, Hausfater (1972) reports that although male rhesus macaques on Cayo Santiago were injured less often than females during intergroup encounters, their wounds were more likely to be severe.

When intergroup encounters escalate to involve adult females and the defense of the range or a particular resource, adult males also occasionally injure and even kill females and immatures in other groups. Over a 7-year period in Amboseli, male vervets were observed to wound females and infants in other groups on five occasions, and two cases of cross-group infanticide also occurred. In each case, the adult males involved were alpha males of groups into whose ranges the other group had trespassed. The males made no subsequent attempts to mate with the females whose infants they had killed, which suggests that the infanticides were related to range defense rather than attempts to improve reproductive opportunities. Male wounding of extragroup females has also been observed in Japanese macaques (Kawanaka 1973).

During intergroup encounters, male baboons often herd the females of their own group away from other males (Stoltz and Saayman 1970; Hamilton, Buskirk, and Buskirk 1975a; Cheney and Seyfarth 1977; Packer 1979a). Such herding is particularly likely to involve sexually receptive females and appears to represent attempts to prevent males in other groups from gaining access to females. Male herding of females is less common in less sexually dimorphic species such as Japanese macaques and vervets (Packer and Pusey 1979; Cheney 1981). As with female aggression toward potential migrants, such interspecific variation in the frequency of herding may be related to differences in the ability of females to retaliate against male aggression.

The behavior of young natal males during intergroup encounters contrasts sharply with that of adult males. In most species characterized by male dispersal, young natal males are the most active participants in intergroup encounters; they often initiate encounters by attempting to approach and interact with the members of other groups (e.g., vervets: Cheney 1981; Cheney and Seyfarth 1983; baboons: Hamilton, Buskirk, and Buskirk 1975a; Cheney and Seyfarth 1977; Packer 1979a; rhesus macaques: Hausfater 1972; Japanese macaques: Sugiyama 1976; black-and-white colobus: Oates 1977c; red howlers: Sekulic 1982c). Such interactions precede and potentially facilitate the natal males' subsequent transfer to other groups.

Individual variation in the behavior of males. It has already been noted that females in at least one species exhibit individual differences in behavior during intergroup encounters and that these differences are correlated with differential access to resources. The same appears true for adult males. For example, in one study of baboons, dominant males were more aggressive toward other groups than were subordinate males (Cheney and Seyfarth 1977). High rates of aggression by dominant males may have been related to the greater reproductive benefits experienced by these individuals. In contrast, subordinate male vervets were more likely than dominant males to intitiate affinitive interactions with the members of other groups (Cheney and Seyfarth 1983). As was the case with natal males, such affinitive interactions often preceded transfer to other groups.

VARIATION IN THE NATURE OF INTERGROUP ENCOUNTERS

Thus far, no attempt has been made to consider variation in the relationships of different neighboring groups. Just as there are individual differences in behavior during intergroup encounters, however, so are there marked differences in the responses of groups to particular neighbors.

As mentioned previously, in many species the outcome of intergroup competition for a particular resource is determined by a dominance hierarchy in which the larger group usually supplants the smaller one. The maintenance of intergroup dominance relations,however, does not depend on the physical presence of all members of both groups. Nonhuman primates seem to recognize the members of other groups as individuals, and the arrival of even one member of a dominant group may be sufficient to cause all the members of a more subordinate

group to leave the area (Kawanaka 1973). The particular male membership of a group may also influence its dominance. At La Cueva, Puerto Rico, for example, a group of rhesus macaques was observed to rise in rank over its neighbors following the transfer of a few individual males (Gabow 1972).

Cross-group individual recognition can extend even to animals who never transfer from their natal groups. Playback experiments using the intergroup calls of females have shown that vervets recognize with which group different individuals are associated, even when they interact with those individuals only during intergroup encounters (Cheney and Seyfarth 1982b). The ability to recognize the members of other groups may permit males to monitor changes in the composition and relationships of neighboring groups and to choose when and where to transfer. Such recognition may also allow females to modify their behavior during intergroup encounters, depending perhaps upon previous patterns of male migration and the past history of their group's relations with each of its neighbors (Cheney and Seyfarth 1983).

Among humans, alliance ties between villages or clans are strongly influenced by past patterns of mate exchange, and marriages are often arranged with the express purpose of solidifying or maintaining alliances with other groups (Fox 1967). Intriguingly, a similar relation between migration patterns and reduced intergroup hostility may also exist in some nonhuman primates. Relations between groups of macaques that have fissioned in the recent past are initially unaggressive and characterized by high rates of male exchange, although this relationship usually deteriorates with time (Koyama 1970; Missakian 1973b). Past patterns of association may also influence relations between groups that do not share a recent common ancestry. In the Amboseli vervet population, females were less aggressive toward males who transferred from groups with which their own group had exchanged males in the past than toward males from groups with which male transfer had not previously occurred (Cheney and Seyfarth 1982b; see also chap. 21).

The causal relation between male transfer and reduced female hostility in vervets has not yet been determined. Males may transfer to groups from which they receive little aggression because reduced hostility facilitates integration into a new group. Males and females may also simply be less hostile toward males coming from groups from which males have transferred in the past. Males in such groups will probably be both more familiar and more closely related genetically than individuals in other neighboring groups, and both factors may influence rates of aggression (Cheney and Seyfarth 1982b).

Peer migration has also been observed in a number of other Old World monkeys (chap. 21), although it is not known if such migration patterns affected subsequent intergroup relations. Indeed, the general applicability of the observation that individuals are less hostile toward groups that contain close kin remains to be determined. For example, it would be of interest to know if gibbon groups consistently treat neighboring groups that include sons or daughters differently from those that are comprised of less closely related individuals.

Although mate exchange may occasionally reduce hostility between groups, in no primate species other than the hamadryas baboon have two social units ever been observed to ally themselves against a third. Although groups are often competitors, cooperation rarely extends beyond passive tolerance. In this sense, the changes in intergroup relations that result from nonrandom transfer or repeated association are less sophisticated and have fewer long-term social consequences than is the case among humans. Nevertheless, it is clear that nonhuman primates are capable of differentiating among individuals and relationships outside their own social groups. Across local populations of groups, individuals recognize each other and often share some degree of genetic relatedness, despite the maintenance of otherwise discrete social units.

Summary

Although groups of nonhuman primates are frequently hostile toward one another, the nature of intergroup interactions varies greatly. Groups that range over relatively small areas are usually territorial, while those that range over larger areas are not. In the latter case, intergroup interactions are mediated by the relative dominance of each group. In species characterized by female dispersal, females tend to avoid intergroup interactions and rarely respond aggressively to females in other groups. In contrast, when females remain in their natal groups or live in monogamous family groups, they are usually hostile toward extragroup females. The relative aggressiveness of females in such species appears to be influenced by a number of factors, including the defensibility of food resources, the degree of sexual dimorphism, and the females' dominance ranks. In all primate species, males typically respond aggressively to males in other groups, perhaps because such individuals represent potential competition for mates. In species characterized by male dispersal, dominant adult males are usually more aggressive than subordinate males or individuals who have not yet emigrated from their natal group. Social groups of primates appear to distinguish among adjacent groups and do not respond similarly to each of their neighbors. Past interactions, in particular previous patterns of dispersal, appear to have an important influence on the relative aggressiveness of intergroup relations.

23 | Evolution of Social Structure

Richard W. Wrangham

There is no consensus as to how primate social organization evolves, but a variety of reasons suggest that ecological pressures bear the principal responsibility for species differences in social behavior. In particular, environmental factors clearly have strong effects on individual survival and reproduction (chaps. 19, 20), and they are often known to differ between species or populations with different social systems (e.g., chap. 17). There are firm theoretical reasons for thinking that aspects of social organization such as average group size can be adaptive responses to problems of obtaining food or avoiding predation (Pulliam and Caraco 1984). Finally, there are striking examples of convergence in the social systems of distantly related species with at least superficial similarities in their ecological adaptations, such as chimpanzees and spider monkeys, or talapoins and squirrel monkeys (chap. 7). These points suggest that ecological variables shape primate societies, just as they appear to do in other animals (Krebs and Davies 1984; Rubenstein and Wrangham 1986). The problem, and this is true for most animal societies, is that we do not know exactly what the relevant ecological pressures are, or which aspects of social life they most directly affect, or how. In this chapter I review the principal hypotheses proposed to explain the effect of ecological pressures on primate social organization.

Ecological pressures are not the only influences on social evolution. Social systems could in theory have significant "cultural" components unrelated to ecological adaptation, but in practice local traditions are known only in the signals that individuals use, rather than in their relationships (chap. 37). More important, the social system could be influenced so strongly by a species' evolutionary history that it might include aspects slow to respond to selective pressures and therefore strongly correlated with phylogeny (Struhsaker 1969). For example, patas monkeys and forest guenons are closely related and have a similar and rather unusual mating system (chap. 9). Yet their habitats, ranging patterns, and feeding behavior differ markedly, which raises the possibility that ecological pressures have little effect on this aspect of their social systems.

Phylogenetic inertia of traits that are subject to divergent ecological pressures could in theory occur in at least two ways. First is the retention of nonadaptive behavior. No examples are known, however, and this is unlikely to be common because major aspects of social organization are known to be labile throughout the primate order. In a few species, for example, some populations form cohesive groups while others forage in unstable parties (brown lemur: Tattersall 1982; red colobus: Starin 1981; muriqui: chap. 7). In savanna baboons, females, but not males, choose to live with their kin, whereas in the closely related hamadryas baboons it is principally males that do so (chap. 21). The chapters in part 1 give numerous cases of intraspecific variation in group size and sex ratio. Social organization is therefore often capable of rapid change, presumably in response to new selection pressures.

A more probable mechanism is that related species share special traits that constrain the way social organization can evolve (E. O. Wilson 1975). This can happen when the shared trait changes the costs and benefits of following particular strategies. For example, it is characteristic of lemuriforms, but of no other superfamily of primates, that the sexes are similar in size and that females routinely dominate males (chap. 3). In other primates female dominance appears due partly to the sexes being similar in size (Packer and Pusey 1979). The unusual nature of lemuriform intersexual relationships therefore appears due to the lack of sexual dimorphism and might accordingly be attributable to phylogenetic inertia in this characteristic. Similarly, constraints on social organization could be imposed by cognitive abilities (chap. 36), which could limit the development of social relationships involving both cooperative and competitive elements (Wrangham 1983), or by communicative abilities, possibly relevant in allowing group living at night.

It cannot be assumed, however, that particular features of social organization are the result of phylogenetic constraints merely because they are similar in closely related species. For instance, ecological pressures may be found to account for the low degree of sexual dimorphism among lemuriforms (chap. 3). Similar arguments can be applied widely to cases where phylogenetic inertia is invoked. An ecological analysis of social organization is therefore critical to understanding phylogenetic constraints on the evolution of social behavior.

ECOLOGICAL AND SOCIAL FACTORS IN SOCIAL EVOLUTION

The relationship between ecology and social organization has been difficult to establish for several reasons, three of which I discuss here. First, the relationship is clearly not very tight. Long-term studies of baboons and chimpanzees, for example, have shown that changes in group size and intergroup and intragroup dynamics often have an internal momentum determined not by short-term variations in ecology but by long-term shifts in demographic patterns and social relationships (Altmann and Altmann 1979; Goodall 1983; Dunbar 1984b). This means that the pattern of social organization is determined not only by ecological pressures but also by recent history. Long-term studies are therefore needed to show which features of social organization are consistent over time. For example, it has become clear only within the last fifteen years that the pattern of cooperative and competitive social relationships within groups is generally more consistent than features of social structure such as group size and sex ratio (e.g., chaps. 6, 11).

Second, it is notoriously difficult to measure several of the ecological parameters believed to be most important, such as predator pressure and patterns of food distribution (chaps. 19, 17). This means that in practice quantitative analyses have been forced to use less appropriate variables, such as habitat type instead of predator pressure or diet type instead of food distribution. When primate socioecology was first developing, it was hoped that grouping patterns would prove to be correlated rather simply with these easily detected ecological categories (e.g., frugivore versus omnivore) (Crook and Gartlan 1966). Over the next decade, however, a variety of studies involving over 100 primate species showed that there are few tight relationships involving these kinds of variables.

For example, the principal results of efforts to understand group size in multifemale groups are that groups tend to be larger (1) in terrestrial species than arboreal species (Clutton-Brock and Harvey 1977a); (2) among frugivores than among folivores (this emerges only when closely related species are compared; Clutton-Brock and Harvey 1977a); (3) in larger than smaller species (Clutton-Brock and Harvey 1977a); and (4) when home ranges are larger (Milton and May 1976). As Clutton-Brock and Harvey (1977a) note, there are numerous exceptions to these trends. This is not surprising given the uncertain significance of the ecological variables, let alone the substantial variation in group size found within species and populations (e.g., chaps. 6, 11), but it means that the results are difficult to interpret. Both predator pressure and food distribution may be responsible for (1), for instance, whereas diet and food distribution are more likely to be responsible for (2) through (4) (Milton

and May 1976; Clutton-Brock and Harvey 1977a). The clearest relationship found between these kinds of variables and aspects of social organization is that nocturnal species live in smaller groups than almost all diurnal species. However, many aspects of social behavior are not explained by these correlations, particularly those concerned with social relationships.

Third, an individual's social behavior is influenced by "social" as well as ecological pressures, that is, the strategies of conspecifics. For example, grouping can in theory be favored in response to the aggressive behavior of conspecifics as well as in response to predators. The relative importance of these two pressures varies between species, as discussed below. Furthermore, their combined effect is complex because the nature of the social pressures depends on the ecological pressures. For instance, coalitions of conspecifics would not be a threat if ecological pressures forced all individuals to forage solitarily. The complexity of the interaction between ecological and social pressures probably explains why species with few ecological differences, such as savanna baboons and hamadryas baboons (Nagel 1973), may have radical differences in their social systems.

The existence of these two kinds of influence (ecological and social pressures) has been recognized for a long time, but different schools have viewed their effects in different ways depending on whether ecological influences are considered separately for each sex. The more traditional approach begins by considering the determinants of group size and dispersion without taking into account the possibility that there are differences in the selective pressures on females and males. Types of social relationship within groups are then considered to be derived, mainly by sexual selection, from the social structure that environmental pressures have generated (e.g., Crook and Gartlan 1966; Crook 1970; Clutton-Brock and Harvey 1977a; Terborgh 1983; van Schaik and van Hooff 1983). The social relationships explained by this procedure are principally those between males and between males and females. These models have not, however, paid much attention to how ecological pressures influence relationships between groups and among females within groups (see chap. 14 for an exception). This kind of analysis appears most appropriate when ecological pressures have equal significance for females and males, for example, when adult males and females are vulnerable to predation in similar ways.

The second approach uses ecological pressures to account for the spatial distribution and social relationships of females, and then views social structure (group characteristics) as emerging out of the interaction between female distribution and male mating strategies (e.g., Orians 1969; Bradbury and Vehrencamp 1977; Emlen and Oring 1977; Wrangham 1980; Clutton-Brock, Guinness, and Albon 1982). This type of analysis is more inclusive than the first because it pays more attention to relationships among adult females, within and between groups. The justification for the initial focus on females is that the behavior of males is generally adapted to maximizing mating success, which depends on the distribution and behavior of females (Trivers 1972). Female behavior, by contrast, often appears to be adapted more directly to ecological pressures because female fitness tends to be limited by environmental pressures such as food availability (chaps. 20, 26). In a similar way female body size is hypothesized to be adapted more closely than male body size to ecological pressures (Gaulin and Sailer 1985).

As a final complexity it should be noted that the impact of male mating competition on female social relationships can vary between species as a result of differences in the ecological pressures acting on males. For instance, Popp (1983) argued that ecologically imposed differences in sexual dimorphism among baboon species are responsible for differences in their social organization. This is an important possibility whose implications have been investigated little to date.

Although the two approaches have different conceptual foundations, many studies use both, and eventually they will probably be united. Here I review theories and data from both approaches, but the sexes are treated separately so as to include hypotheses concerned with the social relationships of both females and males. Accordingly, I begin by considering the nature of ecological pressures on the distribution and social relationships of females. Subsequently male relationships are discussed in terms of (1) the distribution of mating and rearing opportunities (Goss-Custard, Dunbar, and Aldrich-Blake 1972), and (2) their effects on the social organization of females.

Solitary, monogamous or polyandrous, and group-living species are considered separately. Within this traditional classification, variations occur even in major aspects of social organization. Some of the variations are considered here, but inevitably much of the diversity is ignored (fig. 23-1). The appendix indicates which species fall into different categories of social organization and hints at some of the variability.

ECOLOGICAL EFFECTS ON FEMALE RELATIONSHIPS
Species in Which Females Forage Solitarily

Nocturnal Prosimians. At least 15 species of nocturnal prosimians, in 6 subfamilies, forage solitarily (chaps. 2, 3; table 23-1). The only exceptions known are some of the species with a diurnal ancestry, which are monogamous (chaps. 2, 3). Adult females are normally not accompanied even by their offspring, and in some species, such as bushbabies and lorisines, they defend territories against other females. In others, such as the cheirogaleids, it is not yet clear whether females defend foraging territories (chap. 3). In bushbabies and possibly other species, maternally related females tolerate each other and form "matriarchal" sets that share overlapping ranges (chap. 2). Cooperative territory defense by females within matriarchies has not been seen.

The conjunction of solitary foraging and intrasexual territoriality clearly suggests that females distribute themselves so as to defend access to food from other females, but this does not explain why females forage alone. In theory they could be accompanied by offspring or an adult male, and in the species where female relatives share their territories they could forage together. Little attention has been given to this problem, which applies to many nocturnal mammals, but both antipredator strategies and feeding competition have been suggested as explanations for solitary foraging in primates (Eisenberg, Muckenhirn, and Rudran 1972; Clutton-Brock and Harvey 1977a).

The predation theory relies on evidence that small group size reduces the rate at which predators detect cryptic, immobile prey (van Schaik and van Hooff 1983). This seems likely to apply to primates, but there is no direct supportive evidence. Moreover, as Charles-Dominique (1977a) points out, prosimians vary greatly in their reliance on crypsis. Lorisines move slowly and silently and have no loud calls, whereas bushbabies have long night-ranges, move rapidly, and are highly vocal. Bushbabies are therefore less cryptic than lorises.

The insect diets of many prosimians are considered to favor solitary foraging because groups would tend to disturb potential prey (Clutton-Brock and Harvey 1977a). Indirect support for this idea is provided by bushbabies, in which the more frugivorous species are more gregarious than the more insectivorous species, even though all are predominantly solitary (chap. 2). The leaf-eating sportive lemur *Lepilemur mustelinus* is also solitary, however, so disturbance of prey is not the only factor favoring lone foraging. An alternative possibility is that individuals benefit by minimizing travel costs when foraging alone (Alcock 1984).

The economics of resource defense are not well understood. Charles-Dominique (1974a) and Hladik (1975) proposed that territoriality functions to ensure a sufficient food supply during periods of scarcity. This is supported by data on *Lepilemur mustelinus leucopus* living in an area with so few food species and competitors that the total food production could be estimated. Their territories during a lean period were estimated to contain only 1.6 times the minimum amount of necessary food (Hladik 1975). Alternative explanations of territoriality have not yet been investigated.

Territoriality is intriguing in nocturnal prosimians because there is substantial variation in the apparent defensibility of ranges. For instance, territoriality is normally expected to occur where day ranges are long in relation to the size of the home range so that ranges are economically defensible (Mitani and Rodman 1979; Davies and Houston 1984). Although bushbabies can visit their boundaries easily, residents in some species of lorisines are clearly unable to find and repel intruders quickly. In these species individuals apparently maintain exclusive areas by avoiding each other's scent marks (Charles-Dominique 1977a). Other mechanisms are possible, however (Gosling 1982). For example, bushbabies and lorises can emit ultrasonic signals (Zimmerman 1981), although their function is unknown. Again, the occurrence of matriarchies is unexplained. Matriarchies are

an extension of the tendency for female philopatry (females living in or near their mother's range), which is widespread but unexplained in solitary mammals (Waser and Jones 1983). Causes that have been suggested for similar forms of range sharing in solitary carnivores are particularly large food patches (Kruuk 1978) or unusually high rates of resource renewal (Waser 1981).

Frugivorous Great Apes and Spider Monkeys. The second set of relatively solitary female foragers are the large frugivorous apes (orangutans, chimpanzees, and the more gregarious bonobos) and spider monkeys (which have been argued to resemble apes in several aspects of their morphology [Mittermeier 1978]). In some popu-

FIGURE 23-1. The diversity of primate social organization is too great to be captured adequately by any simple scheme, yet traditional classifications remain useful for broad comparisons. Shown here are (*clockwise from top*): a large group of mantled howler monkeys; a solitary bushbaby (uncharacteristically active, in this fantasy, by day); a monogamous pair of indris; a multimale group of Japanese monkeys; a small multimale group of ring-tailed lemurs; part of a patas monkey group, including the resident male on the left; and a mother-offspring pair of orangutans. (Drawing: Teryl Lynn)

lations of muriqui, which are closely related to spider monkeys, females also forage independently (Milton 1984; chap. 7). These species differ from nocturnal prosimians in allowing their juveniles to accompany them—even after weaning, thus forming small maternal groups—and in having widely overlapping core areas, which are not defended as territories. They are excellent subjects for analyzing the proximate effects of ecological pressures because females (mothers plus young) spend some time alone and some time with other adults. This variable grouping pattern offers opportunities for estimating the costs and benefits of grouping.

These species appear little bothered by predators (chap. 19), which raises the possibility that they travel individually simply because they do not need the protection afforded by groups (van Schaik and van Hooff 1983). However, this would not explain why group size varies seasonally. In chimpanzees, bonobos, and spider monkeys, there are consistent indications that food shortage causes smaller groups, which suggests that individuals forage alone when it is important to maximize food intake (Wrangham 1977; Ghiglieri 1984; Kuroda 1980; van Roosmalen 1980). Similarly in orangutans, occasional periods of food abundance are said to favor group foraging (van Schaik and van Hooff 1983). The hypothesis that lone foraging increases food intake is supported by data on female chimpanzees, whose feeding time when they forage alone averages 67% of their activity budget, up from 50% when they forage with others. Exploitation competition at small food sources may account for this effect, which has been argued to explain why female chimpanzees are regularly solitary (Wrangham and Smuts 1980). Similar reasoning is applicable to the other species.

This raises two problems. First, do any features of their ecology prevent these species from forming the cohesive groups typical of most large primates? The frugivorous apes and spider monkeys have unusual diets compared to other large anthropoids because they do not forage for dispersed insects, eat little leaf in relation to their body weight, and appear relatively intolerant of toxic compounds (Wrangham 1980; Temerin and Cant 1983). They are therefore restricted to high-quality diets, especially ripe fruit, thought to be distributed in discrete patches. Accordingly I have suggested that during periods of food shortage, when large patches are rare, they are forced to use small patches, where competition penalizes group foraging (Wrangham 1980). Most large primates, on the other hand, can exploit alternative foods such as grasses, leaves, or dispersed insects and can thereby afford to travel in groups even when ripe fruits are confined to small patches. Data on food distribution are needed to test this hypothesis. A particularly sensitive test will be provided by comparisons of muriqui living in fission-fusion communities with those living in cohesive groups. Milton (1984) found that community-

living muriqui ate a high proportion of leaves (as much as 80%). Leaves are normally thought to occur in rather large patches, but as Milton (1984) noted, this depends on the particular species eaten. She suggested that the distribution of the muriqui's key foods favors solitary foraging in the same way as in chimpanzees and spider monkeys.

Second, why do females not defend their home ranges as territories? Female ranges have been argued to be indefensible because the observed foraging area is too large to allow regular visits to boundaries (Wrangham 1979a; Rodman 1984). This is consistent with the fact that in relation to their size, great apes have large home ranges and low population densities, probably because of their restriction to relatively high-quality diets (Clutton-Brock and Harvey 1979). Although females are not territorial, female orangutans and chimpanzees avoid each other by using dispersed core areas. This suggests that they distribute themselves so as to minimize range overlap, perhaps because lone travel maximizes foraging efficiency (Wrangham 1979a). As expected, their social relationships are neither cooperative nor strikingly competitive (chap. 15).

If their dietary specialization and large home ranges are responsible for the pattern of female dispersion, what explains species differences in grouping patterns? There are no clear answers. Bonobos appear to have larger food patches than chimpanzees do, which could explain why they are able to live in comparatively large and stable groups (chap. 15). Ecological comparisons therefore indicate that grouping patterns are constrained by the abundance and dispersal of food, though they do not suggest why large groups are preferred.

Females Foraging in Monogamous or Polyandrous Groups

Regular monogamy (including systems that are sometimes polyandrous) is known or suspected in nine subfamilies of primates (appendix) and 22% of species (table 23-1). It occurs in both diurnal and nocturnal species. No sex biases have yet been found in the age at which dispersal occurs, the distance that offspring disperse, or the probability of inheriting the parental territory (chap. 21).

There are three main classes of explanation for the occurrence of monogamy in primates. First are the predation theories reviewed in chapter 19. The predation theories are concerned largely with explaining small group size rather than patterns of social relationships. Second are the food defense theories, discussed later as the gibbon model. Third are theories, concerned with paternal care, suggesting that female relationships are determined by competition over males.

In this section we are concerned with relationships between females rather than with reasons females travel with males. Because males are often more aggressive than females in intergroup encounters, the nature of female-female relationships is not immediately obvious. In gibbons, however, which are the best-studied monogamous primates, there is abundant evidence that monogamy is maintained at least partly by female-female aggression. In particular, female calls are an important component of intergroup aggression and elicit aggressive responses from females in neighboring groups. Females also chase members of other groups, especially females (chap. 12, 22). As Leighton demonstrates in chapter 12, these observations are consistent with a model in which females defend access to individual territories in order to maximize net food intake. The precise features of resource dispersal responsible for the small groups and high defensibility of gibbon territories are still uncertain, but the importance of female-female aggression for keeping females apart is clear.

The gibbon model has been applied widely to monogamous primates that are less well studied. For instance, van Schaik and van Hooff (1983) suggested that all Old World species of monogamous primates share the key feature of females having defensible ranges, and that this feature is responsible for their being territorial and hence monogamous. Many monogamous species are indeed consistent with this hypothesis because there is clear evidence of female-female aggression, such as females having specialized loud calls and participating in intergroup encounters (e.g., indri: Pollock 1979b); spectral tarsiers: MacKinnon and KacKinnon 1984; titi monkeys: chap. 5; see chap. 22).

Yet the gibbon model does not fit all species so well. Monogamous primates include various patterns of social behavior, some of which are inconsistent with the idea that females benefit by excluding other females from their ranges. For example, in de Brazza's monkey monogamous groups sometimes meet without aggression, and no loud call is known among females (Gautier-Hion and Gautier 1978). De Brazza's monkey has been studied too little to understand why its social system differs so radically from other Cercopithecus species, but the lack of female-female hostility makes it unlikely that the species owes its monogamy to female competition for feeding ranges. Again, peaceful aggregations of pairs are sometimes found among mongoose lemurs (Tattersall 1982) and owl monkeys (chap. 5), both of which are nocturnal. Female simakobu also lack intergroup calls. Simakobus are monogamous only at low population densities (chap. 19), where ranges are presumably larger than normal and therefore harder to defend; this makes it particularly unlikely that their monogamy is favored by territoriality. Since the distribution of monogamy appears poorly correlated with predator pressure in these species (chap. 19), the adaptive significance of their social systems presents an interesting puzzle.

TABLE 23-1   Grouping Patterns of Nonhuman Primates

	Number of Species	Lone	Monogamy/ Polyandry	Multifemale Group	Community	?
Prosimians	38	15 (6)	3 (5)	5	0	4
Daubentoniidae	1	1	0	0	0	0
Cheirogaleidae	7	3 (2)	0	0	0	2
Indriidae	4	0	1 (2)	1	0	0
Lemuridae	7	0	1 (2)	4	0	0
Lepilemuridae	3	1	0 (1)	0	0	1
Galaginae	8	7 (1)	0	0	0	0
Lorisinae	5	2 (3)	0	0	0	0
Tarsiidae	3	1	1	0	0	1
New World monkeys	47	0	9 (7)	15 (2)	5	9
Callimiconidae	1	0	0	(1)	0	0
Callitrichidae	15	0	1 (7)	0	0	7
Alouattinae	6	0	0	4	0	2
Atelinae	7	0	0	1 (1)	5	0
Cebinae	6	0	0	6	0	0
Pitheciinae	12	0	8	4	0	0
Old World monkeys	68	0	2 (2)	46 (10)	0	8
Cercopithecinae	41	0	1 (2)	28 (6)	0	4
Colobinae	27	0	1	18 (4)	0	4
Apes	13	1	9	1	2	0
Hylobatidae	9	0	9	0	0	0
Pongidae	4	1	0	1	2	0
ALL PRIMATES	166	16 (6)	23 (14)	67 (12)	7	21
%	—	10 (4)	13 (8)	40 (7)	4	13

SOURCE: Appendix.
NOTES: Figures show number of species for which the grouping pattern is confirmed, with additional suspected cases in parentheses.

Paternal care theories have been applied most explicitly to the small (< 1.5 kg) monogamous species of the New World, all of which show high levels of parental care. These are the callitrichids (some or all of which have variable social systems; chap. 4), owl monkey, and titi monkeys (chap. 5). Van Schaik and van Hooff (1983) suggested that these species are derived from larger ancestors living in multifemale groups. They argued that females need paternal care because litter weight is relatively high. Accordingly, females compete for males by showing intolerance to other females.

Litter weight is indeed high in some of these species (table 16-5), females do participate in intergroup encounters, and suppression of other females' breeding by the dominant female is well documented in captivity (chap. 4). However, the argument fails to explain how a relatively high litter weight evolved without paternal care in the first place, why monogamous (or polyandrous) groups travel as separate foraging units, or why tarsiers (with much higher relative litter weight) show less extensive male care. Moreover, paternal care can be viewed equally well as a consequence rather than as a cause of monogamy (see chap. 28). As a result, the importance of female-female competition for male helpers is still unclear.

Multifemale Groups

Multifemale groups are defined here as those in which at least two adult females normally travel together all day. They occur in almost half of primate species and in 10 subfamilies (table 23-1). Multifemale groups are semiclosed (i.e., individuals are intolerant of members of other groups), have at least one adult male, and occur only in diurnal species. In every other respect they occur in a startling diversity of forms. The number of females per group averages between 2 (sifaka, hamadryas baboon) and over 20 (talapoin, some savanna baboons); patterns of intergroup transfer, group cohesion, and social relationships within and between groups vary widely (tables 23-2, 23-3; chaps. 21, 22). Some species with multifemale groups also form single-female (monogamous) groups either in certain parts of their range (e.g., Guatemalan howler, simakobu) or at low frequencies in many populations (e.g., gorilla). In all species there is substantial variation in group size within local populations (chap. 11). Adult sex ratios also vary widely both between and within species.

Three principal kinds of hypothesis have been proposed to explain why groups might be favored purely through ecological pressures. These hypotheses differ in the predictions they make about the economics of group size. The predation hypotheses state that larger groups inflict inevitable costs on female reproductive rate because of feeding competition, but that compensating benefits come from the increased survival of young (Al-

exander 1974; van Schaik 1983; chap. 19). The food defense hypothesis accepts that larger groups may have somewhat greater feeding competition within groups. However, it proposes that females in large groups nevertheless have high fitness because their social dominance over small groups gives them increased access to preferred foods (Wrangham 1980). The "low-cost" hypotheses argue that group living has neither large positive nor large negative effects on foraging efficiency and that even trivial benefits can therefore favor group living (Moore 1984). None of these predictions has yet been generally confirmed because only a few studies have attempted to measure the costs and benefits of different group sizes.

The effect of group size on reproductive rate has been estimated in two ways, using survey data or birth rates. First, van Schaik (1983) assembled data on the size and composition of groups in 14 primate species. He found that in 27 censuses, females in larger groups on average had fewer infants (22 cases) and fewer juveniles (19 cases) than those in smaller groups. Van Schaik's data indicate that females in larger groups often have reduced reproductive rates, but as he notes, they are derived from censuses whose accuracy is unclear. Second, Meikle, Tilford, and Vessey (1984) showed that among four groups of rhesus macaques on La Parguera, Puerto Rico, the reproductive rate per female over a 12-year period was consistently higher in larger and more dominant groups. Furthermore, females in dominant groups had a higher proportion of sons than those in subordinate groups, which indicates that they contributed relatively large numbers of migrants to other groups (Meikle, Tilford, and Vessey 1984). This study indicates a benefit to being in large groups, but its significance may be affected by the fact that the macaques were provisioned at artificial feeding stations. Provisioning can sometimes exaggerate the ordinary effects of social dominance on differences in reproductive rate between individuals (e.g., Japanese macaques: Sugiyama and Ohsawa 1982b). It could therefore have the same effect between groups, so provisioned populations are not necessarily useful for understanding the typical effects of group size on reproductive rate or fitness. This is a key issue for understanding the adaptive significance of multifemale groups.

The predation hypothesis has been proposed as a general explanation for all multifemale groups. It is discussed in detail in chapter 19, which concludes that adequate tests of its effects on the evolution of multifemale groups have yet to be made.

The food defense hypothesis has been proposed only for "female-bonded" groups, that is, those in which females usually remain in their natal groups (Wrangham 1980). Table 23-2 indicates that this pattern of female philopatry is suspected in 36 out of the 48 species (living in groups or communities) with data available and is thus the predominant primate trend. The hypothesis proposes that the preferred foods of these species occur in rare patches that are economically defensible by groups; that competition at the preferred patches favors those individuals who have allies helping to supplant others; and that the need for reliable allies therefore favors the formation of stable groups of kin (Wrangham 1982). Smaller, subordinate groups are supplanted, and their females therefore obtain less of the preferred foods or expend more energy to obtain them elsewhere. According to this idea, territorial or dominance relationships should occur between groups, females should participate actively in intergroup interactions when their group dominance is threatened, and females in dominant groups should have higher fitness than those in subordinate groups. The food defense hypothesis is so far the only one to propose an explanation for the distribution of female philopatry, female involvement in intergroup interactions, and substantial variation in group size within local populations (Wrangham 1980). It assumes that male dispersal is a consequence of the fact that females are philopatric (chap. 21).

Certain predictions of the food defense hypothesis are supported. First, female-bonded groups either defend territories or, where they are nonterritorial, typically have stable dominance relationships with other groups, based largely on group size (chap. 22). Second, in competitive intergroup interactions females are much more likely to be active participants in species with female philopatry than in those with female transfer (chap. 22). Third, the outcome of intergroup interactions commonly determines access to preferred resources (chap. 22),

which in the case of many female-bonded species are ripe fruits (Wrangham 1980).

The prediction that females in large groups have higher fitness than those in small groups was challenged by van Schaik's (1983) census data showing higher reproductive rates in smaller groups. Subsequently Moore (1984) argued that van Schaik's test was invalid because only 2 of the 14 species in the sample were female bonded. However, Moore's criteria for classifying female-bonded species was particularly restrictive because he excluded species with regular male transfer if female transfer had been recorded even occasionally. When Moore's criteria are relaxed to allow the inclusion of *Propithecus verreauxi, Macaca fascicularis, Papio ursinus,* and *Presbytis entellus,* 12 data sets are available. Females in smaller groups had more infants or juveniles in 8 of these. Thus females in larger groups did appear to have lower reproductive rates, in contrast to the prediction of the food defense hypothesis.

Several further points raise questions about the food defense hypothesis. First, the benefits from successful intergroup competition have not been measured. It is not clear whether they are typically large enough to overcome the costs of intragroup competition, which are known to increase consistently with group size in some species, for example, long-tailed macaques (van Schaik et al. 1983). Second, the argument that the preferred patches of female-bonded species are economically defensible is largely circular. It depends primarily on observations that in some species they are defended effectively, though also on the general preference of female-bonded species for ripe fruit, which often occurs in rare,

TABLE 23-2 Patterns of Intergroup Transfer in Primates Living in Multifemale Groups or Communities

	Number of Species	Intergroup Transfers Mainly by			
		Males	Females	Both	?
Prosimians	5	2	0	0	3
Indriidae	1	1	0	0	0
Lemuridae	4	1	0	0	3
New World monkeys	20 (2)	1 (3)	(5)	2	11
Callimiconidae	(1)	0	0	0	1
Alouattinae	4	0	0	2	2
Atelinae	6 (1)	0	(5)	0	2
Cebinae	6	1 (3)	0	0	2
Pitheciinae	4	0	0	0	4
Old World monkeys	46 (10)	15 (15)	2	0	24
Cercopithecinae	28 (6)	12 (11)	1	0	10
Colobinae	18 (4)	3 (4)	1	0	14
Apes	3	0	2 (1)	0	0
Pongidae	3	0	2 (1)	0	0
ALL PRIMATES	74 (12)	18 (18)	4 (6)	2	38
%	—	21 (21)	5 (7)	2	44

SOURCE: Appendix, especially based on chapter 21.

NOTES: Figures show number of species for which the pattern of transfer is known, with additional suspected cases in parentheses.

Families or subfamilies in which no species have been confirmed to live in groups or communities are excluded as are species living in single-female groups.

discrete patches. Third, it should be noted that the food defense hypothesis is not applicable to all female-bonded species since at least two species with female philopatry do not defend food resources in intergroup competition (gelada, guereza) (Wrangham 1980). (However, a possible explanation for gelada groups concerns the fact that geladas commonly spend several minutes digging to reach grass roots: female allies may provide protection against theft of root holes by females in other groups.)

The third ("low-cost") hypothesis proposes that group living has only trivial effects on foraging efficiency and that it occurs because of other kinds of benefit. The logic applies to species that use resources distributed in large and temporary patches, as many primates do. In these circumstances individuals foraging alone would need large home ranges, so they suffer little harm from forming groups (Moore 1984). The benefits of group living may therefore be concomitantly small, such as the ability to capitalize on the food-finding memories of old individuals (Moore 1984), to cue each other to foods that are hard to locate (Milton and May 1976), or to reduce the rate of spread of disease (Freeland 1976). No way has yet been presented to test such ideas, nor have they been used to explain species differences in social organization. Another possibility is that the benefit of grouping in low-cost species is a result of male strategies.

None of the hypotheses discussed have proposed testable explanations for species differences in average group size or for the pattern of variance in group size within species. The size and renewal rate of resource patches, intensity of within-group feeding competition, and degree of predation pressure are considered important variables (Clutton-Brock and Harvey 1977a; Wrangham 1980; Terborgh 1983; van Schaik 1983). In mantled howler monkeys the size of temporary subgroups has been shown to be related to food-patch size (Leighton and Leighton 1982). However, the effect of food-patch size on the size of stable groups has not been investigated, and it is not even clear what characteristics of patch distribution are most important, such as the minimum, average, or maximum size or some aspect of the rate of renewal. It certainly seems likely that species using larger patches tend to live in larger groups (Terborgh 1983), but whether this can account for the distribution of group sizes is unknown. The distribution of food within patches also appears to have effects on social organization within groups. For instance, the relationship between the diet type and intragroup dominance relationships is discussed in chapter 17.

MALE CONTRIBUTIONS TO SOCIAL ORGANIZATION
Species with Lone Females

Nocturnal Prosimians. Among nocturnal prosimians, large males compete to dominate access to areas containing one or more female ranges, as Bearder describes for *Galago senegalensis* (chapter 2). Male territories appear not to be adapted to resource distribution because they are often substantially larger than those of females. By contrast, there is no evidence that female behavior is adapted to male strategies. Females do not follow or wait for males when foraging, and their range use appears determined by relationships with other females (Charles-Dominique 1977a). Thus in these species males apparently have little effect on the social system beyond the superimposition of their ranges and their infrequent interactions with females.

These societies are therefore explicable by separate analyses of the distribution of females (with respect to food) and males (with respect to females). This leaves several problems unanswered, however. For example, in species where a male shares his territory with a single female, why doesn't he accompany her? What explains the size of male ranges? Why don't males cooperate in defending access to females, as chimpanzees do? We are forced to take as a given that the pattern of male distribution and relationships represents what theory suggests it should, namely the consequence of male-male competition for mating opportunities. Similar problems apply even more acutely in many of the examples considered below.

Frugivorous Great Apes and Spider Monkeys. Males have more obvious effects on female behavior in the frugivorous great apes, which show a sequence of increasing complexity of male strategies and their effects on females. First, the distribution of male orangutans resembles the system in nocturnal prosimians (Rodman 1973; chap. 13), and similar conclusions can therefore be drawn. Large males occupy separate ranges and have mutually hostile relationships. Males and females have infrequent contact and travel together only during occasional sexual consortships. The occurrence of forced copulations by subadult and adult males certainly means that females are at times directly affected by males. However, there is no evidence that male strategies affect the ordinary dispersal or social relationships of females.

In chimpanzees it is groups of males, rather than lone males, that occupy separate ranges and have mutually hostile relationships. This has been attributed to the ability of chimpanzees to forage in small parties, which orangutans appear unable to do because the foraging costs are too high (Rodman 1984; Wrangham 1986). According to this theory, communities evolved from a hypothetical solitary-male system because males could afford to travel in small parties, even though the optimal foraging strategy was to travel alone; they were forced to do so because lone males therefore became vulnerable to attacks by pairs. As a result, selection favored mutually protective bonds among males, leading by escalation to the formation of mutually hostile male communities. This analysis relies on evidence that lone chimpanzees are indeed vulnerable to attack by small parties (Goodall et al. 1979) and that when food is scarce males increase the time they spend foraging alone (Wrangham 1977).

When female chimpanzees were discovered to be dispersed in a similar manner to female orangutans, it was uncertain whether their social relationships were affected by the distribution of male communities (Wrangham 1979b). It is now clear that they are (Nishida et al. 1985). Although females living in the border area between two communities may associate with males from each community, border areas are generally avoided, and females show fear responses to less familiar males (Nishida et al. 1985). Females travel throughout the range of "their" males and associate substantially more with females in their own than in neighboring communities. The chimpanzee community thus emerges as a bisexual society formed as a result of male mating strategies. The same may be true of spider monkeys and the muriqui.

The evolutionary impact of males on female relationships may be even greater in bonobos than in chimpanzees. Bonobos form larger and more stable foraging groups than chimpanzees, apparently because their foods occur in larger clumps (chap. 15). No ecological benefits to grouping have been proposed. However, female bonobos have prolonged sexual swellings compared to those of chimpanzees, which suggest that they benefit by attracting males. A possible benefit is that familiar males protect them from harassment by others (Wrangham 1986).

Since the community organization of chimpanzees was discovered, it has been explained entirely in terms of male-male competition. Male strategies appear particularly important for the existence of communities in chimpanzees and spider monkeys because in these species males spend less time alone than females do. This exceedingly unusual pattern (compared to other mammals and birds) is associated with well-developed bonds between males, whereas it is unclear whether females typically form bonds with other females in any of these species. It is difficult to imagine what direct ecological pressures could exist that would favor grouping in males but not in females. It seems more likely that these species exemplify the direct importance of mating competition on social structure. Ecological pressures limit the kinds of groups that can be formed, while mating competition creates the social unit.

Monogamous and Polyandrous Groups

The gibbon model of monogamy suggests that the distribution of males depends on the distribution of females that are already territorial. Males presumably benefit from accompanying females by guarding their mates from rivals, ensuring future mating opportunities, and protecting their offspring. Females gain male assistance in the defense of resources and protection from infanticide or other dangers (chap. 12). The relative importance of these benefits to each sex is unknown, nor is it clear why males are unable to patrol more than one female's territory.

It appears ironic that a system of female-female competition for resources should lead to females sharing their territories with a male and therefore losing precious resources to him. However, although males must impose feeding costs on females, they defer to their mates in ways that reduce feeding competition: females lead at least half of the group progressions, and in feeding contexts they are codominant or dominant to males in gibbons, owl monkeys, titi monkeys, indris, and mongoose lemurs (chaps. 5, 12; Pollock 1979a). This is in contrast to the monogamous de Brazza's monkey, in which males lead females and appear clearly dominant (Gautier-Hion and Gautier 1978). The comparison suggests that there are at least two different types of monogamous species. In territorial species, females are followed by males that assist in the maintenance of the territory. In nonterritorial species, females follow males that protect them from either predators, as Gautier-Hion and Gautier (1978) suggest for de Brazza's monkey, or other males (as known, for instance, in dabbling ducks: Wittenberger and Tilson 1980).

The gibbon model does not explain why territorial females are accompanied by males in some species (i.e., the monogamous species) and not in others (i.e., nocturnal prosimians: chap. 2). If ecological pressures force nocturnal species to travel alone, as discussed earlier, the difference is explicable. However, the distribution of monogamy appears strongly affected also by the species' evolutionary history because the few nocturnal species that are monogamous (in four different families) are all considered unusual in being descended from diurnal primates (i.e., *Lemur mongoz*—partly nocturnal, partly diurnal; *Hapalemur griseus*—crepuscular; *Tarsius spectrum; Aotus trivirgatus*). Eisenberg, Muckenhirn, and Rudran (1972) suggested that only a diurnal ancestry can generate the communicative abilities necessary for coordination of a monogamous group at night. If this proves correct, it would be an example of phylogenetic inertia forcing solitariness in most nocturnal prosimians. Solution of this problem will help explain both the function of male consorts and the reasons most nocturnal prosimians are solitary.

Compared to species that conform to the gibbon model, male-female relationships in the Callitrichidae have two oddities. First, in at least some species, polyandry and polygyny occur at significant levels associated with helping by adult males. Second, relationships are comparatively brief, a result of high rates both of mortality and intergroup transfer (chap. 4; Sussman and Kinzey 1984).

Both aspects reflect greater variation in the social system than occurs in gibbons. In a similar way, some groups of callitrichids are territorial and others are not.

This puzzling variability in social organization appears to be associated with an equivalent diversity of ecological conditions. Unlike gibbons, callitrichids experience high variation in range size and population density (compare tables 4-1 and 12-1). Callitrichids have been argued to be patchily distributed because they are specialized at colonizing temporarily productive habitat (Sussman and Kinzey 1984, but see chap. 4 this volume). Possibly this leads to a variation in range quality that underlies the variability in their social organization.

There is no satisfactory explanation for the distribution of "helpers" (chap. 4). Adult males help most in the small New World species where they carry infants to safety from sudden attacks by raptors. This suggests that predator pressure is responsible for helping in these species (Wright 1984). But it is puzzling that helpers do not occur in other monogamous primates since the "ecological constraints" theory would suggest that helpers are likely to be found in stable populations, such as gibbons, where breeding opportunities arise rarely (Emlen 1984). Furthermore, gibbon groups with older offspring are more successful in territorial defense (Carpenter 1940), which indicates that there are potential benefits from the retention of helpers. If feeding competition within groups is more intense in the larger monogamous species, the costs of retaining old offspring may be prohibitively high. However, feeding competition is generally expected to be greater in smaller species (Wrangham 1979a). The problem is therefore not easily solved.

Multifemale Groups

Unlike some mammals with stable multifemale groups (e.g., coati: Russell 1983), multifemale groups of primates are almost always accompanied by males. Variation in the number of males in the group is most commonly argued to result from species differences in patterns of male mating competition (Goss-Custard, Dunbar, and Aldrich-Blake 1972; Clutton-Brock and Harvey 1976, 1977a). Explanations in terms of predator pressure have not proved instructive to date (chap. 19). Male contributions to the social system differ clearly between species in which females are philopatric and those in which females regularly transfer. The distinction between these two categories is not always clear-cut (chap. 21; Moore 1984), but it is certainly adequate for the purposes of this discussion. Female-bonded species are considered first.

Female-Bonded Species. The number of males per group varies widely between species (table 23-3) and within species. However, when closely related species are compared, the number of males per group increases rather steadily with group size (Andelman 1986; Terborgh 1983). This may account for the fact that, since group size varies in the same way, single-male groups tend to be found in forest while multimale groups tend to occur in open country (Crook 1970; Clutton-Brock and Harvey 1977a). Functional explanations for species differences in adult sex ratio have been discussed in terms of the costs and benefits both to females and males.

The hypothesis that the adult sex ratio is influenced by female interests is supported by observations of females

TABLE 23-3   Distribution of Single-Male and Multimale Groups among Primates

| | Number of Species | Group Size | | | | | |
| | | <5 Females | | | >5 Females | | |
		s1	V	M	s1	V	M
Prosimians	4	0	0	3	0	0	1
Indriidae	1	0	0	1	0	0	0
Lemuridae	3	0	0	2	0	0	1
New World monkeys	10 (8)	1 (1)	2 (1)	2 (1)	0	1	4 (5)
Callimiconidae	(1)	(1)	0	0	0	0	0
Alouattinae	3 (1)	0	2 (1)	0	0	0	1
Atelinae	2 (1)	0	0	(1)	0	0	2
Cebinae	5 (1)	1	0	2	0	1	1 (1)
Pitheciinae	(4)	0	0	0	0	0	(4)
Old World monkeys	29 (3)	11 (2)	4	0	2	3	9 (1)
Cercopithecinae	16 (1)	4	0	0	2	1	9 (1)
Colobinae	13 (2)	7 (2)	4	0	0	2	0
Apes	1	0	1	0	0	0	0
Pongidae	1	0	1	0	0	0	0
ALL PRIMATES	44 (11)	12 (3)	7 (1)	5 (1)	2	4	14 (6)
%	—	22 (5)	13 (2)	9 (2)	4	7	25 (11)

NOTES: 1 = only one breeding male resides in the group; V = groups may have one or more resident adult males; M = groups typically contain more than one resident adult male. Only multifemale groups are included.

acting friendly or hostile to males attempting to immigrate (chaps. 21, 31). In multimale groups, females can benefit reproductively from the presence of extra males, as shown by an increased number of infants per female in groups of savanna baboons with more males (Dunbar and Sharman 1983). One interpretation is that in groups with few males, females suffer from the consequences of competition for access to males (which are needed as mates or protectors in intragroup competition) (Dunbar and Sharman 1983; chap. 31). Another is that the presence of extra males benefits the group as a whole by raising its dominance status in relation to other groups (Wrangham 1980). A third is that extra males help reduce the costs to females of male-male competition, such as infanticide (Wrangham 1980; Hrdy 1981b). The relative importance of these potential benefits has not been evaluated.

The most important consideration from the male point of view is whether other males can be excluded (Clutton-Brock and Harvey 1977a; Terborgh 1983; van Schaik and van Hooff 1983). Both the defensibility of females and the competitive abilities of males are thought to be important. The importance of the defensibility of females has received only anecdotal support so far. For example, van Schaik and van Hooff (1983) argued that populations of hanuman langurs with relatively compact groups have single males, whereas groups that disperse widely during foraging tend to be multimale. The implication is that large, widely dispersed groups are not economically defensible by a single male. Males therefore compete for mating access to individual females within the group rather than for control of the group as a whole. This idea is likely to be helpful when combined with an analysis of female interests.

The effect of habitat differences on male competitive abilities was examined by Popp (1983) among four subspecies of baboons in 17 populations. He found that in areas of higher rainfall, populations had consistently greater sexual dimorphism in body size as well as a trend toward higher sex ratios (more adult females per adult male in breeding groups). He suggested that in areas of higher rainfall, males could afford to expend more energy on excluding other males from groups, but that habitat quality is correlated only weakly with the sex ratio because female strategies also affect grouping patterns.

Male relationships with each other are largely antagonistic, although male-male coalitions occur in several species living in multimale groups (chap. 31). In some species, males and females interact at relatively low rates (e.g., vervet monkeys), whereas in others, close bonds between particular males and females are evident (e.g., macaques and savanna baboons: Takahata 1982a; Chapais 1983a; Smuts 1985). The factors responsible for species differences in these characteristics are unknown (chap. 31).

Female-Transfer Species. Whatever the reason for females living in kin groups, in female-bonded species the fact that a particular set of females lives together is generally affected little by male strategies. In three female-transfer species, by contrast, the proximate causes of group composition are female attraction to particular males (gorilla, hamadryas baboon, and possibly red colobus), and male herding of females (hamadryas baboon) (chaps. 21, 31). This suggests that in these species the existence of social units depends on females being attracted to males rather than to each other. Furthermore, females pay more attention to males than to females within the group: unlike female-bonded groups, the timing and direction of travel is determined by males, only males participate actively in intergroup interactions, and cooperative relationships among females are developed poorly if at all. Three kinds of theory have been proposed to explain why females form a stronger long-term bond with a male than with females. These theories may also be applicable to howler monkeys with female transfer (chap. 6).

The first theory is that females are attracted to males because they provide protection from predators. This has been argued most clearly for gorillas (chap. 14). An antipredator function is supported by the fact that males defend their groups against potential predators, that females rarely travel alone, and that even unmated males sometimes travel in groups. However, males might be expected to defend their groups regardless of the ultimate reasons for grouping, and the tendency for both females and males to avoid traveling alone can be attributed not only to their fear of predators but also to their fear of aggression from other gorillas. These observations are therefore explicable both by the predation hypothesis and by the male harassment hypothesis. A potential test of the importance of predation is that in areas without predators, female-transfer species should not live in cohesive groups.

Second, females could be attracted to males who provide the best resources, as occurs among some polygynous territorial birds (Greenwood 1980). Present data do not support this idea because males have not been observed to defend food resources. However, it is possible that groups with more dominant males deter other groups from using the same area.

Third, females may form a bond with males because lone females are vulnerable to harassment, including infanticide, by "courting" males (Wrangham 1979a, 1982). According to this male harassment theory, female-transfer groups are possible because ecological pressures are weak, creating neither significant costs nor significant benefits for group living. In these circumstances males can afford to harass attractive females, and the best protection that females can obtain is to form a bond with a male who minimizes harassment by other males. This theory is supported by the fact that males do at-

a.

FIGURE 23-2. A comparison of savanna baboons (*a*) and hamadryas baboons (*b*) shows that very closely related species may have striking differences in their social organization. Activities typical for savanna baboons, but rare in hamadryas baboons, are (*clockwise from top right*): a mother initiates troop movement by leading a subgroup of females; a female grooms an adult male "friend"; an adult female grooms her adult sister; a consort pair is harassed by a male attempting to take over as the female's temporary mate; and two females form an alliance to protect access to a small food source.

tempt to harass females and that females seek protection from males. In hamadryas baboons, for example, females line up in the "attack shadow" of their males (Kummer 1968). This theory is challenged by the fact that in female-bonded species, females also have strategies to protect themselves from male harassment, including the formation of coalitions with female kin and the development of special relationships with males (chaps. 31, 32). However, these strategies appear to provide only temporary protection, depending on the availability of an appropriate ally at the time of the encounter. The male harassment theory predicts that females in female-transfer species, by following a male or males in a permanent association, receive fewer attacks from males than in female-bonded species. A comparison of gorillas and chimpanzees, or savanna baboons and hamadryas baboons, would make an appropriate test.

Whether preferred resource patches are indeed not economically defensible, as this theory requires (Wrangham 1980), remains to be proven. However, it is clear that food defense is unusual both between and within groups. The argument that indefensibility of food resources is responsible for the failure to form female-bonded groups is strongest for hamadryas baboons, which meet savanna baboons along a zone of hybridization in Ethiopia (fig. 23-2). The habitats of the two spe-

cies are similar in many respects, but unlike savanna baboons, hamadryas baboons have little access to trees containing ripe fleshy fruits, and they eat fruits rarely. Instead, hamadryas baboons forage principally from abundant *Acacia* bushes (Nagel 1973). This suggests that their adaptation to a desert environment deprives hamadryas baboons of economically defensible resources. Hence, the benefits of forming female-bonded groups are less for hamadryas baboons than for savanna baboons (Wrangham 1980). Among red colobus, ripe fruit is not a preferred food (Wrangham 1980), and availability of the young leaves that most attract red colobus groups does not appear to be affected by intergroup aggression (Struhsaker and Leland 1979).

The male harassment theory suggests that social bonds between males and females may be formed purely as a consequence of the need for protection against conspecifics and that in these species ecological pressures merely permit, but do not actively generate, groups. The same may be true for bonds between males. Chimpanzees offered one example, and hamadryas baboons offer another. As Kummer (1968) stressed, hamadryas society is based on two bonds—those between a male and his females and those between males of different one-male groups. Because females are willing to transfer between groups, breeding males are exposed to the possibility

b.

Equally, a number of typical hamadryas interactions are rare among savanna baboons. Illustrated are (*center and right*): the males of two one-male groups yawn aggressively at each other, while (*right-hand group*) females line up in their male's "attack shadow"; (*center left*) a male bites his female on the neck after she strayed too far from him; and (*upper left*) a male turns to stare at his females as a signal for them to follow.

The reduction of female social power in hamadryas baboons (compared to savanna baboons) may have occurred because hamadryas food resources are less easily defensible (see text).

that other breeding males will "steal" females. Accordingly, breeding males have developed strong defensive alliances within which there is intense inhibition against mutual aggression. This makes an instructive contrast with gelada baboons, where females do not transfer between groups, and breeding males are therefore not threats to each other. Among gelada, breeding males form defensive alliances only against bachelor males living in all-male groups (chap. 10).

Sex ratios in female-transfer groups are determined ultimately by females choosing whether to join groups with few or many males. The basis of female choice is not well understood. In theory, data on female choice offer opportunities to test hypotheses by showing, for example, whether the criteria are the degree of protection against predators, the quality of food resources, or the degree of protection from other males. In practice such data are hard to obtain (see Harcourt 1978b; Marsh 1979a), nor have different hypotheses yet generated different predictions about adult sex ratios.

SOCIOECOLOGY OF PRIMATES COMPARED TO OTHER ANIMALS

Only occasional attempts have been made to compare primate socioecology directly with the socioecology of other animals (e.g., E. O. Wilson 1975; Clutton-Brock

and Harvey 1978; Richard 1985). This is unfortunate because the fundamental principles underlying social evolution are presumably the same in all animals. Comparisons across taxa may therefore provide tests of theories that are applied to a particular set of species, such as the primates. A comparative socioecological analysis is difficult, however, partly because primates include social systems with no exact analogues in other taxa (e.g., female-bonded groups: Wrangham 1983) and partly because theories explaining the ecological basis of social organization are still poorly developed for most animals. Nevertheless, the processes governing primate social evolution appear broadly similar to those operating in other taxa. If so, to what extent are primate social systems unusual compared to other animals, and what is responsible for the features typical of primates?

Solitary and monogamous primates have grouping patterns typical of many other species. There are clear similarities between the social organization of bushbabies and solitary carnivores such as the white-tailed mongoose *Ichneumia albicauda*, for example, in which "matriarchies" (clans) can share ranges (Waser and Waser 1985). Monogamous primates are also comparable to species in a variety of other taxa (Kleiman 1977; Wittenberger and Tilson 1980), although they differ from many monogamous species because all members of the group travel together. This feature also differentiates female-bonded groups of primates from other animals with female philopatry and aggressive relationships between neighboring groups of females. For example, among lions (*Panthera leo*) and spotted hyenas (*Crocuta crocuta*) related females jointly defend their shared ranges, yet females travel independently (Packer 1986). Among female-transfer species, the social structure of gorillas and hamadryas is similar to that of Burchell's zebra (*Equus burchelli*) because they form cohesive groups in which males defend a few females from other males; intergroup relationships are different in all three species, however (Klingel 1974). As a final case, the social structure exhibited by chimpanzees and spider monkeys, in which related males defend access to an area containing independent females, has not yet been described in other animals. Many other taxa include species with unique kinds of social system, however, so the fact that some primate systems are not found elsewhere need not surprise us.

In general, the kinds of explanation proposed for primate social systems are similar to those proposed for comparable systems in other species. For example, the factors thought to favor solitary and monogamous systems in carnivores are similar to those discussed for primates (MacDonald 1983), and there is evidence that group living among larger carnivores is adapted more to competition for food resources or mates than to the benefits of group hunting, as it was more often thought to be previously (Lamprecht 1981; Waser and Waser 1985; Packer 1986). Furthermore, female-bonded groups show defense of food resources (e.g., lions: Packer 1986), whereas male-bonded groups rarely interact over food (African wild dog *Lycaon pictus:* Frame et al. 1979). Nevertheless there are general tendencies in the social organization of primates that await explanation. For example, all the gregarious species form semiclosed social networks, like carnivores (MacDonald 1983) but unlike the loose flocks or groups of many birds and mammals. Differences between species with semiclosed networks and those without could be due to either ecological factors, such as the high quality of primate (and carnivore) foods, which are therefore likely to be worth defending, or to phylogenetic constraints, such as the cognitive abilities to recognize many individuals and act accordingly. Again, no primates have a social system in which males defend food resources while females visit the territories of a number of different males. This system is widespread in other species, such as ungulates (Owen-Smith 1977). Presumably primate food resources are not economically defensible by males, but the conditions under which male primates should adopt this strategy are unknown. Solutions to problems such as these are important for a better understanding of primate socioecology.

CONCLUSION AND SUMMARY

Ecological influences on primate social organization are still poorly understood, but studies are beginning to suggest which variables are relevant and how they affect social life. Three ecological factors appear particularly important for determining both the degree of gregariousness and the pattern of social relationships. First, the defensibility of resources seems to determine whether females benefit by isolating themselves from other females in individual territories, by forming long-term coalitions with other females, or by not attempting to defend resources. Second, the spatial and temporal distribution of food patches may determine how large a group individuals can afford and whether groups must periodically fission. Third, the degree of predator pressure influences whether individuals should travel alone, whether males are needed to protect young, and whether it is worth traveling in groups so large that the individual's net rate of food intake falls. Rigorous investigation of these and other similar ideas are now needed. They will be facilitated by the use of optimality theories (e.g., Davies and Houston 1984; Pulliam and Caraco 1984).

An important problem is why patterns of social relationships differ radically, even between species whose groups have similar size and sex ratio. One possibility is that two different kinds of selective pressure have been responsible for favoring multifemale groups. Where ecological pressures favor grouping, they have created social groups of cooperating females that compete against other groups. Where ecological pressures permit but do not necessarily favor groups, on the other hand, sexual selection has created aggregations of females attracted to males. Among monogamous species, a similar distinction may be found between groups formed because males are attracted to females and groups formed because females are attracted to males. The primates offer excellent opportunities for further understanding of the ecological basis of species differences in social relationships.

PART III

Group Life

The chapters in part 3 view primate societies from within by examining the social dynamics of group life. Each of the first eight chapters (chaps. 24–31) focuses on a different topic and describes the general patterns of behavior that emerge from a comparison of all of the species discussed in part 1. Inevitably, these reviews draw heavily on observations of the best-studied species, especially apes and terrestrial cercopithecines. It is important to bear in mind that the patterns of behavior in other less-well-known species may prove in some cases to be different from those that are emphasized here.

In contrast to other chapters in this volume, the last three chapters in this section (chaps. 32–34) make no attempt to provide systematic reviews of particular topics. Instead, they offer somewhat personal treatments of issues of long-standing interest to social scientists: sex differences in behavior (chap. 32), the relevance of non-human primate studies to understanding human behavior (chap. 33), and mechanisms promoting social cohesion within groups (chap. 34).

The chapters echo several themes that are central to an understanding of primate societies. One theme involves the complex interplay between competition and cooperation. Animals live in stable groups at least in part because of the benefits of associating with particular partners. Yet, precisely because they live together, group companions are often one another's most serious competitors for resources that affect reproductive success. Within the group, affinitive interactions such as grooming, protection, and alliances are common, particularly among close matrilineal relatives, as described by Gouzoules and Gouzoules in chapter 24 on kinship. Yet in chapter 25 on conflict and cooperation, when Walters and Seyfarth analyze the distribution of these behaviors, they find that individuals build cooperative relationships in order to compete more effectively against others. As a result, cooperation between some individuals often leads to competition between others.

This tension between cooperation and competition is a reflection of the selection pressures that underlie social life. As Silk points out in chapter 26 on social behavior in an evolutionary perspective, formation of cooperative relationships is sometimes the most effective way to increase individual reproductive success. However, because the genetic interests of individuals are not identical (unless they are clones), conflicts of interest perpetually endanger the survival of cooperative relationships. This point is well illustrated by Nicolson (chap. 27 on females and infants), who shows that even the most intimate associates, such as mothers and offspring, interact in both affiliative and competitive ways. In chapter 34 on social dynamics, de Waal discusses the social mechanisms that allow resolution of conflicts of interest within the group. Because these mechanisms have developed to solve problems fundamental to all social life, the study of social competition and cooperation in nonhuman primates is likely to contribute to an understanding of human societies as well.

A second theme highlighted in part 3 concerns the crucial significance of social relationships within primate societies. In all of the best-studied species, animals treat one another as individuals and develop a series of multifaceted, dynamic relationships with each of their partners. The formation of distinctive relationships begins soon after birth and is strongly influenced by the mother's relationships with others. This point is illustrated by Nicolson (chap. 27), who describes interactions between infants and adult females, and by Whitten (chap. 28), who describes interactions between infants and adult males. Walters (chap. 29 on the transition to adulthood) shows that, by the time the young primate is a juvenile, its social relationships, like those of adults, differ as a function of sex, rank, and other individual characteristics.

Many primate social relationships appear to rest on social reciprocity. For example, in chapter 25 Walters and Seyfarth describe how grooming among vervet monkeys increases the likelihood that the recipient will later attend to the groomer's vocal solicitations for support during agonistic encounters. Similarly, Smuts (chap. 31 on sexual competition and mate choice) reviews observations indicating that females sometimes develop distinctive mating preferences for particular males. In savanna baboons, these preferences are often based on long-term relationships characterized by male protection of the female and her offspring from aggression by other baboons. Through such mutually advantageous exchanges of benefits, social relationships can contribute to individual reproductive success.

At the same time, primatologists have come to realize that in order to understand primate social relationships, they must consider not just their selective advantages but also the goals and motivations of the animals who create them. Attempting to see the world from the perspective of a marmoset, vervet monkey, or chimpanzee is a challenging and difficult enterprise. It relies on the development of ingenious methods for determining what behaviors are most salient to the animals, such as field playbacks of vocalizations or carefully designed naturalistic experiments (e.g., Bachmann and Kummer 1980; chaps. 34, 36). It involves long hours of patient observation and a commitment to recording, in meticulous detail, the most subtle nuances of posture, gesture, touch, and sound. Finally, it may even benefit from the use of intuitive skills to gain insight into the behavior of another species, as de Waal suggests in chapter 34 on social dynamics.

A third theme linking chapters in part 3 concerns the flexibility of primate social behavior. Many aspects of primate behavior are learned, and primates demonstrate an extraordinary capacity to modify behavior depending on the context in which they find themselves. For example, in chapter 30 Hrdy and Whitten show that, in contrast to most other mammalian females, female nonhuman primates often exhibit sexual receptivity at times when they are unlikely to conceive. In some cases, sexual receptivity is clearly a response to social factors, such as the entry of a new male into a group. The importance of social context is also emphasized by Smuts (chap. 32 on gender, aggression, and influence), who argues that to understand sex differences in behavior, it is useful to focus on differences in the situations that males and females encounter as they pursue sex-specific reproductive strategies rather than on inherent temperamental differences.

Many of the behaviors described in these chapters will remind readers of aspects of human behavior. In chapter 33, Hinde discusses the strengths and weaknesses of different methods of relating nonhuman primate behavior to that of humans. He concludes that the most useful approach is one that derives general principles from a broad-based comparison of many different species.

Taken together, the chapters in part 3 highlight one clear message emerging from recent studies of primate behavior: we have hardly begun to grasp the richness and complexity of these animals' social lives. As study after study reveals unforeseen complexities, the challenge to explain the behavior becomes more compelling, and the likelihood grows that the explanations we discover will hold relevance for our own species.

24 | Kinship

Sarah Gouzoules and Harold Gouzoules

Kinship was recognized as an important determinant of primate social behavior even before its relevance to the evolution of behavior was fully realized. From the longitudinal studies of Japanese researchers on provisioned troops of Japanese macaques, it became evident that a remarkable number of behavioral interactions and social relationships, particularly those among females, were influenced by kinship. Since these original studies, kinship has been shown to be an important correlate of behavior in many primate species that have been studied for any appreciable length of time.

Behavior that is "kin correlated" is distributed among individuals on the basis of some measure of genetic relatedness (S. Gouzoules 1984 provides a review of the nonhuman primate literature relating to kin-correlated behavior). In virtually all primate field studies, knowledge of genealogical relatedness has been restricted to that through female lines. In some species (especially among the Old World monkeys), subgroups of individuals related through maternal lines have been identified and referred to as lineages, matrilines, or matrilineal genealogies.

The most extensive data on primate kin-correlated behavior are available from the Japanese macaque and the rhesus macaques of Cayo Santiago, Puerto Rico (Japanese macaques: Yamada 1963; Koyama 1967, 1970; Kurland 1977; rhesus macaques: Sade 1965, 1972a; Missakian 1972, 1974; Miller, Kling, and Dicks 1973). Long-term study of several other species, both in the wild and in captivity, have more recently begun to yield information on the ways kinship determines interactions among group members (chimpanzees at Gombe National Park and the Mahale Mountains, Tanzania: Goodall 1968, 1983; Nishida 1979; Pusey 1980, 1983; gorillas in the Virunga Mountains, Rwanda: Fossey 1979, 1983; chap. 14; yellow baboons in Amboseli National Park, Kenya: Altmann 1980; Walters 1980; Hausfater, Altmann, and Altmann 1982; vervet monkeys in Amboseli: Cheney 1983a; Cheney and Seyfarth 1983; captive pigtailed macaques: Massey 1977; Defler 1978; captive bonnet macaques: Defler 1978; Silk, Samuels, and Rodman 1981; Silk 1982). Although, as noted above, much of the data on primate kin-correlated behavior come from studies of provisioned populations, no evidence suggests that the patterning of behavior in these populations has been al-

tered by feeding. Japanese macaques have been studied under many different provisioning regimes, and indeed under natural conditions, and the influence of kinship on behavior in this species has remained constant. Provisioning may alter the degree to which various behavior patterns are expressed by increasing the likelihood of survival of different classes of relatives. This is discussed later in more detail.

For the vast majority of nonhuman primate species, however, insufficient information precludes confident statements about the role of kinship as a major determinant of behavior outside the immediate mother-offspring "matrifocal unit." In particular, no field studies of any New World species or any prosimian provide information on kin relationships among individuals other than females and immature offspring. Since some researchers (e.g., Moynihan 1976) have suggested that considerable differences in social structure may exist between New World and the more familiar Old World primates, broad generalizations about the role of kinship across the order Primates are inappropriate. The aim of this chapter is to review how kinship influences behavior in those species where it has been studied and also to consider how kinship interacts with other important social variables such as dominance. Our discussion will focus on behavior such as grooming, aggression, and alliance formation that have been particularly well studied. Lastly, we will consider how primates come to recognize their kin and what processes are likely to constrain kin recognition.

MATRILINEAL KIN-CORRELATED BEHAVIOR, WITH SPECIAL REFERENCE TO OLD WORLD MONKEYS
Affinitive Behavior

Spatial Proximity. Related individuals in species of Old World monkeys that typically form multimale groups (primarily the cercopithecines, see chaps. 9, 11) are often found near and in contact with one another and may travel, feed, and sleep together (e.g., pigtailed macaques: Rosenblum 1971; Rosenblum, Kaufman, and Stynes 1966; Japanese macaques: Yamada 1963; Kurland 1977; chimpanzees: Goodall 1968; Pusey 1983; yellow baboons: Altmann 1980). Mothers and their immediate offspring are particularly spatially cohesive, collateral kin (i.e., nonlineal kin such as aunts/uncles–nieces/

nephews, cousins) to a lesser extent. However, distant kin often associate more than would be expected on the basis of availability (Kurland 1977; Grewal 1980a). In rhesus and Japanese macaques, groups of matrilineally related individuals may also consistently associate with particular males of the group (Grewal 1980b; Chapais 1983a).

Grooming. Social grooming is one of the most widely used measures of affinitive interaction in studies of social relationships (fig. 24-1). In many species of Old World monkeys, grooming has been found to be distributed preferentially among related individuals (Japanese macaques: Yamada 1963; Oki and Maeda 1973; Mori 1975; Kurland 1977; rhesus macaques: Sade 1965, 1972b; Missakian 1974; pigtailed macaques: Defler 1978; bonnet macaques: Silk, Samuels, and Rodman 1981; yellow baboons: Walters 1981; chimpanzees: Goodall 1968; Pusey 1983; patas monkeys: Rowell and Olsen 1983). In several species of macaques, the majority of grooming occurs among kin, most of it between mothers and offspring. In fact, Oki and Maeda (1973) reported that 52% of all grooming interactions in a group of Japanese monkeys occurred among mothers and their offspring. Missakian (1974) found a similar amount of grooming (57%) between mothers and offspring in a rhesus monkey group. Grooming between siblings is also common in several species (e.g., rhesus monkeys: Colvin 1983a; chimpanzees: Goodall 1968; Riss and Goodall 1977), and grooming has also been reported between more distant kin, for instance grandmothers and grandoffspring, and aunts and nieces (Kurland 1977).

The distribution of grooming among related and unrelated individuals varies to some extent between species. For example, Defler (1978) compared the frequency of grooming among related and unrelated individuals in captive pigtailed and bonnet macaques. Groups of both species were large and contained well-developed matrilines. Defler found that female pigtailed macaques directed significantly more grooming to matriline ("clan") members than did female bonnet macaques. Silk (1982) also noted that female bonnet macaques directed a substantial amount of grooming toward unrelated adult females.

Grooming among related individuals also varies with the sex and age of partners. For instance, in Japanese and rhesus macaques, daughters usually groom their mothers more frequently than do sons (Missakian 1974; Kurland 1977). In contrast, Pusey (1983) found that immature male and female chimpanzees groomed their mothers for approximately equal amounts of time. In Japanese macaques, mothers groom their offspring more often than they receive grooming from them, and this is also true of grandmothers (Kurland 1977). Infants generally receive more grooming than do older offspring.

FIGURE 24-1. A group of female gray langurs engaged in grooming interactions. (Photo: Sarah Blaffer Hrdy/Anthrophoto)

Yet another pattern of grooming has been suggested for gelada baboons. Among females suspected of being related, individuals appear to restrict their grooming to only one member of their "matriline," and if this partner dies she may not be replaced (Dunbar 1983a).

Despite the unarguable importance of relatedness to the distribution of grooming in many species of Old World monkeys, grooming partners are not always close kin. For instance, special relationships that include grooming behavior between adult females and unrelated adult males have been reported for several species (gelada baboons: Dunbar 1983b; chacma baboons: Seyfarth 1978b; olive baboons: Smuts 1983a; Japanese macaques: Grewal 1980b; Takahata 1982a; rhesus macaques: Chapais 1983a). Moreover, as mentioned above, grooming between less closely related adult females also occurs. Both Oki and Maeda (1973) and Mori (1975) have described differences in the way females approach and groom related and unrelated individuals. In approaching unrelated individuals, for example, females often give specific vocalizations, lip smacks, and head shakes. Oki and Maeda (1973) note that "these movements probably indicate friendliness" (p. 151) and may not be necessary in the initiation of grooming between kin.

Dominance relationships appear to have important effects on the direction of grooming among unrelated adult females in a number of species (chacma baboons: Seyfarth 1976, 1977; vervet monkeys: Seyfarth 1980, Fairbanks, 1980; bonnet macaques: Silk 1982). In these species, females tend to be groomed by females lower ranking than themselves and attempt to groom females that are higher ranking. Immature individuals, particularly females, have also been reported to direct grooming toward higher-ranking, less closely related females (chacma baboons: Cheney 1978a; bonnet macaques: Silk, Samuels, and Rodman 1981). One suggested explanation for this behavior is that individuals attempt to establish relationships that may result in future benefits, such as aid during agonistic interactions (see chap. 25).

Dominance Rank and the Formation of Alliances

Females of several species of Old World monkeys, for example, macaques, baboons, and vervet monkeys, form conspicuous dominance hierarchies in which matrilineally related individuals tend to occupy adjacent ranks. In these species, offspring acquire dominance ranks just below those of their mothers, and female siblings rank in reverse order of their ages, so a female's youngest daughter is the highest ranking of her offspring, her next youngest daughter is the next highest ranking, and so on (Kawai 1965b; Kawamura 1965a; chaps. 25, 11). As a result, each matriline can also be ranked as a unit (yellow baboons: Hausfater, Altmann, and Altmann 1982; chacma baboons: Cheney 1977a; bonnet macaques: Silk, Samuels, and Rodman 1981; long-tailed macaques: de Waal 1977). In all cases, such "rules" of dominance are more variable for males, which usually emigrate from their natal troop around the time of sexual maturity (chap. 21).

Although a number of studies have noted differences in the way infants are treated by other group members, depending upon the rank of their mothers (e.g., H. Gouzoules 1975; Cheney 1978a; Berman 1983a), the acquisition, and probably the maintenance, of dominance rank is primarily dependent upon agonistic support from other individuals, notably matrilineal kin. During agonistic encounters, individuals previously uninvolved in the fight are recruited by combatants (chap. 25). As might be expected, individuals are most commonly supported by their relatives (rhesus macaques: Kaplan 1978; Bernstein and Ehardt 1985; pigtailed macaques: Massey 1977; Japanese macaques: Kurland 1977; Watanabe 1979; bonnet macaques: Silk 1982; long-tailed macaques: de Waal 1977; chacma baboons: Cheney 1977a) (fig. 24-2). Aid is more likely to come from closely related individuals than from more distantly related ones (table 24-1).

As with grooming behavior, kinship is certainly not the sole determinant of alliance formation. A number of studies have reported that unrelated individuals also

FIGURE 24-2. A subadult female Japanese macaque aids her younger sibling against another juvenile. (Photo: Harold Gouzoules)

TABLE 24-I Agonistic Aiding and Matrilineal Relatedness in Three Macaque Species

$r =$	1/2	1/4	1/8	1/16	1/32	0
Pigtailed macaque	———	———	—————————————		· · · · · · · · · · · · · · · · · ·	
Rhesus macaque	———	———	—————————————	————		
Japanese macaque	———	———	—————————————	· · · · · · ·		————

SOURCES: Pigtailed macaque: Massey 1979; Rhesus macaque: Kaplan 1978; Japanese macaque: Kurland 1977. See original papers for details on methods of data collection.

NOTES: Breaks in lines indicate significant differences in frequency of aiding behavior between individuals of different degrees of relatedness (r), where r is defined as the probability of sharing identical genes by virtue of common descent (see chap. 26). Here r is calculated "minimally" by assuming maternal siblings are always half-siblings. Dotted lines indicate data are unavailable.

form coalitions and aid one another (yellow baboons: Walters 1980; rhesus macaques: Kaplan 1978; Datta 1983a; bonnet macaques: Silk 1982; chacma baboons: Cheney 1977a). Kaplan (1978) noted that aid given to unrelated individuals accounted for nearly one-fifth of all female interference. Such alliance formation among unrelated individuals may function to reinforce existing dominance relations, particularly among females.

Some studies have reported that alliances among kin differ from those involving unrelated individuals in that kin are more likely than nonkin to aid one another against higher-ranking animals during agonistic contests that can involve more serious aggression, such as biting attacks (e.g., Walters 1980; Silk 1982; Datta 1983a; Bernstein and Ehardt 1985). Recently, Gouzoules, Gouzoules, and Marler (1984) found that rhesus monkeys, depending on the dominance rank and matrilineal relatedness of their opponent, use different vocalizations in agonistic encounters. These calls function to recruit allies. When tape-recorded vocalizations were played back to group members, close kin responded most strongly to the calls associated with high-ranking, unrelated opponents and agonistic situations that frequently involved contact aggression.

Kinship and Male Behavior

While matrilineal kinship is most often discussed as a determinant of female behavior in multimale groups of Old World monkeys, it is also important with respect to some aspects of male behavior. For instance, related males sometimes emigrate together from their natal troops (rhesus macaques: Boelkins and Wilson 1972; Meikle and Vessey 1981; Colvin 1983b; vervet monkeys: Cheney and Seyfarth 1983). Male rhesus and vervet monkeys are also known to transfer preferentially into a group that a male relative has previously entered (reviewed in Cheney and Seyfarth 1983). Thus kinship may influence both the timing and distribution of male transfer.

In chimpanzees, where males generally remain in their natal community rather than emigrate from it, kinship may be important in adult male behavior such as maintenance of proximity, grooming, and alliance formation (Riss and Goodall 1977; Bygott 1979; Nishida 1979; chap. 15). For instance, Riss and Goodall (1977) de

scribed how one adult male, Figan, was able to rise to the status of alpha male with the aid of his brother, Faben. They also suggested that other high-ranking male coalitions were probably composed of siblings. Nishida (1979) points out that chimpanzee social structure "has as a nucleus a cluster of strongly bonded adult males" (p. 118). These males often cooperate in finding food and in territorial defense, and it is likely that kinship is one important determinant of male behavior in chimpanzees. Male hamadryas baboons are also likely to remain in their natal "clans," and some authors have suggested that a "patrilineal" system of kinship might exist in this species as well (see Sigg et al. 1982; chap. 10).

Inbreeding Avoidance

The avoidance of inbreeding is likely to constitute a strong selective pressure for the recognition of related individuals, and the most important mechanism for the avoidance of inbreeding in primates is dispersal from kin (reviewed in Cheney and Seyfarth 1983). In some cases, however, males may remain in their natal group for at least some portion of their adult lives. Under these circumstances, it is clear that mate choice is influenced by kinship since closely matrilineally related males and females rarely mate. In rhesus and Japanese macaques, close matrilineal kin clearly avoid consanguineous mating (Sade 1968; Missakian 1973a; Enomoto 1974; Takahata 1982b). Even less closely related kin, such as aunts and nephews and cousins, mate less often than would be predicted by chance (e.g., Sade 1968; Takahata 1982b). More distant categories of matrilineal kin, however, have been found to mate at rates that would be predicted to occur through random choice of partner (Takahata 1982b). Young female chimpanzees have also been found to avoid mating with their brothers when they become sexually mature (Goodall 1968; Pusey 1980).

In some species, such as baboons, males may avoid mating with their own daughters by tending to avoid all females that might be offspring. For example, Packer (1979a) reported that male olive baboons consorted with females that "could" have been their daughters (i.e., were conceived after the males had entered the group) significantly less than with females that could not have been their daughters. In other species, such as rhesus macaques, males do not appear to discriminate in this manner, and the avoidance of inbreeding is probably achieved through other means, most notably male transfer after a period of residency in a group or changes in male dominance rank (Smith 1982).

Behavioral Development

In many species of nonhuman primates, infants and juveniles regularly interact with relatives other than their mothers (rhesus macaques: Berman 1982a, 1983a; Japanese macaques: H. Gouzoules 1980b, in prep.; yellow baboons: Altmann 1980; chap. 27; chimpanzees: Goodall 1968; gorillas: Fossey 1979; vervets: Lee 1983a). Older siblings are usually the first and most persistent individuals other than the mother to interact with infants. For instance, Fossey (1979) reported that siblings were more consistently near infant gorillas than any other class of individuals except mothers, and this proximity relationship was maintained throughout the first 3 years of an infant's life. In chimpanzees, older siblings interact a great deal with infants, touching, grooming, protecting, playing, and even briefly "kidnapping" them (Goodall 1968). Most interestingly, several cases of adoption of infants by older siblings after the death of the mother were recorded at Gombe. Orphaned infants without siblings were not adopted by other chimpanzees in the community.

In contrast, Altmann (1980) found a considerable amount of variation in the time juveniles spent near their mothers and infant siblings in yellow baboons. Those immatures that were attracted to infants seemed to be attracted to nonkin as well as kin. Kinship, in general, was not a good predictor of differences in the number of individuals usually found in proximity to various mothers and infants.

In Japanese and rhesus macaques, a great deal of the variability in the early experiences of infants can be attributed to differences in the number of kin available to them (H. Gouzoules 1980b, in prep.; Berman 1982a, 1983a). Berman (1982a) noted that infant rhesus macaques spent more time near close kin than distant kin or unrelated individuals. She suggested that, like adults, infants form stronger relationships with their matrilineal kin than with other individuals. Infant rhesus monkeys with more close female relatives were more likely to be protected from attack, especially when the infants tended to spend time near these relatives. H. Gouzoules (1980b, in prep.) found that infant Japanese macaques with relatively more close kin developed greater independence from their mothers at an earlier age. Such independence was especially likely when the infant's mother was low ranking or old.

Some of the most elaborate interactions between infants and older siblings have been reported in the New World marmosets and tamarins. In these species older offspring may help their parents with the rearing of younger siblings by carrying infants and by sharing food with them, much as in some species of cooperatively breeding birds (chap. 4).

A phenomenon related to behavioral development, the learning of novel behavior patterns, appears to proceed more rapidly along kin lines. Examples of the acquisition of new habits come from studies of provisioned troops of Japanese monkeys, where techniques for handling introduced food items spread among troop members primarily along kin lines (chap. 38).

LIMITS OF KIN-CORRELATED BEHAVIOR

Although matrilineal kinship is an important correlate of social behavior in many primate species, kin-correlated behavior is not equally apparent throughout the primate order. Variation in the importance of kinship can be attributed largely to species differences in demographic variables such as dispersal and mortality as well as in social organization. These factors limit the extent to which one or more genetically related individuals will have opportunities to interact during their lifetime.

Dispersal and Social Organization. In the macaques and baboons discussed thus far, females generally remain in their natal group throughout their lives, while males emigrate from their natal group upon reaching sexual maturity (chap. 21). As a result, females often have the opportunity to interact with a large number of kin while males do not. In contrast, in several other primate species females disperse from the natal group: examples are hamadryas baboons, red colobus monkeys, gorillas, and chimpanzees (chaps. 8, 10, 14, and 15). In these species, females have less opportunity to interact with kin other than their immature offspring, and bonds among matrilineal kin are likely to be less well developed than in the baboons and macaques or absent altogether. Similarly, in monogamous species such as titi monkeys (chap. 5) and gibbons (chap. 12), extended kin networks such as those found in baboons and macaques are unlikely to develop.

In gibbons, there is evidence that parents sometimes help offspring acquire a territory adjacent to their own (Tilson 1981), although high rates of hunting in the area of this study may have led to unusual numbers of available territories and thus to this phenomenon (chap. 12). However, if parental assistance in territory acquisition is found to be widespread and neighboring gibbon groups contain related individuals, this could have important effects on intergroup relations within a local area (chap. 22). Similarly, one-male groups of hamadryas baboons and gorillas usually contain females that are not close genetic relatives. However, long-term data have shown that, as adults, close female kin do occasionally live in the same group (chaps. 10, 14). In gorillas, when close female relatives live in the same group they exhibit higher levels of spatial proximity and grooming than do less closely related females (Harcourt, in prep.). Even in the well-studied baboons and macaques, patterns of dispersal may have led to an overemphasis on kin relations among females, which do not disperse, as opposed to males, which do. As noted earlier, male rhesus macaques and vervets often disperse in the company of brothers or transfer into groups that a male relative has previously joined. Finally, the need for long-term records in the study of kinship is illustrated particularly in chapter 2, which describes the importance of matrilineal relationships in bushbabies—prosimians that spend most of their time alone.

Thus it is important to distinguish two questions when evaluating the relation between matrilineal kinship and social organization in primates. First, do females interact with kin more often in species where male dispersal, as opposed to female dispersal, predominates? Here the answer is certainly yes, despite the exceptions noted above. Second, given opportunities to interact with both kin and nonkin, do individuals in some species show a greater preference for kin? Here data are available primarily for the macaques, baboons, and vervets discussed earlier (all of which form multimale and female groups), and the answer for females in these species is clearly yes (see S. Gouzoules 1984 for more discussion). Data for males, and for other species, however, are only beginning to become available, and the question of whether biases in favor of kin are pervasive among primates remains open.

Mortality. Just as differences in social organization between species can affect the opportunity for kin-correlated behavior, differences in mortality can have a similar effect between groups of the same species. As Altmann and Altmann (1979) have noted, most free-ranging primate populations are characterized by relatively lower birth rates and higher mortality rates than found in captive or provisioned groups. As a result, large matrilines of the sort described for Cayo Santiago rhesus and some groups of Japanese macaques are unlikely to occur at every study site (Melnick and Kidd 1983; chap. 20). Indeed, Melnick and Kidd (1983) hypothesize that, in stable or declining populations, more individuals may actually be patrilineally rather than matrilineally related.

Mechanisms of Kin Recognition

How do animals recognize their kin? During the past few years a variety of mechanisms have been proposed, most notably developmental association and "phenotype matching." These putative mechanisms are illustrated here by studies conducted on ground squirrels, pigtailed macaques, rhesus macaques, and baboons (for detailed review see Holmes and Sherman 1983; S. Gouzoules 1984).

Holmes and Sherman (1982) conducted a laboratory study in which, immediately after birth, ground squirrel pups were separated into four groups for purposes of rearing: littermates (siblings) reared together, littermates reared apart, nonlittermates reared together, and nonlittermates reared apart. After several months, individuals from different groups were paired together and their social behavior observed. Not surprisingly, the highest rates of affinitive behavior and lowest rates of aggression were shown by littermates reared together, while nonlittermates reared apart showed the opposite extreme. In addition, nonlittermates reared together showed significantly less aggression than did nonlittermates reared apart, which indicates that association during development is an important factor affecting bonds between matrilineally related individuals. Finally, and most interesting, evidence showed that kin reared apart were less aggressive toward one another than were nonkin reared apart. To explain this result, Holmes and Sherman (1982) postulated a mechanism called "phenotype matching," in which individuals compare certain genetically controlled traits in other individuals against a learned "template." If such a mechanism were eventually found to exist in primates, it would have important implications for the study of nonhuman primate kinship because, for example, it would offer a means by which male primates could recognize their offspring even in the absence of any association during development.

To test the idea of paternal kin recognition, laboratory studies have presented young pigtailed macaques with a "choice" between stimulus individuals that were either related or unrelated to the subject through the paternal line. One study (Wu et al. 1980) found evidence for recognition of paternal kin, while an attempt to replicate this study (Fredrickson and Sackett 1984) did not.

Using quite different methods, Stein (1984a) looked for evidence in wild yellow baboons for a relationship between probable paternity (based on copulations around the time the mother conceived) and male relationships with infants. Probable fathers were indeed more likely than other males to develop an affinitive bond with an infant, but Smuts (1985) showed that, in olive baboons, such relationships were nearly always the result of a long-term bond between the male and the infant's mother. Probable fathers that did not share a bond with the mother were unlikely to develop a close bond with her infant, and males that were not observed mating with the mother, but that did have a long-term relationship with her, did form a close bond with the infant. These findings from the wild are consistent with results from a study of captive rhesus macaques, where true paternity was determined by analysis of genetic markers (Berenstein,

Rodman, and Smith 1981). The rhesus fathers showed a slight but significant tendency to associate with their own offspring, but "the effect of paternity disappeared when maternal association with males was controlled" (p. 1061). These studies of baboons and macaques cast considerable doubt on the ability of males to recognize their own infants independent of close behavioral association (for further discussion see S. Gouzoules 1984; chap. 28).

Recognizing the Kin Relationships of Others

Because many primates discriminate among conspecifics on the basis of the closeness of their relatedness to them, primates may be said to "classify" social companions on the basis of kinship (as well as certain other criteria, such as dominance rank). It is also possible that individuals may recognize the kin relationships of other group members. For example, the rhesus monkey recruitment screams described earlier convey information about the caller's kinship and dominance relationship to its opponent. Monkeys hearing and responding to these screams apparently recognize the caller's network of relationships (for similar results see Cheney and Seyfarth 1980, 1982b; chap. 36). In a study of pigtailed macaque agonistic behavior, Judge (1983) reported that individuals involved in aggressive interactions later "reconciled" with the relatives of their opponents. These relatives had not been involved in the earlier dispute. Judge's study, as well as others (de Waal 1982; Smuts 1985), indicates that some primates are indeed aware of the social relationships, including kin relations, of other group members.

Summary

Kinship is among the most important determinants of behaviors such as grooming and agonistic aiding in social groups of several species of Old World monkeys and chimpanzees. Kinship also influences the ontogeny of social behavior in infants as well as the acquisition and maintenance of dominance rank and mate choice. Although kinship is generally more crucial to female than male behavior, at least in those species in which males emigrate from their natal group, it also may influence patterns of male migration. The role of kinship in the social behavior of other species of primates largely remains unknown because of insufficient data, but variation among primate species in the importance it plays is likely to exist due to differences in demographic patterns such as dispersal and mortality as well as social organization. The most important mechanism whereby kin are recognized in primates is probably association during development.

25 | Conflict and Cooperation

Jeffrey R. Walters and Robert M. Seyfarth

Most primates live in groups. There are numerous potential benefits to sociality, including increased protection from predators, cooperative defense of food resources, and collective rearing of offspring. Group life also provides an opportunity for long-term cooperative relationships. At the same time, however, sociality entails a number of costs, since group-living animals must compete with one another for scarce resources such as food, water, and mates.

Life within a primate group is thus delicately balanced between competition and cooperation (Crook 1970). In this chapter we provide an overview of competitive and cooperative interactions within groups, the mechanisms underlying these behaviors, and their short-term effects. In the first part of this chapter, we provide some broad descriptions of the kinds of competition and cooperation that occur among primates. In the second part, we argue that an understanding of the significance of these behaviors depends crucially on their patterns: who interacts with whom, in what contexts, and how different types of behavior are related to each other. Since the distribution of behavior among particular individuals has only been studied in detail among the cercopithecines and some apes, these species are the focus of the second part of the chapter. The evolutionary forces that appear to have shaped such interactions will be considered in chapter 26.

DESCRIPTION OF AGGRESSIVE AND COOPERATIVE BEHAVIORS

Aggression

The most obvious manifestation of competition within primate groups is aggressive behavior. As in many birds and nonprimate mammals, aggression often takes the form of displays: conspicuous signals exchanged between two or more animals that rarely lead to physical contact but usually result in one animal gaining a particular resource. Aggressive displays among primates are extremely varied. Male ring-tailed lemurs (*Lemur catta*) smear a pheromone on their tails and wave them in the air when competing over access to females (Jolly 1972); a common squirrel monkey (*Saimiri sciureus*) male threatens others by displaying his erect penis (Ploog 1967); baboons (*Papio cynocephalus*) "flash" their eyelids, revealing a patch of white skin (Hall and DeVore

1965); gorillas (*Gorilla gorilla*) stand bipedally, beat their chests, and charge (Schaller 1963). Virtually all primate species accompany such visual threats with vocalizations that can also signal aggression when given alone.

In addition to displays, primate aggressive behavior also includes actions such as staring, jerking the head, chasing, lunging, shaking branches, and slapping the ground (fig. 25-1). These may be followed by physical contact in the form of hitting, grappling, holding down, and biting. Such aggression frequently results in injury, but rarely death. In contrast to the diversity of displays, these contact or near-contact behaviors are remarkably uniform in their motor patterns across primate species.

Although aggressive behavior usually reflects competition between two or more individuals, competition is not always overtly aggressive. Perhaps the most common competitive interaction among many primates is what Rowell (1966b) called an "approach-retreat" interaction, in which one animal simply approaches another, shows no obvious posture or threatening expression, and the other moves away. An even more subtle manifestation of competition has been termed "competitive exclusion" (e.g., Hardin 1960). It occurs, for example, when a dominant male chimpanzee (*Pan troglodytes*) feeds uninterrupted on a large fruit while others sit nearby watching. Though no interactions occur, the animals are nonetheless competing, as shown by the eagerness of the others to pick up the fruit once the male has left. Competitive exclusion and approach-retreat interactions are common in species with well-established dominance relations and have important effects on social interactions in many primate groups.

Primates also exhibit behaviors of submission or appeasement. Such actions presumably reduce either the frequency or intensity of aggression directed toward the submissive animal (chap. 34). Appeasement behavior includes screaming, grunting, cowering, and grimacing. Like many aggressive behaviors, many appeasement gestures are rather uniform across species. For example, hitting and cowering appear to be employed similarly by humans and nonhuman primates (Goodall and Hamburg 1975). In contrast, although the human smile resembles the grimace of nonhuman primates, the smile has many more diverse functions (Eibl-Eibesfeldt 1975).

Contexts of Aggression

Some primate aggression is linked directly to the acquisition of resources. Individuals supplant others from food items and resting or sleeping sites, and they defend their resting or feeding sites against intrusion. A brief exchange of overtly aggressive behavior may be involved, or one individual may defer to another before aggression actually occurs.Such interactions are common, but severe aggression in these circumstances is rare.

One context in which aggression can be especially common and severe is in competition between adult males over access to reproductive females. Some such competitive interactions escalate to involve physical contact and wounding (see chap. 31).

Male-male competition does not always involve aggression toward other males. Some forms of infanticide, for example, are thought to be an indirect form of male-male competition (Hrdy 1977, 1979; chap. 8). In seasonally breeding species, male aggression toward females increases during the mating season (Enomoto 1981; see also chap. 32). This aggression sometimes functions to keep a female spatially separated from other males, and it can thus be interpreted as a form of male-male competition for mates. Males also aggressively

herd their females to keep them away from males in other groups (hamadryas baboons [*Papio hamadryas*]: Kummer 1968; chacma baboons: Stoltz and Saayman 1970; Hamilton, Buskirk, and Buskirk 1975a; Cheney and Seyfarth 1977).

Much of the aggression within primate groups is not directly related to resource competition but instead appears to involve the establishment and maintenance of dominance relationships. Such aggression may be an indirect form of resource competition, as dominance is often related to access to resources (chaps. 20, 26). In a 14-month study of competition among adult female vervet monkeys (*Cercopithecus aethiops*), an average of 10.5% of all aggressive interactions occurred over access to food or water, 19.7% occurred over access to a social partner, and 73.9% occurred in circumstances in which the cause was not apparent (Seyfarth 1980; see also Lindburg 1971; Seyfarth 1976; Deag 1977 for similar results among rhesus macaques [*Macaca mulatta*], chacma baboons, and Barbary macaques [*Macaca sylvanus*], respectively). Dominant individuals in many species frequently approach subordinates, apparently causing them to withdraw, or they may actively harass subordinates by directing aggression at them without apparent provocation, thus eliciting submissive behavior (e.g., chim-

FIGURE 25-1. A juvenile long-tailed macaque, lunging at another juvenile, stares at her opponent with a wide open mouth. (Photo: H. van Beek)

panzees: de Waal and Hoekstra 1980; bonnet macaques [*Macaca radiata*]: Silk, Samuels, and Rodman 1981; Japanese macaques [*Macaca fuscata*]: Kurland 1977; savanna baboons: Wasser and Barash 1983). Occasionally, such aggression can be severe, but usually harassment does not involve physical contact. However, when low-ranking animals contest the status of higher-ranking individuals, or when animals support a close genetic relative who has already been attacked, serious aggression can result (e.g., chimpanzees: Goodall 1971; gorillas: Harcourt 1979b, 1979c; bonnet macaques: Silk 1982).

In summary, primate aggression takes a variety of forms, from subtle stares to outright killing. Like many other group-living mammals, however, primates do not consistently fight over resources, but rather exhibit both ritualized displays and relatively stable dominance relations that allow most disputes to be resolved without physical violence. Although the cause of aggression can sometimes be deduced from the circumstances in which it occurs, aggressive behavior more often arises for reasons that are not obvious to the observer. In these cases animals may be concerned primarily with the maintenance of existing dominance relationships.

Performers of Aggression

The only individuals not involved in aggression within primate groups are young infants. Excepting infanticide, young infants in most species neither perform aggression nor receive aggression from other group members (e.g., rhesus macaques: Berman 1980a; yellow baboons: Altmann 1980; vervets: Horrocks and Hunte 1983). This period of amnesty usually ends before infancy does, however, and by the time they are juveniles youngsters are actively involved in aggressive interactions. In fact, juveniles in some species exhibit particularly high rates of aggression, correlated with their acquisition of adult dominance ranks.

Rates of aggression between particular age-sex classes vary widely depending on species, social organization, study site, and social context. For example, although squirrel monkeys live throughout the year in multimale, multifemale groups, outside the breeding season the sexes are spatially segregated. During this time males rarely interact with females and direct little or no aggression toward them, even though they continue to behave aggressively toward other males (Baldwin 1969; DuMond 1968). In species with one-male groups, rates of male aggression toward females are highly variable. Although male hamadryas baboons actively herd their females to keep them near (Kummer 1968), herding in gelada baboons (*Theropithecus gelada*) is less common; males instead use affinitive behavior to influence female movements (Dunbar and Dunbar 1975; U. Mori 1979b). Similarly, Hall and DeVore (1965) found high rates of male-male aggression in open-country, savanna groups of ba-

boons, whereas Rowell (1972) saw many fewer fights among males in forest-dwelling groups. Among immature baboons, Cheney (1977b) found higher rates of aggressive interactions between juvenile females than between juvenile females and juvenile males. This was not because juvenile females were inherently more aggressive, but because juvenile females frequently competed with each other over access to infants.

The frequency of aggression between two individuals does not necessarily reflect the overall aggressiveness of their social relationship. Since rates of aggression are to some extent dependent on the frequency with which individuals come into contact, animals that associate regularly may appear to be more aggressive toward each other than ones that spend less time together. For example, same-sexed peers and kin often show high rates of aggression. However, these individuals usually interact less aggressively than others when the frequency of association is taken into account (e.g., toque macaques [*Macaca sinica*]: Dittus 1977; Japanese macaques: Kurland 1977). Similarly, in species with one-male groups, male-male aggression may occur relatively infrequently simply because there is little opportunity for such interaction. When males do come into contact, however, their interactions are often extremely hostile (e.g., blue monkeys [*Cercopithecus mitis*]: chap. 9; redtail monkeys [*Cercopithecus ascanius*]: Cords 1984a; chap. 9; black-and-white colobus [*Colobus guereza*]: Oates 1977c; gray langurs [*Presbytis entellus*]: Hrdy 1977). In multimale groups of baboons, aggression between adult males does not always occur at high rates, but since aggression often accounts for an extremely high proportion of their interactions, their relationships appear to be relatively hostile.

To characterize the aggressiveness of social relationships, one must therefore consider the frequency of aggression relative to other kinds of interaction as well as its absolute frequency. One must also consider the intensity of aggression. For example, in one group of olive baboons, adult females were attacked by other females four times as often as they were attacked by males, but only male attacks resulted in serious wounds (Smuts 1985). As mentioned above, relatives frequently engage in aggressive interactions, but these tend to involve mild "bickering" rather than serious aggression (Kurland 1977).

Finally, simple frequencies of aggression may not be useful in describing competitive relationships because different individuals compete in different ways. Adult females, for example, generally compete for mates at lower rates than they compete for food, whereas among males competition for mates is much more common. Because a single food item presumably has less effect on an animal's fitness than does access to a mate, competition among females usually involves less high-intensity aggression than that among males (chap. 32). In one group of baboons, for instance, the ratio of approach-retreat interactions to bouts of overt aggression was 20:1 for females and 1.7:1 for males (Seyfarth 1976). Despite such differences in the pattern of competitive interactions, however, competition among same-sex adults may have just as clear an effect on female fitness as it does on male fitness (Hrdy 1981b; see also chap. 26).

Cooperation

As is the case with aggressive behavior, cooperative behavior within primate groups is extraordinarily diverse. Perhaps the most common form of affinitive behavior in primates is grooming, in which one animal, picking through the fur of another, removes ectoparasites and cleanses wounds. In addition to its utilitarian functions (e.g., Hutchins and Barash 1976), grooming is also used to develop cooperative relationships, as discussed later. Other cooperative behaviors include warning calls that signal the presence of predators (reviewed in chap. 36), collective defense against predators (chap. 19), collective defense of a home range (chaps. 22, 23), food sharing (chap. 4), mutual tolerance at food sites (or "cofeeding": Kawai 1965; Stein 1984b), and the formation of alliances.

Alliances occur whenever a third individual intervenes in an aggressive interaction between two others, to aid one of the antagonists in either attack or defense (fig. 25-2) (such "third parties" may also influence aggressive interactions in other, more subtle ways discussed in chapter 34).

Triadic, quadratic, and even more complex interactions are common among primates. In Deag's (1977) study of Barbary macaques, for example, 219 of 531 aggressive interactions (41%) involved more than two par-

ticipants. In Walters's (1980) study of juvenile female baboons, 16% of the aggressive interactions included alliance formation. The pervasiveness and importance of alliances among primates is also suggested by the fact that many species exhibit stereotyped solicitation behaviors, directed at third parties by antagonists (e.g., long-tailed macaques [*Macaca fascicularis*]: de Waal, van Hooff, and Netto 1976; baboons: Packer 1977; Walters 1980; chimpanzees: de Waal 1982; Nishida 1983b). In baboons, the soliciting animal repeatedly looks toward a third party and then vigorously turns its head to gaze briefly at its antagonist. The solicitor continues to look back and forth between potential ally and antagonist, often simultaneously threatening the latter. Screaming may also function to bring about intervention, and information about an angatonist may even be conveyed by the type of scream produced (Gouzoules, Gouzoules, and Marler 1984).

Despite the diversity of their cooperative behaviors, primates do not necessarily exhibit more cooperation than other mammals. The cooperative hunts of lions, for example (e.g., Schaller 1972), have no parallel among primates (Busse 1978), nor is primate food-sharing as widespread as that found in many social carnivores. Nevertheless, cooperative social interactions are one of the most conspicuous features of life in a primate group; they take up a considerable amount of the animals' time.

Participants and Frequencies of Cooperative Behavior

The frequency of cooperative interactions depends, among other factors, on the species, habitat, individual characteristics such as age or reproductive condition,

FIGURE 25-2. Four screaming chimpanzees band together to threaten a displaying subadult male (*right*). The coalition was formed after the male attacked the female second from left. (Photo: Frans de Waal)

and particular sort of cooperative behavior involved. In virtually all primate species, the most common grooming and alliance partners are mother and offspring. Mothers and offspring are also the individuals who most often feed next to each other (chap. 27) and who most often exhibit food sharing (e.g., Goodall 1968; Silk 1979; but see also chap. 4, in which food sharing between tamarin males and infants is described). The factors affecting cooperative behavior between males and infants are discussed in chapter 28 and will not be considered here.

In cercopithecine groups where females remain together throughout their lives (chap. 11), grooming and alliances are also common among other female kin, such as adult sisters (chap. 24). In addition, females groom and form alliances with unrelated individuals, although usually at lower rates. In contrast, among species such as red colobus monkeys (*Colobus badius*), hamadryas baboons, chimpanzees, and gorillas, in which the female members of a group are usually unrelated, female-female grooming and alliances occur at much lower rates (chaps. 8, 10, 14, 15). Thus the pattern of cooperative behavior within any species depends in part on its social structure (chap. 32).

Grooming and alliances between adult males and females in some species occur at higher rates when females are sexually receptive than at other times (rhesus macaques: Kaufmann 1967; hamadryas baboons: Kummer 1968; yellow baboons: Hausfater 1975; vervets: Andelman 1985), whereas in others reproductive changes appear to have less effect (e.g., bonnet macaques: Simonds 1965; geladas: Dunbar 1983b). In many species females groom males more often than vice versa, while in spider monkeys (*Ateles fusciceps*) the reverse is true (Eisenberg 1976). Harcourt (1979c) studied two adjacent gorilla groups that illustrate intraspecific variation in male-female grooming relations. In one group the male groomed the females, while in another group the reverse was true.

Ecological conditions may also affect the frequency of cooperative interactions. When baboons (Oliver and Lee 1978) and vervets (Lee 1983c) were studied in different habitats, those in ecologically richer areas were found to spend less time searching for food and more time in cooperative social behavior. Interestingly, however, all cooperative behaviors were not equally affected. In both species, juveniles in dry habitats played at lower rates than juveniles in wetter habitats, but groomed equally often (Oliver and Lee 1978; Lee 1981, 1983c). Similarly, while forest-dwelling redtail and red colobus monkeys spend quite different amounts of time searching for food, they devote an equal amount of time to grooming (Struhsaker 1980). These observations emphasize the importance of grooming to primates and suggest that only extreme ecological conditions will cause animals to cease grooming altogether.

AGGRESSIVE AND COOPERATIVE INTERACTIONS AND THE STUDY OF SOCIAL RELATIONSHIPS

Dominance

Dominance in primates has been defined in many ways, but it is most commonly measured in terms of the direction of approach-retreat interactions (e.g., Rowell 1966b; Seyfarth 1976) or the direction of submissive and aggressive behaviors in agonistic interactions (Sade 1967; Koyama 1967; Missakian 1972; Hausfater 1975). When such aggressive or submissive behaviors are used to define winners and losers, the outcome of interactions between individuals is often highly consistent over time. Following Schjelderup-Ebbe (1935), who first described such predictable interactions among chickens, the winner in such disputes is called the dominant animal. Dyadic dominance relationships defined in this way often form a linear, transitive dominance hierarchy: that is, A will be dominant to all other individuals, B will be dominant to all but A, C will be dominant to all but A and B, and so on.

Not all aggressive behaviors yield equally clear-cut dominance relationships, however. A chase, for example, often fails to predict the eventual winner of an aggressive bout because during such a bout the subordinate may briefly chase the dominant before retreating (e.g., Angst 1975). Aggressive behaviors involving physical contact, and submissive behaviors give more consistent results.

While dominance relationships are usually defined in terms of one type of competitive interaction, the term *dominance* is most useful if it is relevant to more than one aspect of behavior (Hinde 1978, 1983c). For example, in one study of Japanese macaques (A. Mori 1979a), dominant females not only had priority of access to food but also reproduced more often than subordinates, had heavier babies, and had babies that were more likely than others to survive their first year. In this case, dominance as originally defined in terms of competitive interactions had considerable predictive value in other contexts.

In contrast, in one population of olive baboons, males who were particularly successful in competitive interactions over meat or females were relatively unsuccessful in other circumstances (Strum 1982; Smuts 1982, 1985; Manzolillo 1982). This lack of correlation between dominance in different contexts does not, as some have suggested (Gartlan 1968; Rowell 1974b), argue against the usefulness of the dominance concept (e.g., Bernstein 1981), particularly given the many close correlations between dominance and other aspects of behavior. It does, however, argue against using priority of access to resources to define dominance (Syme 1974; Richards 1974).

Two further points concerning dominance are worth

mentioning. First, dominant animals are not necessarily the most aggressive members of their group. Dominance is generally defined in terms of a consistent direction of aggressive behavior between individuals and does not necessarily predict the rate at which aggression occurs. Indeed, long-term studies have shown that an alpha female in a group of macaques, for example, can maintain high dominance rank while only rarely being aggressive to others (Yamada 1963; but see also Riss and Busse 1977; Horrock and Hunte 1983; and chap. 7, which present data showing that high rank was positively correlated with rates of aggression in male chimpanzees, female vervets, and brown capuchins [*Cebus apella*], respectively). Second, in species with dominance hierarchies, it is not yet clear whether such rank orders are merely inventions of the observer (Altmann 1981; Bernstein 1981) or whether relative ranks are recognized by the individuals themselves. Seyfarth (1981) and Kummer (1982) provide evidence suggesting that monkeys can recognize each other's ranks, but this has not yet been conclusively demonstrated.

While stable, long-term dominance hierarchies are characteristic of many primates, particularly cercopithecines (chaps. 11, 24, 26), there are also many primate species in which dominance hierarchies are unclear, ambiguous, or apparently nonexistent (e.g., black-and-white colobus: Dunbar and Dunbar 1976; patas (*Erythrocebus patas*): Rowell and Olsen 1983; redtail monkeys: chap. 9; blue monkeys: chap. 9; female squirrel monkeys: Baldwin and Baldwin 1981). Furthermore, dominance relations may be complex even in species characterized by clear hierarchies. For instance, alliances may complicate dominance relationships, particularly in large groups. Chimpanzee males form clear, linear dominance hierarchies in small communities (Nishida 1979), but in larger groups only classes of high-, medium-, and low-ranking males can be distinguished (Bygott 1979). Similarly, it is common for the outcome of dominance interactions involving immatures to vary depending on the proximity of potential allies. Dominance relationships are considered in detail here both because we know a great deal about them and because they have important effects on other aspects of social behavior and reproduction (see also chap. 26). This emphasis on dominance should not, however, be taken to mean that such predictable competitive interactions are typical of all primate species.

In the following sections we focus on how dominance ranks are acquired, how males and females differ in the maintenance of dominance relationships, and how dominance affects other aspects of social behavior within groups.

Acquisitions and Maintenance of Rank

One might expect that dominance rank would be a func-

tion of attributes such as size, strength, or agility that clearly affect performance in aggressive interactions. While this may often be true for males (olive baboons: Hall and DeVore 1965; Packer 1979a, 1979b; chimpanzees: Bygott 1979; Nishida 1983b), it is certainly less often true for females. In fact, it is not unusual in many cercopithecine species to see a small juvenile female regularly supplant individuals twice her size, or a small, crippled adult maul a much larger, healthier one. How are these relationships established?

Females. Research on baboons, macaques, and vervets has shown that the development of dominance relationships among females in these species is best explained in terms of several stages. As infants, females are subordinate to all adult females, though infants of high-ranking mothers are treated differently from the infants of low-ranking mothers (Berman 1980a; chap. 27).

Juveniles remain subordinate to females who are dominant to their mother. However, as the infant gradually gains her independence, relationships between herself and adult females subordinate to her mother go through a gradual transition from the elder being dominant to the younger being dominant. In long-tailed macaques (de Waal 1977), yellow baboons (Walters 1980), and rhesus macaques (Datta 1983b), the young juvenile female is initially submissive and the adult female dominant. Then follows a stage in which the juvenile directs both aggressive and submissive behaviors at the adult and eventually ceases performing submissive ones. Finally, the adult begins to direct submissive behaviors toward the juvenile and then ceases being aggressive, so the relationship becomes like that of adults. In vervet monkeys the process is similar, but submission by the juvenile and aggression by the adult are less frequent (Horrocks and Hunte 1983; Lee 1983b).

In all species studied, dominance relationships during these transition periods are inconsistent and can be affected by many different factors. In rhesus monkeys, the timing of an immature female's aggression against those that rank above her is determined largely by the immature's age and size (Datta 1983b). In vervets, social context is important (Lee 1983b). Juveniles can supplant adult females when competing for the opportunity to groom the juvenile's mother or siblings, but not when competing for access to other social partners. In many species, interventions (or the threat of interventions) by the mother and adult female kin play an important role in the acquisition of female dominance ranks (Cheney 1977a; Berman 1980a). Juvenile females typically behave more aggressively (and less submissively) toward adults who rank below the juveniles' mothers when the mother or close kin are nearby than when they are not (Japanese macaques: Kawai 1965b; vervets: Lee 1983b; Horrocks and Hunte 1983). A juvenile female may also be able to elicit deference from adults in the presence of her mother or other close kin long before she can in a one-to-one interaction.

Clearly, however, there is more to female rank acquisition than mothers bullying subordinates into accepting their daughters as dominants. Juvenile macaques, baboons, and vervets sometimes acquire the ranks their mothers occupied at the juvenile's birth even if the mother dies or falls in rank before the process of rank acquisition is completed (Walters 1980; Hasegawa and Hiraiwa 1980; Hamilton, Busse, and Smith 1982; Lee 1983d). In addition, juveniles invariably challenge adults who rank below the juveniles' mothers but rarely challenge adults who rank above the juveniles' mothers (Datta 1983b). It appears that an immature learns about its expected dominance relations with others at a very early age. It may do so by observing how others behave while handling it as an infant (e.g., Cheney 1978a; Berman 1980b), by observing interactions between its mother and others, and by the aggression it receives in later infancy from adults (Gouzoules 1975; Altmann 1980; Berman 1980a; Horrocks and Hunte 1983; Datta 1983b).

Within families, juvenile females usually assume ranks immediately below those of their mothers, but daughters occasionally rise in rank over their mothers in groups where genealogies are small (Chikazawa et al. 1979; Hausfater, Altmann, and Altmann 1982). Rank among female siblings is usually inversely related to age, with younger sisters rising in rank above older sisters, but exceptions occur regularly (e.g., Missakian 1972; Sade 1972a; Kurland 1977; Hausfater, Altmann, and Altmann 1982; Horrocks and Hunte 1983). The proximate cause of this rank reversal seems to be differential maternal support, since mothers usually aid their younger offspring against their older ones (Kurland 1977; Chapais and Schulman 1980), and exceptions are especially frequent when the mother is deceased (e.g., Walters 1980). Chapais and Schulman have argued that, in functional terms, it may be more beneficial for mothers to aid younger offspring because their reproductive value exceeds that of their older siblings. In contrast, Horrocks and Hunte (1983) hypothesize that such differential aiding of younger offspring is a strategy adopted by mothers to prevent daughters from forming coalitions against them.

Thus far we have emphasized dominance ranks and the process of rank acquisition among female cercopithecines, particularly baboons, macaques, and vervets. Some data are available, however, from species in which the characteristics of female dominance differ. Among mantled howlers (*Alouatta palliata*), gray langurs, and chimpanzees, female dominance appears to have an age component, with younger females generally dominant to older females (Hrdy and Hrdy 1976; Dunbar 1980b; Jones 1980b; chap. 15). This effect seems to result from the greater aggressiveness of younger females. However, in howlers and langurs not all young females become

dominant to older ones, and the longitudinal data necessary to determine the role, if any, of maternal rank and such other factors as fighting ability, or to characterize precisely the effect of age, are not yet available.

Capuchins (Izawa 1980; chap. 7), and gorillas (chap. 14) represent another class of species in which dominance relations are known to exist among females, but their mechanisms of acquisition and maintenance are unknown. In most gorilla groups females are unrelated to each other, female-female aggression occurs at very low rates, and there is no consistent direction to approach-retreat interactions. Dominance ranks have only been detected in one group, where many females failed to emigrate and ranks were related to age (chap. 14).

Once established, female dominance relations in cercopithecines are often stable over the adult lives of the individuals. However, adult females occasionally fall in rank. Individual loss of rank sometimes occurs at an advanced age (e.g., Hausfater, Altmann, and Altmann 1982). In other cases, entire matrilines fall in rank as a result of persistent, aggressive challenges from coalitions of previously lower-ranking females (e.g., Koyama 1970; Chance, Emory, and Payne 1977; Smuts 1980). These rare but important events indicate that, at least in baboons and macaques, females are motivated to increase dominance status whenever the opportunity arises.

Males. The development of dominance relationships among immature males is quite different from that among immature females (Angst 1975). In cercopithecines, among male peers of similar size, maternal rank is an important determinant of immature rank, but maternal rank becomes progressively less important with age (baboons: Lee and Oliver 1979; Pereira and Altmann 1985; rhesus macaques: Datta 1983a; vervets: Horrocks and Hunte 1983; Lee 1983b). In Japanese macaques, mothers gradually cease to groom their sons and support them in alliances as the sons grow older, even though the same mothers continue to groom and support their daughters (Oki and Maeda 1973; see also Missakian 1972 for rhesus macaques; Silk, Samuels, and Rodman 1981 for bonnet macaques; Fairbanks and McGuire 1985 for vervets; Pereira and Altmann 1985 for baboons). As young males grow older, their dominance ranks appear to depend increasingly on size, strength, and other determinants of fighting ability. In many species males are larger than females, and they become dominant to female peers and to adult females. Males may not become dominant to all adult females until adolescence or even early adulthood, depending on the species' degree of sexual dimorphism in size. Several studies indicate that in macaques in which adult females weigh 75 to 80% as much as adult males, sons of high-ranking females are more likely than other males to attain high rank as adults, both in their natal groups and following transfer (Japanese macaques: Koyama 1970; Sugiyama 1976; rhesus macaques: Koford 1963; Chapais 1983d; Meikle, Tilford, and Vessey 1984). However, in baboons, in which adult females weigh only about 50% as much as adult males, there is no evidence that maternal rank affects the rank of adult males.

In rhesus and Japanese macaques, males that have recently entered a group are usually low ranking, and rank typically increases over time; in both species, rank is positively correlated with age and length of time in the group (rhesus macaques: Drickamer and Vessey 1973; Tilford 1982; Japanese macaques: Norikoshi and Koyama 1975; Sugiyama 1976). In baboons and vervet monkeys, however, the pattern is very different. Males often achieve very high ranks after transfer, and ranks gradually decline as males age (vervets: Henzi and Lucas 1980; baboons: Packer 1979a, 1979b; Rasmussen 1980; Busse and Hamilton 1981; Strum 1982; Smuts 1985). These interspecific differences remain unexplained. Baboon males form coalitions in competition with other males, but these same-sex coalitions (typically among low-ranking individuals) do not appear to affect dominance rank (chap. 31). In macaques, however, male-male coalitions, which often involve close kin, are occasionally used to increase rank (e.g., Chapais 1983c). In one group of olive baboons (Packer 1979a), and in several groups of toque macaques (Dittus 1977), there was a significant positive correlation between male rank and weight, but it is unclear whether the males achieved high rank because they were larger or were able to obtain more food because of their high rank. In another population of olive baboons, there was no correlation between male weight and rank (Smuts 1982, 1985).

As with females, much less is known about the acquisition and maintenance of male dominance ranks outside the cercopithecines. Among chimpanzees and bonobos (*Pan paniscus*), males form dominance hierarchies, and early in life they may be supported by their mothers in aggressive interactions (Bygott 1979; Kano, in press). Maternal support seems unimportant in interactions among adult males, but, at least in chimpanzees, coalitions with other males are often critical in determining the outcome of contests for alpha status (chap. 15). In howler monkeys, male-male coalitions can result in expulsion of rivals from the group and changes in dominance relationships (C. B. Jones 1982; Sekulic 1983a). Although male ring-tailed lemurs (Budnitz and Dainis 1975), sifakas (*Propithecus verreauxi:* Richard 1974b), red colobus (Struhsaker and Leland 1979), capuchins (chap. 7), and spider monkeys (van Roosmalen 1980) also form dominance hierarchies, little is known about the acquisition or maintenance of dominance in these animals.

Although dominance relationships between males may be stable over short periods of time, a male's rank typically changes many times over the course of his lifetime. Male rank is therefore less stable than female rank. Over a 4-year period in three groups of vervets, adult females changed ranks at the rate of 0.11 ranks per female per year, whereas the comparable figure for males was 0.75 (Cheney 1983a). In a troop of yellow baboons, no agonistically induced (as opposed to demographically induced) rank changes occurred among adult females in a 400-day period (18,884 dyad-days), whereas 19 occurred among adult males (1 per 528 dyad-days) (Hausfater 1975). Among males, dominance reversals appear to result primarily from dyadic agonistic interactions (e.g., Hausfater 1975).

Distribution of Competitive and Cooperative Interactions

Competitive and cooperative behaviors within primate groups are not distributed at random. To cite just one example, during a 14-month study of adult female vervet monkeys, the most frequently groomed individual received four times more grooming than the least frequently groomed individual, and there was also a fourfold difference in the amount of aggression received (Seyfarth 1980). To understand how these differences occur, it is necessary to examine in detail the distribution of competitive and cooperative interactions and to consider how one type of social interaction can affect another.

Distribution of Interactions among Females and Immatures. 1. The Effects of kinship. The pervasive effects of kinship on all aspects of social behavior are reviewed in chapter 24. Summarizing that chapter's main points, interactions between close genetic relatives—particularly matrilineal relatives—account for a substantial portion of the cooperative behavior observed within primate groups. Kin also tend to engage in serious aggression less often than nonkin, especially in relation to their frequency of association. Given the detailed discussion of primate kinship in chapter 24, we concentrate here on other factors that influence the distribution of competitive and cooperative interactions.

2. The Effects of dominance rank. Dominance hierarchies among adult female and immature cercopithecines can affect social interactions in two ways. First, dominance can affect the frequency of interaction. High-ranking animals are sometimes more attractive social partners than are low-ranking animals. Thus dominance as defined in terms of aggressive interactions is often strongly correlated with partner preferences in other, more cooperative behaviors. Second, because females and immatures may compete for social partners, either overtly or through competitive exclusion, dominance relations can affect who is able to interact with whom. Frequently, the behavior and apparent social preferences of low-ranking animals are more constrained than the behavior and social preferences of high-ranking animals (Vaitl 1978).

In many species with adult female dominance hier-archies, it appears that, in addition to their kin, individu-als are preferentially attracted to the members of high-ranking families. For example, in Sade's (1972b) study of grooming among rhesus macaque females, the total amount of grooming an individual received was more closely correlated with her rank than with the size of her family (see also Chapais 1983b). High-ranking females also receive more total grooming than others in capu-chins (chap. 7), hamadryas baboons (Kummer 1968; Stammbach 1978), Japanese macaques (Oki and Maeda 1973), chacma baboons (Seyfarth 1976), vervets (Sey-farth 1980; Fairbanks 1980), and bonnet macaques (Silk, Samuels, and Rodman 1981). Among baboons and ver-vets, females compete for the opportunity to groom one another, and when the confounding effects of repro-ductive state are held separate, high-ranking females are the most attractive grooming partners (Seyfarth 1976, 1980). In other studies, however, no attraction of high rank is found, which suggests that this effect is less uni-versal than that of kinship (Rowell 1966b; Altmann 1980; Walters 1986).

Effects of rank on social interaction other than groom-ing are less clear. For example, in some studies (Lee 1983a; chap. 27), but not others (Altmann 1980), high-ranking infant cercopithecines are handled more than low-ranking infants. High-ranking juvenile baboons (Cheney 1978b) and rhesus macaques (French 1981; Tar-tabini and Dienske 1979) play more often and are more attractive play partners than low-ranking juveniles. And when adult males groom juveniles other than their own offspring, they most often groom juveniles from high-ranking families (rhesus macaques: Missakian 1974; Japanese macaques: Kurland 1977; baboons: Cheney 1978a; bonnet macaques: Silk, Samuels, and Rodman 1981). The complex nature of dominance effects, how-ever, may preclude evaluating its role based on simple correlations between rank and frequency of interaction. For example, in the case of infant handling, an individual subordinate to the infant's mother may have to wait until the infant strays to handle it, whereas one dominant to the mother may simply pull the infant away from her. Rank effects on infant handling may therefore depend on which behaviors are measured.

Alliances provide another example. In many species high-ranking infants and juveniles are more likely than their lower-ranking peers to receive support in alliances from animals outside their immediate kin group (ba-boons: Cheney 1977a; Walters 1980; Japanese macaques: Watanabe 1979; rhesus macaques: Berman 1980a; bon-net macaques: Silk, Samuels, and Rodman 1981; ver-vets: Cheney 1983a). However, when two unrelated ani-mals form an alliance, it is generally against an opponent that ranks lower than both (Japanese macaques: Watanabe 1979; baboons: Walters 1980; vervets: Cheney 1983a). An individual therefore has many more opportunities to

join a high-ranking, rather than a low-ranking, juvenile in such an alliance. Only one study has shown that high-ranking juveniles are more attractive alliance partners given equal opportunity (Cheney 1983a).

The exact role of rank-based attractiveness in social interaction, and how this varies between and within spe-cies, is not yet clear. Current knowledge suggests that kin-based attractiveness explains much of the distribution of social interactions, and rank-based attractiveness may explain a varying proportion of the remainder (e.g., Seyfarth 1977, 1980; Chapais and Schulman 1980). The relative importance of rank-based attractiveness may be determined at least in part by demography. In popula-tions in which females remain in their natal groups throughout their lives and infant survivorship is high (e.g., Cayo Santiago rhesus or provisioned Japanese macaques), many female kin are present and grooming occurs primarily between these individuals. It is in popu-lations where females disperse (e.g., hamadryas ba-boons), or where group size is small and infant sur-vivorship is low (e.g., vervet monkeys in Amboseli, Kenya), so that female kin are not always present, that rank-based attractiveness is reported (see also chap. 20). Still, rank-based attractiveness is not apparent in all such populations.

It is often difficult to separate rank effects from those of other factors, such as kinship. For example, because close kin occupy similar ranks and are attracted to each other, interactions are expected to occur at high rates among animals of similar dominance status. However, if all individuals are attracted to the highest-ranking ani-mals, competition for social partners would also lead to high rates of interaction among those of adjacent ranks (Seyfarth 1977). High-ranking animals might mo-nopolize access to others of high rank, thus preventing middle-ranking animals from gaining access to high-ranking animals. Middle-ranking animals would then compromise by interacting with those of middle rank, while low-ranking animals could only interact with each other. Thus both kin-based and rank-based attractiveness may act to promote grooming between animals of similar status. Where data permit separation of kinship and rank effects, a significant rank effect among unrelated indi-viduals was found in bonnet macaques (Silk 1982) and vervet monkeys (Fairbanks 1980), whereas no effect be-yond that attributable to kinship was found in rhesus ma-caques (Chapais 1983b). A rank effect is also implied by frequent grooming among those of adjacent rank in newly formed captive groups composed of unrelated fe-males (e.g., rhesus monkeys: Seyfarth 1977; stump-tailed macaques [*Macaca arctoides*]: Rhine 1972; ge-ladas: Bramblett 1970; Kummer 1975). Rank-based attractiveness might also explain why in many cases relationships are more cohesive among the members of high-ranking families than among the members of low-ranking families. Data supporting this generalization come from Japanese macaques (Yamada 1963; French 1981), baboons (Cheney 1977a), long-tailed macaques (Fady 1969), rhesus macaques (Berman 1980a), and vervets (Fairbanks, in prep.). The absence of rank-based attractiveness in low-ranking families could account for the difference in cohesion.

3. The Effects of reproductive state. Changes in reproductive state among females have important effects on the pattern of their interaction with others. In some species, for example, when a female gives birth she immediately becomes more attractive to both immature and adult females. She receives more grooming from these animals, and her aggressive interactions are more likely to be followed by "reconciliation" (chap. 34). The attractiveness of such females declines as their infants grow older (rhesus monkeys: Rowell, Hinde, and Spencer-Booth 1964; patas monkeys: Hall and Mayer 1967; vervets: Struhsaker 1971; Japanese macaques: Sugiyama 1971; baboons: Seyfarth 1976; Altmann 1980; see also chap. 27).

Like the arrival of infants, the onset of sexual receptivity also produces marked changes in the pattern of female interactions (chap. 30). When a female is sexually receptive, she tends to groom other females less often and receive more grooming from adult males (savanna baboons: Rowell 1968; hamadryas baboons: Kummer and Kurt 1965). Sexually receptive females may also be supported by males in alliances more often than at other times (rhesus monkeys: Carpenter 1942; Lindburg 1971; baboons: Hall and DeVore 1965; Seyfarth 1978a; see also Tutin 1979 for chimpanzees). Apparently as a consequence of this support, for relatively brief periods sexually receptive females may be able to ignore aggression from higher-ranking females (Seyfarth 1978a), or they may actually win aggressive bouts with dominant individuals (e.g., rhesus monkeys: Altmann 1962; pigtailed macaques [*Macaca nemestrina*]: Tokuda and Jensen 1969). This change in female status is invariably temporary, however, and ends along with other behavioral changes at the end of sexual receptivity.

It should be emphasized that not all interactions between adult males and females are characterized by behavioral changes related to reproduction. Smuts (1985) reviews data from several species in which there are stable, long-term bonds between individual males and females that persist throughout the female's reproductive cycle.

Distribution of interactions among males. 1. Features in which males and females are similar. Some of the above generalizations concerning kinship, dominance rank, and reproductive condition also apply to adult males. For example, although males in most cercopithecine species transfer from their natal group at around sexual maturity (chap. 21), cooperative interactions between male kin occur at high rates both before transfer and even after

transfer, when related males migrate to the same group (rhesus monkeys: Miller, Kling, and Dicks 1973; Meikle and Vessey 1981; Colvin 1983b; vervets: Cheney and Seyfarth 1983). Data in chapter 28 suggest that, in some groups, males preferentially support infants likely to be their offspring. In hamadryas baboons, where one-male groups join together to form clans, the males within a clan appear to be related, and they cooperate to defend each other against less closely associated males (Abegglen 1984). In chimpanzees, where related males live together as adults, high rates of male-male grooming (Nishida 1979), greeting (Hasegawa, unpub.), food sharing (Teleki 1981), and tolerance during copulation (chap. 15) are observed.

There is also evidence for the attractiveness of high-ranking individuals in interactions among males. In chimpanzees and capuchins, dominant males receive more grooming than others (Simpson 1973b; chap. 7). In rhesus monkeys, subordinate males groom dominants more often than vice versa, and grooming most commonly occurs between animals of adjacent rank (Kaufmann 1967; Colvin 1983c).

Finally, male-male interactions, like those of females, can be affected by changes in reproductive state. In rhesus monkeys, male testes regress outside the mating season (Sade 1964), when rates of male-male aggression and the frequency of male rank changes also decline. In many seasonally breeding species, male aggression is higher in the breeding season than outside it (e.g., toque macaques: Dittus 1977), while in non–seasonally–breeding species, rates of aggression may be strongly influenced by the number of sexually receptive females present (e.g., baboons: Packer 1979b).

2. Features in which males and females differ. Although the distribution of interactions among males resembles that among females and immatures in some respects, in other ways interactions involving males are markedly different. Male aggressive and cooperative behavior are perhaps best understood in light of the alternative "strategies" pursued by males in apparent attempts to maximize their reproductive success. As data in chapter 26 indicate, many species with multimale groups exhibit a positive correlation between male dominance rank and mating activity. One strategy of males is thus to maintain high rank for as long as possible, usually through aggressive competition and the formation of alliances. The relation between rank and mating success is not always a close one, however (chap. 26), and males are often unable to maintain high rank for long periods of time (e.g., baboons: Hausfater 1975; chimpanzees: Nishida 1983b). Hence males also employ other behavioral strategies to gain access to females, despite being low ranked.

To this end, male cercopithecines are especially prominent as defenders of females and immatures. Such defense seems to have two consequences that benefit the male. First, in the short term, the females and immatures that a male defends tend to cooperate with him during aggressive encounters with other males (Packer 1980; Strum 1983; Smuts 1985). Such cooperation often allows a male to direct aggression against other males while at the same time diminishing the likelihood that he will receive aggression in return. For instance, "agonistic buffering," in which a male carries an infant during an aggressive interaction with another male (Deag and Crook 1971), appears in some cases to be adaptive because it reduces the severity of aggression against the carrying male (chap. 28). Second, over longer periods of time, defense of adult females and their young may be one way in which males can form long-term, "special relationships" (Smuts 1983a) with those adult females, which may increase chances of mating with the females (baboons: Seyfarth 1978b; Stein 1984a; Smuts 1985; see also chap. 28).

Unlike adult females, adult males frequently form alliances that run counter to existing dominance relationships. This occurs, for example, when two males challenge a third who outranks them both (squirrel monkeys: Baldwin and Baldwin 1981; baboons: Hall and DeVore 1965; Packer 1977; Rasmussen 1980; Collins 1981; Smuts 1985; Japanese macaques: Stephenson 1975; chimpanzees: Goodall 1968; de Waal 1982; Nishida 1983b). Although such alliances often involve unrelated males, they may consistently involve the same individuals (e.g., Packer 1977; Saayman 1971a; Nishida 1983b). Males also consistently defend females and immatures against higher-ranking females. Because females are reluctant to intervene counter to the nepotistic female dominance hierarchy, males may be important in controlling aggression among the female lineages within a group.

Social Relationships

In many primate species, different types of competitive and cooperative behaviors have similar distributions. For example, individuals that groom each other most often are least aggressive and most often feed together, play, and form alliances. This phenomenon is particularly well documented among kin (chap. 24), but it also occurs between unrelated individuals.

For example, in baboons, females form special relationships with one or two males that persist through all phases of the female reproductive cycle (Ransom and Ransom 1971; Seyfarth 1978b; Altmann 1980; Smuts 1983a, 1983b, 1985). These relationships involve frequent grooming, spatial proximity, and reciprocal alliances. The male frequently develops a protective, long-lasting relationship with the female's infants (Seyfarth 1978b; Altmann 1980; Stein 1984b; Johnson 1984; Smuts 1985), and when the female resumes estrous cycles she often prefers to mate with a previously affiliated male (Seyfarth 1978a, 1978b; Rasmussen 1980, 1983b; Smuts

1983b, 1985). Special male-female relationships that persist beyond sexual cycling also occur in mountain gorillas (Harcourt 1979c), gelada baboons (Dunbar 1983b), hamadryas baboons (Kummer 1968; Sigg 1980), rhesus macaques (Kaufmann 1967; Chapais 1983a), and Japanese macaques (Itani 1959; Fedigan and Gouzoules 1978; Takahata 1982a, 1982b). The relationship between the male and the female's offspring may also persist for longer periods. In yellow baboons, relationships between 4-year-old juvenile females and the males with which they had associated as infants were characterized by frequent grooming, frequent alliance formation, and unusually low levels of aggression (Walters 1986; see also Stein 1984b; Johnson 1984; chap. 28). The juveniles were not necessarily related to the males.

Among juvenile vervets, frequent grooming partners are also frequent play partners (Lee 1983e), both among related and unrelated individuals. Similar bonds among both kin and unrelated individuals exist among juvenile male rhesus monkeys (Colvin 1983c). Among unrelated adult male baboons, Packer (1977) found evidence for reciprocal support in alliances. Each male had a favorite alliance partner, whom he supported more than any other individual. If A's favorite partner was B, B's favorite partner was generally A—suggesting that one coalition between two individuals increased the probability of their subsequent cooperative behavior. Similar long-term, cooperative relations occur among both related and unrelated male chimpanzees (Riss and Goodall 1977; Nishida 1981, 1983b; de Waal 1982; chap. 34).

Supporting these correlational data, playback experiments on vervet monkeys provide further support for the hypothesis that cooperative behavior of one type can affect cooperative behavior of another type, even among unrelated animals. In experiments involving two subjects, A and B, the observers played A's threat vocalization within earshot of B roughly 90 minutes after A had groomed B. B's response was filmed to measure its readiness to answer A's solicitation. On another day, as a control, B heard A's solicitation after a 2-hour period when A had not groomed B. Results showed clearly that prior grooming between A and B increased B's willingness to attend to A's solicitation for support (Seyfarth and Cheney 1984a).

Competitive and cooperative interactions may thus affect one another and may have long-term consequences for the individuals involved. Thus, the competitive and cooperative behavior of primates is often best understood not in terms of single interactions, but in terms of the pattern of interactions between two individuals over a relatively long period of time. This pattern of interactions defines the animals' *social relationship* (Hinde 1976). As the editors point out in their introduction, there are many situations in which an interaction between two primates, analyzed on its own, makes little sense. If the interaction is seen as part of a long-term social relationship, however, its function becomes more apparent. Why would two female macaques join together in an alliance against a third who ranked lower than both? Why would a male spend half an hour grooming a pregnant female? In both of these, and many other cases, the beneficial consequences of behavior are perhaps best understood if each interaction is seen as an incremental attempt to reaffirm or establish a close social relationship that itself may have long-term adaptive consequences for the individuals involved.

It is clear that frequencies of cooperative and competitive behaviors are related in predictable ways in certain social relationships. The relationships of kin, for example, are characterized by frequent grooming and alliances, and infrequent severe aggression (chap. 24). Other examples of positive relationships, such as between males and between immatures and their adult male associates, were described earlier. Often, however, the frequency of grooming is not correlated with the frequency of alliances among unrelated individuals within a group (Fairbanks 1980; Silk 1982; Walters 1986). The just-described playback experiments on vervets suggest short-term reciprocity in exchanges of cooperative behavior among unrelated individuals, but the long-term consequences of such interactions remain an important research question. Do the associations frequently observed among unrelated immatures form the basis of especially positive relationships in adulthood? Are special, positive relationships among nonkin restricted to a relatively small number of particular types, so that cooperation is based only on a short-term reciprocity? Can reciprocity account for rank-based attractiveness, in that grooming by low-ranking individuals is repaid, perhaps at a very low rate, by support in alliances from high-ranking ones? The answers to such questions may soon emerge from the several longitudinal field studies of primates now being conducted.

SUMMARY

Social behavior in primate groups is delicately balanced between competition and cooperation. Competition takes a variety of forms, from subtle approach-retreat interactions to dangerous physical violence, but it is most commonly expressed in the form of gestures or behaviors that involve no physical contact. The most common forms of cooperation are grooming and the formation of alliances.

Old World Monkeys, particularly baboons, macaques, and vervets, provide the most detailed data on the distribution of competitive and cooperative interactions. In these species—but not in all others—the outcome of competitive interactions is often highly predictable, and individuals can be ranked in a linear dominance hierarchy. Among females, dominance rank is determined primarily by maternal rank, and rank is relatively stable throughout a female's lifetime. Among adult males, dominance rank appears to be determined primarily by age, sex, fighting ability, length of tenure in a group, and the presence of alliance partners. Dominance ranks often change during a male's lifetime, and rank may affect access to some resources but not others.

In all species studied thus far, grooming, alliances, and other cooperative behavior occur primarily among kin. In addition, evidence from a number of species indicates that high-ranking animals may sometimes be attractive social partners. Several kinds of cooperative bonds have also been described between unrelated group members.

Competitive and cooperative interactions affect one another and can have long-term consequences for the individuals involved. Because of this, the cooperative and competitive behavior of primates is often best understood not in terms of single interactions, but in terms of the pattern of interactions between two individuals over a relatively long period of time. This pattern of interactions defines the animals' social relationship.

26 | Social Behavior in Evolutionary Perspective

Joan B. Silk

The authors of the preceding chapters have described the form and frequency of a wide range of altruistic, competitive, and aggressive behaviors, and the formation of dominance hierarchies within primate groups. The purpose of this chapter is to consider two related questions. What forces have shaped the patterns of behaviors that we observe? What are the adaptive consequences of these behaviors?

To understand how evolution may have shaped the patterns of particular kinds of behavior, such as grooming or aggressive interactions, it is necessary to understand in a general way how evolutionary forces may influence social behavior. It is also important to understand how different evolutionary forces may favor behaviors that have very different effects upon the actors, such as cooperative and competitive behaviors. These issues are considered in the first section of the chapter.

Evolutionary biologists categorize behaviors in terms of their effects upon the genetic fitness of the participants. For example, altruistic interactions are those in which the fitness of the actor is decreased and the fitness of the recipient is increased. Hence, to determine whether a given behavior is altruistic, it is necessary to assess its adaptive consequences. In the second section of this chapter I consider the adaptive consequences of aggressive and competitive behaviors and the adaptive costs and benefits associated with the maintenance of high and low dominance rank. In the final sections, I consider whether the patterns of certain affiliative behaviors, which are usually considered to be altruistic, are consistent with predictions derived from the theories of kin selection and reciprocal altruism.

The reader should be aware from the outset that questions about the evolution of social behavior are very difficult to resolve empirically, especially among long-lived primates. It will become clear that some important variables, such as the effects of particular acts upon genetic fitness, are very difficult to measure with precision. In other cases, there may be no adequate theoretical models to account for the detailed patterns of behavior we observe. Finally, most of the available empirical evidence comes from a relatively small number of primate genera, namely species of macaques, baboons, and vervet monkeys. Nevertheless, it is useful to summarize what we now know about the adaptive consequences of primate social behavior and reasonable to speculate or hypothesize cautiously about how evolution may have shaped the patterns of social behavior within primate groups.

EVOLUTION OF SOCIAL BEHAVIOR

Evolution is defined as a change in the genetic composition of a population. Some genetic processes, such as drift and mutation, randomly alter the frequency of alternative alleles at particular genetic loci. However, natural selection is the only genetic process that produces a directional evolutionary change and causes the distribution of genotypes to shift in a systematic manner. For natural selection to occur, three conditions must be met: (*a*) organisms within populations must vary, and this variation must affect their ability to survive and reproduce; (*b*) these variations must be transmitted from parent to offspring; and (*c*) population growth must be limited by the availability of environmental resources. When variations are transmitted genetically, natural selection changes the distribution of alleles in the population. This occurs because individuals who possess certain traits are able to reproduce more successfully than others, and they transmit the genes that contribute to those traits to their offspring. Thus, the frequency of alleles that produce traits that contribute to reproductive success will increase in the next generation.

Evolutionary biologists frequently use the terms *genetic fitness* and *adaptation* when they describe the action of natural selection. Genetic fitness is a measure of an individual's relative genetic contribution to the gene pool of the next generation. Adaptive traits are those that confer reproductive advantages upon individuals that possess them. When fitness varies, the relative frequency of alleles that produce adaptations will tend to increase in the next generation. Thus, evolution by natural selection produces adaptation.

While the theory of natural selection predicts that evolution will act to increase the frequency of genetically transmitted traits that improve an individual's chances of reproducing more successfully than other members of its population, these are not the only kinds of traits that can evolve. Other evolutionary processes can influence the distribution of genotypes and favor the evolution of other kinds of behavioral traits. Later in this chapter, I consider two evolutionary processes, kin selection and reciprocal altruism, that may favor the evolution of altruistic behaviors that would not be expected to evolve through individual selection.

ADAPTIVE CONSEQUENCES OF COMPETITION

Much of what animals do can be interpreted as the result of the action of individual selection. Individual behaviors that enhance foraging efficiency, reduce vulnerability to predation, improve health and vigor, reduce harassment by conspecifics, or increase longevity and fecundity are likely to increase individual fitness and may thus be favored by individual selection. Some social behaviors that involve interactions among two or more individuals may also result from the action of individual selection. Perhaps the most obvious example of such behaviors are competitive interactions that occur when two or more individuals attempt to monopolize access to a common resource.

It is well documented that in many species of primates, individuals actively compete over a variety of resources, including food, water, grooming partners, mates, resting spots, and membership in social groups.

Animals are more likely to compete over resources than to share or yield access to them if the resources are limited in supply, defendable, and equally valuable to the competitors (Clutton-Brock and Harvey 1976; *Cercopithecus aethiops:* Wrangham 1981; Whitten 1983; *Macaca sinica:* Dittus 1979). While the benefits of winning a competitive encounter might seem obvious, it is usually difficult to quantify the effect of a single encounter upon an individual's fitness. There is very little direct evidence that resources gained through competitive encounters directly enhance fitness. However, a number of lines of evidence indirectly suggest that competitive success is related to reproductive success and thereby influences genetic fitness.

Adaptive Consequences of Resource Competition among Females

For primate females, nutritional status influences several different parameters of reproductive performance (see chap. 20). When free-ranging rhesus and Japanese macaques are provisioned (fig. 26-1), females grow faster, mature earlier, survive longer, and produce infants at

FIGURE 26-1. Rhesus monkeys on Cayo Santiago gathered around a feeding hopper. In several nonhuman primate populations, provisioning results in rapid population growth, which indicates that female nutritional status affects reproductive rates. (Photo: Barbara Smuts/Anthrophoto)

shorter intervals than under natural conditions (*Macaca fuscata:* A. Mori 1979a; Takahata 1980; Sugiyama and Ohsawa 1982b; *Macaca mulatta:* Drickamer 1974). These changes result in the rapid increases in population size characteristic of provisioned populations (*M. fuscata:* Fujii 1975; Koyama, Norikoshi, and Mano 1975; Masui et al. 1975; Sugiyama and Ohsawa 1982; Kurland 1977; *M. mulatta:* Koford 1965). Similarly, improvements in the quality or quantity of food provided to captive tamarins have resulted in consistent increases in litter size (*Saguinus oedipus:* Kirkwood 1983). These data suggest that the limited availability of nutritional resources constrains the reproduction of primate females under natural conditions.

Competition over Resources and Dominance Rank

If food is limited, nutritional status may be related to success in competitive encounters over food. Animals with high agonistic dominance ranks often obtain greater access to preferred, limited, or clumped foods or water (*C. aethiops:* Wrangham 1981; Wrangham and Waterman 1981; Whitten 1983; Fairbanks and McGuire 1984; *M. sinica:* Dittus 1977), or feed more efficiently (*Papio cynocephalus cynocephalus:* Post, Hausfater, and McCuskey 1980) than their low-ranking peers. Given this fact, it is not surprising to find that nutritional status is also correlated with dominance status in some cases. For example, in one group of wild vervet monkeys, high-ranking females weighed more than lower-ranking females of similar body size (Whitten 1983). The amount of body fat was positively correlated with dominance rank among captive rhesus macaque females who were fed ad libitum (Small 1981).

Other Benefits of High Rank

High dominance rank may confer additional benefits upon individuals. For example, high-ranking individuals sometimes receive more grooming than do low-ranking individuals (*C. aethiops:* Seyfarth 1980; *M. mulatta:* Sade 1972b). In several groups, the offspring of high-ranking females are harassed less often and supported more frequently than the offspring of low-ranking females (*C. aethiops:* Cheney 1983a; Horrocks and Hunte 1983; *Macaca radiata:* Silk et al. 1981; *M. mulatta:* Berman 1980a; *Papio cynocephalus ursinus:* Cheney 1977a).

Costs of High Rank

There may also be distinct costs associated with high dominance rank. Cheney, Lee, and Seyfarth (1981), and Cheney et al. (in press) observed that among free-ranging vervet monkeys, the members of high-ranking families were more likely to be preyed upon than were the members of low-ranking families. Higher vulnerability to pre-

dation among the members of high-ranking families was partially offset, however, by relatively lower rates of mortality due to illness within the same families.

Dominance Rank and Reproductive Success among Females

If the benefits associated with high rank offset the costs of occupying high ranks, a positive relationship between dominance rank and reproductive success is expected to exist. The components of female reproductive success include the age at first reproduction, the length of the interval between successive births, the number of infants produced, the fraction of infants surviving to reproductive age, and the length of a female's reproductive career. Thus, we must consider the relationship between dominance rank and each of these parameters.

In the discussion that follows, data from both captive and free-ranging populations will be cited. It might seem unlikely that captive populations that have unlimited access to food and no predators would demonstrate a relationship between dominance rank and reproductive success. However, even in captive situations competition is seen, and females form dominance hierarchies characteristic of conspecifics in the wild. Moreover, if we were to exclude data from captive or provisioned populations, we would eliminate a substantial fraction of the available information about the relationship between dominance rank and reproductive success.

Age at First Birth. In one group of rhesus macaques, the daughters of high-ranking females reached sexual maturity and gave birth for the first time at earlier ages than did the offspring of lower-ranking females (Drickamer 1974). Among the members of one group of Japanese macaques moved to a large enclosure in Texas, there was a positive correlation between dominance rank and the age of reproductive maturity (Gouzoules, Gouzoules, and Fedigan 1982). In a population of free-ranging yellow baboons in Amboseli National Park, Kenya, the daughters of high-ranking females reach reproductive maturity at earlier ages and conceive their first infant approximately 200 days sooner than do the daughters of low-ranking females (Altmann, Altmann, and Hausfater, in press).

Other studies have failed to document differences among high- and low-ranking females in the age of sexual maturity, first conception, or first birth. In two populations of macaques, all females, gave birth for the first time at approximately the same ages (*M. radiata:* Silk, Samuels, and Rodman 1981; *M. fuscata:* Wolfe 1984a). Among wild vervet monkeys, there were no differences within groups in the age at reproductive maturity, but differences in habitat quality were associated with differences between groups in the age at reproductive maturity (Cheney et al., in press).

Interbirth Intervals. In some groups, high-ranking females produced infants at shorter intervals than did low-ranking females (*C. aethiops:* Fairbanks and McGuire 1984; *M. fuscata:* Sugiyama and Ohsawa 1982b; *M. mulatta:* Drickamer 1974), but in others no association was observed (*M. fuscata:* Gouzoules, Gouzoules, and Fedigan 1982; *M. radiata:* Silk, Samuels, and Rodman 1981; *P. c. cynocephalus:* Altmann, Altmann, and Hausfater, in press).

Female Fecundity and Infant Survivorship. In some cases, dominance rank has been found to be positively associated with the total number of infants produced (*C. aethiops:* Whitten 1983; Fairbanks and McGuire 1984; *M. fuscata:* Takahata 1980; Gouzoules, Gouzoules, and Fedigan 1982; *M. mulatta:* Drickamer 1974; Wilson, Gordon, and Bernstein 1978), and with the number that survive (*C. aethiops:* Whitten 1983; *P. c. ursinus:* Busse 1982; *Theropithecus gelada:* Dunbar and Dunbar 1977; *M. fuscata:* A. Mori 1979a; Sugiyama and Ohsawa 1982b; *M. radiata:* Silk et al. 1981; *M. sinica:* Dittus 1979; *M. mulatta:* Wilson, Gordon, and Bernstein 1978).

In other cases there was no significant relationship between dominance rank and fecundity or infant mortality (*C. aethiops:* Cheney, Lee, and Seyfarth 1981; Cheney et al., in press; *M. fuscata:* Wolfe 1984a; *Presbytis entellus:* Dolhinow, McKenna, and Vonder Haar Laws 1979). Among the yellow baboons of Amboseli, there is no evidence that dominance rank affects female fecundity or overall infant survival, but low-ranking females do have slightly higher rates of miscarriage and tend to give birth after shorter gestation periods than do high-ranking females. There is some evidence that the latter reduces the probability that infants will survive (Altmann, Altmann, and Hausfater, in press).

In one case, dominance rank was negatively related to infant survivorship, but positively related to fecundity (*M. fuscata:* Gouzoules, Gouzoules, and Fedigan 1982). Dominance rank was positively related to the rate at which the membership of matrilineages increased on Cayo Santiago Island (*M. mulatta:* Sade et al. 1976), and to infant growth rates in provisioned groups in Japan (*M. fuscata:* Sugiyama and Ohsawa 1982b). In one group of macaques, declines in dominance rank were associated with delays in conception (*M. mulatta:* Wilson, Gordon, and Bernsteon, 1978).

Sex Ratio of Offspring. In several groups, the proportion of sons and daughters produced by high- and low-ranking females differ (reviewed by Hrdy, in press). In three of these groups, high-ranking females produce as many or more female infants than male infants (*P. c. cynocephalus:* Altmann 1980; Altmann, Altmann, and Hausfater, in press; *M. radiata:* Silk et al. 1981; Silk 1983; *M. mulatta:* Simpson and Simpson 1982). In two other groups, high-ranking females appear to produce more male than female infants, and low-ranking females produce as many or more female than male infants (*M. mulatta:* Meikle, Tilford, and Vessey, 1984; *Macaca sylvanus:* Paul and Thommen 1984). Finally, in one group of Japanese macaques the dominance ranks of the mothers of female and male infants do not differ (*M. fuscata:* Noyes 1982).

In two of the populations in which a significant relationship between maternal rank and infant sex has been detected, the bias in the sex ratio reflects differences in the probability that the sons and daughters of high- and low-ranking females will survive. High-ranking female yellow baboons produce more daughters than sons, and their daughters have a higher probablity of surviving than do their sons. At the same time, low-ranking females produce more sons than daughters, and their sons have a higher probability of surviving than do their daughters (*P. c. cynocephalus:* Altmann, Altmann, and Hausfater, in press). Similar patterns occur among captive bonnet macaques (*M. radiata:* Silk 1983). Comparable information is not available for the two cases in which high-ranking females produce more sons than daughters and low-ranking females produce as many or more daughters than sons (*M. sylvanus:* Paul and Thommen 1984; *M. mulatta:* Meikle, Tilford, and Vessey 1984). None of these studies provides precise information about the reproductive success of the male and female offspring of high- and low-ranking females. Clearly, evidence regarding the relationship between maternal rank, infant sex, and infant survival is still quite fragmentary, and it is too early to determine whether facultative adjustment of sex ratios contributes to variation in female reproductive success.

Female Longevity. Little information about factors that influence female longevity is now available. Hence, the relationship between female dominance rank, longevity, and reproductive success cannot be assessed.

Is Dominance Rank Correlated with Female Reproductive Success? While these data indicate that among cercopithecine primate females a positive association often exists between dominance rank and various measures of reproductive success, there are also cases in which dominance rank is unrelated or negatively related to reproductive parameters. It is noteworthy that significant positive correlations between dominance rank and measures of reproductive success outnumber significant negative correlations between these variables. Taken together, these data suggest that dominance rank explains some, but by no means all, of the variance in reproductive success among cercopithecine females. The data also suggest that dominance rank is more clearly associated with reproductive success in some populations than in others.

However, variation in the strength of the relationship between rank and reproductive success does not seem clearly related to environmental conditions, as positive associations between rank and some measures of reproductive success have been found in some provisioned populations and not others, and in both captive and free-ranging populations.

The lack of consistency among these results has created considerable controversy about the significance of the relationship between dominance rank, reproductive success, and genetic fitness among females. This is partly because we cannot fully assess the relationship between various components of reproductive success. All other things equal, successful females are likely to be those that begin to reproduce at early ages, continue to reproduce until later ages, give birth at shorter intervals, produce more infants, or produce healthier infants. However, many of these parameters are likely to be influenced by each other. For example, in both seasonally and non–seasonally breeding species, the average length of the interval following the birth of a surviving infant is longer than the interval following the birth of nonsurviving infants (*C. aethiops:* Cheney et al., in press; *P. c. cynocephalus:* Altmann, Altmann, and Hausfater 1978; *M. radiata:* Silk, Samuels, and Rodman 1981; *M. fuscata:* Tanaka, Tokuda, and Kortera 1970). Thus, we might expect to find that infant survivorship is inversely correlated with interbirth intervals and with the total number of infants produced during a female's lifetime. Moreover, a negative correlation between rank and one component of reproductive success may be balanced by a positive correlation between rank and other components of reproductive success. Gouzoules, Gouzoules, and Fedigan (1982) demonstrated that high-ranking Japanese macaque females produced their first infants at younger ages, and produced a larger number of infants, than did lower-ranking females. Although the infants of high-ranking females were significantly more likely to die before the age of 1 year than were the infants of low-ranking females, 90% of all infants survived their first year. To determine whether high-ranking females achieve higher genetic fitness than lower-ranking females, we need to know whether the effects of accelerated maturation and greater fecundity in high-ranking females offset the slightly higher infant mortality experienced by high-ranking females. Since that is the case among this group of Japanese macaques, it might be reasonable to conclude that high dominance rank confers an overall selective advantage upon high-ranking females in this population.

Adaptive Consequences of Mate Competition among Males

Darwin (1971) was aware that selection pressures upon males may differ from selection pressures upon females, and may thereby create different sources of competition among males than among females. Males must obtain sufficient resources to survive and may therefore benefit from monopolizing access to limited resources such as vertebrate kills (*P. c. ursinus:* Hamilton and Busse 1982). However, in multimale groups the major factor that appears to determine a male's genetic fitness is his ability to obtain access to receptive females within his social group (see chap. 31). Just as dominance rank is generally correlated with access to other kinds of resources, we might also expect to find that dominance rank is correlated with access to receptive females and related to male reproductive success.

Male Rank and Reproductive Success. The reproductive success of males in multimale primate groups is considerably more difficult to assess than that of females. Nevertheless, there are several different means of assessing paternity in multimale groups. The most precise measures of paternity are obtained from immunological analyses that indicate whether or not a given male could have sired a given infant. To date, genetic paternity data are available only for the members of captive macaque groups. In the first study using genetic paternity data, Duvall, Bernstein, and Gordon (1976) found no correlation between dominance rank and the number of live-born rhesus macaque infants sired in the first year after the study group was formed, but did find a significant correlation in the same group during the next year. Smith (1980, 1981) found that in one year high-ranking rhesus macaque males in six different captive groups fathered significantly more offspring than lower-ranking males. Dominance rank was also positively associated with the number of live-born infants sired in a group of captive Barbaray macaques over a 4-year period (Witt, Schmidt, and Schmidt 1981). Finally, during an 8-year period in a captive group of rhesus macaques (Curie-Cohen et al. 1983), the three highest-ranking males produced significantly more offspring than lower-ranking males. However, the alpha male fathered fewer infants than the second- and third-ranking males in all but one year of the study. These data indicate that high-ranking males generally sired a larger number of live-born infants than did lower-ranking males, especially in stable social groups. However, they do not provide information about other components of male fitness, such as the survivorship of infants to reproductive age.

The bulk of information about the relationship between male rank and reproductive success comes from behavioral studies. Such studies are based upon the assumption that males who monopolize access to females, or copulate most frequently with females on the days they are most likely to conceive, are most likely to have inseminated them. In many cases it is possible to assess the female's reproductive condition at the time observations are made or to determine it retrospectively. It is known, for example, that female baboons are most likely to conceive 1 to 3 days before the first day of detumescence of the female's sex skin (Hendrickx and Kraemer 1969; Shaikh et al. 1982). In other cases, the average gestation length is known, and by counting back from the day of birth it is possible to establish a reasonably short period in which conception is likely to have occurred.

Male dominance rank is sometimes correlated with behavioral measures of reproductive activity near the times of ovulation and probable conception. Hausfater (1975) found that among one group of yellow baboons, the highest-ranking male did not always monopolize access to receptive females. Nevertheless, male dominance rank accounted for 56% of the variance in the number of copulations observed near the time of ovulation. Longitudinal studies of members of the same population indicates that male rank may change frequently during their residence in the group (Hausfater 1975; Altmann, Altmann, and Hausfater, in press), but males are likely to be highest ranking and sire the largest number of infants during their first year in the group. Positive relationships between male rank and behavioral measures of reproductive success have also been found in other studies (*P. c. ursinus:* Saayman 1970; Seyfarth 1978a; Busse and Hamilton 1981; *Papio cynocephalus anubis:* Packer 1979b; Ransom 1981; *Pan troglodytes:* Tutin 1979; *M. radiata:* Samuels, Silk, and Rodman 1984; *M. fuscata:* Stephenson 1975; *M. mulatta:* Chapais 1983c, 1983e).

Other studies have revealed no consistent relationship between male rank and reproductive success. In one population of anubis baboons, where males did not form linear dominance hierarchies, dominance rank was not positively correlated with reproductive activity (Strum 1982; Smuts 1982, 1985). Similar data have been obtained for Barbary macaques (Taub 1980b) and chimpanzees (McGinnis 1979).

The accuracy of behavioral measures of reproductive success is difficult to verify for several reasons. First, the use of inappropriate sampling procedures may bias the results obtained (Drickamer 1974; Estep, Johnson, and Gordon 1981). Second, we know little about the nature of sperm competition or the precise timing of ovulation and conception. Thus, if more than one male associates with a female near the time she conceives, paternity is ambiguous. Finally, broadly defined measures of consort behavior and sexual activity may not be good indicators of paternity. In one study where sexual activity was defined to include both complete and incomplete mounts, and consort activity included a variety of affiliative behaviors, behavioral measures of sexual activity were found to be uncorrelated with genetic measures of paternity (*M. mulatta:* Stern and Smith 1984).

Behavioral studies in which the reproductive status of the females is not known or is not taken into account in the analysis are not useful sources of information about the relationship between male rank and reproductive success. The reproductive state of the female influences the probability that conception will occur. Thus, frequent sexual activity with nonreceptive females is not likely to result in insemination, while even infrequent sexual activity with receptive females may do so. Thus, it is not surprising to find that correlations between male dominance rank and male sexual activity are much more variable than correlations based upon more accurate measures of reproductive success (reviewed in Fedigan 1983; Curie-Cohen et al. 1983).

Is Dominance Rank Correlated with Male Reproductive Success? As for females, there is considerable variation in the consistency of the relationship between male rank and reproductive success. Again, it is reasonable to conclude that dominance rank explains some but not all of the variation in reproductive success among males. This conclusion is consistent with evidence that male dominance rank is correlated with reproductive success among a wide range of vertebrate species outside the primate order (reviewed by Dewsbury 1982a). At the same time, other factors, such as male age, length of a male's residence in the group, female choice, and male investment in offspring may influence male reproductive success and mediate the effects of dominance rank. (These issues are discussed more fully in chap. 31.) Moreover, Hausfater (1975) has pointed out that males often change rank several times during their reproductive careers and has argued that it is necessary to assess the reproductive success of males over their entire lifetimes in order to estimate variance in male reproductive success. The debate about the consistency and significance of the relationship between male dominance rank and reproductive success will not be resolved until genetic paternity data are available and lifetime reproductive success of males has been assessed in a number of wild populations.

Evolution and Dominance Rank

At this point, we may ask several very different questions about the evolution of dominance rank: (1) What are the adaptive consequences of high and low dominance rank? (2) How has evolution shaped the formation of transitive dominance relationships and linear dominance hierarchies? (3) Why are dominance interactions prominent elements of the social behavior of some species and not others? We can only suggest very incomplete answers to these questions.

The fact that significant positive correlations between dominance and reproductive success outnumber both significant negative correlations and negative trends suggests that on average high-ranking individuals obtain a reproductive advantage by virtue of their rank. If the maintenance of dominance rank contributes to heritable genetic variation in fitness, selection is likely to favor those behaviors that allow individuals to acquire and maintain high rank. Even if the magnitude of the difference in fitness among high- and low-ranking individuals is very small (and even if it does not achieve statistical significance), the adaptive consequences of dominance rank may be important in shaping behavior.

It is much less clear how evolution has shaped the formation of transitive dominance relationships and linear dominance hierarchies. Transitive dominance relationships are those in which individual A dominates individual B, individual B dominates individual C, and individual A also dominates individual C. When all possible triadic relationships are transitive, linear dominance hierarchies are formed (Chase 1980). In a model that uses the logic of game theory, Maynard Smith and Parker (1976) have shown that animals may adopt a conventional asymmetric character to resolve contests without physical conflict. Such asymmetries may be based upon characters that are related to physical fighting ability, such as size or aggressiveness, or upon an arbitrary rule, such as "resident wins." If asymmetries are based upon a metric character that individuals can assess accurately (e.g., "bigger guy wins"), repeated encounters among individuals will yield a linear dominance hierarchy, even if animals do not recognize their opponents or remember the outcome of previous encounters. However, if the asymmetry is based upon an arbitrary character that is not inherent in the contestants (e.g., "first come, first served"), a linear dominance hierarchy will not necessarily result. At present, it is unclear why dominance hierarchies in most primate species are transitive.

Although observers of cercopithecine primates are usually able to establish the dominance relationships between all pairs of individuals within the groups they study, observers of other species sometimes find that interactions are so rare that they cannot do so (e.g., *Gorilla gorilla beringei:* Harcourt 1979b). We know very little about why dominance relationships play a prominent and apparently important role in the social behavior of some well-studied species and not others. If competitive strategies are influenced by the distribution and quality of resources, then variation in ecological conditions may be associated with differences in the prevalence of dominance interactions. Moreover, if the genetic relationships among group members influence the costs and benefits of competition among them (as discussed later), differences in the structure of relatedness within groups may influence the extent of competition within them. We will need more complete information about the importance of dominance relationships, the role of distribution of resources, and the structure of relatedness within primate groups before this question can be answered.

KIN SELECTION AND ALTRUISTIC BEHAVIOR

Because individual selection generally does not favor individuals who reduce their own fitness to increase the fitness of others, processes other than individual selection must be invoked to explain the evolution of altruistic behaviors. The theory of kin selection is based upon the insight that an individual's fitness has two components: (1) fitness gained through the replication of its own genetic material through reproduction, and (2) inclusive fitness gained from the replication of copies of its own genes carried in others as a result of its own actions (Hamilton 1964). When an actor behaves altruistically toward its kin, fitness benefits to kin also benefit the actor, but the actor's benefits are devalued by the coefficient of relatedness (r) between actor and kin. The coefficient of relatedness represents the probability that two individuals will obtain copies of the same gene through common descent from a single ancestor.

The precise genealogical relationship among kin determines the probability that both will share the same gene through common descent. In diploid species, identical twins share all genetic material ($r = 1.0$), while parents share exactly one-half of their genetic material with their offspring ($r = 0.5$), and share on average one-quarter of their genetic material with their grandchildren ($r = 0.25$). Hamilton (1964) was the first to show that altruistic acts toward kin increase the inclusive fitness of the actor only if the increment to the recipient's fitness (b) weighted by the coefficient of relatedness between them (r) is greater than the decrement to the actor's fitness (c), or $b \times r > c$. Hamilton's rule is sometimes expressed in an equivalent form as $K > 1/r$, where K equals b/c. Under a specified set of conditions, altruistic behaviors are expected to conform to Hamilton's rule. (See Michod 1982 and Grafen 1984 for more detailed discussions of these concepts.)

For altruistic interactions to be favored by kin selection, the conditions of Hamilton's rule must be met. If an actor's behavior decreases his or her own fitness by two units ($c = 2.0$), but increases the fitness of a full sibling ($r = 0.5$) by five units ($b = 5.0$), then the ratio of b/c ($5.0/2.0 = 2.5$) will exceed $1/r$ ($1/0.5 = 2.0$). All other things equal, a mother is expected to behave altruistically toward her offspring ($r = 0.5$) only if the benefits to her offspring are greater than twice the costs of her altruism ($b > 2c$). The same female is expected to behave altruistically toward her first cousin ($r = 0.125$) only if the benefits to her cousin are greater than eight times her own costs ($b > 8c$). Thus, altruism is expected to be selectively directed toward kin, and close kinship is expected to facilitate costly altruism.

For altruistic traits to evolve, individuals with altruistic genotypes must be more likely to interact with individuals who have similar genotypes than would be expected based upon the frequency of their common genotype in the population. For this to occur, they must recognize individuals with similar genotypes, or live among them. Kin, who are likely to have similar genotypes, from the core of many primate groups, and kin recognition through the maternal line is well documented (chap. 24). Moreover, groups may persist as cohesive social units over many generations. In such cases, the average coefficient of relatedness among the natal members of social groups will increase slowly over time. The degree of relatedness among them will increase further if emigrants are drawn nonrandomly from neighboring groups (Cheney and Seyfarth 1983), if groups divide along kinship lines (Chepko-Sade and Sade 1979; Melnick and Kidd 1983), or if variance occurs in reproductive success among females or males. Under such conditions, natal individuals will be surrounded by kin. Even if primates do not recognize all of their relatives, or selectively direct all classes of altruistic behavior toward kin, the structure of relatedness within many primate groups provides the necessary conditions for the evolution of altruistic behaviors (Aoki and Nozawa 1984; Silk 1984).

In practice, however, altruistic interactions are often difficult to identify, and predictions derived from the theory are difficult to evaluate empirically. In most cases, it is difficult to measure the effects of behavioral acts upon the genetic fitness of the participants. Furthermore, exact degrees of relatedness between the participants are often unknown (Fedigan 1982). Thus, it is often difficult to be sure that a given behavior is actually altruistic ($c > 0$), and even more difficult to calculate a numerical solution for the cost-benefit equation. However, in many cases we can make reasonable assumptions about the relative costs and benefits of particular behaviors and obtain minimum estimates of the degrees of relatedness among individuals. We can then ask whether the patterns of observed behaviors are consistent with qualitative predictions derived from kin selection theory.

Coalition Formation

Among cercopithecine females, intervention in aggressive encounters appears to be one form of altruism. While animals who intervene on behalf of others must expend energy and may risk injury, the recipients of aid benefit if intervention terminates aggression or causes aggression to be diverted to other targets. There is overwhelming evidence that females form coalitions on behalf of maternal kin more often than on behalf of other individuals (*P. troglodytes:* de Waal 1978a, 1982; *C. aethiops:* Cheney 1983a; *M. mulatta:* Kaplan 1977, 1978; Berman 1983c; Chapais 1983b; Datta 1983a, 1983b; *M. radiata:* Silk 1982; *M. fuscata:* Kurland

1977; Watanabe 1979; *Macaca nemestrina:* Massey 1977). Moreover, females appear to incur greater risks when they intervene on behalf of close maternal kin than when they intervene on behalf of others. When a female intervenes on behalf of an unrelated female, the opponent is nearly always lower ranking than herself. In contrast, when a female intervenes against an opponent higher ranking than herself, the intervention is much more likely to be on behalf of one of her immediate relatives than on behalf of a distant relative or nonrelative (*M. mulatta:* Chapais 1983b; *M. radiata:* Silk 1982; *M. fuscata:* Kurland 1977). Support of kin against higher-ranking monkeys is risky because coalitions against higher-ranking individuals are more likely to result in retaliatory attacks than are coalitions against lower-ranking animals (*M. mulatta:* Kaplan 1978). Thus, cercopithecine females selectively defend and support maternal kin and reserve costly aid for close kin.

Although some of the coalitions that females form are consistent with predictions derived from the theory of kin selection, the theory does not predict that coalitions will be formed among nonkin. Coalitions among females from different matrilines have been observed in every one of the studies cited above. Many of these coalitions are directed against lower-ranking females and may involve little direct risk of injury to the participants (Chapais 1983b), but all involve the expenditure of time and energy. Thus, predictions derived from the theory of kin selection do not account for all instances of coalition formation. Other explanations for the formation of these coalitions are discussed below.

Social Grooming

It seems plausible that individuals benefit from the removal of ectoparasites and dirt from their skin and hair, and that those who perform such services for others incur some cost. Although the costs and benefits of grooming are difficult to measure, it is known that animals selectively groom the parts of the body that individuals cannot reach themselves (Hutchins and Barash 1976) and that animals spend appreciable amounts of time grooming each other. Thus, it is reasonable to assume that social grooming is altruistic (e.g., Kurland 1977).

If grooming is altruistic, animals should groom relatives more often than other members of their groups. It is more difficult to predict exactly how grooming should be distributed among kin (Altmann 1979a; Weigel 1981; Reiss 1984). To see why this is true, consider the following situations:

(*a*) Suppose that each minute of grooming increases the fitness of the recipient by one unit and decreases the fitness of the actor by one-tenth of one unit. Each actor can devote 100 minutes to grooming each day. Individuals who devote all their grooming to their closest kin

FIGURE 26-2. After persistent begging (*above*), a chimpanzee mother shares part of her banana with her infant (*facing page*). (Photos: Toshisada Nishida)

have a higher ratio of benefits to costs, and thus achieve greater fitness advantages than individuals who divide their grooming among kin of different degrees of relatedness (all grooming to kin: $b = (100$ minutes $\times 1$ unit/minute) $\times .5 = 50$; $c = 100$ minutes $\times .1$ unit/minute $= 10$; $b/c = 5$; grooming divided among full siblings and half-siblings: $b = (50$ minutes $\times 1$ unit/minute) $\times .5 + (50$ minutes $\times 1$ unit/minute) $\times .25 = 37.5$; $c = 10$; $b/c = 3.75$).

(*b*) Now, suppose that the benefits of being groomed decrease over time, and after the first 50 minutes each minute of grooming increases the fitness of the recipient by only one-quarter of one unit. Then it makes sense to switch after 50 minutes from grooming close kin to more distant kin (all grooming toward closest kin: $b = (50$ minutes $\times 1$ unit/minute) $\times .5 + (50$ minutes $\times .25$ unit/minute) $\times .5 = 31.25$; $c = 10$; $b/c = 3.125$; divide grooming among full siblings and half-siblings: $(50$ minutes $\times 1$ unit/minute) $\times .5 + (50$ minutes $\times 1$ unit/minute) $\times .25 = 37.5$; $c = 10$; $b/c = 3.75$)

(*c*) By similar reasoning, if the costs of grooming increase over time, it makes sense to switch from more distant kin to closer kin.

In nearly every group in which patterns of grooming have been studied, grooming is concentrated upon maternal kin (chap. 24). The uniformity of this result strongly suggests that kin selection has shaped the evolution of grooming among primates. However, individuals do not devote all their grooming to their closest relatives and sometimes groom nonrelatives. Thus, the distributions of grooming and coalition formation among kin and nonkin follow a similar pattern. While these behaviors are directed primarily to kin, interactions among nonkin are also observed.

Food Sharing

Food sharing involves the direct transfer of food items from one individual to another. Spontaneous transfer of food items is common in only a few species (*P. troglodytes:* Goodall 1968, 1971; McGrew 1975; Silk 1978, 1979; *Leontopithecus rosalia:* Brown and Mack 1978), although there are isolated reports of such behavior in several more species (Kavanagh 1972; Brown and Mack 1978; McGrew 1975). Similar interactions among tamarins are described in chapter 4. In some species, individuals call others to the site of rich resources (*P. troglodytes:* Wrangham 1975; *M. sinica:* Dittus 1984). More commonly, however, food is transferred when individuals are forced to relinquish access to it when they are displaced by older, more dominant, or stronger animals.

Female chimpanzees sometimes share plant material with their offspring (Goodall 1968; McGrew 1975). Some transfers are spontaneously initiated by the mother, while others are initiated by persistent solicitations from offspring (fig. 26-2). Mothers are more likely to share items that their infants are unable to procure or process than they are to share readily accessible foods (Silk 1978, 1979). In an analysis of the distribution of provisioned bananas at Gombe, McGrew (1975) found that 86% of all transfers were among kin, and virtually all of these were from mother to offspring. Most of the remaining transfers were from adult males to unrelated adult females. While food sharing among mother and infant chimpanzees is consistent with predictions derived from kin selection theory, sharing among unrelated adults is not.

In monogamous captive golden marmosets, donors initiate transfers of food by vocalizing, assuming specific postures, and establishing eye contact with the potential recipient (Brown and Mack 1978). Food is shared with newly weaned offspring, and adults sometimes share with their mates. Juveniles share with their siblings and help to provision their parents after the birth of younger siblings. Sharing food so freely within marmoset family groups is consistent with predictions derived from kin selection theory.

While food sharing among mother and infant chimpanzees and within marmoset families seems altruistic, the relative costs and benefits of food sharing are more ambiguous in other contexts. Male titi monkeys share fruit with their infants (*Callicebus moloch:* Starin 1978) in response to persistent solicitations. There is some evidence that males consume fruits more quickly when they allow their infants to share small portions of the fruit than when they resist the infants' solicitations. Similarly, adult male chimpanzees (Wrangham 1977) may share meat only because it is less costly to share than to defend access to their kills. Male chimpanzees (McGrew 1975) and orangutans (Galdikas and Teleki 1981) may also derive benefits when they share with adult females if such acts increase the probability that sexually receptive females will mate with them. Individual selection or sexual selection may have favored the evolution of food sharing in these contexts.

Has Altruism in Primate Groups Evolved through Kin Selection?

Grooming, coalition formation, and food sharing are not the only examples of altruism among primates. Infants and juveniles whose mothers disappear are sometimes adopted by kin (*P. troglodytes:* Goodall 1971) or by adults to whom they are probably not closely related (*P.*

c. ursinus: Hamilton, Busse, and Smith 1982; *M. fuscata:* Hasegawa and Hiraiwa 1980). Extensive care of infants by females other than their mothers is common among some species, particularly the colobines (reviewed by McKenna 1979a; Hrdy 1976). There is some debate about the quality of care that caretakers provide (Hrdy 1978). Male baboons also develop close affiliative bonds with specific infants and may groom, carry, and defend them (Altmann 1980; Packer 1980; Stein 1984a; Busse and Hamilton 1981; Smuts 1985). In each of these cases, both kin and nonkin participate in interactions that seem to benefit the recipient at some cost to the actor.

The fact that kin are responsible for much of the altruism deployed in primate groups suggests that kin selection has shaped the evolution of all of these altruistic behaviors. Nonetheless, in all of the cases considered here, individuals perform altruistic acts on behalf of individuals from different matrilines and individuals born in different groups. While the members of different matrilines may be paternal kin (Altmann 1979), individuals born in different groups are unlikely to be closely related. Thus, it seems unlikely that kin selection provides a unitary explanation for the evolution and deployment of coalition formation and social grooming.

In addition to kin selection, there are a number of other plausible explanations of such interactions. Some explanations are based upon the theory of reciprocal altruism and suggest that individuals are more likely to receive aid from others if they have previously groomed them (Cheney 1977a, 1983a; Seyfarth 1977, 1980; Seyfarth and Cheney 1984a). Individuals may also benefit directly from grooming others if this reduces the risk of being harassed by higher-ranking group members (Bertram 1982; Silk et al. 1981; Silk 1982). These explanations may be complementary to explanations based upon kin selection and need not be considered competing hypotheses (Seyfarth 1983). A single form of behavior under different circumstances may confer different costs and benefits upon the participants. While kin selection might favor grooming toward kin, reciprocal altruism or individual selection might favor grooming toward others as well. Factors other than kin selection are likely to be particularly important determinants of behavior in free-ranging groups where social or demographic factors decrease the opportunities for kin to interact (chap. 20).

Finally, as pointed out in the introduction to part 3, it may not always be appropriate to assess the adaptive consequences of particular behaviors solely in terms of their immediate consequences upon genetic fitness. Individual acts do not occur independently; over relatively long periods of time, each altruistic act may contribute in an almost imperceptible way to the formation of a particular relationship, which itself has adaptive consequences for the participants (Hinde 1974a, 1983a; chap. 33).

RECIPROCAL ALTRUISM AND ALTRUISTIC BEHAVIOR

The theory of reciprocal altruism predicts that selection will favor those individuals who help only those who will help them in return (Trivers 1971). For this conditional strategy to evolve in any given population, three conditions must be met: (1) pairs of individuals must have the opportunity to interact repeatedly; (2) the benefits to the recipient must be much greater than the costs to the altruist; and (3) individuals must be able to remember their interactions. Axelrod and Hamilton (1981) have shown that under these conditions reciprocal altruism can evolve. It should be emphasized that participants need not be related for reciprocal altruism to be favored by selection. Reciprocal altruism is favored by selection because, over time, the participants in reciprocal acts obtain benefits that outweigh the costs of their interactions.

Like other forms of altruism, reciprocal altruism is difficult to identify in nature. Moreover, the theory of reciprocal altruism has not been developed fully enough to stipulate whether altruistic acts must be reciprocated in kind (Bertram 1982), to designate the time scale over which reciprocation must occur, or to predict the pattern of altruistic interactions among more than two individuals. Although it is therefore difficult to establish the importance of reciprocal altruism in the evolution of primate social behavior, several examples of altruistic behavior among unrelated individuals seem to fit the predictions of the theory (*P. troglodytes:* de Waal 1982; *C. aethiops:* Fairbanks 1980; Seyfarth 1980; Seyfarth and Cheney 1984a; *P. c. anubis:* Packer 1977; *M. mulatta:* Seyfarth 1977).

Seyfarth and Cheney (1984a) conducted a series of elegant field experiments on vervet monkeys to test the hypothesis that grooming among nonkin increases the probability of future support in aggressive encounters. In each experiment, two unrelated individuals, A and B, were selected for one pair of matched trials. A tape-recorded vocalization used by animal A to solicit support from others in a coalition was played near animal B in one of the following conditions: (1) after A had groomed B, or (2) after a fixed period of time during which no grooming had occurred between A and B. The duration of B's visual response to the vocalization was compared under these two conditions. The results of 10 such experiments clearly indicated that "grooming between unrelated individuals increases the probability that they will subsequently attend to each others' solicitations for aid" (Seyfarth and Cheney 1984a). An identical series of experiments among pairs of maternal kin ($r \geq 0.25$) indicated that grooming had no measurable effect upon their responses to solicitations for aid.

This study demonstrates both the usefulness of the notion of reciprocal altruism to describe patterns of altruistic behavior among unrelated individuals and the difficulties in verifying predictions derived from the theory. The data very clearly indicate that animals remember those who have behaved altruistically toward them in the past and adjust their future cooperative behavior accordingly. The data also show that one form of altruism (attention to solicitations for aid) is contingent upon another form of altruism (grooming). Although the data evaluate and support the qualitative predictions of the theory of reciprocal altruism, they do not provide a quantitative test of the predictions because they do not establish (1) whether the benefits to the recipient of being groomed and aided are equivalent, or (2) whether the benefits gained by the recipient of altruism are always greater than the costs incurred by the actor.

CONCLUSION

The data cited in this review provide suggestive evidence that individual selection has shaped the pattern of competition and aggression among primate groups and that kin selection and reciprocal altruism have favored the evolution of altruistic interactions. In general, successful competitors seem to obtain reproductive advantages over their less successful peers, and altruism is selectively directed toward kin or reciprocating partners. Nonetheless, few firm conclusions can be reached about the adaptive consequences of these behaviors. Much of the evidence that we would need to adequately evaluate the adaptive consequences of altruism, competition, and aggression has not yet been collected. Moreover, much of the evidence that has been cited in support of predictions derived from evolutionary theories could also be interpreted differently. As evolutionary theories are developed more fully, and empirical techniques are refined, our knowledge of the evolution of primate social behavior will become more complete. In the meantime, it is critical to understand how evolution may have shaped the patterns of the behavior we observe and to consider the possible adaptive advantages associated with those behaviors. As Darwin noted more than a hundred years ago "without speculation there is no good and original observation" (in a letter to Alfred Russell Wallace in 1857, cited in Brent 1981).

SUMMARY

Primates perform a wide range of different kinds of social behaviors. Some of their interactions with other members of their groups are altruistic or cooperative, while other interactions are competitive or aggressive. The primary purpose of this chapter is to explain how evolutionary processes, such as individual selection, kin selection, and reciprocal altruism, may have shaped the patterns of social behaviors such as grooming, competing over resources and mates, giving aid, and sharing food.

These evolutionary theories generate quantitative and qualitative predictions about the kinds and patterns of social behavior likely to be observed. A review of competitive interactions over resources, reproductive consequences of dominance interactions, and patterns of altruistic behaviors among primates indicates that many forms of behavior are consistent with qualitative predictions derived from evolutionary theory. However, this conclusion must be qualified because it is very difficult to derive or evaluate quantitative predictions or to identify a single adaptive function for any particular kind of social behavior. Thus, evolutionary theory appears to be a profoundly important clue in understanding the function and form of primate social behavior, but it is not yet possible to determine how all the pieces of the adaptive puzzle fit together.

27 | Infants, Mothers, and Other Females

Nancy A. Nicolson

MOTHER-INFANT RELATIONSHIPS
Introduction: A Historical Overview

The nature of the relationship between mother and infant has fascinated primatologists from the beginnings of systematic research. Primate infants are born in a relatively helpless, altricial state, and, compared to other mammals, they are dependent on adults for an unusually long time (Gould 1977). Primates tend, moreover, to be extremely social creatures, capable of forming strong and long-lasting emotional bonds, of which the mother-infant relationship is thought to be a prototype. These parallels between human and nonhuman primates, together with their shared evolutionary history, have sparked the interest of comparative psychologists, anthropologists, and zoologists alike. The marked inter- and intraspecific variability in patterns of mother-infant interaction, as well as the possibility of manipulating the relationship experimentally, gave researchers hope of unraveling the complex effects of early social experience on later behavior.

Pioneering efforts to describe primate mother-infant relations were made as early as the 1920s and 1930s by Yerkes and his contemporaries observing captive chimpanzees (Bingham 1927; Yerkes and Tomilin 1935) and macaques (Tinklepaugh and Hartman 1932). In the 1950s, Harlow and co-workers began a series of experimental investigations designed to test then-current theories of the significance of the human infant's relationship with its mother (Harlow 1958). In addition to the countless observations and experiments in laboratories and captive colonies that followed (usually involving macaques and the great apes), much information on mother-infant relations has been obtained from field studies, which have focused mainly on the more terrestrial and open-country species. Much less is known about mother-infant interaction in the wild in arboreal Old World monkeys, New World monkeys, and prosimians.

Until recently, the unifying theme of laboratory and field studies has been psychodynamic, with the closest theoretical ties to issues in human social development. Bowlby's (1969) concept of attachment (see chap. 33) found an operational definition in terms of contact- and proximity-seeking behaviors; gradually, the development of infant independence has become almost synonymous

with a decline in mother-infant contact. With the expansion of evolutionary theory concerned with parental investment patterns and parent-offspring relations (e.g., Trivers 1972, 1974; Alexander 1974; Parker and Mac-Nair 1978), however, new interest has arisen in socio-ecological and life-historical aspects of maternal and infant behavior. As a result, recent field studies have integrated ecological and demographic data into descriptions of mother-infant relations and have tended to interpret behavioral interactions in terms of the costs and benefits to each member of the dyad (see, for example, Altmann 1980; Lee 1981; Nicolson 1982; P. L. Whitten 1982). Predictions derived from sociobiological theories of parent-offspring relations have proven difficult to test (see Altmann 1980), but efforts made to collect the relevant information have led to a broader understanding of the significance of maternal care and the process through which independence is achieved.

This review focuses on mother-infant relations in free-ranging populations, while drawing from the vast literature on captive mothers and infants to fill in some of the gaps. (For recent summaries that include studies conducted in captivity, see McKenna 1979a; Swartz and Rosenblum 1981; Hinde 1983c.)

Transitions from Dependence on the Mother to Independence

Transition to Independent Feeding. A period of nutritional dependence on parents is a universal feature of mammalian development, which also occurs in most bird species, some insects, and a few spiders (E. O. Wilson 1975). Lactation is an exclusively mammalian trait, and it has been a major factor in the evolution of infant morphology and behavior (Pond 1977). It is also an energetically expensive form of parental care. For primate females, energy requirements are raised somewhere between 20 and 50% during lactation (Portman 1970; Buss and Voss 1971). Nursing is one of the clearest forms of parental investment, as defined by Trivers (1972, 1974), in that it both increases the current offspring's chance of survival and decreases the parent's ability to produce and care for future offspring. For this reason, nutritional weaning is the most commonly cited example of potential parent-offspring conflict. Indeed, for all wild primate species in which the transition from nutritional dependence to independence has been described, conflict is evident; the severity and duration of the weaning period, however, vary markedly both within and between species.

Quantitative data on suckling behavior are available for wild vervet monkeys (Struhsaker 1971; P. L. Whitten 1982), baboons (Altmann 1980; Nash 1978; Nicolson 1982), Japanese macaques (Hiraiwa 1981), provisioned rhesus macaques (Berman 1980b), chimpanzees (Goodall 1968; Clark 1977), and gorillas (Fossey 1979). In each case, the amount of time infants spent at the nipple each day decreased at a fairly steady rate from birth through weaning. Several studies report a decrease in the frequency of suckling bouts in older infants, accompanied by an increase in the length of an average bout, especially during periods of intense weaning conflict (Clark 1977; Hiraiwa 1981; Nicolson 1982). Weaning is apparently accomplished not by the absolute denial of milk, but by a gradual shifting of the conditions under which the infant is granted the nipple. Although suckling takes place more or less on demand in the early months, mothers later begin to restrict suckling to times when it will not interfere with their own activities, such as while grooming or resting (gray langurs: Jay 1963a; baboons: Altmann 1980; Nicolson 1982; vervets: P. L. Whitten 1982).

As nursing becomes less contingent upon infant demands, conflict between mother and infant is often observed. Maternal aggression is rare in wild primates, but threats, hitting, and nipping do occur. More often the mother simply denies access to the breast by turning or running away, blocking the chest with an arm, lying down, or holding the infant at a distance. Infants respond with signs of frustration and distress: whimpering, moaning, and occasional tantrums. Infants soon learn to monitor their mothers' activities until opportune moments for suckling arise; one strategy among baboon infants, for example, was to wait until the mother sat down to defecate, then to suckle for the few moments she was incapacitated (Nicolson 1982).

To understand the dynamics of mother-infant interaction during the weaning period, it is useful to examine what alternatives an infant has, at different ages, to dependence on its mother (see Davies 1976, 1978). It is clear that infants supplement their diet of milk with solid foods long before weaning is completed, but little is actually known about the development of independent feeding in primates. In olive baboons, the time infants spent feeding increased steadily as suckling decreased, until at 14 months infants were spending as much time feeding as were adults (Nicolson 1982). There was considerable variability among infants of a given age in suckling versus feeding time, however, and those who spent less time suckling consistently spent more time

feeding. A plausible explanation is that mothers, by refusing to nurse on demand, forced infants to adopt the alternative strategy of feeding themselves for more time each day.

The primate mother's role in infant nutrition does not end with nursing; by tolerating the infant's proximity during feeding, she further facilitates the development of independent feeding. For example, for much of the first year, baboon infants spent a disproportionate percentage of their total feeding time sharing a feeding site with their mothers (Nicolson and Demment 1982). There are several possible benefits to an infant of co-feeding with its mother. First, by watching her, an infant probably learns what to eat and how to eat it. Second, when the mother is preparing difficult foods (e.g., underground corms or tubers) for herself, the infant can often obtain a free share. Finally, feeding close to the mother may prevent others from supplanting the infant.

In the few species in which mothers actively share food with offspring, it appears that mothers are sensitive to the offspring's developing abilities to fend for themselves. Wild chimpanzee mothers show a willingness to share food in proportion to the age of the offspring and the degree of strength or dexterity required to prepare difficult items, such as hard-shelled fruits (Silk 1978). Wild titi monkeys share food items that are impossible for offspring to obtain independently (Starin 1978).

Transition to Locomotor Independence. Extensive transport of the young is rare among mammals; it occurs only in marsupials, a handful of creatures such as anteaters, tree sloths, and pangolins, and the primates. Although some prosimians (e.g., galagos, mouse lemurs) leave their young in nests or clinging to branches and carry them short distances by mouth (Klopfer and Boskoff 1979), the vast majority of the primates carry their infants on their bodies throughout the day. With a few exceptions, this task falls to mothers. Infant carriage probably never entails as great an energy expenditure as lactation, but it is certainly not a negligible form of parental investment (see P. L. Whitten 1982) for estimates of the energetic costs of infant transport in primates).

The transition from riding to the adult form of locomotion is often speeded along by the mother, with attendant resistance from the young, and for this reason it is appropriate to speak of weaning from riding. Relatively little is known about this process, in part because most studies of mother-infant relations have been conducted in captivity, where long-distance locomotion is not a major developmental issue. In natural habitats, however, it is critical that infants remain near their mothers or other group members during sometimes extensive day journeys. Riding on the mother has a similar, but not identical, time course to that of suckling. Although mothers occasionally carry older infants over difficult passages or during alarm situations, the riding period is virtually over by 6 months in wild vervets (P. L. Whitten 1982), 1 to 1.5 years in olive and yellow baboons (Nash 1978; Nicolson 1982), 2 to 3 years in orangutans (Horr 1977), and 4 to 8 years in chimpanzees (Clark 1977). It is extremely rare for an older sibling to ride after the birth of a new infant.

Data from wild baboons shed some light on the process through which locomotor independence is achieved (Nicolson 1982). Mothers provided all of their infants' transportation in the first 8 weeks. Thereafter, infants spent increasing amounts of time traveling on their own, as well as being carried short distances by nonmothers, especially subadult and juvenile females and siblings. While a few infants achieved total locomotor independence as early as 7 or 8 months, most infants continued riding on some occasions (during troop alarms or rapid progressions, for example) throughout the first year and, in rare cases, beyond 15 months. Rejection from riding was often abrupt and apparently at least as traumatic as rejection from the breast. At all ages throughout the first year, infants who were carried by their mothers least spent the most time in independent locomotion. Infants weaned relatively early from riding spent more time running than later-weaned infants, especially between 6 and

9 months. In general, running time peaked around 12 months, although it remained high compared to adults in the oldest infants observed (18 months). Early weaning from riding thus appears to entail extra energetic costs for infants and perhaps, in addition, an increased risk of predation. The advantage to mothers of refusing to carry their offspring will depend on how well their infants are able to make use of other sources of nutrition and alternate caretakers.

Attachment and the Transition to Social Independence. The early work of Harlow (1958, 1971) established the importance of physical contact to the developing monkey; the mother's body was shown to provide warmth, reassurance, and a safe base from which to explore. As the infant grows older and contact declines, the continuing bond with the mother is shown in part by their spatial association. Thus, developmental changes in contact and proximity can provide insight into how the infant's world gradually expands to include individuals and experiences beyond its primary relationship with the mother.

Good estimates of the amount of time mother and infant spend in physical contact over the period of dependency are available for only a few free-ranging populations (see fig. 27-1). Some generalizations can be made, however, for other species. The range in contact is

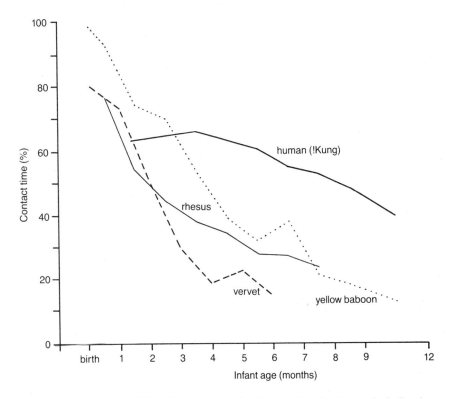

FIGURE 27-1. Mother-infant time in contact for four species of primates, including humans. SOURCES: Yellow baboons: Altmann 1980; vervet monkeys: Struhsaker 1971; rhesus macaques: Berman 1978; human hunter-gatherers (!Kung): Konner 1976. Figure redrawn from Altmann 1980, fig. 60, with permission of Harvard University Press.

exemplified at the lower extreme by prosimian species in which mothers leave their young in nests or clinging to branches (e.g., lorises, galagos, mouse lemurs: Klopfer and Boskoff 1979) and at the upper extrme by the great apes, where the first break in mother-infant contact occurs when the infant is already 4- to 6-months old (Goodall 1968; Horr 1977; Fossey 1979) (fig. 27-2).

Interspecific variations in patterns of contact are largely accounted for by differences in rates of growth and maturation, which are in turn closely related to life-history parameters such as adult body and brain sizes (chap. 16). Several other factors may also contribute. Based on observations of closely related cercopithecine species in captivity, Chalmers (1972) speculated that prolonged mother-infant contact might be particularly advantageous in arboreal primates, due to the risk of infants falling out of trees. It should be noted, however, that although infants of the more arboreal species Chalmers studied were indeed more reluctant to break contact, there were no differences in the restrictiveness of arboreal versus terrestrial mothers. Climate is another environmental variable that influences patterns of contact. Cold weather or rain increases contact between baboon mothers and infants (Nicolson 1982) and may explain why Japanese macaque infants, who must endure extremely cold winters, are carried longer and weaned later than rhesus macaques (Hiraiwa 1981). Lastly, differences among species in the extent to which individuals other than the mother participate in infant care are clearly reflected in mother-infant contact patterns (see "Allomaternal Care").

Measures of contact and proximity, by themselves, provide little insight into the dynamics of the development of independence. Hinde and co-workers have presented quantitative methods for determining the relative contributions of mother and infant to the maintenance of contact and proximity (Hinde and Spencer-Booth 1967; Hinde and Atkinson 1970; Hinde and White 1974). If the infant's responsibility, relative to the mother's, for decreasing proximity exceeds its responsibility for increasing proximity, it can be said to play a greater role than the mother in maintaining mutual proximity. In rhesus macaques (Berman 1980b) and baboons (Altmann 1980; Nicolson 1982), the major responsibility for staying close together, as measured by Hinde's index, shifted from mother to infant by 3 to 4 months. Older infants appeared to play an especially large role in the maintenance of proximity to their mothers in these free-ranging groups, compared to captive rhesus macaques (Hinde and Spencer-Booth 1967).

Roles of Mother and Infant in the Development of Independence

The issue of whether mother or infant plays the more important role in the development of independence has gen-

erated considerable debate. Some researchers maintain that rejection by the mother facilitates the normal development of independent functioning (e.g., Hansen 1966; Jensen, Bobbitt, and Gordon 1967; Hinde 1969, 1974b; Rowell 1972). Others argue that rejection is likely to result in greater, rather than lesser, dependence on the mother (e.g., Rosenblum 1971; Kaufman 1974). The latter view deemphasizes the mother's role in promoting independence and stresses instead the importance of the infant's maturing motor and social skills, which gradually allow the infant access to the world beyond its mother (see Kaufman 1966; Rheingold and Eckerman 1970; Bolwig 1980). The arguments on both sides have been summarized by Suomi (1976). As he points out, the available data are often difficult to interpret; one major problem is that the behaviors of individual mothers and their infants have been separated during analysis, thereby obscuring contingencies between maternal acts and infant responses. Lack of comparability in social groupings, housing conditions, and infant ages from one study to the next has further complicated the picture. Whether or not rejection will increase or decrease infant contact and proximity depends to a large extent on just such variables (see, for example, Jensen, Bobbitt, and Gordon 1969).

In short, most observers agree that conflict between mother and infant occurs, but disagree over the developmental significance of this conflict for the growth of independence. Theories of the evolutionary significance of parent-offspring conflict have shed new light on this controversy. According to Trivers (1974), underlying differences in their genetic self-interest lead parent and offspring to disagree over the length of the period of parental care and the amount of care an infant receives at any particular time. All other things equal, a parent is expected to divide its resources equally among all current and future offspring. An individual offspring, however, is more closely related to itself than to its siblings and will thus benefit from obtaining more parental investment than the parent is selected to give. Rejection and protest can thus be seen as strategies that have evolved to bias the amount of investment given in favor of mother or infant, respectively. Whether mother or infant is more responsible for independence has become less compelling than how specific life-historical, social, and environmental factors can influence the costs and benefits of parental care. Answers to the latter questions will lead to a better understanding of the marked intraspecific variability in weaning, conflict, and development of independence.

Variability in Mother-Infant Relations

One of the most striking features of the primate mother-infant relationship is the extent to which it varies from one pair to another. Individual differences in the fre-

FIGURE 27-2. Mother-infant contact in four nonhuman primates: (*a*) olive baboon; (*b*) indri; (*c*) red howler monkey; (*d*) mountain gorilla. (Photos: (*a*) Barbara Smuts/Anthrophoto; (*b*) Russ Mittermeier/Anthrophoto; (*c*) John Robinson; (*d*) Kelly Stewart and Alexander Harcourt/Anthrophoto)

c.

d.

quency of maternal rejection and the time course of infant independence can be related in some cases to characteristics of the mother. Findings from field studies are shown in table 27-1. These data indicate that high maternal rank is associated with permissiveness and early weaning in some wild cercopithecines (fig. 27-3). The reported effects of maternal age and experience are more difficult to summarize; the findings for macaques, for example, are completely opposite to those for baboons. One possibility is that early weaning may have greater advantages for older mothers in seasonal breeding species such as rhesus and Japanese macaques than in nonseasonal breeders such as baboons (see Altmann 1980; P. L. Whitten 1982). Hiraiwa (1981) speculated that older Japanese macaque mothers, by weaning as soon as possible, were able to reproduce again in the next season, whereas this strategy would have been too costly in terms of infant mortality for primiparas, who tended to delay weaning and never gave birth in successive years. In contrast, Trivers (1974) predicted that older mothers, because they are unlikely to produce many more offspring, should invest more heavily than young mothers in current offspring. A better understanding of the relationship between maternal rank, age, parity, and reproductive performance may help explain how these variables influence patterns of maternal care.

Among infant characteristics that might influence mother-infant relations, only gender has been systematically examined. Although this variable has been associated with differences in both maternal rejection and the development of spatial independence in captive groups (Mitchell 1968; Jensen, Bobbitt, and Gordon 1968; Rosenblum 1974; Simpson 1983), no major sex differences in mother-infant interactions have been found in studies of wild groups when sample sizes were large

enough to permit a quantitative comparison (e.g., Altmann 1980; Nicolson 1982).

Implications of Individual Differences

Consequences for Mothers. A number of recent studies indicate that variations in patterns of maternal care in primates can affect the timing of the mother's future reproduction. In a study of olive baboons, multiparous females who spent a relatively long time nursing each day in the first 32 weeks experienced longer periods of postpartum anestrus and longer total interbirth intervals than females who nursed less (Nicolson 1982). A study of yellow baboons also demonstrated a relationship between maternal care and the length of postpartum anestrus, but in this case females who weaned early had no shorter interbirth intervals than females who weaned late (Altmann, Altmann, and Hausfater 1978). In species that reproduce seasonally, females who wean their infants earlier are more likely than late weaners to conceive and reproduce in successive years (Japanese macaques: Hiraiwa 1981; vervet monkeys: P. L. Whitten 1982; captive rhesus macaques: Simpson et al. 1981; in this last study, lower suckling frequencies were associated with shorter interbirth intervals only for mothers of female infants). Shorter interbirth intervals can contribute to a female's reproductive success either by increasing the total number of offspring she produces in a lifetime or by decreasing generation time, a major determinant of population growth (Cole 1954).

Consequences for Infants. Many studies have documented differences among mothers in how protective, or restrictive, they are of their infants. There is some evidence that permissive mothering is associated with higher infant morbidity and mortality in wild baboons

TABLE 27-1   Maternal Characteristics That May Contribute to Variability in Mother-Infant Relations

Characteristic	Effects	Sources
Dominance rank		
Yellow baboons	Permissive ("laissez-faire") mothers ranked higher than restrictive mothers.	Altmann 1980
Vervet monkeys	Infants of higher-ranking mothers had more contact, lower suckling rates, and earlier weaning in one troop, but not in another.	P. L. Whitten 1982
Olive baboons	No significant relationship evident after the first 8 weeks of infant's life.	Nicolson 1982
Age and parity		
Provisioned rhesus macaques	Mothers older than 6 years were more rejecting and sought less contact and proximity with their infants than younger mothers.	Berman 1982b
Provisioned Japanese macaques	Older multipares more aggressive and weaned infants earlier than younger mothers.	Hiraiwa 1981
Olive baboons	Older mothers spent more time in contact and proximity and were less likely to interrupt suckling bouts than younger mothers.	Nicolson 1982
	Primiparous mothers rejected infants earlier than older multipares.	Ransom and Rowell 1972

FIGURE 27-3. Adult female Japanese macaque preventing her infant from moving away from her. In macaques and baboons, maternal styles vary individually, and some mothers show protective behavior of this type much more often than others. (Photo: Sarah and Harold Gouzoules)

(Altmann 1980, 134). The incompetence of some first-time mothers may also contribute to offspring mortality, especially in the earliest weeks of life. In olive baboons, for example, the mortality rate in the first 6 months for full-term firstborn infants was more than seven times that of later-born infants (43 versus 6%) (Nicolson 1982). Similar relationships between maternal parity and infant mortality have been reported in provisioned rhesus (Drickamer 1974), captive rhesus (Wilson, Gordon, and Bernstein 1978), captive squirrel monkeys (Taub 1980a), and wild mantled howler monkeys (Glander 1980), possibly reflecting differences in maternal behavior in addition to physical factors such as lower birth weights or lactation failure.

The psychosocial consequences of different patterns of mother-infant interaction remain largely unknown. Although experimental perturbations of the relationship can have long-term effects on the offspring's social behavior (reviewed in Hinde 1974a; Suomi 1982), the implications of normal variations are much more elusive. From studies of captive monkeys, it appears that maternal punishment in infancy can have a lasting influence on social responses in stressful situations (Mitchell, Arling, and Moller 1967) and may lead to increased offspring aggressiveness or withdrawal later in life (Chamove 1980). Less is known, however, about how differences in early interactions between mother and infant influence the social development of wild primates. Preliminary

data suggest some continuity in mothering style across generations of yellow baboon females (Altmann 1980). Long-term research at several study sites has the potential to address this and other issues, such as the relationship between mother-infant interaction and the persistence of family ties and kin-directed altruism.

Interactions between Siblings and Infants

Next to its mother, an infant's siblings will probably be its most frequent social partners. An older sibling can be both a rival for the mother's attention and an ally in times of need. Both negative and positive interactions between siblings can, in turn, influence aspects of the mother-infant relationship. In a variety of species, juveniles have been seen to threaten and bite their infant siblings, pull them off the mother's nipple, and interrupt mother-infant grooming sessions. In contrast to these occasional minor assaults, which may temporarily increase maternal protectiveness (see Hooley and Simpson 1983), older siblings provide many beneficial services to infants. Particularly in species with a social organization that precludes infant peer groups (e.g., chimpanzees, monogamous primates), siblings are often playmates. In rhesus macaques (Sade 1965), Japanese macaques (Kurland 1977), baboons (Cheney 1978b; Nicolson 1982), and several other species, siblings are responsible for a disproportionate share of the grooming an infant receives from individuals other than its mother. Older siblings often attempt to carry their younger brothers or sisters and in some cases may provide valuable help with infant transport. To the extent that these activities free the mother's time and relieve some of the burden of parental care, siblings may serve as "helpers at the nest." The most striking example of this phenomenon in primates occurs in marmosets and tamarins, where older sibs may carry infants (usually twins) more often than does the mother (chap. 4). The presence of older offspring in the group may also facilitate early maternal rejection and infant independence (rhesus macaques: Berman 1982b; Japanese macaques: Hiraiwa 1981).

When infants are orphaned, siblings often attempt to care for them; in some cases, such care may make a crucial difference to infant survival (chacma baboons: Hamilton, Busse, and Smith 1982; chimpanzees: Goodall 1983; Pusey 1983). Berman (1983b) has described a case of sibling adoption in rhesus monkeys on Cayo Santiago. By examining measures of typical patterns of mother-infant interaction, she found that the interactions between the orphaned infant and a 3-year-old nulliparous sister developed over a period of 2 months, through adjustments by each, into a relationship very similar to that of a mother and a similar-aged infant. Even if an older sibling does not actually function as a substitute mother, the affiliative relationship between the siblings can be intensified by the mother's death (Lee 1983d).

Persistence of the Mother-Offspring Bond after Weaning

The social structure of various primate species determines which classes of offspring will have the opportunity for an extended relationship with the mother. For example, in species in which males transfer, such as baboons, macaques, and vervets, male offspring will generally have few opportunities to interact with their mothers after emigrating from the natal group. In contrast, mother-daughter bonds in these species often persist into adulthood (chaps. 11, 24). Chimpanzees of both sexes may retain close bonds with their mothers even as adults (Pusey 1983; Goodall 1971). Gorilla mothers and mature daughters also show a persistent bond if they remain in the same group (chap. 14). There is little information about relationships between mothers and grown offspring in other species because few long-term studies of recognized individuals have been conducted (see chap. 24).

How does the relationship between a mother and her weaned offspring differ from that of other kinds of dyads? Preferential affiliative interactions between mother and older offspring have frequently been described, but more often than not maternity has been inferred rather than known. In a yellow baboon troop in which juveniles had been observed from birth, Walters (1981) compared the relationships of 3-year-old females with all adult females

and found that the juveniles' mothers, if still living, could easily be identified by behavioral data. Mother-offspring dyads could be differentiated from other adult female-juvenile dyads on the basis of grooming, presenting for grooming, and interventions during agonistic events. When the mother was not living, however, juveniles formed a strong relationship with another adult female that could be distinguished from a mother-offspring relationship only by using multivariate techniques. In summary, these results suggest that there is indeed a unique bond between mothers and their offspring of all ages, but that similar relationships can develop to compensate for loss of the mother. Although such relationships have been described in several primate species (e.g., captive gray langurs: Dolhinow and DeMay 1982; free-ranging rhesus macaques: Berman 1983b), they appear to be less reciprocal and less stable over time than true mother-offspring relationships.

RELATIONS BETWEEN INFANTS AND FEMALES OTHER THAN THEIR MOTHERS

Characteristic features of the primate mother-infant relationship are highlighted by contrasting maternal behavior with the behaviors of females toward infants who are not their own offspring. Among wild primates, a wide range of female-infant interactions has been observed, from care and even adoption to overt aggression, culminating under rare circumstances in infanticide. These observations and related data from captive groups are reviewed by Hrdy (1976) and McKenna (1979a, 1979b). For the sake of simplicity, female-infant interactions can be divided into two main categories—primarily affinitive (allomaternal care) and primarily aggressive (abuse). This distinction is not always clear-cut; allomaternal care can give rise to infant distress, especially when allomothers are clumsy. Separating care from abuse nevertheless provides a framework within which to discuss the various functional interpretations of female-infant interactions that have been proposed.

Allomaternal Care

Patterns. Lumped under the terms *allomothering* or *aunting* are such behaviors as cuddling, embracing, grooming, carrying, and protecting infants. (Note that, in spite of these terms, females or males of any age can be allomothers. Male participation in allomothering is discussed in chaps. 28, 29.) As a rule, nursing behavior is limited to the mother-infant relationship, although rare instances of allomaternal nursing have been reported in several species and may occur with some regularity in gray (Jay 1963a) and Nilgiri langurs (Poirier 1968). Allomaternal care has been described in free-ranging groups of ring-tailed lemurs, gray langurs, black-and-white colobus monkeys, vervet monkeys, patas monkeys, Japanese macaques, rhesus macaques, savanna baboons, gelada baboons, howler monkeys, gorillas, and chimpanzees, as well as in other species in captivity (see Hrdy 1976 and McKenna 1979a for references). Several studies, especially the more recent ones, provide quantitative data, which allow some generalizations to be made:

1. Allomaternal behavior is much more extensive among some species than others. Table 27-2 gives examples from studies that calculated the amount of time infants spent with allomothers (almost always females, in these cases). Based on qualitative descriptions and reported frequencies of allomaternal acts, vervet monkeys (Struhsaker 1971; P. L. Whitten 1982) and wedge-capped capuchins (Robinson, cited in Mack 1979) can be added to those species with frequent and prolonged allomaternal care. In savanna baboons, rhesus macaques, and Japanese macaques, however, although allomaternal interest in infants is high, the actual amount of time infants spend in contact with females other than their

TABLE 27-2   Some Species Differences in Amount of Allomaternal Contact

Species	Estimated Time Infants Spent in Contact with Allomothers	Sources
Gray langur	35–57% of total time observed (first 4 weeks)	Hrdy 1977; McKenna 1981 (captive group)
Common squirrel monkey	40% of total time observed	Baldwin 1969 (seminatural environment)
Red howler monkey	Less than 6% of total time carried	Mack 1979
Chimpanzee	0.7–8.0% of total time observed	Nishida 1983a

TABLE 27-3   Factors That May Influence the Frequency of Allomaternal Behavior Shown by Females

Factor	Type of Influence	Species and Sources
Infant's age	Very young infants preferred; interest from allomothers declines after first few months	Several colobine species: Hrdy 1977; McKenna 1979b, 1981; vervet monkey: P. L. Whitten 1982
	Mother restricts allomaternal contacts for first month or longer; thereafter, allomaternal interest sustained for several months	Squirrel monkey: Baldwin 1969; wedge-capped capuchin: Robinson, cited by Mack 1979; red howler: Mack 1979; Japanese macaque: Hiraiwa 1981; rhesus macaque: Breuggeman 1973; olive baboon: Nicolson 1982
Mother's rank	Infants of high-ranking mothers receive more attention from allomothers (allomothers rarely abusive)	Chacma baboon: Cheney 1978a; vervet monkey: P. L. Whitten 1982; Lee 1983a
	Infants of low-ranking mothers receive more attention from allomothers (allomothers often abusive)	Captive bonnet macaques: Silk 1980b
	Maternal rank not an important factor	Gray langur: Hrdy 1977; chimpanzee: Nishida 1983a
Relatedness of allomother to mother and infant	Siblings (especially females) participate more than nonrelatives in allomothering	Rhesus macaque: Breuggeman 1973; Japanese macaque: Hiraiwa 1981; vervet monkey: Lee 1983a; baboon: Cheney 1978a; Nicolson 1982; chimpanzee: Nishida 1983a
	Siblings participate less than nonrelatives	Japanese macaque: Kurland 1977
	Kinship not an important factor	Gray langur: Hrdy 1977
Age and maternal experience of the allomother	Subadult and juvenile nulliparous females most frequently involved in allomothering	Gray langur: Hrdy 1977; squirrel monkey: DuMond 1968; bonnet macaque: Rahaman and Parthasarathy 1962; Japanese macaque: Hiraiwa 1981; vervet monkey: Lancaster 1972; olive baboon: Nicolson 1982; chimpanzee: Nishida 1983a
	Adult females most frequently involved in allomothering	Ring-tailed lemur: Jolly 1966a; Nilgiri langur: Poirier 1968; red howler: Mack 1979; wedge-capped capuchin: Robinson, cited by Mack 1979; urban vervet monkey: Kriege and Lucas 1974

mothers is low (unlikely to exceed 5% of time observed). Mothers are also more likely to restrict access to their infants in these species.

Interspecific variations in the extent of allomaternal behavior are not well understood. McKenna (1979b), for example, has argued that differences in dietary adaptations that affect adult female relations have allowed the evolution of early and extensive infant transfer among the colobines, while selecting against such behavior among the cercopithecines. The paucity of quantitative data on female relations and allomaternal behavior in wild colobines other than the gray langur, however, make this hypothesis difficult to evaluate. Moreover, the large amount of allomaternal care displayed by vervet monkeys does not fit the pattern McKenna (1979b) suggests.

2. Within a species, the distribution of allomaternal behavior is not random, but is likely to involve specific classes of female actors and infant recipients. As shown in table 27-3, allomaternal behavior is influenced by the infant's age, its mother's rank, kinship between the mother-infant dyad and potential allomothers, and the age and reproductive experience of the allomothers. These rather disparate results require some comment. First, regarding infant age, one might speculate that the youngest infants are invariably the most attractive to allomothers, but access to them is prevented by mothers in some species. However, in those species in which early infant handling is permitted, allomaternal interest wanes much more quickly than in those species with delayed contact, which suggests that more is involved than interspecific differences in maternal restrictiveness. Second, high maternal rank can contribute both to the attractiveness of an infant and to its mother's ability to restrict the approaches of lower-ranking allomothers. Whether a mother does indeed restrict access will depend in part on how positive she perceives the attentions of the allomother to be. In the case of the captive bonnet macaques listed in table 27-3, for example, female approaches to infants were more aggressive than those described in the vervets and baboons. Third, the greater involvement of siblings and other close female relatives in allomaternal care, as reported in the majority of studies, may reflect the mother's (and perhaps the infant's) greater tolerance of those individuals who are least likely to do harm. Lastly, young nulliparous females (those who have never given birth) displayed allomaternal behavior more frequently than any other classes of females in most studies that provide quantitative data (fig. 27-4). In other cases, all or most allomaternal acts were performed by adult females. More information is needed on the reproductive experience and current status of these females, as well as on the exact nature of their involvement in infant care, before such reported differences can be interpreted.

FIGURE 27-4. Juvenile female gray langur attempts to take infant away from adult female. In langurs and many other species, young, nulliparous females like this one are the most active allomothers. (Photo: Sarah Blaffer Hrdy/ Anthrophoto)

Interpretations. The existence of allomothering poses a problem for evolutionary theory: although such behavior bears some resemblance to patterns of maternal behavior toward own offspring, allomothers are less closely related than mothers to their infant charges. The questions then arise: what benefits (relative to costs) does an individual derive from exhibiting allomaternal behavior? What benefits accrue to the mother who allows allomothering to occur? And finally, what benefits, if any, does the infant gain from such activities? The various interpretations of allomaternal behavior have been thoroughly reviewed by Hrdy (1976, 1977), Quiatt (1979), and McKenna (1979b). The major theories can be summarized briefly.

1. Mothering practice for the allomother. First proposed by Lancaster (1972), this "learning-to-mother" hypothesis remains the leading, though not the exclusive, explanation for allomaternal behavior, based on two lines of supporting evidence. First, the widespread occurrence of such behavior in nulliparous females, in striking contrast to the indifference of many parous adult females to infants other than their own, suggests that allomothering has a specific developmental function in females. Second, captive females who have not had earlier contact with infants often show inadequate maternal behavior (Suomi 1982), and even under natural conditions, first-time mothers are sometimes awkward or incompetent (e.g., chimpanzees: Goodall 1971; savanna baboons: Altmann 1980; Nicolson 1982)—observations that underscore the importance of learning in the development and expression of primate maternal behavior. This theory does not explain why allomothers interact with certain infants rather than others.

2. Enhanced status for the allomother. According to this hypothesis, the possession of an infant, particularly if its mother is high-ranking, gives the allomother certain privileges (e.g., reduced aggression, increased access to higher-ranking individuals and to resources). Cheney (1978a) has suggested that regular contact with unrelated adult females, via interactions with their infants, may also facilitate the integration of immature females into the adult female social structure. This hypothesis is supported by the observations of increased allomaternal attention to infants of high-ranking mothers in vervet monkeys and chacma baboons (see table 27-3).

3. Benefits to the mother. The only direct benefit to mothers that has been proposed is increased foraging efficiency while the infant is being cared for by an allomother. Such a benefit would help explain why mothers in at least some species (e.g., gray langurs, vervets) readily give up their infants to other females. P. L. Whitten (1982) found that vervet monkey females had longer feeding bouts and higher rates of food intake when their infants were at a distance than when they were in contact or nearby. In other species, however, mothers actively resist having their infants taken by allomothers (e.g., chimpanzees: Nishida 1983a).

4. Benefits to the infant. It has also been suggested that allomaternal behavior improves the chances that an infant will be adopted if its own mother dies. Although true cases of adoption are rare among primates (Quiatt 1979), a few well-documented cases do exist, often involving older siblings or unrelated males (Hamilton, Busse, and Smith 1982; see also earlier section, "Interactions between Siblings and Infants"). Dolhinow and DeMay (1982) reported that captive gray langur females frequently adopt rejected or orphaned infants (at the infant's initiative), but evidence that unrelated females are willing and able to adopt orphaned infants in free-ranging groups remains weak.

5. Kin benefits. Allomaternal behavior could have evolved through the process of kin selection if mothers or infants derive some benefit (as discussed above) from the care given by relatives or if the benefit gained by related females from practicing mothering skills outweighs the cost to infants. Given the difficulty of assessing the actual costs and benefits of allomaternal care, these hypotheses have not yet been tested. It is important to note, however, that allomaternal care often involves only distantly related individuals. Moreover, the fact that kin benefits may accrue from allomaternal care does not exclude the possibility that allomothering evolved as a purely selfish behavior (Quiatt 1979; McKenna 1979b).

Finally, it is important to note that these interpretations are not mutually exclusive. Patterns of allomothering vary from species to species, and a given explanation may fit one species or environment better than another.

Female Abuse of Infants

Patterns. When we examine female-infant relations from the perspective of costs and benefits to the interactants, it becomes extremely difficult to say at what point infant care becomes infant use, and at what point use becomes abuse. Nevertheless, several types of female-infant interaction have been observed that, unlike the allomaternal behaviors described above, have few conceivable benefits for infants and definite costs. Again, many of these patterns have been compiled by Hrdy (1976). The behaviors that have been observed include threats, biting, rough handling of infants (e.g., hitting, throwing, dragging, or sitting on infants), aggressive removal of infants from the mother's ventrum (often followed by "kidnapping"), and direct attacks. Kidnapping can have fatal consequences when a female refuses to return an unweaned infant to its mother ("aunting-to-death," in Hrdy's 1976 terms). This phenomenon has been observed in gray langurs, Campbell's guenons, rhesus and Japanese macaques (reviewed by Quiatt 1979), and savanna baboons (Strum 1975; Collins, Busse, and Goodall 1984).

Female aggression toward infants was correlated with lower infant survival in captive bonnet macaques (Silk 1980b). Infanticide by an adult female has been witnessed in wild chimpanzees (J. Goodall 1977) and is suspected in wild gorillas (Fossey 1983). In contrast to allomaternal care, aggression is most commonly displayed by older, multiparous females (e.g., gray langurs: Hrdy 1977; Japanese macaques: Hiraiwa 1981; chimpanzees: Nishida 1983a), by females higher ranking than the mother (e.g., olive baboons: Nicolson, pers. obser.; captive bonnet macaques: Silk 1980b), and by females not closely related to mother or infant (Silk 1980b).

Interpretations. How might females benefit from abusing or killing the infants of other females? The available evidence suggests that such behavior is an aspect of female-female competition. Mild aggression from high-ranking females may serve to reinforce the status quo, most importantly in those species in which daughters inherit rank from their mothers and remain in their natal groups through adulthood. The infant soon learns which females rank above its mother and can mistreat it with impunity and which females are lower ranking and relatively harmless. Another possible function of severe aggression and infanticide is a reduction in the number of competitors for the perpetrator and her own offspring. This interpretation is more plausible when females are not closely related, as in chimpanzees and gorillas. Whether "aunting-to-death" fits with either explanation or may simply be explained as overzealous allomothering (benefiting the actor in terms of improved maternal skills rather than decreased competition) is less clear.

Conclusion

In summary, female-infant interactions have effects on at least three individuals—the infant, its mother, and the participant female—and can be examined from the perspective of each. Although in most species studied young females show greater interest in infants than do young males, it is arguable whether female-infant interactions benefit infants as much as do some male-infant interactions (chap. 28). Perhaps the most striking feature of interactions between unrelated females and infants is their selfish quality. In sharp contrast to true mothers, allomothers generally expend little energy monitoring or retrieving infants from danger, risk little in defending them, and soon lose interest in their charges altogether (Hrdy 1977; Dolhinow and DeMay 1982). Because such interactions are often neither reciprocal nor persistent over time, they do not always constitute a "relationship" (see Hinde 1976), certainly not in the same sense that mother-infant interactions do. The behavior of female relatives (especially siblings) is intermediate, although intermittent, affiliative interactions involving grooming, carrying, and defending form the basis for enduring relationships. The aggressive behavior of females toward infants is noteworthy because it demonstrates that infants do not invariably serve as releasing stimuli for female nurturance. Whether an infant will receive care, hostility, or utter indifference depends on the complex interplay between species-specific adaptations, infant characteristics, and the female's own life history.

SUMMARY

Recently, a number of detailed field studies have helped place the mother-infant relationship, traditionally a focus of psychodynamic research, in its natural context. Interspecific variations in patterns of behavioral development are largely accounted for by species differences in rates of growth and maturation. Intraspecific variations are related to factors such as maternal rank and parity. The timing of infant development affects both the mother's future reproduction and the infant's survival. Conflicts of interest between mothers and infants are particularly evident in two contexts: the development of independent feeding (weaning) and the development of independent locomotion. In many nonhuman primates, a special relationship between mother and infant persists long after weaning. Aside from the mother, infants usually interact most frequently with maternal siblings who often groom, protect, play, and sometimes carry infants. In many species, adult and immature females frequently attempt to inspect, handle, and carry infants. In some species mothers tolerate such attentions, but in others they do not. Other females sometimes abuse infants, and in general, allomothering seems to be a selfish, rather than an altruistic, behavior.

28 | Infants and Adult Males

Patricia L. Whitten

Primatologists have long been aware of the extensive interactions of nonhuman primate males and infants. Authors described the parental behaviors of captive marmoset males as early as the eighteenth century (Hershkovitz 1977), and Itani (1959) described caretaking by males among wild Japanese macaques in one of the first modern studies of primates. Subsequent work demonstrated that male-infant interactions were not only common but also extraordinarily diverse (Mitchell and Brandt 1972). The theoretical significance of these interactions, however, was given little attention. Because individuals were thought to act for the good of the group, male care of infants was an expected and unremarkable phenomenon.

The demise of group selection theory and the revival of interest in individual selection turned male care into a problem worthy of explanation. Male care was intriguing for several reasons. First, kin selection theory predicted that, when benefits and costs were held constant, individuals should invest preferentially in close relatives (Hamilton 1964). Primate males, however, were always less certain of the identity of their offspring than were females. Second, sexual selection theory predicted that males should be fundamentally biased toward the production of many offspring and against substantial investment in any single offspring (Trivers 1972). Finally, new data indicated tremendous interspecific variation in male-infant interactions, ranging from extensive maternal-like care in marmosets to infanticide in langurs (Hrdy 1976). These studies also confirmed Itani's (1959) finding that, even within a single group, males varied widely in the types of relationships that they had with infants. Friendly association, therefore, could not be dismissed simply as a general primate characteristic.

These considerations suggested that the diversity of male-infant relationships among primates could be explained by examining mating systems and genetic relationships. Subsequent reviews therefore attempted to map patterns of male-infant relations onto primate social systems (Redican 1976). These efforts were not wholly successful, however, and authors have since suggested other relevant factors such as the costs of maternal care (Kleiman 1977), female choice (Wittenberger and Tilson 1980), and the balance between male costs and infant benefits in these relationships (Kurland and Gaulin 1984). Although these approaches provide a promising beginning, a coherent explanation of the distribution of male care has yet to be realized.

This chapter examines the types of male-infant relationships observed in primates and considers hypotheses that have been developed to explain these findings. Discussion is restricted to those behaviors that involve actual interactions between males and infants; behaviors such as predator warnings or troop defense, which are not directed specifically toward infants, are not included. To facilitate discussion, relationships have been divided into five, somewhat arbitrary categories: (1) *intensive caretaking:* males spend a large part of the day engaged in infant caretaking; although the actual extent of male participation varies, males predictably perform some parental duties for all infants; (2) *affiliation:* males spend part of the day engaged in affiliative interactions with one or more specific infants; most males interact affiliatively with at least one infant; (3) *occasional affiliation:* males occasionally interact affiliatively with one or more infants, but these associations are not characteristic of all males nor of any single male all of the time; (4) *tolerance:* males permit infants to be near them but otherwise interact rarely with infants; (5) *use and abuse:* males interact with infants in ways that benefit the male and sometimes harm the infant. These categories are heuristic devices; it is important to note that within a group, more than one category can apply (e.g., savanna baboon males show both affiliation and use and abuse). For the purposes of this chapter, an infant will be defined as any immature animal still dependent on its parent(s) for milk or transport.

INTENSIVE CARETAKING

Hidden among the tangled branches and vines of the Neotropical forests are the most paternal of primate fathers. The monogamously mated males of the small titi (*Callicebus* sp.) and night monkeys and the tiny Callimiconidae and Callitrichidae are unique in the intensity and duration of their relations with infants (Redican 1976; Wright 1984; chap. 4). Males of these species share all parental duties except nursing, and although the extent of participation is quite variable within species, they are generally the major caretakers of infants (see table 28-1).

Males in these species are often strongly attracted to infants. Immediately after birth, they have been observed trying to sniff, touch, or hold the still-bloody newborn, and they sometimes even lick off the covering birth fluids (common marmoset and cottontop tamarin: Hershkovitz 1977). Within hours of birth, males carry infants on their backs, groom them, and protect them (Redican 1976; Vogt 1984). Large portions of a male's day are devoted to infant care, and the most devoted fathers return their infants to the mother only to suckle (Hershkovitz 1977).

Males also permit infants to take food from their hands and mouths (Hershkovitz 1977; see table 28-1). In captive lion tamarins and Goeldi's marmosets, males reportedly initiate food sharing with food calls heard only in this context (Brown and Mack 1978; Hershkovitz 1977). In the wild, however, food sharing has been reported primarily as a response to infant begging (saddleback tamarin: chap. 4; black-and-red tamarin: Izawa 1978a; Goeldi's marmoset: Masakata 1981; yellowhanded titi monkey: Starin 1978; dusky titi: Wright 1984). The food items shared are those that infants have difficulty obtaining or processing themselves, such as large mobile insects or hard-shelled fruit (Starin 1978; chap. 4).

Fiercely protective, males will defend infants against any real or imagined threat. In captivity, tiny lion tamarin males have flung themselves against intruders as intimidating as woolly monkeys, macaques, and humans (Hershkovitz 1977). Wild tamarin males have charged and threatened observers in order to rescue wounded offspring (white-footed tamarin, red-handed tamarin, black-and-red tamarin: Hershkovitz 1977). Sometimes, males also defend infants against unrelated conspecifics. In two captive saddleback tamarin groups, the dominant male acted as a "watchman," attacking and chasing unrelated adults whenever they attempted to touch the newborn infants (Epple 1975). In one of these groups, the dominant male himself was unrelated to the infant he so actively defended.

Although in most cases males care for their own offspring, they sometimes care for unrelated infants. In captive saddleback tamarin groups, the dominant male cares for infants regardless of paternity; adult males introduced to pregnant females before delivery assume full parental duties (Epple 1975). In Peru, feral saddleback tamarin groups contain two males, both of whom mate with the adult female. Presumably, neither male knows who fathered the offspring, but both males provide infant care (chap. 4). In one captive common marmoset group, an unrelated male actually spent more time caring for the infants than either of the two parents (Box 1977). In a captive Goeldi's marmoset group, a new male introduced after a female had given birth initially ignored her infants

but became an excellent caretaker after he had impregnated their mother (Hershkovitz 1977).

The duration of male caretaking varies among species, but it generally ends when the infant is capable of independent locomotion. Friendly associations between males and infants, however, can continue even after the infant achieves independent locomotion. In a feral yellow-handed titi monkey group, a male and an unweaned but older infant fed, traveled, and rested together, sometimes grooming or sitting with tails entwined (Kinzey et al., 1977).

Among these small New World monkeys, infant care can be costly for a male. Carriage of the heavy infants, weighing 7 to 27% of adult weight (Kleiman 1977; Vogt 1984), can add as much as 17% to locomotor costs (P. L. Whitten 1982). Moreover, infant care can interfere with foraging ability, as has been recorded among adult males in feral dusky titi monkey groups (Wright 1984). Slowed down by their heavy burdens, males entered fruit trees later than other troop members and frequently found the fruit supply already depleted. Males spent less time feeding and had lower rates of feeding on fruit and lower success in foraging for insects while caring for infants than at comparable times when not caring for infants.

The actual cost of caretaking should vary with participation, and male participation varies among species, among groups, and even among offspring of the same male and female. Among seven captive saddleback tamarin groups, the male's contribution ranged from less than 30 to over 95% of infant care (Epple 1975). Within groups, male participation also varies from year to year. In one common marmoset group, the male performed as little as 7% of infant carriage and as much as 80% in different years (Box 1977). This extensive individual variation in caretaking patterns makes it difficult to identify any species-specific patterns, but the available evidence suggests that species vary in the typical time of onset of male care (Kleiman 1977; chap. 4). This interspecific variability appears to relate to the relative size of the infant: males assume primary responsibility earlier in species in which the infant is relatively heavier and thus more costly for a mother to carry (Kleiman 1977). For example, in *Saguinus* and *Aotus*, where infants weigh 12 to 18% of maternal weight, maternal participation is substantial in the first week, and male participation increases following the first week (Epple 1975; Cebul and Epple 1984; Dixson and Fleming 1981; Wright 1984). In contrast, in *Callimico* the single infant weighs only 7 to 10% of the mother's weight, and males do not assume primary responsibility for the infant until after the third week (Vogt 1984).

Observed variation within species may be due in part to the availability of other helpers. During a longitudinal study of captive common marmoset groups, male involvement with infants declined substantially once juvenile offspring were available to care for infants (Box

1975a, 1977). After the first week, male caretaking was minimal, and mothers actually spent more time carrying infants than did males. However, in another study of the same species, males did not always reduce their participation when juvenile caretakers were available (Ingram 1977).

Despite this variability, monogamous, and sometimes polyandrous (see chap. 4), New World primates are unsurpassed in the extent of their male-infant relations. Two Old World primate species, however, deserve a place among the intensive caretakers: the monogamous siamangs and the promiscuous Barbary macaques. Although males in these species are not the primary infant caretakers, they are co-caretakers for at least a portion of the period of infancy (Taub and Redican 1984).

All gibbons are monogamous, but only the siamang exhibits active male participation in infant care (Wittenberger and Tilson 1980; chap. 12). The largest of the hylobatids, siamangs are an exception to the general rule that male care is most elaborate in the monogamous species that are smallest and that have the largest infants: siamang neonates weigh only 5 to 6% of adult body weight (Kleiman 1977), while the infants of most other monogamous intensive caretakers weigh 12 to 22% of adult female weight. However, in siamangs, mothers are solely responsible for infant care during the first year (Chivers 1974); when male care begins in the second year, the infant's weight relative to the mother's weight is well within the range of other monogamous primates (see table 28-4). The male takes over infant care when the mother begins to reject her infant, generally at 12 to 16 months of age (Chivers 1974). However, males have been observed carrying infants as early as 6 to 8 months (Chivers 1974; Chivers and Raemaekers 1980). The male grooms and carries the infant during the day, spending as much as 78% of the day in infant care (Chivers 1974).

Barbary macaques provide an interesting exception to the general rule that male care is most elaborate in monogamous primates (Redican 1976). This species is notable both for the harshness of its habitat and the intensity of its male-infant interactions (Taub 1978; 1984). Barbary macaques are found only in the mountains of Morocco and Algeria and on the island of Gibraltar (Taub 1978). On Gibraltar the animals are provisioned, but in the high mixed cedar forests of Morocco and Algeria, they extract a large portion of their diet from cedar and oak trees (Burton 1972; Deag 1983; Fa 1984). When heavy snow blankets the rocky slopes, they subsist almost entirely on the bitter needles of cedar trees (Taub 1978). In this harsh environment, reproduction is seasonal, and infants are born in the spring (Taub 1978).

The interactions of Barbary macaque males with infants are neither as frequent nor as consistent as those observed in siamangs or the monogamous New World primates. Nevertheless, they account for a substantial portion of infant caretaking (see table 28-1). Males begin

to interact with infants a few days after birth (Burton 1972; Deag 1974). Approaching the mother of the newborn, a male sits next to her, looks closely at the infant, teeth chatters, and touches it. As soon as infants are able to move independently, males begin interacting with them in a variety of ways (Taub 1978, 1984). They solicit an infant's approach by teeth chattering or by presenting a lowered rump or shoulder. The most common interaction is infant carriage. Males carry infants during feeding, during rapid travel, in response to predators, and during social interactions. Males often sit with an infant clinging to their backs or cradled in their laps. They also groom, nuzzle, and mouth infants, lick and smell them, manipulate their genitalia, and teeth chatter at them. Sometimes males simply sit and watch an infant intently. Alarm barks or distress calls from an unattended infant bring a prompt approach and retrieval. These interactions are most pronounced in the early months of infant life and begin to wane during the following mating season when infants are 5 to 7 months old (Taub 1978). After the first year, caretaking interactions cease altogether.

Although each male interacts to some extent with all or most infants, he interacts primarily with only one or two infants, and most adult males direct over half of their caretaking activities toward only one infant (MacRoberts 1970; Taub 1984). Each infant receives care from several different males, but the majority of care comes from only two or three males. Individual males also vary in their total involvement with infants, and some infants receive more male attention than others.

Observations of mating suggest that these caretaking relationships, unlike those observed in monogamous species, are not closely tied to paternity (Taub 1978). During estrus, each female mates daily with most of the males in the group (Taub 1980b, 1984). Females seek out new males after each ejaculatory copulation and do not allow any male sexual monopoly. Therefore, although each male is a possible father of any given infant, no male is a likely father. Furthermore, the most active caretakers are subadult males that mate with estrous females less often than do fully adult males (Taub 1984). Instead, caretaking relationships may arise from other forms of kinship. The apparently low rate of migration and the link between triadic interactions and shared interest in a single infant (described later) suggest males have ample opportunity to learn the identities of their maternal kin (Taub 1984). Since the mating system seems to preclude knowledge of paternity, any selection for investment in kin would be likely to favor siblings (and maternal sisters' offspring) over own offspring.

AFFILIATION

Affiliative relationships of males and infants are not limited to monogamous primates or to primates that live in difficult habitats. Strong male-infant associations have been documented in primates as diverse as black howler monkeys (*Alouatta pigra*: Bolin 1981), stump-tailed macaques (Estrada 1984a), savanna baboons (Altmann 1980; Ransom and Ransom, 1971; Hamilton 1984), and mountain gorillas (Fossey 1979, 1983). These associations usually lack extensive caretaking, but they often involve strong and enduring relationships. Average rates of male-infant interaction are often low, but much higher rates of interaction occur within specific male-infant pairs (Bolin 1981; Smith and Peffer-Smith 1984; Estrada 1984a; Stein 1984a, 1984b; Hamilton 1984; Smuts 1985).

The most obvious feature of these relationships is frequent spatial proximity (Altmann 1980; Fossey 1979; Bolin 1981; Estrada 1984a) (fig. 28-1). Infants are strongly attracted to males and follow them and forage and rest near them (mountain gorillas: Fossey 1979; black howlers: Bolin 1981; stump-tailed macaques: Estrada 1984a; savanna baboons: Nicolson 1982; Stein 1984a). Some mountain gorilla and black howler infants actually spend more time near or interacting with males than with their own mothers (mountain gorillas: Harcourt 1979c; howlers: Bolin 1981). Friendly contacts are common. Associated males hold, cuddle, nuzzle, examine, and groom infants, and infants turn to these males in times of distress (black howlers: Bolin 1981; savanna baboons:

FIGURE 28-1. Mountain gorilla infant resting with the silver-backed male. (Photo: Kelly Stewart/Anthrophoto)

Stein 1984b; mountain gorillas: Fossey 1983; stump-tailed macaques: Estrada 1984a). Males are tolerant of the play and proximity of youngsters and allow infants to tumble or climb on them, jump on their backs, and even swing on their tails (black howlers: Bolin 1981; mountain gorillas: Fossey 1983). Male associates also participate occasionally in brief periods of maternal-like caretaking. Before going off to forage, a black howler often left her infant with her mate, who, holding it in his arms, hugged and nuzzled the infant until the mother returned (Bolin 1981). Baboon male associates also "baby-sit" while the infant's mother is away, comforting and sometimes grooming or carrying the infant for up to 30 minutes at a time, several times a day (Ransom and Ransom 1971). Gorilla males often groom, cuddle, and nest with their 3- and 4-year-old offspring after the mother has given birth again, died, or transferred to a new troop (Fossey 1979, 1983).

Such associations provide important benefits to infants. One of the most important benefits appears to be protection from conspecifics (fig. 28-2). Savanna baboon males buffer new mothers and their infants from the curiosity and stressful pulling and handling of other group members (Altmann 1980; Stein 1984b). Individuals are more hesitant to approach the infant when a male associate is nearby, and males sometimes actively threaten others away by using threats and attacks to "punish" individuals who harass infants (Altmann 1980; Stein 1984b; Hamilton 1984). Baboon male associates also behave protectively toward infants in the vicinity of immigrant males, whose proximity often elicits distress from mothers and infants (Packer 1980; Busse and Hamilton 1981; Stein 1984b; Busse 1984b; Smuts 1985). Mountain gorilla males also protect infants from the attentions of other group members, and they intervene when juveniles play too roughly with infants or when adult females pig-grunt at unruly youngsters (Fossey 1983). Black howler males hurry to the side of infants left alone when the mother moves away (Bolin 1981).

Proximity to males can also give infants greater access to food. Baboon males share feeding sites with infants, which allows them to feed on seeds from discarded fruits and on grass corms that the males uproot (Stein 1984b; Nicolson 1982; Hamilton 1984). Like the foods shared by the monogamous New World primates (Wright 1984; chap. 4), these are items that infants are unable to process themselves. Unlike the New World primates, however, baboon males "share" food items by tolerating infant proximity at feeding sites rather than by bringing food to the infant or by allowing infants to take food from them directly.

For mountain gorilla and baboon infants, affiliation with a male can also bring benefits in later life. Male gorillas who associate closely with a silverback male in infancy continue to associate with him in adolescence and

FIGURE 28-2. A rhesus monkey adult male (*far left*) protects an infant (*to the right of the male and in contact with him*) from aggression by several adult females and juveniles. When the others began to threaten and chase the infant, it ran to the adult male. Although the females continued to threaten the infant with open mouth threats and lunges, they did not attack it as long as it was near the adult male. (Photo: Barbara Smuts/Anthrophoto)

are more likely eventually to inherit leadership of the group, whereas adolescent males who did not associate with the currently dominant silverback in infancy interact infrequently with him and are more likely eventually to leave the group (Harcourt and Stewart 1981). Among savanna baboons, some male-infant relationships persist for at least 3 to 4 years; these older juveniles continue to receive protection from their male associates (Stein 1984b; Johnson 1984).

Such male-infant associations are often accompanied by a close bond between the male and the infant's mother. Unlike other howler monkeys, the black howlers of Belize live in monogamous groups (Bolin 1981). The gorilla males who affiliate with infants are the dominant silverbacks and thus the major sexual consorts and affiliates of the females in their groups (Fossey 1979, 1983; Harcourt 1979c). Savanna baboon males associate pri-

marily with the infants of females with whom they have a special relationship (Ransom and Ransom 1971; Altmann 1980; Smuts 1985), and one factor affecting stump-tailed macaque affiliations is an association with the infant's mother (Estrada and Sandoval 1977; Smith and Peffer-Smith 1984).

Other factors, however, also influence male-infant relationships. These include male rank, dyadic preferences, and infant needs. Yellow baboon males sometimes compete with one another to be near newborn infants, and high-ranking males follow and herd new mothers, grooming them often and threatening away other males (Altmann 1980). Perhaps as a result, the high-ranking males in one particular troop spent more time with neonates than did lower-ranking males, and this greater proximity affected their later associations with infants (Stein 1984b). During the first 6 weeks of life when infant baboons seldom leave their mothers, their associations reflect maternal preferences, but after the second month, mutual male-infant attraction produces associations even when the mother is not nearby (Stein 1984b; Nicolson 1982). In some cases, associations appear to arise or intensify as a consequence of the infant's loss of, or neglect by, its mother (stump-tailed macaques: Smith and Peffer-Smith 1984; baboons: Ransom and Ransom 1971; Nicolson 1982; mountain gorillas: Fossey 1983).

The inclusion of humans among the affiliative species in table 28-1 may be surprising to some readers. Human males are generally considered unusual in the extent of their infant care (e.g., Alexander and Noonan 1979; Lovejoy 1981). Available data on the frequency and pattern of male-infant interactions, however, place human males squarely within the affiliative rather than the intensive caretaking species (see table 28-1). First, rates of interaction with infants are generally low. In modern industrial societies, some fathers spend as little as 45 minutes each week in interaction with infants, and even the most active spend no more than 3 hours a day (Lamb 1984). Of the 80 cultures included in a world survey of paternal relationships, in 20% fathers were rarely or never near their infants (West and Konner 1976). In the remaining cultures, fathers were occasionally or frequently near their infants, but only 4% of the cultures were characterized by a close father-infant relationship. The !Kung San, a hunting and gathering group in Botswana, were one such culture. However, even here fathers spent only 14% of the time interacting with infants, about the same amount of time spent by the most active fathers in industrial societies (Konner 1976; Lamb 1984). Second, fathers rarely assume major responsibility for child care, and actual caretaking is rare. The !Kung fathers accounted for no more than 6% of infant caretaking even though they were free of subsistence activities for at least half of the week (West and Konner 1976). Instead, in both industrial and nonindustrial societies, the major form of male-infant interaction is play (Lamb 1984). Thus in both frequency of interactions and involvement in caretaking, human males more closely resemble the males in "affiliative" nonhuman primate species than the males in the "intensive caretaking" species. Only in their provision of food do human males resemble the intensive caretakers. However, since subsistence is typically provided to the mother rather than directly to the offspring, even this paternal investment is inextricably linked to mating efforts and interests.

OCCASIONAL AFFILIATION

In addition to these consistent patterns of affiliation, short-term associations of some males and infants occur in several other species (see table 28-1). In these species, males vary in their relationships with infants. Often they are indifferent, but under certain circumstances they af-

filiate with particular infants. The factors influencing such associations and the benefits that males provide to infants are often similar to those already discussed for affiliative species. For example, among rhesus and Japanese macaques, male-infant associations are often an outgrowth of the male's bond with the infant's mother, and males protect these infants from aggression by other troop members (Japanese macaques: Takahata 1982a; rhesus macaques: Kaufman 1967; Breuggeman 1973). Sifaka males groom and associate with infants in the context of a multimale group (Jolly 1966a). In several species, intense bonds sometimes develop after infants are orphaned (rhesus macaques: Berman 1983b; Vessey and Meikle 1984; Japanese macaques: Hiraiwa 1981; hamadryas baboons: Kummer 1968; lar gibbons: Carpenter 1940). As in the mountain gorillas and savanna baboons discussed above, males in such cases take on many of the duties of the mother, such as holding, grooming, comforting, and protecting the infants, and occasionally even carrying and sleeping with them. Because of these similarities, it is more appropriate to view the species classified here under *affiliation* and *occasional affiliation* as points on a continuum rather than as two distinct groups.

Among Japanese macaques, male-infant relationships vary between groups, between males, and over time. In several groups, males comforted and protected weanlings whose mothers were preoccupied with new infants (Itani 1959; Alexander 1970; Hiraiwa 1981), but in another group, male-infant affiliations involved mainly young infants not yet weaned (H. Gouzoules 1984). In free-ranging groups, associations involved mainly older, higher-ranking males (Itani 1959; Hiraiwa 1981), but in a captive group, males of all ranks interacted with infants (Alexander 1970).

Male interest in infants can vary from year to year, and in some cases associations wane during the mating season but are renewed the following year (Itani 1959; Hasegawa and Hiraiwa 1980; H. Gouzoules 1984). Paternity does not seem to explain such variation in male-infant relationships since, in the one group for which data are available, males did not bias their caretaking toward the infants of former consorts (H. Gouzoules 1984).

Young adult hamadryas baboon males also develop close associations with weanlings as a first step in the establishment of a breeding unit (Kummer 1968; chap. 10). Young adult males who do not yet have a unit herd recently weaned females away from their mothers and gradually condition them to follow, even carrying them when they are frightened or having difficulty navigating the steep sleeping cliffs. As males acquire additional females, they behave less paternally toward them, but they still occasionally carry the black infants of the females in their unit.

Dominant red howler males also associate closely with new mothers, possibly protecting likely offspring from

TABLE 28-1   Male-Infant Interactions in Well-Known Primate Species

Species	Frequency of Interactions (% obser. time)	Caretake	Carry	Protect	Share Food	Co-feed	Groom	Play	Contact	Proximity	Onset	Peak	End	Infant Sex	Sources
INTENSIVE CARETAKING															
Callithrix jacchus (c)	1–80%	x	x	x	x		x	x	x	x	1st hr	0–1 wk	17 wk	M > F	Box 1975b, 1977; Epple 1970; Hershkovitz 1977
Cebuella pygmaea (c)		x	x				x		x	x	1st hrs		4 wk		Hershkovitz 1977
Saguinus fuscicollis (c, f)	(18–96% of carrying)	x	x	x	x		x	x	x	x	Carry: day 1	1–2 wk	10 wk	M > F	Epple 1975; Cebul and Epple 1984
											Other: Day 1	3–6 mo	9 mo		Chap. 4
S. oedipus (c)	80%	x	x	x	x				x	x	Day 1	1–6 wk	30 wk		Hershkovitz 1977; Vogt 1984
Leontopithecus rosalia (c)	7–50%	x	x	x	x		x		x	x	1–2 wk	4–12 wk	16 wk	M > F	Ditmars 1933; Hoage 1978
Callimico goeldii (c, f)		x	x	x	x	x	x		x	x	2–3 wk	3–6 wk	12 wk		Hershkovitz 1977; Masakata 1981
Aotus trivirgatus (c, f)	51–81%	x	x	x					x	x	1 wk	4–8 wk	5 mo		Wright 1984; Dixson and Fleming 1981
(c)	10–60%	x	x				x		x	x	1 wk	2–6 wk			
Callicebus torquatus (f)		x	x	x	x	x	x	x	x	x	?	?	4 mo		Kinzey et al. 1977; Kinzey 1981; Starin 1978
C. moloch (f)	1–92%	x	x		x		x	x	x	x	Day 1	0–3 wk	19 wk		Wright 1984; Vogt 1984
Hylobates syndactylus (f)	20–78%	x	x	x			x	x	x	x	6–16 mo	8–12 mo	24 mo		Chivers 1974; Chivers and Raemakers 1980
Macaca sylvanus (f)	20–50% (all other animals)	x	x	x			x	x	x	x	1 wk	0–4 wk	5–12 mo		Deag 1974; Taub 1978, 1984; Burton 1972
(p)	3–8 hr/day (subadults)	x	x	x					x	x	Ad: 1 wk	0–2 wk	?		
											SA: 2 wk	2–3 wk	6 mo		
(c)	8%	x	x	x			x		x	x	1 wk				Lahiri and Southwick 1966
AFFILIATION															
Alouatta pigra (f)	7% interaction time (3–49%)	x		x		x		x	x	x	2 mo	10–20 mo	?	M > F	Bolin 1981
Macaca arctoides (p)	(4/hr)	x	x	x	x		x	x	x			0–6 mo	12 mo	M > F	Estrada and Sandoval 1977; Estrada 1984a
(c)	(18–50/hour)	x					x	x	x	x					Smith and Peffer-Smith 1984
Gorilla gorilla beringei (f)	In proximity 10–80% of time out of arm's reach of mother	x	x				x	x	x	x	1 wk	3–4 yr	Adolescence or beyond if male		Fossey 1979, 1983; Harcourt 1979c; Harcourt and Stewart 1981
Homo sapiens	7 min–3 hr/day (0–14%)	x	x	x	x	x	x	x	x	x	1 wk	Over 30 mo	Adolescence or beyond	M > F	Lamb 1984; West and Konner 1976
Papio cynocephalus anubis (f)	In proximity 20–100% with specific infants (x̄ = 5–10%)	x	x	x			x	x	x	x	1 mo	2–4 mo	?		Ransom and Ransom 1971; Packer 1980; Nicolson 1982; Smuts 1985
P. c. cynocephalus (f)	In proximity 15–33%; contact 1–3%			x			x	x	x	x	1 wk	1–2 mo	8–13 mo	F > M in 1st months; later M > F	Stein 1984a, 1984b

TABLE 28-1 *(continued)*

Species	Frequency of Interactions (% obser. time)	Caretake	Carry	Protect	Share Food	Co-feed	Groom	Play	Contact	Proximity	Onset	Peak	End	Infant Sex	Sources
OCCASIONAL AFFILIATION																
Macaca fuscata (t)		x	x	x		<u>x</u>	x	<u>x</u>	x		12 mo		≥ 15 mo	F > M	Itani 1959; Hasegawa and Hiraiwa 1980; Hiraiwa 1981
(p)	(.06–.17/hr)	x	x		x	<u>x</u>		x		<u>x</u>	10–15 wk	3–7 mo	?	F > M	H. Gouzoules 1984
(c)	(0–.4/male/hr)	<u>x</u>				<u>x</u>	<u>x</u>	x	x		12 mo	12–16 mo	Increases with age		Alexander 1970
Theropithecus gelada (f)				<u>x</u>				x	x			Unweaned infants		M > F	U. Mori 1979a; Dunbar 1984a
Papio hamadryas (f)			x					x	x			0–6 mo; 18 mo			Kummer 1967, 1968
Propithecus verreauxi (f)						<u>x</u>		x	x		2 wk	?	?		Jolly 1966a
Cebus albifrons (f)			x					x	<u>x</u>	<u>x</u>					Defler 1979b
Cercocebus albigena (f)			x					x	x		10 wk	Varies with male	Varies with male		Chalmers 1968a
Macaca radiata (f)			x	x				x	x	x	6 wk				Simonds 1965, 1974; Sugiyama 1971
Cercopithecus aethiops (f)				x	x	x			<u>x</u>		~6 mo				Whitten, pers. obs.
Pan troglodytes (f)	0.4%		x	x		<u>x</u>	x	<u>x</u>	x						Nishida 1983a
Alouatta seniculus (f)									x		1 mo	0–3 mo	?		Sekulic 1983b
Macaca mulatta	<1%	x	x	x		x	x	x	x	<u>x</u>	1 mo	6–12 mo	?		Breuggeman 1973; Taylor et al. 1978; Vessey and Meikle 1984

NOTE: c = captivity; p = provisioned but free ranging; f = feral.

a. The most common interactions are underlined. Caretake: Male has sole responsibility for infant while mother is away foraging or otherwise engaged, and he performs some maternal duties such as guarding, carrying, holding, comforting, etc. Protect: Male defends infant from conspecifics and potential predators (including observer). Co-feed: Male and infant feed at the same feeding site. Proximity: Male and infant are within arm's reach, for most species; exceptions: *Cercopithecus aethiops; Macaca mulatta; Papio cynocephalus anubis* (Nicolson 1982; Smuts 1985); *P. c. cynocephalus:* within 2 m; *Gorilla gorilla beringei; Alouatta seniculus:* within 3 m; *P. c. anubis* (Packer 1980): within 5 m.

b. Based on currently available information; future research may indicate that some of these species are more properly placed with affiliative species.

infanticide (Sekulic 1983a). Subadult hamadryas baboon males sometimes carry or handle infants for short periods, but in these cases the mother follows and watches intently while the infant squeals and struggles to return to her (Kummer 1967, 1968). Similar scrutiny by mothers has been observed during the occasional interactions between chimpanzee males and infants (Nishida 1983a).

Brief affiliations also occur in conjunction with changes or attempted changes in male leadership. In the one-male breeding units of gelada baboons, older males who had lost their units to a younger male but who had remained in their former unit became extremely solicitous of their young offspring and defended them when the infants were threatened by the new leader or by other males (Dunbar and Dunbar 1975; Dunbar 1984a). Young "follower males" seeking to establish their own units sometimes groomed and carried an infant in order to initiate a relationship with its mother (U. Mori 1979a).

Thus, even in species in which male caretaking is not common, a wide variety of different types of male-infant relationships can develop, depending on the attributes of the male, his relationship with the infant's mother, his relationships with other males, and the needs of the infant.

TOLERANCE

In the remaining primate species, males are generally indifferent to infants but tolerate their occasional proximity (see table 28-2). In many cases, little is known about the associations of males or infants, and consequently, for some species this classification may reflect the current

TABLE 28-2 Species in Which Males are Tolerant of, But Rarely Interact with, Infants

Species	Group Structure	Sources
Tarsius bancanus (c)	Monogamy	Vogt 1984
Galago senegalensis (c)	Solitary/sleeping groups	Ibid.
Varecia variegata (f)	Monogamy	Ibid.
Lemur macaco (f)	Monogamy	Ibid.
L. fulvus (f)	Monogamy	Ibid.
Saimiri sciureus (f)	Multimale	Ibid.
Saguinus nigricollis (f)	Monogamy and larger assemblies of monogamous units	Izawa 1978a
Macaca fascicularis (c)	Multimale	Mitchell and Brandt 1972
M. nemestrina (c)	Multimale	Ibid.
Presbytis entellus (f)	Unimale/multimale	Hrdy 1976
P. melalophos (f)	Unimale/multimale	Curtin 1980
P. obscura (f)	Unimale/multimale	Ibid.
P. potenziani (f)	Monogamy	Wittenberger and Tilson 1980
Nasalis concolor (f)	Monogamy	Ibid.
Cercopithecus neglectus	Monogamy	Gautier-Hion and Gautier 1978
C. diana (c)	Unimale	Byrne, Conning, and Young 1983
C. campbelli (f)	Unimale	Galat-Luong and Galat 1979

NOTE: c = captive; f = feral.

lack of knowledge more than the absence of male attention to infants. Many studies of captive prosimians, for example, indicate that male involvement with infants is limited to tolerance and occasional grooming or sniffing (Vogt 1984). When caged alone with infants, however, male lemurs and galagos sometimes hold, groom, and play with infants (Vogt 1984). In most cases, data on male-infant interactions among wild prosimians are lacking.

USE AND ABUSE

Prominent among baboons and some other Old World cercopithecines are interactions in which a male carries or holds an infant during an interaction with another male (Hrdy 1976). These interactions are probably the most controversial of all male-infant interactions among primates. Although almost all primatologists would agree that intensive caretakers are actually caretaking, no agreement exists about what exactly baboon males are doing when they carry infants in agonistic encounters. Such interactions have been variously termed "agonistic buffering" (Deag and Crook 1971), "kidnapping" (Popp 1978), "infant use" (Strum 1984), "infant carrying" (Busse and Hamilton 1981), "countercarrying" (Hamilton 1984), "tripartite relations" (Kummer 1967), and "triadic male-infant interactions" (Taub 1980c). Of all these labels, the latter seems to be the most objective and descriptively appropriate and will be used here.

A variety of contexts surrounded these interactions, but the most common is tension between two or more males (savanna baboons: Strum 1984; bonnet macaques: Silk and Samuels 1984; gelada baboons: Dunbar 1984a) or between a male and an infant (savanna baboons: Busse and Hamilton 1981; sooty mangabeys: Busse and Gor-

don 1983; gelada baboons: Dunbar 1984a). The approach or proximity of another male to the carrier male or to the infant usually precipitates these interactions, but sometimes they are initiated by active aggression between the two males (fig. 28-3).

Males usually handle infants carefully during these interactions (Busse 1984b; Strum 1984). Savanna baboon males grunt and lip smack to the infant, and sometimes it grunts in concert (Stein 1984b; Strum 1984). Infants usually cling willingly, and generally neither infants nor their mothers show resistance (savanna baboons: Busse and Hamilton 1981; Busse 1984b; Packer 1980; Stein 1984b; sooty mangabeys: Busse and Gordon 1983). In spite of this apparent cooperation, distresslike vocalizations often occur during triadic interactions. For example, yellow baboon infants characteristically gecker in a highly stylized manner, similar but qualitatively different from distress signals in other contexts (Stein 1984b). Sometimes, if the infant does not gecker, the male hits, pinches, bites, or squeezes it until it does.

During the triadic episodes of Barbary macaques, both males interact with one another and with the infant in a series of exaggerated and stylized gestures (Taub 1978, 1980c). The males sit facing one another and clutch at each other's sides or genitalia while teeth chattering at one another and at the infant. Holding the infant between them, the males then touch, mouth, and pet it. Then, teeth chattering and shaking the heads from side to side, they hold the infant above their heads and lick or nuzzle its anogenital area. Infants are handled gently and never give signs of distress.

Infants and their mothers do not always cooperate with males, however (savanna baboons: Popp 1978; Packer 1980; Stein 1984b; Strum 1984). In one troop, over a

FIGURE 28-3. Triadic male-infant interaction in olive baboons. The high-ranking male (*left*), Orpheus, initiated the interaction by threatening the low-ranking male (*right*), Alex. Alex ran to a close female associate who was feeding nearby, gathered her infant, moved toward Orpheus, and sat, holding the infant ventral. As shown in the photograph, Orpheus continued to threaten Alex by exposing his canines, but within a few seconds he ceased doing so and walked away. (Photo: Barbara Smuts/Anthrophoto)

third of yellow baboon male solicitations for infant contact were unsuccessful (Stein 1984b). Usually these males did not attempt to carry infants that resisted, but some olive baboon males dragged infants along by an arm or a leg if they were unable to get them to cling (Popp 1978; Packer 1980).

All studies have demonstrated that males tend to carry only particular infants in these interactions (Strum 1984). In most studies, the majority of males who carried infants had been resident in the troop since the infant's conception or birth and therefore were potential fathers of the infants that they carried. In some troops, recent immigrants rarely carried infants (Packer 1980; Busse and Hamilton 1981), but among yellow baboons, newcomer males carried infants in 14% of triadic interactions (Stein 1984b). If unsuccessful in carrying an infant, these males either forced infants to participate or instead used large juveniles or even estrous females as "buffers" (see also Strum 1984). Olive baboon newcomers mainly carried infants that had been weaned early or who were neglected by their mothers (Strum 1984).

Although male choice of an infant is quite specific, it is not consistently related to paternity. Some studies have found that males predominantly carry infants who are likely offspring or putative siblings (chacma baboons: Busse and Hamilton 1981; Busse 1984b; sooty mangabeys: Busse and Gordon 1983), but others have found

no evidence of male preference for likely offspring (olive baboons: Packer 1980; Strum 1984; yellow baboons: Stein 1984b).

Among gelada baboons, two types of triadic interactions have been observed. Unit leaders and old followers who were former leaders handled and groomed their putative infants during takeover attempts or other aggressive interactions with unrelated males (Dunbar 1984a). During aggressive encounters, the likely father touched, groomed, and mounted an infant while threatening an opponent not related to the infant. Infants often cooperated by running to the father, leaping onto his back, and screaming as soon as the two males began to interact. Deposed leaders also protected their infants from aggression by the new leader. In the second type of interaction, young followers attempting to form units of their own carried the leader's infants against him. The infant used by the follower was the offspring of his closest female associate, and, again, infants cooperated with the follower as described above. When this occurred, the follower's attitude changed from submissive to aggressive, and the leader desisted.

A strong affiliation between the male and the infant that he carries is, in fact, the most common feature of triadic interactions across all studies. Males carry infants with whom they are affiliated even if they are unlikely fathers of the infant (savanna baboons: Packer

1980; Stein 1984b; Strum 1984; Smuts 1985; gelada baboons: Dunbar 1984a). Among Barbary macaques, both males involved in the interaction typically share a common caretaking relationship with the infant (Taub 1980c).

A number of explanations have been offered for these interactions. One set of explanations views the interactions as basically exploitative or manipulative in nature. The term *agonistic buffering* was first coined to describe interactions in which low-ranking animals handled infants in order to reduce the likelihood of aggression from higher-ranking animals (Deag and Crook 1971). More recently, the term *passport* (after Itani 1959) has been suggested for interactions in which an animal carries an infant in order to approach or to supplant another (Strum 1984). The most extreme form of the exploitative explanation argued that males held hostage the infants of their opponents in order to force their acquiescence (Popp 1978), but subsequent studies have not confirmed this hypothesis. Aside from infant carrying by young gelada baboon followers, there is no evidence that opponents are normally any more likely to be the infant's father than the carrier, and in fact, several studies (including Popp's) indicate that the reverse is often true (Stein 1984b). Consequently, later studies proposed a more mutualistic form of interaction in which males cared for infants in order to ensure their cooperation (savanna baboons: Packer 1980; Stein 1984b; Strum 1984).

The benefits of these interactions to competing males, however, are uncertain. Among olive baboons, participants in triadic interactions were successful about one-third of the time, when success was defined as one or more of the following: stopping the opponent's aggression, supplanting a previous supplanter, reversing a supplant, becoming more aggressive in approaching an opponent, or inducing the opponent's avoidance (Strum 1984). When the rate of aggression received while carrying an infant was compared to that received at other times, some studies found that males were less likely to be threatened (olive baboons: Packer 1980; bonnet macaques: Silk and Samuels 1984) while another found no difference (yellow baboons: Stein 1984b). However, male baboons carrying infants did appear to engage in riskier interactions; males initiated more aggression and retreated less when carrying an infant than in other circumstances (Packer 1980; Stein 1984b).

An alternative explanation for triadic interactions is that males are protecting offspring from infanticide by immigrant males (Busse and Hamilton 1981). Some of the available data also support this hypothesis. Infants are often afraid of immigrant males, and infant attacks and infanticides have been observed in some savanna baboon populations (Packer 1980; Busse and Hamilton 1981; Busse 1984b; Collins, Busse, and Goodall 1984). Moreover, newcomers are often the opponents in triadic interactions, while the initiators are sometimes likely fa-

thers and usually potential fathers due to their longer residence in the group. However, newcomers did carry infants in some populations, and infants sometimes clung to both males in the same interaction (Stein 1984b).

Recently, Stein (1984b) and Dunbar (1984a) suggested yet a third alternative: carrying the infant of a female associate might be a means of enlisting that female's support against the other male. Among gelada baboons, the mother sometimes joins the male carrying the infant in threatening the opponent (Dunbar 1984a). Among savanna baboons, the aggressive responses of opponents appear to depend on the identity of the carrier and infant, carrying males sometimes give submissive responses to mothers, and mothers and other group members occasionally attack males when infants give indications of real distress (Stein 1984b; Strum 1984). Little attention has been given to the behavior of mothers and to the long-term consequences of triadic interactions, but it is certainly possible that triadic interactions could have as much to do with the long-term relationships of males and females as with the short-term relationships of contesting males (Smuts 1985).

At this point, far too little is known, in spite of a wealth of data, about the causes of triadic interactions. The most general conclusion possible is that a single explanation cannot account for all of these interactions (Smuts 1985). There appear to be several different types of interactions that vary in frequency between species and groups. Moreover, these different types of interactions may even coexist within a single group or single male-infant pair (Altmann 1980; Smuts 1985).

EVOLUTION OF MALE-INFANT RELATIONSHIPS

Why males interact with infants is a question that evolutionary biologists have tackled with much fanfare and considerable relish. At the heart of this problem lie such fundamental issues as the importance of kin selection and the origin and evolution of sex differences (Hamilton 1964; Trivers 1972). As the two major paradigms of sociobiology, kin selection and sexual selection have proven to be powerful tools for understanding the diversity of social behavior. Nowhere do these two forces appear so strongly opposed as in paternal care, which makes male-infant relations a popular testing ground for these theories.

As a consequence of this interest, there are many hypotheses about the evolution of paternal investment. These hypotheses are of two basic types (Kurland and Gaulin 1984). One subset proposes that natural selection has altered female reproductive physiology and behavior to produce the level of paternity certainty eliciting her optimal level of paternal investment. The other subset proposes that other unrelated reproductive adaptations have indirectly affected paternity certainty, thereby making paternal investment more or less likely. Both types of hypotheses, however, implicitly assume that paternal investment arises directly from the likelihood of male-infant relatedness.

However, patterns of primate male-infant affiliation do not accord with simple predictions of paternal investment based on mating systems (Snowdon and Suomi 1982; Smuts 1985). A male should be most certain of the identity of his offspring in monogamous groups and least certain in multimale groups. In one-male groups, extra-group matings may reduce paternity certainty (chap. 9), but the rarity of such matings indicates that males in one-male groups are no less certain of paternity, and probably are often more certain, than are males in multimale groups (Smuts 1985). It is true that the most intense affiliations occur in monogamous groups. Yet males in most one-male groups do not interact at all with infants, while males in some multimale groups form very strong affiliations with infants.

Typological classifications of mating systems may obscure the more finely tuned tactics that individuals utilize in particular situations (Kurland and Gaulin 1984). Therefore, expectations of paternity certainty based simply on mating systems may be invalid. For example, males in particular multimale groups could be much more certain of the identity of their offspring than their mating system would suggest since, for example, females sometimes mate with only one or two males around the time of probable conception (e.g., Hausfater 1975; Chapais 1983a; Smuts 1985). Moreover, such classifications fail to take account of the relative costs and benefits of parental investment; ecological factors such as body size can alter the benefits of parental effort (Kurland and Gaulin 1984). However, even within groups of a single species, male affiliations do not closely match paternal relationships. Males in multimale species often affiliate with infants likely to be their own, but they affiliate just as often with infants who are unlikely offspring (table 28-3). In fact, male-infant affiliations are most pronounced among Barbary macaques, the multimale species in which paternity appears to be the least certain. Even in the Callitrichidae, males do not always interact intensively with their own infants, and sometimes males care for infants that they have not sired (table 28-3). Moreover, infants do not always benefit from interactions with likely fathers. Males who carry infants in triadic interactions are generally potential fathers. The baboon infant studiously protected from pulls and pinches at birth may be thrust into the center of aggression by the same male a few months later.

Genetic relationship seems to be most important in the distribution of abusive interactions. The data clearly indicate that infants are at risk from males who are unlikely fathers. In almost all cases of infanticide, males killed infants who were unlikely or impossible offspring (chap. 8, but see chap. 15 for exceptions).

These data indicate that paternity is neither a necessary nor a sufficient condition for primate male affiliation with infants. Thus, primate males do not appear to follow the dictates of parental investment theory. That may be because most of the interactions of primate males with infants are not parental investment. By definition, parental investment involves some reproductive cost to the parent and some positive effects on the infant's survival and future reproduction (Trivers 1972). What can be observed and measured in the field and laboratory, however, are simply the types and frequencies of interactions between males and infants. The extent to which these interactions can be considered parental investment is unclear. The activities of the intensive caretakers provide some clear benefits to infants: carriage provides protection and reduces energetic expense, and food sharing provides the infants with access to high-quality foods not otherwise available. Moreover, these parental duties also impose some significant costs on the male, if night monkeys are typical.

In other primate species, the benefits and costs of male-infant interactions are more uncertain. Many types of interactions with infants may have no costs to future male reproductive success. Grooming and carrying are relatively uncommon among primates other than the intensive caretakers. Food sharing is also quite limited. Although baboons may willingly share the fruits of their labor with infants, they apparently are unwilling to share the fruits that they intend to eat. Proximity, huddling, and even baby-sitting with infants probably require little energetic expenditure. Infant presence could reduce male foraging efficiency, but no data are available to test this possibility. Mating activity may not be compromised by interactions with infants. Seasonally breeding primates such as Barbary macaques and Japanese macaques cease affiliative interactions with infants at the onset of the mating season. Among non–seasonally breeding primates, there is no evidence that males forego mating activity in order to interact with infants, although, admittedly, no study has ever attempted to test precisely this prediction.

However, there is considerable evidence that affiliations with infants may actually enhance a male's mating success. The data on triadic interactions indicate that males sometimes use infant relationships directly to gain competitive advantages over other males. More commonly, however, benefits may be gained more indirectly. Throughout the primate order, male affiliations with infants are commonly accompanied by strong bonds with the infants' mothers (Hershkovitz 1977; U. Mori 1979a; Takahata 1982a; Strum 1984; Smuts 1985; see table 28-3). An affiliation may be a means of establishing a re-

TABLE 28-3   Indications of Paternity for Males Involved in Affiliative Interactions with Infants

Species	Group Structure	Length of Residency	Rank	Relationship with Infant's Mother			Sources
				Prior Mating	Subsequent Mating	Special Relation	

				INTENSIVE CARETAKING			
Callithrix jacchus (c)	Monogamy	Variable	Variable	Often but not essential	Postpartum	Pair bond or recent addition	Box 1977; Epple 1970; Hershkovitz 1977
Cebuella pygmaea (c)	Monogamy				3–6 wk postpartum	Pair bond	Hershkovitz 1977
Saguinus fuscicollis (c)	Monogamy	Variable	Dominant more active	Often but not essential	Postpartum	Pair bonded male more active	Epple 1975; Cebul and Epple 1984
(f)	Polyandry	Variable	All males	Yes	?	Each male has a pair bond with the female	Chap. 4
Leontopithecus rosalia (c)	Monogamy			Yes	Yes	Pair bond	Hoage 1978
Callimico goeldii (c)	Monogamy	Variable		Often but not essential	Postpartum	Pair bond	Hershkovitz 1977
Aotus trivirgatus (c, f)	Monogamy			Yes	Yes	Pair bond	Wright 1981, 1984
Callicebus torquatus (f)	Monogamy			Yes but extrapair consorts	Yes	Pair bond	Kinzey 1981
C. moloch (f)	Monogamy			Yes	?	Pair bond	Wright 1984
Hylobates syndactylus (f)	Monogamy			Yes	Yes	Pair bond	Chivers 1974
Macaca sylvanus (f)	Multimale	Long (transfer rare)	All ranks	Yes, all males	Yes, all males	?	Taub 1978, 1984
(p)	Only 1–2 males/ troop	Long	Subordinate only when dominant absent	Yes	Yes	Close association	MacRoberts 1970; Burton 1972

				AFFILIATION			
Alouatta pigra (f)	Mostly monogamous	?		?	?	Pair bond?	Bolin 1981
Macaca arctoides (p, c)	Multimale	Long (transfer restricted)	All ranks	In some cases	?	Close association in some cases	Estrada and Sandoval 1977; Smith and Peffer-Smith 1984
Gorilla gorilla beringei (f)	Unimale, occasionally multimale	Usually long	Dominant	Likely	Usually	Mothers increase association with silverback	Harcourt 1979c; Fossey 1983
Homo sapiens	Unimale and monogamous; very rarely polyandrous			Often but not required	Often but not required	Pair bond or kin to mother or her husband; sometimes no relation	Goody 1976; Silk 1980a
Papio cynocephalus anubis (f)	Multimale	Immigrated prior to infant's conception	All ranks	In 50% of affiliations	Increased likelihood	Close association during pregnancy and lactation	Ransom and Ransom 1971; Packer 1980; Nicolson 1982; Smuts 1985
P. c. cynocephalus (f)	Multimale	Immigrated prior to infant's conception	All ranks, but high more active	In 50–70% of associations	?	Often a close association	Altmann 1980; Stein 1984a, 1984b
P. c. ursinus (f)	Multimale	Immigrated prior to infant's conception	High-ranking at infant's conception	54% of infant-carrier pairs in triadic episodes	?	Protect infants of female associates	Busse and Hamilton 1981; Hamilton 1984; Busse 1984b

TABLE 28-3 *(continued)*

Species	Group Structure	Length of Residency	Rank	Relationship with Infant's Mother			Sources
				Prior Mating	Subsequent Mating	Special Relation	
OCCASIONAL AFFILIATION							
Macaca fuscata (p)	Multimale	Long	High	?	?	Protect infants of female associates	Itani 1959; Hiraiwa 1981; A. Mori 1979a
(p)	Multimale	Long	High	In 33% of associations	?	?	H. Gouzoules 1984
(c)	Multimale	Long	All ranks	?	?	?	Alexander 1970
Theropithecus gelada (f)	Unimale in multimale group	Present at infant's conception if old follower	Low (follower) but high at infant conception if old	If old follower	Not if old; possible if young follower	Close association if young follower	U. Mori 1979a; Dunbar 1984a
Papio hamadryas (f)	Unimale in multimale group	Variable		Variable	With mother or "infant"	With mother or "infant"	Kummer 1968
Propithecus verreauxi	Multimale			Yes		Pair bond	Jolly 1966a
Cebus albifrons	Multimale	?	All ranks	?	?	?	Defler 1979b
Cercocebus albigena	Multimale	?					Chalmers 1968a
Macaca radiata	Multimale	?	All ranks	?	?	?	Simonds 1965
Cercopithecus aethiops	Multimale	Prior to infant birth	All ranks	Usually	?	Feeding subgroup	Whitten, pers. obser.
Pan troglodytes	Multimale	Natal	All ranks	Often	Likely	?	Nishida 1983a
Alouatta seniculus	Multimale	Present at infant conception	Dominant more active	Yes	?	?	Sekulic 1983b

NOTE: c = captive; p = provisioned, free ranging; f = feral.

TABLE 28-4 Relative Infant Weight and the Timing of Male Caretaking

Species	No. of Young	Relative Litter Weight (% maternal weight)			Sources
		Onset	Peak	End	
Callithrix jacchus	2	21	21–38	100–120	Kleiman 1977; Hershkovitz 1977
Cebuella pygmaea	2	22	22–200	200	Kleiman 1977; Hershkovitz 1977
Saguinus oedipus	2	14–18	28–40	128	Kleiman 1977; Hershkovitz 1977
Callimico goeldii	2	15–21	21–32	42–59	Kleiman 1977; Hershkovitz 1977
Aotus trivirgatus	1	12–17	22–28	44	Hall, Beattier, and Wychoff 1979
Hylobates syndactylus	1	16[a]			Chivers 1974
Macaca sylvanus	1	6			Roberts 1978

a. Based on a growth rate of 3.8 g/day for infant gibbons (Martin et al. 1979) lationship with the infant's mother, or it may be an extension of an already established relationship. At least in baboons, the advantages gained from such close relationships with females can enhance a male's future mating success (Seyfarth 1978b; Smuts 1985).

Even in monogamous primates, caretaking males may have at stake future as well as current reproductive success. The universal occurrence of postpartum estrus in the Callitrichidae (Hershkovitz 1977) means that females are often simultaneously investing in two sets of offspring. The combination of pregnancy and costs of nursing and carrying two relatively large infants places an excessive burden on the mother that probably could not be sustained without help. By mitigating the costs of maternal care for his pregnant mate, a male is also increasing the chances of survival of his as-yet-unborn offspring.

These data suggest that male-infant affiliations in the primates could be a product of female choice (Smuts 1985). Female choice of mating partners on the basis of male interactions with infants has not yet been distinguished clearly from mating advantages that males gain simply due to their familiarity with close female associates. The hypothesis of female choice, however, is supported by a relationship between the costs of maternal caregiving and the intensity of male-infant relationships. Affiliations are most pronounced in primate species in which infants are 16% or more of maternal weight, and onset of caretaking also is earlier in the species with relatively larger infants (see tables 28-1, 28-4). Reproduction as a whole also is more costly in these species, which are among the smallest primates. Smaller primates have faster growing infants and produce milk that is higher in protein than that of larger primates (Schwartz and Rosenblum 1983). Moreover, males in monogamous species participate more in infant care when immature helpers are unavailable. These patterns of affiliation are not predicted by the hypothesis that the mating advantages that males gain by affiliation with infants are due simply to familiarity with the infant's mother. These patterns are predicted, however, by the hypothesis that females preferentially mate with males who aid in infant care.

Although paternal relations are sometimes important, as a whole the primate data suggest that male affiliations with infants have more to do with maintaining relationships with particular females, and thereby increasing mating success, than with paternity. If so, these behaviors represent mating effort rather than parental investment (West and Konner 1976; Kurland and Gaulin 1984; Smuts 1985). Thus, the relationships of primate males with infants are not simply a consequence but rather are an integral part of male reproductive strategies.

SUMMARY

The relationships of primate males and infants can be divided into the following categories: intensive caretaking, affiliation, occasional affiliation, tolerance, and use and abuse. Males of the monogamous New World primate species exhibit the most intensive caretaking, but the monogamous siamangs and the multimale Barbary macaques also interact intensively with infants. Males in these species look after infants for a large portion of the day, carry them, and even share food with them. Males in many Old World and a few New World primate species have less intensive affiliations with infants. Males in these species carry infants infrequently but sometimes look after them for brief periods. These relationships are most characterized by frequent proximity and close friendly contact. The triadic interactions of baboons and a few macaque species are less easily interpreted. In these interactions, one male carries an infant to or near another male. Triadic interactions occur in a variety of circumstances, but the most common denominator seems to be a tense situation between two males or between an infant and a male. Both protection and manipulation of infants may be involved. All of these interactions of primate males and infants are quite specific: particular males interact only with particular infants. Although male associates of infants are often likely fathers, paternity is neither a necessary nor a sufficient condition for affiliation with infants. A more common denominator seems to be a close relationship between the male and the infant's mother. There is evidence that males gain mating advantages by associating with infants. Moreover, the distribution of male-infant affiliations suggests that these associations may be a product of female choice.

29 | Transition to Adulthood

Jeffrey R. Walters

Primates are characterized by a long period of pre-reproductive development. This period includes not only infancy, when the individual is physically dependent on its mother, but also a longer interval when it is independent but not yet adult. The latter period is the focus of this chapter. Unfortunately, the behavior of juveniles and adolescents has received much less study than the behavior of infants or adults. As a result, much still remains to be learned about the behavior of immature primates.

DEVELOPMENTAL STAGES

Although the terms *infant, juvenile, adolescent,* and *subadult* have been variously used to refer to progressively older nonreproductive animals, only recently have some studies accumulated sufficient long-term data to permit a precise description of these developmental sequences in terms of known ages (e.g., Goodall 1983; Pereira and Altmann 1985; table 29-1). As yet, however, no consistent attempts have been made to use the same definitions across study sites and, in particular, across species.

In most studies, infancy is defined as the period when a youngster is physically dependent on its mother for survival. Although intuitively satisfying, this definition provides no clear marker of infancy's end. The age at which an individual is able to survive its mother's death is a possible criterion, but this may also vary between individuals (Goodall 1983; Pusey 1983) and is difficult to determine on a species-wide basis. Pereira and Altmann (1985) suggest using interbirth interval to approximate the duration of infancy. Although this constitutes a reasonable upper limit, interbirth intervals are affected by food supply, and they often vary greatly across habitats (see chaps. 11, 20). Moreover, interbirth intervals are usually greater than the age at which youngsters can survive their mothers' death. For example, in one population of savanna baboons (*Papio cynocephalus*) where the interbirth interval was 23 months, animals as young as 13 months were able to survive their mother's death (Altmann et al. 1977; Altmann 1980). Similarly, the age of weaning, a frequently used criterion, does not correspond closely with the age at which youngsters are capable of surviving independently (e.g., Pusey 1980) and is highly variable. In one group of baboons, for example, some mothers weaned their infants completely by 13 months of age, whereas others continued to suckle their infants for almost 2 years (Nicolson 1982).

The interval between infancy and the onset of puberty is usually called the juvenile period. Puberty may be defined as all the events, beginning with altered hormonal function, leading to reproductive maturation. These processes are broadly similar in all primates. Puberty begins with a surge of luteinizing hormone that triggers dramatic rises in levels of sex steroids such as testosterone and estrogen (Pereira and Altmann 1985). In many primates, rapid testicular growth and descent are external indicators of the beginning of puberty in males. External signs of puberty in females are often less obvious; in some species the onset of puberty can be detected by the beginning of perineal swellings (e.g., baboons and chim-

TABLE 29-1 Primate Developmental Stages

Stage	Definition	Duration in Females (mo)[a]			
		Lesser Bushbaby[b]	Common Marmoset[b]	Yellow Baboon	Chimpanzee
Infant	Birth until able to survive death of mother	2[c]	5[c]	13[c]	48[c]
Juvenile	From time able to survive death of mother until onset of puberty	6.5[d]	5	41[e]	54
Adolescent	Onset of puberty until reproduction occurs	6.5[d]	10	18[e]	53

SOURCES: Bushbabies (*Galago senegalensis*): Doyle 1979; marmosets (*Callithrix jacchus*): Hearn 1978; baboons (*Papio cynocephalus*): Altmann et al. 1977; chimpanzees (*Pan troglodytes*): Pusey 1983.

a. Species are listed in order of increasing size from left to right.

b. These estimates are derived from captive individuals; durations of these stages are likely longer in nature.

c. Estimated by interbirth interval.

d. The data were not sufficient to separate the juvenile and adolescent periods; the estimates were therefore derived by halving the interval from infancy to adulthood.

e. Assumes a 6-month interval between the onset of puberty and menarche.

panzees [*Pan troglodytes*]: see chap. 30). Menarche is an obvious sign that puberty is underway, but the event occurs well after puberty's onset.

The interval from the onset of puberty to the beginning of effective reproduction is termed *adolescence* by Pereira and Altmann (1985). Females are considered to be adult when they are able to carry a pregnancy to term. In cercopithecines, apes, and humans, menarche is usually followed by a series of irregular, infertile cycles, during which a female is physiologically incapable of maintaining implantation due to immature luteal function (Altmann et al. 1977; Pusey 1983; Scott 1984; Pereira and Altmann 1985). For females, the life stage of adolescence thus corresponds with the process of puberty as defined above.

In many species males reach adulthood later than females, both because puberty may begin later (e.g., baboons: Altmann et al. 1977) and because the interval between the onset of puberty and the start of reproduction may be considerably longer than among females. In sexually dimorphic species such as baboons and chimpanzees, males do not reach adult size until well after they become capable of inseminating females (see references in table 29-1). The transition to adulthood is not completed until individuals become socially, as well as physically, capable of reproduction, and males are often not able to compete successfully for females until they have reached full adult size. By this criterion, male lesser bushbabies (*Galago senegalensis*) reach adulthood 9 months later, and male baboons 2 to 4 years later, than females (Altmann et al. 1977; Doyle 1979).

SIGNIFICANCE OF THE EXTENDED PREPRODUCTIVE PERIOD

Among animals in general, large species live longer and have longer prereproductive periods than small species. The relation between body size and the length of the prereproductive period does not exist because of direct selection for longer life in larger animals but because of the inevitable physiological consequences of body size, known as scaling laws (Lindstedt and Calder 1981; Western and Ssemakula 1982; chap. 16). Even compared to mammals of equivalent size, however, primates devote an inordinately large portion of their life spans to prereproductive development.

The lengthy period of prereproductive development in primates has been explained in several ways. One of these derives from life-history theory, which interprets reproductive and developmental schedules of organisms as adaptations that optimize lifetime reproductive success (Stearns 1977). For example, the prereproductive period may be especially long in primates due to the dependence on learning: the long interval provides an opportunity to learn about the physical and social environment prior to attempting reproduction (e.g., Poirier and

Smith 1974). Thus the beneficial effects of learning during the prereproductive period on adult reproductive success may justify delaying reproduction.

Delayed maturation may also be a physiological consequence of neonatal brain size and fetal growth patterns (Sacher 1982; chap. 16). According to this hypothesis, large brain size at birth may place physiological and energetic constraints on postnatal growth. The long prereproductive period of development in primates is therefore not adaptive, but is a necessary consequence of large brain size, which is adaptive. The learning that occurs during this period may thus be an effect, rather than a cause, of prolonged prereproductive life (Martin 1983; Pereira and Altmann 1985).

For whatever reason, a large proportion of a primate's life is taken up by the period between infancy and adulthood. Whether as a member of a large social group or in the company primarily of its mother, the young primate is closely associated with some conspecifics throughout most or all of this period. With only a few exceptions (e.g., marmosets [*Callithrix* spp.] and tamarins [*Saguinus* spp.]: chap. 4), the immature primate is not directly assisted by group members in obtaining food after infancy. Although it may profit indirectly from the experiences of other group members by traveling to suitable foraging areas with its group and observing what others are eating, the young primate must make its own living after weaning. Typically, mortality of young juveniles is high (Dittus 1977; Pereira and Altmann 1985; chap. 20), whereas that of older juveniles and adolescents is much lower, which suggests that the ability to cope with the physical environment improves with experience. Unfortunately, however, there are few data on how skills associated with obtaining food, traveling with the social unit, and avoiding predators are acquired (but see chap. 36). These important aspects of development deserve more study.

Instead, this chapter focuses primarily on the development of social behavior. After infancy, the social interactions of young primates shift increasingly from their mothers to other kin, peers, and unrelated members of the group. Some of these interactions, such as play, often seem to be without immediate functional benefit, but immatures are also effective participants in their group's adult social network. This chapter attempts to describe these complex and diverse interactions and to assess their effects on the survival and future reproductive success of immature animals.

PLAY

As in many other species, play is an important feature of the behavior of immature primates. Every reader has a reasonable idea of what is meant by the term *play*, and most, without prior experience, could recognize and distinguish it from nonplay. Yet play is notoriously difficult to define. The consensus that emerges from the scores of definitions is that play incorporates many physical components of adult behavior patterns, such as those used in aggression, but without their immediate functional consequences. Play is also more exaggerated, repetitious, and variable than corresponding nonplay behavior (Fagen 1981).

Play occurs in almost all mammals and is also common in birds (see reviews by Smith 1978; Symons 1978a; Fagen 1981; Pereira and Altmann 1985). It generally occurs in three forms. The first is object play, in which an individual manipulates an object in a repetitious manner. Although common in some mammals (for example, cats), object play is relatively rare in primates and apparently occurs at high rates only in apes and humans (Fagen 1981). The second type of play is solitary locomotor play, of which the frolicking of a colt in a pasture is a familiar example. Although such play also occurs in primates, it again is less common than in most other mammals (Fagen 1981). Instead, primates can be distinguished from other group-living animals by the fact that most of their play involves two or more individuals. Such social play is not merely a by-product of group life, since

among mammals the degree to which play is social is not closely correlated with a species' social system (Fagen 1981).

In its physical form, social play is quite uniform across primate species. It consists mostly of chasing and wrestling and is usually accompanied by a relaxed open-mouthed expression (the "play face") that is peculiar to play (fig. 29-1). Although play includes many of the physical components of aggressive behavior, it lacks the visual and vocal signals associated with aggression, such as fear grimaces and screams

Immature primates spend much of their nonfeeding time engaged in social play. For example, young juvenile vervet monkeys (*Cercopithecus aethiops*) in Amboseli, Kenya, engaged in 5 to 10 play bouts per hour (Lee 1983e), whereas adults rarely played. In most species studied thus far, play begins in infancy, becomes much more frequent during the juvenile period, and then declines steadily during adolescence (Fig. 29-2).

Function of Play

Defining play, despite the difficulties involved, has proved easier than determining its function. There are a multi-

FIGURE 29-1. Two juvenile baboons display the "play face" while wrestling with each other. (Photo: Barbara Smuts/Anthrophoto)

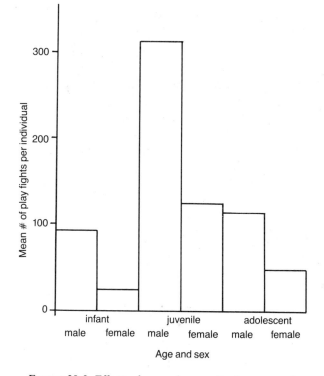

FIGURE 29-2. Effects of age and sex on the frequency of play-fighting in immature rhesus macaques. SOURCE: Symons 1978a, based on 300-hr observation of one group.

tude of theories about the function of play. Many of these propose that play somehow facilitates social development (e.g., Poirier and Smith 1974). For example, through play juveniles may establish dominance relationships, learn social and communication skills, learn to control aggression, learn to distinguish the appropriate recipients of social behavior, and test the strengths of others. Some of these theories propose benefits that may in fact not occur. For example, Symons (1978b) was unable to find any relation between dominance and play-fighting ability in young rhesus macaques, presumably because the ranks of immatures were determined by their mothers' dominance ranks rather than by individual competitive abilities.

Some of the other proposed benefits probably are important, however, and juveniles surely learn things in play that are useful in both current and future social interactions. But play is much more frequent than seems necessary for the acquisition of social skills. Young primates deprived of contact with conspecifics do not develop normal social behavior (Harlow 1969), but factors other than play appear to be responsible. In the wild, play often disappears or becomes infrequent in times of nutritional or social stress (Fagen 1981; Lee 1983c), but animals that are thereby deprived nevertheless develop normal social abilities (Baldwin and Baldwin 1973; Lee 1983c, 1983e). This suggests that the development of so-

cial skills is a fortuitous benefit that may not have been important in the evolution of play. Another possible social benefit of play, social bond formation, is considered below.

The function of play may also be related to physical development. Play may serve as a context for rehearsing motor skills that are too risky to practice in their usual functional context, such as those involved in predator avoidance or intraspecific fighting (Smith 1978; Symons 1978a; Fagen 1981) (fig. 29-3). Much of the play of animals in general and primates in particular is consistent with this hypothesis. Both solitary locomotory play and social play typically involve behaviors that occur in predator avoidance and intraspecific fighting. Symons's (1978a) detailed study of play, for example, showed that the design of play-fighting in rhesus macaques promoted the acquisition of the particular fighting skills employed by adults.

Sex and species differences in play also seem consistent with the hypothesis that play functions to develop motor skills that cannot be practiced in their appropriate context. For example, baboons, macaques, and vervets are some of the many mammalian species in which aggressive competition among adult males for females is preceded ontogenetically by higher rates of play-fighting among immature males than among immature females (Fagen 1981). This sex difference is expected if play rehearses fighting and if fighting skills are more important to male than to female reproductive success. In fact, male dominance rank, which is sometimes correlated with mating activity in cercopithecines, depends more on fighting ability than does female dominance rank (chaps. 25, 26). If the hypothesis is correct, fighting skills should be equally important to the reproductive success of adult males and females in species such as marmosets (*Callithrix jacchus*), which lack sex differences in play (Box 1975a).

Fagen (1981) has proposed a more general function of play as physical training. He views play as a behavioral mediator between the environment and the organism, which enables the developing individual to modify its physiology to match the particular environment in which it was born. Physical training related to predator avoidance and fighting is but a part of this, since the exercise involved in play is hypothesized to produce modifications in bones and muscles according to the particular experiences of the individual. Brain development may similarly be modified through play, thus resulting in more flexible and novel behavior than would occur without such modification. This hypothesis is supported by a strong correlation between brain complexity and elaborateness of play (Fagen 1981), which suggests that play may be particularly important in primates. Enduring responses to exercise in skeletal muscle, bone, and the cardiovascular system are well known, and in mammals

FIGURE 29-3. Two juvenile gray langurs engaged in play-fighting. (Photo: Irven DeVore/Anthrophoto)

are most pronounced in immatures (Fagen 1981). Furthermore, experiments on rodents have demonstrated changes in cerebral growth and brain chemistry in response to play (Rosenzweig and Bennett 1978; Fagen 1981).

Play may therefore provide various benefits both during development and in adult life. However, play also carries potential immediate costs in terms of both the risk of injury and the expenditure of time and energy. The energetic costs of play have never been measured, but play consumes between 1 and 10% of an immature primate's time in the wild, and even more in captivity (data from 13 species in Fagen 1981, table 5-1). Serious injuries during play have also been observed in several species (e.g., chimpanzees: Goodall 1968). We can only speculate that these costs are outweighed by whatever benefits are derived through play.

Play Partners

Although primate play is structurally similar across species, the identity of play partners is more variable. In particular, the play partners available to a young primate are often limited by the species' social structure. In solitary species and species living in family groups, the social unit typically contains only one reproductive female,

and same-age play partners are not usually available unless twinning occurs. Occasionally, juveniles leave their social unit to play with peers (e.g., orangutans [*Pongo pygmaeus*]: Galdikas 1979). More often, in species such as bushbabies and gibbons (*Hylobates* spp.), youngsters play with different-aged siblings or parents, and solitary play is also common (Chivers 1974; Charles-Dominique 1977a; Doyle 1979; Gittins and Raemaekers 1980; chap. 12). When play partners are thus unevenly matched, the older individual usually refrains from employing all of its physical ability, a phenomenon known as self-handicapping (Fagen 1981) (fig. 29-4). Similarly, in chimpanzees both mother-offspring and sibling play occurs at high rates, perhaps because chimpanzee females frequently range alone (Goodall 1968; Pusey 1983). It is not yet known whether play patterns are affected by more subtle differences in social structure, such as the relative lack of affiliative bonds among adult females in species in which females normally disperse (e.g., gorillas [*Gorilla gorilla*]).

Many primates grow up in larger social units containing potential play partners of many different ages. In most cercopithecine groups, there are marked sex differences in play quality, quantity, and partner preferences. Young males usually play more often (fig. 29-2) and play

FIGURE 29-4. Although adults in most primate species play less than juveniles, adult play is not unusual. Here, a silver-backed male gorilla plays with a subadult female (*top*) and an adult male baboon plays with a juvenile male (*bottom*). (Photos: Alexander Harcourt/Anthrophoto and Jim Moore/Anthrophoto)

more with older animals than do females (e.g., rhesus macaques [*Macaca mulatta*: Symons 1978a; rhesus and bonnet macaques [*M. radiata*]: Caine and Mitchell 1979; Japanese macaques [*M. fuscata*]: Hayaki 1983; baboons: Owens 1975; Cheney 1978b; vervets: Fedigan 1972; Lee 1983e; gray langurs [*Presbytis entellus*]: Hrdy 1977; hamadryas baboons [*Papio hamadryas*]: Abegglen 1984). When female rhesus macaques are treated prenatally with androgens, they become predisposed to develop malelike play patterns, which suggests a hormonal influence on the frequency and quality of play (Goy and Resko 1972). However, the social environment may often override this influence, since juvenile female bonnet macaques also show an increase in play when only their companions are treated with gonadal hormones (Rosenblum and Bromley 1978).

When a variety of different play partners is available, immature primates tend to play most with others of comparable size and strength (see earlier references; see also Goodall 1968 for chimpanzees). As a result, siblings are not always the most preferred partners, although siblings are generally preferred over nonsiblings of comparable age and sex (Fedigan 1972; Cheney 1978b; Caine and Mitchell 1979; see also chap. 14 for gorillas). Association with relatives is less pronounced in play, however, than in other types of social interaction.

The social relationships and dominance ranks of adult females may also affect the choice of play partners by immatures. Cheney (1978b) reports that juvenile female baboons preferred the offspring of high-ranking females as play partners. Other studies have also reported that high-ranking juveniles play more than the offspring of low-ranking females (rhesus macaques: Tartabini and Dienske 1979; Japanese macaques: French 1981). Although it is unclear how consistent this preference is (Caine and Mitchell 1979), the suggestion that females may use play to develop bonds with high-ranking individuals is a provocative one. Individuals certainly vary in their choices of play partners, but whether play influences the subsequent relationships of individuals is not known. This is a part of the general question of whether social bonds formed in the prereproductive period benefit individuals during adulthood.

The expression of play-partner preferences may sometimes be constrained by demographic factors. In small groups, potential partners do not always include relatives, same-sex peers, or other preferred individuals. For example, few sex differences in play were evident in the small groups of baboons and vervet monkeys studied by Cheney (1978b) and Lee (1983e), respectively. In both cases, there were few immature females in the groups, and females consequently played more with males and with partners of different ages than they might have had more partners been available.

SOCIAL RELATIONSHIPS OF JUVENILES AND ADOLESCENTS

Although immature primates spend much of their social time playing, they also interact with others in adultlike ways. Such interactions have often been viewed as practice for adulthood in that they provide opportunity for improving motor and social skills. Although performance of many behaviors clearly improves with experiences (e.g., sexual behavior in macaques and baboons: Hanby and Brown 1974; Owens 1976), most are employed in an adultlike manner long before sexual maturity. It seems more likely that the social interactions of immatures function primarily to acquire resources and to develop social bonds, some of which may persist into adulthood. Such interactions fall into four general classes of activity: grooming, infant handling, agonistic behavior, and sexual behavior. Because the most detailed information on social development comes from studies of savanna baboons, macaques, and vervets, the social interactions of immatures in these species are considered first. Their behavior is then contrasted with interactions in species of different social structure.

Macaques, Savanna Baboons, and Vervets

Female macaques, baboons, and vervets usually remain in their natal groups throughout their lives, whereas males emigrate to other groups as adolescents or young adults (chap. 11). The natal group is therefore the social unit for both sexes throughout all or most of the transition to adulthood, but it is the future social unit only for females. This basic dichotomy is consistent with a number of differences in the social relationships of immature males and females. The social development of immature males is characterized by increasing peripheralization from the adult female social network. In contrast, immature females continue to interact at high rates not only with their peers but also with the adult females of their group. Some of the social relationships developed by females during the prereproductive period persist into adulthood, including dominance relationships and affiliative bonds with female kin. The long-term significance of some of their social relationships, however, is less clear. It is not yet known, for example, whether the bonds formed among unrelated juvenile female peers persist into adulthood as special relationships or if they are merely temporary associations specific to the prereproductive period.

Grooming. Sex differences in behavior are particularly evident in grooming. In many ways, females, long before reaching maturity, perform like adults in their grooming interactions. Data on the grooming relationships of immature baboons (Cheney 1978a; Walters 1981; Johnson

1984; Pereira and Altmann 1985), bonnet macaques (Silk, Samuels, and Rodman 1981), rhesus macaques (Sade 1965; Missakian 1974), Japanese macaques (Yamada 1963; Oki and Maeda 1973; Mori 1975; Sugiyama 1976; Kurland 1977), and vervets (Fairbanks and McGuire 1985) all indicate that juvenile females groom other group members, particularly adult females and female peers, at higher rates than do males. Throughout development, the juvenile females' mothers continue to be their primary groomers, and the mother-daughter grooming relationship becomes more reciprocal with age (Missakian 1974; Pereira and Altmann 1985; Fairbanks and McGuire 1985). Immature females also maintain close grooming bonds with other members of their immediate family.

When grooming less closely related individuals, immature females of some species direct much of their grooming toward the members of high-ranking families, as is true for adults (Cheney 1978a; Walters 1981; Silk, Samuels, and Rodman 1981; Johnson 1984; see also chaps. 24–26). It has been hypothesized that such interactions represent attempts to establish bonds with high-ranking individuals from which the females may later benefit (Cheney 1978a; see also chap. 26). Perhaps because the dominance ranks and social relationships of young females are still more flexible than those of adults, immature females often seem particularly active in their attempts to interact with other animals. For example, they often initiate more grooming interactions with unrelated females than do adult females (Pereira and Altmann 1985; see also Cheney 1977b).

In contrast, males become increasingly isolated from the female grooming network as they mature. It is unlikely that most of the bonds formed between immature males and adult females persist into adulthood simply because males usually emigrate from the natal group. Perhaps as a result, the grooming of immature males is focused primarily on the immediate family. Immature males, however, usually groom their mothers less than do immature females, even though they may receive as much grooming. Grooming between mothers and sons also decreases in frequency with age and never becomes reciprocal (Sade 1965; Oki and Maeda 1973; Missakian 1974; Cheney 1978a; Pereira and Altmann 1985; Fairbanks and McGuire 1985). There is some evidence from rhesus (Colvin 1983b) and Japanese macaques (Itoigawa 1975) that males groomed at low rates by their mothers emigrate from the natal group at younger ages than males whose mothers groom them often. Similarly, immature male vervets with no close female kin emigrate at younger ages than males whose relatives are still alive (Cheney 1983b). Eventually, however, almost all males emigrate regardless of the closeness of their bonds with female kin (chap. 21).

Immature males groom unrelated adult females at lower rates than do immature females, but when they do groom such females they tend to groom low-ranking, rather than high-ranking, females; they also groom sexually receptive females more than lactating or pregnant females (Cheney 1978a; Silk, Samuels, and Rodman 1981; Johnson 1984).

The different patterns of grooming by immature males and females suggest that although both sexes derive immediate benefits from parasite removal, the establishment of grooming relationships with adult and immature females is likely to be of long-term importance only for females. In support of this suggestion, adolescent male Japanese and rhesus macaques often groom brothers, peers, and young adult males more than unrelated females (Miller, Kling, and Dicks 1973; Sugiyama 1976; Colvin 1983a). The bonds established through such interactions seem to contribute to and aid in the eventual emigration of males from the natal group, and may even persist following transfer. Male macaques and vervets frequently emigrate in the company of brothers or natal group peers (see chap. 21), and there is some evidence from rhesus macaques that alliances developed in the natal group may enable males to attain high dominance rank in their adopted group (Meikle and Vessey 1981).

Infant Handling Like adult females, juvenile and adolescent females are attracted to infants and often compete with each other to handle or groom them (chap. 27). In one group of baboons, for example, juvenile females interacted with each infant in the group at an average rate of once per hour (Walters 1986). Similar observations have been made in other studies of baboons (Cheney 1978a; Altmann 1980), rhesus macaques (Berman 1982a), and vervets (Lancaster 1972; Lee 1983a), as well as in a number of other Old World monkey species (reviewed in Hrdy 1976; chap. 27). Strong preferences for particular infants are less conspicuous than partner preferences in grooming or other social interactions. However, immature females do not handle all infants at equal rates. For example, they usually interact more with their infant siblings than with other infants, at least in part because their mothers are usually more tolerant than less closely related adult females. Similarly, in at least some studies, immature females have been observed attempting to handle the infants of high-ranking females more often than the infants of low-ranking females (Cheney 1978a; Berman 1982a; Lee 1983a). In these species, interactions with infants may be another means by which immature females attempt to establish bonds with the members of high-ranking families. Interestingly, abusive kidnappings of infants by immature female bonnet macaques are most likely to involve the infants of low-ranking females (Silk 1980b).

In contrast to females, immature males seldom interact with unrelated infants except in play. Although infant handling is observed, it occurs at far lower rates than among females. This is true even for male savanna baboons, who as adults, often form special protective relationships with infants (chap. 28). It is not yet clear if the few cases of special relationships between infants and unrelated immature males reported in some baboons (Ransom 1981; Hamilton, Busse, and Smith 1982) represent bonding similar to that which occurs between adult males and infants. Close bonds often do develop, however, between males and their siblings (chap. 27). In some cases, such bonds may persist into adulthood, if males later migrate to the same group (see chap. 21).

In two species of Old World monkeys, immature males interact at high rates with unrelated infants. In Barbary macaques (*Macaca sylvanus*), the use of infants appears to play an important role in mediating adult male interactions, and both immature and adult males frequently handle infants (Taub 1978). It is possible that infant handling by immature males either provides practice or aids in the development of bonds that will prove beneficial during adulthood. In hamadryas baboons, immature males are the primary handlers of infants (Kummer 1968; Abegglen 1984). Such behavior is the first manifestation of sexual bonding between males and females since young adult males acquire mates by forming protective relationships with young juvenile and infant females from other one-male units (Kummer 1968; chap. 10). In the case of male infants, interactions with juvenile and subadult males may aid in their integration into the natal clan (Abegglen 1984).

Agonistic behavior Data on the agonistic behavior of immatures are available for a variety of species, including vervets (Horrocks and Hunte 1983; Cheney 1983a; Lee 1983b), long-tailed macaques (*Macaca fascicularis*: Angst 1975; de Waal 1977; de Waal, van Hooff, and Netto 1976), rhesus macaques (Sade 1967; Missakian 1972; Kaplan 1977; Berman 1980a; Datta 1983a, 1983b, 1983c), bonnet macaques (Silk, Samuels, and Rodman 1981), stump-tailed macaques (*M. arctoides*: Gouzoules 1975; Massey 1977), Japanese macaques (Kawai 1958; Koyama 1967), and baboons (Cheney 1977a; Lee and Oliver 1979; Walters 1980; Hausfater, Altmann, and Altmann 1982). In all these species, adult females form stable, linear dominance hierarchies in which rank is not consistently related to any physical attribute such as size or fighting ability. Offspring acquire ranks similar to thsoe of their mothers and eventually become dominant to all adult females subordinate to their mothers. Even from infancy, the offspring of high-ranking females receive more affinitive approaches, less aggression, and more agonistic support from both related and unrelated

individuals than do the offspring of low-ranking females (see also chap. 25).

Like adults, immature females are both the recipients and the initiators of aggression related to competition over food and social partners. In fact, the daughters of low-ranking females often receive considerably more aggression from adult females than do their sons (Cheney 1977b; Silk, Samuels, and Rodman 1981; Simpson 1983). In proximate terms, this aggression may occur in part because females of all ages compete for similar social partners. Ultimately, it may occur because females represent more long-term competition for scarce resources than do males, who eventually emigrate from the group. Immature females are also especially active in the formation of alliances, probably because alliances are important in the acquisition of rank (chap. 25). Like adult females, immature females give most of their aid to their close relatives, but they also form many alliances with unrelated individuals, especially high-ranking ones (Massey 1977; Kurland 1977; Cheney 1977a, 1983a; Walters 1980; Datta 1983a, 1983c). Such alliances seem to be related both to rank acquisition and the formation of affiliative bonds (chap. 25).

In most species, males become dominant to adult females when they reach adult size (e.g., Sugiyama 1976; Lee and Oliver 1979; Lee 1983b; Scott 1984; Pereira and Altmann 1985). Before puberty, however, maternal rank strongly influences the agonistic interactions of immature males with females of all ages (see earlier references). In their agonistic interactions with each other, the dominance relationships of immature males are more variable than those of females. Although maternal rank accurately predicts the outcome of aggressive interactions among peers, ranks between males of more disparate ages are usually determined by age and size (e.g., baboons: Cheney 1977a; Japanese macaques: Alexander and Bowers 1969; rhesus macaques: Missakian 1972). Available data suggest that adult male ranks also depend more on such factors as age, fighting ability, size, and the presence of allies than on maternal rank (chaps. 11, 25, 26). There is some evidence from Japanese and rhesus macaques, however, that the sons of high-ranking females are more likely than other males to attain high rank as adults, both in their natal groups and following transfer (Sugiyama 1976; Chapais 1983d; Meikle, Tilford, and Vessey 1984).

Like immature females, immature males form alliances primarily with close relatives and the members of high-ranking families, although they usually form alliances less frequently (Cheney 1977a, 1983; Chapais 1983d). They are also involved in aggressive competition for resources with all group members. Although the frequency of their aggressive interactions is often the same as or even lower than that of females, the relatively low rate of their affinitive interactions often makes them appear more aggressive than females, particularly during adolescence.

Sexual behavior Both male and female cercopithecines engage in sexual behavior long before they are capable of reproduction. In rhesus and Japanese macaques, juvenile males direct most of their sexual activity at their mothers and other close relatives, but cease to do so as they grow older (Missakian 1973a; Hanby and Brown 1974). Mating between males and their mothers or other close matrilineal kin is rare once puberty begins (see also chap. 21). Among baboons, as males grow older they increasingly direct their sexual behavior toward estrous females (Owens 1976; Cheney 1978a), and they may engage in fertile copulations as adolescents (Hausfater 1975). In most cases, however, males do not continue to mate with these same females as adults simply because most males transfer to other groups. Whether males that remain in their natal groups continue to mate with the same females with whom they copulated as adolescents is not known. They do, however, continue to avoid mating with close matrilineal kin (Packer 1979a; Walters, in press; chap. 21).

Among baboons, the sexual behavior of juvenile females occurs at low rates prior to puberty and occurs initially with male peers. Female sexual activity increases markedly in adolescence and becomes directed increasingly at adult males. Typically, however, adult males show little interest in adolescent females until they have completed several estrous cycles (Altmann et al. 1977; Scott 1984). Since most males emigrate from the natal group, it is unlikely that juvenile females form sexual bonds with peers which persist into adulthood. Sexual inhibitions, however, do appear to develop at an early age. Both as adolescents and adults, females copulate at lower rates with their former male associates (who are often close relatives) than with less familiar males (Packer 1979a; Scott 1984). This lack of sexual attraction to familiar individuals appears to be a mechanism for inbreeding avoidance and may be a major cause of male emigration (Packer 1979a; Walters, in press; chap. 21). Males are not forced from their natal group by aggression from adult males or other group members. Prior to emigrating, they often attain high ranks in the natal group and will emigrate even when there are no other adult males (Sugiyama 1976; Packer 1979a).

In many species of monkeys and apes, juveniles and subadults harass copulating pairs by touching, threatening, or vocalizing at them (e.g., gray langurs: Jay 1965; Hrdy 1977; baboons: Rowell 1967; stump-tailed macaques: Gouzoules 1975; vervets: Lee 1981; chimpanzees: Goodall 1968; de Waal 1982; Adang 1984; gray-cheeked mangabeys [*Cercocebus albigena*]: Struhsaker and Le-

land 1979; see also Neimeyer and Anderson 1983 for a review). Males usually harass mating pairs more than do females, and adult males tend to be their primary targets (see fig. 9-4). The function of sexual harassment is not known, but its occurrence in infants may represent a form of parent-offspring conflict (see chap. 27). Such harassment may also enable subordinate juveniles to approach, examine, and threaten dominant males at a time when they are unlikely to retaliate.

Other Species

A number of colobines, New World monkeys, and other cercopithecines are also characterized by female philopatry and male emigration. Although social development in most such species has not been intensively studied, present evidence suggests patterns in the behavior of immature males and females similar to those described above (e.g., gray langurs: Hrdy 1976, 1977; gelada baboons [*Theropithecus gelada*]: Dunbar and Dunbar 1975; talapoins [*Cercopithecus talapoin*]: Wolfheim 1977b).

While the general pattern of development may be similar, however, some differences have also been reported. For example, in many species that typically live in one-male groups, the relationship between adult and immature males is more hostile than in macaques or baboons, and males are sometimes evicted from the natal group (e.g., gray langurs: Sugiyama 1967; purple-faced langurs [*Presbytis senex*]: Rudran 1973a; see also chap. 21). The development of dominance relations among females also differs in some species. Specifically, dominance has been reported to depend more on age than maternal rank in gray langurs (Hrdy 1977) and mantled howlers (*Alouatta palliata*: Jones 1980b), whereas in other species female dominance hierarchies are not readily detected (see chap. 26). Clearly, agonistic relationships among immatures in these species can be expected to differ markedly from those that are found in macaques or baboons.

Species in Which Females Disperse

A number of species live in multimale, multifemale groups that, although superficially similar to cercopithecine groups, are characterized by female dispersal. As a result, a number of differences in social development are apparent. Species for which some data on social development are available include red colobus (*Colobus badius*: Struhsaker 1975; Struhsaker and Leland 1979; chap. 8), chimpanzees (Goodall 1968; Nishida 1979; Bygott 1979; Pusey 1980, 1983; chap. 15), hamadryas baboons (Kummer 1968; chap. 10), and gorillas (Harcourt 1979b, 1979c; chap. 14). Apparently because female kin often reside in different groups as adults, adult female relationships in these species appear to be less affinitive than among female cercopithecines. In chimpanzees, hamadryas baboons, and gorillas, the grooming interactions of immature females are focused primarily on the immediate family (Kummer 1968; Harcourt 1979b; Goodall 1968; Pusey 1983), and grooming with other adult females is infrequent. Interactions with infants, however, are not always similarly restricted. In one population of chimpanzees, for example, both unrelated immigrant females and sisters handled infants at high rates (Nishida 1983a). Unrelated females interacted preferentially with the infants of high-ranking females and were more likely than close relatives to groom mothers before interacting with their infants.

Occasionally, however, female kin end up in the same group as adults (chap. 24). In such cases, they exhibit strong grooming relations and other forms of affiliation (gorillas: Harcourt 1979b; chap. 14; chimpanzees: Pusey 1983; hamadryas baboons: Sigg et al. 1982). Whether unrelated females also develop strong bonds that cause them to remain in or migrate to the same group as adults is not known.

In some species characterized by female dispersal, males often remain in their natal group or clan. The social interactions of males in such species include frequent interactions with unrelated adults as well as with peers and close maternal relatives. Among male hamadryas baboons, for example, there is considerable affiliation between adult and immature males, and the bonds developed by immature members of the same clan are manifested during adulthood through mutual support against males from other clans or bands (Kummer 1968; Abeggelen 1984). Similarly, immature male chimpanzees seek affiliative contact with the adult male members of their community (Pusey 1983), and in some groups of red colobus, adolescent males groom adult females at high rates (Struhsaker and Leland 1979). Even among male gorillas, who only sometimes remain in their natal groups, there are high rates of grooming and play between adult and immature members (Harcourt 1979a, 1979c). Indeed, Harcourt and Stewart (1981) found that immature male gorillas who developed a close bond with the resident silverback were more likely to remain in the natal group as adults.

The sexual relations of immature males and females in female-dispersing species are in many respects similar to those in macaques and baboons. Many of the sexual interactions of immature chimpanzees involve siblings or other close relatives. When they become sexually mature, however, they avoid sexual interactions with kin (Pusey 1983). Similar sexual inhibitions have been observed in gorillas (Harcourt, Stewart, and Fossey 1976), although incestuous matings do occur in this species (chap. 14). Male chimpanzees appear to be particularly sexually precocious; the sexual behavior of males ap-

pears to be fully developed by 1 year of age, even though they do not reach puberty until 7 or 8 (Goodall 1968).

Species Living in Family Groups

Gibbons, marmosets, tamarins, and titi monkeys (*Callicebus* spp.) all live in family groups containing only one breeding female (chaps. 4, 7, 12). Although juveniles in such groups develop close bonds with their relatives, they almost never interact with peers other than their siblings. Immatures are thus relatively free from aggression, but may lack play partners. A number of studies of gibbons (chap. 12) and marmosets (Box 1975a, 1977) have described grooming and play among immatures and their parents and siblings. At least in gibbons, however, play occurs at relatively low rates, apparently because of the lack of suitable partners (chap. 12). Little is known about the long-term social relationships of immatures, but since the members of both sexes disperse at sexual maturity, individuals probably seldom develop relationships that persist into adulthood.

The best-studied aspects of social development in species that live in family groups concern the maturation and dispersal of adolescents. In gibbons, increasing aggression from the same-sex parent forces maturing offspring to become spatially and behaviorally peripheral to the family group and usually culminates in dispersal (Chivers 1974; Chivers and Raemaekers 1980; Gittins and Raemaekers 1980; Tilson 1981; chap. 12). Parent-offspring conflicts may be intensified by the fact that the scarcity of suitable breeding habitats often delays dispersal. Indeed, adolescents have been observed to remain in their parents' territory and to mate with their opposite-sexed parents following their mothers' or fathers' deaths (chap. 12).

Perhaps because offspring represent potential mate competition, in some callitrichids mothers may suppress their daughters' sexual development (saddleback tamarins [*Saguinus fuscicollis*]: Epple and Katz 1984; marmosets: Evans and Hodges 1984). Similar effects have been documented in rodents (Vandenbergh 1983). The behavioral events associated with dispersal in tamarins and marmosets are not known. In some cases, mature offspring remain in their natal groups with their parents and aid their parents in rearing subsequent offspring, which suggests that parents do not force dispersal (chap. 4). To date, infant care by immatures has been studied only in captivity (Box 1977; Hoage 1978; Ingram 1977). In such circumstances, both juvenile males and females play an important role in rearing infants, and groom and carry their siblings at high rates. It is not yet known whether helping behavior in marmosets and tamarins is associated with the saturation of suitable breeding habitats and limited opportunities for dispersal, as is the case in many cooperatively breeding birds and mammalian carnivores (Vehrencamp and Emlen 1983).

Solitary Species

In some primates, adults range by themselves, and immatures have little opportunity to interact with conspecifics other than their mother and occasionally a sibling. The best studied of these species are lesser bushbabies (chap. 2) and orangutans (chap. 13). In each case, the relationship between mother and offspring remains unaggressive throughout development, and adolescent dispersal appears to be voluntary. Given the limited opportunities for social interaction in these species, it is interesting that immatures sometimes leave their mothers to associate with other animals. Immature orangutans form temporary play groups (MacKinnon 1979), and immature male bushbabies have been observed to follow adult males that range through their mother's territory (Charles-Dominique 1977a). There are no data on what sorts of social bonds immatures develop or whether there are sex differences in social development.

Similarly, it is not yet known whether immature orangutans form long-term bonds during play or during the occasional periods when their mothers associate (Galdikas 1979). Adult female bushbabies, however, continue to associate in sleeping groups with their mothers, other females, and immatures, although they forage alone (Charles-Dominique 1977a; chap. 2). Some of the bonds developed by immature females therefore appear to persist into adulthood (see also Clark 1978b). In contrast, males disperse over further distances and probably do not continue to associate with males or close kin.

SUMMARY

At first glance, the period between infancy and adulthood appears to be a time of play and freedom from the energetic costs of reproduction. Perhaps as a result, it is a period that has also been ignored by most behavioral scientists. It is becoming increasingly clear, however, that the transition to adulthood is a time when individuals acquire skills and develop relationships that may be of both immediate and long-term benefit. Many of the sex differences evident in the play, grooming, and other social interactions of immature animals reflect differences in the adult behavior of each sex. While both males and females develop close bonds with their close maternal relatives, only that sex which typically remains in the natal group following adolescence interacts at high rates with unrelated adults. Among most species of Old World monkeys, males play more than females, whereas females groom other females and interact with infants at higher rates than males. In contrast, in species such as hamadryas baboons, chimpanzees, and gorillas, where females normally disperse, the interactions of juvenile females are more restricted to the immediate family. It is hoped that research over the next decade will provide better descriptions of social development in a number of species as well as answers to the many remaining questions about the significance of prereproductive behavior.

30 | Patterning of Sexual Activity

Sarah Blaffer Hrdy and Patricia L. Whitten

The word *estrus* comes from the Greek for gadfly, an insect that deposits its ova—eggs that are shortly to metamorphose into irritating larvae—in the skin of cattle. The image conveyed is of creatures driven to distraction by this temporary itching in their system. The phrase *females in estrus* thus implies the transformation of sedate foragers into active solicitors of male attention.

This chapter reviews the behaviors associated with this transformation. It focuses on the distribution of estrous behavior over time and the morphological and behavioral signals that primates use to communicate sexual availability and sexual interest. This information is summarized in table 30-1, which serves as a reference for much of the following discussion. We also consider the functional significance of particular patterns of sexual availability and sexual communication among primates. The review begins with a definition of estrus and a discussion of the great diversity in patterns of estrus among nonhuman primates.

DEFINING ESTRUS

The term *estrus* has traditionally referred to a discrete period around ovulation that is characterized by three related changes (Beach 1976):

1. an increase in *attractivity*, defined in terms of the female's value as a sexual stimulus to males. Attractivity is measured in terms of male behaviors such as the frequency of approaches and the frequency of attempts to mount.

2. An increase in *proceptive* behaviors, appetitive activities shown by females in response to the presence and behavior of males. Proceptive behaviors include general affiliative behaviors such as approaching to sit near the male as well as acts that solicit sexual contact more directly, such as presenting the hindquarters to the male.

3. An increase in *receptive* behaviors, acts designed to facilitate copulation itself, such as adoption of a stance that permits penile insertion and maintenance of the appropriate posture long enough to allow ejaculation.

In what follows, *receptivity* generally will be used to refer to both proceptive and receptive behaviors, and *estrous behavior* will refer to an increase in receptivity, so defined. Male responses to these behaviors—a measure of female attractivity—will be considered when data are available.

Among most mammals, the period of time during which a female is proceptive, receptive, and attractive is brief. Rats, for example, will position themselves to copulate for only a few hours during a 4- to 5-day estrous cycle. A cow comes into heat for about 6 hours every 3 weeks. In many species, ranging from guinea pigs to prosimians such as the galago, mating outside of estrus is physically impossible since the vagina is fused shut or covered with an epithelial membrane (Butler 1974).

For many mammals, estrus is not only confined to a brief portion of the reproductive cycle, but it is strictly seasonal as well. Reproductive cycling occurs for only a few weeks of the year among Madagascar prosimians such as the sifakas (chap. 3). In some New World monkeys, such as squirrel monkeys, week-long sexual cycles occur only during 3 months of the year, and copulations are confined to 1 or 2 days each cycle. Changes in male squirrel monkey morphology and behavior are also evident during the breeding season. The testicular regression characteristic of males out of season is reversed, and in addition to producing active sperm, males also become larger, partly through the buildup of subcutaneous fat deposits in their shoulders. As with sifaka males, squirrel monkey males become much more assertive and aggressive during the brief breeding season (DuMond 1968).

Many catarrhine primates, however, depart from this generalized mammalian pattern of strictly circumscribed estrous periods (Keverne 1981). Apes, humans, and many monkeys have reproductive cycles that differ in two ways from traditional estrous cycles: first, the cycles include menstruation, a cyclical sloughing of the uterine lining; second, there is greater flexibility in the timing of proceptive and receptive behaviors and a longer duration of estrus. In some species, sexual receptivity throughout all or much of the cycle is commonly exhibited under natural conditions (e.g., humans, pygmy chimpanzees, common marmosets: Thompson-Handler, Malenky, and Badrian 1984; Hearn 1978), whereas in others, extended receptivity is observed primarily under artificial conditions in captivity (common chimpanzees: Lemmon and Allen 1978). These species and many other monkeys and apes exhibit a capacity for flexible receptivity that is quite distinctive and not typically found in other mammals. At the same time, all primates retain a tendency to concentrate matings at midcycle. This tendency may be pronounced, as in savanna baboons (Hausfater 1975) and gorillas (Graham 1981), or slight, as in tamarins (Brand and Martin 1983) or humans (Adams, Gold, and Burt 1978).

Because of the capacity for such flexible receptivity in catarrhine primates, use of the term *estrus* does not necessarily imply the specific physiological conditions surrounding ovulation, although in some cases (such as gorillas) estrus and ovulation do coincide (Mitchell et al. 1982). Increasingly, then, the adjective *estrous* is being redefined by primatologists to denote actively libidinous behavior by a female who permits copulations regardless of whether or not she is near ovulation (e.g., Loy 1970; Wolfe 1984b). Since it is rarely possible to confirm hormone levels or ovulation under field conditions (unless pregnancy is detected subsequently), a behavioral definition of estrus measured by female proceptive and receptive behaviors is useful.

THE MENSTRUAL CYCLE

Apes and Old World monkeys exhibit menstrual cycles ranging from 25 to 35 days (table 30-1). On the basis of laboratory studies of a few species, the endocrine events in these cyles resemble (with minor variations) the well-documented human pattern (see Graham 1981 for great apes; Hess and Resko 1973 for rhesus macaques). In humans, the first phase of the cycle from the end of menstruation until ovulation is called the follicular phase. During this phase, a small pouch or follicle containing an egg matures in an ovary. As the ovarian follicle matures, circulating levels of estrogenic hormones (particularly estradiol) gradually rise, peaking at around day 15 of the cycle. This high level of estrogens causes the pituitary to produce a surge of luteinizing hormone; this is the immediate stimulus for ovulation. Other hormones, including testosterone, also peak at midcycle (reviewed in Chalmers 1980). Morphological changes in the female that accompany these endocrine changes are discussed below.

Once the egg is released, the ovarian follicle is transformed into a solid body called a corpus luteum, which secretes progesterone during this luteal phase of the cycle. Stimulated by progesterone, the lining of the uterus (or endometrium) prepares to receive the fertilized egg. If fertilization does not occur, levels of progesterone fall, the lining of the uterus disintegrates, and menstruation follows. Bleeding is often detectable among humans, apes, and Old World monkeys for a period ranging from 1 to 8 days (reviewed in table 30-1). Vaginal bleeding has never been reported in prosimians, but tiny amounts of blood occasionally have been noted in a few genera of New World monkeys (Hershkovitz 1977, 26).

ESTROUS BEHAVIOR DURING PREGNANCY

If implantation takes place, progesterone levels in the blood remain high until the end of pregnancy and normal menstrual cycles cease. Pregnancy may be signaled by changes in the color of the labia or paracallosal skin, as in orangutans and savanna baboons, species that typically do not mate when pregnant (see table 30-1). In other species, pregnant females may continue to exhibit periods of sexual receptivity. Postconception estrus can be either cyclical (e.g., cottontop and lion tamarins) or very irregular in occurrence (e.g., rhesus and pigtailed macaques, chimpanzees, sooty mangabeys). Postconception solicitations are sometimes more subdued (e.g., the head-shaking solicitations of pregnant gray langurs: Hrdy 1977) or last for shorter periods than the estrous bouts of cycling females (e.g., 11 days comapred to 14 days among Japanese macaques: Takahata 1980), before tapering off to a virtual absence of sexual receptivity in the later stages of pregnancy. In a few cases, however, receptivity actually appears to increase. For example, captive female lion tamarins solicit males more at midpregnancy than at other times (Kleiman and Mack 1977), captive chimpanzees do so during the early months of pregnancy (Wallis 1982), and human females also report enhanced libido during the second trimester (Masters and Johnson 1966). Among wild vervet monkeys, females were more receptive, and attempted copulations were more likely to succeed during the first half of pregnancy than during the 2½ months of the breeding season just preceding conception (Andelman 1986).

Clearly, sexual receptivity during pregnancy cannot be an adaptation to enhance the probability of conception. Alternative explanations include the possibility that enhanced receptivity during pregnancy is (*a*) an artifact of endocrinological changes or (*b*) an adaptation functioning to influence the behavior of males (Hrdy 1979). The latter hypothesis proposes that males are more likely to assist or tolerate infants born to former mates, an idea that is developed further later in this chapter.

MENOPAUSE

A postreproductive life phase has traditionally been thought of as uniquely human, but recent findings suggest that declining fertility with age may be more pronounced among monkeys and apes than previously recognized, to the point that some females cease to reproduce altogether before they die. In the wild, only a handful of seemingly postreproductive old females have been described (e.g., gray-cheeked mangabeys: Waser 1978; toque macaques: Dittus 1975). Their failure to breed might be due either to increasingly long birth intervals terminated by death or to actual cessation of ovulation. Based on captive data, it is now apparent that the end of the female primate life cycle may be marked by irregular and lengthened menstrual cycles, reduced levels of estrogens, longer and longer birth spacing, and in a few cases (e.g., some chimpanzees), cessation of ovulation

TABLE 30-1 Interspecific Variation in Sexual Signaling and Behavior among Primates

Species	Cycle Length (days)	Duration Menstrual Flow (days)	Duration of Receptivity (synonomous with estrus in some but not all species)	Proceptive Behaviors (solicitations)	Visual Cues	Olfactory Cues	Other Indicators	Behavioral Solicitations	Mount Pattern	Cycling	Pregnant	Breeding System/Consort Patterns	Sources
Ring-tailed lemur	39	None	2–10 hr	Present hindquarters or crouch with lordosis	Confined to vulval swelling and pinkening	Genital and urine secretions		Follow, lick female's genitalia	Single (6–35 thrusts; 2–3 min)			Multimale; consorts several hours or less	Evans and Goy 1968; Jolly 1966a
Black lemur		None			Confined to vulval swelling and pinkening							Multimale	Bogart, Komanoto, and Lasley 1977
Ruffed lemur	40	None	4–24 hr	Rubs body against male, then rapidly retreats, slaps male	Confined to vulval swelling and pinkening	Urine marking	Cyclic opening and closing of vagina	Lick female's genitalia, rapid chewing movements of jaw, follow	Multiple?			Multimale	Shideler and Lasley 1982; Foerg 1982b
Western fat-tailed dwarf lemur	50	None	1 day		Confined to vulval swelling and pinkening; may be whitening as labia open		Cyclic opening and closing of vagina	Follow, sniff genitals	Single (2–3 min every 10 min)	20/hr		Solitary	Foerg 1982a
Mouse lemur	48–55	None	3 days		Confined to vulval swelling and pinkening		Cyclic opening and closing of vagina					Female associations	Doyle 1974
White sifaka		None	12–36 hr		Confined to vulval swelling	Genital and urine secretions		Approach, sniff and mark female	Single (40 thrusts; 4 min)	Single		Multimale	Richard 1974b
Angwantibo	36–45	None	A few hrs	Suspends body upside down									Doyle 1974
Potto	37–38	None	2 days	Suspends body upside down	Confined to vulval swelling and reddening	Scent marks						Solitary	Ibid.; Hafez 1971
Slow loris	42	None	1–2.5 days	Suspends body upside down									Doyle 1974
Slender loris	29–40		2 days	Suspends body upside down	Vulval swelling and reddening	Urine marking	Vaginal plug apparent after 67% of ejaculations			6 copulatory bouts/day			Izard and Rasmussen 1985
Lesser bushbaby	30–37	None	1–3 days		Confined to vulval swelling and reddening	White vaginal secretion	Cyclic opening and closing of vagina	Approach, sniff, and mark female	Single (10 sec–7 min)			Solitary	Doyle 1974
Greater bushbaby	34–44	None	2–10 days		Confined to labial swelling and pinkening	Vaginal secretion	Cyclic opening and closing of vagina	Approach, inspect	Single (5–6 thrusts)	0.78/min	0	Solitary	Ibid.; Eaton 1973

TABLE 30-I (continued)

| Species | Cycle Length (days) | Duration Menstrual Flow (days) | Duration of Receptivity (synonomous with estrus in some but not all species) | Signals of Female Receptivity and Factors Affecting Female Attractiveness (as detected by human observers) | | | | Male Sexual Behavior and Ejaculatory Pattern | | Mating Frequency | | Breeding System/Consort Patterns | Sources |
				Proceptive Behaviors (solicitations)	Visual Cues	Olfactory Cues	Other Indicators	Behavioral Solicitations	Mount Pattern	Cycling	Pregnant		
Tarsier	24	None	24 hrs; swelling lasts up to 9 days		*Labial swelling and reddening	Vaginal secretion	Cyclic opening and closing of vagina; opaque, sperm-filled secretion similar to vaginal plug	Approach, lick female's genitalia	Single (2 min)			Monogamous or solitary (?)	Wright, Izard, and Simons 1986
Common marmoset	15–17	Not detectable	Copulates throughout cycle	Arch back, piloerection crouch, and tongue flick	Slight swelling, usually not detectable	Genital and urine marking		Arch back, piloerection, tongue flick	Single (7–10 thrusts, 2–18 sec)			Monogamous or polyandrous? (chap. 4)	Hershkovitz 1977
Pygmy marmoset				Present hindquarters, strut, stroke and nuzzle male				Strut, tongue flick, jaw quiver				Monogamous? (chap. 4)	Hershkovitz 1977
Saddleback tamarin	18								Single (6 sec–5 min)			Monogamous or polyandrous? (chap. 4)	Hershkovitz 1977; Epple and Katz 1984
Cottontop tamarin	23		Copulates throughout cycle	Tongue flicks	None			Dance around and lick female	Single (3–20 pelvic thrusts)	1–3/day	1–3/day	Monogamous or polyandrous? (chap. 4)	Brand and Martin 1983; French, Abbott, and Snowdon 1984; Hershkovitz 1977
Lion tamarin	14–21		3–5 days	Tongue flicks		Scent marking declines		Approach, sniff female	Serial?	1.8/hr	.2/hr	Monogamous? (chap. 4)	Kleiman and Mack 1977; Kleiman 1978a
Goeldi's marmoset	21–24		7 days	Faces male and drops head; raises rear; shows lordosis; tongue flicks		Scent marking increases		Embraces female, licks vagina, shows tongue flicking	Single (20–40 pelvic thrusts)			Monogamous? (chap. 4)	Heltne, Wojcik, and Pook 1981; Hershkovitz 1977
Night monkey	16		Copulates throughout cycle						Single (3–4 thrusts)			Monogamous	Dixson 1982
White-nosed saki				Present hindquarters	Anogenital region reddens		Nipples and anogenital area redden in pregnancy	Stamps feet				Multimale	Van Roosmalen, Mittermeier, and Milton 1981
Red uakari				Present hindquarters		Genital and urine marking			Serial			Multimale or age-graded	Fontaine 1981
Dusky titi				Genital investigation may include touching	No visible change				Single			Monogamous but may mate sometimes with neighboring males	Mason 1966

TABLE 30-1 (continued)

Species	Cycle Length (days)	Duration Menstrual Flow (days)	Duration of Receptivity (synonymous with estrus in some but not all species)	Signals of Female Receptivity and Factors Affecting Female Attractiveness (as detected by human observers)				Male Sexual Behavior and Ejaculatory Pattern		Mating Frequency		Breeding System/Consort Patterns	Sources
				Proceptive Behaviors (solicitations)	Visual Cues	Olfactory Cues	Other Indicators	Behavioral Solicitations	Mount Pattern	Cycling	Pregnant		
Mantled howler	11–24	Present but rarely visible externally	3–4 days	Tongue flicks, crouch, and present hind-quarters	Slight pinkening of labia minora			Tongue flicks	Serial (8–25 thrusts/mt)		Observed once	Multimale	Glander 1980
Red howler	16–27		A few days, but extended estrus observed in case of a solitary female		No visible change					.33/hr	Never seen	Primarily unimale	Sekulic 1981; Sekulic 1982a; Crockett and Sekulic 1984
Brown capuchin	16–20	Present but rarely visible externally	3 days	Chase or dance, lip pucker, present hind-quarters					Serial (1–6 mt up to 2 min/mt)			Multimale	Freese and Oppenheimer 1981
White-fronted capuchin	16–20				Vulva swells								Hafez 1971
Squirrel monkey	7–25	Never seen	hr–2 days	Display erect clitoris, present hind-quarters and look at male	Vulva swells sometimes	Aliphatic acids present, may scent mark through sneezing		Paces, leaps, silent "calls"	Serial (2–25 mt; 1–15 thrust/mt)				Eisenberg 1978; Baldwin and Baldwin 1981; Hennessy et al. 1980
Brown-headed spider monkey	24–27								Single (6–35 min/mt)			Multimale	Eisenberg 1978
Humboldt's woolly monkey	23–36	?							Single (10 min/mt)				Hafez 1971; Eisenberg 1978
Gray langur	27	2–4	5–7 days, but may last several weeks when new males enter troops	Shakes head, presents hind-quarters	None				Single and serial mounts		May be common in the presence of extra-troop males	Unimale or multimale, females occasionally mate with extratroop males	Hrdy 1977
Purple-faced langur				Shakes head, presents hind-quarters	None							Unimale or multimale	Rudran 1973b
Silver leaf monkey				Shakes head, presents hind-quarters	None			None	Serial	.02/hr		Unimale or multimale	Bernstein 1968
Dusty leaf monkey					Vulva swells at estrus							Unimale or multimale	Roonwal and Mohnot 1977; Curtin 1980
Nilgiri langur					Clitoris reddens; related to estrus?				Serial	Rare		Unimale or multimale	Poirier 1970; Michael and Zumpe 1971
Mentawi Island langur												Monogamous	Tilson and Tenaza 1976
Black-and-white colobus					None				Serial?	.01/hr		Unimale	Struhsaker and Leland 1979

TABLE 30-I (continued)

Species	Cycle Length (days)	Duration Menstrual Flow (days)	Duration of Receptivity (synonomous with estrus in some but not all species)	Signals of Female Receptivity and Factors Affecting Female Attractiveness (as detected by human observers)				Male Sexual Behavior and Ejaculatory Pattern				Breeding System/Consort Patterns	Sources
				Proceptive Behaviors (solicitations)	Visual Cues	Olfactory Cues	Other Indicators	Behavioral Solicitations	Mount Pattern	Mating Frequency			
										Cycling	Pregnant		
Red colobus		Not obvious externally			*Perineum swells/cycle				Serial	.14/hr	? Some swellings	Multimale	Ibid.; Struhsaker, pers. comm.
Olive colobus					*Perineum swells/cycle								
Black colobus					*Perineum swells/cycle								Chap. 8
Barbary macaque	31	3–4		Grab and jerk male's head; present hindquarters	*Perineum swells/cycle				Single		Occurs	Multimale; brief consorts (1–93 min)	Taub 1978; Turckheim and Merz 1984
Toque macaque	29	1–4			None	Strong-smelling vaginal discharge		Male usually initiates mount	Single			Multimale	
Bonnet macaque	25–36	10	Throughout cycle; copulations peak midcycle	Present hindquarters	Scarcely detectable perineal swelling	Strong-smelling vaginal discharge	Perineum swells in pregnancy slightly	Rapid tongue flicks while open/close jaws, grin	Single (14–23 thrusts/mt)	3.5 mt/hr	Occurs	Multimale; consorts (hrs–10 day)	Glick 1980; Nadler and Rosenblum 1969
Lion-tailed macaque	40	2.5	Throughout cycle; copulations peak midcycle		*Perineum swells/cycle				90% serial mounts			Unimale or multimale	Green 1981; Shideler and Lasley 1982; Shiveley et al. 1982
Pigtailed macaque	32	Not obvious externally	Throughout cycle; copulations peak midcycle	Presents hindquarters	*Perineum swells/cycle	Aliphatic acids present		Flehmen face	Multiple (8–23 mt, 12 thrusts/mt)	7.5 mt/hr	Occurs	Multimale; consorts (hrs–10 days)	Eaton 1973; Michael, Bonsall, and Warner 1975b
Crab-eating macaque	31	2–7	Occasionally for periods of weeks	Presents hindquarters	Reddening and swelling most pronounced among adolescents; degree and frequency decreases with age (+)	Aliphatic acids present			Single (14–21 thrusts/mt) or serial (2–6 mt, 4–29 thrust/mt, 2–85 min)		May be periods of intense sexual activity	Multimale; consortships last hours to more than 1 week	Shiveley et al. 1982; Michael, Bonsall, and Warner 1975b; Wheatley 1978; Deputte and Gustard 1980; L. Berenstain, pers. comm.; van Noordwijk 1986
Rhesus macaque	29	2.6	8–11 days	Presents hindquarters	Perineum reddens/season and cycle to some extent	Aliphatic acids peak midcycle	Perineum incandescent in pregnancy		Serial (1–20 mt, 1–15 thrusts/mt)	18 mt/hr	In the wild females exhibit 100% estrus in first 12 weeks	Multimale	MacDonald 1971; Hafez 1971; Baulu 1976; Nadler and Rosenblum 1969; Lindberg 1983
Stump-tailed macaque	31	2–4, but not obvious externally	Throughout cycle; copulations peak midcycle	Presents hindquarters	Slight perineal swelling, rare (+)	Aliphatic acids present		Touches or lifts female's hips	Single (1–170 thrusts)			Multimale; no consorts	Bertrand 1969; MacDonald 1971; Eaton 1973; Michael, Bonsall, and Warner 1975b; Murray, Bour, and Smith 1985

TABLE 30-I (continued)

Species	Cycle Length (days)	Duration Menstrual Flow (days)	Duration of Receptivity (synonymous with estrus in some but not all species)	Proceptive Behaviors (solicitations)	Visual Cues	Olfactory Cues	Other Indicators	Behavioral Solicitations	Mount Pattern	Cycling	Pregnant	Breeding System/Consort Patterns	Sources
					Signals of Female Receptivity and Factors Affecting Female Attractiveness (as detected by human observers)			Male Sexual Behavior and Ejaculatory Pattern		Mating Frequency			
Japanese macaque	28	3.5	1–92 days, x̄ = 13.6; copulation peaks midcycle	Presents hindquarters	In older females, perineum and face redden/season	"Sharp-smelling" secretion	Conspicuous swelling in young females only	"Sexual attack," present; touch	Serial (2–30 mt, 1–8 thrusts/mt)	.50/day	.32/day	Multimale; consorts (hrs–days)	Eaton 1973; Aso et al. 1977; Enomoto, Seiki, and Haruki 1979; Takahata 1980
Celebes macaque	34				*Perineum swells/cycle				Serial				Dixson 1977
Formosan macaque	30	3			*Perineum swells/cycle								
Assamese macaque					None		Perineum reddens in pregnancy						
Vervet monkey	33	3.6, but not obvious externally	7–13 days or throughout	Presents hindquarters	Scarcely visible perineal pinkening		Perineum stays pink if pregnant	Male initiates copulation; grabs female's hips	Single (25 thrusts/mt)	.09/hr	.06/hr	Multimale	Gartlan 1969; Rowell 1970; P. L. Whitten 1982; Andelman 1985, 1986
Blue monkey	30	1–4, but not obvious externally	4–5 days, but not related to cycle	Approaches male	None				Serial	.004/hr	Occurs	Unimale	Rowell 1970; Struhsaker and Leland 1979
Spot-nosed monkey	28				Vulva pinkens and swells slightly				Serial				Gautier-Hion and Gautier 1976; Struhsaker 1969
De Brazza's monkey			Throughout cycle										
Redtail monkey		Not obvious externally	1–28 days, x̄ = 6.9 days		None				Serial	.009/hr	Semi-continuous receptivity in captivity	Unimale, but promiscuous mating when extratroop males enter group	Cords 1984a; Struhsaker and Leland 1979; Struhsaker, pers. comm.; Wallis, pers. comm.
Talapoin	33	2–6	Throughout, but copulations peak midcycle	Crouches and presents hindquarters, may look back at male	*Perineum swells/cycle		Perineum reddens in pregnancy	Examine perineum; may "grin"	Single			Multimale; no consorts	Scruton and Herbert 1970; Rowell and Dixson 1975
Allen's swamp monkey					*Perineum swells/cycle								
Patas monkey	32	3	1–66 days, x̄ = 12 days	Crouching gait, blows into cheek pouches, drools, tail curled up	Vulval reddening/cycle	Aliphatic acids present	Facial hair whitens in pregnancy		Single and serial mount reported (10 thrusts to ejaculation in 12 sec)	.14 mt/female hr	.11 mt/female hr	Unimale, but nonresident males may mate	Loy 1975, 1981; Michael, Bonsall, and Warner 1975b; Palmer, London, and Brown 1981; Chism, Rowell, and Olson 1984; Sly et al. 1983

TABLE 30-1 *(continued)*

Species	Cycle Length (days)	Duration Menstrual Flow (days)	Duration of Receptivity (synonomous with estrus in some but not all species)	Proceptive Behaviors (solicitations)	Visual Cues	Olfactory Cues	Other Indicators	Behavioral Solicitations	Mount Pattern	Cycling	Pregnant	Breeding System/ Consort Patterns	Sources
Gray-cheeked mangabey	30	Not obvious externally	5 days at maximum swelling	Presents hind-quarters	*Perineum swells/cycle		Perineum and nipples dark red in pregnancy		Serial or single			Multimale	Gautier-Hion and Gautier 1976; Rowell and Chalmers 1970; Struhsaker and Leland 1979
White-collared mangabey	32	Not obvious externally	5 days		*Perineum swells/cycle			Serial				Multimale	Gautier-Hion and Gautier 1976
Sooty mangabey	31				*Perineum swells/cycle						Occurs	Multimale	Gordon and Busse 1984
Yellow baboon	32	3		Presents hind-quarters	*Perineum swells/cycle		Paracallosal skin reddens in pregnancy			1–6 hr	None	Multimale; consorts	Hausfater 1975; Collins 1981
Olive baboon	31–35	3	15–20 days	Presents hind-quarters	*Perineum swells/cycle	Aliphatic acids present	Perineum dark red in pregnancy		Single (6 thrusts)		None	Multimale; consorts	Altmann 1973; Smuts 1985
Guinea baboon	31–35	3		Presents hind-quarters	*Perineum swells/cycle								
Chacma baboon	31–35	3	Throughout but copulations peak midcycle	Presents hind-quarters; raise eyebrows with ears flattened	*Perineum swells/cycle		Paracallosal skin reddens in pregnancy		Serial (1–20 thrusts/mt)		None	Multimale; consorts	Saayman 1970; Michael, Bonsall, and Warner 1975b
Hamadryas baboon	31–35	3		Presents hind-quarters	*Perineum swells/cycle				Serial	7–12.2 hr		Unimale in multimale group	Kummer 1968
Drill	33	Not obvious externally			*Perineum swells/cycle		Perineum reddens in pregnancy				Occurs	Unimale in multimale group	Hadidan and Bernstein 1979
Mandrill					*Perineum swells/cycle		Perineum reddens in pregnancy					Unimale in multimale group	
Gelada baboon	35	Not obvious externally	1–25 days; copulations peak at midcycle	Presents hind-quarters; "solicit call"	Blisters on chest and para-callosal skin swell/cycle, to some extent	Males smell chest patch	Paracallosal skin reddens in pregnancy	Approach, inspect, lip smack	Single (14 thrusts/mt, 11 sec)	.1–1.3/hr	When new males usurp troop	Unimale in multimale group	Alvarez 1973; Dunbar and Dunbar 1974; A. Mori 1979b
Lar gibbon	30	2–5	Several days	May crouch	Vulva may redden but relation to cycle unclear			Thrusts against female's back	Single (30 thrusts/mt)	<1/day	Occasionally	Monogamous pair	Chivers 1974; Ellefson 1968a
Siamang					Vulva reddens but relation to cycle unclear			Shuffling gait	Single	.5– 6/day		Monogamous pair	Chivers 1974
Orangutan	31	3–4; not obvious externally	Forcible matings may occur throughout	Touch or mouth male's genitalia; hang above and present perineum			Scarcely perceptible labial swelling and slight whitening in pregnancy	Long call?	Single (3–29 min)		Rare	Solitary; consorts or forced copulations	Graham 1981; Galdikas 1981; Mitani 1985b; 1985d

TABLE 30-1 (*continued*)

Species	Cycle Length (days)	Duration Menstrual Flow (days)	Duration of Receptivity (synonomous with estrus in some but not all species)	Signals of Female Receptivity and Factors Affecting Female Attractiveness (as detected by human observers)				Male Sexual Behavior and Ejaculatory Pattern		Mating Frequency		Breeding System/Consort Patterns	Sources
				Proceptive Behaviors (solicitations)	Visual Cues	Olfactory Cues	Other Indicators	Behavioral Solicitations	Mount Pattern	Cycling	Pregnant		
Common chimpanzee	37	3	Wild: 7–17 days; captivity; throughout cycle, copulations peak at midcycle	Crouch and present hind-quarters	*Perineum swells/cycle	Aliphatic acids present		Shakes branch, erects hair, gazes with erect penis	Single (3–30 thrusts, 7 sec)	Female initiated .04/hr; male initiated .07/hr (captive)	Female initiated .03/hr; male initiated .17/hr (captive)	Multimale; consorts and opportunistic mating	Graham 1981; Tutin and McGinnis 1981; Tutin 1975; Fox 1982; Wallis 1982
Pygmy chimpanzee	28–37		Throughout; copulations peak at midcycle	Presents hind-quarters in crawl position	*Perineum swells/cycle				Single (2–54 thrusts/mt)			Multimale	Savage-Rumbaugh and Wilkerson 1978; Graham 1981
Mountain gorilla	28	Not obvious externally	1–3 days	Slow and hesitant approach	Scarcely detectable swelling of the vulva/cycle (only seen in captivity)				Single (15 sec–20 min)	.2–1.2/hr		Unimale, occasionally multimale	Harcourt et al. 1980; Graham 1981; Veit, pers. comm.
Lowland gorilla	31	Not obvious externally	3–4 days	In captivity "very assertive," backs forcefully into male, rubs genitalia, "fluttering" vocalizations	Labial tumescence (captivity)			"Unobtrusive" approaches; may touch female with hand				Unimale (?)	Nadler 1975
Humans	28	2–8	Throughout cycle, but may peak at midcycle	Highly variable; often includes eye contact	Labial tumescence and flushing of vulva during intercourse, may be permanent darkening of sexual skin with parity	Aliphatic acids peak midcycle in some females		Very variable; involves touching, holding	Single (2–10 min)	2.5/week	Common in first and second trimester, diminishes last trimester	Unimale and monogamous; very rarely polyandrous	Adams, Gold, and Burt 1978; Masters and Johnson 1966; Eibl-Eibesfeldt 1975; Kinsey et al. 1953; Michael, Bonsall, and Warner 1975a, 1975b; Stern and Leiblum 1984; Udry and Morris 1968

SOURCES: Butler 1974, Michael and Zumpe 1971, and Dixson 1983 provide general reviews of sexual signaling in nonhuman primates. In addition, see sources in last column.

NOTES: This summary is intended to provide some sense of the range and magnitude of interspecific differences. Readers are cautioned, however, against using any entry in the table as if it represented a "species norm." A tabular format was chosen for brevity and convenience, but it cannot deal adequately with situational and interindividual variation within a species nor resolve problems generated by data drawn from diverse sources. Plus (+) indicates swelling only apparent in some females; asterisk (*) indicates species with conspicuous swelling easily visible at a distance. Species are classified according to whether males ejaculate after a single mount with intromission and thrusting or only after multiple mounts.

altogether (Gould, Flint, and Graham 1981; Graham, Kling, and Steiner 1979; Small 1984). Patterns of vaginal bleeding and serum hormone profiles of macaques in the third decade of life are similar to those described for peri- and postmenopausal women (Hodgen et al. 1977).

Such findings raise new questions. Provided they live long enough, do nonhuman female primates experience menopause? If so, is this nonreproductive period in any sense functional (Hrdy 1981a)? Mayer (1982) has argued that mothers who terminate reproduction prior to death and direct their attention instead toward investment in extant descendants leave more offspring on average than do mothers who breed right up until death and then leave an orphan. Others stress the steady decline in fertility (Small 1981) and regard menopause as a by-product of physical degeneration (Jolly 1985).

SIGNALING RECEPTIVITY
Olfactory Cues

Nonhuman primate females indicate readiness to mate through a variety of olfactory, morphological, and visual signals. Of these different types of signals, olfactory cues have received the most systematic investigation. Male primates frequently approach and smell the perineal region of a female and may even touch her vagina, collecting secretions on the finger to sniff or taste. Such behavior suggests that primates, like other mammals, use smell to monitor female reproductive state. Some progress has been made toward identifying pheromones, the specific hormone-related substances that males find attractive. In the case of macaques (and perhaps also humans: Michael, Bonsall, and Warner, 1975a, 1975b), the pheromones appear to be volatile fatty acids produced by bacteria that grow in the vaginal mucus secreted in response to rising estrogen levels at midcycle (Keverne 1976). When mixtures containing these aliphatic acids from the vaginas of intact females were applied to the sexual skin of ovariectomized females, the number of male mating attempts increased (Keverne 1976).

As Keverne points out, however, both the compositions of the secretions and their effectiveness in promoting sexual behavior are highly variable. Whether or not copulations occur depends on social circumstances and on which individuals are involved, since both males and females show strong partner preferences. For example, in one study, each of two ovariectomized macaques was given the same dose of estradiol, a hormone that presumably altered smell as well as sexual skin color. Invariably, males preferred one female over the other (Everitt and Herbert 1969). The male's attraction to a particular female may be so strong that she is preferred over another female, even when the preferred female is not in estrus and the other female is stimulated by estrogen treatment (Herbert 1970). Furthermore, macaque males continue to copulate and to demonstrate cyclical fluctua-

tions in rates of copulation after their ability to detect odors is experimentally blocked (Goldfoot et al. 1978, 1979). Thus, it appears that olfactory cues are supplemented by other signals that strongly influence sexual preferences (Goldfoot 1981).

Vocalizations

At present there is little evidence to indicate that vocalizations play a major role in signaling readiness to mate. It has been suggested that orangutan males may use the long call to locate females, but the data are equivocal (Mitani 1985d). Among gelada baboons, females emit solicitation calls throughout the cycle, but they become more frequent just prior to ovulation (Dunbar 1984b). However, in many primate species mating itself is accompanied by a wide variety of grunts, screams, and chatters, as well as more elaborate calls emitted by both sexes (table 30-1). In a comparative study of three species, Hamilton and Arrowood (1978) found that calls given toward the end of copulation by promiscuously mating savanna baboons were structurally more complex than those given by either monogamous pairs of gibbons or by human couples. Among baboons there were also interesting differences depending upon the identity of the copulating partners. Juvenile male baboons mated silently, and female calls were typically shorter and less complex when they were mating with subadult males than when they were mating with adult males. Adult males almost always gave long, low-frequency vocalizations while copulating. Although female baboons are more likely to give copulatory calls around the time of ovulation, in species with postconception estrus there may be no difference in the frequency of calling by cycling versus pregnant copulating females (e.g., crab-eating macaques: Deputte and Goustard 1980).

Behavioral Solicitations

In many primates, readiness to mate is indicated by a variety of distinctive postures, gestures, and facial expressions. Female sexual solicitations include crouching behavior (resembling lordosis) in lemurs and marmosets, flicks of the tongue among tamarins and howler monkeys, puckering of the lips in capuchins and patas, shuddering of the head in langurs (fig. 30-1), and displays of the clitoris in squirrel monkeys (selectively reviewed in table 30-1). In many species, females combine such behaviors with presentation of the rump (fig. 30-2). Male sexual solicitations also vary widely between species. For example, male chimpanzees often invite females to mate by shaking branches and raising an arm, as if beckoning the female to approach. During the invitation the male typically spreads his legs to reveal a bright red, erect penis that stands out against the black scrotum (Tutin and McGinnis 1981). Male baboons and rhesus monkeys may invite an estrous female to mate by lip smacking and

FIGURE 30-1. A langur female solicits a male by presenting and frenetically shuddering her head. There is no conspicuous visual signal at ovulation. The possibility of olfactory cues has not yet been examined. (Photo: Daniel Hrdy/Anthrophoto)

FIGURE 30-2. A female hamadryas baboon solicits her male by presenting her sexual swelling. (Photo: Hans Kummer)

making a friendly face (fig. 30-3) or simply by giving a sitting female a gentle shove to make her stand up (Ransom 1981).

Morphological Cues and the Function of Sexual Swellings

In many nonhuman primates, female morphology and coloration vary with different phases of the menstrual cycle (cataloged in table 30-1). In some species, females exhibit a slight—often scarcely detectable—swelling or pinkening of the vulva at midcycle. In other species, however, morphological changes are very conspicuous, even at a distance. These changes include reddening of the face and/or perineum (e.g., rhesus macaques, mandrills), swelling of small blisters on the chest (gelada baboons), and large, perineal swellings ranging in color from bright pink to deep burgundy. At least 24 species in diverse genera (marked by an asterisk in table 30-1) advertise sexual receptivity with perineal swellings.

Based on the current distribution of midcycle morphological changes among catarrhines (Dixson 1983; table 30-1), it seems likely that sexual swellings have evolved independently at least three times: among the cercopithecines, the colobines, and the great apes. Figure 30-4 shows variation in the perineal sexual swellings of several Old World monkeys and one ape.

Male sexual interest seems to be related to cyclical changes in coloration and sexual swellings. In the case of the well-studied savanna baboon, the "sexual skin" around the vulva begins to swell during or just after menstruation. The size of the swelling increases gradually for 2 to 3 weeks, reaching peak size a few days before ovulation. Immediately after ovulation, the swelling rapidly decreases in size and subsides into the "flat" phase of the cycle. Thus in baboons, as in other species with sexual swellings, the female exhibits a dramatic visual indicator of sexual receptivity many days before ovulation occurs.

Early in the cycle, as the skin begins to swell, low-ranking juvenile and subadult males mate with the female. Later in the cycle, as the size of the swelling increases and the female nears ovulation, fully adult males actively compete to monopolize mating opportunities by forming exclusive "consort" relationships with females (Hall and DeVore 1965; Hausfater 1975). In a typical cycle, consortships with adult males begin about a week before ovulation and continue until the swelling begins to rapidly decrease in size (detumescence). Several studies have shown that the males that are most successful in monopolizing access to estrous females consort most often about 2 to 3 days before detumescence begins (Hausfater 1975; Packer 1979b; Smuts 1985), the time when ovulation is most likely to occur, according to laboratory studies (Hendrickx and Kraemer 1969; Wildt et al. 1977; Shaikh et al. 1982). However, despite the obvious importance of sexual swellings in attracting males, this cue, like olfactory cues in macaques, is not the only determinant of female attractiveness. Males show strong partner preferences (Packer 1979b; Rasmussen 1980, 1983b; Collins 1981; Smuts 1985), and there is no simple relationship between the maximum size of a female's sexual swelling and her attractiveness (Scott 1984).

Most of the species that show prominent midcycle sexual swellings live in multimale groups (Clutton-Brock

FIGURE 30-3. (*Left*) A male rhesus monkey waiting to make eye contact with an estrous female. (*Right*) As soon as he catches her eye, he invites her to mate by thrusting out his chin and puckering his lips. (Photos: Barbara Smuts/Anthrophoto)

and Harvey 1976). The association between a multimale breeding system and sexual swellings is particularly striking in the family Colobinae (chap. 8). Of 24 species in the subfamily, only 3 are characterized by breeding systems in which troops typically contain several adult males, and these are also the only 3 in which females exhibit sexual swellings. Similarly, in the superfamily Hominoidea, only common chimpanzees and pygmy chimpanzees live in multimale groups, and only these species exhibit prominent sexual swellings. However, many species that live in multimale groups do not have sexual swellings, including several species of lemurs and New World monkeys. Even within the subfamily in which sexual swellings are most frequent, the cercopithecines, some multimale species, such as vervet monkeys and some macaques, do not show sexual swellings.

Macaques exhibit the greatest variations in sexual skin of all primate genera. Several macaque species, including the Barbary macaque and pigtailed macaque, exhibit pronounced swellings, much like those of savanna baboons, while other macaques, including the Japanese macaque and rhesus macaque, show only minimal swelling of the anogenital region and highly variable color changes in the face and perineum (Dixson 1983). Stumptailed macaques copulate through the cycle and exhibit no external sign of ovulation (Nieuwenhuijsen 1985).

There are no convincing explanations for the distribution of sexual swellings among nonhuman primates. The intriguing association between swellings and multi-

male breeding systems has, however, led to at least three different hypotheses to explain why females might benefit from sexual swellings. The first explanation, the "best male" hypothesis, argues that swellings function to incite competition among troop males for sexual access to the female, thus increasing the chances that the "best" male will be the one that fathers the female's offspring (Clutton-Brock and Harvey 1976). The second explanation, the "many male" hypothesis, argues that sexual swellings function to increase a female's opportunities to mate with several different males (Hrdy 1981b). If male treatment of infants is related to the male's previous mating activity, then by mating with many males, a female may be able to increase the total amount of male care her infant receives (Taub 1980b) or decrease the chances that her infant will be a victim of male infanticide (Hrdy 1979). Hamilton (1984) has recently proposed a third hypothesis. He argues that sexual swellings serve to pinpoint the timing of ovulation, thereby increasing paternity confidence and paternal investment by one or two males. In contrast to the "many male" hypothesis where paternity is obscured, Hamilton's "obvious ovulation" hypothesis assumes that sexual swellings function to increase the information available to males about paternity.

Information on the relationship between prior mating activity and male-infant interactions does not allow rejection of any of these explanations (chap. 28). Furthermore, some observers assume that sexual swellings obscure ovulation, while others assume just the opposite.

The hypotheses discussed above cannot be evaluated until it is clear which assumption is correct; it is likely that this will require experimental studies.

SITUATION-DEPENDENT RECEPTIVITY

Situation-dependent receptivity has been reported most often among species in which female receptivity is indicated primarily by changes in female behavior, rather than by conspicuous changes in coloration or the presence of sexual swellings. These species tend to live either in monogamous or polyandrous units (e.g., humans, tamarins, see chap. 4) or in groups in which one male typically does most of the mating (e.g., gray langurs, red howler monkeys, gelada baboons, and patas monkeys). While humans and tamarins are receptive throughout the cycle with peaks in sexual behavior during the early phases of pair-bonding, gray langurs, red howlers, gelada, and patas typically exhibit behavioral signs of receptivity cyclically, only shifting to situation-dependent receptivity under certain circumstances.

The most frequently reported circumstance precipitating receptivity by females who are not in the middle phase of the cycle (determined in the field by monitoring cyclical changes in receptivity over several months) is an encounter with strange or at least relatively unfamiliar males. In such cases, females not previously showing estrous behavior (including pregnant females) may begin to exhibit proceptivity and receptivity within days after new males invade their troop (captive patas: Loy 1975; gray langurs: Hrdy 1977; gelada baboons: A. Mori 1979c;

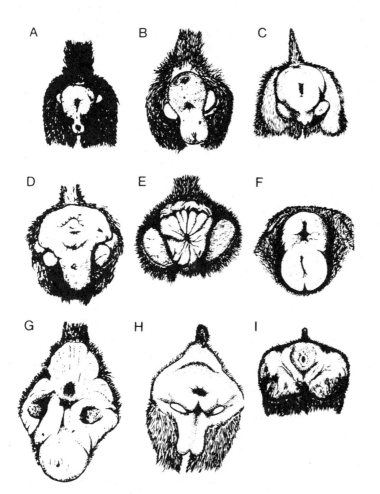

FIGURE 30-4. Conspicuous coloration and/or swelling at midcycle evolved in the lines leading to such cercopithecine genera as *Macaca, Papio, Cercocebus, and Theropithecus* (itemized in table 30-1), in the ancestors of *Colobus badius* and *Colobus verus,* and again in the line leading to *Pan troglodytes* and *Pan paniscus.* As shown in these drawings, the pattern of swellings and the precise tissues implicated vary enormously. (*a*) *Cercocebus albigena* (from Chalmers 1980); (*b*) *Cercocebus atys; (c) Papio sphinx; (d) Papio cynocephalus ursinus* (from a photograph by Dr.Craig Bielert); (*e*) *Colobus verus* (from Hill 1952); (*f*) *Pan troglodytes; (g) Cercopithecus talapoin; (h) Macaca nemestrina; (i) Macaca nigra.* The drawings are not to the same scale. SOURCE: Dixson 1983, fig. 2, with permission.

wild redtail monkeys: Cords 1984a; chap. 9). In the wild, such periods of receptivity are often sporadic, with little evidence of the cyclicity shown at other times. Sekulic (1982a) reports the extreme case of a lone red howler female who was receptive whenever she encountered males from nearby troops.

Situation-dependent receptivity has also been reported in captivity among species that are not known to exhibit it in the wild (e.g., chimpanzees: Lemmon and Allen 1978). Colony-dwelling female Japanese macaques and stump-tailed macaques sometimes exhibit receptivity in "runs" that last for several weeks and that cannot be predicted according to the phase of the menstrual cycle (Wolfe 1984b; Nieuwenhuijsen 1985). Similar runs have been reported for wild vervets (Struhsaker 1967b; Andelman 1985), wild redtails (Cords 1984a), captive blue monkeys (Rowell 1974a), and wild long-tailed macaques (van Noordwijk, in press).

Little is known about the physiological mechanisms underlying noncyclical receptivity or the social and environmental features that affect its occurrence. In an experiment with rhesus macaques, Wallen (1982) found that the sexual behavior of 15 male-female pairs was more cyclical in large compounds than in small cages, but it is not known why this was the case. Similarly, the relationship between postconception estrus (which is often cyclical) and situation-dependent receptivity (which is, by definition, not cyclical) is not clear.

Nevertheless, even at this early stage of knowledge it may be useful to speculate about the functional significance of situation-dependent receptivity. In species with multimale breeding systems, females compress their sexual activity into discrete, well-advertised intervals during which they are able to mate with many or most males in the troop (e.g., savanna baboons, Barbary macaques, common chimpanzees, and other species starred in table 30-1). In polygynous species without conspicuously advertized receptivity, females appear to be trading the efficiency of discrete estrous intervals for greater flexibility in the timing of copulations, thus permitting females to mate opportunistically with males encountered occasionally. In both cases, however, the outcome for the female is essentially the same: she has the opportunity to mate with a range of partners. Whether this is best explained by the "best male" or the "many male" hypothesis or, indeed, by some completely different hypothesis remains to be determined through systematic attempts to evaluate alternative explanations.

PATTERNING OF MALE SEXUALITY
Behavior

Male primates do not have regular cycles of sexual behavior, but males may track patterns of female sexual receptivity. Among seasonally breeding rhesus macaques, for example, males only produce active sperm during the months when females are actually fertile. However, if paired with ovariectomized females artificially brought into estrus out of season, males resume sperm production (Conaway and Sade 1965). Female cyclicity and sexual activity also influence a variety of male behaviors, such as the frequency of intergroup transfer (chap. 21) and the intensity of male-male aggression (chap. 25).

Several aspects of male sexuality vary between species. For example, many primates, including humans, ejaculate after a single intromission. In contrast, in pigtailed macaques and roughly a dozen other species (see table 30-1), males typically mount, thrust, and dismount several times before an ejaculatory mount occurs. The adaptive significance of this behavior is not known, and it appears, at first glance, to be maladaptive, given male-male competition for mates, or else somehow imposed upon males by the evolution of female attributes that require multiple mounts for insemination to occur. Among pigtailed macaques, however, male copulatory behavior is readily modified by the social context. Under normal conditions, the average male copulatory series covers 18.5 minutes and includes 8 intromissions with a 43-minute interval between ejaculations. But if the male sees another male mount his partner, his postejaculatory interval can be reduced by as much as 60% of the normal interval so that the male mates again within just 17.5 minutes (Busse and Estep 1984).

This study and others suggest that some aspects of male sexuality are more responsive to social circumstances and less constrained by physiological limitations than previously thought. For example, in some species, such as gorillas, male libido seems generally low, and females initiate most copulations (Fossey 1983), whereas in other species, such as chimpanzees, male sexual motivation generally seems high, and most copulations occur after the male solicits the female (Tutin 1980). However, when extra-group males follow a gorilla group containing an estrous female, the normally passive silverback takes a much more active role in initiating copulations with his female (Fossey 1983). An experiment with captive rhesus monkeys provides another interesting example. In general, old rhesus males copulate much less often than young adult males in controlled tests with ovariectomized females receiving hormone treatment; researchers have attributed this difference to a decline in the physical capacity for frequent copulations with age. However, when old and young males were placed with a favorite female partner, there was no difference in the copulation rates of the two groups (Chambers and Phoenix 1984). Older males were not less capable of frequent copulations; they were just more discriminating about which females motivated peak performance. Copulation frequency and many other aspects of male sexuality have not received detailed study, and this is a promising area for future research.

Morphology

One aspect of male morphology—relative testes size—has been investigated systematically (Popp 1978; Collins 1978; Harcourt et al. 1981). Species living in multimale breeding systems generally have heavier testes in relation to body weight than either monogamous species or species living in one-male breeding units (Harcourt et al. 1981); a similar pattern has recently been reported for ruminants (Hogg 1984). Harcourt and his coauthors speculate that this correlation is due to the importance of sperm competition in species in which a female often mates sequentially with different males, such as chimpanzees or savanna baboons, in which the testes weigh .27% and .26% of body weight, respectively. In contrast, monogamous primates and species living in groups in which, typically, only one male breeds, have relatively small testes (e.g., .10% and .02% of body weight for gibbons and gorillas, respectively). Initially, the most glaring exception to this pattern was the supposedly monogamous cottontop tamarin whose testes size (.65% of body weight) was more consistent with a multi-male breeding structure. The recent finding that tamarins are sometimes polyandrous (chap. 4) has resolved this inconsistency.

PATTERNING OF HUMAN FEMALE SEXUAL ACTIVITY

Anthropologists have long been concerned with explaining why human females are continuously sexually receptive and why ovulation is concealed rather than advertized—indeed, concealed to such a degree that most women do not seem to know when they are ovulating. Over a dozen different hypotheses have been proposed (e.g., Morris 1967; Benshoof and Thornhill 1979; Symons 1979; Lovejoy 1981; Alexander and Noonan 1979; Strassman 1981; Turke 1984), but none has involved a systematic review of the patterning of sexual activity among nonhuman primates.

A comparative approach that incorporates data on nonhuman primates can make an important contribution to this debate. Most discussions have assumed a dichotomy between humans and other primates: humans are viewed as continuously receptive, while all other primates are assumed to confine sexual activity to discrete periods around the time of ovulation. This dichotomy is an oversimplification. On the one hand, human females

are clearly not receptive all the time; most human females probably mate only a few times a week at most, and in many human societies, women are expected to forego copulation for many months after giving birth. On the other hand, the information reviewed above and summarized in table 30-1 shows that many nonhuman primate females exhibit sexual receptivity outside the brief period right around ovulation. Thus, rather than asking, Why have humans lost estrus? we might better ask, Under what conditions do primates generally shift away from circumscribed to noncyclical or situation-dependent receptivity, and how do these changes contribute to female reproductive success? Attempts to answer this question are likely to yield general principles that will help explain the patterning of sexual activity to a wide range of species, including humans.

SUMMARY

In most mammals, females are sexually receptive during discrete intervals known as estrus, which are signaled by changes in behavior, smells, and morphology. Among apes and Old World monkeys, however, there has been a general trend toward more flexible patterns of female receptivity. Females typically exhibit peaks of sexual activity around the middle of the menstrual cycle when ovulation occurs, but in many species females show a capacity to solicit males and to copulate at other times as well. Females solicit male sexual attention through a wide variety of visual, olfactory, and behavioral cues; the importance of different types of cues varies among different species. Males also show great variability in patterns of sexual behavior and sexual morphology, including interspecific differences in relative testes size that appear to be related to the breeding system. In general, females living in multimale breeding systems advertize receptivity with morphological changes, and they tend to mate over an extended, but discrete, period of time that brackets ovulation. Females in monogamous and one-male breeding systems, in contrast, rarely possess conspicuous morphological indicators of receptivity, and they tend to exhibit greater flexibility in the timing of sexual activity. There exist, however, notable exceptions to all of these generalizations, and systematic assessment of the different hypotheses to explain variations in the patterning of female sexual activity remains a task for the future.

31 | Sexual Competition and Mate Choice

Barbara B. Smuts

Across a wide variety of primates, certain patterns recur: males fight over females, females selectively respond to male sexual solicitations, and strong preferences for particular mating partners often emerge in both sexes. This chapter explores these patterns in nonhuman primates within the framework of sexual selection theory.

Darwin (1871) developed the theory of sexual selection to explain the evolution of traits that did not appear to contribute to an individual's survival, traits such as the enlarged canines of male baboons or the songs of male birds. He argued that these traits evolved because they increased an individual's mating success through either intrasexual competition for mates or intersexual choice of mates. Darwin pointed out that this usually involved males competing for females and females choosing some males over others as mates. These patterns have since been explained in terms of differential parental investment by the two sexes (Trivers 1972). In most species, females invest more in offspring than do males, a tendency that follows from the much greater initial energy expenditure required to produce eggs compared with sperm (Bateman 1948). Because females invest more in each offspring, they function as a limiting resource for male reproductive success. This means that males can normally increase the number of offspring they produce by mating with additional fertile females. Females, in contrast, generally cannot increase the number of offspring they produce by mating with many males. Instead, females are expected to discriminate among possible mates by choosing to mate with those males that contribute most, either behaviorally or genetically, to the female's reproductive success (Trivers 1972).

This chapter focuses on these two primary components of sexual selection: male-male competition for mates and female mate choice. At the end, it considers complementary behaviors that have received less study, female-female competition for mates and male mate choice, and concludes that in primates all four processes are important. The data show that there is substantially more flexibility, variability, and complexity in primate mating tactics than might be expected on the basis of orthodox sexual selection theory.

In what follows, behaviors that appear to increase male mating opportunities at the expense of other males are considered examples of male mating competition.

However, it is important to point out that this convention frequently involves two assumptions. First, it often ignores the issue of proximate causation. For example, dominance status is frequently correlated positively with mating access, so we tend to consider competition for status a form of intrasexual mate competition. Sometimes, however, aggression may be a response to proximate stimuli such as food or close proximity of another male, rather than the presence of an estrous female. Second, in many cases, the reproductive consequences of behaviors that are assumed to reflect intrasexual mate competition have not been conclusively demonstrated. In some instances, the function of male-male competition is probably not increased access to mates but something else, such as increased resources for a mate (e.g., gibbon male territorial aggression) or increased survival of own offspring (e.g., some observations of males protecting infants against aggression by other males). Thus, the functional significance of many of the patterns described below is not certain. Similar cautions apply to the subsequent discussion of female choice. First, behaviors that appear to express sexual preferences, such as the frequency of presenting by estrous females to males, may sometimes be a response to "nonsexual" proximate stimuli, such as the presence of an unfamiliar male that arouses fear. Second, in all cases, the links between female mating preferences and evolutionary consequences remain to be demonstrated.

EFFECTS OF THE SOCIAL SYSTEM ON MALE-MALE COMPETITION FOR MATES

Gross patterns of male-male competition differ according to the species' social system (chap. 23). These patterns are reviewed briefly in order to provide a general background for the detailed discussion of particular behaviors that follows.

In the solitary nocturnal prosimians, females disperse into small territories that they defend against other females. In these species and in the solitary orangutan, males compete for females primarily by aggressively excluding rivals from an area that includes the territories of several females (chaps. 2, 13). Typically, the largest males control the most space and mate with the most females (Charles-Dominique 1977a; Galdikas 1981).

Little is known about how males in monogamous species acquire a mate, but it may sometimes involve acquisition of a high-quality territory that will attract a female (chap. 12). Once a mated pair is established in a territory, males attempt to exclude other males through displays and chases at territory boundaries (chaps. 12, 5).

Species in which two or more adult females live together can be divided into two broad categories: those in which there is typically only one breeding male in the group at a time (e.g., hamadryas baboons, many forest-dwelling cercopithecines, some populations of gray langurs) and those in which there are two or more breeding males (e.g., ring-tailed lemurs, squirrel monkeys, macaques, chimpanzees). The competitive tactics of males in these two types of species are often quite different: in one-male species, males usually compete intensively at intervals for the breeding position within the group, whereas in multimale species, males typically compete within the group for dominance rank, which often mediates access to fertile females. However, as indicated in chapter 9, the sharp distinction between the two types of social system and the associated male competitive tactics is often blurred. For example, in some "one-male" species, several males may invade the group and mate with the females (e.g., redtails and patas monkeys: chap. 9), while in others, a second, subordinate male may share breeding rights with the dominant male for years (e.g., gorillas: Fossey 1983; gelada baboons: Dunbar 1984b). On the other hand, in some multimale groups, the dominant male is so successful at monopolizing matings that the group has, in essence, only one breeding male (e.g., brown capuchins: Janson 1984; mantled howlers: Clarke 1983). Despite these complications, in many cases it is possible to distinguish between a strategy aimed at monopolizing sexual access to an entire group of females and one that simply achieves increased access to a proportion of the group's females at optimal times for fertilization.

TAKING RISKS FOR SEX: MALE FIGHTING AND MOBILITY

In many primates, fighting with other males and moving long distances in search of mating opportunities are two of the most obvious ways in which males compete for females. These tactics are often dangerous, but in primates, as in many other animals (Trivers 1972), males will apparently risk high costs when reproductive benefits are also high. This may be why from puberty on male primates often experience significantly higher mortality than do females (chap. 20). This appears to be true even for humans (Daly and Wilson 1983).

Fighting

In most species, the frequency and intensity of male-male aggression increases when estrous females are

present (sifakas: Richard 1974b; ring-tailed lemurs: Jolly 1966a; red howlers: Sekulic 1983a; brown capuchins: Janson 1984; squirrel monkeys: Baldwin 1968; gray langurs: Boggess 1980; Mohnot 1984; vervets: Henzi and Lucas 1980; rhesus macaques: Vandenbergh and Vessey 1968; Wilson and Boelkins 1970; savanna baboons: Bercovitch 1983; orangutans: Galdikas 1985; chimpanzees: Nishida 1983b). Male aggression over females often produces injuries. For example, in one group of redtails, fights over females were common and "often resulted in severe wounds" (Cords 1984a, 319). Of 16 male gelada baboons observed fighting over control of female groups, 4 suffered severe injuries (Dunbar 1984b). Similarly, in a group of ring-tailed lemurs, by the end of the breeding season all 5 males had wounds 4 to 5 cm long on thighs, arms, or sides (Jolly 1966a). In one population of wild bonnet macaques, "fights resulting in severe injuries among males are almost an everyday affair" (Singh, Akram, and Pirta 1984, 15).

Although male-male aggression over females is therefore common in many primates, its frequency varies both within (Nagel and Kummer 1974) and between species (table 31-1; Janson 1984). Berenstain and Wade (1983) reviewed data indicating that male savanna baboons tend to fight over estrous females more often than do male macaques (figs. 31-1, 31-2). They suggested

a.

FIGURE 31-1. Adult male olive baboons fighting. Males fight more often when competing for estrous females, and injuries from the long, sharp canines are not uncommon. (Photo: Irven DeVore/Anthrophoto)

c.

FIGURE 31-2. Although male macaques do not fight over estrous females as often as savanna baboon males do, high-ranking macaque males frequently disrupt the consortships of lower-ranking males by approaching mating couples. (*a*) The third-ranking male sits close to his consort partner during a mount series; (*b*) he mounts the female; (*c*) the second-ranking male immediately approaches, and the consorting male is forced to move away. If the female follows him (as she did in this case), mating will resume after the rival leaves the vicinity. (Photos: Barbara Smuts/Anthrophoto)

b.

two reasons for this difference. First, mating competition in baboons may be more intense because fewer estrous females are available simultaneously since baboons, in contrast to most macaques, breed throughout the year (chap. 11). Second, less sexual dimorphism in body size in macaques may result in increased female influence over consort partner selection, which, in turn, could reduce the probability that the winner of a male-male fight will gain access to a contested female. Janson (1984) presented a similar argument for the relatively low rates of male-male aggression in a group of brown capuchins. Neither hypothesis, however, explains high rates of male-male aggression among ring-tailed lemurs (table 31-1), which are characterized by seasonal breeding and female dominance over males (chap. 3). Systematic investigation of factors affecting the rate and intensity of male-male competition over females remains to be done.

Researchers usually observe injurious fights among males more often than among females, and in a wide variety of species, wounds are more common among males (table 32-2). The important role of physical combat in male-male competition for mates is the most common evolutionary explanation for the greater body size and larger canines of males in many primate species (Clutton-Brock, Harvey, and Rudder 1977; Harvey, Kavanagh, and Clutton-Brock 1978b; chap. 19). However, we do not know why sexual dimorphism in body size and weaponry is much greater in some species than others (Fedigan 1982; Gaulin and Sailer 1984).

Mobility

In many species, males travel farther than females, often apparently in order to obtain or protect access to mates. For example, among Japanese macaques, as many as half of the adult males travel for years as solitaries, moving from troop to troop during the breeding season in search of mating opportunities (Nishida 1966). In orangutans (chap. 14) and some solitary prosimians, such as lesser bushbabies (chap. 2), dominant males range over a much larger area than individual females. There is no evidence

that males need to travel this far in order to obtain sufficient food, so observers have suggested that males range widely in order to control sexual access to several females. Although data are scarce, it is likely that mobility increases male mortality through increased exposure to disease, predation, and aggressive conspecifics (chap. 21).

MALE DOMINANCE

In most groups containing more than one adult male, some males consistently win agonistic interactions against others; the males can be ranked in a linear dominance hierarchy (chap. 25). The relationship between male dominance and male mating activity has probably received more attention than any other aspect of primate social behavior. The results are reviewed by Robinson (1982a), Dewsbury (1982a), Fedigan (1983), Berenstain and Wade (1983), and chapters 25 and 26. A large number of studies indicate that high rank is often associated with increased mating activity. For example, in rhesus monkeys, Chapais (1983e) reported a high correlation ($r_s = .79$) between male dominance rank and time spent consorting with estrous females near the time of ovulation ($N = 10$ to 17 males; the number varied over the mating season). The importance of high rank is particularly clearly demonstrated by studies showing that an increase or decrease in male rank is quickly followed by a dramatic increase or decrease in mating activity (see fig. 31-3; also chap. 15 and Janson 1984).

Dominance status appears to influence mating activity in two major ways. First, in some groups, dominant males can appropriate estrous females with little or no resistance from subordinate males, presumably because subordinates have learned that they will probably lose any fight that might occur. Examples include chimpanzees (chap. 15), rhesus macaques (Chapais 1983e), and brown capuchins (Janson 1984). Second, in groups in which subordinate males attempt to mate, dominant males often interfere through persistent shadowing, chasing, or aggressive disruption of copulations (e.g., rhesus

TABLE 31-1 Interspecific Variation in Frequencies of Male-Male Aggression over Females

Species	Number of Dyadic Aggressive Encounters per Male-Male Pair/hr	Time Period	Sources
Brown capuchins	0.0007	Days when at least one estrous female was present	Janson 1984
Japanese macaques	0.018	Breeding season	Enomoto 1981
Savanna baboons	0.240	Days when at least one estrous female was present	Hausfater 1975
Ring-tailed lemurs	0.385	Breeding season	Jolly 1967

NOTE: All values derived from calculations performed by Janson 1984.

macaques: Lindburg 1973; Japanese macaques: Hanby, Robertson, and Phoenix 1971; Takahata 1982b; orangutans: Mitani 1985b; Galdikas 1985; savanna baboons: Hall and DeVore 1965; gray langurs: Boggess 1980). (It should be noted that subordinate males sometimes also interfere with the copulations of more dominant males; e.g., savanna baboons: Smuts 1985; gray-cheeked mangabeys: Wallis 1983; ring-tailed lemurs: Jolly 1966a).

Thus, it is clear that in some groups at some times, achieving high rank represents an effective means of competing for females. However, a number of studies have failed to find a positive relationship between male dominance and mating activity. Among ring-tailed lemurs and sifakas, for example, the male dominance hierarchy established outside of the breeding season apparently has no effect on mating success during the breeding season (Jolly 1966a; Richard 1974b). Similarly, middle- and low-ranking male vervet monkeys in Amboseli copulated as often as high-ranking males during the week when conception was most likely to occur (Andelman 1985).

Repeated studies of the same species, such as baboons or macaques, clearly show that the relationship between male dominance rank and mating success varies within species and even within groups. For example, although genetic paternity exclusions for one group of captive rhesus macaques showed that the most dominant male did not father the most offspring (Duvall, Bernstein, and Gordon 1976), a similar study in a different group of captive rhesus produced the opposite result (Smith 1981).

Similarly, while a number of studies have reported a significant positive correlation between male rank and consort activity in savanna baboons (Hausfater 1975; Popp 1978; Packer 1979a, 1979b; Rasmussen 1980; Collins 1981; Sapolsky 1983), several other studies using similar methods have found no such correlation (Saayman 1971a; Strum 1982; Manzolillo 1982; Smuts 1983b, 1985). Furthermore, even in groups that show a positive relationship between rank and mating success, important exceptions frequently occur. Packer (1979b), for example, found that although rank and consort activity were highly correlated among Gombe baboons, old males past their physical prime usually consorted at higher rates than expected based on their dominance ranks. Several other baboon studies report similar, unexpectedly high consort activity by older males (DeVore 1965a; Saayman 1971a; Rasmussen 1980; Collins 1981; Strum 1982; Manzolillo 1982; Smuts 1985).

Thus, dominance rank alone clearly does not determine male mating activity in multimale groups. Three additional factors influencing male access to females are discussed below: male-male coalitions, alternative competitive tactics, and female choice.

MALE-MALE COALITIONS
Types of Coalitions

Male-male coalitions are rare in solitary and monogamous primates. Observations of gibbons suggest a possible important exception: fathers have been seen help-

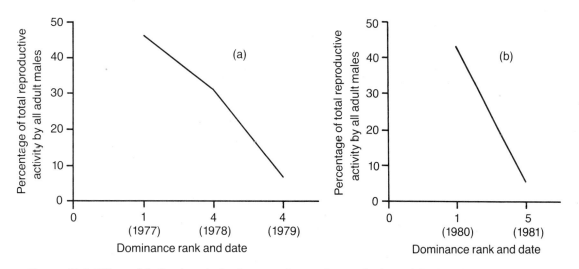

FIGURE 31-3. Effects of decline in male dominance rank on male reproductive activity. (*a*) The reproductive activity of a captive male bonnet macaque is plotted against his dominance rank at three different times. There were 7 adult males in the group in 1977–78 and 8 in 1979. Reproductive activity was based on all observed copulations with females that had not yet conceived (from Samuels, Silk, and Rodman 1984; male M99). (*b*) The reproductive activity of a wild savanna baboon male is plotted against his dominance rank at two different times. There were 14 adult males in the group. Reproductive activity was based on frequency of copulations and time spent in sexual consortships; both measures were weighted by the probability of conception on a given day of the female estrous cycle and then combined (from Sapolsky 1983; male 257).

ing their sons compete against males from neighboring groups in order to establish a territory (chap. 12). Male-male coalitions occur in many primates living in multi-female groups. These coalitions fall into four main types, depending upon their effects:

1. Coalitions used to expel a breeding male from a one-male group and take over the group. This has been reported for gray langurs (Hrdy 1977; Moore 1985), but it appears to be uncommon in other species living in one-male groups (e.g., gelada baboons: Dunbar 1984b). In langurs, one of the allied males expels all of the others after the takeover occurs (chap. 21).

2. Coalitions used to repel outside males. This has been reported for species living in "one-male" groups, in which the coalition involves only two males (e.g., go-rillas: chap. 14; hamadryas: chap. 10; gelada baboons: Dunbar 1984b). It has also been observed in species living in multimale groups in which the coalition in-volves several males (e.g., chimpanzees: chap. 15; red colobus monkeys: chap. 8; spider monkeys: Fedigan and Baxter 1984). Although male savanna baboons occasion-ally jointly threaten extragroup males (Packer 1979a), male-male coalitions against extragroup males seem to occur mainly in species in which males typically remain in their natal groups.

3. Coalitions used to increase dominance rank. Since such coalitions, by definition, must involve at least three males (two allies and an opponent), they can occur only in multimale groups. They have been reported in a variety of macaques (Japanese macaques: Koyama 1967; rhesus macaques: Chapais 1983d; Tilford 1982; captive Barbary macaques: Witt, Schmidt, and Schmidt 1981; captive stump-tailed macaques: Bernstein 1980), mantled howl-ers (Jones 1980b), red howlers (Sekulic 1983a), chim-panzees (chap. 15), and captive gray langurs (Curtin 1982). The absence of such coalitions has been noted in wild multimale bisexual groups of gray langurs (Bog-gess 1980; but see Moore 1985 for coalitions in all-male bands), white-throated capuchins (Oppenheimer 1968), brown capuchins (Janson 1984), and ring-tailed lemurs (Jolly 1966a). Although it is possible that males do not form coalitions in some species that live in multimale groups, it seems more likely that males in all of these species have the potential to form coalitions and that cir-cumstances determine their relative frequency. Kinship, for example, appears to influence coalition formation. In free-living macaques, coalitions usually involve matri-lineal relatives (Meikle and Vessey 1981; Chapais 1983d; Tilford 1982), and in the red howlers mentioned above, the coalition partners might have been father and son (Sekulic 1983a).

In macaques and chimpanzees, male-male coalitions sometimes influence dominance relationships, which in turn affect mating activity, and in howlers, males use coalitions to force other males to the periphery or out of the group. In none of these species, however, have males been observed using coalitions to compete directly for estrous females.

4. Coalitions used to acquire estrous females from other males. In this case, males cooperate to compete di-rectly for females, but such coalitions do not appear to affect dyadic dominance relationships. This pattern is common in savanna baboons (Packer 1977; Smuts 1985; see later section "Alternative Competitive Strategies") and perhaps also in captive sooty mangabeys (Bernstein, Williams, and Ramsay 1983), but it has not been re-ported for other species.

Effect of Coalitions on Relationship between Dominance and Mating Activity

In chimpanzees, the existence of type 3 coalitions some-times reduces the mating advantages of high rank because dominant males can be forced to tolerate a subordinate's mating activity in exchange for the subordinate's coali-tionary support (chap. 15; de Waal 1982). There is no evidence, however, that type 3 coalitions have this effect in macaques (Chapais 1983e) or howlers. In savanna ba-boons, type 4 coalitions clearly reduce the mating advan-tages of high rank, since low-ranking males typically use coalitions to take estrous females away from more domi-nant males (DeVore 1965a; Rasmussen 1980; Collins 1981; Smuts 1985).

The distinction between type 3 and 4 coalitions has not been explicitly recognized, and species' differences in male coalition formation present an important prob-lem for future research.

ALTERNATIVE MALE COMPETITIVE STRATEGIES

Data from many animals indicate that males can acquire mates in more than one way; this is probably true for most nonhuman primates. Table 31-2 summarizes infor-mation on alternative male competitive strategies in seven species. These examples suggest several general principles.

First, alternative strategies are often related to age. This has been demonstrated, for example, in savanna ba-boons (Smuts 1985). In one troop, young adult males (about 8 to 12 years of age) were dominant to older males (about 13 years old and older), but during the day older males repeatedly used coalitions with other older males to take consort partners away from younger males. Older males also used coalitions to protect their own consort-ships from younger males. At night, poor visibility and the difficulty of maneuvering on the sleeping cliffs appar-ently reduced the effectiveness of the older males' co-alitions, and young males were able to take consorting females away from other males through individual ag-gressive challenges. Both sets of males showed statisti-cally significant tendencies to challenge males from the

TABLE 31-2 Examples of Alternative Male Competitive Tactics

Species	Tactic 1	Tactic 2	Sources
Chimpanzees	Achieve alpha rank; monopolize estrous females through possessive behavior within groups	Nonalpha; mate opportunistically within groups and form prolonged sexual consortships away from other males	Tutin 1979; Hasegawa and Hiraiwa-Hasegawa 1983
Orangutans	Resident adult males: longer sexual consortships with females likely to conceive and aggressive interference with consortships of subadult males and nonresident adult males; forced copulations rare	Subadult males: brief, opportunistic associations with females, often involving forced copulations usually at times when conception is not likely	Galdikas 1985; Mitani 1985b; chap. 13
Mountain gorillas	Remain in natal group and inherit females from aging silverback (probably one's father)	Leave natal group and acquire females through aggressive "raids" on other groups	Harcourt and Stewart 1981; chap. 14
Savanna baboons	Young adult males: take estrous females away from older males through solo aggressive challenges, usually at night	Middle-aged and older males: take estrous females away from younger males through aggressive coalitions formed with other older males, usually during the day	Rasmussen 1980; Collins 1981; Smuts 1985
Gelada baboons	Take over a "one-male" group from another male and become the breeding male	Enter a "one-male" unit as a follower, gradually establish bonds with some of the females, and eventually leave with them to form a new group.	Dunbar 1984b
Redtails, patas, langurs, Japanese macaques	Long-term resident in bisexual group; attempt to prevent extragroup males from mating	Haunt or invade groups when females are in estrus and copulate with them; travel alone or in all-male bands at other times, searching for opportunities to invade groups	Chap. 9; Nishida 1966; Hrdy 1977; Moore 1985; Mohnot 1984
Hamadryas baboons	"Adopt" juvenile females and care for them until they mature	Take over adult females by challenging resident male in other groups	Kummer 1968; Sigg et al. 1982; chap. 10

opposite age class and to avoid challenging males from their own age class. Perhaps of greatest interest, the strategies of individual males changed over time as they grew older. Age also clearly influences competitive strategies in orangutans, chimpanzees, and geladas (see table 31-2).

Second, demographic and life-history features of the population or group can influence the effectiveness of different strategies. For example, Dunbar (1984b) showed that in geladas the costs and benefits of the "takeover" and "follower" strategies were determined by a complex interplay of several factors, including male age, the size and age structure of the female group, male life expectancies, and the number of extragroup males (see also chap. 10). In any species living in one-male groups, the number of extragroup males probably influences the frequency with which resident males are challenged and, thus, the costs and benefits of being a resident male as opposed to one who enters bisexual groups only when females are in estrus (Butynski 1982b; Leland, Struhsaker, and Butynski 1984). In savanna baboons, coalitions may be ineffective in small groups with few males or in groups with high rates of male turnover in which stable alliances do not have time to develop (Smuts 1985).

Third, in some cases alternative strategies are not equally effective. This is particularly evident when one strategy is adopted only by smaller, younger males who cannot compete effectively against older, larger males who pursue a more successful strategy (e.g., orangutans: table 31-2). In other cases, however, different strategies appear to be genuine alternatives, that is, they may produce roughly equal reproductive benefits. For example, using a computer simulation, Dunbar (1984b) showed that the expected reproductive benefits for gelada follower males are almost exactly the same as those of males who aggressively take over reproductive units.

Observations of primates suggest two additional generalizations about male competitive options. First, accidents of history and random demographic processes often provide opportunities for creative tactics. For example, in one group of wild chimpanzees containing only three adult males, the relationship between the two top males came to depend entirely on the third. This male took advantage of their dependence on him to increase his own mating success (chap. 15). In species in which coalitions with male kin are an important competitive tactic, availability of a brother, father, or son of the right age influences a male's competitive tactics (e.g., chimpanzees: Riss and Goodall 1977; gorillas: Harcourt and Stewart 1981). In gelada baboons, a random increase in the number of males born into the population over a brief period can influence male competitive tactics several years later by increasing the frequency of adult male attempts to take over one-male units (Dunbar 1979a, 1984b). In general, primatologists have probably underestimated the importance of past events in shaping competitive strategies because studies are usually too brief to provide a historical perspective.

Second, even for animals of the same age, competitive options can depend on individual attributes such as temperament, size, and intelligence. Goodall, for example, describes how Mike, a small, low-ranking, but unusually innovative male chimpanzee, achieved alpha status by incorporating empty kerosene cans into his displays (Goodall 1971). In a group of savanna baboons, male capacity for remaining calm in the presence of rivals correlated significantly with the frequency of sexual consortships, and in two different periods the males who consorted the most shared an extraordinary capacity for ignoring persistent harassment by other males (Smuts 1985). Informal conversations among primatologists indicate a keen awareness of the fact that individual personality traits influence competitive tactics, but such factors have received little formal attention because they defy easy quantification (see Stevenson-Hinde 1983a, 1983b for an exception).

The alternative tactics discussed so far all relate to male-male competition. In order to gain a more complete understanding of the options available to males, however, we must also consider a different kind of tactic: behaving in ways that increase a male's attractiveness to females. This is discussed later, after reviewing the evidence for female mate choice in primates.

EVIDENCE FOR FEMALE CHOICE

Striking anecdotal evidence for female choice has been available since the early 1960s, but only recently has it received systematic attention. Although the study of female choice is still young, recent research indicates that it represents a major force in the evolution of primate societies.

Female primates exert mate choice in two main ways: directly, through responses to opportunities to copulate,

and indirectly, by influencing male group membership. Both appear important in a wide variety of primates.

Direct Female Choice

Initiating Sex. Females have been observed soliciting copulations in virtually every primate species that has been studied, and, in some species, females initiate the majority of copulations (e.g., mountain gorillas: Fossey 1983; dusky leaf monkey: Bernstein 1968; captive slender loris: Izard and Rasmussen 1985; brown capuchins: Janson 1984; gelada baboons: Dunbar 1984b). Sometimes, female solicitations are fairly subtle (at least to the observer), involving gestures such as tongue flicks, headshakes, or simple approaches (see table 30-1). In other cases, however, females indicate their eagerness to mate in more dramatic ways. For example, Janson (1984) describes elaborate female courtship in a group of wild brown capuchins. An estrous female will shadow the dominant male for days. At frequent intervals she approaches him closely, grimaces at him while giving a distinctive vocalization, pushes him on the rump, and shakes branches at him. When she is ready to copulate, she charges him, he runs away, she follows, and when he stops running, they mate (Janson 1984). Female Japanese macaques frequently initiate sex by approaching the male, huddling against his back, and then mounting him (Hanby, Robertson, and Phoenix 1971). In one study, 31% of completed copulations began with the female mounting the male and rubbing her perineum against his back. Those females who mounted males had significantly more sexual partners than females who did not (Wolfe 1979). Captive talapoin females also commonly initiate sex by mounting and thrusting on the male (Wolfheim and Rowell 1972).

Female signaling of willingness to mate can be important even when it appears that the male is the one who initiates and controls copulation. For example, Goldfoot (1982) showed that, in pairs of captive rhesus monkeys, male approaches and attempts to mount were likely to lead to copulation only when the female had already indicated her willingness to mate through subtle glances and postures. Previous observers had failed to detect these signals and had erroneously assumed that the female's behavior had little effect on the frequency of male mounts.

Refusing Sex. Like female sexual solicitations, female refusals of male attempts to copulate have been observed in nearly all nonhuman primates. With the exception of orangutans (chap. 13; Mitani 1985b; Galdikas 1985), forced copulations have not been seen in wild nonhuman primates, which indicates that female cooperation is necessary for mating (Lindburg 1983). In many species, females refuse copulations by simply walking away or sitting down (e.g., Japanese macaques: Stephenson 1975;

rhesus macaques: Lindburg 1983; captive patas monkeys: Hall and Mayer 1967; savanna baboons: Smuts 1985). In other cases, refusals are more dramatic. Female bushbabies (Charles-Dominique 1974a), fat-tailed dwarf lemurs (Foerg 1982a), vervet monkeys (Andelman 1985), talapoin monkeys (Wolfheim and Rowell 1972), and gelada baboons (Dunbar 1984b) sometimes aggressively discourage an unwanted suitor. Vervet females in one group refused slightly more than half of all male copulation attempts, and 11 to 17% of their refusals involved escalated aggression (hits, bites, and chases) (Andelman 1985).

Indirect Female Choice: Group Membership

In many species females influence male group membership (chaps. 21, 22, 32). When they do so, they help determine which males will be available as mates. Gelada baboons provide the clearest example (Dunbar 1984b; Mori and Dunbar 1985). A male attempting to replace the resident male tries to get close to the group's females, despite the resident male's aggression. If the rival succeeds in getting the females to interact with him, the resident male simply gives up and the rival takes over the unit. Dunbar (1984b) notes that the male that appears to be winning the fight is not necessarily the one who ends up with the females because females do not always choose the winner. In one case, a female continued to prefer the unit's "follower" male, who remained in the group after two successive takeovers. Each time, she attacked the new resident male fiercely and drove him away from her mate, even though the male was much larger than she was (Dunbar 1984b, 75). Since male geladas are unable to breed unless they achieve membership in a female group, females have a great deal of influence over male reproductive success.

In a variety of primate species, females have been observed temporarily leaving their groups to interact in friendly ways (including copulation) with outside males (savanna baboons: Cheney and Seyfarth 1977; Packer 1979a; Japanese macaques: Nishida 1966; Wolfe 1984b; rhesus macaques: Brereton 1981; gray langurs: Hrdy 1977; Mohnot 1984). These interactions may influence subsequent male attempts to enter the female's group. For example, an estrous female savanna baboon was observed entering an adjacent group and luring one of its males into her own group, where he remained for many years (Smuts 1985).

In species in which females emigrate from their natal groups (e.g., hamadryas baboons, gorillas, chimpanzees, red colobus monkeys; see chap. 21), females choose their mate(s) by entering and leaving groups. Although males, through aggression and affiliative behaviors, can discourage females from leaving, only for hamadryas is there evidence that they sometimes succeed in keeping females against their will (Kummer 1968).

BASES FOR FEMALE CHOICE

Which males do females prefer as mates? This is a very difficult question to answer for several reasons. First and most important, only a handful of studies have focused on this issue, and very few quantitative data exist. Second, we expect the bases for female choice to depend not only on the species and its social system, but also on such female attributes as age, dominance rank, and presence or absence of a young infant vulnerable to infanticide. Third, female choice will often be constrained by male competitive tactics, making it difficult to know whether a particular act represents the female's "ideal" choice or a compromise between her interests and those of males. Because of these difficulties, much of the following section remains speculative. My main goal is to highlight the need for studies designed to evaluate specific hypotheses.

Resources

In a wide variety of animals, females choose mates based on male ability to provide resources ranging from the nuptial gifts of insects to the feeding territories of birds and some mammals (Searcy 1982; Bateson 1983), which the female can invest in her offspring. There is little direct evidence, however, to indicate that male resource defense is an important basis for female choice in nonhuman primates (Robinson 1982a). In monogamous primates in which females and males defend home ranges against conspecifics, females may prefer males that demonstrate an ability for effective territorial defense (e.g., gibbons: Tilson 1981), but since almost nothing is known about the process of pair-bonding in monogamous primates, this remains merely a plausible hypothesis (chap. 12). In many species in which females live with female relatives in multifemale groups, females defend territories or particular food sources against other groups (chap. 22). Although males sometimes participate in these intergroup encounters, we do not know whether females recruit males on the basis of their willingness to do so. It is possible that females sometimes favor particular males as mates in order to obtain aid in feeding competition within groups, as Janson (1984) has suggested for brown capuchins.

Male Care of Infants

Although direct paternal care is relatively rare in most primates, it does occur in the monogamous siamangs, owl monkeys, and titi monkeys, and in monogamous/ polygynous callitrichids (chap. 28). It seems likely that female choice of mates in these species is based at least in part on male ability to invest in offspring, but this hypothesis cannot be tested with available data.

In a number of species living in multimale groups, males appear to contribute to offspring survival through baby-sitting, protection, occasional carrying, and other affiliative behaviors. Male parental investment does not fully explain these behaviors because the infants are in many cases unlikely to be the male's own offspring (chap. 28). Seyfarth (1978a, 1978b) and Smuts (1983b, 1985) suggested that, by developing an affiliative relationship with an infant, male savanna baboons might be able to improve their chances of mating with the mother in the future. If this hypothesis is correct, then male care of infants sometimes represents mating effort rather than parental investment (Smuts 1985; chap. 28).

Protection from Male Aggression

In many nonhuman primates males show aggression toward females and infants (see, for example, discussions of male infanticide in chap. 8 and male aggression toward females in chap. 32). Several types of evidence suggest that female mate choice sometimes functions to reduce female and infant vulnerability to male aggression.

Mating with Unfamiliar Males

In many species, females show physiological and behavioral adaptations that lead to copulation with more males over a longer period than is necessary to ensure fertilization (chap. 30). For example, savanna baboon females exhibit sexual swellings for 2 to 3 weeks of the 35-day

cycle, and during the period when ovulation is most likely to occur (the week preceding detumescence of the sexual swelling), they consort on average with 4 to 5 different adult males (Rasmussen 1980; Smuts 1985). Barbary macaques represent an extreme example: estrous females initiate and terminate brief sexual consortships with an average of six males per day, and during a single estrous cycle they consort on average with 64 to 73% of all adult and subadult males in the group (Taub 1980b). Taub (1980b) suggested that by mating with many males, a female can confuse paternity and thus increase the probability that several males will invest in her infant; Hrdy (1979) and Wrangham (1980) suggested that this same tactic reduces the probability of infanticide (see also chap. 30).

These two hypotheses are not mutually exclusive. The second one, however, predicts that females will be particularly interested in mating with males who are most likely to commit infanticide (Hrdy 1979), namely, unfamiliar extragroup males or males that have only recently entered a group (chap. 8). In several well-studied species, females seem to be attracted to unfamiliar males (Clutton-Brock and Harvey 1976; Harcourt 1978b) (fig. 31-4). For example, in one troop of Japanese macaques, 30% of the females were observed leaving the troop to

FIGURE 31-4. When a rhesus monkey female mates with an extragroup male, she usually tries to do so in secret, since males from her troop are likely to interfere if they spot her. Just before these pictures were taken, the male sexually solicited the female and then moved quickly away from the troop. After a brief pause, the female followed him. Before mating, the female checks to make sure that none of the adult males in her troop is watching (*left*); confident that they are hidden from view, she mates with the extragroup male (*right*). The wound on the male's skull was inflicted by another male a few days before. (Photos: Barbara Smuts/Anthrophoto)

copulate with extragroup males, and others probably did so without being detected (Wolfe 1981). Similarly, female rhesus macaques have been seen to leave their groups for several days to consort with adult males from other groups (Brereton 1981). Vervet monkey females solicit copulations from immigrant males soon after they enter a group (Henzi and Lucas 1980), and female savanna baboons prefer to mate with unfamiliar immigrant males compared with males born in their own group (Packer 1979a). It is important to note, however, that although these behaviors might function to reduce infanticide, they might also be related to inbreeding avoidance (chap. 21).

Attraction to extragroup males is also apparent among females living in one-male groups, where infanticide frequently follows invasion by unfamiliar males (chap. 8). For example, female gray langurs copulate and even develop consort relationships with extratroop males (Hrdy 1977). When the resident male is trying to fight off males from an all-male band, females sometimes take advantage of this distraction to mate with males from the all-male band or other extragroup males (Mohnot 1984). If challenges by extragroup males signal the resident male's vulnerability to a takeover by one of these males, then females should be particularly motivated to copulate with outside males at these times. Copulation with extragroup males has also been observed in patas monkeys, redtails, and blue monkeys (chap. 9).

Finally, as noted in chapter 30, after takeover of a one-male group, females sometimes exhibit sexual solicitations and copulations even though they are already pregnant. Hrdy (1977) argued that such "situation-dependent estrus" benefits the female by increasing the chances that the new male will treat her next offspring as if it were his own.

Mating with Males Most Likely to Protect Females and Offspring

If females and their offspring are indeed vulnerable to aggression from males, then females might, in some circumstances, benefit from mating with the male or males most able to defend them against other males. For females living in multimale groups, dominant males might be the best defenders since they have demonstrated their superior competitive ability against other males (Wrangham 1979a). Females sometimes seem to prefer dominant males (reviewed by Robinson 1982a; Silk and Boyd 1983; and Berenstain and Wade 1983). For example, Rasmussen (1980) reported that estrous female savanna baboons presented to higher-ranking males significantly more often than to lower-ranking males. In one brown capuchin group females initiated sex with the dominant male much more often than with subordinate males (Janson 1984), and sifaka females have been observed mating only with males that have demonstrated their com-

petitive superiority by fighting off rivals (Richard 1974b).

However, abundant evidence also suggests that females do not always prefer to mate with the highest-ranking males. For example, in one group of vervet monkeys, females refused more copulation attempts by dominant males than by subordinate males (Andelman 1985). Female macaques occasionally resist consortships with alpha males (rhesus monkeys: Lindburg 1983) and sometimes show strong sexual preferences for subordinate males (captive bonnet macaques: Glick 1980).

It is perhaps not surprising that studies fail to reveal consistent preferences for dominant males since other characteristics attractive to females do not always correlate with high rank. For example, in macaques, males that have recently transferred into a group are usually low ranking (chap. 25). To the extent that macaque females show strong preferences for novel males, this will counteract preferences for high-ranking males. Indeed, in one troop of Japanese macaques, a newly transferred male, lowest ranking of all fully adult males, mounted more different females than did any other male (Takahata 1982b).

In other cases, females might achieve more consistent protection for themselves and their offspring by forming a special relationship with one or two particular males. For example, in savanna baboons, females appear to mate preferentially with particular males in exchange for protection for themselves and their offspring against other males (Smuts 1985; see "Affiliative Relationships and Individual Mating Preferences"). It is important to emphasize, however, that female dependence on male protection varies considerably across species. In particular, in species such as vervets, talapoin monkeys, and rhesus and Japanese macaques, in which females are almost as large as males, the effectiveness of female-female coalitions against males probably reduces the need for male protection (chap. 32).

Mate choice in female-transfer species may also sometimes reflect female need for protection against male aggression. Observations of one population of red colobus monkeys suggested that females sometimes emigrated in order to avoid a potentially infanticidal male that had recently entered the group (Marsh 1979b). In other cases, males may use aggression and even infanticide in order to attract mates. For example, many emigrations by female mountain gorillas occurred following harassment (and in a few cases, infanticide) by extragroup males, and females sometimes transferred to the harassing male (chap. 14). This pattern makes sense if the outside male's success in harassing females and infants indicates that he is a stronger competitor and thus a more effective protector than the female's current mate (Wrangham 1979a). Similar factors may influence transfer in female chimpanzees, who are also vulnerable to aggression by males from neighboring groups (chap. 15). Nishida et al. (1985) reported that several males disappeared from one community, probably as a result of aggression from neighboring males. When only one male was left, all but two of the females of the dwindling community deserted en masse to join the neighboring group.

Mating with Males of Superior Genetic Quality

There are two possible ways in which a male's genetic makeup could influence female mate choice. First, females could avoid mating with closely related males to reduce the deleterious effects of inbreeding. This issue is discussed in chapter 21. Second, females could choose to mate with males of superior genetic quality that will pass useful traits on to the females' offspring. No firm evidence indicates that female primates choose males on this basis. Evolutionary biologists disagree about whether female choice for genes occurs in any species, and research directed toward this question constitutes one of the most active areas in modern behavioral ecology (reviewed in Bateson 1983). Evaluation of the hypothesis of mate choice for genes among primates probably must await resolution of this controversy within biology as a whole.

EFFECTS OF FEMALE CHOICE ON MALE BEHAVIOR
Effects on Male-Male Competition

By cooperating with some potential mates and resisting others, females alter the costs and benefits of competition among males and thus influence the form and frequency of male-male competition. For example, in experiments with captive hamadryas baboons, Bachmann and Kummer (1980) found that unmated males were significantly more likely to fight for access to a female when the female exhibited a low preference for her current mate. Similarly, in savanna baboons, males are more reluctant to challenge consortships involving a pair that has a long-term affiliative relationship (Seyfarth 1978a, 1978b; Packer 1979b; Collins 1981; Smuts 1985). Janson (1984) noted an association in brown capuchins between consistent female preferences for the alpha male and very low rates of male-male aggression. He argued that subordinate males do not bother to challenge the alpha male for access to females because the females would be unlikely to allow them to mate even if they won. Dunbar (1984b) discusses still another way in which female choice can influence male-male competition. As the number of adult females in a gelada one-male unit increases, the resident male finds it increasingly difficult to groom with all of his females, and the females become increasingly willing to desert him for another male. This makes it easier for rival males to take over the unit, and,

indeed, takeover attempts directed toward large units occur more often than expected on the basis of their frequency in the population.

Effects on Male Behavior Toward Females

When females exert mate choice, males can increase their reproductive success not only through successful competition with other males but also by behaving in ways that make them more attractive to females. Perhaps because female primates are so often vulnerable to male aggression, they sometimes prefer to mate with males who are unaggressive toward them. This preference has been demonstrated most clearly in laboratory tests in which female macaques control male entry into their cages (Eaton 1973; Michael, Bonsall, and Zumpe 1978). Chapais (1983f) also reported that, in free-ranging rhesus monkeys, the males who mated the most often were the least aggressive toward females. However, as hypothesized earlier, when females rely on males to protect them from other males, they may sometimes choose to mate with aggressive males (e.g., mountain gorillas: Wrangham 1979a). Thus, the relationship between male aggression and female choice seems to vary, depending on the circumstances. Better understanding of these circumstances requires considerably more information about male aggression toward females and female responses than is currently available.

In addition to refraining from aggression, males may attract females by behaving in friendly ways. For example, Tutin's (1979) study of wild chimpanzees showed that sexual consortships, which require female cooperation, most commonly involved males who associated with females, groomed them, and shared meat with them. Similarly, captive female rhesus monkeys preferred males who groomed them the most (Michael, Bonsall, and Zumpe 1978).

As noted above and in chapter 28, males sometimes contribute benefits to infants that are unlikely to be their own, and this may increase their attractiveness to females. In gelada baboons, for example, a young male trying to enter a group as a follower may begin by establishing an affiliative relationship with an infant. Once he has succeeded in doing so, it is easier for him to develop a bond with the infant's mother and subsequently with other females (U. Mori 1979a; Dunbar 1984b). Male Barbary macaques develop unusually intimate relationships with infants (chap. 28), and subordinate males try to attract the attention of an estrous female by marching up and down before her while carrying an infant (Taub 1980b). It is tempting to speculate that the male uses this particular display (which has not been observed in other species) to signal the female that he will take good care of her infant if she mates with him. Finally, in savanna baboons, males contribute a variety of benefits to females and their offspring in the context of long-term, special relationships, and females often prefer these males as mates (Smuts 1985; see "Affiliative Relationships and Individual Mating Preferences").

FEMALE-FEMALE COMPETITION FOR MATES

As noted in the introduction to this chapter, male-male competition for mates and female mate choice are the most common results of sexual selection. But females sometimes compete for mates, and males sometimes exhibit mate choice.

Evidence for female-female competition over males is strongest for callitrichids, species in which males contribute a great deal of infant care (chap. 28). For example, in captive saddleback tamarins, females aggressively prevent friendly interactions between their mate and other females (Epple, Alveario, and Katz 1982), and in captive common marmosets, the mere presence of a dominant female inhibits reproductive activity in subordinate females, even if they are caged separately (Evans and Poole 1983).

Females sometimes compete for sexual access to the breeding male in species living in multifemale, one-male groups. Females have been seen harassing copulating pairs in patas monkeys (Loy and Loy 1977), gray langurs (Yoshiba 1968), and gelada baboons (A. Mori 1979b). Dunbar (1980b, 1984b) found that harassment by dominant female gelada baboons made it difficult for subordinate estrous females to approach the male, and this may have been a cause of their lower conception rates. Laboratory studies of talapoins (Bowman, Dilley, and Keverne 1978) and common marmosets (Abbott and Hearn 1978) have shown hormonally mediated reproductive suppression in subordinate females.

In several species that live in multifemale, multimale groups, females have been observed using aggression to interrupt sexual interactions between males and subordinate females (mantled howlers: Young 1981; red howlers: Sekulic 1983a; captive black howlers: C. B. Jones 1983; savanna baboons: Hall 1962b; Seyfarth 1976, Wasser 1983; captive pigtailed macaques: Gouzoules 1980c). Lindburg (1971) reported that in wild rhesus monkeys, females harass sexually consorting females, sometimes winning the consort male for themselves. In other groups, subordinate females avoid sexual contact with dominant males in the presence of dominant females, even when overt aggression between females over males does not occur (e.g., brown capuchins: Janson 1984; captive pigtailed macaques: Goldfoot 1971). Note, however, that female competition over males is sometimes related to the development of long-term social relationships rather than to sexual access per se, as indicated by Seyfarth's (1976, 1978b) observations of savanna baboons.

MALE MATE CHOICE

As described above, male-male competition often involves severe risks (e.g., injuries from fights) and large expenditures of time and energy (e.g., searching for estrous females; protecting consort partners from rivals). Furthermore, recent evidence indicates that frequent copulation can result in significant sperm depletion (Dewsbury 1982b). When the costs of mating competition are sufficiently high, males are expected to compete selectively in order to maximize the benefits gained. In particular, males should (1) compete most intensively for females who prefer them, (2) distribute their copulations in ways that will ensure fertilization, and (3) prefer to mate with females who are more likely to conceive and to produce surviving offspring (Robinson 1982a; Dewsbury 1982b; Halliday 1983; Berenstain and Wade 1983). The first prediction has already been discussed; the second and third are considered briefly below.

In several species living in multimale groups, males whose mating activity is greatest overall concentrate their copulations around the time when ovulation is most likely to occur (rhesus monkeys: Chapais 1983e; savanna baboons: Hall and DeVore 1965a; Hausfater 1975; Packer 1979b; Smuts 1985; chimpanzees: Tutin 1979; Hasegawa and Hiraiwa-Hasegawa 1983). Similarly, Rasmussen (1980) and Collins (1981) reported that aggressive challenges to sexual consortships by male savanna baboons were less common at times when ovulation was unlikely to occur (i.e., the first estrous cycle after postpartum amenorrhea and early stages of sexual swelling).

Males in many species show distinct mating preferences and aversions. The most widespread of these is relatively low sexual interest in adolescent, nulliparous females compared to older females who have borne at least one offspring (orangutans: Galdikas 1985; chimpanzees: Hasegawa and Hiraiwa-Hasegawa 1983; captive bonnet macaques: Glick 1980; Japanese macaques: Takahata 1982b; savanna baboons: Rasmussen 1983a; Scott 1984). The relative disinterest in adolescent females may reflect the fact that these females typically exhibit lower fertility (chap. 20) and less adequate mothering (chap. 27) than older females. In many groups, males also show mating preferences for high-ranking females (reviewed by Robinson 1982a; Silk and Boyd 1983; and Berenstain and Wade 1983), who often have higher reproductive success than lower-ranking females (chap. 26). However, males do not always prefer high-ranking females (e.g., captive patas monkeys: Loy 1981; rhesus macaques: Loy 1971; captive rhesus: Small and Smith 1982; savanna baboons: Packer 1979b), which indicates that other factors also influence male mate choice. Rasmussen (1980) found that male savanna baboons preferred to consort with females who showed high sexual motivation toward males in general. If sexually aroused females tend to be more cooperative mates, then males might find consortships with these females less costly (and more pleasureable) than consortships with less aroused females.

Male primates, like females, often show strong mating preferences for particular partners. Such preferences are common in multimale groups (e.g., captive crab eating macaques: Deputte and Goustard 1980; captive bonnet macaques: Glick 1980; rhesus macaques: Rowell 1963; Loy 1971; Lindburg 1971; Japanese macaques: Fedigan and Gouzoules 1978; Wolfe 1979; Takahata 1982a; savanna baboons: Seyfarth 1978a, 1978b; Packer 1979b; Rasmussen 1983b; Collins 1981; Smuts 1985). Several studies of captive macaques reveal remarkably strong individual preferences in the absence of competition from other males (Everitt and Herbert 1969; Goldfoot 1971; Perachio, Alexander, and Marr 1973; Phoenix 1973; Slob et al. 1978). In some cases, males continue to prefer a favorite female even when she is not near peak estrus and other females are (captive chimpanzees: Allen 1981; captive rhesus macaques: Herbert 1970). Chambers and Phoenix (1982) found, as had others, that old rhesus males were less sexually active than young adult males. However, when old males were allowed to interact with a preferred female partner, they copulated as often as did young adult males who were also placed with a favorite partner. Similarly, Dunbar (1984b) found that the likelihood of ejaculation per mount was significantly higher (45 versus 30%) for the male gelada's favorite social partner than for other females. Findings like these have led experienced researchers to conclude that "in both sexes individual preferences are as important [to sexual activity] as hormonal requirements" (Phoenix 1973, 369).

AFFILIATIVE RELATIONSHIPS AND INDIVIDUAL MATING PREFERENCES

In many of the studies described above, the observers had no idea why males found particular females much more attractive than others (in several cases, different males preferred different females). Similarly, partner preferences in wild groups often cannot be explained in terms of easily quantifiable characteristics like age or rank. In some of these cases, mating preferences probably reflect the pair's history of social interactions—in other words, their social relationship. Since social relationships are embedded in a larger social context involving the conflicting interests of many individuals, it is likely that individual partner preferences sometimes reflect all four processes discussed in this chapter: male competition for mates, female choice, female competition for mates, and male choice. This point is illustrated below with findings on savanna baboons.

Savanna baboon females and males form long-term special relationships or "friendships" that are distin-

guished from ordinary male-female relationships by more frequent grooming, proximity, and the relaxed, friendly quality of most of their social interactions (Ransom and Ransom 1971; Seyfarth 1978b; Altmann 1980; Smuts 1985) (fig. 31-5). Behavioral evidence suggests that both sexes compete for these relationships (e.g., both males and females sometimes threaten potential rivals away from their friends). Both males and females also tend to choose friends with particular characteristics. For instance, older adult males who have lived in the group for several years and middle-aged and old females form friendships with one another more often than expected. The same is true for adolescent males and females (Smuts 1985).

Elsewhere I have argued that these friendships increase the reproductive success of both females and males by facilitating reciprocal exchanges of benefits (Smuts 1983b, 1985). Females and their offspring are protected by male friends from conspecifics (and possibly predators). The male friend may be particularly important in reducing female and infant vulnerability to aggression (including infanticide) by other males. The male, in turn, is favored by the female as a mate. This hypothesis is supported by two types of data. First, in one group, males consorted with their female friends significantly more often than expected based on their overall consort activity (Smuts 1983b, 1985). Second, in another group, when a female preferred a particular male as a social partner outside of sexual consortship (as revealed by frequencies of approaches), she was a more cooperative sexual consort partner (Rasmussen 1980, 1983b).

Intrasexual competition over access to opposite-sex partners, possessive behavior, and strong mutual attraction are shown by baboon males and females at all phases of the female reproductive cycle, including pregnancy and lactation (Collins 1981; Smuts 1985). This is true not just for baboons, but for some other species as well (e.g., rhesus macaques: Chapais 1981; 1983f; Japanese macaques: Fedigan 1982; Takahata 1982a). Although such behaviors are often ultimately related to mate choice, they can occur independently of immediate motivations to copulate (e.g., a pregnant female baboon is unlikely to be available as a mate for up to 2 years).

Attraction to members of the opposite sex at times when sex itself is unlikely, strong individual partner preferences, and long-term special relationships between males and females indicate that in some nonhuman primates, as in humans, mate choice is intimately related to social bonds. For other species, however, such as vervet monkeys, there is no evidence for long-term male-female bonds or strong mating partner preferences (Andelman 1985; Cheney and Seyfarth, pers. comm.). This is puzzling because vervet social organization is very similar to

FIGURE 31-5. An adult male and female olive baboon, who have developed a long-term, special relationship, asleep on the baboons' sleeping cliffs. Such couples associate during the day and groom frequently. The male protects the female and her offspring from other baboons, and the female, in turn, often prefers the male as a mate. (Photo: Barbara Smuts/Anthrophoto)

that of macaques and savanna baboons (chap. 11). This contrast highlights the need for further research on the causes and consequences of male-female bonds and individual mating preferences among nonhuman primates.

SUMMARY

Male-male competition for access to females frequently involves risky behaviors such as fighting with other males or moving long distances in search of mates. Male dominance rank often, but not invariably, correlates with mating activity. In many species males use coalitions to compete more effectively against other males. Males pursue different competitive tactics, depending upon factors such as age, dominance rank, and demographic features of the population. Access to females is also influenced by female mating preferences.

Females choose mates directly, by facilitating or refusing copulations, and indirectly, by influencing male group membership or, in female-transfer species, by choosing which groups to join. Little is known about the bases for female mate choice in primates, but in several species females appear to distribute copulations in ways that reduce their vulnerability to male aggression, including infanticide. Female-female competition for mates and male mate choice also influence mate selection in primates. In some species, such as savanna baboons, females and males often develop strong preferences for particular mates in the context of long-term social relationships, but other species, such as vervet monkeys, show no evidence of individual mating preferences based on social bonds.

32 | Gender, Aggression, and Influence

Barbara B. Smuts

Nonhuman primates often exhibit striking sex differences in behavior. Individual differences in behavior are widespread among primates, and, as chapters in this volume illustrate, they involve many characteristics, including age, status, and kinship as well as sex. Sex differences in behavior, however, often carry special significance because of controversy about the origins and meaning of human sex differences. The belief that observations of animals can illuminate this controversy has stimulated a search for generalizations about behavioral sex differences, especially in nonhuman primates. Some of the generalizations that have emerged reflect cultural stereotypes: male primates are more aggressive than females; males dominate females; females are more passive than males (e.g., Tiger 1969). Others represent attempts to counteract such cultural biases: female primates are more influential than males; females are more likely than males to form affiliative social bonds (e.g., French 1985).

Widespread acceptance of these types of generalizations has two unfortunate consequences. First, inter- and intraspecific variation in male and female behavior are often ignored. Second, insufficient attention is paid to the functional significance of sex differences in behavior. This chapter stresses the importance of these two issues for understanding behavioral sex differences in nonhuman primates. It makes no attempt to review sex differences in all aspects of behavior but focuses instead on how gender relates to social bonds, aggression, and influence (for discussion of the development of sex differences see chap. 29; for systematic reviews of the relationship between gender and behavior among nonhuman primates see Mitchell 1979; Hrdy 1981b; Fedigan 1982; and Lancaster 1984). The first part of the chapter examines contextual factors that influence sex differences in social behavior. The second part considers how males and females influence the behavior of members of the opposite sex through both aggressive and nonaggressive means.

BEHAVIORAL SEX DIFFERENCES
Impact of the Social System

Among nonhuman primates, patterns of intergroup transfer (chap. 21) appear to have important effects on intra-sexual relationships within groups (Wrangham 1980; Harcourt and Stewart 1983; Smuts, in press). First consider female-female relationships. In species in which females usually remain in their natal groups, females develop strong bonds with other females, revealed by frequent association and grooming. Examples include macaques, vervets, and savanna baboons (chap. 11); gelada baboons (chap. 10; Dunbar 1984b); gray langurs and Nilgiri langurs (chap. 8; Poirier 1970); squirrel monkeys and capuchins (chap. 7); and sifakas and ring-tailed lemurs (chap. 3). In some of these species, females develop highly differentiated dominance relationships; in macaques, vervet monkeys, and baboons, these relationships are based largely on coalitions among female kin (chaps. 24, 25).

In species in which females usually transfer out of their natal groups, female-female relations lack such cohesion. Females rarely rest in close proximity or groom one another. Dyadic agonistic interactions and female-female coalitions against other females are uncommon, and female dominance relationships remain ill defined. Examples include red colobus monkeys, gorillas, and chimpanzees (chaps. 8, 14, 15). Female hamadryas baboons also fit this pattern, except that female dominance relationships appear more clearly defined in this species (chap. 10).

Similarly, male-male relationships also reflect transfer patterns. In chimpanzees and red colobus, males remain in the natal group and form strong bonds characterized by grooming, proximity, and coalitionary support (chaps. 8, 15). Among hamadryas baboons, males born in the same band associate and groom until they acquire their own females. After this they no longer groom or sit together, but coordinated travel, respect for one another's females, and mutual support against males from other bands indicate their continuing bonds with one another (chaps. 10, 22). In all of these species, male-male bonds appear to be stronger than female-female bonds (with the exception of mother-daughter relationships).

Conversely, in species in which males typically transfer, males rarely associate closely or groom one another. Examples include savanna baboons, gelada baboons, squirrel monkeys, capuchins, and ring-tailed lemurs. There are some exceptions to these generalizations, however. In macaques and vervets, brothers sometimes transfer together and maintain close relationships (Meikle and Vessey 1981; Cheney and Seyfarth 1983), and in seasonally breeding species, male-male relations are typically more relaxed outside the mating season (chap. 25). Nevertheless, male-male bonds are generally weaker in these species than female-female bonds.

Thus, in male-transfer species, female-female bonds are most apparent, whereas in female-transfer species male-male bonds are more prominent (Wrangham 1980). This general finding vividly demonstrates the value of a broad-based, comparative approach to the study of sex differences (Hrdy 1981b; Fedigan 1982; Harcourt and Stewart 1983; Smuts, in press). If, for example, data were available only for macaques, vervets, and savanna baboons, one would conclude that female nonhuman primates are more affiliative than males; if data were available only for chimpanzees, red colobus, and hamadryas one would conclude just the opposite. Although we can now clearly see the pitfalls of any investigation of sex differences that fails to include species with different transfer patterns, the importance of this variable went unrecognized for many years due to the lack of data. There are probably other social or ecological variables of which we are not yet aware that are equally critical for understanding sex differences in behavior. For this reason, the ideal method for investigating any aspect of behavioral sex differences involves comparison of many species that differ in a wide variety of ways (e.g., social system, ecology, demography, phylogeny, size, and so on), both because this reduces the chances of invalid generalizations and because it is likely to reveal new and important factors influencing gender-related behaviors.

Even within species, female-female and male-male relationships can vary dramatically, depending on the context. For example, in some male-transfer species, such as gray langurs, extragroup males who are not competing directly over females form stable friendly associations that involve proximity, grooming, and coalitional support (Moore 1985). However, once these males succeed in replacing the breeding male of a bisexual group, they fight among themselves until only one male remains with the females. Similarly, in experiments with captive macaques, the introduction of females severely disrupts male-male affiliative relationships (Bernstein, Gordon, and Rose 1974; Coe and Rosenblum 1984).

Another example of intraspecific variation in social relationships comes from de Waal's study of captive female chimpanzees. In the wild, adult females spend most of their time alone and rarely associate with unrelated females (chap. 15). When brought into prolonged, close proximity in captivity, however, females form close, stable bonds and support one another through agonistic interventions (de Waal 1982).

Several behaviors, however, appear to vary by sex in consistent ways across a wide variety of nonhuman primate species with different social systems. As illustrated later these behaviors are closely tied to fundamental differences in the reproductive strategies of males and females.

Sex Differences in Aggression

Since the first systematic field studies of nonhuman primates, researchers have often claimed that males are more aggressive than females (e.g., Tiger 1969; Nagel and Kummer 1974). However, this hypothesis has received little systematic attention since relatively few studies have provided quantitative comparisons of aggression in the two sexes. The available data suggest that, in terms of the frequency of agonistic interactions (supplants, threats, chases, and fights), no consistent sex differences exist. In some studies, males engaged in agonistic interactions more often than did females, but in others the reverse was true (table 32-1).

A few studies, however, suggest intriguing sex differences in the quality of agonistic interactions. During encounters with other males, male baboons, vervet monkeys, and macaques use ritualized threats such as yawns and teeth grinding that normally are not shown by females (macaques: Blurton-Jones and Trollope 1968; vervets: Durham 1969; baboons: Smuts, pers. obser.), and male chimpanzees routinely use a ritualized charging display to intimidate rivals, whereas females do so only occasionally (Bygott 1979). In addition, several observers have noted that ritualized threats nearly always precede male attacks on other males, whereas females often attack other females without warning (captive patas monkeys and captive Sykes's monkeys: Nagel and Kummer 1974; captive chimpanzees: de Waal 1982; savanna baboons: Smuts, pers. obser.; vervet monkeys: Cheney and Seyfarth, pers. comm.). One possible explanation for these differences is that males attempt to settle conflicts through ritualized signals more often than do females because the risks of injury through physical combat are greater for males.

Since observers rarely witness infliction of wounds, it is difficult to test directly the hypothesis that males injure one another during fights more often than do females. Instead, researchers have relied on comparisons of the frequency in males and females of fresh wounds or scars indicating previous wounds (table 32-2). What can such comparisons tell us? In most species, observations of fights indicate that most, and in some cases all, male wounds are inflicted by other males. Thus, when males have more wounds than females, it usually indicates higher levels of severe aggression among males than among females. However, since males injure females, if females show more wounds than males, we cannot as-

TABLE 32-1   Sex Differences in the Frequency of Agonistic Interactions of Adults

Species	Measure Used	Sex Showing More Agonism	Sources
Cebus apella (w)	Frequency of threats[a]	F	Izawa 1980
Presbytis johnii (w)	Frequency of aggressive encounters[a]	M	Poirier 1974
P. entellus (w)	Frequency of agonistic interactions[a]	M	Boggess 1980
Macaca mulatta (c)	Number of fights[a]	F	Loy et al. 1984
M. nemestrina (c)	Frequency of all agonistic interactions[b]	F	Bernstein 1972
	Frequency of damaging aggression[b]	M	
M. arctoides (c)	Frequency of threats and bites[b]	F	Bernstein 1980
	Frequency of chases and contact aggression[b]	F = M	
	Frequency of initiation of agonistic interactions[b]	M	Whitten and Smith 1984
M. sinica (w)	Frequency of threats given, corrected for age-sex composition of group[b]	M	Dittus 1977
Papio cynocephalus anubis (w)	Time spent in agonistic interactions[b]	F	Bercovitch 1983
P. c. ursinus (w)	Mean number approach-retreat interactions per dyad per hour[a]	F	Seyfarth 1976
	Mean number aggressive interactions per dyad per hour[a]	M	Ibid.
Pan troglodytes (w)	Frequency of aggression[b]	M	Bygott 1979

NOTE: w = wild; c = captive.
a. Includes only agonism between adults of the same sex.
b. Includes agonism between adult and all other age-sex classes.

sume that this indicates higher levels of severe aggression among females than among males. In all but two of the comparisons shown in table 32-2, males have more wounds than females, and thus we can conclude that severe aggression is more common among males.

Why should males injure one another more often than females? First, sex differences in rates of wounding may reflect, at least in part, sexual dimorphism in canine size and shape. In all of the species listed in table 32-2, adult males have longer, sharper canines than do females (Harvey, Kavanagh, and Clutton-Brock 1978b). However, because female teeth can also inflict serious wounds, this is unlikely to be the whole explanation (it also, of course, begs the question of why males have evolved longer, sharper canines to begin with).

A second hypothesis argues that the potential payoffs of escalated aggression (i.e., aggression in which individuals risk injury) are greater for males than for females because of differences in the resources over which they compete (Clutton-Brock, Guinness, and Albon 1982). Among males, the outcome of a single interaction or of a series of interactions over brief periods of time can result in great differences in access to females and thus in reproductive success (e.g., olive baboons: Sapolsky 1983; rhesus macaques: Chapais 1983e; chimpanzees: Riss and Goodall 1977; gray langurs: Hrdy 1977; mantled howlers: Clarke 1983). Among females, however, the outcome of a single interaction rarely leads to large variations in reproductive success because female reproductive performance depends mainly on the ability to sustain investment in offspring over long periods of time.

Also, females may have more to lose from defeat in an escalated encounter since in many cases they risk not only injury to themselves but also harm to a fetus or an infant. Thus, although intrasexual competition plays a critical role in the reproductive success of both sexes, in females relatively small payoffs accumulate over long periods of time so that competition is generally low key and chronic, while for males the payoffs are often large but short term so that competition tends to be intense and episodic.

This does not mean, however, that females will always be more circumspect and males always more belligerent. It suggests, instead, that the contexts that provoke aggression will differ for each sex, and that given particularly high and immediate stakes, females will fight intensely and risk injury as much as will males. The following examples illustrate this point.

First, although females seldom initiate aggression against higher-ranking females, they will sometimes do so in order to protect vulnerable offspring (e.g., Kaplan 1978; Silk 1980b; Bernstein and Ehardt 1985; de Waal 1982). Females have even been known to attack adult males twice their size to protect close relatives. For example, Collins, Busse, and Goodall (1984) describe how an old female baboon attacked an adult male when he bit her grandchild.

Second, although female rank reversals rarely occur in cercopithecines, when they do occur, they often involve intense female-female fights that sometimes result in injuries (e.g., Koyama 1970; Marsden 1968; Witt, Schmidt, and Schmidt 1981). High rank sometimes

TABLE 32-2 Sex Differences in the Frequency of Wounds in Adults

Species	Measure Used	Sex with More Wounds	Sources
Arctocebus calabarensis (w)	Traces of serious wounds in captured animals	M	Charles-Dominique 1974a
Perodicticus potto (w)			
Galago elegantulus (w)			
G. alleni (w)			
G. demidovii (w)			
Propithecus verreauxi (w)	Fresh wounds during breeding season	M	Sussman and Richard 1974
Lemur catta (w)	Fresh wounds during breeding season	M	Jolly 1966a
Alouatta seniculus (w)	Lip injuries, scars, and missing fingers	M	Sekulic 1983a
A. palliata (w)	Scars and freshly torn lips	M	Chivers 1969
Presbytis entellus (w)	Fresh wounds	M	Hrdy 1977
Cercopithecus aethiops (w)	Fresh wounds	M	Henzi and Lucas 1980
Macaca mulatta (p)	Fresh wounds	M	Drickamer 1975; Vandenbergh and Vessey 1968
(p)	Fresh wounds during breeding season	M	Wilson and Boelkins 1970
(p)	Fresh wounds: posterior	F	Hausfater 1972
	Fresh wounds: anterior	M	
(w)	Fresh wounds	M	Lindburg 1971
(c)	Fresh wounds	F	Loy et al. 1984
(c)	Fresh wounds	M	Bernstein, Williams, and Ramsay 1983
M. arctoides (c)	Fresh wounds	M	Ibid.; Whitten and Smith 1984
M. nemestrina (c)	Fresh wounds	M	Bernstein, Williams, and Ramsay 1983
Cercocebus atys (c)	Fresh wounds	M	Bernstein 1971; Bernstein, Williams, and Ramsay 1983
Papio cynocephalus anubis (w)	Fresh wounds	M	Smuts, unpub. obser.

NOTE: w = wild; p = provisioned; c = captive.

leads to higher reproductive success for females, and, since females usually inherit ranks from their mothers in these species, the benefits of high rank can accumulate through more than one generation (chap. 26). Thus, in the rare instances in which low-ranking females have an opportunity to raise their dominance status, the stakes may be high, and we might expect females to engage in escalated aggression. In one troop of wild baboons undergoing a series of dramatic shifts in female dominance relationships, females showed three types of behavior that were otherwise observed exclusively among males (Smuts and Nicolson, unpub. obser.). First, females fought face to face and inflicted facial lacerations similar to the wounds males inflict on one another, although these were less deep. Second, to intimidate opponents, females used ritualized threats such as repeated circling, tooth grinding, and deep yawns that exposed the canines. Third, females solicited aid from one another through a combination of rapid head flagging, hip grasping, and mounting of potential allies—all tactics employed in male-male coalitions.

Finally, females in several female-bonded species engage in escalated aggression against females from other groups during intergroup encounters (chap. 22). For example, during intergroup encounters in olive baboons and rhesus monkeys, females from one group repeatedly hit, pushed, jumped on, and bit females from the other group (Nicolson and Smuts, unpub. obser.; fig. 32-1). The outcome of such encounters can have long-term reproductive consequences for females in different groups (chap. 23).

The three examples discussed above concern cases in which the payoffs of intense aggression seemed to be unusually high. A fourth example concerns situations in which the costs of intense aggression appear to be unusually low: gang attacks on single females, often by a group of closely allied female relatives. These attacks occur occasionally in female-bonded species in the wild (e.g., Wasser 1983), but they are more common in captivity when the victim can neither escape nor effectively fight back once she is cornered (e.g., de Waal 1978b; Alexander and Bowers 1969). In the wild, such attacks may be more costly both because they often require repeated chasing and because a male relative or long-term associate of the victim may intervene and attack the perpetrators (e.g., Kaplan 1978; Watanabe 1979; Smuts

FIGURE 32-1. Female rhesus monkeys slapping at females from another group during an intergroup encounter. Notice that the two females in the foreground are carrying infants. (Photo: Barbara Smuts/Anthrophoto)

1985). Because group membership is often unstable in captive groups, male intervention on behalf of relatives and associates may be less common than in the wild.

These examples indicate that the lower frequency of intense intrasexual aggression among females than among males depends, at least in part, on differences in the context of male-male versus female-female conflicts rather than on inherent sex differences in the potential for violent combat.

Sex Differences in Types of Agonistic Interventions

In addition to sex differences in the context and frequency of aggression, researchers have also considered sex differences in patterns of aggression. One such pattern involves interference by third parties in dyadic disputes (chap. 25). It takes two forms: (1) intervention on behalf of a victim (loser support), and (2) support for the aggressor (winner support). In a study of captive chimpanzees, de Waal (1984b) reported that adult male interventions involved a significantly higher proportion of winner support than those of adult females, who generally supported losers. Furthermore, females usually

supported relatives and friends, whereas males sometimes intervened against their closest associates. Based on these quantitative findings and qualitative observations of the outcomes of interventions, de Waal concluded that males used interventions to form opportunistic alliances that might increase status, whereas females intervened to protect vulnerable group members and to maintain peaceful relations within the group.

Although these sex differences are intriguing, no evidence indicates that they hold for primates in general. Both males and females show winner support and loser support, depending on the circumstances. For example, young adult male baboons sometimes support winners in an opportunistic manner, but older males tend to support their stable coalition partners, whether or not the partner is winning or losing a dispute (Smuts 1985). Similarly, male macaques tend to support winners (Bernstein and Ehardt 1985) except when they intervene to protect matrilineal kin who are the victims of aggression (e.g., Kaplan 1978). In baboons and macaques, adult males consistently support losers when the victim is a close female associate or one of her offspring (e.g., Kurland

1977; Chapais 1983f; Seyfarth 1978b; Smuts 1985). Finally, even among the chimpanzees studied by de Waal, patterns of male intervention varied depending on the circumstances. When vying for rank, males supported winners, but once a male had firmly established the alpha position, he shifted toward loser support (de Waal 1984b).

Females, like males, may support either winners or losers, depending on the context. In cercopithecines, females intervene on behalf of relatives most often, and when they do so, they support both winners and losers (e.g., Datta 1983a, 1983c). However, females also support unrelated females against other females. When this occurs, they generally support higher-ranking aggressors. These alliances with higher-ranking females appear to help females maintain or enhance their own status (chap. 25). Finally, as noted below, females sometimes support unrelated females and immatures when they are victims of aggression by adult males.

Thus, both males and females are capable of opportunistic alliances and defense of victims, depending on the circumstances. It is therefore inappropriate to conclude that males in general are more concerned with winning than are females or that females are more concerned with protecting friends and relatives than are males.

Observed Behavior versus Behavioral Potentials

The examples discussed earlier clearly demonstrate, in both females and males, a potential for agonistic behaviors typically shown by the opposite sex. Observations of nonhuman primates suggest that, as among humans, expression of behavioral potentials may be inhibited in the presence of members of the opposite sex and released in their absence (Goldfoot and Neff, in press). Two sets of findings illustrate this point.

First, experiments with captive macaques and squirrel monkeys have shown that, in the absence of females, males exhibit a wide variety of infant caretaking behaviors that rarely occur when females are present (Vaitl 1977b; Snowdon and Suomi 1982). For instance, when male and female rhesus macaques were individually exposed to a distressed infant, both responded with equal solicitude. When the experiment was repeated with male-female pairs, the females showed caretaking behaviors but the males did not, and some actively avoided the infants (Gibber 1981, cited in Goldfoot and Neff, in press). Second, in the absence of males, some females in all-female groups adopt behavior typical of the male or males in bisexual groups. Examples include wild rhesus females who displayed at strange males (Neville 1968a), immature rhesus females who showed the typical male foot-clasp mount (Goldfoot, Wallen, and Neff 1984), and female hamadryas who adopted the centripetal social role normally taken by the single breeding male (Stamm-

bach 1978). Thus, the fact that a behavior has been observed only among males or only among females does not necessarily imply that members of the opposite sex are unable to perform the behavior or lack the motivation to do so.

Conclusion: The Study of Sex Differences

There is a widespread tendency to interpret sex differences in terms of qualities or traits inherent to males and females. So, for example, if males are observed to injure one another more often than females, and if females are observed to support victims more often than males do, this difference is attributed to the fact that males are more aggressive than females and that females are more protective than males. In other words, intrinsic sex differences in temperament are used to explain sex differences in behavior. This is circular reasoning, analogous to "explaining" a behavior in terms of instinct, which, to use a memorable phrase, "is just another name for an inscrutable behavior" (Itani and Nishimura 1973, 23).

The examples discussed earlier suggest that, in studying behavioral sex differences, it is useful to shift attention away from intrinsic properties of individuals and toward social interactions and the contexts in which they take place (see also Fedigan 1982; Goldfoot and Neff, in press). By examining how, and in what situations, the behaviors of females and males are similar and different, primatologists can begin to develop testable hypotheses about why these patterns occur.

DOMINANCE, AGGRESSION, AND INFLUENCE AMONG FEMALES AND MALES
Introduction

Nonhuman primates influence the behavior of others in two principal ways (Fedigan 1982; Noë, de Waal, and van Hooff 1980). The first involves superior competitive ability or "control that is backed by aggressive sanctions" (Noë, de Waal, and van Hooff 1980, p. 108). In other cases, individuals influence one another's behavior through control over benefits that cannot be taken by force, such as affiliative behaviors or support by a third party during agonistic interactions. Nonhuman primates exploit both sources of influence in their interactions with members of the opposite sex, and which sex exerts more influence depends on the situation and the individuals involved. This is not widely recognized, in part because most researchers interested in conflicts of interest tend to focus on the use of aggressive power, which is more obvious and easier to study than other, nonagonistic forms of influence. From this perspective, males often—but not always—have a competitive edge, as indicated by their tendency to show aggression toward females and their ability to dominate females physically.

In the following section I briefly discuss male aggression toward females and dominance relationships between the sexes. The rest of the chapter considers agonistic and nonagonistic sources of female influence vis-à-vis males.

Male Aggression toward Females

Male aggression toward females is common in many nonhuman primates, but the situations in which it occurs and its effect on females have received little systematic attention. What follows is therefore incomplete, and its main purpose is to emphasize the need for further research.

Male aggression toward females often occurs in a sexual context. Males occasionally attack females who refuse to mate with them, and in macaques and baboons, estrous females usually receive more aggression and more wounds from males than do anestrous females (Tokuda 1961; Enomoto 1981; Kurland 1977; Hausfater 1975; Fedigan 1982; but see Loy et al. 1984). During intergroup encounters, male baboons and vervets chase and sometimes attack females from their own group in an attempt to minimize contact between them and males from

other groups (chap. 22); male hamadryas baboons use a neck bite to discourage females from straying too close to other males (fig. 32-2). Male chimpanzees and gorillas attack females from other groups, and these attacks sometimes result in female transfer to the attacking male's group (Goodall et al. 1979; Fossey 1983). In all of these cases, male aggression may represent sexual harassment, that is, its function may be to increase the probability, by increasing the costs of noncooperation, that a female will mate with the aggressor or join his group (and thus become a potential mate). Unfortunately, few data exist to test this hypothesis.

The most extreme example of male aggression apparently designed to increase mating opportunities is, of course, infanticide, which frequently involves attacks on females as well as infants (chap. 8).

Although male sexual harassment appears to be common in many species, it is important to stress three points. First, in some species, male aggression toward females has not been observed in a sexual context (e.g., spider monkeys: Fedigan and Baxter 1984). Second, al-

FIGURE 32-2. Male hamadryas baboon using a neck bite to punish a female who has strayed too close to other males. (Photo: Hans Kummer)

though males often use aggression to influence female sexual availability and mate choice, they do not force females to mate, with the exception of orangutans (chap. 13). It is possible that forced copulations do not occur in other monkeys and apes because of potential interference by other animals, both males and females (Smuts 1985). Third, males show aggression toward females in many contexts other than mating. For example, in rhesus monkeys (Chapais 1983f) and olive baboons (Smuts 1985), males routinely threaten and attack anestrous females. In one olive baboon troop, anestrous females were victims of male aggression on average once every 17 daylight hours; 25% of the aggression involved attacks (Smuts 1985). Males inflicted severe wounds during approximately 2% of these attacks, and each anestrous female, on average, received one severe wound per year. Male aggression toward anestrous females occurred primarily in the following contexts: (1) during feeding competition; (2) in defense of a close female associate or one of her offspring; (3) in order to "punish" a female for previous aggression toward a female associate or her offspring; and (4) during a dispute with another male (redirected aggression). Male-female aggression thus often involved protection or manipulation of relationships with others.

Although females are vulnerable to male aggression, they have means of protecting themselves. These are discussed first in the context of dyadic dominance relationships and then in terms of other sources of female influence.

Dominance

Dyadic dominance relationships between adult male and female nonhuman primates can be divided into five major types:

1. Species in which sexual dimorphism in body size is slight (with females sometimes slightly larger than males) and in which females are clearly dominant to males. This group includes many (perhaps all) lemuriforms (chap. 3) but no other species, as far as is known.

2. Species in which sexual dimorphism is slight and the sexes are "codominant," that is, agonistic interactions are uncommon; when they do occur, neither sex consistently dominates the other. This group includes callitrichids, monogamous New World monkeys, and monogamous gibbons (e.g., Evans and Poole 1984; Chivers 1974; Tilson 1981).

3. Species in which males and females are similar in size and in which males consistently dominate females through male-male coalitions. As far as is known, spider monkeys are the only primates that fit this pattern (Klein 1974; Fedigan and Baxter 1984).

4. Species in which males are larger than females and in which females nevertheless sometimes dominate males, often through female-female coalitions. This group, discussed further later on, includes squirrel monkeys, talapoins, vervets, macaques, and possibly patas monkeys and Sykes's monkeys.

5. Species in which males are larger than females and in which females rarely, if ever, dominate males. This group includes gelada, hamadryas, and savanna baboons (Dunbar 1984b; Kummer 1968; Hausfater 1975), all great apes in the wild (Kuroda 1980; Bygott 1979; Harcourt 1979c; chap. 13), mantled howlers (Jones 1980b), and gray langurs (Hrdy 1977).

For many species, insufficient information precludes definitive classification. This is particularly true for many cebids, colobines, and nocturnal prosimians, although in at least some nocturnal prosimians in which sexual dimorphism in body size is slight, females use aggression, including biting of the testicles, in order to rebuff mating attempts (Charles-Dominique 1974a; Foerg 1982a).

In type 1, 3, and 5 species, heterosexual dominance relations are clear-cut, and in type 2 species they appear relatively unimportant. In species grouped under type 4, however, dominance relationships between the sexes are complex, in part because of the effects of female-female coalitions against males.

Female-Female Coalitions against Males

Females join forces against males in a wide variety of nonhuman primate species, including species in which males always dominate single females (e.g., chimpanzees, baboons) and species in which single females sometimes dominate males (e.g., talapoins, macaques). Female coalitions are most common in three contexts: repulsion of strange males who have entered or are attempting to enter a group, protection of an adult female attacked by a male, and, probably the most common situation of all, protection of infants (table 32-3).

Protection of infants occurs during infanticidal attacks, but it is also common at other times. Females in many species are extraordinarily sensitive to male proximity to infants, and the faintest hint of infant distress can provoke explosive gang attacks on a nearby male even if he has shown no aggression toward the infant (e.g., Ransom 1981; Jay 1963b). This extreme female vigilance may explain why males often avoid infants and why they sometimes even exhibit fearful responses when an infant approaches (macaques: Itani 1959; Kurland 1977; Simonds 1974; de Waal 1977; Alexander 1970; vervets: Lancaster 1975; olive baboons: Ransom 1981; squirrel monkeys: Vaitl 1977b). It also suggests that the potential for male harm to infants has been a potent selective force operating on the behavior of females, even in species, such as Japanese macaques, in which infanticide has not been observed in the wild (Smuts 1985).

TABLE 32-3 Contexts in Which Females Form Aggressive Coalitions against Males

Context	Species	Sources
Male attempting to enter group	*Colobus badius* (w)	Starin 1981
	Erythrocebus patas (p)	Kaplan and Zucker 1980
	Cercopithecus talapoin (c)	Rowell 1974a
	C. ascanius (w)	Cords 1984a
	C. aethiops (w)	Cheney 1983a, 1983b
	Macaca mulatta (w, p)	Neville 1968a; Vessey 1971
	M. fuscata (p)	Packer and Pusey 1979
Male attacks, herds, or frightens female	*M. mulatta* (c, p)	Bernstein and Ehardt 1985; Chapais 1981
	M. fuscata (p)	Watanabe 1979
	Papio cynocephalus anubis (w)	Smuts, unpub. obser.
	P. c. ursinus (w)	Hall 1962b
	Theropithecus gelada (w)	Dunbar 1984b
	Pan troglodytes (c)	de Waal 1982
	Presbytis cristata (w)	Bernstein 1968
Male attempts to harm infant or is nearby when infant shows distress	*P. entellus* (w)	Jay 1963a; Hrdy 1977; Boggess 1979
	Cercopithecus mitis (w)	Butynski 1982b
	C. ascanius (w)	Struhsaker 1977
	C. aethiops (w)	Lancaster 1972
	Erythrocebus patas (c, w)	Hall and Mayer 1967; Hall 1968a
	Macaca mulatta (c, w)	Bernstein and Ehardt 1985; Lindburg 1971
	M. fuscata (p)	Watanabe 1979; Kurland 1977
	M. fascicularis (c)	Chance, Emory, and Payne 1977
	Papio cynocephalus anubis (w)	Ransom 1981; Smuts 1985
Females attempt to drive male to periphery or expel him from group	*Cebus apella* (w)	Robinson 1981a; Izawa 1980
	Erythrocebus patas (c)	Hall 1967
	Cercopithecus talapoin (c)	Rowell 1974a
	Macaca fuscata (p)	Kawamura 1967; Koyama 1970
Male approaches females and/or young	*Saimiri oerstedii* (w)	Baldwin and Baldwin 1972b
	S. sciureus (c)	Baldwin 1968
Male attempts to mate with unwilling female	*Cercopithecus aethiops* (w)	Andelman 1985
	Erythrocebus patas (c)	Hall 1967
	Pan troglodytes (c)	de Waal 1982
Male attempts to inspect female perineum	*Saimiri sciureus* (c)	Baldwin 1968
Competition for dominance or leadership	*Macaca fuscata* (p)	Yamada 1963
	M. arctoides (c)	Bernstein 1980
Context not specified	*Cercopithecus talapoin* (c)	Wolfheim and Rowell 1972

NOTE: w = wild; p = provisioned; c = captive.

Although information on kinship among females who form coalitions against males is often lacking, in at least some cases coalitions involve females from different matrilines (rhesus monkeys: Bernstein and Ehardt 1985; squirrel monkeys: Baldwin and Baldwin 1972b; gray langurs: Hrdy 1977; vervets: Cheney 1983a; olive baboons; Smuts, pers. obser.). Why should females protect females or infants to whom they are not closely related? One explanation suggests that female mobbing of males provides benefits similar to those proposed for mobbing of predators (Harvey and Greenwood 1978): because mobbing serves to warn the "predator" (in this case, a male) that hostility toward females and infants is costly, it may reduce the chances that the mobbing female (or one of her relatives) will be the victim of male aggression in the future. Other instances of female coalitions against males (e.g., preventing males from entering groups or chasing them away from feeding sites) may reflect the mutual benefits females receive from reduced feeding competition with males (Cheney 1983a) and reduced sexual harassment. Finally, Cheney (1983a) suggested that low-ranking females might support a higher-ranking female against a male in exchange for later support by the higher-ranking female in another context (e.g., during female-female agonistic interactions).

Usually female coalitions simply force males to withdraw or run away, but occasionally females injure males (vervets: Cheney 1981), sometimes severely (Japanese macaques: Packer and Pusey 1979; see also chap. 22).

The widespread occurrence of female-female coalitions against males makes it difficult to interpret observations of smaller females dominating larger males (type 4 species, table 32-4). First, in many reports, authors do not differentiate between cases in which a female dominated a male as a member of a coalition and cases in which single females dominated males in dyadic encounters. Second, even in those cases in which single females have been observed to dominate males, it is difficult to know the extent to which this reflects the ever-present possibility that, if the male does not give way, the female will be supported by other females.

The same issue arises in studying intrasexual dominance relationships since, in many species, rank is influenced or even determined by coalitionary support (chap. 25). In these species, however, subordinates use ritualized submissive gestures to clearly indicate "formal" acknowledgment of status differences; such gestures are common during intrasexual interactions and when females submit to males (chap. 34). It is unclear, however, how often males who appear individually subordinate to females show similar formal acknowledgment of status differences. Most reports describe males giving way to females or running away from them but do not mention submissive gestures. In at least two studies, males never showed submission to aggressive females (crab-eating macaques: de Waal 1977; patas monkeys: Kaplan and Zucker 1980). A systematic study of the frequency of adult male submissive gestures toward adult females would help resolve this issue.

In many of the species listed in table 32-4, females influence male group membership, rank, and mating success, which is another reason why interpreting observations of females dominating males is difficult. In these species, males may be willing to defer to females during agonistic encounters in order to foster or maintain affiliative relationships with influential females. For example, in one group of rhesus macaques, the alpha female dominated several central, high-ranking males, even though she ranked below peripheral, lower-ranking males. Chapais (1983d, 1983f) argued that the high-ranking males' deference toward this female reflected their reliance on her support in coalitions against other males rather than a true dominant-subordinate relationship. This hypothesis is consistent with the fact that the alpha female's son dominated her; he could presumably depend on her support without trading favors.

Finally, in heterosexual dominance interactions, as in dominance interactions in general (chap. 25), the outcome can depend on the context. For example, in a group of captive chimpanzees, males deferred to females in situations involving competition over space or objects but dominated females in other contexts (Noë, de Waal, and van Hooff 1980).

However one interprets observations of individual females dominating males, it is clear that in many species, female coalitions represent an important source of influence over male behavior. For example, in a group of rhesus monkeys, every time a low-ranking adult male

was observed attacking an adult female, a coalition of females interfered by chasing him away (Chapais 1981). In a group of Japanese macaques, six of nine adult males received more aggression from adult females (per unit time spent within 3 m) than from other adult males (Packer and Pusey 1979).

Female Control over Benefits That Cannot Be Acquired by Force

Female Influence over Male Group Membership. In a number of species, females unite to chase strange males away from their groups (table 32-2; chaps. 21, 22). Hall (1967), Packer and Pusey (1979), and Cheney (1981) concluded that in patas monkeys, Japanese macaques, and vervets, respectively, female aggression is a major determinant of which males achieve group membership.

Females also use coalitionary aggression to drive resident males to the periphery of the group or to exclude them altogether (captive patas monkeys: Hall 1967; captive talapoins: Rowell 1974a; Japanese macaques: Kawamura 1967; wedge-capped capuchins: Robinson 1981a). In some cases, females continue to show frequent aggression toward new males long after resident males have ceased to do so (e.g., gray langurs: Boggess 1980).

Females influence male troop membership through affiliative behaviors as well as aggression. In a captive gorilla group, females, by allying with one male against the other, determined the outcome of aggressive competition between males vying for control of the group (Tilford and Nadler 1978). In gelada baboons, female preferences, demonstrated by allowing males to mount and other affiliative behaviors, strongly influence the outcome of challenges to the resident male (Dunbar 1984b; Mori and Dunbar 1985). In savanna baboons and macaques, male immigrants are usually peripheral initially; in order to move into the main body of the troop they must first succeed in forming affiliative relationships with several adult females (Japanese macaques: Fedigan 1976; rhesus macaques: Breuggeman 1973; Kaplan 1978; olive baboons: Strum 1982; Smuts 1985). Some males fail to gain such acceptance, and even when they are able to dominate the males in the group, they do not achieve integration (Smuts 1985). Finally, females sometimes actively entice males into their groups through affiliative behaviors (Cheney and Seyfarth 1977; Packer 1979a; Smuts 1985).

Relationships with females can continue to affect male group membership after a male has achieved initial acceptance. In gelada baboons, males that are unable to groom with each of their females are more vulnerable to takeovers by another male (chap. 20; Dunbar 1984b). In one troop of olive baboons, adult males whom females avoided the least subsequently remained in the group the longest (Smuts 1985).

TABLE 32-4 Species in Which Adult Females May Sometimes Dominate Larger Males

Species	Adult Female Weight as Percentage of Adult Male Weight	Type of Evidence	Sources
Saimiri sciureus (c)	75[a]	3 adult males had priority of access to food over 3 adult females	Green et al. 1972
S. oerstedii (w)	—	Females submissive to males during dyadic encounters; females dominate males only when they form a coalition with another female	Baldwin and Baldwin 1972b
Cebus apella (w)	79[b]	Adult males and females are interspersed in the hierarchy; alpha female dominant to all males except alpha. Based on dyadic agonistic encounters, including feeding supplants and fights, but excluding submission and avoidance in the absence of threats or a contested resource	Janson 1984
Cercopithecus talapoin (c)	81[c]	Based on outcomes of dyadic approach-retreat interactions, 2 females ranked above all 3 adult males, and 2 other females ranked above 1 adult male	Wolfheim 1977a
(c)	81[c]	All adult females ranked above all adult males	Dixson et al. 1973
C. mitis (c)	72[d]	Based on outcome of approach-retreat interactions, all adult females ranked higher than single adult male, and females took food away from male	Rowell 1974a
C. aethiops (c)	80[e]	Some adult females dominated some adult males; females from high-ranking lineages dominated their brothers	Bramblett et al. 1982
(c)	80[e]	Based on outcome of approach-retreat interactions, 1 adult female ranked above single adult male	Rowell 1974a
Erythrocebus patas (c)	56[b]	Females defeated the male 47 times; males defeated females 41 times (defeats defined as threat followed by flight). Author does not specify whether female victories involved single females or female coalitions	Loy 1981
(c)	56[b]	Male has priority of access to resources and usually holds ground when attacked by single female	Hall and Mayer 1967
(p)	56[b]	Single females attacked male, but male never showed submission to females	Kaplan and Zucker 1980

Female Influence on Male Dominance Rank. In macaques, female agonistic support can influence, or even determine, male dominance ranks vis-à-vis other males (rhesus macaques: Chapais 1983d; Japanese macaques: Kawamura 1967; Gouzoules 1980a; captive pigtailed macaques: Bernstein 1969; captive bonnet macaques: Glick 1980). Female agonistic support can also influence male dominance relationships in captive vervets (Raleigh, McGuire, and Brammer 1982) and chimpanzees (de Waal 1982).

Female Mate Choice. As noted earlier, forced copulations do not occur in wild nonhuman primates (with the exception of orangutans), and female preferences affect male mating opportunities and male behavior in many—perhaps all—nonhuman primate species, as discussed in detail in chapter 31.

Power and Reciprocity between the Sexes

In many primates, females control benefits important to reproductive success of males. Males also control benefits important to females, including the capacity to protect females and young from other males and to aid females in competition with other females. Nonhuman primates exploit this situation by forming heterosexual relationships that facilitate the exchange of benefits. Female chimpanzees prefer to consort with males that spend time with them, groom them, and share meat with them (Tutin 1979). Female macaques (and, in captivity, female chimpanzees) help males acquire high rank; these males in turn protect females from aggression by other troop members (Chapais 1983f; de Waal 1982). Vervet monkey females allow certain males to join their groups, and the males aid females in territorial defense (Cheney 1981). Male baboons form long-term, protective rela-

TABLE 32-4 *continued*

Species	Adult Female Weight as Percentage of Adult Male Weight	Type of Evidence	Sources
Macaca arctoides (c)	—	Alpha female showed aggression toward second-ranking male; in 10 of 12 months, she won more of their agonistic interactions than he did	Bernstein 1980
M. fascicularis (c)	69[b]	1 old female was dominant to "one or two" of the adult males	Bunnel, Gore, and Perkins 1980
(c, w)	69[b]	In captivity, females occasionally dominated young adult males, but in the wild, all adult males dominated all adult females	Angst 1975
M. mulatta (c)	83[f]	"Most matrilines are dominant to all but the highest-ranking non-natal adult males"	Bernstein and Ehardt 1985, 48
(c)	83[f]	Some females dominate some males; females are more likely to dominate males as adults when raised in groups with mothers present for first year of life	Goldfoot, Wallen, and Neff 1984
(c)	83[f]	Several instances of adult females dominating adult males. Female victory rate = 44% of male victory rate in all heterosexual fights	Loy et al. 1984
(p)	83[f]	Alpha female dominant to all high-ranking males except her son; same female subordinate to low-ranking males	Chapais 1983d
M. fuscata (c)	84[g]	Adult males tended to cluster at the top and bottom of the dominance hierarchy. During the birth season, males dominated on average 36% of the adult females; during the breeding season, they dominated 55% of the females. Based on dyadic agonistic interactions in which the recipient of aggression was clearly submissive	Johnson, Modahl, and Eaton 1982

NOTE: c = captive; p = provisioned; w = wild.
a. Jorde and Spuhler 1974.
b. Napier and Napier 1967.
c. Gautier-Hion 1975.
d. Coelho 1974.
e. Fedigan 1982.
f. Rawlins, Kessler, and Turnquist 1984.
g. Sugiyama and Ohsawa 1982b.

tionships with anestrous females and their offspring; when the female resumes estrous cycles, she often prefers to mate with her close associate (Smuts 1985). Thus, in studying power relations between the sexes, it is important to focus not only on aggression and dominance, but also on the ways in which females and males use benefits that cannot be taken by force to influence one another's behavior (Noë, de Waal, and van Hooff 1980).

CONCLUSION

Behavioral variations between the sexes in nonhuman primates cannot easily be reduced to inherent gender differences. The relative costs and benefits of different reproductive strategies influence the overall behavioral dispositions of males and females, but particular behaviors reveal great variation within these general parameters: In most nonhuman primates, males fight more intensely than females and injure one another more frequently. Yet in many species, females supplant, threaten, and chase one another as often as do males. Which sex is more aggressive? In most nonhuman primates, females are smaller than males and usually give way to males in dyadic contests. Yet female nonhuman primates routinely form coalitions against males, sometimes driving them from the group. Which sex is more influential?

As the observations described in this chapter indicate, there are no simple answers to these questions for primates as a whole. Better understanding of sex differences in nonhuman, as well as human, primates awaits a clearer appreciation and investigation of the complex social environments in which these differences find their varied articulation.

SUMMARY

Aggressive and affiliative behaviors of male and female primates vary depending on the species, the social context, and the individual. In species in which females remain in their natal groups, female-female bonds are more prominent than male-male bonds, but male-male bonds are more prominent in female-transfer species. Among primates as a whole, females show aggressive behavior as often as do males, but males generally wound one another more often than do females. These sex differences are best understood in terms of sex-specific reproductive strategies rather than as reflections of intrinsic temperamental differences between the sexes. In most nonhuman primates, males dominate females individually, but in several species females dominate males, sometimes individually and sometimes as the result of female-female coalitions. Females use coalitions with other females to gain protection against male aggression and to influence male group membership. For both females and males, control of benefits that cannot be taken by force, such as affiliative behaviors, represents a source of leverage in interactions with others.

33 | Can Nonhuman Primates Help Us Understand Human Behavior?

Robert A. Hinde

Studies of nonhuman primates form an important part of a scientific enterprise that is giving us increased understanding of the world about us, of the evolutionary processes that gave rise to it, and of our place within it. Beyond that, however, nonhuman species, and especially nonhuman primates, provide an important source of data for understanding many aspects of human behavior and physiology in terms of causation, developmental processes, function, and evolution. However, comparisons between human and nonhuman behavior are not to be undertaken lightly, for they can give rise to misleading conclusions. In this chapter five ways in which studies of nonhuman primates have been used to enhance understanding of our own species are discussed, the primary aim being to specify where comparisons are and are not useful. The fivefold division is, of course, merely a heuristic device: there is no implication that the list is exhaustive, nor that any one study cannot contribute in more than one way. The chapter concentrates on the relevance of nonhuman primates to aspects of human psychology and psychopathology. I have drawn on data from both field and laboratory, the former often aiding interpretation of the latter, and used work on animals other than primates where it seemed appropriate.

DIRECT BEHAVIORAL PARALLELS BETWEEN PARTICULAR NONHUMAN SPECIES AND PARTICULAR HUMAN CULTURES

Attempting to draw direct parallels in behavior between human and nonhuman species is a dangerous pastime. Since there are about two hundred different species of nonhuman primates, and an even larger number of recognizably distinct human societies, it is not difficult to find parallels to prove whatever one wishes. In any one case the use of a different species or a different culture could produce a very different perspective (e.g., Lehrman 1974). For example, earlier chapters in this volume show that nonhuman primates may live solitarily, monogamously, in one-male, or in multimale groups. Humans usually live in monogamous, polygynous, or (rarely) polyandrous societies: conclusions over matters that relate to sociosexual arrangements could easily mislead if based on simple comparisons. While the writings of authors such as Morris (e.g., 1967) may have been useful in raising consciousness about our biological ori-

gins, conclusions about the human condition drawn from simple parallels with monkeys or apes should be treated with great caution. We shall see later, however, that the application of *principles* drawn from a range of species can be very fertile.

NONHUMAN PRIMATES AS MODELS

Important advances in the study of learning, motivation, physiological psychology, drug effects, and so on have come from the experimental use of monkeys and apes (e.g., Serban and Kling 1976). Innumerable examples are available, but only one will be mentioned here. Rhesus monkeys brought up in various degrees of social deprivation show aberrant behavior and are socially inept in various ways. For example, isolation-reared monkeys placed with socially reared age-mates were overwhelmed by the age-mates' playfulness and aggressiveness. Suomi and Harlow (1972) suggested that the rehabilitation of such monkeys might be achieved by exposing them to younger socially reared animals who approached them in a nonthreatening manner. Such proved to be the case: 6-month-old isolates paired with 3-month-old socially reared "therapists" gradually lost their aberrant behavior patterns and subsequently developed near-normal social behavior. On the basis of this work, the same principle has been applied with success to the rehabilitation of socially withdrawn children (Furman, Rahe, and Hartup 1979).

In using nonhuman primates as models, which species is chosen for study may be crucial. For example, studies of the effects of mother-infant separation on the behavioral development of the infant have been carried out with a number of species of monkeys. Among the earlier studies, rhesus and pigtailed macaque infants showed marked phases of "protest" and "despair" or depression after separation, but bonnet macaques did not show the latter to anything like the same extent. This is perhaps related to a difference in social structure between the species: bonnet infants received more attention from other females during the mothers' absences than did rhesus and pigtailed infants (Kaufman and Rosenblum 1967). If only one of these macaques were studied, the implications for the human case might seem very different according to the species selected (Lehrman 1974): study of several species helped clarify the factors leading in the short-term to infant distress and in the longer term to distortions of development.

An important criterion for choosing an animal species for studying a problem in human behavior or physiology is often its similarity to our own species. Apes are thought of as more "relevant" than monkeys, and monkeys more relevant than rats. However, several issues relating to this criterion of similarity must be borne in mind. The first is a general one: the value of models in science lies in part in their differences from the object, species, or problem of central interest. Models are useful because, by virtue of availability, simplicity, or manipulability, they pose questions, suggest relations, or can be used in experiments in a way not possible with the original. If a model comes too close to being a replica, it loses some of its raison d'être (Craik 1943).

The second point relates to this and concerns a common implicit reason for using nonhuman primates as models, namely, it is legally permissible to manipulate them in a way that would be considered quite improper with human subjects. In general, doubts about the propriety of using animals decrease, the greater the difference in their cognitive capacities from our own. But the converse of this must not be forgotten: ethical considerations must be more important the closer the model is to our own species. No animals should be used in experiments unless careful consideration indicates that the probable gains are sufficient to justify the suffering caused, and then higher mammals, monkeys, and especially apes should never be used if other species will do.

Third, the very act of comparing may lead us to underestimate the complexity of the human case: overt similarities between certain aspects of nonhuman and human behavior do not mean that concepts or principles derived from the behavior of animals are wholly adequate to account for our own. The study of aggression is a case in point. Physiological studies of animals can tell us much about the neural and hormonal mechanisms of aggressive behavior. Behavioral studies of animals can tell us something about the role of conflict in human aggression, about its development in the individual, and about the situations that induce it (e.g., Hamburg and Trudeau 1981). But that does not mean that lessons drawn from the study of aggression in animals are wholly adequate to account for the complexities of aggression in humans, and animal data must not cause us to underestimate those complexities. Biological principles might contribute a little toward an understanding of the sexual jealousy leading to the death of Desdemona, but they would be stretched in explaining the mixture of vengeance, jealousy, misplaced compassion, and self-destruction that led Medea to murder her sons.

Fourth, the choice of a primate model often depends on comparable independent or dependent variables in human and nonhuman primates. Sometimes a particular human symptom or syndrome, such as depression, is to be studied. An animal model is found in which similar symptoms can be induced, and their aetiology in the animal is studied. Alternatively, the effects of a particular independent variable with known pathological consequences in man, such as crowding, are studied in an animal. In either case it is necessary to remember that pathological symptoms have multiple causes and involve a complex nexus of events: we may be able to identify one point in the nexus in the animal model (e.g., an independent or dependent variable) with a similar point in the human case, but as we move away from that point, similarity becomes increasingly less probable. Thus social isolation in a vertical chamber may be an effective way of inducing symptoms in rhesus monkeys that resemble those of human depression (cf. McKinney, Suomi, and Harlow 1971), but the biochemical intermediaries may or may not be the same as those of human clinical depression. Resemblances to reactive depression are perhaps more probable than resemblances to endogenous depression, but even in the former case the precipitating factors are surely very different from those operating in the human case.

HUMAN ETHOLOGY

Ethologists were formerly concerned primarily with nonhuman species. In recent years, however, there has developed a discipline of human ethology in which techniques developed for the study of animals, and concepts and principles derived from the study of animals, are applied to our own species. Its achievements, which have been considerable, and its limitations have been assessed recently by von Cranach et al. (1979). One product of its development has been an increase in the number of studies of child behavior that use observational techniques. Although such techniques had been much used by developmental psychologists in the 1930s (reviewed by Hinde 1983b), they fell into disuse as problems became more focused and other techniques were developed. The influence of primate studies has been important in the renewed emphasis given to the recording of behavior as it occurs in real-life situations. For example, until recently a high proportion of the studies of children's friendships and popularity relied on sociometric techniques—individuals were asked who their friends were and so on. However, such methods have their limitations. For instance, it became apparent that preschool children often give answers that do not distinguish between their actual friends and those with whom they would like to be friends: the use of an observational approach enables this issue to be resolved and also facilitates study of why some children are better than others at making friends, entering groups, and so forth (Asher and Gottman 1981; Foot 1980).

Another field in which human ethology has been outstandingly successful is the study of communication (e.g., Eibl-Eibesfeldt 1975; Lockard 1980). Ethologists' earlier clashes of opinion with more traditionally minded psychologists, in which both sides overstated their cases, led to a fruitful synthesis. The ethologists at first tended to overemphasize the cross-cultural generality and independence from experience of human signal movements (e.g., Eibl-Eibesfeldt 1975), while others (e.g., Birdwhistell 1967) emphasized their cultural specificity and dependence on early learning. Rapprochement was due largely to the work of the psychologist Ekman, who classified human signal movements into categories that indicated just where each of these two approaches was more relevant (Ekman and Friesen 1969). In some cases comparative data from nonhuman primates can throw light on the nature of human expressive movements: thus van Hooff's (1972) suggestion that the human smile and laugh are related to the fear grin and play face of nonhuman primates does much to resolve the problem of the relation between smiling and laughing. Of course comparative data can never fully explain all aspects of human nonverbal communication.

We shall meet other examples of the value of human ethology later, but in general it seems likely that an ethological approach will prove more valuable in the understanding of human behavior if it can be married to techniques already in the field and not isolated as a new discipline of human ethology. The application of ethologically derived observation techniques to the study of human behavior, in isolation from other methods of data collection, can lead to the study of rather sterile problems (e.g., an excessive emphasis on dominance hierarchies in children). The methods of primatology are more valuable if applied in parallel with those of experimentation, questionnaire, and interview already in use in this field. For example, in a recent study of preschool children's home relationships, observational techniques were combined with maternal interviews and questionnaires. Comparisons between data sets showed, for instance, that mothers scoring high on scales of irritability and anxiety tended to have children rated high on the temperamental characteristics moody and intense; and that while girls rated high on the characteristic shy had less tense relationships with their mothers (assessed by observation) than nonshy girls, shy boys had more tense relationships than nonshy boys. The interview data suggested that this sex difference could be explained in terms of maternal adherence to social norms (Simpson and Stevenson-Hinde 1985; Hinde and Stevenson-Hinde 1986). Such relations would not be revealed by the sole use of either observational or interview and questionnaire techniques.

In general, in studies of both nonhuman and human primates, what an individual did at a given moment may be less important for understanding what he does next than what he thought he did or hoped to do. The nonhuman primatologist is handicapped because information about motives and beliefs is not easily available: the human ethologist must not ignore the unique opportunities made possible by the existence of human language.

ABSTRACTION OF PRINCIPLES FROM STUDIES OF NONHUMAN PRIMATES

While direct comparisons between particular nonhuman primate models and the human case are often profitable, the scientist is often on surer ground if she or he can abstract principles from studies of other species. The applicability of these principles to the human case can then be assessed.

Sometimes such principles impress themselves upon us just because of the differences between nonhuman and human primates. For example, although mothers were well aware that human babies could be soothed by being held, pediatric practice in the 1940s and 1950s underplayed the importance of contact comfort with the mother. This came about in part because mother-child interaction was interpreted largely in terms of reinforcement theory, and food and warmth were seen as primary. The demonstration by Harlow and Zimmerman (1959) of the important part played by contact comfort in the development of rhesus monkeys facilitated recognition of its lesser but real importance in the human case (e.g., Bowlby 1969).

Another example concerns the relative roles of mother and infant in determining the nature of the mother-infant relationship. Perhaps because they can tell mothers what to do, while babies pay no heed to their admonitions, pediatricians were formerly wont to emphasize the extent to which parents control the behavior of their infants and to disregard the extent to which infants control their parents. In part because in many animal species parental care is more obviously controlled by sign stimuli from the young, a more balanced view had been present in the animal literature; its relevance to our own species is now increasingly accepted (Bell and Harper 1977). This has opened the way for a more sensitive analysis of mother-child interaction, with due attention to the ways in which each affects the actions of the other (Stern 1977).

Another type of difference lies in the relative simplicity of the behavior of nonhuman primates. Although complicated enough, as every primatologist knows, the lack of verbal language leads to the highlighting of issues that are too easily shrouded in the complexity of the human case. Indeed, understanding of the influence of language on culture or on human cognitive abilities, or the influence of cultural factors on human relationships, may be enhanced by comparison with nonhumans, where such influences are less pervasive. We will consider some examples in the following paragraphs.

One example is provided by the manner in which studies of nonhuman primates have led to a clear recognition of successive levels of complexity in social phenomena, namely, (*a*) the behavior of individuals; (*b*) interactions; (*c*) relationships involving series of interactions between individuals known to each other; (*d*) social structure involving the patterning of dyadic and higher-order relationships; and (*e*) (in the human case) sociocultural structure involving the interrelations between the institutions of the society and the associated beliefs and so forth (fig 33-1) (Hinde 1979, 1984b). Each of the items in this sequence influences and is influenced by its neighbors: while ethologists and primatologists have been mainly interested in influences from less to more complex levels, anthropologists have focused mainly on how institutions influence relationships and interactions. Full understanding can be achieved only by recognition of the mutual influences at each level. While with some honorable exceptions ethologists and behavioral ecologists have tended to underestimate the influences of cultural factors on individual behavior, social anthropologists have underestimated the influences of individual characteristics on culture. An example is provided by the differences in behavior between men and women. While social scientists have emphasized the ubiquitous influence of socialization in producing these differences, they have underestimated the importance of sex differences in reproductive strategies in shaping many social practices (e.g., Alexander et al. 1979; Dickemann 1979; Hinde 1984b).

Recognition of these levels of increasing complexity in social phenomena, and of the successive dialectics between them, has also facilitated a clear separation of the data language used to describe social phenomena from the theory languages used to explain it—a separation not always easy in the social sciences. For example the term *institution* is used sometimes in a descriptive sense (e.g., to refer to the common features of village councils in a group of villages) and sometimes in an explanatory sense (e.g., to explain why the members of such councils behave as they do) (Hinde 1979). Similar confusions occur in primatology (e.g., the use of the term *dominance*; see Bernstein 1981 and discussants), but should be easier to put right.

In other cases insights have come from principles derived by comparative study of a range of animal species. For example, the suckling frequency of mammals is inversely related to the protein and fat content of the milk. Human milk has a relatively low protein and fat content, which suggests that neonates should be fed much more frequently than the once-every-four-hour feeds formerly common in many maternity hospitals (table 33-1). In harmony with this, mammalian babies that are suckled infrequently suck faster, and for a shorter period, than those that are suckled frequently: human babies suck slowly. Thus our own species could be placed on a spectrum derived from comparative study of other species, with the clear implication that a four-hour schedule is not appropriate (Blurton Jones 1972). This has facilitated a more natural approach to breast-feeding.

It is important to note that principles must sometimes be abstracted at a level other than that of the empirical data. Experiments on mother-infant separation again provide an example (Hinde 1974a). In experiments on rhesus monkeys living in small groups, separation was achieved by removing the mother from the pen and leaving the infant otherwise in the social situation to which it was accustomed ("mother removed paradigm"). In another series of experiments, the infant was removed and the mother left behind ("infant-removed"). In the second case the infant suffered exposure to a strange environment as well as the trauma of maternal separation. Surprisingly, however, such infants were less disturbed after reunion than those tested under the "mother-removed" procedure. Although no hard data exist, it is common ex-

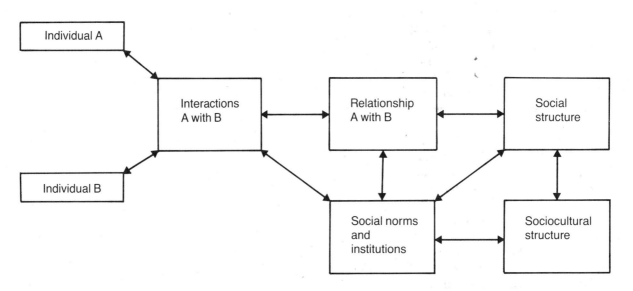

FIGURE 33-1. Successive levels of complexity in social phenomena.

TABLE 33-1 Milk Content and Suckling Interval in Various Mammals

Species	Fat Content	Suckling Interval (approx. hr)
Tupaia	Very high	48
Rabbits	High	24
Ungulates (Cached young)	High/moderate	4
Ungulates (Following young)	Moderate	2–3
Apes	Low	<1
Humans	Low	?

SOURCE: Based on Blurton Jones 1972.

perience that young children who go away to a strange place for a while are more disturbed after reunion than children whose mothers go, leaving them in a familiar place (Bowlby 1973). The animal and human data thus seem to conflict. However, at a different level of abstraction, there is a parallel between the two sets of data. The effects of a separation experience in children are related to the resilience of, and the disturbance caused to, the mother-child relationship. For two reasons, in the monkey experiments the mother-infant relationship was disturbed more by the mother-removed procedure. First, for reasons that are not understood, infants left in the home environment go through a phase of protest and into despair more quickly than infants in a strange environment; they are thus less effective in eliciting maternal behavior from their mothers on reunion. Second, rhesus mothers who have been temporarily excluded from their home group must, on return, reestablish their relationships with their group companions; they thus have less to spare for their infants. The mother-removed infants are therefore more disturbed after reunion because maternal attention is more difficult for them to obtain, and they are less effective in obtaining it. Thus to find the parallel between monkey and human we must move to the level of the mother-infant relationship (Hinde 1983b). In each case the infant shows more distress in the circumstances in which the mother-infant relationship is more disturbed.

Special mention must be made of the use of principles derived from the study of nonhuman primates in the development of attachment theory. Children's fears of the dark, of being alone, and of strange places and people were once commonly referred to as "irrational fears of childhood." Bowlby (1969), using evidence derived from studies of monkeys and apes, argued that, in the environment of evolutionary adaptedness of the human species, such fears were highly adaptive. Safety for the child lay in proximity to the mother, and both child and mother must have been adapted to maintain it. Although the adaptedness of proximity seeking is no longer so obvious under modern conditions, the behavior persists.

This biological view permitted a new approach to the mother-child relationship in which proximity to and interaction with a sensitive parent were recognized as equally important for child development as food or warmth.

Subsequent development of attachment theory has continued to be influenced by ethological studies, especially those of nonhuman primates. For example, a baby can use several types of behavior to maintain proximity with, or gain the attention of, the parent. These are seen as controlled by an "attachment behavioral system"—the concept of behavioural system being borrowed from studies of nonhuman species. The concept helped to explain why the nature of the relationship with the parent was not necessarily closely related to the intensity of any particular type of "attachment behavior" shown by the infant, but involved a patterning of responses tuned to the situation and especially to the nature of the parent.

More recently studies of nonhuman primates have highlighted other issues of crucial importance to attachment theorists (see chapters in Parkes and Stevenson-Hinde 1982). First, studies of rhesus monkeys had led to the recognition that the behavior of individuals within a relationship reflects the characteristics of the relationship rather than of one or other individual (reviewed by Hinde 1979). Attachment theorists, however, were prone to classify babies as "securely attached," "anxiously attached," and so forth. These assessments were made on the basis of episodes in which the mother-infant pair were first alone, then joined by a stranger, then the mother left, and so on. Since the mother was instructed how to behave and had an essentially passive role, it was natural to think that the behavior shown by the infant was a reflection of the infant's own characteristics. But when it was shown that infants can be "securely attached" to one parent and not the other, it became apparent that the procedure assessed the parent-child relationship and not the infant; the importance of ethological data became apparent.

Second, attachment theorists assumed an ideal mother-child relationship, based on sensitive responsiveness to the child's needs. Primatological data, however, indicate that the maternal style optimal for the mother may not be optimal for the child, and that what is best for either may depend on the social and environmental situation. Thus baboon mothers vary in the extent to which they restrict their infants' excursions or are "laissez-faire." Altmann (1980) has argued that the relative advantages of these strategies vary with circumstances. For example, although the infants of restrictive mothers may be less susceptible to accident or predation, they are probably less able to cope on their own if orphaned than those of laissez-faire mothers. The best maternal style also depends on the social status of the mother. Restrictive mothering helps protect infants from harassment by other group members. Since high-status mothers are less

a.

FIGURE 33-2. Weaning conflict between rhesus monkey mother and infant. (*a*) The infant reaches under the mother to touch the nipple, but she is prevented from nursing by the mother's position; (*b*) the infant rounds its mouth to give a soft "coo," a common distress vocalization in young macaques. (Photos: Barbara Smuts/Anthrophoto)

susceptible to harassment than low-status ones, they can more easily afford to be laissez-faire mothers.

As a third example, considerations of behavioral consequences, comparable to those used by primatologists, have advanced understanding of some of the less usual types of behavior in the mother-child relationship. For instance, some infants, reunited with their mothers after a brief period of separation, show only half-hearted attempts to regain close contact and some degree of avoidance of the mother. Such behavior seems inappropriate, but makes sense if we see it as necessary for the preservation of the child's behavioral organization in the face of a somewhat rejecting mother who is also often a little restricted in emotional expression (Main and Weston 1982).

MAKING SENSE OF THE HUMAN CONDITION

In the development of attachment theory, Bowlby made use of functional arguments—the fears of childhood made good functional sense in the environments in which our earlier ancestors lived. Functional arguments,

however, can be dangerous; while some find it fun to sit back in an armchair and speculate about functional parallels between the behavior of nonhuman species and our own, especially when the speculations are unverifiable, such an occupation is not science. The dangers arise in part from the fact, referred to above, that the diversity of primate species and the diversity of human cultures make it easy to find functional parallels that will support almost any hypothesis. This is especially the case if functional parallels are used as a basis for conclusions about causation: similar ends can be reached by different means in different species and vice versa, so great caution is necessary. In attachment theory it is the functional end of gaining proximity to mother that provides the basis for comparison between monkey infant and human infant, not the causal mechanism by which proximity is achieved.

This danger is less evident in another use of the functional approach, namely, to integrate diverse facts about human behavior and physiology. For example, we know a great deal about human infancy, but a new perspective on the human condition is gained if we can integrate the

b.

scattered facts into a coherent whole. The argument depends on the concept of the adaptive complex. The characters of a species, its anatomy, physiology, and behavior, can be seen as forming a coadapted complex fitting it for its way of life. Evolutionary change in any one aspect is liable to produce effects ramifying through the whole. Applying this to the mother-child relationship, we see that many facts about infant behavior, including aspects of sucking behavior, their poor thermo-regulating ability, Moro reflex, and vestigial grasping reflex, are compatible with a close mother-infant relationship in which the infant was carried rather than cached. Mothers would have been predisposed to look after their own infants in preference to those of others, and under some conditions they may have been selected to behave with hostility to the infants of others. It would thus have been in the infants' interests to show fear of strangers—and it is probably no coincidence that in human infants this develops at about the same time as the capacity for independent locomotion—and to form a special bond with the mother. Hence the infant smile, the games that parent and infant play, and other activities that cement the bond between them can be seen as making good biological sense (Hinde 1984a).

It is possible to extend this argument to embrace many other aspects of the mother-child relationship. For instance, mothers tend to move their children on from breast to solid food, from cot to bed, and from home to school before the child, if left to itself, would take these steps. Given the importance of maternal care, one might ask whether pushing an infant on before it is apparently ready to go could be in the infant's interests. However, monkey mothers behave in a similar way, and their behavior is exactly that predicted from evolutionary principles if one takes the behavior of both mother and infant into account. Natural selection promotes parental care because the offspring carries the parents' genes, and gene survival depends on the offspring's well-being. At the same time, the parent expends energy and other resources in caring for the infant, which can reduce the parent's chances of subsequently rearing further young. The relative importance of these two issues changes with time. At first, parental care is essential for infant survival. As the infant grows, it demands more milk and thus the cost of parental care (measured in terms of the parent's inclusive fitness) increases. Over the same period parental care becomes less essential to the offspring, so the benefits it confers on the offspring (measured in terms of the offspring's inclusive fitness) decrease. There thus comes a point at which the costs to the mother of a parental act exceed the benefit to her in terms of its effects on the young. However, the infant will continue to benefit from maternal care in its own right, and the cost to the mother will affect the infant's inclusive fitness less than the mother's since only some of their genes are shared. Thus continued maternal care favors the infant's evolutionary fitness and diminishes the mother's. On this basis it has been suggested that there must come a point when it is in the mother's interests to halt maternal care and in the infant's to elicit it (Trivers 1974). (In addition, continued suckling may delay the mother's next pregnancy, and thus the arrival of a sibling who will compete with the infant.) The infant is thus selected to elicit more parental investment than the parent is selected to give. Weaning conflict is thus based firmly in the processes of evolution (see chap. 27; fig. 33-2).

Numerous other aspects of the parent-child relationship, features that we mostly take for granted, such as the tendencies of parents to treat male and female children differently (Hinde 1984a), make sense as an integrated whole when seen in the evolutionary perspective provided by study of our primate relations.

CONCLUSION

Studies of nonhuman primates contribute to an understanding of the human species in two main ways. As experimental animals, they have long been used to advance knowledge in anatomy, physiology, psychology, pathology, and related areas. More recently, as knowledge of their behavior in the field has increased, it has become increasingly possible to derive principles, often based on comparative study of a range of species, that throw light on the human condition and whose detailed applicability to the human case can be assessed. In either case there are, of course, dangers, and in this chapter I have tried to point to some of them—dangers stemming from inappropriate selection of data, too narrow a perspective, or lack of humility. But the need to use data with circumspection in no way detracts from the value of those data. Nor does it lessen the need to pursue further studies of nonhuman primates with vigor, and to conserve primate species so that future generations can observe them and, in doing so, learn new lessons and reflect further on their own place in nature.

34 | Dynamics of Social Relationships

Frans B. M. de Waal

The basic tendencies governing interactions among primates are competition, social attraction, and cooperation. The objective of this chapter is to describe and discuss the way in which these different and often conflicting tendencies are integrated into a cohesive system of social relationships. The emphasis lies on *proximate* explanations, that is, explanations based on immediate causes and goals, and on the social histories, experience, and intelligence of the individuals involved. This approach asks how social mechanisms work rather than, as in evolutionary biology, why they came into existence.

One major problem in studies at the proximate level is an underdeveloped vocabulary to describe and think about subtle characteristics of social relationships. This problem is partly due to the combination of great variability in primate personalities and relationships and a reluctance to borrow from the richest source of social concepts: human language. In primatology we are used to investigating social components, such as the submissiveness, the affinity, or the supportiveness of one individual to another. We run out of words, however, if these different components form a single complex, that is, if they are closely interlinked and simultaneously expressed. A person who is treated with a similar combination of sympathy, submission and corroboration is said to have respectful and loyal colleagues, but these terms are considered inappropriate when applied to animals. It may be difficult, though, to avoid such so-called anthropomorphisms completely if we wish to move from analyses of aspects of social relationships to more encompassing levels.

The dilemma of a primatologist may be compared to that of a pianist listening to a record of a classical piano concert. He is unable to distance himself from the process by which the music is produced. Instead of "pure" enjoyment of a series of patterned sound waves, he automatically imagines a grand piano and feels the chords and melodies, so to speak, in his own fingertips. Similarly, scientists cannot completely distance themselves from primate behavior. Almost everything they see reminds them, consciously or not, of their own experiences and feelings.

Rather than regarding this as a disadvantage, or, worse, denying it, we should exploit the situation (Menzel 1979). Piano players undoubtedly listen more carefully and analytically to a piano concert than the average listener. In the same way, our background as social beings provides us with a depth of intuitive insight into social relationships that is bound to guide our thinking and theorizing when studying primates. Allowing for this influence is not the same as uncritically giving in to it. The tradition of rigorous quantification is now so well established in the field of primate behavior that there is hardly any danger of a descent to the level of pet lover's talk. In the history of ethology, references to human behavior have been of great heuristic value, and large parts of our present vocabulary (e.g., threat, appeasement, bonding) started out as anthropomorphisms (Asquith 1984).

New, easily recognizable terms for patterns of interaction or particular types of relationships are not to be confused with an understanding of the mechanisms involved. Such understanding usually comes later, after much research, and may as a matter of fact lead to the rejection of the initial terms. What new metaphors may do, though—and this is essential in science—is to reorganize our views and provide new frameworks for observations and experiments.

A complicating factor in the study of dynamics of primate groups is that the integration and balancing of competitive and cooperative tendencies cannot be understood at a dyadic level, that is, without considering the influence of third individuals or the group as a whole. Some studies indicate, for example, that the expression of interindividual preferences, as measured in dyadic tests, may be inhibited in the group context (Kummer 1975; Vaitl 1977a; Stammbach 1978). Without experimental procedures it is difficult to specify such influences, but Seyfarth (1977) has developed a detailed model that can be tested in the field.

Triadic influences also play a role in dominance relationships, that is, A's dominance over B may depend on C (e.g., Kawai 1958; Kawamura 1958; Hall and DeVore 1965; Varley and Symmes 1966). Recently, a number of observational studies have focused on the role of supportive relationships in structuring and changing the rank order; primates recruit allies to maintain or improve their social position. In this paper I will review some of my own observations of such dominance processes in macaques and chimpanzees. The same themes are recognizable in the work of Dunbar and Dunbar (1975), Chance,

Emory, and Payne (1977), Cheney (1977a), Walker Leonard (1979), Walters (1980), and Datta (1983a).

DOMINANCE AND SOCIAL INTEGRATION

The highly structured group life of most primate species did not evolve through disappearance of competitive and aggressive tendencies, but through the development of powerful mechanisms of conflict resolution. The conflict management of group-living species requires reconciliation and tolerance, which allows losers to live together with winners without provoking further violence. Among other things, losers must have a way to indicate their willingness, at least for the moment, to refrain from behaviors that pose a threat to the life, territory, or social position of winners.

The solution to this problem, widely found in the animal kingdom, is to provide losers with a sort of white flag and winners with an understanding of its meaning. The resulting harmonization, and the increased predictability of the direction of possible further conflicts, is often referred to as a dominance relationship.

It does not matter what form the white flag takes. Teeth baring and high-pitched screaming in many monkey species, low panting grunts in chimpanzees, licking of the other's mouth corners in canids, posture freezing in rats, and in general, any behavior that makes one look small and vulnerable can serve to signal submission. Fear may be an important motivational component of these behaviors, but it rarely seems the only motivation involved. In group-living species the subordinate's fear is often mixed with social attraction. Thus, submissive signals, often appearing to take the form of a greeting or the paying of respects, may be given while actually approaching the dominant. This notion of submissive behavior is expressed most succinctly by Schenkel (1967, 319) regarding wolves and dogs: "Submission is the effort of the inferior to attain friendly or harmonic social integration."

We know relatively little about socially positive aspects of dominance relationships. Traditionally most attention has gone to inequalities resulting from the relationship: the dominant's priority of access to limited resources (chap. 26). Although it is logical, from an evolutionary standpoint, to analyze the phenomenon in terms of competition and reproductive benefits, evolutionary analyses may be limited because they tend to isolate dominance from its social context. In baboons and macaques this problem is not immediately apparent because of a fairly close link between social expressions of dominance and priority of access to resources (chap. 25). Dominants go as far as forcefully removing food from a subordinate's cheek pouches. But even in these species, individuals known to be capable of claiming a resource may fail to do so either for lack of motivation or because of a special relationship with the subordinate involved.

If such special relationships become the rule, the problem with priority-of-access criteria of dominance becomes serious. In the large chimpanzee colony of Arnhem Zoo (Netherlands), some females are more successful than adult males in claiming objects or places to sit. It is quite common for these females simply to take the leaves on which a male was feeding out of his hands. At the same time, the adult males in this colony clearly win most of their fights against the same females, and all females show submission toward adult males (Noë, de Waal, and van Hooff 1980). How can we explain this? Are the females dominant or are the males tolerant?

Social integration and peaceful coexistence are important aspects of relationships between dominant and subordinate animals—often as important as the outcome of resource competition. By regarding dominance relationships as a compromise between inevitable antagonistic tendencies and a need for life in cohesive groups, one gets a feeling for the precarious equilibrium that exists between the two. A greater emphasis on the social side of dominance relationships does not necessarily (and should not) conflict with an evolutionary viewpoint. Vehrencamp's (1983) balanced evolutionary model of dominance, resulting from her assumptions concerning the benefits of group living, is an improvement upon the strictly competitive picture emerging from previous models (e.g., Popp and DeVore 1979).

STATUS MECHANISMS

The hierarchical organization of primates depends on the following social mechanisms: formalization, conditional reassurance, and status striving.

Formalization

In many macaque species, facial expressions with baring of the teeth (fig. 34-1) are exclusively shown by the lower ranking of two partners in a relationship. The occurrence of simultaneous teeth baring by two monkeys to each other is extremely rare, and the most dominant individual never bares his teeth in agonistic situations, or, if he does so, it is a sure sign he is on the verge of losing his position (Angst 1975; de Waal 1977).

In chimpanzees, in contrast, mutual teeth baring between individuals is not uncommon during agonistic encounters, and a leader with a stable position may also show this behavior. Exposure of the teeth among chimpanzees appears to signal fear, nervousness, and hesitation, but not submission. At the same time, the chimpanzee has another set of signals, very different in form, to express status differences. The subordinate approaches with bowing movements while uttering a series of soft panting grunts, whereas the dominant makes himself bigger by standing up with his hair erect (fig. 34-2). The dominant may also perform a "bluff-over," which is a passing charge in which one arm is raised and moved over the crouching subordinate; sometimes this display

FIGURE 34-1. A juvenile rhesus macaque responds with submissive teeth baring to the approach of an adult male.
(Photo: Frans de Waal)

FIGURE 34-2. Chimpanzees communicate their dominance relationships in ritualized encounters. The alpha male (*left*) has his hair on end and draws himself up while approaching another adult male. The subordinate bends down and utters a series of submissive pant-grunts. (Photo: Frans de Waal)

even takes the form of a spectacular jump over the subordinate (Bygott 1974; Noë, de Waal, and van Hooff 1980; de Waal 1982).

These almost ritual encounters confirm the assymmetrical state of the relationship. Dominance relationships that have reached this stage of communication are called *formalized* here. This term allows for a contrast with nonformal dominance. We all realize that in our own society there often is a difference between a person's formal position and his or her de facto power (for example, the influence of some elder statesmen). This discrepancy is also discernible in nonhuman primates. Coalition networks and other mutual dependencies seem to allow skilled and experienced individuals to exert greater influence than one would expect on the basis of their formal ranks (see "Power versus Rank").

By their extreme consistency of direction, status signals, such as the macaque's teeth baring or the chimpanzee's pant-grunting, contrast sharply with other measures of dominance. Priority of access to resources, aggression, flight, and avoidance may occasionally be in favor of the subordinate party. Since exceptional outcomes can

generally be ascribed to variations in social context (for example, the presence/absence of supporters), the absence of similar reversals in the direction of formal status communication suggests that it is "context free." These signals seem to communicate the overall state of the relationship rather than transient outcomes of encounters. They might, in that sense, be interpreted as *meta*communication (Altmann 1962).

Conditional Reassurance

Our studies have produced evidence for reconciliation behavior in both chimpanzees and rhesus monkeys (de Waal and van Roosmalen 1979; de Waal and Yoshihara 1983). The phenomenon is likely to exist in other species as well (e.g., McKenna 1978; Seyfarth 1976). The evidence for reconciliation can be summarized as follows: (1) During a limited period after an aggressive incident there is an increased probability of contact between former opponents. (2) The contact increase is selective, that is, adversaries seek contact with one another rather than with bystanders. (3) The behaviors during these post-conflict contacts differ from those during normal contacts; rhesus monkeys show a higher frequency of embracing and lip smacking, whereas for chimpanzees the mouth-mouth kiss is the most characteristic contact pattern.

There is likely to be some relation between well-established dominance relationships and the tendency to reconcile after conflicts. The hypothetical link is that of "conditional reassurance." According to this hypothesis, dominant individuals are prepared to reconcile only with subordinates who clearly and regularly demonstrate that they recognize their position. For example, Kummer (1975) found that fights between male gelada baboons stopped after reaching a decisive outcome. The winner approached the loser with appeasing gestures such as presenting and lip smacking. The animals then proceeded to mounting and grooming, and finally relaxed. Also, Maxim (1976), in experiments on rhesus monkeys, found a link between the establishment of dominance and the development of a more friendly relationship. Conversely, observations by Bernstein (1969) on pigtailed males in a captive troop indicate that violence may persist, even to a fatal point, if the loser of a dominance struggle fails to submit. This is not to say that fights always stop short after submission by the loser; these mechanisms may only work between individuals with good reasons to maintain a relationship.

If peaceful coexistence depends on formalization of dominance relationships, this is presumably due to the dominant reading the signs of submission as indications that his or her position is safe. At this point conditional reassurance comes into play. During the final weeks of dominance struggles among male chimpanzees, for example, if it is becoming evident that male A generally has the upper hand in his confrontations with B, reconciliations will become increasingly rare. This is due to A's systematic rejections of overtures by B. Each time B seeks contact after one of their aggressive encounters, A calmly walks away, avoiding the contact. This may happen many times a day. Only when B starts to utter the first submissive panting grunts, not heard for months between the two males, will A accept B's attempts to reconcile. It seems as if A blackmails B by withholding friendly interactions: kissing and grooming contact can only be obtained through a formal acknowledgment of the outcome of the dominance struggle (de Waal 1982, in press).

Status Striving

It is difficult to explain why such an old and valuable concept as Maslow's (1937) *dominance drive* has been taboo for such a long time. An important factor undoubtedly was that behaviorism, until recently a very influential school of thought, dictated that since we cannot know whether animals have intentions, it is better to describe their behavior as having specific consequences rather than as aiming at this or that goal. Thus, scientists might speak of an animal's rise in rank, but would avoid speaking of aspirations or striving for status.

That there is little or no purposefulness (in the cognitive sense) in animal behavior is only an assumption. We seem to be reaching a point at which this assumption is starting to hamper further research. Unless we open our eyes to the possibility of intentionality in animals, it seems almost impossible, especially in the case of primates, to make sense of the complex group processes accompanying and producing rank reversals. The number of studies concerning dominance strategies is on the increase, and the result is a cautious change in outlook and language over recent years.

Let me summarize two relatively simple quantitative results relating to aggression in the Arnhem chimpanzee colony. These results can easily be explained if we assume active status striving, whereas they would demand separate and more difficult explanations if status reversals were to be more or less unintended, passive processes of change.

Aggressive actions by females and immatures usually occur in response to particular events in the group (e.g., in order to protect an infant), whereas a much higher proportion of aggressive actions by adult males occur without any obvious reason (de Waal and Hoekstra 1980). It can also be demonstrated that both the intensity and frequency of male aggression decreases after the establishment of formal dominance relationships (de Waal, in press). The most dramatic illustration occurred when a young adult male, Nikkie, started provoking fights with all eight females in the colony. After months of daily confrontation the females, one by one, began to defer to Nikkie by approaching him while uttering pant-grunts of submission. Over the same period, Nikkie's tendency to direct provocative bluff displays at females diminished significantly.

All dominance reversals observed over the years involved adult males. The greater "spontaneity" of their aggression, and the inhibitory effect of their targets' subordination, may be explained by the use of aggressive behavior for the pursuit of a goal that is not immediately visible to us. This hypothetical goal is to maintain or achieve the subordination of as many other individuals as possible.

Retaining the Dominance Concept

More attention to the ways in which ranks are achieved and maintained may solve some of the much discussed problems with the dominance model (see Bernstein 1981 for an overview of the debate). A central problem has been the correlation, or rather the lack of correlation, between different measures of dominance. Is dominance a unitary concept? Can it be used as an intervening variable? The idea behind these questions was that the explanatory value of the model depended on whether a rank order based on one variable, say aggression, provides any information about rank orders based on other variables, say priority in competitive situations. Note, however, the narrowness of this interpretation of explanatory value, being limited to correlations between simultaneously existing distributions of behavior.

Actually, lack of agreement between different measures of dominance is something to be expected, if dominance is a dynamic phenomenon, because rank relationships are likely to change gradually, stage by stage. Thus in long-tailed macaques an almost perfect correlation exists between four agonistic rank criteria as applied to adult group members, but a very poor correlation if juveniles and infants are considered (de Waal 1977). Data suggested that, whereas adults had well-established ranks, youngsters were still in the middle of a long-lasting process of rank acquisition. At a certain point in its ontogeny a particular juvenile might already be dominant over another individual according to one or two criteria, but still be subordinate to the same individual according to other criteria (fig. 34-3; see also chap. 25).

In spite of all criticism, there is no way to discard the dominance model. Dominance relationships, and their transitive arrangement into a hierarchical structure, are too obvious to be ignored. One step necessary to revitalize the concept is its integration with other aspects of group life. The explanatory value of such an extended

model is evident from long-term research on Old World monkeys. These studies demonstrate that accurate predictions of the direction and outcome of dominance processes involving young females can be based on knowledge of their mothers' positions in the hierarchy: daughters usually reach ranks close to their mothers' (chap. 11).

Thus, the dominance model may be more useful in combination with other factors, such as kinship, than in isolation. Previous sections of this chapter suggest the following additional ways to improve the model: (*a*) distinction between formal dominance relationships, as expressed in ritualized communication, and success in competitive contexts; (*b*) attention to socially positive aspects of relationships between dominants and subordinates, for example, reconciliation and tolerance; (*c*) reconsideration of assumptions inherited from behaviorism; and (*d*) incorporation of the factor *time* in our analyses.

Finally, it should be noted that the dominance model's usefulness may strongly depend on the species and sex under study. Whereas male chimpanzees seem much more dominance oriented than females (Bygott 1974; Nishida 1979; de Waal 1982), this sex difference does not seem to hold for many Old World monkeys where females are just as involved in dominance strategies as males (e.g., Chance, Emory, and Payne 1977; de Waal 1977; Walters 1980; Hrdy 1981b; chap. 25).

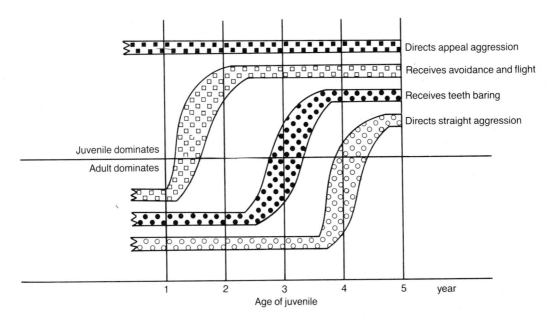

FIGURE 34-3. Young macaques usually achieve dominance over all adults ranking under their mother. This process takes several years. In the beginning the adult still dominates by directing straight aggression to, and receiving flight behavior and submissive teeth baring from, the juvenile. The juvenile dominates in terms of appeal aggression only. In subsequent years the adult starts to respond with avoidance and flight. Eventually, the direction of teeth baring also reverses. After this formal acknowledgement of the new relationship, the juvenile's aggressive behavior may change toward straight aggression. This model of rank acquisition is based on comparisons between age classes (de Waal 1977).

COOPERATION
Side-Directed Communication

If we regard interactions between adversaries as the main feature of behavior during agonistic interactions, patterns aimed at third individuals can be taken together as *side-directed* behavior. This is one of the most variable categories of communication in primates. Through side-directed behavior, individuals make their aggressive encounters "public," that is, open to all kinds of influence by the group. Chimpanzees may do so by begging outsiders, with an outstretched hand, for protection (de Waal and van Hooff 1981), hamadryas females by presenting their hindquarters to a male while threatening another female (Kummer's 1957, 1967, "protected threat"), and rhesus monkeys by different scream types (Gouzoules, Gouzoules, and Marler 1984). For other descriptions see Hall and DeVore, (1965) and Wolfheim and Rowell (1972).

In addition to screaming in defense, macaques use a special form of threat behavior for the recruitment of support. The ears are laid against the head, the eyebrows pulled up, the chin usually pointing upward, and a series of loud grunts is uttered to the opponent while repeatedly looking around at potential supporters with jerky turns of the head. Since this pattern demonstrably increases the probability of receiving support, it was labeled "appeal-aggression" (de Waal, van Hooff, and Netto 1976). It is typically shown by young monkeys when challenging dominant group members. If they succeed in reversing ranks, a process that may take years, young monkeys will start using a different form of threat behavior. "Straight-aggression" refers to the self-confident, silent threat with wide open mouth and staring eyes used by well-established dominants. Thus aggressive macaques seem to use different sets of signals depending on whether they are trying to increase their dominance rank or whether they are reaffirming existing rank positions (de Waal 1977; see also fig. 34-1).

Types of Intervention

Interventions by third individuals can be peaceful, disruptive, or aggressive. For example, in the chimpanzee colony of Arnhem Zoo it is not uncommon for females to "confiscate" a displaying male's weapons (sticks or stones) by removing them from his hands or to placate him by grooming. In macaques, dominant males often give branch-shaking (or cage-shaking) demonstrations during aggressive incidents between others. Such displays may serve as a warning to the combatants that intervention is at hand. But the most common and perhaps the most effective form is the aggressive intervention. This form has received the most attention and is known under various names such as fight interference, agonistic aiding, alliance or coalition formation, and support; see, for example, Kaplan (1977), Cheney (1977a), de Waal (1977, 1978b), and Walters (1980). Recently I developed a classification of aggressive interventions based on an analysis of several thousand instances observed in the Arnhem chimpanzee colony (de Waal 1984b):

Bond-dependent Interventions. The side an individual takes in aggressive encounters depends on the social bonds with the two combatants. The closer the bond, in terms of friendly contact and grooming, the more likely this individual will be given support against attackers. The main function of these interventions seems to be the protection of friends and relatives.

Scapegoating. Irritations and tensions among high-ranking group members may lead to redirection by jointly threatening or attacking a bottom-ranking individual. For example, we found evidence that crowding stimulates the occurrence of such aggressive alliances. Their function seems to be the maintenance of peace at the top of the hierarchy.

Exploitative Coalitions. In the Arnhem chimpanzee colony, adult male coalitions are, to a certain degree, independent of previously existing social bonds. This is in contrast to the largely bond-dependent interventions of adult females. Qualitative observations strongly indicate that male coalitions serve status competition and are opportunistic in the sense that they are not based on stable preferences for particular individuals, but on the usefulness and willingness of partners, at a given place and time, to contribute to beneficial conflict outcomes (de Waal 1982). This leads to a testable hypothesis: if the independence between male coalitions and social bonds is related to status competition, it is expected to be greatest during periods of hierarchical instability. Our data confirmed this prediction.

Breaking Up Fights. Since the performance of this type of intervention is usually restricted to a single, high-ranking member of the group, it is often considered part of a role pattern: the so-called control role. It involves protective and impartial interventions. The goal seems to be to stop aggression rather than to help particular individuals. This behavior definitely benefits weaker group members, but it may also serve the performer himself. Among the Arnhem chimpanzees, there are indications that subordinates who are regularly protected by a performer may mount massive support to prevent his overthrow if his position is challenged (de Waal 1978a). Breaking up fights might, therefore, be part of a strategy aimed at status stabilization.

The extent to which this classification of intervention types is useful for other primate species remains to be tested. However, indications for each type can be found in the literature on macaques and baboons: influence of (kinship) bonds on interventions (e.g. Massey 1977; Kaplan 1978), scapegoating (e.g., Kawai 1960; de Waal 1977), exploitative coalitions (e.g., Packer 1977; Smuts 1985), and breaking up fights (e.g., Bernstein and Sharpe 1966).

Ambivalence

The principal aim of a coalition is to gain advantage over other competitors in a situation in which all parties compete. This means that there is also a potential for disagreement among the cooperating parties themselves. To illustrate the extremely tense relationships resulting from such intracoalitional competition, let me summarize the situation among three adult males in the Arnhem chimpanzee colony (de Waal 1982).

The oldest male had lost his alpha position to a coalition of two other males. The younger of these two males, however, started to compete with his former coalition partner, the new alpha, over access to the fallen leader. Both tried to sit and groom with him and to prevent the other from doing so. After about 1 year the old male began to develop a preference for the young male. Thus, the youngest and least experienced of the three males

was made alpha male and depended completely on the "old fox." Their coalition lasted several years, but was not free of tense incidents.

In figure 34-4 the young alpha male, in the center, grins and pants while holding out a hand to his coalition partner. This scene illustrates his dependence on the old male. A few minutes before, the two coalition partners had been chasing each other, loudly screaming in a conflict over access to an estrous female. Such conflicts within the ruling coalition created a very unstable situation because there was no one to keep the impressive third male from giving his charging displays and terrorizing the group. This rival male, visible on the left of the picture, is watching how the alpha male hurries to make up with his ally. Only after the reconciliation could alpha reconfirm his position by bluffing over the third male.

Our steadily growing knowledge of dominance processes among chimpanzees in their natural habitat indicates a similar important role for male coalitions as observed in the Arnhem colony (chap. 15). Male baboons also may show exploitative coalition formation (Hall and DeVore 1965; Packer 1977), but for adult male macaques there is much less evidence. It almost seems as if the am-

FIGURE 34-4. Under pressure from his displaying rival (*left*), the alpha male (*center*) tries to mend the breach in his coalition by begging his partner (*right*) for reconciliation. (Photo: Frans de Waal)

bivalence discussed above, that is, tension and competition within the coalition, cannot be overcome by male macaques. Part of the reason may be that many macaque species have a mating season, thus concentrating sexual competition in a few months of the year. Under such circumstances, coalition maintenance may demand more efficient mechanisms of tension reduction than those present in macaques (Tilford 1982).

Power versus Rank

One of the most intriguing problems in recent research on dominance is that of restraints on the dominant's control and discrepancies between power and rank. The close interaction between cooperation and competition seems to create plenty of opportunities for individuals to exert greater influence, at least occasionally, than their formal and agonistic ranks would suggest.

Strum (1982) found that during competition over females or meat resident male baboons were more successful than newly immigrated males, in spite of the obvious dominance of new males during aggressive confrontations. She suggests that intimate knowledge of the troop and a well-established set of social connections favor resident males. Other studies of sexual competition among baboons have indicated that males known to be capable of challenging another male do not always do so, even if they have the opportunity (chap. 31). Dependence on supportive relationships may explain this apparent restraint, as it is shown especially toward frequent coalition partners (Rasmussen 1980; Smuts 1985).

Similarly, in small isosexual groups of rhesus monkeys, I found in both male and female groups that the first-ranking monkey was not more successful than the second-ranking one in obtaining and keeping an apple piece that was thrown into their pen (de Waal 1984a). Second-ranking monkeys were able to keep the piece because, although they might be threatened, they were not attacked by alphas. Against lower-ranking monkeys this inhibition was lacking. Since in all groups the two top monkeys formed a coalition against the rest, the selective tolerance may have reflected the need for alpha monkeys to maintain a good relationship with their supporter (see also chaps. 25, 26, 31).

The relationship between the sexes is complex in a large group of captive chimpanzees. Males may depend on females for agonistic support, for reassurance, or for mediation in reconciliations with adversaries. The fact that some females can take priority over adult males may well be related to the importance of these female behaviors for males (Noë, de Waal, and van Hooff 1980; de Waal 1982). Another example of subordinate control concerns the tactic of low-ranking males to regularly change sides in disputes between dominant males over access to estrus females. Since each dominant needs the subordinate's support in order to have any chance at sexual contact, the subordinate has created a very powerful key position for himself. Apparently, this playing off of one male against the other brings sexual advantages to the subordinate. Nishida (1983b) observed the tactic in the wild and labeled it "allegiance fickleness," while I observed it in captivity and described it as the manipulation of sexual jealousy between others (de Waal 1982). In both cases priority of access was decided by the power balance rather than by the formal rank order.

The concept of power in primate societies is still poorly developed, and all this may sound vague and paradoxical. Why do primates strive for formal dominance if there are other ways to gain certain advantages? What sometimes prevents the most powerful and influential individuals from also attaining formal dominance? We do not have the answers yet, but we should keep in mind that primates, just like humans, may live in double-layered societies with considerable room for influence behind the scenes.

SUMMARY

The dominance concept remains central to the explanation of primate social organization, but three major modifications of the classical concept seem underway: (1) dominance relationships are anything but static; we need to study the proximate mechanisms through which they are established, maintained, and changed; (2) dominance is not a unitary concept; we need to distinguish between, for example, status communication, enforcement of positions, and benefits associated with dominance; and (3) the evolutionary approach has put too much emphasis on competitive aspects; we also need to study dominance in the context of reconciliation, social tolerance, and group cohesiveness. These new approaches are illustrated with selected examples from the literature and from the author's observations on captive macaques and chimpanzees. Special attention is paid to the role of cooperation in agonistic situations and its attenuating effect on power differences between dominants and subordinates.

PART IV

Communication and Intelligence

For centuries, we have been both fascinated and repelled by the similarities between ourselves and nonhuman primates. In zoos, primates are often among the most taunted and harassed of any animals on display, at least in part because their appearance and behavior seem sufficiently humanlike to invite both curiosity and embarrassed ridicule. But how similar really are primates to ourselves? We implicitly assume that primate behavior and intelligence are more like our own than like those of any other animals, yet this assumption is rarely based on systematic evidence. We have some reason to suspect that primates are indeed more intelligent than other animals since primates have bigger brains than animals of comparable body size (chap. 16) and their brains also contain proportionately more neocortex (Passingham 1982). At present, however, the relation among brain size, neocortical volume, and intelligence is poorly understood.

The assumption that primates are somehow the animals most like ourselves coexists uneasily in our minds with a second, equally pervasive view, which holds that nonhuman primates differ fundamentally from humans because they lack language. Some have argued that the lack of language affects not only the capacity for speech itself but also the capacity for abstract thought (e.g., Premack 1983b). Thus although nonhuman primates may be more similar to ourselves than any other animals, they also differ from us in significant respects. This second assumption has persisted primarily because until recently it was never tested.

We think of human speech as distinct from nonhuman vocal communication for a number of reasons. First, human language is considered to be under voluntary, cortical control. Human speech sounds are also semantic, in the sense that they represent or refer to objects or events in the external world and convey a meaning that can be relatively independent of the context in which they occur. In contrast, the vocalizations of nonhuman primates have traditionally been seen as involuntary, referring only to the signaler's internal emotional state. Second, humans are able to create distinct words that differ in meaning from other words with similar acoustic features because we perceive a continuous array of speech sounds in terms of relatively discrete categories (e.g., Studdert-Kennedy 1975). Monkeys and apes, however, have tradi-

tionally been described as having a graded vocal repertoire without discrete perceptual boundaries.

As described in chapter 36, however, a variety of field and laboratory studies now provide convincing evidence that these dichotomies may be too rigid and simplistic. Not only do the vocalizations of monkeys and apes often function in a rudimentary semantic manner, but it also appears that such vocalizations can be given or withheld depending on social circumstances. Some striking differences between human speech and primate vocalizations remain, however. Field and laboratory studies have yet to demonstrate convincingly that the vocal repertoires of nonhuman primates, like those of humans or birds, are learned or modified during development, although this remains a distinct possibility. Moreover, considerable controversy remains about whether language-trained captive apes or any other primates are capable of grammar or syntax. Nevertheless, existing data suggest that nonhuman primates' vocalizations are considerably more complex and convey considerably more information than was previously believed. Chapter 35 describes further the rich olfactory and visual communicative abilities of nonhuman primates and emphasizes that we are just beginning to understand the complexities of primate olfaction as well as the information that can be obtained through a detailed study of facial expressions.

Specifying the nature of primate intelligence has posed one of the most elusive problems since people first began to study monkeys and apes in the late nineteenth century. The difficulty stems in part from the fact that field and laboratory research on intelligence have traditionally used different methods to address fundamentally different questions. Laboratory tests have focused on learning paradigms and cross-species comparisons of problem-solving abilities using arbitrary objects as test stimuli. While these experiments have the virtue of being tightly controlled and systematic, they have been divorced from the question of function, and their relevance to the animal's natural social behavior has never been clear. Indeed, the laboratory psychologists' concern for methodological rigor has instead often led to a kind of intellectual rigor mortis in which precise results are obtained but the diversity and complexity of intelligence are lost.

In contrast, while field investigations have at least considered the function of intelligence, with some notable exceptions they have too often relied on anecdotal accounts rather than on systematic observations or experiments. As chapter 37 emphasizes, some of the most intriguing information on the extent and limitations of primate intelligence emerges when field and laboratory studies are linked. Increasingly, laboratory investigators are attempting to design experiments that test the skills that have evolved among their subjects in the wild, whereas field observers are beginning to realize that their richly detailed accounts of social behavior provide data directly related to laboratory tests. In particular, there are compelling speculations that the intelligence of nonhuman primates may be most highly developed in the social domain, since some of the tasks that monkeys and apes solve only with great difficulty in the laboratory appear to be more easily mastered when the animals are dealing with one another.

Many of the problems surrounding the study of intelligence ultimately stem from the difficulty of defining intelligence, cognition, and abstract reasoning in nonverbal animals. Human intelligence is to a large extent expressed verbally, and tests of intelligence and cognition generally rely at least to some degree on language. If intelligence is defined in terms of uniquely human attributes such as speech or the skills associated with language, then other species will obviously be found wanting. Moreover, even if we adopt the more interesting and evolutionary logical hypothesis that the antecedents of human intelligence, cognition, and self-awareness may be sought in nonhuman primates, investigators are still confronted with a fundamental methodological problem. As Griffin (1984) has pointed out, even if a nonlinguistic animal were capable of abstract thought or self-awareness, how would we know? Nonhuman primates cannot be interviewed, and in the absence of verbal proof of complex reasoning abilities, what data would constitute satisfactory evidence? It is clear that the investigation of nonhuman-primate intelligence is a field that will require, almost more than anything else, imaginative and careful experiments, preferably conducted under naturalistic conditions and using wherever possible stimuli that reflect the particular environment in which each species has evolved.

The assumption that nonhuman primates are more intelligent than other animals suggests that they may also depend more on learning and that information can be transmitted across generations by cultural rather than simply genetic means. We think of ourselves as the ultimate primates in this respect, with an ability to manipulate the environment that has largely freed us from many of the selective pressures that confronted our early ancestors. It therefore seems logical to search for evidence of cultural transmission in nonhuman primates. Chapter 38 provides a thorough review of known cases of cultural transmission in primates and emphasizes how difficult it often is to distinguish genuinely cultural behavior from behavior that may have arisen by other means. For example, unless the investigator is lucky enough to be present when a new technology is "invented," it will never be clear if a pattern of behavior that is now shared by most or all of the individuals in a social group was originally acquired through individual learning, observation, or genetic means. Inevitably, therefore, accounts of cultural transmission in natural populations are anecdotal. Such anecdotes, however, should not be dismissed lightly because they suggest that the exchange of information among nonhuman primates may be far more sophisticated than we suspect, and they challenge investigators to design further observations and experiments that will specify more precisely the interaction between cultural and genetic factors in the evolution of primate adaptations.

35 | Communication by Sight and Smell

Anne C. Zeller

The complexity of primate communication, and its integration into all aspects of life, makes communication a very difficult term to define. In bypassing the morass of a formal definition, I have chosen to regard communication from a functional viewpoint, in an effort to see what influences it has on the lives of animals. These influences are divided into the proximate and ultimate functions of communication.

At the proximate level, communication aids an individual in meeting the needs of everyday life. The information on food location, approaching predators, and interspecific interactions that the animal gains from social companions assists it in moving through the web of relationships that compose its daily life. The ultimate function of an individual's use of communication is to increase its reproductive success. Territorial defense, the socialization of young, the formation of alliances, and sexual interactions are all facilitated by communication and serve to increase survival and reproduction. In primates, as well as in other group-living animals, such communication may result in "traditions" of habitat use, food preparation, and social structure that persist in a given area for generations (chap. 38).

Primate communication is frequently divided into the auditory/vocal, tactile, chemical, and visual systems. The chemical and visual systems are the focus of attention in this chapter, although most primate messages are received through a multitude of channels. The chemical communication system can be divided into olfactory and gustatory senses, although in general "the chemicals are perceived by olfaction rather than gustation" (Shorey 1976, 7).

OLFACTORY COMMUNICATION
Introduction

Olfactory communication is used in many primate species, primarily during sexual interactions. The use of olfaction in sexual behavior is discussed in chapter 30; the present chapter is limited to a discussion of olfactory communication outside sexual contexts.

Methods of Marking

Some primates have specialized glandular structures that produce highly volatile, strong-smelling chemicals called pheromones, which are selectively emitted. In such spe-

cies, the odor of the pheromone alone can allow recognition of an individual's identity, species, age, and sex (for reviews see Epple, Alveario, and Katz 1982; Evans 1980; Harrington 1974; Schilling 1974, 1979; Shorey 1976). Anatomical and physiological specialization for pheromonal communication are most common in the prosimians, the callitrichids, and some members of the Cebidae. Even within each group, however, the presence and structure of pheromone-producing glands may vary. Glandular formations are generally absent in the lorises, who nonetheless scent mark, while some animals, such as female woolly lemurs (*Avahi laniger*), have scent glands but do not use them (Schilling 1979). All members of the genus *Galago* practice chest rubbing, although there is some controversy over whether they all possess sternal glands (Dixson 1976). The glands that produce pheromones can be on the head, sternal region, abdomen, or forelegs, but they are most commonly found in the anogenital region.

Olfactory communication may also involve urine or feces, which are not used for their instrinsic smell, but because some glands have duct openings in the urethra or anal canal that scent these substances with glandular secretions (Shorey 1976; Schilling 1979).

Scent marking may occur in what Shorey (1976) calls a passive, that is, nonspecific, liberation as the animal moves, or in active marking, where the animal selects the time and place, and performs particular bodily maneuvers in applying the urine, feces, or glandular area of the skin to the substrate. This active marking is seen particularly in the ring-tailed lemur (*Lemur catta*), where individuals may leave a resting group to go and mark in a stand of trees that is a particular target of attention (Schilling 1974). The male ring-tailed lemur has a sebaceous brachial gland on the upper arm and an antebrachial gland on the wrist, surmounted by a horny spur. He wipes the antebrachial gland across the brachial gland, and then draws the forearm and spur obliquely across tree trunks. This produces both a scent mark and a scar in the bark caused by the spur. Such behavior may continue for periods of 2 to 20 minutes before the animal returns to its group (Schilling 1974).

Urine washing is another specialized form of olfactory signaling. It is a very stereotyped behavior, quite distinct from functional urination and characterized by different movement patterns in different species (Andrew and Klopman 1974). In general, the hand and foot on one side of the animal are raised and placed under the urine flow, then wiped on the branch. This is repeated on the other side of the body, with a side-to-side rocking motion. Tail-shaking and lowering of the anogenital region may accompany the behavior, thus drawing attention to its occurrence. Capuchins (*Cebus*), night monkeys (*Aotus*), and squirrel monkeys (*Saimiri*) have their own distinctive movements for urine washing. In capuchins, for example, the hands are rubbed up the flanks, thus marking the animal's own body rather than the substrate. Urine washing may alternate with, or be replaced by, a rhythmic micturition in some species, or by another behavior, such as genital scratching, in the potto (*Perodicticus potto*) (Andrew and Klopman 1974).

Function of Olfactory Communication

Surprisingly little is known about the precise function of olfactory communication in primates. Some studies, however, are suggestive, and these are reviewed briefly.

Territory defense. Epple, Alveario, and Katz (1982) suggest that copious scent marking of an environment may inhibit the approach of conspecifics. The dwarf galago, Allen's galago, and pottos urine mark four times more frequently near the periphery of their ranges than in the center; the marking may serve as a territorial deterrent (Schilling 1979). Intensive marking of range edges occurs in many territorial mammals, but the clearest experimental evidence that pheromones alone can deter territorial invasion comes from studies of mice (Shorey 1976). Ring-tailed lemurs also mark intensively at the edges of their territories and visually reinforce this olfactory cue by waving their long black-and-white-striped tails over their heads when they discover other groups approaching their boundaries. Since their tails are frequently impregnated with scent from being rubbed across their brachial glands, additional odors are broadcast by tail waving (Schilling 1974).

Aggregation. On the other hand, some marking situations foster aggregation of group members. After a frenzied explosion of marking (Andrew and Klopman 1976), galagos may mob a predator in response to its presence. An estrous female lemur who performs genital marking may attract a number of males, who overmark her scent and tail wave or brachial mark at each other (Schilling 1974). Aggregation by scent may be an important force in binding mothers and young together (Shorey 1976). Mother mouse lemurs with olfactory bulb ablations show very disturbed maternal behavior; if their young are removed and then returned, the mothers will frequently not accept them (Schilling 1979). Another aggregative function may be the use of olfactory cues to retain proximity to the group at night or to locate frequently used arboreal trails (Schilling 1979). Lemurs who wander away from their groups are frequently marked by other group members upon return (Schilling 1979).

Alarm. In many fish, insects, and mammals, alarm odors appear to have been derived from defensive secretions (Lewis and Gower 1980). As noted above, galagos often scent mark at high rates immediately before mobbing a predator, and scent marking in this case may attract others for the purpose of mobbing, thus communicating alarm. Brown lemurs accompany scent marking with tail swinging and shrieks when alarmed by a human observer or another group, and this behavior may escalate to mobbing (Harrington 1974). Silverback male gorillas emit a fear scent when engaged in fights or frightened by humans (Fossey 1983). Manley (1974) notes that slow loris (*Nycticebus coucang*) lack alarm vocalizations, but produce a pheromone that may serve this function. Manley has also observed copious glandular secretions when angwantibos are stressed. He suggests that the instant immobility or clinging response seen in young angwantibos when their mothers are alarmed may be a response to an olfactory cue (see also chap. 2).

Aggression. Olfactory signals are also used in aggressive, within-group interactions such as those involved in competition for dominance rank or mates. Male saddleback tamarins respond more frequently and aggressively to scent marks of same-sex conspecifics than to those of other species and are attracted to female tamarin marks. They seem to prefer marks of animals who are subordinate to them (Epple, Alveario, and Katz 1982). Harrington (1974) notes that brown lemurs also respond more strongly to scents of animals of the same sex. Although the point is controversial, Shorey (1976) indicates than dominant males in many mammalian species secrete more pheromones, have larger scent glands, and mark more frequently than subordinate animals. Among ring-tailed lemurs, subordinate males show less genital marking than dominant ones, although the amount of brachial marking is equivalent (Schilling 1974). The tail-waving display of ring-tailed lemurs, called a "stink fight" by Jolly (1972), is frequently associated with aggressive displays, especially during competition over estrous females.

Individual Recognition. Evidence for individual recognition by pheromonal cues is rare for at least two reasons. First, it is often difficult to separate the effects of odor from those of vision and hearing. Second, recognition of individuals often occurs in conjunction with recognition of other variables such as familiarity, age, status, sex, and reproductive condition. Nonetheless, there is evidence that mice can recognize up to 18 individuals (Shorey 1976), and Schilling (1979) argues that, while olfactory signals may convey general physiological and social information, they will always be strongly influenced by individual components. Harrington (1974) showed that brown lemurs could discriminate between individuals of the same sex and subspecies. This observation is supported by Epple, Alveario, and Katz (1982), who state that responses to the odors of known tamarins vary depending on whether previous encounters were friendly or antagonistic, thus suggesting a recognition of particular individuals (for a description of recognition by auditory cues see chap. 36).

Discussion

Olfactory communication may have originated as a long-lasting form of communication for relatively nocturnal, solitary animals, but in addition to these original functions it has developed into a system for influencing immediate social interactions among both nocturnal and diurnal group-living species. Although many of the motor patterns of urine washing and anal marking are both species-specific and appear fully formed in young animals, other forms of marking seem to require experience for effective use. For example, social learning and imitation appear to be important among lemur species for the development of anal rubbing, and brachial, head, and tail marking (Schilling 1979). In addition, the use of olfactory cues to promote aggregation and dispersion or aggressive interactions would also seem to require learning before animals are able to use the appropriate signal in the appropriate context (see chap. 36 for similar examples of development in vocal communication). Finally, the ability of animals to recognize individual conspecifics, if it occurs, almost certainly indicates that the development of olfactory communication is affected by experience. Olfactory communication may be phylogenetically old in the primate order, but its potential as a sophisticated system of information transfer has only recently been recognized. A great deal of future work will be necessary to clarify the interrelationship between pheromones and their environmental and social consequences.

VISUAL COMMUNICATION
Introduction and Historical Background

Visual communication occurs when one animal gains information from another by looking at it. Perhaps because their visual system is so much like our own, the visual communication of nonhuman primates is more easily studied and better understood than their olfactory communication, and primatologists have been able to uncover great complexity in the range of information that can be drawn from another individual's spacing, body postures, and facial signals.

While spacing and body postures are used throughout the primate order, the occurrence of complex facial signals is most evident in monkeys and apes. By contrast, among prosimians, longer noses, tethered lips, and reduced facial enervation limit the variety and complexity of facial signals that can be produced. The noctural activity of many prosimians also makes them less suitable subjects for the analysis of facial movements. Thus in many respects the enhanced visual communication of diurnal higher primates seems to fulfill a function analogous to the olfactory communication of nocturnal prosimians.

The ethological approach to visual communication developed from the work of Lorenz (1950) and Tinbergen (1951), who characterized communication as a series of genetically coded units of behavior ("fixed action patterns"). Because such behavioral units were assumed to be both stereotyped and species-specific, analysis of communicative gestures could lead to the discovery of phylogenetic relationships between species. By comparing expressions of more than a dozen genera and species, Andrew (1963a, 1963b, 1963c) and van Hooff (1962, 1967) utilized this approach in their pioneering studies describing primate facial gestures. Lists of behavioral elements were assembled by observation and filming of paired animals in small cages. Elements were then used to describe gestures, and these gestures were related to emotional states such as aggression or fear. For example, van Hooff (1967, 29) described the "frowning bared teeth scream face" as follows: "The eyes are closed, or opened to only a small degree . . . the animal looks away . . . the eyebrows are lowered in a frown. The ears are retracted. The mouth is opened. The lips are vertically retracted and the mouth corners are fully retracted thus baring all of the teeth. The body posture is . . . crouched. Often the animal presses its belly against the ground . . . The vocalization is usually a loud high-pitched screaming sound like 'aach' or 'eech.'" This expression is used in circumstances where frightened or submissive animals have just lost an aggressive interaction.

Other studies, such as Hinde and Rowell's (1962) work on rhesus macaques, focused on a single species in an attempt to obtain a complete description of its communication. As with Andrew and van Hooff, the emphasis was on normative behavior patterns: the average expression by members of a given species, rather than differences in communication among individuals. As more studies were conducted, the descriptive approach was expanded to include a quantitative association of face and body movements with functional systems of behavior such as play or submission (e.g., Altmann 1965; Berdicio and Nash 1981; Sade 1973; van Hooff 1973). For example, among chimpanzees the "relaxed open mouth face" is frequently seen when animals are playing together. It is characterized by a lack of tension and extremely quick and unpredictable body movements. The open mouth is sometimes closed on the body of the other individual in a playful "gnaw wrestle" that is more gentle than the struggling and biting characteristic of aggressive encounters. Relaxed open mouth faces and gnaw wrestling are much more frequently associated with each other than either is with tense, agonistic charges and attacks. By analyzing communicative elements according to their temporal relations, it becomes possible to associate expressions arising from the same motivational state and to predict subsequent behavior from subtle communicative cues.

Variation in the Use of Visual Signals

To further investigate the subtleties inherent in primate visual communication, recent studies have focused on individual differences in facial and gestural communication. One study, for example, used frame-by-frame film analysis to investigate individual differences in the aggressive and friendly expressions of Barbary macaques (*Macaca sylvanus*) in Gibralter (Zeller 1980). "Components" were defined as the movement of a particular portion of the face in a specific direction, such that, for example, "eyebrows raised" was distinguished from "eyebrows lowered." Different individuals used from 7 to 22 of the 33 components available in the repertoire. Facial "gestures" consisted of a group of components used either concurrently or serially in the course of a communicative episode. The entire transmission of a gesture was considered a "unit" of interaction.

Three results of interest emerged from film analysis. First, aggressive and friendly gestures shared a surprisingly large number of components. The 4 most frequently used components in aggressive gestures (eyes staring, eyebrows raised, eyebrows lowered, and piloerection) were among the 7 most frequently used components in friendly gestures (Zeller 1982). This suggests that components in a gesture act collectively to produce a message and that the information value of components may vary with the context of the interaction. Second, components that consistently occurred in both threatening and friendly gestures were concentrated in the nonmouth regions of the face, such as the eyes and eyebrows. Components that were centered on the mouth region were much less consistent from individual to individual, and their inclusion may have been influenced by the context of the interaction (Zeller 1982). This emphasizes the special importance of the mouth, and regions around the mouth, when interpreting the visual signals of macaques, and probably of many other primates.

Third, animals could not be distinguished *individually* by their use of particular components, and thus facial gestures did not seem to involve individual "signatures" as has been shown in primate vocalizations (chap. 36). The use of components in facial gestures, however, did differ significantly across age-sex classes and across kin groups. In particular, the four matrilines were clearly distinguishable from each other as groups, and one animal, unrelated to the rest, was also clearly distinguishable (Zeller 1980). Thus a male Barbary macaque, transferring into a new social group (chap. 21), might be able to acquire information about matrilineal relatedness in the group simply by observing facial expressions.

Function of Visual Communication

Questions about the function of visual signals in nonhuman primates overlap in many respects with questions about the function of primate vocalizations (chap. 36). How, for example, are visual signals used in competitive and cooperative social interactions? How is the use of visual signals affected by experience or by prior social interactions? Can primates use visual signals to convey information about objects in their environment? Can they conceal information from or deceive others? A review of all these issues is inevitably frustrating because, while some areas are richly documented (competition and cooperation: see chaps. 25, 34), others are little studied and highly speculative.

Visual Signals in Competitive and Cooperative Interactions. Data summarized in chapters 25 and 34 clearly demonstrate that nonhuman primates regularly maintain and manipulate social relations through the judicious use of visual communication (fig. 35-1). Threats can be used to coerce another individual into performing a particular activity. Male hamadryas baboons threaten and neck-bite their females to encourage them to follow closely (chap. 10), and de Waal (1982) observed one clear case of a female chimpanzee who used threats to force another old female to suckle her infant. Attempts to gain the support and cooperation of another may also involve gestures of reassurance and conciliation, such as friendly approaches or grooming. During a status takeover attempt among wild chimpanzees, Nishida (1983b) observed that the third-ranking male appeared to manipulate the two main competitors by siding first with one then with the other so that the contestants would vie for his support with friendly gestures. De Waal (1982) observed a similar incident of social maneuvering among captive chimpanzees.

Successful solicitation of aid in agonistic situations frequently depends on the relationships established between participants. Mothers usually assist their offspring, members of a matriline tend to support each other, and close associates ("friends") are also a frequent source of aid. Assistance is usually solicited by threatening and sometimes vocalizing toward the antagonist, while simultaneously retreating toward the source of support with frequent quick looks back and forth between supporter and antagonist. If an individual of low dominance rank, with few kin nearby, directs aggression toward a higher-ranking individual who is near its relatives, the antagonist may find four or five individuals threatening it from all sides. Thus, although communicative gestures transmit messages, the effectiveness of these messages frequently varies, depending on the social relations of the signaling animal.

Factors Affecting Development of Visual Signals. The many experiments of Harlow and colleagues (summarized in Hinde 1974a) demonstrate that, in cases of extreme deprivation, a young primate's social experience can exert major effects on its ability to use visual signals. The effect of experience and social learning of the development of complex social strategies has been investigated in experiments by Anderson and Mason (1978). Subjects in these experiments were two groups of young rhesus macaques, one containing animals who had been reared normally, the other containing animals who had been reared either without their mother or without peers. In both groups, animals formed dominance ranks based on the direction of dyadic interactions, and such ranks were correlated with many other sorts of social interaction. Dominance, however, was less predictive (and dominance ranks changed more often) in the group containing animals reared normally. This was apparently due to the social strategies observed among experienced ani-

mals, who formed coalitions at high rates and could thereby influence the behavior of higher-status members to obtain their objectives. The development of such complex "triadic" and "quadratic" interactions only among experienced animals clearly indicated that "the development of higher orders of social cognition is dependent on early social experience" (Anderson and Mason 1978, 289).

Experiments conducted by Miller (1967, 1971, 1975) on captive rhesus macaques indicate that the ability of two individuals to communicate cooperatively, and to "read" each other's facial expressions, may depend on their prior social interactions. Miller first trained monkeys to respond in different ways to two different stimuli. If a monkey saw stimulus A, it learned to pull a bar, otherwise it received a shock. If the monkey saw stimulus B, it learned to press a second bar which released a food reward. Having trained monkeys individually, Miller placed them in a situation where cooperation was encouraged. One monkey (the "stimulus monkey") could see the lighted panel where one of the two stimuli would appear, but this animal had no levers with which to respond. A second monkey (the "responder") had the necessary levers, but could not see the stimuli that indicated which bar to use. The responder could only see the face of the stimulus monkey, projected onto a television

FIGURE 35-1. With eye stares and raised eyebrows, two adult female baboons, mother and daughter, threaten a common opponent. (Photo: Barbara Smuts/Anthrophoto)

screen. If the responder pressed the correct lever (presumably on the basis of the stimulus animal's face, and his own knowledge of the experiment), both animals received food. If an incorrect response occurred, both animals received a mild shock.

Miller found that animals unknown to each other fared poorly, whereas cagemates responded correctly on 80 to 90% of all tasks. Certain kin pairs, such as mothers and sons, performed better than nonkin, although no detailed comparison among classes of kin was made. Analysis of efficient senders and receivers indicated that few animals were effective at both. The animals most effective in both areas were those of high dominance status, whose behavior in many natural settings has the strongest influence on other group members (e.g., chap. 25).

Information Contained in Visual Signals. As noted earlier, visual signals are often used by observers as clues to the motivations underlying primate social interactions. This does not necessarily mean, however, that when one primate gestures to another the only information conveyed concerns the motivational state of the signaler. Savage-Rumbaugh, Wilkerson, and Bakeman (1977), for example, found that gestures used by bonobos (*Pan paniscus*) before copulation referred both to the signaler's emotional state and to aspects of the environment external to the signaler. The most common gestures observed between bonobos involved directional contact in which a particular form of body contact directed the receiver to take up a new position. For example, one animal would touch the outside of the shoulder and push the torso of the partner away gently to indicate that the partner should turn her back to facilitate dorso-ventral copulation. A more complex gesture occurred when one animal touched the partner's shoulder, then passed an arm across his or her own body, apparently indicating dorso-ventral copulation. Still more complex were cases in which one animal pointed in the direction of desired movement or flipped the hand upward at the wrist to indicate that the partner should stand bipedally in preparation for ventro-ventral copulation (for similar observations among bonobos in the wild, see Kano, unpub.; chap. 15). Such gestures are intriguing because, in addition to revealing the signaler's emotions, they appear to represent a location to be occupied, or a particular activity to be performed, by both signaler and recipient.

Menzel (1969, 1971, 1973b, 1975) was one of the first to examine the ability of nonhuman primates to communicate about their environment. He maintained four to six young chimpanzees in a large outdoor enclosure that was also connected to a smaller holding cage. He restrained all but one animal in the holding cage, while showing this chosen "leader" the hidden location of either an amount of food or an aversive stimulus such as a stuffed snake. The leader was then returned to the holding cage, and the whole group was released. According to Menzel's reports, the variable behavior of the animals indicated that they "seemed to know approximately where the object was, and what sort of object it was, long before the leader reached the spot where it had been hidden" (1975, 112). If the goal was food, they ran ahead looking in possible hiding places; if it was a stuffed alligator or snake, they emerged from their cage showing piloerection and staying close to their companions. If the hidden item was an alligator or snake, they became very cautious in their approach and often mobbed the area, hooting in the direction of the hidden item and hitting at it with sticks. If the hidden item was food, the animals searched the area intensively and showed little fear or distress. The behaviors occurred even if the aversive stimulus had been removed before the animals were released from the holding cage, so it was not the item itself that produced these reactions.

In the food tests, one male (Rocky) began to monopolize the food supply when it was located. When Belle, a female, served as leader, she attempted to avoid indicating the location of the food cache, but Rocky could often extrapolate from her line of orientation and find the food. If Belle were shown two caches, one large and one small, she would lead Rocky to the small one and, while he was busy eating, run to the larger one which she would share with other individuals. Menzel concluded that chimpanzees could communicate the direction, amount, quality, and nature of the goal, as well as attempt to conceal at least some of this information, but precisely how chimpanzees achieve such communication is still not known (Menzel 1969, 1971, 1973b, 1975).

Deception. Primate gestures often reveal an animal's intentions or imminent behavior. For individuals competing with one another, this is not always an asset, and there are occasions when one can imagine a primate "wanting" to withold information about its intentions. The experiments of Miller and Menzel described above indicate that the use of visual signals can be modified by experience: can primates also modify their signals to conceal or give false information?

In de Waal's captive chimpanzee group at Arnhem Zoo, Netherlands, he saw animals manipulating their faces with their hands and pushing their lips together, apparently in an effort to cover up expressions of fear so that rival animals would not infer their emotional state. In another case a lower-ranking male, who had been making approaches to an estrous female, suddenly became aware of the alpha male's presence. The lower-ranking male covered his erection with his hands and turned his back on the approaching male until his erection had subsided. One female often approached others with gestures of reconciliation, only to bite when she got close enough.

Sometimes a lower-ranking chimpanzee male who wished to mate with an estrous female would catch her eye and look toward some bushes at the edge of the enclosure. Then he would unobtrusively disappear, followed later by the female, and copulation would occur out of sight of animals who might intervene, such as the alpha male. In some cases de Waal saw a third chimpanzee observe the clandestine mating and rush up to the alpha male, barking and grabbing his arm, and leading him to the spot where the infraction was occurring (de Waal 1982). De Waal's observations indicate that primates use visual communication not only in relatively straightforward aggressive or cooperative interactions, but also in more complex social exchanges, where competition masquerades as cooperation and deception is the animal's aim.

SUMMARY

Although primates, compared with other mammals, rely more on vision and touch than on smell, olfactory communication nevertheless plays an important role in primate social behavior. Outside of sexual contexts, olfactory signals are most common in prosimians and some New World Monkeys, which employ stereotyped behavior and specialized scent-marking glands in territorial defense, aggregation, alarm, and individual recognition.

Careful analyses of visual communication underlie much of the research currently being conducted on behavioral development and social relationships. Although visual signals within a species often appear quite similar on first examination, laboratory studies have shown that both their form and their function are strongly affected by social experience and by an animal's prior interactions with specific individuals. As a result, visual signals may be culturally transmitted, leading to local "traditions" within a particular area (chap. 38).

Examination of the components used in visual signals allows observers to compare species, to compare kin groups within species, to describe social interactions quantitatively, and to attribute "motives" and "strategies" to their subjects. Both Menzel (1975) and de Waal (1982; chap. 34) suggest that nonhuman primates use visual signals not only to signal their intentions but also to communicate about the direction, amount, and quality of external objects. Examples of chimpanzees attempting to deceive one another are particularly striking. In this respect, the study of visual signals, like the study of vocalizations (chap. 36), may eventually provide an insight into how primates think.

36 | Vocal Communication and Its Relation to Language

Robert M. Seyfarth

INTRODUCTION: EARLY THEORIES CONCERNING NONHUMAN PRIMATE VOCALIZATIONS

There has always been a temptation to compare the behavior of nonhuman primates with human behavior and to compare nonhuman primate vocalizations with human language. With this latter comparison foremost in mind, scientists who first studied the vocalizations of monkeys and apes described their results in terms of a series of dichotomies, each indicating a fundamental difference between humans and their nonhuman primate relatives. While human language was assumed to be under voluntary, higher cortical control, and human words assumed to represent objects or events in the external world, observers described the vocalizations of monkeys and apes as occurring only in highly emotional circumstances, such as fights or encounters with predators. It was concluded that nonhuman primate vocalizations were relatively involuntary, under limited higher cortical control, and that such calls could not be interpreted as "representing" anything other than an individual's internal emotional state (e.g., Lancaster 1975; Washburn 1982). Supporting neurophysiological evidence came from laboratory work on squirrel monkeys (*Saimiri sciureus*), where it was shown that seemingly normal vocalizations could be elicited through electrical stimulation of subcortical areas in the brain, such as the anterior limbic cortex, known to be associated with emotional expression in humans (Jurgens 1979; Ploog 1981). Morton (1977), specifying the link between emotional state and the physical structure of vocalizations more precisely, argued that birds and mammals give harsh, low-frequency sounds when hostile, and higher-frequency, less harsh sounds when frightened, appeasing, or behaving in a friendly manner.

Primate ethologists also compared their work with research on language perception, where it had been shown that, with certain consonants, humans perceived an acoustically graded continuum of speech sounds as a series of discrete categories. Many linguists believed that such "categorical" perception was a necessary precursor of language. On the basis of their own experience and acoustic analyses using sound spectrographs, field biologists noted that monkeys and apes also had "discrete" and "graded" calls in their repertoires. Discrete calls were easy to distinguish by ear in the field and acoustically different from one another. Other call types graded into one another and were generally difficult to distinguish, either by ear or spectographically (e.g., Marler 1965; Struhsaker 1967a). Discrete vocal repertoires were thought to be common in many forest monkeys, where visibility was poor and long-distance calls predominated. Such calls allowed animals to communicate unambiguously, even in the absence of supporting visual signals. In contrast, monkeys and apes living in open country were described as having a more acoustically continuous array of vocal signals and were thought to use these signals in combination with other visual cues (Marler 1965). Marler (1976a) hypothesized that a crucial step in the evolution of language occurred when early humans began to perceive graded vocal signals in a discrete manner.

A final and important dichotomy between nonhuman primate vocalizations and human language concerned learning. Humans obviously learn many aspects of language, and language learning may be severely disrupted by early deafness or social isolation. In contrast, the few early studies conducted on nonhuman primate vocalizations showed neither gradual development nor dependence on environmental cues. Winter et al. (1973) found that squirrel monkeys could apparently produce their entire species' vocal repertoire, without anything resembling practice, 6 days after birth. Similarly, Talmadge-Riggs et al. (1972) found no vocal abnormalities in squirrel monkeys deafened shortly after birth or in animals raised with muted mothers.

In recent years, these clear (and simplistic) dichotomies between human and nonhuman primate vocalizations—voluntary versus involuntary, semantic versus affective, graded versus discrete, learned versus unmodifiable—have been challenged by research in three interrelated areas. First, in the well-known "ape-language" projects, scientists have achieved some success in their attempts to teach elements of human language to chimpanzees, gorillas, and orangutans. Second, studies of the natural vocalizations of monkeys have suggested that there are rudimentary parallels between the way monkeys use vocalizations and the ways humans use words. The notion that nonhuman primate vocalizations are involuntary and provide information only about an animal's internal emotional state has largely been disproven. Finally, just as psychologists and anthropologists have long used language as a means of studying how humans think, primatologists are beginning to discover that the communication of monkeys and apes can reveal much about how these animals classify and think about objects in the world around them. Although primate vocalizations are not language, they can, like language, provide primatologists with a "mirror on the mind" of their subjects. The aim of this chapter is to review this research and to discuss the relation between vocal communication, cognition, and social behavior in nonhuman primates.

APE-LANGUAGE PROJECTS
Introduction

The earliest attempts to teach human language to nonhuman primates aimed, quite literally, to teach spoken English to chimpanzees (e.g., Hayes 1951). These efforts were failures, largely because the vocal apparatus of the chimpanzee prevents it from articulating many of the sounds used in human speech (Lieberman 1975). More recent projects, using a variety of methods, have employed as subjects either chimpanzees, gorillas, or orangutans. Despite differences in species and techniques, all of these studies began with similar goals: to decompose language into its constituent parts, to develop training procedures for each constituent, and to establish whether a nonhuman species could be taught all, or any part of, human language (for a detailed review, see Ristau and Robbins 1982).

The two most basic constituents of language are referential signals and the ability to generate sentences. As many have emphasized, true linguistic reference (or semantics) requires more than just the use of signs to name objects. An animal must understand the *relation*, "X is the sign for Y," and be able to distinguish such referential relations from other sorts of relations, such as, for instance, "X is the cause of Y" (e.g., Premack 1976; Savage-Rumbaugh et al. 1980). Compare, for example, reference in human language with the dance "language" of the honey bee, where different components of the bee's dance represent the relative size and location of a food source (reviewed in Gould 1982). As Premack (1976, 8) points out, although a bee can encode specific information in its dance, and a second bee can decode this information, it is unlikely that bees are aware of precisely how the components of the dance represent, or "stand for," different aspects of the food source. The bee's dance may function as a representational signaling system, but bees, unlike humans, probably do not understand the relations between signs and objects that underlie their communication (Gould and Gould 1982). Attempts to establish whether nonhuman primates understand the relations that underlie representational signal-

ing have comprised a major part of the ape-language projects, as well as other work on nonhuman primate communication.

The vocalizations of many animals, including nonhuman primates, are often given in strings that include a number of different call types (e.g., Beer 1976; Marler 1972; Deputte 1982). In comparing such sequences with human sentences, it is important to distinguish two sorts of signal combinations—one simple and the other far more complex (Marler 1977). In simpler sequences, new signals are formed through the combination of existing sounds. There is, however, no "rule" underlying the construction of such combinations, and there is no apparent functional relation between different strings. In contrast, when humans use words to generate sentences, they do so according to identifiable rules, called grammar (e.g., Chomsky 1957). This formal structure specifies the relation between words in a string and between strings. It is grammar that gives language its extraordinary communicative power and allows humans to create novel messages and to generate an infinite number of sign combinations that are readily comprehended by others. In the search for grammar among nonhuman primates, research has focused on the sequences of signs used by apes in captivity. Does the ordering of signs itself convey different meanings? Are these signs used in predictable orders, from which rules for sequence formation can be abstracted? If a rule-based sequencing system exists, does it emerge with little or no training, as Chomsky (1957) believes is the case with human infants?

Methods

The most common training method in the ape-language projects involves use of a form of American Sign Language for the Deaf (AMESLAN). Using one of two methods, apes are trained to associate a unique hand signal with a particular object. In one method, an object or a picture of an object is held in front of the ape while the instructor repeatedly demonstrates the appropriate sign (fig. 36-1a). Alternatively, the instructor may actively mould or shape the subject's hands until they form the correct sign. Records of the animal's progress, and tests of the animal's competence, are conducted using either observational sampling, videotaping, or elaborate "double-blind" observations. In some cases, double-blind tests involve experienced AMESLAN signers as judges. Studies using AMESLAN include the work of Gardner and Gardner (1969), Fouts (1973) and Terrace et al. (1979) on chimpanzees, Patterson (1978b) on gorillas, and Miles (1983) on orangutans.

Two other projects explore the linguistic abilities of chimpanzees using artificial lexicons. At the Yerkes Primate Research Center, Rumbaugh, Savage-Rumbaugh, and associates have used a system of signs linked to a computer to study the communication first of a single chimpanzee, Lana, and more recently of two chimpanzees, Sherman and Austin (Rumbaugh 1977; Savage-Rumbaugh et al. 1980). The animals are housed in one or more rooms containing a computer console with many large, lighted keys (fig. 36-1b). On the keys are printed lexigrams in "Yerkish," an artificial set of symbols. Chimpanzees are rewarded for pressing the appropriate keys at the appropriate time in the appropriate sequence, and their total communicative output can be stored in the computer for future analysis. Research at Yerkes is of particular note because of the methods used and because it has been one of the few projects systematically to study communication *between* two or more nonhuman primates.

Premack (1976, 1983b) uses an artificial lexicon of plastic chips to study communication and intelligence in chimpanzees. The chips are differently shaped pieces of colored plastic that adhere when placed against a magnetic board. A chip may represent an object, such as *apple,* it may represent a concept such as *same, different,* or *name of,* or a chip may represent a more abstract descriptor such as *color.* Sequences of words are placed in vertical order on the magnetic board. Premack's research is noteworthy because, in contrast to the original goals of other ape-language studies, its aim has been from the start to understand chimpanzee cognition rather than to teach the animals a version of human language.

Results

From the beginning, the ape-language projects have demonstrated that apes can use a large number of signs in a manner strikingly similar to the way humans use words. Washoe, the chimpanzee studied by Gardner and Gardner (1969, 1975), reportedly mastered over 200 signs, and similar vocabularies are reported for Nim, the chimpanzee studied by Terrace et al. (1979), Lana, one of the chimps studied by Rumbaugh (1977), and Sarah, a chimpanzee studied by Premack (1976). Koko, the gorilla studied by Patterson, is reported to have mastered a considerably larger vocabulary (Patterson and Linden 1981). All of these animals used signs correctly when shown an object, or a picture of an object, and asked "What this?" They also used signs correctly in sequences of other signs. Terrace (1979, 3), for example, describes one of his associates walking outdoors with Nim when a robin flew overhead. "Bird there," Nim signs.

"Who there?" the teacher answers.

"Bird." Then Nim pauses, and looks in another direction, at some flowers. "Bug, flower there," he signs.

"Yes, many things see," the teacher replies.

Descriptions of this sort raise two crucial questions. First, what do animals actually mean when they use signs in this way; second, is their use of signs sufficient proof that the animals actually understand the labeling relation between a sign and the thing for which it stands? For example, the chimpanzee Lana required 1,600 trials before she correctly answered the question "? what name of this," as it applied to the signs for banana and M&M candy. Extensive training was necessary despite the fact that she had already used these words hundreds of times in stock phrases such as "Please machine give M&M" (Rumbaugh and Gill 1977; Ristau and Robbins 1982).

FIGURE 36-1. Two methods of teaching artificial "languages" to captive chimpanzees: (*a*) A trainer holds up a picture of an orange while simultaneously making the sign for *orange* using American Sign Language for the Deaf, and the chimpanzee Nim imitates; (*b*) Sherman (*on the left*), uses a computer keyboard to ask another chimpanzee for one of several food items on the tray. (Photos: (*a*) H. Terrace; (*b*) E. Rupert, courtesy of D. Rumbaugh and E. Savage-Rumbaugh)

Thus, although Lana could use words correctly to request food, she apparently had merely formed an association between particular food rewards and a specific sequence of key punches. Additional training appeared to be necessary before she could use such words in a way that suggested she understood the relation between a sign and the object for which it stands.

Different sorts of data can provide insight into what a particular sign means to an ape. One of the most striking occurs in "overgeneralization," when an animal spontaneously uses a sign in a way that extends its original meaning. Washoe, for example, was originally taught the sign for flower in the presence of a real flower. She seems to have interpreted the sign as meaning smell, however, since she subsequently generalized it to pipe tobacco and kitchen fumes (Gardner and Gardner 1969, 1975; Ristau and Robbins 1982). Similarly, the gorilla Koko was originally taught that *straw* meant drinking straw. She later used the word to refer to plastic tubing, clear plastic hose, cigarettes, a pen, and a car radio antenna. All these objects were long and thin, and were objects for which Koko had no word in her vocabulary (Patterson and Linden 1981; Ristau and Robbins 1982).

The meaning of a sign can also be revealed by its spontaneous use, together with other signs, in novel situations. The most famous example is Washoe's signing *water bird* when she first saw a swan (Fouts 1974). Patterson (1978b) reports that Koko called a ring *finger bracelet,* a stale cake *cookie rock,* and a Pinocchio doll *elephant baby.* From these data, and from those on overgeneralization, it is hard to escape the conclusion that the animals attached a specific meaning to their signs.

The problem with such observations, however, is that too often they were not recorded by human trainers in a way that can be evaluated systematically. In most cases, when an animal overgeneralized, or used a novel combination of words in a way that seemed insightful, a detailed record was made. However, when the overgeneralization or novel use made no sense, the data were generally discarded. Thus while such observations do give some clue to the meaning attached to words, it is impossible to tell whether the ape was showing impressive imagination or was simply a "random sign generator" (Seidenberg and Pettito 1979).

A more systematic analysis of meaning was conducted by Premack (1976), who investigated whether the chimpanzee Sarah could perform a feature of analysis of an object and its label. Sarah's first test was to describe the features of an actual apple: was it red, was it round, did it have a stem? Then Sarah was asked the same questions about the *sign* for an apple, in Premack's lexicon a blue plastic triangle. Similarly, Sarah was asked whether the sign for a caramel candy was cube- or disk-shaped, white or brown, smooth or crumpled. In both cases Sarah used the same features to describe both the object and the sign that represented it.

Premack then reversed the question and asked Sarah to begin with an object and then describe the name for that object. Shown an apple, for example, Sarah correctly answered that the sign for this object was triangular not round, blue not green, and small not big (Premack 1976). These results show clearly that, for Sarah, a small blue triangle was a sign for an apple.

Further evidence in support of this claim comes from work by Savage-Rumbaugh et al. (1980, 1983). They taught the chimpanzees Sherman and Austin to sort objects such as keys, cakes, and sticks into two groups: food and tools. The chimps were then taught that a specific button on a keyboard represented all "food" while another button represented all "tools." Given this training, Sherman and Austin could not only correctly classify objects they had never seen before, but they could also classify photographs and lexigrams that represented different objects (Savage-Rumbaugh et al. 1980).

Considerable data, then, indicate that captive apes can learn to use signs to represent objects and relations, and that in doing so they understand the relation between a sign and its referent. Despite these semantic abilities, however, it seems unlikely that apes can combine such words into anything resembling human sentences.

The most thorough test of an ape's syntactic abilities is that of Terrace et al. (1979). Terrace and colleagues used observational data and videotapes to examine over 19,000 of the chimpanzee Nim's multisign combinations. They first asked whether Nim's sign combinations exhibited any regularities in sign order. There were regularities in two-sign combinations, but no similar regularities appeared in three- or four-sign utterances, and combinations of five or more signs were too rare to permit analysis. In addition, where Nim did exhibit predictably ordered two-sign combinations, his signs were frequently repetitions of what the trainer had just signed to him. Thus it was impossible to determine whether Nim's sign order was his own product or whether it simply emerged because Nim was copying the regular, ordered signs of his trainers.

Also arguing against Nim's sentence-generating abilities was the fact that most of his three- and four-sign combinations involved repetitions. For example, the most frequent three-sign combinations were "Play me Nim," "Eat me Nim," "Eat Nim eat," and "Tickle me Nim." "Eat drink eat drink" and "Eat Nim eat Nim" were the most common four-sign combinations (Terrace et al. 1979; Ristau and Robbins 1982). Terrace also searched for any indication that Nim systematically varied word order in different circumstances. For example, he might have used "Tickle me" whenever he solicited tickling from another, and "Me tickle" whenever he was about to tickle someone else. There was little evidence for regularities of this sort (Terrace 1979; Terrace et al. 1979).

Terrace's results have not yet been convincingly refuted, although they have been challenged on a number of grounds (e.g., Patterson and Linden 1981). To conclude this brief review of the ape-language projects, I discuss two issues, raised by Terrace's results, that have implications for all of the projects dealing with captive apes.

Terrace's work has been criticized because of both the way Nim was trained and the way Nim's communicative abilities were tested. Unlike many apes, who were trained by a small number of caretakers, Nim was exposed to 60 trainers over a 4-year period. This frequent turnover in instructors, it is argued, prevented Nim from forming close bonds with his human caretakers and may have impeded his communicative development (Ristau and Robbins 1982). Given data on ways in which the natural communicative development of primates can be affected by disruption of close social bonds (e.g., chap. 27), this criticism seems well founded. In addition to having many trainers, both Nim's training and the testing of his abilities were carried out in rather formal sessions, in an empty classroom from which most objects had been removed (Terrace et al. 1979). Such conditions are by no means optimal for either communicative development or test performance. Nim's motivation to interact may therefore have been reduced, with the result that his full communicative abilities were underestimated.

Indeed, it is generally true of the ape-language projects that they cannot distinguish between a failure of ability and a failure of motivation. When a captive animal fails to exhibit a particular behavior, this may indicate either limited cognitive skills or an unwillingness to perform under artificial conditions. To overcome the problems of motivation, Savage-Rumbaugh et al. (1983) and Premack (1983b) have begun working with a number of captive chimps simultaneously, while others have placed renewed emphasis on the communicative skills exhibited by monkeys and apes in their natural habitat.

NATURAL VOCALIZATIONS OF MONKEYS AND APES
Introduction

Early studies of vocalizations in free-ranging monkeys and apes were largely descriptive. Observers tape-recorded sounds, displayed them on sound spectrographs, noted the social contexts in which they were given, and described the function of vocal signals in much the same way they described the function of visual signals such as bared teeth or piloerection (e.g., Rowell and Hinde 1962; Goodall 1968; Fossey 1974b). The ape-language studies, however, raised major questions about this approach. Work with captive apes suggested that these animals could signal to each other *about* things, and they hinted that a chimpanzee's scream, for example, might mean more to nearby chimpanzees than human observers had previously imagined.

Predator Alarm Calls and the Possibility of Referential Signaling

In 1967, T. T. Struhsaker reported that East African vervet monkeys gave different-sounding alarm calls to at least three different predators: leopards, eagles, and snakes. Each alarm elicited a different, apparently adaptive response from other vervets nearby. Struhsaker's observations were important because they suggested that nonhuman primates might in some cases use different sounds to designate different objects or types of danger in the external world (Struhsaker 1967a).

To test this hypothesis, Seyfarth, Cheney, and Marler (1980a, 1980b), working with the same population of vervet monkeys originally studied by Struhsaker, carried out playback experiments. They began by tape-recording alarm calls given by vervets in actual encounters with leopards, eagles, and snakes. Then they played tape-recordings of alarm calls in the absence of predators and filmed the monkeys' responses.

Monkeys responded to playback of leopard alarms by running into trees, to eagle alarms by looking up in the air or running into bushes, and to snake alarms by looking down in the grass around them. Results also showed that varying the length and amplitude of different alarm types had little effect on the different responses they evoked. The only features of both necessary and sufficient to evoke the different responses were the different acoustic features of the alarm calls themselves (Seyfarth, Cheney, and Marler 1980a).

Acoustically distinct alarm calls have also been reported in a number of other nonhuman primates (cotton-top tamarins: Cleveland and Snowdon 1982; pygmy marmosets: Pola and Snowdon 1975; Goeldi's monkey: Masataka 1983; red colobus: Struhsaker 1975; gibbons: Tenaza and Tilson 1977), as well as in many nonprimate mammals and birds (references in Seyfarth, Cheney, and Marler 1980a). Dittus (1984) describes acoustically distinct vocalizations given by toque macaques in the presence of different foods. Although such calls have not always been studied experimentally, descriptions of them suggest that the generalizations applied to vervet alarm calls are likely to be true for many other species.

When interpreting data on primate alarm calls and comparing them with laboratory research, it is important to bear in mind the distinction between a behavior's function (its long-term consequences for survival and reproduction) and a behavior's causation (the short-term social or physiological changes that cause it to occur). In a functional sense, the alarm calls of primates (and other animals) have some intriguing parallels with human words. Like words, they provide nearby individuals with accurate information about the presence of different kinds of danger, and, at least in the case of vervet alarm calls, they do so in a way that is relatively independent of the caller's arousal state. In playback experiments, vervet alarms satisfy one requirement of representational signals: they accurately replace (i.e., elicit the same response as) the object for which they stand, even when that object is not itself present (e.g., Hockett 1960).

Playback experiments, however, tell us little about the causation of animal vocalizations because such experiments measure only the responses a call evokes, not the underlying processes (mental or otherwise) of the individual who is vocalizing. Thus experiments offer no proof that primates in the wild recognize the relationship between a vocalization and its referent, and care must be taken before drawing close parallels between the production of calls by free-ranging primates and the conscious intention to communicate that underlies some instances of communication in captive apes.

At the same time, a variety of evidence indicates that the vocalizations of primates are under some voluntary control, that they are not simply obligatory manifestations of certain arousal states, and that they provide information about events other than the signaler's emotional state. For example, rhesus monkeys (Sutton et al. 1973; Sutton, Trachy, and Lindeman 1981) and ring-tailed lemurs (W. A. Wilson 1975) in captivity can learn to vocalize only upon presentation of a certain stimulus, while adult female vervet monkeys, if exposed to an artificial predator, give more alarm calls in the presence of their juvenile offspring than in the presence of an unrelated juvenile (Cheney and Seyfarth 1985c). Similarly, in the field experiments on vervet alarm calls, subjects invariably responded to playbacks without giving calls themselves. Thus, monkeys can regulate the production of calls depending on a variety of stimuli, and they can respond to calls without vocalizing themselves.

Vocalization Specialized for Long-Distance Transmission

When one imagines a primate vulnerable to a variety of different predators with different hunting methods, it is easy to imagine why natural selection would have favored the evolution of acoustically distinct, representational alarm calls. In an analogous manner, forest-dwelling primates that use vocalizations to defend territories or to increase distance between groups face the problem of sound transmission through dense vegetation. Acoustic analysis indicates that such calls have indeed been selected for long-distance transmission through rain forest environments.

Experimental studies of loud vocalizations used in intergroup spacing have been conducted on both New World and Old World Monkeys. Waser (1977b) demonstrated that the "whoop-gobble" of the gray-cheeked mangabey mediated intergroup avoidance. When the members of a group heard playback of a neighboring male's whoop-gobble, they moved away from the sound source if it was closer than 400 m, but ignored it if it was farther away. If the members of a group heard playback of their own male's whoop-gobble, they approached the sound source. Waser (1977b, 1982b) has shown that the initial whoop of the call is relatively similar across males, and he suggests that its function is to draw attention to the caller. The gobble that immediately follows, however, differs markedly from one male to another and seems to provide acoustic cues for individual recognition. Similarly, Robinson (1979a) showed that, in monogamous titi monkeys, the male gives loud calls and moves toward a territorial border whenever he hears the calls of other males. As the male approaches the border, his mate joins him, and they countercall (or duet) as they approach a neighboring pair (see also Kinzey and Robinson 1983).

In tropical forests, low-frequency sounds travel for longer distances than do high-frequency sounds of the same amplitude (Marten and Marler 1977; Wiley and Richards 1978). Sounds between 500 and 2,500 Hz show the least attenuation of amplitude with distance (Waser and Waser 1977), and virtually all of the long-distance calls used by primates have energy within this range (cottontop and Geoffroy's tamarins: Moyniham 1970; Cleveland and Snowdon 1982; langurs: Horwich 1976; mangabeys: Waser 1982b; gibbons: Deputte 1982). Further, Waser and Waser (1977) have shown that the greatest sound attenuation occurs during midday, with less attenuation in the afternoon and the least attenuation at dawn. A number of primates that use long-distance vocalizations call at the highest rates during the early morning hours (indris: Oliver and O'Connor 1980; howler monkeys: Sekulic 1982b; langurs: Horwich 1976; gibbons: Marler and Tenaza 1977). Finally, detailed study of the vocal repertoire of mangabeys has shown that calls not used for intergroup communication have acoustic features that cause them to be less well transmitted through the environment (Waser and Waser 1977). Thus there is a direct relation between the function of a call and its acoustic properties.

The acoustic constraints imposed by forest environments are not, however, the only factors that have affected the evolution of long-distance primate vocalizations. For example, Waser (1982b) reviewed the acoustic features of loud calls given by a variety of baboons and mangabeys that inhabit environments ranging from rain forest to open savanna. He found that although environmental factors had clearly played a role in shaping the acoustic structure of some calls, other variables were also important, and in some cases appeared to override environmental considerations. For example, calls with homologous acoustic features could be found in many of the species studied despite wide variation in habitat. Moreover, these species appeared to use acoustically similar calls in similar social circumstances. Such data have led some to suggest that the acoustic structure of vocalization is, compared with other features such as coat color or cranial characters, phylogenically conservative. As a result, species that have recently occupied different habitats may nevertheless exhibit acoustically similar vocalizations, and these calls may be used to establish their evolutionary relationship (Struhsaker 1981b; Oates and Trocco 1983).

Vocalizations Used in Close-Range Social Interactions

Predator alarm calls and calls used in intergroup spacing comprise only a small part of primate vocal repertoires, and there is no a priori reason to believe that they are representative of primate vocal abilities. Indeed, if one is interested in comparing the natural vocalizations of monkeys with the "linguistic" communication of captive apes, the most important data would seem to come from the many calls exchanged among free-ranging primates in relaxed social situations, when animals are resting, grooming, foraging, or playing. In the past, such vocalizations were described simply as "contact calls." Recent research has revealed greater complexity, however, and has suggested parallels with language in both physiology and vocal function.

"Coos" of Japanese Macaques. Green (1975a), studying the vocalizations of Japanese macaques, concentrated on the class of calls he labeled "coo" vocalizations. Macaques gave coos to each other in a variety of situations, and on first hearing all coos sounded more or less the same to the human ear. With experience, however, differences began to emerge: certain coos, for example, had a frequency peak near the start of the call, with steadily falling frequency thereafter ("smooth early highs"), while others had a steadily rising frequency, reaching a peak in the call's latter half ("smooth late highs"—fig. 36-2). Green found, moreover, that the Japanese macaques used different coos in different social situations. Smooth early highs, for example, were most commonly used by infants sitting apart from their mothers, while smooth late highs were used most frequently by sexually receptive females.

Green's results had an important influence on subsequent research for a number of reasons. First, they indicated that the vocal repertoires of nonhuman primates are likely to be larger than the human ear initially perceives them to be. Second, because humans may underestimate the size of primate vocal repertoires, they may also underestimate the extent to which specific signals are used selectively in particular social situations. Third, Green's results suggested that dividing nonhuman primate vocal repertoires into "graded" and "discrete" classes may be misleading since, like humans, the animals may often perceive a graded continuum of sounds in an apparently discrete manner.

When humans discriminate between phonemes, two phenomena are apparent. First, they divide acoustic stimuli into categories by using some physical features to set category boundaries while at the same time ignoring others (e.g., Ladefoged 1975). Second, for most humans the perception of many linguistically relevant sounds is specialized in the left cerebral hemisphere. As a result, subjects commonly show a right-ear advantage in tests of discrimination (reviewed in Bradshaw and

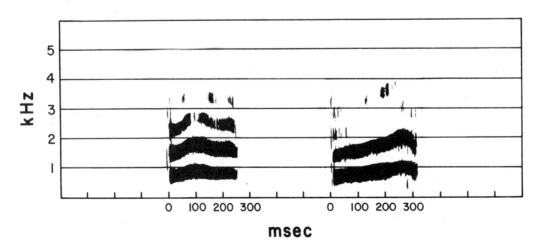

FIGURE 36-2. Two "coo" calls recorded from Japanese macaques. The Y-axis is measured in frequency, the X-axis in time. The call on the left is a "smooth early high," and the call on the right is a "smooth late high." Calls recorded by S. Green and used with permission.

Nettleton 1981). Following Green's research, the following experiment was designed to test for similar phenomena in nonhuman primates.

First, a number of smooth early and smooth late high coos were selected as test stimuli. One task of primates in the laboratory was to divide calls into two classes on the basis of peak position, a feature that Green's work had suggested was biologically meaningful for Japanese macaques. At the same time, coo stimuli also differed in terms of onset frequency: some had a relatively high and others a relatively low initial frequency. A second task of subject animals was to divide stimuli into two classes on the basis of initial frequency, presumably a feature that was less biologically important than peak position, at least for Japanese macaques.

Intriguingly, when subjects were rewarded for discriminating coos on the basis of peak position, Japanese macaques learned to do so significantly faster than control species (a rhesus macaque, two bonnet macaques, and a vervet). In contrast, when a second group of subjects was rewarded for discriminating coos on the basis of initial frequency, Japanese macaques learned to do so at a slower rate (Zoloth et al. 1979).

Moreover, throughout these experiments Japanese macaques exhibited a right-ear advantage while other species did not (Petersen et al. 1978, 1984; Hefner and Hefner 1984). The neural lateralization shown by Japanese macaques is particularly important because it parallels human behavior and because it contradicts earlier studies that failed to find evidence of nonhuman primate lateralization (e.g., Hamilton 1977; see also chap. 1).

Other Studies. Data from a variety of species indicate that Japanese macaques are by no means the only primate to make subtle acoustic discriminations within a broad class of vocalizations. Similar descriptive findings have been reported for talapoins (Gautier 1974), spider monkeys (Eisenberg 1976), and cottontop tamarins (Cleveland and Snowdon 1982).

Many studies have gone beyond the description of subtle acoustic variation, have used playback experiments to verify that such variation is perceived by the animals themselves, and have examined how such call types function. For example, in pygmy marmosets, trills whose acoustic features make them most easy to locate are used when animals are most widely separated, and those whose features make them difficult to locate are used when animals are close together (Snowdon and Pola 1978; Snowdon and Hodun 1981). In cebus monkeys, the "huh" call is given frequently throughout the day and appears to counteract the animals' tendency to clump together, especially at food sources. "Heh" calls also function to separate animals spatially, while "arrawhs" bring them together (Robinson 1982b).

Among vervet monkeys, playback experiments have shown that animals distinguish at least four different grunts and that certain grunt types can, like the vervets' alarm calls, function to convey information about external events. For example, one grunt type is given when a vervet first spots the members of another group. If this grunt is played to monkeys from a concealed loudspeaker, they respond by looking toward the horizon in the direction the speaker is pointed. In a manner reminiscent of the signs of captive apes, the grunt alone provides specific information about external events, without requiring support from other cues such as postures or facial expressions (Cheney and Seyfarth 1982a).

Nonhuman primates thus make subtle acoustic discrimination between vocalizations, and the use of their calls reveals some rudimentary functional and physiological parallels with the use of human words. As noted earlier, however, care must be taken when drawing such parallels because we still know very little about the neurological and cognitive basis of primate vocal signals. It is entirely possible, for example, that the different grunts of vervet monkeys are closely associated with different levels of arousal, and that all vervets "agree" on which social situations are most arousing. Alternatively, vervet grunts may result from some more cognitive process that is less dependent upon arousal and that shows more parallels with the production of human speech.

Clarifying the relative importance of intention, cognition, and affect in primate vocalizations will not be easy because even in the well-studied human case it has proved difficult to separate the cognitive and affective components of human speech. The fact that apparently normal vocalizations of squirrel monkeys, for example, can be elicited in the laboratory through electrical stimulation of certain areas in the limbic system (e.g., Jurgens 1979; Ploog 1981) does not necessarily mean that higher cortical areas play no role in the normal production of calls, since animals are selective about whom they vocalize to and under what conditions. The squirrel monkey "chuck," a common affinitive vocalization, is given only between females who, by other behavioral measures, can be described as close "friends" (Smith, Newman, and Symmes 1982).

Similarly, classifying primate vocalizations according to whether they are harsh, low-frequency sounds signaling aggression or tonal, high-frequency sounds signaling appeasement (Morton 1977) is often misleading because to do so omits much of the specific information conveyed by different calls. Japanese macaques use many different variants on the same tonal "coo" pattern and do so in situations ranging from affinity to distress (Green 1975a). Vervet monkeys show similarly broad variation in the circumstances under which they give acoustically quite similar grunts (Seyfarth and Cheney 1984b).

In summary, simplistic dichotomies separating the "affective" signals of primates from the "cognitive" signals of humans (e.g., Washburn 1982) are misleading, largely because they ignore recent studies suggesting that primate vocalizations are under voluntary control and may *function* to designate objects and events in the external world.

VOCAL COMMUNICATION AND COGNITION

As our understanding of primate communication improves, it is becoming increasingly clear that we can use vocalizations to study how individuals classify and associate objects, including other primates, in the world around them. The most obvious illustration occurs in tests for individual recognition; but there are other, more complex examples that in some cases allow us to study how primates think and in others offer an intriguing glimpse of primate social structure from the animal's point of view.

Individual Recognition

A number of studies have selected a particular call type within a species' repertoire, demonstrated consistent acoustic differences between the calls of different individuals (pygmy marmosets: Snowdon and Cleveland 1980; squirrel monkeys: Smith, Newman, and Symmes 1982; stump-tailed macaques: Lillehei and Snowdon 1978; gibbons: Tenaza 1976; chimpanzees: Marler and Hobbett 1975), and shown through playback experiments that animals respond selectively to the calls of certain individuals (squirrel monkeys: Kaplan, Winship-Ball, and Sim (1978); mangabeys: Waser 1977b; rhesus macaques: Hansen 1976). Studies of individual recognition in primates parallel those from many other species of birds and mammals (reviewed in Green and Marler 1979).

More Elaborate Recognition and Classification

In a test for individual recognition in vervet monkeys, the scream of a 2-year-old juvenile was played from a concealed loudspeaker to three adult females, one of whom was the juvenile's mother. Mothers responded more strongly than control females, which demonstrated that they recognized the screams of their offspring. More intriguingly, however, control females responded to playbacks by looking at the mother, often before the mother herself had responded (Cheney and Seyfarth 1980, 1982b). Control females behaved as if they were able to associate particular screams with particular juveniles, and these juveniles with their mothers. This result is of interest because it indicates that monkeys not only distinguish their own offspring from others, but they also recognize the associations that exist among other group members. Such information can only be obtained by observing the interactions of other group members and making appropriate deductions.

Further evidence of primate social knowledge comes from experiments on rhesus macaques. Gouzoules, Gouzoules, and Marler (1984) noticed that the screams of juveniles varied in their acoustic features (fig. 36-3), and that certain screams were given almost exclusively in particular social situations. Some screams, for example, were acoustically noisy, and were given most often when the vocalizer was interacting with an individual higher ranking than itself and in situations involving physical contact. "Arched" screams (fig. 36-3) were given to lower-ranking individuals in the absence of physical contact. Both "tonal" and "pulsed" screams were given more often than expected to relatives, and "undulated" screams were given almost exclusively to higher-ranking individuals when no physical contact occurred. Observation revealed no relation between the different screams and the vocalizer's subsequent behavior, thus arguing against the hypothesis that screams were simply manifestations of arousal, linked in some physiological way to the animal's imminent activity. Playback experiments, however, showed clearly that mothers responded differently to the different types of scream from their offspring. Mothers responded most strongly to playback of noisy screams (physical contact with higher-ranking opponents), and next most strongly to playback of arched (lower-ranking opponents), tonal, and pulsed screams (genetically related opponents) in that order. This suggests that acoustic differences between screams function to convey information about different external referents, specifically the type of opponent a vocalizer faces and the severity of aggression involved.

These results are important because they provide an insight into rhesus social organization from the monkey's point of view. When it screams, a juvenile effectively classifies its opponent according to kinship, dominance rank, and the severity of aggression (see also Seyfarth and Cheney 1984a). By her selective response, an adult female reveals knowledge of both her offspring's voice and network of social relations. Results suggest that concepts such as kinship and dominance rank, devised by human observers to explain and organize data on nonhuman primate behavior, exist not only in the minds of observers but also in the minds of their subjects.

Deceit

The interrelation between communication and cognition is nowhere more apparent than in the case of deceit. For one animal to deceive another, it must be aware that its communcation transmits information, it must recognize that other animals know that its behavior is informative, and it must select, from a set of alternatives, communication that provides, suppresses, or distorts information (Woodruff and Premack 1979). Given the elaborate ways in which primates use behaviors such as grooming and alliances to manipulate social relations (e.g., chap. 34), we should always be open to the possibility of deception in primate communication. Thus far, however, active deception has only been demonstrated in the laboratory, where it emerged after considerable training.

Woodruff and Premack (1979) showed a chimpanzee two containers, one with food hidden inside. The chimpanzee was then introduced to two different trainers, neither of whom knew the location of the food. One was a "cooperative" trainer: if the chimpanzee signaled which container held the food, this trainer collected the food and shared it with the chimp. The second trainer was "uncooperative": when shown the location of the food, he ate

the food himself. Over time, chimpanzees were tested with each trainer, in trials where chimp and trainer alternatively served as sender and recipient of information.

When interacting with the cooperative trainer, chimps from the very beginning were able to produce and comprehend accurate cues about the location of food. When interacting with the uncooperative trainer and playing the role of sender, chimps learned, after many trials, first to withold information and then to mislead the recipient. In the role of recipient, they learned to discount or controvert the sender's misleading cues (Woodruff and Premack 1979).

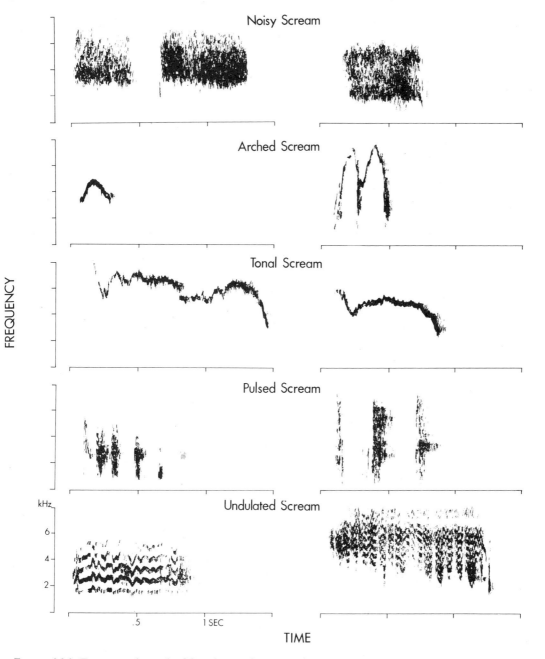

FIGURE 36-3. Two examples each of five classes of screams given by juvenile rhesus macaques. The Y-axis is measured in frequency, the X-axis in time. From Gouzoules, Gouzoules, and Marler 1984, with permission.

Although deception in animal communication has only begun to be studied, three sorts of evidence distinguish these data on chimpanzees from other examples of deceit, such as the "broken-wing display" of birds such as the killdeer (*Charadrius vociferus*) (Skutch 1976). First, in Woodruff and Premack's experiments, deception emerged gradually over the course of many trials. Unlike distraction displays in birds, chimpanzee deception appeared less "hard-wired" and more easily modified as the result of memory and experience. Second, chimpanzees were able to vary their deceptive behavior depending on the specific individuals with whom they were interacting. In contrast, there are as yet no data indicating that birds, for example, have sufficient control over their distraction displays that they can withold such displays depending on the individual identities of specific predators. Third, our knowledge of chimpanzees under other circumstances further suggests that their deceit is based on conscious reasoning. For example, at the Gombe Stream Reserve in Tanzania, the subadult male Figan was unable to obtain a share of the bananas provided by researchers, due to competition from other chimpanzees. He solved this problem by marching purposefully out of camp in a manner that caused everyone present to follow him. Shortly thereafter he abandoned his companions, circled back, and ate the bananas himself (Goodall 1971; see also de Waal 1982 for further examples). Thus while there may be many examples of behavior in animals that has the consequence of deceiving others, so far only deception in chimpanzees appears to be based on the same sort of cognitive "strategies" familiar in human interactions.

UNRESOLVED ISSUES IN PRIMATE VOCAL COMMUNICATION

Despite considerable progress in the past 10 years, there are still major gaps in our understanding of nonhuman primate vocalizations. For example, while apes have been taught striking communicative skills in the laboratory, almost nothing is known about the vocalizations of chimpanzees, gorillas, and orangutans in their natural habitat. Some New and Old World monkeys have been well studied in the wild, but psychologists have only recently begun to employ natural vocalizations as stimuli in laboratory tests of the same species. Although laboratory chimps have failed to show evidence of syntactic ability, rule-governed ordering of natural vocalizations remains a possibility (Robinson 1984; Snowdon and Cleveland 1984).

Perhaps the most obvious gap in our knowledge, however, concerns vocal development. As noted earlier, there is little evidence for developmental modifiability in nonhuman primate vocalizations. Instead, the only evidence for learned, modifiable vocalizations in species other than humans comes from songbirds (Marler 1981).

Research on primate vocal development must deal with three areas: vocal production, vocal usage, and the response to vocalizations by others. Work on vocal production concerns the development of species-typical calls, and it is here that scientists face their most serious methodological problems. Unlike birds, nonhuman primates cannot be raised in isolation without developing severe behavioral pathologies. As a result, it is difficult to separate the relative roles of genetic and environmental factors affecting development. Newman and Symmes (1982) review data suggesting that, in both rhesus macaques and squirrel monkeys, normal vocal ontogeny depends on both hearing conspecifics and maternal interaction. Snowdon, French, and Cleveland (in press) describe an early stage of "babbling" or vocal practice in marmosets and tamarins. These studies conflict with data from squirrel monkeys (e.g., Winter et al. 1973) and gibbons (Brockelman and Schilling 1984), which suggest that crucial observations and experiments remain to be done.

The ontogeny of vocal usage concerns the ways in which animals develop an ability to give a particular vocalization in a specific circumstance. Here there is some evidence that both genetic and environmental factors are important. For example, while adult vervet monkeys restrict their eagle alarm calls to a small number of genuine avian predators, infants give alarm calls to many different species, some of which present no danger. Eagle alarms given by infants, however, are not entirely random and are restricted to objects flying in the air (fig. 36-4). From a very early age, therefore, infants seem predisposed to divide external stimuli into different classes of danger. This general predisposition is then sharpened with experience, as infants learn which of the many birds they encounter daily pose a threat to them (Seyfarth and Cheney 1980, 1986).

Similarly, learning plays a role in an infant vervet's development of appropriate responses to alarm calls. When a 10- to 16-week-old infant hears an alarm call, its response does not differentiate among alarms to leopards, eagles, or snakes. Only gradually, over the next 14 weeks, do infants come to respond differently upon hearing each alarm call type. During this period the infants have many opportunities to learn "correct" responses by observing other group members, and experiments have shown that infants are more likely to respond correctly if they first look at an adult (Seyfarth and Cheney 1986).

A final issue relevant to vocal ontogeny concerns dialects. Strictly defined, dialects in animals are differences in vocal signals between neighboring populations of potentially interbreeding individuals (e.g., Baker and Cunningham 1985). As such they are both learned and potentially modifiable. Where true dialects are found in nature, they set the stage for observations or experiments that could reveal the extent of developmental modifiability

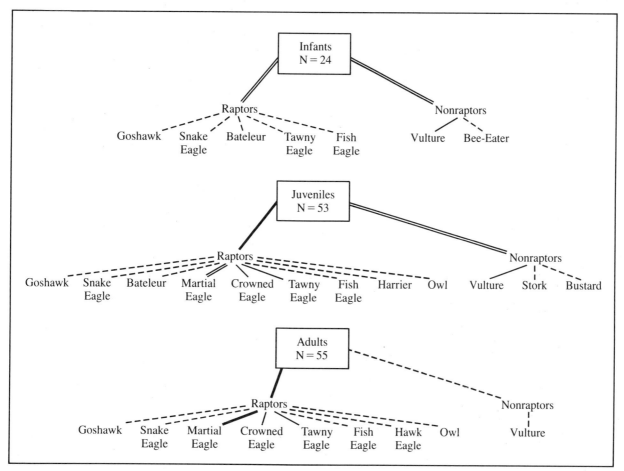

FIGURE 36-4. The stimuli that elicited eagle alarm calls from vervet monkeys of different ages. Data collected over 18 months. Infants are animals less than one-year old, juveniles are 1-4 years old, and adults are over 4. *N* = number of alarm calls from animals in each age class. Broken lines indicate five or fewer alarms, single lines 6-10, double lines 11-15, and thick solid lines more than 15.

in primate vocalizations. Green (1975b) and Hodun, Snowdon, and Soini (1981) suggest that dialects may well occur in Japanese macaques and saddleback tamarins, respectively, but to date the necessary studies to confirm these observations have not yet been conducted.

SUMMARY

Early studies drew sharp distinctions—in neural control, development, and function—between nonhuman primate vocalizations and human language. Subsequent research paints a more complicated picture. While nonhuman primates do not possess humanlike language, captive apes can learn large numbers of communicative signs and genuinely recognize the referential relation between a sign and the thing for which it stands. Free-

ranging monkeys also use calls in a manner that effectively represents objects in their environment.

The vocal repertoires of nonhuman primates are far larger, and contain more specific information, than initially believed. Many studies indicate that vocalizations are under volitional control. Primates make subtle acoustic discriminations when distinguishing between calls, and in one case there is evidence for left hemispheric specialization in vocal perception.

Through observations and experiments, primate vocalizations can reveal how monkeys perceive and classify objects, including other monkeys. Vocalizations are thus not only of interest in themselves, but also provide a tool for studying social structure from the animal's point of view.

37 | Intelligence and Social Cognition

Susan Essock-Vitale and Robert M. Seyfarth

The complexities of their social behavior suggest that primates are intelligent. They recognize each other as individuals, distinguish kin from nonkin (chap. 24), and behave differently toward those of different dominance ranks (chap. 25). Experiments indicate that these social skills are learned and that they develop with experience (e.g., Anderson and Mason 1974). Moreover, primates can remember past interactions, seem able to predict the behavior of others on the basis of prior observations (Kummer 1982), and discriminate among their own and other individuals' close associations. Among vervets, for example, grooming between two unrelated animals increases the likelihood that they will subsequently support each other in an alliance. Close kin, however, support each other at high rates regardless of prior interactions (Seyfarth and Cheney 1984a). Finally, although primates are not qualitatively different from other animals in this respect, they do seem outstanding in their ability to maintain simultaneously many different kinds of relationships, each finely tuned to the individual characteristics of the participants (e.g., Kummer 1982; de Waal 1982; Hinde 1983c; Dasser 1985).

We humans believe that our complex relationships and social organizations are strongly linked to our intelligence. It seems logical, therefore, to assume that this is also true for nonhuman primates. But is it? To find out, we would need (1) a means of recognizing "intelligent" behavior and distinguishing it from other activities; (2) an objective way to measure intelligence, and to compare it across different behaviors and species; and (3) an objective way to measure the complexity of social interactions.

Intuitively, there seems little doubt that a species' intelligence must affect many aspects of its social behavior. At present, however, the relation between intelligence and social behavior is poorly understood. In this chapter, therefore, we discuss, first, the attempts that have been made to measure and compare intelligence across primate species, second, the various ways that primates seem to use intelligence under natural or near-natural conditions, and third, the possible functional and evolutionary links between intelligence and social behavior.

ATTEMPTS TO MEASURE AND COMPARE INTELLIGENCE ACROSS SPECIES

Cross-Modal Perception

Investigators have long wondered whether animals can make information received in one sensory modality (for example, touch) available for performance using another sensory modality (for example, vision). Davenport and Rogers (1970) were able to show that apes (chimpanzees and orangutans) who were first allowed to feel but not see an object, and were then forced to choose between the original object and another, were able to identify by sight the object they had felt. After about 3 months' training, the apes were able to pick the correct object about 77% of the time, counting only the first presentation of each stimulus. In contrast, rhesus macaques studied under the same conditions and trained by the same staff performed only slightly above chance (50% correct), even after more than 3 years of training (Davenport, pers. comm.). Many scientists concluded that these data showed that apes were smarter than monkeys and claimed that only apes could recognize in one sensory modality information that had been presented in another.

These claims were dashed, however, by Cowey and Weiskrantz (1975), who demonstrated with a novel experimental design that monkeys could show beautiful transfer from touch to vision when given a task that suited them. Cowey and Weiskrantz ground up monkey chow, made it into a paste, and reformed it into various shapes, lacing some of the shapes with an odorless, bitter substance. In the dark, rhesus macaques were given many of the "good" and "bad" shapes. They quickly learned to sort good from bad using touch or taste (a quick lick), as evidenced by piles of unbitten shapes on the floor of the cages. Then, in the light, they were offered the same shapes, and they immediately selected the good shapes using vision rather than touch or taste. These results not only demonstrate cross-modal perception, but they also illustrate the methodological point that it is very difficult to prove than an animal cannot do something. Negative results might mean only that a different experimental design is needed.

Discrimination Learning and Learning Sets

Learning means that information gained at one point in time can subsequently be used at a later time. In the traditional study of primate learning, subjects who solve new problems by making use of what they have already learned are described as having "learned to learn." Animals that have learned to learn are also described as having formed a "learning set" (Harlow 1949; for a summary see Schrier 1984). The classic example of learning-set formation involves exposing subjects to, for instance, two differently shaped objects. On the first trial, the subject is rewarded when it picks up one of the objects (say, a square). If it picks up the other object the subject is not rewarded. Then, on the second, third, fourth, and later trials, the subject is presented with the same two objects, with their locations randomly determined. On each trial the subject is rewarded only if it picks up the square object.

On such a "two-choice object discrimination problem," performance on the first trial is governed by chance since the subject has no way of knowing how to identify the correct object. The subject's task is to learn, as quickly as possible, which object is the correct one. In other words, the subject must learn that its experience on trial 1 provides the information needed to solve trial 2 and all of the subsequent trials of that problem. Of course, the information is not always that "the square one is correct," but varies from one problem to another. The subject's improvement, measured over many problems by the proportion of correct responses on trial 2, indicates how well the subject is learning to learn. Harlow (1949, 51) proposed that "this learning-to-learn transforms the organism from a creature that adapts to a changing environment by trial and error to one that adapts by seeming hypothesis and insight."

When the results of learning set tasks were first presented, they strongly suggested that performance was closely related to taxonomy. Apes and Old World monkeys were quantitatively superior to nonprimate mammals (Warren 1965), and, within the primate order, apes performed better than Old World monkeys, Old World monkeys performed better than New World monkeys, and New World monkeys performed better than prosimians. More recently, these conclusions have been strongly criticized (e.g., Hodos and Campbell 1969; Warren 1973; Hodos 1982) for a variety of reasons.

First, one cannot be certain that all species were tested under equivalent conditions (Kintz et al. 1969). For example, were they all equally motivated to perform the task? Rhesus monkeys will form learning sets faster if they are rewarded with two, four, or eight food pellets as opposed to one, and learning-set tests have not always used comparable rates of reward (Schrier and Harlow 1956; Schrier 1958).

Second, comparative data on learning-set performance were drawn from small numbers of subjects in each case, while later work has demonstrated considerable individual variation within species. For example, Rumbaugh (1968) found that some squirrel monkeys performed as well as the best macaques and apes, while others performed far worse than many nonprimates. Warren (1973) reports similar individual variation among rhesus macaques, while in another "standardized" test (the transfer index; Rumbaugh and McCormack 1967), "bright" chimpanzees and orangutans performed better than other primates, while "dull" ones performed only as well as talapoin monkeys and ringtailed lemurs (reviewed in Essock and Rumbaugh 1978). Thus among primates, as well as among other animals (Warren 1973), within-species differences are often so great that they exceed the variation between species, and even mean performance scores do not consistently differentiate between species.

Third, considerable evidence argues against any simple correlation between learning-set performance and either phylogenetic status or cortical complexity. For example, Doty, Jones, and Doty (1967) found that minks and ferrets performed better on such tests than New World monkeys, and roughly as well as rhesus macaques or chimpanzees. Chickens performed about as well as raccoons, cats, and marmosets (Plotnik and Tallarico 1966), porpoises performed less well than chickens and many mammals (Herman et al. 1969), while in a comparison between two species that are relatively similar compared with other species pairs, gray langurs performed far better than rhesus macaques (Manocha 1967; see also Fragaszy 1981).

Fourth, the great majority of research on learning sets has asked subjects to respond to the visual characteristics of stimulus objects ("pick the square one"), while ignoring spatial characteristics ("pick the one on the left"). Species differ, however, in their tendency to respond to visual, as opposed to spatial, stimuli (Warren 1973). Thus a ranking of species on the basis of learning-set performance may more accurately reflect the species' different use of a specific sensory modality rather than their cortical complexity or "intelligence."

The limited value of learning-set data in comparisons among species is worth emphasizing because it illustrates the problems that investigators face whenever they attempt to formulate a single measure or set of measures that compares intelligence in different species. As Tinbergen (1951, 12) has noted: "one should *not* use identical experimental techniques to compare two species, because they would almost certainly not be the same to *them*."

Tests That Can Only Be Solved Using Abstract Hypotheses

Members of many primate species are able to solve problems that require the association of a particular response with a particular stimulus, hereafter called associational learning. At the same time, animals may also be able to base their responses on abstract hypotheses such as "if the previous selection was rewarded, pick it up again; otherwise pick up the other object." Although discriminating between relatively simple associational learning and more sophisticated hypothesis formation might allow us to assess species differences, in practice the two sorts of behavior are often difficult to distinguish. For example, correct trial 2 performance in an object discrimination study may result either when the subject has formed a "win-stay, lose-shift" hypothesis or when associations between particular stimuli are made and broken very quickly (Bessemer and Stollnitz 1971). Learning-set performance alone, then, is not proof of hypothesis-based behavior. However, hypothesis-based tests are possible.

Tests Based on Oddity. Consider what degree of cognitive sophistication would be necessary to solve the following task: On each trial, the subject is presented with three objects, two of which are the same and one of which is different. A food reward lies under the different object only. If a small pool of stimulus objects were recombined trial after trial, the subject could learn specific stimulus combinations (for example, if blue circles are on right and center and red square is on left, pick left). Above-chance performance could thus be attained through associational learning. But what if new stimuli were used for each problem, and each set of stimuli were presented for only one trial? Then an animal could solve the problem only by using an abstract hypothesis such as, pick the odd object. The hypothesis is called abstract because *odd* does not refer to any specific stimulus dimension, as does *red* or *square*. Rather, oddity is a concept that specifies a relationship between objects independent of specific stimulus attributes.

Other examples of tasks requiring abstract hypotheses for correct trial 1 performance when new stimuli are used are matching-from-sample and Weigl oddity. In a matching-from-sample task (also called matching-to-sample), the subject is shown one stimulus (the sample) and two other stimuli, one of which is the same as the sample and the other different. The subject is rewarded for picking the object that is the same as the sample. In a Weigl oddity problem, a single stimulus (such as the color of the tray on which the stimulus objects sit) specifies which of two dimensions (for example, color or shape) is relevant in choosing the correct response on that trial. On a typical trial, there may be three stimulus objects on the tray, two of the same color and two of the same shape; for instance, red circle, red square, blue square. If the tray were, say, pink, then the odd-colored object would be correct; if the tray were green then the odd-shaped object would be correct.

Which species have demonstrated an ability to solve such tasks? Using a simple oddity paradigm, Strong and Hedges (1966) found that cats and raccoons failed to reach a criterion of 90% correct over 48 trials, even after 4,800 tests, while three rhesus macaques and three chimpanzees all learned the task. The chimpanzees showed less variability in their performance than the rhesus macaques, but some rhesus performed as well as the best chimpanzees.

In another oddity test administered to eight primate species, spot-nosed monkeys (*Cercopithecus nictitans*) failed to perform above chance, while ring-tailed lemurs and squirrel monkeys obtained only 60% correct over 12 trials after 120 tests. Cebus and woolly monkeys performed as well as or better than many of the rhesus, pig-tailed, and stump-tailed macaques (Davis et al. 1967; see also Bernstein 1961; Strong, Drash, and Hedges 1968). Two further results from the Davis et al. study are of interest. First, as a group, Old World monkeys were more likely than New World monkeys to use color as a cue, and they performed better than New World monkeys in tests requiring color discrimination. This is not surprising since color vision is better developed in Old World monkeys and apes than it is in monkeys of the New World (Passingham 1982). Second, woolly monkeys were outstanding in their perception of oddity based on height; in these tests no other species came close to equaling their performance. Along with squirrel monkeys, woolly monkeys are the most arboreal of the eight primates tested (chap. 7). Both of these results illustrate once again the difficulty of standardizing tests across a variety of species whose morphology, behavior, and, presumably, problem-solving skills have evolved in different habitats.

Tests That Examine How Species Solve Problems. Rumbaugh (1971) and Essock-Vitale (1978) designed tests for a variety of Old World monkeys and apes that could be solved either by simple associative learning or by abstract hypothesis formation. An added feature was that the investigator could distinguish between these two learning methods: after the animals had learned a specific discrimination, they were given a new problem designed in such a way that those using associative learning would perform worse, while those using an abstract hypothesis would perform equally well (for details see Essock-Vitale 1978).

In Rumbaugh's (1971) study, talapoin monkeys used associative learning to solve the problem at hand, while lowland gorillas appeared to use an abstract hypothesis (see also Gill and Rumbaugh 1974). This led Rumbaugh to conclude that there might be "monkeylike" and "ape-like" ways of solving the same problem. But monkeylike need not mean associational. In Essock-Vitale's study, orangutans used an abstract hypothesis from the outset, while rhesus and stump-tailed macaques began by solving the same problem through association. The macaques' percentage of correct scores, however, was almost always higher than the apes', and the monkeys eventually also used an abstract hypothesis (Essock-Vitale 1978). These results emphasize that hypothesis formation, though it may subjectively seem more intelligent to us, is not necessarily superior in problem-solving tasks to more simple learning methods. Results also underscore the need to test subjects until their performance stabilizes. Had Essock-Vitale's experiment ended earlier than it did, one might erroneously have concluded that apes, but not monkeys, could generate a hypothesis to solve the task.

Stages of Intellectual Development

Drawing on his observations of Swiss children, Piaget (1963) hypothesized that human intellectual development proceeds in stages. Each stage is characterized by specific cognitive deficits and abilities, with the order of development from one stage to the next being highly predictable. For example, by the time they have reached the sensorimotor stage (birth to 2 years) and the early preoperational stage (2 to 4 years), children have begun to explore and manipulate, with some sophistication, objects in their environment. They fail, however, to recognize that liquids retain the same volume even when moved from one container to another (tests of conservation) or that, if object A is larger than object B, and object B is larger than object C, then A must be larger than C (tests of transitivity).

Piaget's research has had a major influence on studies of human development and has also been extended to investigations of the intellectual development of nonhuman primates (e.g., capuchins: Parker and Gibson 1977; gray langurs: Chevalier-Skolnikoff 1982; stump-tailed macaques: Parker 1977; lowland gorillas: Redshaw 1978; chimpanzees: Mathieu 1978; Chevalier-Skolnikoff 1982; orangutans: Chevalier-Skolnikoff 1982). In general, authors have concluded that nonhuman primates pass through developmental stages in approximately the same order as do humans (e.g., Redshaw 1978). They have also suggested that primate species differ in the extent to which they complete the developmental sequence that Piaget described for humans. Parker and Gibson (1979), for example, suggest that while prosimians exhibit only the earlier stages of sensorimotor intelligence, Old World monkeys and especially apes reach higher levels of development.

The application of Piaget's developmental stages to nonhuman primates, like its application to humans, has been criticized on a number of grounds (Brainerd 1978). For example, experimenters have been criticized for imposing discrete stages onto what may in fact be a more continuous developmental process. In Chevalier-Skolnikoff's (1982) study, stages were defined in terms of object manipulation and accepted uncritically as having clear boundaries and transitions. Data analysis, however, revealed considerable evidence of incomplete stages, vague boundaries, and overlap between stages. Thus a strictly Piagetian approach may not be biologically accurate or relevant.

Equally important, the Piagetian approach accepts failure on one, two, or a few tests as proof that the subject lacks a *general* cognitive skill, such as the ability to recognize conservation of volume. As we have already mentioned, however, such negative evidence can be misleading since a change in the design of an experiment can reveal abilities that were previously thought not to exist (e.g., Cowey and Weiskrantz 1975). Even in the case of children, many recent studies have shown that Piaget underestimated his subjects' abilities at different ages (Gelman and Gallistel 1978). In the case of nonhuman primates, concerns about both the application of discrete stages and the limited value of negative evidence are particularly important since sample sizes for each species are small and only a limited variety of tests have been performed.

Languagelike Performance

For over a decade, several groups of scientists have been trying to teach apes to use abstract codes to communicate with humans or with other apes (chap. 36). Whether learning these abstract codes is synonymous with learning language is irrelevant to our concerns here. Instead, we present two examples of the sorts of problem-solving abilities that can be investigated with the help of language training.

Abstract Concepts. Rumbaugh and colleagues (Rumbaugh 1977) have trained subjects to press keys on a specially designed keyboard (fig. 36-1b). Each key displays a geometric figure; for example, a circle with a dot in it on a red background. The figures are called lexigrams and represent a single word or concept. Pressing keys causes duplicates of the lexigrams to be illuminated in a row above the keyboard, thereby forming a "sentence." The experimenter also has a keyboard, with English equivalents, and can make strings of lexigrams appear above the subject's keyboard.

The first subject of Rumbaugh's project, a chimpanzee named Lana, demonstrated that she could respond accurately on trial 1 to questions such as "What color of ball?" when six different objects, each of a different color, were presented (Essock, Gill, and Rumbaugh 1977). The task required that Lana identify what information was being requested, identify the target object, determine the answer to the question based on the physical characteristics of the target object, and finally code the answer into lexigrams. Hence Lana demonstrated that she could mentally manipulate abstract concepts and use the information to perform a goal-directed visual search (see also Dohl 1966, 1968). Inadvertent cuing by the experimenter was precluded by having the experimenter arrange the stimulus objects before he or she knew what question would be asked.

Analogical Reasoning. Premack (1976) has taught captive chimpanzees the use of a languagelike communication system based on arbitrary plastic shapes that the animals can manipulate to form or complete "sentences." As part of this work, chimpanzees have learned to make same/different judgments and to reason analogically.

In the initial stages of such tests, the subject is presented with two objects that either are or are not the same (say, two apples, or an apple and a pear), and is required to apply the plastic word *same* in one case and the plastic word *different* in another (Premack 1976, 1983b). Subjects are trained with a small set of stimuli, then tested with many more, unfamiliar stimuli in order to determine whether they have learned the meaning of *same* or *different*. A number of steps are taken to reduce the likelihood of cuing by the experimenters (e.g., Premack and Premack 1983). In further tests of analogical reasons, the chimpanzee is presented with two pairs of items arranged in the form A/A′ and B/B′. In simple tests, the objects involved are physically similar, for example, a large triangular piece of plastic (A) above a smaller triangular piece (A′), and a large crescent-shaped piece (B) above a smaller crescent-shaped piece (B′). The subject is then required to judge whether the relation exhibited by one pair is the same as the relation exhibited by the other (Premack 1983b). In more difficult tests, the objects share no obvious physical similarity. For example, the subject is asked "lock is to key as closed paint can is to _____," where the options for completing the analogy are a can opener (the correct response) and a paint brush. Note that although both of the alternatives can be associated with a paint can, only one correctly completes the analogy. Sarah, the only chimpanzee given such analogy tests, learned to solve them in both the simpler and the more complex formats (Premack 1976; Gillan, Premack, and Woodruff 1981).

Concept Formation

Abstract hypotheses used to categorize stimuli are often called concepts. For example, subjects that can solve oddity problems are thought to have a concept of oddity because they can respond differentially (categorize) on the basis of an abstract relation between stimuli, independent of the particular stimuli involved.

Herrnstein and Loveland (1964) showed slides to pigeons, and rewarded the pigeons for pecking only when they saw a slide containing one or more trees. When the pigeons had achieved a certain level of performance, they were tested with hundreds of new slides. They found that a great variety of stimuli—tall trees, short trees, leafless trees, and parts of trees—were all more likely to receive pecks than were stimuli without

trees. The experiments concluded that pigeons could form the natural concept *tree*. Similar results were obtained when pigeons were differentially rewarded for responding to pictures with and without people, and with and without bodies of water (Herrnstein 1979). Monkeys in similar tests can discriminate slides with humans, or monkeys, or various typescripts of the letter *A* (D'Amato and Salmon 1983; Schrier, Angarella, and Povar, in press). As Premack (1983a), and Schrier, Angarella, and Povar (in press) have argued, however, it is unclear exactly what these results show. They do not reveal, for example, how any species might form a concept, or to which aspects of the stimuli the subjects are attending.

Concept of Self

Some laboratory experiments suggest that great apes can recognize themselves. Chimpanzees, for example, can learn to recognize their own image in a mirror and can use mirrors to groom otherwise inaccessible parts of their bodies. In more controlled tests, experimenters anesthetized subjects and marked part of the eyebrow with a tasteless, odorless dye. When the animals awoke and looked into the mirror, instances of touching the dyed body parts were much higher than before the dye had been applied (Gallup 1970; Gallup and McClure 1971; Gallup, McClure, and Bundy 1971). In contrast to the performance of normal chimpanzees, rhesus macaques and isolate-reared chimpanzees responded to mirror images as if they were other animals. Similar experiments have demonstrated self-recognition in orangutans (Suarez and Gallup 1981), but not in gorillas (Ledbetter and Basen 1982; Suarez and Gallup 1981; but see Patterson 1978a). In a study involving marmosets, however, Eglash and Snowdon (1983) demonstrated what they termed "early stages of mirror-image recognition." When placed in front of a mirror, pygmy marmosets rapidly ceased to make threats at their own image. It is unlikely that this was due simply to habituation, since the animals then used the mirror to locate otherwise unseen animals in other groups, and then directed threats toward the actual location of these animals.

Spontaneous Displays of Intelligence

Just as humans do not always display their greatest intellectual feats during tests, some of the most striking examples of intelligence in nonhuman primates come from accounts of apparently spontaneous behavior.

Insight

Between 1913 and 1917, Wolfgang Kohler conducted observations and experiments on the intelligence of chimpanzees at a field station in North Africa. In one study a male chimpanzee, Sultan, was led into a room where a banana had been tied to a string and suspended from the ceiling in a corner. A large wooden box had also been placed in the center of the room, open side up. Sultan first tried to reach the fruit by jumping, but this quickly proved futile. He then "paced restlessly up and down, suddenly stood still in front of the box, seized it, tipped it . . . straight towards the objective . . . began to climb up it . . . and springing upwards with all his force, tore down the banana" (Kohler [1925] 1959, 38). A few days later Sultan was taken into a room with a much higher ceiling, where again there was a suspended banana, as well as a wooden box and a stick. After failing to get the banana with the stick alone, Sultan sat down "with an air of fatigue . . . gazed about him, and scratched his head." He then stared at the boxes, suddenly leaped up, seized a box and a stick, pushed the box underneath the banana, reached up with the stick and knocked the fruit down. Kohler was struck with the apparently thoughtful period that preceded Sultan's solution, as well as with his sudden and directed performance. Such "insightful" behavior apparently constrasted with other forms of learning, which develop gradually and depend on reinforcement (Epstein et al. 1984).

Since Kohler's original study, numerous examples of apparent insight in nonhuman primates have been reported. Kawamura (1959) describes the sudden appearance, in one female Japanese macaque, of new techniques for preparing food. The novel behavior then rapidly spread throughout the group (for a full account see chap. 38). Menzel (1972, 1973a) observed the spontaneous invention of ladders in a group of eight immature chimpanzees. Initially, one individual learned to stand a pole upright, rapidly climb up it, then leap off before the pole lost its balance. Later, the same individual (a young male) leaned the pole against a wall and climbed up. Ladder-making behavior rapidly spread throughout the group. The chimpanzees first used ladders to break into an elevated observation booth, where they obviously enjoyed themselves, and later used ladders to escape over a fence, leaving their enclosure entirely.

Regardless of whether insight is fundamentally different from associative or trial-and-error learning, three features are worth noting. First, insight depends on the circumstances in which the problem is presented. In Kohler's experiments, chimpanzees were more likely to exhibit insightful behavior if the necessary objects were close to one another or if boxes had been arranged in a pile. Second, once an insightful conclusion has emerged, it readily appears again. Finally, skills acquired through insight can often be generalized to other situations, as shown by Menzel's observation of ladder making.

Tool Use

In terms of both frequency and variety, tool use is substantially greater among nonhuman primates than among all other animal groups. Although there are no reported cases of tool use by prosimians in the wild, tool use has

been reported for 18 species of monkeys and apes (Beck 1980). The many ways in which primates use tools are summarized in table 37-1.

Wild primates use tools in three main contexts (Passingham 1982), perhaps the most common of which occurs when threatening or attacking intruders. All apes, and many forest-dwelling monkeys, drop twigs or branches from trees onto human observers, while gorillas often throw vegetation as part of their chest-beating display (chap. 14). In most cases it is difficult to determine whether such "tools" are deliberately aimed. When chacma baboons dislodged stones from the top of a cliff onto humans (Hamilton, Buskirk, and Buskirk 1975b), they did so only from a point directly above the observers, but they also dislodged stones when the observers were too far away from the edge of the cliff to be struck. At Gombe, Tanzania, when chimpanzees and baboons competed for provisioned food, Goodall (1968) saw chimpanzees throwing objects at baboons on 20 occasions. Thrown objects seemed to intimidate the baboons on 15 occasions, and in 3 cases the baboons were hit. Baboons were never seen to throw objects at chimpanzees.

Tool use also occurs in the acquisition and preparation of food. Chimpanzees use tools to obtain termites, honey, and ants, and this use involves at least five different types of tool (chap. 38). In the Ivory Coast, the use of hammers and anvils by chimpanzees to break open nuts (Boesch and Boesch 1983, 1984) illustrates clearly the animals' foresight and care in selecting appropriate materials. At the start of a typical nut-smashing session, the chimpanzee will collect as many nuts of a particular species as it can and then carry these to a broad, flat rock or surface root that serves as an anvil. Usually a root can be found within 30 m of the tree where the nuts were gathered, but if the nuts are *Panda oleosa*, a particularly hard species, the chimp will travel more than 30 m to find a stone anvil. The animal's route suggests it has a "mental map" of the area (Boesch and Boesch 1984). Often a wooden club or stone (the hammer) will already be lying next to the anvil, but on other occasions the chimp will first locate a hammer and then carry it to the anvil along with the nuts. For *Panda oleosa* nuts, a stone hammer is almost always used (Boesch and Boesch 1983, 1984; see also Sugiyama 1981 for a description of Guinean chimpanzees using stone hammers and stone anvils to open oil-palm nuts).

A third context of tool use involves bodily care. A baboon has been seen wiping blood from its face with a maize kernel (Goodall, Van Lawick, and Packer 1973); chimpanzees wipe blood and feces from their hair with leaves (McGrew and Tutin 1973); orangutans construct crude shelters to protect themselves from rain (MacKinnon 1974).

In two respects, data on tool use support the conclusions already drawn about species differences in intelli-

TABLE 37-1   Tool Use by Anthropoid Primates in the Wild and Captivity

Action	Tool	Aim	Chim-panzee	Baboon	Macaque	Cebus	Other Apes	Other Monkeys
Drop	Branches, etc.	Hit or scare intruder	W	W	W	W	2W	6W
Throw	Stones, etc.	Hit or scare intruder	W	W	C	W	2C	2W, 2C
Club	Sticks	Hit or scare intruder	W			C	1C	
Pound	Stones	Open fruit, nuts	W	W	W	W		
Dig	Sticks	Open nests, dig up roots	W, C	W				2W
Lever	Sticks	Open food container	W	C			1C	
Insert	Twigs	Probe for insects, honey	W				1C	
Sponge up	Leaves, rope	Collect water	W				2C	
Wipe	Stones, leaves	Clean self	W	W	W		1W	
Wash	In water	Clear food			W			
Reach out	Sticks	Touch object to investigate	W					
Rake in	Sticks	Reach and draw in	W	C	C	C	2C	1C
Prop up	Branches	Provide a ladder	C				1C	
Stack	Boxes	Provide a "staircase"	C					

SOURCES: Adopted from Passingham 1982; data from Beck 1975, 1980, and Warren 1976.

NOTES: W = observed in the wild; C = observed in captivity. Numbers indicate number of different species.

gence among primates. First, just as prosimians generally perform worse than other primates on intelligence tests, they also exhibit much less tool use under natural conditions. Second, just as apes almost always do at least as well, and often better, than other primates on intelligence tests, they also are the primates most commonly observed using tools. This is particularly true when one bears in mind the important distinction between tool using and tool making. Chimpanzees that fish for termites and use hammers to crack open nuts are the only primates that select particular objects as tools, modify them appropriately, and do so in a way that shows foresight.

Comparisons of tool use in different primate species are not always straightforward, however. For example, tool use is observed much more commonly in terrestrial than arboreal species (table 37-1; see also Beck 1980). This could occur because arboreal species have been studied less intensively than terrestrial ones, because the rudimentary thumb of many arboreal species makes them less skillful manipulators, or because there are fewer materials available for tool use in the tops of trees. Thus data should not be taken as an indication that arboreal species are less intelligent than terrestrial ones. To cite one example, while reports of tool using and tool making among chimpanzees are numerous, there are few such reports for the arboreal orangutan. Orangutans perform as well as chimpanzees on laboratory tests of learn-

ing abilities, however, and no one who has worked with orangutans in captivity would deny their extraordinary skill at manipulating objects (e.g., Wright 1972).

SOCIAL INTELLIGENCE

Many observers have suggested that the skills shown by primates in manipulating objects are rudimentary compared with the skills they exhibit when interacting with each other. Early in the study of primate behavior, Chance (1961) and Kummer (1968) were struck by interactions in which one individual appeared to "use" the behavior of another to achieve its own ends. They described this as "social tool use," and noted that in most species studied the use of social companions as tools was far more common than the use of objects (see also Kummer 1982; de Waal 1982; chap. 34). Jolly (1966b) and Humphrey (1976) further speculated that the intelligence of primates originally evolved to solve social problems and that their ability to use tools or to solve laboratory tests is largely a result of natural selection acting on individuals in the social domain.

The goal of this chapter is to examine the variety of research on nonhuman primate intelligence and, if possible, to link field and laboratory approaches in a way that clarifies the relation between intelligence and social behavior. With this in mind, we begin this concluding section by considering the hypothesis that intelligence among primates is most richly illustrated in their social behavior. We then consider whether there are levels of complexity in social interactions that might permit comparisons across or within species. Finally, we discuss whether the intelligence shown in social interactions is in any way similar to the features of intelligence measured in the laboratory.

Performance in Social and Nonsocial Domains

The idea that intelligence may be manifested more in some spheres than in others is not unusual (e.g., Rozin 1976). The various intelligence tests described earlier show clearly that different species have evolved in such a way that each performs well on some tests and poorly on others. Similarly, there is empirical support for the idea that primates, as a group, may be particularly sensitive to social, as compared with nonsocial, stimuli. Human infants, for example, are particularly sensitive to speech sounds as opposed to other acoustic stimuli (Eimas et al. 1971) and particularly sensitive to faces as opposed to other visual stimuli (e.g., Sherrod 1981). Moreover, when human infants first begin to express an understanding of causality, they do so not by talking about causal events involving objects but by discussing the intentions and motivations of people (Hood and Bloom 1979; see also Gelman and Spelke 1981; Hoffman 1981; MacNamara 1982).

Do nonhuman primates, like human infants, also exhibit unusual skills in social interactions? Thus far the issue has been tested directly only on vervet monkeys. When vervets are presented with logically similar problems, some of which use social, and other nonsocial, stimuli, they perform much better on tests that use social stimuli (Cheney and Seyfarth 1985a, 1985b). For example, when interacting with each other vervets, like other primates, are able to form complex associations between individuals. Within a local population they can both recognize individuals and associate them with particular groups (Cheney and Seyfarth 1982b). Within their own groups they can recognize individuals' relative dominance ranks, as well as the close associations among kin (chap. 24), and there is a suggestion that monkeys may recognize the relation between their own kinship bonds and the kinship bonds of others. Outside the social domain, however, the monkeys' performance is less impressive. Although they recognize the alarm calls of other species, they have not been shown to recognize the visual cues associated with predators, such as a gazelle carcass left in a tree by the leopard or a fresh track in the dirt left by a python. Similarly, vervets seem unable to recognize common features of other species' behavior, such as the fact that hippopotamuses and black-winged stilts habitually live near water.

Nonhuman primates appear to use social (as opposed to physical) stimuli in their interactions with others. For example, cooperation and reciprocity are familiar features of primate social groups (chap. 26), but reports of behavior using nonsocial currency (for example, food sharing) are relatively rare. Similarly, while data on primate tool use are easy to list and summarize, accounts of complex social behavior, including what appear to be elaborate, long-term strategies, are too numerous to count. Thus despite an enormous increase in field research on primates since their ideas were originally presented, Chance's and Kummer's original point about social and object tool use still appears to apply today.

Problem Solving Using Behavioral Stimuli

No studies have yet been designed that use social behavior to compare intelligence in different species. There are some data, however, that seem to reveal different levels of social intelligence within the same species, either between normally and abnormally reared juveniles or between juveniles and adults. In a study of captive rhesus monkeys, for example, Anderson and Mason (1974) noted that normally reared juveniles seemed to recognize the dominance relationship between two other animals. When a dominant juvenile, A, threatened a subordinate individual, B, B would often "redirect" aggression toward a third, previously uninvolved, individual. Regardless of their dominance relationship with B, these third animals were always subordinate to A and never one of A's preferred social partners (Anderson and Mason 1974). B's ability to recognize relations between A and other animals contrasted with the behavior of abnormally reared immatures, who showed no similar social skills (Anderson and Mason 1974; Mason 1978).

Redirected aggression also occurs in free-ranging vervet monkeys and suggests greater social knowledge in adults than in immatures. Rhesus macaques (Judge 1982), baboons (Smuts 1985), and vervet monkeys frequently redirect aggression toward animals who are close relatives or associates of their original antagonist. In vervet monkeys, such behavior is common among both adults and juveniles (Cheney and Seyfarth 1985b, 1986), which suggests that animals of all ages recognize the social relationships of others.

Data further indicate that an individual is more likely than expected to threaten another if that animal's close kin and its own close kin have previously been involved in a fight. Thus, for example, in a group that included two sisters as well as an adult female and her offspring, a fight between one sister and the offspring early in the day significantly increased the likelihood that the other sister would later threaten the offspring's mother. Interestingly, this generalization is true for adult animals, but not for juveniles under the age of 3 years (Cheney and Seyfarth 1985b, 1986). This suggests that while adults can recognize the relationship between their own kinship bonds and the kinship bonds of others, juvenile vervets may not. Rather than learning about the relationships of others only through direct experience or observation, adult vervets may be able to infer properties of social relationships (Dasser 1985). Studies that suggest different degrees of social intelligence are important because they suggest that in future it may be possible to compare quantitatively the degrees of social complexity in different species.

Intelligence in the Laboratory and in the Field

If the study of primate intelligence is to be productive, we must find ways both to measure the complexities of primate social behavior and to link its research with laboratory studies. Are the kinds of intelligence studied in the laboratory similar to those used in naturally occurring social behavior? If so, might we eventually be able to use different performances on laboratory tests to predict qualitative differences in social relationships? At present there are few links between field and laboratory research on primate intelligence. Although our arguments are both preliminary and speculative, we offer below three examples of how such links might be forged.

Analogical Reasoning. As noted earlier, Premack (1976, 1983b) has demonstrated that, with some training, chimpanzees can learn to reason analogically in the laboratory. Are there circumstances in which free-ranging primates might need to make use of analogical reasoning? One possible example is suggested by the need for individuals to recognize the different sorts of social relationships that exist among others in their group. Consider, for instance, the data presented earlier on redirected aggression in vervet monkeys. Although it is entirely possible that the sequences described occur because monkeys have formed relatively simple associations between other individuals in their group, we nevertheless cannot rule out the possibility that some form of analogical reasoning is at work and that vervet monkeys understand that certain types of associations—between mothers and offspring and between siblings—share the same characteristics. This sort of observation suggests that species differences in analogical reasoning in the laboratory may eventually direct our attention to similar differences in social behavior in the wild.

Concept Formation. It is possible to imagine ways in which free-ranging nonhuman primates might benefit from an ability to form abstract categories of some sorts. Consider, for example, the problems facing a young male monkey who emigrates from his natal group. The male leaves behind a familiar environment, where social relations are well known, and attempts to join a new group, where animals are new and where acceptance appears to depend on his ability to recognize the relationships that exist among others. He must avoid certain males when the allies are near, challenge others when they are alone, and quickly learn the dominance ranks and kinship bonds of females and immatures. Of course, this could all be done through associational learning: by memorizing individuals and forming particular associations between all possible pairs of individuals in his new group. The male's task would be much easier, however, if he had an abstract concept such as *closely bonded* that allowed him to categorize different group members. The example of redirected aggression described above suggests that monkeys may be capable of such behavior, but more direct evidence is needed to substantiate this hypothesis.

Observations of male takeover attempts in hamadryas and gelada baboon one-male groups provide another example of circumstances in which free-ranging primates might benefit from the use of an abstract concept. In the wild, adult male hamadryas and gelada baboons challenge male leaders of one-male groups in an attempt to take possession of their females (Kummer 1968; Dunbar and Dunbar 1975; chap. 10). Among geladas, takeover attempts are less likely to succeed if there are close bonds between a male unit leader and his females. Among hamadryas baboons, experiments have shown that one male may refrain from challenging another if the male in possession of a female is strongly preferred by her (Bachmann and Kummer 1980). This suggests that male challengers are able to assess a female's preference for her current partner, and one may speculate that such assessments might be facilitated if male baboons, having observed only one or two interactions, were capable of forming the abstract concepts *strongly bonded* and *weakly bonded.*

Concept of Self. How might a concept of self be advantageous to a free-ranging primate? Observing a large colony of chimpanzees, de Waal (1982) recorded the following observation: One male chimpanzee, Luit, was dominant to all others but was regularly being challenged by another male, Nikki. Although Luit generally won such encounters, both males were in their prime, and there was considerable danger of injury and nervousness whenever they met. During one fight between the two, Luit was supported by two females and Nikki was driven into a tree. As Luit sat at the bottom of the tree, he nervously fear-grinned, baring his teeth. He then quickly turned away from Nikki and the females, put his hand over his mouth, and pressed his lips together to hide this sign of submission. These actions were repeated a second and a third time, but only after the third attempt, when Luit apparently succeeded in wiping the fear grin from his face, did he once again turn around to face Nikki. Luit's actions suggest that he was aware of his own nervousness, of the external manifestation of his fear, and of the need to hide this sign from his rival. It is difficult to imagine how this would be possible without some, at least rudimentary, concept of the effect of his own behavior on others. But as a timely caution about the ease of anthropomorphism, especially when dealing with apes, consider how oddly this paragraph would read were *it* to be substituted for *he*. Perhaps the concept of self is more likely to be bestowed upon that which we perceive to be like ourselves.

Among rhesus macaques, as among many other terrestrial cercopithecines, adult females can be ranked in a linear dominance hierarchy, and immature animals acquire ranks immediately below those of their mothers (chap. 11). Thus adult female dominance rank is the major factor influencing the dominance ranks of immatures. As Datta (1983b) has shown, however, there are cases when a younger, smaller juvenile from a high-ranking family is subordinate to an older, larger juvenile from a lower-ranking family. When this occurred in Datta's study, a significant number of juveniles challenged those who "inappropriately" outranked them based on family membership. They did so by initiating, joining, or returning aggression against these rivals. Such challenges were usually successful and accounted for 30 of 32 rank reversals observed during a 16-month period. Immatures behaved as if they had a clear idea of whom, on the basis of family rank, they "ought" to outrank in the group (Datta 1983b). Immature monkeys thus seem able not only to rank the others in their group (Seyfarth 1981; Kummer 1982), but also to recognize their own appropriate place within such rank orders. Again, while this degree of social knowledge could result from simple associate learning, it seems possible that it reflects a concept of self.

SUMMARY

We generally assume that our complex social behavior is due, at least in part, to our intelligence. The social behavior of nonhuman primates is complex, but as yet we do not understand the extent to which this is due to intelligence. We lack an objective means of measuring and comparing intelligence across species as well as a method for assessing the complexity of different social interactions.

Laboratory tests that aim to compare intelligence across primate species provide some support for the hypothesis that prosimians perform worse than other primates and that apes perform better. However, these tests are fraught with both conceptual and methodological problems. They reveal considerable overlap in performance between species and wide variation in performance from one type of test to another.

Other laboratory studies have examined single species—often the chimpanzee—and have shown impressive skills, such as the ability to manipulate symbols, to reason analogically, and to recognize oneself. This performance on tests is paralleled by displays, again among chimpanzees, of both insight and tool use under more natural conditions.

Data from many species suggest that primates display great intelligence in social interactions. Although at present this idea is speculative, we suggest a number of ways in which studies of primate intelligence and studies of their social behavior may be linked.

38 | Local Traditions and Cultural Transmission

Toshisada Nishida

For individual animals, there are three, sometimes interdependent, means of information acquisition: genetic transmission, individual learning, and cultural transmission (see Galef 1976; Kummer 1971; Washburn and Hamburg 1965).

Genetic transmission of information is advantageous particularly for relatively simple behaviors that have direct and immediate consequences for survival. The advantage of learning is that it provides animals with far more detailed information and that it can replace old, outmoded information with newer, more effective information in circumstances where the environment has changed.

Cultural transmission usually has further advantages over individual learning. Kummer (1971) summarizes these as follows.

1. Not every individual is equally inventive. Moreover, different members of the society may be skilled at different types of learning. Tradition can pool the individual achievements.

2. Experimenting directly with the environment may be dangerous, as with poisonous food plants or predators. In these cases, tradition is a safer way of acquiring information.

3. Some environmental situations such as drought are too rare to permit direct experience for every group member. In this case, an experienced elder may be the only animal in the group that has the relevant information.

Social learning probably originated early in our evolutionary history, particularly after the emergence of birds and mammals, when parent and offspring interacted directly and offspring spent an increasingly long period of time dependent on the mother.

Earlier experiments by psychologists such as Thorndike (1901) or Watson (1914) failed to show social learning in monkeys, mainly because of limitations in the experiments. Scientists used too few subjects, knew little about their subjects' natural social behavior, and created inhibiting experimental settings where social forms of learning would be least likely to occur (Hall 1963a).

Just before the onset of the modern era in primate field studies, Imanishi (1952) proposed the epoch-making idea that many species with a continuing group life must have their own cultures. By *culture* he meant the "so-cially admitted," adjustable behavior that exists under natural conditions. A rush of field studies of Japanese monkeys followed. Imanishi (1957a, 1957b, 1960) classified the behavior of Japanese monkeys into two categories: "individual oriented" and "troop oriented." The former concept included maintenance activities and sexual behavior, and was similar to selfish behavior in modern terms. The latter concept resembled altruistic behavior in modern terms and included behaviors that appeared contradictory to an individual's biological interest, such as warning calls and leaders' "control" behavior. Imanishi implicitly assumed that troop-oriented behavior could not be genetically transmitted because it presumably lowered an individual's fitness; he pointed out that it could not be acquired by the usual means of learning because there was no obvious reinforcement or reward for imitating such behavior. Therefore, Imanishi proposed "identification" as the mechanism for learning troop-oriented behavior. For example, an infant monkey who was close to a leader male might identify himself with the leader and thus incorporate the leader's personality into his own.

Now that troop-oriented, or altruistic, behavior can be understood to have evolved through natural selection (chap. 26), the importance of Imanishi's argument may be unclear. However, Imanishi's unprejudiced perspective has had an enormous impact upon the worldwide study of primate learning and culture. In this chapter, not short-lived individually inherited traditions but only long-lived group-inherited traditions are included in culture. Cultural behavior is thus defined here as behavior that is (*a*) transmitted socially rather than genetically, (*b*) shared by many members within a group, (*c*) persistent over generations, and (*d*) not simply the result of adaptation to different local conditions.

INDIVIDUAL LEARNING

Learning does not necessarily lead to cultural transmission. A monkey infant must learn to recognize the members of his natal group. Such knowledge is crucial to his survival since it enables him to discriminate his relatives from nonrelatives and group members from strangers. As he matures, he must establish and learn gradually his own status within the group and the social relationships that exist between other members. This is critically important to an individual's reproductive success in higher primates because individuals influence the status of one another. For example, in macaques daughters come to occupy dominance ranks immediately below that of their mother (Kawamura 1958), and the formation of coalitions tends to determine the reproductive success of adult males in macaques, baboons, and chimpanzees (chap. 26). Since such knowledge is of a particular individual's personal relationships with others, however, it is lost when the individual dies.

CONFOUNDING FACTORS THAT COMPLICATE THE ANALYSIS OF CULTURAL BEHAVIOR

The most readily observable result of social transmission processes would be the existence of different modes of behavior within different geographic subpopulations of a species, uncorrelated with gene or resource distribution (Galef 1976). However, even if differences are found in behavior between local populations of the same species, and genetic differences are improbable, the differences may not be necessarily cultural. Many local differences that appear cultural at first sight can be more parsimoniously explained by other factors. One such factor is the flexibility of a species responding to variable local environments; another can be described as observational biases caused by differences in sampling methods. These are pitfalls in which field-workers are easily trapped.

Ecological Factors

Yoshiba (1968) found conspicuous variation in the behavior of hanuman langurs from one area to another. The differences include the size, adult sex ratio, and composition of troops, types of social units, relationships among troop members, time spent on the ground, and even the age of weaning. The study sites differed in such environmental factors as vegetation, climate, and density of predators. At least some of the variations may be ascribed to environmental differences, although many others remain obscure.

It had long been puzzling that chimpanzees in the Mahale Mountains were not observed to fish for termites, although they were seen to fish for arboreal ants (Nishida 1973), and many termites were present. At Gombe, termite fishing was discovered very early in the study (Goodall 1967). The puzzle was solved when *Macrotermes,* the genus for which Gombe's chimpanzees fish, was found to be absent in the study area of the habituated chimpanzees of Mahale (Nishida and Uehara 1980). Termites of the genus *Odontotermes* are present at Mahale and as easy to fish for (at least by human observers) as those of *Macrotermes.* However, they are not fished for by habituated chimpanzees, perhaps because they produce a more distasteful defensive secretion than those of *Macrotermes.*

The Mahale chimpanzees were observed fishing for termites of *Psudacanthotermes* only once because fishing opportunities are seasonally very limited on account of the termites' special life-style (Uehara 1982).

The chimpanzees of Mt. Assirik, Senegal, prey upon nocturnal primates such as *Galago* and *Potto,* but neglect *Cercopithecus* monkeys and ungulates, which are the commonest prey of eastern chimpanzees. McGrew (1983) suggests that the abundance of competing predators and the extremely open habitat of Mt. Assirik might be major factors contributing to the lack of these species in the diet of Mt. Assirik chimpanzees.

These ecological factors, which are related to the origin of the local differences, might lead to cultural differences. For example, young chimpanzees of Mt. Assirik might come habitually to prey on prosimians as a result of observing their seniors' similar habits, although ultimately the behavior was ecologically influenced. The problem is that field-workers can only rarely determine whether or not a particular behavioral pattern was learned socially.

Sampling Methods

Primate behavior changes seasonally and annually. Behavior changes ontogenetically (e.g., Casimir 1979), and sex differences often occur (see Mitchell 1979 for review). Sampling data during a limited period, or from only a limited number of subjects, might suggest local differences that had in fact only been produced by observational bias.

Nishida et al. (1983) reported that unripe seeds of *Saba florida* (Apocynaceae) were neglected by Mahale chimpanzees, while Gombe chimpanzees ate them often. However, in 1983 (after 18 years of study) Mahale chimpanzees were observed for the first time to feed on them. Even a 4-year fecal analysis had not revealed this habit. The fact that the ripening of *S. florida* was much delayed in 1983 may explain this unusual variation in feeding. One hypothesis is that the chimpanzees invented the habit in that year. Alternatively (and this is more plausible), Mahale chimpanzees had fed on them before, but this had been overlooked by observers because it occurred only rarely. This episode illustrates that we must be careful when drawing conclusions about supposed cultural differences in primate diet and behavior.

CULTURAL TRANSMISSION

As stated earlier, it is often quite difficult to determine, by field observation alone, whether local behavioral differences are truly cultural. The examples given below are the most plausible cases of cultural transmission in primates, but readers should keep in mind that some of them need to be confirmed in more controlled conditions.

Space Utilization

The home ranges of groups of primates have been known to change little over many years, despite the turnover of individuals.

For example, many troops of Japanese monkeys in the wild have occupied virtually the same areas for at least 20 years. A group's range is the area its members have learned to know, and boundary lines appear to be limited psychologically (i.e., by familiarity) as well as ecologically (by the availability of food). Washburn and Hamburg (1965, 616) documented an interesting episode: "Hall tried to drive a group of baboons beyond its usual range. As long as the animals were within the area they knew, they were easily driven, but on reaching the edge of their range they turned back."

Food Selection

Local differences in diet selection have been extensively discussed for Japanese monkeys by Kawamura (1965b). For example, monkeys of two troops of Minoo, Osaka, regularly dig and feed on the roots of *Dioscorea* and lilies, while monkeys of Takasakiyama were never observed to do so, although the plants were available.

In chimpanzees, many differences have been recognized between the populations of Mahale and Gombe (Nishida et al. 1983). For example, chimpanzees at Gombe eat fruits, pith of leaf fronds, flowers, and wood of the oil palm (*Elaeis guineensis*), whereas those at Mahale do not eat any part of this species. Chimpanzees in West Africa eat both pulp and kernel of the oil palm, although Gombe chimpanzees reject the kernel. The palm tree is a relatively recent introduction to East Africa.

Chimpanzees of Mahale feed on the spiny leaves of *Blepharis buchneri* (Acanthaceae). Since the leaves occasionally hurt the mouth and lips, chimpanzees grimace when chewing them. Chimpanzees of Gombe have never been observed to feed on these leaves. Conspicuous differences exist in regard to insect eating (Nishida and Hiraiwa 1982; McGrew 1983). Gombe chimpanzees habitually eat *Dorylus* ants, rarely eat *Crematogaster* ants, and reject *Camponotus* ants, while Mahale chimpanzees eat the latter two habitually and reject the former completely. Similar local food differences have been reported for Nilgiri langurs (Poirier 1970) and the mountain gorilla (Casimir 1975).

There may be many different explanations for such local differences in diet: relative availability of the food, human interaction, and characteristics of the food type itself. Availability limits opportunities to experiment with (e.g., inspect and taste) potential food items. A long history of contact with humans often provides the animals with more opportunity for experimental feeding. Such a food type as leaves of *Blepharis* is likely to discourage an animal who tastes it for the first time and so will be more likely to be ignored by the local group. Pre-

cisely how a group first begins to exploit a new food resource is discussed later in greater detail.

Food Processing and Tool Use

The details of food manipulation may also be largely acquired through the accumulations of daily observational learning.

There is some evidence that chimpanzees differ in their processing of the same food items (Nishida et al. 1983). Very subtle differrences in manipulation occur in feeding on rough-surfaced leaves and on a particular type of pod. Second, in opening hard-shelled fruits, the chimpanzees of Gombe bang them against tree trunks or rocks. At Mahale, the same kind of fruits are always bitten open. Only the chimpanzees of West Africa use stones in opening hard nuts (e.g., oil-palm nuts) to obtain the kernels inside (Sugiyama and Koman 1979b). These differences cannot be ascribed to the presence or absence of either tools or hard nuts because Gombe chimpanzees eat the pulps of the oil palm and use stones in social display (Goodall 1970). The "hammer-and-anvil technique" may be a regional culture of the far western chimpanzees. "Fishing" for termites or ants has been reported for chimpanzees from various parts of Africa (McGrew, Tutin, and Baldwin 1979a). It would be reasonable to assume that this technique is transmitted from mother to infant, for infants watch their mothers' activities intently and try to copy their behavior (Goodall 1970; Uehara 1982). The raw materials of tools are selected in terms of the quality (Nishida 1973).

The chimpanzees of Okorobiko, Rio Muni, do not fish for termites in the same way as the others; circumstantial evidence suggests that they use relatively large sticks to "perforate" the termite mound, then pick out the termites by hand (Sabater Pi 1974; McGrew, Tutin, and Baldwin 1979a).

A chimpanzee of Gombe puts a long stick or branch into a swarm of *Dorylus* ants. Then, hundreds of ants stream up it. The chimpanzee watches their progress, and when ants have almost reached its hand, the tool is quickly withdrawn. Immediately, the opposite hand sweeps the length of the tool catching the ants in a mass. These are popped into the open mouth (McGrew 1974; fig. 38-1). This "ant-dipping" technique was never observed among Mahale chimpanzees. In both ant fishing and ant dipping, a great amount of trial-and-error learning appears to be needed for tool manipulation, especially for tactics against the ants' antipredator responses (McGrew 1977; Nishida and Hiraiwa 1982).

Occasionally Gombe chimpanzees crush leaves and use them as sponges to soak up water for drinking from holes in trees. This technique of "leaf sponging" has never been seen in the chimpanzees of Mahale, although they put their hands into such holes and lick them. This difference is puzzling because the physiognomy and distribution of water courses differ little between Gombe and Mahale (Nishida 1980a).

Brown capuchins (*Cebus apella*) of La Macarena National Park, Colombia, employ ingenious techniques of exploiting palm nuts (*Astrocaryum chambira*), according to the ripeness of the palm (Izawa and Mizuno 1977; Izawa 1979; Struhsaker and Leland 1977). When the juice becomes sweet and sticky, and the coco begins to be yogurtlike in the husk, capuchins penetrate a large eyehole of the palm with their canine: they first drink the juice and then strike the fruit against the guadua bamboo (*Bambusa guadua*) just above the joint and lick the yogurtlike coco oozing out through the hole. When the coco becomes harder and will not exude through the eyeholes, capuchins hold the fruit with both hands and strike it against the tree joint. After cracking the husk, they take out some part of hardened coco inside the husk with their teeth.

The capuchins also recover old fruits from the ground, most of which are infected by bruchid beetles. They first take a fruit in the palm and inspect it (gaze, wave up and down, or shake right to left). After selecting an infection-free fruit, they climb a tree with the fruit and smash it against the tree. This latter behavior has also been observed by Terborgh (1983) for *Cebus albifrons*.

Capuchins also acquire frogs in a unique way (Izawa 1978b). They gnaw the guadua bamboo at the fringe of a slit in the internode (the slit was probably made by insects or birds) and pull the epidermis and cortex strongly downward or upward with their teeth. They then strip the epidermis and cortex off with their hands. After repeating this several times, they put their hands into the enlarged slit to capture frogs. In this way they also obtain drinking water, termites, and ants (Izawa 1979).

Judging from their sophistication, these techniques may very probably be cultural behaviors.

Attitude toward Humans

Kawamura (1959) reported considerable differences in the ease or difficulty of provisioning among different troops of Japanese monkeys. Normally, monkeys were habituated after a few months. However, some troops were habituated in just a week, but others not even after several years. There are also great differences among various troops in their speed of acquiring new food habits. Such differences across monkey troops in their atti-

FIGURE 38-1. An adult female chimpanzee at Gombe, Tanzania, "ant dipping" as her infant looks on. (Photo: Jim Moore/Anthrophoto)

tude to human beings and novel food items may be explained by historical differences in contacts with human neighbors. In some places, but not in others, monkeys were hunted intensively by humans.

Gestural Communication

Among the nonhuman primates, gestural dialects appear to be known only in chimpanzees. The chimpanzees of Mahale (McGrew and Tutin 1978) and Kibale Forest, Uganda (Ghiglieri 1984), display the same stereotypic pattern of mutual grooming, the "grooming hand-clasp."

Two chimpanzees sitting face to face each simultaneously extend an arm overhead; then one clasps the other's wrist or hand or both clasp each other's hand. The other hand engages in social grooming of the underarm area revealed by the upraised limb. Either both raise their right arms and groom with their left, or vice versa (McGrew and Tutin 1978). This behavior has never been observed at Gombe. In similar contexts, each simply holds a bough overhead with one hand and does not clasp the other's hand or wrist. No ecological explanation appears probable for this difference, and therefore McGrew and Tutin called it a "social custom."

Another behavioral pattern peculiar to Mahale chimpanzees is the "leaf-clipping display." A chimpanzee picks one to five stiff leaves, grasps the petiole between the thumb and the index finger (fig. 38-2, *top*), repeatedly pulls it from side to side while removing the leaf blade with the incisors, and thus bites the leaf to pieces (fig. 38-2, *bottom*). In removing the leaf blade, a ripping sound is conspicuously and distinctly produced. When only the midrib with tiny pieces of the leaf blade remains, it is dropped and another sequence of ripping a new leaf is often repeated. This occurs most commonly (23 of 41 observations) in sexual contexts, such as "herding" behavior, or as a courtship display. Otherwise it occurs when the chimpanzee seems frustrated (Nishida 1980b).

Leaf clipping has been observed only twice at Gombe, both times in the context of frustration of young females (Goodall, pers. comm.). Sugiyama (1981) reported that a similar pattern was often observed in chimpanzees at Bossou, Guinea. There it occurred mostly (41 of 44 observations) as a display, in apparent frustration, or during play (often in the presence of humans), and it did not appear to be ritualized as a courtship signal. Both leaf clipping and grooming hand-clasping are shown by many chimpanzees of two habituated unit-groups (K- and M-group).

Recently, another style of courtship display was discovered among Mahale chimpanzees. Typically, a male sitting on the ground or in the tree, faces an estrous female and makes a crude day bed or cushion (often bending two to four shrubs down to the ground), which he sits on. Then he stamps with one foot. This behavior may be

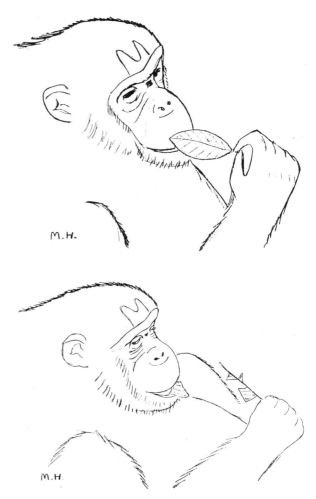

FIGURE 38-2. The leaf-clipping display in chimpanzees of the Mahale Mountains, Tanzania. (Drawing: Mariko Hasegawa from a videotape)

derived from the branch shaking in courtship, which is commonly observed in the chimpanzees of Gombe and Mahale. Interestingly, this pattern has not been observed for K-group's chimpanzees, but has been recorded only for about 10 (mostly immature) males of M-group. Therefore, it is plausible that this is a newly acquired behavioral pattern that has been transmitted neither to all members of M-group not to the local population; Thus, it might be "incipient" culture. Newly invented behavioral patterns can be "exported" to other groups when females acquire them because, in the chimpanzee, females, more often than males, transfer (chap. 15). Although the leaf-clipping display is shown by both males and females, the cushion-making display has not been seen in any females of M-group.

MECHANISMS OF CULTURAL TRANSMISSION
Three Ways of Information Transfer

There are three ways of transmitting cultural behavior in a society of wild nonhuman primates.

1. *Propagation,* or "one-to-many" cultural transmission, defined as the spread of new information introduced by one individual to other members of a group (or society). Since the innovator often is a juvenile, the new information tends to flow from younger to older generations.

2. *Tradition* or "mother-to-child" transmission, defined as the spread of established information from older (often mothers) to younger individuals (often infants). This is the usual course of cultural transmission in a conservative society and plays a major part in the process of so-called socialization of individuals.

3. *Enculturation,* or "many-to-one" tranmission, defined as the spread of established information from many group members to one or a few newly immigrated individuals. The information tends to flow from one to another adult individual.

It took many years to habituate most of Mahale's chimpanzees to human observers. However, after most of them were habituated, newly immigrated females from unhabituated unit-groups could be approached within 10 m 1 month or even 2 weeks after the first contact. This amazing speed of habituation is explicable only by observational learning on the part of the immigrants, who usually were close to adult males who were the least shy of human observers. By contrast, some peripheral females who were rarely observed with adult males remain shy even after 18 years of study.

Innovation, Propagation, and Tradition

It is usually impossible to elucidate the process of propagation of new information from one individual to another because innovation is rarely observed in the wild except after human intervention.

When Japanese monkeys in Koshima Islet came in contact with human observers as a result of provisioning, innovation and propagation were observed (Kawamura 1959; Kawai 1965a; Itani and Nishimura 1973). In September 1953, Imo, a 1.5-year-old female (fig. 38-3), took her sweet potato and washed the sand off it in a small brook, thereafter continuing to do so regularly. Other monkeys started doing the same (figs. 38-4, 38-5).

After Imo, the next individual to learn potato washing was Imo's playmate, who did so in October. Imo's mother and another male peer began to wash in January 1954. In subsequent years (1955 and 1956), three of Imo's lineage (younger brother, elder sister, and niece) and four animals from other lineages (two were a year younger and two were a year older than Imo) started to do so. Thus, with the exception of her mother, all the individuals that learned potato washing quickly were either peers or young close relatives of Imo.

By March 1958, 2 of the 11 adults (18.1%), and 15 of 19 monkeys (78.9%) aged betwen 2 and 7 years had acquired the behavior. In August 1962, 36 out of 46

FIGURE 38-3. The Japanese macaque, Imo, at age 19, 3 months before her death. (Photo: Umeyo Mori)

(73.4%) monkeys above 2 years of age washed potatoes. Out of 11 monkeys older than 12, however, only 2 females had acquired the behavior. Thus, the rate of adults' acquisition of the behavior was very low.

After 1959, features of information transfer changed. Sweet-potato washing was no longer a new mode of behavior: when infants were born, they found most of their mothers and elders washing potatoes and learned this behavior from them as they learned the group's usual food repertoire. Infants are taken to the edge of water during the period when they are dependent on mothers' milk. While their mothers wash potatoes, infants watch carefully and put into their mouths pieces of potatoes that mothers drop in the water. Most of the infants acquire potato washing around 1 to 2.5 years old (Kawai 1965a).

Thus, in the first period of propagation (1953–58, the period of "individual propagation" in Kawai's term), sex and age were important factors facilitating information transfer. But in the second period (1959–present, the period of "precultural propagation"), acquisition of potato washing occurred independent of sex and age. During

FIGURE 38-4. Two Japanese macaques wash sweet potatoes in the ocean. (Photo: Umeyo Mori)

the second period, virtually all individuals (16 of 19 to date) acquired this habit through their mothers or playmates when they were infants or juveniles.

Meanwhile, the monkeys were fed wheat as well as sweet potatoes on the sandy beach, and the wheat was difficult to separate from the sand. In 1956, when Imo was 4-years old, she took a handful of mixed wheat and sand to the brook. When it was dropped on the water, the sand sank and the floating wheat could be skimmed off the water's surface, now clean again. This "placer-mining" technique was also adopted by some of the other monkeys, and soon more and more animals learned it (fig. 38-5).

Compared to potato washing, placer mining was quite slow to propagate. The first 3 individuals to learn placer mining from Imo did so only in 1958, after more than 1 year had elapsed. They were a 1-year-older male from the lineage other than Imo's and Imo's 2 sisters. In 1959, Imo's son, her niece, and two adolescent females younger than Imo started to do placer mining. In 1961, 5 other monkeys began to do so. Thus, within 5 years only 12 individuals had successfully acquired this behavior: 2 sisters, 1 son, 1 niece, and 1 nephew from Imo's lineage and 1 adult female, 1 peer male, and 5 younger females from lineages other than Imo's. By 1962, 19 of 49 monkeys (38.7%) above 2-years old did placer mining.

The process of the propagation of placer mining was similar to that of potato-washing behavior, that is, through the inventor's playmates and lineage. In 1962, 2 1-year-

old infants (the only infants born in 1961) were seen several times scratching sand with their fingers while their mothers were doing placer mining. Five 2-year-old monkeys were occasionally observed picking up wheat in the water, although they did not do placer mining. Thus, around that year placer mining was slowly developing as the tradition of the Koshima troop (Kawai 1965a).

The age of acquisition of placer mining during the period of propagation was between 2- and 4-years old, while that of sweet-potato washing was 1.5- to 2-years old (fig. 38-6). This difference appears due to the relative difficulty of acquisition. Potato-washing behavior is sometimes observed in other troops as individual behavior and seems to be acquired easily. Placer-mining behavior, however, is never observed in other troops. Placer mining appears to require more understanding of complex relations between objects and may be particularly difficult to learn because a monkey must "discard" his food first, while in potato washing he can keep the potato from the beginning to the end.

Itani (1958) studied the propagation of caramel eating among the monkeys of Takasakiyama. In May 1953, only one adult male accepted a caramel, wrapped in cellophane. From about 1 year after that, Itani checked the acquisition of the new habit six times during 14 months.

Figure 38-7 compares the rates of acquisition in different age-sex groups, using the data from the third and sixth tests (Itani and Nishimura 1973). The results are as follows.

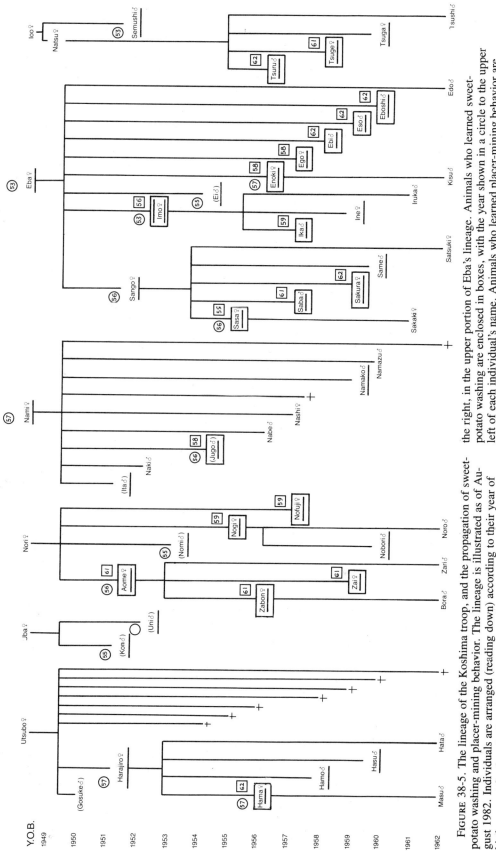

FIGURE 38-5. The lineage of the Koshima troop, and the propagation of sweet-potato washing and placer-mining behavior. The lineage is illustrated as of August 1982. Individuals are arranged (reading down) according to their year of birth, except for females born before 1949. Solitary monkeys are enclosed in parentheses (see fig. 38-3) can be found at the right, in the upper portion of Eba's lineage. Animals who learned sweet-potato washing are enclosed in boxes, with the year shown in a circle to the upper left of each individual's name. Animals who learned placer-mining behavior are underlined, with the year shown in a square to the upper right of their name. Imo, the originator of both behaviors (see fig. 38-3) can be found at the right, in the upper portion of Eba's lineage.
SOURCE: Kawai 1965a

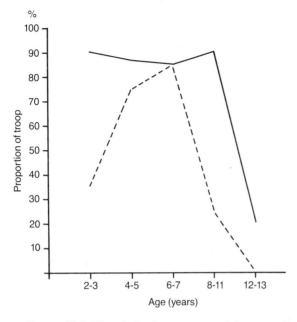

FIGURE 38-6. The relation between age and the proportion of animals that acquired sweet-potato washing (solid line) and placer-mining (broken line) in the Koshima troop. SOURCE: Kawai 1965a.

1. One-year-old infants, both male and female, showed a 100% acquisition rate at the sixth test.

2. Two- and 3-year-old infants, both male and female, showed the next highest acquisition rate.

3. Conspicuous sex differences, however, emerged when monkeys reached adolescence: adolescent and young adult females showed roughly a 75% acquisition rate, while males of comparable ages showed less than a 25% acquisition rate. Thus above adolescence, females generally showed higher acquisition rates than males.

As Itani and Nishimura (1973) stated, these patterns suggest that the acquisition of this behavior resulted not only from "invention" based on the trial and error of individuals but also from "imitation" of those who had already acquired the behavior by those who had not. For adults of both sexes, Itani stressed the importance of the play group, which served as a "pool for propagation," largely because infants and juveniles were so innovative and showed the highest rate of acquisition. Females, for example, maintained close social relationships with and had ample opportunities to observe the behaviors of the infants who showed the highest ability to acquire new traits. The same can be said of some males. Older males showed "paternal care" (Itani 1959) and had many opportunities to interact with and observe infants. In contrast, younger males who spent most of their time in the peripheral part of the troop had few chances to interact with infants. Perhaps as a result, adult female acquisition rate was higher than that for adult males, and the rate for males was directly related to age. The only clear exception to these generalizations was the oldest (leader) males, who showed the lowest acquisition rate.

From studies of Koshima and Takasakiyama, Itani and Nishimura (1973) concluded that there are several channels through which new adaptive behavior spreads through a monkey troop: (1) via playmates, (2) via matrilines (i.e., from infant to mother and between siblings), (3) via paternal care (i.e., from infant to its male protector). Youngsters were the least conservative regarding a new habit. This tendency has been confirmed for many other primates. It is occasionally argued that infants and juveniles are most curious and approach a new object most unhesitatingly because they are the most "expendable" if it turns out to be dangerous (Kummer 1971; Jolly 1972). However, this is a group selectionist explanation and seems unlikely to be true. The youngsters' exploratory tendency is part of the primate life-cycle strategy in which an individual youngster must learn most of its survival techniques from observational learning and trial and error, while coping with complicated social and physical environments.

There is an interesting sequel to the monkey culture at Koshima. In 1972, in order to limit population growth and to avoid disturbances from tourists, provisioning was stopped except on rare occasions when monkeys were given a small number of items, and these mostly away from the sandy beach. Thus, most infants born after 1972 had no chance to observe their mothers washing potatoes or wheat. In 1980, Kawai and his colleagues (Kawai, pers. comm.) started to feed monkeys again along the beach, to check for retention of the monkeys' culture. Most of the juvenile monkeys of the most dominant lineage (led by alpha female Satsuki, Imo's niece) washed potatoes in the sea, but very few youngsters of the other lineages did so. The placer-mining habit was almost unknown to youngsters. Only two youngsters, (again, from Satsuki's lineage), took wheat to the brook. It appeared that, during the period when limited provisioned food was available, mothers of the dominant lineage had monopolized what food was available, and thus were the only females who could transmit their tradition to their offspring. This episode illustrates the importance of dominance rank and how adaptive acquired responses rapidly vanish if the environment changes. Of course the older generation of any lineage still knows the technique, and if researchers were to put out food regularly on the beach from now on, the monkey culture of Koshima would not become extinct.

Propagation and Dominance Rank

Imo was a member of a lineage of high dominance rank. Did her age and dominance status influence the speed of propagation in the Koshima troop? Because she was the inventor in both potato-washing and placer-mining behaviors, no comparative data allow us to examine the relation between these parameters and the speed of propagation.

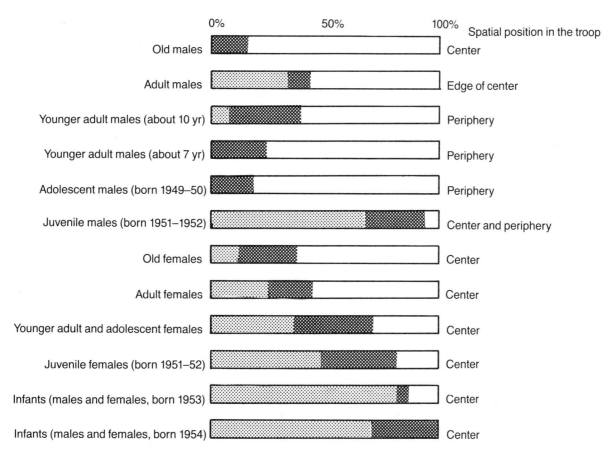

FIGURE 38-7. The acquisition of candy eating in the Takasakiyama troop. The figure shows the proportion of animals in each class that acquired the candy-eating habit in the third test (hatched bars), sixth test (dark bars), or not at all (light bars). SOURCE: Itani and Nishimura 1973.

A study of the acquisition of wheat eating among Japanese monkeys of Minoo-B troop (Yamada 1957) is suggestive in this respect. The Minoo-B troop consisted of 17 individuals: 2 adult males, 5 adult females, 3 young females, 4 juveniles, and 3 infants. One of the adult males (the second-ranking) had previously left the troop and acquired the habit of wheat eating. To examine the order in which this new behavior spread through the group, observers placed two boxes containing wheat at intervals of 10 m on the feeding ground. At the beginning of the test, only the second-ranking male was eating wheat. Several minutes later, however, the alpha male began to do so. Within roughly an hour, the alpha female and her two daughters and another adult female of the central part began to feed on wheat. Within 4 hours, all the troop members except infants began to feed on wheat.

The acquisition of wheat feeding in this troop was much faster than that of caramel feeding in Takasakiyama. Moreover, the older individuals and individuals in higher status were generally earlier in acquisition. Yamada stated that propagation passed from elder to younger monkeys. However, caution is needed before accepting this conclusion. Because only two food boxes were used in the test, it seems likely that dominant animals had priority of access to the food. Therefore, the possibility cannot be excluded that those monkeys acquired the wheat-feeding habit through simultaneous individual innovation rather than through propagation. Itani and Nishimura (1973) attributed this rapid acquisition to the history of the Minoo-B troop, in particular, to the fact that this troop had been accustomed to various artificial objects in the park and had scarcely any resistance for taking new food.

It is tempting to argue that if an inventor is an adult occupying high social status rather than a juvenile, propagation would be rapid. However, there are yet no decisive data supporting this hypothesis.

Mechanism of Social Learning

Propagation, tradition, and enculturation are based on communication between conspecifics: usually the communication is one-sided and described as imitation or observational learning. When the "demonstrator" behaves positively toward a naïve animal, such a pattern of apparently intentional transmission is called teaching.

Observational Learning. Social facilitation is the performance of a behavioral pattern, already in an individual's repertoire, which is released by the same behavior in other individuals (Thorpe 1963). Hall (1968b) showed that a previously positive response to a box by two young patas monkeys changed to strong avoidance after seeing their mothers startled by it. The reaction was induced in the mothers by a live snake that was visible to them when the lid of the box was lifted. It was not visible to the young patas, in a cage nearby. Such avoidance learning may be an example of social facilitation (Strayer 1976).

Local enhancement is a special form of social facilitation and means an increased tendency to respond to a particular object as a result of another individual's response to it. It usually combines with trial-and-error learning (Thorpe 1963). Imitative feeding behavior, such as the acquisition of traditional food items by infants from mothers (Kawamura 1959; Jay 1963a; Hall 1963a; Goodall 1973), should be classed as local enhancement, since it entails trial and error.

True imitation is the copying of a novel or otherwise improbable act for which there is clearly no instinctive tendency (Thorpe 1963). The propagation of placer mining of wheat provides the best example of true imitation among feral monkeys (Strayer 1976). Fishing for termites and ants, and ant dipping may be examples of true imitation among wild chimpanzees.

Teaching. Teaching can be divided into discouragement and encouragement. Teaching by discouragement can be distinguished from other behaviors that are learned through punishment (e.g., behaviors associated with deference to high-ranking animals) in that the behavior being transmitted has consequences primarily for the learner. Discouragement plays a part in avoidance learning in monkeys and apes. Macaque mothers pull their offspring away from a novel object (Kawamura 1959; Menzel 1966). Chimpanzee mothers and allomothers were observed to pull infants away from a novel object or to snatch parts of plants not included in the group's food repertoire, thus discouraging the infants' exploratory activities (Goodall 1973; Wrangham 1977; Nishida 1983b).

When subadult and juvenile chacma baboons displayed interest in fruits drugged with cynalin, they were usually threatened by a high-ranking male who had tested and rejected them. Thus, avoidance response to the fruits spread through the troop (Fletemeyer 1978).

PRIMATE AND NONPRIMATE CULTURE

The significance of cultural behavior lies primarily in the fact that it opens a new ecological niche not exploited by other animals. *Cebus* monkeys, for example, can obtain important nutrients (palm nuts, frogs, and insects) that could not easily be obtained without apparently cultural knowledge and techniques. Chimpanzees fool safari ants and wood-boring ants by a traditional technique that outwits the ants' instinctive defensive strategies. Placer

mining of wheat spares monkeys time and energy and provides them with pure grain that might otherwise make their teeth unnecessarily worn.

Avoidance learning by tradition is critically important for the survival of animals, for they learn what to avoid without having to experience events directly themselves (Washburn and Hamburg 1965). Learning not to take avoidance responses is also important in sparing time and energy expended in useless flight from unharmful novel objects or animals. Naïve animals learn more quickly by tradition, than by individual learning, to accept novel objects.

Nonprimate mammals also have great learning powers. For example, young meerkats failed to recognize a novel food as edible and only began to eat it when their mother did so (Ewer 1963). Examples of cultural transmission in nonprimate mammals are reviewed by Galef (1976), Mainardi (1980), and Bonner (1980); a few of them are introduced later.

The size and position of the territories of prairie dog coteries remain essentially unchanged through complete population turnovers (King 1955). Similarly, female mountain sheep inherit the home ranges of the female group that raised them (Geist 1971).

Some clans of Norway rats feed extensively on bivalves, while other clans do not prey on mollusks despite their ready availability. Moreover, the specific mode of opening the shells of these prey differ from colony to colony and is also socially transmitted (Gandolfi and Parisi 1973).

In 1919 at the request of farmers in South Africa, an attempt was made to annihilate a group of about 140 elephants. A famous hunter killed elephants one by one. Survivors perhaps associated the sound of a shot and the hunter's smell with their family members' deaths. Within a year there were only 16 to 30 animals left alive, but by then the remaining elephants had become extremely wary and never came out of the thicket until after dark. As a result, the hunter gave up the attempt. In 1930 the elephants were granted a sanctuary, but the behavior of these survivors has changed very little. "Few if any of those shot at in 1919 can still be alive, so it seems that their defensive behavior has been transmitted to their offspring, now adult, and even to calves of the third and fourth generations, not one of which had itself suffered attack from man" (Douglas-Hamilton and Douglas-Hamilton 1975).

As in the case of primates, the above examples of social learning in nonprimate mammals may be explained by social facilitation and local enhancement. Is there any difference, then, in cultural transmission between primate and nonprimate mammals?

First, true imitation has rarely been demonstrated in nonprimate mammals. Certainly, cetaceans and carnivores may be exceptions. Cats that had had the opportunity to watch another cat obtain a food reward by a simple manipulation, such as lever pressing, were themselves significantly quicker at learning the task than were controls (Thorpe 1963). Bottle-nosed dolphins spontaneously learn complex behavioral sequences by observation and imitate a wide variety of previously unfamiliar motor patterns and sounds, such as the comfort and sleeping posture of a Cape fur seal and activities and sounds of human drivers (Taylor and Saayman 1973).

Second, "teaching" may be more common among nonhuman primates than among other animals. As mentioned earlier, teaching by discouragement was observed for monkeys and apes. However, the use of punishment in the attempt to train their young in anything other than avoidance seems exclusively limited to humans (Barnett 1968). No nonprimate mammals are known to teach by discouragement.

There are some indications of teaching by encouragement, at least in Old World monkeys and apes. A baboon mother takes a few steps away from her infant, pauses, and looks back at it, encouraging it to move toward her (Altmann 1980). Similar behavior is reported for caged rhesus mothers (Hinde and Simpson 1975).

When infant vervet monkeys give alarm calls, they often make "mistakes" and alarm at harmless species such as pigeons. If an infant gives the "eagle alarm" to a pigeon, adults nearby look up, but then do nothing else. In contrast, if an infant is the first group member to alarm-call at a martial eagle (a true predator), adults will look up and then give alarm calls themselves. This behavior seems to help the infants learn which are their predators, and they quickly cease giving alarm calls to harmless creatures (chap. 36). These may be borderline cases of encouragement because it is not clear whether the behaviors of adults are goal directed and because the infant would probably learn such behavior anyway, even if it were not taught.

Under captive conditions, a chimpanzee mother stretches her infant's limbs by holding it up as if to make it walk. She encourages her young to walk by dragging it with one hand or by crouching in front of it and calling (Yerkes and Tomilin 1935). Similar maternal behavior is reported for a captive gorilla (Passingham 1982). However, there has been no evidence of such tuition in the wild.

Ewer (1969, 698) proposed another type of teaching by encouragement in nonprimate mammals, namely, "the creation by the parent of a situation in which the responses of the young automatically lead to their learning." She mentioned as the most striking examples the introduction of the young, in carnivores such as cats, tigers, cheetahs, otters, and caracals to the killing and eating of prey. It is likely that in these animals modes of learning and teaching are innately programmed.

Third, social structures in primates are often diversified and complex than those in many nonprimate mammals, where adult males and females may live separately except in breeding season. Channels for cultural transmission are thus limited. Monkeys and apes usually form bisexual groups and may learn socially from adult males and females, elders, peers, and particularly infants, as well as from their mothers.

Differentiation of social roles by age, sex, and rank is obvious in many primates (see reviews in part 1). In one-male units of hamadryas baboons, a central female behaves as social coordinator while a peripheral female tends to be the earliest to detect food, water, and potential danger (Sigg 1980). Although enough studies have not yet been done, "attention structure" (Chance 1961) in primate groups might have important consequences for cultural transmission (Cambefort 1981).

Social roles and many aspects of social behavior in primates are likely to be learned from elders and culturally transmitted, although this is virtually impossible to document without experimentation. Many of the complex "social strategies" in chimpanzees documented by de Waal (1982, 48) may be learned: for example, an intact older male simulated limping pitifully only when he was being watched by the stronger competitor and thus effectively avoided being attacked by him. The older male "may have learned from incidents in the past in which he had been seriously wounded that his rival was less hard on him during periods when he was (of necessity) limping." Similar strategies might differ between groups in manners that we cannot detect.

At our present stage of knowledge, it would be premature to draw any rigid distinction between nonhuman primate and nonprimate mammal culture. However, the three features outlined above—imitation, "teaching," and the effects of different social "roles"—are presently more well documented in primates than in any other mammalian order (see also Mainardi 1980).

Human culture is asserted to be different from nonhuman culture in that the former depends on language (Washburn and Benedict 1979). However, one tends to forget the great importance of observational learning even in human culture. Children are likely to learn most social behavior, social relationships, and attitudes by observing family members, elders, and peers rather than by verbal tuition. A Japanese proverb says, "Children do not do as parents tell them to do, but do as parents do."

SUMMARY

Cultural behavior is behavior that is transmitted socially rather than genetically, shared by many members within a group, persistent over generations, and not simply the result of adaptation to local conditions. Although it is often difficult to determine by field observation alone which behaviors are truly cultural, primates exhibit a variety of such activities, including food processing, tool use, and gestural communication. The mechanisms underlying cultural transmission are well documented for sweet-potato washing and wheat cleaning among Japanese macaques. Primate culture differs in four respects from similar behavior in other animals: primate culture is more widespread; it is often characterized by genuine imitation; it commonly involves behavior that resembles teaching; and the social organization of primates allows behavior to be transmitted through many different channels in addition to that transmitted between mothers and infants.

PART V

The Future

39 | Conservation of Primates and Their Habitats

Russell A. Mittermeier and Dorothy L. Cheney

As this volume and many other recent publications clearly indicate, interest in nonhuman primates has increased tremendously over the past two decades, with researchers from many disciplines in both developed and developing countries focusing on this unique and diverse order. Ironically, however, these years have also seen serious declines in wild primate populations. Although difficult to express in quantitative terms, these declines threaten the survival of a growing number of the world's approximately two hundred primate species. At present, over 50% of all primate species are in some jeopardy (Wolfheim 1983), and this figure will continue to increase in the remaining years of this century.

The goal of this chapter is to review the factors contributing to the disappearance of primates and to provide an overview of primate conservation efforts around the world. The chapter begins with a discussion of the major threats to wild primate populations, which is followed by a brief review of primate conservation priorities in the four major regions where primates occur. Finally, we discuss how conservation efforts might best be focused to ensure that these animals do not disappear from the face of the earth.

MAJOR THREATS TO PRIMATE POPULATIONS

The major threats to wild primate populations fall into three broad categories: habitat destruction, hunting, and live capture for export or local trade. The relative importance of these factors varies across species and regions, but one or more of them influence almost all existing primate populations.

Habitat Destruction

By far the most important factor contributing to the decline of the world's nonhuman primates is habitat destruction (fig. 39-1). More than 90% of all primate species occur in the tropical forests of Asia, Africa, and South and Central America, and as these forests disappear, so too do the animals that depend on them for survival. Tropical forests are destroyed for a variety of reasons, among them conversion for agriculture or ranching, demand for charcoal and cooking fuel, poor management of industrial logging, and construction of massive hydroelectric projects. Throughout the world, tropical forests are disappearing at a rate of 10 to 20 million hectares per year (U.S. Department of State 1980). The Office of Technology Assessment (1984) cites a more conservative figure of 11.3 million hectares per year. This is roughly the equivalent of an area the size of West Germany disappearing every 2 years.

The ultimate cause of forest destruction is the human population explosion, which is greater in the poor, underdeveloped countries of the tropics. Growth rates in almost all tropical countries are far higher than in the developed world and average over 3% per year. For example, Brazil's annual growth rate is 2.3%, Nigeria's 3.3%, and Kenya's over 4%, compared with 0.7% for the United States and −0.2% for West Germany. World population, now at five billion, is expected to increase to 8.086 billion by the year 2020, and most of this growth will be in the same countries where most nonhuman primates now occur. Over the next 35 years, the population of underdeveloped countries in the Third World will double, from 2.561 billion to 5.191 billion. In contrast, population in the developed world will rise from 1.166 to only 1.35 billion, and that of China from 1.034 to 1.545 billion (Population Reference Bureau, cited in Raven 1984).

Each year this population growth adds many millions of landless people to the already overburdened developing countries. Existing agricultural lands are unable to absorb most of these people for a number of reasons. First, intensified agricultural production, which could substantially increase yields, requires sophisticated technology that is often unavailable to many farmers and is certainly more expensive than simply expanding the area under cultivation. Moreover, inequitable land ownership systems in many parts of the world keep existing farmland from absorbing as many people as it could. Third, trends toward cash crops, ranching, and urbanization occupy areas that could be used by small farms. At present, major urban areas are being overwhelmed by increasing numbers of migrants. With no place else to go, more and more people are settling in the remaining undisturbed forest areas (Office of Technology Assessment 1984; U.S. Department of State 1980). As a result, the trend toward large-scale forest loss seems irreversible.

The extent and effects of habitat destruction vary considerably. In the Ivory Coast, 6.5% of the remaining forest is being destroyed every year, whereas in Bolivia it is

FIGURE 39-1. Forest destruction in Bahia, Brazil, for an agricultural project. (Photo: Russell Mittermeier)

presently only 0.2%. In Madagascar and the Atlantic forest region of eastern Brazil, so much has already been destroyed that any further loss of forest constitutes a grave threat to the remnant primate populations, many of which are already endangered. Similarly, if present trends continue, many of the forests of West Africa and Central America, as well as almost all of the lowland forests in peninsular Malaysia, will disappear by the early 1990s (Myers 1984).

The rapid rate of forest destruction in many parts of the world emphasizes the vulnerability of those regions that still include huge tracts of forest. For example, since the advent of Europeans, the Atlantic forest of Brazil has been reduced to between 1 to 5% of its original size. Since most of this destruction has occurred in the last 20 or 30 years, it is easy to imagine how even more vast expanses of forest such as those found in Amazonia could be vulnerable to equally rapid devastation (Mittermeier et al. 1982). Indeed, if present trends continue, it is estimated that by the year 2020 the forests of Amazonia and the Zaire-Congo river basin will have been reduced by 50% and 25%, respectively, and that all other accessible forests outside parks and reserves will have been completely destroyed (Skorupa 1985).

Conversion of tropical forest for agricultural or ranching purposes is the single most important cause of forest destruction, whether as a result of traditional shifting "slash-and-burn" subsistence farming or large-scale commercial agricultural projects. In productive, fertile areas such as Malaysia and the flood plains of West Africa, most forests have already been replaced by permanent farms or plantations (Wolfheim 1983). In many other areas, forests have been replaced with subsistence farming by people who have neither the capital nor the expertise to invest in land maintenance technology that might permit more sustained, less destructive agriculture. Such farms often operate in upland areas or regions of low soil fertility (e.g., much of Amazonia) that can support only a few seasons of intensive farming before erosion and overuse deplete the nutrients in the soil. As long as cleared areas are small, and land use is followed by a fallow period, such farming can be ecologically sound in some areas of the tropics. However, population growth is leading to inadequate fallow periods and overuse of land in many cases, with the result that degraded plots are abandoned and new areas of forest cleared to begin the cycle of destruction anew (Office of Technology Assessment 1984; U.S. Department of State, 1980).

A second major cause of forest destruction in Central and South America is cattle ranching. In contrast to subsistence farming, however, cattle ranching often involves foreign-owned corporations and active government support. Brazil, for example, encourages cattle ranching in Amazonia as part of its campaign to extend its control over the interior and to develop its export market. Large ranching consortiums in Brazil include such multinational corporations as Volkswagen, which holds a concession for 1,400 km² of Amazonian forest (Myers 1984). Ironically, cattle ranching in tropical forest areas is much less efficient than on existing subtropical pastures, but the terms offered by the government to encourage the development of Amazonia are so attractive that there is little inducement for better land management (Myers 1979, 1984).

The cutting down of trees for fuel is also a significant and often underestimated threat. It accounts for about 80% of all wood removed from tropical forests, and in some countries it is the major cause of deforestation (e.g., India, Pakistan, Java, much of Thailand, northern Sumatra, eastern Madagascar, parts of West Africa and Central America; Myers 1984). At least 1.5 billion people in the Third World are estimated to suffer some scarcity of fuelwood, and forests are being subjected to increasingly demanding harvesting rates (Myers 1984). The need for firewood and charcoal is usually thought to have only a local impact, but in fact there is also a large international trade in charcoal (primarily to Middle Eastern countries) that is having a global effect on tropical forests (Office of Technological Assessment 1984).

Industrial logging for both local use and export accounts for about 20% of the total volume of wood removed from tropical forests. Such logging depletes or destroys the forest as a resource for local people, with the benefit going to foreign consumers, national governments, or large companies. Although much commercial extraction of wood is selective, even such selective deforestation can adversely affect species that depend on mature primary forest (e.g., *Cercopithecus diana, Cercocebus albigena, Ateles paniscus; Lagothrix lagotricha*; Wolfheim 1983; Skorupa 1986). Moreover, there is growing evidence that the productivity of the remaining trees is severely reduced by deforestation (Skorupa 1986), especially since the use of heavy equipment usually results in secondary damage that kills many of the unfelled trees. Moreover, roads built by logging companies to remove target species also provide access to farmers, who complete the process of deforestation. Finally, although companies are usually only granted licenses to remove portions of the forest or particular species, local corruption or lack of enforcement can result in considerable abuse of contracts.

New technologies that allow whole wood harvesting and pulp production are making it profitable both to exploit the forest more fully and to use species that formerly had little or no commercial value. The result has been increasing exploitation of tropical forests (Office of Technological Assessment 1984). Between 1950 and 1970, world output of paper products increased from 40

to 130 million metric tons, and the demand has continued to rise (Myers 1979). Prior to the mid-1970s, most paper was produced from softwoods, but new technologies now permit hardwood to be converted easily and quickly into pulp. Japan is particularly dependent on foreign sources of pulp and relies increasingly on Southeast Asian and South American hardwood trees for pulpwood (Myers 1984).

In addition to removing trees, forest destruction dramatically increases soil erosion, and the effects of soil depletion are soon felt well beyond the immediate area of land use (Myers 1979, 1984). Without trees, the land absorbs less water. As a result, the rainy season runoff is released in a flood rather than at a steady and low rate and is followed by months of drought. The runoff carries away great quantities of topsoil, not only reducing fertility in the immediate area but also ruining lowland crops and clogging streams, reservoirs, and hydroelectric projects. The land is converted into a dry, nutrient-depleted area no longer capable of supporting farms, and rivers dry up for large portions of the year. Ultimately, the repercussions of tropical forest destruction extend even to the developed countries of North America, Europe, and Japan, since forest destruction not only eliminates crucial genetic resources that are important to agricultural and biomedical research, but also may even alter climatic and rainfall patterns in the Northern Hemisphere (Myers 1979, 1984).

Although replanting and management of valuable forest trees could make sustained utilization of forests possible, in practice such activities are uncommon and carried out on only a small scale when they do occur (U.S. Department of State 1980; Office of Technological Assessment 1984). Moreover, throughout Asia, Africa, and South America, forests are frequently replanted with exotic species such as pine or eucalyptus that are regarded as fast growing and of more commercial use than indigenous species. Such exotic species, however, usually cannot sustain indigenous birds or mammals.

Finally, hydroelectric projects also pose a threat to tropical forests. Although their overall effect is far less than that of the factors already discussed, they often destroy some of the least disturbed and most important areas of tropical forest, including those occurring in parks or reserves. Such "protected" areas are often chosen as project sites both because they are relatively uninhabited and because they are already in government hands. In some cases, the areas of forest to be destroyed can be significant. The Tucurui Project in Amazonian Brazil, for example, will flood an area of approximately 3,000 km² (A. Johns, pers. comm.).

Hunting

Primates are hunted for many different reasons, the most important being as a source of food (fig. 39-2). As with habitat destruction, the effects of hunting vary consider-

FIGURE 39-2. Spider monkey (*Ateles belzebuth belzebuth*) shot for food in Colombia. (Photo: F. Medem)

FIGURE 39-3. Black-and-white colobus monkey (*Colobus guereza*) rugs for sale in a Nairobi tourist shop in 1973. (Photo: Russell Mittermeier)

ably from region to region and across species. Hunting is not very important, for example, in India, where primates are linked to the monkey god Hanuman, who occupies an important role in the Hindu religion. Hunting of primates for food is a major threat, however, in the Amazonian region of South America and in both West and central Africa. Thousands of primates in these regions are killed each year for immediate consumption or sale in local markets. In some places (e.g., the interior of Suriname), primates can account for as much as 25% of the meat intake of forest people (Mittermeier 1977). Primate hunting is often prohibited by law, but enforcement of such legislation is rare or nonexistent in the remote areas where most hunting occurs (Mittermeier and Coimbra-Filho 1977).

Even where hunting is common, it does not affect all species equally. In Amazonia, for example, larger species such as woolly monkeys (*Lagothrix* spp.), spider monkeys (*Ateles* spp.), howler monkeys (*Alouatta* spp.), and capuchins (*Cebus* spp.) are heavily hunted, while smaller ones such as squirrel monkeys (*Saimiri sciureus*) and tamarins (*Saguinus* spp.) are rarely persecuted since they barely recompense the hunter for his shotgun shell. In areas of heavy hunting pressure, the killing of pri-

mates for food can result in local extinctions, even where there is suitable forest habitat. This appears to be the case, for example, for *Lagothrix* and *Ateles* in many parts of Brazilian and Peruvian Amazonia (Mittermeier and Coimbra-Filho 1977; Soini 1982c). Primates are also sometimes killed or maimed by traps set for other species, as occurs with chimpanzees and gorillas in parts of Africa (Fossey 1983; Ghiglieri 1984).

Primates are also killed to obtain skins or other body parts for human ornamentation. The best example of this is the use of black-and-white colobus (*Colobus guereza*) skins to make cloaks and headdresses for native Africans and rugs and coats for Europeans, Americans, and Japanese (fig. 39-3). In the last two decades of the nineteenth century, some 2.5 million skins reached Europe, and colobus skin rugs were still common in tourist shops in East Africa as late as 1978 (Mittermeier 1973; Oates 1977a). For some species, the tourist trade remains a serious threat, as is the case with the slaughter of mountain gorillas in Rwanda and Zaire to obtain hands and skulls for sale to European tourists (Fossey 1983). Similarly, in Amazonia, various tribes use monkey teeth, skulls, and skins for ornamentation, and all of these items occasionally find their way into tourist shops. By and large, however, such uses of primates are a by-product of food hunting.

Sport and trophy hunting must also be considered, although they are relatively unimportant in the case of most primate species. Sport hunting has generally focused on the large, spectacular species, and, judging from the number of books that appeared on gorilla hunting in the first half of this century (e.g., Du Chaillu 1930), gorillas were the most attractive targets. Even today there is occasional interest in shooting gorillas and other primates as trophies, as shown by a 1978 permit application (not granted) from the Safari Club International to the U.S. Endangered Species Scientific Authority to hunt gorillas, orangutans, black colobus (*Colobus satanas*), and Zanzibar red colobus (*Colobus badius kirkii*), and a 1981 advertisement for gorilla-hunting safaris organized by a French company.

Finally, primates are hunted as agricultural pests. One of the best examples of such pest control comes from Sierra Leone, where the government considered primate crop raiding so serious a threat that it sponsored periodic "monkey drives" from 1948 to 1962 to eliminate these animals from agricultural areas. According to government records, an average of 19,000 primates were destroyed each year by these drives (Tappen 1964). Although this example is an extreme one, it emphasizes that primates are considered a serious menace to agriculture in some parts of the world, the major offenders being semiterrestrial species such as baboons (*Papio* spp.) and vervets (*Cercopithecus aethiops*) in Africa and macaques (*Macaca* spp.) in Asia. It is difficult to assess the damage inflicted by primates on crops in different parts of the world and equally difficult to determine the impact of pest control on global primate decline. However, the Sierra Leone data and anecdotal reports from other countries indicate that the killing of primates as crop pests may result in more losses than is recognized. As primate habitats continue to be encroached upon, it is likely that the more adaptable species will rely increasingly on crops. The unfortunate result will be heightened conflict between humans and nonhuman primates, which in turn will exacerbate the difficulty of gaining the sympathy of local people for primate conservation.

Live Capture of Primates

The third major threat to primate populations is live capture to serve the pet or research market. At the height of the international primate trade in the 1950s and 1960s, the export of primates for these purposes involved hundreds of thousands of animals. These were mainly rhesus monkeys (*Macaca mulatta*), which were exported from India for biomedical research, and squirrel monkeys (*Saimiri sciureus*), which were exported from Colombia and Peru primarily for the pet trade. Dozens of other species were also affected, although usually in much smaller numbers. The decline of this trade can be attributed to several factors, among them the imposition of export bans by source countries, import restrictions by user countries for health reasons, and decreased demand for live-caught primates in biomedical research (Mack and Mittermeier 1984).

Although the international trade in primates is not as significant as habitat destruction in the decline of primate populations, it can have a profound effect on the most heavily traded species, especially those that are already endangered or vulnerable (e.g., cottontop tamarins, *Saguinus oedipus;* chimpanzees, *Pan troglodytes*). The international trade in primates has been curtailed significantly since the drafting of the Convention of Trade in Endangered Species of Wild Flora and Fauna (CITES) in 1973 and the imposition of a number of export bans by producing countries. Parties to CITES agree to ban commercial trade in endangered species and to monitor trade in others that may become endangered. To date, 87 countries have ratified or acceded to CITES, although some (e.g., Belgium) did not do so until 1984 and continued to serve as conduits for the importation of primates well into the 1980s. Although CITES provides valuable guidelines for the control of wildlife trade, it is only as effective as the countries that implement it, and some countries follow its requirements more carefully than others. Loopholes that permit, for example, the legal sale of captive-born animals or the importation of animals captured in specific countries have contributed to the illegal trade through bribery, corruption, and misidentification of species (Wolfheim 1983; IUCN Traffic Bulletin 1984).

Although the CITES convention is occasionally violated, it has caused a marked reduction in the numbers of nonhuman primates imported to the developed countries each year. In 1968, before the ratification of the CITES accord, the United States imported 113,714 monkeys and apes; by 1983, this number had dropped to 13,148 (Wolfheim 1983; Mack and Mittermeier 1984). Efforts have also been made to set guidelines for the use of endangered and wild-caught primates in biomedical research. In 1981, for example, the IUCN Primate Specialist Group formulated a policy on the use of primates in biomedical research that specifically recommends that endangered, vulnerable, and rare species be used for medical research only when they are obtained from existing, self-sustaining captive breeding colonies. Although there is increasing compliance with these guidelines, they are still occasionally ignored. For example, in 1983 Japan received 30 chimpanzees from Sierra Leone for use in hepatitis research despite the fact that Japan is a party to the CITES convention (Stevenson 1984).

The local trade in primates within source countries is also potentially significant. Some developing countries, such as Brazil and India, are beginning to use indigenous species in their own biomedical research programs. Several countries also have a large internal pet market aimed at urban areas. In Brazil, for example, hundreds of marmosets (*Callithrix* spp.) are sold each year in illegal pet markets in the large cities of Rio de Janeiro and São Paulo. The Sunday Market in Bangkok, Thailand, provides a similar market for young gibbons and macaques. Many of these primates eventually find their way to the international market, and pet dealers in such countries as Japan continue to receive primates from this trade (IUCN Traffic Bulletin 1984).

In areas such as Amazonia, where primates are already hunted for food, live capture of primates as pets is usually just a by-product of meat hunting. For some of these persecuted species (e.g., *Lagothrix lagotricha*), however, even the small added incentive of the pet trade can compound the other pressures the animals face. Woolly monkeys are both the favorite primate meat source and the most desirable primate pets, and as a result a hunter will usually attempt to shoot females with infants in preference to other group members. The effects of such selective hunting on the reproduction rates of a slow-breeding species can clearly be disastrous.

The trade in young apes is a particularly serious problem. The great apes have now become so valuable that the incentives for both local people and unscrupulous dealers to trade in them are great. A lowland gorilla, for example, is worth at least $75,000 in the zoo trade, and chimpanzees are still considered to be important for biomedical research. Despite the CITES accord, zoos still occasionally purchase apes from traders, on the assumption that animals that have already been captured are otherwise likely to die. This was the justification used by the Burger's Zoo in the Dutch city of Arnhem when, under the auspices of the Netherlands IUCN Committee, it purchased seven young gorillas from Cameroon in 1984 (IUCN Traffic Bulletin 1984). Although such purchases do save the lives of the individuals in question, they also provide continued incentive for future poaching and are likely to hinder efforts to curtail the international trade in endangered species.

A final point to emphasize is that live capture of any kind usually involves the death of many individuals for every one that survives. Captured animals are usually handled by a number of middlemen before they reach their final destination, and handling and transport often occur under stressful and inhumane conditions. There are countless reports of dead or starving monkeys and apes arriving at their European or American destinations in small, inadequate crates. Few accurate estimates exist for the number of primates that are lost each year during transport, but it is certain that the trade figures that are usually cited provide a gross underestimate of the numbers of primates actually removed from wild populations to support the trade.

A REGIONAL VIEW OF GLOBAL PRIMATE CONSERVATION PRIORITIES
Central and South America

Sixteen genera and 64 species of New World monkeys are currently recognized (Mittermeier and Coimbra-Filho 1981; Hershkovitz 1983), of which 27 species and subspecies are already listed as either endangered, vulnerable, rare, or indeterminate in the IUCN *Red Data Book*. (The *Red Data Book* is the major reference work on endangered species and is produced by the Conservation Monitoring Centre of the International Union for Conservation of Nature and Natural Resources.)

Using the criterion of taxonomic uniqueness, the most critically endangered New World monkeys are members of the genera *Brachyteles* and *Leontopithecus*. Both occur only in the tiny remnant forests of the Atlantic forest region in eastern Brazil, and both are on the verge of extinction. The muriqui (*Brachyteles arachnoides*) is the only member of its genus and is the largest South American primate. Remaining populations may total as few as 300 to 400 individuals. Similarly, the lion tamarin genus, *Leontopithecus*, includes three species that are all in a very precarious position. All three occur in only a few remnant forest patches, and the reserves created for them are in some cases inadequate and occupied by squatters (Mittermeier et al. 1982). (Although some taxonomists regard *Leontopithecus* as a single species, here we follow Rosenberger and Coimbra-Filho 1984 in recognizing three *Leontopithecus* species; see also table 39-1.)

At the generic level, the next most endangered New World monkeys are the spider monkeys (*Ateles*) and the woolly monkeys (*Lagothrix*). Though both genera still have large ranges, they are heavily hunted and require undisturbed forest for their survival. Moreover, as is true for most large primates, females in both genera reproduce at relatively slow rates (in *Ateles* interbirth intervals are about 3 years; Eisenberg 1978) and therefore recover slowly from any kind of exploitation. As a result, *Ateles* and *Lagothrix* have disappeared from many parts of their former range, even in areas of still suitable primary forest (Mittermeier and Coimbra-Filho 1977).

The highest priority area in Central or South America is without doubt the Atlantic forest region of eastern Brazil. This region is distinct from Amazonia and is the agricultural, industrial, and population center of the country. Only a tiny proportion of the original forest cover remains, and three-quarters of the 21 species and subspecies of primates (80% of which occur nowhere else) are already endangered.

Taken as a whole, Amazonia must be considered the next highest primate conservation priority. It is by far the largest tropical forest region in the world and is home to the majority of New World monkey species. In some areas, there are as many as 13 sympatric species. However, this vast region is shared by nine different coun-

TABLE 39-1   World's Most Endangered Primates

Genus/Species	Range	Comments
ENDANGERED FAMILIES		
Daubentoniidae *Daubentonia madagascariensis*	Formerly the eastern forests of Madagascar	A monotypic family; highly endangered by habitat destruction and killed on sight by villagers as evil omen; no population estimates available
ENDANGERED MONOTYPIC GENERA		
Brachyteles *Brachyteles arachnoides*	Formerly the southeastern portion of Brazil's Atlantic forest region	A monotypic genus; highly endangered by habitat destruction and poaching; 300–400 remain
Allocebus *Allocebus trichotis*	Eastern forests of Madagascar	A monotypic genus; known from four specimens, three of them collected in the last century; possibly extinct?
Indri *Indri indri*	Eastern forests of Madagascar	A monotypic genus; endangered by habitat destruction within its small and patchy range; no population estimates available
Varecia *Varecia variegata*	Eastern forests of Madagascar	A monotypic genus; endangered by habitat destruction and hunting; no population estimates available
Simias *Simias concolor*[a]	Mentawai Islands off Sumatra, Indonesia	A monotypic genus; endangered by habitat destruction and hunting and the most endangered of the four Mentawai primates; population estimated at 19,000 (World Wildlife Fund 1980)
ENDANGERED POLYTYPIC GENERA		
Leontopithecus[b] *Leontopithecus rosalia* *Leontopithecus chrysomelas* *Leontopithecus chrysopygus*	Southeastern portion of Brazil's Atlantic forest region in states of Rio de Janeiro, Bahia, and Saõ Paulo	All three species highly endangered by habitat destruction and illegal live capture within their very small ranges; only a few hundred *rosalia* and *chrysopygus* survive in the wild; *chrysomelas* also depleted, but no recent population estimates available
Rhinopithecus *Rhinopithecus roxellanae* (including *roxellanae*, *brelichi* and *bietii* as subspecies) *Rhinopithecus avunculus*	China and Vietnam	All four taxa endangered by habitat destruction and perhaps also hunting. *R. roxellanae brelichi* down to 500 individuals; *R. roxellanae bietii* to about 200; *R. roxellanae* to 3,700–5,700 (Poirier 1983); *R. avunculus* known from only a handful of museum specimens, possibly extinct?
ENDANGERED SPECIES AND SUBSPECIES		

Species/Subspecies Numbering Only in the Hundreds

Genus/Species	Range	Comments
Hapalemur	Now known only from the humid forest east of Fianarantsoa on the east coast of Madagascar	Although no population estimates are available, this species has been considered extremely rare and on the brink of extinction
Propithecus diadema (including *diadema, candidus edwardsi, holomelas* and *perrieri* as subspecies)	Eastern forests of Madagascar	Although no population estimates are available, the species as a whole is considered highly endangered because of habitat destruction and perhaps some hunting, and all subspecies must exist at low population sizes; *P. d. perrieri* is the rarest (Tattersall 1982)
Macaca silenus	Western Ghats of south India	Highly endangered by habitat destruction and occasional hunting; population estimated at 670–2,000 (Ali 1982)
Colobus badius gordonorum	Uzungwa Mountains and Magombera Forest Reserve in Tanzania	Endangered by habitat destruction and heavy hunting pressure; Magombera Forest Reserve, thought to be its main stronghold, has a population of only 300 animals (Rodgers 1981)

TABLE 39-1   *(continued)*

Genus/Species	Range	Comments

ENDANGERED SPECIES AND SUBSPECIES

Genus/Species	Range	Comments
Gorilla gorilla beringei	Virunga Volcanoes of Zaire, Rwanda, and Uganda; Bwindi Forest in Uganda	Highly endangered by habitat encroachment and poaching; total population in the Virungas 255 animals, in Bwindi 95–130 (Harcourt et al. 1983)

Species/Subspecies Numbering in the Low Thousands

Genus/Species	Range	Comments
Macaca nemestrina pagensis	Mentawai Islands off Sumatra, Indonesia	Endangered by habitat destruction and hunting in very restricted range; population estimated at 39,000 (World Wildlife Fund 1980)
Cercocebus galeritus galeritus	Tana River, Kenya	Endangered, very small population in fragmented habitat; total numbers estimated at 1,200–1,700 (Marsh 1978)
Colobus badius rufomitratus	Tana River, Kenya	Endangered, very small population in fragmented habitat; total numbers estimated at 1,400–2,000
Colobus badius kirkii	Zanibar	Very small population in fragmented habitat; total numbers estimated at 1,700 (Silkiluwasha 1981)
Colobus badius preussi	Cameroon	Endangered by habitat destruction and hunting; population estimated at less than 8,000 (Gartlan, pers. comm.)
Presbytis potenziani	Mentawai islands off Sumatra, Indonesia	Endangered by habitat destruction and hunting; population estimated at 46,000 (World Wildlife Fund 1980)
Hylobates klossi	Mentawai Islands off Sumatra, Indonesia	Endangered by habitat destruction and hunting in very restricted range; population estimated at 36,000 (Ibid.)
Hylobates moloch	West Java	Endangered by habitat destruction in already fragmented habitat; total numbers estimated at 2,400–7,900 (Kappeler 1981)
Pan troglodytes verus	West Africa from southern Senegal to western Nigeria; now largely restricted to Guinea, Sierra Leone, Liberia, and Ivory Coast	Endangered by habitat destruction, hunting, and live capture for export; total in known habitats estimated at 1,500; in potential habitats 15,700 (Teleki and Baldwin 1979)
Pan paniscus	Central Zaire south of the Zaire River	At risk from habitat destruction and hunting in some parts of its range; total in known habitats about 2,200; in potential habitats 13,000 (Teleki and Baldwin 1979)
Pongo pygmaeus (including *pygmaeus* and *abelii* as subspecies)	Borneo and Sumatra	Threatened by habitat destruction and occasional live capture; 5,000–15,000 still thought to exist in Sumatra, 3,500 in Sabah, and about 250 in Sarawak (Rijksen 1978; Davies & Payne 1982); no published estimate for Kalimantan; recent unpublished information indicates that the total population may be higher than previously believed
Gorilla gorilla graueri	Eastern Zaire	Endangered by habitat destruction and hunting; total estimate at 4,000
Gorilla gorilla gorilla	West Africa (including Cameroon, Gabon, Central African Republic, Congo Republic, Equatorial Guinea, and Angola)	Threatened by habitat destruction and hunting; total numbers unknown, but recent estimate for Gabon 35,000 + 7,000 indicates that total population may be higher than previously believed (Tutin and Fernandez 1984)

TABLE 39-1   *(continued)*

Genus/Species	Range	Comments
SPECIES/SUBSPECIES FOR WHICH POPULATION ESTIMATES NOT AVAILABLE, BUT KNOWN TO BE OF CONSERVATION CONCERN		
Mirza coquereli	Forests of western Madagascar	May be endangered
Lepilemur mustelinus (including *ruficaudatus*, *leucopus*, and *dorsalis* as subspecies)	Forests of western and southern Madagascar	May be endangered
Lemur macaco	Coastal areas of northwestern Madagascar and islands of Nosy Be and Nosy Komba	Probably endangered or vulnerable
Lemur mongoz	Northwestern Madagascar and Moheli and Ndzouani in the Comores	Endangered both on the Madagascan mainland and the Comores (Tattersall 1982)
Lemur rubriventer	Sparsely distributed in eastern forests of Madagascar	Poorly known, but probably endangered
Callithrix flaviceps[b]	Atlantic forest region of eastern Brazil in southeastern Espirito Santo and adjacent parts of Minas Gerais	Endangered by widespread habitat destruction
Callithrix aurita[b]	Atlantic forest region of eastern Brazil in Saõ Paulo and adjacent parts of Minas Gerais and Rio de Janeiro	Endangered by widespread habitat destruction
Callithrix geoffroyi[b]	Atlantic forest region of eastern Brazil in Espirito Santo and Minas Gerais	Endangered by widespread habitat destruction
Saguinus oedipus	Northwestern Colombia	Endangered by widespread habitat destruction and live capture for export
Saguinus bicolor bicolor	Vicinity of Manaus in Brazilian Amazonia	Endangered by habitat destruction in a very small range
Callimico goeldii	Widely but very sparsely distributed in upper Amazonia	A rare species about which very little is known
Callicebus personatus (including *personatus*, *melanochir*, and *nigrifrons* as subspecies)	Atlantic forest region of eastern Brazil from southern Bahia to Saõ Paulo	Endangered by widespread habitat destruction and hunting in some areas
Saimiri oerstedii	Western Panama and southern Costa Rica	Endangered by widespread habitat destruction
Chiropotes satanas satanas	Lower Brazilian Amazonia	Endangered by habitat destruction and heavy hunting pressure in its restricted range
Cacajao calvus calvus	Upper Brazilian Amazonia, between the Rio Japura, the Rio Solimoes, and the Rio Auati-Parana	Possibly endangered
Cebus apella xanthosternos	Atlantic forest region of eastern Brazil, restricted to small area in southern Bahia	Endangered by widespread habitat destruction and heavy hunting pressure
Alouatta fusca (including *fusca* and *clamitans* as subspecies)	Atlantic forest region of eastern Brazil from southern Bahia to Rio Grande do Sul	Endangered by widespread habitat destruction and hunting; the northern subspecies *fusca* may be close to extinction
Lagothrix flavicauda	Cloud forest region of the northern Peruvian Andes	Endangered by habitat destruction and hunting
Ateles geoffroyi azuerensis	Azuero Peninsula of Panama	Endangered by habitat destruction and hunting; possibly on the verge of extinction?
Ateles fusciceps fusciceps	Pacific slope forests of northern Ecuador	Endangered by habitat destruction and hunting; possibly on the verge of extinction

TABLE 39-1   *(continued)*

Genus/Species	Range	Comments
SPECIES/SUBSPECIES FOR WHICH POPULATION ESTIMATES NOT AVAILABLE, BUT KNOWN TO BE OF CONSERVATION CONCERN		
Ateles belzebuth hybridus	Disjunct distribution in northern Colombia and northern Venezuela	Endangered by habitat destruction and hunting
Cercocebus galeritus sanjei	Uzungwa Mountains, Tanzania	A recently discovered subspecies quite restricted in range (Homewood and Rodgers 1981)
Cercopithecus erythrogaster	Southern Nigeria	Endangered by habitat destruction and hunting (Oates and Anadu 1982)
Cercopithecus erythrotis erythrotis	Fernando Poo in Equatorial Guinea	Thought to be endangered
Cercopithecus l'hoesti preussi	Mountains of western Cameroon and eastern Nigeria	Endangered by habitat destruction and a very restricted range
Cercopithecus l'hoesti l'hoesti	Eastern Zaire, Uganda, Rwanda, and Burundi	Rare everywhere except Nyungwe Forest; appears to be sensitive to even moderate habitat destruction (J. Skorupa, pers. comm.)
Papio leucophaeus	Cameroon, eastern Nigeria, and Fernando Poo	Endangered by heavy hunting pressure
Colobus satanas	Cameroon, Equatorial Guinea (including Fernando Poo), Gabon, and Congo Republic	Endangered by habitat destruction
Colobus badius bouvieri	Congo Republic	Possibly endangered
Colobus badius pennanti	Fernando Poo	Possibly endangered
Presbytis aygula	West Java	Endangered by habitat destruction
Pygathrix nemaeus	Laos, Cambodia, Vietnam, perhaps Hainan in China	A monotypic genus; poorly known and possibly endangered or vulnerable
Hylobates pileatus	Thailand, Laos, and Cambodia	Endangered by habitat destruction and hunting in Thailand; status unknown in Laos and Cambodia

NOTE: Species are ranked according to taxonomic uniqueness and then according to degree of depletion of wild populations.

a. Alternative name, as shown in the appendix, is *Nasalis concolor.*

b. Hershkovitz 1977 and others (see chap. 4) consider the three *Leontopithecus* to be only subspecies and *Callithrix aurita, C. fla-* *viceps,* and *C. geoffroyi* as subspecies of *C. jacchus.* Here, the authors follow Rosenberger and Coimbra-Filho 1984 and Coimbra-Filho and Mittermeier 1974 in recognizing the three *Leontopithecus* and the three *Callithrix* in this table as distinct species.

tries, with the result that conservation programs must be developed on a country-by-country basis. Primate conservation in Amazonia is currently less urgent than in other areas of the world, but certain species (e.g., *Saguinus bicolor bicolor, Ateles belzebuth marginatus, Chiropotes satanas satanas*) are already endangered, and high rates of forest destruction in parts of Amazonia suggest that this region will not long remain unthreatened.

Elsewhere in Central and South America (e.g., the forests of Mexico and Central America, the cloud forests of northern Peru, and the drier forest formations of Argentina, Paraguay, and Bolivia), the status of primate populations and habitats varies greatly. In some areas, the habitat is as undisturbed as the most remote areas of Amazonia, but more frequently the situation approaches that of the Atlantic forest region. The top priority in these non-Amazonian regions is to focus on highly endangered species (e.g., the yellow-tailed woolly monkey, *Lagothrix flavicauda,* from the cloud forests of northern Peru; or the cottontop tamarin, *Saguinus oedipus,* from

northern Colombia), and to attempt to set aside parks and reserves of appropriate size wherever possible.

Africa (Excluding Madagascar)

Ten to 15 genera and approximately 55 primate species are currently recognized from mainland Africa and nearby islands, there being no general agreement among taxonomists about the exact numbers involved. These include 3 species of great apes, 40 to 45 species of monkeys, and 8 to 13 lorises and galagos (Thorington and Groves 1970; Dandelot 1971; Petter and Petter-Rousseaux 1979; Olson 1979). Of these, 14 species and subspecies are listed in the IUCN *Red Data Book,* including all of the great apes. Although this represents less than one-quarter of all African primates, African primates have not yet been thoroughly reviewed; it is likely that many more will be reclassified as endangered when such a revision occurs.

The single most endangered African primate is probably the mountain gorilla (*Gorilla gorilla beringei*), which lives only in the Virunga Volcanoes on the Rwanda-Zaire-Uganda borders and in the Bwindi Forest of Uganda. This subspecies is down to about 255 animals in the Virungas and 95 to 120 in the Bwindi Forest (Harcourt et al. 1983), and continues to be threatened by poaching, habitat destruction, and one of the highest human population densities in Africa. The reserves set aside for the gorillas have been severely encroached upon in the last decade, and there is continuing political pressure to convert the remaining areas to farming or commercial use (Harcourt et al. 1983; Wolfheim 1983). The low reproductive rate of gorillas also dramatically diminishes their ability to recover quickly from poaching.

Other endangered African species and subspecies include the red-bellied monkey (*Cercopithecus erythrogaster*) from southwest Nigeria (Oates and Anadu 1982), l'Hoest's monkey (*Cercopithecus l'hoesti preussi*) from the mountains of western Cameroon and eastern Nigeria, the drill (*Papio leucophaeus*) from Cameroon, eastern Nigeria, and Equatorial Guinea, the new mangabey subspecies (*Cercocebus galeritus sanjei*) from the Uzungwa Mountains in Tanzania (Homewood and Rodgers 1981), the Tana River mangabey (*Cercocebus galeritus galeritus*) and red colobus monkey (*Colobus badius rufomitratus*) from Kenya (Marsh 1978), the Zanzibar red colobus monkey (*Colobus badius kirkii*: Silkiluwasha 1981), the Uhehe red colobus (*Colobus badius gordonorum*) from Tanzania (Struhsaker and Leland 1980; Rodgers 1981), Preuss's red colobus (*Colobus badius preussi*) from Cameroon (Gartlan, pers. comm.), and the black colobus (*Colobus satanas*) from Cameroon (Gartlan, pers. comm.). Also of great conservation concern are all of the great apes, including the three chimpanzee (*Pan troglodytes*) subspecies, the bonobo (*Pan paniscus*), and the two other gorilla subspecies (*Gorilla gorilla gorilla* and *G. g. graueri*). Very few data are available on populations of most of these species, and much further research on them is needed.

Although terrestrial, open-country species are an important part of the African primate fauna, the tropical forest species are of the greatest conservation concern. As the remaining forests continue to be encroached upon, those species that are restricted to a forest habitat will approach extinction, while savanna-dwelling species such as the baboon may continue to survive on pastureland and areas unsuited to farming. Of particular importance, both for the diversity of their primates and the extent of habitat destruction that has already occurred, are the forests of West Africa (especially Cameroon and Equatorial Guinea) and the small remaining forest islands of East Africa and Zanzibar. Zaire is also of great conservation importance because its forests harbor the greatest diversity of species in Africa (14 genera and 33 species). Most of Zaire's forests remain undisturbed, but they will be put under increasing pressure in the remaining years of this century.

TABLE 39-2   Recently Extinct Madagascar Primates

Family	Genus and Species
Lemuridae	*Varecia insignis*
	Varecia jullyi
Lepilemuridae	*Hapalemur galieni*
	Megaladapis madagascariensis
	Megaladapis edwardsi
	Megaladapis grandidieri
Indriidae	*Mesopropithecus pithecoides*
	Mesopropithecus globiceps
	Archaeolemur majori
	Archaeolemur edwardsi
	Hadropithecus stenognathus
	Palaeopropithecus ingens
	Archaeondris fontoynonti
Daubentoniidae	*Daubentonia robusta*
Cheirogaleidae	*Allocebus trichotis*[a]

SOURCE: Tattersall 1982.
NOTE: See chapter 3 for further discussion.
a. Possibly extinct.

Madagascar

With 13 genera and 28 species of living lemurs (Petter *et al.* 1977; Tattersall 1982), Madagascar has a unique primate fauna that merits consideration apart from the African mainland. Indeed, the island's *entire* primate fauna must be considered the highest primate conservation priority in the world, with four endangered genera (*Daubentonia, Allocebus, Indri, Varecia*) and several additional species and subspecies on the verge of extinction (*Hapalemur simus, Propithecus diadema perrieri*). Madagascar is also home of the aye-aye (*Daubentonia madagascariensis*), the one living member of its family and the only endangered primate taxon above the generic level (table 39-1). Widespread habitat destruction is the primary threat to these primates, as it has been since the arrival of humans approximately 1,500 years ago. Six genera and at least 14 species have already disappeared from the island due to hunting and habitat destruction (Tattersall 1982; table 39-2), and the remaining population is severely threatened.

While all of Madagascar must be considered a high-priority area, the eastern forest region probably requires the most attention. This is the most diverse and least known part of the island, and it harbors the most endangered lemur species, including *Daubentonia madagascariensis, Allocebus trichotis, Indri indri, Varecia variegata, Hapalemur simus,* and *Propithecus diadema perrieri* (A. Richard and J. Pollock, pers. comm.).

Asia

Nine to 16 genera and 50 to 56 species of Asian primates are currently recognized, there being again no general agreement among taxonomists about the exact numbers involved (Thorington and Groves 1970; Napier 1981; Honacki, Kinman, and Koeppl 1982). These include 1 species of great ape, all 9 species of lesser apes (or gibbons), 34 to 40 species of monkeys, 3 species of lorises, and at least 3 species of tarsiers. Of these, 16 species and subspecies are currently listed in the IUCN *Red Data Book,* among them the 1 great ape, 4 gibbons, 9 monkey species, and 2 tarsier species. Although these endangered species represent about 30% of all Asian species, a number of others are being considered for inclusion.

One of the most endangered Asian species is the lion-tailed macaque (*Macaca silenus*) from South India, which is down to between 670 to 2,000 individuals in the wild (Ali 1982). However, this species is receiving considerable attention, both within India and internationally, and a number of measures are being taken to ensure its survival in captivity and in the wild.

Of perhaps greater concern are the four members of the snub-nosed monkey genus *Rhinopithecus,* the most endangered Asian primate genus. Two species and four subspecies are recognized, and all are at low population levels. The most recent population estimates of the golden monkey (*Rhinopithecus roxellanae roxellanae*) from China are between 3,700 to 5,700 individuals (Poirier 1983), and the other three members of the genus are far rarer. Indeed, the Vietnamese snub-nosed monkey (*Rhinopithecus avunculus*) from Tonkin may already be extinct. It is known from only a handful of museum specimens collected earlier in this century, and there are no recent reports of it from the wild.

The four primate species endemic to the tiny Mentawai Islands off the west coast of Sumatra are also of great concern. Kloss's gibbon (*Hylobates klossi*), the Mentawai langur (*Presbytis potenziani*), the pigtailed langur (*Simias concolor*), and the Mentawai macaque (*Macaca pagensis* or a subspecies of *M. nemestrina*) are all listed in the *Red Data Book,* and the monotypic genus *Simias* is probably the second most endangered Asian genus after *Rhinopithecus.*

Other species of conservation concern in Asia are listed in table 39-1. Of special note is the only Asian great ape, the orangutan (*Pongo pygmaeus*), as well as several of the more threatened gibbons (e.g., *Hylobates moloch, Hylobates pileatus, Hylobates concolor*).

Indonesia is probably the most important Asian country for primate conservation in Asia, since it has by far the highest primate diversity (8 genera and 27 to 30 species). Commercial loggers, however, have already exploited the majority of its lowland forests (Myers 1979, 1984), and large areas of remaining forest continue to be converted each year. Indonesia includes a number of the top-priority conservation areas, including the Mentawai Islands. Elsewhere in Southeast Asia, peninsular Malaysia and Thailand are also of great concern, both for the diversity of their species and the large-scale forest destruction that has already occurred. In China, the ranges of all *Rhinopithecus* monkeys are of importance,

and on the subcontinent, South India and Assam are also high priorities.

OUTLOOK FOR THE FUTURE

Although the planet's expanding human population and the increasing global pressure for development make it inevitable that a proportion of the world's forests and primates will disappear, the role of conservationists is to limit this loss wherever possible. This can be accomplished by the following: (1) protecting areas for particularly endangered and vulnerable species; (2) creating large national parks and reserves in areas of high primate diversity or abundance; (3) maintaining parks and reserves that already exist and enforcing protective legislation in them; (4) creating public awareness of the need for primate conservation and the importance of primates as both a national heritage and a resource, especially in the countries where they occur; (5) determining ways in which people and other primates can coexist in multiple-use areas.

While the creation of parks and reserves and the maintenance of those that already exist clearly offer the greatest protection to endangered species, only a small portion of the world's primate populations occur in areas that have been designated as protected. Elsewhere, ways to encourage the coexistence of human and nonhuman primates will have to be developed, although this may prove difficult in many agricultural areas. In both cases, it will be essential to educate people in both source countries and the developed world that the conservation of habitats and animal species is of practical as well as aesthetic importance (Myers 1984).

Clearly, the ultimate needs of human and nonhuman primates are not as divergent as they first appear, since the survival of both depends on the conservation of forests. The long-term implications of tropical forest destruction have not always been understood, however, in either the developing or developed countries. Within developing countries, tropical forest conservation is essential for agriculture, clean and reliable water supplies, and hydroelectric power. With good management, forests also provide a long-term source of foreign currency. To the developed world, tropical forests offer an irreplaceable genetic repository for medicinal and food plants. There is also increasing evidence that the destruction and burning of tropical forests are altering global weather patterns and that the resulting changes in rainfall and wind currents will seriously damage agriculture in Europe and North America (Myers 1984).

If tropical forests are to be saved, efforts will have to be directed as much toward education as toward the protection of existing parks and reserves because, in the absence of the former, the latter will not survive. Parks are frequently an important source of foreign currency for many countries, especially those in East Africa. Unfortunately, however, parks located in tropical forests are usually less popular tourist attractions than those situated in the savannas, at least in part because forest-dwelling animals are more difficult to observe and often also less dramatic and photogenic than many of the large African savanna species. The survival of forest parks and reserves cannot, therefore, depend solely on their importance to tourism.

In the developing countries where most of the world's tropical forests are found, there is evidence of increasing realization that agriculture, reliable water supplies, and hydroelectric projects all ultimately depend on forest conservation, and it would be incorrect to imply that developing countries have done little to conserve their environment. Indeed, many developing countries have committed a far larger proportion of their land to wildlife reserves and national parks than have the richer developed countries. For example, over 9% of Tanzania has been set aside for wildlife reserves, the equivalent of the United States setting aside California, Oregon, and Washington combined (Myers 1979). Large areas of Indonesia, Brazil, and other source countries have also been dedicated to wildlife conservation. In practice, however, many of these areas are simply "paper parks," where exploitation of timber is permitted, either legally or informally. For example, in the Kutai Reserve of East Kalimantan, Indonesia, an area with a large orangutan population, 1,000 out of 3,000 km^2 had been illegally logged by 1977 (Myers 1979). Recently, however, Indonesia has taken more concrete action to conserve its forests, and there is evidence that the rate of forest destruction has slowed (Myers 1984).

Unfortunately, forest conservation is seldom a high economic or political priority since it does not translate easily into immediate benefits. The temptation, for example, to improve foreign trade imbalances through the export of timber is great, and short-term economic incentives can overwhelm long-term considerations even when the consequences of forest destruction are acknowledged. Moreover, the growing human population imposes political pressures that make it difficult to prohibit human expansion into unsettled areas.

The developed countries in temperate regions of the world also benefit from the conservation of tropical forests, however, and if tropical forests are to survive, these countries must acknowledge both their dependence on and their responsibility for forest conservation. It is the modern, developed countries, for example, that have benefited most from advances in medical research and increases in crop production derived from the germ plasm of wild plant strains from the tropics. At present, the conservation of the world's tropical forests relies disproportionately on the willingness and ability of underdeveloped source countries to pay the economic price

of conservation for the benefit of all nations. Given their many other economic problems, however, this will be difficult or impossible to continue without outside assistance.

Conservation ultimately depends heavily on money to enforce current restrictions on habitat destruction, to provide economic incentives against further exploitation, and to educate people about the importance of conservation. Given the poverty of most of the countries in which tropical forests are found, such money must ultimately come from the developed countries of the world. Aid from external sources is important not only in itself, but also because it often provides host countries with an incentive to meet such aid with matching funds.

At present, only a fraction of the money provided each year by institutions such as the World Bank or U.S. AID go to tropical forest conservation, although such institutions are beginning to recognize the interdependence of developing countries' economies and forest conservation. For example, between 1968 and 1977 the World Bank loaned only $13 million for tropical forestry. Since then, there has been a tenfold increase in funding, with half of all programs devoted to conservation (e.g., watershed management) rather than production (Myers 1984).

While the World Bank and AID still direct only a small proportion of their funds to forest conservation, their expenditures are huge compared to those of conservation organizations such as the World Wildlife Fund, which rely largely on private donations. Between 1978 and 1984, the WWF-U.S. Primate Program spent an average of $250,000 per year on more than 80 primate conservation projects in 25 countries. Although these expenditures represent a substantial increase over previous years, they do not begin to cover even the highest conservation priorities. Unless the developed countries acknowledge their responsibility and provide large-scale support for the conservation of wildlife and tropical forests in the developing world, we can expect forests not only to continue to disappear, but to do so at an ever-increasing rate.

Conservation efforts should not, however, be limited only to those countries in which tropical forests or nonhuman primates are found. Primate conservation can also be furthered by (6) establishing conservation-oriented captive breeding programs for endangered species; (7) ending all illegal and otherwise destructive traffic in primates; (8) ensuring that research institutions that use primates are aware of the conservation status of the species they use, that they use primates as prudently as possible, and that they make every effort to breed in captivity the primates that they need.

The first of these points recognizes that some highly endangered species may not survive in the wild due to the destruction of their remaining habitat. Although breeding programs offer the only hope for these species, and always offer the possibility of eventual reintroduction, they should not replace or even be regarded as equivalent to efforts to conserve wild populations.

The seventh and eighth points emphasize that the trade in wild primates can only be curtailed if the developed countries refuse to trade in them. In the countries where it occurs, the economic incentives for the capture of primates are often high, and as long as individuals, zoos, and medical institutions are willing to trade in wild primates, primates will continue to be removed from the wild. As already noted, conventions such as CITES have done much to reduce the international primate trade, but they must continue to be enforced if they are to be effective. Finally, it seems not unreasonable to require research institutions that benefit from the use of primates to be aware of the status of their subjects and, whenever possible, to restrict their use of primates to individuals that have been bred in captivity. Such measures not only aid primate conservation efforts, but also improve the quality of scientific research, since background data on experimental subjects are then available. Research institutions should also consider substantially increasing their financial support of conservation efforts because they benefit from the maintenance of wild populations.

Since 1977, the IUCN Primate Specialist Group has attempted to coordinate primate conservation activities on a global basis, both to determine the highest international priorities and to ensure that the limited funds available for primate conservation are put to the best possible use. The IUCN Group has tried to bring the problems of primate conservation to the attention of governments, funding organizations, and the general public in the hope of increasing the resources available for essential programs. As a result of these efforts, it has been possible to set up a communication network that links researchers interested in carrying out conservation-oriented projects with both local conservation organizations that need technical assistance and sources of funding that can support such work. These efforts have also led to a substantial increase in interest and funding for primate conservation activities and permitted the establishment of a special Primate Program by World Wildlife Fund–United States in 1979. Many other private organizations have also played an increasingly important role in primate conservation efforts. The most significant of these has been the New York Zoological Society, which continues to fund many primate conservation projects. They also include the Wildlife Preservation Trust International, the Fauna and Flora Preservation Society, the National Geographic Society, the African Wildlife Foundation, the L. S. B. Leakey Foundation, and numerous other groups.

The bulk of this volume has been devoted to the investigation of primate ecology and behavior and has been concerned with issues that may seem esoteric or trivial compared with the urgency of conservation efforts. In closing, we would like to emphasize that although scientific research derives obvious benefits from primate conservation efforts, it also makes important contributions to conservation. Field research is of importance both in establishing where and at what densities primates occur and in determining the size of reserves required to maintain viable breeding populations. Behavioral and ecological field research also provides a scientific presence that in itself offers a degree of protection and prestige to a given protected area. In some cases, it has even provided an incentive for the establishment of national parks or reserves where none existed previously. For example, Mahale Mountain National Park in Tanzania was established as a direct result of behavioral research on chimpanzees, and there are similar examples from other countries. Perhaps more important, such research has piqued an interest about the behavior of our closest living relatives that provides a powerful incentive for conservation.

Although the outlook for primates and their habitats is gloomy, it must not be regarded as hopeless. The global community of primate researchers can make a major contribution by always incorporating a conservation component into scientific research programs and by helping to increase public awareness of both the plight and importance of these animals. In so doing, we can increase the hope that we will enter the next century with a full complement of the primate species with which we now share our planet.

40 | Future of Primate Research

Dorothy L. Cheney, Robert M. Seyfarth, Barbara B. Smuts, and Richard W. Wrangham

DeVore's *Primate Behavior* (1965b) was one of the first collections of research on wild primates. In the concluding chapter, Washburn and Hamburg (1965) emphasized the importance of field studies for understanding behavior as a product of natural selection and stressed the potential significance of primate research for understanding aspects of human behavior, especially group-living tendencies. At the same time, they noted that "investigations of the quantity and quality [needed] to build reliable theories of primate behavior have yet to be made" (p. 608) and looked forward to a time when this would no longer be true. How far have we come, and where will we go in the next few decades?

DESCRIPTIVE BASE

We have seen a dramatic increase in our knowledge of wild primates over the last 20 years. Two decades ago, data on group composition, ranging, feeding, and social organization were available for less than a dozen species, compared to more than a hundred now. This larger data base has allowed us to make systematic interspecific comparisons, and these in turn make it possible to generate testable theories, such as those that explain the distribution of infanticide or patterns of intergroup competition. Similarly, a number of long-term field studies are now an order of magnitude longer than those of the 1960s. They have shown that demographic accidents such as a temporarily skewed sex ratio or rare events such as a single aggressive intergroup interaction can have long-lasting and biologically significant effects on behavior. Long-term field studies have also allowed observers to document both general tendencies and occasional rare events, such as high rates of male migration and occasional transfer of females among Japanese macaques. Perhaps most significantly, they have revealed the remarkable importance of long-term social relationships for patterning social interactions, such as those involving grandmothers and grandoffspring or long-established "friendships."

These achievements are a source of satisfaction, but every chapter in this book reminds us of frustrating gaps in our knowledge. Some loom especially large; in some cases we do not know even a species' grouping pattern. In the appendix, 21 species are listed as having unknown grouping patterns, and others are the objects of thin

guesses. Unfortunately, some species, such as the aye-aye or the hairy-eared dwarf lemur, may never be studied in the wild because they are close to extinction. Others, such as the Brazilian bare-faced tamarin, the red-handed howler, the uakaris, Allen's swamp monkey, the owl-faced monkey, and the douc langur, should eventually attract resolute observers prepared to risk time on a path-breaking venture. Recent investigations of little-studied species have revealed unexpected patterns of social organization, including polyandry in supposedly monogamous tamarins, monogamy in supposedly polygynous cercopithecines, and the possibility of multilevel social organization in several lemuriforms and bonded pairs within saki groups. Thus, opportunities for new, theoretically significant discoveries abound.

Of equal concern for those seeking an understanding of primate societies is the comparatively shallow descriptive base for many species, even where the outline of their social organization is known. Many of our generalizations about group-living primates still are derived from observations of a small number of well-studied species, especially rhesus and Japanese macaques, savanna baboons, vervets, gray langurs, chimpanzees, and mountain gorillas. These species are far from representative of the primate order as a whole. They are well-studied because they share an unusual set of characteristics that make them easy to observe, including relatively large size, terrestrial locomotion, and the use of fairly open habitats. Many other species are less well known because they are smaller or mainly arboreal animals living in heavily forested areas. Such species are almost certain to broaden our concepts of primate behavior. A striking example comes from recent studies of howler monkeys, whose complex patterns of intergroup migration show that primates cannot be divided simply into male-transfer or female-transfer species.

Finally, Washburn and Hamburg (1965) noted the need to investigate species adaptations in several different localities. Progress has clearly been made in this respect. Too often, however, conclusions must be tempered with caution because of the restricted number of study sites. Our detailed knowledge of chimpanzee society, for example, comes largely from the shores of Lake Tanganyika, in the extreme southeast of the species' distribution. The behavior of wild gorillas is known almost wholly from the Virunga Volcanoes, since the first group of lowland gorillas has yet to be habituated. Even for species that have been studied at a number of different sites, such as baboons and vervets, the difficulties of making comparisons between study sites have been so great that explanations for intraspecific variation have rarely emerged. It is still not clear, for example, what factors influence variation in group size in these species. Nevertheless, replicate studies have been invaluable in showing which aspects of behavior tend to be uniform within species (such as patterns of intergroup dispersal) and which tend to vary (such as the presence or absence of territorial defense).

EXPLAINING BEHAVIORAL DIVERSITY

Considering the growth in the data base, it is remarkable how many striking problems in primate social evolution remain unanswered. Why do vervet, gray langur, and squirrel monkey females readily allow other females to handle their infants, whereas baboon, macaque, and chimpanzee females are usually more restrictive? Why do baboon males use coalitions to compete for access to estrous females, while macaque and vervet males do so only rarely? Why do Mentawai langurs and some populations of simakobus and de Brazza's monkeys live in monogamous groups, while nearly all other members of their respective genera live in multifemale groups? Why do female baboons, macaques, and vervets develop clearcut dominance relationships, while female blue and redtail monkeys apparently do not? What accounts for species differences in the presence or absence of female sexual swellings, color signals of pregnancy, or malefemale "friendships"? The answers to many questions such as these are little more than speculation. Even such basic problems as the distribution of multifemale and multimale groups have not been clearly solved. There is no room for complacency in the modern study of primate behavior, whatever the advances in our knowledge.

We draw attention to these analytical gaps not from despair or cynicism but in order to stress the need for new methods to describe and explain behavior. It could once be said that primatology suffered by comparison with studies of other animals because it had not incorporated the principles of ecology and evolutionary biology. Doubtless primatologists still have much to learn from studies of other species, but theoretical concepts no longer act as barriers between primatology and other areas of behavioral biology. Equivalent gaps in understanding social behavior loom fully as large in the study of other organisms. Research on many primate species will always be plagued by ethical and practical difficulties that interfere, for example, with conducting field experiments, obtaining sufficiently large samples, and monitoring complex environments. This may mean that in some respects primate studies will continue to be more often collections of case studies than those of more easily studied birds and mammals where larger samples can be readily obtained. Nevertheless a comparison of this book and DeVore's (1965b) *Primate Behavior* shows clearly that rapid advances are possible; we fully expect that in 20 years we will have firm answers to many of the problems that puzzle us today. In the remainder of this chapter we speculate freely about where and how a few of these advances might come.

RELATION BETWEEN ECOLOGY AND SOCIAL BEHAVIOR

In the past, studies of primate social behavior were often conducted without regard to what animals ate and how they obtained it. In part this reflected the fact that the most detailed data on social behavior came from provisioned populations, such as those in Japan and on Cayo Santiago. Conversely, ecological studies typically led to broad, cross-species comparisons that categorized species into classes with little concern for intraspecific variation and the ways in which ecological factors and intragroup social behavior affect one another. Typical questions in socioecology were broad and crude (e.g., Why do grouping patterns differ?), and the answers were plausible but untested assertions that failed to acknowledge numerous complexities arising from the nature of selective processes and the intricacy of the system. For example, predator pressure was said to favor large groups, without consideration of the different interests of the two sexes. These broadly phrased questions have now been expanded and focused to include new issues such as dispersal patterns and female social relationships. Hypotheses are more sophisticated than before because they incorporate progress in evolutionary theory, and they are more credible because attempts to test them are more frequent and more thorough. But all too often the answers are not firm because they fail fully to explain exceptions to general tendencies and patterns of variation.

The chapters in part 2 of this book give many examples of these problems and also indicate directions for finding solutions. They emphasize the difficulty of relating behavior to ecology when the tools of ecological analysis are imprecise and involve poorly defined concepts such as predator pressure, food quality, patch size and distribution, and defensibility. Efforts must be made to find measures of such variables that are not only comparable across studies but that are also relevant to social behavior. Predator pressure, for example, must be measured in terms of the effects of predators on individual behavior, rather than simply estimated as high or low in relation to some arbitrary standard. Similarly, food quality needs to be measured not in terms of a nutritionist's recommendations but in relation to animal feeding strategies and effects on reproduction and survival. Recent efforts to estimate patch size (e.g., Terborgh

1983) need to be pursued broadly in combination with attempts to measure individual rates of food intake in different circumstances, such as in small or large groups. None of these problems is easily solved, but the last 20 years have clearly shown that interspecific comparisons provide valuable insights when the right variables are used. As primate behavioral ecology focuses increasingly on food distribution and predator pressure as the major ultimate sources of species differences in social behavior, it becomes correspondingly important to identify and quantify the relevant variables and their effects.

While such advances may make it possible to provide ecological explanations for the general outlines of a given species' social system, we do not know whether ecological variables can explain the intricacies of intragroup behavior. For example, in both mountain gorillas and Burchell's zebras (*Equus burchelli*), females disperse from their natal groups to live with a single dominant male and a number of unrelated females (chap. 14; Klingel 1974). Since both species also feed on evenly distributed, widely abundant food, it is tempting to hypothesize that each species' mating system has evolved in response to similar selective pressures exerted by similar ecological factors. Behavior within gorilla groups, however, is characterized by a complexity of social interactions that seems unmatched by zebras. Why do these differences exist? Cognitive abilities are likely to be important here, and they may also help explain differences in social behavior among primates. More refined ecological analyses will therefore help generate new hypotheses, some of which may diminish the importance of the environment as the dominant influence on social evolution.

DEVELOPMENT AND FUNCTION OF SOCIAL RELATIONSHIPS

In the 1950s, psychologists began to use nonhuman primates as experimental models for the study of human emotional disturbances (e.g., Harlow 1958). This research revealed striking similarities between the social bonds of human and nonhuman primates, including the profound emotional trauma induced by separation from a loved one and the importance of contact comfort for the development of bonding. It has thus been clear for many years that nonhuman primates and humans share capacities for the development and expression of social relationships. However, with few exceptions, primatologists have only recently begun to apply systematic, quantitative methods to the detailed study of primate social relationships in the wild. The results thus far indicate that this is likely to become a particularly exciting area of research, producing information that will capture the attention of a wide variety of social scientists.

The future of research in this area depends on the development of new and ingenious methods for describing relationships. In the past, primatologists have generally been content to describe social relationships in terms of fairly gross distinctions, such as differences in dominance status or degree of relatedness. While these distinctions are useful for some purposes, they fail to reveal—and sometimes actually obscure—the much-finer-grained distinction that the animals themselves make in their interactions with one another. For example, female baboons tend to retreat from the approaches of all "unrelated" (e.g., from different matrilines) higher-ranking females. However, individuals tend consistently to avoid the approaches of some females even when they are far away, while others are avoided only after the approach is completed (Smuts, in prep.). There are comparable differences in the way females respond to other females' attempts to touch their infants or to groom them. When combined with other detailed measures of behavior, these observations indicate that each female develops two or three distinctive types of relationships with "unrelated" females. Dominance rank is only one of several as-yet-undetermined factors influencing the form that these relationships take.

Thus, future studies of primate social relationships must rely even more on meticulous records of interactions at a fine-grained level. Although well-established methods can often serve as a useful starting point for such investigations, it is becoming increasingly clear that we may sometimes need to create our own methods, uniquely tailored to the species we are studying and the particular questions we are asking. In deciding what to look for and how to organize the observations we record, the animals themselves may often be our most valuable source of inspiration and insight, for they know better than we the significance of behaviors that initially go unnoticed by human observers. We can take advantage of the animals' knowledge in several ways, including the use of open-minded, unstructured observations to increase our sensitivity to behavioral nuances and the use of carefully planned playbacks of vocalizations to "crack the code" of primate communication. It is through a combination of such diverse methods that we will continue to discover complex and intriguing features of relationships, such as reconciliation, redirected aggression toward relatives and friends of the initial target, and the exchange of grooming for agonistic support.

Studies of primate social relationships are likely to generate important new advances in understanding the evolution of social behavior. Field research has already begun to force a revision and refinement of evolutionary theories that were too simple to predict the complex and flexible behavior of wild primates. Consider, for example, male-infant relationships. Sexual selection theory predicted that in polygynous species, male investment in young would be minimal because paternity certainty would normally be low. However, several recent studies of male-infant relationships in baboons and macaques show that males often develop long-term, affiliative relationships with infants who, in some cases, are unlikely to be their own offspring (chap. 28). Usually, the males have or are in the process of developing special relationships with the infants' mothers. In many cases, males seem to provide females and their infants with benefits, such as protection from conspecifics, in exchange for benefits that the females provide them, such as enhanced mating opportunities in the future or agonistic support (Smuts 1985).

Knowledge of evolutionary theory alone neither predicts nor explains these relationships. In order to make sense of them, we also need to understand the immediate needs and motivations of female baboons and macaques that, in turn, affect the reproductive tactics available to males. Future studies may show that, in many other cases, evolutionary theories cannot predict the form that social relationships take because the theories have not yet taken into account the important constraints and opportunities created by proximate factors, such as the effects of some social relationships on others, variations in cognitive skills, and even individual differences in emotional makeup due to early social experiences. If the study of primate social relationships remains firmly committed to the analysis of both ultimate and proximate causation, we may help foster the integration between these two levels of analysis that is needed to create the next "revolution" in behavioral biology.

Finally, the study of primate societies offers the possibility of extraordinary insights into the biological processes underlying human social relationships. We say "extraordinary" to convey our surprise at the apparent emotional intensity of primate social bonds and the complex strategies primates use to develop and change their relationships. The discovery of long-term, affiliative bonds, not just among mothers and offspring but also among siblings, mates, and friends, opens up exciting new areas for the study of attachment and other important aspects of long-term relationships, including especially the mechanisms by which reciprocity is initiated and maintained between individuals. Similarly, research on patterns of competition, alliances, and reconciliation in nonhuman primates may eventually provide insights into the nature of human power and conflict. The possibility of such important interactions with the social sciences again emphasizes the need for firm, testable theories based on clear and comparable measures.

EVOLUTION, COMPLEX SOCIAL BEHAVIOR, AND COGNITIVE SCIENCE

At present, one of the most exciting areas of research in the behavioral sciences lies at the interface between linguistics, computer science, cognitive psychology, and philosophy. The ultimate object of interest is the human brain. How does the brain represent and process information, particularly words and rules of grammar? Can we find a nonhuman species, build a machine, or write a computer program that duplicates the human brain's performance? What would it mean if we could?

The growth of interest and interdisciplinary research in the field now broadly known as "cognitive science" has made it essential for linguists to learn computer programming, computer scientists to learn philosophy, and philosophers to inform themselves about research in cognitive psychology. At the same time, some studies in cognitive psychology that have looked at animal subjects have typically studied them under controlled, captive conditions; there has been little recognition of the theoretical importance of evolutionary theory or natural social behavior.

We believe, however, that there are potentially important links to be forged between evolutionary theory, cognitive science, and the study of animals, particularly nonhuman primates, under natural conditions. The ultimate goal is a fusion of these disciplines that reveals not only how primates think but also why they think as they do. Consider, for example, some of the questions that Hamilton's (1964) theory of kin selection raises about the nature of kin recognition—the mechanism that underlies cooperative behavior among genetic relatives.

Although kin recognition is used as a general term to describe preferential association among related individuals, it is in fact possible to distinguish a number of different types of kin recognition, each with a different level of complexity and functional significance. Many animals seem capable of distinguishing between kin and nonkin and, within the kin class, of distinguishing among individuals of different degrees of relatedness (reviewed by Holmes and Sherman 1983). Discrimination may also extend beyond one's own relationships to include those of other individuals. For example, animals may recognize that certain other individuals associate at high rates. This kind of "nonegocentric" kin recognition requires that animals attend to interactions in which they themselves are not involved. Few studies have documented this sort of kin recognition, although it is suggested by some data on nonhuman primates (Judge 1982; Cheney and Seyfarth 1986). Finally, at its most complex, individuals may not only discriminate among the associations of others, but also recognize that other individuals' relationships are similar to their own. This more abstract form of kin recognition demands that animals be able to understand, based on generalizations derived from specific interactions, that certain types of social bonds share the same characteristics independent of the individuals involved. Humans are clearly capable of this form of kin recognition, and there is some evidence that nonhuman primates may also be capable of such generalization, although in a more rudimentary form. For example, vervet monkeys are more likely to threaten a particular individual when that individual's close kin and their own close kin have recently been involved in a fight (Cheney and Seyfarth, in press; chap. 37). These data suggest that monkeys recognize that many individuals share the same types of relationships.

These observations emphasize the need for additional research on the nature of kin recognition. Such investigations are not just exercises in cataloging the species that show more complex behavior than others. Instead, they allow us to ask questions about why an animal might need to distinguish a more abstract class of kin and how it does so. Such an ability seems beneficial primarily because it allows individuals to make inferences about the behavior of others without having to experience directly or observe interactions with every possible combination of individuals (Dasser 1985). If it is shown, for example, that some nonhuman primate species are capable of this form of kin recognition while others are not, or that nonhuman primate capacities for kin recognition differ from those of other mammals, then we can begin to investigate the adaptive significance of different forms of kin recognition and to identify the proximate social and environmental factors that allow complex forms of kin recognition in some species but not in others. For example, Sherman and Holmes (1985) have indicated that differences in kin recognition abilities in ground squirrels are related to species differences in dispersal, nepotism, and mating patterns. Identifying such factors in primates is no less important than—and indeed complements—research in more traditional areas of behavioral ecology.

Vocal communication provides a second example of the potential interaction between evolutionary theory and cognitive science. With the exception of studies of alarm calls, most studies of vocal communication in nonhuman primates have been conducted without attention to the ways in which vocalizations might contribute to individual fitness. Similarly, studies concerned with the function of behavior have tended to ignore vocalizations and to concentrate on behaviors such as fighting and mating, whose effects on fitness are easier to measure. To the extent that vocalizations are considered at all, they are often simply assumed to reflect variation in the vocalizer's emotional state. Recent research, however, is beginning to reveal the deficiency of divorcing communicative studies from more functional considerations of behavior.

There is growing evidence that many, if not all, nonhuman primates are capable of a rudimentary form of semantic signaling in which subtle variation in acoustic signals can be used to communicate about objects or events in the external world (chap. 36). These findings are of interest not only because they suggest rudimentary parallels with human speech, but also because they force us to take a less simplistic approach to the functional sig-

nificance of vocalizations. How might individuals benefit by semantic signaling? More puzzling, if we can imagine situations in which there might be benefits to such forms of communication, then why are the representational calls of monkeys and apes (apparently) so few in number? Vervet monkeys, for example, seem to possess fewer than ten different grunts that signal different social or environmental events, and their other classes of vocalizations, though more elaborate than previously imagined, seem to be similarly restricted. Why might a species that had developed some ability to signal semantically be limited in expanding its vocal repertoire to include many thousands of "words"? If representational signaling is so obviously adaptive, what sets an upper limit to an organism's semantic skills, and why should such a limit exist? Finally, since the adaptive advantages of grammar are also obvious, why has syntax not yet been demonstrated in the natural vocalizations of any nonhuman primate? These are questions for which we currently lack answers, largely because the questions are seldom asked.

In sum, cognitive science today focuses on the nature of the human mind. One research strategy, pursued in a variety of forms, has been to investigate "almost minds," such as the minds of children or the "minds" of computer programs, to see what makes them different and what would be needed to elevate them to fully human status. As the chapters in this volume illustrate, nonhuman primates provide an extraordinary diversity of "almost minds" that, in their social interactions with one another, promise to provide unique insights into the study of intelligence. As subjects, nonhuman primates are not as easy to find as children, nor will they wait patiently in labs like computer programs. In their natural habitats, however, primates are uniquely poised to reveal how, in the first instance, some minds gained an advantage over others.

CONSERVATION AND PRIMATE RESEARCH

We have attempted to provide a brief and by no means exhaustive sketch of some of the promising areas of research in primate behavior. No doubt a few hours of armchair speculation would reveal many more. Therefore, it is especially frustrating to admit that many questions will not only remain unanswered but also unasked, simply because we are rapidly losing our ability to gather the necessary data. Human population growth and habitat destruction have now become irreversible in many areas of the world, and the next two decades will surely see the extinction of large numbers of plant and animal species. The ramifications of habitat destruction and species extinction are described in chapter 39. Here we would simply reiterate the urgency of research on the less well-studied primate species. Such studies are our last opportunity to learn about the behavior of species that may not survive this century, and they will also provide knowledge critical to conservation efforts. Behavioral and ecological studies provide information vital to establishing reserves and delineating their necessary size and boundaries. Equally important, they provide the opportunity for local people to learn about their indigenous animals and thus they can spur local interest in conservation. As indicated in chapter 39, such interest and support are necessary if reserves and national parks are to be established and maintained. For example, only a few hundred muriqui (*Brachyteles arachnoides*) survive in the Atlantic forest of eastern Brazil. In recent years, however, publicity stemming from scientific research has spurred local efforts to preserve this species, to the point that the muriqui has become a symbol of conservation efforts throughout Brazil (fig. 40-1).

Tragically, it may prove impossible to save some species from extinction, and thus it is tempting to ask if the effort is worth the trouble. However, the disappearance of even a few species serves as a potent reminder of the grave consequences of habitat destruction on the future environmental and economic health of our own species. Attempts to save them offer a glimmer of hope. However, for these attempts to become more effective, there must be increased efforts by all concerned with the future survival of primate species in particular and genetic diversity in general. The academician and theoretician are especially able to contribute to this effort by demonstrating the unique value of primates and by making these values known to policymakers and the general public. After all, conservation is not the sole responsibility of any single group or profession.

CONCLUSION

Washburn and Hamburg (1965) concluded that the most important single contribution of the field studies summarized in DeVore's *Primate Behavior* was the demonstration that it is possible to collect reliable behavioral data in the wild. Perhaps the most important result of the last 20 years has been the demonstration that it is possible to collect not just reliable data, but the kinds of data needed to test interesting hypotheses. We can now look forward to a steady growth in the standardization of methods and concepts, in the ability to replicate results, and in the power of explanations.

We hope that *Primate Societies* has stimulated your interest in the complexities of primate social behavior. We also hope that many of you will pursue your interest and contribute both to future empirical and theoretical insights and efforts to save endangered primates before it is too late.

FIGURE 40-1. Adult female muriqui (or woolly spider monkey) from the Atlantic forest in Brazil. (Photo: Andrew Young)

APPENDIX   The Order Primates: Species Names and a Guide to Social Organization

The following taxonomy is not meant to be definitive. It is presented as a guide to Latin and common names and to provide readers with an overview of the evolutionary relationships among different species. In some cases where relationships and taxonomic names have recently been changed or are in dispute, alternative forms are given. Subspecies are listed only in those cases where the behavior of different subspecies has been discussed in the text.

The summaries of grouping types and intergroup transfer patterns are also presented merely as a guide. They should not be used to test hypotheses without checking further because in many species there is a substantial amount of variation in social organization that could not be included for reasons of space. Furthermore, the standards of evidence vary between species, depending on how many studies have been conducted. The aim of these tables is to indicate significant gaps in knowledge and to give a broad overview of the distribution of types of social organization. Three summary tables are given in chapter 23.

TABLE A-I   Taxonomy and Social Organization of Living Primates.

		Grouping Pattern	Intergroup Transfer	Sources
	PROSIMIANS			
Prosimii				
Lemuriformes				
Daubentonioidea				
Daubentoniidae				
Daubentonia madagascariensis	Aye-aye	Sol		Tattersall 1982
Lemuroidea				
Cheirogaleidae				
Allocebus trichotis	Hairy-eared dwarf lemur	?		Tattersall 1982
Cheirogaleus major	Eastern dwarf lemur	Sol?		Tattersall 1982
Cheirogaleus medius	Western fat-tailed dwarf lemur	Sol?		Tattersall 1982
Microcebus murinus	Western gray mouse lemur	Sol		Chap. 3
Microcebus rufus	Eastern brown mouse lemur	?		Tattersall 1982
Mirza coquereli	Coquerel's mouse lemur	Sol		Tattersall 1982
Phaner furcifer	Forked lemur	Sol		Chap. 3
Indriidae				
Avahi laniger	Woolly lemur	Mon?		Chap. 3
Indri indri	Indri	Mon		Chap. 3
Propithecus diadema	Diademed sifaka	Mon?		Chap. 3
Propithecus verreauxi	White or Verreaux's sifaka	Sm-M, C?	M	Chap. 3
Lemuridae				
Lemur catta	Ring-tailed lemur	Lg-M	M	Chap. 3
Lemur coronatus	Crowned lemur	Sm-M?, Lg-M?	?	Tattersall 1982
Lemur fulvus	Brown lemur	Sm-M, C?	?	Tattersall 1982; Chap. 3
Lemur macaco	Black lemur	Sm-M, C?	?	Tattersall 1982; Chap. 3
Lemur mongoz	Mongoose lemur	Mon, C?		Chap. 3
Lemur rubriventer	Red-bellied lemur	Mon?		Chap. 3
Varecia variegata	Ruffed or variegated lemur	Mon?		Chap. 3
Lepilemuridae				
Hapalemur griseus	Gray gentle lemur	Mon?		Chap. 3
Hapalemur simus	Broad-nosed gentle lemur	?		Tattersall 1982
Lepilemur mustelinus	Sportive lemur	Sol		Chap. 3
Lorisiformes				
Lorisoidea				
Lorisidae				
Galaginae				
Galago alleni	Allen's bushbaby	Sol		Chap. 2
Galago crassicaudatus	Thick-tailed bushbaby	Sol		Chap. 2
Galago demidovii	Demidoff's or dwarf bushbaby	Sol		Chap. 2
Galago (Euoticus) elegantulus	Needle-clawed bushbaby	Sol		Chap. 2
Galago garnettii	Greater bushbaby	Sol		Chap. 2
Galago inustus	Lesser needle-clawed bushbaby	Sol?		Chap. 2
Galago senegalensis	Lesser bushbaby	Sol		Chap. 2
Galago zanzibaricus	Zanzibar bushbaby	Sol		Chap. 2
Lorisinae				
Arctocebus calabarensis	Angwantibo	Sol		Chap. 2
Loris tardigradus	Slender loris	Sol?		Chap. 2
Nycticebus coucang	Common slow loris	Sol?		Chap. 2
Nycticebus pygmaeus	Pygmy slow loris	Sol?		Chap. 2
Perodicticus potto	Potto	Sol		Chap. 2
Tarsiiformes				
Tarsioidea				
Tarsiidae				
Tarsiinae				
Tarsius bancanus	Borneo tarsier	Sol		Chap. 2
Tarsius spectrum	Spectral tarsier	Mon		Chap. 2
Tarsius syrichta	Philippine tarsier	?		Chap. 2

TABLE A-1   *(continued)*

		Grouping Pattern	Intergroup Transfer	Sources
	NEW WORLD MONKEYS			
Platyrrhini				
Ceboidea				
Callimiconidae				
Callimico goeldii	Goeldi's marmoset	Sm-1?	?	Chap. 4
Callitrichidae				
Callithrix argentata	Silvery bare-ear marmoset	?		Chap. 4
Callithrix humeralifer	Tassel-ear marmoset	Mon/Pol?		Chap. 4
Callithrix jacchus	Common or tufted-ear marmoset	?		Chap. 4
Cebuella pygmaea	Pygmy marmoset	Mon/Pol?		Chap. 4
Leontopithecus rosalia	Lion tamarin	Mon/Pol?		Chap. 4
Saguinus bicolor	Brazilian barefaced tamarin	?		Chap. 4
Saguinus fuscicollis	Saddleback tamarin	Mon/Pol, (Sm-V)		Chap. 4
Saguinus imperator	Emperor tamarin	Mon/Pol?		Chap. 4
Saguinus inustus	Mottled-face tamarin	?		Chap. 4
Saguinus labiatus	Red-chested or white-lipped tamarin	Mon/Pol?		Chap. 4
Saguinus leucopus	White-footed tamarin	?		Chap. 4
Saguinus midas	Red-handed tamarin	?		Chap. 4
Saguinus mystax	Moustached tamarin	Mon/Pol?		Chap. 4
Saguinus nigricollis	Black-and-red tamarin	?		Chap. 4
Saguinus oedipus	Cottontop tamarin	Mon/Pol?		Chap. 4
Cebidae				
Alouattinae				
Alouatta belzebul	Red-handed howler	?		Chap. 6
Alouatta caraya	Black howler	Sm-V?	?	Chap. 6
Alouatta fusca (guariba)	Brown howler	?		Chap. 6
Alouatta palliata (villosa)	Mantled howler	Lg-M	B	Chap. 6
Alouatta pigra (palliata, villosa pigra)	Guatemalan or black howler	Sm-V, (Mon)	?	Chap. 6
Alouatta seniculus	Red howler	Sm-V	B	Chap. 6
Atelinae				
Ateles belzebuth	Long-haired spider monkey	C-i	F?	Chap. 7
Ateles fusciceps	Brown-headed spider monkey	C-i	F?	Chap. 7
Ateles geoffroyi	Black-handed spider monkey	C-i	F?	Chap. 7
Ateles paniscus	Black spider monkey	C-i	F?	Chap. 7
Brachyteles arachnoides	Muriqui or woolly spider monkey	C-i, (Lg-M)	F?	Chap. 7
Lagothrix flavicauda	Yellow-tailed woolly monkey	Sm-M?	?	Chap. 7
Lagothrix lagothricha	Humboldt's, or common woolly monkey	Lg-M	?	Chap. 7
Cebinae				
Cebus albifrons	White-fronted capuchin	Sm-M	M	Chap. 7
Cebus apella	Brown or tufted capuchin	Sm-M	M?	Chap. 7
Cebus capucinus	White-throated capuchin	Sm-1	?	Chap. 7
Cebus olivaceus (nigrivittatus)	Wedge-capped capuchin	Lg-V	M?	Chap. 7
Saimiri oerstedii	Red-backed squirrel monkey	Lg-M?	?	Chap. 7
Saimiri sciureus	Common squirrel monkey	Lg-M	M?	Chap. 7
Pitheciinae				
Aotus trivirgatus	Night or owl monkey	Mon		Chap. 5
Cacajao calvus	White or red uakari	Lg?	?	Chap. 5
Cacajao melanocephalus	Black uakari	Lg?	?	Chap. 5
Callicebus moloch	Dusky titi	Mon		Chap. 5
Callicebus personatus	Masked titi	Mon		Chap. 5
Callicebus torquatus	Yellow-handed titi	Mon		Chap. 5
Chiropotes albinasus	White-nosed saki	Lg?, Mon?	?	Chap. 5
Chiropotes satanas	Black saki	Lg?	?	Chap. 5
Pithecia albicans	Buffy saki	Mon		Chap. 5
Pithecia hirsuta	Black-bearded saki	Mon		Chap. 5
Pithecia monachus	Red-bearded saki	Mon		Chap. 5
Pithecia pithecia	White-faced or Guianan saki	Mon		Chap. 5

TABLE A-1   *(continued)*

		Grouping Pattern	Intergroup Transfer	Sources
	OLD WORLD MONKEYS			
Catarrhini				
Cercopithecoidea				
Cercopithecidae				
Cercopithecinae				
Allenopithecus (Cercopithecus) nigroviridis	Allen's swamp monkey	Mon?		T. Butynski, pers. comm.
Cercocebus albigena	Gray-cheeked mangabey	Lg-M	M?	Chap. 11
Cercocebus aterrimus	Black mangabey	?		Wolfheim 1983
Cercocebus atys	Sooty mangabey	Sm?, (Lg?)	M?	Chap. 11
Cercocebus galeritus	Agile mangabey	Lg-M	?	Homewood 1978
Cercocebus torquatus	White-collared or cherry-crowned mangabey	Lg?	?	Wolfheim 1983
Cercopithecus aethiops	Vervet monkey	Lg-M, (Lg-V)	M	Chap. 11
Cercopithecus ascanius	Redtail monkey	Lg-1	M	Chap. 9
Cercopithecus campbelli	Campbell's guenon	Sm-1, (Lg-1)	M	Chap. 9; R. Kluberdanz, pers. comm.
Cercopithecus cephus	Moustached guenon	Sm-1	M?	Chap. 9; T. Butynski, pers. comm.
Cercopithecus diana	Diana monkey	Lg-1	M?	Chap. 9; J. Oates, pers. comm.
Cercopithecus erythrogaster	Red-bellied monkey	?		Wolfheim 1983
Cercopithecus erythrotis	Red-eared nose-spotted monkey	Lg?	?	Wolfheim 1983
Cercopithecus hamlyni	Owl-faced or Hamlyn's monkey	Mon?, (Sm?)		J. Hart, pers. comm.; T. Butynski, pers. comm.
Cercopithecus l'hoesti	L'Hoest's monkey	Lg-1, (Sm-1)	?	Chap. 9; T. Butynski, pers. comm.
Cercopithecus mitis	Blue or Sykes's monkey	Lg-1	M	Chap. 9
Cercopithecus mona	Mona monkey	Sm?	?	T. Butynski, pers. comm.
Cercopithecus neglectus	De Brazza's monkey	Mon, (Sm?)		Chap. 9
Cercopithecus nictitans	Spot-nosed guenon	Lg-1	M?	Chap. 9
Cercopithecus petaurista	Lesser spot-nosed monkey	Lg?	?	Chap. 9; T. Butynski, pers. comm.
Cercopithecus pogonias	Crowned guenon	Lg-1	M?	Chap. 9; T. Butynski, pers. comm.
Cercopithecus salongo	Salongo monkey	?		Chap. 9
Cercopithecus (Miopithecus) talapoin	Talapoin	Lg-M	M?	Chap. 11
Erythrocebus patas	Patas monkey	Lg-1	M	Chap. 9
Macaca arctoides	Stump-tailed macaque	Lg-M	M?	Chap. 11; Wolfheim 1983
Macaca assamensis	Assamese macaque	Lg-M?	?	Wolfheim 1983
Macaca cyclopis	Formosan or Taiwan macaque	?	?	Wolfheim 1983
Macaca fascicularis	Long-tailed, cynomolgus, or crab-eating macaque	Lg-M	M	Chap. 11
Macaca fuscata	Japanese macaque	Lg-M	M	Chap. 11
Macaca mulatta	Rhesus macaque	Lg-M	M	Chap. 11
Macaca nemestrina	Pigtailed macaque	Lg-V	M?	Chap. 11
Macaca (Cynopithecus) nigra	Celebes macaque	Lg?, (Sm?)	?	Wolfheim 1983
Macaca radiata	Bonnet macaque	Lg-M	M?	Chaps. 11, 26
Macaca silenus	Lion-tailed macaque	Lg-M	M?	Chap. 11
Macaca sinica	Toque macaque	Lg-M	M	Chap. 11
Macaca sylvanus	Barbary macaque	Lg-M	M	Chap. 11
Papio cynocephalus anubis	Olive or anubis baboon[a]	Lg-M	M	Chap. 11

TABLE A-1   *(continued)*

		Grouping Pattern	Intergroup Transfer	Sources
OLD WORLD MONKEYS				
Papio cynocephalus cynocephalus	Yellow baboon[a]	Lg-M	M	Chap. 11
Papio cynocephalus papio	Guinea baboon	Lg-M	M	Chap. 11
Papio cynocephalus ursinus	Chacma baboon[a]	Lg-M	M	Chap. 11
Papio hamadryas	Hamadryas baboon	Sm-1, C-g	F	Chap. 10
Papio leucophaeus	Drill	Lg-1, C-g?	?	Chap. 10
Papio sphinx	Mandrill	Lg-1, C-g?	M?	Chap. 10
Theropithecus gelada	Gelada baboon	Sm-1, C-g	M	Chap. 10
Colobinae				
Colobus angolensis	Angolan black-and-white colobus	Sm?, Lg?	?	Chap. 8; Wolfheim 1983
Colobus (Piliocolobus) badius	Red colobus	Lg-V, (C-i)	F	Chap. 8
Colobus guereza	Guereza, or black-and-white colobus	Sm-1	M	Chap. 8
Colobus polykomos	Western black-and-white colobus	Sm-V?, (Lg-V?)	?	Wolfheim 1983; J. Oates, pers. comm.
Colobus satanas	Black colobus	Sm-V	M?	Chap. 8; Wolfheim 1983
Colobus vellerosus	Geoffroy's black-and-white colobus	Sm-V, Lg-V	?	D. Olson, pers. comm.
Colobus (Procolobus) verus	Olive colobus	Sm-V	?	J. Oates, pers. comm.
Nasalis (Simias) concolor	Simakobu, or pigtailed langur	Sm-1, (Mon)	?	Chap. 8
Nasalis larvatus	Proboscis monkey	Sm-1?, (Lg-M?)	?	Chap. 8
Presbytis aygula	Sunda Island leaf monkey	Sm-1, (Mon)	?	Chap. 8
Presbytis cristata	Silver leaf monkey	Sm-1	M?	Chap. 8
Presbytis entellus	Gray or hanuman langur	Lg-V	M	Chap. 8
Presbytis francoisi	Francois's langur	?		Wolfheim 1983
Presbytis frontata	White-fronted leaf monkey	?		Wolfheim 1983
Presbytis geei	Golden leaf monkey	Sm?, Lg?	?	Wolfheim 1983
Presbytis johnii	Nilgiri langur	Sm-1	M?	Chap. 8
Presbytis melalophos	Banded leaf monkey	Sm-V	M?	Chap. 8
Presbytis obscura	Dusky leaf monkey	Sm-V	?	Chap. 8
Presbytis phayrei	Phayre's leaf monkey	Sm?, Lg?	?	Wolfheim 1983
Presbytis pileata	Capped leaf monkey	Sm-1?	?	Chap. 8
Presbytis potenziani	Mentawai langur	Mon		Chap. 8
Presbytis rubicunda	Maroon leaf monkey	Sm?	?	Wolfheim 1983
Presbytis senex	Purple-faced langur	Sm-1	M	Chap. 8
Presbytis thomasi	Thomas's langur	Sm-1	?	Chap. 8
Pygathrix nemaeus	Douc langur	?		Wolfheim 1983
Rhinopithecus avunculus	Tonkin snub-nosed langur	?		Wolfheim 1983
Rhinopithecus roxellanae	Golden snub-nosed langur	Lg-1, C-g?	?	Chap. 8
APES AND HUMANS				
Hominoidea				
Hylobatidae				
Hylobates agilis	Agile gibbon	Mon		Chap. 12
Hylobates concolor	Black- or white-cheeked gibbon	Mon		Chap. 12
Hylobates hoolock	Hoolock gibbon	Mon		Chap. 12
Hylobates klossii	Kloss's gibbon	Mon		Chap. 12
Hylobates lar	Whitehanded or lar gibbon	Mon		Chap. 12

TABLE A-1 *(continued)*

		Grouping Pattern	Intergroup Transfer	Sources
APES AND HUMANS				
Hylobates moloch	Silvery or moloch gibbon	Mon		Chap. 12
Hylobates muelleri	Mueller's gibbon	Mon		Chap. 12
Hylobates pileatus	Pileated gibbon	Mon		Chap. 12
Hylobates (Symphalangus) syndactylus	Siamang	Mon		Chap. 12
Pongidae				
Gorilla gorilla beringei	Mountain gorilla	Sm-V	F	Chap. 14
Gorilla gorilla gorilla	Western lowland gorilla	Sm-V	?	Chap. 14
Gorilla gorilla graueri	Eastern lowland gorilla	Sm-V	?	Chap. 14
Pan paniscus	Bonobo or pygmy chimpanzee	C-i	F?	Chap. 15
Pan troglodytes	Chimpanzee	C-i	F	Chap. 15
Pongo pygmaeus	Orangutan	Sol		Chap. 13
Hominidae				
Homo sapiens	Human	C-i	F, B	Fox 1967

NOTES:

Grouping Pattern

C-i = community—closed social network (containing several breeding females and males) within which individuals forage partly independently.

C-g = community—closed social network within which breeding groups forage partly independently.

Lg = large groups (averaging 6 or more adult females, foraging primarily as a single unit); number of males unknown.

Lg-1 = large groups, with one resident adult male.

Lg-V = large groups, having variable number of resident adult males, from one to many.

Lg-M = large groups, having more than one resident adult male.

Mon = monogamous pairs, traveling as a family group.

Mon/Pol = monogamous pairs and polyandrous groups occur in the same population.

Sm = small groups (averaging 2–5 adult females, foraging primarily as a single unit); number of males unknown.

Sm-1 = small groups, with one resident adult male.

Sm-V = small groups, having variable number of resident adult males, from one to many.

Sm-M = small groups, having more than one resident adult male.

Sol = individuals forage solitarily.

Where two grouping patterns are given without parentheses, they are found together in the same population. Parentheses indicate alternative grouping patterns found in some populations only.

Intergroup Transfer

M = primarily males transfer between breeding units.

F = primarily females transfer between breeding units.

B = both sexes show a strong tendency to transfer between breeding units.

Data are given only for groups (or communities) having more than one breeding female. The primary source is chap. 21.

Taxonomy

Subfamily names (ending in *-inae*) are given only where a family (ending in *-idae*) contains more than one subfamily.

a. These species are also known as savanna baboons.

Contributors

SIMON K. BEARDER
Department of Social Studies
Oxford Polytechnic,
Headington, Oxford OX3 OBP
England

DOROTHY L. CHENEY
Department of Anthropology
University of Pennsylvania
Philadelphia, Pennsylvania 19104

T. H. CLUTTON-BROCK
Large Animal Research Group
34A Storey's Way
Cambridge CB3 ODT
England

MARINA CORDS
Department of Zoology
University of California
Berkeley
California 94720

CAROLYN M. CROCKETT
National Zoological Park
Smithsonian Institution
Washington, D.C. 20008
and
Departments of Anthropology and Psychology
University of Washington
Seattle, Washington 98195

R. I. M. DUNBAR
Department of Zoology
University of Liverpool
P.O. Box 147
Liverpool L69 3BX
England

JOHN F. EISENBERG
Department of Natural Sciences
Florida State Museum
University of Florida
Gainesville, Florida 32611

SUSAN ESSOCK-VITALE
Department of Psychiatry and Biobehavioral Sciences
University of California
Los Angeles, California 90024

ANNE WILSON GOLDIZEN
Department of Biology
University of Michigan
Ann Arbor, Michigan 48109

HAROLD GOUZOULES
Department of Psychology
Emory University
Atlanta, Georgia 30322

SARAH GOUZOULES
Yerkes Regional Primate Research Center Field Station
2409 Collins Hill Road
Lawrenceville, Georgia 30245

ALEXANDER H. HARCOURT
Department of Applied Biology
Cambridge University
Cambridge CB2 3DX
England

PAUL H. HARVEY
Department of Zoology
University of Oxford
South Parks Road
Oxford OX1 3PS
England

ROBERT A. HINDE
MRC Unit on the Development and Integration of
    Behaviour
Cambridge University
Madingley, Cambridge CB3 8AA
England

MARIKO HIRAIWA-HASEGAWA
Department of Anthropology
University of Tokyo
Hongo, Bunkyo-ku
Tokyo 113
Japan

SARAH BLAFFER HRDY
Department of Anthropology
University of California
Davis, California 95616

CHARLES H. JANSON
   Department of Ecology and Evolution
   State University of New York
   Stony Brook, New York 11794

WARREN G. KINZEY
   Department of Anthropology
   City University of New York
   New York, New York 10031

DONNA ROBBINS LEIGHTON
   Center for Field Research
   Earthwatch
   P.O. Box 403
   Watertown, Massachusetts 02172

LYSA LELAND
   Kibale Forest Project
   P.O. Box 409
   Fort Portal
   Uganda

ROBERT D. MARTIN
   Department of Anthropology
   University College
   Gower Street
   London W1
   England

DON J. MELNICK
   Department of Anthropology
   Columbia University
   New York, New York 10027

JOHN C. MITANI
   Rockefeller University
   Field Research Center
   Tyrrel Road
   Millbrook, New York 12545

RUSSELL A. MITTERMEIER
   World Wildlife Fund
   1255 23rd Street, Northwest
   Washington, D.C. 20037
   and
   Department of Anatomical Sciences
   Health Sciences Center
   State University of New York
   Stony Brook, New York 11794

NANCY A. NICOLSON
   Human Biology
   University of Limburg
   P.O. Box 616
   6200 MD Maastricht
   The Netherlands

TOSHISADA NISHIDA
   Department of Anthropology
   University of Tokyo
   Hongo, Bunkyo-ku
   Tokyo 113
   Japan

JOHN F. OATES
   Department of Anthropology
   Hunter College
   City University of New York
   695 Park Avenue
   New York, New York 10021

CRAIG PACKER
   Department of Ecology and Behavioral Biology
   University of Minnesota
   318 Church Street Southeast
   Minneapolis, Minnesota 55455

MARY C. PEARL
   New York Zoological Society
   Bronx Park
   185th Street and Southern Boulevard
   Bronx, New York, 10460

ANNE E. PUSEY
   Department of Ecology and Behavioral Biology
   University of Minnesota
   318 Church Street Southeast
   Minneapolis, Minnesota 55455

ALISON F. RICHARD
   Department of Anthropology
   Yale University
   New Haven, Connecticut 06520

JOHN G. ROBINSON
   Department of Wildlife and Range Sciences
   University of Florida
   Gainesville, Florida 32611

PETER S. RODMAN
   Department of Anthropology and California Primate
      Research Center
   University of California
   Davis, California 95616

ROBERT M. SEYFARTH
   Department of Psychology
   University of Pennsylvania
   Philadelphia, Pennsylvania 19104

JOAN B. SILK
Department of Anthropology
Emory University
Atlanta, Georgia 30322
and
California Primate Research Center
University of California
Davis, California 95616

BARBARA B. SMUTS
Departments of Psychology and Anthropology
University of Michigan
580 Union Drive
Ann Arbor, Michigan 48109

EDUARD STAMMBACH
Institute of Zoology
Ethology and Wildlife Research
University of Zurich
Winterthurerstrasse 190
CH-8057 Zurich
Switzerland

KELLY J. STEWART
Sub-Department of Animal Behavior
Cambridge University
Madingley, Cambridge CB3 8AA
England

THOMAS T. STRUHSAKER
Kibale Forest Project
P.O. Box 409
Fort Portal
Uganda

FRANS B. M. DE WAAL
Wisconsin Regional Primate Research Center
University of Wisconsin
1223 Capitol Court
Madison, Wisconsin 53715-1299

JEFFREY R. WALTERS
Department of Zoology
North Carolina State University
Campus Box 7617
Raleigh, North Carolina 27695-7617

PETER M. WASER
Department of Biological Sciences
Purdue University
West Lafayette, Indiana 47097

PATRICIA L. WHITTEN
Department of Obstetrics and Gynecology
Yale University School of Medicine
P.O. Box 3333
New Haven, Connecticut 06510

RICHARD W. WRANGHAM
Department of Anthropology
University of Michigan
Ann Arbor, Michigan 48109

PATRICIA C. WRIGHT
Duke University Primate Center
Duke University
Durham, North Carolina 27706

ANNE C. ZELLER
Department of Anthropology
University of Waterloo
Waterloo, Ontario
Canada N2L 3G1

Bibliography

Abbott, D. M., and Hearn, J. P. 1978. Physical, hormonal, and behavioral aspects of sexual development in the marmoset monkey, *Callithrix jacchus*. *J. Reprod. Fertil.* 53:155–66.

Abegglen, J. J. 1984. *On socialization in hamadryas baboons.* Cranbury, N.J.: Associated University Presses.

Adams, D. B.; Gold, A. R.; and Burt, A. D. 1978. Rise in female initiated sexual activity at ovulation and its suppression by oral contraceptives. *N. Eng. J. Med.* 299:1146–48.

Adang, O. M. J. 1984. Teasing in young chimpanzees. *Behaviour* 88:98–122.

Aguirre, A. C. 1971. O mono *Brachyteles arachnoides*. Situaco atual da especie no Brasil. Rio de Janeiro: Academia Brasileira de Ciencias.

Albrecht, H., and Dunnett, S. C. 1971. *Chimpanzees in West Africa.* Munich: Piper Verlag.

Alcock, J. 1984. *Animal behavior.* 3d ed. Sunderland, Mass.: Sinauer Associates.

Aldrich-Blake, F. P. G. 1970. The ecology and behaviour of the blue monkey *Cercopithecus mitis stuhlmanni*. Ph.D. diss., University of Bristol.

Aldrich-Blake, F. P. G.; Bunn, T. K.; Dunbar, R. I. M.; and Headley, P. M. 1971. Observations on baboons, *Papio anubis*, in an arid region in Ethiopia. *Folia Primatol.* 15:1–35.

Alexander, B. K. 1970. Parental behavior of adult male Japanese monkeys. *Behaviour* 36:270–85.

Alexander, B. K., and Bowers, J. M. 1969. Social organization of a troop of Japanese monkeys in a two-acre enclosure. *Folia Primatol.* 10:230–42.

Alexander, R. D. 1974. The evolution of social behavior. *Ann. Rev. Ecol. Syst.* 5:325–83.

Alexander, R. D.; Hoogland, J. L.; Howard, R. D.; Noonan, K. M.; and Sherman, P. W. 1979. Sexual dimorphism and breeding systems in pinnipeds, ungulates, primates, and humans. In *Evolutionary biology and human social behavior: An anthropological perspective,* ed. N. A. Chagnon and W. Irons. N. Scituate, Mass.: Duxbury Press.

Alexander, R. D., and Noonan, K. M. 1979. Concealment of ovulation, paternal care, and human social evolution. In *Evolutionary biology and human social behavior,* ed. N. A. Chagnon and W. Irons. N. Scituate, Mass.: Duxbury Press.

Ali, R. 1981. The ecology and social behaviour of the Agastyamali bonnet macaque (*Macaca radiata diluta*). Ph.D. diss., University of Bristol.

———. 1982. Update on the status of India's lion-tailed macaque. *IUCN/SSC Primate Specialist Group Newsletter* 2:21.

Allen, G. M. 1939. A checklist of African mammals. *Bull. Museum Comp. Zool.* 83:173–81.

Allen, M. 1981. Individual copulatory preference and the "strange female effect" in a captive group-living male chimpanzee (*Pan troglodytes*). *Primates* 22:221–36.

Altmann, J. 1974. Observational study of behavior: Sampling methods. *Behaviour* 49:227–65.

———. 1979. Age cohorts as paternal sibships. *Behav. Ecol. Sociobiol.* 6:161–69.

———. 1980. *Baboon mothers and infants.* Cambridge: Harvard University Press.

Altmann, J.; Altmann, S. A.; and Hausfater, G. 1978. Primate infant's effects on mother's future reproduction. *Science* 201:1028–30.

———. In press. Determinants of reproductive success in savannah baboons (*Papio cynocephalus*). In *Reproductive success,* ed. T. H. Clutton-Brock. Chicago: University of Chicago Press.

Altmann, J.; Altmann, S. A.; Hausfater, G.; and McCuskey, S. 1977. Life histories of yellow baboons: Physical development, reproductive parameters, and infant mortality. *Primates* 18:315–30.

Altmann, S. A. 1962. A field study on the sociobiology of rhesus monkeys (*Macaca mulatta*). *Ann. N.Y. Acad. Sci.* 102:338–435.

———. 1965. Sociobiology of rhesus monkeys, 2: Stochastics of social communication. *J. Theoret. Biol.* 8:490–522.

———. 1973. The pregnancy sign in savanna baboons. *J. Zool. Anim. Med.* 4:8–12.

———. 1979a. Altruistic behaviour: The fallacy of kin deployment. *Anim. Behav.* 27:958–59.

———. 1979b. Baboon progressions: Order or chaos? A study of one-dimensional group geometry. *Anim. Behav.* 27:46–80.

———. 1981. Dominance relationships: The Cheshire cat's grin? *Behav. Brain Sci.* 4:430–31.

Altmann, S. A., and Altmann, J. 1970. *Baboon ecology.* Chicago: University of Chicago Press.

———. 1979. Demographic constraints on behavior and social organization. In *Primate ecology and human origins,* ed. I. S. Bernstein and E. O. Smith. New York: Garland Press.

Alvarez, F. 1973. Periodic changes in the bare skin areas of *Theropithecus gelada*. *Primates* 14:195–99.

Andelman, S. J. 1983. The adaptive significance of prolonged sexual receptivity and concealed ovulation in vervet monkeys. Paper presented at Animal Behavior Society Meetings, Bucknell University, Lewisburg, Pa.

———. 1985. Ecology and reproductive strategies of vervet monkeys (*Cercopithecus aethiops*) in Amboseli National Park, Kenya. Ph.D. diss., University of Washington.

———. 1986. Ecological and social determinants of cercopithecine mating patterns. In *Ecology and social evolu-*

tion: Birds and mammals, ed. D. I. Rubenstein and R. W. Wrangham. Princeton: Princeton University Press.

Anderson, C. M. 1981. Intertroop relations of chacma baboons (*Papio ursinus*). *Int. J. Primatol.* 2:285–310.

Anderson, C. O., and Mason, W. A. 1974. Early experience and complexity of social organization in groups of young rhesus monkeys. *J. Comp. Physiol. Psychol.* 87:681–90.

———. 1978. Competitive social strategies in groups of deprived and experienced rhesus monkeys. *Dev. Psychobiol.* 11:289–99.

Anderson, J. R.; Williamson, E. A.; and Carter, J. 1983. Chimpanzees of Sapo Forest, Liberia: Density, nests, tools and meat-eating. *Primates* 24:594–601.

Andrew, R. J. 1963a. Evolution of facial expression. *Science* 2:1034–41.

———. 1963b. The origin and evolution of the calls and facial expressions of the primates. *Behaviour* 20:1–109.

———. 1963c. Trends apparent in the evolution of vocalization in the Old World monkeys and apes. *Symp. Zool. Soc. Lond.* 10–89–101.

Andrew, R. J., and Klopman, R. B. 1974. Urine washing: Comparative notes. In *Prosimian biology,* ed. R. D. Martin, G. A. Doyle, and A. C. Walker, pp. 303–10. London: Duckworth.

Angst, W. 1975. Basic data and concepts in the social organization of *Macaca fascicularis.* In *Primate behavior,* vol. 4, ed. L. A. Rosenblum. New York: Academic Press.

Angst, W., and Thommen, D. 1977. New data and a discussion of infant killing in Old World monkeys and apes. *Folia Primatol.* 27:298–29.

Aoki, K., and Nozawa, K. 1984. Average coefficient of relationship within troops of the Japanese monkey and other primate species with reference to the possibility of group selection. *Primates* 25:171–84.

Ardito, G. 1976. Checklist of the data on the gestation length of primates. *J. Hum. Evol.* 5:213–22.

Armstrong, E. 1983. Relative brain size and metabolism in mammals. *Science* 220:1302–04.

Asher, S. R., and Gottman, J. M., eds. 1981. *The development of children's friendships.* Cambridge: Cambridge University Press.

Aso, T.; Tominaga, T.; Osima, K.; and Matsubayashi, K. 1977. Seasonal changes of plasma estradiol and progesterone in the Japanese monkey (*Macaca fuscata fuscata*). *Endocrinology* 100:745–50.

Asquith, P. 1984. The inevitability and utility of anthropomorphism in descriptions of primate behaviour. In *The meaning of primate signals,* ed. V. Reynolds and R. Harré. Cambridge: Cambridge University Press.

Axelrod, R., and Hamilton, W. D. 1981. The evolution of cooperation. *Science* 211:1390–96.

Ayres, J. M. 1981. Observações sobre a ecologia e o comportamento dos cuxiús (*Chiropotes albinasus* e *Chiropotes satanas.* Cebidae. Primates). Manaus, Brazil: Instituto Nacional de Pesquisas da Amazonia (INPA).

Ayres, J. M., and Nessimian, J. L. 1982. Evidence of insectivory in *Chiropotes satanas. Primates* 23:458–59.

Bachmann, C., and Kummer, H. 1980. Male assessment of female choice in hamadryas baboons. *Behav. Ecol. Sociobiol.* 6:315–21.

Badrian, A., and Badrian, N. 1984. Social organization of *Pan paniscus* in the Lomako Forest, Zaire. In *The pygmy chimpanzee: Evolutionary biology and behavior,* ed. R. L. Susman. New York: Plenum Press.

Badrian, N.; Badrian, A.; and Susman, R. L. 1981. Preliminary observations on the feeding behavior of *Pan paniscus* in the Lomako forest of central Zaire. *Primates* 22:173–81.

Badrian, N., and Malenky, R. 1984. Feeding ecology of *Pan paniscus* in the Lomako Forest, Zaire. In *The pygmy chimpanzee: Evolutionary biology and behavior,* ed. R. L. Susman. New York: Plenum Press.

Bailey, R. C. 1985. Socioecology of Efe pygmies in northeastern Zaire. Ph.D. diss., Harvard University.

Baker, M. C., and Cunningham, M. A. 1985. The biology of bird song dialects. *Behav. Brain Sci.* 8:85–133.

Baker, R. R. 1978. *The evolutionary ecology of animal migration.* London: Hodder and Stoughton.

Baldwin, J. D. 1968. The social behavior of adult male squirrel monkeys (*Saimiri sciureus*) in a seminatural environment. *Folia Primatol.* 9:281–314.

———. 1969. The ontogeny of social behaviour of squirrel monkeys (*Saimiri sciureus*) in a seminatural environment. *Folia Primatol.* 11:35–79.

Baldwin, J. D., and Baldwin, J. I. 1972a. Population density and use of space by howling monkeys (*Alouatta villosa*) in southwestern Panama. *Primates* 13:371–79.

———. 1972b. The ecology and behavior of squirrel monkeys (*Saimiri oerstedi*) in a natural forest in western Panama. *Folia Primatol.* 19:161–84.

———. 1973. The role of play in social organizations: Comparative observations on squirrel monkeys (*Saimiri*). *Primates* 14:369–81.

———. 1976a. Primate populations in Chiriqui, Panama. In *Neotropical primates: Field studies and conservation,* ed. R. W. Thorington, Jr., and P. G. Heltne. Washington D.C.: National Academy of Sciences.

———. 1976b. The vocalizations of howler monkeys (*Alouatta palliata*) in southwestern Panama. *Folia Primatol.* 26:81–108.

———. 1981. The squirrel monkeys, genus *Saimiri.* In *Ecology and behavior of Neotropical primates,* vol. 1, ed. A. F. Coimbra-Filho and R. A. Mittermeier. Rio de Janeiro: Academia Brasileira de Ciencias.

Baldwin, P. J.; McGrew, W. C.; and Tutin, C. E. G. 1982. Wide-ranging chimpanzees at Mt. Assirik, Senegal. *Int. J. Primatol.* 3:367–85.

Barnett, S. A. 1968. The "instinct to teach." *Nature* 220:747–49.

Barnicot, N. A.; Jolly, C. J.; and Wade, P. T. 1967. Protein variations and primatology. *Am. J. Phys. Anthropol.* 27:343–55.

Barrett, E. 1981. The present distribution and status of the slow loris in peninsular Malaysia. *Malasian Appl. Biol.* 10:205–11.

———. 1984. The ecology of some nocturnal, arboreal mammals in the rainforest of peninsular Malaysia. Ph.D. diss., Cambridge University.

Barrit, D. 1985. Raging monkeys take revenge—trap terrified family in home. *National Enquirer,* February 1985.

Bateman, A. J. 1948. Intra-sexual selection in *Drosophila. Heredity* 2:349–68.

Bateson, P. P. G., ed. 1983. *Mate choice.* Cambridge: Cambridge University Press.

Bauchop, T. 1978. Digestion of leaves in vertebrate arboreal folivores. In *The ecology of arboreal folivores,* ed. G. G. Montgomery. Washington, D.C.: Smithsonian Institution Press.

Baulu, T. 1976. Seasonal sex skin coloration and hormonal fluctuations in free-ranging and captive monkeys. *Horm. Behav.* 7:481–94.

Beach, F. A. 1976. Sexual activity, proceptivity, and receptivity in female mammals. *Horm. Behav.* 7:105–38.

Bearder, S. K. 1984. The relevance of field studies to the captive management of bushbabies (Primates: Lorisidae). *Proc. Symp. Assoc. Brit. Wild Anim. Keepers* 8:29–48.

Bearder, S. K., and Doyle, G. A. 1974. Ecology of bushbabies *Galago senegalensis* and *Galago crassicaudatus,* with some notes on their behaviour in the field. In *Prosimian biology,* ed. R. D. Martin, G. A. Doyle, and A. C. Walker. London: Duckworth.

Bearder, S. K., and Martin, R. D. 1979. The social organization of a nocturnal primate revealed by radio tracking. In *A handbook on biotelemetry and radio tracking,* ed. C. J. Amlaner, Jr., and D. W. Macdonald. Oxford: Pergamon Press.

———. 1980. Acacia gum and its use by bushbabies, *Galago senegalensis* (Primates: Lorisidae). *Int. J. Primatol.* 1:103–28.

Beck, B. B. 1975. Primate tool behavior. In *Primate socioecology and psychology,* ed. R. H. Tuttle. The Hague: Mouton.

———. 1980. *Animal tool behavior.* New York: Garland Press.

Beer, C. G. 1976. Some complexities in the communication behavior of gulls. *Ann. N. Y. Acad. Sci.* 280:413–32.

Bell, R. Q., and Harper, L. V. 1977. *Child effects on adults.* Hillsdale, N.J.: Lawrence Erlbaum Associates.

Bengtsson, B. O. 1978. Avoiding inbreeding: At what cost? *J. Theoret. Biol.* 73:439–44.

Bennett, E. L. 1983. The banded langur: Ecology of a colobine in West Malaysian rain-forest. Ph.D. diss., University of Cambridge.

Bennett, P. M., and Harvey, P. H. 1984. Why mammals are not bird-brained. *New Scient.* 1450:16–17.

———. 1985. Brain size, development, and metabolism in birds and mammals. *J. Zool. (Lond.)* 207:491–509.

Benshoof, L., and Thornhill, R. 1979. The evolution of monogamy and concealed ovulation in humans. *J. Biol. Soc. Struct.* 2:95–106.

Bercovitch, F. B. 1983. Time budgets and consortships in olive baboons (*Papio anubis*). *Folia Primatol.* 41:180–90.

Berdicio, S., and Nash, L. T. 1981. Chimpanzee visual communication. Anthro. Papers No. 26. Tempe: Arizona State University.

Berenstain, L.; Rodman, P. S.; and Smith, D. G. 1981. Social relationships between fathers and offspring in a captive group of rhesus monkeys (*Macaca mulatta*). *Anim. Behav.* 29:1057–63.

Berenstain, L., and Wade, T. D. 1983. Intrasexual selection and male mating strategies in baboons and macaques. *Int. J. Primatol.* 4:201–35.

Berman, C. M. 1978. Social relationships among free-ranging infant rhesus monkeys. Ph.D. diss., Cambridge University.

———. 1980a. Early agonistic experience and rank acquisition among free-ranging infant rhesus monkeys. *Int. J. Primatol.* 1:153–70.

———. 1980b. Mother-infant relationships among free-ranging rhesus monkeys on Cayo Santiago: A comparison with captive pairs. *Anim. Behav.* 28:860–73.

———. 1982a. The ontogeny of social relationships with group companions among free-ranging infant rhesus monkeys, 1: Social networks and differentiation. *Anim. Behav.* 30:149–62.

———. 1982b. The roles of maternal age and the presence of close kin on mother-infant interaction among free-ranging monkeys on Cayo Santiago (abstract). *Int. J. Primatol.* 3:261.

———. 1983a. Early differences in relationships between infants and other group members based on the mothers' status: Their possible relationship to peer-peer rank acquisition. In *Primate social relationships: An integrated approach,* ed. R. A. Hinde. Oxford: Blackwell.

———. 1983b. Effects of being orphaned: A detailed case study of an infant rhesus. In *Primate social relationships: An integrated approach,* ed. R. A. Hinde. Oxford: Blackwell.

———. 1983c. Matriline differences and infant development. In *Primate social relationships: An integrated approach,* ed. R. A. Hinde. Oxford: Blackwell.

———. 1984. Variation in mother-infant relationships: Traditional and nontraditional factors. In *Female primates: Studies by women primatologists,* ed. M. F. Small. New York: Alan R. Liss.

Bernstein, I. S. 1961. The utilization of visual cues in dimension-abstracted oddity by primates. *J. Comp. Physiol. Psychol.* 54:243–47.

———. 1967a. A field study of the pigtail monkey (*Macaca nemestrina*). *Primates* 8:217–28.

———. 1967b. Intertaxa interactions in a Malayan primate community. *Folia Primatol.* 7:198–207.

———. 1968. The lutong of Kuala Selangor. *Behaviour* 32:1–16.

———. 1969. Spontaneous reorganization of a pigtail monkey group. In *Proceedings of the Second Congress of the International Primatological Society,* vol. 1, ed. C. R. Carpenter. Basel: S. Karger.

———. 1971. Activity profiles of primate groups. In *Behavior of nonhuman primates,* vol. 3, ed. A. M. Schrier and F. Stollnitz. New York: Academic Press.

———. 1972. Daily activity cycles and weather influences on a pigtail monkey group. *Folia Primatol.* 18:390–415.

———. 1976a. Activity patterns in a sooty mangabey group. *Folia Primatol.* 26:185–206.

———. 1976b. Dominance, aggression, and reproduction in primate societies. *J. Theoret. Biol.* 60:459–72.

———. 1980. Activity patterns in a stumptail macaque group (*Macaca arctoides*). *Folia Primatol.* 33:20–45.

———. 1981. Dominance: The baby and the bathwater. *Behav. Brain Sci.* 4:419–58.

Bernstein, I. S., and Ehardt, C. L. 1985. Agonistic aiding: Kinship, rank, age, and sex influences. *Am. J. Primatol.* 8:37–52.

Bernstein, I. S.; Gordon, T. P.; and Rose, R. M. 1974. Aggression and social controls in rhesus monkey (*Macaca mulatta*) groups revealed in group formation studies. *Folia Primatol.* 21:81–107.

Bernstein, I. S., and Sharpe, L. 1966. Social roles in a rhesus monkey group. *Behaviour* 26:91–103.

Bernstein, L. S.; Williams, L.; and Ramsay, M. 1983. The expression of aggression in Old World monkeys. *Int. J. Primatol.* 4:113–25.

Bertram, B. C. R. 1976. Kin selection in lions and in evolution. In *Growing Points in Ethology,* ed. P. P. G. Bateson and R. A. Hinde. Cambridge: Cambridge University Press.

———. 1978. Living in groups: Predators and prey. In *Behavioral ecology: An evolutionary approach,* ed. J. R. Krebs and N. B. Davies. Sunderland, Mass.: Sinauer Associates.

———. 1982. Problems with altruism. In *Current problems in sociobiology,* ed. Kings' College Sociobiology Group, Cambridge University. Cambridge: Cambridge University Press.

———. 1983. Kin selection and altruism. In *Advances in the study of mammalian behavior,* ed. J. F. Eisenberg and D. G. Kleiman. Special Publication No. 7. American Society of Mammalogists.

Bertrand, M. 1969. *The behavioral repertoire of the stumptail macaque.* Basel: S. Karger.

Bessemer, D. W., and Stollnitz, F. 1971. Retention of discrimination and analysis of learning set. In *Behavior of nonhuman primates,* vol. 4, ed. A. M. Schrier and F. Stollnitz. New York: Academic Press.

Bingham, H. C. 1927. Parental play of chimpanzees. *J. Mammal.* 8:77–89.

Birdwhistell, R. L. 1967. Communication without words. In *L'aventure humaine,* ed. P. Alexandre. Paris: Societé d'Etudes Littéraires et Art.

Blaxter, K. L. 1971. The comparative biology of lactation. In *Lactation,* ed. I. R. Falconer. London: Butterworth.

Blurton-Jones, N. G. 1972. Comparative aspects of mother-child contact. In *Ethological studies of child behaviour,* ed. N. G. Blurton-Jones. Cambridge: Cambridge University Press.

Blurton-Jones, N. G., and Trollope, J. 1968. Social behavior of stump-tail macaques in captivity. *Primates* 9:365–94.

Bodmer, W. F., and Cavalli-Sforza, L. L. 1968. A migration model for the study of random genetic drift. *Genetics* 59:565–92.

Boelkins, R. C., and Wilson, A. P. 1972. Intergroup social dynamics of the Cayo Santiago rhesus (*Macaca mulatta*) with special reference to changes in group membership by males. *Primates* 13:125–40.

Boesch, C., and Boesch, H. 1981. Sex differences in the use of natural hammers by wild chimpanzees: A preliminary report. *J. Hum. Evol.* 10:585–93.

———. 1983. Optimisation of nut-cracking with natural hammers by wild chimpanzees. *Behaviour* 83:265–86.

———. 1984. Mental map in wild chimpanzees: An analysis of hammer transports for nut cracking. *Primates* 25:160–70.

Bogart, M. H.; Komanoto, A. T.; and Lasley, B. C. 1977. A comparison of the reproductive cycles of three species of *Lemur. Folia Primatol.* 28:134–43.

Boggess, J. 1979. Troop male membership changes and infant killing in langurs (*Presbytis entellus*). *Folia Primatol.* 32:65–107.

———. 1980. Intermale relations and troop male membership changes in langurs *Presbytis entellus* in Nepal. *Int. J. Primatol.* 1:233–74.

———. 1984. Infant killing and male reproductive strategies in langurs (*Presbytis entellus*). In *Infanticide: Comparative and evolutionary perspectives,* ed. G. Hausfater and S. B. Hrdy. Hawthorne, N.Y.: Aldine.

Bolin, I. 1981. Male parental behavior in black howler monkeys (*Alouatta palliata pigra*) in Belize and Guatemala. *Primates* 22:349–60.

Bolwig, N. 1980. Early social development and emancipation of *Macaca nemestrina* and species of *Papio. Primates* 21:357–75.

Bond, J., and Vinacke, W. 1961. Coalitions in mixed-sex triads. *Sociometry* 24:61–75.

Bongaarts, J. 1982. Malnutrition and fertility. *Science* 215:1273–74.

Bonner, J. T. 1980. *The evolution of culture in animals.* Princeton: Princeton University Press.

Bonney, R. C.; Dixson, A. F.; and Fleming, D. 1980. Plasma concentration of oestradial-17-beta, estrone, progesterone, and testosterone during the ovarian cycle of the owl monkey (*Aotus trivirgatus*). *J. Reprod. Fertil.* 60:101–7.

Bourlière, F.; Hunkeler, C.; and Bertrand, M. 1970. Ecology and behaviour of Lowe's guenon (*Cercopithecus campbelli lowei*) in the Ivory Coast. In *Old World monkeys,* ed. J. R. Napier and P. H. Napier. New York: Academic Press.

Bowlby, J. 1969. *Attachment and loss,* vol. 1: *Attachment.* New York: Basic Books.

———. 1973. *Attachment and Loss,* vol. 2: *Separation.* London: Hogarth Press.

Bowman, L. A.; Dilley, S.; and Keverne, E. B. 1978. Suppression of oestrogen-induced LH surges by social subordination in talapoin monkeys. *Nature* 275:56–58.

Box, H. O. 1975a. A social developmental study of young monkeys (*Callithrix jacchus*) within a captive family group. *Primates* 16:419–35.

———. 1975b. Quantitative studies of behaviour within captive groups of marmoset monkeys (*Callithrix jacchus*). *Primates* 16:155–74.

———. 1977. Quantitative data on the carrying of young captive monkeys (*Callithrix jacchus*) by other members of their captive groups. *Primates* 18:475–84.

Bradbury, J. W., and Vehrencamp, S. L. 1977. Social organisation and foraging in emballonurid bats. *Behav. Ecol. Sociobiol.* 2:1–17.

Bradshaw, J. L., and Nettleton, N. C. 1981. The nature of hemispheric specialization in man. *Behav. Brain Sci.* 4:51–91.

Brainerd, C. J. 1978. The stage question in cognitive-developmental theory. *Behav. Brain Sci.* 1:173–213.

Bramblett, C. 1970. Coalitions among gelada baboons. *Primates* 11:327–34.

———. 1983. Incest avoidance in socially living vervet monkeys. *Am. J. Phys. Anthropol.* 60:176–77.

Bramblett, C.; Bramblett, S. S.; Bishop, D. A.; and Coelho, A. M. 1982. Longitudinal stability in adult status hierarchies among vervet monkeys (*Cercopithecus aethiops*). *Am. J. Primatol.* 2:43–51.

Brand, H. M., and Martin, R. D. 1983. The relationship between female urinary estrogen secretion and mating behavior in cotton-topped tamarins, *Saguinus oedipus oedipus*. *Int. J. Primatol.* 4:275–90.

Braza, F.; Alvarez, F.; and Azcarate, T. 1981. Behaviour of the red howler monkey (*Alouatta seniculus*) in the llanos of Venezuela. *Primates* 22:459–73.

Brennan, E. J. 1985. De Brazza's monkey (*Cercopithecus neglectus*) in Kenya: Census, distribution, and conservation. *Am. J. Primatol.* 8:269–77.

Brent, P. 1981. *Charles Darwin: A man of enlarged curiosity*. London: Hienemann.

Brereton, A. 1981. Inter-group consorting by a free-ranging female rhesus monkey (*Macaca mulatta*). *Primates* 22:417–23.

Brett, F. L.; Jolly, C. J.; Socha, W.; and Weiner, A. S. 1977. Human-like ABO blood groups in wild Ethiopian baboons. *Yrbk. Phys. Anthropol.* 20:276–89.

Breuggeman, J. A. 1973. Parental care in a group of free-ranging rhesus monkeys (*Macaca mulatta*). *Folia Primatol.* 20:178–210.

Brockelman, W. Y., and Schilling, D. 1984. Inheritance of stereotyped gibbon calls. *Nature* 312:634–36.

Brockelman, W. Y., and Srikosamatara, S. 1984. Maintenance and evolution of social structure in gibbons. In *The lesser apes: Evolutionary and behavioral biology,* ed. H. Preuschoft, D. Chivers, W. Brockelman, and N. Creel. Edinburgh: Edinburgh University Press.

Brockmann, H. J.; Grafen, A.; and Dawkins, R. 1979. Evolutionary stable nesting in a digger wasp. *J. Theoret. Biol.* 77:473–96.

Brown, J. L. 1964. The evolution of diversity in avian territorial systems. *Wilson Bull.* 76:160–69.

———. 1978. Avian communal breeding systems. *Ann. Rev. Ecol. Syst.* 9:123–55.

Brown, J. L., and Langley, L. H. 1979. Reevaluation of level of genic heterozygosity in natural populations of *Drosophila melanogaster* by two-dimensional electrophoresis. *Proc. Nat. Acad. Sci. USA* 76:2381–84.

Brown, K., and Mack, D. S. 1978. Food sharing among captive *Leontopithecus rosalia*. *Folia Primatol.* 29:268–90.

Buchanan, D. B. 1978. Communication and ecology of pithecine monkeys, with special reference to *Pithecia pithecia*. Ph.D. diss., Wayne State University.

Buchanan, D. B.; Mittermeier, R. A.; and van Roosmalen, M. G. M. 1981. The saki monkeys, genus *Pithecia*. In *Ecology and behavior of Neotropical primates*, vol. 1, ed. A. F. Coimbra-Filho and R. A. Mittermeier. Rio de Janeiro: Academia Brasileira de Ciencias.

Budnitz, N., and Dainis, K. 1975. *Lemur catta*: Ecology and behavior. In *Lemur biology*, ed. I. Tattersall and R. W. Sussman. New York: Plenum Press.

Buettner-Janusch, J.; Olivier, T. J.; Ober, C. C.; and Chepko-Sade, B. D. 1983. Models for lineal effects in rhesus group fissions. *Am. J. Phys. Anthropol.* 61:347–53.

Bunnel, B. N.; Gore, W. T.; and Perkins, M. N. 1980. Performance correlates of social behavior and organization: Social rank and reversal learning in crab-eating macaques *M. fascicularis*. *Primates* 21:376–88.

Burton, F. D. 1972. The integration of biology and behavior in the socialization of *Macaca sylvana* of Gibraltar. In *Primate socialization*, ed. F. E. Poirier. New York: Random House.

Burton, F. D., and Fukuda, F. 1981. On female mobility: The case of the Yugawara-T group of Macaca fuscata. *J. Hum. Evol.* 10:381–86.

Buss, D. H., and Voss, W. R. 1971. Evaluation of four methods for estimating the milk yield of baboons. *J. Nutr.* 101:901–10.

Busse, C. D. 1976. Chimpanzee predation as a possible factor in the evolution of red colobus monkey social organization. *Evolution* 31:907–11.

———. 1978. Do chimpanzees hunt cooperatively? *Am. Nat.* 112:767–70.

———. 1980. Leopard and lion predation upon chacma baboons living in the Moremi Wildlife Reserve. *Botswana Notes Rec.* 12:15–21.

———. 1982. Social dominance and offspring mortality among female chacma baboons (abstract). *Int. J. Primatol.* 3:267.

———. 1984a. Spatial structure of chacma baboon groups. *Int. J. Primatol.* 5:247–62.

———. 1984b. Triadic interactions among male and infant chacma baboons. In *Primate paternalism*, ed. D. M. Taub. New York: Van Nostrand Reinhold.

Busse, C. D., and Estep, D. Q. 1984. Sexual arousal in male pigtailed monkeys (*Macaca nemestrina*): Effects of serial matings by two males. *J. Comp. Psychol.* 98:227–31.

Busse, C. D., and Gordon, T. P. 1983. Infant carrying by adult male mangabeys (*Cercocebus atys*). *Am. J. Primatol.* 6:133–41.

Busse, C. D., and Hamilton, W. J., III. 1981. Infant carrying by male chacma baboons. *Science* 212:1281–83.

Butler, H. 1974. Evolutionary trends in primate sex cycles. In *Contributions to Primatology*, vol. 3. Basel: S. Karger.

Butynski, T. M. 1982a. Blue monkey (*Cercopithecus mitis stuhlmanii*) predation on galagos. *Primates* 23:563–66.

———. 1982b. Harem-male replacement and infanticide in the blue monkey (*Cercopithecus mitis stuhlmanni*) in the Kibale Forest, Uganda. *Am. J. Primatol.* 3:1–22.

Bygott, J. D. 1974. Agonistic behavior and social relationships among adult male chimpanzees. Ph.D. diss., Cambridge University.

———. 1979. Agonistic behaviour, dominance, and social structure in wild chimpanzees of the Gombe National Park. In *The great apes*, ed. D. A. Hamburg and E. R. McCown. Menlo Park, Calif.: Benjamin/Cummings.

Bygott, J. D.; Bertram, B. C. R.; and Hanby, J. P. 1979. Male lions in large coalitions gain reproductive advantages. *Nature* 282:839–41.

Byles, R. H., and Sanders, M. F. 1981. Intertroop variation in the frequencies of ABO alleles in a population of olive baboons. *Int. J. Primatol.* 2:35–46.

Byrne, R. W. 1981. Distance vocalizations of Guinea baboons (*Papio papio*) in Senegal: An analysis of function. *Behaviour* 78:288–313.

Byrne, R. W.; Conning, A. M.; and Young, J. 1983. Social relationships in a captive group of Diana monkeys (*Cercopithecus diana*). *Primates* 24:360–70.

Caine, N. G. 1984. Visual scanning by tamarins: A description of the behavior and tests of two derived hypotheses. *Folia*

Primatol. 43:59–67.

Caine, N. G., and Mitchell, G. D. 1979. The relationship between maternal rank and companion choice in immature macaques (*Macaca mulatta and M. radiata*). *Primates* 20:583–90.

Caldecott, J. O. 1981. Findings on the behavioural ecology of the pigtailed macaque. *Malaysian Appl. Biol.* 10:213–20.

Cambefort, J. P. 1981. A comparative study of culturally transmitted patterns of feeding habits in the chacma baboons *Papio ursinus* and the vervet monkeys *Cercopithecus aethiops*. *Folia Primatol.* 36:243–63.

Candland, D. K.; Blumer, E. S.; and Mumford, M. D. 1980. Urine as a communicator in a New World primate, *Saimiri sciureus*. *Anim. Learn. Behav.* 8:468–80.

Cant, J. G. H. 1977. Ecology, locomotion, and social organization of spider monkeys (*Ateles geoffroyi*). Ph.D. diss., University of California, Davis.

———. 1980. What limits primates: *Primates* 21:538–44.

Caro, T. M. 1976. Observations on the ranging behaviour and daily activity of lone silverback mountain gorillas (*Gorilla gorilla beringei*). *Anim. Behav.* 24:889–97.

Carpenter, C. R. 1934. A field study of the behavior and social relations of howling monkeys (*Alouatta palliata*). *Comp. Psychol. Monogr.* 10:1–168.

———. 1940. A field study in Siam of the behavior and social relations of the gibbon (*Hylobates lar*). *Comp. Psychol. Monogr.* 6:1–212.

———. 1942. Sexual behaviour of free-ranging rhesus monkeys, *Macaca mulatta*, 1: Specimens, procedures, and behavioural characteristics of estrus. *J. Comp. Psychol.* 33:113–42.

Cartmill, M. 1972. Arboreal adaptations and the origin of the order Primates. In *The functional and evolutionary biology of primates*, ed. R. Tuttle. Chicago: Aldine.

———. 1974a. *Lemur catta*: Ecology and behavior. In *Lemur biology*, ed. I. Tattersall and R. W. Sussman. New York: Plenum.

———. 1974b. Pads and claws in arboreal locomotion. In *Primate locomotion*, ed. F. A. Jenkins, Jr. New York: Academic Press.

———. 1974c. Rethinking primate origins. *Science* 184:436–43.

———. 1974d. *Daubentonia, Dactylopsila*, woodpeckers, and klinorhynchy. In *Prosimian Biology*, ed. R. D. Martin, G. A. Doyle, and A. C. Walker. London: Duckworth.

Casimir, M. J. 1975. Feeding ecology and nutrition of an eastern gorilla group in the Mt. Kahuzi region (Republique du Zaire). *Folia Primatol.* 24:81–136.

———. 1979. An analysis of gorilla nesting sites of the Mt. Kahuzi Region (Zaire). *Folia Primatol.* 32:290–308.

Castro, N.; Revilla, J.; and Neville, M. 1975. "Carne de monte" como una fuente de proteínas en Iquitos, con referencia especial a monos. *Revista Forestal de Perú* 5:19–32.

Caughley, G. 1977. *Analysis of vertebrate populations.* Chichester: John Wiley.

Cebul, M. S., and Epple, G. 1984. Father-offspring relationships in laboratory families of saddle-back tamarins (*Saguinus fuscicollis*). In *Primate paternalism*, ed. D. M. Taub. New York: Van Nostrand Reinhold.

Chalmers, N. R. 1968a. The social behavior of free-living mangabeys in Uganda. *Folia Primatol.* 8:263–81.

———. 1968b. Group composition, ecology, and daily activities of free living mangabeys in Uganda. *Folia Primatol.* 8:247–62.

———. 1972. Comparative aspects of early infant development in some captive cercopithecines. In *Primate socialization*, ed. F. E. Poirier. New York: Random House.

———. 1980. *Social behaviour in primates.* Baltimore: University Park Press.

Chalmers, N. R., and Rowell, T. E. 1971. Behavior and female reproductive cycles in a captive group of mangabeys. *Folia Primatol.* 14:1–14.

Chambers, K. C., and Phoenix, C. M. 1982. Sexual behavior in old male rhesus monkeys: Influence of familiarity and age of female partners. *Arch. Sex. Behav.* 11:299–308.

———. 1984. Restoration of sexual performance in old rhesus macaques paired with a preferred female partner. *Int. J. Primatol.* 5:287–98.

Chamove, A. S. 1980. Nongenetic induction of acquired levels of aggression. *J. Abnorm. Psychol.* 89:469–88.

Chance, M. 1961. The nature and special features of the instinctive social bond of primates. In *Social life of early man*, ed. S. L. Washburn. New York: Viking Fund Publications.

Chance, M.; Emory, G.; and Payne, R. 1977. Status referents in long-tailed macaques (*Macaca fascicularis*): Precursors and effects of a female rebellion. *Primates* 18:611–32.

Chapais, B. 1981. The adaptiveness of social relationships among adult rhesus monkeys. Ph.D. diss., University of Cambridge.

———. 1983a. Autonomous, bisexual subgroups in a troop of rhesus monkeys. In *Primate social relationships: An integrated approach*, ed. R. A. Hinde. Oxford: Blackwell.

———. 1983b. Dominance, relatedness, and the structure of female relationships in rhesus monkeys. In *Primate social relationships: An integrated approach*, ed. R. A. Hinde. Oxford: Blackwell.

———. 1983c. Male dominance and reproductive activity in rhesus monkeys. In *Primate social relationships: An integrated approach*, ed. R. A. Hinde. Oxford: Blackwell.

———. 1983d. Matriline membership and male rhesus reaching high ranks in their natal troops. In *Primate social relationships: An integrated approach*, ed. R. A. Hinde. Oxford: Blackwell.

———. 1983e. Reproductive activity in relation to male dominance and the likelihood of ovulation in rhesus monkeys. *Behav. Ecol. Sociobiol.* 12:215–28.

———. 1983f. Structure of the birth season relationship among adult male and female rhesus monkeys. In *Primate social relationships: An integrated approach*, ed. R. A. Hinde. Oxford: Blackwell.

Chapais, B., and Schulman, S. 1980. An evolutionary model of female dominance relations in primates. *J. Theoret. Biol.* 82:47–89.

Charles-Dominique, P. 1974a. Aggression and territoriality in nocturnal prosimians. In *Primate aggression, territoriality, and xenophobia*, ed. Ralph L. Holloway. New York: Academic Press.

———. 1974b. Ecology and feeding behaviour of five symmpatric lorisids in Gabon. In *Prosimian biology*, ed. R. D. Martin, G. A. Doyle, and A. C. Walker. London: Duckworth.

———. 1975. Nocturnality and diurnality: An ecological interpretation of these two modes of life by an analysis of the higher vertebrate fauna in tropical forest ecosystems. In *Phylogeny of the primates: A multidisciplinary approach,* ed. W. P. Luckett and F. S. Szalay. New York: Plenum Press.

———. 1977a. *Ecology and behaviour of nocturnal prosimians.* London: Duckworth.

———. 1977b. Urine marking and territoriality in *Galago alleni.* A field study by radio-telemetry. *Z. Tierpsychol.* 43: 113–38.

———. 1978. Solitary and gregarious prosimians: Evolution of social structures in primates. In *Recent advances in primatology,* vol. 3: *Evolution,* ed. D. J. Chivers and K. A. Joysey. London: Academic Press.

Charles-Dominique, P., and Bearder, S. K. 1979. Field studies of lorisid behavior: Methodological aspects. In *The study of prosimian behavior,* ed. G. A. Doyle and R. D. Martin. New York: Academic Press.

Charles-Dominique, P.; Cooper, H. M.; Hladik, A.; Hladik, C. M.; Pages, E.; Pariente, G. F.; Petter-Rousseaux, A.; Petter, J. J.; and Schilling, A., eds. 1980. *Nocturnal Malagasy primates: Ecology, physiology, and behavior.* New York: Academic Press.

Charles-Dominique, P., and Petter, J. J. 1980. Ecology and social life of *Phaner furcifer.* In *Nocturnal Malagasy primates: Ecology, physiology, and Behaviour,* ed. P. Charles-Dominique, H. M. Cooper, A. Hladik, E. Pages, G. F. Pariente, A. Petter-Rousseaux, J. J. Petter, and A. Schilling. New York: Academic Press.

Charlesworth, B. 1980. *Evolution in age-structured populations.* Cambridge: Cambridge University Press.

Chase, I. D. 1980. Social process and hierarchy formation in small groups: A comparative perspective. *Am. Sociol. Rev.* 45: 905–24.

Cheney, D. L. 1977a. The acquisition of rank and the development of reciprocal alliances among free-ranging immature baboons. *Behav. Ecol. Sociobiol.* 2: 303–18.

———. 1977b. *The social development of immature baboons.* Ph.D. diss., University of Cambridge.

———. 1978a. Interactions of immature male and female baboons with adult females. *Anim. Behav.* 26: 389–408.

———. 1978b. The play partners of immature baboons. *Anim. Behav.* 26: 1038–50.

———. 1981. Inter-group encounters among free-ranging vervet monkeys. *Folia Primatol.* 35: 124–46.

———. 1983a. Extra-familial alliances among vervet monkeys. In *Primate social relationships: An integrated approach,* ed. R. A. Hinde. Oxford: Blackwell.

———. 1983b. Proximate and ultimate factors related to the distribution of male migration. In *Primate social relationships: An integrated approach,* ed. R. A. Hinde. Oxford: Blackwell.

Cheney, D. L.; Lee, P. C.; and Seyfarth, R. M. 1981. Behavioral correlates of non-random mortality among free-ranging adult female vervet monkeys. *Behav. Ecol. Sociobiol.* 9: 153–61.

Cheney, D. L., and Seyfarth, R. M. 1977. Behaviour of adult and immature male baboons during inter-group encounters. *Nature* 269: 404–6.

———. 1980. Vocal recognition in free-ranging vervet monkeys. *Anim. Behav.* 28: 362–67.

———. 1981. Selective forces affecting the predator alarm calls of vervet monkeys. *Behaviour* 76: 25–61.

———. 1982a. How vervet monkeys perceive their grunts: Field playback experiments. *Anim. Behav.* 30: 739–51.

———. 1982b. Recognition of individuals within and between groups of free-ranging vervet monkeys. *Am. Zool.* 22: 519–29.

———. 1983. Non-random dispersal in free-ranging vervet monkeys: Social and genetic consequences. *Am. Nat.* 122: 392–412.

———. 1985a. Social and non-social knowledge in vervet monkeys. *Phil. Trans. Roy. Soc. Lond. B* 308: 187–201.

———. 1985b. The social and non-social world of non-human primates. In *Social relationships and cognitive development,* ed. R. A. Hinde, A. Perret-Clermont, and J. Stevenson Hinde. Oxford: Oxford University Press.

———. 1985c. Vervet monkey alarm calls: Manipulation through shared information? *Behaviour* 94: 150–66.

———. 1986. The recognition of social alliances among vervet monkeys. *Anim. Behav.* 34: 1722–31.

Cheney, D. L.; Seyfarth, R. M.; Andelman, S. J.; and Lee, P. C. In press. Reproductive success in vervet monkeys. In *Reproductive success,* ed. T. H. Clutton-Brock. Chicago: University of Chicago Press.

Chepko-Sade, B. D., and Olivier, T. J. 1979. Coefficient of genetic relationship and the probability of intragenealogical fission in *Macaca mulatta. Behav. Ecol. Sociobiol.* 5: 263–78.

Chepko-Sade, B. D., and Sade, D. S. 1979. Patterns of group splitting within matrilineal kinship groups: A study of social group structure in *Macaca mulatta* (Cercopithecidae: Primates). *Behav. Ecol. Sociobiol.* 5: 67–87.

Chevalier-Skolnikoff, S. 1982. A cognitive analysis of facial behavior in Old World monkeys, apes, and humans. In *Primate communication,* ed. C. Snowdon, C. H. Brown, and M. Petersen. Cambridge: Cambridge University Press.

Cheverud, J. M.; Buettner-Janusch, J.; and Sade, D. S. 1978. Social group fission and the origin of intergroup genetic differentiation among the rhesus monkeys of Cayo Santiago. *Am. J. Phys. Anthropol.* 49: 449–56.

Chikazawa, D.; Gordon, T.; Bean, C.; and Bernstein, I. 1979. Mother daughter dominance reversals in rhesus monkeys (*Macaca mulatta*). *Primates* 20: 301–5.

Chism, J., and Olson, D. 1982. Paper presented at Tenth Congress of the International Primatological Society, Atlanta.

Chism, J.; Rowell, T. E.; and Olson, D. K. 1984. Life history patterns of female patas monkeys. In *Female primates: Studies by women primatologists,* ed. M. F. Small. New York: Alan R. Liss.

Chivers, D. J. 1969. On the daily behavior and spacing of howling monkey groups. *Folia Primatol.* 10: 48–102.

———. 1974. The siamang in Malaya. In *Contributions to Primatology,* vol. 4. Basel: S. Karger.

———. 1977. The lesser apes. In *Primate conservation,* ed. H. R. H. Prince Rainier III and G. H. Bourne. New York: Academic Press.

———. 1984. Feeding and ranging in gibbons: A summary. In *The lesser apes: Evolutionary and behavioral biology,* ed. H. Preuschoft, D. J. Chivers, W. Brockelman, and N. Creel.

Edinburgh: Edinburgh University Press.

————, ed. 1980. *Malayan forest primates: Ten years' study in tropical forest.* Plenum Press: New York.

Chivers, D. J., and Hladik, C. M. 1980. Morphology of the gastrointestinal tract in primates: Comparisons with other mammals in relation to diet. *J. Morphol.* 166:337–86.

Chivers, D. J., and MacKinnon, J. 1977. On the behavior of siamang after playback of their calls. *Primates* 18:943–48.

Chivers, D. J., and Raemaekers, J. J. 1980. Long-term changes in behaviour. In *Malayan forest primates: Ten years' study in tropical rain forest,* ed. D. J. Chivers. New York: Plenum Press.

Chivers, D. J.; Wood, B. A.; and Bilsborough, A., eds. 1984. *Food acquisition and processing in primates.* New York: Plenum Press.

Chomsky, N. 1957. *Syntactic structures.* The Hague: Mouton.

Clark, A. B. 1978a. Olfactory communication, *Galago crassicaudatus,* and the social life of prosimians. In *Recent advances in primatology,* vol. 3, ed. D. J. Chivers and K. A. Joysey. New York: Academic Press.

————. 1978b. Sex ratio and local resource competition in a prosimian primate. *Science* 201:163–65.

————. 1984. Prolonged copulation and mate guarding in galagos. Typescript.

Clark, C. B. 1977. A preliminary report on weaning among chimpanzees of the Gombe National Park, Tanzania. In *Primate bio-social development,* ed. S. Chevalier-Skolnikoff and F. E. Poirier. New York: Garland Press.

Clark, C. W., and Mangel, M. 1984. Foraging and flocking strategies: Information in an uncertain environment. *Am. Nat.* 123:626–41.

Clarke, M. R. 1983. Infant-killing and infant disappearance following male takeovers in a group of free-ranging howling monkeys (*Alouatta palliata*) in Costa Rica. *Am. J. Primatol.* 5:241–47.

Clarke, M. R., and Glander, K. E. 1984. Female reproductive success in a group of free-ranging howling monkeys (*Alouatta palliata*) in Costa Rica. In *Female primates: Studies by women primatologists,* ed. M. Small. New York: Alan R. Liss.

Clegg, E. J., and Harrison, G. A. 1971. Reproduction in human high altitude populations. *Hormones* 2:13–25.

Cleveland, J., and Snowdon, C. T. 1982. The complex vocal repertoire of the adult cotton-top tamarin (*Saguinus oedipus oedipus*). *Z. Tierpsychol.* 58:231–70.

Clutton-Brock, T. H. 1974. Activity patterns of red colobus (*Colobus badius tephrosceles*). *Folia Primatol.* 21:161–87.

————. 1977. Some aspects of intraspecific variation in feeding and ranging behaviour in primates. In *Primate ecology: Studies of feeding and ranging behaviour in lemurs, monkeys, and apes,* ed. T. H. Clutton-Brock. London: Academic Press.

————. 1982a. Sexual dimorphism in primates. Paper presented at the Ninth Congress of the International Primatological Society, Atlanta.

————. 1982b. Sons and daughters. *Nature* 298:11–13.

————. 1985. Size, sexual dimorphism, and polygymy in primates. In *Size and scaling in primate biology,* ed. W. L. Junger. New York: Plenum Press.

Clutton-Brock, T. H.; Albon, S. D.; Gibson, R. M.; and Guiness, F. E. 1979. The logical stag: Adaptive aspects of fighting in red deer (*Cervus elaphus L.*). *Anim. Behav.* 27:211–25.

Clutton-Brock, T. H.; Guinness, F. E.; and Albon, S. D. 1982. *Red deer: Behavior and ecology of two sexes.* Chicago: University of Chicago Press.

————. 1983. The costs of reproduction to red deer hinds. *J. Anim. Ecol.* 52:367–83.

Clutton-Brock, T. H., and Harvey, P. H. 1976. Evolutionary rules and primate societies. In *Growing points in ethology,* ed. P. P. G. Bateson and R. A. Hinde. Cambridge: Cambridge University Press.

————. 1977a. Primate ecology and social organization. *J. Zool., Lond.* 183:1–39.

————. 1977b. Species differences in feeding and ranging behaviour in primates. In *Primate ecology,* ed. T. H. Clutton-Brock. London: Academic Press.

————. 1978. Mammals, resources, and reproductive strategies. *Nature* 273:191–95.

————. 1979. Home range size, population density, and phylogeny in primates. In *Primate ecology and human origins,* ed. I. S. Bernstein and E. O. Smith. New York: Garland Press.

————. 1980. Primates, brains, and ecology. *J. Zool., Lond.* 190:309–23.

————. 1983. The functional significance of variation in body size among mammals. In *Recent advances in the study of mammalian behavior,* ed. J. F. Eisenberg and D. G. Kleiman. Special Publication of the American Society of Mammalogists, no. 7. Shippensburg, Penn.

————. 1984. Comparative approaches to investigating adaptation. In *Behavioural ecology: An evolutionary approach,* ed. N. B. Davies and J. R. Krebs. Oxford: Blackwell.

Clutton-Brock, T. H.; Harvey, P.; and Rudder, B. 1977. Sexual dimorphism, socionomic sex ratio, and body weight in primates. *Nature* 269:797–99.

Cody, M. L. 1971. Finch flocks in the Mojave Desert. *Theoret. Pop. Biol.* 2:141–58.

Coe, C. L.; Connolly, A. C.; Kraemer, H. C.; and Levine, S. 1979. Reproductive development and behavior of captive female chimpanzees. *Primates* 20:571–82.

Coe, C. L., and Rosenblum, L. A. 1974. Sexual segregation and its ontogeny in squirrel monkey social structure. *J. Hum. Evol.* 3:551–61.

————. 1984. Male dominance in the bonnet macaque: A malleable relationship. In *Social cohesion: Essays toward a sociophysiological perspective,* ed. P. Barchar and S. P. Mendoza. Westport, Conn.: Greenwood Press.

Coelho, A. M. 1974. Socio-bioenergetics and sexual dimorphism in primates. *Primates* 15:263–69.

Coelho, A. M., and Bramblett, C. A. 1981. Sexual dimorphism in the activity of olive baboons (*Papio cynocephalus anubis*) housed in mono-sexual groups. *Arch. Sex. Behav.* 10:79–91.

Coimbra-Filho, A. F., and Mittermeier, R. A. 1973. Distribution and ecology of the genus *Leontopithecus* Lesson, 1840 in Brazil. *Primates* 14:47–66.

————. 1974. New data on the taxonomy of the Brazilian marmosets of the genus *Callithrix* Erxleben, 1777. *Folia Primatol.* 20:241–64.

————. 1978. Tree-gouging, exudate-eating, and the "short-tusked" condition in *Callithrix* and *Cebuella*. In *The biology and conservation of the Callitrichidae,* ed. D. G. Kleiman. Washington, D.C.: Smithsonian Institution Press.

Cole, L. C. 1954. The population consequences of life history phenomena. *Quart. Rev. Biol.* 29:103–37.

Colillas, O., and Coppo, J. 1978. Breeding *Alouatta caraya* in Centro Argentino de Primates. In *Recent advances in primatology,* vol. 2: *Conservation,* ed. D. J. Chivers and W. Lane-Petter. New York: Academic Press.

Collias, N., and Southwick, C. 1952. A field study of population density and social organization in howling monkeys. *Proc. Am. Philos. Soc.* 96:143–56.

Collins, D. A. 1978. Why do some baboons have red bottoms? *New Scient.* 78:12–14.

———. 1981. Social behaviour and patterns of mating among adult yellow baboons (*Papio c. cynocephalus* L. 1766). Ph.D. diss., Cambridge University.

———. 1984. Spatial patterns in a troop of yellow baboons, *Papio cynocephalus*) in Tanzania. *Anim. Behav.* 32:536–53.

Collins, D. A.; Busse, C. D.; and Goodall, J. 1984. Infanticide in two populations of savannah baboons. In *Infanticide: Comparative and evolutionary perspectives,* ed. G. Hausfater and S. B. Hrdy. Hawthorne, N.Y.: Aldine.

Colvin, J. 1983a. Description of sibling and peer relationships among immature male rhesus monkeys. In *Primate social relationships: An integrated approach,* ed. R. A. Hinde. Oxford: Blackwell.

———. 1983b. Influences of the social situation on male emigration. In *Primate social relationships: An integrated approach,* ed. R. A. Hinde. Oxford: Blackwell.

———. 1983c. Rank influences rhesus male peer relationships. In *Primate social relationships: An integrated approach,* ed. R. A. Hinde. Oxford: Blackwell.

Colwell, R. W., and Fuentes, E. R. 1975. Experimental studies of the niche. *Ann. Rev. Ecol. Syst.* 6:281–310.

Conaway, C., and Sade, D. 1965. The seasonal spermatogenic cycle in free ranging rhesus monkeys. *Folia Primatol.* 3:1–12.

Connell, J. H. 1983. On the prevalence and relative importance of interspecific competition: Evidence from field experiments. *Am. Nat.* 122:661–96.

Corbet, G. B., and Hill, J. E. 1980. *A world list of mammalian species.* London: British Museum (Natural History).

Cords, M. 1984a. Mating patterns and social structure in redtail monkeys (*Cercopithecus ascanius*). *Z. Tierpsychol.* 64:313–29.

———. 1984b. Mixed-species groups of *Cercopithecus* monkeys in the Kakamega Forest, Kenya. Ph.D. diss., University of California Berkeley.

———. In press. Interspecific and intraspecific variation in diet of two forest guenons, *Cercopithecus ascanius* and *C. mitis. J. Anim. Ecol.*

Cowey, A., and Weiskrantz, L. 1975. Demonstration of cross-model matching in rhesus monkeys, *Macaca mulatta.* Neuropsychologia 13:117–20.

Cox, C., and Le Boeuf, B. 1977. Female incitation of male competition: A mechanism in sexual selection. *Am. Nat.* 111:317–35.

Craik, K. J. W. 1943. *The nature of explanation.* Cambridge: Cambridge University Press.

von Cranach, M.; Foppa, K.; Lepenies, W.; and Ploog, D. 1979. *Human ethology: Claims and limits of a new discipline.* Editions de la Maison des Sciences de l'Homme. Cambridge: Cambridge University Press.

Crespo, J. A. 1982. Ecología de la comunidad de mamíferos del Parque Nacional Iguazu, Misiones. *Revista de Museo Argentino de Ciencias Naturales "Bernadino Rivadavia".*

Crockett, C. M. 1984. Emigration by female red howler monkeys and the case for female competition. In *Female primates: Studies by women primatologists,* ed. M. F. Small. New York: Alan R. Liss.

———. 1985. Population studies of red howler monkeys (*Alouatta seniculus*). *Nat. Geogr. Res.* 1:264–73.

———. 1986. Diet, dimorphism, and demography: Perspectives from howlers to hominids. In *Primate models for the evolution of human behavior,* ed. W. G. Kinzey. New York: SUNY Press.

Crockett, C. M., and Sekulic, R. 1982. Gestation length in red howler monkeys. *Am. J. Primatol.* 3:291–94.

———. 1984. Infanticide in red howler monkeys (*Alouatta seniculus*). In *Infanticide: Comparative and evolutionary perspectives,* ed. G. Hausfater and S. B. Hrdy. Hawthorne, N.Y.: Aldine.

Crockett, C. M., and Wilson, W. L. 1980. The ecological separation of *Macaca nemestrina* and *M. fascicularis* in Sumatra. In *The macaques,* ed. D. Lindburg. New York: Van Nostrand Reinhold.

Crompton, R. H. 1983. Age differences in locomotion of two subtropical Galaginae. *Primates* 24:241–59.

———. 1984. Foraging, habitat structure, and locomotion in two species of galago. In *Adaptations for foraging in nonhuman primates,* ed. P. S. Rodman and J. G. H. Cant. New York: Columbia University Press.

Crook, J. H. 1965. The adaptive significance of avian social organization. *Symp. Zool. Soc. Lond.* 14:181–218.

———. 1970. Social organization and environment: Aspects of contemporary social ethology. *Anim. Behav.* 18:197–209.

———. 1972. Sexual selection, dimorphism, and social organization in the primates. In *Sexual selection and the descent of man,* ed. B. G. Campbell. Chicago: Aldine.

———. 1980. *The evolution of human consciousness.* Oxford: Clarendon Press.

Crook, J. H.; Ellis, J. E.; and Goss-Custard, J. D. 1976. Mammalian social systems: Structure and function. *Anim. Behav.* 24:261–74.

Crook, J. H., and Gartlan, J. S. 1966. Evolution of primate societies. *Nature* 210:1200–1203.

Cubicciotti, D. D., III, and Mason, W. A. 1975. Comparative studies of social behavior in *Callicebus* and *Saimiri:* Male-female emotional attachments. *Behav. Biol.* 16:185–97.

Curie-Cohen, M.; Yoshihara, D.; Luttrell, L.; Bentorado, K.; MacCluer, J.; and Stone, W. H. 1983. The effects of dominance on mating behavior and paternity in a captive troop of rhesus monkeys (*Macaca mulatta*). *Am. J. Primatol.* 5:127–38.

Curtin, R. 1982. Females, male competition, and grey langur troop structure. *Folia Primatol.* 37:216–27.

Curtin, R., and Dolhinow, P. 1978. Primate social behavior in a changing world. *Am. Scient.* 66:468–75.

Curtin, S. H. 1980. Dusky and banded leaf monkeys. In *Ma-*

layan forest primates: Ten years' study in tropical rain forest, ed. D. J. Chivers. New York: Plenum Press.

Daly, M., and Wilson, M. 1983. *Sex, evolution, and behavior.* 2d ed. Boston: Willard Grant Press.

D'Amato, M., and Salmon, D. 1983. Person concept in cebus monkeys. Paper presented at the annual meeting of the American Psychological Association, Anaheim, Calif.

Dandelot, P. 1971. *The mammals of Africa: An identification manual.* Washington, D.C.: Smithsonian Institution Press.

Dare, R. 1974. The social behavior and ecology of spider monkeys, *Ateles geoffroyi,* on Barro Colorado Island. Ph.D. diss., University of Oregon.

Darling, F. F. 1937. *A herd of red deer.* Oxford: Oxford University Press.

Darwin, C. 1871. *The descent of man and selection in relation to sex.* London: J. Murray.

Dasser, V. 1985. Cognitive complexity in primate social relationships. In *Social relationships and cognitive development,* ed. R. A. Hinde, A. Perret-Clermont and J. Stevenson Hinde. Oxford: Oxford University Press.

Datta, S. B. 1981. Dynamics of dominance among rhesus females. Ph.D. diss., Cambridge University.

———. 1983a. Patterns of agonistic interference. In *Primate social relationships: An integrated approach,* ed. R. A. Hinde. Oxford: Blackwell.

———. 1983b. Relative power and the maintenance of dominance. In *Primate social relationships: An integrated approach,* ed. R. A. Hinde. Oxford: Blackwell.

———. 1983c. Relative power and the acquisition of rank. In *Primate social relationships: An integrated approach,* ed. R. A. Hinde. Oxford: Blackwell.

Davenport, R. K., and Rogers, C. M. 1970. Intermodal equivalence of stimuli in apes. *Science* 168:179–80.

Davies, G., and Payne, J. 1982. *A faunal survey of Sabah.* IUCN/WWF Project No. 1692. World Wildlife Fund—Malaysia, Kuala Lumpur.

Davies, N. B. 1976. Parental care and the transition to independent feeding in the young spotted flycatcher (*Muscicapa striata*). *Behaviour* 59:280–95.

———. 1978. Parental meanness and offspring independence: An experiment with hand-reared great tits, *Parus major. Ibis* 120:509–14.

Davies, N. B., and Houston, A. I. 1984. Territory economics. In *Behavioral ecology: An evolutionary approach,* ed. J. R. Krebs and N. B. Davies. Sunderland, Mass.: Sinauer Association.

Davis, R. T.; Leary, R. W.; Stevens, D. A.; and Thompson, R. 1967. Learning and perception of oddity problems by lemurs and seven species of monkeys. *Primates* 8:311–22.

Dawkins, R. 1976. *The selfish gene.* Oxford: Oxford University Press.

Dawson, G. A. 1978. Composition and stability of social groups of the tamarin, *Saguinus oedipus geoffroyi,* in Panama: Ecological and behavioral implications. In *The biology and conservation of the Callitrichidae,* ed. D. G. Kleiman. Washington, D.C.: Smithsonian Institution Press.

———. 1979. The use of time and space by the Panamanian tamarin, *Saguinus oedipus. Folia Primatol.* 31:253–84.

Deag, J. M. 1973. Intergroup encounters in the wild Barbary macaque *Macaca sylvanus.* In *Comparative ecology and behaviour of primates,* ed. R. P. Michael and J. H. Crook. New York: Academic Press.

———. 1974. A study of the social behaviour and ecology of the wild Barbary macaque, *Macaca sylvanus* L. Ph.D. diss., Bristol University.

———. 1977. Aggression and submission in monkey societies. *Anim. Behav.* 25:465–74.

———. 1983. Feeding habits of *Macaca sylvanus* (Primates: Cercopithecinae) in a commercial Moroccan cedar forest. *J. Zool.* 201:570–74.

Deag, J. M., and Crook, J. M. 1971. Social behaviour and "agonistic buffering" in the wild Barbary macaque *Macaca sylvanus* L. *Folia Primatol.* 15:183–200.

Defler, T. R. 1978. Allogrooming in two species of macaque (*Macaca nemestrina* and *Macaca radiata*). *Primates* 19:153–67.

———. 1979a. On the ecology and behavior of *Cebus albifrons* in northern Colombia, 1: Ecology. *Primates* 20:475–90.

———. 1979b. On the ecology and behavior of *Cebus albifrons* in eastern Colombia, 2: Behavior. *Primates* 20:491–502.

———. 1981. The density of *Alouatta seniculus* in the eastern llanos of Colombia. *Primates* 22:564–69.

———. 1983. Some population characteristics of *Callicebus torquatus lugens* (Humbolt, 1812) (Primates: Cebidae) in eastern Colombia. *Lozania (Acta Zoologica Colombiana)* 38:1–19.

Demment, M. W. 1983. Feeding ecology and the evolution of body size of baboons. *Afr. J. Ecol.* 21:219–33.

Deputte, B. 1982. Duetting in male and female songs of the white-cheeked gibbon (*Hylobates concolor leucogenys*). In *Primate communication,* ed. C. T. Snowdon, C. H. Brown, and M. Petersen. Cambridge: Cambridge University Press.

Deputte, B. L., and Goustard, M. 1980. Copulatory vocalizations of female macaques (*Macaca fascicularis*): Variability factors analysis. *Primates* 21:83–99.

DeVore, I. 1963. Mother-infant relations in free-ranging baboons. In *Maternal behavior in mammals,* ed. H. L. Reingold. New York: John Wiley.

———. 1965a. Male dominance and mating behavior in baboons. In *Sex and behavior,* ed. F. A. Beach. New York: John Wiley.

———, ed. 1965b. *Primate behavior: Field studies of monkeys and apes.* New York: Holt, Rinehart and Winston.

DeVore, I., and Hall, K. R. L. 1965. Baboon ecology. In *Primate behavior: Field studies of monkeys and apes,* ed. I. DeVore. New York: Holt, Rinehart, and Winston.

DeVore, I., and Washburn, S. L. 1963. Baboon ecology and human evolution. In *African ecology and human evolution,* ed. F. C. Howell and F. Bourlière. Chicago: Aldine.

DeVos, A., and Omar, A. 1971. Territories and movements of Sykes monkeys (*Cercopithecus mitis kolbi* Neuman) in Kenya. *Folia Primatol.* 16:196–205.

Dewar, R. E. 1984. Recent extinctions in Madagascar: The loss of the sub-fossil fauna. In *Quarternary extinctions,* ed. P. S. Martin and R. Klein. Tucson: University of Arizona Press.

Dewsbury, D. A. 1982a. Dominance rank, copulatory behavior, and differential reproduction. *Quart. Rev. Biol.* 57:135–59.

———. 1982b. Ejaculate cost and male choice. *Am. Nat.* 119:601–10.

Dickemann, M. 1979. Female infanticide, reproductive strategies, and social stratification: A preliminary model. In *Evolutionary biology and human social behavior: An anthropological perspective,* ed. N. A. Chagnon and W. Irons. N. Scituate, Mass.: Duxbury Press.

Ditmars, R. L. 1933. Development of the silky marmoset. *Bull. N.Y. Zool. Soc.* 36:175–76.

Dittus, W. 1975. Population dynamics of the toque monkey, *Macaca sinica.* In *Socioecology and psychology of primates,* ed. R. H. Tuttle. The Hague: Mouton.

———. 1977. The social regulation of population density and age-sex distribution in the toque monkey. *Behaviour* 63: 281–322.

———. 1979. The evolution of behavior regulating density and age-specific sex ratios in a primate population. *Behaviour* 69:265–302.

———. 1984. Toque macaque food calls: Semantic communication concerning food distribution in the environment. *Anim. Behav.* 32:470–77.

Dixson, A. F. 1976. Effects of testosterone on the sternal cutaneous glands and genitalia of the male greater galago (*Galago crassicaudatus crassicaudatus*). *Folia Primatol.* 26:207–13.

———. 1977. Observations on the displays, menstrual cycles, and sexual behaviour of the "Black Ape" of Celebes. *J. Zool., Lond.* 182:63–84.

———. 1981. *The natural history of the gorilla.* London: Weidenfeld and Nicolson.

———. 1982. Some observations on the reproductive physiology and behavior of the owl monkey. *Int. Zool. Yrbk.* 22: 115–19.

———. 1983. Observations on the evolution and behavioral significance of "sexual skin" in female primates. In *Advances in the study of behavior,* ed. J. S. Rosenblatt, R. A. Hinde, C. Beer, and M.-C. Busnel, 13:63–106. New York: Academic Press.

Dixson, A. F.; Everitt, B.; Herbert, J.; Rugman, S. M.; and Scruton, D. M. 1973. Hormonal and other determinants of sexual attractiveness and receptivity in rhesus and talapoin monkeys. In *Primate reproductive behavior,* vol. 2, ed. C. H. Phoenix. Basel: S. Karger.

Dixson, A. F., and Fleming, D. 1981. Parental behaviour and infant development in owl monkeys (*Aotus trivirgatus griseimembra*). *J. Zool., Lond.* 194:25–39.

Dixson, A. F., and van Horn, R. N. 1977. Comparative studies of morphology and reproduction in two subspecies of the greater bushbaby, *Galago crassicaudatus crassicaudatus* and *G. c. argentatus. J. Zool., Lond.* 183:517–26.

Dobson, F. S. 1982. Competition for mates and predominate juvenile dispersal in mammals. *Anim. Behav.* 30:1183–92.

Döhl, J. 1966. Manipulability and "insightful" behavior of a chimpanzee from complicated behavior chains. *Z. Tierpsychol.* 23:77–113.

———. 1968. The ability of a female chimpanzee to overlook problems with intermediate goals. *Z. Tierpsychol.* 25: 89–103.

Dolhinow, P. 1977. Normal monkeys? *Am. Scient.* 65:266.

Dolhinow, P., and DeMay, M. G. 1982. Adoption: The importance of infant choice. *J. Hum. Evol.* 11:391–420.

Dolhinow, P.; McKenna, J.; and Vonder Haar Laws, J. 1979. Rank and reproduction among female langur monkeys: Aging and improvement (they're not just getting older, they're getting better). *J. Aggr. Behav.* 5:19–30.

Dorst, J., and Dandelot, P. 1969. *A field guide to the larger mammals of Africa.* Boston: Houghton Mifflin.

Doty, B. A.; Jones, C. N.; and Doty, L. A. 1967. Learning set formation by mink, ferrets, skunks, and cats. *Science* 155:1579–80.

Douglas-Hamilton, I., and Douglas-Hamilton, O. 1975. *Among the elephants.* New York: Viking Press.

Downhower, J. F., and Armitage, K. B. 1971. The yellow-bellied marmot and the evolution of polygamy. *Am. Nat.* 105:355–70.

Doyle, G. A. 1974. Behavior of prosimians. In *Behavior of nonhuman primates,* vol. 4, ed. A. M. Schrier and F. Stollnitz. New York: Academic Press.

———. 1979. Development of behavior in prosimians with special reference to the lesser bushbaby, *Galago senegalensis moholi.* In *The study of prosimian behavior,* ed. G. A. Doyle and R. D. Martin. New York: Academic Press.

Doyle, G. A., and Bearder, S. K. 1977. The galagines of South Africa. In *Primate conservation,* ed. H. R. H. Prince Rainier III, and G. H. Bourne. New York: Academic Press.

Dracopoli, N.; Brett, F. L.; Turner, T. R.; and Jolly, C. J. 1983. Patterns of genetic variability in the serum proteins of the Kenyan vervet monkey (*Cercopithecus aethiops*). *Am. J. Phys. Anthropol.* 61:39–49.

Drickamer, L. C. 1974. A ten-year summary of reproductive data for free-ranging *Macaca mulatta. Folia Primatol.* 21: 61–80.

———. 1975. Quantitative observation of behavior in free-ranging *Macaca mulatta:* Methodology and aggression. *Behaviour* 55:209–36.

Drickamer, L. C., and Vessey, S. 1973. Group changing in free-ranging male rhesus monkeys. *Primates* 14:359–68.

Du Chaillu, P. 1930. *Paul Du Chaillu: Gorilla hunter.* New York: Harper Bros.

Duggleby, C. 1978. Blood group antigens and the population genetics of *Macaca mulatta* on Cayo Santiago, 1: Genetic differentiation of social groups. *Am. J. Phys. Anthropol.* 48.35–40.

DuMond, F. V. 1968. The squirrel monkey in a semi-natural environment. In *The squirrel monkey,* ed. L. Rosenblum and R. W. Cooper. New York: Academic Press.

DuMond, F. V., and Hutchinson, T. C. 1967. Squirrel monkey reproduction: The "fatted" male phenomenon and seasonal spermatogenesis. *Science* 158:1067–70.

Dunbar, R. I. M. 1974. The reproductive cycle of the gelada baboon. *Anim. Behav.* 22:203–10.

———. 1976. Some aspects of research design and their implications in the observational study of behaviour. *Behaviour* 58:79–98.

———. 1977. Feeding ecology of gelada baboons: A preliminary report. In *Primate ecology: Studies of feeding and ranging behaviour in lemurs, monkeys, and apes,* ed. T. H. Clutton-Brock. London: Academic Press.

———. 1979a. Population demography, social organization, and mating strategies. In *Primate ecology and human ori-*

gins, ed. I. S. Bernstein and E. O. Smith. New York: Garland Press.

———. 1979b. Structure of gelada baboon reproductive units, 1: Stability of social relationships. *Behaviour* 69:72–87.

———. 1980a. Demographic and life history variables of a population of gelada baboons (*Theropithecus gelada*). *J. Anim. Ecol.* 49:485–506.

———. 1980b. Determinants and evolutionary consequences of dominance among female gelada baboons. *Behav. Ecol. Sociobiol.* 7:253–65.

———. 1983a. Structure of gelada baboon reproductive units, 2: Social relationships between reproductive females. *Anim. Behav.* 31:556–64.

———. 1983b. Structure of gelada baboon reproductive units, 3: The male's relationship with his females. *Anim. Behav.* 31:565–75.

———. 1983c. Structure of gelada baboon reproductive units, 4: Integration at group level. *Z. Tierpsychol.* 63:265–82.

———. 1983d. Relationships and social structure in gelada and hamadryas baboons. In *Primate social relationships: An integrated approach,* ed. R. A. Hinde. Oxford: Blackwell.

———. 1984a. Infant-use by male gelada in agonistic contexts: Agonistic buffering, progeny protection, or soliciting support? *Primates* 25:28–35.

———. 1984b. *Reproductive decisions.* Princeton: Princeton University Press.

Dunbar, R. I. M., and Dunbar, E. P. 1974. The reproductive cycle of the gelada baboon. *Anim. Behav.* 22:203–10.

———. 1975. *Social dynamics of gelada baboons.* Basel: S. Karger.

———. 1976. Contrasts in social structure among black and white colobus monkey groups. *Anim. Behav.* 24:84–92.

———. 1977. Dominance and reproductive success among female gelada baboons. *Nature* 266:351–52.

Dunbar, R. I. M., and Nathan, M. F. 1972. Social organization of the Guinea baboon, *Papio papio. Folia Primatol.* 17:251–52.

Dunbar, R. I. M., and Sharman, M. 1983. Female competition for access to males affects birth rate in baboons. *Behav. Ecol. Sociobiol.* 13:157–59.

Dunford, C. 1977. Kin selection for ground squirrel alarm calls. *Am. Nat.* 111:782–85.

Durham, N. M. 1969. Sex differences in visual threat displays of West African vervets. *Primates* 10:91–95.

———. 1975. Some ecological, distributional, and group behavioral patterns of Atelinae in southern Peru: With comments on interspecific relations. In *Socioecology and psychology of primates,* ed. R. H. Tuttle. The Hague: Mouton.

Duvall, S. W.; Bernstein, I. S.; and Gordon, T. P. 1976. Paternity and status in a rhesus monkey group. *J. Reprod. Fertil.* 47:25–31.

Easley, S. P. 1982. Ecology and behavior of *Callicebus torquatus,* Cebidae, Primates. Ph.D. diss., Washington University.

Easley, S. P., and Kinzey, W. 1985. Territorial shift in the yellow-handed titi monkey. Typescript.

Eaton, G. G. 1973. Social and endocrine determinates of sexual behavior in simian and prosimian females. In *Primate reproductive behavior: Symposium of the Fourth Congress of the International Primatological Society,* vol. 2, ed. C. H. Phoenix. Basel: S. Karger.

Eberhart, J. A.; Keverne, E. B.; and Meller, R. E. 1980. Social influences on plasma testosterone in male talapoin monkeys. *Horm. Behav.* 14:247–66.

Economos, A. C. 1980. Brain-life span conjecture: A re-evaluation of the evidence. *Gerontology* 26:82–89.

Eglash, A. R., and Snowdon, C. T. 1983. Mirror-image responses in pygmy marmosets. *Am. J. Primatol.* 5:211–19.

Eibl-Eibesfeldt, I. 1975. *Ethology,* 2d ed. New York: Holt, Rinehart, and Winston.

Eimas, P. D.; Siqueland, P.; Jusczyk, P.; and Vigorito, J. 1971. Speech perception in infants. *Science* 171:303–6.

Eisenberg, J. F. 1973. Reproduction in two species of spider monkeys, *Ateles fusciceps* and *Ateles geoffroyi. J. Mammal.* 54:955–57.

———. 1976. Communication mechanisms and social integration in the black spider monkey (*Ateles fusciceps robustus*), and related species. Smithsonian Contributions to Zoology, No. 213. Washington, D.C.: Smithsonian Institution Press.

———. 1978. Comparative ecology and reproduction of New World monkeys. In *The biology and conservation of the Callitrichidae,* ed. D. G. Kleiman. Washington D.C.: Smithsonian Institution Press.

———. 1979. Habitat, economy, and society: Some correlations and hypotheses for the Neotropical primates. In *Primate ecology and human origins,* ed. I. S. Bernstein and E. O. Smith. New York: Garland Press.

———. 1981. *The mammalian radiations.* Chicago: University of Chicago Press.

Eisenberg, J. F., and Kuehn, R. E. 1966. The behavior of *Ateles geoffroyi* and related species. *Smithsonian Miscellaneous Collections* 151:1–63.

Eisenberg, J. F.; Muckenhirn, N. A.; and Rudran, R. 1972. The relation between ecology and social structure in primates. *Science* 176:863–74.

Eisenberg, J. F., and Thorington, R. W. 1973. A preliminary analysis of a Neotropical mammal fauna. *Biotropica* 5:150–61.

Ekman, P., and Friesen, W. V. 1969. The repertoire of nonverbal behavior: Categories, origins, usage, and coding. *Semiotica.* 1:49–98.

Ellefson, J. O. 1968a. A natural history of white-handed gibbons in the Malay Peninsula. In *Gibbon and siamang,* vol. 3, ed. D. M. Rumbaugh. Basel: S. Karger.

———. 1968b. Territorial behavior in the common white-handed gibbon, *Hylobates lar* Linn. In *Primates: Studies in adaptation and variability,* ed. P. C. Jay. New York: Holt, Rinehart, and Winston.

———. 1974. A natural history of white-handed gibbons in the Malayan peninsula. In *Gibbon and siamang,* vol. 3, ed. D. M. Rumbaugh. Basel: S. Karger.

Elliott, M. E.; Sehgal, P. D.; and Shalifoux, L. V. 1976. Management and breeding of *Aotus trivirgatus. Lab. Anim. Sci.* 26:1037–44.

Elliot, O., and Elliot, M. 1967. Field notes on the slow loris in Malaya. *J. Mammal.* 48:497–98.

Emlen, J. M. 1973. *Ecology: An evolutionary approach.* Reading, Mass.: Addison-Wesley.

Emlen, S. T. 1982. The evolution of helping, 1: An ecological constraints model. *Am. Nat.* 119:29–39.

———. 1984. Cooperative breeding in birds and mammals. In

Behavioral ecology: An evolutionary approach, ed. J. R. Krebs and N. B. Davies. Sunderland, Mass.: Sinauer.

Emlen, S. T., and Oring, L. W. 1977. Ecology, sexual selection, and the evolution of mating systems. *Science* 197: 215–23.

Emmons, L. H.; Gautier-Hion, A.; and Dubost, G. 1983. Community structure of the frugivorous-folivorous mammals of Gabon. *J. Zool., Lond.* 199:209–22.

Emory, G. R. 1975. Comparison of spatial and orientational relationships as manifestations of divergent modes of social organisation in captive groups of *Mandrillus sphinx* and *Theropithecus gelada. Folia Primatol.* 24:293–314.

———. 1976. Aspects of attention, orientation, and status hierarchy in mandrills (*Mandrillus sphinx*) and gelada baboons. *Behaviour* 59:70–87.

Enomoto, T. 1974. The sexual behavior of Japanese monkeys. *J. Hum. Evol.* 3:351–72.

———. 1978. On social preference in sexual behavior of Japanese monkeys (*Macaca fuscata*). *J. Hum. Evol.* 7:283–93.

———. 1981. Male aggression and sexual behavior of Japanese monkeys. *Primates* 22:15–23.

Enomoto, T.; Seiki, K.; and Haruki, Y. 1979. On the correlation between sexual behavior and ovarian hormone level during the menstrual cycle in captive Japanese monkeys. *Primates* 20:563–70.

Epple, G. 1970. Maintenance, breeding, and development of marmoset monkeys (Callitrichidae) in captivity. *Folia Primatol.* 12:56–76.

———. 1975. Parental behavior in *Saguinus fuscicollis* spp. (*Callithricidae*). *Folia Primatol.* 24:221–38.

Epple, G.; Alveario, M. C.; and Katz, Y. 1982. The role of chemical communication in aggressive behavior and its gonadal control in the tamarin (*Saguinus fuscicollis*). In *Primate communication,* ed. C. T. Snowdon, C. H. Brown, and M. R. Peterson. Cambridge: Cambridge University Press.

Epple, G., and Katz, V. 1984. Social influences on estrogen excretion and ovarian cylicity in saddle back tamarins (*Saguinus fuscicollis*). *Am. J. Primatol.* 6:215–28.

Epple, G., and Lorenz, R. 1967. Vorkommen, Morphologie, und Funcktion der Sternaldruse bei den Platyrrhini. *Folia Primatol.* 7:98–126.

Epple, G., and Moulton, D. 1978. Structural organization and communicatory functions of olfaction in nonhuman primates. In *Sensory systems of primates,* ed. C. R. Noback. New York: Plenum Press.

Epstein, R.; Kirschnit, C.; Lanza, R.; and Rubin, L. 1984. "Insight" in the pigeon: Antecedents and determinants of an intelligent performance. *Nature* 308:61–62.

Essock, S. M.; Gill, T. V.; and Rumbaugh, D. 1977. Language relevant object- and color-naming tasks. In *Language learning by a chimpanzee: The Lana project,* ed. D. M. Rumbaugh. New York: Academic Press.

Essock, S. M., and Rumbaugh, D. M. 1978. Development and measurement of cognitive capabilities in captive non-human primates. In *The behavior of captive wild animals,* ed. H. Markowitz and V. Stevens. New York: Academic Press.

Essock-Vitale, S. M. 1978. Comparison of ape and monkey modes of problem solution. *J. Comp. Physiol. Psychol.* 92: 942–57.

Estep, D. Q.; Johnson, M. E.; and Gordon, T. P. 1981. The effectiveness of sampling methods in detecting copulatory behaviour in *Macaca arctoides. Am. J. Primatol.* 1:453–55.

Estrada, A. 1982. Survey and census of howler monkeys (*Alouatta palliata*) in the rain forest of "Los tuxtlas," Veracruz, Mexico. *Am. J. Primatol.* 2:363–72.

———. 1984a. Male-infant interactions among free-ranging stumptail macaques. In *Primate paternalism,* ed. D. M. Taub. New York: Van Nostrand Reinhold.

———. 1984b. Resource use by howler monkeys (*Alouatta palliata*) in the rain forest of Los Tuxtlas, Veracruz, Mexico. *Int. J. Primatol.* 5:105–31.

Estrada, A., and Sandoval, J. M. 1977. Social relations in a free-ranging troop of stumptail macaques (*Macaca arctoides*): Male-care behaviour. *Primates* 18:793–813.

Eudey, A. 1980. Pleistocene glacial phenomena and the evolution of Asian macaques. In *The macaques,* ed. D. G. Lindburg. New York: Van Nostrand Reinhold.

Evans, C. S. 1980. Disomic responses to scent signals in *Lemur catta.* In *Chemical signals,* ed. H. Muller-Schwarze and J. Silverstein. New York: Plenum Press.

Evans, C. S., and Goy, R. W. 1968. Social behaviour and reproductive cycles in captive ringtailed lemurs (*Lemur catta*). *J. Zool., Lond.* 156:181–97.

Evans, S. 1983. The pair-bond of the common marmoset, *Callithrix jacchus jacchus:* An experimental investigation. *Anim. Behav.* 31:651–58.

Evans, S., and Hodges, J. K. 1984. Reproductive status of adult daughters in family groups of common marmosets (*Callithrix jacchus jacchus*). *Folia Primatol.* 42:127–33.

Evans, S., and Poole, T. B. 1983. Pair-bond formation and breeding success in the common marmoset *Callithrix jacchus jacchus. Int. J. Primatol.* 4:83–97.

———. 1984. Long-term changes and maintenance of the pair-bond in common marmosets *Callithrix jacchus jacchus. Folia Primatol.* 42:33–41.

Everitt, B. J., and Herbert, J. 1969. The role of ovarian hormones in the sexual preference of rhesus monkeys. *Anim. Behav.* 17:738–46.

Ewer, R. F. 1963. The behaviour of the meerkat, *Suricata suricatta* (Schreber). *Z. Tierpsychol.* 20:570–607.

———. 1969. The "instinct to teach." *Nature* 222:698.

Fa, J. E. 1984. Habitat distribution and habitat preference in Barbary macaques (*Macaca sylvanus*). *Int. J. Primatol.* 5:273–86.

Fady, J. C. 1969. Les jeux sociaux: Le compagnon de jeux chez les jeunes. Observatis chez *Macaca iris. Folia Primatol.* 11:134–43.

Fagen, R. 1981. *Animal play behavior.* Oxford: Oxford University Press.

Fairbanks, L. A. 1980. Relationships among adult females in captive vervet monkeys: Testing a model of rank-related attractiveness. *Anim. Behav.* 28:853–59.

———. 1984. Predation by vervet monkeys in an outdoor enclosure: The effect of age, rank, and kinship on prey capture and consumption. *Int. J. Primatol.* 5:263–72.

Fairbanks, L. A., and McGuire, M. T. 1984. Determinants of fecundity and reproductive success in captive vervet monkeys. *Am. J. Primatol.* 7:27–38.

———. 1985. Relationships of vervet monkeys with sons and daughters from one through three years of age. *Anim. Be-*

hav. 33:40–50.

Falconer, D. S. 1961. Introduction to quantitative genetics. New York: Ronald Press.

Farabaugh, S. M. 1982. The ecological and social significance of duetting. In *Acoustic communication in birds,* vol. 2, ed. D. E. Kroodsma and E. H. Miller. New York: Academic Press.

Fedigan, L. 1972. Social and solitary play in a colony of vervet monkeys. *Primates* 13:347–64.

————. 1976. A study of roles in the Arashiyama monkeys (*Macaca fuscata*). In *Contributions to primatology,* vol. 9. Basel: S. Karger.

————. 1982. *Primate paradigms: Sex roles and social bonds.* Montreal: Eden Press.

————. 1983. Dominance and reproductive success in primates. *Yrbk. Phys. Anthropol.* 26:91–129.

Fedigan, L., and Baxter, M. J. 1984. Sex differences and social organization in free-ranging spider monkeys (*Ateles geoffroyi*). *Primates* 25:279–94.

Fedigan, L., and Gouzoules, H. 1978. The consort relationship in a troop of Japanese monkeys, 1: Partner selection. In *Proceedings of the Symposium of the Sixth Congress of the International Primatological Society,* ed. R. Goy. Basel: S. Karger.

Fisher, R. A. 1958. *The genetical theory of natural selection.* New York: Dover.

Fix. A. G. 1978. The role of the kin-structured migration in genetic microdifferentiation, *Ann. Hum. Genet. (Lond.)* 41:239–39.

Fleagle, J. G. 1976. Locomotion and posture of the Malayan siamang and implications for hominoid evolution. *Folia Primatol.* 26:245–69.

Fleagle, J. G., and Mittermeier, R. A. 1980. Locomotor behavior, body size, and comparative ecology of seven Surinam monkeys. *Am. J. Phys. Anthropol.* 52:301–14.

Fletemeyer, J. R. 1978. Communication about potentially harmful foods in free-ranging chacma baboons, *Papio ursinus. Primates* 19:223–26.

Foerg, R. 1982a. Reproduction in *Cheirogaleus medius. Folia Primatol.* 39:49–62.

————. 1982b. Reproductive behavior in *Varecia variegata. Folia Primatol.* 38:108–21.

Fonseca, G. A. B. 1983. The role of deforestation and private reserves in the conservation of the woolly spider monkey (*Brachyteles arachnoides*). M. A. thesis, University of Florida.

Fonseca, G. A. B.; Mittermeier, R. A.; Nishimura, A.; and Valle, C. M. In press. The muriqui, genus *Brachyteles.* In *Ecology and behavior of Neotropical primates,* vol. 2, ed. A. F. Coimbra-Filho and R. A. Mittermeier. Rio de Janeiro: Academia Brasileira de Ciencias.

Fontaine, R. 1981. The uakaris, genus *Cacajao.* In *Ecology and behavior of Neotropical primates,* ed. A. F. Coimbra-Filho and R. A. Mittermeier. Rio de Janeiro: Academia Brasileira de Ciencias.

Fooden, J. 1971. Report on primates collected in western Thailand, January–April 1967. *Fieldiana Zool.* 59:1–62.

————. 1980. Classification and distribution of the living macaques. In *The macaques: Studies in ecology, behavior,*

and evolution, ed. D. Lindburg. New York: Van Nostrand Reinhold.

Foot, H. C.; Chapman, A. J.; and Smith, J. R. 1980. *Friendship and social relations in children.* Chichester: John Wiley.

Fossey, D. 1974a. Observations on the home range of one group of mountain gorillas *Gorilla gorilla beringei. Anim. Behav.* 22:568–81.

————. 1974b. Vocalizations of the mountain gorilla (*Gorilla gorilla*). *Anim. Behav.* 20:36–53.

————. 1979. Development of the mountain gorilla (*Gorilla gorilla beringei*): The first thirty-six months. In *The great apes,* ed. D. A. Hamburg and E. R. McCown. Menlo Park, Calif.: Benjamin/Cummings.

————. 1983. *Gorillas in the mist.* Boston: Houghton Mifflin.

————. 1984. Infanticide in mountain gorillas (*Gorilla gorilla beringei*) with comparative note on chimpanzees. In *Infanticide: Comparative and evolutionary perspectives,* ed. G. Hausfater and S. B. Hrdy. Hawthorne, N.Y.: Aldine.

Fossey, D., and Harcourt, A. H. 1977. Feeding ecology of free-ranging mountain gorillas (*Gorilla gorilla beringei*). In *Primate ecology,* ed. T. H. Clutton-Brock. London: Academic Press.

Foster, R. B. 1982. Famine on Barro Colorado Island. In *The ecology of a tropical forest: Seasonal rhythms and longterm changes,* ed. E. G. Leigh, A. S. Rand, and D. M. Windsor. Washington, D.C.: Smithsonian Institution Press.

Foster, R. B., and Brokow, N. V. L. 1982. Structure and history of the vegetation of Barro Colorado Island. In *The ecology of a tropical forest: Seasonal rhythms and longterm changes,* ed. E. G. Leigh, A. S. Rand, and D. M. Windsor. Washington, D.C.: Smithsonian Institution Press.

Fouts, R. F. 1973. Acquisition and testing of gestural signs in four young chimpanzees. *Science* 180:978–80.

————. 1974. Capacities for language in great apes. In *Proceedings of the Eighteenth International Congress of Anthropological and Ethnological Sciences,* 1–20. The Hague: Mouton.

Fox, G. J. 1982. Potentials for pheromones in chimpanzee vaginal fatty acids. *Folia Primatol.* 37:255–66.

Fox, R. 1967. *Kinship and marriage.* Harmondsworth, England: Penguin.

Fragaszy, D. M. 1978. Contrasts in feeding behavior in squirrel and titi monkeys. In *Recent advances in primatology,* vol. 1, ed. D. J. Chivers and J. Herbert. New York: Academic Press.

————. 1981. Comparative performance in discrimination learning tasks in two New World primates (*Saimiri sciureus* and *Callicebus moloch*). *Anim. Learn. Behav.* 9:127–34.

Fragaszy, D. M.; Schwartz, S.; and Schinosaka, D. 1982. Longitudinal observation of care and development of infant titi monkeys (*Callicebus moloch*). *Am. J. Primatol.* 2:191–200.

Frame, L. H.; Malcolm, J. R.; Frame, G. W.; and van Lawick, H. 1979. Social organization of African wild dogs (*Lycaon pictus*) on the Serengeti plains, Tanzania, 1967–1978. *Z. Tierpsychol.* 50:225–49.

Frankie, G. W.; Baker, H. G.; and Opler, P. A. 1974. Comparative phenological studies of trees in tropical wet and dry forests in the lowlands of Costa Rica. *J. Ecol.* 62:881–919.

Fredrickson, W. T., and Sackett, G. P. 1984. Kin preferences in primates (*Macaca nemestrina*): Relatedness or familiarity? *J. Comp. Psychol.* 98:29–34.

Freeland, W. J. 1976. Pathogens and the evolution of primate sociality. *Biotropica* 8(1):12–24.

Freese, C. H. 1976. Censusing *Alouatta palliata, Ateles geoffroyi,* and *Cebus capucinus* in the Costa Rican dry forest. In *Neotropical primates: Field studies and conservation,* ed. R. W. Thorington and P. G. Heltne. Washington, D.C.: National Academy of Science.

Freese, C. H.; Heltne, P. G.; Castro R. N.; and Whitesides, G. 1982. Patterns and determinants of monkey densities in Peru and Bolivia, with notes on distributions. *Int. J. Primatol.* 3:53–90.

Freese, C. H., and Oppenheimer, J. R. 1981. The capuchin monkeys, genus *Cebus*. In *Ecology and behavior of Neotropical primates,* vol. 1, ed. A. F. Coimbra-Filho and R. A. Mittermeier. Rio de Janeiro: Academia Brasileira de Ciencias.

French, J. A. 1981. Individual differences in play in *Macaca fuscata:* The role of maternal status and proximity. *Int. J. Primatol.* 2:237–46.

French, J. A.; Abbott, D. H.; and Snowdon, C. T. 1984. The effect of social environment on estrogen excretion, scent marking, and sociosexual behavior in tamarins (*Saguinas oedipus*). *Am. J. Primatol.* 6:155–67.

French, M. 1985. *Beyond power: On women, men, and morals.* New York: Summit.

Frisch, R. E. 1982. Malnutrition and fertility. *Science* 215:1271–73.

Frisch, R. E., and McArthur, J. W. 1974. Menstrual cycles: Fatness as a determinant of minimum weight for height necessary for their maintenance or onset. *Science* 185:949–51.

Froehlich, J. W., Jr.; Thorington, R. W., Jr.; and Otis, J. S. 1981. The demography of howler monkeys (*Alouatta palliata*) on Barro Colorado Island, Panama. *Int. J. Primatol.* 2:207–36.

Fujii, H. 1975. A psychological study of the social structure of a free-ranging group of Japanese monkeys in Katsuyama. In *Contemporary primatology: Proceedings of the Fifth Congress of the International Primatological Society,* ed. S. Kondo, M. Kawai, and A. Ehara. Basel: S. Karger.

Furman, W.; Rahe, D. F.; and Hartup, W. W. 1979. Rehabilitation of socially withdrawn preschool children through mixed-age and same-age socialization. *Child Development* 50:915–22.

Furuya, Y. 1957. Grooming behavior in the wild Japanese monkeys. *Primates* 1:47–68.

———. 1965. Social organization of the crab-eating monkey. *Primates* 6:285–337.

———. 1969. On the fission of troops of Japanese monkeys, 2: General view of troop fission of Japanese monkeys. *Primates* 10:47–69.

Gabow, S. L. 1972. Dominance order reversal between two groups of free-ranging rhesus monkeys. *Primates* 14:215–23.

Galat-Luong, A. 1975. Notes preliminaires sur l'écologie de *Cercopithecus ascanius schmidti* dans les environs de Bangui (R.C.A.) *Terre et Vie* 122:288–97.

Galat-Luong, A., and Galat, G. 1979. Conséquences comportementales des perturbations sociales repetées sur une troupe

de Mones de Lowe *Cercopithecus campbelli lowei* de Côte d'Ivoire. *Terre et Vie* 33:4–57.

Galdikas, B. M. F. 1978. Orangutan adaptation at Tanjung Puting Reserve, Central Borneo. Ph.D. diss., University of California, Los Angeles.

———. 1979. Orangutan adaptation at Tanjung Puting Reserve: Mating and ecology. In *The great apes,* ed. D. A. Hamburg and E. R. McCown. Menlo Park, Calif.: Benjamin/Cummings.

———. 1981. Orangutan reproduction in the wild. In *Reproductive biology of the great apes,* ed. C. E. Graham. New York: Academic Press.

———. 1983. The orangutan long call and snag crashing at Tanjung Puting Reserve. *Primates* 24:371–84.

———. 1984. Adult female sociality among wild orangutans at Tanjung Puting Reserve. In *Female primates: Studies by women primatologists,* ed. M. F. Small. New York: Alan R. Liss.

———. 1985. Subadult male orangutan sociality and reproductive behavior at Tanjung Puting. *Am. J. Primatol.* 8:87–99.

Galdikas, B. M. F., and Teleki, G. 1981. Variations in subsistence activities of male and female pongids: New perspectives on the origins of human labor divisions. *Curr. Anthropol.* 22:241–56.

Galef, B. G. 1976. Social transmission of acquired behavior: A discussion of tradition and social learning in vertebrates. In *Advances in the study of behavior,* vol. 6, ed. J. S. Rosenblatt, R. A. Hinde, E. Shaw, and C. Beer. New York: Academic Press.

Gallup, G. G. 1970. Champanzees: Self-recognition. *Science* 167:86–87.

Gallup, G. G., and McClure, M. 1971. Preference for mirror-image stimulation in differentially-reared rhesus monkeys. *J. Comp. Physiol. Psychol.* 78:403–7.

Gallup, G. G.; McClure, M.; and Bundy, R. 1971. Capacity of self recognition in differentially-reared champanzees. *Psychol. Rec.* 21:69–74.

Gandini, G.; and Baldwin, P. J. 1978. An encounter between chimpanzees and a leopard in Senegal. *Carnivore* 1:107–9.

Gandolfi, G., and Parisi, V. 1973. Ethological aspects of predation by rats, *Rattus norvegicus* (Berkenhout), on bivalves *Unio pictorum,* L., and *Cerastoderma lamarcki* (Reeve). *Bollettino di Zoologia* 40:69–74.

Garber, P. A. 1980. Locomotor behavior and feeding ecology of the Panamanian tamarin (*Saguinus oedipus geoffroyi,* Callitrichidae, Primates). *Int. J. Primatol.* 1:185–201.

———. 1984. Proposed nutritional importance of plant exudates in the diet of the Panamanian tamarin, *Saguinus oedipus geoffroyi. Int. J. Primatol.* 5:1–15.

Garber, P. A.; Moya, L.; and Malaga, C. 1984. A preliminary field study of the moustached tamarin monkey (*Saguinus mystax*) in northeastern Peru: Questions concerned with the evolution of a communal breeding system. *Folia Primatol.* 42:17–32.

Gardner, R. A., and Gardner, B. T. 1969. Teaching sign language to a chimpanzee. *Science* 165:664–72.

———. 1975. Early signs of language in child and chimpanzee. *Science* 187:752–53.

Gartlan, J. S. 1968. Structure and function in primate society. *Folia Primatol.* 8:89–120.

———. 1969. Sexual and maternal behavior of the vervet monkey, *Cerocopithecus aethiops. J. Reprod. Fertil. Suppl.* 6:137–50.

———. 1970. Preliminary notes on the ecology and behavior of the drill. In *Old World monkeys,* ed. J. R. Napier and P. H. Napier. New York: Academic Press.

———. 1975. Adaptive aspects of social structure in *Erythrocebus patas.* In *Proceedings from the Symposia of the Fifth Congress of the International Primatological Society,* ed. S. Kondo, M. Kawai, A. Ehara, and S. Kawamura. Tokyo: Japan Science Press.

Gartlan, J. S., and Brain, C. K. 1968. Ecology and social variability in *Cercopithecus aethiops* and *C. mitis.* In *Primates: Studies in adaptation and variability,* ed. P. Jay. New York: Holt, Rinehart, and Winston.

Gartlan, J. S., and Struhsaker, T. T. 1972. Polyspecific associations and niche separation of rain forest anthropoids in Cameroon, West Africa. *J. Zool. Soc. Lond.* 168:221–66.

Gatinot, B. L. 1978. Characteristics of the diet of West African red colobus. In *Recent advances in primatology,* vol. 1, ed. D. J. Chivers and J. Herbert. London: Academic Press.

Gaulin, S. J. C., and Gaulin, C. K. 1982. Behavioral ecology of *Alouatta seniculus* in Andean cloud forest. *Int. J. Primatol.* 3:1–32.

Gaulin, S. J. C.; Knight, D. H.; and Gaulin, C. K. 1980. Local variance in *Alouatta* group size and food availability on Barro Colorado Island. *Biotropica* 12:137–43.

Gaulin, S. J. C., and Sailer, L. 1984. Sexual dimorphism in weight among the primates: The relative impact of allometric and sexual selection. *Int. J. Primatol.* 5:515–35.

———. 1985. Are females the ecological sex? *Am. Anthropol.* 87:111–19.

Gauthreaux, S. A. 1978. Ecological significance of behavioral dominance. In *Perspectives in ethology,* vol. 3, ed. P. P. G. Bateson and P. H. Klopfer. New York: Plenum Press.

Gautier, J.-P. 1974. Field and laboratory studies of the vocalizations of talapoin monkeys (*Miopithecus talapoin*). *Behaviour* 51:209–73.

Gautier, J.-P., and Gautier-Hion, A. 1969. Les associations poly-specifiques chez les Cercopithecidae du Gabon. *Terre et Vie* 23:164–201.

———. 1983. Comportement vocal des males adultes et organisation dans les troupes polyspecifiques de cercopitheques. *Folia Primatol.* 40:161–74.

Gautier-Hion, A. 1971a. L'écologie du talapoin du Gabon, *Miopithecus talapoin. Terre et Vie* 4:427–90.

———. 1971b. L'organisation sociale d'un bande de talapoins (*Miopithecus talapoin*) dans le nord-est du Gabon. *Folia Primatol.* 12:116–41.

———. 1973. Social and ecological features of talapoin monkeys: Comparisons with sympatric cercopithecines. In *Comparative ecology and behaviour of primates,* ed. R. P. Michael and J. H. Crook. New York: Academic Press.

———. 1975. Dimorphisme sexuelle et organisation sociale chez les Ceropithecines forestiers africains. *Mammalia* 39: 365–74.

———. 1978a. Food niches and coexistence in sympatric primates in Gabon. In *Recent advances in primatology,* vol. 1, ed. D. J. Chivers and J. Herbert. New York: Academic Press.

———. 1980. Seasonal variations of diet related to species and sex in a community of *Cercopithecus* monkeys. *J. Anim. Ecol.* 49:237–69.

Gautier-Hion, A.; Emmons, L. H.; and Dubost, G. 1980. A comparison of the diets of three major groups of primary consumers of Gabon (primates, squirrels, and ruminants). *Oecologia* 45:182–89.

Gautier-Hion, A., and Gautier, J.-P. 1974. Les associations polyspecifiques de Cercopitheques du Plateau de M'passa (Gabon). *Folia Primatol.* 22:134–77.

———. 1976. Croissance, maturité sexuelle et sociale, reproduction chez les cercopithecines forestiers africains. *Folia Primatol.* 22:134–77.

———. 1978. Le singe de Brazza: Une stratégie originale. *Z. Tierpsychol.* 46:84–104.

———. 1979. Niche écologiques et diversite des espèces sympatriques dans le genre *Cercopithecus. Terre et Vie* 33: 493–507.

Gautier-Hion, A.; Quris, R.; and Gautier, J.-P. 1983. Monospecific vs. polyspecific life: A comparative study of foraging and antipredatory tactics in a community of *Cercopithecus* monkeys. *Behav. Ecol. Sociobiol.* 12:325–35.

Geist, V. 1971. *Mountain sheep: A study in behavior and evolution.* Chicago: University of Chicago Press.

Gelman, R., and Gallistel, C. R. 1978. *The child's understanding of number.* Cambridge: Harvard University Press.

Gelman, R., and Spelke, E. 1981. The development of thoughts about animate and inanimate objects. In *Social cognitive development,* ed. J. H. Flavell and L. Ross. Cambridge: Cambridge University Press.

Ghiglieri, M. P. 1984. *The chimpanzees of Kibale Forest.* New York: Columbia University Press.

———. 1985. The social ecology of chimpanzees. *Sci. Am.* 252:102–13.

Gibber, J. R. 1981. Infant-directed behaviors in male and female rhesus monkeys. Ph.D. diss., University of Wisconsin–Madison.

Gibber, J. R., and Goy, R. W. 1985. Infant-directed behavior in young rhesus monkeys: Sex differences and effects of prenatal androgens. *Am. J. Primatol.* 8:225–37.

Gill, T., and Rumbaugh, D. M. 1974. Learning processes of bright and dull apes. *J. Ment. Defic.* 78:683–87.

Gillan, D. J.; Premack, D.; and Woodruff, G. 1981. Reasoning in the chimpanzee, 1: Analogical reasoning. *J. Exp. Psychol.: Anim. Behav. Proc.* 7:1–17.

Gittins, S. P. 1980. Territorial behavior in the agile gibbon. *Int. J. Primatol.* 1:381–99.

———. 1982. Feeding and ranging in the agile gibbon. *Folia Primatol.* 38:39–71.

———. 1984. Territorial advertisement and defence in gibbons. In *The lesser apes: Evolutionary and behavioural biology,* ed. H. Preuschoft, D. J. Chivers, W. Y. Brockelman, and N. Creel. Edinburgh: Edinburgh University Press.

Gittins, S. P., and Raemaekers, J. J. 1980. Siamang, lar, and agile gibbons. In *Malayan forest primates: Ten years' study in tropical rain forest,* ed. D. J. Chivers. New York: Plenum Press.

vervet monkeys in Uganda. In *The baboon in medical research*, vol. 1, ed. H. Vagtborg. Austin: University of Texas Press.

———. 1967. Social interactions of the adult male and adult females of a patas monkey group. In *Social communication among primates*, ed. S. A. Altmann. Chicago: Chicago University Press.

———. 1968a. Behavior and ecology of the wild patas monkey, *Erythrocebus patas* in Uganda. In *Primates: Studies in adaptation and variability*, ed. P. C. Jay. New York: Holt, Rinehart, and Winston.

———. 1968b. Social learning in monkeys. *Primates: Studies in adaptation and variability*, ed. P. C. Jay. New York: Holt, Rinehardt, and Winston.

Hall, K. R. L., and DeVore, I. 1965. Baboon social behavior. In *Primate behavior*, ed. I. DeVore. New York: Holt, Rinehardt, and Winston.

Hall, K. R. L., and Gartlan, J. S. 1965. Ecology and behaviour of the vervet monkey, *C. aethiops*, Lolui Island, Lake Victoria. *Proc. Zool. Soc. Lond.* 145:37–56.

Hall, K. R. L., and Mayer, B. 1967. Social interactions in a group of captive patas monkeys. *Folia Primatol.* 5:213–36.

Hall, R. D.; Beattie, R. J.; Wychoff, G. H., Jr. 1979. Weight gains and sequence of dental eruptions in infant owl monkeys (*Aotus trivirgatus*). In *Nursery care of nonhuman primates*, ed. G. C. Ruppenthal. New York: Plenum Press.

Halliday, T. R. 1983. The study of mate choice. In *Mate choice*, ed. P. P. G. Bateson. Cambridge: Cambridge University Press.

Haltenorth, T., and Diller, H. 1980. *A field guide to the mammals of Africa*. London: Collins.

Hamburg, D. A., and Trudeau, M. B., eds. 1981. *Biobehavioral aspects of aggression*. New York: Alan R. Liss.

Hamilton, A. C. 1982. *Environmental history of East Africa: A study of the quaternary*. London: Academic Press.

Hamilton, C. R. 1977. An assessment of hemispheric specialization in monkeys. *Ann. N.Y. Acad. Sci.* 299–332.

Hamilton, W. D. 1964. The genetical evolution of social behavior. *J. Theoret. Biol.* 7:1–51.

Hamilton, W. D., and Zuk, M. 1982. Heritable true fitness and bright birds: A role for parasites? *Science* 218:384–87.

Hamilton, W. J. 1984. Significance of paternal investment by primates to the evolution of male-female associations. In *Primate paternalism*, ed. D. M. Taub. New York: Van Nostrand Reinhold.

Hamilton, W. J., and Arrowood, P. C. 1978. Copulatory vocalizations of chacma baboons (*Papio ursinus*), gibbons (*Hylobates hoolock*), and humans. *Science* 200:1405–9.

Hamilton, W. J.; Buskirk, R. E.; and Buskirk, W. H. 1975a. Chacma baboon tactics during intertroop encounters. *J. Mammal.* 56:857–70.

———. 1975b. Defensive stoning by baboons. *Nature* 256:488–89.

———. 1976. Defense of space and resources by chacma (*Papio ursinus*) baboon troops in an African desert and swamp. *Ecology* 57:1264–72.

———. 1978. Omnivory and utilization of food resources by chacma baboons, *Papio ursinus*. *Am. Nat.* 112:911–24.

Hamilton, W. J., and Busse, C. D. 1982. Social dominance and predatory behavior of chacma baboons. *J. Hum. Evol.* 11:567–74.

Hamilton, W. J.; Busse, C.; and Smith, K. S. 1982. Adoption of infant orphan chacma baboons. *Anim. Behav.* 30:29–34.

Hamilton, W. J., and Tilson, R. L. 1980. Solitary male chacma baboons in a desert canyon. *Am. J. Primatol.* 2:149–58.

Hampton, S. H., and Hampton, J. K., Jr. 1978. Detection of reproductive cycles and pregnancy in tamarins (*Saguinus* spp.). In *The biology and conservation of the Callitrichidae*, ed. D. G. Kleiman. Washington, D.C.: Smithsonian Institution Press.

Hanby, J. P. 1976. Sociosexual development in primates. In *Perspectives in ethology*, vol. 2, ed. P. P. G. Bateson and P. H. Klopfer. New York: Plenum Press.

Hanby, J. P., and Brown, C. E. 1974. The development of sociosexual behaviours in Japanese macaques, *Macaca fuscata*. *Behaviour* 49:152–96.

Hanby, J. P.; Robertson, L. T.; and Phoenix, C. 1971. The sexual behavior of a confined troop of Japanese macaques. *Folia Primatol.* 16:123–43.

Hansen, E. W. 1966. The development of maternal and infant behavior in the rhesus monkey. *Behaviour* 27:109–49.

———. 1976. Selective responding by recently separated juvenile rhesus monkeys to the calls of their mothers. *Dev. Psychobiol.* 9:83–88.

Happel, R. 1982. Ecology of *Pithecia hirsuta* in Peru. *J. Hum. Evol.* 11:581–90.

Harcourt, A. H. 1978a. Activity periods and patterns of social interaction: A neglected problem. *Behaviour* 66:121–35.

———. 1978b. Strategies of emigration and transfer by primates, with particular reference to gorillas. *Z. Tierpsychol.* 48:401–20.

———. 1979a. Contrasts between male relationships in wild gorilla groups. *Behav. Ecol. Sociobiol.* 5:39–49.

———. 1979b. Social relationships among female mountain gorillas. *Anim. Behav.* 27:251–64.

———. 1979c. Social relationships between adult male and female mountain gorillas in the wild. *Anim. Behav.* 27:325–42.

———. 1984. Conservation of the Virunga gorillas. *IUCN/SSC Primate Specialist Group Newsletter* 4:36–37.

Harcourt, A. H.; Fossey, D.; and Sabater Pi, J. 1981. Demography of *Gorilla gorilla*. *J. Zool., Lond.* 195:215–33.

Harcourt, A. H.; Fossey, D.; Stewart, K.; and Watts, D. P. 1980. Reproduction in wild gorillas and some comparisons with chimpanzees. *J. Reprod. Fertil. Suppl.* 28:59–70.

Harcourt, A. H.; Harvey, P. H.; Larson, S. G.; and Short, R. V. 1981. Testis weight, body weight, and breeding system in primates. *Nature* 293:55–57.

Harcourt, A. H.; Kineman, J.; Campbell, G.; Yamagiwa, J.; Redmond, I.; Aveling, C.; and Condiotti, M. 1983. Conservation and the Virunga gorilla population. *Afr. J. Ecol.* 21:139–42.

Harcourt, A. H., and Stewart, K. J. 1981. Gorilla male relationships: Can differences during immaturity lead to contrasting reproductive tactics in adulthood? *Anim. Behav.* 29:206–10.

———. 1983. Interactions, relationships, and social structure: The great apes. In *Primate social relationships: An integrated approach*, ed. R. A. Hinde. Oxford: Blackwell.

———. 1984. Gorillas' time feeding: Aspects of methodology, body size, competition and diet. *Afr. J. Ecol.* 22:

207–15.

Harcourt, A. H.; Stewart, K. J.; and Fossey, D. 1976. Male emigration and female transfer in wild mountain gorilla. *Nature* 263:226–27.

Harcourt, C. S. 1980. Behavioral adaptations in South African galagos. M. S. diss., University of the Witwaterand, Johannesburg.

———. 1981. An examination of the function of urine washing in *Galago senegalensis*. *Z. Tierpsychol.* 55:119–28.

———. 1984. The behaviour and ecology of galagos in Kenyan coastal forest. Ph.D. diss., Cambridge University.

Harcourt, C. S., and Nash, L. T. In press. Social organization of galagos in Kenyan coastal forests, 1: *Galago zanzibaricus*. *Am. J. Primatol.*

Hardin, G. 1960. The competitive exclusion principle. *Science* 131:1292–97.

Harding, R. S. O. 1976. Ranging patterns of a troop of baboons (*Papio anubis*) in Kenya. *Folia Primatol.* 25:143–85.

———. 1977. Patterns of movement in open country baboons. *Am. J. Phys. Anthropol.* 47:349–54.

———. 1980. Agonism, ranking, and the social behavior of adult male baboons. *Am. J. Anthropol.* 53:203–16.

———. 1981. An order of omnivores: Nonhuman primate diets in the wild. In *Omnivorous primates: Gathering and hunting in human evolution,* ed. R. S. O. Harding and G. Teleki. New York: Columbia University Press.

Harlow, H. F. 1949. The formation of learning sets. *Psychol. Rev.* 56:51–65.

———. 1958. The nature of love. *Am. Psychol.* 13:673–85.

———. 1969. Age-mate or peer affectional systems. In *Advances of the Study of Behavior,* vol. 2., ed. D. Lehrman, R. A. Hinde, and E. Shaw. New York: Academic Press.

———. 1971. *Learning to love.* New York: Ballantine Books.

Harlow, H. F., and Zimmermann, R. R. 1959. Affectional responses in the infant monkey. *Science* 130:421–32.

Harrington, J. 1974. Olfactory communication in *Lemur fulvus*. In *Prosimian biology,* ed. R. D. Martin, G. A. Doyle, and A. C. Walker. London: Duckworth.

———. 1975. Field observations of social behavior of *Lemur fulvus fulvus* E. geoffroy 1812. In *Lemur biology,* ed. I. Tattersall and R. Sussman. New York: Plenum Press.

Harrison, M. J. S. 1983a. Patterns of range use by the green monkey, *Cercopithecus sabaeus,* at Mt. Assirik, Senegal. *Folia Primatol.* 41:157–79.

———. 1983b. Territorial behaviour in the green monkey, *Cercopithecus sabaeus:* Seasonal defense of local food supplies. *Behav. Ecol. Sociobiol.* 12:85–94.

Harvey, P. H., and Bennett, P. M. 1983. Brains size, energetics, ecology, and life history patterns. *Nature* 306:314–15.

Harvey, P. H., and Clutton-Brock, T. H. 1985. Life history variation in primates. *Evolution* 39:559–81.

Harvey, P. H., and Greenwood, P. J. 1978. Anti-predator defense strategies: Some evolutionary problems. In *Behavioural ecology: An evolutionary approach,* ed. J. R. Krebs and N. B. Davies. Sunderland, Mass.: Sinauer Associates.

Harvey, P. H.; Kavanagh, M.; and Clutton-Brock, T. H. 1978a. Canine tooth size in female primates. *Nature* 276:817–18.

———. 1978b. Sexual dimorphism in primate teeth. *J. Zool. Soc. Lond.* 186:475–85.

Harvey, P. H., and Mace, G. M. 1982. Comparisons between taxa and adaptive trends. In *Current problems in sociobiology,* ed. King's College Sociobiology Group. Cambridge: Cambridge University Press.

Hasegawa, T., and Hiraiwa M. 1980. Social interactions of orphans observed in a free-ranging troop of Japanese monkeys. *Folia primatol.* 33:129–58.

Hasegawa, T. and Hiraiwa-Hasegawa, M. 1983. Opportunistic and restrictive matings among wild chimpanzees in the Mahale Mountains, Tanzania. *J. Ethol.* 1:75–85.

Hasegawa T.; Hiraiwa-Hasegawa, M.; Nishida, T.; and Takasaki, H. 1983. New evidence of scavenging behavior of wild chimpanzees. *Curr. Anthropol.* 24:231–32.

Hauser, M. D.; Cheney, D. L.; and Seyfarth, R. M. In press. Group extinction and fusion in free-ranging vervet monkeys. *Am. J. Primatol.*

Hausfater, G. 1972. Intergroup behavior of free-ranging rhesus monkeys (*Macaca mulatta*). *Folia Primatol.* 18:78–107.

———. 1975. Dominance and reproduction in baboons (*Papio cynocephalus*). In *Contributions to primatology,* vol. 7. Basel: S. Karger.

———. 1976. Predatory behavior of yellow baboons. *Behaviour* 56:44–68.

———. 1981. Long-term consistency of dominance relations in baboons (*Papio cynocephalus*). *Am. J. Primatol.* 1:310.

Hausfater, G.; Altmann, J.; and Altmann, S. 1982. Long-term consistency of dominance relations among female baboons (*Papio cynocephalus*). *Science* 217:752–55.

Hausfater, G., and Hrdy S. B., eds. 1984. *Infanticide: Comparative and evolutionary perspectives.* Hawthorne, N.Y.: Aldine.

Hausfater, G.; Saunders, C. D.; and Chapman, M. 1981. Some applications of computer models to the study of primate mating and social systems. In *Natural selection and social behavior,* ed. R. D. Alexander and D. W. Tinkle. New York: Chiron Press.

Hausfater, G., and Vogel, C. 1982. Infanticide in langurs (*Presbytis entellus*): Recent research and a review of hypotheses. In *Advanced views in primate biology,* ed. A. B. Chiarelli and R. S. Corruccini. Berlin: Springer-Verlag.

Hawkes, K.; Hill, K.; and O'Connell, J. F. 1982. Why hunters gather: Optimal foraging and the Aché of eastern Paraguay. *Am. Ethnol.* 9:379–98.

Hayaki, H. 1983. The social interactions of juvenile Japanese monkeys on Koshima Islet. *Primates* 24:139–53.

Hayes, C. 1951. *The ape in our house.* New York: Harper.

Hearn, J. P. 1978. The endocrinology of reproduction in the common marmoset, *Callithrix jacchus.* In *The biology and conservation of the Callitrichidae,* ed. D. G. Kleiman. Washington, D.C.: Smithsonian Institution Press.

Hefner, H. E., and Hefner, R. S. 1984. Temporal lobe lesions and perception of species-specific vocalizations by macaques. *Science* 226:75–76.

Helfman, G. S. 1984. School fidelity in fishes: The yellow perch pattern. *Anim. Behav.* 32:663–73.

Heltne, P. G.; Turner, D. C.; and Scott, N. J., Jr. 1976. Comparison of census data on *Alouatta palliata* from Costa Rica and Panama. In *Neotropical primates: Field studies and conservation,* ed. R. W. Thorington, Jr., and P. G. Heltne. Washington, D.C.: National Academy of Sciences.

Heltne, P. G.; Wojcik, T. K.; and Pook, A. G. 1981. Goeldi's monkey, genus *Callimico*. In *Ecology and behavior of neotropical primates,* vol. 1, ed. A. F. Coimbra-Filho and R. A. Mittermeier. Rio de Janeiro: Academia Brasileira de Ciencias.

Hendrickx, A. G., and Kraemer, D. C. 1969. Observation of the menstrual cycle, optimal mating time, and preimplantation embryos of the baboon, *Papio anubis* and *Papio cynocephalus*. *J. Reprod. Fertil. Suppl.* 6:119–28.

Hennessy, M. B.; Mendoza, S. P.; Coe, C. L.; Lowe, E. L.; and Levine, S. 1980. Androgen-related behavior in the squirrel monkey: An issue that is nothing to sneeze at. *Behav. Neural Biol.* 30:103–8.

Henzi, S. P. 1981. Causes of testis-adduction in vervet monkeys *Cercopithecus aethiops pygerythrus*). *Anim. Behav.* 29:961.

Henzi, S. P., and Lucas, J. W. 1980. Observations on the intertroop movement of adult vervet monkeys (*Cercopithecus aethiops*). *Folia Primatol.* 33:220–35.

Herbert, J. 1970. Hormones and reproductive behavior in rhesus and talapoin monkeys. *J. Reprod. Fert. Suppl.* 11:119–40.

Herman, L. M.; Beach, F.; Pepper, R. L.; and Stalling, R. B. 1969. Learning set formation in the bottlenose dolphin. *Psychonom. Sci.* 14:98–99.

Hernández-Camacho, J., and Cooper, R. W. 1976. The nonhuman primates of Colombia. In *Neotropical primates: Field studies and conservation,* ed. R. W. Thorington, Jr., and P. G. Heltne. Washington, D.C.: National Academy of Sciences.

Herrnstein, R. 1979. Acquisition, generalization, and discrimination reversal of a natural concept. *J. Exp. Psychol.: Anim. Behav. Proc.* 5:116–29.

Herrnstein, R., and Loveland, D. 1964. Complex visual concepts in the pigeon. *Science* 146:549–51.

Hershkovitz, P. 1977. *Living New World monkeys (Platyrrhini),* vol. 1. Chicago: University of Chicago Press.

———. 1979. The species of sakis, genus *Pithecia* (Cebidae, Primates), with notes on sexual dichromatism. *Folia Primatol.* 31:1–22.

———. 1983. Two new species of night monkeys, genus *Aotus* (Cebidae, Platyrrhini): A preliminary report on *Aotus* taxonomy. *Am. J. Primatol.* 4:209–43.

———. 1984. Taxonomy of squirrel monkeys, genus *Saimiri* (Cebidae, Platyrrhini): A preliminary report with description of a hitherto unnamed form. *Am. J. Primatol.* 6:257–312.

———. In press. A preliminary taxonomic review of the South American bearded saki monkeys, genus *Chiropotes* (Cebidae, Platyrrhini), with the description of a new form. *Fieldiana.*

Hess, D. L., and Resko, J. A. 1973. The effects of progesterone on the patterns of testosterone and estradiol concentrations in the systemic plasma of the female rhesus monkey during the intermenstrual period. *Endocrinology* 92:446–53.

Hick, U. 1968a. Erstmalig gelungene Zucht eines Bartsakis (Vater: Rotruckensaki, *Chiropotes chiropotes* (Humbolt, 1811), Mutter: Weissnasensaki, *Chiropotes albinasus* (Geoffroy et Deville 1848) im Kolner Zoo. *Freunde des Kolner Zoo* 11:35–41.

———. 1968b. The collection of saki monkeys at Cologne Zoo.

Int. Zool. Yrbk. 8:192–94.

———. 1973. Wir sind umgezogen. *Z. Kolner Zoo* 16:127–45.

Hill, A. 1976. Non-aggressive tactile interactions of *Hippopotamus amphibius* Linn. with *Syncerus caffer* (Sparrman). *Mammalia* 40:161–72.

Hill, J. L. 1974. *Peromyscus:* Effect of early pairing on reproduction. *Science* 186:1042–44.

Hill, W. C. O. 1952. The external and visceral anatomy of the olive colobus monkey (*Procolobus verus*). *Proc. Zool. Soc. London* 122:127–86.

———. 1962. *Primates: Comparative anatomy and taxonomy,* vol. 5: *Cebidae,* part B. New York: Wiley Interscience.

———. 1966. *Primates: Comparative anatomy and taxonomy,* 6: *Cercopithecinae.* Edinburgh: Edinburgh University Press.

———. 1970. *Primates: Comparative anatomy, and taxonomy,* vol. 8: *Cynopithecinae.* Edinburgh: Edinburgh University Press.

Hinde, R. A. 1969. Analysing the roles of the partners in a behavioural interaction: Mother-infant relations in rhesus macaques. *Ann. N.Y. Acad. Sci.* 159:651–67.

———. 1973. On the design of check-sheets. *Primates* 14:393–406.

———. 1974a. *Biological bases of human social behaviour.* New York: McGraw-Hill.

———. 1974b. Mother/infant relations in rhesus monkeys. In *Ethology and psychiatry,* ed. N. F. White. Toronto: University of Toronto Press.

———. 1976. Interactions, relationships, and social structure. *Man* 11:1–17.

———. 1978. Dominance and role: Two concepts with dual meanings. *J. Soc. Biol. Struct.* 1:27–38.

———. 1979. *Towards understanding relationships.* London: Academic Press.

———. 1983a. A conceptual framework. In *Primate social relationships: An integrated approach,* ed. R. A. Hinde. Oxford: Blackwell.

———. 1983b. Ethology and child development. In *Handbook of child psychology;* vol. 2, ed. P. H. Mussen. New York: John Wiley.

———, ed. 1983c. *Primate social relationships: An integrated approach.* Oxford: Blackwell.

———. 1984a. Biological bases of the mother-child relationships. In *Frontiers of infant psychiatry,* vol. 2, ed. J. Call, E. Galensen, and R. L. Tyson. New York: Basic Books.

———. 1984b. Why do the sexes behave differently in close relationships? *J. Soc. Pers. Relation.* 1:471–501.

Hinde, R. A., and Atkinson, S. 1970. Assessing the roles of social partners in maintaining mutual proximity, as exemplified by mother-infant relations in rhesus monkeys. *Anim. Behav.* 18:169–76.

Hinde, R. A., and Rowell, T. E. 1962. Communication by postures and facial expressions in the rhesus monkey (*Macaca mulatta*). *Proc. Zool. Soc. London* 138:1–21.

Hinde, R. A., and Simpson, M. J. A. 1975. Qualities of mother-infant relationships in monkeys. In *Parent-infant interaction.* Ciba Foundation Symposium 33, Amsterdam.

Hinde, R. A., and Spencer-Booth, Y. 1967. The behaviour of socially living rhesus monkeys in their first two and a half years. *Anim. Behav.* 15:169–96.

Hinde, R. A., and Stevenson-Hinde, J. 1986. Relationships, personality, and the social situation. In *Key issues in interpersonal relationships,* ed. R. Gilmour and S. Duck. Hillsdale, N.J.: Erlbaum.

————, eds. 1973. *Constraints on learning: Limitations and predispositions.* New York: Academic Press.

Hinde, R. A., and White, L. E. 1974. Dynamics of a relationship: Rhesus mother-infant ventro-ventral contact. *J. Comp. Physiol. Psychol.* 86:8–23.

Hiraiwa, M. 1981. Maternal and alloparental care in a troop of free-ranging Japanese monkeys. *Primates* 22:309–29.

Hiraiwa-Hasegawa, M.; Hasegawa, T.; and Nishida, T. 1984. Demographic study of a large-sized unit-group of chimpanzees in the Mahale Mountains, Tanzania: A preliminary report. *Primates* 25:401–13.

Hladik, A. 1978. Phenology of leaf production in rain forest of Gabon: Distribution and composition of food for folivores. In *The ecology of arboreal folivores,* ed. G. G. Montgomery. Washington, D.C.: Smithsonian Institution Press.

Hladik, A., and Hladik, C. M. 1969. Rapports trophiques entre végétation et Primates dans la forêt de Barro Colorado (Panama). *Terre et Vie* 23:25–117.

Hladik, C. M. 1975. Ecology, diet, and social patterns in Old and New World primates. In *Socioecology and psychology of primates,* ed. R. H. Tuttle. The Hague: Mouton.

————. 1977. Chimpanzees of Gabon and chimpanzees of Gombe: Some comparative data on diet. In *Primate ecology,* ed. T. H. Clutton-Brock. New York: Academic Press.

————. 1978. Adaptive strategies of primates in relation to leaf-eating. In *The ecology of arboreal folivores,* ed. G. G. Montgomery. Washington, D.C.: Smithsonian Institution Press.

————. 1979. Diet and ecology of prosimians. In *The study of prosimian behavior,* ed. G. A. Doyle and R. D. Martin. New York: Academic Press.

Hladik, C. M., and Charles-Dominique, P. 1974. The behaviour and ecology of the sportive lemur (*Lepilemur mustelinus*) in relation to its dietary peculiarities. In *Prosimian biology,* ed. R. D. Martin, G. A. Doyle, and A. C. Walker. London: Duckworth.

Hladik, C. M.; Charles-Dominique, P.; and Petter, J. J. 1980. Feeding strategies of five nocturnal prosimians in the dry forest of the west coast of Madagascar. In *Nocturnal Malagasy primates,* ed. P. Charles-Dominique, H. M. Cooper, A. Hladik, C. M. Hladik, E. Pages, G. F. Pariente, A. Petter-Rousseaux, and A. Schilling. New York: Academic Press.

Hladik, C. M., and Hladik, A. 1972. Disponibilités alimentaires et domaines vitaux des primates à Ceylan. *Terre et Vie* 26:149–215.

Hoage, R. J. 1978. Parental care in *Leontopithecus rosalia rosalia:* Sex and age differences in carrying behavior and the role of prior experience. In *the biology and conservation of the Callitrichidae,* ed. D. G. Kleiman. Washington, D.C.: Smithsonian Institution Press.

Hockett, C. F. 1960. Logical considerations in the study of animal communication. In *Animal sounds and communicaton,* ed. W. E. Lanyon and W. N. Tavolga. Washington, D.C.: American Institute of Biological Sciences.

Hodges, T. K., and Eastman, S. A. K. 1984. Monitoring ovarian function in marmosets and tamarins by the measurement of urinary estrogen metabolites. *Am. J. Primatol.* 6:187–97.

Hodos, W. 1982. Some perspectives on the evolution of intelligence and the brain. In *Animal mind—Human mind,* ed. D. R. Griffin. New York: Springer-Verlag.

Hodos, W., and Campbell, C. B. G. 1969. Scala naturae: Why there is no theory in comparative psychology. *Psychol. Rev.* 76:337–50.

Hodun, A.; Snowdon, C. T.; and Soini, P. 1981. Subspecific variation in the long calls of the tamarin, *Saguinus fuscicollis Z. Tierpsychol.* 57:97–110.

Hoffman, M. L. 1981. Perspectives on the difference between understanding people and understanding things: The role of affect. In *Social cognitive development,* ed. J. H. Flavell and L. Ross. Cambridge: Cambridge University Press.

Hogden, G. D.; Goodman, A. K.; O'Conner, A.; and Johnson, D. K. 1977. Menopause in rhesus monkeys: Model for study of disorders in the human climacteric. *Am. J. Obstet. Gynec.* 127:581–84.

Hogg, J. T. 1984. Mating in bighorn sheep: Multiple creative male strategies. *Science* 225:526–29.

Holmes, R. J., and Pitelka, F. A. 1968. Food overlap among coexisting sandpipers on northern Alaskan tundra. *Syst. Zool.* 17:305–18.

Holmes, W. G., and Sherman, P. W. 1982. The ontogeny of kin recognition in two species of ground squirrels. *Am. Zool.* 22:491–517.

————. 1983. Kin recognition in animals. *Am. Sci.* 71:46–55.

Homewood, K. M. 1978. Feeding strategy of the Tana mangabey—*Cercocebus galeritus galeritus. J. Zool., Lond.* 186:375–92.

Homewood, K. M., and Rodgers, W. A. 1981. A previously undescribed mangabey from southern Tanzania. *Int. J. Primatol.* 2:47–55.

Honacki, J. H.; Kinman, K. E.; and Koeppl, J. W. 1982. *Mammal species of the world: A taxonomic and geographic reference.* Lawrence, Kans.: Allen Press and the Association of Systematic Collections.

Hood, L. and Bloom, L. 1979. What, when, and how about why: A longitudinal study of early expressions of causality. *Monogr. Soc. Res. Child Dev.* 44:1–47.

van Hooff, J. A. R. A. M. 1962. Facial expressions in higher primates. *Symp. Zool. Soc. Lond.* 8:97–125.

————. 1967. The facial displays of the catarrhine monkeys and apes. In *Primate ethology,* ed. D. Morris. London: Weidenfeld and Nicolson.

————. 1972. A comparative approach to the phylogeny of laughter and smiling. In *Non-verbal communincation,* ed. R. A. Hinde. Cambridge: Cambridge University Press.

————. 1973. A structural analysis of the social behaviour of a semi-captive group of chimpanzees. In *Social communication and movement,* ed. M. von Cranach and I. Vine. London: Academic Press.

Hoogland, J. L. 1981. Nepotism and cooperative breeding in the black-tailed prairie dog (Sciuridae: *Cynomys ludovicianus*). In *Natural selection and social behavior,* ed. R. D. Alexander and D. W. Tinkle. New York: Chiron Press.

————. 1983. Nepotism and alarm-calling in the black-tailed prairie dog (*Cynomys ludovicianus*). *Anim. Behav.* 31:472–79.

Hooley, J. M., and Simpson, M. J. A. 1983. Influence of siblings on the infant's relationship with the mother and others. In *Primate social relationships: An integrated approach,* ed. R. A. Hinde. Oxford: Blackwell.

Hopkins, B. 1974. *Forest and savanna.* 2d ed. London: Heinemann.

Horn, A. D. 1980. Some observations on the ecology of the bonobo chimpanzee (*Pan paniscus* Schwarz 1929) near Lake Tumba, Zaire. *Folia Primatol.* 34:145–69.

Horn, H. S. 1978. Optimal tactics of reproduction and life history. In *Behavioural ecology: An evolutionary approach,* ed. J. R. Drebs and N. B. Davies. Sunderland, Mass.: Sinauer Associates.

van Horn, R. N and Eaton, G. G. 1979. Reproductive physiology and behavior in prosimians. In *The study of prosimian behavior,* ed. G. A. Doyle and R. D. Martin. New York: Academic Press.

Horr, D. A. 1975. The Borneo orangutan: Population structure and dynamics in relationship to ecology and reproductive strategy. In *Primate behavior: Developments in field and laboratory research,* vol. 4, ed. L. A. Rosenblum. New York: Academic Press.

———. 1977. Orangutan maturation: Growing up in a female world. In *Primate bio-social development,* ed. S. Chevalier-Skolnikoff and F. E. Poirier. New York: Garland Press.

Horrocks, J., and Hunte, W. 1983. Maternal rank and offspring rank in vervet monkeys: An appraisal of the mechanisms of rank acquisition. *Anim. Behav.* 31:772–82.

Horwich, R. H. 1976. The whooping display in Nilgiri langurs: An example of daily fluctuations superimposed on a general trend. *Primates* 17:419–31.

———. 1983a. Breeding behaviors in the black howler monkey (*Alouatta pigra*) of Belize. *Primates* 24:222–30.

———. 1983b. Species status of the black howler monkey, *Alouatta pigra,* of Belize. *Primates* 24:288–89.

Horwich, R. H., and Gebhard, K. 1983. Roaring rhythms in black howler monkeys (*Alouatta pigra*) of Belize. *Primates* 24:290–96.

Hoshino, J.; Mori, A.; Kudo, H.; and Kawai, M. 1984. A preliminary report on the grouping of mandrills (*Mandrillus sphinx*) in Cameroon. *Primates* 25:295–307.

Howard, W. E. 1960. Innate and environmental dispersal of individual vertebrates, *Am. Mid. Nat.* 63:152–61.

Hrdy, S. B. 1974. Male-male competition and infanticide among the langurs (*Presbytis entellus*) of Abu, Rajasthan. *Folia Primatol.* 22:19–58.

———. 1976. The care and exploitation of nonhuman primate infants by conspecifics other than the mother. In *Advances in the study of behavior,* vol. 6, ed. J. Rosenblatt, R. Hinde, E. Shaw, and C. Beer. New York: Academic Press.

———. 1977. *The langurs of Abu.* Cambridge: Harvard University Press.

———. 1978. Allomaternal care and abuse of infants among Hanuman langurs. In *Recent advances in primatology,* vol. 1, ed. D. J. Chivers and P. Herbert, New York: Academic Press.

———. 1979. Infanticide among animals: A review, classification, and examination of the implications for the reproductive strategies of females. *Ethol. Sociobiol.* 1:13–40.

———. 1981a. "Nepotists" and "altruists": The behavior of old females among macaques and langur monkeys. In *Other*

ways of growing old, ed. P. Amoss and S. Harrell. Stanford: Stanford University Press.

———. 1981b. *The woman that never evolved.* Cambridge: Harvard University Press.

———. 1984. Assumptions and evidence regarding the sexual selection hypothesis: A reply to Boggess. In *Infanticide: Comparative and evolutionary perspectives,* ed. G. Hausfater and S. B. Hrdy. Hawthorne, N.Y.: Aldine.

———. In press. Sex-biased parental investment among primates and other mammals: A critical evaluation of the Trivers-Willard hypothesis. In *Offspring abuse and neglect in biosocial perspective,* ed. R. Gelles and J. Lancaster. Hawthorne, N.Y.: Aldine.

Hrdy, S. B., and Hausfater, G. 1984. Comparative and evolutionary perspectives on infanticide: An introduction and overview. In *Infanticide: Comparative and evolutionary perspectives,* ed. G. Hausfater and S. B. Hrdy. Hawthorne, N.Y.: Aldine.

Hrdy, S. B., and Hrdy, D. B. 1976. Hierarchical relations among female hanuman langurs (Primates: Colobinae, *Presbytis entellus*). *Science* 193:913–15.

Huck, U. W.; Soltis R. L.; and Coopersmith C. B. 1982. Infanticide in male laboratory mice: Effects of social status, prior sexual experience, and basis for discrimination between related and unrelated young. *Anim. Behav.* 30:1158–65.

Humphrey, N. K. 1976. The social function of intellect. In *Growing points in ethology,* ed. P. P. G. Bateson and R. A. Hinde. Cambridge: Cambridge University Press.

Hunkeler, C.; Bourlière, F.; and Bertrand, M. 1972. Le comportement social de la Mone de Lowe (*Cercopithecus campbelli lowei*). *Folia Primatol.* 17:218–36.

Hunter, J.; Martin, R. D.; Dixson, A. F.; and Rudder, B. C. C. 1979. Gestation and interbirth intervals in the owl monkey (*Aotus trivirgatus griseimembra*). *Folia Primatol.* 31:165–75.

Hutchins, M., and Barash, D. P. 1976. Grooming in primates: Implications for its utilitarian function. *Primates* 17:145–50.

Imanishi, K. 1952. Evolution of humanity. In *Man,* ed. K. Imanishi, Tokyo: Mainichi-Shinbunsha.

———. 1957a. Identification: A process of enculturation in the subhuman society of *Macaca fuscata. Primates* 1:1–29.

———. 1957b. Learned behavior of Japanese monkeys. *Jap. J. Ethnol.* 21:185–89.

———. 1960. Bird, monkey, and man: Is it possible to build a general theory to support "identification"? *Jinbun Gakuho* 10:1–24.

Ingram, J. C. 1977. Interactions between parents and infants, and the development of independence in the common marmoset (*Callithrix jacchus*). *Anim. Behav.* 25:811–27.

Isbell, L. A. 1984. Daily ranging behavior of red colobus (*Colobus badius*) in Kibale Forest, Uganda. *Folia Primatol.* 41:34–48.

Islam, M. I., and Husain K. Z. 1982. A preliminary study on the ecology of the capped langur. *Folia Primatol.* 39:145–59.

Itani, J. 1958. On the acquisition and propagation of new food habits in the troop of Japanese monkeys at Takasakiyama. *Primates* 1:84–98.

———. 1959. Paternal care in the wild Japanese monkey, *Macaca fuscata fuscata. Primates* 2:61–93.

———. 1972. A preliminary essay on the relationship between

social organisation and incest avoidance in nonhuman primates. In *Primate socialization,* ed. F. E. Poirier. New York: Random House.

———. 1980. Social structures of African great apes. *J. Reprod. Fertil. Suppl.* 28:33–41.

Itani, J., and Nishimura, A. 1973. The study of infrahuman culture in Japan. In *Symposia of the Fourth Congress of the International Primatological Society,* vol. 1: *Precultural primate behavior,* ed. E. Menzel. Basel: S. Karger.

Itoigawa, N. 1975. Variables in male leaving a group of Japanese macaques. In *Proceedings of the Fifth Congress of the International Primatological Society,* ed. S. Kondo, M. Kawai, A. Ehara, and S. Kawamura. Tokyo: Japan Science Press.

IUCN. 1982a. Conservation status of the great apes. Cambridge: IUCN. Unpublished report.

IUCN. 1982b. *Proceedings of the third meeting of the conference of the parties, New Delhi, India,* vol. 2, Secretariat of the Convention. Gland, Switzerland: IUCN.

IUCN mammal red data book, part 1. 1982. Gland, Switzerland: IUCN.

IUCN Wildlife Trade Monitoring Unit Traffic Bulletin. 1984. vol. 6, no. 2.

Iwamoto, T. 1979. Feeding ecology, ecological, and sociological studies of gelada baboons. In *Contributions to primatology,* vol. 16: *Ecological and sociological studies of gelada baboons,* ed. M. Kawai. Basel: S. Karger.

———. 1982. Food and nutritional condition of Japanese monkeys on Koshima Islet during winter. *Primates* 23:153–70.

Iwamoto, T., and Dunbar, R. I. M. 1983. Thermoregulation, habitat quality, and the behavioural ecology of gelada baboons. *J. Anim. Ecol.* 52:357–66.

Izard, M. K., and Rasmussen, D. T. 1985. Reproduction in the slender loris (*Loris tardigradus malabaricus*). *Am. J. Primatol.* 8:153–65.

Izawa, K. 1970. Unit-groups of chimpanzees and their nomadism in the savannah woodland. Primates 11:1–46.

———. 1975. Foods and feeding behavior of monkeys in the upper Amazon basin. *Primates* 16:295–316.

———. 1978a. A field study of the ecology and behavior of the black-mantle tamarin (*Saguinus nigricollis*). *Primates* 19:241–74.

———. 1978b. Frog-eating behavior of wild black-capped capuchin (*Cebus apella*). *Primates* 19:633–42.

———. 1979. Foods and feeding behavior of wild black-capped capuchin (*Cebus apella*). *Primates* 21:57–76.

———. 1980. Social behavior of the wild black-capped capuchin (*Cebus apella*). *Primates* 21:443–67.

Izawa, K., and Bejarano, G. 1981. Distribution ranges and patterns of nonhuman primates in western Pando, Bolivia. *Kyoto University Overseas Research Reports of New World Monkeys* 2:1–11.

Izawa, K.; Kimura, K.; and Nieto, A. S. 1979. Grouping of the wild spider monkey. *Primates* 20:503–12.

Izawa, K., and Mizuno, A. 1977. Palm-fruit cracking: Behavior of wild black-capped capuchin (*Cebus apella*). *Primates* 18:773–92.

Janson, C. H. 1975. Ecology and population densities of primates in a Peruvian rainforest. B.A. thesis, Princeton University.

———. 1984. Female choice and mating system of the brown capuchin monkey *Cebus apella* (Primates: Cebidae). *Z. Tierpsychol.* 65:177–200.

Janzen, D. H. 1975. *Ecology of plants in the tropics.* London: Edward Arnold.

———, ed. 1983. *Costa Rican natural history.* Chicago: University of Chicago Press.

Jarman, P. J. 1974. The social organisation of antelope in relation to their ecology. *Behaviour* 48:215–67.

Jay, P. 1963a. Mother-infant relations in langurs. In *Maternal behavior in mammals,* ed. H. L. Reingold. New York: John Wiley.

———. 1963b. The Indian langur monkey (*Presbytis entellus*). In *Primate social behavior,* ed. C. H. Southwick. New York: Van Nostrand.

———. 1965. The common langur of North India. In *Primate behavior,* ed. I. DeVore. New York: Holt, Rinehart, and Winston.

Jensen, G. D.; Bobbitt, R. A.; and Gordon, B. N. 1967. The development of mutual independence in mother-infant pigtailed monkeys, *Macaca nemestrina.* In *Social communication among primates,* ed. S. Altmann. Chicago: University of Chicago Press.

———. 1968. Sex differences in the development of independence in infant monkeys. *Behaviour* 30:1–14.

———. 1969. Patterns and sequences of hitting behavior in mother and infant monkeys (*Macaca nemestrina*). *J. Psychol. Res.* 7:55–61.

Jerison, H. J. 1973. *The evolution of the brain and intelligence.* New York: Academic Press.

Jewell, P. A. 1966. The concept of home range in mammals. *Symp. Zool. Soc. Lond.* 18:85–110.

Johns, A. D. 1981. The effects of selective logging on the social structure of resident primates. *Malaysian Appl. Biol.* 10:221–26.

———. 1983. Ecological effects of selective logging in a west Malaysian rain-forest. Ph.D. diss., University of Cambridge.

Johnson, D. F.; Modahl, K. B.; and Eaton, G. 1982. Dominance status of adult male Japanese macaques: Relationship to female dominance status, male mating behavior, seasonal changes, and developmental changes. *Anim. Behav.* 30:383–92.

Johnson, J. A. 1984. Social relationships of juvenile olive baboons. Ph.D. diss., University of Edinburgh.

Johnson, S. C. 1981. Bonobos: Generalized hominid prototypes or specialized insular dwarfs? *Curr. Anthropol.* 22:363–75.

Jolly, A. 1966a. *Lemur behavior.* Chicago: University of Chicago Press.

———. 1966b. Lemur social behavior and primate intelligence. *Science* 153:501–6.

———. 1967. Breeding synchrony in wild *Lemur catta.* In *Social communication among primates,* ed. S. A. Altmann. Chicago: University of Chicago Press.

———. 1972. *The evolution of primate behavior.* New York: Macmillan.

———. 1984. The puzzle of female feeding priority. In *Female primates: Studies by women primatologists,* ed. M. Small. New York: Alan R. Liss.

———. 1985. *The evolution of primate behavior,* 2d ed. New

York: Macmillan.

Jolly, A.; Gustafson, H.; Oliver, W. L. R.; and O'Connor, S. M. 1982. *Propithecus verreauxi* population and ranging at Berenty, Madagascar, 1975 and 1980. *Folia Primatol.* 39:124–44.

Jones, C. B. 1980a. Seasonal parturition, mortality, and dispersal in the mantled howler monkey, *Alouatta paliata* Gray. *Brenesia* 17:1–10.

———. 1980b. The functions of status in the mantled howler monkey, *Alouatta palliata* Gray: Intraspecific competition for group membership in a folivorous Neotropical primate. *Primates* 21:389–405.

———. 1981. The evolution and socioecology of dominance in primate groups: Theoretical formulation, classification, and assessment. *Primates* 22:70–83.

———. 1982. A field manipulation of spatial relations among male mantled howler monkeys. *Primates* 23:130–34.

———. 1983. Social organization of captive black howler monkeys (*Alouatta caraya*): "Social competition" and the use of non-damaging behavior. *Primates* 24:25–39.

Jones, C. B., and Sabater Pi, J. 1968. Comparative ecology of *Cercocebus albigena* and *Cercocebus torquatus* in Rio Muni, West Africa. *Folia Primatol.* 9:99–113.

———. 1971. Comparative ecology of *Gorilla gorilla* (Savage and Wyman) and *Pan troglodytes* (Blumenback) in Rio Muni, West Africa. *Biblio. Primatol.* 13:1–96.

Jones, D. M. 1982. Veterinary aspects of the maintenance of orangutans in captivity. In *The orangutan: Its biology and conservation,* ed. L. de Boer. The Hague: W. Junk.

Jones, K. C. 1983. Inter-troop transfer of *Lemur catta* males at Berenty, Madagascar. *Folia Primatol.* 40:145–60.

Jorde, L. B., and Spuhler, J. N. 1974. A statistical analysis of selected aspects of primate demography, ecology, and social behavior. *J. Anthropol. Res.* 30:199–224.

Jouventin, P. 1975a. Les roles des colorations du mandrill (*Mandrillus sphinx*). *Z. Tierpsychol.* 39:455–62.

———. 1975b. Observations sur la socioécologie du mandrill. *Terre et Vie* 29:493–532.

Judge, P. 1982. Redirection of aggression based on kinship in a captive group of pigtail macaques (abstract). *Int. J. Primatol.* 3:301.

———. 1983. Reconciliation based on kinship in a captive group of pigtail macaques (abstract). *Am. J. Primatol.* 4:346.

Jungers, W. L., and Susman, R. L. 1984. Body size and skeletal allometry in African apes. In *The pygmy chimpanzee: Evolutionary biology and behavior,* ed. R. L. Susman. New York: Plenum Press.

Jurgens, U. 1979. Neural control of vocalization in non-human primates. In *Neurobiology of social communication in primates,* ed. H. D. Steklis and M. J. Raleigh. New York: Academic Press.

Kano, T. 1972. Distribution and adaptation of the chimpanzee on the eastern shore of Lake Tanganyika. *Kyoto University African Studies* 7:37–129.

———. 1980. The social behavior of wild pygmy chimpanzees (*Pan paniscus*) of Wamba: A preliminary report. *J. Hum. Evol.* 9:243–60.

———. 1982a. The social group of pygmy chimpanzees (*Pan pansiscus*) of Wamba. *Primates* 23:171–88.

———. 1982b. The use of leafy twigs for rain cover by pygmy chimpanzees of Wamba. *Primates* 23:453–57.

———. 1983. An ecological study of the pygmy chimpanzees (*Pan paniscus*) of Yalosidi, Republic of Zaire. *Int. J. Primatol.* 4:1–25.

———. 1984. Distribution of pygmy chimpanzees (*Pan paniscus*) in the central Zaire Basin. *Folia Primatol.* 43:36–52.

———. In press. Social regulation for individual coexistence of wild bonobos. In *Status and rank: Missing variables in the analysis of war,* ed. D. McGuinnes. New York: Paragon House Press.

Kano, T., and Mulavwa, M. 1984. Feeding ecology of the bonobos (*Pan paniscus*) of Wamba. In *The pygmy chimpanzee: Evolutionary biology and behavior,* ed. R. L. Susman. New York: Plenum Press.

Kaplan, J. N.; Winship-Ball, A.; and Sim L. 1978. Maternal discrimination of infant vocalizations in the squirrel monkey. *Primates* 19:187–93.

Kaplan, J. R. 1977. Patterns of fight interference in free-ranging rhesus monkeys. *Am. J. Phys. Anthropol.* 47:279–88.

———. 1978. Fight interference and altruism in rhesus monkeys. *Am. J. Phys. Anthropol.* 47:241–49.

Kaplan, J. R., and Zucker, E. 1980. Social organization in a group of free-ranging patas monkeys. *Folia Primatol.* 34:196–213.

Kappeler, M. 1981. The Javan silvery gibbon (*Hylobates lar moloch*), habitat, distribution, numbers. Ph.D. diss., part 1, University of Basel.

———. 1984a. Diet and feeding behavior of the moloch gibbon. In *The lesser apes: Evolutionary and behavioral biology,* ed. H. Preuschoft, D. J. Chivers, W. Brockelman, and N. Creel. Edinburgh: Edinburgh University Press.

———. 1984b. Vocal bouts and territorial maintenance in the moloch gibbon. In *The lesser apes: Evolutionary and behavioral biology,* ed. H. Preuschoft, D. J. Chivers, W. Brockelman, and N. Creel. Edinburgh: Edinburgh University Press.

Kaufman, I. C. 1974. Mother/infant relations in monkeys and humans: A reply to Professor Hinde. In *Ethology and psychiatry,* ed. N. F. White. Toronto: University of Toronto Press.

Kaufman, I. C., and Rosenblum, L. A. 1967. The reaction to separation in infant monkeys: Anaclitic depression and conservation-withdrawal. *Psychosom. Med.* 29:648–75.

Kaufmann, J. H. 1965. A three-year study of mating behavior in a free-ranging band of rhesus monkeys. *Ecology* 46:500–512.

———. 1966. Behaviour of infant rhesus monkeys and their mothers in a free-ranging band. *Zoologica* 51:17–28.

———. 1967. Social relations of adult males in a free-ranging band of rhesus monkeys. In *Social communication among primates,* ed. S. A. Altmann. Chicago: University of Chicago Press.

Kavanagh, M. 1972. Food sharing behavior in a group of douc langurs (*Pygathrix nemaeus*). *Nature* 239:406–7.

———. 1978. The diet and feeding behaviour of *Cercopithecus aethiops tantalus*. *Folia Primatol.* 30:30–63.

———. 1981. Variable territoriality among tantalus monkeys in Cameroon. *Folia Primatol.* 36:76–98.

Kavanagh, M., and Dresdale, L. 1975. Observations on the

woolly monkey (*Lagothrix lagothricha*) in northern Colombia. *Primates* 16:285–94.

Kavanagh, M., and Laursen, E. 1984. Breeding seasonality among long-tailed macaques, *Macaca fascicularis*, in peninsular Malaysia. *Int. J. Primatol.* 5:17–30.

Kawabe, M., and Mano, T. 1972. Ecology and behavior of the wild proboscis monkey, *Nasalis larvatus* (Wurmb), in Sabah, Malaysia. *Primates* 13:213–28.

Kawai, M. 1958. On the system of social ranks in a natural group of Japanese monkeys. *Primates* 1:11–48.

———. 1960. A field experiment on the process of group formation in the Japanese monkey (*Macaca fuscata*), and the releasing of the group at Ohirayama. *Primates* 2:181–253.

———. 1965a. Newly acquired pre-cultural behavior of the natural troop of Japanese monkeys on Koshima Islet. *Primates* 1:1 30.

———. 1965b. On the system of social ranks in a natural troop of Japanese monkeys, 1: Basic rank and dependent rank. In *Japanese monkeys*, ed. K. Imanishi and S. A. Altmann. Atlanta: Emory University Press.

———, ed. 1979. *Contributions to primatology*, vol. 16: *Ecological and sociological studies of gelada baboons*. Basel: S. Karger.

Kawai, M.; Azuma, S.; and Yoshiba, K. 1967. Ecological studies of reproduction in Japanese monkeys (*Macaca fuscata*), 1: Problems of the birth season. *Primates* 8:35–74.

Kawai, M.; Dunbar, R. I. M.; Ohsawa, H.; and Mori, U. 1983. Social organization of gelada baboons: Social units and definitions. *Primates* 24:13–24.

Kawamoto, Y.; Ischak, T. M.; and Supriatna, J. 1984. Genetic variation within and between troops of the crab-eating macaque (*Macaca fascicularis*) on Sumatra, Java, Bali, Lombok, and Sumbawa, Indonesia. *Primates* 25:131–59.

Kawamura, S. 1958. Matriarchal social order in the Minoo-B Group: A study on the rank system of Japanese macaques. *Primates* 1:149–56.

———. 1959. The process of sub-culture propagation among Japanese macaques. *Primates* 2:43–60.

———. 1965a. Matriarchial social ranks in the Minoo-B troop: A study of the rank system of Japanese monkeys. In *Japanese Monkeys*, ed. K. Imanishi and S. A. Altmann. Atlanta: Emory University Press.

———. 1965b. Sub-culture among Japanese macaques. In *Monkeys and apes: Sociological studies*, ed. S. Kawamura and J. Itani. Tokyo: Chuokoronsha.

———. 1967. Aggression as studied in troops of Japanese monkeys. *Univ. Calif. Forum Med. Sci.* 7:195–223.

Kawanaka, K. 1973. Intertroop relations among Japanese monkeys. *Primates* 14:113–59.

———. 1981. Infanticide and cannibalism in chimpanzees—with special reference to the newly observed case in the Mahale Mountains. *Afr. Stud. Monogr.* 1:69–99.

———. 1984. Association, ranging, and the social unit in chimpanzees of the Mahale Mountains, Tanzania. *Int. J. Primatol.* 5:411–34.

Kay, R. F., and Cartmill, M. 1977. Cranial morphology and adaptations of *Palaechthon nacimienti* and other Paramomgidae (Plesiadapoidea, ? Primates), with a description of a new genus and species. *J. Hum. Evol.* 6:13–53.

Kern, J. A. 1964. Observations on the habits of the proboscis monkey, *Nasalis larvatus* (Wurmb), made in the Brunei Bay area, Borneo. *Zoologica* 49:183–92.

Keverne, E. B. 1976. Sexual receptivity and attractiveness in the female rhesus monkey. *Advances in the study of behavior*, vol. 7, ed. J. S. Rosenblatt, R. A. Hinde, E. Shaw, and C. Beer. New York: Academic Press.

———. 1981. Do Old World primates have oestrus? *Malaysian Appl. Biol.* 10:119–26.

———. 1982. Olfaction and reproductive behavior of nonhuman primates. In *Primate communication*, ed. C. T. Snowdon, C. H. Brown, and M. R. Petersen. Cambridge: Cambridge University Press.

Kihlström, J. E. 1972. Period of gestation and body weight in some placental mammals. *Comp. Biochem. Physiol.* 43:673–79

King, J. 1955. Social behavior, social organization, and population dynamics in a black-tailed prairie dog town in the Black Hills of South Dakota. *Contributions from the Lab. Vert. Biol. Univ. Mich.* 67.

Kingdon, J. S. 1974. *East African mammals*. Vol. 1. Chicago: University of Chicago Press.

———. 1980. The role of visual signals and face patterns in African forest monkeys (guenons) of the genus *Cercopithecus*. *Trans. Zool. Soc. Lond.* 35:425–75.

Kingsley, S. 1982. Causes of non-breeding and the development of the secondary sexual characteristics in the male orangutan: A hormonal study. In *The orangutan: Its biology and conservation*, ed. L. de Boer. The Hague: W. Junk.

Kinsey, A. C.; Pomeroy, W. B.; Martin, C. E.; and Gebhard, P. H. 1953. *Sexual behavior in the human female*. Philadelphia: W. B. Saunders.

Kintz, B. L.; Foster, M. S.; Hart, J. O.; O'Malley, J. J.; Palmer, E. L.; and Sullivan, S. L. 1969. A comparison of learning sets in humans, primates, and subprimates. *J. Gen. Psychol.* 80:189–204.

Kinzey, W. G. 1972. Canine teeth of the monkey, *Callicebus moloch*: Lack of sexual dimorphism. *Primates* 13:365–69.

———. 1976. Positional behavior and ecology in *Callicebus torquatus*. *Yrbk. Phys. Anthropol.* 20:468–80.

———. 1977. Diet and feeding behaviour of *Callicebus torquatus*. In *Primate ecology*, ed. T. H. Clutton-Brock. London: Academic Press.

———. 1978. Feeding behaviour and molar features in two species of titi monkey. In *Recent advances in primatology*, vol. 1, ed. D. J. Chivers and J. Herbert. London: Academic Press.

———. 1981. The titi monkeys, genus *Callicebus*. In *Ecology and behavior in Neotropical primates*, vol. 1, ed. A. F. Coimbra-Filho and R. A. Mittermeier. Rio de Janeiro: Academia Brasileira de Ciencias.

———. 1982. Distribution of primates and forest refuges. In *Biological diversification in the tropics*, ed. G. T. Prance. New York: Columbia University Press.

Kinzey, W. G., and Becker, M. 1983. Activity pattern of the masked titi monkey, *Callicebus personatus*. *Primates* 24:337–43.

Kinzey, W. G., and Gentry, A. H. 1979. Habitat utilization in two species of *Callicebus*. In *Primate ecology: Problem-oriented field studies*, ed. R. W. Sussman. New York: John Wiley.

Kinzey, W. G., and Robinson, J. G. 1983. Intergroup loud calls, range size, and spacing in *Callicebus torquatus. Am. J. Phys. Anthropol.* 60:539–44.

Kinzey, W. G.; Rosenberger, A. L.; Heisler, P. S.; Prowse, D. L.; and Trilling, J. S. 1977. A preliminary field investigation of the yellow-handed titi monkey, *Callicebus torquatus torquatus,* in northern Peru. *Primates* 18:159–81.

Kinzey, W. G., and Wright, P. C. 1982. Grooming behavior in the titi monkey. *Callicebus torquatus. Am. J. Primatol.* 3:267–75.

Kirkwood, J. K. 1983. The effects of diet on health, weight, and litter size in captive cotton-top tamarins *Saguinus oedipus oedipus. Primates* 24:515–20.

Kitamura, K. 1983. Pygmy chimpanzee association patterns and ranging. *Primates* 24:1–12.

Kleiber, M. 1961. *The fire of life: An introduction to animal energetics.* New York: John Wiley.

———. 1975. *The fire of life.* 2d ed. New York: Krieger.

Kleiman, D. G. 1977. Monogamy in mammals. *Quart. Rev. Biol.* 52:39–69.

———. 1978a. Characteristics of reproduction and sociosexual interaction in pairs of lion tamarins (*Leontopithecus rosalia*) during the reproductive cycle. In *The biology and conservation of the Callitrichidae,* ed. D. G. Kleiman. Washington, D.C.: Smithsonian Institution Press.

———. 1978b. Parent-offspring conflict and sibling competition in a monogamous primate. *Am. Nat.* 194:753–60.

———. 1981. Correlations among life history characteristics of mammalian species exhibiting two extreme forms of monogamy. In *Natural selection and social behavior: Recent research and theory,* ed. R. D. Alexander and D. W. Tinkle. New York: Chiron Press.

Kleiman, D. G., and Mack, D. S. 1977. A peak in sexual activity during mid-pregnancy in the golden lion tamarin, *Leontopithecus rosalia* (Primates: Callithrichidae). *J. Mammal.* 58:657–60.

Klein, L. L. 1971. Observations on copulation and seasonal reproduction of two species of spider monkeys, *Ateles belzebuth* and *A. geoffroyi. Folia Primatol.* 15:233–48.

———. 1974. Agonistic behavior in Neotropical primates. In *Primate aggression, territoriality, and xenophobia,* ed. R. L. Holloway. New York: Academic Press.

Klein, L. L., and Klein, D. J. 1971. Aspects of social behaviour in a colony of spider monkeys *Ateles geoffroyi* at San Francisco Zoo. *Int. Zool. Yrbk.* 11:175–81.

———. 1973. Observations on two types of Neotropical primate intertaxa associations. *Am. J. Phys. Anthropol.* 38:649–54.

———. 1975. Social and ecological contrasts between four taxa of Neotropical primates. In *Socioecology and psychology of primates,* ed. R. Tuttle. The Hague: Mouton.

———. 1976. Neotropical primates, aspects of habitat usage, population density, and regional distribution in La Macarena, Colombia. In *Neotropical primates: Field studies and conservation,* ed. R. W. Thorington, Jr., and P. G. Heltne. Washington D.C.: National Academy of Sciences.

———. 1977. Feeding behavior of the Colombian spider monkey, *Ateles belzebuth.* In *Primate ecology.* ed. T. H. Clutton-Brock. New York: Academic Press.

Klingel, H. 1974. A comparison of the social behavior of the Equidae. In *The behaviour of Ungulates and its relation to management,* ed. V. Geist and F. Walther. Morges, Switzerland: International Union for the Conservation of Nature and Natural Resources.

Klopfer, P. H., and Boskoff, K. J. 1979. Maternal behavior in prosimians. In *The study of prosimian behavior,* ed. G. A. Doyle and R. D. Martin. New York: Academic Press.

Klopfer, P. H., and Jolly, A. 1970. The stability of territorial boundaries in a lemur troop. *Folia Primatol.* 12:199–208.

Koford, C. B. 1963. Rank of mothers and sons in bands of rhesus monkeys. *Science* 141:356–57.

———. 1965. Population dynamics of rhesus monkeys on Cayo Santiago. In *Primate behaviour: Field studies of monkeys and apes,* ed. I. DeVore. New York: Holt, Rinehart, and Winston.

———. 1966. Population changes in rhesus monkeys: Cayo Santiago, 1960–64. *Tulane Stud. Zool.* 13:1–7.

Kohler, W. [1925] 1959. *The mentality of apes.* 2d ed. New York: Viking.

Konner, M. J. 1976. Maternal care, infant behavior, and development among the !Kung. In *Kalahari hunter-gatherers: Studies of the !Kung San and their neighbors,* ed. R. B. Lee and I. DeVore. Cambridge: Harvard University Press.

Kortlandt, A. 1972. *New perspectives on ape and human evolution.* Amsterdam: Stichting voor Psychobiologie.

———. 1980. How might early hominids have defended themselves against large predators and food competitors? *J. Hum. Evol.* 9:79–112.

———. 1983. Marginal habitats of chimpanzees. *J. Hum. Evol.* 12:231–78.

Koyama, N. 1967. On dominance rank and kinship of a wild Japanese monkey troop in Arashiyama. *Primates* 8:189–216.

———. 1970. Changes in dominance rank and division of a wild Japanese monkey troop in Arashiyama. *Primates* 11:335–90.

Koyama, N.; Norikoshi, K.; and Mano, T. 1975. Population dynamics of Japanese monkeys at Arashiyama. In *Contemporary primatology: Proceedings of the Fifth Congress of the International Primatological Society,* ed. M. Kawai, S. Kondo, and A. Ehara. Basel: S. Karger.

Krebs, J. R. 1978. Optimal foraging: Decision rules for predators. In *Behavioural ecology: An evolutionary approach,* ed. J. R. Krebs and N. B. Davis. Oxford: Blackwell.

Krebs, J. R., and Davies, N. B. 1981. *An introduction to behavioral ecology.* Sunderland, Mass.: Sinauer Associates.

———, eds. 1984. *Behavioural ecology: An evolutionary approach.* Sunderland, Mass.: Sinauer Associates.

Kriege, P. D., and Lucas, J. W. 1974. Aunting behavior in an urban troop of *Cercopithecus aethiops. J. Behav. Sci.* 2:55–61.

Kruuk, H. 1972. *The spotted hyena: A study of predation and social behavior.* Chicago: University of Chicago Press.

———. 1978. Foraging and spatial organization in the European badger, *Meles meles* L. *Behav. Ecol. Sociobiol.* 4:75–89.

Kummer, H. 1957. *Soziales Verhalten einer Mantelpavian Gruppe.* Bern: Huber.

———. 1967. Tripartite relations in hamadryas baboons. In *Social communication among primates,* ed. S. A. Altmann. Chicago: University of Chicago Press.

———. 1968. *Social organization of hamadryas baboons.* Chicago: University of Chicago Press.

———. 1971. *Primate societies.* Chicago: Aldine.

———. 1975. Rules of dyad and group formation among captive gelada baboons (*Theropithecus gelada*). In *Proceedings from the Symposia of the Fifth Congress of the International Primatological Society,* ed. S. Kondo, M. Kawai, A. Ehara, and S. Kawamura. Tokyo: Japan Science Press.

———. 1982. Social knowledge in free-ranging primates. In *Animal mind—Human mind,* ed. D. R. Griffin. Berlin: Springer-Verlag.

Kummer, H.; Götz, W.; and Angst, W. 1974. Triadic differentiation: An inhibitory process protecting pair bonds in baboons. *Behaviour* 49:62–87.

Kummer, H.; Banaja, A. A.; Abo-Khatwa, A. N.; and Ghandour, A. M. 1981. A survey of hamadryas baboons in Saudi Arabia. *Fauna of Saudi Arabia* 3:441–71.

Kummer, H., and Kurt, F. 1965. A comparison of social behavior in captive and wild hamadryas baboons. In *The baboon in medical research,* ed. H. Vagtborg. Austin: University of Texas Press.

Kurland, J. A. 1973. A natural history of kra macaques (*Macaca fascicularis* Raffles, 1821) at the Kutai Reserve, Kalimantan Timur, Indonesia. *Primates* 14:254–62.

———. 1977. Kin selection in the Japanese monkey. In *contributions to Primatology,* vol. 12. Basel: S. Karger.

Kurland, J. A., and Gaulin, S. J. C. 1984. The evolution of male parental investment: Effects of genetic relatedness and feeding ecology on the allocation of reproductive effort. In *Primate paternalism,* ed. D. M. Taub. New York: Van Nostrand Reinhold.

Kuroda, S. 1979. Grouping of the pygmy chimpanzees. *Primates* 20:161–83.

———. 1980. Social behavior of the pygmy chimpanzees. *Primates* 21:181–97.

———. 1982. *Pygmy chimpanzees.* Tokyo: Chikumashobo.

———. 1984. Interaction over food among pygmy chimpanzees. In *The pygmy chimpanzee: Evolutionary biology and behavior,* ed. R. L. Susman. New York: Plenum.

Lack, D. 1968. *Ecological adaptations for breeding in birds.* London: Methuen.

Ladefoged, P. 1975. *A course in phonetics.* New York: Harcourt, Brace, Jovanovich.

Lahiri, R. K., and Southwick, C. H. 1966. Paternal care in *Macaca sylvana. Folia Primatol.* 4:257–64.

Lamb, M. E. 1984. Observational studies of father-child relationships in humans. In *Primate paternalism,* ed. D. M. Taub. New York: Van Nostrand Reinhold.

Lamprecht, J. 1981. The function of social hunting in larger terrestrial carnivores. *Mammal. Rev.* 11:169–79.

Lancaster, J. B. 1972. Play-mothering: The relations between juvenile females and young infants among free-ranging vervets. In *Primate socialization,* ed. F. E. Poirier. New York: Random House.

———. 1975. *Primate behavior and the emergence of human culture.* New York: Holt, Rinehart, and Winston.

———. 1979. Sex and gender in evolutionary perspective. In *Human sexuality,* ed. H. A. Katchadourian. Berkeley: University of California Press.

———. 1984. Evolutionary perspectives on sex differences in the higher primates. In *Gender and the life course,* ed. A. S. Rossi. Hawthorne, N.Y.: Aldine.

Lancaster, J. B., and Lee, R. B. 1965. The annual reproductive cycle in monkeys and apes. In *Primate behavior: Field studies of monkeys and apes,* ed. I. DeVore. New York: Holt, Rinehart, and Winston.

Laws, J. W., and Vonder Haar Laws, J. 1984. Social interaction among adult male langurs (*presbytis entellus*) at Rajaji Wildlife Sanctuary. *Int. J. Primatol.* 5:31–50.

Le Boeuf, B. J. 1974. Male-male competition and reproductive success in elephant seals. *Am. Zool.* 14:163–76.

Ledbetter, D., and Basen, J. 1982. Failure to demonstrate self-recognition in gorillas. *Am. J. Primatol.* 2:307–10.

Lee, P. C. 1981. Ecological and social influences on the development of vervet monkeys (*Cercopithecus aethiops*). Ph.D. Diss., Cambridge University.

———. 1983a. Caretaking of infants and mother-infant relationships. In *Primate social relationships: An integrated approach,* ed. R. A. Hinde. Oxford: Blackwell.

———. 1983b. Context-specific unpredictability in dominance interactions. In *Primate social relationships: An integrated approach,* ed. R. A. Hinde. Oxford: Blackwell.

———. 1983c. Ecological influences on relationships and social structures. In *Primate social relationships: An integrated approach,* ed. R. A. Hinde. Oxford: Blackwell.

———. 1983d. Effects of the loss of the mother on social development. In *Primate social relationships: An integrated approach,* ed. R. A. Hinde. Oxford: Blackwell.

———. 1983e. Play as a means for developing relationships. In *Primate social relationships: An integrated approach,* ed. R. A. Hinde. Oxford: Blackwell.

Lee, P. C., and Oliver, J. I. 1979. Competition, dominance, and the acquisition of rank in juvenile yellow baboons (*Papio cynocephalus*). *Anim. Behav.* 27:576–85.

Le Gros Clark, W. E. 1971. *The antecedents of man.* 3d ed. Edinburgh: Edinburgh University Press.

Lehrman, D. S. 1974. Can psychiatrists use ethology? In *Ethology and psychiatry,* ed. N. F. White. Toronto: University of Toronto.

Leigh, E. G.; Rand, A. S.; and Windsor, D. M. 1982. *The ecology of a tropical forest.* Washington, D.C.: Smithsonian Institution Press.

Leighton, M.; and Leighton, D. R. 1982. The relationship of size of feeding aggregate to size of food patch: Howler monkeys (*Alouatta palliata*) feeding in *Trichilia cipo* fruit trees on Barro Colorado Island. *Biotropica* 14(2):81–90.

Leland, L.; Struhsaker, T. T.; and Butynski, T. M. 1984. Infanticide by adult males in three primate species of the Kibale Forest, Uganda: A test of hypotheses. In *Infanticide: Comparative and evolutionary perspectives,* ed. G. Hausfater and S. B. Hrdy. Hawthorne, N.Y.: Aldine.

Lemmon, W. B., and Allen, M. L. 1978. Continual sexual receptivity in the female chimpanzee (*Pan troglodytes*). *Folia Primatol.* 30:80–88.

Leo Luna, M. 1981. First field study of the yellow-tailed woolly monkey. *Oryx* 15:386–89.

———. 1982. Estudio preliminar sobre la biología y ecología del mono chorro de cola amarilla, *Lagothrix flavicauda,* Homboldt 1812. Thesis, Universidad Nacional Agraria, La Molina, Perú.

Leutenegger, W. 1973. Maternal-fetal weight relationships in primates. *Folia Primatol.* 20:280–93.

———. 1976. Allometry of neonatal size in eutherian mammals. *Nature* 263:229–30.

———. 1979. Evolution of litter size in primates. *Am. Nat.* 114:525–31.

Leutenegger, W., and Kelly, J. T. 1977. Relationship of sexual dimorphism in canine size and body size to social, behavioral, and ecological correlates in anthropoid primates. *Primates* 18:117–36.

Lewis, D. B., and Gower, D. M. 1980. *Biology of communication*. London: Blackie Press.

Lidicker, W. Z., Jr. 1962. Emigration as a possible mechanism permitting the regulation of population density below carrying capacity. *Am. Nat.* 96:23–29.

Lieberman, P. 1975. *On the origins of language*. New York: Macmillan.

Lillegraven, J. A.; Kielan-Jaworowska, Z.; and Clemens, W. A. 1980. *Mesozoic mammals*. Berkeley: University of California Press.

Lillehei, R., and Snowdon, C. T. 1978. Individual and situational differences in the vocalizations of young stumptail macaques (*Macaca arctoides*). *Behaviour* 64:270–81.

Lindburg, D. G. 1969. Rhesus monkeys: Mating season mobility of adult males. *Science* 166:1176–78.

———. 1971. The rhesus monkey in northern India: An ecological and behavioral study. In *Primate behavior*, vol. 2, ed. L. A. Rosenblum. New York: Academic Press.

———. 1973. Grooming behavior as a regulator of social interactions in rhesus monkeys. In *Behavioral regulators of behavior* ed. C. R. Carpenter. Lewisburg, Pa.: Bucknell University Press.

———. 1977. Feeding behaviour and diet of rhesus monkeys (*Macaca mulatta*) in a Siwalik forest in North India. In *Primate ecology*, ed. T. H. Clutton-Brock. London: Academic Press.

———. 1983. Mating behavior and estrus in the Indian rhesus monkey. In *Perspectives in primate biology*, ed. P. K. Seth. New Dehli: Today and Tomorrow.

Lindstedt, S. L., and Calder, W. A., III 1981. Body size, physiological time, and longevity of homeothermic animals. *Quart. Rev. Biol.* 56:1–16.

Lockard, J. S. 1980. *The evolution of human social behavior*. New York: Elsevier.

Lorenz, K. Z. 1950. The comparative method of studying innate behavior patterns. *Symp. Soc. Exp. Biol.* 4:221–68.

Lott, D. 1984. Intraspecific variation in the social systems of wild vertebrates. *Behaviour* 88:266–325.

Lovejoy, C. O. 1981. The origin of man. *Science* 211:341–50.

Loy, J. 1970. Peri-menstrual behavior among rhesus macaques. *Folia Primatol.* 13:286–97.

———. 1971. Estrous behavior of free-ranging rhesus monkeys (*M. mulatta*). *Primates* 12:1–31.

———. 1975. The copulatory behavior of adult male patas monkeys, *Erythrocebus patas*. *J. Reprod. Fertil.* 45:193–95.

———. 1981. The reproductive and heterosexual behaviours of adult patas monkeys in captivity. *Anim. Behav.* 29:714–26.

Loy, J., and Loy, K. 1977. Sexual harassment among captive patas monkeys (*Erythrocebus patas*) *Primates* 18:691–99.

Loy, J.; Loy, K.; Keifer, G.; and Conoway, C. 1984. The behav-

ior of gonadectomized rhesus monkeys. In *Contributions to Primatology*, vol. 20. Basel: S. Karger.

McCann, C. 1933. Notes on the colouration and habits of the white-browed gibbon or hoolock (*Hylobates hoolock* Harl.). *J. Bombay Nat. Hist. Soc.* 36:395–405.

McConkey, E. H.; Taylor, B. J.; and Phan, P. 1979. Human heterozygosity: A new estimate. *Proc. Nat. Acad. Sci.* 76:6500–04.

MacDonald, D. W. 1982. Notes on the size and composition of groups of proboscis monkey, *Nasalis larvatus*. *Folia Primatol.* 37:95–98.

———. 1983. The ecology of carnivore social behaviour. *Nature* 301:379–84.

MacDonald, G. J. 1971. Reproductive patterns of three species of macaques. *Fertil. Steril.* 22:373–77.

Mace, G. M. 1979. The evolutionary ecology of small mammals. Ph.D. diss., University of Sussex.

McGinnis, P. R. 1979. Sexual behaviour in free-living chimpanzees: Consort relationships. In *The great apes*, ed. D. A. Hamburg and E. R. McCown. Menlo Park, Calif.: Benjamin/Cummings.

McGrew, W. C. 1974. Tool use by wild chimpanzees in feeding upon driver ants. *J. Hum. Evol.* 3:501–8.

———. 1975. Patterns of plant food sharing by wild chimpanzees. In *Contemporary primatology: Proceedings of the Fifth Congress of the International Primatological Society*, ed. M. Kawai, S. Kondo, and A. Ehara. Basel: S. Karger.

———. 1977. Socialization and object manipulation of wild chimpanzees. In *Primate bio-social development*, ed. S. Chevalier-Skolnikoff and F. E. Poirier. New York: Garland.

———. 1979. Evolutionary implications of sex differences in chimpanzee predation and tool use. In *The great apes*, ed. D. A. Hamburg and E. R. McCown. Menlo Park, Calif.: Benjamin/Cummings.

———. 1981. The female chimpanzee as a human evolutionary prototype. In *woman the gatherer*, ed. F. Dahlberg. New Haven: Yale University Press.

———. 1983. Animal foods in the diets of wild chimpanzees (*Pan troglodytes*): Why cross-cultural variation? *J. Ethol.* 1:46–61.

McGrew, W. C.; Baldwin, P. J.; and Tutin, C. E. G. 1981. Chimpanzees in a hot, dry and open habitat: Mt. Assirik, Senegal. *J. Hum. Evol.* 10:227–44.

McGrew, W. C., and Tutin, C. E. G. 1973. Chimpanzee tool use in dental grooming. *Nature* 241:477–78.

———. 1978. Evidence for a social custom in wild chimpanzees? *Man* 13:234–51.

McGrew, W. C.; Tutin, C. E. G.; and Baldwin, P. J. 1979a. Chimpanzees, tools, and termites: Cross-cultural comparisons of Senegal, Tanzania, and Rio Muni. *Man* 14:185–215.

———. 1979b. New data on meat-eating by wild chimpanzees. *Curr. Anthropol.* 20:238–39.

Mack, D. 1979. Growth and development of infant red howling monkeys (*Alouatta seniculus*) in a free ranging population. In *Vertebrate ecology in the northern Neotropics*, ed. J. F. Eisenberg. Washington, D.C.: Smithsonian Institution Press.

Mack, D., and Mittermeier, R. A. 1984. *The primate trade*. Washington, D.C.: World Wildlife Fund—United States.

McKenna, J. J. 1978. Biosocial function of grooming behavior among the common langur monkey (*Presbytus entellus*).

Am. J. Phys. Anthropol. 48:503–10.

———. 1979a. Aspects of infant socialization, attachment, and maternal caregiving patterns among primates: A cross-disciplinary review. *Yrbk. Phys. Anthropol.* 22:250–86.

———. 1979b. The evolution of allomothering behavior among colobine monkeys: Function and opportunism in evolution. *Am. Anthropol.* 81:818–40.

———. 1981. Primate infant caregiving behavior. In *Parental care in mammals*, ed. D. J. Gubernick and P. H. Klopfer. New York: Plenum Press.

McKey, D. B. 1978. Plant chemical defenses and the feeding and ranging behavior of *Colobus* monkeys in African rain forests. Ph.D. diss., University of Michigan.

McKey, D. B.; Gartlan J. S.; Waterman P. G.; and Choo, G. M. 1981. Food selection by black colobus monkey (*Colobus satanas*) in relation to plant chemistry. *Biol. J. Linn. Soc.* 16:115–46.

McKey, D. B., and Waterman P. G. 1982. Ranging behavior of a group of black colobus (*Colobus satanas*) in the Douala-Edea Reserve, Cameroon. *Folia Primatol.* 39:264–304.

McKinney, W. T.; Suomi, S. J.; and Harlow, H. F. 1971. Depression in primates. *Am. J. Psychol.* 127:1313–20.

MacKinnon, J. R. 1974. The ecology and behaviour of wild orangutans (*Pongo pygmaeus*). *Anim. Behav.* 22:3–74.

———. 1977. A comparative ecology of Asian apes. *Primates* 18:747–72.

———. 1979. Reproductive behavior in wild orangutan populations. In *The great apes*, ed. D. A. Hamburg and E. R. McCown. Menlo Park, Calif.: Benjamin/Cummings.

MacKinnon, J. R., and MacKinnon, K. S. 1978. Comparative feeding ecology of six sympatric primates in West Malaysia. In *Recent advances in primatology*, vol. 1, ed. D. J. Chivers and J. A. Herbert. London: Academic Press.

———. 1980a. Niche differentiation in a primate community. In *Malayan forest primates: Ten years' study in tropical rain forest*, ed. D. J. Chivers. New York: Plenum Press.

———. 1980b. The behaviour of wild spectral tarsiers. *Int. J. Primatol.* 1:361–79.

———. 1984. Territoriality, monogamy, and song in gibbons and tarsiers. In *The lesser apes: Evolutionary and behavioral biology*, eds. H. Preuschoft, D. J. Chivers, W. Brockelman, and N. Creel. Edinburgh: Edinburgh University Press.

MacMillan, C., and Duggleby, C. 1981. Interlineage genetic differentiation among rhesus macaques on Cayo Santiago. *Am. J. Phys. Anthropol.* 56:305–12.

MacNamara, J. 1982. *Names for things.* Cambridge: MIT Press.

MacRoberts, M. H. 1970. The social organization of Barbary apes (*Macaca sylvana*) on Gibraltar. *Am. J. Phys. Anthropol.* 33:83–100.

Magnanini, A. 1978. Progress in the development of Poco Das Antas Biological Reserve for *Leontopithecus rosalia rosalia* in Brazil. In *The biology and conservation of the Callitrichidae*, ed. D. G. Kleiman. Washington, D.C.: Smithsonian Institution Press.

Main, M., and Weston, D. R. 1982. Avoidance of the attachment figure in infancy: Descriptions and interpretation. In *The place of attachment in human behavior*, ed. C. Murray Parkes and J. Stevenson Hinde. New York: Basic Books.

Mainardi, D. 1980. Tradition and the social transmission of behavior in animals. In *Sociobiology: Beyond nature/nurture?*, ed. G. W. Barlow and J. Silverberg. Boulder, Colo.: Westview Press.

Makwana, S. C. 1978. Field ecology and behaviour of the rhesus macaque (*Macaca mulatta*), 1: Group composition, home range, roosting sites, and foraging routes in the Asarori Forest. *Primates* 19:483–92.

Makwana, S. C., and Advani, R. 1981. Social changes in the hanuman langur, *Presbytis entellus*, around Jodphur. *J. Bombay Nat. Hist. Soc.* 78:152–54.

Manley, G. H. 1974. Functions of the external genital glands of Periodictus and Arctocebus. In *Prosimian biology*, ed. R. D. Martin, G. A. Doyle, and A. C. Walker. London: Duckworth.

———. 1984. Through the territorial barrier: Harem accretion in *Presbytis senex*. *Int. J. Primatol.* 5:360.

Manocha, S. N. 1967. Discrimination learning in langurs and rhesus monkeys. *Percep. Motor Skills* 24:805–6.

Manzolillo, D. L. 1982. Intertroop transfer by adult male *Papio anubis*. Ph.D. diss., University of California, Los Angeles.

———. 1984. Intertroop transfer by male *Papio anubis* (abstract). *Int. J. Primatol.* 5:360.

Marais, E. 1939. *My friends the baboons.* Capetown: Human and Rousseau.

Marler, P. 1965. Communication in monkeys and apes. In *Primate behavior*, ed. I. DeVore. New York: Holt, Rinehart, and Winston.

———. 1969. *Colobus guereza:* Territoriality and group composition. *Science* 163:93–95.

———. 1972. Vocalizations of East African monkeys, 2: Black and white colobus. *Behaviour* 42:175–97.

———. 1976a. An ethological theory of the origin of vocal learning. *Ann. N.Y. Acad. Sci.* 280:386–95.

———. 1976b. Social organization, communication, and graded signals: The chimpanzee and the gorilla. In *Growing points in ethology*, ed. P. P. G. Bateson and R. A. Hinde. Cambridge: Cambridge University Press.

———. 1977. Primate vocalizations: Affective or symbolic? In *Progress in ape research*, ed. G. H. Bourne. New York: Academic Press.

———. 1981. Birdsong: The acquisition of a learned motor skill. *Trends in Neurosciences* 4:88–94.

Marler, P., and Hobbett, L. 1975. Individuality in a long-range vocalization of wild chimpanzees. *Z. Tierpsychol.* 38:97–109.

Marler, P., and Tenaza, R. 1977. Signalling behavior of apes with special reference to vocalization. In *How animals communicate*, ed. T. Sebeok. Bloomington: Indiana University Press.

Marsden, H. M. 1968. Agonistic behavior of young rhesus monkeys after changes induced in social rank of their mothers. *Anim. Behav.* 16:38–44.

Marsh, C. W. 1978. Problems of primate conservation in a patchy environment along the lower Tana River, Kenya. In *Recent advances in primatology*, vol. 2, ed. D. J. Chivers and W. Lane-Petter. London: Academic Press.

———. 1979a. Comparative aspects of social organization in the Tana River red colobus, *Colobus badius rufomitratus*. *Z. Tierpsychol.* 51:337–62.

————. 1979b. Female transference and mate choice among Tana River red colobus. *Nature* 281:568–69.

————. 1981a. Diet choice among red colobus (*Colobus badius rufomitratus*) on the Tana River, Kenya. *Folia Primatol.* 35:147–78.

————. 1981b. Ranging behaviour and its relation to diet selection in Tana River red colobus (*Colobus badius rufomitratus*). *J. Zool. Soc. Lond.* 195:473–92.

Marsh, C. W., and Wilson, W. L. 1981a. *A survey of primates in peninsular Malaysian forests.* Universiti Kebangsaan Malaysia.

————. 1981b. Effects of natural habitat disturbances on the abundance of Malaysian primates. *Malaysian Appl. Biol.* 10:227–49.

Marten, K., and Marler, P. 1977. Sound transmission and its significance for animal vocalizations, 1: Temperate habitats. *Behav. Ecol. Sociobiol.* 3:271–90.

Martin, D. E. 1981. Breeding great apes in captivity. In *Reproductive biology of the great apes,* ed. C. Graham. New York: Academic Press.

Martin, D. P.; Golway, P. L.; George, M. J.; and Smith, J. A. 1979. Care of the infant and juvenile gibbon (*Hylobates lar*). In *Nursery care of nonhuman primates,* ed. G. C. Ruppenthal. New York: Academic Press.

Martin, R. D. 1972. A preliminary field-study of the lesser mouse lemur (*Mocrocebus murinus* J. F. Miller, 1777). *Z. Tierpsychol. Suppl.* 9:43–89.

————. 1973a. A review of the behaviour and ecology of the lesser mouse lemur (*Microcebus murinus* J. F. Miller 1777). In *Comparative ecology and behaviour of primates,* ed. R. P. Michael and J. H. Crook. London: Academic Press.

————. 1973b. Comparative anatomy and primate systematics. *Symp. Zool. Soc. Lond.* 33:301–37.

————. 1975. The bearing of reproductive behaviour and ontogeny on strepsirhine phylogeny. In *Phylogeny of the primates: A multidisciplinary approach,* ed. W. P. Luckett and F. S. Szalay. New York: Plenum Press.

————. 1978. Major features of prosimian evolution: A discussion in the light of chromosomal evidence. In *Recent advances in primatology,* vol. 3, ed. D. J. Chivers and K. A. Joysey. London: Academic Press.

————. 1979. Phylogenetic aspects of prosimian behaviour. In *The study of prosimian behaviour,* ed. G. A. Doyle and R. D. Martin. New York: Academic Press.

————. 1981a. Field studies of primate behaviour. *Symp. Zool. Soc. Lond.* 46:287–336.

————. 1981b. Relative brain size and basal metabolic rate in terrestrial vertebrates. *Nature* 293:56–60.

————. 1983. Human brain evolution in an ecological context. *Fifty-Second James Arthur Lecture on the Evolution of the Human Brain.* New York: American Museum of Natural History.

————. 1984a. Body size, brain size, and feeding strategies. In *Food acquisition and processing in primates,* ed. D. J. Chivers, B. A. Wood, and A. Bilsborough. New York: Plenum Press.

————. 1984b. Scaling effects and adaptive strategies in mammalian reproduction. *Symp. Zool. Soc. Lond.* 51:81–117.

Martin, R. D., and Harvey, P. H. 1984. Brain size allometry: Ontogeny and phylogeny. In *Size and scaling in primate biology,* ed. W. L. Jungers. New York: Plenum Press.

Martin, R. D., and MacLarnon, A. M. 1985. Gestation period, neonatal size, and maternal investment in placental mammals. *Nature.*

Masataka, N. 1981. A field study of the social behavior of Goeldi's monkeys (*Callimico goeldii*) in North Bolivia, 1: Group composition, breeding cycle, and infant development. Kyoto University Overseas Research Reports of New World Monkeys 2:23–41.

————. 1983. Categorical responses to natural and synthesized alarm calls in Goeldi's monkeys (*Callimico goeldii*). *Primates* 24:40–51.

Maslow, A. 1937. The role of dominance in social and sexual behavior of infrahuman primates, 4: The determinants of a hierarchy in pairs and in a group. *J. Genet. Psychol.* 49:161–98.

Mason, W. A. 1966. Social organization of the South American monkey, *Callicebus moloch:* A preliminary report. *Tulane Stud. Zool.* 13:23–28.

————. 1968. Use of space by callicebus groups. In *Primates: Studies in adaptation and variability,* ed. P. C. Jay. New York: Holt, Rinehart, and Winston.

————. 1971. Field and laboratory studies of social organization in *Saimiri* and *Callicebus.* In *Primate behavior,* vol. 2, ed. L. A. Rosenblum. New York: Academic Press.

————. 1974. Comparative studies of social behavior in *Callicebus* and *Saimiri:* Behavior of male-female pairs. *Folia Primatol.* 22:1–8.

————. 1975. Comparative studies of social behavior in *Callicebus* and *Saimiri:* Strength and specificity of attraction between male-female cagemates. *Folia Primatol.* 23:113–23.

————. 1978. Ontogeny of social systems. In *Recent advances in primatology,* vol. 1, ed. D. J. Chivers and J. Herbert. New York: Academic Press.

Massey, A. 1977. Agonistic aids and kinship in a group of pigtail macaques. *Behav. Ecol. Sociobiol.* 2:31–40.

————. 1979. Reply to Kurland and Gaulin. *Behav. Ecol. Sociobiol.* 6:81–83.

Masters, W., and Johnson, V. 1966. *Human sexual response.* Boston: Little, Brown.

Masterson, R. B., and Imig, T. J. 1984. Neural mechanisms for sound localizations. *Ann. Rev. Physiol.* 46:275–87.

Masui, K.; Sugiyama, Y.; Nishimura, A.; and Ohsawa, H. 1975. The life table of Japanese monkeys at Takasakiyama. In *Contemporary primatology: Proceedings of the Fifth Congress of the International Primatological Society,* ed. M. Kawai, S. Kondo, and A. Ehara. Basel: S. Karger.

Mathieu, M. 1978. Piagetian assessment of cognitive development in primates. Paper presented at the Forty-seventh Meeting of the American Anthropological Association, Los Angeles, Calif.

Maxim, P. E. 1976. An interval scale for studying and quantifying social relations in pairs of rhesus monkeys. *J. Exp. Psychol. (Gen.)* 105:123–47.

Maxim, P. E., and Buettner-Janusch, J. 1963. A field study of the Kenya baboon. *Am. J. Phys. Anthropol.* 21:165–80.

May, R. M., and Rubenstein, D. I. 1983. Reproductive strategies. In *Reproduction fitness,* 2d ed., vol. 4, ed. R. Short and M. V. Austin. Cambridge: Cambridge University Press.

Mayer, P. J. 1982. Evolutionary advantage of the menopause.

Hum. Ecol. 10:477–93.

Maynard Smith, J. 1965. The evolution of alarm calls. *Am. Nat.* 100:637–50.

———. 1982. *Evolution and the theory of games.* Cambridge: Cambridge University Press.

Maynard Smith, J., and Parker, G. A. 1976. The logic of asymmetric contests. *Anim. Behav.* 24:159–75.

Maynard Smith, J., and Price, G. R. 1973. The logic of animal conflicts. *Nature* 246:15–18.

Medway, Lord. 1970. The monkeys of Sundaland: Ecology and systematics of the cercopithecids of a humid equatorial environment. In *Old World monkeys: Evolution, systematics, and behavior,* ed. J. R. Napier and P. H. Napier. New York: Academic Press.

Meikle, D. B.; Tilford, B. L.; and Vessey, S. H. 1984. Dominance rank, secondary sex ratios, and reproduction of offspring in polygynous primates. *Am. Nat.* 124:173–88.

Meikle, D. B., and Vessey, S. H. 1981. Nepotism among rhesus monkey brothers. *Nature* 294:160–61.

Mellen, J.; Littlewood, A. P.; Barrow, B. C.; and Stevens, V. J. 1981. Individual and social behavior in a captive group of mandrills (*Mandrillus sphinx*). *Primates* 22:206–20.

Melnick, D. J. 1981. Microevolution in a population of Himalayan rhesus monkeys (*Macaca mulatta*). Ph.D. diss., Yale University.

———. 1982. Are social mammals really inbred? *Genetics* 100:46.

———. 1983. Genetic diversity in wild rhesus monkeys. *Am. J. Primatol.* 4:348–49.

Melnick, D. J.; Jolly, C. J.; and Kidd, K. K. 1984. The genetics of a wild population of rhesus monkeys (*Macaca mulatta*), 1: Genetic variability within and between social groups. *Am. J. Phys. Anthropol.* 63:341–60.

Melnick, D. J., and Kidd, K. K. 1981. Social group fission and paternal relatedness. *Am. J. Primatol.* 1:333–34.

———. 1983. The genetic consequences of social group fission in a wild population of rhesus monkeys (*Macaca mulatta*). *Behav. Ecol. Sociobiol.* 12:229–36.

———. 1985. Genetic and evolutionary relationships among Asian macaques. *Int. J. Primatol.* 6:123–60.

Melnick, D. J.; Pearl, M. C.; and Richard A. F. 1984. Male migration and inbreeding avoidance in wild rhesus monkeys. *Am. J. Primatol.* 7:229–43.

Menzel, E. W. 1966. Responsiveness to objects in free-ranging Japanese monkeys. *Behaviour* 26:130–49.

———. 1969. Naturalistic and experimental approaches to primate behavior. In *Naturalistic viewpoints in psychology,* ed. E. Willems and H. Raush. New York: Holt, Rinehart, and Winston.

———. 1971. Communication about the environment in a group of young chimpanzees. *Folia Primatol.* 15:220–32.

———. 1972. Spontaneous invention of ladders in a group of young chimpanzees. *Folia Primatol.* 17:87–106.

———. 1973a. Further observations on the use of ladders in a group of young chimpanzees. *Folia Primatol.* 19:450–57.

———. 1973b. Leadership and communication in young chimpanzees. In *Symposia of the Fourth Congress of the International Primatological Society,* vol. 1: *Precultural behavior,* ed. E. Menzel. Basel: S. Karger.

———. 1975. Communication and aggression in a group of young chimpanzees. In *Nonverbal communication of aggression,* ed. P. Pliner, L. Krames, and T. Alloway. New York: Plenum Press.

———. 1979. General discussion of the methodological problems involved in the study of social interactions. In *Social interaction analysis,* ed. M. Lamb, S. Suomi, and G. Stephenson. Madison: University of Wisconsin Press.

Michael, R. P.; Bonsall, R. W.; and Warner, P. 1975a. Human vaginal secretions: Volatile fatty acids content. *Science* 186:1217–19.

———. 1975b. Primate sexual pheromones. In *Olfaction and taste,* ed. D. A. Denton and T. P. Coghlan. New York: Academic Press.

Michael, R. P.; Bonsall, R. W.; and Zumpe, D. 1978. Consort bonding and operant behavior by female rhesus monkeys. *J. Comp. Physiol. Psychol.* 92:837–45.

Michael, R. P.; Wilson, M.; and Plant, T. M. 1973. Sexual behavior of male primates and the role of testosterone. In *Comparative ecology and behaviour of primates,* ed. R. P. Michael and J. H. Crook. New York: Academic Press.

Michael, R. P., and Zumpe, D. 1971. Patterns of reproductive behavior. In *Comparative reproduction of nonhuman primates,* ed. E. S. E. Hafez. Springfield, Ill.: C. C. Thomas.

Michod, R. E. 1982. The theory of kin selection. *Ann. Rev. Ecol. Syst.* 13:23–55.

Miles, L. 1983. Apes and language: The search for communicative competence. In *Language in primates: Perspectives and implications,* ed. J. de Luce and T. H. Wilder. New York: Springer Verlag.

Millar, J. S. 1977. Adaptive features of mammalian reproduction. *Evolution* 31:370–86.

———. 1982. Pre-partum reproductive characteristics of eutherian mammals. *Evolution* 36:1149–93.

Miller, M.; Kling, A.; and Dicks, D. 1973. Familial interactions of male rhesus monkeys in a semi-free-ranging troop. *Am. J. Phys. Anthropol.* 38:605–12.

Miller, R. E. 1967. Experimental approaches to the physiological and behavioral concomitants of affective communication in rhesus monkeys. In *Social communication among primates,* ed. S. A. Altmann. Chicago: University of Chicago Press.

———. 1971. Experimental studies of communication in the monkey. In *Primate behavior,* vol. 2, ed. L. A. Rosenblum. New York: Academic Press.

———. 1975. Nonverbal expressions of aggression and submission in social groups of primates. In *Nonverbal communication of aggression,* ed. P. Pliner, L. Krames, and T. Alloway. New York: Plenum Press.

Miller, R. S. 1968. Pattern and process in competition. *Rec. Adv. Ecol. Res.* 4:1–74.

Milton, K. 1978. Role of the upper canine and p2 in increasing the harvesting efficiency of *Hapalemur griseus* Linn., 1795. *J. Mammal.* 59:188–90.

———. 1980. *The foraging strategy of howler monkeys.* New York: Columbia University Press.

———. 1981. Distribution patterns of tropical plant foods as an evolutionary stimulus to primate mental development. *Am. Anthropol.* 83:534–48.

———. 1982. Dietary quality and demographic regulation in a howler monkey population. In *The ecology of a tropical for-*

est: Seasonal rhythms and long-term changes, ed. E. G. Leigh, Jr., A. S. Rand, and D. M. Windsor. Washington, D.C.: Smithsonian Institution Press.

———. 1984. Habitat, diet, and activity patterns of free-ranging woolly spider monkeys (*Brachyteles arachnoides* E. Geoffroy 1806). *Int. J. Primatol.* 5:491–514.

Milton, K., and de Lucca, C. 1984. Population estimate for *Brachyteles* at Fazenda Barreiro Rico, Saõ Paulo State, Brazil. *IUCN/SSC Primate Specialist Group Newsletter* 4:27–28.

Milton, K., and May, M. L. 1976. Body weight, diet, and home range area in primates. *Nature* 259:459–62.

Milton, K.; Windsor, D. M.; Morrison, D. W.; and Estribi, M. A. 1982. Fruiting phenologies of two Neotropical *Ficus* species. *Ecology* 63:752–62.

Missakian, E. A. 1972. Genealogical and cross-genealogical dominance relations in a group of free-ranging rhesus monkeys (*Macaca mulatta*) on Cayo Santiago. *Primates* 13:169–80.

———. 1973a. Genealogical mating activity in free-ranging groups of rhesus monkeys (*Macaca mulatta*) on Cayo Santiago. *Behaviour* 45:224–40.

———. 1973b. The timing of fission among free-ranging rhesus monkeys. *Am. J. Phys. Anthropol.* 38:621–24.

———. 1974. Mother-offspring grooming relations in rhesus monkeys. *Arch. Sex. Behav.* 3:135–41.

Mitani, J. C. 1984. The behavioral regulation of monogamy in gibbons (*Hylobates muelleri*). *Behav. Ecol. Sociobiol.* 15:225–29.

———. 1985a. Gibbon song duets and intergroup spacing behavior. *Behaviour* 92:59–96.

———. 1985b. Mating behavior of male orangutans in the Kutai Reserve, East Kalimantan, Indonesia. *Anim. Behav.* 33:392–402.

———. 1985c. Responses of gibbons (*Hylobates muelleri*) to self, neighbor, and stranger song duets. *Int. J. Primatol.* 6:193–200.

———. 1985d. Sexual selection and adult male orangutan long calls. *Anim. Behav.* 33:272–83.

Mitani, J. C., and Rodman, P. S. 1979. Territoriality: The relation of ranging patterns and home range size to defendability, with an analysis of territoriality among primate species. *Behav. Ecol. Sociobiol.* 5:241–51.

Mitchell, G. D. 1968. Attachment differences in male and female infant monkeys. *Child Dev.* 39:612–20.

———. 1979. *Behavioral sex differences in nonhuman primates.* New York: Van Nostrand Reinhold.

Mitchell, G. D.; Arling, G. L.; and Moller, G. W. 1967. Long-term effects of maternal punishment on the behaviour of monkeys. *Psychonom. Sci.* 8:209–10.

Mitchell, G. D., and Brandt, E. M. 1972. Paternal behavior in primates. In *Primate socialization,* ed. F. E. Poirier. New York: Random House.

Mitchell, U. R.; Presely, S.; Czekala, N. M.; and Lasley, B. S. 1982. Urinary immunoreactive estrogen and pregnanediol-3-glucuronide during the normal menstrual cycle of the female lowland gorilla (*Gorilla gorilla*). *Am. J. Primatol.* 2:167–75.

Mittermeier, R. A. 1973. Colobus monkeys and the tourist trade. *Oryx* 12:113–17.

———. 1977. Distribution, synecology, and conservation of Surinam primates. Ph.D. diss., Harvard University.

———. 1978. Locomotion and posture in *Ateles geoffroyi* and *Ateles paniscus. Folia Primatol.* 30:161–93.

Mittermeier, R. A., and Coimbra-Filho, A. F. 1977. Primate conservation in Brazilian Amazonia. In *Primate conservation,* ed. H. R. H. Prince Rainier III and G. H. Bourne. New York: Academic Press.

———. 1981. Systematics: Species and subspecies. In *Ecology and behavior of Neotropical primates,* vol. 1, ed. A. F. Coimbra-Filho and R. A. Mittermeier. Rio de Janeiro: Academia Brasileira de Ciencias.

Mittermeier, R. A.; Coimbra-Filho, A. F.; Constable, I. D.; Rylands, A. B.; and Valle, C. 1982. Conservation of primates in the Atlantic forest region of eastern Brazil. *Int. Zoo Yrbk.* 22:2–17.

Mittermeier, R. A., and van Roosmalen, M. G. M. 1981. Preliminary observations on habitat utilization and diet in eight Surinam monkeys. *Folia Primatol.* 36:1–39.

Mizuhara, H. 1964. Social changes of Japanese monkey troops in the Takasakiyama. *Primates* 5:21–52.

Mohnot, S. M. 1971. Some aspects of social changes and infant-killing in the hanuman langur (*Presbytis entellus*) (Primates: Cercopithecidae) in western India. *Mammalia* 35:175–98.

———. 1978. Peripheralization of weaned male juveniles in *Presbytis entellus.* In *Recent advances in primatology,* vol. 1, ed. D. J. Chivers and J. Herbert. New York: Academic Press.

———. 1980. Inter-group infant kidnapping in hanuman langurs. *Folia Primatol.* 34:259–77.

———. 1984. Langur interactions around Jodhpur (*Presbytis entellus*). In *Current primate researches,* ed. M. L. Roonwal, S. M. Mohnot, and N. S. Rathore. Jodhpur: University of Jodhpur.

Moore, J. 1982. Coalitions in langur all-male bands. *Int. J. Primatol.* 3:314.

———. 1984. Female transfer in primates. *Int. J. Primatol.* 5:537–89.

———. 1985. Demography and sociality in primates. Ph.D. diss., Harvard University.

Moore, J., and Ali, R. 1984. Are dispersal and inbreeding avoidance related? *Anim. Behav.* 32:94–112.

Moreno-Black, G. S., and Bent E. F. 1982. Secondary compounds in the diet of *Colobus angolensis. Afr. J. Ecol.* 20:29–36.

Mori, A. 1975 Signals found in the grooming interactions of wild Japanese monkeys of the Koshima troop. *Primates* 20:371–98.

———. 1979a. Analysis of population changes by measurement of body weight in the Koshima troop of Japanese monkeys. *Primates* 20:371–97.

———. 1979b. Reproductive behaviour. In *Contributions to primatology,* vol. 16: *Ecological and sociological studies of gelada baboons,* ed. M. Kawai. Basel: S. Karger.

———. 1979c. Unit formation and the emergence of a new leader. In *Contributions to primatology,* vol. 16: *Ecological and sociological studies of gelada baboons,* ed. M. Kawai. Basel: S. Karger.

Mori, U. 1979a. Development of sociability and social status.

In *Contributions to primatology*, vol. 16: *Ecological and sociological studies of gelada baboons*, ed. M. Kawai. Basel: S. Karger.

———. 1979b. Inter-unit relationships. In *Contributions to primatology*, vol. 16: *Ecological and sociological studies of gelada baboons*, ed. M. Kawai. Basel: S. Karger.

Mori, U., and Dunbar, R. I. M. 1985. Changes in the reproductive condition of female gelada baboons following the takeover of one-male units. *Z. Tierpsychol.* 67:215–24.

Morris, D. 1967. *The naked ape*. London: Cape.

Morse, D. H. 1969. Ecological aspects of some mixed-species foraging flocks of birds. *Ecol. Monogr.* 40:119–68.

Morton, E. S. 1977. On the occurrence and significance of motivation-structural rules in some bird and mammal sounds. *Am. Nat.* 111:855–69.

Moynihan, M. 1964. Some behavior patterns of platyrrhine monkeys, 1: The night monkey *Aotus trivirgatus*. *Smithson. Misc. Coll.* 146:1–84.

———. 1966. Communication in the titi monkey *Callicebus*. *J. Zool. Soc. Lond.* 150:77–127.

———. 1970. Some behavioral patterns of platyrrhine monkeys, 2: *Saguinus geoffroyi* and some other tamarins. *Smithson. Contrib. Zool.* 28:1–27.

———. 1976. *The New World primates*. Princeton: Princeton University Press.

Müller, E. F.; Kamau, J. M. Z.; and Maloiy, G. M. O. 1983. A comparative study of basal metabolism and thermoregulation in a folivorous (*Colobus guereza*) and an omnivorous (*Cercopithecus mitis*) primate species. *Comp. Biochem. Physiol.* 74A:319–22.

Müller, H. U. 1980. Variations of social behavior in a baboon hybrid zone (*Papio anubis* and *Papio hamadryas*) in Ethiopia. Ph.D. diss., University of Zurich.

Murray, R. D.; Bour, E. S.; and Smith, E. O. 1985. Female menstrual cyclicity and sexual behavior in stumptail macaques (*Macaca arctoides*). *Int. J. Primatol.* 6:101–13.

Murray, R. D., and Smith, E. O. 1985. The role of dominance and intrafamilial bonding in the avoidance of close inbreeding. *J. Hum. Evol.* 12:481–86.

Myers, N. 1979. *The sinking ark*. New York: W. W. Norton.

———. 1984. *The primary source*. New York: W. W. Norton.

Myers, R. H., and Shafer, D. A. 1979. Hybrid ape offspring of a gibbon and siamang. *Science* 205:308–10.

Nacci, P., and Tedeschi, J. 1976. Liking and power as factors affecting coalition choices in the triad. *Soc. Behav. Personal.* 4:27–32.

Nadler, R. D. 1975. Sexual cyclicity in captive lowland gorillas. *Science* 189:813–14.

———. 1977. Sexual behavior of captive orangutans. *Arch. Sex. Behav.* 6:457–75.

Nadler, R. D., and Rosenblum, L. A. 1969. Sexual behavior of male bonnet monkeys in the laboratory. *Brain Behav. Evol.* 2:482–97.

Nagel, U. 1973. A comparison of anubis baboons, hamadryas baboons, and their hybrids at a species border in Ethiopia. *Folia Primatol.* 19:104–65.

Nagel, U., and Kummer, H. 1974. Variation in cercopithecoid aggressive behavior. In *Primate aggression, territoriality, and xenophobia*, ed. R. L. Holloway. New York: Academic Press.

Napier, P. H. 1976. *Catalogue of primates in the British Museum (Natural History)*, part 1: *Families Callitrichidae and Cebidae*. London: British Museum (Natural History).

———. 1981. *Catalogue of the primates in the British Museum (Natural History) and elsewhere in the British Isles*, part 2: *Family Cercopithecidae, subfamily Cercopithecinae*. London: British Museum (Natural History).

Napier, J. R., and Napier P. H. 1967. *A handbook of living primates*. London: Academic Press.

Nash, L. T. 1976. Troop fission in free-ranging baboons in the Gombe Stream National Park, Tanzania. *Am. J. Phys. Anthropol.* 44:63–78.

———. 1978. The development of the mother-infant relationship in wild baboons (*Papio anubis*). *Anim. Behav.* 26:746–59.

———. 1982. Influence of moonlight level on travelling and calling patterns in two sympatric species of *Galago* in Kenya. In *Selected Proceedings of the Ninth Congress of the International Primatological Society*, ed. F. A. King and D. Taub. New York: Van Nostrand Reinhold.

———. 1983. Reproductive patterns in galagos (*Galago zanzibaricus* and *Galago garnettii*) in relation to climatic variability. *Am. J. Primatol.* 3:181–96.

Nathan, M. 1973. The social life of geladas. *Animals*, May, 208–14.

Neel, J. V., and Salzano, F. M. 1967. Further studies of the Xavante Indians, 10: Some hypotheses-generalizations resulting from these studies. *Am. J. Hum. Genet.* 19:554–74.

Nei, M. 1977. F-statistics and analysis of gene diversity in subdivided populations. *Ann. Hum. Genet. (Lond.)* 41:225–33.

Neimeyer, C. L., and Anderson, J. R. 1983. Primate harassment of matings. *Ethol. Sociobiol.* 4:205–20.

Neville, M. K. 1968a. A free-ranging rhesus monkey troop lacking males. *J. Mammal.* 49:771–78.

———. 1968b. Ecology and activity of Himalayan foothill rhesus monkeys (*Macaca mulatta*). *Ecology* 49:110–23.

———. 1972a. Social relations within troops of red howler monkeys (*Alouatta seniculus*). *Folia Primatol.* 18:47–77.

———. 1972b. The population structure of red howler monkeys (*Alouatta seniculus*) in Trinidad and Venezuela. *Folia Primatol.* 17:56–86.

———. 1976. The population and conservation of howler monkeys in Venezuela and Trinidad. In *Neotropical primates: Field studies and conservation*, ed. R. W. Thorington and P. G. Heltne. Washington, D.C.: National Academy of Science.

———. 1983. Social affinities in the Riverbanks howler monkeys. In *Perspectives in primate biology*, ed. P. K. Seth. New Delhi: Today and Tomorrow's Printers.

Newman, J. D., and Symmes, D. 1982. Inheritance and experience in the acquisition of primate acoustic behavior. In *Primate communication*, ed. C. T. Snowdon, C. H. Brown, and M. Petersen. New York: Cambridge University Press.

Newton, P. N. 1984. Infanticide and social change in forest grey langurs, *Presbytis entellus*, in Kanha Tiger Reserve, India. *Int. J. Primatol.* 5:366.

Neyman, P. F. 1978. Aspects of the ecology and social organization of free-ranging cotton-top tamarins (*Saguinus oedipus*) and the conservation status of the species. In *Biology and conservation of the Callitrichidae*, ed. D. G. Kleiman. Washington, D.C.: Smithsonian Institution Press.

Nicolson, N. 1982. Weaning and the development of independence in olive baboons. Ph.D. diss., Harvard University.

Nicolson, N., and Demment, M. W. 1982. The transition from suckling to independent feeding in wild baboon infants. *Int. J. Primatol.* 3:318.

Niemitz, C. 1979. An outline of the behaviour of *Tarsius bancanus*. In *The study of prosimian behavior*, ed. G. A. Doyle and R. D. Martin. New York: Academic Press.

——., ed. 1984. *Biology of tarsiers*. Stuttgart: Fischer Verlag.

Nieuwenhuijsen, C. 1985. Sex hormones and behaviour in stumptail macaques (*Macaca arctoides*). Ph.D. diss., Erasmus University, Rotterdam, The Netherlands.

Nieuwolt, S. 1977. *Tropical climatology*. Chichester: John Wiley.

Nishida, T. 1966. A sociological study of solitary male monkeys. *Primates* 7:141–204.

——. 1968. The social group of wild chimpanzees in the Mahali Mountains. *Primates* 9:167–224.

——. 1970. Social behavior and relationship among wild chimpanzees of the Mahali Mountains. *Primates* 11:47–87.

——. 1973. The ant-gathering behaviour by the use of tools among wild chimpanzees of the Mahale Mountains. *J. Hum. Evol.* 2:357–70.

——. 1976. The bark-eating habits in primates, with special reference to their status in the diet of wild chimpanzees. *Folia Primatol.* 25:277–87.

——. 1979. The social structure of chimpanzees of the Mahale Mountains. In *The great apes*, ed. D. A. Hamburg and E. R. McCown. Menlo Park, Calif.: Benjamin/Cummings.

——. 1980a. Local differences in response to water among wild chimpanzees. *Folia Primatol.* 33:189–209.

——. 1980b. The leaf-clipping display: A newly-discovered expressive gesture in wild chimpanzees. *J. Hum. Evol.* 9:117–28.

——. 1981. *The world of wild chimpanzees*. Tokyo: Chuokoronsha.

——. 1983a. Alloparental behavior in wild chimpanzees of the Mahale Mountains, Tanzania. *Folia Primatol.* 41:1–33.

——. 1983b. Alpha status and agonistic alliance in wild chimpanzees (*Pan troglodytes schweinfurthii*). *Primates* 24:318–36.

——. 1986. Social structure and dynamics of the chimpanzee: A review. In *Perspectives in primate biology*, ed. P. K. Seth and S. Seth. New Delhi: Today and Tomorrow's Printers and Publishers.

Nishida, T., and Hiraiwa, M. 1982. Natural history of a tool-using behaviour by wild chimpanzees in feeding upon wood-boring ants. *J. Hum. Evol.* 11:73–99.

Nishida, T., and Hiraiwa-Hasegawa, M. 1984. Behaviour of an adult male in a one-male unit group of chimpanzees in the Mahale mountains, Tanzania. *Int. J. Primatol.* 5:367.

——. 1985. Responses to a stranger mother-son pair in the wild chimpanzee: A case report. *Primates* 26:1–13.

Nishida, T.; Hiraiwa-Hasegawa, M.; Hasegawa, T.; and Takahata, Y. 1985. Group extinction and female transfer in wild chimpanzees in the Mahale Mountains. *Z. Tierpsychol.* 67:284–301.

Nishida, T., and Kawanaka, K. 1985. Within-group cannibalism

by adult male chimpanzees. *Primates* 26:274–84.

Nishida, T., and Uehara, S. 1980. Chimpanzees, tools, and termites: Another example from Tanzania. *Curr. Anthropol.* 21:671–72.

——. 1983. Natural diet of chimpanzees (*Pan troglodytes schweinfurthii*): Long-term record from the Mahale Mountains, Tanzania. *Afr. Stud. Monogr.* 3:109–30.

Nishida, T.; Wrangham, R. W.; Goodall, J.; and Uehara, S. 1983. Local differences in plant-feeding habits of chimpanzees between the Mahale Mountains and Gombe National Park, Tanzania. *J. Hum. Evol.* 12:467–80.

Nishimura, A., and Izawa, K. 1975. The group characteristics of woolly monkeys (*Lagothrix lagothrica*) in the upper Amazonian basin. In *Proceedings from the Symposia of the Fifth Congress of the International Primatological Society*, ed. S. Kondo, M. Kawai, A. Ehara, and S. Kawamura. Tokyo: Japan Science Press.

Nissen, H. W., and Yerkes, R. M. 1943. Reproduction in the chimpanzee: Report on forty-nine births. *Anat. Rec.* 86:567–78.

Noë, R.; de Waal, F.; and van Hooff, J. 1980. Types of dominance in a chimpanzee colony. *Folia Primatol.* 34:90–110.

van Noordwijk, M. A. 1986. Sexual behaviour of Sumatran long-tailed macaques (*Macaca fascicularis*). *Z. Tierpsychol.*

van Noordwijk, M. A., and van Schaik, C. P. 1985. Male migration and rank acquisition in wild long-tailed macaques *Macaca fascicularis*. *Anim. Behav.* 33:849–61.

Norikoshi, K. 1982. One observed case of cannibalism among wild chimpanzees of the Mahale Mountains. *Primates* 23:66–74.

Norikoshi, K., and Koyama, N. 1975. Group shifting and social organization among Japanese monkeys. In *Proceedings from the Symposia of the Fifth Congress of the International Primatological Society*, ed. S. Kondo, M. Kawai, A. Ehara, and S. Kawamura. Tokyo: Japan Science Press.

Nowak, R. M., and Paradiso, J. L., eds. 1983. *Walker's mammals of the world*, 4th ed. Baltimore: Johns Hopkins University Press.

Noyes, M. J. S. 1982. The association of maternal attributes with infant gender in a group of Japanese monkeys (abstract). *Int. J. Primatol.* 3:320.

Nozawa, K.; Shotake, T.; Kawamoto, Y.; and Tanabe, Y. 1982. Population genetics of Japanese monkeys, 2: Blood protein polymorphisms and population structure. *Primates* 23:252–71.

Oates, J. F. 1977a. The guereza and man. In *Primate conservation*, ed. H. R. H. Prince Rainier III and G. H. Bourne. New York: Academic Press.

——. 1977b. The guereza and its food. In *Primate ecology: Studies of feeding and ranging behaviour in lemurs, monkey, and apes*, ed. T. H. Clutton-Brock. London: Academic Press.

——. 1977c. The social life of a black-and-white colobus monkey (*Colobus guereza*). *Z. Tierpsychol.* 45:1–60.

——. 1982. In search of rare forest primates in Nigeria. *Oryx* 16:431–36.

Oates, J. F., and Anadu, P. A. 1982. Report on a survey of rainforest primates in southwest Nigeria. *IUCN/SSC Primate Specialist Group Newsletter* 2:17.

Oates, J. F.; Gartlan, J. S.; and Struhsaker, T. T. 1982. A framework for planning rain-forest primate conservation. *IPS News.* 1982:4–9.

Oates, J. F., and Trocco, T. F. 1983. Taxonomy and phylogeny of black-and-white colobus monkeys: Inferences from an analysis of loud call variation. *Folia Primatol.* 40:83–113.

Ober, C.; Olivier, T. J.; and Buettner-Janusch, J. 1980. Genetic aspects of migration in a rhesus monkey population. *J. Hum. Evol.* 9:197–203.

O'Brien, S. J.; Goldman, D.; Merril, C. R.; Bush, M.; and Wildt, D. E. 1983. The cheetah is depauperate in biochemical genetic variation. *Science* 221:459–62.

O'Brien, S. J.; Roelke, M. E.; Marker, L.; Newman, A.; Winkler, C. A.; Meltzer, D.; Colly, L.; Evermann, J. F.; Bush, M.; and Wildt, D. E. 1985. A genetic basis for species vulnerability in the cheetah. *Science* 227:1428–34.

Office of Technology Assessment. 1984. Technologies to sustain tropical forest resources. Congress of the United States. Washington, D.C.: Government Printing Office.

Ohsawa, H., and Dunbar, R. I. M. 1984. Variations in the demographic structure and dynamics of gelada baboon populations. *Behav. Ecol. Sociobiol.* 15:231–40.

Oki, J., and Maeda, Y. 1973. Grooming as a regulator of behavior in Japanese macaques. In *Behavioral regulators of behavior in primates,* ed. C. R. Carpenter. Lewisburg, Pa.: Bucknell University Press.

Oliver, J. I., and Lee, P. C. 1978. Comparative aspects of the behavior of juveniles in two species of baboons in Tanzania. In *Recent advances in primatology,* vol. 1, ed. D. J. Chivers and J. Herbert. New York: Academic Press.

Oliver, W., and O'Connor, S. 1980. Circadian distribution of *Indri indri* group vocalizations: A short period sampling at two study sites near Perinet, Eastern Madagascar. *Dodo* 17:19–27.

Olson, T. R. 1979. Studies on aspects of the morphology and systematics of the genus *Otolemur* Coquerel, 1859 (Primates: Galagidae). Ann Arbor, Mich.: University Microfilms International.

Omar, A., and DeVos, A. 1971. The annual reproductive cycle of an African monkey (*Cercopithecus mitis kolbi* Neuman). *Folia Primatol.* 16:206–15.

Oppenheimer, J. R. 1968. Behavior and ecology of the white-faced monkey *Cebus capucinus* on Barro Colorado Island. Ph.D. diss., University of Illinois, Urbana.

———. 1969. Changes in forehead patterns and group composition of the white-faced monkey (*Cebus capucinus*). In *Proceedings of the Second Congress of the International Primatological Society,* vol. 1, ed. C. R. Carpenter. Basel: S. Karger.

Orians, G. H. 1969. On the evolution of mating systems in birds and mammals. *Am. Nat.* 108:589–603.

Otis, J. S.; Froehlich, J. W.; and Thorington, R. W., Jr. 1981. Seasonal and age-related differential mortality by sex in the mantled howler monkey, *Alouatta palliata. Int. J. Primatol.* 2:197–205.

Owens, N. W. 1975. Social play behavior in free-living baboons, *Papio anubis. Anim. Behav.* 23:387–408.

———. 1976. The development of sociosexual behavior in free-living baboons, *Papio anubis. Behaviour* 57:241–59.

Owen-Smith, N. 1977. On territoriality in ungulates and an evolutionary model. *Quart. Rev. Biol.* 52:1–38.

Packer, C. 1977. Reciprocal altruism in olive baboons. *Nature* 265:441–43.

———. 1979a. Inter-troop transfer and inbreeding avoidance in *Papio anubis. Anim. Behav.* 27:1–36.

———. 1979b. Male dominance and reproductive activity in *Papio anubis. Anim. Behav.* 27:37–45.

———. 1980. Male care and exploitation of infants in *Papio anubis. Anim. Behav.* 28:512–20.

———. 1985. Dispersal and inbreeding avoidance. *Anim. Behav.* 33:676–78.

———. 1986. The ecology of sociality in felids. In *Ecology and social evolution: Birds and mammals,* ed. D. I. Rubenstein and R. W. Wrangham. Princeton: Princeton University Press.

Packer, C. and Pusey, A. E. 1979. Female aggression and male membership in troops of Japanese macaques and olive baboons. *Folia Primatol.* 31:212–18.

———. 1982. Cooperation and competition within coalitions of male lions: Kin selection or game theory? *Nature* 296:740–42.

———. 1983. Adaptations of female lions to infanticide by incoming males. *Am. Nat.* 121:716–28.

———. 1984. Infanticide in carnivores. In *Infanticide: Comparative and evolutionary perspectives,* ed. G. Hausfater and S. B. Hrdy. Hawthorne, N.Y.: Aldine.

Pages, E. 1980. Ethoecology of *Microebus coquereli* during the dry season. In *Nocturnal Malagasy primates, ecology, physiology, and behaviour,* ed. P. Charles-Dominique, H. M. Cooper, A. Hladik, E. Pages, G. F. Pariente, A. Petter-Rousseaux, J. J. Petter, and A. Schilling. New York: Academic Press.

Palmer, A. E.; London, W. T.; and Brown, R. L. 1981. Color changes in the haircoat of patas monkeys (*Erythrocebus patas*). *Am. J. Primatol.* 1:371–78.

Pariente, G. F. 1979. The role of vision in prosimian behavior. In *The study of prosimian behavior,* ed. G. A. Doyle and R. D. Martin. New York: Academic Press.

Parker, G. A. 1984. Evolutionarily stable strategies. In *Behavioral ecology: An evolutionary approach,* 2d ed., ed. J. R. Krebs and N. B. Davies. Sunderland, Mass.: Sinauer Associates.

Parker, G. A., and MacNair, M. R. 1978. Models of parent-offspring conflict, 1: Monogamy. *Anim. Behav.* 26:97–110.

Parker, S. T. 1977. Piaget's sensorimotor period series in the infant macaque: A model for comparing intelligence and feeding adaptation in cebus monkeys and great apes. *J. Hum. Evol.* 6:623–41.

Parker, S. T., and Gibson, K. R. 1977. Object manipulation, tool use, and sensorimotor intelligence as feeding adaptations in cebus monkeys and great apes. *J. Hum. Evol.* 6:623–41.

———. 1979. A developmental model for the evolution of language and intelligence in early hominids. *Behav. Brain Sci.* 2:367–408.

Parkes, C. M., and Stevenson-Hinde, J., eds. 1982. *The place of attachment in human behavior.* New York: Basic Books.

Passingham, R. E. 1982. *The human primate.* San Francisco: W. H. Freeman.

Patterson, F. G. 1978a. Conversations with a gorilla. *Nat. Geogr.* 154:438–65.

————. 1978b. The gestures of a gorilla: Language acquisition in another pongid. *Brain Lang.* 5:72–97.

Patterson, F. G., and Linden, E. 1981. *The education of Koko.* New York: Holt, Rinehart, and Winston.

Paul, A., and Kuester, J. 1985. Inter-group transfer and incest avoidance in semi-free-ranging Barbary macaques (*Macaca sylvanus*) at Salem (FRG). *Am. J. Primatol.* 8:317–22.

Paul, A., and Thommen, D. 1984. Timing of birth, female reproductive success, and infant sex ratio in semifree-ranging Barbary macaques (*Macaca sylvana*). *Folia Primatol.* 42:2–16.

Pearl, M. C. 1982. Networks of social relations among Himalayan rhesus monkeys (*Macaca mulatta*). Ph.D. diss., Yale University.

Pearl, M. C., and Melnick, D. 1983. Inferring maternity from the behavior and genetics of wild rhesus monkeys. *Am. J. Primatol.* 4:351.

Pearl, M. C., and Schulman, S. R. 1983. Techniques for the analysis of social structure in animal societies. In *Advances in the study of behavior,* vol. 13, ed. J. S. Rosenblatt, R. A. Hinde, C. Beer, and M. C. Busnel. New York: Academic Press.

Perachio, A. A.; Alexander, M.; and Marr, L. D. 1973. Hormonal and social factors affecting evoked sexual behavior in rhesus monkeys. *Am. J. Phys. Anthropol.* 38:227–32.

Pereira, M. E., and Altmann, J. 1985. Development of social behavior in free-living non-human primates. In *Non-human primate models for human growth and development,* ed. E. S. Watts. New York: Alan R. Liss.

Perret, M. 1982. Influence du groupement social sur la reproduction de la femelle de *Microcebus murinus* (Miller, 1777). *Z. Tierpsychol.* 60:47–65.

Petersen, M. R.; Beecher, M. D.; Zoloth, S. R.; Green, S.; Marler, P.; Moody, D. B.; and Stebbins, W. C. 1984. Neural lateralization of vocalizations by Japanese macaques: Communicative significance is more important than acoustic structure. *Behav. Neurosci.* 98:779–90.

Petersen, M. R.; Beecher, M. D.; Zoloth, S. R.; Moody, D. B.; and Stebbins, W. C. 1978. Neural lateralization of species-specific vocalizations by Japanese macaques (*Macaca fuscata*). *Science* 202:324–27.

Petter, J. J. 1962. Recherches sur l'écologie et l'éthologie des lemuriens malgaches. *Mem. Mus. Nat. Hist. Natur.* (Paris) 27:1–146.

————. 1977. The aye-aye. In *Primate conservation,* ed. H. R. H. Prince Rainier III and G. H. Bourne. New York: Academic Press.

Petter, J. J.; Albignac, R.; and Rumpler, Y. 1977. *Faune de Madagascar 44: Mammifères lémuriens (primates prosimiens).* Paris: ORSTOM/CNRS.

Petter, J. J., and Charles-Dominique, P. 1979. Vocal communication in prosimians. In *The study of prosimian behavior,* ed. G. A. Doyle and R. D. Martin. New York: Academic Press.

Petter, J. J., and Petter-Rousseaux, A. 1979. Classification of the prosimians. In *The study of prosimian behavior,* ed. G. A. Doyle and R. D. Martin. New York: Academic Press.

Petter, J. J., and Peyrieras, A. 1970. Observations écoéthologiques sur les lémuriens malgaches du genre *Hapalemur. Terre et Vie* 117:356–82.

Petter-Rousseaux, A. 1980. Seasonal activity rhythms, repro-

duction, and body weight variations in five sympatric nocturnal prosimians, in simulated light and climatic conditions. In *Nocturnal Malagasy primates,* ed. P. Charles-Dominique, H. M. Cooper, A. Hladik, E. Pages, G. F. Pariente, A. Petter-Rousseaux, J. J. Petter, and A. Schilling. New York: Academic Press.

Phillips, M. J., and Mason, W. A. 1976. Comparative studies of social behavior in *Callicebus* and *Saimiri:* Social looking in male-female pairs. *Bull. Psychonom. Soc.* 7:55–56.

Phoenix, C. M. 1973. Ejaculation by male rhesus as a function of female partner. *Horm. Behav.* 4:365–70.

Piaget, J. 1963. *The origins of intelligence in children.* New York: W. W. Norton.

Pianka, E. R. 1974. *Evolutionary ecology.* New York: Harper and Row.

Pilbeam, D. A. 1984. The descent of hominoids and hominids. *Sci. Am.* 250:84–96.

Pires, J. M. 1973. Tipos y vegetaçâo da Amazônia. *Publiçaes avulsas do Museo Paraensis Emilio Goeldi* 20:179–202.

Ploog, D. 1967. The behavior of squirrel monkeys (*Saimiri sciureus*) as revealed by sociometry, bioacoustics, and brain stimulation. In *Social communication among primates,* ed. S. A. Altmann. Chicago: University of Chicago Press.

————. 1981. Neurobiology of primate audio-visual behavior. *Brain Res. Rev.* 3:35–61.

Plotnick, R. J., and Tallarico, R. B. 1966. Object-quality learning set formation in the young chicken. *Psychonom. Sci.* 5:195–96.

Poirier, F. E. 1968. Nilgiri langur (*Presbytis johnii*) territorial behavior. In *Proceedings of the Second Congress of the International Primatological Society,* vol. 1, ed. C. R. Carpenter. Basel: S. Karger.

————. 1969. The Nilgiri langur (*Presbytis johnii*) troop: Its composition, structure, function, and change. *Folia Primatol.* 10:20–47.

————. 1970. The Nilgiri langur (*Presbytis johnii*) of South India. In *Primate behavior: Developments in field and laboratory research,* vol. 1, ed. L. A. Rosenblum. New York: Academic Press.

————. 1974. Colobine aggression: A review. In *Primate aggression, territoriality, and xenophobia: A comparative perspective,* ed. R. L. Holloway. New York: Academic Press.

————. 1983. The golden monkey in the People's Republic of China. *IUCN/SSC Primate Specialist Group Newsletter* 3: 31–32.

Poirier, F., and Smith, E. O. 1974. Socializing function of primate play. *Am. Zool.* 14:275–87.

Pola, Y. V., and Snowdon, C. T. 1975. The vocalizations of pygmy marmosets (*Cebuella pygmaea*). *Anim. Behav.* 23: 826–42.

Pollock, J. I. 1975. Field observations on *Indri indri:* A preliminary report. In *Lemur biology,* ed. I. Tattersall and R. W. Sussman. New York: Plenum Press.

————. 1979a. Female dominance in *Indri indri. Folia Primatol.* 31:143–64.

————. 1979b. Spatial distribution and ranging behavior in lemurs. In *The study of prosimian behavior,* ed. G. A. Doyle and R. D. Martin. New York: Academic Press.

Pond, C. M. 1977. The significance of lactation in the evolution of mammals. *Evolution* 31:177–99.

Pook, A. G., and Pook, G. 1981. A field study of the socio-ecology of the Goeldi's monkey (*Callimico goeldii*) in northern Bolivia. *Folia Primatol.* 35:288–312.

———. 1982. Polyspecific association between *Saguinus fuscicollis, Saguinus labiatus, Callimico goeldii* and other primates in north-western Bolivia. *Folia Primatol.* 38:196–216.

Poole, J. H., and Moss, C. J. 1981. Musth in the African elephant, *Loxodonta africana*. *Nature* 292:830–31.

Popp, J. L. 1978. Male baboons and evolutionary principles. Ph.D. diss., Harvard University.

———. 1983. Ecological determinism in the life histories of baboons. *Primates* 24:198–210.

Popp, J., and DeVore, I. 1979. Aggressive competition and social dominance theory: A synopsis. In *The great apes,* ed. D. Hamburg and E. R. McCown. Menlo Park, Calif.: Benjamin/Cummings.

Portman, O. 1970. Nutritional requirements (NRC) of nonhuman primates. In *Feeding and nutrition of nonhuman primates,* ed. R. S. Harris. New York: Academic Press.

Post, D. G. 1982. Feeding behavior of yellow baboons (*Papio cynocephalus*) in the Amboseli National Park, Kenya. *Int. J. Primatol.* 3:403–30.

Post, D. G.; Hausfater, G.; and McCuskey, S. A. 1980. Feeding behavior of yellow baboons (*Papio cynocephalus*): Relationship to age, gender, and dominance rank. *Folia Primatol.* 34:170–95.

Prakash, I. 1962. Group organization, sexual behavior, and breeding season of certain Indian monkeys. *Jap. J. Ecol.* 12:83–86.

Premack, D. 1976. *Intelligence in ape and man.* Hillsdale, N.J.: Lawrence Erlbaum Associates.

———. 1983a. Social cognition. *Ann. Rev. Psychol.* 34:351–62.

———. 1983b. The codes of man and beast. *Behav. Brain Sci.* 6:125–67.

Premack, D., and Premack, A. 1983. *The mind of an ape.* New York: W. W. Norton.

Preuschoft, H.; Chivers, D. J.; Brockelman, W.; and Creel, N. 1984. *The lesser apes: Evolutionary and behavioral biology.* Edinburgh: Edinburgh University Press.

Prouty, L. A.; Buchanan, P. D.; Pollitzer, W. S.; and Mootnick, A. R. 1983. A presumptive new hylobatid subgenus with thirty-eight chromosomes. *Cytogenet. and Cell Genet.* 35:141–42.

Pulliam, H. R., and Caraco, T. 1984. Living in groups: Is there an optimal group size? In *Behavioural ecology: An evolutionary approach,* 2d ed., ed. J. R. Krebs and N. B. Davies. Sunderland, Mass.: Sinauer Associates.

Pusey, A. E. 1978. The physical and social development of wild adolescent chimpanzees (*Pan troglodytes schweinfurthii*). Ph.D. diss., Stanford University.

———. 1979. Intercommunity transfer of chimpanzees in Gombe National Park. In *The great apes,* ed. D. A. Hamburg and E. R. McCown, Menlo Park, Calif.: Benjamin/Cummings.

———. 1980. Inbreeding avoidance in chimpanzees. *Anim. Behav.* 28:543–82.

———. 1983. Mother-offspring relationships in chimpanzees after weaning. *Anim. Behav.* 31:363–77.

Pyke, G. H.; Pulliam, H. R.; and Charnov, E. L. 1977. Optimal foraging: A selective review of theory and tests. *Quart. Rev. Biol.* 52:137–53.

Quiatt, D. D. 1979. Aunts and mothers: Adaptive implications of allomaternal behavior of nonhuman primates. *Am. Anthropol.* 81:311–19.

Quris, R. 1975. Ecologie et organisation sociale de *Cercocebus galeritus agilis. Terre et Vie* 29:337–98.

———. 1976. Données comparatives sur la socioécologie de huit espèces de Cercopithecidae vivant dans une même zone de forêt primitive periodiquement inondée (nord-est du Gabon). *Terre et Vie* 30:193–209.

Quris, R.; Gautier, J.-P.; Gautier, J.-Y.; and Gautier-Hion, A. 1981. Organisation spatio-temporelle des activités individuelles et sociales dans une troupe de *Cercopithecus cephus. Rev. Ecol. (Terre et Vie)* 35:37–53.

Raemaekers, J. J. 1979. Ecology of sympatric gibbons. *Folia Primatol.* 31:227–45.

———. 1980. Causes of variation between months in the distance traveled daily by gibbons. *Folia Primatol.* 34:46–60.

Raemaekers, J. J.; and Raemaekers, P. M. 1985. Field playback of loud calls to gibbons (*Hylobates lar*): Territorial, sex-specific, and species-specific responses. *Anim. Behav.* 33:481–93.

———. In press. Vocal interactions between two gibbons (*Hylobates lar*). *Nat. Hist. Bull. Siam Soc.*

Raemaekers, J. J., Raemaekers, P. M.; and Haimoff, E. 1984. Loud calls of the gibbon (*Hylobates lar*): Repertoire, organization, and context. *Behavior* 91:146–89.

Rahaman, H., and Parthasarathy, M. D. 1962. Studies of the social behavior of bonnet monkeys. *Primates* 10:149–62.

———. 1969. The home range, roosting places, and the day ranges of the bonnet macaque *Macaca radiata. J. Zool.* 157:267–76.

Raleigh, M. J.; McGuire, M. T.; and Brammer, G. L. 1982. Female choice and male dominance in captive vervet monkeys (*Cercopithecus aethiops sabaeus*). Paper presented at the Ninth Congress of the International Primatological Society, Atlanta.

Ralls, K. 1976. Mammals in which females are larger than males. *Quart. Rev. Biol.* 51:245–76.

Ralls, K., and Ballou, J. 1982a. Effect of inbreeding on infant mortality in captive primates. *Int. J. Primatol.* 3:491–505.

———. 1982b. Effect of inbreeding on juvenile mortality in some small mammal species. *Lab. Anim.* 16:159–66.

Ralls, K.; Brugger, K.; and Ballou, J. 1979. Inbreeding and juvenile mortality in small populations of ungulates. *Science* 206:1101–3.

Ramaswami, L. S. 1975. Some aspects of the reproductive biology of the langur monkey *Presbytis entellus entellus* Dufresne. In *Proceedings of the Indian National Science Academy,* vol. 41, part B: *Biological Sciences.* New Delhi: Indian National Science Academy.

Ramirez, M. In press. The woolly monkey, genus *Lagothrix.* In *Ecology and behavior of Neotropical primates,* vol. 2, ed. A. F. Coimbra-Filho and R. A. Mittermeier. Rio de Janiero: Academia Brasileira de Ciencias.

Ramirez, M. F.; Freese, C. H.; and Revilla, J. 1978. Feeding ecology of the pygmy marmoset, *Cebuella pygmaea,* in northeastern Peru. In *The biology and conservation of the Callitrichidae,* ed. D. G. Kleiman. Washington, D.C.: Smithsonian Institution Press.

Ramshaw, J. A. M.; Coyne, J. A. and Lewontin, R. C. 1979. The sensitivity of gel electrophoresis as a detector of genetic variation. *Genetics* 93:1019–37.

Ransom, T. W. 1981. *Beach troop of the Gombe.* East Brunswick, N.J.: Associated University Presses.

Ransom, T. W., and Ransom, B. S. 1971. Adult male-infant relations among baboons (*Papio anubis*). *Folia Primatol.* 16:179–95.

Ransom, T. W. and Rowell, T. E. 1972. Early social development of feral baboons. In *Primate socialization,* ed. F. E. Poirier. New York: Random House.

Rasmussen, D. R. 1979. Correlates of patterns of range use of troop of yellow baboons *Papio cynocephalus,* 1: Sleeping sites, impregnable females, births, and male emigrations and immigrations. *Anim. Behav.* 27:1098–112.

———. 1981. Communities of baboon troops (*P. cynocephalus*) in Mikumi National Park, Tanzania. *Folia Primatol.* 36:232–42.

Rasmussen, K. L. R. 1980. Consort behavior and mate selection in yellow baboons (*Papio cynocephalus*). Ph.D. diss., University of Cambridge.

———. 1983a. Age-related variation in the interactions of adult females with adult males in yellow baboons. In *Primate social relationships: An integrated approach,* ed. R. A. Hinde. Oxford: Blackwell.

———. 1983b. Influences of affiliative preferences upon the behaviour of male and female baboons during consortships. In *Primate social relationships: An integrated approach,* ed. R. A. Hinde. Oxford: Blackwell.

Rathbun, C. D. 1979. Description and analysis of the arch display in the golden lion tamarin, *Leontopithecus rosalia rosalia. Folia Primatol.* 32:125–48.

Raven, P. 1984. Knockdown-dragout on the global future. Paper presented at the AAAS Meetings, New York, May 25, 1984.

Rawlins, R. G.; Kessler, M. J.; and Turnquist, J. E. 1984. Reproductive performance, population dynamics, and anthropometrics of the free-ranging Cayo Santiago rhesus macaques. *J. Med. Primatol.* 13:247–59.

Redican, W. K. 1976. Adult male-infant interactions in nonhuman primates. In *The role of the father in child development,* ed. M. E. Lamb. New York: John Wiley.

Redshaw, M. 1978. Cognitive development in human and gorilla infants. *J. Hum. Evol.* 7:133–41.

Reiss, M. J. 1982. Functional aspects of reproduction: Some theoretical considerations. Ph.D. diss., University of Cambridge.

———. 1984. Kin selection, social grooming, and the removal of ectoparasites: A theoretical investigation. *Primates* 25:185–91.

Rettig, N. 1978. Breeding behavior of the harpy eagle (*Harpia Harpyja*). *Auk* 629–43.

Reynolds, V., and Reynolds, F. 1965. Chimpanzees in the Budongo Forest. In *Primate behavior,* ed. I. DeVore. New York: Holt, Rinehart, and Winston.

Rheingold, H. L., and Eckerman, C. D. 1970. The infant separates himself from his mother. *Science* 168:78–83.

Rhine, R. J. 1972. Changes in the social structure of two groups of stump-tail macaques. *Primates* 13:181–94.

Rhine, R. J., and Westlund, B. J. 1981. Adult male positioning in baboon progressions: Order and chaos revisited. *Folia Primatol.* 35:77–116.

Richard, A. 1970. A comparative study of the activity patterns and behavior of *Alouatta villosa* and *Ateles geoffroyi. Folia Primatol.* 12:241–63.

———. 1974a. Intra-specific variation in the social organization and ecology of *Propithecus verreauxi.* In *Primate ecology,* ed. R. W. Sussman. New York: John Wiley.

———. 1974b. Patterns of mating in *Propithecus verreauxi verreauxi.* In *Prosimian biology,* ed. R. D. Martin, G. A. Doyle, and A. C. Walker. London: Duckworth.

———. 1977. The feeding behaviour of *Propithecus verreauxi.* In *Primate ecology,* ed. T. H. Clutton-Brock. London: Academic Press.

———. 1978. *Behavioral variation: Case study of a Malagasy Lemur.* Lewisburg, Pa.: Bucknell University Press.

———. 1981. Changing assumptions in primate ecology. *Am. Anthropol.* 83:517–33.

———. 1985. *Primates in nature.* New York: W. H. Freeman.

———. In press. Social boundaries in a Malagasy prosimian, the sifaka (*Propithecus verreauxi*). *Int. J. Primatol.*

Richard, A., and Heimbuch, R. 1975. An analysis of the social behavior of three groups of *Propithecus verreauxi.* In *Lemur biology,* ed. I. Tattersall and R. W. Sussman. New York: Plenum.

Richards, P. W. 1952. *The tropical rain forest: An ecological study.* Cambridge: Cambridge University Press.

Richards, S. 1974. The concept of dominance and methods of assessment. *Anim. Behav.* 22:914–30.

Ricklefs, R. E. 1979. Patterns of growth in birds, 5: A comparative study of development in the starling, common tern, and Japanese quail. *Auk* 96:10–30.

Rijksen, H. D. 1978. *A field study on Sumatran orang-utans (Pongo pygmaeus abelii, Lesson 1827): Ecology, behaviour, and conservation.* Wageningen, The Netherlands: H. Veenman and Zonen.

Rijksen, H. D., and Rijksen-Graatsma, A. G. 1975. Orangutan reserve work in north Sumatra. *Oryx* 13:63–73.

Ripley, S. 1965. The ecology and social behavior of the Ceylon gray langur, *Presbytis entellus thersites.* Ph.D. diss., University of California, Berkeley.

———. 1967. Intertroop encounters among Ceylon gray langurs (*Presbytis entellus*). In *Social communication among primates,* ed. S. A. Altmann. Chicago: University of Chicago Press.

———. 1970. Leaves and leaf-monkeys: The social organization of foraging in gray langurs *Presbytis entellus thersites.* In *Old World monkeys: Evolution, systematics, and behavior,* ed. J. R. Napier and P. H. Napier. New York: Academic Press.

Riss, D. C., and Busse, C. D. 1977. Fifty-day observation of a free-ranging adult male chimpanzee. *Folia Primatol.* 28:283–97.

Riss, D. C., and Goodall, J. 1977. The recent rise to the alpha rank in a population of free-living chimpanzees. *Folia Pri-*

matol. 27:134–51.

Ristau, C., and Robbins, D. 1982. Language in the great apes: A critical review. In *Advances in the study of behavior,* vol. 12, ed. J. Rosenblatt, R. A. Hinde, C. Beer, and M. C. Busnel. New York: Academic Press.

Roberts, M. S. 1978. The annual reproductive cycle of captive *Macaca sylvana. Folia Primatol.* 29:229–35.

Robinson, J. G. 1977. The vocal regulation of spacing in the titi monkey, *Callicebus moloch.* Ph.D. diss., University of North Carolina.

———. 1979a. An analysis of the organization of vocal communication in the titi monkey *Callicebus moloch. Z. Tierpsychol.* 49:381–405.

———. 1979b. Correlates of urine washing in the wedge-capped capuchin *Cebus nigrivittatus.* In *Vertebrate ecology in the northern Neotropics,* ed. J. F. Eisenberg. Washington, D.C.: Smithsonian Institution Press.

———. 1979c. Vocal regulation of use of space by groups of titi monkeys, *Callicebus moloch. Behav. Ecol. Sociobiol.* 5:1–15.

———. 1981a. Spatial structure in foraging groups of wedge-capped capuchin monkeys *Cebus nigrivittatus. Anim. Behav.* 29:1036–56.

———. 1981b. Vocal regulation of inter- and intragroup spacing during boundary encounters in the titi monkey, *Callicebus moloch. Primates* 22:161–72.

———. 1982a. Intrasexual competition and mate choice in primates. *Am. J. Primatol. Suppl.* 1:131–44.

———. 1982b. Vocal systems regulating within-group spacing. In *Primate communication,* ed. C. T. Snowdon, C. H. Brown, and M. Petersen. Cambridge: Cambridge University Press.

———. 1984. Syntactic structures in the vocalizations of wedge-capped capuchin monkeys *Cebus nigrivittatus. Behaviour* 90:46–79.

———. In press *a.* Expected benefits determine area defense: Experiments with capuchin monkeys. *Nat. Geogr. Soc. Res. Rep.*

———. In press *b.* Seasonal variation in use of time and space by the wedge-capped capuchin monkey *Cebus olivaceus:* Implications for foraging theory. *Smithson. Contrib. Zool.*

Robinson, J. G., and Ramirez, J. 1982. Conservation biology of Neotropical primates. In *Mammalian biology in South America,* ed. M. A. Mares and H. Genoways. Pittsburgh, Pa.: University of Pittsburgh Press.

Rodgers, W. A. 1981. The distribution and conservation status of colobus monkeys in Tanzania. *Primates* 22:33–45.

Rodman, P. S. 1973. Population composition and adaptive organisation among orangutans of the Kutai Reserve. In *Comparative ecology and behaviour of primates,* ed. R. P. Michael and J. H. Crook. London: Academic Press.

———. 1977. Feeding behaviour of orangutans of the Kutai Nature Reserve, East Kalimantan. In *Primate ecology: Studies of feeding and ranging behaviour in lemurs, monkeys, and apes,* ed. T. H. Clutton-Brock. London: Academic Press.

———. 1978. Diets, densities, and distributions of Bornean primates. In *The ecology of arboreal folivores,* ed. G. Montgomery. Washington, D.C.: Smithsonian Institution Press.

———. 1979. Individual activity patterns and the solitary nature of orangutans. In *The great apes,* ed. D. A. Hamburg and E. R. McCown. Menlo Park, Calif.: Benjamin/Cummings.

———. 1984. Foraging and social systems of orangutans and chimpanzees. In *Adaptations for foraging in nonhuman primates: Contributions to an organismal biology of prosimians, monkeys, and apes,* ed. P. S. Rodman and J. G. H. Cant. New York: Columbia University Press.

Rodman, P. S. and Cant, J. G. H., eds. 1984. *Adaptations for foraging in nonhuman primates.* New York: Columbia University Press.

Roonwal, M. L., and Mohnot, S. M. 1977. *Primates of South Asia: Ecology, sociobiology, and behavior.* Cambridge: Harvard University Press.

van Roosmalen, M. G. M. 1980. Habitat preferences, diet, feeding strategy, and social organization of the black spider monkey (*Ateles p. paniscus* Linnaeus 1758) in Surinam. Arnhem, The Netherlands: Rijksinstituut voor Natuurbeheer.

van Roosmalen, M. G. M.; Mittermeier, R. A.; and Milton, K. 1981. The bearded sakis, genus *Chiropotes.* In *Ecology and behavior of Neotropical primates,* vol. 1, eds. A. F. Coimbra-Filho and R. A. Mittermeier. Rio de Janeiro: Academia Brasileira de Ciencias.

Rosenberger, A. L. 1971. Kinship interaction patterns in pigtail and bonnet macaques. In *Proceedings of the Third International Congress of Primatology, Zurich, 1970,* vol. 3, ed. H. Kummer. Basel: S. Karger.

———. 1981. Systematics: The higher taxa. In *Ecology and behavior of Neotropical primates,* vol. 1, eds. A. F. Coimbra-Filho and R. A. Mittermeier. Rio de Janeiro: Academia Brasileira de Ciencias.

Rosenberger, A. L.; and Coimbra-Filho, A. F. 1984. Morphology, taxonomic status, and affinities of the lion tamarins, *Leontopithecus* (Callitrichinae, Cebidae). *Folia Primatol.* 42:149–79.

Rosenblum, L. A. 1971. The ontogeny of mother-infant relations in macaques. In *The ontogeny of vertebrate behavior,* ed. H. Moltz. New York: Academic Press.

———. 1974. Sex differences, environmental complexity, and mother-infant relations. *Arch. Sex. Behav.* 3:117–28.

Rosenblum, L. A., and Bromley, L. J. 1978. The effects of gonadal hormones on peer interactions. In *Recent advances in primatology,* vol. 1, ed. D. J. Chivers and J. Herbert. New York: Academic Press.

Rosenblum, L. A.; Kaufman, I. C.; and Stynes, A. J. 1966. Some characteristics of adult social and autogrooming patterns in two species of macaque. *Folia Primatol.* 4:438–51.

Rosenzweig, M. R., and Bennett, E. L. 1978. Experiential influences on brain anatomy and brain chemistry in rodents. In *Studies on the development of behavior and the nervous system,* vol. 4: *Early influences,* ed. G. Gottlieb. New York: Academic Press.

Rowell, T. E. 1963. Behaviour and female reproductive cycles of macaques. *J. Reprod. Fertil.* 6:193–203.

———. 1966a. Forest living baboons in Uganda. *J. Zool. Soc. Lond.* 147:344–64.

———. 1966b. Hierarchy in the organization of a captive baboon group. *Anim. Behav.* 14:430–43.

———. 1967. A quantitative comparison of the behaviour of a

wild and caged baboon troop. *Anim. Behav.* 15:499–509.

———. 1968. Grooming by adult baboons in relation to reproductive cycles. *Anim. Behav.* 16:585–88.

———. 1969. Long-term changes in a population of Ugandan baboons. *Folia Primatol.* 11:241–54.

———. 1970. Reproductive cycles of two *Cercopithecus* monkeys. *J. Reprod. Fertil.* 22:321–38.

———. 1972. *Social behaviour of monkeys.* Baltimore: Penguin Books.

———. 1973. Social organization of wild talapoin monkeys. *Am. J. Phys. Anthropol.* 38:593–98.

———. 1974a. Contrasting adult male roles in different species of nonhuman primates. *Arch. Sex. Behav.* 3:143–49.

———. 1974b. The concept of social dominance. *Behav. Biol.* 11:131–54.

———. 1977. Variation in age at puberty in monkeys. *Folia Primatol.* 27:284–96.

Rowell, T. E., and Chalmers, N. R. 1970. Reproductive cycles of the mangabey *Cercocebus albigena. Folia Primatol.* 12:264–72.

Rowell, T. E., and Dixson, A. F. 1975. Changes in social organization during the breeding season of wild talapoin monkeys. *J. Reprod. Fertil.* 43:419–34.

Rowell, T. E., and Hartwell, K. M. 1978. The interaction of behaviour and reproductive cycles in patas monkeys. *Behav. Biol.* 24:141–67.

Rowell, T. E., and Hinde, R. A. 1962. Vocal communication by the rhesus monkey (*Macaca mulatta*). *Proc. Zool. Soc. Lond.* 138:279–94.

Rowell, T. E.; Hinde, R. A.; and Spencer-Booth, Y. 1964. Aunt-infant interactions in captive rhesus monkeys. *Anim. Behav.* 12:219–26.

Rowell, T. E., and Olson, D. K. 1983. Alternative mechanisms of social organization in monkeys. *Behaviour* 86:31–54.

Rowell, T. E., and Richards, S. M. 1979. Reproductive strategies of some African monkeys. *J. Mammal.* 60:58–69.

Rozin, P. 1976. The evolution of intelligence and access to the cognitive unconscious. In *Progress in psychology,* vol. 6, ed. J. N. Sprague and A. N. Epstein. New York: Academic Press.

Rubenstein, D. I. 1978. On predation, competition, and the advantages of group living. *Perspectives in ethology,* vol. 3, ed. P. P. G. Bateson and P. F. Klopfer. New York: Plenum Press.

———. 1980. On the evolution of alternative mating strategies. In *Limits to action: The allocation of individual behavior,* ed. J. E. R. Staddon. New York: Academic Press.

Rubenstein, D. I., and Wrangham, R. W. 1986. Socioecology: Origins and trends. In *Ecology and social evolution: Birds and mammals,* ed. D. I. Rubenstein and R. W. Wrangham. Princeton: Princeton University Press.

Rudder, B. C. C. 1979. *The allometry of primate reproduction parameters.* Ph.D. diss., University of London.

Rudran, R. 1973a. Adult male replacement in one-male troops of purple-faced langurs (*Presbytis senex senex*) and its effect on population structure. *Folia Primatol.* 19:166–92.

———. 1973b. The reproductive cycles of two subspecies of purple-faced langurs (*Presbytis senex*) with relation to environmental factors. *Folia Primatol.* 19:41–60.

———. 1978. Socioecology of the blue monkeys (*Cercopithe-*

cus mitis stuhlmanni) of the Kibale Forest, Uganda. *Smithson. Contrib. Zool.* 249:1–88.

———. 1979. The demography and social mobility of a red howler (*Alouatta seniculus*) population in Venezuela. In *Vertebrate ecology in the northern Neotropics,* ed. J. F. Eisenberg. Washington, D.C.: Smithsonian Institution Press.

Ruhiyat, Y. 1983. Socio-ecological study of *Presbytis aygula* in West Java. *Primates:* 24:344–59.

Rumbaugh, D. 1968. The learning and sensory capacities of the squirrel monkey in phylogenetic perspective. In *The squirrel monkey,* ed. L. A. Rosenblum and R. W. Cooper. New York: Academic Press.

———. 1971. Evidence of qualitative differences in learning processes among primates. *J. Comp. Physiol. Psychol.* 76:250–55.

———. ed. 1977. *Language learning by a chimpanzee: The Lana project.* New York: Academic Press.

Rumbaugh, D., and Gill, T. 1977. Lana's acquisition of language skills. In *Language learning by a chimpanzee: The Lana project,* ed. D. Rumbaugh. New York: Academic Press.

Rumbaugh, D., and McCormack, C. 1967. The learning skill of primates: A comparative study of apes and monkeys. In *Progress in primatology,* ed. D. Starck, R. Schneider, and H. Kuhn. Stuttgart: Fischer Verlag.

Russell, E. M. 1982. Patterns of parental care and parental investment in marsupials. *Biol. Rev.* 57:1423–86.

Russell, J. K. 1983. Altruism in coati bands: Nepotism or reciprocity? In *Social behavior of female vertebrates,* ed. S. K. Wasser. New York: Academic Press.

Russell, R. J. 1977. The behavior, ecology, and environmental physiology of a nocturnal primate, *Lepilemur mustelinus* (Strepsirhini, Lemuriformes, Lepilemuridae). Ph.D. diss., Duke University.

Rutberg, A. T. 1983. The evolution of monogamy in primates. *J. Theoret. Biol.* 104:93–112.

Ruvolo, M. 1983. Genetic evolution in the African guenon monkeys (Primates, Cercopithecinae). Ph.D. diss., Harvard University.

Rylands, A. B. 1981. Preliminary field observations on the marmoset, *Callithrix humeralifer intermedius* (Hershkovitz, 1977) at Dadanelos, Rio Aripuana, Mato Gross. *Primates* 22:46–59.

vom Saal, F. S., and Howard, L. S. 1982. The regulation of infanticide and parental behavior: Implications for reproductive success in male mice. *Science* 215:1270–72.

Saayman, G. S. 1968. Oestrogen, behaviour, and permeability of a troop of chacma baboons. *Nature* 220:1339–40.

———. 1970. The menstrual cycle and sexual behaviour in a troop of free-ranging chacma baboons (*Papio ursinus*). *Folia Primatol.* 15:36–57.

———. 1971a. Behavior of the adult males in a troop of free-ranging chacma baboons. *Folia Primatol.* 15:36–57.

———. 1971b. Grooming behaviour in a troop of free-ranging chacma baboons (*Papio ursinus*). *Folia Primatol.* 16:161–78.

Sabater Pi, J. 1972. Contribution to the ecology of *Mandrillus sphinx,* Linnaeus 1758 of Rio Muni (Republic of Equatorial Guinea). *Folia Primatol.* 17:304–19.

———. 1973. Contribution to the ecology of *Colobus poly-*

komos satanas (Waterhouse, 1838) of Rio Muni, Republic of Equatorial Guinea. *Folia Primatol.* 19:193–207.

———. 1974. An elementary industry of the chimpanzees in the Okorobiko Mountains, Rio Muni (Republic of Equatorial Guinea), West Africa *Primates.* 15:351–64.

———. 1977. Contribution to the study of alimentation of lowland gorillas in the natural state, in Rio Muni, Republic of Equatorial Guinea (West Africa). *Primates* 18:183–204.

———. 1979. Feeding behavior and diet of chimpanzees (*Pan troglodytes troglodytes*) in the Okorobiko Mountains of Rio Muni (West Africa) *Z. Tierpsychol.* 50:265–81.

Sacher, G. A. 1959. Relationship of lifespan to brain weight and body weight in mammals. In *C.I.B.A. Foundation Symposium on the Lifespan of Animals,* ed. G. E. W. Wolstenholme and M. O'Connor. Boston: Little, Brown.

———. 1982. The role of brain maturation in the evolution of the primates. In *Primate brain evolution,* ed. E. Armstrong and D. Falk. New York: Plenum Press.

Sacher, G. A., and Staffeldt, E. F. 1974. Relation of gestation time to brain weight for placental mammals. *Am. Nat.* 108:593–616.

Sackett, G. P. 1974. Sex differences in rhesus monkeys following varied rearing experiences. In *Sex differences in behavior,* ed. P. C. Friedman, R. M. Richart, and R. L. Van de Weile. New York: John Wiley.

Sade, D. S. 1964. Seasonal cycle in size of testes of free-ranging *Macaca mulatta. Folia Primatol.* 2:171–80.

———. 1965. Some aspects of parent-offspring and sibling relations in a group of rhesus monkeys, with a discussion of grooming. *Am. J. Phys. Anthropol.* 23:1–18.

———. 1967. Determinants of dominance in a group of free-ranging rhesus monkeys. In *Social communication among primates,* ed. S. Altmann. Chicago: University of Chicago Press.

———. 1968. Inhibition of mother-son mating among free-ranging rhesus monkeys. *Sci. Psychoanal.* 12:18–38.

———. 1972a. A longitudinal study of social behavior of rhesus monkeys. In *The functional and evolutionary biology of primates,* ed. R. H. Tuttle. Chicago: Aldine.

———. 1972b. Sociometrics of *Macaca mulatta,* 1: Linkages and cliques in grooming matrices. *Folia Primatol.* 18:196–223.

———. 1973. An ethogram for rhesus monkeys, 1: Antithetical contrasts in posture and movement. *Am. J. Phys. Anthropol.* 38:537–42.

Sade, D. S.; Cushing, K.; Cushing, P.; Dunaif, J.; Figueroa, A.; Kaplan, J. R.; Laurer, C.; Rhodes, D.; and Schneider, J. 1976. Population dynamics in relation to social structure on Cayo Santiago. *Yrbk. Phys. Anthropol.* 20:253–62.

Sadleir, R. M. S. 1969. *Ecology of reproduction in wild and domestic mammals.* London: Methuen.

Samuels, A.; Silk, J. B.; and Rodman, P. S. 1984. Changes in the dominance rank and reproductive behaviour of male bonnet macaques (*Macaca radiata*). *Anim. Behav.* 32:994–1003.

San José, J. J., and Fariñas, M. R. 1983. Changes in tree density and species composition in a protected *Trachypogon* savanna, Venezuela. *Ecology* 64:447–53.

Sapolsky, R. M. 1983. Endocrine aspects of social instability in the olive baboon (*Papio anubis*). *Am. J. Primatol.* 5:365–79.

Sassenrath, E. N.; Mason, W. A.; Fitzgerald, R. C.; and Kenney, M. D. 1980. Comparative endocrine correlates of reproductive states in *Callicebus* (titi) and *Saimiri* (squirrel) monkeys. *Anthropol. Contemp.* 3:265.

Savage, T. S., and Wyman, J. 1847. Notice of the external characters and habits of *Troglodytes gorilla,* a new species of orang from the Gabon River. *Boston J. Nat. Hist.* 5:28–43.

Savage-Rumbaugh, E. S.; Pate, J. L.; Lawson, J.; Smith, S. T.; and Rosenbaum, S. 1983. Can a chimpanzee make a statement? *J. Exp. Psychol. (Gen.)* 112:457–92.

Savage-Rumbaugh, E. S.; Rumbaugh, D. M.; Smith, S. T.; and Lawson, J. 1980. Reference: The linguistic essential. *Science* 210:922–25.

Savage-Rumbaugh, E. S.; and Wilkerson, B. J. 1978. Sociosexual behavior in *Pan paniscus* and *Pan troglodytes:* A comparative study. *J. Hum. Evol.* 7:327–44.

Savage-Rumbaugh, E. S.; Wilkerson, B. J.; and Bakeman, R. 1977. Spontaneous gestural communication among conspecifics in the pygmy chimpanzee. In *Progress in ape research,* ed. G. H. Bourne. New York: Academic Press.

van Schaik, C. P. 1983. Why are diurnal primates living in groups? *Behaviour* 87:120–44.

van Schaik, C. P., and van Hooff, J. 1983. On the ultimate causes of primate social systems. *Behaviour* 85:91–117.

van Schaik, C. P.; van Noordwijk, M. A.; de Boer, R. J.; and den Tonkelaar, I. 1983. The effect of group size on time budgets and social behaviour in wild long-tailed macaques (*Macaca fascicularis*). *Behav. Ecol. Sociobiol.* 13:173–81.

Schaller, G. B. 1963. *The mountain gorilla: Ecology and behavior.* Chicago: University of Chicago Press.

———. 1965. Behavioral comparisons of the apes. In *Primate Behavior: Field studies of monkeys and apes,* ed. I. DeVore. New York: Holt, Rinehart, and Winston.

———. 1972. *The Serengeti lion.* Chicago: University of Chicago Press.

Schenkel, R. 1967. Submission: Its features and function in the wolf and dog. *Am. Zool.* 7:319–29.

Schilling, A. 1974. A study of marking behavior in *Lemur catta.* In *Prosimian biology,* ed. R. D. Martin, G. A. Doyle, and A. C. Walker. London: Duckworth.

———. 1979. Olfactory communication in prosimians. In *The study of prosimian behavior,* ed. G. A. Doyle and R. D. Martin. New York: Academic Press.

Schjelderup-Ebbe, T. 1935. Social behavior of birds. In *Handbook of social psychology,* ed. C. Murchison. Worcester, Mass.: Clark University Press.

Schlichte, H.-J. 1978a. A preliminary report on the habitat utilization of a group of howler monkeys (*Alouatta villosa pigra*) in the National Park of Tikal, Guatemala. In *The ecology of arboreal folivores,* ed. G. G. Montgomery. Washington, D.C.: Smithsonian Institution Press.

———. 1978b. The ecology of two groups of blue monkeys, *Cercopithecus mitis stuhlmanni,* in an isolated habitat of poor vegetation. In *The ecology of arboreal folivores,* ed.

G. G. Montgomery. Washington, D.C.: Smithsonian Institution Press.

Schön Ybarra, M. A. 1984. Locomotion and postures of red howlers in a deciduous forest-savanna interface. *Am. J. Phys. Anthropol.* 63:65–76.

Schrier, A. M. 1958. Comparison of two methods of investigating effects of amount of reward on performance. *J. Comp. Physiol. Psychol.* 51:725–31.

———. 1984. Learning how to learn: The significance and current status of learning set formation. *Primates.* 25:95–102.

Schrier, A. M.; Angarella, R.; and Povar, M. L. In press. Studies of concept formation by stumptail monkeys: Concepts, humans, monkeys, and letter A. *J. Exp. Psychol.*

Schrier, A. M., and Harlow, H. 1956. Effect of amount of incentive on discrimination learning by monkeys. *J. Comp. Physiol. Psychol.* 49:117–22.

Schultz, A. H. 1940. Growth and development of the orangutan. *Contrib. Embryol.* 28:57–110.

Schurmann, C. 1982. Mating behaviour of wild orangutans. In *The orangutan: Its biology and conservation*, ed. L. de Boer. The Hague: W. Junk.

Schwartz, C. G., and Rosenblum, L. A. 1983. Allometric influences on primate mothers and infants. In *Symbiosis in parent-offspring interactions*, ed. L. A. Rosenblum and H. Moltz. New York: Plenum Press.

Schwartz, J. H. 1984. The evolutionary relationships of man and orang-utans. *Nature* 308:501–6.

Scollay, P. A., and Judge, P. 1981. The dynamics of social organization in a population of squirrel monkeys *Saimiri sciureus* in a seminatural environment. *Primates* 22:60–69.

Scott, L. M. 1984. Reproductive behavior of adolescent female baboons (*Papio anubis*) in Kenya. In *Female primates: Studies by women primatologists*, ed. M. F. Small. New York: Alan R. Liss.

Scott, N. J.; Malmgren, L. A.; and Glander, K. E. 1978. Grouping behavior and sex ratio in mantled howler monkeys. In *Recent advances in primatology*, vol. 1, ed. D. J. Chivers and J. Herbert. New York: Academic Press.

Scruton, D. M. and Herbert, T. 1970. The menstrual cycle and its effect on behaviour in the talapoin monkey (*Miopithecus talapoin*). *J. Zool. Soc. Lond.* 162:419–36.

Searcy, W. A. 1982. The evolutionary effects of mate selection. *Ann. Rev. Ecol. Syst.* 13:57–85.

Seidenberg, M. S., and Pettito, L. A. 1979. Signing behavior in apes: A critical review. *Cognition* 7:177–215.

Seim, E., and Saether, B.-E. 1983. On rethinking allometry: What regression model to use? *J. Theoret. Biol.* 104:161–68.

Sekulic, R. 1981. The significance of howling in the red howler monkey (*Alouatta seniculus*). Ph.D. diss., University of Maryland, College Park.

———. 1982a. Behavior and ranging patterns of a solitary female red howler (*Alouatta seniculus*). *Folia Primatol.* 38:217–32.

———. 1982b. Daily and seasonal patterns of roaring and spacing in four red howler (*Alouatta seniculus*) troops. *Folia Primatol.* 39:22–48.

———. 1982c. The function of howling in red howler monkeys (*Alouatta seniculus*). *Behaviour* 81:38–54.

———. 1983a. Male relationships and infant deaths in red howler monkeys (*Alouatta seniculus*). *Z. Tierpsychol.* 61:185–202.

———. 1983b. Spatial relationships between recent mothers and other troop members in red howler monkeys (*Alouatta seniculus*). *Primates* 24:475–85.

Sekulic, R., and Eisenberg, J. F. 1983. Throat-rubbing in red howler monkeys. In *Chemical signals in vertebrates*, vol. 3, ed. D. Muller-Schwarze and R. M. Silverstein. New York: Plenum Press.

Selander, R. K. 1972. Sexual selection and sexual dimorphism in birds. In *Sexual selection and the descent of man, 1871–1971*, ed. B. G. Campbell. Chicago: Aldine.

Seligman, M. E. P., and Hager, J. L. 1972. *Biological boundaries of learning.* New York: Appleton Century Crofts.

Serban, G., and Kling, A., eds. 1976. *Animal models in human psychobiology.* New York: Plenum Press.

Serpell, J. A. 1981. Duetting in birds and primates: A question of function. *Anim. Behav.* 29:963–65.

Seyfarth. R. M. 1976. Social relationships among adult female baboons. *Anim. Behav.* 24:917–38.

———. 1977. A model of social grooming among adult female monkeys. *J. Theoret. Biol.* 65:671–98.

———. 1978a. Social relationships among adult male and female baboons, 1: Behaviour during sexual consortship. *Behaviour* 64:204–26.

———. 1978b. Social relationships among adult male and female baboons, 2: Behaviour throughout the female reproductive cycle. *Behaviour* 64:227–47.

———. 1980. The distribution of grooming and related behaviours among adult female vervet monkeys. *Anim. Behav.* 28:798–813.

———. 1981. Do monkeys rank each other? *Behav. Brain Sci.* 4:447–48.

———. 1983. Grooming and social competition in primates. In *Primate social relationships: An integrated approach*, ed. R. A. Hinde. Oxford: Blackwell.

Seyfarth, R. M., and Cheney, D. L. 1980. The ontogeny of vervet monkey alarm-calling behavior: A preliminary report. *Z. Tierpsychol.* 54:37–56.

———. 1984a. Grooming, alliances, and reciprocal altruism in vervet monkeys. *Nature* 308:541–43.

———. 1984b. The acoustic features of vervet monkey grunts. *J. Acoust. Soc. Am.* 75:1623–28.

———. 1986. Vocal development in vervet monkeys. *Anim. Behav.* 34:1450–68.

Seyfarth, R. M.; Cheney, D. L.; and Hinde, R. A. 1978. Some principles relating social interaction and social structure among primates. In *Recent advances in primatology*, vol. 1, ed. D. J. Chivers and J. Herbert. New York: Academic Press.

Seyfarth, R. M.; Cheney, D. L.; and Marler, P. 1980a. Monkey responses to three different alarm calls: Evidence for predator classification and semantic communication. *Science* 210:801–3.

———. 1980b. Vervet monkey alarm calls: Semantic communication in a free-ranging primate. *Anim. Behav.* 28:1070–94.

Shaikh, A. A.; Celaya, C. L.; Gomez, I.; and Shaikh, S. A. 1982. Temporal relationships of hormonal peaks to ovulation and sex skin deturgescence in the baboon. *Primates* 23:444–52.

Sharman, M., and Dunbar, R. I. M. 1982. Observer bias in selection of study group in baboon field studies. *Primates* 23:567–73.

Sheine, W. 1979. The effect of variations in molar morphology on masticatory effectiveness and digestion of cellulose in prosimian primates. Ph.D. diss., Duke University.

Shepher, J. 1971. Mate selection among second generation kibbutz adolescents and adults: Incest avoidance and negative imprinting. *Arch. Sex. Behav.* 1:293–307.

Sherman, P. W. 1977. Nepotism and the evolution of alarm calls. *Science* 197:1246–53.

———. 1980. The limits of ground squirrel nepotism. In *Sociobiology: Beyond nature/nurture?* ed. G. W. Barlow and J. Silverberg. Boulder, Colo.: Westview Press.

———. 1981. Reproductive competition and infanticide in Belding's ground squirrels and other animals. In *Natural selection and social behavior,* ed. R. D. Alexander and D. W. Tinkle. New York: Chiron Press.

Sherman, P. W., and Holmes, W. G. 1985. Kin recognition: Issues and evidence. In *Experimental behavioral ecology,* ed. B. Hölldobler and M. Lindauer. Stuttgart: Fischer Verlag.

Sherrod, L. R. 1981. Issues in cognitive-perceptual development: The special case of social stimuli. In *Infant social cognition,* ed. M. E. Lamb and L. R. Sherrod. Hillsdale, N.J.: Lawrence Erlbaum.

Shideler, S. E., and Lasley, B. L. 1982. A comparison of primate ovarian cycles. *Am. J. Primatol. Suppl.* 1:171–80.

Shields, W. M. 1982. *Philopatry, Inbreeding, and the Evolution of Sex.* Albany: State University of New York Press.

Shiveley, C.; Clarke, S.; King, N.; Schapiro, S.; and Mitchell, G. 1982. Patterns of sexual behavior in male macaques. *Am. J. Primatol.* 2:373–84.

Shoemaker, A. H. 1978. Observations on howler monkeys, *Alouatta caraya,* in captivity. *Der Zoologishe Garten* 48:225–34.

———. 1979. Reproduction and development of the black howler monkey *Alouatta caraya* at Columbia Zoo. *Int. Zoo Yrbk.* 19:150–55.

———. 1982. Fecundity in the captive howler monkey, *Alouatta caraya. Zoo Biol.* 1:149–56.

Shopland, J. M. 1982. An intergroup encounter with fatal consequences in yellow baboons (*Papio cynocephalus*). *Am. J. Primatol.* 3:263–66.

Shorey, H. H. 1976. *Animal communication by pheromones.* London: Academic Press.

Short, R. V. 1979. Sexual selection and its component parts, somatic and genital selection, as illustrated by man and the great apes. In *Advances in the study of behaviour,* vol. 9, ed. J. S. Rosenblatt, R. A. Hinde, C. Beer, and M. C. Busnel. New York: Academic Press.

Siegel, S. 1956. *Nonparametric statistics for the behavioral sciences.* New York: McGraw-Hill.

Sigg, H. 1980. Differentiation of female positions in hamadryas one-male-units. *Z. Tierpsychol.* 53:265–302.

Sigg, H., and Falett, J. 1985. Experiments on respect of possession and property in hamadryas baboons (*Papio hamadryas*). *Anim. Behav.* 33:978–84.

Sigg, H., and Stolba, A. 1981. Home range and daily march in a hamadryas baboon troop. *Folia Primatol.* 36:40–75.

Sigg, H.; Stolba, A.; Abegglen, J. J.; and Dasser, V. 1982. Life history of hamadryas baboons: Physical development, infant mortality, reproductive parameters, and family relationships. *Primates* 23:473–87.

Silk, J. B. 1978. Patterns of food sharing among mother and infant chimpanzees at Gombe National Park, Tanzania. *Folia Primatol.* 29:129–41.

———. 1979. Feeding, foraging, and food sharing behavior in immature chimpanzees. *Folia Primatol.* 31:12–42.

———. 1980a. Adoption and kinship in Oceania. *Am. Anthropol.* 82:799–820.

———. 1980b. Kidnapping and female competition among captive bonnet macaques. *Primates* 21:100–110.

———. 1982. Altruism among female *Macaca radiata:* Explanations and analysis of patterns of grooming and coalition formation. *Behaviour* 79:162–88.

———. 1983. Local resource competition and facultative adjustment of sex ratios in relation to competitive abilities. *Am. Nat.* 121:56–66.

———. 1984. Measurement of the relative importance of individual selection and kin selection among females of the genus *Macaca. Evolution* 38:553–59.

Silk, J. B., and Boyd, R. 1983. Cooperation, competition, and mate choice in matrilineal macaque groups. In *Social behavior of female vertebrates,* ed. S. Wasser. New York: Academic Press.

Silk, J. B.; Clark-Wheatley, C. B.; Rodman, P. S.; and Samuels, A. 1981. Differential reproductive success and facultative adjustment of sex ratios among captive female bonnet macaques (*Macaca radiata*). *Anim. Behav.* 29:1106–20.

Silk, J. B., and Samuels, A. 1984. Triadic interactions among *Macaca radiata:* Passports and buffers. *Am. J. Primatol.* 6:373–76.

Silk, J. B.; Samuels, A.; and Rodman, P. 1981. The influence of kinship, rank, and sex on affiliation and aggression between adult female and immature bonnet macaques (*Macaca radiata*). *Behaviour* 78:111–77.

Silkiluwasha, F. 1981. The distribution and conservation status of the Zanzibar red colobus. *Afr. J. Ecol.* 19:187–94.

Simonds, P. E. 1965. The bonnet macaque in South India. In *Primate behavior: Field studies of monkeys and apes,* ed. I. DeVore. New York: Holt, Rinehart, and Winston.

———. 1973. Outcast males and social structure among bonnet macaques *Macaca radiata. Am. J. Phys. Anthropol.* 38:599–604.

———. 1974. Sex differences in bonnet macaque networks and social structure. *Arch. Sex. Behav.* 3:151–66.

Simpson, A. E. and Stevenson-Hinde, J. 1985. Temperamental characteristics of three- to four-year-old boys and girls and child-family interactions. *J. Child Psychol. Psychiat.* 26:43–53.

Simpson, M. J. A. 1973a. Social displays and the recognition of individuals. In *Perspectives in ethology,* vol. 1, ed. P. P. G. Bateson and P. H. Klopfer. New York: Plenum Press.

———. 1973b. Social grooming of male chimpanzees. In

Comparative ecology and behavior of primates, ed. J. H. Crook and R. P. Michael. New York: Academic Press.

———. 1983. Effect of the sex of an infant on the mother-infant relationship and the mother's subsequent reproduction. In *Primate social relationships: An integrated approach,* ed. R. A. Hinde. Oxford: Blackwell.

Simpson, M. J. A., and Simpson, A. E. 1982. Birth sex ratios and social rank in rhesus monkey mothers. *Nature* 300: 440–41.

Simpson, M. J. A.; Simpson, A. E.; Hooley, J.; and Zunz, M. 1981. Infant-related influences on birth intervals in rhesus monkeys. *Nature* 290:49–51.

Singh, M; Akram, N.; and Pirta, R. S. 1984. Evolution of demographic patterns in the bonnet monkey (*Macaca radiata*). In *Current primate researches,* ed. M. L. Roonwal, S. M. Mohnot, and N. S. Rathore. Jodhpur: University of Jodhpur.

Skorupa, J. 1986. Effects of selective timber harvesting on rain forest primates in Kibale forest, Uganda. Ph.D. diss., University of California, Davis.

Skutch, A. F. 1976. *Parent birds and their young.* Austin: University of Texas Press.

Slatkin, M., and Hausfater, G. 1976. A note on the activities of a solitary male baboon. *Primates* 17:311–22.

Slob, A. K.; Wiegand, S. J.; Goy, R. W.; and Robinson, J. A. 1978. Heterosexual interactions in laboratory-housed stumptail macaques (*Macaca arctoides*): Observations during the menstrual cycle and after ovariectomy. *Horm. Behav.* 10: 193–211.

Sly, D. L.; Harbough, S. W.; London, W. T.; and Rice, J. M. 1983. Reproductive performance in a laboratory breeding colony of patas monkeys (*Erythrocebus patas*). *Am. J. Primatol.* 4:23–32.

Small, M. F. 1981. Body fat, rank, and nutritional status in a captive group of rhesus macaques. *Int. J. Primatol.* 2: 91–96.

———. 1984. Aging and reproductive success in female *Macaca mulatta.* In *Female primates: Studies by women primatologists,* ed. M. F. Small. New York: Alan R. Liss.

Small, M. F., and Smith, D. G. 1981. Interactions with infants by full siblings, paternal half-siblings, and nonrelatives in a captive group of rhesus macaques (*Macaca mulatta*). *Am. J. Primatol.* 1:91–94.

———. 1982. The relationship between maternal and paternal rank in rhesus macaques (*Macaca mulatta*) *Anim. Behav.* 30:626–33.

Smith, C. C. 1977. Feeding behavior and social organization in howling monkeys. In *Primate ecology: Studies of feeding and ranging behaviour in lemurs, monkeys, and apes,* ed. T. H. Clutton-Brock. London: Academic Press.

Smith, D. G. 1980. Paternity exclusion in six captive groups of rhesus monkeys (*Macaca mulatta*). *Am. J. Phys. Anthropol.* 53:243–49.

———. 1981. The association between rank and reproductive success of male rhesus monkeys. *Am. J. Primatol.* 1:83–90.

———. 1982. Inbreeding in three captive groups of rhesus monkeys. *Am. J. Anthropol.* 58:447–51.

Smith, E. O., ed. 1978. *Social play in primates.* New York: Academic Press.

Smith, E. O., and Peffer-Smith, P. G. 1984. Adult male-immature interactions in captive stumptail macaques (*Macaca arctoides*). In *Primate paternalism,* ed. D. M. Taub. New York: Van Nostrand Reinhold.

Smith, H. J.; Newman, J. D.; and Symmes, D. 1982. Vocal concomitants of affiliative behavior in squirrel monkeys. In *Primate communication,* ed. C. T. Snowdon, C. H. Brown, and M. Petersen. New York: Cambridge University Press.

Smith, J. D. 1970. The systematic status of the black howler monkey. *Alouatta pigra* Lawrence. *J. Mammal.* 51:358–69.

Smuts, B. B. 1980. Effects on social behavior of loss of high rank in wild adult female baboons (*Papio anubis*). Paper presented at the Annual Meeting of the Animal Behavior Society, Fort Collins, Colo.

———. 1982. Special relationships between adult male and female olive baboons (*Papio anubis*). Ph.D. diss., Stanford University.

———. 1983a. Dynamics of "special relationships" between adult male and female olive baboons. In *Primate social relationships: An integrated approach,* ed. R. A. Hinde. Oxford: Blackwell.

———. 1983b. Special relationships between adult male and female olive baboons: Selective advantages. In *Primate social relationships: An integrated approach,* ed. R. A. Hinde. Oxford: Blackwell.

———. 1985. *Sex and friendship in baboons.* Hawthorne, N.Y.: Aldine.

———. In press. Sisterhood is powerful: Aggression, competition, and cooperation in nonhuman primate societies. In *The aggressive female,* ed. D. Benton and P. Brain. Montreal: Eden Press.

Snowdon, C. T., and Cleveland, J. 1980. Individual recognition of contact calls in pygmy marmosets. *Anim. Behav.* 28:717–27.

———. 1984. "Conversations" among pygmy marmosets. *Am. J. Primatol.* 7:15–20.

Snowdon, C. T.; French, J. A.; and Cleveland, J. In press. Ontogeny of primate vocalizations: Models from birdsong and human speech. In *Current perspectives in primate social behavior,* ed. D. Taub and F. E. King. New York: Van Nostrand Reinhold.

Snowdon, C. T., and Hodun, A. 1981. Acoustic adaptations in pygmy marmoset contact calls: Locational cues vary with distance between conspecifics. *Behav. Ecol. Sociobiol.* 9: 295–300.

Snowdon, C. T., and Pola, Y. 1978. Interspecific and intraspecific responses to synthesized pygmy marmoset vocalizations. *Anim. Behav.* 26:192–206.

Snowdon, C. T., and Suomi, S. J. 1982. Paternal behavior in primates. In *Child nurturance,* vol. 3: *Studies of development in nonhuman primates,* ed. H. Fitzgerald, J. Mullins, and P. Gage. New York: Plenum Press.

van Soest, P. J. 1982. *Nutritional ecology of the ruminant.* Corvallis, Wash.: O and B Books.

Soini, P. 1982a. Distribución geográfica y ecología poblacional de *Saguinus mystax.* Report to the Peruvian Minister of Agriculture.

———. 1982b. Ecology and population dynamics of the pygmy marmoset, *Cebuella pygmaea. Folia Primatol.* 39:1–21.

———. 1982c. Primate conservation in Peruvian Amazonia.

Int. Zoo Yrbk. 22:37–47.

Sokal, R., and Rohlf, F. 1981. *Biometry.* San Francisco: W. H. Freeman.

Southwick, C. H., ed. 1963. *Primate social behavior.* Princeton: Van Nostrand.

Southwick, C. H.; Beg, M. A.; and Siddiqi, M. R. 1965. Rhesus monkeys in North India. In *Primate behavior,* ed. I. DeVore. New York: Holt, Rinehart, and Winston.

Southwick, C. H., and Siddiqi, M. F. 1977. Population dynamics of rhesus monkeys in northern India. In *Primate conservation,* ed. H. R. H. Prince Rainier III and G. H. Bourne. New York: Academic Press.

Srikosamatara, S. 1980. Ecology and behavior of the pileated gibbon (*Hylobates pileatus*) in Khao Soi Dao Wildlife Sanctuary, Thailand. M.A. thesis, Mahidol University.

———. 1984. Ecology of pileated gibbons in south-east Thailand. In *The lesser apes: Evolutionary and behavioral ecology,* ed. H. Preuschoft, D. J. Chivers, W. Brockelman, and N. Creel. Edinburgh: Edinburgh University Press.

Stallings, J. 1984. Status and conservation of Paraguayan primates. M.A. thesis, University of Florida.

Stammbach, E. 1978. On social differentiation in groups of captive female hamadryas baboons. *Behaviour* 67:322–38.

Starin, E. D. 1978. Food transfer by wild titi monkeys (*Callicebus torquatus torquatus*). *Folia Primatol.* 30:145–51.

———. 1981. Monkey moves. *Nat. Hist.* 90:36–43.

Stearns, S. C. 1977. The evolution of life history traits: A critique of the theory and a review of the data. *Ann. Rev. Ecol. Syst.* 8:145–71.

———. 1983. The influence of size and phylogeny on patterns of covariation among life-history traits in the mammals. *Oikos* 41:173–87.

Stein, D. M. 1981. The nature and function of social interactions between infant and adult male yellow baboons (*Papio cynocephalus*). Ph.D. diss., University of Chicago.

———. 1984a. Ontogeny of infant-adult male relationships during the first year of life for yellow baboons (*Papio cynocephalus*). In *Primate paternalism,* ed. D. M. Taub. New York: Van Nostrand Reinhold.

———. 1984b. *The sociobiology of infant and adult male baboons.* Norwood, N.J.: Albex.

Stephenson, G. R. 1975. Social structure of mating activity in Japanese macaques. In *Proceedings from the Symposia of the Fifth Congress of the International Primatological Society,* ed. S. Kondo, M. Kawai, A. Ehara, and S. Kawamura. Tokyo: Japan Science Press.

Stern, B. R., and Smith, D. G. 1984. Sexual behavior and paternity in three captive groups of rhesus monkeys (*Macaca mulatta*). *Anim. Behav.* 32:23–32.

Stern, D. 1977. *The first relationship: Infant and mother.* London: Fontana/Open Books.

Sterne, J. M., and Leiblum, S. 1984. Postpartum resumption of sexual activity in American women: Effects of absence or frequency of breast feeding. *Am. J. Primatol.* 5:381.

Stevenson, M. F. 1984. Conservation working party report: May 1984. *Primate Eye* 23:16–19.

Stevenson-Hinde, J. 1983a. Consistency over time. In *Primate social relationships: An integrated approach,* ed. R. A. Hinde. Oxford: Blackwell.

———. 1983b. Individual characteristics: A statement of the problem. In *Primate social relationships: An integrated approach,* ed. R. A. Hinde. Oxford: Blackwell.

Stevenson-Hinde, J. and Simpson, A. E. 1985. Temperamental characteristics of three- to four-year-old boys and girls and child-family interactions. *J. Child Psychol. Psychiat.* 26:43–53.

Stewart, K. J. 1981. Social development of wild mountain gorillas. Ph.D. diss., University of Cambridge.

Stolba, A. 1979. Entscheidungsfindung in Verbänden von Papio hamadryas. Ph.D. diss., University of Zurich.

Stoltz, L. P., and Saayman, G. S. 1970. Ecology and behavior of baboons in the northern Transvaal. *Ann. Transvaal Mus.* 26:99–143.

Strassmann, B. 1981. Sexual selection, parental care, and concealed ovulation in humans. *Ethol. Sociobiol.* 2:3–40.

Strayer, F. F. 1976. Learning and imitation as a function of social status in macaque monkeys (*Macaca nemstrina*). *Anim. Behav.* 24:835–48.

Strong, P. N.; Drash, P.; and Hedges, M. 1968. Solution of dimension abstracted oddity as a function of species, experience, and intelligence. *Psychonom. Sci.* 11:337–38.

Strong, P. N., and Hedges, M. 1966. Comparative studies in simple oddity learning, 1: Cats, raccoons, monkeys, and chimpanzees. *Psychonom. Sci.* 5:13–14.

Struhsaker, T. T. 1967a. Auditory communication among vervet monkeys (*Cercopithecus aethiops*). In *Social communication among primates,* ed. S. A. Altmann. Chicago: University of Chicago Press.

———. 1967b. Behavior of vervet monkeys (*Cercopithecus aethiops*). *Univ. Calif. Pub. Zool.* 82:1–74.

———. 1967c. Ecology of vervet monkeys (*Cercopithecus aethiops*) in the Masai-Amboseli Game Reserve, Kenya. *Ecology* 48:891–904.

———. 1967d. Social structure among vervet monkeys. *Behaviour* 29:83–121.

———. 1969. Correlates of ecology and social organization among African cercopithecines. *Folia Primatol.* 11:80–118.

———. 1970. Phylogenetic implications of some vocalizations of *Cercopithecus* monkeys. In *Old World monkeys,* ed. J. R. Napier and P. H. Napier. New York: Academic Press.

———. 1971. Social behavior of mother and infant vervet monkeys (*Cercopithecus aethiops*). *Anim. Behav.* 19:233–50.

———. 1973. A recensus of vervet monkeys in the Masai-Amboseli Game Reserve, Kenya. *Ecology* 54:930–32.

———. 1975. *The red colobus monkey.* Chicago: University of Chicago Press.

———. 1976. A further decline in numbers of Amboseli vervet monkeys. *Biotropica* 8:211–14.

———. 1977. Infanticide and social organization in the redtail monkey (*Cercopithecus ascanius schmidti*) in the Kibale Forest, Uganda. *Z. Tierpsychol.* 45:75–84.

———. 1978. Food habits of five monkey species in the Kibale Forest, Uganda. In *Recent advances in primatology,* vol. 1, ed. D. J. Chivers and J. Herbert. New York: Academic Press.

———. 1980. Comparison of the behaviour and ecology of red colobus and redtail monkeys in the Kibale Forest, Uganda. *Afr. J. Ecol.* 18:33–51.

———. 1981a. Polyspecific association among tropical rainforest primates. *Z. Tierpsychol.* 57:268–304.

———. 1981b. Vocalizations, phylogeny, and paleogeography

of red colobus monkeys (*Colobus badius*). *Afr. J. Ecol.* 19:265–84.

Struhsaker, T. T., and Gartlan, J. S. 1970. Observations on the behaviour and ecology of the patas monkey (*Erythrocebus patas*) in the Waza Reserve, Cameroon. *J. Zool. Soc. Lond.* 161:49–63.

Struhsaker, T. T., and Leland, L. 1977. Palmnut smashing by *Cebus a. apella* in Colombia. *Biotropica* 9:124–26.

———. 1979. Socioecology of five sympatric monkey species in the Kibale Forest, Uganda. In *Advances in the study of behavior,* vol. 9, ed. J. S. Rosenblatt, R. A. Hinde, C. Beer, and M. C. Busnel. New York: Academic Press.

———. 1980. Observations on two rare and endangered populations of red colobus monkeys in East Africa: *Colobus badius gordonorum* and *Colobus badius kirkii. Afr. J. Ecol.* 18:191–216.

———. 1985. Infanticide in a patrilineal society of red colobus monkeys. *Z. Tierpsychol.* 69:89–132.

Struhsaker, T. T., and Oates, J. F. 1975. Comparison of the behavior and ecology of red colobus and black-and-white colobus monkeys in Uganda: A summary. In *Socioecology and psychology of primates,* ed. R. H. Tuttle. The Hague: Mouton.

Strum, S. C. 1975. Life with the pumphouse gang: New insights into baboon behavior. *Nat. Geogr.* 147:672–91.

———. 1981. Processes and products of change: Baboon predatory behavior at Gilgil, Kenya. In *Omnivorous primates: Gathering and hunting in human evolution,* ed. R. S. O. Harding and G. Teleki. New York: Columbia University Press.

———. 1982. Agonistic dominance in male baboons: An alternative view. *Int. J. Primatol.* 3:175–202.

———. 1983. Use of females by male olive baboons (*Papio anubis*). *Am. J. Primatol.* 5:93–109.

———. 1984. Why males use infants. In *Primate paternalism,* ed. D. M. Taub. New York: Van Nostrand Reinhold.

Strum, S. C., and Western, D. 1982. Variations in fecundity with age and environment in olive baboons (*Papio anubis*). *Am. J. Primatol.* 3:61–76.

Studdert-Kennedy, M. 1975. Speech perception. In *Contemporary issues in experimental phonetics,* ed. N. J. Lass. Springfield, Ill.: Charles Thomas.

Suarez, S., and Gallup, G. G. 1981. Self-recognition in chimpanzees and orangutans, but not gorillas. *J. Hum. Evol.* 10:175–88.

Sugawara, K. 1979 Sociological study of a wild group of hybrid baboons between *Papio anubis* and *P. hamadryas* in the Awash Valley, Ethiopia. *Primates* 20:21–56.

Sugiyama, Y. 1965. On the social change of hanuman langurs (*Presbytis entellus*) in their natural conditions. *Primates* 6:213–47.

———. 1966. An artificial social change in a hanuman langur troop (*Presbytis entellus*). *Primates* 7:41–72.

———. 1967. Social organization of hanuman langurs. In *Social communication among primates,* ed. S A. Altmann. Chicago: University of Chicago Press.

———. 1968. Social organization of chimpanzees in the Budongo Forest, Uganda. *Primates* 9:225–58.

———. 1971. Characteristics of the social life of bonnet macaques (*Macaca radiata*). *Primates* 12:247–66.

———. 1976. Life history of male Japanese monkeys. In *Advances in the study of behavior,* vol. 7, ed. J. S. Rosenblatt, R. A. Hinde, E. Shaw, and C. Beer. New York: Academic Press.

———. 1981. Observations on the population dynamics and behavior of wild chimpanzees at Bossou, Guinea, in 1979–1980. *Primates* 22:435–44.

———. 1984. Population dynamics of wild chimpanzees at Bossou, Guinea, between 1976 and 1983. *Primates* 25:391–400.

Sugiyama, Y., and Koman, J. 1979a. Social structure and dynamics of wild chimpanzees at Bossou, Guinea. *Primates* 20:323–39.

Sugiyama, Y., and Koman, J. 1979b. Tool-using and making behaviour in wild chimpanzees at Bossou, Guinea. *Primates* 20:513–24.

Sugiyama, Y., and Ohsawa, H. 1975. Life history of male Japanese macaques at Ryozenyama. In *Contemporary primatology: Proceedings of the Fifth Congress of the International Primatological Society,* ed. S. Kondo, M. Kawai, and A. Ehara. Basel: S. Karger.

———. 1982a. Population dynamics of Japanese macaques at Ryozenyama, 3: Female desertion of the troop. *Primates* 23:31–44.

———. 1982b. Population dynamics of Japanese monkeys with special reference to the effect of artificial feeding. *Folia Primatol.* 39:238–63.

Sugiyama, Y.; Yoshiba, K.; and Parthasarathy, M. D. 1965. Home range, mating season, male group, and intertroop relations in hanuman langurs (*Presbytis entellus*). *Primates* 6:73–106.

Suomi, S. J. 1976. Mechanisms underlying social development: A reexamination of mother-infant interactions in monkeys. In *Minnesota symposium on child psychology,* vol. 10, ed. A. Pick. Minneapolis: University of Minnesota Press.

———. 1982. Abnormal behavior and primate models of psychopathology. In *Primate behavior,* ed. J. Fobes and J. King. New York: Academic Press.

Suomi, S. J., and Harlow, H. F. 1972. Social rehabilitation of isolate-reared monkeys. *Dev. Psychobiol.* 6:487–96.

Susman, R. L.; Badrian, N. L.; and Badrian, A. J. 1980. Locomotor behavior of *Pan paniscus* in Zaire. *Am. J. Phys. Anthropol.* 53:69–80.

Sussman, R. W. 1974. Ecological distinctions of sympatric species of *Lemur.* In *Prosimian biology,* ed. R. D. Martin, G. A. Doyle, and A. C. Walker. London: Duckworth.

Sussman, R. W., and Kinzey, W. G. 1984. The ecological role of the Callitrichidae: A review. *Amer. J. Phys. Anthropol.* 64:419–49.

Sussman, R. W., and Richard, A. F. 1974. The role of aggression among diurnal prosimians. In *Primate aggression, territoriality, and xenophobia,* ed. R. L. Holloway. New York: Academic Press.

Sussman, R. W.; Richard, A. F.; and Ravelojaona, G. 1985. Madagascar: Current projects and problems in conservation. *Primate Conservation* 5:53–59.

Sussman, R. W., and Tattersall, I. 1976. Cycles of activity, group composition, and diet of *Lemur mongoz mongoz* Linnaeus, 1766 in Madagascar. *Folia Primatol.* 26:270–83.

Sutton, D.; Larson, C.; Taylor, E. M.; and Lindeman, R. C. 1973. Vocalizations in rhesus monkeys: Conditionability. *Brain Res.* 52:225–31.

Sutton, D.; Trachy, R. E.; and Lindeman, R. C. 1981. Vocal and nonvocal discrimination performance in monkeys. *Brain Lang.* 14:93–105.

Suzuki, A. 1965. An ecological study of wild Japanese monkeys in snowy area, focussed on their food habits. *Primates* 6:31–72.

———. 1969. An ecological study of chimpanzees in a savanna woodland. *Primates* 10:103–48.

Svare, B., and Mann, M. 1981. Infanticide: Genetics, developmental and hormonal influences in mice. *Physiol. Behav.* 27:921–27.

Sved, J. A., and Latter, B. H. 1977. Migration and mutation in stochastic models of gene frequency change, 1: The island model. *J. Math. Biol.* 5:61–73.

Swartz, K. B., and Rosenblum, L. A. 1981. The social context of parental behavior: A perspective on primate socialization. In *Parental behavior in mammals,* ed. D. J. Gubernick and P. H. Klopfer. New York: Plenum Press.

Syme, G. J. 1974. Competitive orders as measures of social dominance. *Anim. Behav.* 22:931–40.

Symons, D. 1978a. *Play and aggression: A study of rhesus monkeys.* New York: Columbia University Press.

———. 1978b. The question of functions: Dominance and play. In *Social play in primates,* ed. E. O. Smith. New York: Academic Press.

———. 1979. *The evolution of human sexuality.* Oxford: Oxford University Press.

Szalay, F. S., and Delson, E. 1979. *Evolutionary history of the primates.* New York: Academic Press.

Takahata, Y. 1980. The reproductive biology of a free-ranging troop of Japanese monkeys. *Primates* 21:303–29.

———. 1982a. Social relations between adult males and females of Japanese monkeys in the Arashiyama B troop. *Primates* 23:1–23.

———. 1982b. The socio-sexual behavior of Japanese monkeys. *Z. Tierpsychol.* 59:89–108.

Takahata, Y.; Hasegawa, T.; and Nishida, T. 1984. Chimpanzee predation in the Mahale Mountains from August 1979 to May 1982. *Int. J. Primatol.* 5:213–33.

Takasaki, H. 1981. Troop size, habitat quality, and home range area in Japanese macaques. *Behav. Ecol. Sociobiol.* 9:277–81.

———. 1984. A model for relating troop size and home range area in a primate species. *Primates* 25:22–27.

Talmadge-Riggs, G.; Winter, P.; Ploog, D.; and Mayer, W. 1972. Effect of deafening on the vocal behavior of the squirrel monkey (*Saimiri sciureus*). *Folia Primatol.* 17:404–20.

Tanaka, J. 1965. Social structure of Nilgiri langurs. *Primates* 6:107–22.

Tanaka, T.; Tokuda, K.; and Kortera, S. 1970. Effects of infant loss on the interbirth interval of Japanese monkeys. *Primates* 11:113–17.

Tappen, N. 1964. Primate studies in Sierra Leone. *Curr. Anthropol.* 5:339–40.

Tartabini, A., and Dienske, H. 1979. Social play and rank order in rhesus monkeys. *Behav. Proc.* 4:375–83.

Tattersall, I. 1978. Behavioural variation in *Lemur mongoz* (= *L. m. mongoz*). In *Recent advances in primatology,* vol. 3, ed. D. J. Chivers, and K. A. Joysey. London: Academic Press.

———. 1979. Patterns of activity in the Mayotte lemur, *Lemur fulvus mayottensis. J. Mammal.* 60:314–23.

———. 1982. *The primates of madagascar.* New York: Columbia University Press.

Taub, D. M. 1977. Geographic distribution and habitat diversity of the Barbary macaque (*Macaca sylvanus*). *Folia Primatol.* 27:108–33.

———. 1978. Aspects of the biology of the wild Barbary macaque (Primates, Cercopithecinae, *Macaca sylvanus* L. 1758): Biogeography, the mating system, and male-infant associations. Ph.D. diss., University of California, Davis.

———. 1980a. Age at first pregnancy and reproductive outcome among colony-born squirrel monkeys (*Saimiri sciureus* Brazilian). *Folia Primatol.* 33:262–72.

———. 1980b. Female choice and mating strategies among wild Barbary macaques (*Macaca sylvanus* L.). In *The macaques: Studies in ecology, behavior, and evolution,* ed. D. G. Lindburg. New York: Van Nostrand Reinhold.

———. 1980c. Testing the "agonistic buffering" hypothesis, 1: The dynamics of participation in the triadic interaction. *Behav. Ecol. Sociobiol.* 6:187–97.

———. 1984. Male caretaking behavior among wild Barbary macaques (*Macaca sylvanus*). In *Primate paternalism,* ed. D. M. Taub. New York: Van Nostrand Reinhold.

Taub, D. M., and Redican, W. K. 1984. Adult male-infant interactions in Old World Monkeys and apes. In *Primate paternalism,* ed. D. M. Taub. New York: Van Nostrand Reinhold.

Taylor, C. K., and Saayman, G. S. 1973. Imitative behaviour by Indian Ocean bottlenose dolphins (*Tursiops aduncus*) in captivity. *Behaviour* 44:286–98.

Taylor, H.; Teas, J.; Richie, T.; and Southwick, C. 1978. Social interactions between adult male and infant rhesus monkeys in Nepal. *Primates* 19:343–51.

Teas, J.; Richie, T.; Taylor, H.; and Southwick, C. 1980. Population patterns and behavioral ecology of rhesus monkeys (*Macaca mulatta*) in Nepal. In *The macaques: Studies in ecology, behavior, and evolution,* ed. D. G. Lindburg. New York: Van Nostrand Reinhold.

Teleki, G. 1973a. The omnivorous chimpanzee. *Sci. Am.* 228:32–42.

———. 1973b. *The predatory behavior of chimpanzees.* Lewisburg, Pa.: Bucknell University Press.

———. 1974. Chimpanzee subsistence technology: Materials and skills. *J. Hum. Evol.* 3:575–94.

———. 1977. Spatial and temporal dimensions of routine activities performed by chimpanzees in Gombe National Park, Tanzania: An ethnological study of adaptive strategy. Ph.D. diss., Pennsylvania State University.

———. 1981. The omnivorous diet and eclectic feeding habits of chimpanzees in Gombe National Park, Tanzania. In *Omnivorous primates,* ed. R. S. O. Harding and G. Teleki. New York: Columbia University Press.

Teleki, G., and Baldwin, L. 1979. Known and estimated distributions of extant chimpanzee populations (*Pan troglodytes* and *Pan paniscus*) in Equatorial Africa. Special report to the IUCN/SSC Primate Specialist Group.

Teleki, G.; Hunt, E. E.; and Pfifferling, J. H. 1976. Demographic observations (1963–1973) on the chimpanzees of Gombe National Park, Tanzania. *J. Hum. Evol.* 5:559–96.

Temerin, L. A., and Cant, J. H. 1983. The evolutionary divergence of Old World monkeys and apes. *Am. Nat.* 122:335–51.

Tenaza, R. R. 1975. Territory and monogamy among Kloss' gibbons (*Hylobates klossii*) in Siberut Island, Indonesia. *Folia Primatol.* 24:68–80.

———. 1976. Songs and related behavior of Kloss' gibbon (*Hylobates klossii*) in Siberut Island, Indonesia. *Z. Tierpsychol.* 40:37–52.

Tenaza, R., and Tilson, R. 1977. Evolution of long-distance alarm-calls in Kloss' gibbon. *Nature* 268:233–35.

———. 1985. Human predation and Kloss's gibbon (*Hylobates klossii*) sleeping trees in Siberut Island, Indonesia. *Am. J. Primatol.* 8:299–308.

Terborgh, J. 1983. *Five New World primates: A study in comparative ecology.* Princeton: Princeton University Press.

Terborgh, J., and Janson, C. H. 1983. Ecology of primates in southeastern Peru. *Nat. Geogr. Soc. Res. Reports* 15:655–62.

Terborgh, J., and Wilson Goldizen, A. 1985. On the mating system of the cooperativcly breeding saddle-backed tamarin (*Saguinus fuscicollis*). *Behav. Ecol. Sociobiol.* 16:293–99.

Terrace, H. S. 1979. *Nim.* New York: Knopf.

Terrace, H. S.; Pettito, L. A.; Sanders, R. J.; and Bever, T. G. 1979. Can an ape create a sentence? *Science* 206:891–902.

Thompson-Handler, N.; Malenky, R. K.; and Badrian, N. 1984. Sexual behavior of *Pan paniscus* under natural conditions in the Lomako Forest, Equateur, Zaire. In *The pygmy chimpanzee: Evolution, biology, and behavior,* ed. R. L. Susman. New York: Plenum Press.

Thorington, R. W., Jr. 1968. Observations of squirrel monkeys in a Colombian forest. In *The squirrel monkey,* ed. L. A. Rosenblum and R. W. Cooper. New York: Academic Press.

Thorington, R. W., Jr., and Groves, C. P. 1970. An annotated classification of the Cercopithecoidea. In *Old World monkeys: Evolution, systematics, and behavior,* ed. J. R. Napier and P. H. Napier. New York: Academic Press.

Thorington, R. W., Jr.; Rudran, R.; and Mack, D. 1979. Sexual dimorphism of *Alouatta seniculus* and observations on capture techniques. In *Vertebrate ecology in the northern Neotropics,* ed. J. F. Eisenberg. Washington, D.C.: Smithsonian Institution Press.

Thorington, R. W., Jr.; Ruiz, J. C.; and Eisenberg, J. F. 1984. A study of a black howler monkey (*Alouatta caraya*) population in northern Argentina. *Am. J. Primatol.* 6:357–66.

Thorington, R. W., Jr., and Vorek, R. E. 1976. Observations on the geographic variations and skeletal development of *Aotus. Lab. Anim. Sci.* 26:1006–21.

Thornback, J., and Jenkins, M. 1982. *The IUCN mammal red data book,* part 1: *Threatened mammalian taxa of the Americas and the Australian zoogeographica region (excluding Cetacea).* Gland, Switzerland: IUCN.

Thorndike, E. L. 1901. The mental life of the monkeys. *Psycol. Rev. Monogr. Suppl.* 3, no. 5.

Thorpe, W. 1963. *Learning and instinct in animals.* London: Methuen.

———. 1979. *The origins and rise of ethology,* London: Heinemann.

Tiger, L. 1969. *Men in groups.* New York: Vintage Books.

Tilford, B. 1982. Seasonal rank changes for adolescent and subadult natal males in a free-ranging group of rhesus monkeys. *Int. J. Primatol.* 3:483–90.

Tilford, B., and Nadler, R. D. 1978. Male parental behavior in a captive group of lowland gorillas (*Gorilla gorilla gorilla*). *Folia Primatol.* 29:218–28.

Tilson, R. L. 1977. Social organization of Simakobu monkeys (*Nasalis concolor*) in Siberut Island, Indonesia. *J. Mammal.* 58:202–12.

———. 1979. Behavior of hoolock gibbons (*Hylobates hoolock*) during different seasons in Assam, India. *J. Bombay Nat. Hist. Soc.* 76:1–16.

———. 1981. Family formation strategies of Kloss's gibbons. *Folia Primatol.* 35:259–87.

Tilson, R. L., and Hamilton, W. J. III, In prep. Evolution of social systems among Mentaiwai Island colobines.

Tilson, R. L., and Tenaza, R. R. 1976. Monogamy and duetting in an Old World monkey. *Nature* 263:320–21.

———. 1977. Social organization of simakobu monkeys (*Nasalis concolor*) in Siberut island, Indonesia. *J. Mammal.* 58:202–12.

Tinbergen, N. 1951. *The study of instinct.* New York: Oxford University Press.

Tinklepaugh, O. L, and Hartman, C. G. 1932. Behavior and maternal care of the newborn monkey (*Macaca mulatta-M. rhesus*). *J. Genet. Psychol.* 40:257–86.

Tokuda, A., and Jensen, G. 1969. Determinants of dominance hierarchy in a captive group of pigtailed monkeys (*Macaca nemestrina*). *Primates* 10:227–36.

Tokuda, K. 1961. A study on the sexual behavior in the Japanese monkey troop. *Primates* 3:1–40.

Trivers, R. L. 1971. The evolution of reciprocal altruism. *Quart. Rev. Biol.* 46:35–57.

———. 1972. Parental investment and sexual selection. In *Sexual selection and the descent of man, 1871–1971,* ed. B. Campbell. Chicago: Aldine.

———. 1974. Parent-offspring conflict. *Am. Zool.* 14:249–64.

———. 1985. *Social evolution.* Menlo Park, Calif.: Benjamin/Cummings.

Tsingalia, H. M.; Cords, M.; Mitchell, B.; and Rowell, T. 1984. Mating behavior of blue monkeys. *Int. J. Primatol.* 5:388.

Tsingalia, H. M., and Rowell, T. E. 1984. The behaviour of adult male blue monkeys. *Z. Tierpsychol.* 64:253–68.

Tucker, V. A. 1975. The energetic cost of moving about. *Am. Scient.* 63:413–19.

de Turckheim, G., and Merz, E. 1984. Breeding Barbary macaques in outdoor open enclosures. In *The Barbary macaque,* ed. J. Fa. New York: Plenum.

Turke, P. W. 1984. Effects of ovulatory concealment and synchrony on protohominid mating systems and parental roles. *Ethol. Sociobiol.* 5:33–44.

Turner, T. 1981. Blood protein variation in a population of Ethiopian vervet monkeys (*Cercopithecus aethiops aethiops*). *Am. J. Phys. Anthropol.* 55:225–32.

Tutin, C. E. G. 1975. Sexual behaviour and mating patterns in community of wild chimpanzees (*Pan troglodytes schwein-furthii*). Ph.D. diss., University of Edinburgh.

———. 1979. Mating patterns and reproductive strategies in a community of wild chimpanzees (*Pan troglodytes schwein-furthii*). *Behav. Ecol. Sociobiol.* 6:29–38.

———. 1980. Reproductive behavior of wild chimpanzees in the Gombe National Park, Tanzania. *J. Reprod. Fertil. Supp.* 28:43–57.

Tutin, C. E. G., and Fernandez, M. 1983. Gorillas feeding on termites in Gabon, West Africa. *J. Mammal.* 64:530–31.

———. 1984. Nationwide census of gorilla (*Gorilla g. gorilla*) and chimpanzee (*Pan t. troglodytes*) populations in Gabon. *Am. J. Primatol.* 6:313–36.

Tutin, C. E. G., and McGinnis, P. R. 1981. Chimpanzee reproduction in the wild. In *Reproductive biology of the great apes*, ed. C. E. Graham. New York: Academic Press.

Tutin, C. E. G.; McGrew, W. C.; and Baldwin, P. 1983. Social organization of savanna-dwelling chimpanzees, *Pan troglodytes verus*, at Mt. Assirik, Senegal. *Primates* 24:154–73.

Udry, J. R., and Morris, N. 1968. The distribution of coitus in the menstrual cycle. *Nature* 200:593–96.

Uehara, S. 1982. Seasonal changes in the techniques employed by wild chimpanzees in the Mahale Mountains, Tanzania, to feed on termites (*Pseudacanthotermes spiniger*). *Folia Primatol.* 37:44–76.

———. 1984. Sex differences in feeding on *Camponotus* ants among wild chimpanzees in the Mahale Mountains, Tanzania. *Int. J. Primatol.* 5:389.

U.S. Department of State. 1980. *The world's tropical forests: A policy, strategy, and program for the United States.* Department of State Publication no. 9117, Washington, D.C.

Vaitl, E. 1977a. Experimental analysis of the nature of social context in captive groups of squirrel monkeys (*Saimiri sciureus*). *Primates* 18:849–59.

———. 1977b. Social context as a structuring mechanism in captive groups of squirrel monkeys (*Saimiri sciureus*). *Primates* 18:861–74.

———. 1978. Nature and implications of the complexly organized social system in non-human primates. In *Recent advances in primatology*, vol. 1, ed. D. J. Chivers and J. Herbert. New York: Academic Press.

Vandenbergh, J. G. 1983. Pheromonal regulation of puberty. In *Pheromones and reproduction in mammals*, ed. J. G. Vandenbergh. New York: Academic Press.

Vandenbergh, J. G., and Vessey, S. H. 1968. Seasonal breeding of free-ranging rhesus monkeys and related ecological factors. *J. Reprod. Fertil.* 15:71–79.

Varley, M., and Symmes, D. 1966. The hierarchy of dominance in a group of macaques. *Behaviour* 27:54–75.

Vehrencamp, S. 1983. A model for the evolution of despotic versus egalitarian societies. *Anim. Behav.* 31:667–82.

Vehrencamp, S., and Emlen, S. T. 1983. Cooperative breeding strategies among birds. In *Perspectives in ornithology*, ed. A. H. Brush and G. A. Clarke. Cambridge: Cambridge University Press.

Verheyen, W. N. 1963. New data on the geographical distribution of *Cercopithecus* (*Allenopithecus*) *nigroviridis* Pocock 1907. *Rev. Zool. Bot. Afr.* 68:393–96.

Vessey, S. H. 1968. Interactions between free-ranging groups of rhesus monkeys. *Folia Primatol.* 8:228–39.

———. 1971. Free-ranging rhesus monkeys: Behavioral effects of removal, separation, and reintroduction of group members. *Behaviour* 40:216–27.

Vessey, S. H., and Meikle, D. B. 1984. Free-living rhesus monkeys: Adult male interactions with infants and juveniles. In *Primate paternalism*, ed. D. M. Taub. New York: Van Nostrand Reinhold.

Vogel, C. 1975. Intergroup relations of *Presbytis entellus* in the Kumaon Hills and in Rajasthan (North India). In *Proceedings from the Symposia of the Fifth Congress of the International Primatological Society*, ed. S. Kondo, M. Kawai, A. Ehara, and S. Kawamura. Tokyo: Japan Science Press.

Vogel, C., and Loch, H 1984. Reproductive parameters, adult-male replacements, and infanticide among free-ranging langurs (*Presbytis entellus*) at Jodhpur (Rajasthan), India. In *Infanticide: Comparative and evolutionary perspectives*, ed. G. Hausfater and S. B. Hrdy. Hawthorne, N.Y.: Aldine.

Vogt, J. L. 1984. Interactions between adult males and infants in prosimians and New World Monkeys. In *Primate paternalism*, ed. D. M. Taub. New York: Van Nostrand Reinhold.

Vogt, J. L.; Carlson, H.; and Menzel, E. 1978. Social behavior of a marmoset (*Saguinus fuscicollis*) group, 1: Parental care and infant development. *Primates* 19:715–26.

de Waal, F. B. M. 1976. Straight-aggression and appeal-aggression in *Macaca fascicularis*. *Experientia* 32:1268–70.

———. 1977. The organization of agonistic relations within two captive groups of Java-monkeys (*Macaca fascicularis*). *Z. Tierpsychol.* 44:225–82.

———. 1978a. Exploitative and familiarity-dependent support strategies in a colony of semi-free-living chimpanzees. *Behaviour* 66:268–312.

———. 1978b. Join-aggression and protective aggression among captive *Macaca fascicularis*. In *Recent advances in primatology*, vol. 1., ed. D. J. Chivers and J. Herbert. London: Academic Press.

———. 1982. *Chimpanzee politics*. New York: Harper and Row.

———. 1984a. Coping with social tension: Sex differences in the effect of food provision to small rhesus monkey groups. *Anim. Behav.* 32:765–73.

———. 1984b. Sex differences in the formation of coalitions among chimpanzees. *Ethol. Sociobiol.* 5:239–55.

———. In press. The reconciled hierarchy: On the integration of dominance and social bonding in primates. *Quart. Rev. Biol.*

de Waal, F. B. M., and Hoekstra, J. 1980. Contexts and predictability of aggression in chimpanzees. *Anim. Behav.* 28:929–37.

de Waal, F. B. M., and van Hooff, J. 1981. Side-directed communication and agonistic interactions in chimpanzees. *Behaviour* 77:164–98.

de Waal, F. B. M.; van Hooff, J.; and Netto, W. 1976. An ethological analysis of types of agonistic interaction in a captive group of Java-monkeys (*Macaca fascicularis*). *Primates* 17:257–90.

de Waal, F. B. M., and van Roosmalen, A. 1979. Reconciliation and consolation among chimpanzees. *Behav. Ecol. Sociobiol.* 5:55–66.

de Waal, F. B. M., and Yoshihara, D. 1983. Reconciliation and redirected affection in rhesus monkeys. *Behaviour* 85:

224–41.

Wade, T. D. 1979. Inbreeding, kin selection, and primate evolution. *Primates* 20:355–70.

Walker, A. 1984. Mechanisms of honing in the male baboon canine. *Am J. Phys. Anthropol.* 65:47–60.

Walker, Leonard, J. 1979. A strategy approach to the study of primate dominance behaviour. *Behav. Processes* 4:155–72.

Wallace, A. R. 1869. *The Malay archipelago.* London: Macmillan.

Wallen, K. 1982. Influence of female hormonal state on rhesus sexual behavior varies with space for social interactions. *Science* 217:375–77.

Wallis, S. J. 1982. Sexual behavior of captive chimpanzees (*Pan troglodytes*): Pregnant versus cycling females. *Am. J. Primatol.* 3:77–88.

———. 1983. Sexual behavior and reproduction of *Cercocebus albigena johnstonii* in Kibale Forest, western Uganda. *Int. J. Primatol.* 4:153–66.

Walter, H. 1973. *Vegetation of the earth in relation to climate and the eco-physiological conditions.* New York: Springer-Verlag.

Walters, J. R. 1980. Interventions and the development of dominance relationships in female baboons. *Folia Primatol.* 34:61–89.

———. 1981. Inferring kinship from behaviour: Maternity determinations in yellow baboons. *Anim. Behav.* 29:126–36.

———. 1986. Determinants of social interaction in juvenile female baboons: Kinship, dominance, personality, and special relationships. Typescript.

———. In press. Kin recognition in nonhuman primates. In *Kin recognition in animals,* ed. D. F. Fletcher and C. D. Michener. New York: John Wiley.

Warren, J. M. 1965. The comparative psychology of learning. *Ann. Rev. Psychol.* 16:95–118.

———. 1973. Learning in vertebrates. In *Comparative psychology: A modern survey,* ed. D. A. Dewesbury and D. A. Rethlingshafer. New York: McGraw Hill.

———. 1976. Tool use in mammals. In *Evolution of brain and behavior in vertebrates,* ed. R. Masterson, M. Bitterman, C. Campbell, and N. Hotton. Hillsdale, N.J.: Erlbaum.

Waser, P. 1974. Inter-group interactions in a forest monkey: The mangabey *Cercocebus albigena.* Ph.D. diss., Rockefeller University.

———. 1975. Monthly variations in feeding and activity patterns of the mangabey. *Cercocebus albigena. E. Afr. Wild. J.* 13:249–63.

———. 1976. *Cercocebus albigena:* Site attachment, avoidance, and intergroup spacing. *Am. Nat.* 110:911–35.

———. 1977a. Feeding, ranging, and group size in the mangabey *Cercocebus albigena.* In *Primate ecology,* ed. T. H. Clutton-Brock. London: Academic Press.

———. 1977b. Individual recognition, intragroup cohesion, and intergroup spacing: Evidence from sound playback to forest monkeys. *Behaviour* 60:28–74.

———. 1978. Postreproductive survival and behavior in a free-ranging female mangabey. *Folia Primatol.* 29:142–60.

———. 1980. Polyspecific associations of *Cercocebus albigena:* Geographic variation and ecological correlates. *Folia Primatol.* 33:57–76.

———. 1981. Sociality or territorial defense? The influence of

resource renewal. *Behav. Ecol. Sociobiol.* 8:231–37.

———. 1982a. Polyspecific associations: Do they occur by chance? *Anim. Behav.* 30:1–8.

———. 1982b. The evolution of male loud calls among mangabeys and baboons. In *Primate communication,* ed. C. T. Snowdon, C. H. Brown, and M. Petersen. New York: Cambridge University Press.

———. 1984. "Chance" and mixed-species associations. *Behav. Ecol. Sociobiol.* 15:197–202.

Waser, P., and Case, T. J. 1981. Monkeys and matrices: On the coexistence of "omnivorous" forest primates. *Oecologia* 49:102–8.

Waser, P., and Jones, W. T. 1983. Natal philopatry among solitary mammals. *Quart. Rev. Biol.* 58:355–90.

Waser, P., and Waser, M. S. 1977. Experimental studies of primate vocalizations: Specializations for long-distance propagation. *Z. Tierpsychol.* 43:239–63.

———. 1985. *Ichneumia albicauda* and the evolution of viverrid gregariousness. *Z. Tierpsychol.* 68:137–51.

Waser, P., and Wiley, R. H. 1979. Mechanisms and evolution of spacing in animals. In *Handbook of behavioral neurobiology,* vol. 3, ed. P. Marler and J. G. Vandenbergh. New York: Plenum Press.

Washburn, S. L. 1982. Language and the fossil record. *Anthropol. UCLA* 7:231–38.

Washburn, S. L., and Benedict, B. 1979. Non-human primate culture. *Man* 14:163–64.

Washburn, S. L., and Hamburg, D. A. 1965. The implications of primate research. In *Primate behavior,* ed. I. DeVore. New York: Holt, Rinehart, and Winston.

Wasser, S. K. 1982. Reciprocity and the trade-off between associate quality and relatedness. *Am. Nat.* 119:720–31.

———. 1983. Reproductive competition and cooperation among female yellow baboons. In *Social behavior of female vertebrates,* ed. S. K. Wasser. New York: Academic Press.

Wasser, S. K., and Barash, D. P. 1983. Reproductive suppression among female mammals: Implications for biomedicine and sexual selection theory. *Quart. Rev. Biol.* 58:513–38.

Watanabe, K. 1979. Alliance formation in a free-ranging troop of Japanese macaques. *Primates* 20:459–74.

———. 1981. Variations in group composition and population density of two sympatric Mentawaian leaf-monkeys. *Primates* 22:145–60.

Waterman, P. G. 1984. Food acquisition and processing as a function of plant chemistry. In *Food acquisition and processing in primates,* ed. D. J. Chivers, B. A. Wood, and A. Bilsborough. New York: Plenum Press.

Watson, J. B. 1914. Imitation in monkeys. *Psychol. Bull.* 5:169–78.

Watts, D. 1985. Relations between group size and composition and feeding competition in mountain gorilla groups. *Anim. Behav.* 33:72–85.

Webster's Third New International Dictionary. 1965. Springfield, Mass.: G. and C. Merriam Co.

Weigel, R. M. 1981. The distribution of altruism among kin: A mathematical model. *Am. Nat.* 118:191–201.

Welker, C.; Pippert, P.; and Witt, C. 1983. Die Kapuzinerkolonie der Universitat Kassel. *Z. Kolner Zoo* 26:115–26.

West, K. 1981. The behavior and ecology of the siamang in Sumatra. M.A. thesis, University of California, Davis.

West, M. M., and Konner, M. J. 1976. The role of the father: An anthropological perspective. In *The role of the father in child development,* ed. M. E. Lamb. New York: Plenum Press.

Western, D. 1979. Size, life-history, and ecology in mammals. *Afr. J. Ecol.* 17:185–204.

Western, D. and Ssemakula, J. 1982. Life history patterns in birds and mammals and their evolutionary interpretation. *Oecologia* 54:281–90.

Wheatley, B. P. 1978. The behavior and ecology of the crab-eating macaque (*Macaca fascicularis*) in the Kutai Native Reserve, East Kalimantan, Indonesia. Ph.D. University of California, Davis.

———. 1980. Feeding and ranging of east Bornean *M. fascicularis.* In *The macaques; Studies in ecology, behavior, and evolution,* ed. D. Lindburg. New York: Van Nostrand Reinhold.

———. 1982. Adult male replacment in *Macaca fascicularis* of East Kalimantan, Indonesia. *Int. J. Primatol.* 3:203–19.

Whitmore, T. C. 1984. *Tropical rain forests of the Far East,* 2d ed. Oxford: Clarendon Press.

Whitten, A. J. 1980. *The Kloss gibbon in Siberut rain forest.* Ph.D. diss., University of Cambridge.

———. 1982. The ecology of singing in Kloss gibbons (*Hylobates klossii* Siberut Island, Indonesia. *Int. J. Primatol.* 3:33–51.

———. 1984. Ecological comparisons between Kloss gibbons and other small gibbons. In *The lesser apes: Evolutionary and behavioral biology,* ed. H. Preuschoft, D. J. Chivers, W. Brockelman, and N. Creel. Edinburgh: Edinburgh University Press.

Whitten, P. L. 1982. Female reproductive strategies among vervet monkeys. Ph.D. diss., Harvard University.

———. 1983. Diet and dominance among female vervet monkeys (*Cercopithecus aethiops*) *Am. J. Primatol.* 5:139–59.

Whitten, P. L., and Smith, E. O. 1984. Patterns of wounding in stumptail macaques (*Macaca arctoides*). *Primates* 25:326–36.

Wiens, J. A. 1977. On competition and variable environments. *Am. Scient.* 65:590–97.

Wildt, D. E.; Doyle, L. L.; Stone, S. C.; and Harrison, R. M. 1977. Correlation of perineal swelling with serum ovarian hormone levels, vaginal cytology, and ovarian follicular development during the baboon reproductive cycle. *Primates* 18:261–70.

Wiley, R. H., and Richards, D. G. 1978. Physical constraints on acoustic communication in the atmosphere: Implications for the evolution of animal vocalizations. *Behav. Ecol. Sociobiol.* 3:69–94.

Williams, G. C., ed. 1971. *Group selection.* Chicago: Aldine.

Williams, L. 1967. Breeding Humboldt's woolly monkey *Lagothrix lagotricha* at Murrayton Woolly Monkey Sanctuary. *Int. Zoo Yrbk.* 7:86–89.

Wilson, A. P., and Boelkins, R. C. 1970. Evidence for seasonal variation in aggressive behavior by *Macaca mulatta. Anim. Behav.* 18:719–24.

Wilson, E. O. 1971. *The insect societies.* Cambridge: Harvard University Press.

———. 1975. *Sociobiology: The new synthesis.* Cambridge: Harvard University Press.

Wilson, M. E.; Gordon, T. P.; and Bernstein, I. S. 1978. Timing of births and reproductive success in rhesus monkey social groups. *J. Med. Primatol.* 7:202–12.

Wilson, M E.; Gordon, T. P.; and Chikazawa, D. 1982. Female mating relationships in rhesus monkeys. *Am. J. Primatol.* 2:21–28.

Wilson, M. E.; Walker, M. L.; and Gordon, T. P. 1983. Consequences of first pregnancy in rhesus monkeys. *Am. J. Phys. Anthropol.* 61:103–10.

Wilson, W. A. 1975. Discriminative conditioning of vocalizations in *Lemur catta. Anim. Behav.* 23:432–36.

Wilson, W. L., and Wilson, C. C. 1975. Species-specific vocalizations and the determination of phylogenetic affinities of the *Presbytis uygula-melaphos* group in Sumatra. In *Contemporary primatology: Proceedings of the Fifth Congress of the International Primatological Society,* ed. S. Kondo, M. Kawai, and A. Ehara. Basel: S. Karger.

Winkler, P.; Loch, H.; and Vogel, C. 1984. Life history of Hanuman langurs (*Presbytis entellus*): Reproductive parameters, infant mortality, and troop development. *Folia Primatol.* 43:1–23.

Winter, P.; Handley, P.; Ploog, D.; and Schott, D. 1973. Ontogeny of squirrel monkey calls under normal conditions and under acoustic isolation. *Behaviour* 47:230–39.

Witt, R.; Schmidt, C.; and Schmidt, J. 1981. Social rank and Darwinian fitness in a multimale group of Barbary macaques (*Macaca sylvanus* Linnaeus, 1758). *Folia Primatol.* 36:201–11.

Wittenberger, J. F. 1980. Group size and polygamy in social mammals. *Am. Nat.* 115:197–222.

———. 1981. *Animal social behavior.* Boston: Duxbury Press.

Wittenberger, J. F., and Tilson, R. L. 1980. The evolution of monogamy: Hypotheses and evidence. *Ann. Rev. Ecol. Syst.* 11:197–232.

Wolf, A. P., and Huang, Chieh-shan. 1980. *Marriage and adoption in China, 1845–1945.* Stanford: Stanford University Press.

Wolf, K. 1980. Social change and male reproductive strategy in silvered leaf-monkeys, *Presbytis cristata,* in Kuala Selangor, Peninsular Malaysia. *Am. J. Phys. Anthropol.* 52:294.

Wolf, K., and Fleagle, J. G. 1977. Adult male replacement in a group of silvered leaf-monkeys (*Presbytis cristata*) at Kuala Selangor, Malaysia. *Primates* 18:949–55.

Wolf, K., and Schulman, S. R. 1984. Male response to "stranger" females as a function of female reproductive value among chimpanzees. *Am. Nat.* 123:163–74.

Wolf, R. H.; Harrison, R. M.; and Martin, T. W. 1975. A review of reproductive patterns in New World monkeys. *Lab. Anim. Sci.* 25:814–21.

Wolfe, L. D. 1979. Behavioral patterns of estrous females of the Arashiyama West troop of Japanese macaques (*Macaca fuscata*). *Primates* 20:525–34.

———. 1981. Display behavior of three troops of Japanese monkeys. *Primates* 22:24–32.

———. 1984a. Female rank and reproductive success among Arashiyama B Japanese macaques (*Macaca fuscata*). *Int. J. Primatol.* 5:133–43.

———. 1984b. Japanese macaque female sexual behavior: A comparison of Arashiyama East and West. In *Female primates: Studies by female primatologists,* ed. M. Small. New

York: Alan R. Liss.

Wolfheim, J. H. 1977a. A quantitative analysis of the organization of a group of captive talapoin monkeys. *Folia Primatol.* 27:1–27.

———. 1977b. Sex differences in behavior in a group of captive juvenile talapoin monkeys (*Miopithecus talapoin*). *Behaviour* 63:110–28.

———. 1983. *Primates of the world: Distribution, abundance, and conservation.* Seattle: University of Washington Press.

Wolfheim, J., and Rowell, T. 1972. Communication among captive talapoin monkeys (*Miopithecus talapoin*). *Folia Primatol.* 18:224–55.

Woodruff, G., and Premack, D. 1979. Intentional communication in the chimpanzee: The development of deception. *Cognition* 7:333–62.

Woolfenden, G. E., and Fitzpatrick, J. W. 1978. The inheritance of territory in group-breeding birds. *Bioscience* 28:104–8.

World Wildlife Fund. 1980. *Saving Siberut: A conservation master plan.* World Wildlife Fund—Indonesia, Bogor, Indonesia.

Wrangham, R. W. 1974. Artificial feeding of chimpanzees and baboons in their natural habitat. *Anim. Behav.* 22:83–94.

———. 1975. The behavioural ecology of chimpanzees in the Gombe National Park, Tanzania. Ph.D. diss., Cambridge University.

———. 1977. Feeding behaviour of chimpanzees in Gombe National Park, Tanzania. In *Primate ecology,* ed. T. H. Clutton-Brock. New York: Academic Press.

———. 1979a. On the evolution of ape social systems. *Soc. Sci. Information* 18:334–68.

———. 1979b. Sex differences in chimpanzee dispersion. In *The great apes,* ed. D. A. Hamburg and E. R. McCown. Menlo Park, Calif.: Benjamin/Cummings.

———. 1980. An ecological model of female-bonded primate groups. *Behaviour* 75:262–300.

———. 1981. Drinking competition in vervet monkeys. *Anim. Behav.* 29:904–10.

———. 1982. Mutualism, kinship, and social evolution. In *Current problems in sociobiology,* ed. King's College Sociobiology Group. Cambridge: Cambridge University Press.

———. 1983. Social relationships in comparative perspective. In *Primate social relationships: An integrated approach,* ed. R. A. Hinde. Oxford: Blackwell.

———. 1986. Ecology and social relationships in two species of chimpanzee. In *Ecology and social evolution: Birds and mammals,* ed. D. I. Rubenstein and R. W. Wrangham. Princeton: Princeton University Press.

Wrangham, R. W., and Smuts, B. B. 1980. Sex differences in the behavioural ecology of chimpanzees in the Gombe National Park, Tanzania. *J. Reprod. Fertil. Suppl.* 28:13–31.

Wrangham, R. W., and Waterman, P. G. 1981. Feeding behaviour of vervet monkeys on *A. tortilis* and *A. xanthophloea:* With special reference to reproductive strategies and tannin production. *J. Anim. Ecol.* 50:715–31.

Wright, E. M., Jr., and Bush, D. E. 1977. The reproductive cycle of the capuchin (*Cebus apella*). *Lab. Anim. Sci.* 27:651–54.

Wright, P. C. 1978. Home range, activity patterns, and agonistic encounters of a group of night monkeys (*Aotus trivirgatus*) in Peru. *Folia Primatol.* 29:43–55.

———. 1981. The night monkeys, genus *Aotus.* In *Ecology and behavior of Neotropical primates,* vol. 1, ed. A. F. Coimbra-Filho and R. A. Mittermeier. Rio de Janeiro: Academia Brasileira de Ciencias.

———. 1984. Biparental care in *Aotus trivirgatus* and *Callicebus moloch.* In *Female primates: Studies by women primatologists,* ed. M. F. Small. New York: Alan R. Liss.

———. 1985. The costs and benefits of nocturnality for *Aotus trivirgatus* (the night monkey). Ph.D. diss., City University of New York.

Wright, P. C.; Izard, M. K.; and Simons, E. L. 1986. Reproductive cycles in *Tarsier banacus.* Typescript.

Wright, R. V. S. 1972. Imitative learning of a flaked-tool technology: The case of an orangutan. *Mankind* 8:296–306.

Wright, S. 1943. Isolation by distance. *Genetics* 28:114–38.

———. 1965. The interpretation of population structure by F-statistics with special regard to systems of mating. *Evolution* 19:395–420.

Wu, H.; Holmes, W. G.; Medina, S. R.; and Sackett, G. P. 1980. Kin preference in infant *Macaca nemestrina. Nature* 285:225–27.

Yamada, M. 1957. A case of acculturation in the subhuman society of Japanese monkeys. *Primates* 1:20–46.

———. 1963. A study of blood-relationship in the natural society of the Japanese macaque: An analysis of co-feeding, grooming, and playmate relationships in Minoo-B troop. *Primates* 4:43–65.

Yamagiwa, J. 1983. Diachronic changes in two eastern lowland gorilla groups (*Gorilla gorilla graueri*) in the Mt. Kahuzi region, Zaire. *Primates* 24:174–83.

Yerkes, R. M. 1941. Conjugal contrasts among chimpanzees. *J. Abnorm. Soc. Psychol.* 36:175–99.

———. 1943. *Chimpanzees: A laboratory colony.* New Haven: Yale University Press.

Yerkes, R. M., and Tomilin, M. I. 1935. Mother-infant relations in chimpanzees. *Comp. Psychol.* 20:321–49.

Yoneda, M. 1981. Ecological studies of *Saguinus fuscicollis* and *Saguinus labiatus* with reference to habitat segregation and height preference. *Kyoto University Overseas Research Reports of New World Monkeys* 2:43–50.

Yoshiba, K. 1967. An ecological study of Hanuman langurs, *Presbytis entellus. Primates* 8:127–54.

———. 1968. Local and intertroop variability in ecology and social behavior of common Indian langurs. In *Primates: Studies in adaptation and variability,* ed. P. C. Jay. New York: Holt, Rinehart, and Winston.

Yost, J. A., and Kelley, P. M. 1983. Shotguns, blowguns, and spears: The analysis of technological efficiency. In *Adaptive responses of native Amazonians,* ed. R. Hames and W. T. Vickers. New York: Academic Press.

Young, A. L. 1983. Preliminary observations on the ecology and behavior of the muriqui and the brown howler monkey. B.A. thesis, Harvard University.

Young, O. P. 1981. Copulation-interrupting behavior between females within a howler monkey troop. *Primates* 22:135–36.

———. 1982. Tree-rubbing behavior of a solitary male howler monkey. *Primates* 23:303–6.

Zeller, A. C. 1980. Primate facial gestures: a study of communication. *Papers in Linguistics: Int. J. Hum. Comm.* 13:

565–606.

⸺. 1982. Speaking of clever apes. *Recherches Semiotiques/Semiotic Inquiry* 2:276–308.

Zimmerman, E. 1981. First record of ultrasound in two prosimian species. *Naturwissenschaften* 68:531–32.

⸺. 1982. First record of ultrasound in two prosimian species. *Naturwissen* 68:531.

Zoloth, S. R.; Petersen, M. R.; Beecher, M. D.; Green, S.; Marler, P.; Moody, D. B.; and Stebbins, W. 1979. Species-specific perceptual processing of vocal sounds by monkeys. *Science* 204:870–72.

Index

All taxonomic terms are italicized for easy location.

Species are indexed by genus and species names the first time they appear in the text (e.g., *Alouatta caraya; Cebus capucinus*). After that, species are usually indexed at level of the genus (e.g., *Alouatta*) or subfamily (e.g., *Cebinae*). Exceptions occur when a single species is discussed extensively; these are typically indexed by genus and species names (e.g., *Cebus apella*, pp. 207–8). Cross references for more general taxonomic categories are given under lower level categories, but not vice versa (e.g., under *Ateles paniscus* you will see a cross reference for *Atelinae*, but under *Atelinae* you will not see cross references for the species in this subfamily). Thus to find all references for a particular species, one must check the entry for that species, its genus, and its subfamily. For example, for rhesus macaques check *Macaca mulatta*, *Macaca*, and *Cercopithecinae*. (For a list of Latin and common names, see the appendix.)

In general, discussions of particular topics have been indexed by both taxon and subject matter. However, to ensure finding all relevant references, check both the taxon and topic. For example, if interested in aggression in rhesus monkeys, check under Agression for references to *Cercopithecinae* and *Macaca*, and check under *Cercopithecinae* and *Macaca* for references to Aggression.

Abduction. *See* Dispersal, voluntary vs. involuntary
Activity budgets
  Alouatta, 56–59
  Atelinae, table 7-1
  Callitrichidae, 36
  Cebinae, 69–71; table 7-1
  diet, 202–4, figures 17-4, 17-5
  foraging behavior, 201–5
  Gorilla, 157, 204–5; figure 17-5
  Hylobates, 136–39; table 12-3
  Pan, 165–66; table 15-1
  Papio, 112–13
  Pitheciinae, 46
  Pongo, 146–47
  Prosimii, 13
Adolescence, 358–59, 364–69
Adoption, 303, 327–28, 338, 341. *See also* Allomaternal care; Male-infant relationships
Affinitive behavior. *See also* Alliances; Cooperation; Food-sharing; Grooming; Reciprocal altruism; Reciprocity
  among kin, 299–303
  Atelinae, 79–80
  Cebinae, 79–80
  Cercopithecinae, 130–32, 299–305, 309–17
  female-female, 299–303
  male-female, 396–97
  male-male, 174–75, 302, 313
  Pan, 174–76
  Pongo, 150
  sex differences, 400–401
Aggression. *See also* Alliances; Competition *under taxonomic headings;* Dominance; Female dominance; Female-female competition; Infanticide; In-

tergroup relationships; Interspecific relationships; Male dominance; Male-male competition; Resource competition
  Atelinae, 73, 77–79
  Callicebus, 51–53
  Callitrichidae, 40
  Cebinae, 73, 77–79
  Cercopithecinae, 118, 129–32, 306–8, 310–11, 315–16, 366–67, 401–12
  Cercopithecoidea, 276–80
  Cercopithecus, 104–5
  Colobinae, 91–93
  communication, displays of, 306–8, 401, 422–24, 434–36; figures 5-2, 14-1, 14-2, 25-1, 25-2
  contexts of, 307–8, 315–16, 401–4; table 32-3
  costs and benefits of, 401–3
  degree of association and, 308
  Erythrocebus patas, 105
  female, 276–79, 308, 311–12, 397; table 32-3
  Gorilla, 157, 163
  Hylobates, 140–41
  injuries, table 32-2; figures 6-2, 20-1, 31-1
  intergroup encounters, 156–57, 274–81
  interspecific, 212–14
  Lemuriformes, 32
  male, 279–80, 308–9, 386–89, 396, 401–2; figures 20-1, 31-1
  male-female, 396, 406–12; tables 32-1, 32-2
  maternal, 331
  Pan, 170–71, 173–74, 404
  Papio, 118, 129–32, 308, 310–11, 402–3; figure 31-1

  Pitheciinae, 51–53
  Pongo, 150
  rates of, 311, 401–4; table 32-1
  relationship to dominance, 311, 388
  sex differences, 401–12; tables 32-1, 32-2
  sexual dimorphism, 388
  sexual harassment, 150, 290–91, 293–94, 406–7; figure 9-4
Agonistic buffering, 316. *See also* Male-infant relationship; Triadic male-infant interactions
Alarm calls, 218, 222, 228–29, 444–45. *See also* Predation on primates; Vocal communication
  Callitrichidae, 37
  Cercopithecus, 228–29; figures 19-1, 36-4
All-Male-Groups (AMG), figure 21-1. *See also* Demography; Social organization *under taxonomic headings*
  adaptive significance, 259
  attraction to, 258–59
  Colobinae, 90
  Erythrocebus patas, 102
  formation of, 254
  Papio, 114, 116, 248
  predation on, 238
  social behavior, 261
Allenopithecus nigroviridis, 99; table 9-1. *See also Cercopithecinae*
Alliances. *See also* Cooperation; Female-female competition; Female-female relationships; Male-male competition; Male-male relationships
  against males, by females, 78, 93, 407–9; table 32-3; figures 8-4, 21-2
  Cebinae, 78

565

Alliances (*continued*)
  Cercopithecinae, 130–32, 174, 301–2,
    309–17, 325, 366, 407–9, 427
  Colobinae, 92–93, 174
  definition of, 309
  dominance rank in, 311–17, 390
  female-female, 78, 93, 311–14, 325,
    366, 407–9; table 32-3; figure 35-1
  greeting behavior, 132
  grooming, 132
  kin vs. non-kin, 132, 301–2, 310–14,
    316–17, 325, 366, 389–90, 408–9
  male expulsion, 389–90, 407–9
  male-male, 78, 92, 132, 174–76, 313,
    315–16, 389–90
  mating strategies, 390
  Pan, 174–76, 302, 309–10, 316, 404–
    5, 427–29
  sex differences, 404–5, 427
  solicitation of, 309, 427
Allomaternal care, 105, 338–42; tables
  27-2, 27-3; figure 27-4. *See also*
  Adoption; Infant care; Infant care
  under taxonomic headings; Infants
Alloparenting. *See* Adoption; Allomaternal
  care; Infant care; Infant care *under*
  taxonomic headings; Infants; Male-
  infant relationships
Alouatta, 54–68
  activity budgets, 56–59
  body size/weight, 54–55; table 6-1
  communication, 55–56, 58–59; figure
    6-1
  competition, 60, 65–68
  conservation, 54
  demography, 61, 63–67; tables 6-5, 6-6,
    6-7
  dispersal, 65–68, 257, 266
  ecology, 54–58; tables 6-2, 6-3
  geographic distribution, 54
  group size and population density, 57–
    58, 60–67; tables 6-3, 6-5, 6-6; figures
    6-3, 6-4
  home range and territoriality, 55–59;
    table 6-3
  infanticide, 65, 348–50; table 8-4
  intergroup interactions, 55, 58–59
  physical characteristics, 54–55
  reproduction, 54, 59–60; tables 6-1, 6-4
  sexual dimorphism, 54–55, 65; table 6-1
  social organization and behavior, 55–56,
    60–68, 346–50; table 8-4; figure 6-4
  taxonomy, 54
Alouatta belzebul, 54
Alouatta caraya, 54
Alouatta fusca, 54
Alouatta palliata, 54–68
Alouatta pigra, 54
Alouatta seniculus, 54–67; figures 6-1, 6-2
Alpha male, 175–76. *See also* Aggression;
  Dominance; Male-male competition;
  Male dominance
Altruism. *See also* Affinitive behavior; Al-
  liances; Cooperation; Food-sharing;
  Grooming; Kin selection; Reciprocal
  altruism

evolution of, 324–29
  toward infants, 327
  Hamilton's rule, 324
Aotus, 22–23, 44–53; tables 5-1, 5-2, 5-3;
  figure 5-4. *See also Pitheciinae*
  communication, 50, 52
  conservation, 47
  dispersal, 49–50
  ecology, 45–47, 52–53, 286; tables 5-1,
    5-2
  geographic distribution, 45
  group size and population density, 46–
    48; figure 5-3; table 5-2
  home range, 46–47
  infant care, 43, 50; figure 5-4
  intergroup interactions, 52–53
  reproduction, 48; table 5-3
  social organization and behavior, 48–50,
    286
  taxonomy, 44
Ape language studies, 441–44, 455–56;
  figure 36-1
Appeasement. *See* Aggression; Dominance;
  Submissive behavior
Arctocebus, 12, 15–17. See also
  Lorisiformes; Prosimii
Ateles belzebuth, 69. See also *Atelinae*
Ateles fusciceps, 69. See also *Atelinae*
Ateles geoffroyi, 69. See also *Atelinae*
Ateles paniscus, 69. See also *Atelinae*
Atelinae, 69–82
  activity budgets, table 7-1
  characteristics, 69
  communication, 73
  compared to other primates, 80–82
  competition, 73, 77–79; table 7-4
  conservation, 73–74
  demography, 76–77
  dispersal, 77; table 7-4
  ecology, 69–73, 80–82; tables 7-1, 7-2
  geographic distribution, 69
  group size and population density, 77,
    81–82; tables 7-2, 7-4; figure 7-2
  home range, 71, 73, 81–82; table 7-2
  intergroup interactions, 73, 81–82
  reproduction, 74–77; table 7-3
  sexual dimorphism, 69; table 7-4
  social organization and behavior, 77–82;
    table 7-4
  taxonomy, 69
Attachment, vii, 330, 332–33, 417–20.
  See also Mother-infant relationships
Avahi, 22; See also *Lemuriformes; Prosimii*

Birth rate, 18, 241, 244-47; table 20-3. *See*
  also Reproduction *under taxonomic*
  headings; Reproductive parameters;
  Reproductive success
Birth seasonality
  Alouatta, table 6-4
  Atelinae, table 7-3
  Cebinae, 74–76; table 7-3
  Cercopithecinae, 113, 126; table 11-4
  Cercopithecus, 102; table 9-3
  Colobinae, 86–87; table 8-3; figure 8-2
  dispersal, 254, 260

Erythrocebus patas, 102, table 9-3
Hylobates, 136–37; table 12-2
Papio, 113; tables 10-1, 11-4
Pitheciinae, table 5-3
Prosimii, 17, 33; table 2-3
Birth weight, 18, 344–45. *See also* Body
  size/weight, neonatal vs. adult
Body size/weight, 181–96; figures 16-1,
  16-4
  allometric analysis, 186–92, tables 16-3,
    16-6; figure 16-3
  Alouatta, 54–55, table 6-1
  Atelinae, 69
  brain size and, 186–89, 193–96; tables
    16-3, 16-6
  Cebinae, 69
  Cercopithecinae, 132; table 10-1
  Colobinae, 83
  demographic variables and, 18, 74, 151–
    52, 181–82
  ecological variables and, 28, 47, 73, 122,
    151–53, 182–85, 194–95, 206–7;
    tables 7-5, 18-1; figures 5-3, 7-2
  Erythrocebus patas, 98
  evolution of, 24, 151–54
  Hylobates, 135
  neonatal vs. adult, 18, 34, 190–93; fig-
    ures 16-3, 16-4
  Pitheciinae, table 5-1
  Pongo, 146–47, 152–54
  predator defense and, 185–86, 231–32,
    234
  Prosimii, 12, 16–18, 20, 24, 28; table 2-1
  social behavior, 20, 132, 313; table 32-4
Border patrols, *Pan,* 170
Brachiation, 135
Brachyteles, 69; figure 40-1. See also
  Atelinae
Brain size
  allometric analysis, 187–88; tables 16-3,
    16-5, 16-6
  body size/weight, and, 186–89, 193–96;
    tables 16-3, 16-6
  demographic variables and, 191–96,
    359; tables 16-5, 16-6
  diet and, 187–88, 194–95
  Homo sapiens, 193–94
  metabolic rate, and, 187, 195
  neonatal, 193–95; figure 16-5

Cacajao calvus, 44–50; figure 5-2. See
  also *Pitheciinae*
Callicebus. See also *Pitheciinae*
  aggression, 51–52
  communication, 50–53
  conservation, 47
  dispersal, 49–50
  ecology, 45–47, 52–53; table 5-1
  geographic distribution, 45–46
  group size and population density, 47;
    table 5-2; figure 5-3
  home range and territoriality, 46, 50–53
  infant care, 50
  intergroup interactions, 50–53
  reproduction, 48–49; table 5-3
  social organization and behavior, 48–50

Callicebus moloch, 46. See also *Callicebus; Pitheciinae*

Callicebus torquatus, 46. See also *Callicebus; Pitheciinae*

Callimico goeldii. See *Callimiconidae*

Callimiconidae, 34–40, 343–45, 368; tables 4-1, 4-2, 4-3

Callithrix humeralifer, 34. See also *Callitrichidae*

Callithrix jacchus, 34–39, 41–43. See also *Callitrichidae*

Callitrichidae, 34–43, 343–45, 368; tables 4-1, 4-2, 4-3
  activity budgets, 36
  characteristics of, 34
  communication, 36, 434
  competition, 40, 397
  conservation, 37
  demography, 39–40; table 4-3
  ecology, 35–37; table 4-1
  food-sharing, 41, 328, 344
  geographic distribution, 34–35
  group size and population density, 37; tables 4-1, 4-3
  home range/territoriality, 37–40, 208–9; table 4-1
  infant care, 40–43, 343–45; table 4-4
  intergroup interactions, 40
  interspecific relationships, 36–37
  reproduction, 37–40, 42–43; table 4-2
  social organization and behavior, 37–43, 368; table 4-3
  taxonomy, 34–35

Cebinae, 69–82
  activity budgets, 69–71; table 7-1
  adaptations of, 69
  body size/weight, 69
  communication, 73
  compared to *Catarrhines,* 81
  competition, 73, 77–79; table 7-4
  conservation, 73–74
  demography, 76–77
  dispersal, 77; table 7-4
  ecology, 69–73, 81, 204, 207–8; tables 7-1, 7-2
  geographic distribution, 69
  group size and population density, 77, 81; tables 7-2, 7-4; figure 7-2
  home range/territoriality, 71–73, 81; table 7-2
  intergroup interactions, 73
  reproduction, 74–77, 81; table 7-3
  sexual dimorphism, 69; table 7-4
  social organization and behavior, 74–80; table 7-4
  taxonomy, 69

Ceboidea
  demographic and life-history variables, 255; table 16-1
  female-female competition, 312, 397
  olfactory signals, 13, 434

Cebuella, 34–40. See also *Callitrichidae*

Cebus albifrons, 69, 204, 207–8, 465. See also *Cebinae*

Cebus apella, 69, 204, 207–8, 396, 465; figure 17-7. See also *Cebinae*

Cebus capúcinus, 69. See also *Cebinae*

Cebus olivaceus, 69; figure 7-1. See also *Cebinae*

Cercocebus albigena, 121. See also *Cercopithecinae*

Cercocebus aterrimus, 121. See also *Cercopithecinae*

Cercocebus atys, 121. See also *Cercopithecinae*

Cercocebus galeritus, 121. See also *Cercopithecinae*

Cercocebus torquatus, 121. See also *Cercopithecinae*

Cercopithecinae, 121–34. See also *Cercopithecus; Cercopithecus aethiops; Macaca; Papio; Papio hamadryas*
  affinitive behavior, 130–32, 309–17
  communication, 119–20, 132, 422–27, 435–38, 444–51
  competition, 126–32, 205–6, 301–2, 306–17, 319–25, 397, 401–11, 425, 428–29
  demography and life-histories, 126–29, 240–49; tables 11-5, 16-1, 20-2; figure 20-3
  dispersal, 115–17, 129, 132–34, 242, 302–4; table 21-1
  ecology, 122–25; table 11-1
  female choice, 392–96, 398–99
  genetic variation in, 132–34
  geographic distribution, 121–22
  grooming, 117, 131–32, 300–301, 313–14, 364–65; figure 11-2
  group size and population density, 123–25, 134; tables 11-1, 11-2, 11-3, 11-5, 22-2
  home ranges, 122–23; tables 11-1, 22-2
  infancy, 303, 330–57, 365–66; table 27-1
  infanticide, 92–93, 279, 353; table 8-4
  intergroup interactions, 104–5, 114, 131, 267–80; table 22-2; figures 22-1, 22-2
  kinship, 129–32, 299–305, 313–14, 364–65
  ontogeny, 314, 358–69
  reproduction, 113, 125–26, 130, 132, 134, 244–46, 322–23, 395; tables 11-4, 20-3
  reproductive success, 130, 132, 320–23, 389
  sex differences, 400–405
  sexual dimorphism, 132
  social behavior and organization, 126–32, 299–317; tables 8-4, 11-1, 11-2
  taxonomy, 112, 121

Cercopithecoidea, 250–66, 278–81

Cercopithecus, 98–111, 212–14, 221, 230–31, 237
  communication, 105, 230, 237
  competition, 104–11
  conservation, 98–99
  demography, 104, 110
  dispersal, 102–4, 109–10
  ecology, 99–101, 212–14, 221, 230–31; tables 9-1, 9-2, 18-2

  geographic distribution, 98–99; table 9-1
  group size and population density, 101, 104; table 9-2
  home range and territoriality, 101, 104–5; table 9-2
  intergroup interactions, 104–5
  interspecific interactions, 212–14, 221, 230–31
  reproduction, 102, 107–11; tables 9-3, 9-5
  social organization and behavior, 102–11
  taxonomy, 98

Cercopithecus aethiops, 121–22; figure 11-1. See also *Cercopithecinae*
  alarm calls, 228–29; figures 19-1, 36-4
  competition, 277–78, 313–14
  ecology, 122, 204
  genetic variation and dispersal, 133–34; figure 22-1
  geographic distribution, 121–22
  home range and territoriality, 122, 267
  intelligence, social, 459, 473
  intergroup relationships, 267, 277–80; figure 22-1
  playback experiments, 7–8, 316, 328–29, 444–45, 447–48
  predation on, 234–35, 238
  reproductive success, 320–22

Cercopithecus ascanius, 99, 111, 202; figure 9-1

Cercopithecus campbelli, 98

Cercopithecus cephus, 99

Cercopithecus diana, 99

Cercopithecus hamlyni, 99

Cercopithecus l'hoesti, 99

Cercopithecus mitis, 98, 110–11; figure 9-3

Cercopithecus mona, 98

Cercopithecus neglectus, 98–99, 104; figure 16-2

Cercopithecus nictitans, 99

Cercopithecus pogonias, 98

Cercopithecus salongo, 9

Cercopithecus talapoin, 121–23, 126. See also *Cercopithecinae*

Cheirogaleus medius, 25, 28–29. See also *Lemuriformes; prosimii*

Chiropotes, 44, 46–49. See also *Pitheciinae*

Coalitions. *See* Alliances

Cognitive capacities. *See also* Intelligence/cognition
  foraging and cognitive maps, 206

Colobinae, 83–97
  body size/weight, 83
  characteristics of, 83
  communication, 83, 87, 90–92
  competition, 91–92
  conservation, 85
  demography, 89–91, table 16-1
  dispersal, 90, 253
  ecology, 83–85, 197, 238; table 8-1
  geographic distribution, 83
  group size and population density, 84, 89–90; tables 8-1, 8-2
  home range and territoriality, 84, 91–92

Colobinae (continued)
  infanticide, 86, 92–97; table 8-4
  intergroup interactions, 90–92, 275; figure 22-1
  interspecific interactions, 221, 230
  reproduction, 85–89; table 8-3
  social organization and behavior, 89–93, 174; table 8-4
  taxonomy, 83
Colobus badius, 83–92, 174, 202, 275, 367; table 8-2; figures 8-1, 22-1
Colobus guereza, 83; figure 8-1
Colobus polykomos, 85
Colobus satanas, 84
Colobus vellerosus, 85
Colobus verus, 85
Communication, 3, 431–61. See also Ape language studies; Communication under taxonomic headings; Olfactory signals; Visual communication; Vocal communication
  Alouatta, 55–56, 58–59; figure 6-1
  ancestral primates, 23
  Atelinae, 73
  Cebinae, 73
  Cercopithecinae, 119–20, 132, 422–27, 435–38, 444–51
  cerebral lateralization and, 2, 446
  Colobinae, 83, 87, 91–92
  Pan, 171–73, 422–25, 427–28, 435–36, 438–44, 448–51
  Pitheciinae, 49–53
  Prosimii, 13, 15, 18, 21–22, 32, 433–34
Competition. See Aggression; Dominance; Female-female competition; Interspecific relationships; Male-male competition; Resource competition
Conflict. See Aggression; Dominance; Female-female competition; Interspecific relationships; Male-male competition; Resource competition
Conservation, vii, 4–5, 9–10, 477–90, 495–96; tables 39-1, 39-2. See also Conservation under taxonomic headings
  Africa, 486–87
  Alouatta, 54
  Asia, 487–88
  Atelinae, 73–74
  Callitrichidae, 37
  Cebinae, 73–74
  Central and South America, 482–86
  Cercopithecus, 98–99
  Colobinae, 85
  future goals, 488–90
  Gorilla, 155–56
  habitat destruction, 477–79; figure 39-1
  hunting, 479–81; figures 39-2, 39-3
  Hylobates, 136
  live capture, 481–82
  Madagascar, 28–29, 487
  Pan, 166
  Papio, 113
  Pitheciinae, 47–48

Pongo, 147
Prosimii, 17, 28–29, 487
Consortships. See also Sexual behavior
  Alouatta, 59
  Atelinae, 77
  Cebinae, 77
  Cercopithecinae, 126, 380, 394–95; figure 31-2
  dominance and, 126
  Pan, 169–70
  Pongo, 148
Cooperation, 309–17, 427–29. See also Affinitive behavior; Alliances; Food-sharing; Grooming; Reciprocity
  Alouatta, 68
  Cercopithecinae, 309–17
  communication and, 436–38
  defense of food, 205–6
  ecological conditions affecting, 309–10
  frequency of, 310
  hunting, 206
  kinds of, 309–10
  kin vs. non-kin, 68, 310, 313
  laboratory studies, 437–38
  male-male, 315
  Pan, 175–76, 315
  participants, 310
  sexual receptivity and, 310, 315
Culture
  benefits of, 462, 472
  Cebus apella, 465
  definition, 462
  ecological influences, 463–65
  Macaca fuscata, 464, 467–72; figures 38-3, 38-4, 38-5, 38-6, 38-7
  mechanisms of transmission, 462–73
  Pan, 463–67, 473
  primates vs. non-primates, 472–73

Daubentonia, 28–29; figure 3-1. See also Lemuriformes; Prosimii
Deception, Pan, 438–39, 448–50
Demography, 242–49. See also Body size/weight; Demography under taxonomic headings; Group size; Life-history; Population density; Reproductive parameters; Sex ratio
Diet, 197, 202–4, 224–25; tables 7-5, 16-2, 18-2–18-7. See also Ecology under taxonomic headings; Folivory; Food distribution; Foraging behavior; Frugivory; Insectivory; Predation by primates; Resource competition
  Alouatta, 56; table 6-2
  Atelinae, 69–71, 81–82; table 7-1
  body size/weight, 73, 195
  brain size, 187–88, 194–95
  Callimiconidae, 36
  Callitrichidae, 36
  Cebinae, 69–73, 81
  Cercopithecinae, 112–13, 122–23
  Cercopithecus, 99–101; table 9-2
  Colobinae, 83–85, 197
  dominance status, 320
  Erythrocebus patas, 99
  fruits, 13, 27–28, 36, 46, 56, 69–70,

81, 101, 112, 122–23, 135, 147, 155, 163, 166, 182–83, 202–3, 206, 220–21, 224–25, 285–86, 290–91
  Gorilla, 155
  gum feeding, 13–15, 28, 36
  home range, 112–13, 202, 217; figure 17-6
  Hylobates, 135–36
  interspecific competition, 212–17
  invertebrates, feeding on, 13, 27–28, 36, 46, 69–70, 81, 101, 112, 122, 155, 166, 182–83, 202–3, 206, 220, 224
  leaves, 13, 27–28, 46, 56, 69–70, 81, 101, 112, 122, 135, 147, 155, 166, 182–83, 202–3, 206, 224
  Pan, 166, 464
  Pitheciinae, 46; table 5-1
  Pongo, 147
  Prosimii, 13–17, 27–28, 204; tables 2-2, 3-1
  reproduction, effects on, 319–20
  seasonal fluctuations in, 216–17
Dispersal, 250–66. See also Emigration; Female transfer; Male transfer; Transfer
  Alouatta, 65–68, 252–53, 257, 266
  Atelinae, 77; table 7-4
  both sexes, 252–54
  Cebinae, 77; table 7-4
  Cercopithecinae, 115–17, 129, 132–34, 242, 253, 303; table 21-1
  Cercopithecus, 102–4, 109–10
  Colobinae, 90, 253
  costs and benefits, 241–42, 263–65
  definition, 250
  effects of provisioning on, 259
  Erythrocebus patas, 102–3, 109–10
  explanations, proximate, 67, 162–64, 254–59, 261–63
  explanations, ultimate, 67–68, 162–64, 262–66
  female, 67–68, 162–64, 169, 252–54, 257–58, 261, 265–66, 274–75, 293–95, 367–68; table 23-2; figures 21-3, 21-4
  genetic variation and, 133–34
  Gorilla, 156–57, 160–64, 253
  Hylobates, 141–42, 254, 257, 303; table 12-4
  inbreeding avoidance and, 162, 262–66
  kinship and, 164, 365
  male, 250–58, 260–62, 265, 276–78, 292–93; tables 21-2, 23-2; figures 21-3, 21-4
  natal, 250–53, 258–60; table 21-1
  Pan, 169, 253, 261, 265; table 15-3
  philopatry, 250, 265–66
  Pitheciinae, 49–50
  Pongo, 148–49
  Prosimii, 18–20, 30–31, 253–54
  reproductive influences, 169, 257–59
  secondary, 250, 257
  sex differences, table 21-1; figure 21-3
  social organization and, 162–64, 250–54, 265–66
  voluntary vs. involuntary, 254–57

Dominance. *See also* Alliances; Aggression; Competition *under taxonomic headings;* Female-female competition; Male-male competition; Resource competition
aggression, 310–11
Alouatta, 60, 65–66
Atelinae, 77–79
body size/weight and, 313; table 32-4
Cebinae, 73, 77–79
Cercopithecinae, 118, 126–32, 243, 301, 311–13, 366, 402–3, 409, 422–26, 429
Colobinae, 91–92
costs and benefits, 320
definition, 310
ecological influences, 33
evolution of, 323–24, 422
formal, 422–24
Gorilla, 157–63
grooming, 4, 79–80, 301
hierarchies, 91, 310–11, 323–24
Hylobates, 139–41, 291
intergroup, 268–72, 288
interspecific, 212
kinship, 301, 311–13
matrilineal groups, 118, 301. See also *Cercopithecinae*
models of, 422–29
monogamous primates, 291
mortality rates and, 244
Pan, 170–71, 173–77, 310, 313, 409, 422–29; figure 34-2
Pitheciinae, 50
predator defense and, 229–30
primate awareness of, 311
Prosimii, 18–20, 30–33
rank acquisition and reversals, 129, 248, 311–13, 366, 402–3
reproduction, effects on, 33, 65, 77, 81, 126, 130, 175–76, 245, 288, 320–24, 388–89; figure 31-1
sex differences, 405–11
special relationships and, 422, 429
status striving, 425

Ecological selection, 150–52
Ecology, table 16-2. *See also* Activity budgets; Diet; Ecology *under taxonomic headings;* Foraging behavior; Group size; Habitat; Home range; Interspecific relationships; Population density; Predation by primates; Predation on primates; Territoriality
Alouatta, 54–58; tables 6-2, 6-3
Atelinae, 69–73, 80–82; tables 7-1, 7-2
body size/weight and, figure 17-6
Callimiconidae, 36; table 4-1
Callitrichidae, 35–37; table 4-1
Catarrhines vs. *Platyrrhines,* 95
Cercopithecinae, 122–25; table 11-1
Cercopithecus, 99–101, 212–14, 221, 230–31; tables 9-1, 9-2, 18-2
Colobinae, 83–85, 197, 238; table 8-1
Erythrocebus patas, 99; table 9-2
factors influencing group size, 61–65

Hylobates, 135–36, 143–45, 203, 207; table 12-1
niche divergence, 223–25
nocturnal vs. diurnal, 11, 24
Pan, 165–66, 171, 174, 206, 285–86, 290–91, 464; table 15-1
Papio, 112–13, 119, 122, 205, 229–30, 294–95
Pitheciinae, 46–48; tables 5-1, 5-2
and population growth, 246–47
Prosimii, 13–15, 25–29, 36, 204, 284; tables 2-1, 2-2
and reproduction, 18, 125, 319–20
savanna, 201; figure 17-3
and social organization, 118–19
sympatric species, 15–17, 98–101, 121–22, 135; tables 18-2–18-10
tropical, 197–201; figure 17-2
Emigration, 250–60. *See also* Dispersal *under taxonomic headings*
aggression precipitating, 255–57
definition, 250
extragroup females, 254
extragroup males, 254
natal, 250
secondary, 250
solo vs. peer, 260
Erythrocebus patas. See also Cercopithecinae
communication, 105
competition, 104–10
conservation, 98–99
demography, 101, 110
dispersal, 101
ecology, 99; table 9-2
geographic distribution, 98
group size and population density, 101, 105
home range and territoriality, 101, 105; table 9-2
intelligence, 472
intergroup interactions, 105
reproduction, 102, 108–9; tables 9-3, 9-5; figure 9-4
social organization and behavior, 102, 105–10; figures 9-2, 9-4
taxonomy, 98
Estrous behavior. *See* Sexual behavior
Estrus. *See also* Female choice; Consortships; Sexual behavior; Reproductive parameters; Reproduction *under taxonomic headings*
Alouatta, 59–60; table 6-4
Atelinae, 77; table 7-3
Callimiconidae, 38
Callitrichidae, 38, 43
Catarrhines, 370–71
Cebinae, 77; table 7-3
Cercopithecinae, 113, 126; table 11-4
Cercopithecus, 102
Colobinae, 87; table 8-3
definition, 370–71
effects on social interactions, 315
Erythrocebus patas, 102
external signs of, 113, 126; tables 11-4, 30-1

Gorilla, 156; table 14-1
Hylobates, 137
male aggression during, 406–7
olfactory cues, 17, 29–30, 379
Pan, 167–70; table 15-2
Papio, 113, 126, 391
Pitheciinae, 48; table 5-3
Pongo, 146, 148
Prosimii, 17, 29–30
Ethology, 4, 414–15
Eviction. *See* Dispersal, voluntary vs. involuntary
Evolution. *See also* Natural selection
of altruism, 324–29
ancestral primates, 23–24
definition of, 318
diversity, 9
of dominance, 323–24
of food-sharing, 326–27
of grooming, 325–26
life-history variables, trade-offs, 192–95
of social behavior and organization, 282–96, 318–19
of vocal communication, 445–46
Evolutionary Theory. *See also* Kin selection; Parental investment; Reproductive success; Sexual selection
applications to primate research, 2–3, 492–95
Extinction. *See* Conservation

Fecundity, 241, 244–47, 321. *See also* Birth rate; Reproductive parameters; Reproductive success
Female choice, 343, 385–86, 392–97. *See also* Sexual behavior, partner preferences
Atelinae, 77; table 7-4
bases for, 354–57, 393–96, 398–99
Cebinae, 77; table 7-4
Cebus apella, 396
Cercopithecinae, 392–99; figure 31-4
Cercopithecus, 102, 110
Gorilla, 163
immigrant males, 260–61
male-infant relationships and, 354–57, 393–94
male protection and, 395
males, effects on, 392–93, 396–97, 409–10
matrilineal groups, 392–99
monogamous primates, 393
Pan, 396
signs of, 392–93
special relationships, 395
Female dominance. *See also* Dominance; Female-female competition
alliances, 314
allomaternal care, 338–41
Ceboidea, 312, 397
Cercopithecinae, 118–19, 126–31, 311–15, 366, 402–4, 409
cooperation, 313–14
grooming, 314
infant abuse, 341–42
kinship, 311–12, 314

Female dominance (*continued*)
  Lemuriformes, 30, 32–33
  longevity, 320–21
  maternal styles, 336–37
  over males, 23, 30, 32–33, 78, 409;
    table 32-4
  play, 314
  rank acquisition and reversals, 129, 311–
    12, 402–3
  reproduction, 33, 245, 320–22
  sex ratio of offspring, 321
  sexual behavior, 397
  social attractiveness, 314
Female-female alliances. *See* Alliances,
    female-female
Female-female competition, 308. *See also*
    Aggression, female; Alliances, female-
    female, kin vs. non-kin; Female domi-
    nance; Matrilineal groups; Resource
    competition
  aggression, 286–88
  Alouatta, 67–68
  Callitrichidae, 397
  Ceboidea, 397
  Cercopithecinae, 129–31, 311, 397
  dispersal, 160–62, 263
  feeding, 160–61, 288–90
  Gorilla, 160–62
  infant abuse, 342
  over males, 397
  reproduction, 245, 288, 320–22
Female-female relationships. *See also* Af-
    finitive behavior, female-female; Alliances,
    female-female; Female dominance;
    Female-female competition; Groom-
    ing, female-female; Kinship; Matrilin-
    eal groups
  affinitive behavior, 299–302, 310
  Cercopithecinae, 118–19, 129–31,
    299–302, 310
  ecological influences, 284–90
  Gorilla, 160–64
  kinship, 299–302
  Pan, 168, 291
  social system, effects on, 400–401
Female transfer, 162–63, 169, 252–55,
    258, 367–68; table 32-2. *See also* Dis-
    persal; Emigration
  evolution of, theories, 293–95
  intergroup relationships, 274–75
  juvenile social relationships and, 368
  social organization and, 293–95
Fertility. *See* Reproductive parameters
Field experiments. *See also* Field methods;
    Playback experiments
  Cercopithecus aethiops, 7–8
  Macaca fuscata, 467–72
  Papio, 7, 114
Field methods, 5–8. *See also* Field experi-
    ments; Observational sampling tech-
    niques; Playback experiments
  demographic data, 240–41
  Galago, 18
  habituation of subjects, 5
  nocturnal observations, 11

Pan, 463
  recognition of individuals, 5
Field studies. *See also* Field experiments;
    Field methods; Observational sampling
    techniques
  future research, 491–96
  history of, ix
  importance of, vii, viii
  long-term, 299
  obstacles in conducting, vii–viii, 2–3
  provisioning, 165, 299, 304, 468–71;
    figure 26-1
Fission-fusion
  Atelinae, 81–82
  Cercopithecinae, 243
  Pan, 81–82, 172, 176
Folivory, 122, 202–3, 206, 224; figure
    17-6. *See also* Diet, leaves
  Alouatta, 56
  Atelinae, 69–70
  brain size and, 187–88, 194–95
  Cercopithecus, 99
  Colobinae, 83–84, 197
  Gorilla, 155
  Hylobates, 135
Food distribution, 197–209. *See also* Diet;
    Ecology; Foraging behavior; Habitat;
    Resource competition
  effects on foraging behavior, 206–9
  interspecific resource competition, 215–
    17, 223–25
  primate biomass and, 215–16
  rain forest, 197–201
  savanna, 201
  seasonal fluctuations, 203–4, 207
  territoriality and, 208–9
Food-sharing, 310, 326–27. *See also* Af-
    finitive behavior
  Atelinae, 80
  Callimiconidae, 344
  Callitrichidae, 41, 327, 344
  Cebinae, 80
  evolution of, 327–28
  mothers and infants, 310, 331
  Pan, 172, 175, 315, 326–27; figure 26-2
Foraging behavior, 197–209. *See also* Diet;
    Ecology; Ecology *under taxonomic
    headings*; Food Distribution; Habitat;
    Home range; Predation by primates;
    Resource competition
  activity budgets and, 201–5
  ancestral primates, 23–24
  Atelinae, 69–71; table 7-1
  banqueters vs. foragers, 202–3, 205
  body size/weight and, 206–7
  Callitrichidae, 36–37
  Callimiconidae, 36
  Cebinae, 69–73; table 7-1
  Cebus apella, 465
  Cercopithecinae, 112–13, 122–23
  cognitive maps of home range, 206
  costs of infant care, 344–45
  effects on food distribution, 206, 286
  frugivorous species, 286
  Gorilla, 157

group size, 207–9
  gum-feeding, 13–15
  Hylobates, 135–36, 143–45
  interspecific mutualism, 220–22
  metabolic rate, 206
  monogamous species, 144, 286–87
  nocturnal vs. diurnal, 11, 36, 206, 290
  Pan, 165–66, 171, 286, 290–91, 464;
    figure 38-1
  Papio, 113, 119, 122
  Pitheciinae, 46
  polyandrous primates, 286
  Pongo, 146–47, 152–53, 290
  Prosimii, 13–15, 18, 22, 27–28, 284,
    290
  rain forest, 198–201
  rank effects, 320
  savanna, 201
  seasonal fluctuations, 203–4, 207
  solitary, 11, 284, 290
  temporal patterning, 202–3
  weather, 203
Forced copulation
  absence of, 392–93
  Pongo, 149–50
Friendship. *See* Special relationships
Frugivory, 123, 182, 206, 224–25, 285–
    86, 290–91; figures 17-6, 18-1. *See
    also* Diet, fruits
  Alouatta, 56
  Atelinae, 69–71
  Cebinae, 69–71
  Cercopithecus, 99–101
  Hylobates, 135
  Pongo, 147
  population densities and, 215–16
  social organization, influences on, 285–
    86, 290–91

Galago, 11–24; table 2-4. *See also*
    Lorisiformes; Prosimii
  communication, 13, 15, 18, 21–22, 433
  ecology, 13–17
  gum-feeding, 13–15; figure 2-4
  infants, 22
  nest building, 15
  sex ratio, 245–46; table 2-4
  social organization and behavior, 18–24;
    table 2-4
Galago alleni, 15–16
Galago crassicaudatus, 13, 17–18, 245–
    46
Galago demidovii, 13, 15, 17–18
Galago elegantulus, 13, 18
Galago garnetti, 13, 16
Galago inustus, 13
Galago senegalensis, 13, 17–22. See also
    Galago; Lorisiformes; Prosimii
  body size/weight, 20
  communication, 18, 21–22
  dominance, 18–20
  ecology, 13, 15, 18
  field methods, 18
  home range and territoriality, 18–20, 22;
    figure 2-5

ontogeny, 18–20
reproduction, 17
social organization and behavior, 18–22;
   table 2-4; figure 2-5
Galago zanzibaricus, 13, 17–18, 22
Genetic variation, 132–34
Gentle giant. *See* Gorilla
Geographic distribution
   Alouatta, 54
   Atelinae, 69
   Callimiconidae, 35
   Callitrichidae, 34–35
   Cebinae, 60
   Cercopithecinae, 112, 121–22
   Cercopithecus, 98–99; table 9-1
   Colobinae, 83
   Gorilla, 155
   Hylobates, 135
   Pan, 165
   Pitheciinae, 45–46
   Pongo, 146
   Prosimii, 12; table 2-1
Gestation. *See also* Reproductive parame-
      ters
   Alouatta, 59; table 6-4
   Atelinae, table 7-3
   body size/weight and, 190–91; table
      16-3
   Callimiconidae, 38
   Callitrichidae, 37–38; table 4-2
   Cebinae, table 7-3
   Cercopithecinae, 125; tables 10-1, 11-4
   Cercopithecus, 102; table 9-3
   Colobinae, 87; table 8-3
   Erythrocebus patas, 102; table 9-3
   Gorilla, table 14-1
   Hylobates, table 12-2
   Pan, table 15-2
   Pitheciinae, 48; table 5-3
   Pongo, 147
   Prosimii, 18, 29; tables 2-3, 3-2
Gestural communication. *See* Communica-
      tion; Visual communication
Gorilla, 155–64; figures 14-3, 28-1, 29-4.
      (Refers to *Gorilla gorilla beringei*)
   activity budgets, 157, 204–5; figure 17-5
   competition, 157–63
   conservation, 155–56
   demography, 156; table 14-2
   dispersal, 156–57, 160–64, 253; table
      14-2
   ecology, 155, 163
   female choice, 163
   geographic distribution, 155
   group size and population density, 155–
      56, 162–63
   immatures, 158–59, 368
   infanticide, 163
   intergroup interactions, 156–57, 267,
      275
   kinship, 160–62, 303–4; tables 14-4,
      14-5
   male protection, 293
   reproduction, 156, 383; table 14-1
   sexual dimorphism, 157

silverback male, 157–60
social behavior, 157–64, 316, 346–47;
   tables 14-4, 14-5; figures 14-1, 14-2
taxonomy, 155
Gorilla gorilla beringei, 155–64. *See also*
      Gorilla
Gorilla gorilla gorilla, 155
Gorilla gorilla graueri, 155, 157
Greetings, 132, 315
Grooming. *See also* Affinitive behavior;
      Reciprocity
   Alouatta, 60
   Atelinae, 80
   Cebinae, 80
   Cercopithecinae, 117, 131–32, 300–
      301, 313–14, 364–65; figure 11-2
   Colobinae, 91
   costs and benefits, 4, 325–26
   direction of, 300–301, 310
   dominance, 4, 79, 301, 365
   ecological conditions effecting, 310
   evolution of, 325–26
   female-female, 313–14, 364–65; figure
      24-1
   Gorilla, 158–59
   kin vs. non-kin, 300–301, 310, 364–65
   male-male, 315, 365
   Pan, 172, 174, 315
   partner preference, 364–65
   patrilineal groups, 91
   Pitheciinae, 50
   reciprocity and, 4
   sex differences, 364–65
Group size, 283; table 23-3. *See also* Popu-
      lation density; Social organization
   Alouatta, 57–58, 60–67; tables 6-3, 6-5,
      6-6; figures 6-3, 6-4
   Atelinae, 77, 81–82; tables 7-2, 7-4; fig-
      ure 7-2
   Callimiconidae, table 4-3
   Callitrichidae, table 4-3
   Cebinae, 77, 81; tables 7-2, 7-4
   Cercopithecinae, 123–25, 134; tables
      11-1–11-3, 11-5
   Cercopithecus, 104, table 9-2
   Colobinae, 84, 89–90; tables 8-1, 8-2
   costs and benefits, 288–90
   dispersal, effects on, 162–63
   Erythrocebus patas, 105; table 9-2
   foraging behavior, 207–8
   Gorilla, 155–56, 162–63
   Hylobates, 137; table 12-1
   Papio, 112, 122
   Pitheciinae, 46, 48–49
   predation pressure, 179, 237–38
   Prosimii, tables 2-4, 3-1
   effects on reproductive success, 162–63,
      288–89
   resource competition, 238

Habitat, 223–25; figures 17-1, 17-2, 17-3.
      See also Ecology, *including listings
      under taxonomic headings;* Food dis-
      tribution; Home range
   Alouatta, 54–58; table 6-3

Callitrichidae, 35
Cercocebus, 122
Cercopithecinae, 112–13, 122–23
Cercopithecus, 98–99; table 9-1
Colobinae, 83–85; table 8-1
Gorilla, 155
Hylobates, 135
Pan, 165–66
Pitheciinae, 46
Pongo, 146
Hapalemur griseus, 23, 27, 29. *See also*
      Lemuriformes; Prosimii
Helping behavior, 40–43. *See also* Af-
      finitive behavior; Altruism; Allomaternal care;
      Food-sharing; Infants
Home range, tables 7-5, 8-2–8-9. *See also*
      Diet; Ecology; Foraging behavior;
      Food distribution; Habitat; Population
      density; Territoriality
   Alouatta, 54–59; table 6-3
   Atelinae, 71, 73, 81–82; table 7-2
   body size/weight and, 122
   Callicebus, 52
   Callitrichidae, 37–40; table 4-1
   Cebinae, 71–73, 81; table 7-2; figure
      17-7
   Cercopithecinae, 112–13, 122–23; table
      11-1
   Cercopithecus, 101; table 9-2
   Colobinae, 84
   dietary influences on, 113
   Erythrocebus patas, 101, 105; table 9-2
   food distribution, 268
   Gorilla, 155
   Hylobates, 135–36; table 12-1
   intergroup relationships and, 268–72
   interspecific influences, 216–17
   noctural primates, 17
   Pan, 166, 286; table 15-1
   Pitheciinae, 46–47; table 5-2
   Pongo, 146–47, 148–49, 286
   Prosimii, 17–20, 28, 30–31, 284–85;
      tables 2-3, 3-1
Homo sapiens
   applications of nonhuman primate
      models to, 413–20
   brain size, 195
   female sexuality, 384
   human ethology, 414–15
   intergroup relationships, 280
   male-infant relationships, 348
   mother-infant relationships, 414–20
Hunting. *See* Conservation, hunting; Preda-
      tion by primates; Predation on pri-
      mates
Hylobates, 135–45
   activity budgets, 136–40; table 12-3
   body size/weight, 135
   communication, 136–41, 267, 287; table
      12-3
   competition, 139–41, 287, 291
   conservation, 136
   dispersal, 141–42, 254, 257; table 12-4
   ecology, 135–36, 143–45, 203, 207;
      table 12-1

Hylobates (*continued*)
  geographic distribution, 135
  group size, 137; table 12-1
  home range and territoriality, 135–45,
    267, 291, 303; table 12-1
  intergroup interactions, 137, 140–41,
    267, 287
  playback experiments, 136, 140
  reproduction, 136–37; table 12-2
  sexual dimorphism, 139; figure 16-2
  social organization and behavior, 137–
    45, 287, 291, 303, 345, 362, 368;
    tables 12-1, 12-2
  taxonomy, 135
Hylobates agilis, 135
Hylobates concolor, 135
Hylobates hoolock, 135
Hylobates klossii, 135
Hylobates lar, 135; figure 12-1
Hylobates moloch, 135
Hylobates muelleri, 135
Hylobates pileatus, 135
Hylobates syndactylus, 135
Hyoid bone, *Alouatta*, 54–55; table 6-1

Inbreeding avoidance. *See also* Dispersal
  Alouatta, 67–68
  Cercopithecinae, 132–34, 263, 302–3,
    367
  Cercopithecoidea, 262–65
  dispersal, 262–65
  Gorilla, 162
  Hylobates, 141–42
  Pan, 302
Incest avoidance. *See* Inbreeding avoidance
Inclusive fitness. *See* Kin selection
Indri indri, 25; figure 3-1. See also
    Lemuriformes; Prosimii
Infant care, 327, 330–41. *See also* Allo-
    maternal care; Helping behavior; In-
    fanticide; Infants; Male-infant relation-
    ships; Mother-infant relationships;
    Weaning
  Aotus, 43, 50; figure 5-4
  Atelinae, 74, 80; figure 7-4
  Callitrichidae, 40–43, 288, 292, 343–
    45; table 4-4
  Cebinae, 74, 80
  Cercopithecinae, 303, 330–41, 365–66;
    table 27-1
  Gorilla, 158–59, 303
  Hylobates, 140
  Pitheciinae, 50
  Prosimii, 22; table 2-3
Infanticide, 233; table 8-4
  Alouatta, 65
  Cercopithecinae, 92–93, 279, 353
  circumstances of, 92–93; table 8-4
  Colobinae, 92–97
  during intergroup encounters, 279
  evidence for, 92
  by females, 342
  genetic relatedness of perpetrator and
    victim, 93
  Gorilla, 163

hypotheses, 92–97
  and interbirth intervals, 92–93
  male-male competition, 307
  by males, 92–97, 170–71
  Pan, 93, 97, 170–71
  Papio, 93
  paternity confidence, 354
  Ponginae, 92
  protection against, 93–96, 394–95, 407;
    figure 8-4
  social organization and, 92–97; table 8-4
Infant-male relationships. *See* Male-infant
    relationships
Infants. *See also* Allomaternal care; Help-
    ing behavior; Infanticide; Infant care;
    Male-infant relationships; Mother-
    infant relationships; Weaning
  adoption, 303, 328
  aggression toward, 308, 341–42
  definition of, 358
  diet of, 330–33
  female dominance and, 320–21, 341–42
  handling of by adult females, 338–42;
    tables 27-2, 27-3
  handling of by immatures, 365–66; fig-
    ure 27-4
  locomotion, 331–32
  and maternal rank effects, 313–14, 321
  and maternal style, 336–37
  effects on mother's social interactions,
    315
  sex ratio, 321
  sibling relationships, 50, 337–38
  survivorship, 320–21
  transition to independence of, 331–33
Insectivory, 182. *See also* Diet, insects
  Callimiconidae, 36
  Callitrichidae, 36
  Cebinae, 69; table 7-1
  Cercopithecus, 99
  Lemuriformes, 27–28
  Pitheciinae, 46
Intelligence/cognition, 431–32, 448–61.
    See also Ape language studies; Com-
    munication; Deception; Culture; Social
    intelligence; Tool use
  abstractions and analogical reasoning,
    455–56, 460
  Cercopithecinae, 448, 459–61
  comparisons across species, 452–61
  concepts and categories, 456, 460
  cross-modal perception, 452–53
  deception, 448–50
  discrimination learning, 452–53
  insight, 456–57
  interdisciplinary approaches to, 494–95
  laboratory studies of, 452–61
  ontogeny of, 455
  Pan, 448–50, 457–58, 460–61, 473
  Piagetian stages, 455
  problem solving strategies, 454–55
  tool use, 457–58
  vocal communication and, 448–51
Interbirth intervals, table 5-3. *See also* Re-
    productive parameters

Alouatta, 59; tables 6-4, 8-4
Atelinae, table 7-3
  and body size/weight, 191; table 16-3
Callimiconidae, table 4-2
Callitrichidae, table 4-2
Cebinae, 74–76; table 7-3
Cercopithecinae, 126, 358; tables 8-4,
    10-1, 11-4
Cercopithecus, 102; table 9-3
Colobinae, 87–89; tables 8-3, 8-4; figure
    8-3
  dominance rank, 321
  ecological influences on, 87–89, 126
Erythrocebus patas, 102; table 9-3
Hylobates, table 12-2
  nursing effects on, 336
Pongidae, 147–48, 173; tables 8-4, 14-1
Prosimii, 17–18, 29; tables 2-3, 3-2
  social influences on, 92–93, 126; table
    8-4
Intergroup relationships, 267–80; table
    22-1
  affinitive, 272
  aggression, 274–80; figures 22-1, 22-2,
    32-1
  Alouatta, 55, 58–59
  Aotus, 52–53
  Atelinae, 73, 81–82
  Callicebus, 50–53
  Callitrichidae, 40
  Cebinae, 73
  Cercopithecinae, 104–5, 114, 131, 267–
    80; table 22-2; figures 22-1, 22-2
  Cercopithecoidea, 278–81
  Cercopithecus, 104–5; figure 22-1
  Colobinae, 90–92, 275; figure 22-1
  definitions of encounters, 267–68
  female-bonded groups, 274, 276–81
  Gorilla, 156–57, 267, 275
  home ranges and territoriality, 268
  Homo sapiens, 280
  Hylobates, 137, 140–41, 267
  intergroup dominance, 268–72
  intragroup dominance, 278–80
  Pan, 170–71, 274
  Pitheciinae, 50–3
  effects on population density, 273; table
    22-2
  reproduction, 288
  resource defense, 205–6, 268, 277–78
  sex differences and, 273–74, 276–80
  effects on social organization, 274–81
  variation in, 272–73, 280–81
  vocal communication, 52–53, 272
Interspecific relationships, 210–26; figures
    18-2, 18-3
  Alouatta, 62
  Callitrichidae, 36–37
  Cercopithecus, 212–14, 221, 230–31
  Colobinae, 221, 230
  communication, 218, 222
  competition, 210–17, 222–26, 457;
    tables 18-2–18-9; figure 18-4
  costs and benefits, 217–22
  evolution of, 225–26

predation, 218, 221–22, 230–31
Prosimii, 15–17
social behavior in, 222; table 18-10
Intrasexual competition. *See* Female-female
competition; Male-male competition

Juveniles, 358–69
aggression, 307–8; figure 25-1
dominance and rank acquisition, 311–
12, 314, 365–67; figure 34-3
grooming, 364–65
infant interactions, 365–66
kinship, 365
mortality rates, 359
play, 314, 359–64
sexual behavior, 367
social relationships, 364–69

Kin selection, 327–29. *See also* Kinship
alarm calling, 229
alliances, 325
altruism, 324–28
coefficient of relatedness, 324–25
food-sharing, 326–27
grooming, 325–26
Hamilton's rule, 324
inclusive fitness, 324–25
kin recognition, 304–5, 494–95
male-infant relationships, 353–57
paternity confidence, 343, 345, 353–54
Kinship, 299–305. *See also* Kin selection
Atelinae, 79
adoption, 303
affinitive behavior, 299–302
alliances, 301–2, 390, 408; figure 35-1
allomaternal care, 341
Cebinae, 79
Cercopithecinae, 129–32, 299–305,
314, 365
cooperation, 313–14
dispersal, effects on, 302–4, 365
female dominance, 314
Gorilla, 160–62, 303–4; tables 14-4,
14-5
grooming, 300–301, 325–26, 364–65
kin recognition, 304–5, 494–95
matrilineal groups, 129–32, 299–304
Pan, 302
protection of kin, 402
rank acquisition, effects on, 311–12
siblings, 303, 337–38
social organization, 303–4
territorial defense, 303–4
!Kung San. See Homo sapiens

Laboratory studies, 2
communication in *Pan,* 438–44
cooperation, 437–38
intelligence/cognition, 452–61
Macaca, 413
mother-infant relationships, 330–38,
413–14
Lactation. *See also* Mother-infant relationships; Weaning
milk composition, table 33-1
parental investment, 330–31

Lagothrix flavicauda, 69. See also *Atelinae*
Lagothrix lagothricha, 69. See also *Atelinae*
Language. *See* Communication
Lemur catta, 25–33; figure 3-1. See also
Lemuriformes; Prosimii
Lemur coronatus, 25–27, 29. See also *Lemuriformes; Prosimii*
Lemur fulvus, 25, 28–31; figure 3-1. See
also *Lemuriformes; Prosimii*
Lemur macaco, 29–30; figure 3-3. See also
Lemuriformes; Prosimii
Lemur mongoz, 25. See also Lemuriformes; Prosimii
Lemuriformes, 25–33. See also prosimii
ancestral species, 25
body size/weight, 28
characteristics of, 25
communication, 13, 30, 32
conservation, 28–29
dominance, 30–33
ecology, 25–29
home range/territoriality, 28, 30–31
monogamy, 30
population densities, 28
reproduction, 29–30, 33
social organization and behavior, 22–23,
30–33
taxonomy, 11
Leontopithecus rosalia, 37, 40–42, 326;
figure 4-1. See also *Callitrichidae*
Lepilemur mustelinus, 25; figure 3-1. See
also *Lemuriformes; Prosimii*
Life-history. *See also* Demography; Reproductive parameters
comparative perspectives, 181–96
life expectancy, 241
life-tables, 240–42
mortality, 241–44; figure 20-1
relationship to body size/weight, 191–
93; tables 16-3, 16-6
relationship to brain size, 191–93; tables
16-5, 16-6
survival rate, 241–44; figure 20-2
variables, 240; tables 16-1, 16-4
Locomotion
Alouatta, 55
Colobinae, 83
Hylobates, 135
infants, 331–32
Papio infants, 331–32
Pongo, 146
Prosimii, 13, 25–27; tables 2-2, 3-1
Long-term relationships, 3–4. *See also* Alliances; Kinship; Reciprocity; Social
behavior; Special relationships
Loris tardigradus, 13; table 2-1. See also
Lorisiformes; Prosimii
Lorisiformes. See also Prosimii
body size/weight, 11
communication, 13, 22
ecology, 13–14
locomotion, 13
reproduction, 17–18
social organization and behavior, 22
taxonomy, 11

Macaca. See also Cercopithecinae
alliances, 409; table 24-1
competition, 310–11, 322, 389, 422–
24, 437
ecology, 122; table 11-1
female-female relationships, 129–31
genetic variation and dispersal, 133–34
geographic distribution, 121
kinship, 303; table 24-1
laboratory studies, 413
reproduction, 320–23, 345, 381, 383,
389, 394–95, 397–98; figure 31-2
social behavior and organization, 316–
17, 336–37, 345–46, 348, 397–98,
422–25
taxonomy, 121
Macaca fascicularis, 122; figure 25-1. See
also *Cercopithecinae*
Macaca fuscata, 122; figures 11-3, 24-2.
See also *Cercopithecinae*
culture, 464, 467–72; figures 38-3–38-7
demographic and life-history variables,
table 20-1
vocal communication, 446–47
Macaca mulatta, 122, 125; figures 11-2,
22-2, 26-1, 28-2, 29-2, 30-3, 31-2,
31-4, 32-1, 33-2, 34-1. See also
Cercopithecinae
dominance, 125, 313, 389, 409, 422–24,
437
laboratory studies, 389, 413, 437–38
playback experiments, 448
vocal communication, 448
Macaca nemestrina, 122. See also *Cercopithecinae*
Macaca nigra, 121. See also
Cercopithecinae
Macaca radiata, 125. See also *Cercopithecinae*
Macaca sinica, 126. See also *Cercopithecinae*
Macaca sylvanus, 125. See also *Cercopithecinae*
birth seasonality, 345
communication, 436
male-infant relationships, 345, 351–52,
366
reproduction, 345
sexual behavior, 394
Malagasy *Prosimii. See Prosimii;
Lemuriformes*
Male choice, 397–99
Male dominance. *See also* Alliances, male-
male; Dominance; Male-male competition
after transfer, 313
body size/weight, 313
Cercopithecinae, 132, 312–13, 322–23
female choice and, 395
ontogeny, *Cercopithecinae,* 366
Pan, 175–77, 313
rank acquisition and reversals, 175–76,
312–13
reproductive success, 132, 388–89; figure 31-1

Male-female relationships. *See also* Consortships; Sexual behavior; Special relationships
    Gorilla, 163
    Pan, 169–70, 315
    Papio, 118–19, 132, 316; figure 31-5
    sexual dimorphism, 132, 277, 388, 407
Male-infant relationships, 40–43, 50, 292, 343–57, 366, 393–94; table 28-2. *See also* Agonistic buffering; Infant care; Infanticide; Infants
    abuse and infanticide, 343, 351–53
    affiliation, 343, 345–50; figure 28-1
    Alouatta, 345–50
    bonds with infant's mother, 354–57
    Callimiconidae, 288, 343–45
    Callitrichidae, 40–43, 288, 292, 343–45; table 4-4
    Cercopithecinae, 316, 345–54, 396; figure 28-1
    costs and benefits, 343–44, 352–57; table 28-4
    evolution of, 353–57
    explanations of, 353
    food-sharing, 344
    Gorilla, 158–59, 346–48; figure 28-1
    Homo sapiens, 348
    Hylobates, 140, 345
    infanticide, protection against, 353
    influence on female choice, 393–94
    intensive caretaking, 343–45
    Macaca sylvanus, 344–45
    monogamous primates, 288, 343–45
    ontogeny, 366
    Pan, 170–71, 350
    paternity and, 394–95; table 28-3
    Pitheciinae, 50
    Prosimii, 351
    triadic interactions, 351–53; figure 28-3
    variation in, table 28-1
Male-male alliances. *See* Alliances, male-male
Male-male competition, 307–9, 315–16, 385–92; figures 31-1, 31-2. *See also* Aggression; Alliances, male-male; Competition; Male dominance; Male-male relationships
    Alouatta, 65–66
    Atelinae, 77–78; table 7-4
    Cebinae, 77–78; table 7-4
    Cercopithecinae, 115–18, 120, 132, 386–92
    Cercopithecus, 104–11
    dispersal, 263, 388
    Erythrocebus patas, 104–11
    over females, 307–9, 381, 385–92; table 31-1; figures 31-1, 31-2
    female choice, 396
    Gorilla, 156–57
    infanticide, 308
    male-infant interactions, 351–53
    Pan, 170–71, 175–77, 290–91, 392
    Pongo, 149–50, 152–54
    Prosimii, 18–20, 30

social organization, 110–11, 279, 283–84; table 31-1
    sperm competition, 384
Male-male relationships. *See also* Alliances, male-male; Grooming, male-male; Male dominance; Male-male competition
    affiliative, 174–77
    all-male groups, 261
    Cercopithecinae, 131–32, 315–16
    Colobus badius, 174
    dispersal, 400–401
    grooming, 174
    Pan, 174–77, 315
    Pongidae, 174
    social system, effects on, 400–401
Male transfer, 90, 115–17, 258, 260–61, 292–93, 313; table 23-2. *See also* Dispersal; Emigration
    intergroup relationships, 274, 276–77, 279–80
Mate-guarding, 17, 53
Mating strategies, 315–16, 323, 390–95; table 31-2. *See also* Reproductive success; Sexual behavior
    alliances, 390–91
    Cercopithecinae, 115–17, 132, 323, 390–91
    Cercopithecus, 109–10
    Erythrocebus patas, 109–10
    Galago, 17
    Pan, 174, 392
    Pongo, 149–50
Matrifocal units, *Pan paniscus,* 172
Matrilineal groups, 125–34. See also *Cercopithecinae;* Kinship; Social organization *under taxonomic headings*
    affinitive behavior, 130–32, 299–305
    Cercopithecinae, 125–34, 299–305, 309–17
    definition, 90
    dominance relationships, 118, 301
    inbreeding avoidance, 262–65, 302
    infanticide, 93
    kinship, 129–32, 299–304
Meat-eating. *See* Predation by primates
Menarche, 358–59. *See also* Reproductive parameters; Adolescence
Menopause, 371–79
Menstruation, 371
Metabolic rate, brain size and, 187, 195
Methodology. *See* Field methods
Microcebus murinus, 23, 25, 28–30, 32; figure 3-1. See also *Lemuriformes; Prosimii*
Miopithecus talapoin. See *Cercopithecus talapoin*
Mirza coquereli, 23, 25, 28. See also *Lemuriformes; Prosimii*
Mixed species associations. *See* Interspecific interactions
Mobbing
    of males, 407–8
    of predators, 15, 222, 227–28

Monogamous primates. See also *Hylobates; Callitrichidae*
    Callitrichidae, 38–40
    Colobinae, 90
    competition, 33, 287, 291, 386
    definition, 142
    dispersal, 254, 257
    ecology, 142–45, 236–37, 287, 291
    evolution of, 142–45, 287
    female choice, 393
    female dominance, 33
    Hylobates, 137–45, 287, 291
    infant care, 142–43, 343–45, 357
    intergroup relationships, 275
    kinship, 303
    Lemuriformes, 30, 33
    Pitheciinae, 48–50
    social organization and behavior, 287, 291, 303, 345, 362, 368
    territoriality and, 52–53, 142–45, 267, 291, 303
    vocal communication, 52–53, 136–41
Mother-infant relationships. *See also* Infants; Infant care; Parental investment
    attachment theory, 330, 332–33, 417–20
    Cercopithecinae, 331–38; table 27-1
    contact, 332–33; figures 27-1, 27-2
    dominance rank, 336
    ecological influences on, 333, 336
    food-sharing, 331
    Galago, 18, 22
    grooming, 300
    human vs. non-human primates, 415–20
    interspecific variation in, 333
    laboratory studies, 330, 332–38, 413–14
    maternal aggression, 331, 333
    maternal styles, 333–37; figure 27-3
    Pongidae, 331–32, 338
    sex differences, 336
    social organization, influences on, 337–38
    variation in, 333–37; table 27-1
    weaning, 330–31, 333
    after weaning, 338; figure 15-1
Multifemale groups, 98–111. *See also* Group size; Sex ratio; Social organization; One-male groups
Multimale groups, 65–66, 121–34, 238; table 23-3. *See also* Group size; Sex ratio; Social organization
Mutualism, 218–22

Nasalis, 83, 237. See also *Colobinae*
Natural selection, 318–19
Nesting behavior, 15, 18, 146
Nocturnal primates, 23–24. See also *Aotus; Prosimii*
    activity budgets, 13
    adaptations of, 11
    Aotus, 44–53
    competition, 385
    ecology, 206, 284, 290
    field methods, 11

home ranges and territoriality, 17, 284–85; figure 2-6
Prosimii, 25, 284–85
social organization and behavior, 20–24, 290–91
Nomascus. See *Hylobates concolor*
Nycticebus coucang, 12, 15, 17; figure 2-2. See also *Lorisiformes; Prosimii*

Observation Sampling Techniques, 5–8, 414–15
Olfactory signals, 379, 433–39. *See also* Communication
aggression, 434
Alouatta, 56, 59
Callitrichidae, 434
Ceboidea, 13, 434
Colobinae, 87
mating, 17, 87, 379; table 30-1
Prosimii, 13, 15, 18, 21–22, 32, 285, 433–34
One-Male Groups, 98–111; table 23-3. *See also* Demography; Polygyny; Social organization *under taxonomic headings*
adaptive significance, 109–11
Cercopithecus, 98–111
male-male competition, 105–11, 279, 386
reproductive success in, 102, 107–10
One-Male Unit (OMU), 112–20. *See also* Demography; Social organization *under taxonomic headings*
adaptive significance of, 118–20
definition of, 112
Papio, 112–20; table 10-4
Ontogeny, 358–69; table 28-4. *See also* Adolescence; Infants; Juveniles; Mother-infant relationships
Cercopithecinae, 364–67
evolutionary significance of prereproductive period, 359

Pan, 165–77; tables 15-1–15-4. *See also* Ape language studies; *Pongidae*
activity budgets, 165–66, 286; table 15-1
communication, 171–73, 422–25, 427–28, 435–36, 438–44, 448–51, 466; table 15-4; figures 15-3, 15-4, 34-2, 34-4, 38-2
competition, 170–71, 173–77, 290–91, 311, 392, 404, 422–29; figures 25-2, 34-2
conservation, 166
cultural transmission, 463–67, 473
deception, 438–39, 448–50
demography, 167, 172, 242; tables 15-3, 20-2
dispersal, 169, 253, 261, 265; table 15-3
ecology, 165–66, 171, 174, 206, 285–86, 290–91, 464; table 15-1
food-sharing, 172, 175, 315, 326–27; figure 26-2
geographic distribution, 165

home range and territoriality, 166, 170–71, 286; table 15-1
intelligence, 448–50, 457–58, 460–61, 473
intergroup relationships, 170–71, 274
laboratory studies of communication, 438–44
male-male relationships, 170, 174–77, 315; figure 15-1
predation by, 166, 206, 463; figure 19-2
reproduction, 166–67, 169, 359, 583–84; tables 15-2, 20-3
sex differences, 168, 404
sexual behavior, 169, 172–75, 392, 396, 466, 583; tables 15-2, 15-4; figure 15-2
sexual dimorphism, 165
social organization and behavior, 81–82, 93, 166–77, 290–91, 302, 309, 315, 326–27, 350, 367–68, 392, 396, 404, 422–49; tables 15-3, 15-5; figure 15-1
taxonomy, 165
tool use, 166, 457–58; figure 38-1
Pan paniscus, 172–74, 438; tables 15-2, 15-4. See also *Pan*
compared to *Pan troglodytes,* 174; tables 15-2, 15-4
distinct features, 165
Pan troglodytes, 165–77. See also *Pan*
Papio, 112–20; figures 28-3, 29-1, 29-4, 31-1, 31-5, 35-1. See also *Cercopithecinae*
communication, 119, 229
competition, 115–16, 118, 126, 129–32, 248, 308, 310–13, 402–3
conservation, 113
dispersal, 114–16, 252–55
ecology, 112–13, 119, 122, 205, 229–30, 294–95
field experiments, 7, 114
geographic distribution, 112, 121
group size and population density, 112–14, 122; tables 11-1, 11-2, 11-3
home range, 112–14, 122
hybrids, 118
intergroup interactions, 114
reproduction, 113, 117, 125–26, 320–23, 358–59, 367, 379–80, 389; tables 10-1, 11-4
sexual dimorphism, 112
social intelligence, 460, 473
social organization and behavior, 93, 113–20, 126–32, 294–95, 302–3, 308, 310–13, 316, 331–40, 346–53, 358–59, 367, 390–95, 398, 402–3; table 10-4; figure 23-2
taxonomy, 112, 121
Papio cynocephalus, 112. See also *Cercopithecinae; Papio*
Papio cynocephalus anubis, 112. See also *Cercopithecinae; Papio*
Papio cynocephalus cynocephalus, 121. See also *Cercopithecinae; Papio*
Papio cynocephalus papio, 121. See also *Cercopithecinae; Papio*

Papio cynocephalus ursinus, 121. See also *Cercopithecinae; Papio*
Papio hamadryas, 112–20. See also *Cercopithecinae; Papio*
communication, 119; figure 30-2
dispersal, 114–16, 252–55, 393
ecology, 112–13, 119, 294–95
intergroup relationships, 114, 275
social organization and behavior, 113–20, 294–95, 302–3, 348–50, 367–68; figures 10-1, 23-2, 32-2
Papio leucophaeus, 112, 114, 118–20. See also *Cercopithecinae; Papio*
Papio sphinx, 112, 114, 118–20. See also *Cercopithecinae; Papio*
Parental care. *See* Allomaternal care; Infant care; Infants; Lactation; Male-infant relationships; Mother-infant relationships; Parental investment; Weaning
Parental investment, 330–31, 385, 419–20
Parent-offspring conflict. *See* Infant care; Infants; Lactation; Mother-infant relationships; Parental investment; Weaning
Paternal care. *See* Male-infant relationships
Paternity confidence, 92–96, 343–45, 354, 381, 394–95; table 28-3
Paternity, assessments of, 322–23
Patrilineal groups, 90–93, 302
Perodicticus potto, 12, 15, 22, 24. See also *Lorisiformes; Prosimii*
Phaner furcifer, 25, 28–30, 32–33. See also *Lemuriformes; Prosimii*
Philopatry. *See* Dispersal
Pithecia hirsuta, 45. See also *Pitheciinae*
Pithecia monachus, 44. See also *Pitheciinae*
Pithecia pithecia, 44; figure 5-1. See also *Pitheciinae*
Pitheciinae, 44–53
body size/weight, table 5-1
communication, 49–53
conservation, 47–48
demography, 48–49; table 5-3
dispersal, 49–50
ecology, 46–48; tables 5-1, 5-2
geographic distribution, 45–46
group size and population density, 46, 48–49; table 5-2, figure 5-3
home range and territoriality, 46–47, 50–53; table 5-2
intergroup interactions, 50–53
physical characteristics, 44–45
reproduction, 48; table 5-3
social organization and behavior, 48–53; table 5-2
taxonomy, 44
Placenta, 190–91
Play, 37, 314, 359–64; figures 29-1–29-4
Playback experiments, 444–45
Callicebus, 52
Cebus, 73
Cercocebus albigena, 445
Cercopithecus aethiops, 7–8, 316, 328–29, 444–45, 447–48

Playback experiments (*continued*)
Hylobates, 136, 140
intergroup relationships, 272
Macaca, 302, 448
Pongo, 150
Polyandrous primates, *Callitrichidae*, 34–43, 291–92
Polygyny, 23–24, 65–66, 90, 265. *See also* One-male groups; Sex ratio; Social organization
Polyspecific associations. *See* Interspecific interactions
Pongidae. See also *Pongo; Gorilla*
ape language studies, 441–44
dispersal, 265–66
infanticide, table 8-4
responses to predators, 227
social organization and behavior, 174, 332, 364; table 8-4
Pongo, 146–54; figures 13-1, 13-2. See also *Pongidae*
activity budgets, 146–47
body size/weight, 146–47, 152–53; table 13-1
communication, 150, 152
conservation, 147
demography, 147–48
dispersal, 148–49
ecology, 146–47, 152–53, 285–86, 290–91
forced copulation, 149–50
geographic distribution, 146
home range and population density, 146–47, 148–49, 286
reproduction, 146–50
sexual dimorphism, 149–54; table 13-1; figure 13-2
sexual selection, 153–54
social organization and behavior, 148–54, 290–91
taxonomy, 146
tool use, 458
Pongo pygmaeus. See Pongo
Population density. *See also* Demography; Geographic distribution; Group size; Home range
Alouatta, 56–57, 62–65; tables 6-2, 6-3, 6-5, 6-6
Atelinae, table 7-2
body size/weight, 73, 122; table 18-1; figures 5-3, 7-2, 18-1
Callitrichidae, 37; table 4-1
Cebinae, table 7-2
Cercopithecinae, tables 11-1, 22-2
Cercopithecus, 101, 104
Colobinae, 84; table 8-1
ecological influences, 246–47
effects on intergroup relationships, 272–73; table 22-2
Erythrocebus patas, 101; table 9-2
Gorilla, 155–56
interspecific resource competition and, 215–17
Pan, table 15-1
Pitheciinae, 47; table 5-2

Pongo, 147
Prosimii, table 5-2
Predation by primates, 224; table 19-1
Callimiconidae, 36
Callitrichidae, 36
Cercopithecus, 101
cooperative hunting, 206
Erythrocebus patas, 101
Pan, 166, 206, 463; figure 19-2
Papio, 112, 122, 206
Prosimii, 13–14
Predation on primates, 227–39. *See also* Conservation, hunting; Alarm calls; Mobbing
Atelinae, 69
behavioral defenses, 15, 222, 227–31, 293, 434
Callitrichidae, 37
Cebinae, 69
Cercopithecinae, 227–32, 234–35, 238
evolution, influence on, 23, 236–39, 286, 288
Gorilla, 163
habitat differences, 235
by humans, 231–33, 237
Hylobates, 143
infanticide, 233
interspecific associations and, 230–31
intragroup spacing and progression order, 229–30
predator types, 231–33
rates of, 231–35
sexual dimorphism, 236
social organization, 233–39, 284, 286
solitary primates, 236–37
Prehensile tail, 55, 69
Presbytis, 83–97, 237, 395, 463. See also *Colobinae*
Presbytis aygula, 84
Presbytis entellus, 83, 86–92, 463; figures 21-1, 22-1, 24-1, 27-4, 29-3, 30-1. See also *Presbytis; Colobinae*
Presbytis geei, 85
Presbytis johnii, 85
Presbytis melalophos, 84
Presbytis obscura, 84
Presbytis pileata, 85
Presbytis senex, 84
Procolobus, 83. See also *Colobinae*
Promiscuity, 77, 108, 345
Propithecus verreauxi, 25, 29–33; figure 3-2. See also *Lemuriformes; Prosimii*
Prosimii, 11–33. See also *Lemuriformes; Lorisiformes; Tarsiiformes*
body size/weight, 12, 16–17, 20, 28; table 2-1
communication, 13, 15, 18, 21–22, 30, 32, 433–34
conservation, 17, 28–29
demographic and life-history variables, 245–46; tables 2-3, 2-4, 16-1
dispersal, 18–20, 30–31, 253–54
ecology, 13–15, 25–29, 36, 204, 284; tables 2-1, 2-2, 2-4, 3-1
geographical distribution of, 12; table 2-1

group size and population density, 28; tables 2-2, 2-4, 3-1
home range and territoriality, 17–20, 28, 30–31, 284–85; tables 2-4, 3-1
interspecific competition, 17
nocturnal primates, 284–85
reproduction, 17–18, 20, 29–30, 33; tables 2-3, 3-2
social organization and behavior, 18–24, 30–33; table 2-4
taxonomy, 11–13; table 2-1
Provisioning. *See* Field studies, provisioning
Pygathrix nemaeus, 83, 85. See also *Colobinae*

Reciprocal altruism, 328–29. *See also* Alliances; Altruism; Cooperation; Reciprocity
Reciprocity, 131, 316–7, 328–29, 410–11. *See also* Alliances; Altruism; Cooperation
Reconciliation, 424–25; figure 34-4
Reproductive parameters, 370–84. *See also* Birth seasonality; Estrus; Fecundity; Gestation; Interbirth intervals; Sexual behavior; Weaning
age at first birth, 320
age of maturity, 125–26; tables 2-3, 3-2, 4-2, 5-3, 7-3, 8-3, 9-3, 10-1, 12-2; figure 16-4
Alouatta, 54, 59–60; tables 6-1, 6-4, 8-4
Atelinae, 74–77; table 7-4
birth rate, 18, 241, 244–47; table 20-3
body size/weight and, 74, 181, 191–92; tables 16-3, 16-6
brain size and, 191–92; tables 16-5, 16-6; figure 16-5
Callimiconidae, 37–38; table 4-2
Callitrichidae, 37–40; table 4-2
Cebinae, 74–77; table 7-4
Cercopithecinae, 113, 125–26; tables 8-4, 10-1, 11-1, 11-2
Cercopithecus, 102
Colobinae, 85–89; table 8-4
concealed ovulation, 43, 384
demographic variables, 191–92; table 16-5
dominance, influence of, 130, 320–24, 388–89
Erythrocebus patas, 102; table 9-3
food availability, effects on, 18, 125, 319–20
Gorilla, 156; table 14-1
Hylobates, 136–37; tables 12-1, 12-2
litter size, 74; table 2-3
net reproductive rate, 241
Pan, 166–67, 169, 173; table 15-2
parity, 336–37, 340
Pitheciinae, 46; table 5-3
Pongo, 146–48
Prosimii, 17–18, 29, 33; tables 2-3, 3-2
reproductive value, 241–42
social interactions, effects on, 315

Reproductive success. *See also* Fecundity; Mating strategies; Reproductive parameters; Sexual behavior
  age at first birth and, 320
  age differences in, 389–91
  Cercopithecinae, 115–17, 130, 132, 320–23, 389–91
  Cercopithecus, 107–11; table 9-5
  dispersal, 265
  dominance, 77, 130–31, 288, 388–92
  ecological effects on, 18, 125, 319–20
  Erythrocebus patas, 108–9; table 9-5
  genetic fitness and, 319
  group size, effects on, 163, 288–89
  interbirth intervals, 18
  male-infant relationships, effects of, 357
  mating strategies, 149, 323, 390–95
  one-male groups, 107–10
  Pongo, 149
  rank, effects on, 320–24
  social organization, 385–86
Resource competition, 31–33, 204–9, 292, 307. *See also* Ecology, sympatric species; Female-female competition; Foraging behavior; Interspecific relationships; Home range; Male-male competition; Territoriality
  adaptive consequences of, 319–20
  among females, 160, 288–90, 320–21
  among males, 322–23
  cooperative defense of food patches, 205–6
  dominance, 20, 33, 205, 320–23
  exploitative, 160, 212–17
  food distribution, 215–16
  group size, 238
  home range, effects on, 20
  interference, 160, 205, 212
  interspecific, 212–17, 223–26
  Lemuriformes, 31–33
  population densities, 216
Rhinopithecus roxellanae, 83. See also *Colobinae*

Saguinus, 34–43, 208, 254; figure 4-2. See also *Callitrichidae*
Saguinus fuscicollis, 34–43; figure 4-2
Saguinus imperator, 34–38, 40
Saguinus labiatus, 34, 36–40
Saguinus mystax, 34, 36, 38–40
Saguinus oedipus, 34–40, 43
Saimiri, 69–82, 204, 209, 237, 370; figure 7-1. See also *Cebinae*
Scent marking. *See* Olfactory signals
Seasonal breeding. *See* Birth Seasonality
Sex differences, 400–412. *See also* Sexual dimorphism
  aggression, 401–12; tables 32-1, 32-2
  alliances, 404–5, 427
  behavioral, 168, 400–412
  dominance, 405–11
  grooming, 364–65
  intergroup relationships, 273–74, 276–80
  Pan, 168, 404

play, figure 29-2
social organization, effects of, 400–404; table 21-1
vocal communication, 272
Sex ratio, 152, 234, 238, 242, 245–46, 248–49, 262, 295, 321, 354; table 23-3. *See also* Demography *under taxonomic headings;* Social organization
  Alouatta, 62–67; tables 6-5, 6-6, 6-7; figure 6-4
  Atelinae, 77
  Callimiconidae, 39–40; table 4-3
  Callitrichidae, 39–40; table 4-3
  Cebinae, 77
  Cercopithecinae, 126, 242, 246; table 11-5, figure 20-3
  Cercopithecus, 104, 110
  Colobinae, 89–91
  Erythrocebus patas, 104, 110
  Galago, 245–46
  Gorilla, 156–57; table 14-2
  Lemuriformes, 29–30
  Pan, 167, 172, 242; table 15-3
  Pongo, 148, 152
Sexual behavior, 87, 370–84; table 30-1. *See also* Estrus; Female choice; Forced copulation; Male choice; Mating strategies
  Alouatta, 59
  Catarrhines, 370
  Cercopithecinae, 126, 322–23, 395
  Cercopithecine immatures, 367
  Cercopithecus, 102, 107–10
  Colobinae, 87
  dominance, 388–89
  Erythrocebus patas, 102, 108–9; figure 9-4
  Gorilla, 156, 383
  Homo sapiens, 384
  Hylobates, 137
  interference during, 91, 388–89; figure 9-4
  Macaca, 126, 383, 394–95
  olfactory cues, 17, 87, 379
  Pan, 169, 172–75, 383, 392, 396, 466; tables 15-2, 15-4; figure 15-2
  partner preferences, 379–80, 383
  Pongo, 149–50
  Prosimii, 17, 20–22, 30
  solicitations, 59, 102, 137, 379–80, 392–93; figures 30-1, 30-2, 30-3
  testes size, 384
  unfamiliar males, 394–95
  vocal communication, 379
Sexual dimorphism
  Alouatta, 54–55, 65; table 6-1
  Atelinae, 69; table 7-4
  canine size, 157, 236
  Cebinae, 69; table 7-4
  Cercopithecinae, 112, 119–20, 132
  dichromatism, 44–45, 54; table 6-1; figure 3-3
  dominance, 407
  evolution of, 150–54

foraging behavior, 153
Gorilla, 157; figures 14-1, 14-2
Hylobates, 139
male-female relationships, 132, 277, 388, 407; figure 16-2
non-primates, table 13-1
Pan, 165
Pitheciinae, 44–45
Pongo, 149–54; figure 13-2
predator defense and, 236
reproduction, age of, 359
sex ratio, 152
Sexual selection
  definition, 385
  infanticide, 96–97
  male-infant relationships, 343
  Pongo, 150–54
  sexual dimorphism, 150–54
Sexual swellings. *See also* Estrus
  adaptive significance, 380–81
  Cercopithecinae, 113, 126
  Cercopithecus, 102
  illustrations of, figure 30-4
  onset, 359
  Pan, 167
Sibling relationships, 337–38, 362–65. *See also* Kinship
Simias. See Nasalis; Colobinae
Sleep. *See* Activity budgets
Social behavior. *See also* Affinitive behavior; Aggression; Allomaternal care; Communication; Cooperation; Dominance; Female-female relationships; Grooming; Kinship; Male-infant relationships; Male-female relationships; Male-male relationships; Mother-infant relationships; Sibling relationships; Social organization and behavior *under taxonomic headings;* Special relationships
  demographic effects on, 248–49
  evolutionary perspectives, 318–29
  ontogeny of, 364–69
Social intelligence, 458–61. *See also* Intelligence/cognition
Social organization, 3; tables 16-2, 23-1, 23-3. *See also* All-male groups; Demography; Dispersal; Group size; Monogamous primates; One-male groups; Polyandrous primates; Polygyny; Sex ratio; Social organization and behavior *under taxonomic headings*
  behavioral influences, 274, 282–83, 338, 362–64, 367–68, 385–86
  demography, 247–48
  diversity of, 3; figures 23-1, 23-2
  ecological influences, 118–19, 233–39, 274, 283–86, 290–95
  evolution of, 282–96, 318–19
  female transfer species, 292–95
  frugivorous species, 285–86, 290–91
  infanticide, 92–97; table 8-4
  kinship, 303–4
  levels of complexity, figure 33-1
  male transfer species, 292–93

Social organization (*continued*)
  monogamy, 287, 291, 303, 345, 362, 368
  multifemale groups, 288–90, 292–95
  nocturnal primates, 20–24, 290–91
  predation, effects on, 233–39, 284, 286
  primates vs. non-primates, 295–96
  reproduction, 283–84; table 30-1
Social relationships, 316–17, 421–29. *See also* Social behavior; Special relationships
Socioecology, 179–80, 182–85, 295–96
Solitary primates, 18, 90, 146–54, 236–37, 254, 368, 385
Special relationships. *See also* Female-female relationships; Male-female relationships; Male-infant relationships; Male-male relationships; Social organization and behavior *under taxonomic headings*
  adaptive significance, 398–99
  dominance, 422
  female choice, 395
  Gorilla, 316
  Macaca, 316
  ontogeny of, 362–65
  Papio, 316, 366, 395, 398–99; figure 31-5
Submissive behavior, 306–8; figure 34-1. *See also* Aggression; Dominance; Female-female competition; Male-male competition
Supplant. *See* Competition; Aggression; Resource competition
Symphalangus. See *Hylobates syndactylus*

Tarsiiformes, 11–18, 22; tables 2-1–2-4; figures 2-1, 2-3. See also *Prosimii*
  body size/weight, 12
  communication, 22
  ecology, 13–14
  home range and territoriality, 22
  locomotion, 13
  reproduction, 22
  social organization and behavior, 22
  taxonomy, 12
Tarsius bancanus, 13–14, 22; figure 2-3. See also *Prosimii; Tarsiiformes*
Tarsius spectrum, 14, 22; figure 2-1. See also *Prosimii; Tarsiiformes*
Taxonomy, 9; tables 0-1, A-1
  Alouatta, 54
  Atelinae, 69
  Callimiconidae, 34
  Callitrichidae, 34
  Cebinae, 69

Cercopithecinae, 112, 121
Cercopithecus, 98
Colobinae, 83
Erythrocebus patas, 98
Gorilla, 155
Hylobates, 135
Pan, 165
Pitheciinae, 44
Pongo, 146
Prosimii, 12–13; table 2-1
Territoriality, 268–72, 284–85, 287, 303, 433–34. *See also* Home range; Intergroup relationships
  Alouatta, 58–59
  Atelinae, 71–73
  Callitrichidae, 37–40, 208–9
  Cebinae, 71–73
  Cercopithecus, 104–5
  Cercopithecus aethiops, 267
  Colobinae, 91–92
  Erythrocebus patas, 105
  Hylobates, 135–45, 267, 291, 303
  Pan, 170–71
  Pitheciinae, 50–53
  Prosimii, 17–20, 284–85; table 2-6
Theropithecus gelada, 112–20. See also *Cercopithecinae; Papio*
  body size/weight, table 10-1
  communication, 117–18
  dispersal, 115–17
  ecology, 112, 119, 203
  intergroup interactions, 114; figure 22-1
  reproduction, 113, 117, 391; table 10-1
  sexual dimorphism, 112
  social organization and behavior, 114–18, 248, 295, 350–53, 391, 393, 396; table 10-4; figures 10-2, 20-1
Tool use, 457–58; table 37-1; figure 38-1
Transfer. *See also* Dispersal; Emigration; Female transfer; Male transfer
  by both sexes, 252–53; table 23-2
  definition, 250
  dominance and, 262
  by females, 160–64, 169, 252–55, 258, 293–95, 367–68; table 23-2
  Lemuriformes, 30–31
  by males, 96, 115–17, 258, 260–61, 292–93, 313; table 23-2
  male attraction to all-male groups, 258
  natal, 250, 253, 258–59
  Pan, 169, 253, 261, 265
  Papio hamadryas, 115–17, 252–55, 393
  proximate causes, 257–58
  reception in and choice of new group, 260–62; figure 21-2

  secondary, 250, 253, 262
  sexual attraction to unfamiliar individuals, 257–58
  sexual status, 257–58
  solo vs. peer, 260
Triadic male-infant interactions, 351–53. *See also* Agonistic buffering; Male-infant relationships
Tumescence. *See* Sexual swellings

Varecia, 29. See also *Lemuriformes; Prosimii*
Visual communication, 90, 119, 435–39, 466; table 30-1; figure 29-1. *See also* Communication
  Alouatta, 55–56
  cultural differences, 466–67
  Gorilla, 157
  Pan, 171–73, 466–67
  Papio, 119
Vocal communication, 440–51. *See also* Playback experiments
  adult males and infants, 351
  Alouatta, 54–55
  Callitrichidae, 36–37
  Cercopithecinae, 119–20, 132, 444–51; figures 36-2, 36-3
  Cercopithecus, 105
  Colobinae, 90
  deception, 448–50
  ecological influences, 445–46
  Erythrocebus patas, 105
  evolution of, 445–46, 495
  Hylobates, 136–41, 229, 267, 287; table 12-3
  individual recognition, 448
  intelligence/cognition, 448–51
  intergroup, 52–53, 272
  interspecific, 36–37, 218, 222
  ontogeny, 450
  Pan, 171–73
  Pitheciinae, 50–53
  Pongo, 150, 152
  Prosimii, 21–22
  referential signaling, 444–45
  right-ear advantage, 2, 446–47
  sex differences, 272
  sexual behavior, 379
  specializations, 445

Weaning, 330–31, 333, 358; table 16-3; figure 33-2. *See also* Lactation; Mother-infant relationships
Wounds. *See* Aggression, injuries

Hawker

8/88